Skills in
Mathematics for
JEE Main & Advanced

Trigonometry

With Sessionwise Theory & Exercises...

Amit M. Agarwal

arihant
ARIHANT PRAKASHAN (Series), MEERUT

ARIHANT PRAKASHAN (Series), MEERUT

卐 Administrative & Production Offices

Regd. Office

'Ramchhaya' 4577/15, Agarwal Road, Darya Ganj, New Delhi -110002
Tele: 011- 47630600, 43518550

卐 Head Office

Kalindi, TP Nagar, Meerut (UP) - 250002 Tel: 0121-7156203, 7156204

卐 Sales & Support Offices

Agra, Ahmedabad, Bengaluru, Bareilly, Chennai, Delhi, Guwahati, Hyderabad, Jaipur, Jhansi, Kolkata, Lucknow, Nagpur & Pune.

卐 ISBN : 978-93-26191-65-4

PO No : TXT-XX-XXXXXXX-X-XX

Published by Arihant Publications (India) Ltd.

For further information about the books published by Arihant, log on to www.arihantbooks.com or e-mail at info@arihantbooks.com

Follow us on

PREFACE

"YOU CAN DO ANYTHING IF YOU SET YOUR MIND TO IT, I TEACH TRIGONOMETRY TO JEE ASPIRANTS BUT BELIEVE THE MOST IMPORTANT FORMULA IS COURAGE + DREAMS = SUCCESS"

It is a matter of great pride and honour for me to have received such an overwhelming response to the previous editions of this book from the readers. In a way, this has inspired me to revise this book thoroughly as per the changed pattern of JEE Main & Advanced. I have tried to make the contents more relevant as per the needs of students, many topics have been re-written, a lot of new problems of new types have been added in etc. All possible efforts are made to remove all the printing errors that had crept in previous editions. The book is now in such a shape that the students would feel at ease while going through the problems, which will in turn clear their concepts too.

A Summary of changes that have been made in Revised & Enlarged Edition

- Theory has been completely updated so as to accommodate all the changes made in JEE Syllabus & Pattern in recent years.
- The most important point about this new edition is, now the whole text matter of each chapter has been divided into small sessions with exercise in each session. In this way the reader will be able to go through the whole chapter in a systematic way.
- Just after completion of theory, Solved Examples of all JEE types have been given, providing the students a complete understanding of all the formats of JEE questions & the level of difficulty of questions generally asked in JEE.
- Along with exercises given with each session, a complete cumulative exercises have been given at the end of each chapter so as to give the students complete practice for JEE along with the assessment of knowledge that they have gained with the study of the chapter.
- Previous Years questions asked in JEE Main & Adv, IIT-JEE & AIEEE have been covered in all the chapters.

However I have made the best efforts and put my all teaching experience in revising this book. Still I am looking forward to get the valuable suggestions and criticism from my own fraternity i.e. the fraternity of JEE teachers.

I would also like to motivate the students to send their suggestions or the changes that they want to be incorporated in this book. All the suggestions given by you all will be kept in prime focus at the time of next revision of the book.

Amit M. Agarwal

CONTENTS

SYLLABUS

JEE MAIN

Trigonometry

Trigonometrical identities and equations. Trigonometrical functions. Inverse trigonometrical functions and their properties. Heights and Distances.

JEE Advanced

Trigonometry

Trigonometric functions, their periodicity and graphs, addition and subtraction formulae, formulae involving multiple and sub-multiple angles, general solution of trigonometric equations.

Inverse trigonometric functions (principal value only) and their elementary properties.

CHAPTER

01

Trigonometric Functions and Identities

Learning Part

Session 1
- Measurement of Angles

Session 2
- Definition of Trigonometric Functions

Session 3
- Application of Basic Trigonometry on Eliminating Variables or Parameters and Geometry

Session 4
- Signs and Graph of Trigonometric Functions

Session 5
- Trigonometric Ratios of any Angle

Session 6
- Trigonometric Ratios of Compound Angles

Session 7
- Sum of Sines/Cosines in Terms of Products

Session 8
- Trigonometric Ratios of Multiples of an Angle

Session 9
- Trigonometric Ratios of sub multiple of an Angle

Session 10
- Trigonometric Ratios of the Sum of Three or More Angles

Session 11
- Maximum and Minimum Values of Trigonometrical Functions

Practice Part

- JEE Type Examples
- Chapter Exercises

Session 1

Measurement of Angles

The word 'Trigonometry' is derived from two Greek words.

(i) trigonon (ii) metron

The word trigonon means a triangle and the word metron mean a measure. Hence, trigonometry means measuring the sides and angle of triangle. The subject was originally develop to solve geometric problems involving triangle.

Angle

In trigonometry, as in case of geometry. Angle is measure of rotation from the direction of one ray about its initial point. The original ray called the initial side and the final position of the ray after rotation is called the terminal side of the angle. The point of rotation is called the vertex. If the direction of rotation is anti-clockwise, the angle is said to be positive and if the direction of rotation is clockwise, then the angle is negative

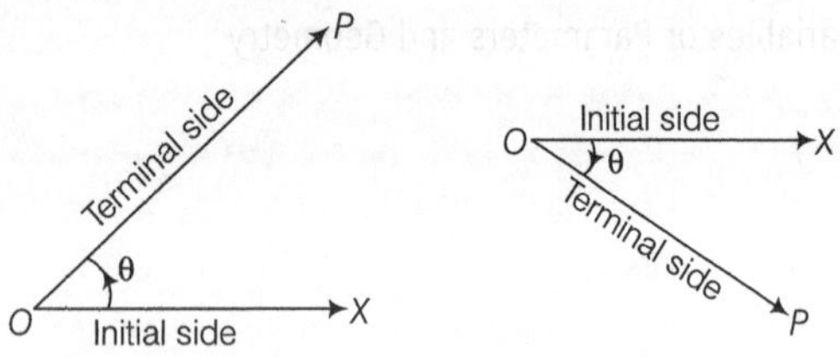

(i) Positive angle (ii) Negative angle

Measurement of Angles

There are three systems used for the measurement of angles.

1. Sexagesimal system or English system (degree)
2. Circular measurement (radian)
3. Centesimal system or French system (grade)

We shall describe the units of measurement of angle which are most commonly used, i.e sexagesimal system (degree measure) and circular measurement (radian measure)

1. **Sexagesimal or Degree measure** If a rotation from the initial side to the terminal side is (1/360)th of a revolution, the angle is said to have a measure of one degree, written as $1°$. A degree is divided into 60 minutes, and a minute is divided into 60 seconds. One sixtieth of a degree is called a minute, written as $1'$; one sixtieth of minute is called a second, written as $1''$. Thus, $1° = 60'$ and $1' = 60''$.

2. **Circular measurement or Radian measure** The angle subtended at the centre of a circle by an arc whose length is equal to the radius of the circle is called a radian and denoted by 1^c.

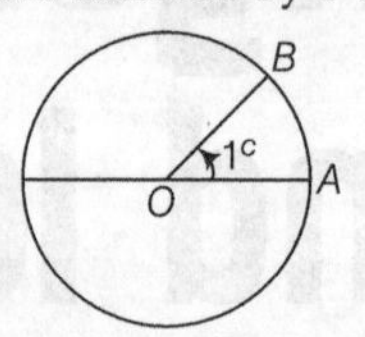

3. **Centesimal or French system** In this system of measurement a right angle is divided into 100 equal parts called **Grades**. Each grade is then divided into 100 equal parts called **minutes** and each minute is further divided into 100 equal parts called **Seconds**.

Thus, right angle $= 100^g$

$$1° = 100'$$

$$1' = 100''$$

Note

Angle of 90° is called a right angle 1′ of centesimal system ≠ 1′ of sexagesimal system 1″ of centesimal system ≠ 1″ of sexagesimal system.

This system of measurement of angles is not commonly used and so here we will not study this system of measurement of angles.

Radian is a Constant Angle

Let ABC be a circle whose centre is O and radius is r. Let the length of arc AB of the circle by equal to r. Then by the definition of radian.

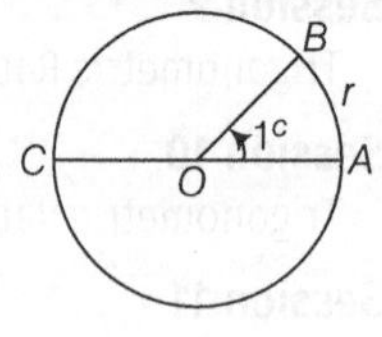

$$\angle AOB = 1 \text{ radian}$$

Produce AO and let it cut the circle at C. Then AC is a diameter of the circle and arc ABC is equal to half the circumference of the circle.

Also $\angle AOC = 2$ right angle $= 180°$

By geometry, we know that angles subtended at the centre of a circle are proportional to the lengths of the arcs which subtend them

$$\therefore \quad \frac{\angle AOB}{\angle AOC} = \frac{\text{arc } AB}{\text{arc } ABC} \text{ or } \frac{1^c}{180^\circ} = \frac{r}{\frac{2\pi r}{2}}$$

[∵ circumference of the circle $= 2\pi r$]

$$\therefore \quad 1 \text{ radian} = \frac{180^\circ}{\pi} = \frac{2 \text{ right angle}}{\pi} = \text{constant}$$

[since a right angle and π are constants]

Relation between Radians and Real Numbers

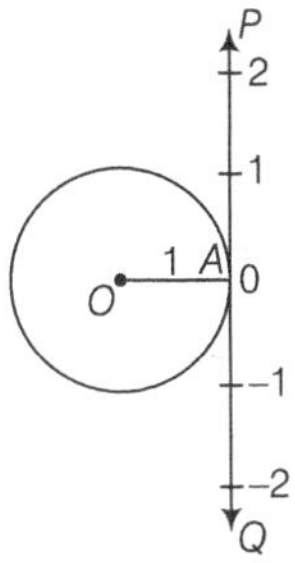

Consider a unit circle with center O. Let A be any point on the circle. Consider OA as the initial side of an angle. Then the length of an arc of the circle gives the radian measure of the angle which the arc subtends at the center of the circle. Consider line PAQ which is tangent to the circle at A. Let point A represents the real number zero, AP represents a positive real number, and AQ represents a negative real number. If we rope line AP in the counter-clockwise direction along the circle, and AQ in the clockwise direction, then every real number corresponds to a radian measure and conversely. Thus, radian measures and real numbers can be considered as one and the same.

Relation between Degree and Radian

It follows that the magnitude in radian of one complete revolution (360 degree) is the length of the entire circumference divided by the radius, or $\frac{2\pi r}{r}$ or 2π.

Therefore, 2π radian $= 360^\circ$

or π radian $= 180^\circ$

or $1 \text{ radian} = \frac{180^\circ}{\pi} = 57^\circ 16'$ (approximately)

Again, $180^\circ = \pi$ radian

$\therefore 1^\circ = \frac{\pi}{180}$ radian $= 0.01746$ radian (approximately)

Thus radian measure of an angle $= \frac{\pi}{180} \times$ degree measure of the angle and degree measure of an angle $= \frac{180}{\pi} \times$ radian measure of the angle.

Thus if the measure of an angle in degrees, and radians be D and C respectively, then

$$\frac{D}{180} = \frac{C}{\pi}$$

The Relation between Degree Measures and Radian Measures of Some Common Angles

Degree	30°	45°	60°	90°	180°	270°	360°
Radians	$\frac{\pi}{6}$	$\frac{\pi}{4}$	$\frac{\pi}{3}$	$\frac{\pi}{2}$	π	$\frac{3\pi}{2}$	2π

Note

(i) Radian is the unit to measure angle and **it does not means that π stands for 180°**, π is a real number. Where as π^c **stands for 180°**.

Remember the relation π **radians = 180 degrees = 200 grade**.

(ii) The number of radians in an angle subtended by an arc of a circle at the centre is equal to $\frac{\text{arc}}{\text{radius}}$.

$$\Rightarrow \quad \theta = \frac{s}{r}$$

Example 1. Convert 40°21′ into radian measure.

Sol. We know that $180^\circ = \pi$ radian.

Hence $\quad 40^\circ 21' = 40\frac{1}{3}$ degree

$$= \frac{\pi}{180} \times \frac{121}{3} \text{ radian} = \frac{121\pi}{540} \text{ radian.}$$

Therefore $40^\circ 21' = \frac{121\pi}{540}$ radian.

Example 2. Express the following angle in degrees.

(i) $\left(\frac{5\pi}{12}\right)^c$ (ii) $-\left(\frac{7\pi}{12}\right)^c$

(iii) $\frac{1^c}{3}$ (iv) $-\frac{2\pi^c}{9}$

Sol. (i) $\left(\frac{5\pi}{12}\right)^c = \left(\frac{5\pi}{12} \times \frac{180}{\pi}\right)^\circ = (5 \times 15)^\circ = 75^\circ$

(ii) $-\left(\frac{7\pi}{12}\right)^c = -\left(\frac{7\pi}{12} \times \frac{180}{\pi}\right)^\circ$

$= -(7 \times 15)^\circ = -105^\circ$

(iii) $\left(\frac{1}{3}\right)^c = -\left(\frac{1}{3} \times \frac{180}{\pi}\right)^\circ = -\left(\frac{60}{\pi}\right)^\circ = 19^\circ\, 5'\, 27''$

(iv) $-\frac{2\pi^c}{9} = -\left(\frac{2\pi}{9} \times \frac{180}{\pi}\right)^\circ = -(2 \times 20)^\circ = -40^\circ$

Example 3. Express the following angle in degrees, minutes and seconds form

$(321.9)^\circ$

Sol. $(321.9)^\circ = 321^\circ + 0.9^\circ$

$= 321^\circ + (0.9^\circ \times 60)'$

$= 321^\circ + 54' = 321^\circ 54'$

Example 4. In ΔABC, $m\angle A = \frac{2\pi^c}{3}$ and $m\angle B = 45°$. Find $m\angle C$ in both the systems.

Sol. $m\angle A = \frac{2\pi^c}{3} = \left(\frac{2\pi}{3} \times \frac{180}{\pi}\right)^\circ = 120°$

$m\angle B = 45°$

$$= \left(45 \times \frac{\pi}{180}\right)^c = \frac{\pi^c}{4}$$

In ΔABC, $m\angle A + m\angle B + m\angle C = 180°$

$\because$ The sum of angles of a triangle is 180°

$\Rightarrow \quad 120° + 45° + m\angle C = 180°$

$\Rightarrow \quad 165° + m\angle C = 180°$

$\Rightarrow \quad m\angle C = 180° - 165°$

$\Rightarrow \quad m\angle C = 15°$

$\Rightarrow \quad m\angle C = \left(15 \times \frac{\pi}{180}\right)^\circ$

$\therefore \quad m\angle C = \frac{\pi^c}{12}$

Example 5. The sum of two angles is $5\pi^c$ and their difference is 60°. Find the angles in degrees.

Sol. Let the angles be x and y in degrees.

Then, $\quad x + y = 5\pi^c \Rightarrow x + y = \left(5\pi \times \frac{180}{\pi}\right)^\circ$

$\therefore \quad x + y = 900° \quad$ ...(i)

$x - y = 60° \quad$ ...(ii)

On adding Eqs. (i) and (ii), we get

$2x = 960°$

$\therefore \quad x = 480°$

On putting $x = 480°$ in Eq. (i), we get

$480° + y = 900°$

$\therefore \quad y = 420°$

$\therefore$ Hence, the angles are 480° and 420°.

Example 6. One angle of a quadrilateral has measure $\frac{2\pi^c}{5}$ and the measures of other three angles are in the ratio 2 : 3 : 4. Find their measures in radians and in degrees.

Sol. One angle $= \frac{2\pi^c}{5} = \left(\frac{2\pi}{5} \times \frac{180}{\pi}\right)^\circ = 72°$

Since, measures of other three angles are in the ratio 2 : 3 : 4. Let the angle be $2k$, $3k$ and $4k$ measured in degree.

$\therefore$ Sum of all angles of quadrilateral = 360°

$\Rightarrow \quad 72° + 2k + 3k + 4k = 360°$

$\Rightarrow \quad 9k = 288° \Rightarrow k = 32°$

$\therefore$ The other three angles are

$2k = 2 \times 32 = 64°$

$3k = 3 \times 32 = 96°$

$4k = 4 \times 32 = 128°$

$\therefore$ The other three angles measured in degree are 64°, 96° and 128°.

The angles in radians are

$$64° = \left(64 \times \frac{\pi}{180}\right)^c = \frac{16\pi^c}{45}$$

$$96° = \left(96 \times \frac{\pi}{180°}\right)^c = \frac{8\pi^c}{15}$$

$$128° = \left(128 \times \frac{\pi}{180}\right)^c = \frac{32\pi^c}{45}$$

$\therefore$ The other three angles measured in radian are

$$\frac{16\pi^c}{45}, \frac{8\pi^c}{15} \text{ and } \frac{32\pi^c}{45}.$$

Example 7. Express the following angles in radians.

(i) 120° (ii) 600°

(iii) − 144°

Sol. (i) $120° = \left(120 \times \frac{\pi}{180}\right)^c = \frac{2\pi^c}{3}$

(ii) $-600° = -\left(600 \times \frac{\pi}{180}\right)^c = -\frac{10\pi}{3}$

(iii) $-144° = \left(-144 \times \frac{\pi}{180}\right) = -\frac{4\pi^c}{5}$

Example 8. If the three angles of a quadrilateral are $60°$, 60^g and $\frac{5\pi}{6}$. Then, find the fourth angle.

Sol. First angle = 60°

Second angle $= 60^g = 60 \times \frac{90}{100}$ degrees $= 54°$

Third angle $= \frac{5\pi}{6}$ radian $= \frac{5 \times 180}{6} = 150°$

$\therefore \quad$ Fourth angle $= 360° - (60° + 54° + 150°) = 96°$

Example 9. In a circle of diameter 40 cm, the length of a chord is 20 cm. Find the length of minor arc corresponding to the chord.

Sol. Let arc $AB = S$. It is given that $OA = 20$ cm and chord $AB = 20$ cm. Therefore, ΔOAB is an equilateral triangle.

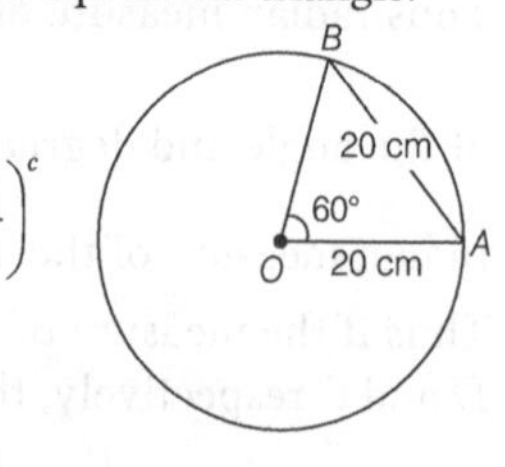

Hence, $\angle AOB = 60°$

$$= \left(60 \times \frac{\pi}{180}\right)^c = \left(\frac{\pi}{3}\right)^c$$

Now, $\quad \theta = \frac{\text{arc}}{\text{radius}}$

$\Rightarrow \quad \frac{\pi}{3} = \frac{S}{20} \Rightarrow S = \frac{20\pi}{3}$ cm

Example 10. In the circle of 5 cm. radius, what is the length of the arc which subtends and angle of 33°15′ at the centre.

Sol. Here, $r = 5$ cm; $15' = \frac{15}{60} = \left(\frac{1}{4}\right)^\circ$

$$\therefore \quad \theta = 33°15' = 33 + \frac{1}{4} = \frac{133}{4} \text{ degrees}$$

$$= \frac{133}{4} \times \frac{\pi}{180} = \frac{133}{4} \times \frac{22}{7 \times 180} = \frac{1463}{2520} \text{ radians}$$

Now, $\theta = \frac{l}{r}$

$$\therefore \quad l = \theta r = \frac{1463}{2520} \times 5 = 2\frac{65}{72} \text{ cm (approx.)}$$

Example 11. The minute hand of a watch is 35 cm long. How far does its tip move in 18 minutes? $\left(\text{use } \pi = \frac{22}{7}\right)$

Sol. The minute hand of a watch completes one revolution in 60 minutes. Therefore the angle traced by a minute hand in 60 minutes = 360° = 2π radians.

$\therefore$ Angle traced by the minute hand in 18 minutes

$$= 2\pi \times \frac{18}{60} \text{ radians} = \frac{3\pi}{5} \text{ radians}$$

Let the distance moved by the tip in 18 minutes be l, then

$l = r\theta$

$$= 35 \times \frac{3\pi}{5} = 21\pi = 21 \times \frac{22}{7} = 66 \text{ cm}$$

Example 12. The wheel of a railway carriage is 40 cm. in diameter and makes 6 revolutions in a second; how fast is the train going?

Sol. Diameter of the wheel = 40 cm

$\therefore$ radius of the wheel = 20 cm

Circumference of the wheel = $2\pi r = 2\pi \times 20 = 40\pi$ cm

Number of revolutions made in 1 second = 6

$\therefore$ Distance covered in 1 second = $40\pi \times 6 = 240\pi$ cm

$\therefore$ Speed of the train = 240π cm/sec.

Example 13. Assuming that a person of normal sight can read print at such a distance that the letters subtend an angle of 5′ at his eye, find the height of the letters that he can read at a distance of 12 metres.

Sol. Let the height of the letters be h metres.

Now, h many be considered as the arc of a circle of radius 12 m, which subtends an angle of 5′ at its centre.

$$\therefore \quad \theta = 5' = \left(\frac{5}{60} \times \frac{\pi}{180}\right) \text{radians} = \left(\frac{\pi}{12 \times 180}\right) \text{radian}$$

and $\quad r = 12$ m

$$\therefore \quad h = r\theta = 12 \times \frac{\pi}{12 \times 180} = \left(\frac{\pi}{180}\right) \text{metres} \approx 1.7 \text{ cm}$$

Exercise for Session 1

1. The difference between two acute angles of a right angle triangle is $\frac{3\pi}{10}$ rad. Find the angles in degree.
2. Find the length of an arc of a circle of radius 6 cm subtending an angle of 15° at the centre.
3. A horse is tied to post by a rope. If the horse moves along circular path always keeping the tight and describes 88 m, when it has traced out 72° at centre, find the length of rope.
4. Find the angle between the minute hand and hour hand of a clock, when the time is 7 : 30 pm.
5. If OQ makes 4 revolutions in 1s, find the angular velocity in radians per second.
6. If a train is moving on the circular path of 1500 m radius at the rate of 66 km/h, find the angle in radian, if it has in 10 second.
7. Find the distance from the eye at which a coin of 2.2 cm diameter should be held so as to conceal the full moon with angular diameter 30′.
8. The wheel of a railway carriage is 40 cm in diameter and makes 7 revolutions in a second, find the speed of train.
9. Assuming that a person of normal sight can read print at such a distance that the letters subtend an angle of 5′ at his eye, find the height of letters that he can read a distance of 12 m.
10. For each natural number k, let C_K denotes the circle with radius k cm and centre at origin. On the circle C_K, a particle moves k cm in the counter-clockwise direction. After completing its motion on C_K, the particle moves on C_{K+1} in the radial direction. The motion of the particle continues in this manner. The particle starts at (1, 0). If the particle crosses the positive direction of the x-axis for the first time on the circle C_n, then n is equal to

Session 2

Definition of Trigonometric Functions

Definition of Trigonometric Functions

An angle whose measure is greater than 0° but less than 90° is called an acute angle.

In a right angled triangle ABC, $\angle CAB = A$ and $\angle BCA = 90^\circ = \pi/2$. AC is the base, BC the altitude and AB is the hypotenuse. We refer to the base as the adjacent side and to the altitude as the opposite side. There are six trigonometric ratios, also called **trigonometric functions or circular functions** with reference to $\angle A$, the six ratio are

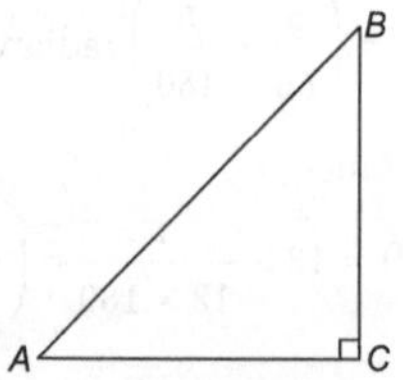

$\dfrac{BC}{AB} = \dfrac{\text{opposite side}}{\text{hypotenuse}}$, is called sine of A, and written as $\sin A$.

$\dfrac{AC}{AB} = \dfrac{\text{adjacent side}}{\text{hypotenuse}}$, is called the cosine of A, and written as $\cos A$.

$\dfrac{BC}{AC} = \dfrac{\text{opposite side}}{\text{adjacent side}}$, is called the tangent of A, and written as $\tan A$.

$\dfrac{AB}{BC} = \dfrac{\text{hypotenuse}}{\text{opposite side}}$, is called cosecant of A, and written as $\text{cosec } A$.

$\dfrac{AB}{AC} = \dfrac{\text{hypotenuse}}{\text{opposite side}}$, is called secant of A, and written as $\sec A$.

$\dfrac{AC}{BC} = \dfrac{\text{adjacent side}}{\text{opposite side}}$, is called cotangent of A, and written as $\cot A$.

Since, the hypotenuse is the greatest side in a right angle triangle, $\sin A$ and $\cos A$ can never be greater than unity and $\text{cosec } A$ and $\sec A$ can never be less than unity.

Hence, $|\sin A| \le 1$, $\cos A| \le 1|$, $|\text{cosec } A| \ge 1$, $|\sec A| \ge 1$, while $\tan A$ and $\cot A$ may have any numerical value lying between $-\infty$ to $+\infty$.

Note

Student must remember the following results

(i) $-1 \le \sin A \le 1$ (ii) $-1 \le \cos A \le 1$

(iii) $\text{cosec } A \ge 1$ or $\text{cosec } A \le -1$ (iv) $\sec A \ge 1$ or $\sec A \le -1$

(v) $\tan A \in R$ (vi) $\cot A \in R$

Some values of Trigonometrical Ratios

Students are already familiar with the values of sin, cos, tan, cot, sec and cosec of angles 0°, 30°, 45°, 60° and 90° which have been given in the following table

	0°	30°	45°	60°	90°
sin	0	$\frac{1}{2}$	$\frac{1}{\sqrt{2}}$	$\frac{\sqrt{3}}{2}$	1
cos	1	$\frac{\sqrt{3}}{2}$	$\frac{1}{\sqrt{2}}$	$\frac{1}{2}$	0
tan	0	$\frac{1}{\sqrt{3}}$	1	$\sqrt{3}$	undefined
cot	undefined	$\sqrt{3}$	1	$\frac{1}{\sqrt{3}}$	0
sec	1	$\frac{2}{\sqrt{3}}$	$\sqrt{2}$	2	undefined
cosec	undefined	2	$\sqrt{2}$	$\frac{2}{\sqrt{3}}$	1

Trigonometric Identities

Trigonometric identities are equalities that involve trigonometric functions that are true for every single value of the occurring variables. In other words, they are equations that hold true regardless of the value of the angles being chosen.

Trigonometric identities are as follows

1. $\sin^2 A + \cos^2 A = 1 \Rightarrow \cos^2 A = 1 - \sin^2 A$

   or $\sin^2 A = 1 - \cos^2 A$

2. $1 + \tan^2 A = \sec^2 A \Rightarrow \sec^2 A - \tan^2 A = 1$

3. $\cot^2 A + 1 = \text{cosec}^2 A$

   $\Rightarrow \text{cosec}^2 A - \cot^2 A = 1$

4. $\tan A = \dfrac{\sin A}{\cos A}$ and $\cot A = \dfrac{\cos A}{\sin A}$
5. Fundamental inequalities: For $0 < A < \pi / 2$;
$$0 < \cos A < \frac{\sin A}{A} < \frac{1}{\cos A}$$
6. It is possible to express trigonometrical ratios in terms of any one of them as,
$$\sin\theta = \frac{1}{\sqrt{1+\cot^2\theta}},$$
$$\cos\theta = \frac{\cot\theta}{\sqrt{1+\cot^2\theta}}, \tan\theta = \frac{1}{\cot\theta},$$
$$\operatorname{cosec}\theta = \sqrt{1+\cot^2\theta}, \sec\theta = \frac{\sqrt{1+\cot^2\theta}}{\cot\theta}$$
i.e. all trigonometrical functions have been expressed in terms of $\cot\theta$.

Similarly, we can express all trigonometric function in other trigonometric ratios.

Example 14. Show that $2(\sin^6 x + \cos^6 x) - 3(\sin^4 x + \cos^4 x) + 1 = 0.$

Sol. $2(\sin^6 x + \cos^6 x) - 3(\sin^4 x + \cos^4 x) + 1$
$$= 2[(\sin^2 x)^3 + (\cos^2 x)^3] - 3(\sin^4 x + \cos^4 x) + 1$$
$$= 2[(\sin^2 x + \cos^2 x)^3 - 3\sin^2 x \cos^2 x$$
$$(\sin^2 x + \cos^2 x) - 3[\sin^2 x + \cos^2 x)^2 - 2\sin^2 x \cos^2 x] + 1$$
$$= 2[1 + 3\sin^2 x \cos^2 x] - 3[1 - 2\sin^2 x \cos^2 x] + 1 = 0$$

Example 15. Show that

(i) $\sin^8 A - \cos^8 A = (\sin^2 A - \cos^2 A)(1 - 2\sin^2 A \cdot \cos^2 A)$

(ii) $\dfrac{1}{\sec A - \tan A} - \dfrac{1}{\cos A} = \dfrac{1}{\cos A} - \dfrac{1}{\sec A + \tan A}$

Sol. (i) L.H.S. $= \sin^8 A - \cos^8 A = (\sin^4 A)^2 - (\cos^4 A)^2$
$$= (\sin^4 A - \cos^4 A)(\sin^4 A + \cos^4 A)$$
$$= (\sin^2 A - \cos^2 A)(\sin^2 A + \cos^2 A)$$
$$[(\sin^2 A + \cos^2 A)^2 - 2\sin^2 A \cos^2 A]$$
$$= (\sin^2 A - \cos^2 A)(1 - 2\sin^2 A \cos^2 A)$$
$$[\because \sin^2 A + \cos^2 A = 1]$$

(ii) Given, $\dfrac{1}{\sec A - \tan A} - \dfrac{1}{\cos A} = \dfrac{1}{\cos A} - \dfrac{1}{\sec A + \tan A}$

or $\dfrac{1}{\sec A - \tan A} + \dfrac{1}{\sec A + \tan A} = \dfrac{1}{\cos A} + \dfrac{1}{\cos A}$

Here, R.H.S. $= \dfrac{2}{\cos A}$

Now L.H.S. $= \dfrac{1}{\sec A - \tan A} \; \dfrac{1}{\sec A + \tan A}$
$$= \frac{\sec A + \tan A + \sec A - \tan A}{(\sec A - \tan A)(\sec A + \tan A)} = \frac{2}{\cos A}$$
Thus, L.H.S. = R.H.S.

Example 16. If $\tan\theta + \sec\theta = 1.5$, find $\sin\theta$, $\tan\theta$ and $\sec\theta$.

Sol. Given, $\sec\theta + \tan\theta = \dfrac{3}{2}$ ...(i)

Now, $\sec\theta - \tan\theta = \dfrac{1}{\sec\theta + \tan\theta} = \dfrac{2}{3}$ ...(ii)

Adding Eqs. (i) and (ii), we get
$$2\sec\theta = \frac{3}{2} + \frac{2}{3} = \frac{13}{6}$$
$\therefore$ $\sec\theta = \dfrac{13}{12}$

$\therefore$ $\tan\theta = \dfrac{5}{12}$

and $\sin\theta = \dfrac{5}{13}$

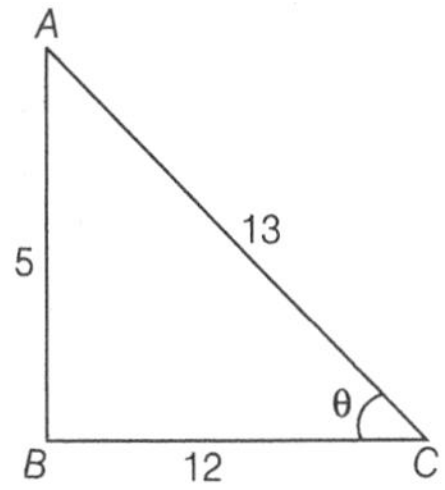

Example 17. If $\dfrac{\cos^4 A}{\cos^2 B} + \dfrac{\sin^4 A}{\sin^2 B} = 1$, then prove that

(i) $\sin^4 A + \sin^4 B = 2\sin^2 A \sin^2 B$

(ii) $\dfrac{\cos^4 B}{\cos^2 A} + \dfrac{\sin^4 B}{\sin^2 A} = 1$

Sol. Given, $\dfrac{\cos^4 A}{\cos^2 B} + \dfrac{\sin^4 A}{\sin^2 B} = 1\,(\cos^2 A + \sin^2 A)$

or $\dfrac{\cos^4 A}{\cos^2 B} - \cos^2 A = \sin^2 A - \dfrac{\sin^4 A}{\sin^2 B}$

or $\dfrac{\cos^2 A(\cos^2 A - \cos^2 B)}{\cos^2 B} = \sin^2 A \dfrac{(\sin^2 B - \sin^2 A)}{\sin^2 B}$

or $\dfrac{\cos^2 A}{\cos^2 B}(\cos^2 A - \cos^2 B) = \dfrac{\sin^2 A}{\sin^2 B}$
$$[(1 - \cos^2 B) - (1 - \cos^2 A)]$$

or $\dfrac{\cos^2 A}{\cos^2 B}(\cos^2 A - \cos^2 B) = \dfrac{\sin^2 A}{\sin^2 B}(\cos^2 A - \cos^2 B)$

or $(\cos^2 A - \cos^2 B)\left(\dfrac{\cos^2 A}{\cos^2 B} - \dfrac{\sin^2 A}{\sin^2 B}\right) = 0$

When $\cos^2 A - \cos^2 B = 0$, we have
$$\cos^2 A = \cos^2 B \quad \text{...(i)}$$

When $\dfrac{\cos^2 A}{\cos^2 B} - \dfrac{\sin^2 A}{\sin^2 B} = 0$, we have
$$\cos^2 A \sin^2 B = \sin^2 A \cos^2 B$$

or $\cos^2 A(1-\cos^2 B) = (1-\cos^2 A)\cos^2 B$

or $\cos^2 A - \cos^2 A\cos^2 B = \cos^2 B - \cos^2 A\cos^2 B$

or $\cos^2 A = \cos^2 B$ ...(ii)

Thus, in both the cases, $\cos^2 A = \cos^2 B$. Therefore,

$\therefore \quad 1-\sin^2 A = 1-\sin^2 B$ or $\sin^2 A = \sin^2 B$ ...(iii)

(i) L.H.S. $= \sin^4 A + \sin^4 B$

$= (\sin^2 A - \sin^2 B)^2 + 2\sin^2 A\sin^2 B$

$= 2\sin^2 A\sin^2 B = $ R.H.S. $[\because \sin^2 A = \sin^2 B]$

(ii) L.H.S. $= \dfrac{\cos^4 B}{\cos^2 A} + \dfrac{\sin^4 B}{\sin^2 A} = \dfrac{\cos^4 B}{\cos^2 B} + \dfrac{\sin^4 B}{\sin^2 B}$

$= \cos^2 B + \sin^2 B = 1 = $ R.H.S.

Example 18. If $\tan^2\theta = 1 - e^2$, prove that $\sec\theta + \tan^3\theta\operatorname{cosec}\theta = (2-e^2)^{\frac{3}{2}}$

Sol. Given, $\tan^2\theta = 1 - e^2$

Now, L.H.S. $= \sec\theta + \tan^3\theta\operatorname{cosec}\theta$

$= \sec\theta\left(1 + \tan^3\theta\,\dfrac{\operatorname{cosec}\theta}{\sec\theta}\right)$

$= \sec\theta(1+\tan^3\theta\cdot\cot\theta) = \sec\theta(1+\tan^2\theta) = \sec\theta\sec^2\theta$

$= \sec^3\theta = (\sec^2\theta)^{\frac{3}{2}} = (1+\tan^2\theta)^{\frac{3}{2}} = (1+1-e^2)^{\frac{3}{2}} = (2-e^2)^{\frac{3}{2}}$

Example 19. For what real values of x and y is the equation $\sec^2\theta = \dfrac{4xy}{(x+y)^2}$ possible?

Sol. Here, $\sec^2\theta = \dfrac{4xy}{(x+y)^2}$

We know $\sec^2\theta \ge 1$ and $\dfrac{4xy}{(x+y)^2} \le 1$ [as AM $\ge$ GM]

$\Rightarrow \quad \sec^2\theta = \dfrac{4xy}{(x+y)^2}$ is only possible if $\sec^2\theta = 1$

i.e. $\dfrac{4xy}{(x+y)^2} = 1, \forall\, x, y \in R^+$

or $4xy = (x+y)^2 \quad \forall\, x, y \in R^+$

$\Rightarrow \quad x^2 + y^2 + 2xy - 4xy = 0, \quad \forall\, x, y \in R^+$

$\Rightarrow \quad (x-y)^2 = 0, \quad \forall\, x, y \in R^+$

or $x = y; \quad \forall\, x, y \in R^+$

Example 20. Show that the equation $\sin\theta = x + \dfrac{1}{x}$ is impossible if x is real.

Sol. Given, $\sin\theta = x + \dfrac{1}{x}$

$\therefore \quad \sin^2\theta = x^2 + \dfrac{1}{x^2} + 2x\cdot\dfrac{1}{x}$

$= x^2 + \dfrac{1}{x^2} + 2 \ge 2$

which is not possible since $\sin^2\theta \le 1$

Exercise for Session 2

1. Prove that $(\operatorname{cosec}\theta - \sin\theta)(\sec\theta - \cos\theta)(\tan\theta + \cot\theta) = 1$.
2. If $\cos^2\alpha - \sin^2\alpha = \tan^2\beta$, then show that $\tan^2\alpha = \cos^2\beta - \sin^2\beta$.
3. If $\sin^6\theta + \cos^6\theta - 1 = \lambda\sin^2\theta\cos^2\theta$, find the value of λ.
4. If $a\cos\theta - b\sin\theta = c$, then find the value of $a\sin\theta + b\cos\theta$.
5. Find the value of $3(\sin x - \cos x)^4 + 6(\sin x + \cos x)^2 + 4(\sin^6 x + \cos^6 x)$.
6. If $\sin\theta + \operatorname{cosec}\theta = 2$, then find the value of $\sin^{20}\theta + \operatorname{cosec}^{20}\theta$.
7. Let $F_k(x) = \dfrac{1}{k}(\sin^k x + \cos^k x)$, where $x \in R$ and $k \ge 1$, then find the value of $F_4(x) - F_6(x)$.
8. If $\dfrac{\sin^4 x}{2} + \dfrac{\cos^4 x}{3} = \dfrac{1}{5}$, then show that $\dfrac{\sin^8 x}{8} + \dfrac{\cos^8 x}{27} = \dfrac{1}{125}$.
9. If $\cot\theta + \tan\theta = x$ and $\sec\theta - \cos\theta = y$, then show that $\sin\theta\cdot\cos\theta = \dfrac{1}{x}$ or $\sin\theta\cdot\tan\theta = y$ or $(x^2y)^{2/3} - (xy^2)^{2/3} = 1$.
10. If $\sin A + \sin^2 A + \sin^3 A = 1$, then find the value of $\cos^6 A - 4\cos^4 A + 8\cos^2 A$.

Session 3

Application of Basic Trigonometry on Eliminating Variables or Parameters and Geometry

Application of Basic Trigonometry on Eliminating Variables or Parameters

As we know, parameter are those values which could vary, e.g. θ if parameter could take any value as;

$$\theta = 0^\circ, 30^\circ, 45^\circ, 60^\circ, 90^\circ, 120^\circ, \ldots$$

Thus, to eliminate these parameter, we have to use basic trigonometric formulae, it could be more clear by some examples :

Example 21. If $\operatorname{cosec}\theta - \sin\theta = m$ and $\sec\theta - \cos\theta = n$, eliminate θ.

Sol. Given, $\operatorname{cosec}\theta - \sin\theta = m$ or, $\dfrac{1}{\sin\theta} - \sin\theta = m$

$$\text{or,} \quad \frac{1-\sin^2\theta}{\sin\theta} = m \text{ or } \frac{\cos^2\theta}{\sin\theta} = m \qquad \ldots\text{(i)}$$

Again $\quad \sec\theta - \cos\theta = n$

$$\text{or} \quad \frac{1}{\cos\theta} - \cos\theta = n$$

$$\text{or} \quad \frac{1-\cos^2\theta}{\cos\theta} = n \text{ or } \frac{\sin^2\theta}{\cos\theta} = n \qquad \ldots\text{(ii)}$$

From Eq. (i) $\sin\theta = \dfrac{\cos^2\theta}{m}$ $\qquad \ldots$(iii)

Putting in (ii), we get $\dfrac{\cos^4\theta}{m^2\cos\theta} = n$ or, $\cos^3\theta = m^2 n$

$\therefore$ $\cos\theta = (m^2n)^{\frac{1}{3}}$ or, $\cos^2\theta = (m^2n)^{\frac{2}{3}}$ $\qquad \ldots$(iv)

From Eq. (iii), $\sin\theta = \dfrac{\cos^2\theta}{m} = \dfrac{(m^2n)^{\frac{2}{3}}}{m}$

$$= \frac{m^{\frac{4}{3}}n^{\frac{2}{3}}}{m} = m^{\frac{1}{3}}n^{\frac{2}{3}} = (mn^2)^{\frac{1}{3}}$$

$$\therefore \quad \sin^2\theta = (mn^2)^{\frac{2}{3}} \qquad \ldots\text{(v)}$$

Adding Eqs. (iv) and (v), we get

$$(m^2n)^{\frac{2}{3}} + (mn^2)^{\frac{2}{3}} = \cos^2\theta + \sin^2\theta$$

$$\text{or,} \quad (m^2n)^{\frac{2}{3}} + (mn^2)^{\frac{2}{3}} = 1$$

Example 22. If $3\sin\theta + 4\cos\theta = 5$, then find the value of $4\sin\theta - 3\cos\theta$.

Sol. Let $4\sin\theta - 3\cos\theta = a$ $\qquad \ldots$(i)

Thus, we want to eliminate θ from both $3\sin\theta + 4\cos\theta = 5$ and $4\sin\theta - 3\cos\theta = a$, i.e. squaring and adding these equations, we get

$$(3\sin\theta + 4\cos\theta)^2 + (4\sin\theta - 3\cos\theta)^2 = 25 + a^2$$

$$9\sin^2\theta + 16\cos^2\theta + 24\sin\theta\cos\theta$$

$$+ 16\sin^2\theta + 9\cos^2\theta - 24\cos\theta\sin\theta = 25 + a^2$$

$$9 + 16 = 25 + a^2 \quad \text{or} \quad a^2 = 0$$

$$a = 0$$

$$\therefore \quad 4\sin\theta - 3\cos\theta = 0$$

Example 23. If $a\sec\alpha - c\tan\alpha = d$ and $b\sec\alpha + d\tan\alpha = c$, then eliminate α from above equations.

Sol. Here, $a\sec\alpha - c\tan\alpha = d$ and $b\sec\alpha + d\tan\alpha = c$ could be written as

$$a = d\cos\alpha + c\sin\alpha \qquad \ldots\text{(i)}$$

$$\text{and} \quad b = c\cos\alpha - d\sin\alpha \qquad \ldots\text{(ii)}$$

On squaring and adding Eqs. (i) and (ii), we get

$$a^2 + b^2 = (d\cos\alpha + c\sin\alpha)^2 + (c\cos\alpha - d\sin\alpha)^2$$

$$\Rightarrow \quad a^2 + b^2 = d^2\cos^2\alpha + c^2\sin^2\alpha + 2dc\cos\alpha\sin\alpha$$

$$+c^2\cos^2\alpha + d^2\sin^2\alpha - 2cd\cos\alpha\sin\alpha.$$

$$= d^2(\cos^2\alpha + \sin^2\alpha) + c^2(\sin^2\alpha + \cos^2\alpha)$$

$$\therefore \quad a^2 + b^2 = c^2 + d^2$$

Example 24. Eliminate θ between the equations $a\sec\theta + b\tan\theta + c = 0$ and $p\sec\theta + q\tan\theta + r = 0$.

Sol. Given $\quad a\sec\theta + b\tan\theta + c = 0$ $\qquad \ldots$(i)

and $\quad p\sec\theta + q\tan\theta + r = 0$ $\qquad \ldots$(ii)

Solving Eqs. (i) and (ii) by cross multiplication method, we have

$$\underset{\text{(i)}}{\frac{\sec\theta}{br - qc}} = \underset{\text{(ii)}}{\frac{\tan\theta}{pc - ar}} = \underset{\text{(iii)}}{\frac{1}{aq - pb}}$$

From Eqs. (i) and (iii), we get

$$\sec\theta = \frac{br - qc}{aq - pb} \qquad \ldots\text{(iii)}$$

From Eqs. (ii) and (iii), we get

$$\tan\theta = \frac{pc - ar}{pc - pb} \qquad \text{...(iv)}$$

$\because \qquad \sec^2\theta - \tan^2\theta = 1$

$$\therefore \qquad \left(\frac{br - qc}{aq - pb}\right)^2 - \left(\frac{pc - ar}{aq - pb}\right)^2 = 1$$

$$\text{or} \qquad (br - qc)^2 - (pc - ar)^2 = (aq - pb)^2$$

Example 25. If $x = \sec\theta - \tan\theta$ and $y = \text{cosec}\,\theta + \cot\theta$, then prove that $xy + 1 = y - x$.

Sol. $xy + 1 = \left(\dfrac{1 - \sin\theta}{\cos\theta}\right)\left(\dfrac{1 + \cos\theta}{\sin\theta}\right) + 1 = \dfrac{1 - \sin\theta + \cos\theta}{\sin\theta\cos\theta}$

$$= \frac{(\sin^2\theta + \cos^2\theta)}{\sin\theta\cos\theta} - \frac{(\sin\theta - \cos\theta)}{\sin\theta\cos\theta}$$

$$= (\tan\theta + \cot\theta) - (\sec\theta - \text{cosec}\,\theta)$$

$$= (\text{cosec}\,\theta + \cot\theta) - (\sec\theta - \tan\theta) = y - x$$

Example 26. If $x = r\sin\theta\cos\phi$, $y = r\sin\theta\sin\phi$ and $z = r\cos\theta$. Find the value of $x^2 + y^2 + z^2$.

Sol. Here,

$$x^2 + y^2 + z^2 = r^2\sin^2\theta\cos^2\phi + r^2\sin^2\theta\sin^2\phi + r^2\cos^2\theta$$

$$= r^2\sin^2\theta(\cos^2\phi + \sin^2\phi) + r^2\cos^2\theta$$

$$= r^2\sin^2\theta + r^2\cos^2\theta$$

$$= r^2$$

$\therefore \; x^2 + y^2 + z^2 = r^2$

Example 27. If $0 < \theta < \dfrac{\pi}{2}$, $x = \displaystyle\sum_{n=0}^{\infty} \cos^{2n}\theta$, $y = \displaystyle\sum_{n=0}^{\infty} \sin^{2n}\theta$ and $z = \displaystyle\sum_{n=0}^{\infty} \cos^{2n}\theta \cdot \sin^{2n}\theta$, then show $xyz = xy + z$.

Sol. Here, $x = \displaystyle\sum_{n=0}^{\infty} \cos^{2n}\theta = 1 + \cos^2\theta + \cos^4\theta + \cos^6\theta + \ldots\infty$

$$= \frac{1}{1 - \cos^2\theta} = \frac{1}{\sin^2\theta}$$

$$\left[\text{using, } S_\infty = \frac{a}{1 - r} \text{ sum of infinite GP}\right]$$

Similarly, $\quad y = \dfrac{1}{1 - \sin^2\theta} = \dfrac{1}{\cos^2\theta}$

and $\quad z = \dfrac{1}{1 - \sin^2\theta \cdot \cos^2\theta}$

$$\therefore \qquad xyz = \frac{1}{\sin^2\theta\cos^2\theta\,(1 - \sin^2\theta\cos^2\theta)}$$

$$= \frac{(1 - \sin^2\theta\cos^2\theta) + (\sin^2\theta\cos^2\theta)}{\sin^2\theta\cos^2\theta\,(1 - \sin^2\theta\cos^2\theta)}$$

$$= \frac{1}{\sin^2\theta\cos^2\theta} + \frac{1}{1 - \sin^2\theta\cos^2\theta}$$

$$= xy + z$$

Application of Basic Trigonometry in Geometry

Example 28. If in given fig, $\tan(\angle BAO) = 3$, then find the ratio $BC : CA$.

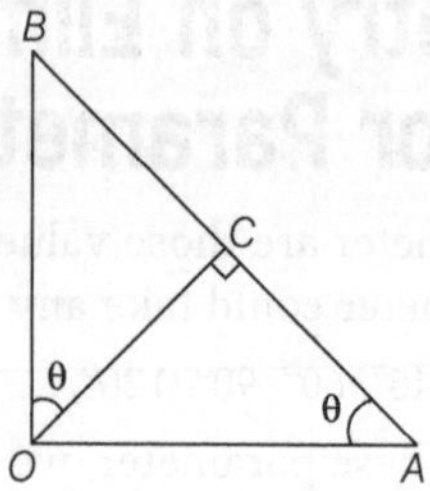

Sol. From Fig., we have

$$\tan\theta = 3$$

In ΔOCA and ΔOCB respectively, we get

$$\therefore \qquad \frac{OC}{AC} = \tan\theta, \frac{OC}{BC} = \cot\theta$$

On dividing, we get

$$\text{or} \qquad \frac{BC}{AC} = \frac{\tan\theta}{\cot\theta} = \tan^2\theta = 9$$

$$\Rightarrow \qquad BC : AC = 9 : 1$$

Example 29. If angle C of triangle ABC is $90°$, then prove that $\tan A + \tan B = \dfrac{c^2}{ab}$ (where, a, b, c are sides opposite to angles A, B, C, respectively).

Sol. Draw ΔABC with $\angle C = 90°$. We have

$$\tan A + \tan B = \frac{a}{b} + \frac{b}{a} = \frac{a^2 + b^2}{ab} = \frac{c^2}{ab}$$

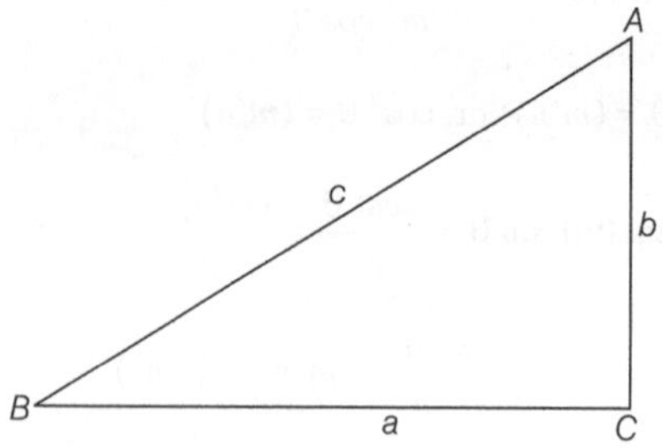

Example 30. In triangle ABC, $BC = 8$, $CA = 6$, and $AB = 10$. A line dividing the triangle ABC into two regions of equal area is perpendicular to AB at point X. Find the value of $\dfrac{BX}{\sqrt{2}}$.

Sol.

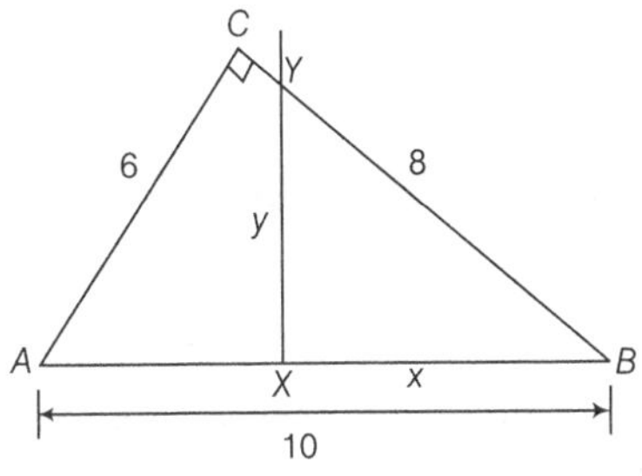

We have area of $\Delta XYB = \frac{1}{2}$ area of ΔABC

$$\therefore \quad \frac{1}{2}(XY)\cdot(XB) = \frac{1}{2}\times\frac{1}{2}\times AC \times BC$$

$$2\left(\frac{x \times y}{2}\right) = \frac{8\times 6}{2} = 24$$

or $\quad x \times x \tan B = 24 \quad [\because y = x \tan B]$

or $\quad x^2 \times \frac{3}{4} = 24 \quad [\because \tan B = \frac{AC}{BC}]$

or $\quad x^2 = 32$ or $x = 4\sqrt{2}$

Example 31. Let PQ and RS be tangents at the extremities of the diameter PR or a circle of radius r. If PS and RQ intersect at a point X on the circumference of the circle, then prove that $2r = \sqrt{PQ \times RS}$.

Sol. From Fig. we have

$$\frac{PQ}{PR} = \tan\left(\frac{\pi}{2} - \theta\right)$$

$$= \cot\theta \text{ and } \frac{RS}{PR} = \tan\theta$$

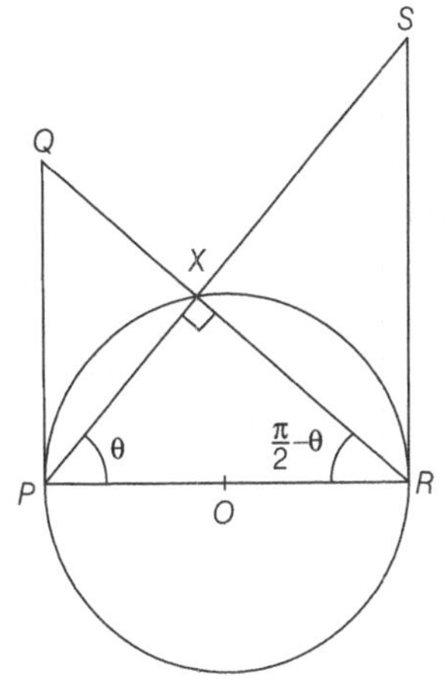

$$\therefore \quad \frac{PQ}{PR}\times\frac{RS}{PR} = 1$$

or $\quad (PR)^2 = PQ \times PS$

or $\quad (2r)^2 = PQ \times PS$

or $\quad 2r = \sqrt{PQ \times PS}$

Exercise for Session 3

1. If $\sec\theta + \tan\theta = k$, find the value of $\cos\theta$.
2. If $x\sin^3\theta + y\cos^3\theta = \sin\theta\cos\theta$ and $x\sin\theta = y\cos\theta$, Find the value of $x^2 + y^2$.
3. If $\sin A + \cos A = m$ and $\sin^3 A + \cos^3 A = n$, prove that $m^3 - 3m + 2n = 0$.
4. If $\sin^2\theta = \frac{x^2 + y^2 + 1}{2x}$. Find the value of x and y.
5. If $\sin\theta - \sqrt{6}\cos\theta = \sqrt{7}\cos\theta$. Prove that $\cos\theta + \sqrt{6}\sin\theta - \sqrt{7}\sin\theta = 0$.
6. If $\sin x + \sin y + \sin z = 3$. Find the value of $\cos x + \cos y + \cos z$.
7. If $\frac{x}{a}\cos\theta + \frac{y}{b}\sin\theta = 1$, $\frac{x}{a}\sin\theta - \frac{y}{b}\cos\theta = 1$, then eliminate θ.
8. If $a\sin^2 x + b\cos^2 x = c$, $b\sin^2 y + a\cos^2 y = d$ and $a\tan x = b\tan y$, then prove that $\frac{a^2}{b^2} = \frac{(d-a)(c-a)}{(b-c)(b-d)}$.
9. If $a + b\tan\theta = \sec\theta$ and $b - a\tan\theta = 3\sec\theta$, then find the value of $a^2 + b^2$.
10. Two circles of radii 4 cm and 1 cm touch each other externally and 8 is the angle contained by their direct common tangents. Find the value of $\sin\frac{\theta}{2} + \cos\frac{\theta}{2}$.

Session 4

Signs and Graph of Trigonometric Functions

Signs of Trigonometric Functions

The signs of the trigonometric ratios of an angle depend on the quadrant in which the terminal side of the angle lies. We always take $OP = r$ to be positive (see figure). Thus the signs of all the trigonometric ratios depend on the signs of x and/or y.

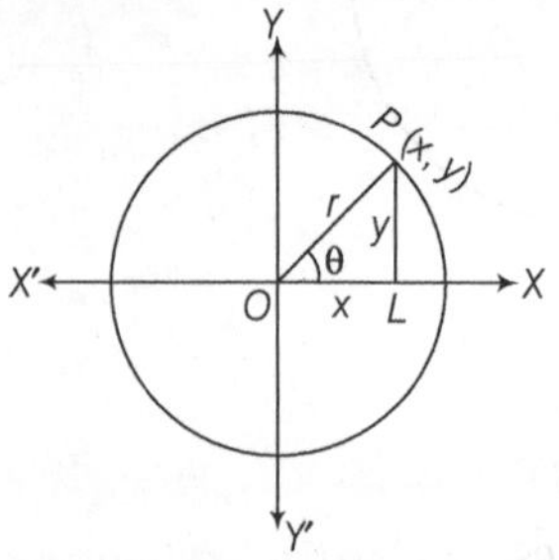

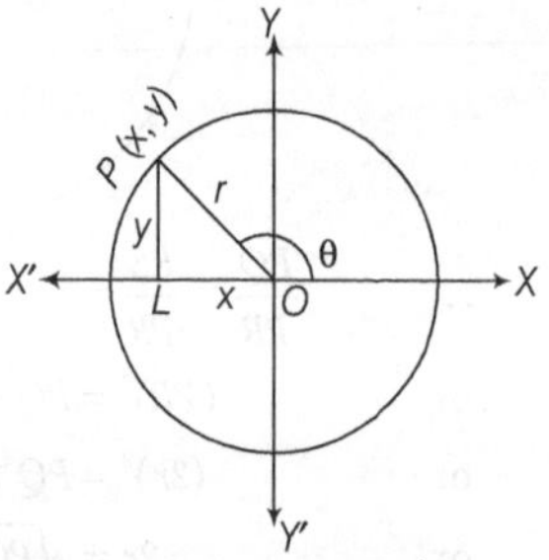

An angle is said to be in that quadrant in which its terminal ray lies

For positive acute angles this definition gives the same result as in case of a right angled triangle since x and y are both positive for any point in the first quadrant and consequently they are the length of base and perpendicular of the angle θ.

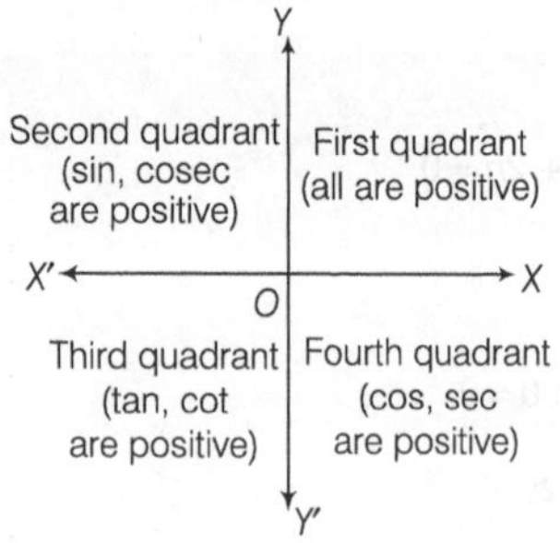

1. Clearly in first quadrant sin θ, cos θ, tan θ, cot θ, sec θ and cosec θ are all positive as x, y are positive.
2. In second quadrant, x is negative and y is positive, therefore, only sin θ and cosec θ are positive.
3. In third quadrant, x and y are both negative, therefore, only tan θ and cot θ are positive.
4. In fourth quadrant, x is positive and y is negative, therefore, only cos θ and sec θ are positive.

Quadrant →	**I**	**II**	**III**	**IV**
sin θ	+	+	–	–
cos θ	+	–	–	+
tan θ	+	–	+	–
cosec θ	+	+	–	–
sec θ	+	–	–	+
cot θ	+	–	+	–

Variation in the Values of Trigonometric Functions in Different Quadrants

We observe that in the first quadrant, as x increases from 0 to $\frac{\pi}{2}$, sin x increases from 0 to 1 and in the second quadrant as x increases from $\frac{\pi}{2}$ to π, sin x decreases from 1 to 0.

In the third quadrant, as x increases from π to $\frac{3\pi}{2}$, sin x decreases from 0 to – 1 and finally, in the fourth quadrant, sin x increases from – 1 to 0 as x increase from $\frac{3\pi}{2}$ to 2π.

Function	**1st quadrant**	**2nd quadrant**	**3rd quadrant**	**4th quadrant**
sin θ	↑ from 0 to 1	↓ from 1 to 0	↓ from 0 to – 1	↑ from – 1 to 0
cos θ	↓ from 1 to 0	↓ from 0 to – 1	↑ from – 1 to 0	↑ from 0 to 1
tan θ	↑ from 0 to ∞	↑ from – ∞ to 0	↑ from 0 to ∞	↑ from – ∞ to 0
cot θ	↓ from ∞ to 0	↓ from 0 to – ∞	↓ from ∞ to 0	↓ from 0 to – ∞
sec θ	↑ from 1 to ∞	↑ from – ∞ to – 1	↓ from – 1 to – ∞	↓ from ∞ to 1
cosec θ	↓ from ∞ to 1	↑ from 1 to ∞	↑ from – ∞ to – 1	↓ from – 1 to – ∞

Note

$+\infty$ and $-\infty$ are two symbols. These are not real numbers. When we say that $\tan\theta$ increases from 0 to ∞ as θ varies from 0 to $\frac{\pi}{2}$, it means that $\tan\theta$ increases in the interval $\left(0, \frac{\pi}{2}\right)$ and it attains arbitrarily large positive values as θ tends to $\frac{\pi}{2}$. This rule applies to other trigonometric functions also.

Graphs of Trigonometric Functions

As in case of algebraic function, we can have some idea about the nature of a trigonometric function by its graph. Graph has many important applications in mathematical problems. We shall discuss the graphs of trigonometrical functions. We know that $\sin x$, $\cos x$, $\sec x$ and $\operatorname{cosec} x$ are periodic functions with period 2π and $\tan x$ and $\cot x$ are trigonometric functions of period π. Also if the period of function $f(x)$ is T, then period of $f(ax+b)$ is $\frac{T}{|a|}$.

Graph and Other Useful Data of Trigonometric Functions

1. $y = f(x) = \sin x$

   Domain $\to R$,

   Range $\to [-1, 1]$

   Period $\to 2\pi$

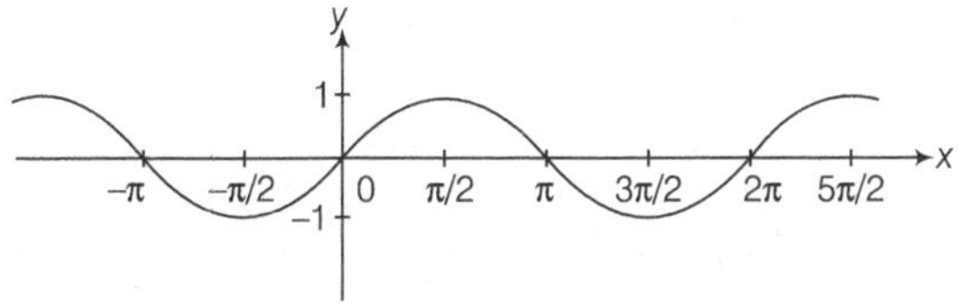

2. $y = f(x) = \cos x$

   Domain $\to R$, Range $\to [-1, 1]$

   Period $\to 2\pi$

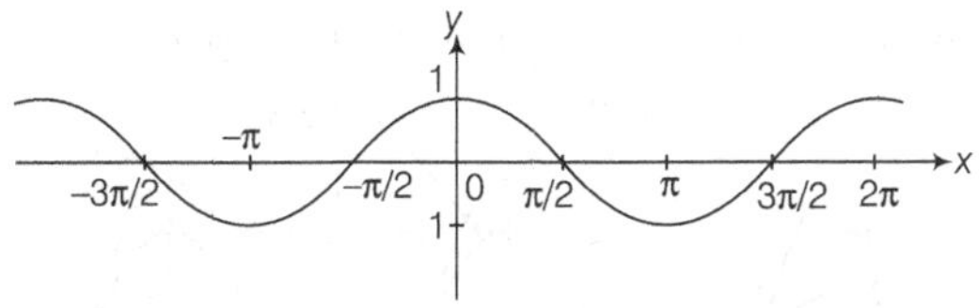

3. $y = f(x) = \tan x$

   Domain $\to R \sim (2n+1)\frac{\pi}{2}, n \in I$

   Range $\to (-\infty, \infty)$

   Period $\to \pi$

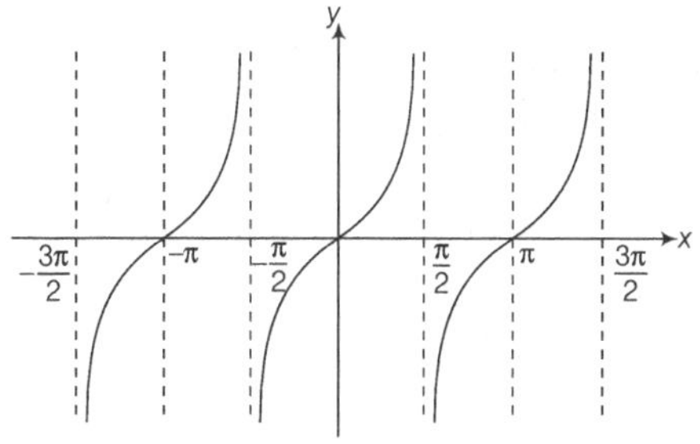

4. $y = f(x) = \cot x$

   Domain $\to R \sim n\pi, n \in I$; Range $\to (-\infty, \infty)$; Period $\to \pi$,

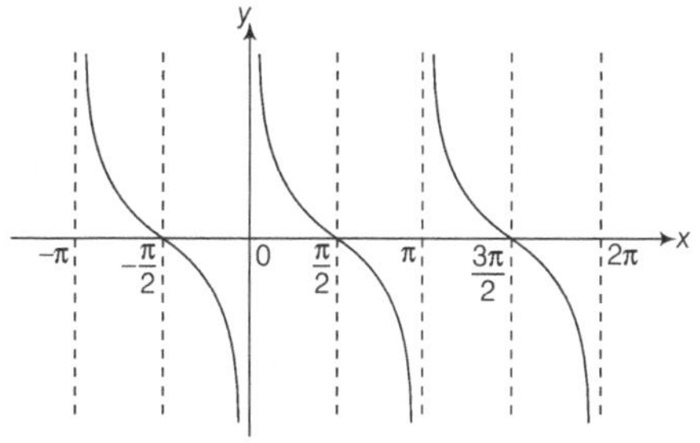

5. $y = f(x) = \sec x$

   Domain $\to R \sim (2n+1)\frac{\pi}{2}, n \in I$

   Range $\to (-\infty, -1] \cup [1, \infty)$

   Period $\to 2\pi$, $\sec^2 x$, $|\sec x| \in [1, \infty)$

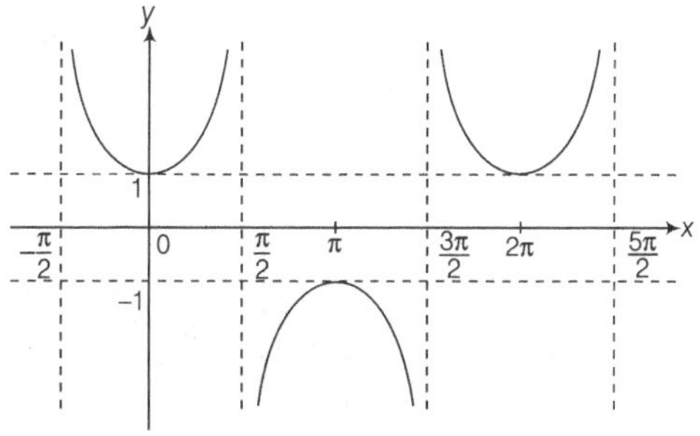

6. $y = f(x) = \operatorname{cosec} x$

   Domain $\to R \sim n\pi, n \in I$;

   Range $\to (-\infty, -1] \cup [1, \infty)$

   Period $\to 2\pi$, $\operatorname{cosec}^2 x$, $|\operatorname{cosec} x| \in [1, \infty)$

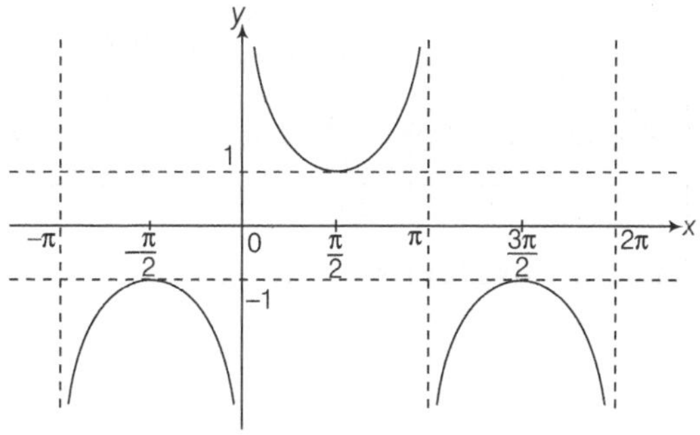

Transformation of the Graphs of Trigonometric Functions

1. To draw the graph of $y = f(x + a); (a > 0)$ from the graph of $y = f(x)$, shift the graph of $y = f(x)$, a units left along the x-axis.

   Consider the following illustration.

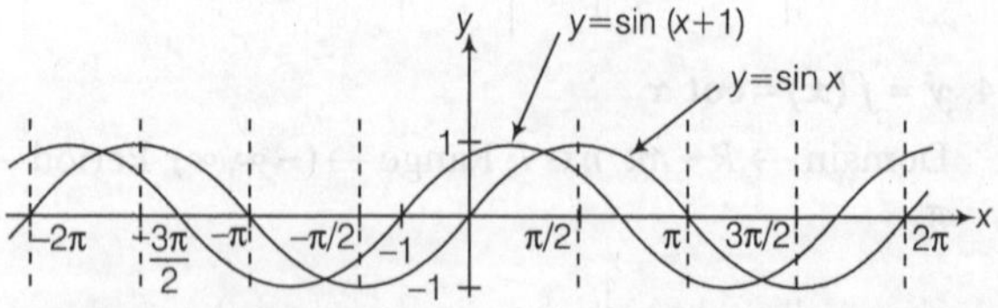

   To draw the graph of $y = f(x - a); (a > 0)$ from the graph of $y = f(x)$, shift the graph of $y = f(x)$, a units right along the x-axis.

   Consider the following illustration.

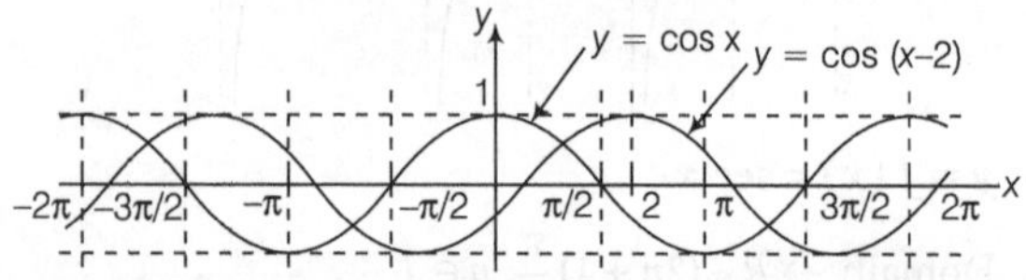

2. To draw the graph of $y = f(x) + a; (a > 0)$ from the graph of $y = f(x)$, shift the graph of $y = f(x)$, a units upwards along the y-axis.

   To draw the graph of $y = f(x) - a; (a > 0)$ from the graph of $y = f(x)$, shift the graph of $y = f(x)$, a units downward along the y-axis.

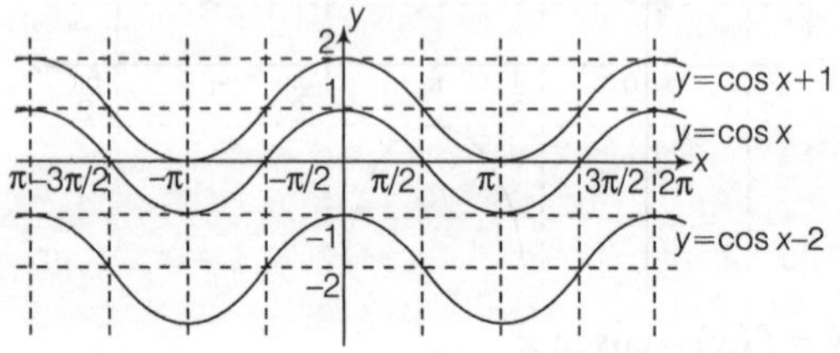

3. If $y = f(x)$ has period T, then period of $y = f(ax)$ is $\frac{T}{|a|}$.

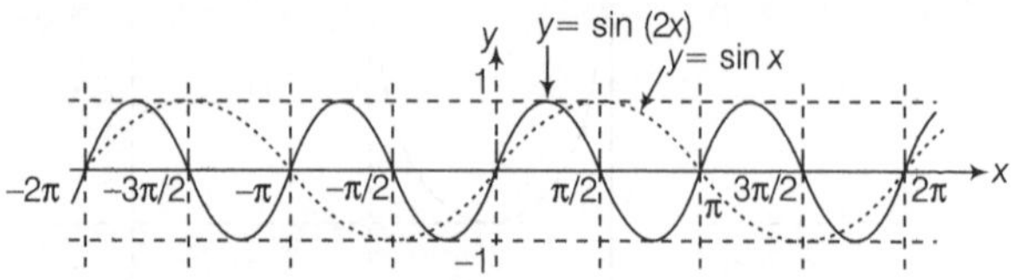

   Period of $y = \sin(2x)$ is $\frac{2\pi}{2} = \pi$

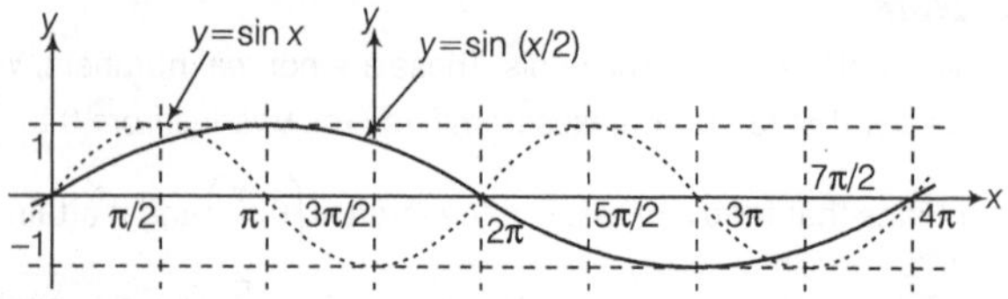

Period of $y = \sin\left(\frac{x}{2}\right)$ is $\frac{2\pi}{\left(\frac{1}{2}\right)} = 4\pi$

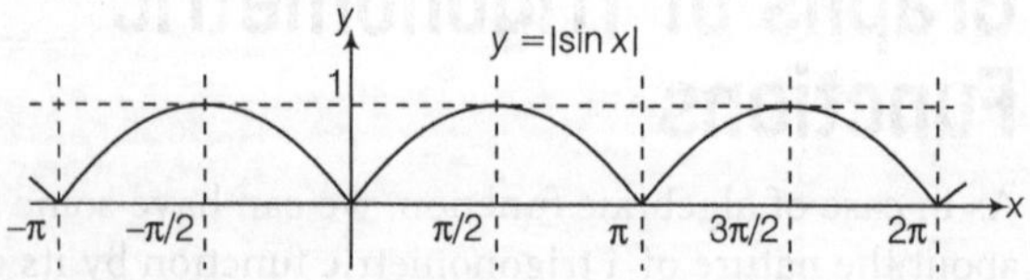

4. Since $y = |f(x)| \geq 0$, to draw the graph of $y = |f(x)|$, take the mirror of the graph of $y = f(x)$ in the x-axis for $f(x) < 0$, retaining the graph for $f(x) > 0$.

   Consider the following illustrations.

   Here, period of $f(x) = |\sin x|$ is π.

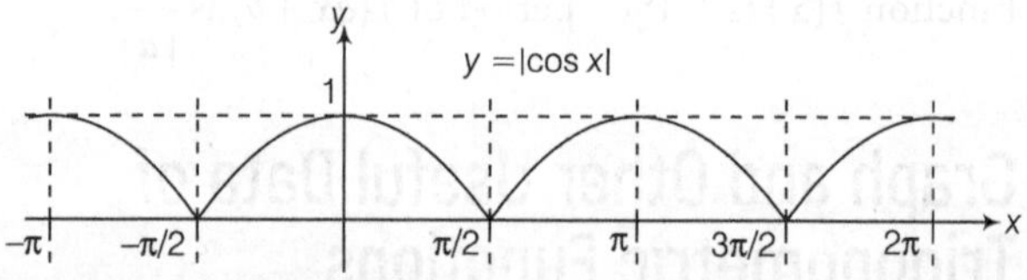

   Here, period of $f(x) = |\cos x|$ is π.

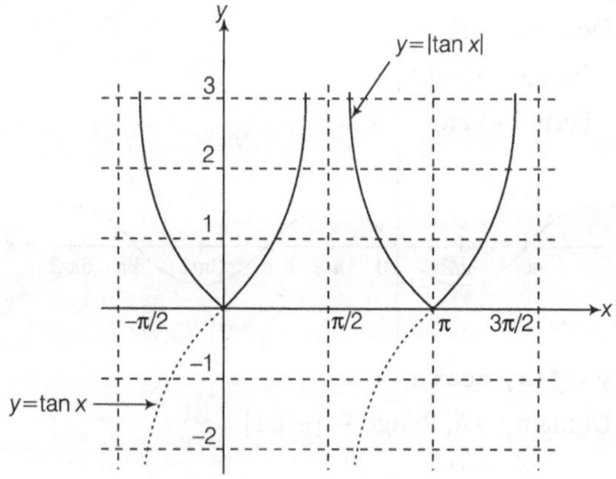

5. Graph of $y = af(x)$ from the graph of $y\ f(x)$

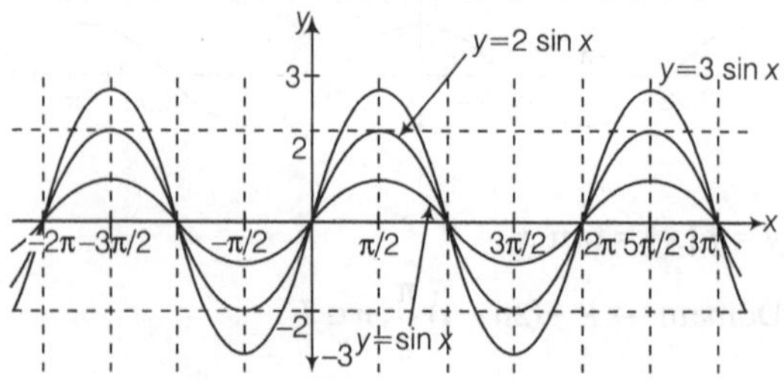

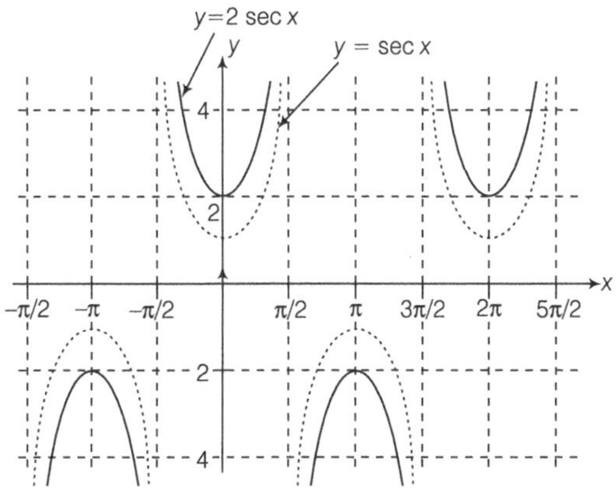

Some Important Graphical Deductions

To find relation between $\sin x$, x and $\tan x$

(i)

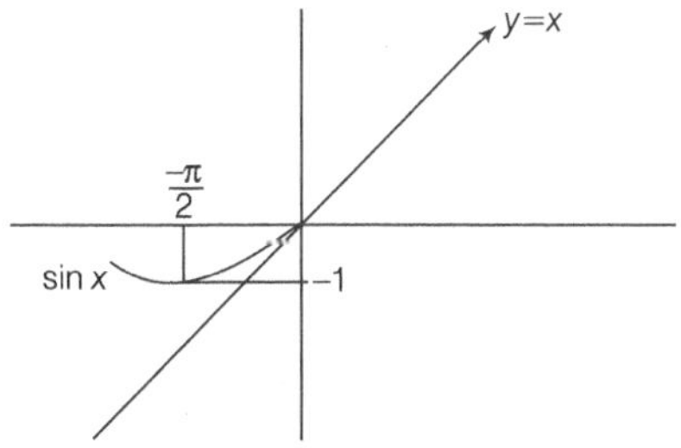

Thus, when $-\infty < x < 0$

$\Rightarrow \sin x > x$

(ii)

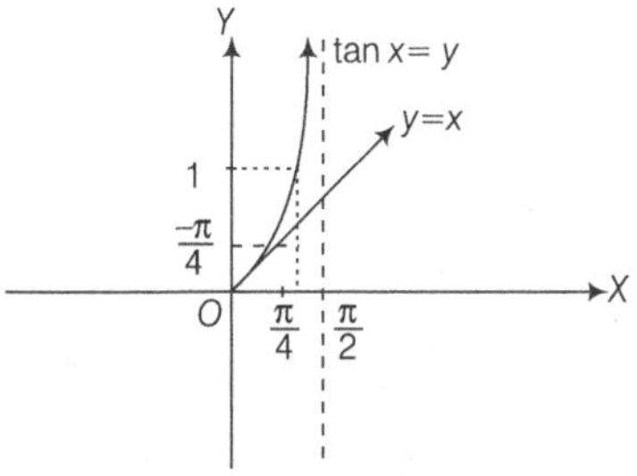

$\therefore \quad \tan x > x$, when $0 > x > \dfrac{\pi}{2}$

(iii) In general,

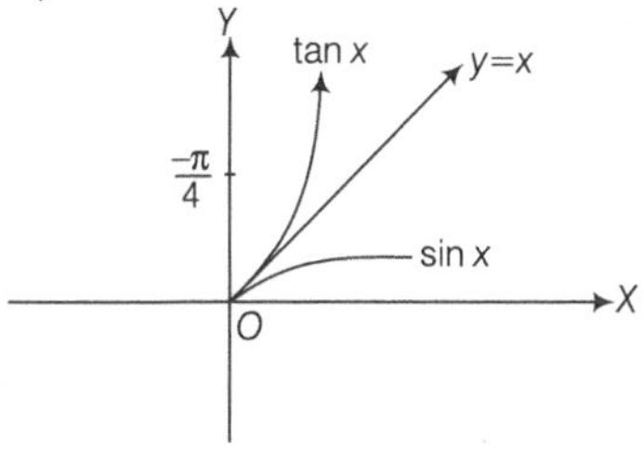

Thus, $\tan x > x > \sin x, \forall\, x \in \left(0, \dfrac{\pi}{2}\right)$

and $\sin x > x > \tan x, \forall\, x \in \left(-\dfrac{\pi}{2}, 0\right)$.

Example 32. Find the values of the other five trigonometric functions in each of the following questions

(i) $\tan\theta = \dfrac{5}{12}$, where θ is in third quadrant.

(ii) $\sin\theta = \dfrac{3}{5}$, where θ is in second quadrant.

Sol. (i) Since θ is in third quadrant,

$\therefore$ Only $\tan\theta$ and $\cot\theta$ are positive

Now, $\tan\theta = \dfrac{5}{12}$

Therefore, $\cot\theta = \dfrac{12}{5}$,

$\sin\theta = -\dfrac{5}{13}$,

$\text{cosec} = -\dfrac{13}{5}$

$\cos\theta = -\dfrac{12}{13}$ and $\sec\theta = -\dfrac{13}{12}$.

(ii) Since θ is in the second quadrant,

$\therefore$ Only $\sin\theta$ and $\text{cosec}\,\theta$ will be positive.

Now, $\sin\theta = \dfrac{3}{5}$.

Therefore,

$\text{cosec}\,\theta = \dfrac{5}{3}, \cos\theta = -\dfrac{4}{5}$,

$\sec\theta = -\dfrac{5}{4}, \tan\theta = -\dfrac{3}{4}$

and $\cot\theta = -\dfrac{4}{3}$.

Example 33. If $\sin\theta = \dfrac{12}{13}$ and θ lies in the second quadrant, find the value of $\sec\theta + \tan\theta$.

Sol. We have $\sin^2\theta + \cos^2\theta = 1$

$\Rightarrow \quad \cos\theta = \pm\sqrt{1-\sin^2\theta}$

In the second quadrant, $\cos\theta$ is negative

$\therefore \quad \cos\theta = -\sqrt{1-\sin^2\theta}$

Now, $\sec\theta + \tan\theta = \dfrac{1}{\cos\theta} + \dfrac{\sin\theta}{\cos\theta} = \dfrac{1+\sin\theta}{\cos\theta}$

$$= \frac{1+\sin\theta}{-\sqrt{1-\sin^2\theta}} = \frac{1+\frac{12}{13}}{-\sqrt{1-\left(\frac{12}{13}\right)^2}}$$

$$= \frac{\frac{25}{13}}{-\sqrt{\frac{25}{169}}} = \frac{\frac{25}{13}}{-\frac{5}{13}} = -5$$

Example 34. Draw the graph of $y = 3\sin 2x$.

Sol. $\sin x$ is a periodic function with period 2π, therefore,

$\sin 2x$ will be a periodic function of period $\frac{2\pi}{|2|} = \pi$

Also $\quad -1 \le \sin 2x \le 1$

$\therefore \quad -3 \le 3\sin 2x \le 3$

In order to draw the graph of $y = 3\sin 2x$, draw the graph of $y = \sin x$ and on X-axis change k to $\frac{k}{2}$, i.e. write $\frac{k}{2}$ wherever it is k. For example, write 15° in place of 30°, 45° in place of 90° etc.

On Y-axis change k to $3k$, i.e. write $3k$ wherever it is k for example, write 3 in place of 1, – 3 in place of – 1, 1.5 in place of 0.5 etc.

The graph of $y = 3\sin 2x$ will be as given in the figure.

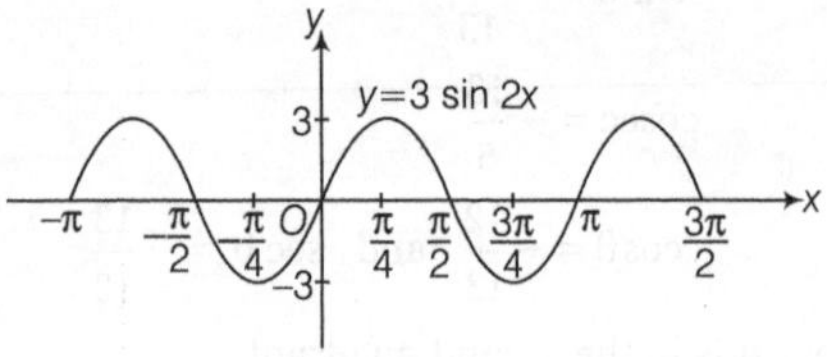

Example 35. Draw the graph of $y = \cos\left(x - \frac{\pi}{4}\right)$

Sol. Given function is $y = \cos\left(x - \frac{\pi}{4}\right)$ ...(i)

Given function is $Y = \cos X$, where

$$X = x - \frac{\pi}{4} \text{ and } Y = y$$

or $\quad Y = 0 \Rightarrow y = 0 \text{ and } X = 0$

$\Rightarrow \quad x - \frac{\pi}{4} = 0 \Rightarrow x = \frac{\pi}{4}$

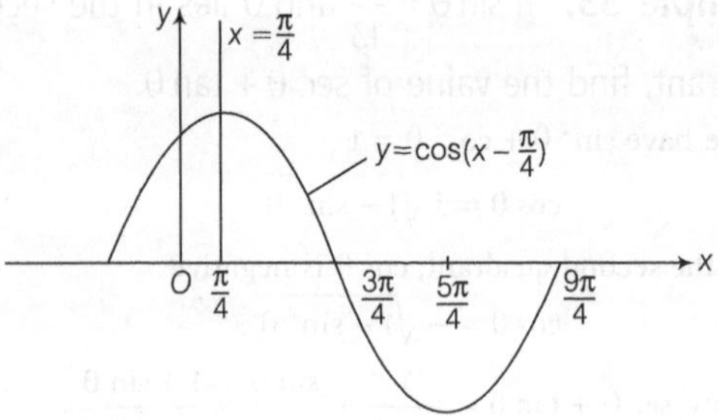

In order to draw the graph of $y = \cos\left(x - \frac{\pi}{4}\right)$, we draw the graph of $y = \cos x$ and shift it on the right side through a distance of $\frac{\pi}{4}$ unit.

Example 36. Which of the following is the least?

(a) sin 3 (b) sin 2

(c) sin 1 (d) sin 7

Sol. (a) $\sin 3 = \sin[\pi - (\pi - 3)] = \sin(\pi - 3) = \sin(0.14)$

$\sin 2 = \sin[\pi - (\pi - 2)]$

$= \sin(\pi - 2) = \sin(1.14)$

$\sin 7 = \sin[2\pi + (7 - 2\pi)]$

$= \sin(7 - 2\pi) = \sin(0.72)$

Now, $\quad 1.14 > 1 > 0.72 > 0.14$

$\Rightarrow \quad \sin(1.14) > \sin 1 > \sin(0.72) > \sin(0.14)$

[as 1.14, 0.72, 0.14 lie in the first quadrant and sine functions increase in the first quadrant]

Hence, among the given values, sin 3 is the least.

Alternate solution

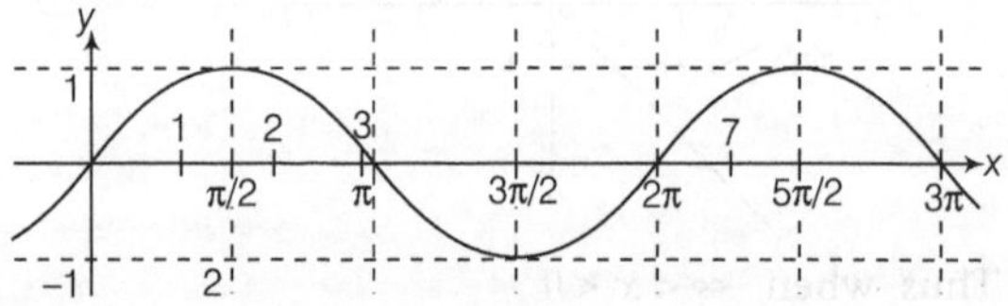

From the graph, obviously sin 3 is the least.

Example 37. Find the value of x for which $f(x) = \sqrt{\sin x - \cos x}$ is defined, $x \in [0, 2\pi]$.

Sol. $f(x) = \sqrt{\sin x - \cos x}$ is defined if $\sin x \ge \cos x$.

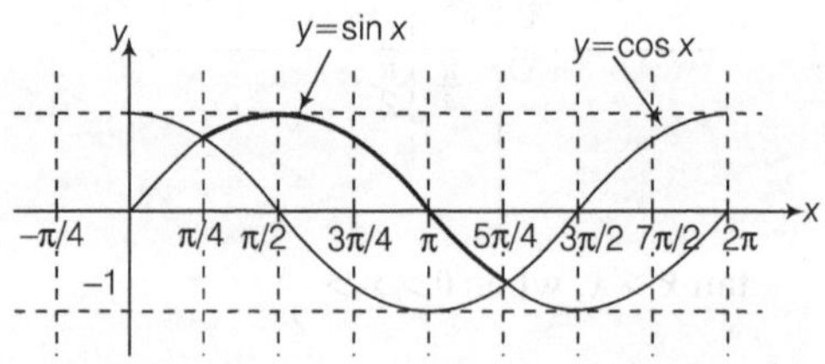

From the graph, $\sin x \ge \cos x$, for $x \in \left[\frac{\pi}{4}, \frac{5\pi}{4}\right]$.

Example 38. Solve $\tan x > \cot x$, where $x \in [0, 2\pi]$.

Sol.

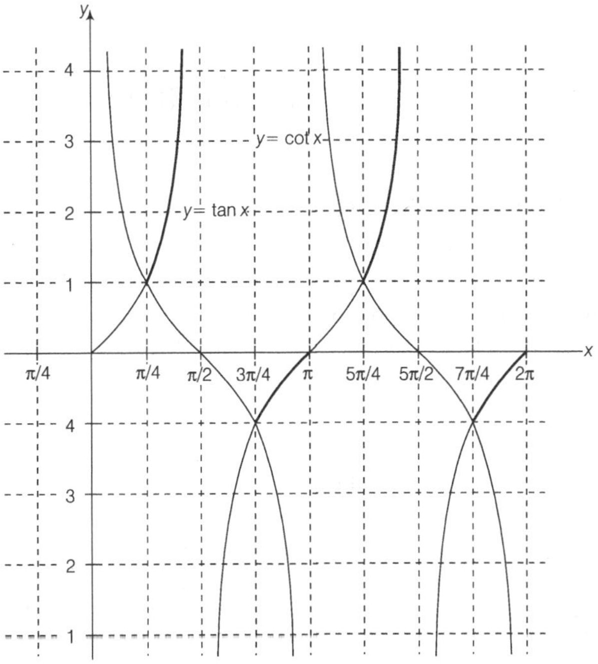

We find that $\tan x \geq \cot x$. Therefore, the values of $\tan x$ are more than the value of $\cot x$.

That is, the value of x for which graph of $y = \tan x$ is above the graph of $y = \cot x$.

From the graph, it is clear that

$$x \in \left(\frac{\pi}{4}, \frac{\pi}{2}\right) \cup \left(\frac{3\pi}{4}, \pi\right) \cup \left(\frac{5\pi}{4}, \frac{3\pi}{2}\right) \cup \left(\frac{7\pi}{4}, 2\pi\right).$$

Exercise for Session 4

1. If $\tan x = -\frac{4}{3}, \frac{3\pi}{2} < x < 2\pi$, find the value of $9\sec^2 x - 4\cot x$.
2. Show that $\sin^2 x = p + \frac{1}{p}$ is impossible if x is real.
3. If $\cos x = \frac{3}{5}$ and x lies in the fourth quadrant find the values of $\text{cosec}\, x + \cot x$.
4. Draw the graph of $y = \sin x$ and $y = \sin \frac{x}{2}$.
5. Draw the graph of $y = \sec^2 x - \tan^2 x$. Is $f(x)$ periodic? If yes, what is its fundamental period?
6. Prove that $\sin \theta < \theta < \tan \theta$ for $\theta \in \left(0, \frac{\pi}{2}\right)$.
7. Find the value of x for which $f(x) = \sqrt{\sin x - \cos x}$ is defined, $x \in [0, 2\pi]$.
8. Draw the graph of $y = \sin x$ and $y = \cos x, 0 \leq x \leq 2\pi$.
9. Draw the graph of $y = \tan(3x)$.
10. If $\cos x = -\frac{\sqrt{15}}{4}$ and $\frac{\pi}{2} < x < \pi$, find the value of $\sin x$.

Session 5

Trigonometric Ratios of any Angle

Trigonometric Ratios of any Angle

Consider the system of rectangular coordinates axis dividing plane into four quadrants. A line OP makes angle θ with the positive x-axis. The angle θ is said to be positive if measured in counter clockwise direction from the positive x-axis and is negative if measured in clockwise direction.

The positive values of the trigonometric ratios in the various quadrants are shown, the signs of the other ratios may be derived.

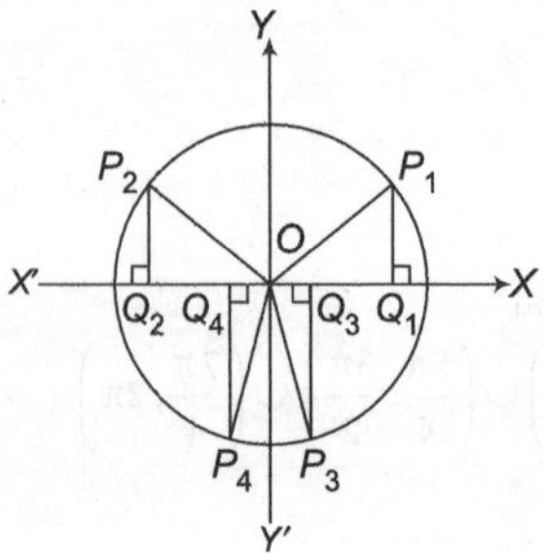

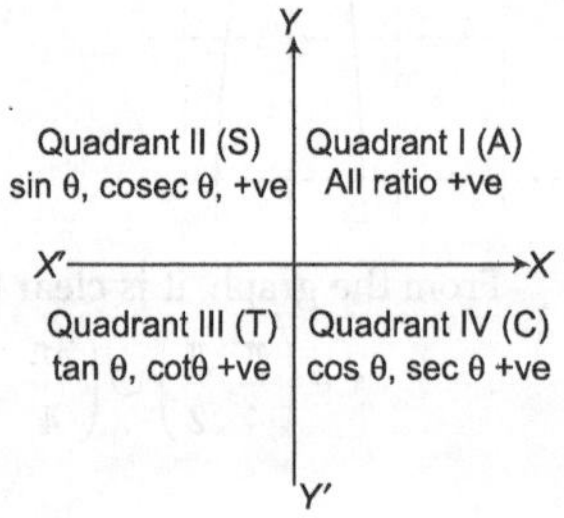

Note that $\angle XOY = \frac{\pi}{2}$, $\angle XOY' = \pi$, $\angle XOY' = \frac{3\pi}{2}$

P_iQ_i is positive if above the x-axis, negative if below the x-axis, OP_i is always taken positive. OQ_i is positive if along x-axis, negative if in oppositive direction.

$$\sin \angle Q_iOP_i = \frac{P_iQ_i}{OP_i}.$$

$$\cos \angle Q_iOP_i = \frac{O_iQ_i}{OP_i}.$$

$$\tan \angle Q_iOP_i = \frac{P_iQ_i}{OQ_i} \qquad [i = 1, 2, 3]$$

Thus, depending on signs of OQ_i and P_iQ_i, the various trigonometrical ratios will have different signs given

$\sin(-\theta) = -\sin\theta$ $\qquad$ $\cos(-\theta) = \cos\theta$

$\sin\left(\frac{\pi}{2} - \theta\right) = \cos\theta$ $\qquad$ $\cos\left(\frac{\pi}{2} - \theta\right) = \sin\theta$

$\tan(-\theta) = -\tan\theta$ $\qquad$ $\cot(-\theta) = -\cot\theta$

$\tan\left(\frac{\pi}{2} - \theta\right) = \cot\theta$ $\qquad$ $\cot\left(\frac{\pi}{2} - \theta\right) = \tan\theta$

$\text{cosec}(-\theta) = -\text{cosec}\,\theta$ $\qquad$ $\sec(-\theta) = \sec\theta$

$\sec\left(\frac{\pi}{2} - \theta\right) = \text{cosec}\,\theta$ $\qquad$ $\text{cosec}\left(\frac{\pi}{2} - \theta\right) = \sec\theta$

$\sin\left(\frac{\pi}{2} + \theta\right) = \cos\theta$ $\qquad$ $\cos\left(\frac{\pi}{2} + \theta\right) = -\sin\theta$

$\sin(\pi - \theta) = \sin\theta$ $\qquad$ $\cos(\pi - \theta) = -\cos\theta$

$\tan\left(\frac{\pi}{2} + \theta\right) = -\cot\theta$ $\qquad$ $\cot\left(\frac{\pi}{2} + \theta\right) = -\tan\theta$

$\tan(\pi - \theta) = -\tan\theta$ $\qquad$ $\cot(\pi - \theta) = -\cot\theta$

$\sec\left(\frac{\pi}{2} + \theta\right) = -\text{cosec}\,\theta$ $\qquad$ $\text{cosec}\left(\frac{\pi}{2} + \theta\right) = \sec\theta$

$\sec(\pi - \theta) = -\sec\theta$ $\qquad$ $\text{cosec}(\pi - \theta) = \text{cosec}\,\theta$

$\sin(\pi + \theta) = -\sin\theta$ $\qquad$ $\cos(\pi + \theta) = -\cos\theta$

$\sin\left(\frac{3\pi}{2} - \theta\right) = -\cos\theta$ $\qquad$ $\cos\left(\frac{3\pi}{2} - \theta\right) = -\sin\theta$

$\tan(\pi + \theta) = \tan\theta$ $\qquad$ $\cot(\pi + \theta) = \cot\theta$

$\tan\left(\frac{3\pi}{2} - \theta\right) = \cot\theta$ $\qquad$ $\cot\left(\frac{3\pi}{2} - \theta\right) = \tan\theta$

$\sec(\pi + \theta) = -\sec\theta$ $\qquad$ $\text{cosec}(\pi + \theta) = -\text{cosec}\,\theta$

$\sec\left(\frac{3\pi}{2} - \theta\right) = -\text{cosec}\,\theta$ $\qquad$ $\text{cosec}\left(\frac{3\pi}{2} - \theta\right) = -\sec\theta$

$\sin\left(\frac{3\pi}{2} + \theta\right) = -\cos\theta$ $\qquad$ $\cos\left(\frac{3\pi}{2} + \theta\right) = \sin\theta$

$\sin(2\pi - \theta) = -\sin\theta$ $\qquad$ $\cos(2\pi - \theta) = \cos\theta$

$\tan\left(\frac{3\pi}{2} + \theta\right) = -\cot\theta$ $\qquad$ $\cot\left(\frac{3\pi}{2} + \theta\right) = -\tan\theta$

$\tan(2\pi - \theta) = -\tan\theta$ $\qquad$ $\cot(2\pi - \theta) = -\cot\theta$

$\sec\left(\frac{3\pi}{2} + \theta\right) = \text{cosec}\,\theta$ $\qquad$ $\text{cosec}\left(\frac{3\pi}{2} + \theta\right) = -\sec\theta$

$\sec(2\pi - \theta) = \sec\theta$ $\qquad$ $\text{cosec}(2\pi - \theta) = -\text{cosec}\,\theta$

$\sin(2\pi + \theta) = \sin\theta$ $\qquad$ $\cos(2\pi + \theta) = \cos\theta$

$\tan(2\pi + \theta) = \tan\theta$ $\qquad$ $\cot(2\pi + \theta) = \cot\theta$

$\sec(2\pi + \theta) = \sec\theta$ $\qquad$ $\text{cosec}(2\pi + \theta) = \text{cosec}\,\theta$

Allied angles (or numbers)

Two angles (or numbers) are called allied iff their sum or difference is a multiple of $\frac{\pi}{2}$. For example, $\frac{\pi}{3}$ and $\frac{\pi}{6}$ are allied, $\frac{5\pi}{6}$ and $-\frac{\pi}{6}$ are allied.

AID TO MEMORY

You must have been overwhelmed by large number of formulae for allied angles (or numbers). Instead of memorising all of them, use the following rules

1. Any trigonometric function of a real number $n\pi \pm x (n \in I)$, treating x as $0 < x < \frac{\pi}{2}$, is numerically equal to the same function of x, with sign depending upon the quadrant in which the arc length (on the unit circle) terminates. The proper sign can be ascertained by 'All – Sin – Tan – Cos' rule. For example, $\sin(\pi + x) = -\sin x$; – *ve* sign was chosen because $\pi + x$ lies in the third quadrant and sin is – ve in the third quadrant.
2. Any trigonometric function of a real number $(2n+1)\frac{\pi}{2} \pm xm$ treating x as $0 < x < \frac{\pi}{2}$, is numerically equal to cofunction of x, with sign depending upon the quadrant in which the arc length (on the unit circle) terminates. Note that sin and cos are cofunctions of each other; tan and cot are cofunctions of each other; sec and cosec are cofunctions of each other. For example, $\sec\left(\frac{\pi}{2} + x\right) = -\operatorname{cosec} x$, – ve sign was chosen because $\frac{\pi}{2} + x$ lies in the second quadrant and sec is – ve in the second quadrant.

I. Method

To prove $\cos\left(\frac{\pi}{2} \pm \theta\right) = \mp \sin\theta$ and $\sin\left(\frac{\pi}{2} \pm \theta\right) = \cos\theta$

Proof

$$e^{i\left(\frac{\pi}{2} \pm \theta\right)} = \cos\left(\frac{\pi}{2} \pm \theta\right) + i\sin\left(\frac{\pi}{2} \pm \theta\right)$$

$$\Rightarrow e^{i\frac{\pi}{2}} \cdot e^{\pm i\theta} = \cos\left(\frac{\pi}{2} \pm \theta\right) + i\sin\left(\frac{\pi}{2} \pm \theta\right)$$

$$\Rightarrow i \cdot e^{i(\pm\theta)} = \cos\left(\frac{\pi}{2} \pm \theta\right) + i\sin\left(\frac{\pi}{2} \pm \theta\right)$$

$$\Rightarrow i \cdot (\cos\theta \pm i\sin\theta) = \cos\left(\frac{\pi}{2} \pm \theta\right) + i\sin\left(\frac{\pi}{2} \pm \theta\right)$$

$$\Rightarrow i \cdot \cos\theta \mp \sin\theta = \cos\left(\frac{\pi}{2} \pm \theta\right) + i\sin\left(\frac{\pi}{2} \pm \theta\right)$$

On comparing real and imaginary part of LHS and RHS, we get

$$\cos\left(\frac{\pi}{2} + \theta\right) = -\sin\theta \qquad \cos\left(\frac{\pi}{2} - \theta\right) = \sin\theta$$

$$\sin\left(\frac{\pi}{2} + \theta\right) = \cos\theta \qquad \sin\left(\frac{\pi}{2} - \theta\right) = \cos\theta$$

II. Method

To prove $\cos(\pi \pm \theta) = -\cos\theta$ and $\sin(\pi \pm \theta) = \mp\sin\theta$

Since, $\quad e^{i(\pi \pm \theta)} = \cos(\pi \pm \theta) + i\sin(\pi \pm \theta)$

$$\Rightarrow e^{i\pi} \cdot e^{i(\pm\theta)} = \cos(\pi \pm \theta) + i\sin(\pi \pm \theta)$$

$$\Rightarrow -(\cos(\pm\theta) + i\sin(\pm\theta)) = \cos(\pi \pm \theta) + i\sin(\pi \pm \theta)$$

On comparing real and imaginary part, we get

$$\cos(\pi + \theta) = -\cos\theta$$
$$\cos(\pi - \theta) = -\cos\theta$$
$$\sin(\pi + \theta) = -\sin\theta$$
$$\sin(\pi - \theta) = \sin\theta$$

Example 39. Prove that

(i) $\sin^2\frac{\pi}{6} + \cos^2\frac{\pi}{3} - \tan^2\frac{\pi}{4} = -\frac{1}{2}$

(ii) $2\sin^2\frac{\pi}{6} + \operatorname{cosec}^2\frac{7\pi}{6}\cos^2\frac{\pi}{3} = \frac{3}{2}$

(iii) $\cot^2\frac{\pi}{6} + \operatorname{cosec}\frac{5\pi}{6} + 3\tan^2\frac{\pi}{6} = 6$

(iv) $2\sin^2\frac{3\pi}{4} + 2\cos^2\frac{\pi}{4} + 2\sec^2\frac{\pi}{3} = 10$

Sol. (i) We have,

$$\sin^2\frac{\pi}{6} + \cos^2\frac{\pi}{3} - \tan^2\frac{\pi}{4} = -\frac{1}{2}$$

$$= \left(\sin\frac{\pi}{6}\right)^2 + \left(\cos\frac{\pi}{3}\right)^2 - \left(\tan\frac{\pi}{4}\right)^2$$

$$= \left(\frac{1}{2}\right)^2 + \left(\frac{1}{2}\right)^2 - (1)^2$$

$$= \frac{1}{4} + \frac{1}{4} - 1 = \frac{1}{2} - 1 = -\frac{1}{2}$$

(ii) We have,

$$2\sin^2\frac{\pi}{6}+\text{cosec}^2\frac{7\pi}{6}\cdot\cos^2\frac{\pi}{3}$$

$$=2\left(\sin\frac{\pi}{6}\right)^2+\left(\text{cosec}\,\frac{7\pi}{6}\right)^2\cdot\left(\cos\frac{\pi}{3}\right)^2$$

$$=2\left(\sin\frac{\pi}{6}\right)^2+\left\{\text{cosec}\left(\pi+\frac{\pi}{6}\right)\right\}^2\left(\cos\frac{\pi}{3}\right)^2$$

$$=2\left(\sin\frac{\pi}{6}\right)^2+\left\{-\text{cosec}\frac{\pi}{6}\right\}^2\left(\cos\frac{\pi}{3}\right)^2$$

$$[\because \text{cosec}(\pi+\theta)=-\text{cosec}\theta]$$

$$=2\left(\frac{1}{2}\right)^2+(-2)^2\times\left(\frac{1}{2}\right)^2=\frac{1}{2}+1=\frac{3}{2}$$

(iii) We have,

$$\cot^2\frac{\pi}{6}+\text{cosec}\frac{5\pi}{6}+3\tan^2\frac{\pi}{6}$$

$$=\left(\cot\frac{\pi}{6}\right)^2+\text{cosec}\left(\pi-\frac{\pi}{6}\right)+3\left(\tan\frac{\pi}{6}\right)^2$$

$$=(\sqrt{3})^2+2+3\left(\frac{1}{\sqrt{3}}\right)^2$$

$$=3+2+1=6$$

(iv) We have,

$$2\sin^2\frac{3\pi}{4}+2\cos^2\frac{\pi}{4}+2\sec^2\frac{\pi}{3}$$

$$=2\left(\sin\frac{3\pi}{4}\right)^2+2\left(\cos\frac{\pi}{4}\right)^2+2\left(\sec\frac{\pi}{3}\right)^2$$

$$=2\left(\sin\frac{\pi}{4}\right)^2+2\left(\cos\frac{\pi}{4}\right)^2-2\left(\sec\frac{\pi}{3}\right)^2$$

$$\left[\because \sin\frac{3\pi}{4}=\sin\left(\pi-\frac{\pi}{4}\right)=\sin\frac{\pi}{4}\right]$$

$$=2\left(\frac{1}{\sqrt{2}}\right)^2+2\left(\frac{1}{\sqrt{2}}\right)^2+2(2)^2$$

$$=1+1+8=10$$

Example 40. Prove that

$$\frac{\cos(90^\circ+\theta)\sec(-\theta)\tan(180^\circ-\theta)}{\sec(360^\circ-\theta)\sin(180^\circ+\theta)\cot(90^\circ-\theta)}=-1.$$

Sol. L.H.S. $\dfrac{\cos(90^\circ+\theta)\sec(-\theta)\tan(180^\circ-\theta)}{\sec(360^\circ-\theta)\sin(180^\circ+\theta)\cot(90^\circ-\theta)}$

$$=\frac{(-\sin\theta)(\sec\theta)(-\tan\theta)}{(\sec\theta)(-\sin\theta)(\tan\theta)}$$

$$=-1$$

= R.H.S.

Example 41. Show that $\tan 1^\circ \tan 2^\circ \ldots \tan 89^\circ = 1$

Sol. L.H.S. $=(\tan 1^\circ \tan 89^\circ)(\tan 2^\circ\cdot\tan 88^\circ)\ldots$

$$=[\tan 1^\circ \tan(90^\circ-1^\circ)]\cdot[\tan 2^\circ\tan(90^\circ-2^\circ)]$$

$$\ldots[\tan 44^\circ \tan(90^\circ-44^\circ)]\tan 45^\circ$$

$$=(\tan 1^\circ\cdot\cot 1^\circ)(\tan 2^\circ\cdot\cot 2^\circ)$$

$$\ldots(\tan 44^\circ\cdot\cot 44^\circ)\tan 45^\circ$$

$$=1 \qquad [\because \tan\theta\cot\theta=1 \text{ and } \tan 45^\circ=1]$$

Example 42. Show that

$$\sin^2 5^\circ+\sin^2 10^\circ+\sin^2 15^\circ+\ldots+\sin^2 90^\circ=9\frac{1}{2}$$

Sol. L.H.S. $=(\sin^2 5^\circ+\sin^2 85^\circ)+(\sin^2 10^\circ+\sin^2 80^\circ)+\ldots+$

$$(\sin^2 40^\circ+\sin^2 50^\circ)+\sin^2 45^\circ+\sin^2 90^\circ$$

$$=(\sin^2 5^\circ+\cos^2 5^\circ)+(\sin^2 10+\cos^2 10^\circ)$$

$$+\ldots+(\sin^2 40^\circ+\cos^2 40^\circ)+\sin^2 45^\circ+\sin^2 90^\circ$$

$$=(1+1+1+1+1+1+1+1)+\left(\frac{1}{\sqrt{2}}\right)^2+1$$

$$=9\frac{1}{2}$$

Example 43. Find the value of

$$\cos^2\frac{\pi}{16}+\cos^2\frac{3\pi}{16}+\cos^2\frac{5\pi}{16}+\cos^2\frac{7\pi}{16}$$

Sol. L.H.S. $=\cos^2\dfrac{\pi}{16}+\cos^2\dfrac{3\pi}{16}+\cos^2\left(\dfrac{\pi}{2}-\dfrac{3\pi}{16}\right)+\cos^2\left(\dfrac{\pi}{2}-\dfrac{\pi}{16}\right)$

$$=\cos^2\frac{\pi}{16}+\cos^2\frac{3\pi}{16}+\sin^2\frac{3\pi}{16}+\sin^2\frac{\pi}{16}$$

$$=\left(\cos^2\frac{\pi}{16}+\sin^2\frac{\pi}{16}\right)+\left(\cos^2\frac{3\pi}{16}+\sin^2\frac{3\pi}{16}\right)$$

$$=1+1=2$$

Exercise for Session 5

1. Find the value of $\tan \frac{19\pi}{3}$.

2. Find the sign of $\sec 2000°$.

3. The value of $\cos 1° + \cos 2° + \cos 3° + \ldots + \cos 180°$.

4. Find the value of $\cos(270° + \theta)\cos(90° - \theta) - \sin(270° - \theta)\cos\theta$.

5. If $S_n = \cos^n \theta + \sin^n \theta$, find the value of $3S_4 - 2S_6$.

6. $\sin^2 \theta = \frac{x^2 + y^2 + 1}{2x}$, then x must be.

7. If $\sin x + \operatorname{cosec} x = 2$, then find the value of $\sin^{10} x + \operatorname{cosec}^{10} x$.

8. $e^{\sin x} - e^{-\sin x} = 4$ then find the number of real solutions.

9. If $\pi < \alpha < \frac{3\pi}{2}$, then find the value of expression $\sqrt{4\sin^4 \alpha + \sin^2 2\alpha} + 4\cos^2\left(\frac{\pi}{4} - \frac{\alpha}{2}\right)$.

10. If $\sum_{i=1}^{n} \cos \theta_i = n$, then the value of $\sum_{i=1}^{n} \sin \theta_i$.

Session 6

Trigonometric Ratios of Compound Angles

Trigonometric Ratios of Compound Angles

Algebraic sum of two or more angles is called a compound angle. If A, B, C are any angles then $A + B$, $A - B$, $A + B + C$, $A - B + C$, $A - B - C$, $A + B - C$, etc., are all compound angles.

Till now, we have learnt the values of trigonometric ratios between 0° to 360°. Now, we are going to learn the values of trigonometric ratios of compound angles.

Note

Trigonometric ratios if i.e. sine, cosine, tan, cot, sec and cosec are not distributed over addition and substraction of 2 angles.

i.e. $\sin(A+B) \neq \sin A + \sin B$

Proof : $A = 60°, B = 30°$

$\sin(90°) \neq \sin 60° + \sin 30°$

The Addition Formula

(i) $\sin(A + B) = \sin A \cos B + \cos A \sin B$

(ii) $\cos(A + B) = \cos A \cos B - \sin A \sin B$

(iii) $\tan(A + B) = \frac{\tan A + \tan B}{1 - \tan A \tan B}$

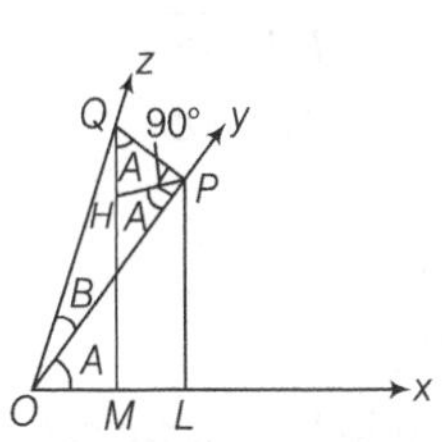

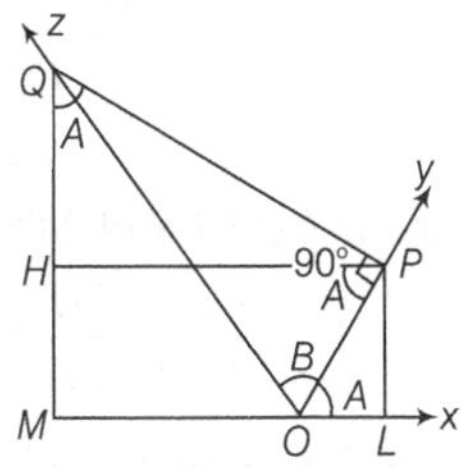

Let the revolving line starting from the position OX describe first $\angle XOY = A$ and then proceed further so as to describe $\angle YOZ = B$ in its position OZ.

Then, $\angle XOZ = A + B$

In figure 6.1 $A + B < 90°$ and in figure 6.2 $A + B > 90°$

Let Q be a point on OZ. From Q draw $QM \perp OX$ and $QP \perp OY$. From P draw $PH \perp QM$.

Now, $\angle HPO = \angle POX = A$

$\because \quad \angle QPO = 90°$

$\therefore \quad \angle QPH = 90° - A$

$\therefore \quad \angle HQP = A$

In ΔQOM,

$$\sin(A+B) = \frac{QM}{OQ} = \frac{QH + HM}{OQ} = \frac{QH + PL}{OQ}$$

$$= \frac{QH}{OQ} + \frac{PL}{OQ} = \frac{QH}{QP} \cdot \frac{QP}{OQ} + \frac{PL}{OP} \cdot \frac{OP}{OQ}$$

$$= \frac{PL}{OP} \cdot \frac{OP}{OQ} + \frac{QH}{QP} \cdot \frac{QP}{OQ}$$

$$= \sin POL \cdot \cos POQ + \cos HQP \cdot \sin POQ$$

$$= \sin A \cos B + \cos A \sin B$$

From figure 6.1, $\cos(A+B) = \dfrac{OM}{Q} = \dfrac{OL - ML}{OQ} = \dfrac{OL - PH}{OQ}$

From figure 6.2, $\cos(A+B) = -\dfrac{OM}{OQ} = -\dfrac{ML - OL}{OQ}$

$$= \frac{OL - ML}{OQ} = \frac{OL - PH}{OQ}$$

$\therefore$ In both cases $\cos(A+B)$

$$= \frac{OL}{OQ} - \frac{PH}{OQ} = \frac{OL}{OP} \cdot \frac{OP}{OQ} - \frac{PH}{QP} \cdot \frac{QP}{OQ}$$

$$= \cos POL \cdot \cos POQ - \sin PQH \cdot \sin POQ$$

$$= \cos A \cos B - \sin A \sin B$$

In both cases

$$\tan(A+B) = \frac{QM}{OM} = \frac{QH + HM}{OL - ML} = \frac{QH + PL}{OL - PH}$$

$$= \frac{\frac{QH}{OL} + \frac{PL}{OL}}{1 - \frac{PH}{OL}} = \frac{\frac{QH}{OL} + \frac{PL}{OL}}{1 - \frac{PH}{PL} \cdot \frac{PL}{OL}} \quad \text{...(i)}$$

From similar ΔQPH and ΔOPL

$$\frac{QH}{OL} = \frac{PH}{PL} = \frac{PQ}{OP} \quad \text{...(ii)}$$

On putting the value from Eq. (ii) in Eq. (i), we get

$$\tan(A+B) = \frac{\frac{PQ}{OP} + \frac{PL}{OL}}{1 - \frac{PQ}{OP} \cdot \frac{PL}{OL}}$$

$$= \frac{\tan B + \tan A}{1 - \tan B \tan A} = \frac{\tan A + \tan B}{1 - \tan A \tan B}$$

$$\left[\because \text{from } \Delta POQ, \frac{PQ}{OP} = \tan B, \text{ from } \Delta POL, \frac{PL}{OL} = \tan A\right]$$

Second Proof of Formulae

1. **$\cos(A+B) = \cos A \cos B - \sin A \sin B$**

Proof Let O be the centre of a unit circle.

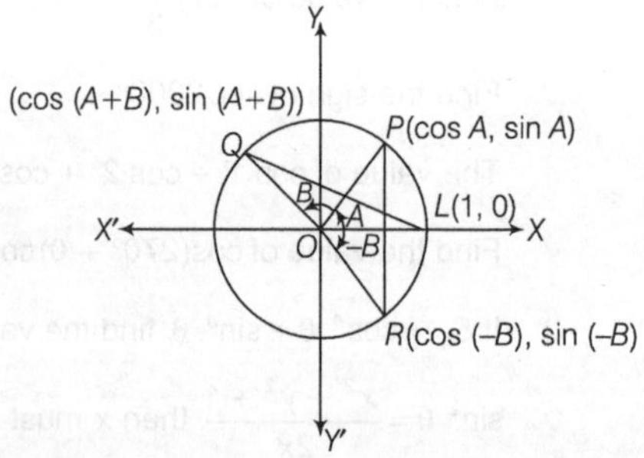

Let $\angle LOP = A$ radian, $\angle POQ = B$ radian, $\angle LOR = -B$ radian

(This angle has been measured in clockwise direction)

Now $\angle LOQ = A + B$ and $\angle ROP = A - B$

Since radius of circle is unity

$\therefore \quad$ arc $LP = A$, arc $PQ = B$, arc $LR = |-B| = B$

[in formulae $\theta = \dfrac{l}{r}$, θ is always taken a positive]

Also as radius of the circle is 1.

$\therefore P \equiv (\cos A, \sin A)$,

$Q \equiv \cos(A+B), \sin(A+B)$,

$R \equiv (\cos(-B), \sin(-B)$ or $R \equiv (\cos B, -\sin B)$

$\therefore \quad \Delta LOQ \cong \Delta POR$

$\therefore \quad LQ = PR$

$\Rightarrow \quad LQ^2 = PR^2$

$\Rightarrow [1 - \cos(A+B)]^2 + [0 - \sin(A+B)]^2$

$= [\cos A - \cos(-B)]^2 + [(\sin A - \sin(-B)]^2$

$\Rightarrow 1 + \cos^2(A+B) - 2\cos(A+B) + \sin^2(A+B)$

$= (\cos A - \cos B)^2 + (\sin A - \sin B)^2$

$\Rightarrow 1 + \cos^2(A+B) + \sin^2(A+B) - 2\cos(A+B)$

$\Rightarrow \quad = \cos^2 A + \cos^2 B - 2\cos A \cos B + \sin^2 A + \sin^2 B + 2\sin A \sin B$

$\Rightarrow \quad 2 - 2\cos(A+B) = (\cos^2 A + \sin^2 A) + (\cos^2 B + \sin^2 B) - 2(\cos A \cos B - \sin A \sin B)$

$\Rightarrow 2 - 2\cos(A+B) = 2 - 2(\cos A \cos B - \sin A \sin B)$

$\Rightarrow \quad \cos(A+B) = \cos A \cos B - \sin A \sin B \quad$ ...(i)

2. **Putting $-B$ in place of B in (1), we get**

$\cos(A-B) = \cos A \cos(-B) - \sin A \sin(-B)$

$= \cos A \cos B + \sin A \sin B \quad$ ...(ii)

3. $\sin(A+B)=\cos\left[\frac{\pi}{2}-(A+B)\right]$

$$=\cos\left[\left(\frac{\pi}{2}-A\right)-B\right]$$

$$=\cos\left(\frac{\pi}{2}-A\right)\cos B+\sin\left(\frac{\pi}{2}-A\right)\sin B$$

$$=\sin A\cos B+\cos A\sin B \quad \text{...(iii)}$$

4. $\tan(A+B)=\frac{\sin(A+B)}{\cos(A+B)}$

$$=\frac{\sin A\cos B+\cos A\sin B}{\cos A\cos B-\sin A\sin B}$$

$$=\frac{\tan A+\tan B}{1-\tan A\tan B} \quad \text{...(iv)}$$

[dividing numerator and denominator by $\cos A\cos B$]

5. Putting $-B$ in place of B in (3), we get

$$\sin(A-B)=\sin A\cos B-\cos A\sin B \quad \text{...(v)}$$

6. Putting $-B$ in place of B in (4), we get

$$\tan(A-B)=\frac{\tan A-\tan B}{1+\tan A\tan B} \quad \text{...(vi)}$$

7. $\cot(A+B)=\frac{\cos(A+B)}{\sin(A+B)}$

$$=\frac{\cos A\cos B-\sin A\sin B}{\sin A\cos B+\cos A\sin B}$$

$$=\frac{\cot A\cot B-1}{\cot B+\cot A} \quad \text{...(vii)}$$

[dividing numerator and denominator by $\sin A\sin B$]

8. Putting $-B$ in place of B in (7), we get

$$\cot(A-B)=\frac{-\cot A\cot B-1}{-\cot B+\cot A}$$

$$=\frac{\cot A\cot B+1}{\cot B-\cot A} \quad \text{...(viii)}$$

Third Proof by Complex Number Method

The result of the sine, cosine and tangent of compound angle can also be derived using the concept of complex numbers as discussed.

$$\cos(A\pm B)+i\sin(A\pm B)=e^{i(A\pm B)}$$

$$=e^{iA}\cdot e^{i(\pm B)}=(\cos A+i\sin A)(\cos(\pm B)+i\sin(\pm B))$$

$$=(\cos A\cos B\pm i\cos A\sin B+i\sin A\cos B\mp\sin A\sin B)$$

$$=(\cos A\cos B\mp\sin A\sin B)+i(\sin A\cos A\pm\cos A\sin B)$$

Comparing real and imaginary parts of the left and right hand side, we get,

$$\cos(A\pm B)=(\cos A\cos B\mp\sin A\sin B)$$

$$\sin(A\pm B)=(\sin A\cos B\pm\cos A\sin B)$$

TWO VERY IMPORTANT IDENTITIES

(a) $\sin(A+B)\cdot\sin(A-B)$

$$=\sin^2 A-\sin^2 B=\cos^2 B-\cos^2 A$$

(b) $\cos(A+B)\cdot\cos(A-B)=\cos^2 A-\sin^2 B$

Proof : (a) $\sin(A+B)\cdot\sin(A-B)$

$$=(\sin A\cos B+\cos A\sin B)(\sin A\cos B-\cos A\sin B)$$

$$=\sin^2 A\cos^2 B-\cos^2 A\sin^2 B$$

$$=\sin^2 A(1-\sin^2 B)-\sin^2 B(1-\sin^2 A)$$

$$=\sin^2 A-\sin^2 B$$

(b) $\cos(A+B)\cdot\cos(A-B)$

$$=(\cos A\cos B-\sin A\sin B)(\cos A\cos B+\sin A\sin B)$$

$$=\cos^2 A\cos^2 B-\sin^2 A\sin^2 B$$

$$=\cos^2 A(1-\sin^2 B)-(1-\cos^2 A)\sin^2 B$$

$$=\cos^2 A-\sin^2 B$$

Example 44. Find the value of $\tan 105°$.

Sol. $\tan 105°=\tan(60°+45°)=\frac{\tan 60°+\tan 45°}{1-\tan 60°\tan 45°}$

$$=\frac{\sqrt{3}+1}{1-\sqrt{3}\cdot 1}=\frac{(\sqrt{3}+1)^2}{1-3}=-(2+\sqrt{3})$$

$\therefore \quad \tan 105°=-(2+\sqrt{3})$

Example 45. Prove that $\tan 70°=\tan 20°+2\tan 50°$.

Sol. $\tan 70°=\tan(20°+50°)=\frac{\tan 20°+\tan 50°}{1-\tan 20°\tan 50°}$

or $\quad \tan 70°-\tan 20°\cdot\tan 50°\cdot\tan 70°=\tan 20°+\tan 50°$

or $\quad \tan 70°=\tan 70°\tan 50°\tan 50°+\tan 20°+\tan 50°$

$$=\cot 20°\tan 50°\tan 20°+\tan 20°+\tan 50°$$

[$\because \tan 70°=\tan(90°-20°)=\cot 20°$]

$$=2\tan 50°+\tan 20°$$

Example 46. If $A+B=45°$, then show that $(1+\tan A)(1+\tan B)=2$.

Sol. $\tan(A+B)=\frac{\tan A+\tan B}{1-\tan A\tan B}$; $1=\frac{\tan A+\tan B}{1-\tan A\tan B}$

[as $A+B=45°$, $\tan(A+B)=1$]

$\therefore \quad \tan A+\tan B+\tan A\tan B=1$

or $\quad 1+\tan A+\tan B+\tan A\tan B=1+1$

[$\because$ adding '1' on both sides]

$\Rightarrow \quad (1+\tan A)+\tan B(1+\tan A)=2$

$\Rightarrow \quad (1+\tan A)(1+\tan B)=2$

Example 47. Find the value of $\dfrac{\tan 495^\circ}{\cot 855^\circ}$

Sol. $\tan 495^\circ = \tan(2.180^\circ + 135^\circ) = \tan 135^\circ = -1$

$$\cot 855^\circ = \cot(4.180^\circ + 135^\circ)$$
$$= \cot 135^\circ = -1 \qquad [\because \cot(4.180^\circ + \theta) = \cot\theta]$$
$$\therefore \quad \frac{\tan 495^\circ}{\cot 855^\circ} = \frac{-1}{-1} = 1$$

Example 48. Evaluate $\sin\left\{n\pi + (-1)^n \dfrac{\pi}{4}\right\}$; where n is an integer.

Sol. $\because \quad \sin(\pi + \theta) = -\sin\theta$

$$\therefore \quad \sin(n\pi + \theta) = (-1)^n \sin\theta \Rightarrow \sin\left\{n\pi + (-1)^n \frac{\pi}{4}\right\}$$
$$= (-1)^n \sin\left\{(-1)^n \frac{\pi}{4}\right\}$$
$$= (-1)^n (-1)^n \sin\frac{\pi}{4} \qquad [\because \sin(-\theta) = -\sin\theta]$$
$$\therefore \sin\{(-1)^n\theta\} = (-1)^n \sin\theta$$
$$= (-1)^{2n} \sin\frac{\pi}{4} = \sin\frac{\pi}{4} = \frac{1}{\sqrt{2}}$$

Example 49. Prove that $\cos 18^\circ - \sin 18^\circ = \sqrt{2}\sin 27^\circ$

Sol. R.H.S. $= \sqrt{2}\sin 27^\circ = \sqrt{2}\sin(45^\circ - 18^\circ)$

$$= \sqrt{2}(\sin 45^\circ \cos 18^\circ - \cos 45^\circ \sin 18^\circ)$$
$$= \sqrt{2}\left(\frac{1}{\sqrt{2}}\cos 18^\circ - \frac{1}{\sqrt{2}}\sin 18^\circ\right)$$
$$= \cos 18^\circ - \sin 18^\circ$$
$$= \text{L.H.S.}$$

Example 50. Show that $\cot\left(\dfrac{\pi}{4} + x\right) \cdot \cot\left(\dfrac{\pi}{4} - x\right) = 1$

Sol. $$\text{L.H.S.} = \frac{\cos\left(\frac{\pi}{4} + x\right)\cos\left(\frac{\pi}{4} - x\right)}{\sin\left(\frac{\pi}{4} + x\right)\sin\left(\frac{\pi}{4} - x\right)}$$

$$= \frac{\cos^2\frac{\pi}{4} - \sin^2 x}{\sin^2\frac{\pi}{4} - \sin^2 x} = \frac{\frac{1}{2} - \sin^2 x}{\frac{1}{2} - \sin^2 x} = 1$$

Example 51. If $\sin\alpha\sin\beta - \cos\alpha\cos\beta + 1 = 0$, Prove that $1 + \cot\alpha\tan\beta = 0$

Sol. Given, $\sin\alpha\sin\beta - \cos\alpha\cos\beta + 1 = 0$

$$\Rightarrow \quad \cos\alpha\cos\beta - \sin\alpha\sin\beta = 1$$
$$\Rightarrow \quad \cos(\alpha + \beta) = 1 \qquad \text{...(i)}$$

Now, $1 + \cos\alpha\tan\beta = 1 + \dfrac{\cos\alpha}{\sin\alpha} \cdot \dfrac{\sin\beta}{\cos\beta}$

$$= \frac{\sin\alpha\cos\beta + \cos\alpha\sin\beta}{\sin\alpha\cos\beta} = \frac{\sin(\alpha + \beta)}{\sin\alpha\cos\beta}$$
$$= \frac{0}{\sin\alpha\cos\beta} = 0$$
$$[\because \sin^2(\alpha + \beta) = 1 - \cos^2(\alpha + \beta) = 1 - 1 = 0]$$

Example 52. Prove that
$$\frac{\sin(B - C)}{\cos B\cos C} + \frac{\sin(C - A)}{\cos C\cos A} + \frac{\sin(A - B)}{\cos A\cos B} = 0$$

Sol. First term of L.H.S.

$$= \frac{\sin(B - C)}{\cos B\cos C} = \frac{\sin B\cos C - \cos B\sin C}{\cos B\cos C}$$
$$= \frac{\sin B\cos C}{\cos B\cos C} - \frac{\cos B\sin C}{\cos B\cos C}$$
$$= \tan B - \tan C$$

Similarly, second term of L.H.S. $= \tan C - \tan A$ and 3rd term of L.H.S. $= \tan A - \tan B$

Now L.H.S. $= (\tan B - \tan C) + (\tan C - \tan A) + (\tan A - \tan B) = 0$

Example 53. Show that $\tan 75^\circ + \cot 75^\circ = 4$.

Sol. $$\tan 75^\circ = \tan(45^\circ + 30^\circ) = \frac{\tan 45^\circ + \tan 30^\circ}{1 - \tan 45^\circ \tan 30^\circ}$$

$$= \frac{1 + \frac{1}{\sqrt{3}}}{1 - \frac{1}{\sqrt{3}}} = \frac{\sqrt{3} + 1}{\sqrt{3} - 1} \qquad \text{...(i)}$$

and $$\cot 75^\circ = \frac{1}{\tan 75^\circ} = \frac{\sqrt{3} - 1}{\sqrt{3} + 1} \qquad \text{...(ii)}$$

Now, L.H.S. $= \tan 75^\circ + \cot 75^\circ$

$$= \frac{\sqrt{3} + 1}{\sqrt{3} - 1} + \frac{\sqrt{3} - 1}{\sqrt{3} + 1} \qquad \text{[from Eqs. (i) and (ii)]}$$
$$= \frac{(\sqrt{3} + 1)^2 + (\sqrt{3} - 1)^2}{(\sqrt{3} - 1)(\sqrt{3} + 1)}$$
$$= \frac{(4 + 2\sqrt{3}) + (4 - 2\sqrt{3})}{3 - 1} = \frac{8}{2} = 4 = \text{R.H.S.}$$

Example 54. If $\tan\beta = \dfrac{n\sin\alpha\cos\alpha}{1 - n\sin^2\alpha}$. Prove that $\tan(\alpha - \beta) = (1 - n)\tan\alpha$.

Sol. $$\tan\beta = \frac{n\sin\alpha\cos\beta}{1 - n\sin^2\alpha}$$

$$= \frac{\frac{n\sin\alpha\cos\alpha}{\cos^2\alpha}}{\frac{1}{\cos^2\alpha} - \frac{n\sin^2\alpha}{\cos^2\alpha}}$$

[dividing numerator and denominator by $\cos\alpha$]

$$= \frac{n \tan \alpha}{\sec^2 \alpha - n \tan^2 \alpha}$$

$$= \frac{n \tan \alpha}{1 + \tan^2 \alpha - n \tan^2 \alpha}$$

$$= \frac{n \tan \alpha}{1 + (1-n) \tan^2 \alpha} \quad \text{...(i)}$$

Now, L.H.S. $= \tan(\alpha - \beta) = \dfrac{\tan \alpha - \tan \beta}{1 + \tan \alpha \tan \beta}$

$$= \frac{\tan \alpha - \dfrac{n \tan \alpha}{1+(1-n)\tan^2 \alpha}}{1 + \tan \alpha \dfrac{n \tan \alpha}{1+(1-n)\tan^2 \alpha}} \quad \text{[from Eqs. (i)]}$$

$$= \frac{\tan \alpha + (1-n)\tan^3 \alpha - n \tan \alpha}{1 + (1-n)\tan^2 \alpha + n \tan^2 \alpha}$$

$$= \frac{(1-n)\tan \alpha + (1-n)\tan^3 \alpha}{1 + \tan^2 \alpha}$$

$$= \frac{(1-n)\tan \alpha (1 + \tan^2 \alpha)}{1 + \tan^2 \alpha}$$

$$= (1-n) \tan \alpha$$

Example 55. Show that $\cos^2 \theta + \cos^2(\alpha + \theta) - 2\cos \alpha \cos \theta \cos(\alpha + \theta)$ in independent of θ.

Sol. $\cos^2 \theta + \cos^2(\alpha + \theta) - 2 \cos \alpha \cos \theta \cos(\alpha + \theta)$

$= \cos^2 \theta + \cos(\alpha + \theta)[\cos(\alpha + \theta) - 2 \cos \alpha \cos \theta]$

$= \cos^2 \theta + \cos(\alpha + \theta)$

$[\cos \alpha \cos \theta - \sin \alpha \sin \theta - 2 \cos \alpha \cos \theta]$

$= \cos^2 \theta - \cos(\alpha + \theta)[\cos \alpha \cos \theta + \sin \alpha \sin \theta]$

$= \cos^2 \theta - \cos(\alpha + \theta) \cos(\alpha - \theta)$

$= \cos^2 \theta - [\cos^2 \alpha - \sin^2 \theta]$

$= \cos^2 \theta + \sin^2 \theta - \cos^2 \alpha$

$= 1 - \cos^2 \alpha$, which is independent of θ.

Example 56. If $3 \tan \theta \tan \phi = 1$, then prove that $2\cos(\theta + \phi) = \cos(\theta - \phi)$.

Sol. Given, $3 \tan \theta \tan \phi = 1$ or $\cot \theta \cot \phi = 3$

or $$\frac{\cos \theta \cos \phi}{\sin \theta \sin \phi} = \frac{3}{1}$$

By componendo and dividendo, we get

$$\frac{\cos \theta \cos \phi + \sin \theta \sin \phi}{\cos \theta \cos \phi - \sin \theta \sin \phi} = \frac{3+1}{3-1}$$

or $$\frac{\cos(\theta - \phi)}{\cos(\theta + \phi)} = 2$$

or $$2 \cos(\theta + \phi) = \cos(\theta - \phi)$$

Example 57. Let A, B, C be the three angles such that $A + B + C = \pi$. If $\tan A \cdot \tan B = 2$, then find the value of $\dfrac{\cos A \cos B}{\cos C}$.

Sol. Given, $\tan A \cdot \tan B = 2$

Let $$y = \frac{\cos A \cos B}{\cos C} = -\frac{\cos A \cdot \cos B}{\cos(A+B)}$$

$[\because \cos C = \cos(\pi - (A+B)) = -\cos(A+B)]$

$$= \frac{\cos A \cdot \cos B}{\sin A \sin B - \cos A \cos B}$$

$$= \frac{1}{\tan A \tan B - 1} = \frac{1}{2-1} = 1$$

Example 58. Prove that $\dfrac{\cos 10° + \sin 10°}{\cos 10° - \sin 10°} = \tan 55°$.

Sol. $$\frac{\cos 10° + \sin 10°}{\cos 10° - \sin 10°} = \frac{1 + \tan 10°}{1 - \tan 10°} = \frac{\tan 45° + \tan 10°}{1 - \tan 45° \tan 10°}$$

$= \tan(45° + 10°) = \tan 55°$ (dividing by $\cos 10°$)

Example 59. If $\sin(A - B) = \dfrac{1}{\sqrt{10}}$, $\cos(A + B) = \dfrac{2}{\sqrt{29}}$, find the value of $\tan 2A$ where A and B lie between 0 and $\dfrac{\pi}{4}$.

Sol. $\tan 2A = \tan[(A+B) + (A-B)]$

$$= \frac{\tan(A+B) + \tan(A-B)}{1 - \tan(A+B)\tan(A-B)} \quad \text{...(i)}$$

Given that, $0 < A < \dfrac{\pi}{4}$ and $0 < B < \dfrac{\pi}{4}$. Therefore,

$$0 < A + B < \frac{\pi}{2}$$

Also, $-\dfrac{\pi}{4} < A - B < \dfrac{\pi}{4}$ and $\sin(A-B) = \dfrac{1}{\sqrt{10}}$

$\therefore$ $$0 < A - B < \frac{\pi}{4}$$

Now, $$\sin(A-B) = \frac{1}{\sqrt{10}}$$

$\Rightarrow$ $$\tan(A-B) = \frac{1}{3} \quad \text{...(ii)}$$

$$\cos(A+B) = \frac{2}{\sqrt{29}}$$

$\Rightarrow$ $$\tan(A+B) = \frac{5}{2} \quad \text{...(iii)}$$

From Eqs. (i), (ii) and (iii), we get

$$\tan 2A = \frac{\dfrac{5}{2} + \dfrac{1}{3}}{1 - \dfrac{5}{2} \times \dfrac{1}{3}} = \frac{17}{6} \times \frac{16}{1} = 17$$

Example 60. Prove that $(1+\tan 1^\circ)(1+\tan 2^\circ)\ldots(1+\tan 45^\circ)=2^{23}$.

Sol. $(1+\tan x^\circ)(1+\tan(45^\circ - x^\circ))$

$$=(1+\tan x^\circ)\left(1+\frac{1-\tan x^\circ}{1+\tan x^\circ}\right)=2$$

$$\therefore (1+\tan 1^\circ)(1+\tan 44^\circ)$$
$$=(1+\tan 2^\circ)(1+\tan 43^\circ)$$
$$=(1+\tan 3^\circ)(1+\tan 42^\circ)$$
$$\ldots$$
$$\ldots$$
$$=(1+\tan 22^\circ)(1+\tan 23^\circ)$$
$$=2$$

$$(1+\tan 1^\circ)(1+\tan 2^\circ)\ldots(1+\tan 45^\circ)=2^{23}$$

(as $1+\tan 45^\circ = 2$)

Example 61. If $\cos(\beta-\gamma)+\cos(\gamma-\alpha)+\cos(\alpha-\beta)=-\frac{3}{2}$, Prove that $\cos\alpha+\cos\beta+\cos\gamma=\sin\alpha+\sin\beta+\sin\gamma=0$

Sol. Given, $\cos(\beta-\gamma)+\cos(\gamma-\alpha)+\cos(\alpha-\beta)=-\frac{3}{2}$

or $3+2\cos(\beta-\gamma)+2\cos(\gamma-\alpha)+2\cos(\alpha-\beta)=0$

or $3+2(\cos\beta\cos\gamma+\sin\beta\sin\gamma)+2(\cos\gamma\cos\alpha+\sin\gamma\sin\alpha)+2(\cos\alpha\cos\beta+\sin\alpha\sin\beta)=0$

or $(\cos^2\alpha+\sin^2\alpha)+(\cos^2\beta+\sin^2\beta)+(\cos^2\gamma+\sin^2\gamma)+2(\cos\beta\cos\gamma+\sin\beta\sin\gamma)+2(\cos\gamma\cos\alpha+\sin\gamma\sin\alpha)+2(\cos\alpha\cos\beta+\sin\alpha\sin\beta)=0$

or $(\cos^2\alpha+\cos^2\beta+\cos^2\gamma+2\cos\alpha\cos\beta+2\cos\beta\cos\gamma+2\cos\gamma\cos\gamma)+(\sin^2\alpha+\sin^2\beta+\sin^2\gamma+2\sin\alpha\sin\beta+2\sin\beta\sin\gamma+2\sin\gamma\sin\alpha)=0$

or $(\cos\alpha+\cos\beta+\cos\gamma)^2+(\sin\alpha+\sin\beta+\sin\gamma)^2=0$

which is possible only when

$\cos\alpha+\cos\beta+\cos\gamma=0$ and $\sin\alpha+\sin\beta+\sin\gamma=0$

Example 62. Find the value of $\frac{\cos 25^\circ+\cot 55^\circ}{\tan 25^\circ+\tan 55^\circ}+\frac{\cot 55^\circ+\cot 100^\circ}{\tan 55^\circ+\tan 100^\circ}+\frac{\cot 100^\circ+\cot 25^\circ}{\tan 100^\circ+\tan 25^\circ}$.

Sol. $$E=\frac{\cot 25^\circ+\cot 55^\circ}{\tan 25^\circ+\tan 55^\circ}+\frac{\cot 55^\circ+\cot 100^\circ}{\tan 55^\circ+\tan 100^\circ}+\frac{\cot 100^\circ+\cot 25^\circ}{\tan 100^\circ+\tan 25^\circ}$$

$$=\frac{1}{\tan 55^\circ\tan 100^\circ}+\frac{1}{\tan 55^\circ\tan 100^\circ}+\frac{1}{\tan 100^\circ\tan 25^\circ}$$

$$=\frac{\tan 25^\circ+\tan 55^\circ+\tan 100^\circ}{\tan 25^\circ\cdot\tan 55^\circ\cdot\tan 100^\circ}$$

Since, $25^\circ+55^\circ+100^\circ=180^\circ$

$\tan 25^\circ+\tan 55^\circ+\tan 100^\circ=\tan 25^\circ\tan 55^\circ\tan 100^\circ$

$\Rightarrow \quad E=1$

Example 63. Prove that $\sum_{k=1}^{100}\sin(kx)\cos(101-k)x=50\sin(101x)$

Sol. Let $S=\sum_{k=1}^{100}\sin(kx)\cos(101-k)x$

$\Rightarrow \quad S=\sin x\cos 100x+\sin 2x\cos 99x+\ldots+\sin 100x\cos x$ ...(i)

$S=\cos x\sin 100x+\cos 2x\sin 99x+\ldots+\sin x\cos 100x$ ...(ii)

(on writing in reverse order)

On adding Eqs. (i) and (ii), we get

$2S=(\sin x\cos 100x+\cos x\sin 100x)+(\sin 2x\cos 99x+\cos 2x\sin 99x)$

...

...

$+(\sin 100x\cos x+\sin x\cos 100x)$

$=\sin 101x+\sin 101x+\ldots+\sin 101x$ (100 times)

Hence, $S=50\sin(101x)$

Example 64. If $A=\frac{\pi}{5}$, then find the value of $\sum_{r=1}^{8}\tan(rA)\cdot\tan((r+1)A)$.

Sol. $$\tan((r+1)A-(rA))=\frac{\tan(r+1)A-\tan(rA)}{1+\tan(r+1)A\cdot\tan(rA)}$$

$$\Rightarrow \quad S=\sum_{r=1}^{8}\tan(rA)\cdot\tan(r+1)A$$

$$=\sum_{r=1}^{8}(-1)+\frac{1}{\tan A}\sum_{r=1}^{8}(\tan(r+1)A-\tan(rA))$$

$$=-8+\frac{1}{\tan A}\cdot(\tan 9A-\tan A)$$

Now, $$\tan 9A=\tan\frac{9\pi}{5}=\tan\left(2\pi-\frac{\pi}{5}\right)=-\tan\frac{\pi}{5}$$

$$\Rightarrow \quad S=-8+\frac{1}{\tan A}(-2\tan A)$$

$$=-8-2=-10$$

Example 65. Prove that

$\sin\theta\cdot\sec 3\theta+\sin 3\theta\cdot\sec 3^2\theta+\sin 3^2\theta\cdot\sec 3^3\theta+\ldots$ upto n terms $=\frac{1}{2}[\tan 3^n\theta-\tan\theta]$

Sol. $\sin\theta\cdot\sec 3\theta+\sin 3\theta\cdot\sec 3^2\theta+\sin 3^2\theta\cdot\sec 3^3\theta+\ldots$ upto n terms

$$=\sum_{r=1}^{n}\sin 3^{r-1}\theta\cdot\sec 3^r\theta$$

$$=\sum_{r=1}^{n}\frac{2\cos 3^{r-1}\theta\sin 3^{r-1}\theta}{2\cos 3^{r-1}\theta\cdot\cos 3^r\theta}$$

$$=\frac{1}{2}\sum_{r=1}^{n}\frac{\sin(2\cdot 3^{r-1}\theta)}{\cos 3^{r-1}\theta\cdot\cos 3^r\theta}$$

$$=\frac{1}{2}\sum_{r=1}^{n}\frac{\sin(3^r\theta-3^{r-1}\theta)}{\cos 3^{r-1}\theta\cdot\cos 3^r\theta}$$

$$\sin 3^r\theta\cdot\cos 3^{r-1}\theta$$

$$=\frac{1}{2}\sum_{r=1}^{n}\frac{-\cos 3^r\theta\cdot\sin 3^{r-1}\theta}{\cos 3^{r-1}\theta\cdot\cos 3^r\theta}$$

$$=\frac{1}{2}\sum_{r=1}^{n}(\tan 3^r\theta-\tan 3^{r-1}\theta)$$

$$=\frac{1}{2}[\tan 3^n\theta-\tan\theta]$$

Example 66. In a triangle ABC, if $\sin A\sin(B-C)=\sin C\sin(A-B)$, then prove that $\cot A$, $\cot B$, $\cot C$ are in AP.

Sol. $\sin A\sin(B-C)=\sin C\sin(A-B)$

$$\Rightarrow\quad \frac{\sin(B-C)}{\sin C\sin B}=\frac{\sin(A-B)}{\sin A\sin B}$$

$$\Rightarrow\frac{\sin B\cos C-\sin C\cos B}{\sin C\sin B}=\frac{\sin A\cos B-\sin B\cos A}{\sin A\sin B}$$

$$\Rightarrow\quad \cot C-\cot B=\cot B-\cot A$$

$$\Rightarrow\quad 2\cot B=\cot A+\cot C$$

$\therefore$ $\cot A$, $\cot B$, $\cot C$ are in A.P.

Example 67. If $0<\beta<\alpha<\pi/4$, $\cos(\alpha+\beta)=3/5$ and $\cos(\alpha-\beta)=4/5$, then evaluate $\sin 2\alpha$.

Sol. We know, $\sin(2\alpha)=\sin\{(\alpha+\beta)+(\alpha-\beta)\}$

$$=\sin(\alpha+\beta)\cdot\cos(\alpha-\beta)+\sin(\alpha-\beta)\cdot\cos(\alpha+\beta).$$

$$=\frac{4}{5}\cdot\frac{4}{5}+\frac{3}{5}\cdot\frac{3}{5}$$

[using $\cos(\alpha+\beta)=3/5$, $\cos(\alpha-\beta)=4/5$
$\Rightarrow\sin(\alpha+\beta)=4/5$, $\sin(\alpha-\beta)=3/5$]

$$=\frac{16+9}{25}=1\Rightarrow\sin 2\alpha=1$$

Example 68. If $\cos\alpha=\frac{1}{2}\left(x+\frac{1}{x}\right)$, $\cos\beta=\frac{1}{2}\left(y+\frac{1}{y}\right)$, then evaluate $\cos(\alpha-\beta)$.

Sol. $\cos\alpha=\frac{1}{2}\left(x+\frac{1}{x}\right)$

$$\Rightarrow x^2-2x\cos\alpha+1=0\Rightarrow x=\frac{2\cos\alpha\pm\sqrt{4\cos^2\alpha-4}}{2}$$

$$\Rightarrow\quad x=\frac{2\cos\alpha\pm 2i\sin\alpha}{2}\qquad\{\text{as }\sqrt{-1}=i\}$$

$\therefore$ $x=\cos\alpha\pm i\sin\alpha$

Similarly, $y=\cos\beta\pm i\sin\beta$

$$\therefore\quad \frac{x}{y}=\frac{\cos\alpha\pm i\sin\alpha}{\cos\beta\pm i\sin\beta}=\cos(\alpha-\beta)\pm i\sin(\alpha-\beta)\quad\ldots(\text{i})$$

$$\text{and}\quad \frac{y}{x}=\frac{\cos\beta\pm i\sin\beta}{\cos\alpha\pm i\sin\alpha}=\cos(\alpha-\beta)\mp i\sin(\alpha-\beta)\quad\ldots(\text{ii})$$

On adding Eqs. (i) and (ii), we get

$$\frac{x}{y}+\frac{y}{x}=2\cos(\alpha-\beta)$$

i.e. $\cos(\alpha-\beta)=\frac{1}{2}\left(\frac{x}{y}+\frac{y}{x}\right)$.

Example 69. If $2\sin\alpha\cos\beta\sin\gamma=\sin\beta\sin(\alpha+\gamma)$. Then, show $\tan\alpha$, $\tan\beta$ and $\tan\gamma$ are in harmonic progression.

Sol. We have, $2\sin\alpha\cos\beta\sin\gamma=\sin\beta\sin(\alpha+\gamma)$

or $2\sin\alpha\cos\beta\sin\gamma=\sin\beta\{\sin\alpha\cos\gamma+\cos\alpha\sin\gamma\}$

$\Rightarrow$ $2\sin\alpha\cos\beta\sin\gamma=\sin\alpha\sin\beta\cos\gamma+\cos\alpha\sin\beta\sin\gamma$

On dividing both sides by $\sin\alpha\sin\beta\sin\gamma$, we get

$$2\cot\beta=\cot\alpha+\cot\gamma\quad\text{or}\quad\frac{2}{\tan\beta}=\frac{1}{\tan\alpha}+\frac{1}{\tan\gamma}$$

i.e. $\frac{1}{\tan\alpha},\frac{1}{\tan\beta},\frac{1}{\tan\gamma}$ are in AP

or $\tan\alpha,\tan\beta,\tan\gamma$ are in HP.

Exercise for Session 6

1. If α lies in II quadrant, β lies in III quadrant and $\tan(\alpha+\beta)>0$, then $(\alpha+\beta)$ lies in .... quadrants.
2. If $3\tan A\tan B=1$, then prove that $\dfrac{\cos(A-B)}{\cos(A+B)}=2$.
3. If $\tan\alpha=\dfrac{m}{m+1}$ and $\tan\beta=\dfrac{1}{2m+1}$, the find the value of $\alpha+\beta$.
4. If $\cos(\alpha+\beta)=\dfrac{4}{5}$, $\sin(\alpha-\beta)=\dfrac{5}{13}$ and $\alpha,\beta\in\left(0,\dfrac{\pi}{4}\right)$, then find the value of $\tan 2\alpha$.
5. If $\alpha+\beta=\dfrac{\pi}{2}$ and $\beta+\gamma=\alpha$, then find the value of $\tan\alpha$.
6. If $\cos(\theta-\alpha)=a$ and $\cos(\theta-\beta)=b$ then the value of $\sin^2(\alpha-\beta)+2ab\cos(\alpha-\beta)$.
7. If $2\cos A=x+\dfrac{1}{x}$, $2\cos\beta=y+\dfrac{1}{y}$ then show that $2\cos(A-B)=\dfrac{x}{y}+\dfrac{y}{x}$.
8. If $y=(1+\tan A)(1-\tan B)$, where $A-B=\dfrac{\pi}{4}$, then find the value of $(y+1)^{y+1}$.

Session 7

Sum of Sines/Cosines in Terms of Products

Converting Product into Sum/ Difference and Vice-Versa

Product into Sum/Difference

1. $2\sin A\cos B=\sin(A+B)+\sin(A-B)$
2. $2\cos A\sin B=\sin(A+B)-\sin(A-B)$
3. $2\cos A\cos B=\cos(A+B)+\cos(A-B)$
4. $2\sin A\sin B=\cos(A-B)-\cos(A+B)$

Proof We know that

$$\sin A\cos B+\cos A\sin B=\sin(A+B) \quad \text{...(i)}$$
$$\sin A\cos B-\cos A\sin B=\sin(A-B) \quad \text{...(ii)}$$
$$\cos A\cos B-\sin A\sin B=\cos(A+B) \quad \text{...(iii)}$$
$$\cos A\cos B+\sin A\sin B=\cos(A-B) \quad \text{...(iv)}$$

Adding Eqs. (i) and (ii), we obtain

$$2\sin A\cos B=\sin(A+B)+\sin(A-B) \quad \text{...(v)}$$

Subtracting Eqs. (ii) from (i), we get

$$2\cos A\sin B=\sin(A+B)-\sin(A-B) \quad \text{...(vi)}$$

Adding Eqs. (iii) and (iv), we get

$$2\cos A\cos B=\cos(A+B)+\cos(A-B) \quad \text{...(vii)}$$

Subtracting Eqs. (iii) from (iv), we get

$$2\sin A\sin B=\cos(A-B)-\cos(A+B) \quad \text{...(viii)}$$

Above four formulas are used to convert product of two sines and cosines into the sum or difference of two sines and cosines.

Sum/Difference into Products

1. $\sin A+\sin B=2\sin\left(\dfrac{A+B}{2}\right)\cos\left(\dfrac{A-B}{2}\right)$
2. $\sin A-\sin B=2\cos\left(\dfrac{A+B}{2}\right)\sin\left(\dfrac{A-B}{2}\right)$
3. $\cos A+\cos B=2\cos\left(\dfrac{A-B}{2}\right)\cdot\cos\left(\dfrac{A+B}{2}\right)$
4. $\cos A-\cos B=2\sin\left(\dfrac{A+B}{2}\right)\cdot\sin\left(\dfrac{B-A}{2}\right)$

Proof (i) Let $A=C+D$ and $B=C-D$, then $C=\dfrac{A+B}{2}$

and $\qquad D=\dfrac{A-B}{2}$

$\therefore$ $\qquad$ L.H.S. $=\sin(C+D)+\sin(C-D)=2\sin C\cos D$

$$=2\sin\frac{A+B}{2}\cos\frac{A-B}{2}=\text{R.H.S}$$

Similarly we proof of (ii), (iii) and (iv).

Some other Useful Results

1. $\tan A + \tan B = \dfrac{\sin(A+B)}{\cos A \cos B}$, where $A, B \neq n\pi + \dfrac{\pi}{2}$
2. $\tan A - \tan B = \dfrac{\sin(A-B)}{\cos A \cos B}$, where $A, B \neq n\pi + \dfrac{\pi}{2}$
3. $\cot A + \cot B = \dfrac{\sin(A+B)}{\sin A \sin B}$, where $A, B \neq n\pi, n \in z$
4. $\cot A - \cot B = \dfrac{\sin(B-A)}{\sin A \sin B}$, where $A, B \neq n\pi, n \in z$

Example 70. Prove that $\cos 55° + \cos 65° + \cos 175° = 0$.

Sol. L.H.S. $= \cos 55° + \cos 65° + \cos 175°$

$$= 2\cos\frac{55°+65°}{2}\cos\frac{55°-65°}{2} + \cos 175°$$

$$= 2\cos 60° \cos(-5°) + \cos 175°$$

$$= 2 \times \frac{1}{2}\cos 5° + \cos(180° - 5°)$$

$$= \cos 5° - \cos 5° = 0$$

Example 71. Prove that $\dfrac{\sin A + \sin 2A + \sin 4A + \sin 5A}{\cos A + \cos 2A + \cos 4A + \cos 5A} = \tan 3A$.

Sol. $\dfrac{\sin A + \sin 2A + \sin 4A + \sin 5A}{\cos A + \cos 2A + \cos 4A + \cos 5A}$

$$= \frac{(\sin 5A + \sin A) + (\sin 4A + \sin 2A)}{(\cos 5A + \cos A) + (\cos 4A + \cos 2A)}$$

$$= \frac{2\sin 3A \cos 2A + 2\sin 3A \cos A}{2\cos 3A \cos 2A + 2\cos 3A \cos A}$$

$$= \frac{2\sin 3A(\cos 2A + \cos A)}{2\cos 3A(\cos 2A + \cos A)} = \tan 3A$$

Example 72. Prove that $(\cos\alpha + \cos\beta)^2 + (\sin\alpha + \sin\beta)^2 = 4\cos^2\left(\dfrac{\alpha-\beta}{2}\right)$.

Sol. L.H.S. $= (\cos\alpha + \cos\beta)^2 + (\sin\alpha + \sin\beta)^2$

$$= \left\{2\cos\left(\frac{\alpha+\beta}{2}\right)\cos\left(\frac{\alpha-\beta}{2}\right)\right\}^2 + \left\{2\sin\left(\frac{\alpha+\beta}{2}\right)\cos\left(\frac{\alpha-\beta}{2}\right)\right\}^2$$

$$= 4\cos^2\left(\frac{\alpha-\beta}{2}\right)\left\{\cos^2\frac{\alpha+\beta}{2} + \sin^2\frac{\alpha+\beta}{2}\right\}$$

$$= 4\cos^2\left(\frac{\alpha-\beta}{2}\right) = \text{R.H.S.}$$

Example 73. If $\sin A = \sin B$ and $\cos A = \cos B$, then prove that $\sin\dfrac{A-B}{2} = 0$.

Sol. We have $\sin A = \sin B$ and $\cos A = \cos B$

or $\quad \sin A - \sin B = 0$ and $\cos A - \cos B = 0$

or $\quad 2\sin\left(\dfrac{A-B}{2}\right)\sin\left(\dfrac{A+B}{2}\right) = 0$

and $\quad -2\sin\left(\dfrac{A-B}{2}\right)\sin\left(\dfrac{A+B}{2}\right) = 0$

or $\sin\dfrac{A-B}{2} = 0$, which is common for both the equations.

Example 74. Prove that $\sin 20° \sin 40° \sin 80° = \dfrac{\sqrt{3}}{8}$

Sol. L.H.S. $= \sin 20° \sin 40° \sin 80° = \dfrac{1}{2}(2\sin 80° \sin 40°)\ \sin 20°$

$$= \frac{1}{2}[\cos(80° - 40°) - \cos(80° + 40°)\sin 20°$$

$$= \frac{1}{2}(\cos 40° - \cos 120°)\sin 20°$$

$$= \frac{1}{4}(2\cos 40° \sin 20° - 2\cos 120° \sin 20°)$$

$$= \frac{1}{4}\left[\sin(40° + 20°) - \sin(40° - 20°) - 2\left(-\frac{1}{2}\right)\sin 20°\right]$$

$$= \frac{1}{4}[\sin 60° - \sin 20° + \sin 20°] = \frac{1}{4}\sin 60°$$

$$= \frac{1}{4}\cdot\frac{\sqrt{3}}{2} = \frac{\sqrt{3}}{8}$$

Example 75. Prove that $\sin A \cdot \sin(60° - A) \cdot \sin(60° + A) = \dfrac{1}{4}\sin 3A$

Sol. L.H.S. $= \sin A \cdot \sin(60° - A) \cdot \sin(60° + A)$

$$= \frac{1}{2}\sin A[2\sin(60° + A)\cdot\sin(60° - A)]$$

$$= \frac{1}{2}\sin A[\cos(60° + A - 60° + A) - \cos(60° + A + 60° - A)]$$

$$= \frac{1}{2}\sin A(\cos 2A - \cos 120°)$$

$$= \frac{1}{4}(2\cos 2A \sin A - 2\cos 120° \sin A)$$

$$= \frac{1}{4}\left[\sin(2A + A) - \sin(2A - A) - 2\left(-\frac{1}{2}\right)\sin A\right]$$

$$= \frac{1}{4}(\sin 3A - \sin A - \sin A) = \frac{1}{4}\sin 3A$$

Exercise for Session 7

1. Show that $\sin x + \sin 3x + \sin 5x + \sin 7x = 4 \sin 4x \cos 2x \cos x$.
2. Show that $\sin A \cdot \sin(B - C) + \sin B \cdot \sin(C - A) + \sin C \cdot \sin(A - B) = 0$.
3. Show that $\cos \alpha + \cos \beta + \cos \gamma + \cos(\alpha + \beta + \gamma) = 4 \cos \frac{\alpha + \beta}{2} \cdot \cos \frac{\beta + \gamma}{2} \cdot \cos \frac{\gamma + \alpha}{2}$.
4. If x and y are acute angles, such that $\cos x + \cos y = \frac{3}{2}$ and $\sin x + \sin y = \frac{3}{4}$, then the value of $\sin(x + y)$.
5. Find the value of expression $2 \cos \frac{\pi}{3} \cos \frac{9\pi}{13} + \cos \frac{3\pi}{13} + \cos \frac{5\pi}{13}$.
6. Find the value of $\left(\frac{\cos A + \cos B}{\sin A - \sin B}\right)^n + \left(\frac{\sin A + \sin B}{\cos A - \cos B}\right)^n$ (where, n is an even)
7. Find the value of $\left(1 + \cos \frac{\pi}{8}\right)\left(1 + \cos \frac{3\pi}{8}\right)\left(1 + \cos \frac{5\pi}{8}\right)\left(1 + \cos \frac{7\pi}{8}\right)$.
8. In a triangle ABC, $\cos 3A + \cos 3B + \cos 3C = 1$, then find any one angle.

Session 8

Trigonometric Ratios of Multiples of an Angle

Trigonometric Ratios of Multiples of an Angle

Definition An angle of the form nA, where n is an integer is called a multiple angle, for example $2A, 3A, 4A, \ldots$ etc. are multiple angles of A.

In this session we shall express trigonometrical ratios of multiple angles of A in terms of trigonometrical ratios of A.

Trigonometrical Ratios of 2*A* in term of Trigonometrical Ratio of *A*

1. $\sin 2A = 2 \sin A \cos A = \frac{2 \tan A}{1 + \tan^2 A}$
2. $\cos 2A = \cos^2 A - \sin^2 A = 1 - 2 \sin^2 A$
$$= 2\cos^2 A - 1 = \frac{1 - \tan^2 A}{1 + \tan^2 A}$$
3. $1 + \cos 2A = 2 \cos^2 A, 1 - \cos 2A = 2 \sin^2 A$

or $\frac{1 + \cos 2A}{2} = \cos^2 A, \frac{1 - \cos 2A}{2} = \sin^2 A$

4. $\tan 2A = \frac{2 \tan A}{1 - \tan^2 A}$, where $A \neq (2n + 1)\frac{\pi}{4}$

Proof $\sin 2A = \sin(A + A) = \sin A \cos A + \cos A \sin A$

[using the formula $\sin(A + B) = \sin A \cos B + \cos A \sin B$]

$$\boxed{= 2 \sin A \cos A}$$

$$\cos 2A = \cos(A + A) = \cos A \cos A - \sin A \sin A$$

$$\boxed{= \cos^2 A - \sin^2 A}$$

$$\tan 2A = \tan (A + A)$$

$$= \frac{\tan A + \tan A}{1 - \tan A \tan A} = \boxed{\frac{2 \tan A}{1 - \tan^2 A}}$$

Trigonometrical Ratios of 3*A* in terms of Trigonometrical Ratio of *A*

1. $\sin 3A = 3 \sin A - 4 \sin^3 A$
$$= 4 \sin(60° - A) \cdot \sin A \cdot \sin(60° + A)$$
2. $\cos 3A = 4 \cos^3 A - 3 \cos A$
$$= 4 \cos(60° - A) \cos A \cos(60° + A)$$
3. $\tan 3A = \frac{3 \tan A - \tan^3 A}{1 - 3 \tan^2 A}$

Proof

1. $\sin 3A = \sin(2A + A) = \sin 2A \cos A + \cos 2A \sin A$

$= 2 \sin A \cos A \cdot \cos A + (1 - 2\sin^2 A) \sin A$

$= 2 \sin A \cos^2 A + \sin A - 2 \sin^3 A$

$= 2 \sin A(1 - \sin^2 A) + \sin A - 2 \sin^3 A$

$= 2 \sin A - 2 \sin^3 A + \sin A - 2 \sin^3 A$

$= \boxed{3 \sin A - 4 \sin^3 A}$

2. $\cos 3A = \cos(2A + A) = \cos 2A \cdot A \cos A - \sin 2A \sin A$

$= (2 \cos^2 A - 1) \cos A - 2 \sin A \cos A \cdot \sin A$

$= 2 \cos^3 A - \cos A - 2 \cos A(1 - \cos^2 A)$

$= 2 \cos^3 A - \cos A - 2 \cos A + 2 \cos^3 A$

$= \boxed{4 \cos^3 A - 3 \cos A}$

3. $\tan 3A = \dfrac{\sin 3A}{\cos 3A} = \dfrac{3 \sin A - 4 \sin^3 A}{4 \cos^3 A - 3 \cos A}$

$= \dfrac{\sin A(3 - 4 \sin^2 A)}{\cos A(4 \cos^2 A - 3)} = \dfrac{\tan A(3 - 4 \sin^2 A)}{4 \cos^2 A - 3}$

On dividing by $\cos^2 A$ numerator and denominator

$= \dfrac{\tan A(3 \sec^2 A - 4 \tan^2 A)}{4 - 3 \sec^2 A}$

$= \dfrac{\tan A(3 + 3 \tan^2 A - 4 \tan^2 A)}{4 - 3 - 3 \tan^2 A}$

$= \dfrac{\tan A(3 - \tan^2 A)}{1 - 3 \tan^2 A} = \dfrac{3 \tan A - \tan^3 A}{1 - 3 \tan^2 A}$

Example 76. If $\sin A = \frac{3}{5}$, where $0° < A < 90°$, find the values of $\sin 2A$, $\cos 2A$, $\tan 2A$ and $\sin 4A$.

Sol. We have, $\sin A = \frac{3}{5}$, where $0° < A < 90°$

$\therefore \quad \cos^2 A = 1 - \sin^2 A$

$\Rightarrow \quad \cos A = +\sqrt{1 - \sin^2 A} = \sqrt{1 - \dfrac{9}{25}} = \dfrac{4}{5}$

$\therefore \quad \tan A = \dfrac{\sin A}{\cos A} = \dfrac{3}{4}$

$\sin 2A = 2 \sin A \cos A = 2 \times \dfrac{3}{5} \times \dfrac{4}{5} = \dfrac{24}{25}$

$\cos 2A = 1 - 2 \sin^2 A = 1 - 2 \times \left(\dfrac{3}{5}\right)^2 = \dfrac{7}{25}$

$\tan 2A = \dfrac{2 \tan A}{1 - \tan^2 A} = \dfrac{2 \times \frac{3}{4}}{1 - \left(\frac{3}{4}\right)^2} = \dfrac{\frac{6}{4}}{1 - \frac{9}{16}} = \dfrac{24}{7}$

$\left[\because \tan A = \dfrac{3}{4}\right]$

$\sin 4A = \sin 2A \cos 2A = 2 \times \dfrac{24}{25} \times \dfrac{7}{25} = \dfrac{336}{625}$

$\left[\because \sin 2A = \dfrac{24}{25} \text{ and } \cos 2A = \dfrac{7}{25}\right]$

Example 77. Prove that $\dfrac{1 + \sin 2\theta}{1 - \sin 2\theta} = \left(\dfrac{1 + \tan\theta}{1 - \tan\theta}\right)^2$.

Sol. L.H.S. $= \dfrac{1 + \sin 2\theta}{1 - \sin 2\theta} = \dfrac{\sin^2\theta + \cos^2\theta + 2\sin\theta\cos\theta}{\sin^2\theta + \cos^2\theta - 2\sin\theta\cos\theta}$

$= \left(\dfrac{\sin\theta + \cos\theta}{\sin\theta - \cos\theta}\right)^2 = \left(\dfrac{1 + \tan\theta}{1 - \tan\theta}\right)^2$

[dividing numerator and denominator by $\cos\theta$]

Example 78. Prove that $\dfrac{1 - \tan^2\left(\frac{\pi}{4} - 4\right)}{1 + \tan^2\left(\frac{\pi}{4} - A\right)} = \sin 2A$.

Sol. $\dfrac{1 + \tan^2\left(\frac{\pi}{4} - A\right)}{1 + \tan^2\left(\frac{\pi}{4} - A\right)} = \dfrac{1 - \tan^2\theta}{1 + \tan^2\theta} \left(\text{where } \dfrac{\pi}{4} - A = \theta\right)$

$= \cos 2\theta = \cos\left(\dfrac{\pi}{2} - 2A\right) = \sin 2A$

Example 79. Prove that $\dfrac{\sec 8\theta - 1}{\sec 4\theta - 1} = \dfrac{\tan 8\theta}{\tan 2\theta}$

Sol. We have, LHS $= \dfrac{\sec 8\theta - 1}{\sec 4\theta - 1}$

$\Rightarrow \quad \text{LHS} = \dfrac{\frac{1}{\cos 8\theta} - 1}{\frac{1}{\cos 4\theta} - 1} = \dfrac{1 - \cos 8\theta}{\cos 8\theta} \cdot \dfrac{\cos 4\theta}{1 - \cos\theta}$

$\Rightarrow \quad \text{LHS} = \dfrac{2\sin^2 4\theta}{\cos 8\theta} \cdot \dfrac{\cos 4\theta}{2\sin^2 2\theta}$

$\left[\because 1 - \cos 8\theta = 2\sin^2\dfrac{8\theta}{2} = 2\sin^2 4\theta \text{ and, } 1 - \cos 4\theta = 2\sin^2\dfrac{4\theta}{2} = 2\sin^2 2\theta\right]$

$\Rightarrow \quad \text{LHS} = \dfrac{(2\sin 4\theta\cos 4\theta)}{\cos 8\theta} \times \dfrac{\sin 4\theta}{2\sin^2 2\theta}$

$\Rightarrow \quad \text{LHS} = \left(\dfrac{2\sin 4\theta\cos 4\theta}{\cos 8\theta}\right) \times \left(\dfrac{2\sin 2\theta\cos 2\theta}{2\sin^2 2\theta}\right)$

$$\Rightarrow \quad LHS = \left(\frac{\sin 2(4\theta)}{\cos 8\theta}\right) \times \left(\frac{\cos 2\theta}{\sin 2\theta}\right)$$

$$= \left(\frac{\sin 8\theta}{\cos 8\theta}\right) \times \left(\frac{\cos 2\theta}{\sin 2\theta}\right) = \tan 8\theta \cot 2\theta$$

$$\Rightarrow \quad LHS = \frac{\tan 8\theta}{\tan 2\theta} = RHS$$

Example 80. Show that

$$\sqrt{2+\sqrt{2+\sqrt{2+2\cos 8}}} = 2\cos\theta$$

Sol. We have, $LHS = \sqrt{2+\sqrt{2+\sqrt{2(1+\cos 8\theta)}}}$

$$\Rightarrow \quad LHS = \sqrt{2+\sqrt{2+\sqrt{2(2\cos^2 4\theta)}}} \quad \left[\because 1+\cos 8\theta = 2\cos^2\frac{8\theta}{2}\right]$$

$$\Rightarrow \quad LHS = \sqrt{2+\sqrt{2+\sqrt{(4\cos^2 4\theta)}}}$$

$$\Rightarrow \quad LHS = \sqrt{2+\sqrt{2+2\cos 4\theta}}$$

$$\Rightarrow \quad LHS = \sqrt{2+\sqrt{2(1+\cos 4\theta)}}$$

$$\Rightarrow \quad LHS = \sqrt{2+\sqrt{2(2\cos^2 2\theta)}} \quad [\because 1+\cos 4\theta = 2\cos^2 2\theta]$$

$$\Rightarrow \quad LHS = \sqrt{2+2\cos 2\theta} = \sqrt{2(1+\cos 2\theta)}$$

$$= \sqrt{2(2\cos^2\theta)} = 2\cos\theta = RHS$$

Example 81. Show that $\sqrt{3}\,\text{cosec}\,20^\circ - \sec 20^\circ = 4$

Sol. We have, $LHS = \sqrt{3}\,\text{cosec}\,20^\circ - \sec 20^\circ$

$$\Rightarrow \quad LHS = \frac{\sqrt{3}}{\sin 20^\circ} - \frac{1}{\cos 20^\circ}$$

$$\Rightarrow \quad LHS = \frac{\sqrt{3}\cos 20^\circ - \sin 20^\circ}{\sin 20^\circ \cos 20^\circ}$$

$$\Rightarrow \quad LHS = \frac{2\left\{\frac{\sqrt{3}}{2}\cos 20^\circ - \frac{1}{2}\sin 20^\circ\right\}}{\sin 20^\circ \cos 20^\circ}$$

$$\Rightarrow \quad LHS = \frac{2(\sin 60^\circ \cos 20^\circ - \cos 60^\circ \sin 20^\circ)}{\sin 20^\circ \cos 20^\circ}$$

$$\Rightarrow \quad LHS = \frac{2\sin(60^\circ - 20^\circ)}{\sin 20^\circ \cos 20^\circ} \Rightarrow LHS = \frac{2\sin 40^\circ}{\sin 20^\circ \cos 20^\circ}$$

$$= \frac{4\sin 40^\circ}{2\sin 20^\circ \cos 20^\circ} = \frac{4\sin 40^\circ}{\sin 40^\circ} = 4 = RHS$$

Example 82. Prove that $\left(1+\cos\frac{\pi}{8}\right)\left(1+\cos\frac{3\pi}{8}\right)\left(1+\cos\frac{5\pi}{8}\right)\left(1+\cos\frac{7\pi}{8}\right) = \frac{1}{8}$.

Sol. We have,

$$\cos\frac{7\pi}{8} = \cos\left(\pi - \frac{\pi}{8}\right) = -\cos\frac{\pi}{8}$$

and $$\cos\frac{5\pi}{8} = \cos\left(\pi - \frac{3\pi}{8}\right) = -\cos\frac{3\pi}{8}$$

$$\therefore LHS = \left(1+\cos\frac{\pi}{8}\right)\left(1+\cos\frac{3\pi}{8}\right)\left(1-\cos\frac{3\pi}{8}\right)\left(1-\cos\frac{\pi}{8}\right)$$

$$\Rightarrow LHS = \left(1-\cos^2\frac{\pi}{8}\right)\left(1-\cos^2\frac{3\pi}{8}\right)$$

$$\Rightarrow LHS = \sin^2\frac{\pi}{8}\sin^2\frac{3\pi}{8}$$

$$\Rightarrow LHS = \frac{1}{4}\left(3\sin^2\frac{\pi}{8}\right)\left(2\sin^2\frac{3\pi}{8}\right)$$

$$\Rightarrow LHS = \frac{1}{4}\left[\left(1-\cos\frac{\pi}{4}\right)\left(1-\cos\frac{3\pi}{4}\right)\right] \quad \left[\because 1-\cos\theta = 2\sin^2\frac{\theta}{2}\right]$$

$$\Rightarrow LHS = \frac{1}{4}\left[\left(1-\frac{1}{\sqrt{2}}\right)\left(1+\frac{1}{\sqrt{2}}\right)\right] = \frac{1}{4}\left(1-\frac{1}{2}\right) = \frac{1}{8} = RHS$$

Example 83. If $\tan^2\theta = 2\tan^2\phi + 1$, prove that $\cos 2\theta + \sin^2\phi = 0$.

Sol. We have, $\cos 2\theta = \dfrac{1-\tan^2\theta}{1+\tan^2\theta}$

$$\Rightarrow \quad \cos 2\theta = \frac{1-(2\tan^2\phi+1)}{1+2\tan^2\phi+1} \quad [\because \tan^2\theta = 2\tan^2\phi + 1]$$

$$\Rightarrow \quad \cos 2\theta = \frac{-2\tan^2\phi}{2+2\tan^2\phi} = \frac{-\tan^2\phi}{\sec^2\phi} = -\sin^2\phi$$

$$\therefore \quad \cos 2\theta + \sin^2\phi = 0$$

Example 84. Prove that $\tan\alpha + 2\tan 2\alpha + 4\tan 4\alpha + 8\cos 8\alpha = \cot\alpha$

Sol. We have, $\cot\theta - \tan\theta = \dfrac{1}{\tan\theta} - \tan\theta$

$$= \frac{1-\tan^2\theta}{\tan\theta} = 2\left\{\frac{1-\tan^2\theta}{2\tan\theta}\right\}$$

$$\Rightarrow \quad \cot\theta - \tan\theta = \frac{2}{\tan 2\theta}$$

$$\Rightarrow \quad \cot\theta - \tan\theta = 2\cot 2\theta \qquad \text{...(i)}$$

We have to prove that

$$\tan\alpha + 2\tan 2\alpha + 4\tan 4\alpha + 8\cot 8\alpha = \cot\alpha$$

or, $\cot\alpha - \tan\alpha - 2\tan\alpha - 4\tan 4\alpha - 8\cot 8\alpha = 0$

Now,

$$LHS = \cot\alpha - \tan\alpha - 2\tan 2\alpha - 4\tan 4\alpha - 8\cot 8\alpha$$

$$\Rightarrow LHS = (\cot\alpha - \tan\alpha) - 2\tan 2\alpha - 4\tan 4\alpha - 8\cot 8\alpha$$

$$\Rightarrow LHS = 2\cot 2\alpha - 2\tan 2\alpha - 4\tan 4\alpha - 8\cot 8\alpha \quad \text{[using (i)]}$$

$$\Rightarrow LHS = 2(\cot 2\alpha - \tan 2\alpha) - 4\tan 4\alpha - 8\cot 8\alpha$$

$\Rightarrow$ LHS = 2(2 cot 4α) − 4 tan 4α − 8 cot 8α

[On replacing θ by 2α in Eq. (i)]

$\Rightarrow$ LHS = 4 cot 4α − 4 tan 4α − 8 cot 8α

$\Rightarrow$ LHS = 4(cot 4α − tan 4α) − 8 cot 8α

$\Rightarrow$ LHS = 4(2 cot 8α) − 8 cot 8α

[On replacing θ by 4α in Eq. (i)]

$\Rightarrow$ LHS = 8 cot 8α − 8 cot 8α $\Rightarrow$ LHS = 0 = RHS

Example 85. Determine the smallest positive value of x (in degrees) for which $\tan(x+100^\circ) = \tan(x+50^\circ)\tan x \tan(x-50^\circ)$

Sol. We have, $\tan(x+100^\circ) = \tan(x+50^\circ)\tan x\tan(x-50^\circ)$

$$\Rightarrow \frac{\tan(x+100^\circ)}{\tan(x+50^\circ)} = \tan(x+50^\circ)\tan x^\circ$$

$$\Rightarrow \frac{\sin(x+100^\circ)\cos(x-50^\circ)}{\cos(x+100^\circ)\sin(x-50^\circ)} = \frac{\sin(x+50^\circ)\sin x}{\cos(x+50^\circ)\cos x}$$

$$\Rightarrow \frac{\sin(x+100^\circ)\cos(x-50^\circ)+\cos(x+100^\circ)\sin(x-50^\circ)}{\sin(x+100^\circ)\cos(x-50^\circ)-\cos(x+100^\circ)\sin(x-50^\circ)} = \frac{\sin(x+50^\circ)\sin x+\cos(x+50^\circ)\cos x}{\sin(x+50^\circ)\sin x-\cos(x+50^\circ)\cos x}$$

$$\Rightarrow \frac{\sin(x+100^\circ+x-50^\circ)}{\sin(x+100^\circ-x+50^\circ)} = \frac{\cos(x+50^\circ-x)}{-\cos(x+50^\circ+x)}$$

$$\Rightarrow \frac{\sin(2x+50^\circ)}{\sin 150^\circ} = \frac{\cos 50^\circ}{-\cos(2x+50^\circ)}$$

$$\Rightarrow \sin(2x+50^\circ)\cos(2x+50^\circ) = -\sin 150^\circ\cos 50^\circ$$

$$\Rightarrow 2\sin(2x+50^\circ)\cos(2x+50^\circ) = -2\cos 60^\circ\cos 50^\circ$$

$[\because \sin 150^\circ = \cos 60^\circ]$

$$\Rightarrow \sin(4x+100^\circ) = \sin(270-50^\circ)$$

$$\Rightarrow \sin(4x+100^\circ) = \sin 220^\circ$$

$$\Rightarrow 4x+100^\circ = 220^\circ \Rightarrow x = 30^\circ$$

Example 86. Prove that

$$\frac{\sin x}{\cos 3x}+\frac{\sin 3x}{\cos 9x}+\frac{\sin 9x}{\cos 27x} = \frac{1}{2}(\tan 27x-\tan x)$$

Sol. We have, $\dfrac{\sin x}{\cos 3x}+\dfrac{\sin 3x}{\cos 9x}+\dfrac{\sin 9x}{\cos 27x}$

$$= \frac{1}{2}\left\{\frac{2\sin x\cos x}{\cos 3x\cos x}+\frac{2\sin 3x\cos 3x}{\cos 9x\cos 3x}+\frac{2\sin 9x\cos 9x}{\cos 27x\cos 9x}\right\}$$

$$= \frac{1}{2}\left\{\frac{\sin 2x}{\cos 3x\cos x}+\frac{\sin 6x}{\cos 9x\cos 3x}+\frac{\sin 18x}{\cos 27x\cos 9x}\right\}$$

$$= \frac{1}{2}\left\{\frac{\sin(3x-x)}{\cos 3x\cos x}+\frac{\sin(9x-3x)}{\cos 9x\cos 3x}+\frac{\sin(27x-9x)}{\cos 27x\cos 9x}\right\}$$

$$= \frac{1}{2}\{(\tan 3x-\tan x)+(\tan 9x-\tan 3x)+(\tan 27x-\tan 9x)\}$$

$$= \frac{1}{2}(\tan 27x-\tan x)$$

Exercise for Session 8

1. This question has statement which is true or false.

   If $\dfrac{\pi}{9}<\theta<\dfrac{\pi}{2}$, then the value of $\sqrt{1-\sin 2\theta} = \cos\theta-\sin\theta$.

2. If $\pi<\theta<\dfrac{3\pi}{2}$, then find the value of $\sqrt{\dfrac{1-\cos 2\theta}{1+\cos 2\theta}}$.

3. If $\tan x = -\dfrac{4}{3}$, x lies in II quadrant, then find the value of $\sin\dfrac{x}{2}$.

4. Prove that $\sin^4\dfrac{\pi}{8}+\sin^4\dfrac{3\pi}{8}+\sin^4\dfrac{5\pi}{8}+\sin^4\dfrac{7\pi}{8} = \dfrac{3}{2}$.

5. If $A = 2\sin^2\theta-\cos 2\theta$ and $A\in[\alpha,\beta]$, then find the values of α and β.

6. If $\sin x+\cos x = \dfrac{1}{5}$, then find the value of $\tan 2x$.

7. If $\tan 3A = \dfrac{3\tan A+k\tan^3 A}{1-3\tan^2 A}$, then k is equal to

8. If $\tan A+2\tan 2A+4\tan 4A+8\cot 8A = k\cot A$, then find the value of k.

9. If $m^2\cos\dfrac{2\pi}{15}\cos\dfrac{4\pi}{15}\cos\dfrac{8\pi}{15}\cos\dfrac{14\pi}{15} = n^2$, then find the value of $\dfrac{m^2-n^2}{n^2}$.

10. If $(2^n+1)\theta = \pi$, then find the value of $2^n\cos\theta\cos 2\theta\cos 2^2\theta\ldots\cos 2^{n-1}\theta$.

Session 9

Trigonometric Ratios of Submultiple of an Angle

Definition

An angle of the form $\frac{A}{n}$, where n is an integer is called a submultiple angle of A.

For example $\frac{A}{2}, \frac{A}{3}, \frac{A}{4} \ldots$ etc., are submultiple angles of A.

In this session we shall express the trigonometric ratios of A in terms of the trigonometric ratios of submultiple angles $\frac{A}{2}, \frac{A}{3}, \ldots$ etc., and *vice-versa.*

Trigonometric Ratios of A in Terms of Trigonometric Ratios of $\frac{A}{2}$

(i) $\sin 2A = 2 \sin A \cos A$. Putting $\frac{A}{2}$ in place of A, we get

$\sin A = 2 \sin \frac{A}{2} \cos \frac{A}{2}$

(ii) $\cos 2A = \cos^2 A - \sin^2 A$. Putting $\frac{A}{2}$ in place of A, we get $\cos A = \cos^2 \frac{A}{2} - \sin^2 \frac{A}{2}$

(iii) $\cos 2A = 2\cos^2 A - 1$. Putting $\frac{A}{2}$ in place of A, we get

$\cos A = 2\cos^2 \frac{A}{2} - 1$

(iv) $\cos 2A = 1 - 2\sin^2 A$. Putting $\frac{A}{2}$ in place of A, we get

$\cos A = 1 - 2\sin^2 \frac{A}{2}$

(v) $\tan 2A = \frac{2 \tan A}{1 - \tan^2 A}$

$\therefore \tan A = \frac{2 \tan \frac{A}{2}}{1 - \tan^2 \frac{A}{2}}$, putting $\frac{A}{2}$ in place of A

(vi) $\sin 2A = \frac{2 \tan A}{1 + \tan^2 A}$ $\therefore \sin A = \frac{2 \tan \frac{A}{2}}{1 + \tan^2 \frac{A}{2}}$, putting $\frac{A}{2}$ in place of A

(vii) $\cos 2A = \frac{1 - \tan^2 A}{1 + \tan^2 A}$ $\therefore \cos A = \frac{1 - \tan^2 \frac{A}{2}}{1 + \tan^2 \frac{A}{2}}$, putting $\frac{A}{2}$ in place of A

(viii) $\cot 2A = \frac{\cot^2 A - 1}{2 \cot A}$ $\therefore \cot A = \frac{\cot^2 \frac{A}{2} - 1}{2 \cot \frac{A}{2}}$, putting $\frac{A}{2}$ in place of A

Trigonometric Ratios of A in Terms of Trigonometric Ratios of $\frac{A}{3}$

(i) $\sin 3A = 3 \sin A - 4 \sin^3 A$. Putting $\frac{A}{3}$ in place of A, we get $\sin A = 3 \sin \frac{A}{3} - 4 \sin^3 \frac{A}{3}$

(ii) $\cos 3A = 4\cos^3 A - 3 \cos A$

$\therefore \cos A = 4\cos^3 \frac{A}{3} - 3 \cos \frac{A}{3}$, putting $\frac{A}{3}$ in place of A

(iii) $\tan 3A = \frac{3 \tan A - \tan^3 A}{1 - 3\tan^2 A}$, putting $\frac{A}{3}$ in place of A

$\therefore \tan A = \frac{3 \tan \frac{A}{3} - \tan^3 \frac{A}{3}}{1 - 3 \tan^2 \frac{A}{3}}$

Values of $\cos \frac{A}{2}$, $\sin \frac{A}{2}$ and $\tan \frac{A}{2}$ in Terms of $\cos A$

(i) $\cos^2 \frac{A}{2} = \frac{1 + \cos A}{2}$ $\therefore \cos \frac{A}{2} = \pm \sqrt{\frac{1 + \cos A}{2}}$

(ii) $\sin^2 \frac{A}{2} = \frac{1-\cos A}{2} \quad \therefore \sin \frac{A}{2} = \pm\sqrt{\frac{1-\cos A}{2}}$

(iii) $\tan^2 \frac{A}{2} = \frac{1-\cos A}{1+\cos A} \quad \therefore \tan \frac{A}{2} = \pm\sqrt{\frac{1-\cos A}{1+\cos A}}$

Note

If $\cos A$ is given, then there will be two values of $\cos \frac{A}{2}$, $\sin \frac{A}{2}$ and $\tan \frac{A}{2}$ but if A is given, then there will be only one value of $\cos \frac{A}{2}$, $\sin \frac{A}{2}$ and $\tan \frac{A}{2}$ because + sign or – sign before the radical sign can be fixed by knowing the quadrant in which $\frac{A}{2}$ lies.

Values of $\sin \frac{A}{2}$ and $\cos \frac{A}{2}$ in Terms of $\sin A$

$$\left(\cos \frac{A}{2} + \sin \frac{A}{2}\right)^2 = \cos^2 \frac{A}{2} + \sin^2 \frac{A}{2} + 2\cos \frac{A}{2} \sin \frac{A}{2}$$
$$= 1 + \sin A$$

$$\therefore \quad \cos \frac{A}{2} + \sin \frac{A}{2} = \pm\sqrt{1+\sin A} \quad \text{...(i)}$$

Similarly, $$\cos \frac{A}{2} - \sin \frac{A}{2} = \pm\sqrt{1-\sin A} \quad \text{...(ii)}$$

Adding Eqs. (i) and (ii), we get

$$\cos \frac{A}{2} = \pm\frac{1}{2}\sqrt{1+\sin A} \pm \frac{1}{2}\sqrt{1-\sin A} \quad \text{...(iii)}$$

Subtracting Eq. (ii) from Eq. (i), we get

$$\sin \frac{A}{2} = \pm\frac{1}{2}\sqrt{1+\sin A} \mp \frac{1}{2}\sqrt{1-\sin A} \quad \text{...(iv)}$$

Note

If $\sin A$ is given, then there will be 4 values of $\sin \frac{A}{2}$ and $\cos \frac{A}{2}$ but if A is given. there will be one and only one value of $\cos \frac{A}{2}$ and $\sin \frac{A}{2}$ because the + or – sign can be fixed before the radical sign in the following way

$$\cos \frac{A}{2} + \sin \frac{A}{2} = \sqrt{2}\left(\frac{1}{\sqrt{2}}\cos \frac{A}{2} + \frac{1}{\sqrt{2}}\sin \frac{A}{2}\right)$$
$$= \sqrt{2}\left(\sin \frac{\pi}{4}\cos \frac{A}{2} + \cos \frac{\pi}{4}\sin \frac{A}{2}\right) = \sqrt{2}\sin\left(\frac{\pi}{4} + \frac{A}{2}\right)$$

Similarly, $\cos \frac{A}{2} - \sin \frac{A}{2} = \sqrt{2}\cos\left(\frac{\pi}{4} + \frac{\pi}{2}\right)$

Thus the sign of $\cos \frac{A}{2} + \sin \frac{A}{2}$ and $\cos \frac{A}{2} - \sin \frac{A}{2}$ can be fixed by knowing the quadrant in which $\frac{\pi}{4} + \frac{A}{2}$ lies.

Values of Trigonometric Ratios of Some Particular Angles

I. (i) Value of sin 18°

Let $\theta = 18°$, then $5\theta = 90° \therefore 2\theta + 3\theta = 90°$

or $2\theta = 90° - 3\theta \ \therefore \ \sin 2\theta = \sin(90° - 3\theta)$

or $\sin 2\theta = \cos 3\theta$ or $2\sin\theta\cos\theta = 4\cos^3\theta - 3\cos\theta$

or $2\sin\theta = 4\cos^2\theta - 3$ [dividing by $\cos\theta$]

or $2\sin\theta = 4(1-\sin^2\theta) - 3 = 1 - 4\sin^2\theta$

or $4\sin^2\theta + 2\sin\theta - 1 = 0$

$$\therefore \sin\theta = \frac{-2\pm\sqrt{4+16}}{8} = \frac{-2\pm2\sqrt{5}}{8} = \frac{-1\pm\sqrt{5}}{4}$$

Thus $\sin\theta = \frac{-1+\sqrt{5}}{4}, \frac{-1-\sqrt{5}}{4}$

$\because \theta = 18°$

$\therefore \sin\theta = \sin 18° > 0$, for 18° lies in the 1st quadrant

$\therefore \sin\theta$ i.e., $\sin 18° = \frac{\sqrt{5}-1}{4}$

(ii) Value of cos 18°

$$\cos^2 18° = 1 - \sin^2 18° = 1 - \left(\frac{\sqrt{5}-1}{4}\right)^2$$
$$= 1 - \frac{5+1-2\sqrt{5}}{16} = 1 - \frac{6-2\sqrt{5}}{16}$$
$$= \frac{16-6+2\sqrt{5}}{16} = \frac{10+2\sqrt{5}}{16}$$

$$\therefore \quad \cos 18° = \frac{1}{4}\sqrt{10+2\sqrt{5}} \quad [\because \cos 18° > 0]$$

(iii) Value of tan 18°

$$\tan 18° = \frac{\sin 18°}{\cos 18°} = \frac{\sqrt{5}-1}{4} \div \frac{\sqrt{10+2\sqrt{5}}}{4}$$
$$= \frac{\sqrt{5}-1}{\sqrt{10+2\sqrt{5}}}$$

(iv) Value of cos 72° and sin 72°

(a) $\cos 72° = \cos(90° - 18°) = \sin 18° = \frac{\sqrt{5}-1}{4}$

(b) $\sin 72° = \sin(90° - 18°) = \cos 18° = \frac{1}{4}\sqrt{10+2\sqrt{5}}$

II. (i) Value of cos 36°

$$\cos 36° = 1 - 2\sin^2 18° = 1 - 2\times\left(\frac{\sqrt{5}-1}{4}\right)^2$$

$$=1-2\times\frac{5+1-2\sqrt{5}}{16}$$

$$=1-\frac{3-\sqrt{5}}{4}$$

$$=\frac{4-3+\sqrt{5}}{4}=\frac{\sqrt{5}+1}{4}$$

Thus, $\cos 36^\circ=\dfrac{\sqrt{5}+1}{4}$

(ii) **Value of sin 36°**

$$\sin^2 36^\circ=1-\cos^2 36^\circ=1-\left(\frac{\sqrt{5}+1}{4}\right)^2$$

$$=1-\frac{6+2\sqrt{5}}{16}=\frac{16-6-2\sqrt{5}}{16}=\frac{10-2\sqrt{5}}{16}$$

$$\therefore\quad \sin 36^\circ=\frac{1}{4}\sqrt{10-2\sqrt{5}}\qquad [\because \sin 36^\circ>0]$$

(iii) **Values of sin 54° and cos 54°**

(a) $\sin 54^\circ=\sin(90^\circ-36^\circ)=\cos 36^\circ=\dfrac{\sqrt{5}+1}{4}$

(b) $\cos 54^\circ=\cos(90^\circ-36^\circ)=\sin 36^\circ=\dfrac{1}{4}(\sqrt{10-2\sqrt{5}})$

III. (i) **Value of $\tan 7\frac{1^\circ}{2}$**

Let $\theta=7\frac{1^\circ}{2}$, then $2\theta=15^\circ$

Now, $\tan\theta=\dfrac{1-\cos 2\theta}{\sin 2\theta}$

$[\because 1-\cos 2\theta=2\sin^2\theta \text{ and } \sin 2\theta=2\sin\theta\cos\theta]$

$$=\frac{1-\cos 15^\circ}{\sin 15^\circ}=\frac{1-\dfrac{\sqrt{3}+1}{2\sqrt{2}}}{\dfrac{\sqrt{3}-1}{2\sqrt{2}}}$$

$$=\frac{2\sqrt{2}-\sqrt{3}-1}{\sqrt{3}-1}=\frac{(2\sqrt{2}-\sqrt{3}-1)(\sqrt{3}+1)}{(\sqrt{3}-1)(\sqrt{3}+1)}$$

$$=\frac{2\sqrt{6}-3-\sqrt{3}+2\sqrt{2}-\sqrt{3}-1}{3-1}$$

$$=\frac{2\sqrt{6}-2\sqrt{3}-4+2\sqrt{2}}{2}$$

$$=\sqrt{6}-\sqrt{3}-2+\sqrt{2}=\sqrt{3}(\sqrt{2}-1)-\sqrt{2}(\sqrt{2}-1)$$

$$=(\sqrt{3}-\sqrt{2})(\sqrt{2}-1)]$$

(ii) **Value of $\cot 82\frac{1^\circ}{2}$**

$$\cot 82\frac{1^\circ}{2}=\cot\left(90^\circ-7\frac{1^\circ}{2}\right)$$

$$=\tan 7\frac{1^\circ}{2}=(\sqrt{3}-\sqrt{2})(\sqrt{2}-1)$$

(iii) **Value of $\cot 7\frac{1^\circ}{2}$**

Let $\theta=7\frac{1^\circ}{2}$, then $2\theta=15^\circ$

Now, $\cot\theta=\dfrac{1+\cos 2\theta}{\sin 2\theta}=\dfrac{1+\cos 15^\circ}{\sin 15^\circ}$

$$=\frac{1+\dfrac{\sqrt{3}+1}{2\sqrt{2}}}{\dfrac{\sqrt{3}-1}{2\sqrt{2}}}=\frac{2\sqrt{2}+\sqrt{3}+1}{\sqrt{3}-1}$$

$$=\frac{(2\sqrt{2}+\sqrt{3}+1)(\sqrt{3}+1)}{(\sqrt{3}-1)(\sqrt{3}+1)}$$

$$=\frac{2\sqrt{6}+3+\sqrt{3}+2\sqrt{2}+\sqrt{3}+1}{3-1}$$

$$=\frac{2\sqrt{6}+2\sqrt{3}+2\sqrt{2}+4}{2}=\sqrt{6}+\sqrt{3}+\sqrt{2}+2$$

$$=\sqrt{3}(\sqrt{2}+1)+\sqrt{2}(\sqrt{2}+1)$$

$$=(\sqrt{3}+\sqrt{2})(\sqrt{2}+1)$$

$$\therefore\quad \cot 7\frac{1^\circ}{2}=(\sqrt{3}+\sqrt{2})(\sqrt{2}+1)$$

(iv) **Value of $\tan 82\frac{1^\circ}{2}$**

$$\tan 82\frac{1^\circ}{2}=\tan\left(90^\circ-7\frac{1^\circ}{2}\right)$$

$$=\cot 7\frac{1^\circ}{2}=(\sqrt{3}+\sqrt{2})(\sqrt{2}+1)$$

IV. (i) **Value of $\cos 22\frac{1^\circ}{2}$**

Let $\theta=22\frac{1^\circ}{2}$, then $2\theta=45^\circ$

Now, $\cos^2 22\frac{1^\circ}{2}=\dfrac{1+\cos 45^\circ}{2}=\dfrac{1+\dfrac{1}{\sqrt{2}}}{2}$

$$=\frac{\sqrt{2}+1}{2\sqrt{2}}=\frac{2+\sqrt{2}}{4}$$

$$\therefore \quad \cos 22\frac{1^\circ}{2} = \frac{1}{2}\sqrt{2+\sqrt{2}} \qquad \left[\because \cos 22\frac{1^\circ}{2} > 0\right]$$

(ii) **Value of sin $22\frac{1^\circ}{2}$**

$$\sin^2 22\frac{1^\circ}{2} = \frac{1-\cos 45^\circ}{2} = \frac{1-\frac{1}{\sqrt{2}}}{2} = \frac{\sqrt{2}-1}{2\sqrt{2}} = \frac{2-\sqrt{2}}{4}$$

$$\therefore \quad \sin 22\frac{1^\circ}{2} = \frac{1}{2}\sqrt{2-\sqrt{2}} \qquad \left[\because \sin 22\frac{1^\circ}{2} > 0\right]$$

(iii) **Value of tan $22\frac{1^\circ}{2}$**

$$\tan 22\frac{1^\circ}{2} > 0$$

$$\therefore \quad \tan 22\frac{1^\circ}{2} = \sqrt{\frac{1-\cos 45^\circ}{1+\cos 45^\circ}}$$

$$\left[\because \tan\theta = \sqrt{\frac{1-\cos 2\theta}{1+\cos 2\theta}}\right]$$

$$= \sqrt{\frac{1-\frac{1}{\sqrt{2}}}{1+\frac{1}{\sqrt{2}}}} = \sqrt{\frac{\sqrt{2}-1}{\sqrt{2}+1}}$$

$$= \sqrt{\frac{(\sqrt{2}-1)^2}{(\sqrt{2}+1)(\sqrt{2}-1)}} = \sqrt{\frac{(\sqrt{2}-1)^2}{2-1}}$$

$$= \sqrt{(\sqrt{2}-1)^2} = \sqrt{2}-1$$

V. **Value of cos $67\frac{1^\circ}{2}$ and sin $67\frac{1^\circ}{2}$**

$$\cos 67\frac{1^\circ}{2} + \sin 67\frac{1^\circ}{2} = \sqrt{2}\left(\frac{1}{\sqrt{2}}\cos 67\frac{1^\circ}{2} + \frac{1}{\sqrt{2}}\sin 67\frac{1^\circ}{2}\right)$$

$$= \sqrt{2}\left(\sin 45^\circ \cos 67\frac{1^\circ}{2} + \cos 45^\circ \sin 67\frac{1^\circ}{2}\right)$$

$$= \sqrt{2}\sin\left(45^\circ + 67\frac{1^\circ}{2}\right) = \sqrt{2}\sin 112\frac{1^\circ}{2} > 0$$

$$\therefore \cos 67\frac{1^\circ}{2} + \sin 67\frac{1^\circ}{2}$$

$$= \sqrt{\left(\cos 67\frac{1^\circ}{2} + \sin 67\frac{1^\circ}{2}\right)^2} = \sqrt{1+\sin 135^\circ}$$

$$= \sqrt{1+\frac{1}{\sqrt{2}}} = \sqrt{\frac{(\sqrt{2}+1)}{\sqrt{2}}} = \sqrt{\frac{4+2\sqrt{2}}{\sqrt{2}\cdot 2\sqrt{2}}}$$

$$= \sqrt{\frac{4+2\sqrt{2}}{4}} = \frac{1}{2}\sqrt{4+2\sqrt{2}} \qquad \text{...(i)}$$

Again, $\cos 67\frac{1^\circ}{2} - \sin 67\frac{1^\circ}{2} = \sqrt{2}\sin\left(45^\circ - 67\frac{1^\circ}{2}\right)$

$$= -\sqrt{2}\sin 22\frac{1^\circ}{2} < 0$$

$$\therefore \cos 67\frac{1^\circ}{2} - \sin 67\frac{1^\circ}{2}$$

$$= -\sqrt{\left(\cos 67\frac{1^\circ}{2} - \sin 67\frac{1^\circ}{2}\right)^2}$$

$$= -\sqrt{1-\sin 135^\circ}$$

$$= \sqrt{1-\frac{1}{\sqrt{2}}} = -\sqrt{\frac{\sqrt{2}-1}{\sqrt{2}}} = \sqrt{\frac{4-2\sqrt{2}}{4}}$$

$$= -\frac{1}{2}\sqrt{4-2\sqrt{2}} \qquad \text{...(i)}$$

Adding Eqs. (i) and (ii), we get

$$2\cos 67\frac{1^\circ}{2} - \frac{1}{2}\left(\sqrt{4+2\sqrt{2}} - \sqrt{4-2\sqrt{2}}\right)$$

$$\therefore \quad \cos 67\frac{1^\circ}{2} = \frac{1}{4}\left(\sqrt{4+2\sqrt{2}} - \sqrt{4-2\sqrt{2}}\right)$$

Similarly, subtracting Eq. (ii) from Eq. (i), we get

$$\sin 67\frac{1^\circ}{2} = \frac{1}{4}\left(\sqrt{4+2\sqrt{2}} + \sqrt{4-2\sqrt{2}}\right)$$

VI. **Value of sin $157\frac{1^\circ}{2}$**

Let $\theta = 157\frac{1^\circ}{2} \quad \therefore 2\theta = 315^\circ$

Now, $\sin\theta = \sqrt{\frac{1-\cos 2\theta}{2}} \qquad \left[\because \sin 157\frac{1^\circ}{2} > 0\right]$

$$= \sqrt{\frac{1-\cos 315^\circ}{2}} = \sqrt{\frac{1-\cos 45^\circ}{2}} = \sqrt{\frac{1-\frac{1}{\sqrt{2}}}{2}}$$

Thus $\quad \sin 157\frac{1^\circ}{2} = \sqrt{\frac{\sqrt{2}-1}{2\sqrt{2}}} = \sqrt{\frac{2-\sqrt{2}}{4}}$

$$= \frac{1}{2}\sqrt{2-\sqrt{2}}$$

Similarly, $\cos 157\frac{1^\circ}{2} = -\sqrt{\frac{1+\cos 2\theta}{2}}$

$$= -\frac{1}{2}\sqrt{2+\sqrt{2}} \qquad \left[\because \cos 157\frac{1^\circ}{2} < 0\right]$$

All these values are tabulated as follows:

	7.5°	15°	18°	22.5°	36°	67.5°	75°
sin	$\frac{\sqrt{8-2\sqrt{6}-2\sqrt{5}}}{4}$	$\frac{\sqrt{3}-1}{2\sqrt{2}}$	$\frac{\sqrt{5}-1}{4}$	$\frac{\sqrt{2-\sqrt{2}}}{2}$	$\frac{\sqrt{10-2\sqrt{5}}}{4}$	$\frac{\sqrt{2+\sqrt{2}}}{2}$	$\frac{\sqrt{3}+1}{2\sqrt{2}}$
cos	$\frac{\sqrt{8+2\sqrt{6}+2\sqrt{5}}}{4}$	$\frac{\sqrt{3}+1}{2\sqrt{2}}$	$\frac{\sqrt{10+2\sqrt{5}}}{4}$	$\frac{\sqrt{2+\sqrt{2}}}{2}$	$\frac{\sqrt{5}+1}{4}$	$\frac{\sqrt{2-\sqrt{2}}}{2}$	$\frac{\sqrt{3}-1}{2\sqrt{2}}$
tan	$(\sqrt{3}-\sqrt{2})(\sqrt{2}-1)$	$2-\sqrt{3}$	$\sqrt{1-\frac{2}{\sqrt{5}}}$	$\sqrt{2}-1$	$\sqrt{5-2\sqrt{5}}$	$\sqrt{2}+1$	$2+\sqrt{3}$
cot	$(\sqrt{3}+\sqrt{2})(\sqrt{2}+1)$	$2+\sqrt{3}$	$\sqrt{(5+2\sqrt{5})}$	$\sqrt{2}+1$	$\sqrt{\left(1+\frac{2}{\sqrt{5}}\right)}$	$\sqrt{2}-1$	$2-\sqrt{3}$

Example 87. Show that

$$\frac{1+\tan\frac{\theta}{2}}{1-\tan\frac{\theta}{2}}=\frac{1+\sin\theta}{\cos\theta}=\tan\left(\frac{\pi}{2}+\frac{\theta}{2}\right).$$

Sol. L.H.S. $=\dfrac{1+\dfrac{\sin\frac{\theta}{2}}{\cos\frac{\theta}{2}}}{1-\dfrac{\sin\frac{\theta}{2}}{\cos\frac{\theta}{2}}}=\dfrac{\cos\frac{\theta}{2}+\sin\frac{\theta}{2}}{\cos\frac{\theta}{2}-\sin\frac{\theta}{2}}$

$$=\frac{\left(\cos\frac{\theta}{2}+\sin\frac{\theta}{2}\right)^2}{\cos^2\frac{\theta}{2}-\sin^2\frac{\theta}{2}}$$

$$=\frac{\cos^2\frac{\theta}{2}+\sin^2\frac{\theta}{2}+2\sin\frac{\theta}{2}\cos\frac{\theta}{2}}{\cos^2\frac{\theta}{2}-\sin^2\frac{\theta}{2}}=\frac{1+\sin\theta}{\cos\theta}$$

Again, $\dfrac{1+\tan\frac{\theta}{2}}{1-\tan\frac{\theta}{2}}=\dfrac{\tan\frac{\pi}{4}+\tan\frac{\theta}{2}}{1-\tan\frac{\pi}{4}\tan\frac{\theta}{2}}=\tan\left(\dfrac{\pi}{4}+\dfrac{\theta}{2}\right)$

Thus $\dfrac{1+\tan\frac{\theta}{2}}{1-\tan\frac{\theta}{2}}=\dfrac{1+\sin\theta}{\cos\theta}=\tan\left(\dfrac{\pi}{4}+\dfrac{\theta}{2}\right)=$ RHS

Example 88. Prove that,

$$(\cos A+\cos B)^2+(\sin A+\sin B)^2=4\cos^2\frac{A-B}{2}.$$

Sol. L.H.S. $=(\cos^2 A+\cos^2 B+2\cos A\cos B)$
$+(\sin^2 A+\sin^2 B+2\sin A\sin B)$

$=(\cos^2 A+\sin^2 A)+(\cos^2 B+\sin^2 B)$
$+2(\cos A\cos B+\sin A\sin B)$

$=2+2\cos(A-B)=2[1+\cos(A-B)]$

$=2\times 2\cos^2\dfrac{A-B}{2}=4\cos^2\dfrac{A-B}{2}$

Example 89. Prove that,

$$\cos^4\frac{\pi}{8}+\cos^4\frac{3\pi}{8}+\cos^4\frac{5\pi}{8}+\cos^4\frac{7\pi}{8}=\frac{3}{2}$$

Sol. L.H.S. $=\left(\cos^4\dfrac{\pi}{8}+\cos^4\dfrac{7\pi}{8}\right)+\left(\cos^4\dfrac{3\pi}{8}+\cos^4\dfrac{5\pi}{8}\right)$

$=\left[\cos^4\dfrac{\pi}{8}+\cos^4\left(\pi-\dfrac{\pi}{8}\right)\right]+\left[\cos^4\dfrac{3\pi}{8}+\cos^4\left(\pi-\dfrac{3\pi}{8}\right)\right]$

$=2\cos^4\dfrac{\pi}{8}+2\cos^4\dfrac{3\pi}{8}$ $\quad[\because \cos(\pi-\theta)=-\cos\theta]$

$=2\left[\left(\cos^2\dfrac{\pi}{8}\right)^2+\left(\cos^2\dfrac{3\pi}{8}\right)^2\right]$

$=2\left[\left(\dfrac{1+\cos\frac{\pi}{4}}{2}\right)^2+\left(\dfrac{1+\cos\frac{3\pi}{4}}{2}\right)^2\right]$

$=\dfrac{1}{2}\left[\left(1+\dfrac{1}{\sqrt{2}}\right)^2+\left(1-\dfrac{1}{\sqrt{2}}\right)^2\right]$

$=\dfrac{1}{2}\left[1+\dfrac{1}{2}+2\cdot\dfrac{1}{\sqrt{2}}+1+\dfrac{1}{2}-2\cdot\dfrac{1}{\sqrt{2}}\right]$

$=\dfrac{3}{2}=$ R.H.S

Example 90. Find the value of $\tan\dfrac{\pi}{8}$.

Sol. Let $\theta=\dfrac{\pi}{8}$, then $2\theta=\dfrac{\pi}{4}$

Now, $\quad\tan 2\theta=\dfrac{2\tan\theta}{1-\tan^2\theta}$

$$\therefore \quad \tan\frac{\pi}{4} = \frac{2\tan\frac{\pi}{8}}{1-\tan^2\frac{\pi}{8}}$$

$$\Rightarrow \quad 1 = \frac{2x}{1-x^2}, \text{ where } x = \tan\frac{\pi}{8}$$

$$\Rightarrow \quad 1 - x^2 = 2x$$

$$\Rightarrow \quad x^2 + 2x - 1 = 0$$

$$\Rightarrow \quad x = \frac{-2 \pm 2\sqrt{2}}{2} = -1+\sqrt{2}, -1-\sqrt{2}$$

$$\therefore \quad x = \tan\frac{\pi}{8} = \sqrt{2}-1 \qquad \left[\because \tan\frac{\pi}{8} > 0\right]$$

Example 91. If $\tan x = -\frac{4}{3}, \frac{\pi}{2} < x < \pi$, then find the value of $\sin\frac{x}{2}$, $\cos\frac{x}{2}$ and $\tan\frac{x}{2}$.

Sol. Here $\frac{\pi}{2} < x < \pi \quad \therefore \frac{\pi}{4} < \frac{x}{2} < \frac{\pi}{2}$

Hence $\sin\frac{x}{2}, \cos\frac{x}{2}, \tan\frac{x}{2}$ will be all positive.

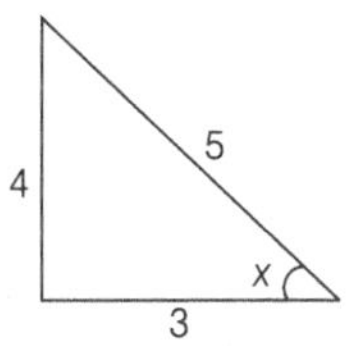

Given, $\tan x = -\frac{4}{3}$ and $\frac{\pi}{2} < x < \pi$

$$\therefore \quad \cos x = -\frac{3}{5}$$

Now, $\sin\frac{x}{2} = \sqrt{\frac{1-\cos x}{2}} \qquad \left[\because \sin\frac{x}{2} > 0\right]$

$$= \sqrt{\frac{1+\frac{3}{5}}{2}} = \frac{2}{\sqrt{5}} = \frac{2\sqrt{5}}{5}$$

$$\cos\frac{x}{2} = \sqrt{\frac{1+\cos x}{2}} \qquad \left[\because \cos\frac{x}{2} > 0\right]$$

$$= \sqrt{\frac{1-\frac{3}{5}}{2}} = \frac{1}{\sqrt{5}} = \frac{\sqrt{5}}{5} \text{ and } \tan\frac{x}{2} = \frac{\sin\frac{x}{2}}{\cos\frac{x}{2}} = 2$$

Example 92. Find the value of $\sin\frac{23\pi}{24}$.

Sol. $\sin\frac{23\pi}{24} = \sin\left(\pi - \frac{\pi}{24}\right) = \sin\frac{\pi}{24} = \sin 7\frac{1^\circ}{2}$

Let $\theta = 7\frac{1^\circ}{2}$, then $2\theta = 15^\circ$

Now, $\sin^2\theta = \frac{1-\cos 2\theta}{2} = \frac{1}{2}(1-\cos 15^\circ)$

$$= \frac{1}{2}\left(1 - \frac{\sqrt{3}+1}{2\sqrt{2}}\right) = \frac{1}{2}\left(\frac{2\sqrt{2}-\sqrt{3}-1}{2\sqrt{2}}\right)$$

$$= \frac{2\sqrt{2}-\sqrt{3}-1}{4\sqrt{2}}$$

$$\therefore \quad \sin\theta = \sin 7\frac{1^\circ}{2} > 0$$

$$\therefore \quad \sin 7\frac{1^\circ}{2} = \sqrt{\frac{2\sqrt{2}-\sqrt{3}-1}{4\sqrt{2}}} = \frac{1}{4}\sqrt{8-2\sqrt{6}-2\sqrt{2}}$$

Example 93. If $\alpha = 112^\circ 30'$, find the value of $\sin\alpha$ and $\cos\alpha$.

Sol. Given, $\alpha = 112^\circ 30'$

$$\therefore \quad 2\alpha = 225^\circ$$

or $\cos 2\alpha = \cos 225^\circ = \cos(180^\circ + 45^\circ)$

$$= -\cos 45^\circ = -\frac{1}{\sqrt{2}}$$

Now, $\sin^2\alpha = \frac{1-\cos 2\alpha}{2}$

Since α lies in the 2nd quadrant $\therefore \sin\alpha$ is positive

$$\therefore \quad \sin\alpha = \sqrt{\frac{1-\cos 2\alpha}{2}} = \sqrt{\frac{1-\left(-\frac{1}{\sqrt{2}}\right)}{2}}$$

$$= \sqrt{\frac{\sqrt{2}+1}{2\sqrt{2}}} = \frac{\sqrt{2+2\sqrt{2}}}{2}$$

Hence, $\sin\alpha = \frac{\sqrt{2+2\sqrt{2}}}{2}$

But $\cos\alpha$ is negative in 2nd quadrant

$$\therefore \quad \cos\alpha = -\sqrt{\frac{1+\cos 2\alpha}{2}}$$

$$= -\sqrt{\frac{1-\frac{1}{\sqrt{2}}}{2}} = -\sqrt{\frac{2-\sqrt{2}}{2}}$$

Example 94. If $\tan\frac{\theta}{2} = \sqrt{\frac{a-b}{a+b}}\tan\frac{\varphi}{2}$, prove that, $\cos\alpha = \frac{a\cos\varphi + b}{a + b\cos\varphi}$.

Sol. Given, $\tan\frac{\theta}{2} = \sqrt{\frac{a-b}{a+b}}\tan\frac{\varphi}{2}$

Now, $\cos\theta = \frac{1-\tan^2\frac{\theta}{2}}{1+\tan^2\frac{\theta}{2}} = \frac{1-\frac{a-b}{a+b}\tan^2\frac{\varphi}{2}}{1+\frac{a-b}{a+b}\tan^2\frac{\varphi}{2}}$

$$= \frac{1 - \frac{a-b}{a+b} \cdot \frac{\sin^2 \frac{\varphi}{2}}{\cos^2 \frac{\varphi}{2}}}{1 + \frac{a-b}{a+b} \cdot \frac{\sin^2 \frac{\varphi}{2}}{\cos^2 \frac{\varphi}{2}}}$$

$$= \frac{(a+b)\cos^2 \frac{\varphi}{2} - (a-b)\sin^2 \frac{\varphi}{2}}{(a+b)\cos^2 \frac{\varphi}{2} + (a-b)\sin^2 \frac{\varphi}{2}}$$

$$= \frac{a\left(\cos^2 \frac{\varphi}{2} - \sin^2 \frac{\varphi}{2}\right) + b\left(\cos^2 \frac{\varphi}{2} + \sin^2 \frac{\varphi}{2}\right)}{a\left(\cos^2 \frac{\varphi}{2} + \sin^2 \frac{\varphi}{2}\right) + b\left(\cos^2 \frac{\varphi}{2} - \sin^2 \frac{\varphi}{2}\right)}$$

$$= \frac{a \cos \varphi + b}{a + b \cos \varphi}$$

Example 95. If $\cos \theta = \frac{\cos \alpha - \cos \beta}{1 - \cos \alpha \cos \beta}$, then prove that one of the values of $\tan \frac{\theta}{2}$ is $\tan \frac{\alpha}{2} \cot \frac{\beta}{2}$.

Sol. $\tan^2 \frac{\theta}{2} = \frac{1 - \cos \theta}{1 + \cos \theta} = \frac{1 - \frac{\cos \alpha - \cos \beta}{1 - \cos \alpha \cos \beta}}{1 + \frac{\cos \alpha - \cos \beta}{1 - \cos \alpha \cos \beta}}$

$$= \frac{1 - \cos \alpha \cos \beta - \cos \alpha + \cos \beta}{1 - \cos \alpha \cos \beta + \cos \alpha - \cos \beta}$$

$$= \frac{(1 - \cos \alpha) + \cos \beta(1 - \cos \alpha)}{(1 + \cos \alpha) - \cos \beta(1 + \cos \alpha)}$$

$$= \frac{(1 - \cos \alpha)(1 + \cos \beta)}{(1 + \cos \alpha)(1 + \cos \beta)} = \tan^2 \frac{\alpha}{2} \cot^2 \frac{\beta}{2}$$

$$\therefore \tan \frac{\theta}{2} = \pm \tan \frac{\alpha}{2} \cot \frac{\beta}{2}$$

Hence one of the values of $\tan \frac{\theta}{2}$ is $\tan \frac{\alpha}{2} \cot \frac{\beta}{2}$.

Example 96. Prove that $\cos 6° \cos 42° \cos 66° \cos 78° = \frac{1}{16}$.

Sol. We have

$$\text{LHS} = \cos 6° \cos 42° \cos 66° \cos 78°$$

$$= \frac{1}{4}(2 \cos 66° \cos 6°)(2 \cos 78° \cos 42°)$$

$$= \frac{1}{4}[\cos(66° + 6°) + \cos(66° - 6°)] \times [\cos(78° + 42°) + \cos(78° - 42°)]$$

$$= \frac{1}{4}(\cos 72° + \cos 60°)(\cos 120° + \cos 36°)$$

$$= \frac{1}{4}\left(\sin 18° + \frac{1}{2}\right)\left(-\frac{1}{2} + \cos 36°\right)$$

$[\because \cos 72° = \cos(90° - 18°) = \sin 18°]$

$$= \frac{1}{4}\left[\frac{(\sqrt{5} - 1)}{4} + \frac{1}{2}\right]\left[-\frac{1}{2} + \frac{(\sqrt{5} - 1)}{4}\right]$$

$\left[\because \sin 18° = \frac{(\sqrt{5} - 1)}{4} \text{ and } \cos 36° = \frac{(\sqrt{5} + 1)}{4}\right]$

$$= \frac{1}{4} \cdot \frac{(\sqrt{5} + 1)}{4} \cdot \frac{(\sqrt{5} - 1)}{4} = \frac{(5 - 1)}{64}$$

$$= \frac{4}{64} = \frac{1}{16} = \text{RHS}$$

Example 97. Prove that $\sin \frac{\pi}{5} \sin \frac{2\pi}{5} \sin \frac{3\pi}{5} \sin \frac{4\pi}{5} = \frac{5}{16}$.

Sol. We have

$$\text{LHS} = \sin \frac{\pi}{5} \sin \frac{2\pi}{5} \sin \frac{3\pi}{5} \sin \frac{4\pi}{5}$$

$$= \sin \frac{\pi}{5} \sin \frac{2\pi}{5} \sin\left(\pi - \frac{2\pi}{5}\right) \sin\left(\pi - \frac{\pi}{5}\right)$$

$$= \sin^2 \frac{\pi}{5} \sin^2 \frac{2\pi}{5} \qquad [\because \sin(\pi - \theta) = \sin \theta]$$

$$= (\sin 36°)^2 \times (\sin 72°)^2$$

$$= (\sin 36°)^2 \times (\cos 18°)^2$$

$[\because \sin 72° = \sin(90° - 18°) = \cos 18°]$

$$= \frac{(10 - 2\sqrt{5})}{16} \times \frac{(10 + 2\sqrt{5})}{16} = \frac{(100 - 20)}{(16 \times 16)}$$

$\left[\because \sin 36° = \frac{\sqrt{10 - 2\sqrt{5}}}{4} \text{ and } \cos 18° = \frac{\sqrt{10 + 2\sqrt{5}}}{4}\right]$

$$= \frac{80}{(16 \times 16)} = \frac{5}{16} = \text{RHS}$$

Example 98. Find the value of

(i) $\sin 22°30'$ (ii) $\cos 22°30'$

(iii) $\tan 22°30'$

Sol. (i) $\sin^2 \theta = \frac{(1 - \cos 2\theta)}{2}$

$$\Rightarrow \quad \sin^2(22°30') = \frac{(1 - \cos 45°)}{2} = \frac{\left(1 - \frac{1}{\sqrt{2}}\right)}{2} = \frac{\sqrt{2} - 1}{2\sqrt{2}}$$

$$\Rightarrow \quad \sin(22°30') = \sqrt{\frac{(\sqrt{2} - 1)}{2\sqrt{2}}}$$

(ii) $\cos^2\theta = \dfrac{(1+\cos 2\theta)}{2}$

$\Rightarrow \quad \cos^2(22°30') \dfrac{(1+\cos 45°)}{2} = \left(\dfrac{1+\dfrac{1}{\sqrt{2}}}{2}\right) = \dfrac{(\sqrt{2}+1)}{2\sqrt{2}}$

$\Rightarrow \quad \cos(20°30') = \sqrt{\dfrac{(\sqrt{2}+1)}{2\sqrt{2}}}$

(iii) $\tan^2(22°30') = \dfrac{\sin^2(22°30')}{\cos^2(22°30')} = \dfrac{(\sqrt{2}-1)}{(2\sqrt{2})} \times \dfrac{(2\sqrt{2})}{(\sqrt{2}+1)}$

$= \dfrac{\sqrt{2}-1}{\sqrt{2}+1} = \dfrac{(\sqrt{2}-1)}{(\sqrt{2}+1)} \times \dfrac{(\sqrt{2}-1)}{(\sqrt{2}-1)}$

$= (\sqrt{2}-1)^2$

$\Rightarrow \quad \tan(20°30') = (\sqrt{2}-1)$

Example 99. If $0 < x < \pi$ and $\cos x + \sin x = \dfrac{1}{2}$, then find the value of $\tan x$.

Sol. Given, $\cos x + \sin x = \dfrac{1}{2}$

$\Rightarrow \quad \dfrac{1-\tan^2\dfrac{x}{2}}{1+\tan^2\dfrac{x}{2}} + \dfrac{2\tan\dfrac{x}{2}}{1+\tan^2\dfrac{x}{2}} = \dfrac{1}{2}$

Let $\quad \tan\dfrac{x}{2} = t$, then $\dfrac{1-t^2}{1+t^2} + \dfrac{2t}{1+t^2} = \dfrac{1}{2}$

$\Rightarrow \quad 3t^2 - 4t - 1 = 0 \Rightarrow t = \dfrac{2 \pm \sqrt{7}}{3}$

$\Rightarrow \quad t = \tan\dfrac{x}{2} = \dfrac{2+\sqrt{7}}{3} \quad \left[\because 0 < \dfrac{\pi}{2} < \dfrac{\pi}{2}, \tan\dfrac{x}{2}\right]$

Now, $\quad \tan x = \dfrac{2\tan\dfrac{x}{2}}{1-\tan^2\dfrac{x}{2}}$

$= \dfrac{2\left(\dfrac{2+\sqrt{7}}{2}\right)}{1-\left(\dfrac{2+\sqrt{7}}{3}\right)^2} = -\dfrac{3(2+\sqrt{7})}{1+2\sqrt{7}} \times \dfrac{1-2\sqrt{7}}{1-2\sqrt{7}}$

$\therefore \quad \tan x = -\left(\dfrac{4+\sqrt{7}}{3}\right)$

Exercise on Session 9

1. If $\tan\left(\dfrac{x}{2}\right) = \operatorname{cosec} x - \sin x$, then find the value of $\tan^2\left(\dfrac{x}{2}\right)$.
2. Find the value of $\cos^4\left(\dfrac{\pi}{8}\right)$.
3. Find the value of expression $\dfrac{1}{\cos 290°} + \dfrac{1}{\sqrt{3}\sin 250°}$.
4. If $x + \dfrac{1}{x} = 2\cos\theta$ then find the value of $x^n + \dfrac{1}{x^n}$.
5. Show that $\sin 47° + \sin 61° - \sin 11° - \sin 25° = \cos 7°$.
6. If α and β be two different roots of equation $a\cos\theta + b\sin\theta = c$, then show that $\sin(\alpha+\beta) = \dfrac{2ab}{a^2+b^2}$.
7. If $\sin\alpha + \sin\beta = a$ and $\cos\alpha + \cos\beta = b$, then show that $\tan\left(\dfrac{\alpha-\beta}{2}\right) = \pm\sqrt{\dfrac{4-a^2-b^2}{a^2+b^2}}$.
8. Show that $\cot 142\dfrac{1°}{2} = \sqrt{2} + \sqrt{3} - 2 - \sqrt{6}$.
9. If $\cos\theta = \dfrac{\cos\alpha + \cos\beta}{1+\cos\alpha\cos\beta}$, then prove that one of values of $\tan\dfrac{\theta}{2}$ is $\tan\dfrac{\alpha}{2}\tan\dfrac{\beta}{2}$.
10. Find the value $\tan\dfrac{\pi}{5} + 2\tan\dfrac{2\pi}{5} + 4\cot\dfrac{4\pi}{5}$.

Session 10

Trigonometric Ratios of the Sum of Three or More Angles

Trigonometric Ratios of the Sum of Three or More Angles

$$e^{i(A+B+C)} = e^{iA} \cdot e^{iB} \cdot e^{iC}$$

$$\Rightarrow \cos(A+B+C) + i\sin(A+B+C)$$
$$= (\cos A + i\sin A)(\cos B + i\sin B)(\cos C + i\sin C)$$
$$\cos A \cos B \cos C + i\Sigma \cos A \cdot \cos B \sin C + i^2$$
$$\Sigma \cos A \cdot \sin B \cdot \sin C + i^3 \cdot \sin A \cdot \sin B \cdot \sin C$$
$$= \cos A \cdot \cos B \cdot \cos C \cdot$$
$$\{1 + i\Sigma \tan A + i^2 \Sigma \tan A \tan B + i^3 \cdot \tan A \tan B \tan C\}$$

Now, equating real and imaginary parts, we get

$$\mathbf{\cos(A+B+C) = \cos A \cdot \cos B \cdot \cos C \cdot \{1 - \Sigma \tan A \tan B\}}$$

and $$\mathbf{\sin(A+B+C) = \cos A \cos B \cos C \{\Sigma \tan A - \tan A \tan B \tan C\}}$$

Generalising Using this Method

$$\cos(A_1 + A_2 + \ldots + A_n) + i\sin(A_1 + A_2 + \ldots + A_n)$$
$$= \cos A_1 \cdot \cos A_2 \ldots \cos A_n \{1 + i \cdot \Sigma \tan A_1 + i^2$$
$$\Sigma \tan A_1 \tan A_2 + i^3 \Sigma \tan A_1 \tan A_2 \tan A_3 + \ldots i^n$$
$$\tan A_1 \cdot \tan A_2 \ldots \tan A_n\}$$

(i) On comparing real and imaginary parts, we get

$$\cos(A_1 + A_2 + \ldots + A_n)$$
$$= \cos A_1 \cdot \cos A_2 \ldots \cos A_n \cdot \{1 - \Sigma \tan A \tan A_2 + \Sigma \tan A_1 \cdot \tan A_2 \tan A_3 \cdot \tan A_4 \ldots\}$$

(ii) $(\sin(A_1 + A_2 + \ldots + A_n)$

$$= \cos A_1 \cdot \cos A_2 \ldots \cos A_n \cdot \{\Sigma \tan A_1 - \Sigma \tan A_1 \cdot \tan A_2 \cdot \tan A_3 + \ldots\}$$

(iii) $$\tan(A_1 + A_2 + \ldots + A_n) = \frac{S_1 - S_3 + S_5 \ldots}{1 - S_2 + S_4 - S_6 \ldots}$$

where, $S_1 = \Sigma \tan A_1$,

$S_2 = \Sigma \tan A_1 \cdot \tan A_2$;

$S_3 = \Sigma \tan A_1 . \tan A_2 . \tan A_3$

Example 100. If $\tan\theta_1, \tan\theta_2, \tan\theta_3, \tan\theta_4$ are the roots of the equation

$$x^4 - (\sin 2\beta)^3 + (\cos 2\beta)x^2 - (\cos\beta)x - \sin\beta = 0$$

Then, $\tan(\theta_1 + \theta_2 + \theta_3 + \theta_4)$, is equal to

Sol. $$\tan(\theta_1 + \theta_2 + \theta_3 + \theta_4) = \frac{S_1 - S_3}{1 - S_2 + S_4}$$

$$= \frac{\sin(2\beta) - \cos\beta}{1 - \cos 2\beta - \sin\beta} = \frac{\cos\beta(2\sin\beta - 1)}{\sin\beta(2\sin\beta - 1)} = \cot\beta$$

Multiple Angle Results in the General Form

Using the De-Moiver's theorem,

$$(\cos n\alpha + i\sin n\alpha) = (\cos\alpha + i\sin\alpha)^n$$
$$= {}^nC_0 \cdot \cos^n\alpha + {}^nC_1 . \cos^{n-1}\alpha\,(i\sin\alpha) + {}^nC_2 \cos^{n-2}\alpha\,(i\sin\alpha)^2 + \ldots + {}^nC_n(i\sin\alpha)^n$$
$$\Rightarrow \cos(n\alpha) + i\sin(n\alpha) = ({}^nC_0 \cos^n\alpha - {}^nC_2 \cos^{n-2}\alpha \sin^2\alpha + {}^nC_4 . \cos^{n-4}\alpha \cdot \sin^4\alpha \ldots)$$
$$+ i({}^nC_1 \cdot \cos^{n-1}\alpha \cdot \sin\alpha - {}^nC_3 \cos^{n-3}\alpha . \sin^3\alpha + {}^nC_5 \cos^{n-5} \cdot \sin^5\alpha \ldots)$$

On comparing real and imaginary part, we get

$$\cos(n\alpha) = \cos^n\alpha(1 - {}^nC_2 \cdot \tan^2\alpha + {}^nC_4 \tan^4\alpha \ldots) \quad \ldots(\text{i})$$

and $$\sin(n\alpha) = \cos^n\alpha({}^nC_1 \cdot \tan\alpha - {}^nC_3 \tan^3\alpha + {}^nC_5 \tan^5\alpha \ldots) \quad \ldots(\text{ii})$$

On dividing Eq. (ii) by Eq. (i), we get,

$$\tan(n\alpha) = \frac{{}^nC_1 - {}^nC_3 \tan^3\alpha + {}^nC_5 \tan^5\alpha \ldots}{1 - {}^nC_2 \tan^2\alpha + {}^nC_4 \tan^4\alpha - {}^nC_6 \tan^6\alpha \ldots}$$

On adding and subtracting Eq. (i) and Eq. (ii), we get

$$\sin(n\alpha) + \cos(n\alpha) = \cos^n\alpha\{1 + {}^nC_1 \tan\alpha + {}^nC_2 \tan^2\alpha - {}^nC_3 \tan^3\alpha + {}^nC_4 \tan^4\alpha + {}^nC_5 \tan^5\alpha \ldots\}$$

and $$\sin(n\alpha) - \cos(n\alpha) = \cos^n\alpha\{-1 + {}^nC_1 \tan\alpha + {}^nC_2 \tan^2\alpha - {}^nC_3 \tan^3\alpha - {}^nC_4 \tan^4\alpha + \ldots\}$$

Example 101. Express $\sin^5\theta$ in term of $\sin(n\theta); n \in N$.

Sol. We know,

$$\sin\theta = \frac{1}{2i}\left(z - \frac{1}{z}\right), \text{ using}$$

$$e^{i\theta} = \cos\theta + i\sin\theta = z$$

$$\therefore \quad \sin^5\theta = \left(\frac{1}{2i}\left(z - \frac{1}{z}\right)\right)^5$$

$$= \frac{1}{32i}\left\{{}^5C_0 \cdot z^5 - {}^5C_1 \cdot z^3 + {}^5C_2 \cdot z - {}^5C_3\left(\frac{1}{z}\right) + {}^5C_4\left(\frac{1}{z}\right)^3 - {}^5C_5\left(\frac{1}{z}\right)^5\right\}$$

$$= \frac{1}{32i}\left\{\left(z^5 - \frac{1}{z^5}\right) - 5\left(z^3 - \frac{1}{z^3}\right) + 10\left(z - \frac{1}{z}\right)\right\}$$

$$= \frac{1}{32i}\{2i\sin 5\theta - 5(2i\sin 3\theta) + 10(2i\sin\theta)\}$$

$$= \frac{1}{16}\{\sin 5\theta - 5\sin 3\theta + 10\sin\theta\}$$

Summation of Series Containing sine and Cosine of Angles Forming an AP

(i) **Sine of angle forming an AP**

Let the series be,

$$S = \sin(\alpha) + \sin(\alpha+\beta) + \sin(\alpha+2\beta) + \ldots + \sin(\alpha+(n-1)\beta)$$

On multiplying and dividing by $2\sin\frac{\beta}{2}$

$$= \frac{1}{2\sin\left(\frac{\beta}{2}\right)}\left\{2\sin\alpha\cdot\sin\left(\frac{\beta}{2}\right) + 2\sin(\alpha+\beta).\sin\left(\frac{\beta}{2}\right) + \ldots + 2\sin(\alpha+(n-1)\beta).\sin\left(\frac{\beta}{2}\right)\right\}$$

$$= \frac{1}{2\sin\left(\frac{\beta}{2}\right)}\cdot\left\{\left(\cos\left(\alpha - \frac{\beta}{2}\right) - \cos\left(\alpha + \frac{\beta}{2}\right)\right) + \left(\cos\left(\alpha + \frac{\beta}{2}\right) - \cos\left(\alpha + \frac{3\beta}{2}\right)\right) + \ldots + \left(\cos(\alpha + \left(n - \frac{3}{2}\right)\beta\right) - \cos\left[(\alpha + \left(n - \frac{1}{2}\right)\beta\right]\right\}$$

$$= \frac{1}{2\sin\left(\frac{\beta}{2}\right)}\left\{\cos\left(\alpha - \frac{\beta}{2}\right) - \cos\left(\alpha + \left(n - \frac{1}{2}\right)\beta\right)\right\}$$

$$= \frac{2\sin\left(\alpha + \left(\frac{n-1}{2}\right)\beta\right)\cdot\sin\left(n\frac{\beta}{2}\right)}{2\sin\frac{\beta}{2}}$$

$\therefore \sin\alpha + \sin(\alpha+\beta) + \sin(\alpha+2\beta) + \ldots + \sin[(\alpha+(n-1)\beta)]$

$$= \frac{\sin\left(n\frac{\beta}{2}\right)\cdot\sin\left[\alpha+(n-1)\frac{\beta}{2}\right]}{\sin\left(\frac{\beta}{2}\right)}$$

(ii) **Cosine of angle forming an AP.**

Let,

$$S = \cos(\alpha) + \cos(\alpha+\beta) + \cos(\alpha+2\beta) + \ldots + \cos[\alpha+(n-1)\beta]$$

On multiplying and dividing by $2\sin\left(\frac{\beta}{2}\right)$, we get

$$S = \frac{1}{2\sin\left(\frac{\beta}{2}\right)}\left\{2\sin\left(\frac{\beta}{2}\right)\cdot\cos(\alpha) + 2\sin\left(\frac{\beta}{2}\right)\cdot\cos(\alpha+\beta) + 2\sin\left(\frac{\beta}{2}\right)\cdot\cos(\alpha+2\beta) + \ldots + 2\sin\left(\frac{\beta}{2}\right)\cos(\alpha+(n+1)\beta\right\}$$

$$= \frac{1}{2\sin\left(\frac{\beta}{2}\right)}\left[\sin\left(\alpha+\frac{\beta}{2}\right) - \sin\left(\alpha-\frac{\beta}{2}\right)\right] + \left[\sin\left(\alpha+\frac{3\beta}{2}\right) - \sin\left(\alpha+\frac{\beta}{2}\right)\right] + \ldots + \left[\sin[\alpha+\left(n-\frac{3}{2}\right)\beta\right] - \sin\left[\alpha+\left(n-\frac{5}{2}\right)\beta\right]\right]$$

$$= \frac{1}{2\sin\left(\frac{\beta}{2}\right)}\cdot\left[\sin\left(\alpha+\left(n-\frac{1}{2}\right)\beta\right) - \sin\left(\alpha-\frac{\beta}{2}\right)\right]$$

$$= \frac{\cos\left(\alpha+(n-1)\left(\frac{\beta}{2}\right)\right)\cdot\sin\left(n\frac{\beta}{2}\right)}{1-\left(\frac{\beta}{2}\right)}$$

$\cos(\alpha) + \cos(\alpha+\beta) + \ldots + \cos(\alpha+(n-1)\beta)$

$$= \frac{\sin\left(n\frac{\beta}{2}\right)}{\sin\left(\frac{\beta}{2}\right)}\cdot\left\{\cos\left(\alpha+(n-1)\left(\frac{\beta}{2}\right)\right)\right\}$$

II Method

Let $S = \sin\alpha + \sin(\alpha+\beta) + \sin(\alpha+2\beta) + \ldots + \sin(\alpha+(n-1)\beta)$ ...(i)

$C = \cos\alpha + \cos(\alpha+\beta) + \cos(\alpha+2\beta) + \ldots + \cos(\alpha(n-1)\beta)$ ...(ii)

$$C + iS = (\cos\alpha + i\sin\alpha) + (\cos(\alpha+\beta) + i\sin(\alpha+\beta)) + (\cos(\alpha+2\beta) + i\sin(\alpha+2\beta)) + \ldots + (\cos(\alpha+(n-1)\beta) + i\sin(\alpha+(n-1)\beta))\, i\,(\alpha+(n-1)\beta)$$

$$= e^{i\alpha} + e^{i(\alpha+\beta)} + e^{i(\alpha+2\beta)} + \ldots + e^{i(\alpha+(n-1)\beta)}$$

$$= e^{i\alpha}\cdot\left(\frac{(e^{i\beta})^n - 1}{e^{i\beta} - 1}\right)$$

$$= e^{i\alpha}\left\{\frac{\cos(n\beta) - 1 + i\sin(n\beta)}{\cos\beta - 1 + i\sin\beta}\right\}$$

$$= e^{i\alpha}\left\{\frac{2i^2\sin^2\left(n\frac{\beta}{2}\right) + 2i\cdot\sin\left(n\frac{\beta}{2}\right)\cdot\cos\left(n\frac{\beta}{2}\right)}{2i^2\cdot\sin^2\left(\frac{\beta}{2}\right) + 2i\cdot\sin\left(\frac{\beta}{2}\right)\cdot\cos\left(\frac{\beta}{2}\right)}\right\}$$

$$= \frac{e^{i\alpha}\cdot 2i\sin\left(n\frac{\beta}{2}\right)}{2i\sin\left(\frac{\beta}{2}\right)}\left(\frac{\cos\left(n\frac{\beta}{2}\right) + i\cdot\sin\left(n\frac{\beta}{2}\right)}{\cos\left(\frac{\beta}{2}\right) + i\cdot\sin\left(\frac{\beta}{2}\right)}\right)$$

$$= \frac{e^{i\alpha}\sin\left(n\frac{\beta}{2}\right)}{\sin\left(\frac{\beta}{2}\right)}\cdot e^{i\left(\frac{n-1}{2}\right)\beta} = \frac{\sin\left(n\frac{\beta}{2}\right)}{\sin\left(\frac{\beta}{2}\right)}\cdot e^{i\left[\alpha+\left(\frac{n-1}{2}\right)\beta\right]}$$

$\therefore$ On comparing real and imaginary part, we get

$$C = \frac{\sin\left(n\frac{\beta}{2}\right)}{\sin\left(\frac{\beta}{2}\right)}\cdot\left\{\cos\left(\alpha+(n-1)\left(\frac{\beta}{2}\right)\right)\right\}$$

and
$$C = \frac{\sin\left(n\frac{\beta}{2}\right)}{\sin\left(\frac{\beta}{2}\right)}\cdot\left\{\sin\left(\alpha+(n-1)\left(\frac{\beta}{2}\right)\right)\right\}$$

1. $\sin(A+B+C) = \sin A\cos B\cos C + \cos A\sin B\cos C + \cos A\cos B\sin C - \sin A\sin B\sin C$

   or $\sin(A+B+C) = \cos A\cos B\cos C(\tan A + \tan B + \tan C - \tan A\tan B\tan C)$

2. $\cos(A+B+C) = \cos A\cos B\cos C - \sin A\sin B\cos C - \sin A\cos B\sin C - \cos A\sin B\sin C$

   $\cos(A+B+C) = \cos A\cos B\cos C(1 - \tan A\tan B - \tan B\tan C - \tan C\tan A)$

3. $\tan(A+B+C) = \dfrac{\tan A + \tan B + \tan C - \tan A\tan B\tan C}{1 - \tan A\tan B - \tan B\tan C - \tan C\tan A}$

In general;

4. $\sin(A_1 + A_2 + \ldots + A_n) = \cos A_1\cos A_2\ldots\cos A_n(S_1 - S_3 + S_5 - S_7 + \ldots)$
5. $\cos(A_1 + A_2 + \ldots + A_n) = \cos A_1\cos A_2\ldots\cos A_n(1 - S_2 + S_4 - S_6 + \ldots)$
6. $\tan(A_1 + A_2 + \ldots + A_n) = \dfrac{S_1 - S_3 + S_5 - S_7 + \ldots}{1 - S_2 + S_4 - S_6 + \ldots}$

where;

$S_1 = \tan A_1 + \tan A_2 + \ldots\ldots + \tan A_n$
= tan sum of the tangents of the separate angles.

$S_2 = \tan A_1\tan A_2 + \tan A_1\tan A_3 + \ldots$
= the sum of the tangents taken two at a time.

$S_3 = \tan A_1\tan A_2\tan A_3 + \tan A_2\tan A_3\tan A_4 + \ldots$
= sum of the tangents three at a time and so on.

If $A_1 = A_2 = \ldots = A_n = A$, then

$S_1 = n\tan A$

$S_2 = {}^nC_2\tan^2 A$

$S_3 = {}^nC_3\tan^3 A, \ldots$

7. $\sin nA = \cos^n A({}^nC_1\tan A - {}^nC_3\tan^3 A + {}^nC_5\tan^5 A\ldots)$
8. $\cos nA = \cos^n A(1 - {}^nC_2\tan^2 A + {}^nC_4\tan^4 A - \ldots)$
9. $\tan nA = \dfrac{{}^nC_1\tan A - {}^nC_3\tan^3 A + {}^nC_5\tan^5 A\ldots}{1 - {}^nC_2\tan^2 A + {}^nC_4\tan^4 A - {}^nC_6\tan^6 A + \ldots}$
10. $\sin nA + \cos nA = \cos^n A(1 + {}^nC_1\tan A - {}^nC_2\tan^2 A - {}^nC_3\tan^3 A + {}^nC_4\tan^4 A + {}^nC_5\tan^5 A - {}^nC_6\tan^6 A\ldots)$
11. $\sin nA - \cos nA = \cos^n A(-1 + {}^nC_1\tan A + {}^nC_2\tan^2 A - {}^nC_3\tan^3 A - {}^nC_4\tan^4 A + {}^nC_5\tan^5 A + {}^nC_6\tan^6 A\ldots)$
12. $\sin(\alpha) + \sin(\alpha+\beta) + \sin(\alpha+2\beta) + \ldots + \sin(\alpha+(n-1)\beta) = \dfrac{\sin\{\alpha+(n-1)(\beta/2)\}\cdot\sin(n\beta/2)}{\sin(\beta/2)}$
13. $\cos(\alpha) + \cos(\alpha+\beta) + \cos(\alpha+2\beta) + \ldots + \cos(\alpha+(n-1)\beta) = \dfrac{\cos(\alpha+(n-1)(\beta/2))\cdot\sin(n(\beta/2))}{\sin(\beta/2)}$

Example 102. Let n be an odd integer. In $\sin n\theta = \sum_{r=0}^{n} b_r \sin^r \theta$, for all real θ. Then, find b_0 and b_1.

Sol. Here, $\sin n\theta = b_0 + b_1 \sin\theta + b_2 \sin^2\theta + \dots$ ...(i)

Putting $\theta = 0$, we get

$$0 = b_0 \quad \text{...(ii)}$$

Again, on differentiating Eq. (i) both sides w.r.t. θ, we get

$$n\cos n\theta = 0 + b_1 \cos\theta + b_2 2\sin\theta \cdot \cos\theta + \dots$$

Again, putting $\theta = 0$, we get

$$n = b_1 \quad \text{...(iii)}$$

$\therefore$ From Eqs. (ii) and (iii),

$$b_0 = 0 \text{ and } b_1 = n$$

Example 103. If $\cos 5\theta = a\cos^5\theta + b\cos^3\theta + c\cos\theta$. Then, find the value of c.

Sol. Here, $\cos 5\theta = a\cos^5\theta + b\cos^3\theta + c\cos\theta$ ...(i)

On differentiating Eq. (i) w.r.t. θ, we get

$$-5\sin 5\theta = a(5\cos^4\theta)(-\sin\theta) + b(3\cos^2\theta)(-\sin\theta) - c\sin\theta$$

Putting $\theta = \frac{\pi}{2}$: $-5\sin\frac{5\pi}{2} = -c\sin\frac{\pi}{2}$ $\left[\text{as } \cos\frac{\pi}{2} = 0\right]$

$\therefore$ $c = 5$

Example 104. If $\sin^3 x \sin 3x = \sum_{m=0}^{n} c_m \cdot \cos mx$ is an identity in x, where c_m's are constants, then find the value of n.

Sol. Here, $\sin^3 x \sin 3x = \frac{3\sin x - \sin 3x}{4} \cdot \sin 3x$

$$= \frac{3}{8}(2\sin x \sin 3x) - \frac{1}{4}(\sin^2 3x)$$

$$= \frac{3}{8}(\cos 2x - \cos 4x) - \frac{1}{8}(1 - \cos 6x)$$

$$= -\frac{1}{8} + \frac{3}{8}\cos 2x - \frac{3}{8}\cos 4x + \frac{1}{8}\cos 6x \quad \text{...(i)}$$

Also, $\sum_{m=0}^{n} c_m \cdot \cos mx = c_0 + c_1\cos x + c_2 \cos 2x + \dots + c_n \cos nx$ ...(ii)

On comparing Eqs. (i) and (ii), we get

$$n = 6.$$

Example 105. Evaluate $\sum_{r=1}^{n-1} \cos^2\left(\frac{r\pi}{n}\right)$.

Sol. Sum $= \frac{1}{2}\sum_{r=1}^{n-1}\left(1 + \cos\frac{2r\pi}{n}\right)$

$$= \frac{1}{2}(n-1) + \frac{1}{2}\left\{\cos\frac{2\pi}{n} + \cos\frac{4\pi}{n} + \dots + \cos\frac{(2n-2)\pi}{n}\right\}$$

$$= \frac{1}{2}(n-1) + \frac{1}{2}\left\{\frac{\sin(n-1)\frac{2\pi}{2n}}{\sin\frac{2\pi}{n.2}} \cdot \cos\left\{\frac{2\left(\frac{2\pi}{n}\right) + (n-2)\frac{2\pi}{n}}{2}\right\}\right\}$$

Using, $\cos\alpha + \cos(\alpha + \beta) + \cos(\alpha + 2\beta) + \dots + \cos[\alpha + (n-1)\beta]$

$$= \frac{\sin\frac{n\beta}{2}}{\sin\frac{\beta}{2}} \cdot \cos\left\{\frac{2\alpha + (n-1)\beta}{2}\right\}$$

$$= \frac{1}{2}(n-1) + \frac{1}{2}\left\{\frac{\sin\frac{(n-1)\pi}{n} \cdot \cos\pi}{\sin(\pi/n)}\right\}$$

$$= \frac{1}{2}(n-1) + \frac{1}{2}\left\{\frac{\left(\sin\frac{\pi}{n}\right)(-1)}{\sin(\pi/n)}\right\}$$

$$= \frac{1}{2}(n-1) - \frac{1}{2} = \frac{n}{2} - 1$$

$$\therefore \quad \sum_{r=1}^{n-1} \cos^2\left(\frac{r\pi}{n}\right) = \frac{n}{2} - 1$$

Example 106. Evaluate $\sin\frac{\pi}{n} + \sin\frac{3\pi}{n} + \sin\frac{5\pi}{n} + \dots$ to n terms.

Sol. $\sin\frac{\pi}{n} + \sin\frac{3\pi}{n} + \sin\frac{5\pi}{n} + \dots$ to n terms

$$= \frac{\sin\frac{n2.\pi}{2.n}}{\sin\frac{2.\pi}{2.n}} \cdot \sin\left\{\frac{2\cdot\frac{\pi}{n} + (n-1)\cdot\frac{2\pi}{n}}{2}\right\}$$

Using, $\sin\alpha + \sin(\alpha + \beta) + \dots + \sin(\alpha + (n-1)\beta)$

$$= \frac{\sin n\beta/2}{\sin\beta/2} \cdot \sin\left\{\frac{2\alpha + (n-1)\beta}{2}\right\}$$

$$= \frac{\sin\pi}{\sin\pi/n} \sin\left\{\frac{2\pi + 2n\pi - 2\pi}{2n}\right\}$$

$$= \frac{\sin^2\pi}{\sin\pi/n} = 0$$

$$\therefore \quad \sin\frac{\pi}{n} + \sin\frac{3\pi}{n} + \sin\frac{5\pi}{n} + \dots \text{ to } n \text{ terms} = 0$$

Exercise for Session 10

1. If $A+B+C=180°$, then prove that $\sin 2A+\sin 2B+\sin 2C=4\sin A\sin B\sin C$.
2. If $A+B+C=180°$, then prove that $\tan^2\frac{\theta}{2}=\tan\frac{B}{2}\tan\frac{C}{2}$. when $\cos\theta(\sin B+\sin C)=\sin A$.
3. If A, B, C are angles of a ΔABC, then prove that $\sin 2A+\sin 2B-\sin 2C=4\cos A\cos B\sin C$.
4. If in a ΔABC, $\frac{\sin A+\sin B+\sin C}{\sin A+\sin B-\sin C}=2\lambda\cot\frac{A}{2}\cot\frac{B}{2}$, then find the value of λ.
5. If $A+B+C=180°$, then find the value of $\frac{\cos A}{\sin B\sin C}+\frac{\cos B}{\sin C\sin A}+\frac{\cos C}{\sin A\sin B}$.
6. In ΔABC, show that $\frac{1-\cos A+\cos B+\cos C}{1-\cos C+\cos A+\cos B}=\tan\frac{A}{2}\cot\frac{C}{2}$.
7. In a ΔABC, if $\tan\left(\frac{B+C-A}{4}\right)\tan\left(\frac{C+A-B}{4}\right)\tan\left(\frac{A+B-C}{4}\right)=1$, then find the value of $\cos A+\cos B+\cos C$.
8. If in a ΔABC, $\cot\frac{A}{2}+\cot\frac{B}{2}+\cot\frac{C}{2}=\lambda\cot\frac{A}{2}\cot\frac{B}{2}\cot\frac{C}{2}$, then find the value of λ.
9. If $A+B+C=\frac{3\pi}{2}$, then show that $\cos 2A+\cos 2B+\cos 2C=1-4\sin A\sin B\sin C$.
10. If $\alpha+\beta+\gamma=2\pi$, then show that $\tan\frac{\alpha}{2}+\tan\frac{\beta}{2}+\tan\frac{\gamma}{2}=\tan\frac{\alpha}{2}\tan\frac{\beta}{2}\tan\frac{\gamma}{2}$.

Session 11

Maximum and Minimum Values of Trigonometrical Functions

Conditional Trigonometrical Identities

We have certain trigonometric identities like, $\sin^2\theta+\cos^2\theta=1$ and $1+\tan^2\theta=\sec^2\theta$ etc. Such identities are identities in the sense that they hold for all value of the angles which satisfy the given condition among them and they are called conditional identities.

If A, B, C denote the angle of a ΔABC, then the relation $A+B+C=\pi$ enables us to establish many important identities involving trigonometric ratios of these angles.

(i) If $A+B+C=\pi$, then $A+B=\pi-C$, $B+C=\pi-A$ and $C+A=\pi-\beta$

(ii) If $A+B+C=\pi$, then $\sin(A+B)$
$=\sin(\pi-C)=\sin C$
Similarly, $\sin(B+C)=\sin(\pi-A)=\sin A$
and $\sin(C+A)=\sin(\pi-B)=\sin B$

(iii) If $A+B+C=\pi$, then $\cos(A+B)$
$=\cos(\pi-C)=-\cos C$
Similarly, $\cos(B+C)=\cos(\pi-A)=-\cos A$
and $\cos(C+A)=\tan(\pi-B)=-\tan B$

(iv) If $A+B+C=\pi$, then $\tan(A+B)$
$=\tan(\pi-C)=-\tan C$
Similarly, $\tan(B+C)=\tan(\pi-A)=-\tan A$
and, $\tan(C+A)=\tan(\pi-B)=-\tan B$

(v) If $A+B+C=\pi$, then $\frac{A+B}{2}=\frac{\pi}{2}-\frac{C}{2}$
and $\frac{B+C}{2}=\frac{\pi}{2}-\frac{A}{2}$ and $\frac{C+A}{2}=\frac{\pi}{2}-\frac{B}{2}$

$$\sin\left(\frac{A+B}{2}\right)=\sin\left(\frac{\pi}{2}-\frac{C}{2}\right)=\cos\left(\frac{C}{2}\right)$$

$$\cos\left(\frac{A+B}{2}\right)=\cos\left(\frac{\pi}{2}-\frac{C}{2}\right)=\sin\left(\frac{C}{2}\right)$$

$$\cos\left(\frac{A+B}{2}\right)=\cos\left(\frac{\pi}{2}-\frac{C}{2}\right)=\sin\left(\frac{C}{2}\right)$$
$$\tan\left(\frac{A+B}{2}\right)=\tan\left(\frac{\pi}{2}-\frac{C}{2}\right)=\cot\left(\frac{C}{2}\right)$$

All problems on conditional identities are broadly divided into the following four types :

(i) Identities involving sines and cosines of the multiple or sub-multiples of the angles involved.

(ii) Identities involving squares of sines and cosines of the multiple or sub-multiples of the angles involved.

(iii) Identities involving tangents and cotangents of the multiples or sub-multiples of the angles involved.

(iv) Identities involving cubes and higher powers of sines and cosines and some mixed identities.

TYPE I ***Identities involving sines and cosines of the multiple or submultiple of the angles involved***

Working Method

***Step* 1** Use C and D formulae.

***Step* 2** Use the given relation $(A+B+C=\pi)$ in the expression obtained in step 1 such that a factor can be taken common after using multiple angles formulae in the remaining term.

***Step* 3** Take the common factor outside.

***Step* 4** Again use the given relation $(A+B+C=\pi)$ within the bracket in such a manner so that we can apply C and D formulae.

Step 5 Find the result according to the given options.

TYPE II ***Identities involving squares of sines and cosines of multiple or sub-multiples of the angles involved.***

***Step* 1** Arrange the terms of the identify such that either $\sin^2 A-\sin^2 B=\sin(A+B)\,.\sin(A-B)$

or $\cos^2 A-\sin^2 B=\cos(A+B)\cos(A-B)$ can be used.

***Step* 2** Take the common factor outside.

***Step* 3** Use the given relations $(A+B+C=\pi)$ within the bracket in such a manner so that we can apply C and D formulae.

***Step* 4** Find the result according to the given options.

Type III ***Identities for tan and cot of the angles***

Working Method

***Step* 1** Express the sum of the two angles in terms of third angle by using the given relation $(A+B+C=\pi)$

***Step* 2** Taking tangent or cotangent of the angles of both the sides.

***Step* 3** Use sum and difference formulae in the left hand side.

***Step* 4** Use cross-multiplication in the expression obtained in the step 3.

***Step* 5** Arrange the terms as per the result required.

Example 107. If $A+B+C=\pi$, then, find

$$\sin 2A+\sin 2B+\sin 2C.$$

Sol. $\sin 2A+\sin 2B+\sin 2C$

$$=2\sin\left(\frac{2A+2B}{2}\right)\cos\left(\frac{2A-2B}{2}\right)+\sin 2C$$
$$=2\sin(A+B)\cos(A-B)+\sin 2C$$
$$=2\sin(\pi-C)\cdot\cos(A-B)+\sin 2C$$

[$\because A+B+C=\pi,\ A+B=\pi-C$
$\therefore \sin(A+B)=\sin(\pi-C)=\sin C$]

$$=2\sin C\cos(A-B)+2\sin C\cos C$$
$$=2\sin C\,[\cos(A-B)+\cos C]$$
$$=2\sin C\,[\cos(A-B)-\cos(A+B)]$$

[$\because \cos(A-B)-\cos(A+B)=2\sin A\sin B$, by C and D formula]

$$=2\sin C\,[2\sin A\sin B]$$
$$=4\sin A\sin B\sin C$$

Example 108. If $A+B+C=\pi$, then, find

$$\tan A+\tan B+B+\tan C$$

Sol. $A+B+C=\pi$

$$A+B=\pi-C\Rightarrow\tan(A+B)=\tan(\pi-C)$$
$$\Rightarrow\quad\frac{\tan A+\tan B}{1-\tan A\tan B}=-\tan C$$
$$\Rightarrow\quad\tan A+\tan B=-\tan C+\tan A\cdot\tan B\cdot\tan C$$
$$\tan A+\tan B+\tan C=\tan A\cdot\tan B\cdot\tan C$$

Maximum and Minimum Values of Trigonometrical Functions

As we have discussed in previous article that $-1\le\sin x\le 1$ and $-1\le\cos x\le 1$.

If there is a trigonometrical function of the form $a\sin x+b\cos x$, then by putting $a=r\cos\theta$, $b=r\sin\theta$, we have

$$a\sin x+b\cos x=r\cos\theta\sin x+r\sin\theta\cdot\cos x$$
$$=r(\cos\theta\sin x+\sin\theta\cos x)$$
$$=r\sin(x+\theta),\text{ where } r=\sqrt{a^2+b^2},\ \tan\theta=\frac{b}{a}$$

Since, $-1 \le \sin(x+\theta) \le 1$

$\therefore \quad -r \le r\sin(x+\theta) \le r$

or $-\sqrt{a^2+b^2} \le a\sin x + b\cos x \le \sqrt{a^2+b^2}$, for all x.

Hence, the maximum and minimum values of trigonometrical functions of the form $a\sin x + b\cos x$ are $\sqrt{a^2+b^2}$ and $-\sqrt{a^2+b^2}$, respectively.

Note $|a\cos A + b\sin A| \le \sqrt{a^2+b^2}$.

Also, $\cos A \pm \sin A = \sqrt{2}\sin\left(\frac{\pi}{4} \pm A\right) = \sqrt{2}\cos\left(A \mp \frac{\pi}{4}\right)$

Example 109. Find the maximum and minimum value of $3\sin 2x + 4\cos 2x + 3$.

Sol. As we know,

$$-\sqrt{a^2+b^2} \le a\sin A + b\cos A \le \sqrt{a^2+b^2}$$

$\Rightarrow \quad -\sqrt{3^2+4^2} \le 3\sin 2x + 4\cos 2x \le \sqrt{3^2+4^2}$

$\Rightarrow \quad -5 \le 3\sin 2x + 4\cos 2x \le 5$

$\therefore \quad -5+3 \le 3\sin 2x + 4\cos 2x + 3 \le 5+3$

$\Rightarrow \quad (3\sin 2x + 4\cos 2x + 3) \in [-2, 8]$.

Example 110. Find the maximum and minimum value of $6\sin x\cos x + 4\cos 2x$.

Sol. We have, $6\sin x\cos x + 4\cos 2x$

$\Rightarrow \quad 3\sin 2x + 4\cos 2x$

$\therefore \quad -\sqrt{3^2+4^2} \le 3\sin 2x + 4\cos 2x \le \sqrt{3^2+4^2}$

$\Rightarrow \quad 3\sin 2x + 4\cos 2x \in [-5, 5]$

Hence, the maximum value is 5 and minimum value is –5.

Example 111. Prove that $-4 \le 5\cos\theta + 3\cos\left(\theta + \frac{\pi}{3}\right) + 3 \le 10$, for all values of θ.

Sol. We have, $5\cos\theta + 3\cos\left(\theta + \frac{\pi}{3}\right)$

$\Rightarrow 5\cos\theta + 3\cos\theta\cos\frac{\pi}{3} - 3\sin\theta\sin\frac{\pi}{3}$

$\Rightarrow \frac{13}{2}\cos\theta - \frac{3\sqrt{3}}{2}\sin\theta \Rightarrow -7 \le \frac{13}{2}\cos\theta - \frac{3\sqrt{3}}{2}\sin\theta \le 7$

$\Rightarrow \quad -7 \le 5\cos\theta + 3\cos(\theta + \pi/3) \le 7$, for all θ.

$\Rightarrow \quad -7+3 \le 5\cos\theta + 3\cos(\theta+\pi/3) + 3 \le 7+3$, for all θ.

$\Rightarrow -4 \le 5\cos\theta + 3\cos(\theta+\pi/3) + 3 \le 10$, for all θ.

Hence proved.

Example 112. Find the maximum value of $1 + \sin\left(\frac{\pi}{4}+\theta\right) + 2\cos\left(\frac{\pi}{4}-\theta\right)$ for all real value of θ.

Sol. We have, $1 + \sin\left(\frac{\pi}{4}+\theta\right) + 2\cos\left(\frac{\pi}{4}-\theta\right)$

$$= 1 + \frac{1}{\sqrt{2}}(\cos\theta + \sin\theta) + \sqrt{2}(\cos\theta + \sin\theta)$$

$$= 1 + \left(\frac{1}{\sqrt{2}} + \sqrt{2}\right)(\cos\theta + \sin\theta)$$

$$= 1 + \left(\frac{1}{\sqrt{2}} + \sqrt{2}\right)\cdot\sqrt{2}\cos\left(\theta - \frac{\pi}{4}\right)$$

$\therefore$ The maximum value of $1 + \left(\frac{1}{\sqrt{2}} + \sqrt{2}\right)\cdot\sqrt{2} = 4$.

Example 113. Find the maximum and minimum value of $\cos^2\theta - 6\sin\theta\cos\theta + 3\sin^2\theta + 2$.

Sol. We have, $\cos^2\theta - 6\sin\theta\cos\theta + 3\sin^2\theta + 2$

$$= (1-\sin^2\theta) - 3\sin 2\theta + 3\sin^2\theta + 2$$

$$= 2\sin^2\theta - 3\sin 2\theta + 3$$

$$= (1-\cos 2\theta) - 3\sin 2\theta + 3$$

$$= 4 - (\cos 2\theta + 3\sin 2\theta) \quad \text{...(i)}$$

As we have, $-\sqrt{10} \le \cos 2\theta + 3\sin 2\theta \le \sqrt{10}$

$\therefore \quad -\sqrt{10} \le -(\cos 2\theta + 3\sin 2\theta) \le \sqrt{10}$

or $\quad 4 - \sqrt{10} \le 4 - (\cos 2\theta + 3\sin 2\theta) \le 4 + \sqrt{10} \quad$ ...(ii)

$\therefore$ From Eqs. (i) and (ii),

$$4 - \sqrt{10} \le \cos^2\theta - 6\sin\theta\cos\theta + 3\sin^2\theta + 2 \le 4 + \sqrt{10}$$

Example 114. The minimum value of $\cos 2\theta + \cos\theta$ for all real values of θ.

Sol. $\cos 2\theta + \cos\theta = 2\cos^2\theta - 1 + \cos\theta$

$$= -1 + 2\left(\cos^2\theta + \frac{1}{2}\cos\theta\right)$$

$$= -1 + 2\left(\cos^2\theta + \frac{1}{2}\cos\theta + \frac{1}{16} - \frac{1}{16}\right)$$

$$= -1 + 2\left(\cos\theta + \frac{1}{4}\right)^2 - \frac{1}{8}$$

$$= -\frac{9}{8} + 2\left(\cos\theta + \frac{1}{4}\right)^2 \ge -\frac{9}{8}$$

So, the minimum value of $\cos 2\theta + \cos\theta$ is $-\frac{9}{8}$.

Example 115. If $f(x) = \frac{\sin 3x}{\sin x}$, $x \ne n\pi$, then find range of $f(x)$.

Sol. $f(x) = \frac{\sin 3x}{\sin x} = \frac{3\sin x - 4\sin^3 x}{\sin x}$

$\Rightarrow \quad f(x) = 3 - 4\sin^2 x. \quad$ ...(i)

We know, $0 < \sin^2 x \le 1 (\sin x \ne 0$ as $x \ne n\pi)$

or $\quad -1 \le -\sin^2 x < 0$

or $\quad -4 \le -4\sin^2 x < 0$ or $3 - 4 \le 3 - 4\sin^2 x < 3$

or $\quad -1 \le 3 - 4\sin^2 x < 3 \Rightarrow -1 \le f(x) < 3$

Hence, range of $f(x) \in [-1, 3)$.

Application on Quadratic Equations

As we know, $ax^2 + bx + c = 0$, represents the quadratic equation whose,

$$\text{sum of roots} = \frac{-b}{a}.$$

$$\text{product of roots} = \frac{c}{a}.$$

$$(\alpha, \beta) \text{ roots} = \frac{-b \pm \sqrt{b^2 - 4ac}}{2a}$$

and if we want to form quadratic equation whose roots are given.

$\Rightarrow$ $x^2 - (\text{sum of the roots})x + (\text{product of the roots}) = 0.$

As above mentioned results are basics for quadratic equations, we discuss certain application on trigonometry.

Example 116. Find $\cos(\alpha + \beta)$, if $\tan\frac{\alpha}{2}$ and $\tan\frac{\beta}{2}$ are roots of the equations $8x^2 - 26x + 15 = 0$.

Sol. It is given $\tan\frac{\alpha}{2}$ and $\tan\frac{\beta}{2}$ are roots of $8x^2 - 26x + 15 = 0$.

$$\Rightarrow \quad \tan\frac{\alpha}{2} + \tan\frac{\beta}{2} = \frac{13}{4} \text{ and } \tan\frac{\alpha}{2}.\tan\frac{\beta}{2} = \frac{15}{8}$$

$$\therefore \quad \cos(\alpha + \beta) = \frac{1 - \tan^2\left(\frac{\alpha + \beta}{2}\right)}{1 + \tan^2\left(\frac{\alpha + \beta}{2}\right)},$$

$$\text{where } \tan\left(\frac{\alpha + \beta}{2}\right) = \frac{\tan\frac{\alpha}{2} + \tan\frac{\beta}{2}}{1 - \tan\frac{\alpha}{2}\tan\frac{\beta}{2}}$$

$$\text{or} \quad \tan\left(\frac{\alpha + \beta}{2}\right) = \frac{\frac{13}{4}}{1 - \frac{15}{8}} = \frac{-26}{7}$$

$$\Rightarrow \quad \cos(\alpha + \beta) = \frac{1 - \left(\frac{676}{49}\right)}{1 + \left(\frac{676}{49}\right)} = \frac{-627}{725}$$

$$\Rightarrow \quad \cos(\alpha + \beta) = \frac{-627}{725}$$

Example 117. If the solutions for θ from the equation $\sin^2\theta - 2\sin\theta + \lambda = 0$ lie in $\bigcup_{n \in Z}\left(2n\pi - \frac{\pi}{6}, (2n + 1)\pi + \frac{\pi}{6}\right)$. Then, find the possible set values of λ.

Sol. $\sin^2\theta - 2\sin\theta + \lambda = 0 \Rightarrow \sin\theta = \frac{2 \pm \sqrt{4 - 4\lambda}}{2} = 1 \pm \sqrt{1 - \lambda}.$

For real values, $1 - \lambda \geq 0$, i.e. $\lambda \leq 1$.

As $-1 \leq \sin\theta \leq 1$, $\sin\theta = 1 - \sqrt{1 - \lambda}$. (neglecting $1 + \sqrt{1 - \lambda}$)

From question, $\sin\theta > -\frac{1}{2}$

$$\text{Thus,} \quad -\frac{1}{2} < 1 - \sqrt{1 - \lambda} \leq 1$$

$$\text{or} \quad -\frac{3}{2} < -\sqrt{1 - \lambda} \leq 0 \Rightarrow \sqrt{1 - \lambda} < \frac{3}{2}$$

$$\Rightarrow \quad 1 - \lambda < \frac{9}{2} \Rightarrow \lambda > -\frac{5}{4}$$

Example 118. If $ABCD$ is a convex quadrilateral such that $4\sec A + 5 = 0$, then find the quadratic equation whose roots are $\tan A$ and $\operatorname{cosec} A$.

Sol. $\sec A = -\frac{5}{4}$. So, $\frac{\pi}{2} < A < \pi$

Hence, $\tan A = -\frac{3}{4}$ and $\operatorname{cosec} A = \frac{5}{3}$

$\therefore$ Required quadratic equation is

$$x^2 - \left(-\frac{3}{4} + \frac{5}{3}\right)x + \left(-\frac{3}{4}\right) \times \frac{5}{3} = 0$$

$$x^2 - \frac{11}{12}x - \frac{5}{4} = 0 \text{ or } 12x^2 - 11x - 15 = 0$$

Example 119. If $\sec\alpha$ and $\operatorname{cosec}\alpha$ are the roots of $x^2 - px + q = 0$, then show $p^2 = q(q + 2)$.

Sol. Since, $\sec\alpha$ and $\operatorname{cosec}\alpha$ are the roots of $x^2 - px + q = 0$

$\therefore$ $\sec\alpha + \operatorname{cosec}\alpha = p$ and $\sec\alpha \operatorname{cosec}\alpha = q$

$\therefore$ $\sin\alpha + \cos\alpha = p\sin\alpha\cos\alpha$ and $\sin\alpha\cos\alpha = \frac{1}{q}$

$$\therefore \quad \sin\alpha + \cos\alpha = \frac{p}{q}$$

On squaring both sides, we get

$$\sin^2\alpha + \cos^2\alpha + 2\sin\alpha\cos\alpha = \frac{p^2}{q^2}$$

$$1 + 2\sin\alpha\cos\alpha = \frac{p^2}{q^2}$$

$$\text{or} \quad 1 + \frac{2}{q} = \frac{p^2}{q^2} \quad \Rightarrow p^2 = q(q + 2)$$

Example 120. Find the number of values of x in the interval $[0, 5\pi]$ satisfying the equation $3\sin^2 x - 7\sin x + 2 = 0$.

Sol. $3\sin^2 x - 7\sin x + 2 = 0$

$$\Rightarrow \quad \sin x = \frac{7 \pm \sqrt{49 - 24}}{6} = \frac{7 \pm 5}{6} = \frac{1}{3}, 2$$

$$\therefore \quad \sin x = \frac{1}{3} \quad \text{(where, 2 is not possible).}$$

$$\sin x = \frac{1}{3} = \sin\alpha; (0 < \alpha < \pi/2)$$

$$\therefore \quad x = \alpha, \pi - \alpha, 2\pi + \alpha, 3\pi - \alpha, 4\pi + \alpha, 5\pi - \alpha$$

Thus, the number of values of x is 6.

Example 121. $0 \le a \le 3$, $0 \le b \le 3$ and the equation, $x^2 + 4 + 3\cos(ax + b) = 2x$ has atleast one solution, then find the value of $(a + b)$.

Sol. $x^2 - 2x + 4 = -3\cos(ax + b)$

$\Rightarrow \quad (x-1)^2 + 3 = -3\cos(ax + b) \quad \text{...(i)}$

As $-1 \le \cos(ax + b) \le 1$ and $(x-1)^2 \ge 0$

$\therefore$ Eq. (i) is only possible if,

$\cos(ax + b) = -1$ and $(x - 1) = 0$.

So, $a + b = \pi, 3\pi, 5\pi, \ldots$

and $3\pi > 6$

where $a + b \le 6$

$\Rightarrow \quad a + b = \pi$

Example 122. Find the values of p if it satisfy;

$\cos\theta = x + \frac{p}{x}$, $x \in R$ for all real values of θ.

Sol. $x^2 - \cos\theta\, x + p = 0$

$$\Rightarrow \quad x = \frac{\cos\theta \pm \sqrt{\cos^2\theta - 4p}}{2}$$

For real x, $\cos^2\theta - 4p \ge 0. \Rightarrow 4p \le \cos^2\theta$

$4p \le \cos^2\theta \le 1.$

$\Rightarrow \quad p \le \frac{1}{4}$ for all values of θ.

Example 123. Find the set of values of $\lambda \in R$ such that $\tan^2\theta + \sec\theta = \lambda$ holds for some θ.

Sol. $\tan^2\theta + \sec\theta = \lambda \Rightarrow \sec^2\theta + \sec\theta - (\lambda + 1) = 0$

$$\therefore \quad \sec\theta = \frac{-1 \pm \sqrt{1 + 4(\lambda + 1)}}{2} = \frac{-1 \pm \sqrt{4\lambda + 5}}{2}$$

For real $\sec\theta$, $4\lambda + 5 \ge 0$,

i.e. $\lambda \ge -\frac{5}{4} \quad \text{...(i)}$

Also, $\sec\theta \ge 1$ or $\sec\theta \le -1$

$$\therefore \quad \frac{-1 \pm \sqrt{4\lambda + 5}}{2} \ge 1$$

or $$\frac{-1 \pm \sqrt{4\lambda + 5}}{2} \le -1.$$

$\Rightarrow \quad -1 \pm \sqrt{4\lambda + 5} \ge 2$

or $-1 \pm \sqrt{4\lambda + 5} \le -2.$

$\Rightarrow \quad 4\lambda + 5 \ge 9$

or $4\lambda + 5 > 1$

$\Rightarrow \quad \lambda \ge 1$

or $\lambda \ge -1 \quad \text{...(ii)}$

From Eqs. (i) and (ii),

$\therefore \quad \lambda \ge -\frac{5}{4}$ and $\lambda \ge 1$

or $\lambda \ge -\frac{5}{4}$ and $\lambda \ge -1$

$\therefore \quad \lambda \ge 1$ or $\lambda \ge -1$

$\therefore \quad \lambda \in [-1, \infty)$.

Proving Trigonometric Inequality

Jensen's Inequality

(i) Suppose that 'f' is a convex function on $[a, b] \in R$, for all $x_1, x_2, x_3 \ldots, x_n \in [a, b]$,

we have

$$f(x_1) + f(x_2) + \ldots + f(x_n) \ge n \cdot f\left(\frac{x_1 + x_2 + \ldots + x_n}{n}\right)$$

Proof If $f(x)$ is concave up,

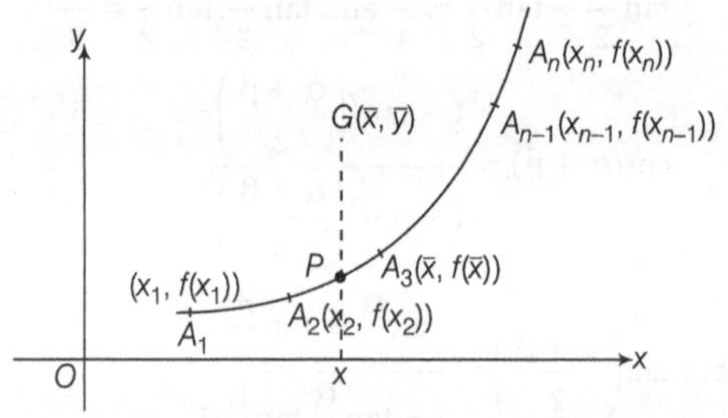

Here, $$G\left(\frac{x_1 + x_2 + \ldots + x_n}{n}, \frac{f(x_1) + f(x_2) + \ldots + f(x_n)}{n}\right)$$

and $$P\left(\frac{x_1 + x_2 + \ldots + x_n}{n}, f\left(\frac{x_1 + x_2 + \ldots + x_n}{n}\right)\right)$$

From figure, ordinate of $G \ge$ ordinate of P.

$$\frac{f(x_1) + f(x_2) + \ldots + f(x_n)}{n} \ge f\left(\frac{x_1 + x_2 + \ldots + x_n}{n}\right)$$

$$\therefore \quad f(x_1) + f(x_2) + \ldots + f(x_n) \ge n \cdot f\left(\frac{x_1 + x_2 + \ldots + x_n}{n}\right)$$

(ii) Similarly, suppose that f is concave function on $[a, b] \in R$, for all $x_1, x_2, x_3, \ldots x_n \in [a, b]$, we have

$$f(x_1) + f(x_2) + \ldots + f(x_n) \le n \cdot f\left(\frac{x_1 + x_2 + \ldots + x_n}{n}\right)$$

Example 124. If $A, B, C, \in \left(-\frac{\pi}{2}, \frac{\pi}{2}\right)$. Then prove that $\cos A \cos B + \cos C \le \frac{3}{2}$.

Sol. Since, for a function which is concave downwards

$$f\left(\frac{x_1 + x_2 + x_3}{3}\right) > \frac{f(x_1) + f(x_2) + f(x_3)}{3}$$

and we know that the graph of $y = \cos x$ is concave downwards for all $x \in \left(-\frac{\pi}{2}, \frac{\pi}{2}\right)$.

Let $P(A, \cos A)$, $Q(B, \cos B)$ and $R(C, \cos C)$ be any three points on $y = \cos x$, then it is clear from the graph $GM \le ML$

$$\Rightarrow \quad \frac{\cos A + \cos B + \cos C}{3} \le \cos\left(\frac{A + B + C}{3}\right)$$

$$\therefore \quad \cos A + \cos B + \cos C \le \frac{3}{2}, \text{ as at } B + C = \pi$$

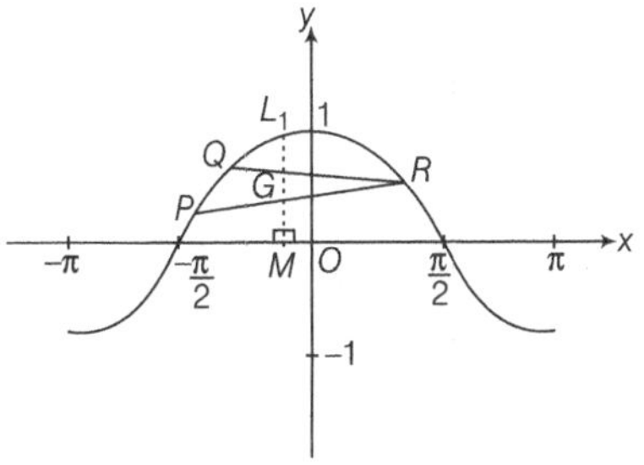

Exercise for Session 11

1. Prove that the minimum value of $3 \cos x + 4 \sin x + 5$ is 0.
2. If $\sin \theta_1 + \sin \theta_2 + \sin \theta_3 = 3$, then find the value of $\cos \theta_1 + \cos \theta_2 + \cos \theta_3$.
3. If $x = r \sin \theta \cos \phi$, $y = r \sin \theta \sin \phi$ and $z = r \cos \theta$, then prove that $x^2 + y^2 + z^2$ is independent of θ and ϕ.
4. Find the least value of $2 \sin^2 \theta + 3 \cos^2 \theta$.
5. α, β, γ are real numbers satisfying $\alpha + \beta + \gamma = \pi$. The find the minimum value of given expression $\sin \alpha + \sin \beta + \sin \gamma$.
6. If $A = \sin^2 \theta + \cos^4 \theta$, then find all real values of θ.
7. Find the minimum value of $\sec^2\theta + \operatorname{cosec}^2\theta - 4$.
8. If $P = \cos(\cos x) + \sin(\cos x)$, then the least and greatest value of P respectively.

   (a) -1 and 1 (b) 0 and 2 (c) $-\sqrt{2}$ and $\sqrt{2}$ (d) 0 and $\sqrt{2}$
9. Let $\theta \in \left(0, \frac{\pi}{4}\right)$ and $t_1 = (\tan \theta)^{\tan \theta}$, $t_2 = (\tan \theta)^{\cot \theta}$, $t_3 = (\cot \theta)^{\tan \theta}$ and $t_4 = (\cot \theta)^{\cot \theta}$, then show that $t_4 > t_3 > t_1 > t_2$.
10. Find the ratio of greatest value of $2 - \cos x + \sin^2 x$ to its least value.

JEE Type Solved Examples :

Single Option Correct Type Questions

● **Ex. 1.** *In the inequality below, the value of the angle is expressed in radian measure. Which one of the inequalities below is true?*

(a) $\sin 1 < \sin 2 < \sin 3$ (b) $\sin 3 < \sin 2 < \sin 1$

(c) $\sin 2 < \sin 1 < \sin 3$ (d) $\sin 3 < \sin 1 < \sin 2$

Sol. (*d*) We have, $\sin 1 - \sin 2$

$$= -2\cos\frac{3}{2}\cdot\sin\frac{1}{2} < 0$$

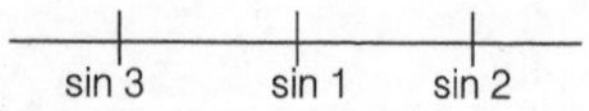

$\therefore$ $\sin 1 < \sin 2$

Similarly $\sin 1 - \sin 3$ ...(i)

$= -2\cos 2 \sin 1 > 0$

$\Rightarrow$ $\sin 1 > \sin 3$ ...(ii)

From Eqs. (i) and (ii), we get

$\sin 3 < \sin 1 < \sin 2$

● **Ex. 2.** *In a ΔABC, $\angle B < \angle C$ and the values of B and C satisfy the equation $2\tan x - k(1+\tan^2 x) = 0$, where $(0 < k < 1)$. Then, the measure of $\angle A$ is*

(a) $\frac{\pi}{3}$ (b) $\frac{2\pi}{3}$

(c) $\frac{\pi}{2}$ (d) $\frac{3\pi}{4}$

Sol. (*c*) $\because k = \frac{2\tan x}{1+\tan^2 x} = \sin 2x$

$\Rightarrow$ $\sin 2C = \sin 2B$

But $\angle C > \angle B$

$\Rightarrow$ $2C = \pi - 2B \Rightarrow B + C = \frac{\pi}{2}$

$\therefore$ $\angle A = \frac{\pi}{2}$

● **Ex. 3.** *If M and m are maximum and minimum value of the function $f(x) = \frac{\tan^2 x + 4\tan x + 9}{1+\tan^2 x}$, then $(M+m)$ equals*

(a) 20 (b) 14

(c) 10 (d) 8

Sol. (*c*) Given, $f(x) = \frac{\tan^2 x + 4\tan x + 9}{1+\tan^2 x}$

$$= \frac{2(2\tan x)}{1+\tan^2 x} + 4\left(\frac{1-\tan^2 A}{1+\tan^2 A}\right) + 5$$

$$= 2\sin 2x + 4\cos 2x + 5$$

$\therefore$ $R_f = [\sqrt{5} - \sqrt{20}, 5 + \sqrt{20}]$

Hence, $(M+m) = 10$.

Alternate Method

$$f(x) = \frac{\tan^2 x}{1+\tan^2 x} + \frac{4\tan x}{1+\tan^2 x} + \frac{9}{1+\tan^2 x}$$

$$= \sin^2 x + 2\sin 2x + 9\cos^2 x$$

$$= 1 + 4(1+\cos 2x) + 2\sin 2x$$

$$= 5 + 2\sin 2x + 4\cos 2x$$

● **Ex. 4.** *The value of $4\cos\frac{\pi}{10} - 3\sec\frac{\pi}{10} - \tan\frac{\pi}{10}$ is equal to*

(a) 1

(b) $\sqrt{5} - 1$

(c) $\sqrt{5} + 1$

(d) zero

Sol. (*d*) We have, $4\cos 18° - \frac{3}{\cos 18°} - 2\tan 18°$

$$= \frac{4\cos^2 18° - 3 - 2\sin 18°}{\cos 18°}$$

$$= \frac{2(1+\cos 36°) - 2\sin 18° - 3}{\cos 18°}$$

$$= \frac{2(1+\cos 36° - \sin 18°) - 3}{\cos 18°}$$

$$= \frac{2\left(1+\frac{1}{2}\right) - 3}{\cos 18°} = 0$$

● **Ex. 5.** *For $0 < A < \frac{\pi}{2}$, the value of*

$$\log_{\frac{1}{2}}\left(\frac{1}{1+2\cos^2 A} + \frac{2}{\sec^2 A + 2}\right)$$ *is equal to*

(a) 1 (b) -1

(c) 2 (d) 0

Sol. (*d*) As, $\left(\frac{1}{1+2\cos^2 A} + \frac{2}{\sec^2 A + 2}\right)$

$$= \left(\frac{1}{1+2\cos^2 A} + \frac{2\cos^2 A}{1+2\cos^2 A}\right)$$

$$= \frac{(1+2\cos^2 A)}{(1+2\cos^2 A)} = 1$$

Hence, $\log_{\frac{1}{2}}(1) = 0$.

Ex. 6. *The sum* $\dfrac{1}{\sin 45^\circ \sin 46^\circ} + \dfrac{1}{\sin 47^\circ \sin 48^\circ} + \dfrac{1}{\sin 49^\circ \sin 50^\circ} + \ldots + \dfrac{1}{\sin 133^\circ \sin 134^\circ}$ *is equal to*

(a) $\sec(1^\circ)$ (b) $\text{cosec}(1^\circ)$
(c) $\cot(1^\circ)$ (d) None of these

Sol. (*b*) $T_1 = \dfrac{1}{\sin 1^\circ}\left[\dfrac{\sin(46^\circ - 45^\circ)}{\sin 45^\circ \sin 46^\circ}\right] = \dfrac{1}{\sin 1^\circ}[\cot 45^\circ - \cot 46^\circ]$

$$T_2 = \frac{1}{\sin 1^\circ}\left[\frac{\sin(48^\circ - 47^\circ)}{\sin 48^\circ \sin 47^\circ}\right]$$

$$= \frac{1}{\sin 1^\circ}[\cot 47^\circ - \cot 48^\circ]$$

$$T_l = \frac{1}{\sin 1^\circ}\left[\frac{\sin(133^\circ - 134^\circ)}{\sin 133^\circ \sin 134^\circ}\right]$$

$$= \frac{1}{\sin 1^\circ}[\cot 133^\circ - \cot 134^\circ]$$

On adding

$$\sum_{r=1}^{l} T_r = \frac{1}{\sin 1^\circ}[\{\cot 45^\circ + \cot 47^\circ + \cot 49^\circ + \ldots + \cot 133^\circ\} - \{\cot 46^\circ + \cot 48^\circ + \cot 50^\circ + \ldots + \cot 134^\circ\}]$$

$= \text{cosec } 1^\circ$

[all terms cancelled except $\cot 45^\circ$ remains]

Ex. 7. *The range of k for which the inequality* $k\cos^2 x - k\cos x + 1 \geq 0 \ \forall\, x \in (-\infty, \infty)$, *is*

(a) $k < \dfrac{-1}{2}$ (b) $k < 4$
(c) $\dfrac{-1}{2} \leq k \leq 4$ (d) $\dfrac{1}{2} \leq k \leq 5$

Sol. (*c*) We have

$$k\cos^2 x - k\cos x + 1 \geq 0 \ \forall\, x \in (-\infty, \infty)$$

$\Rightarrow \quad k(\cos^2 x - \cos x) + 1 \geq 0$

But $\quad \cos^2 x - \cos x = \left(\cos x - \dfrac{1}{2}\right)^2 - \dfrac{1}{4}$

$\Rightarrow \quad -\dfrac{1}{4} \leq \cos^2 x - \cos x \leq 2$

$\therefore$ We have, $\quad 2k + 1 \geq 0$ and $-\dfrac{k}{4} + 1 \geq 0$

Hence, $\quad -\dfrac{1}{2} \leq k \leq 4.$

Ex. 8. *If* $f(\theta) = \dfrac{1 - \sin 2\theta + \cos 2\theta}{2\cos 2\theta}$, *then value of* $f(11^\circ)\cdot f(34^\circ)$ *equals*

(a) $\dfrac{1}{2}$ (b) $\dfrac{3}{4}$ (c) $\dfrac{1}{4}$ (d) 1

Sol. (*a*) $f(\theta) = \dfrac{1 - \sin 2\theta + \cos 2\theta}{2\cos 2\theta}$

$$= \frac{(\cos\theta - \sin\theta)^2 + (\cos^2\theta - \sin^2\theta)}{2(\cos\theta - \sin\theta)(\cos\theta + \sin\theta)}$$

$$= \frac{(\cos\theta - \sin\theta) + (\cos\theta + \sin\theta)}{2(\cos\theta + \sin\theta)}$$

$$= \frac{2\cos\theta}{2(\cos\theta + \sin\theta)} = \frac{1}{1 + \tan\theta}$$

$$f(11^\circ)\cdot f(34^\circ) = \frac{1}{(1 + \tan 11^\circ)}\cdot\frac{1}{(1 + \tan 34^\circ)}$$

$$= \frac{1}{(1 + \tan 11^\circ)}\cdot\frac{1}{(1 + \tan(45^\circ - 11^\circ))}$$

$$= \frac{1}{(1 + \tan 11^\circ)}\cdot\frac{1}{1 + \dfrac{1 - \tan 11^\circ}{1 + \tan 11^\circ}}$$

$$= \frac{1}{(1 + \tan 11^\circ)}\cdot\frac{1 + \tan 11^\circ}{2} = \frac{1}{2}$$

Ex. 9. *The variable 'x' satisfying the equation* $|\sin x \cos x| + \sqrt{2 + \tan^2 x + \cot^2 x} = \sqrt{3}$, *belongs to the interval*

(a) $\left[0, \dfrac{\pi}{3}\right]$ (b) $\left[\dfrac{\pi}{3}, \dfrac{\pi}{2}\right]$
(c) $\left[\dfrac{3\pi}{4}, \pi\right]$ (d) Non-existent

Sol. (*d*) $|\sin x \cos x| + |\tan x + \cot x| = \sqrt{3}$

$\Rightarrow \quad |\sin x \cos x| + \dfrac{1}{|\sin x \cos x|} = \sqrt{3}$

but $\quad |\sin x \cos x| + \dfrac{1}{|\sin x \cos x|} \geq 2$

Hence, no solution.

Ex. 10. *Let* α *be a real number such that* $0 \leq \alpha \leq \pi$. *If* $f(x) = \cos x + \cos(x + \alpha) + \cos(x + 2\alpha)$ *takes some constant number c for any* $x \in R$, *then the value of* $[c + \alpha]$ *is equal to*

Note $[y]$ denotes greatest integer less than or equal to y.

(a) 0 (b) 1 (c) -1 (d) 2

Sol. (*d*) $f(x) = \cos x + \cos(x + 2\alpha) + \cos(x + \alpha)$

$$= 2\cos(x + \alpha)\cos\alpha + \cos(x + \alpha)$$

$$= (2\cos\alpha + 1)\cos(x + \alpha)$$

As $\cos(x + \alpha)$ can take any real value from -1 to 1, $\forall x \in R$

$\therefore \quad f(x)$ is constant, so $(2\cos\alpha + 1) = 0$ must hold.

$\Rightarrow \quad \alpha = \dfrac{2\pi}{3}$ and $c = 0$

Hence, $\quad [c + \alpha] = \left[0 + \dfrac{2\pi}{3}\right] = 2$

Ex. 11. *In a ΔABC, if $4\cos A \cos B + \sin 2A + \sin 2B + \sin 2C = 4$, then ΔABC is*

(a) right angle but not isosceles
(b) isosceles but not right angled
(c) right angle isosceles
(d) obtuse angled

Sol. (c) We have, $4\cos A \cos B + 4\sin A \sin B \sin C = 4$

$$\Rightarrow \quad \sin C = \frac{1-\cos A \cos B}{\sin A \cos B} \le 1$$

$$\Rightarrow \quad 1 \le \sin A \sin B + \cos A \cos B$$

$$\Rightarrow \quad \cos(A-B) \ge 1$$

$$\Rightarrow \quad A = B \text{ and } \sin C = \frac{1-\cos^2 A}{\sin^2 A} = 1$$

$$\therefore \quad C = 90^\circ$$

and $\quad A = B = \frac{\pi}{4}$ (each).

Ex. 12. *For $\theta_1, \theta_2, \ldots, \theta_n \in \left(0, \frac{\pi}{2}\right)$, if $ln(\sec\theta_1 - \tan\theta_1) + ln(\sec\theta_2 - \tan\theta_2) + \ldots + ln(\sec\theta_n - \tan\theta_n) + ln\,\pi = 0$, then the value of $\cos((\sec\theta_1 + \tan\theta_1)(\sec\theta_2 + \tan\theta_2)\ldots\ldots(\sec\theta_n + \tan\theta_n))$ is equal to*

(a) $\cos\left(\frac{1}{\pi}\right)$ (b) -1
(c) 1 (d) 0

Sol. (b) $\ln\{(\sec\theta_1 - \tan\theta_1)(\sec\theta_2 - \tan\theta_2)\ldots(\sec\theta_n - \tan\theta_n)\}$

$$= \ln\left\{\frac{1}{\pi}\right\}$$

[**Note** If $0 < x < \frac{\pi}{2}$, $\sec x - \tan x = \frac{1-\sin x}{\cos x} > 0$]

$$\therefore (\sec\theta_1 - \tan\theta_1)(\sec\theta_2 - \tan\theta_2)\ldots(\sec\theta_n - \tan\theta_n) = \frac{1}{\pi} \quad \ldots(i)$$

Let $(\sec\theta_1 + \tan\theta_1)(\sec\theta_2 + \tan\theta_2)\ldots$

$$(\sec\theta_n + \tan\theta_n) = x \quad \ldots(ii)$$

On multiplying Eqs. (i) and (ii) we get

$$1 = \frac{x}{\pi}$$

$$\therefore \quad x = \pi$$

$$\therefore \cos((\sec\theta_1 + \tan\theta_1)(\sec\theta_2 + \tan\theta_2)\ldots(\sec\theta_n + \tan\theta_n)) = \cos\pi = -1$$

Ex. 13. *If A, B, C are interior angles of ΔABC such that $(\cos A + \cos B + \cos C)^2 + (\sin A + \sin B + \sin C)^2 = 9$, then number of possible triangles is*

(a) 0 (b) 1
(c) 3 (d) infinite

Sol. (d) $(\Sigma \cos A)^2 + (\Sigma \sin A)^2 = 9$

$$\Sigma(\cos^2 A + \sin^2 A) + 2(\Sigma \cos A \cos B + \sin A \sin B)$$

$$3 + 2\Sigma\cos(A-B) \le 3 + 2(3) = 9.$$

Equality holds if $A = B = C$

$\Rightarrow$ ΔABC is equilateral $\Rightarrow$ Infinite many equilateral

[**Note** We can vary side length of equilateral triangle]

Ex. 14. *If $\text{cosec}\frac{\pi}{32} + \text{cosec}\frac{\pi}{16} + \text{cosec}\frac{\pi}{8} + \text{cosec}\frac{\pi}{4} + \text{cosec}\frac{\pi}{2} = \cot\frac{\pi}{k}$, then the value of k is*

(a) 64 (b) 96 (c) 48 (d) 32

Sol. (a) $T_1 = \text{cosec}\,\theta = \dfrac{\sin\left(\theta - \frac{\theta}{2}\right)}{\sin\frac{\theta}{2}\sin\theta}; \theta = \frac{\pi}{32}$

$$T_1 = \cot\frac{\theta}{2} - \cot\theta$$

$$\left.\begin{aligned} T_2 &= \cot\theta - \cot 2\theta \\ T_3 &= \cot 2\theta - \cos 2^2\theta \\ T_4 &= \cot 2^2\theta - \cos 2^3\theta \\ T_5 &= 1 \end{aligned}\right] \text{sum} = 1 + \cot\frac{\theta}{2} - \cot 8\theta$$

$$\text{Sum} = 1 + \cot\frac{\theta}{2} - \cot 8\theta$$

$$= 1 + \cot\frac{\pi}{64} - \cot\frac{\pi}{4} = \cot\frac{\pi}{64} = \cot\frac{\pi}{k} \therefore k = 64$$

Ex. 15. *Let $S = \sum_{r=1}^{5}\cos(2r-1)\frac{\pi}{11}$ and $P = \prod_{r=1}^{4}\cos\left(2^r\frac{\pi}{15}\right)$, then*

(a) $\log_S P = -4$ (b) $P = 3S$
(c) $\text{cosec}\, S > \text{cosec}\, P$ (d) $\tan^{-1}P < \tan^{-1}S$

Sol. (d) We have, $\sum_{r=1}^{5}\cos(2r-1)\frac{\pi}{11}$

$$= \underbrace{\cos\frac{\pi}{11} + \cos\frac{3\pi}{11} + \cos\frac{5\pi}{11} + \cos\frac{7\pi}{11} + \cos\frac{9\pi}{11}}_{\text{cosine series}}$$

$$= \frac{2.\cos\left(\frac{5\pi}{11}\right).\sin\left(\frac{5\pi}{11}\right)}{2\cdot\sin\frac{\pi}{11}} = \frac{\sin\left(\frac{10\pi}{11}\right)}{2\cdot\sin\left(\frac{\pi}{11}\right)} = \frac{1}{2}$$

Also, $\prod_{r=1}^{\pi}\cos\left(2^r\frac{\pi}{15}\right) = \cos\frac{2\pi}{15}.\cos\frac{4\pi}{15}.\cos\frac{8\pi}{15}.\cos\frac{16\pi}{15}$

$$= \frac{\sin\left(\frac{32\pi}{15}\right)}{2^4\cdot\sin\left(\frac{2\pi}{15}\right)} = \frac{\sin\left(2\pi + \frac{2\pi}{15}\right)}{16\cdot\sin\left(\frac{2\pi}{15}\right)}$$

$$= \frac{\sin\left(\frac{2\pi}{15}\right)}{16 \cdot \sin\left(\frac{2\pi}{15}\right)} = \frac{1}{16}$$

Therefore, $\tan^{-1} P < \tan^{-1} S$.

Ex. 16. *Set of values of x lying in* $[0, 2\pi]$ *satisfying the inequality* $|\sin x| > 2\sin^2 x$ *contains*

(a) $\left(0, \frac{\pi}{6}\right) \cup \left(\pi, \frac{7\pi}{6}\right)$ (b) $\left(0, \frac{7\pi}{6}\right)$

(c) $\frac{\pi}{6}$ (d) None of these

Sol. (a) $|\sin x| > 2\sin^2 x$

$\Rightarrow |\sin x|(2|\sin x| - 1) < 0$

$\Rightarrow 0 < |\sin x| < \frac{1}{2}$

$\Rightarrow x \in \left(0, \frac{\pi}{6}\right) \cup \left(\frac{5\pi}{6}, \pi\right) \cup \left(\pi, \frac{7\pi}{6}\right) \cup \left(\frac{11\pi}{6}, 2\pi\right)$

Ex. 17. *The number of ordered pairs* (x, y)*, when* $x, y \in [0, 10]$ *satisfying* $\left(\sqrt{\sin^2 x - \sin x + \frac{1}{2}}\right) \cdot 2^{\sec^2 y} \le 1$ *is*

(a) 0 (b) 16

(c) infinite (d) 12

Sol. (b) $\sqrt{\sin^2 x - \sin x + \frac{1}{2}} = \sqrt{\left(\sin x - \frac{1}{2}\right)^2 + \frac{1}{2}} \ge \frac{1}{2}, \forall x$

and $\sec^2 y \ge 1, \forall y$, so $2^{\sec^2 y} \ge 2$. Hence, the above inequality holds only for those values of x and y for which $\sin x = \frac{1}{2}$

and $\sec^2 y = 1$.

Hence, $x = \frac{\pi}{6}, \frac{5\pi}{6}, \frac{13\pi}{6}, \frac{17\pi}{6}$ and $y = 0, \pi, 2\pi, 3\pi$. Hence, required number of ordered pairs are 16.

Ex. 18. *The least values of* $\operatorname{cosec}^2 x + 25\sec^2 x$ *is*

(a) 0 (b) 26

(c) 28 (d) 36

Sol. (d) $\operatorname{cosec}^2 x + 25\sec^2 x = 26 + \cot^2 x + 25\tan^2 x$

$= 26 + 10 + (\cot x - 5\tan x)^2 \ge 36$

Ex. 19. *If* $x\sin a + y\sin 2a + z\sin 3a = \sin 4a$

$$x\sin b + y\sin 2b + z\sin 3b = \sin 4b$$
$$x\sin c + y\sin 2c + z\sin 3c = \sin 4c$$

Then, the roots of the equation

$t^3 - \left(\frac{z}{2}\right)t^2 - \left(\frac{y+2}{4}\right)t + \left(\frac{z-x}{8}\right) = 0$, $a, b, c \ne n\pi$, *are*

(a) $\sin a, \sin b, \sin c$

(b) $\cos a, \cos b, \cos c$

(c) $\sin 2a, \sin 2b, \sin 2c$

(d) $\cos 2a, \cos 2b, \cos 2c$

Sol. (b) Equation first can be written as

$x\sin a + y \times 2\sin a\cos a + z \times \sin a(3 - 4\sin^2 a)$

$= 2 \times 2\sin a\cos a\cos 2a$

$\Rightarrow x + 2y\cos a + z(3 + 4\cos^2 a - 4)$

$= 4\cos a(2\cos^2 a - 1)$ as $\sin a \ne 0$

$\Rightarrow 8\cos^3 a - 4z\cos^2 a - (2y + 4)\cos a + (z - x) = 0$

$\Rightarrow \cos^3 a - \left(\frac{z}{2}\right)\cos^2 a - \left(\frac{y+2}{4}\right)\cos a + \left(\frac{z-x}{8}\right) = 0$

which shows that $\cos a$ is a root of the equation

$$t^3 = \left(\frac{z}{2}\right)t^2 - \left(\frac{y+z}{4}\right)t + \left(\frac{z-x}{8}\right) = 0$$

Similarly, from second and third equation we can verify that $\cos b$ and $\cos c$ are the roots of the given equation.

Ex. 20. *Let* α *and* β *be any two positive values of x for which* $2\cos x, |\cos x|$ *and* $1 - 3\cos^2 x$ *are in GP. The minium value of* $|\alpha + \beta|$ *is*

(a) $\frac{\pi}{3}$ (b) $\frac{\pi}{4}$

(c) $\frac{\pi}{2}$ (d) None of these

Sol. (d) $\because 2\cos x, |\cos x|, 1 - 3\cos^2 x$ are in GP.

$\therefore \cos^2 x = 2\cos x \cdot (1 - 3\cos^2 x)$

$\Rightarrow 6\cos^3 x + \cos^2 x - 2\cos x = 0$

$\therefore \cos x = 0, \frac{1}{2}, -\frac{2}{3}$

$\therefore x = \frac{\pi}{2}, \frac{\pi}{3}, \cos^{-1}\left(-\frac{2}{3}\right)$ [$\because \alpha, \beta$ are positive]

If $\alpha = \frac{\pi}{2}, \beta = \frac{\pi}{3}$

Then, $|\alpha - \beta| = \frac{\pi}{6}$

Ex. 21. *Let n be an odd integer. If* $\sin n\theta = \sum_{r=0}^{n} b_r \sin^r \theta$ *for all real* θ*, then*

(a) $b_0 = 1, b_1 = 3$ (b) $b_0 = 0, b_1 = n$

(c) $b_0 = -1, b_1 = n$ (d) $b_0 = 0, b_1 = n^2 - 3n - 3$

Sol. (b) Given, $\sin n\theta = \sum_{r=0}^{n} b_r \sin^r \theta = b_0 + b_1 \sin\theta + b_2 \sin^2\theta + \dots + b_n \sin^n \theta$ ...(i)

Putting $\theta = 0$ in Eq. (i), we get $0 = b_0$

Again, Eq. (i) can be written as $\sin n\theta = \sum_{r=0}^{n} b_r \sin^r \theta$

$$\frac{\sin n\theta}{\sin\theta} = \sum_{r=0}^{n} b_r \sin^{r-1}\theta$$

On taking limit as $\theta \to 0$, we get

$$\lim_{\theta\to 0} \frac{\sin n\theta}{\sin\theta} = b_1$$

$$\Rightarrow \quad \lim_{\theta\to 0} n\left(\frac{\sin\theta}{n\theta}\right)\left(\frac{\theta}{\sin\theta}\right) = b_1$$

$$\Rightarrow \quad n = b_1$$

Hence, $\quad b_0 = 0;\ b_1 = n$

Ex. 22. *The minimum and maximum values of $ab\sin x + b\sqrt{(1-a^2)}\cos x + c\ (|a|<1, b>0)$ respectively are*

(a) $\{b-c, b+c\}$ (b) $\{b+c, b-c\}$
(c) $\{c-b, b+c\}$ (d) None of these

Sol. (c) $ab\sin x + b\sqrt{(1-a^2)}\cos x$

Now, $\sqrt{(ab)^2 + (b\sqrt{(1-a^2)})^2}$

$$= \sqrt{a^2b^2 + b^2(1-a^2)}$$

$$= b\sqrt{(a^2+1-a^2)} = b$$

$$\Rightarrow \quad b\{(a\sin x + \sqrt{(1-a^2)}\cos x)\}$$

Let, $\quad a = \cos\alpha,$

$\therefore \quad \sqrt{(1-a^2)} = \sin\alpha$

$\Rightarrow \quad b\sin(x+a)$

$\because \quad -1 \le \sin(x+\alpha) \le 1$

$\therefore \quad c - b \le b\sin(x+\alpha) + c \le b + c$

$\therefore \ b\sin(x+\alpha) + c \in [c-b, c+b]$

Ex. 23. *If* $\dfrac{\sin^3\theta - \cos^3\theta}{\sin\theta - \cos\theta} - \dfrac{\cos\theta}{\sqrt{(1+\cot^2\theta)}}$

$-2\tan\theta\cot\theta = -1, \theta \in [0, 2\pi]$, *then*

(a) $\theta \in \left(0, \frac{\pi}{2}\right) - \left\{\frac{\pi}{4}\right\}$ (b) $\theta \in \left(\frac{\pi}{2}, \pi\right) - \left\{\frac{3\pi}{4}\right\}$

(c) $\theta \in \left(\pi, \frac{3\pi}{2}\right) - \left\{\frac{5\pi}{4}\right\}$ (d) $\theta \in (0, \pi) - \left\{\frac{\pi}{4}, \frac{\pi}{2}\right\}$

Sol. (d) $\dfrac{\sin^3\theta - \cos^3\theta}{\sin\theta - \cos\theta} = \sin^2\theta + \cos^2\theta + \sin\theta\cos\theta,$

$\theta \ne \frac{\pi}{4}, \frac{5\pi}{4}$

$= 1 + \sin\theta\cos\theta$

and $\dfrac{\cos\theta}{\sqrt{(1+\cot^2\theta)}} = \dfrac{\cos\theta}{|\operatorname{cosec}\theta|} = \sin\theta\cos\theta\ \forall\ \theta \in (0,\pi)$

and $-2\tan\theta\cot\theta = -2, \theta \ne \frac{\pi}{2}$

Hence, $\quad$ LHS = RHS

But $\quad \theta \ne \frac{\pi}{4}, \frac{5\pi}{4}, \frac{\pi}{2}$

Hence, $\quad \theta \in (0, \pi) \sim \left\{\frac{\pi}{4}, \frac{\pi}{2}\right\}$

Ex. 24. *If* $\cos x + \sin x = a\left(-\frac{\pi}{2} < x < -\frac{\pi}{4}\right)$, *then* $\cos 2x$ *is equal to*

(a) a^2 (b) $a\sqrt{(2-a)}$
(b) $a\sqrt{(2+a)}$ (d) $a\sqrt{(2-a^2)}$

Sol. (d) $\because -\frac{\pi}{2} < x < -\frac{\pi}{4}\left(-\pi < 2x < -\frac{\pi}{2}\text{, i.e, in III quadrant}\right)$

$$\Rightarrow \quad \cos x + \sin x = a$$

Squaring both sides $\cos^2 x + \sin^2 x + 2\cos x\sin x = a^2$

$$\Rightarrow \quad \sin 2x = (a^2 - 1)$$

$$\cos 2x = \sqrt{1 - (a^2-1)^2}$$

$$= \sqrt{a^2(2-a^2)}$$

$$= a\sqrt{(2-a^2)}$$

Ex. 25. *If* $S = \cos^2\frac{\pi}{2} + \cos^2\frac{2\pi}{n} + \ldots + \cos^2\frac{(n-1)\pi}{n}$, *then S equals*

(a) $\dfrac{n}{2(n+1)}$ (b) $\dfrac{1}{2(n-1)}$

(c) $\dfrac{1}{2(n-2)}$ (d) $\dfrac{n}{2}$

Sol. (c) $S = \cos^2\frac{\pi}{n} + \cos^2\frac{2\pi}{n} + \ldots + \cos^2(n-1)\frac{\pi}{n}$

$$\frac{1}{2}\begin{bmatrix} 1 + \cos\frac{2\pi}{n} + 1 + \cos\frac{4\pi}{n} + 1 + \cos\frac{6\pi}{n} \\ + \ldots + 1 + \cos 2(n-1)\frac{\pi}{n} \end{bmatrix}$$

$$= \frac{1}{2}\left[n - 1 + \sum_{k=1}^{n-1}\cos\frac{2k\pi}{n}\right]$$

$$= \frac{1}{2}[n-1-1] = \frac{1}{2}(n-2)$$

Ex. 26. *If* $\cos 5\theta = a\cos\theta + b\cos^3\theta + c\cos^5\theta + d$, *then*

(a) $a = 20$ (b) $b = -30$
(c) $a + b + c = 2$ (d) $a + b + c + d = 1$

Sol. (d) Put $\theta = \frac{\pi}{2}$ in the given inequality, we get $d = 0$

Put $\theta = 0$ in the given inequality, we get

$$a + b + c + d = 1 \quad \text{...(i)}$$

So, (d) is correct and (c) is not correct.

Now differentiate both sides with respect to θ, we get

$$-5\sin\theta = -a\sin\theta - 3b\cos^2\theta\sin\theta - 5c\cos^4\theta\sin\theta \quad ...(ii)$$

Put $\theta = \frac{\pi}{2}$, then $a = 5$ ...(iii)

Again putting $\theta = \frac{\pi}{4}$ in the given expression or in (2), we get

$$4a + 2b + c = -4 \quad ...(iv)$$

From (i), (iii) and (iv) we have $b = -20$ and $c = 16$

[**Note** We have found correct answer at the second step only however the complete solution is desired for better understanding of the solution.]

Alternates Solution

$$\cos 5\theta = \cos(2\theta + 3\theta) = \cos 2\theta\cos 3\theta - \sin 2\theta\sin 3\theta$$
$$=(2\cos^2\theta - 1)(4\cos^3\theta - 3\cos\theta) - (2\sin\theta\cos\theta)(3\sin\theta - 4\sin^3\theta)$$
$$= 8\cos^5\theta - 10\cos^3\theta + 3\cos\theta - 2(1-\cos^2\theta)\cos\theta\{3 - 4(1-\cos^2\theta)\}$$
$$= 8\cos^5\theta - 10\cos^3\theta + 3\cos\theta - 2(\cos\theta - \cos^3\theta)(4\cos^2\theta - 1) = 16\cos^5\theta - 20\cos^3\theta + 5\cos\theta.$$

Ex. 27. *If A and B are acute positive angles satisfying the equations* $3\sin^2 A + 2\sin^2 B = 1$ *and* $3\sin 2A - 2\sin 3B = 0$, *then* $A + 2B$ *is equal to*

(a) $\frac{\pi}{4}$ (b) $\frac{\pi}{2}$

(c) $\frac{3\pi}{4}$ (d) $\frac{2\pi}{3}$

Sol. (*b*) From the given relations, we have

$\sin 2B = \left(\frac{3}{2}\right)\sin 2A$ and $3\sin^2 A = 1 - 2\sin^2 B = \cos 2B$

so that

$$\cos(A + 2B) = \cos A\cos 2B - \sin A\sin 2B$$
$$= \cos A \cdot 3\sin^2 A - \left(\frac{3}{2}\right)\sin A\sin 2A$$
$$= 3\cos A\sin^2 A - 3\sin^2 A\cos A = 0$$
$$A + 2B = \frac{\pi}{2}$$

Ex. 28. *If* $A = \begin{bmatrix} \cos^2\alpha & \cos\alpha\sin\alpha \\ \cos\alpha\sin\alpha & \sin^2\alpha \end{bmatrix}$ *and* $B = \begin{bmatrix} \cos^2\beta & \cos\beta\sin\beta \\ \cos\beta\sin\beta & \sin^2\beta \end{bmatrix}$

are two matrices such that *AB* is the null matrix, then

(a) $\alpha = \beta$ (b) $\cos(\alpha - \beta) = 0$

(c) $\sin(\alpha - \beta) = 0$ (d) None of these

Sol. (*b*) $AB = 0$

$$\Rightarrow \begin{bmatrix} \cos^2\alpha & \cos\alpha\sin\alpha \\ \cos\alpha\sin\alpha & \sin^2\alpha \end{bmatrix}\begin{bmatrix} \cos^2\beta & \cos\beta\sin\beta \\ \cos\beta\sin\beta & \sin^2\beta \end{bmatrix} = \begin{bmatrix} 0 & 0 \\ 0 & 0 \end{bmatrix}$$

$$\Rightarrow \begin{bmatrix} \cos\alpha\cos\beta\cos(\alpha-\beta) & \cos\alpha\sin\beta\cos(\alpha-\beta) \\ \cos\beta\sin\alpha\cos(\alpha-\beta) & \sin\alpha\sin\beta\cos(\alpha-\beta) \end{bmatrix} = \begin{bmatrix} 0 & 0 \\ 0 & 0 \end{bmatrix}$$

$$\Rightarrow \cos(\alpha - \beta) = 0$$

Ex. 29. *If* $k_1 = \tan 27\theta - \tan\theta$ *and* $k_2 = \frac{\sin\theta}{\cos 3\theta} + \frac{\sin 3\theta}{\cos 9\theta} + \frac{\sin 9\theta}{\cos 27\theta}$ *then,*

(a) $k_1 = k_2$

(b) $k_1 = 2k_2$

(c) $k_1 + k_2 = 2$

(d) $k_2 = 2k_1$

Sol. (*b*) We can write

$$k_1 = \tan 27\theta - \tan 9\theta + \tan 9\theta - \tan 3\theta + \tan 3\theta - \tan\theta$$

But $\tan 3\theta - \tan\theta = \frac{\sin 3\theta\cos\theta - \cos 3\theta\sin\theta}{\cos 3\theta\cos\theta}$

$$= \frac{\sin 2\theta}{\cos 3\theta\cos\theta} = \frac{2\sin\theta}{\cos 3\theta}$$

$$\therefore \quad k_1 = 2\left[\frac{\sin 9\theta}{\cos 27\theta} + \frac{\sin 3\theta}{\cos 9\theta} + \frac{\sin\theta}{\cos 3\theta}\right] = 2k_2$$

Ex. 30. *If* $a^2 - 2a\cos x + 1 = 674$ *and* $\tan\left(\frac{x}{2}\right) = 7$ *then the integral value of a is*

(a) 25 (b) 49

(c) 67 (d) 74

Sol. (*a*) $674 = a^2 - 2a\frac{1-\tan^2\left(\frac{x}{2}\right)}{1+\tan^2\left(\frac{x}{2}\right)} + 1$

$$= a^2 - 2a \times \frac{1-49}{1+49} + 1$$
$$= a^2 + 2a \times \frac{48}{50} + 1$$

$\Rightarrow \quad 25a^2 + 48a - 673 \times 25 = 0$

$\Rightarrow \quad (a - 25)(25a + 673) = 0$

$\Rightarrow \quad a = 25$ (taking the integral value of a).

Ex. 31. *The maximum value of* $(\cos\alpha_1)(\cos\alpha_2)\ldots(\cos\alpha_n)$ *under the restriction* $0\le\alpha_1,\alpha_2,\ldots,\alpha_n\le\frac{\pi}{2}$ *and* $(\cot\alpha_1)(\cot\alpha_2)\ldots(\cot\alpha_n)=1$ *is*

(a) $\frac{1}{2^{\frac{n}{2}}}$ (b) $\frac{1}{2^n}$

(c) $\frac{1}{2n}$ (d) 1

Sol. (*a*) From the given relations we have

$$\prod_{i=1}^{n}(\cos\alpha_i)=\prod_{i=1}^{n}(\sin\alpha_i)$$

$$\Rightarrow \prod_{i=1}^{n}(\cos^2\alpha_i)=\prod_{i=1}^{n}(\cos\alpha_i\sin\alpha_i)=\prod_{i=1}^{n}\left(\frac{\sin 2\alpha_i}{2}\right)$$

Since, $0\le\alpha_i\le\frac{\pi}{2}\Rightarrow 0\le 2\alpha_i\le\pi$

$\therefore \prod_{i=1}^{n}(\cos^2\alpha_i)\le\frac{1}{2^n}$ as max. value of $\sin 2\alpha_i$ is 1 for all *i*.

$$\Rightarrow \prod_{i=1}^{n}(\cos\alpha_i)\le\frac{1}{2^{\frac{n}{2}}}$$

So, the maximum value of the given expression is $\frac{1}{2^{\frac{n}{2}}}$.

Ex. 32. *The value of expression* $\frac{\sin^3 x}{1+\cos x}+\frac{\cos^3 x}{1-\sin x}$ *is/are*

(a) $\sqrt{2}\cos\left[\frac{\pi}{4}-x\right]$ (b) $\sqrt{2}\cos\left[\frac{\pi}{4}+x\right]$

(c) $\sqrt{2}\sin\left[\frac{\pi}{4}-x\right]$ (d) None of these

Sol. (*a*) Let $\frac{\sin^3 x}{1+\cos x}+\frac{\cos^3 x}{1-\sin x}=A$, then

$$A=\frac{(\sin^3 x+\cos^3 x)+(\cos^4 x-\sin^4 x)}{(1+\cos x)(1-\sin x)}$$

or $$A=\frac{\{(\sin^3 x+\cos^3 x)\}+\left\{\begin{matrix}(\cos x+\sin x)(\cos x-\sin x)\\(\cos^2 x+\sin^2 x)\end{matrix}\right\}}{(1+\cos x)(1-\sin x)}$$

or $$A=\frac{(\sin x+\cos x)\{(1-\sin x\cos x)+(\cos x-\sin x)\}}{1+\cos x-\sin x-\sin x\cos x}$$

or $A=\sin x+\cos x$

or $A=\sqrt{2}\left[\frac{1}{\sqrt{2}}\sin x+\frac{1}{\sqrt{2}}\cos x\right]$ ...(i)

or $A=\sqrt{2}\left[\cos\frac{\pi}{4}\sin x+\sin\frac{\pi}{4}\cos x\right]$

$=\sqrt{2}\sin\left[\frac{\pi}{4}+x\right]$

Again, by Eq. (i)

$A=\sqrt{2}\left[\sin\frac{\pi}{4}\sin x+\cos\frac{\pi}{4}\cos x\right]$

$=\sqrt{2}\cos\left[\frac{\pi}{4}-x\right]$

Ex. 33. *Let* $0\le\theta\le\frac{\pi}{2}$ *and* $x=X\cos\theta+Y\sin\theta$, $y=X\sin\theta-Y\cos\theta$ *such that* $x^2+2xy+y^2=aX^2+bY^2$, *where a and b are constant, then*

(a) $a=-1, b=-3$ (b) $\theta=\frac{\pi}{2}$

(c) $a=3, b=-1$ (d) $\theta=\frac{\pi}{3}$

Sol. (*c*) $x^2+y^2=X^2+Y^2$,

$xy=(X^2-Y^2)\sin\theta\cdot\cos\theta-XY(\cos^2\theta-\sin^2\theta)$

$x^2+4xy+y^2=X^2+Y^2+2(X^2-Y^2)\sin 2\theta-2XY\cos 2\theta$

$=(1+2\sin 2\theta)X^2+(1-2\sin 2\theta)Y^2-2\cos 2\theta\cdot XY$

From the question,

$a=1+2\sin 2\theta,\ b=1-2\sin 2\theta,\ \cos 2\theta=0$

$\cos 2\theta=0\Rightarrow\theta=\frac{\pi}{4}$, then

$a=1+2\sin\frac{\pi}{2},\ b=1-2\sin\frac{\pi}{2}$

$\therefore a=3, b=-1$

Ex. 34. *If* $0<x<\frac{\pi}{2}$ *and* $\sin^n x+\cos^n x\ge 1$, *then*

(a) $n\in[2,\infty)$ (b) $n\in(-\infty,2]$

(c) $n\in[-1,1]$ (d) None of these

Sol. (*b*) Since, $0<x<\frac{\pi}{2}$

$\therefore 0<\sin x<1$ and $0<\cos x<1$

when $x=2$, $\sin^n x+\cos^n x=1$

when $n>2$, both $\sin^n x$ and $\cos^n x$ will decreases and hence $\sin^n x+\cos^n x<1$.

when $n>2$, both $\sin^n x$ and $\cos^n x$ will increase and hence $\sin^n x+\cos^n x>1$.

Thus, $\sin^n x+\cos^n x\ge 1$ for $n\le 2$.

Ex. 35. *If* $a = \sin\frac{\pi}{18}\sin\frac{5\pi}{18}\sin\frac{7\pi}{18}$, *and x is the solution the equation* $y = 2[x] + 2$ *and* $y = 3[x-2]$, *where* $[x]$ *denotes the integral part of x, then a is equal to*

(a) $[x]$ (b) $\frac{1}{[x]}$

(c) $2[x]$ (d) $[x]^2$

Sol. (*b*) $a = \sin\frac{\pi}{18}\sin\frac{5\pi}{18}\sin\frac{7\pi}{18}$

$$= \sin 10^\circ \sin 50^\circ \sin 70^\circ$$
$$= \frac{1}{2}[2\sin 70^\circ \sin 10^\circ]\sin 50^\circ$$
$$= \frac{1}{2}[\cos 60^\circ - \cos 80^\circ]\sin 50^\circ$$
$$= \frac{1}{4}\sin 50^\circ - \frac{1}{4}(2\cos 80^\circ \sin 50^\circ)$$
$$= \frac{1}{4}\sin 50^\circ - \frac{1}{4}(\sin 130^\circ - \sin 30^\circ)$$
$$= \frac{1}{4}\sin 50^\circ - \frac{1}{4}\sin 50^\circ + \frac{1}{4}\cdot\frac{1}{2} = \frac{1}{8}$$

$y = 2[x] + 2$ and $y = 3[x-2]$

$\Rightarrow 2[x] + 2 = 3[x-2]$

$= 3[x] + 3[-2] \Rightarrow [x] = 8$

$\therefore \quad a = \frac{1}{[x]}$

Ex. 36. *If the mapping* $f(x) = ax + b$, $a < 0$ *and maps* $[-1, 1]$ *onto* $[0, 2]$, *then for all values of* θ, $A = \cos^2\theta + \sin^4\theta$ *is such that*

(a) $f\left(\frac{1}{4}\right) \le A \le f(0)$ (b) $f(0) \le A \le f(-2)$

(c) $f\left(\frac{1}{3}\right) \le A \le f(0)$ (d) $f(-1) < A \le f(-2)$

Sol. (*a*) Given, $f(x) = ax + b$

$\therefore \quad f'(x) = a$

Since, $a < 0$, $f(x)$ is a decreasing function

$\therefore \quad f(-1) = 2$ and $f(1) = 0$

$\Rightarrow \quad -a + b = 2$ and $a + b = 0$

$\therefore \quad a = -1$ and $b = 1$

Thus, $f(x) = -x + 1$

Clearly, $f(0) = 1$, $f\left(\frac{1}{4}\right) = \frac{3}{4}$, $f(-2) = 3$,

$f\left(\frac{1}{3}\right) = \frac{2}{3}$, $f(-1) = 2$

Also, $A = \frac{1+\cos 2\theta}{2} + \left(\frac{1-\cos 2\theta}{2}\right)^2$

$$= \frac{1}{2} + \frac{1}{2}\cos 2\theta + \frac{1}{4} - \frac{1}{2}\cos 2\theta + \frac{1}{4}\cos^2 2\theta$$
$$= \frac{3}{4} + \frac{1}{4}\left(\frac{1+\cos 4\theta}{2}\right) = \frac{7}{8} + \frac{1}{8}\cos 4\theta$$

$\therefore \quad \frac{3}{4} \le A \le 1 \Rightarrow f\left(\frac{1}{4}\right) \le A \le f(0)$

Ex. 37. *The value of* $\cos\frac{2\pi}{7} + \cos\frac{4\pi}{7} + \cos\frac{6\pi}{7}$ *is equal to*

(a) 1 (b) -1

(b) $\frac{1}{2}$ (c) $-\frac{1}{2}$

Sol. (*d*) $\cos\left(\frac{2\pi}{7}\right) + \cos\left(\frac{4\pi}{7}\right) + \cos\left(\frac{6\pi}{7}\right)$

$$= \text{Re}\left\{e^{\frac{2\pi i}{7}} + e^{\frac{4\pi i}{7}} + e^{\frac{6\pi i}{7}}\right\}$$
$$= \frac{e^{\frac{2\pi i}{7}} + e^{\frac{4\pi i}{7}} + e^{\frac{6\pi i}{7}} + e^{\frac{-4\pi i}{7}} + e^{\frac{-4\pi i}{7}} + e^{\frac{-6\pi i}{7}}}{2}$$
$$= \frac{-1 + \left(1 + e^{\frac{2\pi i}{7}} + e^{\frac{4\pi i}{7}} + e^{\frac{6\pi i}{7}} + e^{\frac{-2\pi i}{7}} + e^{\frac{-4\pi i}{7}} + e^{\frac{-6\pi i}{7}}\right)}{2}$$
$$= \frac{-1 + (\text{Sum of seven roots of unity})}{2}$$
$$= \frac{-1+0}{2} = -\frac{1}{2}$$

Ex. 38. *The number of integral value of k for which the equation* $7\cos x + 5\sin x = 2k + 1$ *has a solution is*

(a) 4 (b) 8

(c) 10 (d) 12

Sol. (*b*) Since, $-\sqrt{a^2 + b^2} \le a\sin x + b\cos x \le \sqrt{a^2 + b^2}$

$\therefore \quad -\sqrt{74} \le 7\cos x + 5\sin x \le \sqrt{74}$

So, $\quad -\sqrt{74} < 2k + 1 < \sqrt{74}$

Therefore, $2k + 1 = \pm 8, \pm 7, \pm 6, \ldots, \pm 1, 0$

So, $\quad k = -4, \pm 3, \pm 2, \pm 1, 0$, so, 8 values of k.

Ex. 39. *If* $y = \frac{\sin^4 x - \cos^4 x + \sin^2 x \cos^2 x}{\sin^4 x + \cos^4 x + \sin^2 x \cos^2 x}$, $x \in \left(0, \frac{\pi}{2}\right)$, *then*

(a) $-\frac{3}{2} \le y \le \frac{1}{2}$ (b) $1 \le y \le \frac{1}{2}$

(c) $-\frac{5}{3} \le y \le 1$ (d) None of these

Sol. (*d*) $y = \frac{\sin^4 x - \cos^4 x + \sin^2 x \cos^2 x}{\sin^4 x + \cos^4 x + \sin^2 x \cos^2 x}$

$$= \frac{(\sin^2 x - \cos^2 x)(\sin^2 x + \cos^2 x) + \sin^2 x \cos^2 x}{(\sin^2 x + \cos^2 x)^2 - \sin^2 x \cos^2 x}$$

$$= \frac{-\cos 2x + \frac{1}{4}\sin^2 2x}{1 - \frac{1}{4}\sin^2 2x}$$

$$= \frac{-4\cos 2x + 1 - \cos^2 2x}{4 - 1 + \cos^2 2x}$$

$$= \frac{1 - 4\cos 2x - \cos^2 2x}{3 + \cos^2 2x}$$

$\Rightarrow \quad (1 + y)\cos^2 2x + 4\cos 2x + 3y - 1 = 0$

Since $\cos 2x$ is real, we have

$$16 - 4(3y - 1)(1 + y) > 0$$

or $\quad 3y^2 + 2y - 5 \le 0$

or $\quad (3y + 5)(y - 1) \le 0 \Rightarrow -\frac{5}{3} \le y \le 1$

But $y = 1$ implies $\cos 2x = -1$ i.e. $x = \frac{\pi}{2}$ which is not permissible.

Ex. 40. *The distance between the two parallel lines is 1 unit. A pont 'A' is chosen to lie between the lines at a distance 'd' from one of them. Triangle ABC is equilateral with B on one line and C on the other parallel line. The length of the side of the equilateral triangle is*

(a) $\frac{2}{3}\sqrt{d^2 + d + 1}$ (b) $2\sqrt{\frac{d^2 - d + 1}{3}}$

(c) $2\sqrt{d^2 - d + 1}$ (d) $\sqrt{d^2 - d + 1}$

Sol. (*b*) From, figure

$$x\cos(\theta + 30^\circ) = d \quad \text{...(i)}$$

and $\quad x\sin\theta = 1 - d \quad$ ...(ii)

Dividing $\sqrt{3}\cot\theta = \frac{1 + d}{1 - d}$, squaring Eq. (ii) and putting the value of $\cot\theta$, we get

$$x^2 = \frac{1}{3}(4d^2 - 4d + 4)$$

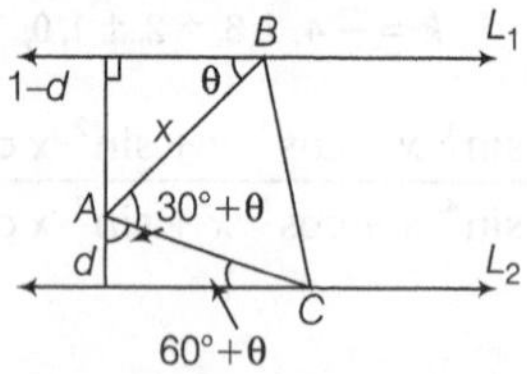

Ex. 41. *If $a\sin x + b\cos(x + \theta) + b\cos(x - \theta) = d$, then the minimum value of $|\cos\theta|$ is equal to*

(a) $\frac{1}{2|b|}\sqrt{d^2 - a^2}$ (b) $\frac{1}{2|a|}\sqrt{d^2 - a^2}$

(c) $\frac{1}{2|d|}\sqrt{d^2 - a^2}$ (d) None of these

Sol. (*a*) $a\sin x + b\cos(x + \theta) + b\cos(x - \theta) = d$

$\Rightarrow \quad a\sin x + 2b\cdot\cos x\cdot\cos\theta = d$

$\Rightarrow \quad |d| \le \sqrt{a^2 + 4b^2\cdot\cos^2\theta}$

$\Rightarrow \quad \frac{d^2 - a^2}{4b^2} \le \cos^2\theta \Rightarrow |\cos\theta| \ge \frac{\sqrt{d^2 - a^2}}{2|b|}$

Ex. 42. *The set of values of $\lambda \in R$ such that $\tan^2\theta + \sec\theta = \lambda$ holds for some θ is*

(a) $(-\infty, 1]$ (b) $(-\infty, -1]$

(c) ϕ (d) $[1, \infty)$

Sol. (*d*) $\because \tan^2\theta + \sec\theta = \lambda$

$\Rightarrow \quad \sec^2\theta + \sec\theta - 1 - \lambda = 0$

$\therefore \quad \sec\theta = \frac{-1 \pm \sqrt{(4\lambda + 5)}}{2}$

for real $\sec\theta$, $\quad 4\lambda + 5 \ge 0$ i.e, $\lambda \ge -\frac{5}{4}$

But we know that

$$\sec\theta \le -1 \text{ and } \sec\theta \ge 1$$

$\therefore \quad \frac{-1 \pm \sqrt{(4\lambda + 5)}}{2} \le -1$ and $\frac{-1 \pm \sqrt{(4\lambda + 5)}}{2} \ge 1$

$\Rightarrow \quad -1 - \sqrt{(4\lambda + 5)} \le -2$ and $-1 + \sqrt{(4\lambda + 5)} \ge 2$

$\Rightarrow \quad \sqrt{(4\lambda + 5)} \ge 1$ and $\sqrt{(4\lambda + 5)} \ge 3$

$\Rightarrow \quad \sqrt{4\lambda + 5} \ge 3$

or $\quad 4\lambda + 5 \ge 9$ or $\lambda \ge 1$

$\therefore \lambda \in [1, \infty)$

Ex. 43. *For $0 < \phi, \frac{\pi}{2}$, if $x = \sum_{n=0}^{\infty}\cos^{2n}\phi$, $y = \sum_{n=0}^{\infty}\sin^{2n}\phi$ and $z = \sum_{n=0}^{\infty}\cos^{2n}\phi\sin^{2n}\phi$, then*

(a) $xyz = xz + y$ (b) $xyz = xy + y$

(c) $xyz = x + y + z$ (d) $xyz = yz + x$

Sol. (*c*) We have,

$$x = \sum_{n=0}^{\infty}\cos^{2n}\phi = 1 + \cos^2\phi + \cos^4\phi + \ldots$$

$$= \frac{1}{1 - \cos^2\phi} = \frac{1}{\sin^2\phi}$$

Similarly, $\quad y = \frac{1}{1 - \sin^2\phi} = \frac{1}{\cos^2\phi}$

and $\quad z = \frac{1}{1 - \sin^2\phi\cos^2 x}$

$\therefore \quad z = \frac{1}{1 - \frac{1}{x}\cdot\frac{1}{y}} = \frac{xy}{xy - 1}$

$\Rightarrow \quad xyz = xy + z$

Ex. 44. *If* $\frac{x}{a}\cos\alpha+\frac{y}{b}\sin\alpha=1$, $\frac{x}{a}\cos\beta+\frac{y}{b}\sin\beta=1$ *and* $\frac{\cos\alpha\cos\beta}{a^2}+\frac{\sin\alpha\sin\beta}{b^2}=0$, *then*

(a) $\tan\alpha\tan\beta=\frac{b^2(x^2+a^2)}{a^2(y^2+b^2)}$

(b) $x^2+y^2=a^2+b^2$

(c) $\tan\alpha\tan\beta=\frac{a^2}{b^2}$

(c) None of the above

Sol. (*b*) $\frac{x}{a}\cos\alpha+\frac{y}{b}\sin\alpha=1$ ...(i)

$$\frac{x}{a}\cos\beta+\frac{y}{b}\sin\beta=1 \quad \text{...(ii)}$$

$$\frac{\cos\alpha\cdot\cos\beta}{a^2}+\frac{\sin\alpha\cdot\sin\beta}{b^2}=0 \quad \text{...(iii)}$$

From Eqs. (i) and (ii), we get

$$x=a\frac{\sin\alpha-\sin\beta}{\sin(\alpha-\beta)}$$

and $$y=b\frac{\cos\beta-\cos\alpha}{\sin(\alpha-\beta)}$$

Now, $$x^2+y^2=\frac{a^2(\sin\alpha-\sin\beta)^2+b^2(\cos\beta-\cos\alpha)^2}{\sin^2(\alpha-\beta)}$$

$$=\frac{a^2(\sin^2\alpha\sin^2\beta)+b^2(\cos^2\alpha+\cos^2\beta)-2(a^2\sin\alpha\sin\beta+b^2\cos\alpha\cos\beta)}{\sin^2(\alpha-\beta)}$$

$$=\frac{a^2(\sin^2\alpha+\sin^2\beta)+b^2(\cos^2\alpha+\cos^2\beta)}{\sin^2(\alpha-\beta)}$$

[from Eq. (iii)]

$$=a^2+b^2+\frac{a^2\{\sin^2\alpha+\sin^2\beta-\sin^2(\alpha-\beta)\}+b^2\{\cos^2\alpha+\cos^2\beta-\sin^2(\alpha-\beta)\}}{\sin^2(\alpha-\beta)}$$

$$=a^2+b^2+\frac{1}{\sin^2(\alpha-\beta)}\left[a^2\left(\begin{array}{l}\sin^2\alpha+\sin^2\beta-\sin^2\alpha\cos^2\beta\\-\cos^2\alpha\sin^2\beta+2\sin\alpha\sin\beta\\\cos\alpha\cos\beta\end{array}\right)+b^2\left\{\begin{array}{l}\cos^2\alpha(1-\sin^2\beta)\\+\cos^2\beta(1-\sin^2\alpha)\\+2\sin\alpha\sin\beta\sin\alpha\\\sin\beta\cos\alpha\cos\beta\end{array}\right\}\right]$$

$$=a^2+b^2+\frac{2a^2\sin\alpha\sin\beta\cos(\alpha-\beta)+2b^2\cos\alpha\cos\beta\cos(\alpha-\beta)}{\sin^2(\alpha-\beta)}$$

$$=a^2+b^2+\frac{2a^2b^2\cos(\alpha-\beta)}{\sin^2(\alpha-\beta)}\left[\frac{\sin\alpha\sin\beta}{b^2}+\frac{\cos\alpha\cos\beta}{a^2}\right] \quad \text{[from Eq. (iii)]}$$

$$=a^2+b^2+0=a^2+b^2$$

Thus, $x^2+y^2=a^2+b^2$

Now, $x^2+y^2=a^2+b^2$

$\Rightarrow$ $x^2-a^2=-(y^2-b^2)$

$\Rightarrow$ $\frac{x^2-a^2}{y^2-b^2}=-1$

$\Rightarrow$ $\frac{b^2(x^2-a^2)}{a^2(y^2-b^2)}=-\frac{b^2}{a^2}$

$=\tan\alpha\tan\beta$ [from Eq. (iii)]

Ex. 45. *If* α,β,γ *are acute angles and* $\cos\theta=\sin\beta/\sin\alpha$, $\cos\phi=\sin\gamma/\sin\alpha$ *and* $\cos(\theta-\phi)=\sin\beta\sin\gamma$, *then the value of* $\tan^2\alpha-\tan^2\beta-\tan^2\gamma$ *is equal to*

(a) -1 (b) 0

(c) 1 (d) 2

Sol. (*b*) From the third relation we get

$$\cos\theta\cos\phi+\sin\theta\sin\phi=\sin\beta\sin\gamma$$

$\Rightarrow$ $\sin^2\theta\sin^2\phi=(\cos\theta\cos\phi-\sin\beta\sin\gamma)^2$

$\Rightarrow$ $\left(1-\frac{\sin^2\beta}{\sin^2\alpha}\right)\left(1-\frac{\sin^2\gamma}{\sin^2\alpha}\right)=\left(\frac{\sin\beta\sin\gamma}{\sin^2\alpha}-\sin\beta\sin\gamma\right)^2$

$\Rightarrow$ $(\sin^2\alpha-\sin^2\beta)(\sin^2\alpha-\sin^2\gamma)=\sin^2\beta\sin^2\gamma(1-\sin^2\alpha)^2$

$\Rightarrow$ $\sin^4\alpha(1-\sin^2\beta\sin^2\gamma)-\sin^2\alpha(\sin^2\beta+\sin^2\gamma-2\sin^2\beta\sin^2\gamma)=0$

$\therefore$ $\sin^2\alpha=\frac{\sin^2\beta+\sin^2\gamma-2\sin^2\beta\sin^2\gamma}{1-\sin^2\beta\sin^2\gamma}$

and $\cos^2\alpha=\frac{1-\sin^2\beta-\sin^2\gamma+\sin^2\beta\sin^2\gamma}{1-\sin^2\beta\sin^2\gamma}$

$\Rightarrow$ $\tan^2\alpha=\frac{\sin^2\beta-\sin^2\beta\sin^2\gamma+\sin^2\gamma-\sin^2\beta\sin^2\gamma}{\cos^2\beta-\sin^2\gamma(1-\sin^2\beta)}$

$$=\frac{\sin^2\beta\cos^2\gamma+\cos^2\beta\sin^2\gamma}{\cos^2\beta\cos^2\gamma}$$

$$=\tan^2\beta+\tan^2\gamma$$

$\Rightarrow$ $\tan^2\alpha-\tan^2\beta-\tan^2\gamma=0$

Ex. 46. *If* $\sqrt{2}\cos A = \cos B + \cos^3 B$, *and* $\sqrt{2}\sin A = \sin B - \sin^3 B$ *then* $\sin(A - B) =$

(a) ± 1 (b) $\pm \frac{1}{2}$

(c) $\pm \frac{1}{3}$ (d) $\pm \frac{1}{4}$

Sol. (c) $\sqrt{2}\cos A = \cos B + \cos^3 B$ ...(i)

and $\sqrt{2}\sin A = \sin B - \sin^3 B$ ...(ii)

$\Rightarrow \sqrt{2}\sin A\cos B - \sqrt{2}\cos A\sin B$

$= (\sin B - \sin^3 B)\cos B - (\cos B + \cos^3 B)\sin B$

$= -\sin B\cos B$

$\Rightarrow \sin(A - B) = \frac{-1}{2\sqrt{2}}\sin 2B$

Now squaring and adding Eqs. (i) and (ii), we get

$$2 = \cos^2 B + \sin^2 B + \cos^6 B + \sin^6 B + 2(\cos^4 B - \sin^4 B)$$

$$\Rightarrow 1 = (\cos^2 A + \sin^2 A)^3 - 3\cos^2 A\sin^2 A(\cos^2 A + \sin^2 A) + 2\cos 2B$$

$$\Rightarrow 1 = 1 - \left(\frac{3}{4}\right)\sin^2 2B + 2\cos 2B$$

$$\Rightarrow -3\sin^2 2B + 8\cos 2B = 0$$

$$\Rightarrow 3\cos^2 2B + 8\cos 2B - 3 = 0$$

$$\Rightarrow \cos 2B = \frac{1}{3}$$

$$\Rightarrow \sin 2B = \pm\frac{2\sqrt{2}}{3}$$

$$\therefore \sin(A - B) = \pm\frac{1}{3}$$

Ex. 47. *If* x_1 *and* x_2 *are two distinct roots of the equation* $a\cos x + b\sin x = c$, *then* $\tan\frac{x_1 + x_2}{2}$ *is equal to*

(a) $\frac{a}{b}$ (b) $\frac{b}{a}$

(c) $\frac{c}{a}$ (d) $\frac{a}{c}$

Sol. (b) $a\cos x + b\sin x = c$

$$\Rightarrow \frac{a\left(1 - \tan^2\frac{x}{2}\right)}{1 + \tan^2\frac{x}{2}} + \frac{2b\tan\frac{x}{2}}{1 + \tan^2\frac{x}{2}} = c$$

$$\Rightarrow (c + a)\tan^2\frac{x}{2} - 2b\tan\frac{x}{2} + c - a = 0$$

$$\Rightarrow \tan\frac{x_1}{2} + \tan\frac{x_2}{2} = \frac{2b}{c + a},$$

$$\tan\frac{x_1}{2}\cdot\tan\frac{x_2}{2} = \frac{c - a}{c + a}$$

Thus, $$\tan\left(\frac{x_1 + x_2}{2}\right) = \frac{\tan\frac{x_1}{2} + \tan\frac{x_2}{2}}{1 - \tan\frac{x_1}{2}\tan\frac{x_2}{2}}$$

$$= \frac{\frac{2b}{c + a}}{1 - \left(\frac{c - a}{c + a}\right)} = \frac{b^2}{a}$$

Ex. 48. *The minimum value of the function*

$$f(x) = \frac{\sin x}{\sqrt{1 - \cos^2 x}} + \frac{\cos x}{\sqrt{1 - \sec^2 x}} + \frac{\tan x}{\sqrt{\sec^2 x - 1}} + \frac{\cot x}{\sqrt{\operatorname{cosec}^2 x - 1}}$$ *whenever it is defined is*

(a) 4 (b) -2

(c) 0 (d) 2

Sol. (b) $$f(x) = \frac{\sin x}{\sqrt{1 - \cos^2 x}} + \frac{\cos x}{\sqrt{1 - \sin^2 x}} + \frac{\tan x}{\sqrt{\sec^2 x - 1}} + \frac{\cot x}{\sqrt{\operatorname{cosec}^2 x - 1}}$$

$$= \frac{\sin x}{|\sin x|} + \frac{\cos x}{|\cos x|} + \frac{\tan x}{|\tan x|} + \frac{\cot x}{|\cot x|}$$

$$= \begin{cases} 4, & x \in \text{1st quadrant} \\ -2, & x \in \text{2nd quadrant} \\ 0, & x \in \text{3rd quadrant} \\ -2, & x \in \text{4th quadrant} \end{cases}$$

$f(x)_{\text{min.}} = -2$

Ex. 49. *If* $0 < \alpha < \frac{\pi}{6}$, *then* $\alpha(\operatorname{cosec}\alpha)$ *is*

(a) less than $\frac{\pi}{6}$ (b) greater than $\frac{\pi}{6}$

(c) less than $\frac{\pi}{3}$ (d) greater than $\frac{\pi}{3}$

Sol. (c) In the graph of $y = \sin x$. Let

$$A \equiv (\alpha, \sin\alpha),\ B = \left(\frac{\pi}{6}, \sin\frac{\pi}{6}\right)$$

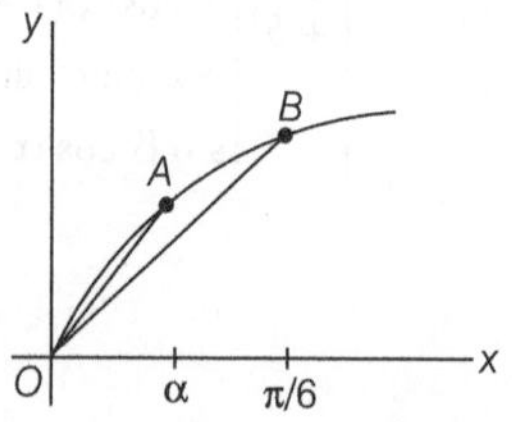

Clearly, slope of $OA >$ slope of OB, so

$$\frac{\sin\alpha}{\alpha} > \frac{\sin\frac{\pi}{6}}{\frac{\pi}{6}} = \frac{3}{\pi} \Rightarrow \frac{\alpha}{\sin\alpha} < \frac{\pi}{3}.$$

Ex. 50. *In which one of the following intervals the inequality* $\sin x < \cos x < \tan x < \cot x$ *can hold good?*

(a) $\left(\frac{7\pi}{4}, 2\pi\right)$ (b) $\left(\frac{3\pi}{4}, \pi\right)$

(c) $\left(\frac{5\pi}{4}, \frac{3\pi}{2}\right)$ (d) $\left(0, \frac{\pi}{4}\right)$

Sol. (*d*) In the second quadrant, $\sin x < \cos x$ is false, as $\sin x$ is positive and $\cos x$ is negative.

In the fourth quadrant, $\cos x < \tan x$ is false, as $\cos x$ is positive and $\tan x$ is negative.

In the third quadrant, i.e. $\left(\frac{5\pi}{4}, \frac{3\pi}{2}\right)$ if $\tan x < \cot x$ then $\tan^2 x < 1$, which is false.

Now, $\sin x < \cos x$ is true in $\left(0, \frac{\pi}{4}\right)$ and $\tan x < \cot x$ is also true.

Further, $\cos x < \tan x$, as $\tan x = \frac{(\sin x)}{(\cos x)}$ and $\cos x < 1$.

JEE Type Solved Examples : More than One Correct Option Type Questions

Ex. 51. *If* $x \in (0, \pi)$ *and* $\cos x + \sin x = \frac{1}{2}$, *then* $\tan x$ *is equal to*

(a) $\frac{4-\sqrt{7}}{3}$ (b) $\frac{4+\sqrt{7}}{3}$

(c) $\frac{-(4+\sqrt{7})}{3}$ (d) $\frac{-4+\sqrt{7}}{3}$

Sol. (*c,d*) Given, $\cos x + \sin x = \frac{1}{2}$

$$\Rightarrow \quad 1 + \sin 2x = \frac{1}{4}$$

$$\sin 2x = -\frac{3}{4} \quad \Rightarrow \quad 2x \in (\pi, 2\pi)$$

$$\Rightarrow \quad x \in \left(\frac{\pi}{2}, \pi\right) \quad \Rightarrow \quad \tan x < 0$$

$$\frac{2t}{1+t^2} = -\frac{3}{4} \quad \Rightarrow \quad 8t = -3 - 3t^2$$

$\Rightarrow 3t^2 + 8t + 3 = 0$, where $t = \tan x$

$$t = \frac{-8 \pm \sqrt{64-36}}{2\cdot 3};$$

$$t = \frac{-8 \pm \sqrt{28}}{2\cdot 3};$$

$$t = \frac{-(4+\sqrt{7})}{3}$$

or $$= \frac{-4+\sqrt{7}}{3}$$

Ex. 52. *The value of the expression* $\tan\frac{\pi}{7} + 2\tan\frac{2\pi}{7} + 4\tan\frac{4\pi}{7} + 8\cot\frac{8\pi}{7}$ *is equal to*

(a) $\operatorname{cosec}\frac{2\pi}{7} + \cot\frac{2\pi}{7}$ (b) $\tan\frac{\pi}{14} - \cot\frac{\pi}{14}$

(c) $\dfrac{\sin\frac{2\pi}{7}}{1-\cos\frac{2\pi}{7}}$ (d) $\dfrac{1+\cos\frac{\pi}{7}+\cos\frac{2\pi}{7}}{\sin\frac{\pi}{7}+\sin\frac{2\pi}{7}}$

Sol. (*a,c,d*) $\tan\frac{\pi}{7} + 2\tan\frac{2\pi}{7} + 4\tan\frac{4\pi}{7} + 8\cot\frac{8\pi}{7} = \cot\frac{\pi}{7}$

$[\tan\theta + 2\tan 2\theta + 4\tan 4\theta + 8\cot 8\theta = \cot\theta$ when $\theta = \frac{\pi}{7}]$

(a) $\operatorname{cosec} 2\theta + \cot 2\theta = \dfrac{1+\cot 2\theta}{\sin 2\theta} = \cot\theta = \cot\frac{\pi}{7}$

(where, $\theta = \frac{\pi}{7}$)

(b) $\tan\frac{\pi}{14} - \cot\frac{\pi}{14} = -2\cot\frac{\pi}{7}$

(c) $\dfrac{\sin\frac{2\pi}{7}}{1-\cos\frac{2\pi}{7}} = \dfrac{2\sin\frac{\pi}{7}\cos\frac{\pi}{7}}{2\sin^2\frac{\pi}{7}} = \cot\frac{\pi}{7}$

(d) $\dfrac{\left(1+\cos\frac{2\pi}{7}\right)+\cos\frac{\pi}{7}}{2\sin\frac{\pi}{7}\cos\frac{\pi}{7}+\sin\frac{\pi}{7}} = \dfrac{2\cos^2\frac{\pi}{7}+\cos\frac{\pi}{7}}{2\sin\frac{\pi}{7}\left(\cos\frac{\pi}{7}+1\right)}$

$$= \cot\frac{\pi}{7}$$

Ex. 53. *Two parallel chords are drawn on the same side of the centre of a circle of radius R. It is found that they subtend and angle of θ and 2θ at the centre of the circle. The perpendicular distance between the chords is*

(a) $2R\sin\frac{3\theta}{2}\sin\frac{\theta}{2}$

(b) $\left(1-\cos\frac{\theta}{2}\right)\left(1+2\cos\frac{\theta}{2}\right)R$

(c) $\left(1+\cos\frac{\theta}{2}\right)\left(1-2\cos\frac{\theta}{2}\right)R$

(d) $2R\sin\frac{3\theta}{4}\sin\frac{\theta}{4}$

Sol. (b,d) $OM = p_1 = R\cos\frac{\theta}{2}$

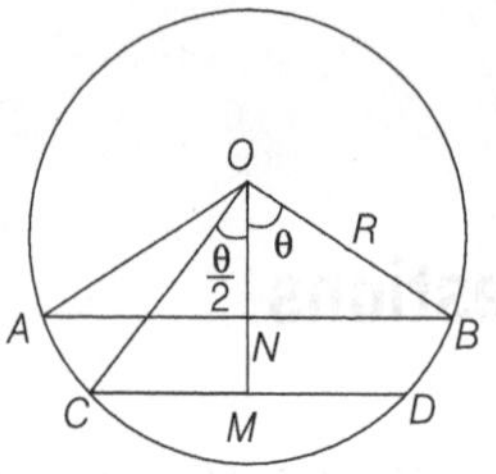

$ON = p_2 = R\cos\theta$

$MN = p_1 - p_2 = R\left(\cos\frac{\theta}{2} - \cos\theta\right)$

$= R\,2\sin\frac{3\theta}{4}\sin\frac{\theta}{4}$ (d)

Again convert $\cos\theta = 2\cos^2\frac{\theta}{2} - 1$ and factorise, we get

$= R(1-\cos\theta/2)^2\,(1+2\cos\theta/2)$

Ex. 54. *If* $2x$ *and* $2y$ *are complementary angles and* $\tan(x+2y)=2$, *then which of the following is(are) correct ?*

(a) $\sin(x+y)=\frac{1}{2}$ (b) $\tan(x-y)=\frac{1}{7}$

(c) $\cot x + \cot y = 5$ (d) $\tan x\tan y = 6$

Sol. (b,c) We have, $2x+2y=\frac{\pi}{2}$

$\Rightarrow \quad x+y=\frac{\pi}{4} \Rightarrow \sin(x+y)=\frac{1}{\sqrt{2}}$

Also, $y=\left(\frac{\pi}{4}-x\right)$,

So, $\tan(x+2y)=\tan\left(x+\frac{\pi}{2}-2x\right)$

$=\tan\left(\frac{\pi}{2}-x\right)=\cot x$

$\therefore \quad 2=\cot x \Rightarrow \tan x=\frac{1}{2}$

Similarly, $\quad x=\left(\frac{\pi}{4}-y\right)$

So, $\quad \tan(x+2y)=\tan\left(\frac{\pi}{4}+y\right)=\frac{1+\tan y}{1-\tan y}$

$\Rightarrow \quad 2=\frac{1+\tan y}{1-\tan y} \Rightarrow \tan y=\frac{1}{3}$

$\cot y = 3$

Also, $\tan(x-y)-\frac{\tan x-\tan y}{1+\tan x\tan y}=\frac{\frac{1}{2}-\frac{1}{3}}{1+\left(\frac{1}{2}\right)\left(\frac{1}{3}\right)}$

$=\left(\frac{1}{6}\right)\left(\frac{6}{7}\right)=\frac{1}{7}$

Now, verify alternatives.

Ex. 55. *If* $2\cos\theta+2\sqrt{2}=3\sec\theta$, *where* $\theta\in(0,2\pi)$, *then which of the following can be correct?*

(a) $\cos\theta=\frac{1}{\sqrt{2}}$ (b) $\tan\theta=1$

(c) $\sin\theta=-\frac{1}{\sqrt{2}}$ (d) $\cot\theta=-1$

Sol. (a,b,c,d) $2\cos\theta+2\sqrt{2}=3\sec\theta$

$\therefore \quad 2\cos^2\theta+2\sqrt{2}\cos\theta-3=0$

$\cos\theta=\frac{-2\sqrt{2}\pm\sqrt{32}}{4}=\frac{-2\sqrt{2}\pm4\sqrt{2}}{4}$

$\therefore \quad \cos\theta=\frac{1}{\sqrt{2}}$ or $\cos\theta=-\frac{3}{\sqrt{2}}$ (rejected)

$\therefore \quad \theta=\frac{\pi}{4}$ or $\frac{7\pi}{4} \Rightarrow \sin\theta=-\frac{1}{\sqrt{2}}$;

$\cot\theta=-1;\ \tan\theta=1$

Ex. 56. *The value of x in* $(0,\pi/2)$ *satisfying the equation,* $\frac{\sqrt{3}-1}{\sin x}+\frac{\sqrt{3}+1}{\cos x}=4\sqrt{2}$ *is*

(a) $\frac{\pi}{12}$ (b) $\frac{5\pi}{12}$

(c) $\frac{7\pi}{24}$ (d) $\frac{11\pi}{36}$

Sol. (a,d) $\frac{\sqrt{3}-1}{2\sqrt{2}\sin x}+\frac{\sqrt{3}+1}{2\sqrt{2}\cos x}=2$

$\sin\frac{\pi}{12}\cos x+\cos\frac{\pi}{12}\sin x=\sin 2x$

$\sin 2x=\sin\left(x+\frac{\pi}{12}\right)$

$\therefore \quad 2x=x+\frac{\pi}{12}$

or $\quad 2x=\pi-x-\frac{\pi}{12}$

$x=\frac{\pi}{12}$

or $\quad 3x=\frac{11\pi}{12}$

$\therefore \quad x=\frac{\pi}{12}$ or $\frac{11\pi}{36}$

Ex. 57. *Which of the following statements are always correct? (where, Q denotes the set of rationals)*

(a) $\cos 2\theta \in Q$ and $\sin 2\theta \in Q \Rightarrow \tan\theta \in Q$ (if defined)
(b) $\tan\theta \in Q \Rightarrow \sin 2\theta$, $\cos 2\theta$ and $\tan 2\theta \in Q$ (if defined)
(c) If $\sin\theta \in Q$ and $\cos\theta \in Q \Rightarrow \tan 3\theta \in Q$ (if defined)
(d) If $\sin\theta \in Q \Rightarrow \cos 3\theta \in Q$

Sol. (a,b,c)

(a) $\tan\theta = \dfrac{1-\cos 2\theta}{\sin 2\theta} \Rightarrow$ (a) is correct

(b) $\sin 2\theta = \dfrac{2\tan\theta}{1+\tan^2\theta}$; $\cos 2\theta = \dfrac{1-\tan^2\theta}{1+\tan^2\theta}$;

$\tan 2\theta = \dfrac{2\tan\theta}{1-\tan^2\theta} \Rightarrow$ (b) is correct

(c) $\tan 3\theta = \dfrac{\sin 3\theta}{\cos 3\theta} \Rightarrow$ (c) is correct

(d) $\sin\theta = \dfrac{1}{3}$ which is rational but

$\cos 3\theta = \cos\theta(4\cos^2\theta - 3)$ which is irrational $\Rightarrow$ (d) is incorrect.

Ex. 58. *In* ΔABC, $\tan B + \tan C = 5$ *and* $\tan A \tan C = 3$, *then*

(a) ΔABC is an acute angled triangle
(b) ΔABC is an obtuse angled triangle
(c) sum of all possible values of $\tan A$ is 10
(d) sum of all possible values of $\tan A$ is 9

Sol. (a,c) $\tan A + \tan B + \tan C = \tan A \tan B \tan C$

$\Rightarrow \quad \tan A + 5 = 3\tan B$

$\Rightarrow \quad 5 + \tan A = 3(5 - \tan C)$

$\Rightarrow \quad 5 + \tan A = 15 - \dfrac{9}{\tan A}$

$\Rightarrow \quad \tan^2 A - 10\tan A + 9 = 0$

$\Rightarrow \quad \tan A = 1$ or $\tan A = 9$

$\Rightarrow \tan B$ and $\tan C$ are 2, 3 or $\dfrac{14}{3}, \dfrac{1}{3}$, respectively

$\Rightarrow \Delta ABC$ is always on acute angled triangle and sum of all possible values of $\tan A$ is 10.

Ex. 59. $(m+2)\sin\theta + (2m-1)\cos\theta = 2m+1$, *if*

(a) $\tan\theta = \dfrac{3}{4}$ (b) $\tan\theta = \dfrac{4}{3}$

(c) $\tan\theta = \dfrac{2m}{(m^2-1)}$ (d) $\tan\theta = \dfrac{2m}{(m^2+1)}$

Sol. (b, c) The given relation can be written as

$(m+2)\tan\theta + (2m-1) = (2m+1)\sec\theta$

$\Rightarrow (m+2)^2\tan^2\theta + 2(m+2)(2m-1)\tan\theta + (2m-1)^2$
$= (2m+1)^2(1+\tan^2\theta)$

$\Rightarrow [(m+2^2) - (2m+1)^2]\tan^2\theta + 2(m+2)\tan\theta + (2m-1)^2 - (2m+1)^2 = 0$

$\Rightarrow 3(1-m^2)\tan^2\theta + (4m^2+6m-4)\tan\theta - 8m = 0$

$\Rightarrow \quad (3\tan\theta - 4)[(1-m)^2\tan\theta + 2m] = 0$

which is true if $\tan\theta = \dfrac{4}{3}$ or $\tan\theta = \dfrac{2m}{(m^2-1)}$.

Ex. 60. *If* $x\cos\alpha + y\sin\alpha = x\cos\beta + y\sin\beta = 2a\left(0 < \alpha, \beta < \dfrac{\pi}{2}\right)$, *then*

(a) $\cos\alpha + \cos\beta = \dfrac{4ax}{x^2+y^2}$

(b) $\cos\alpha\cos\beta = \dfrac{4a^2-y^2}{x^2+y^2}$

(c) $\sin\alpha + \sin\beta = \dfrac{4ay}{x^2+y^2}$

(d) $\sin\alpha\sin\beta = \dfrac{4a^2-x^2}{x^2+y^2}$

Sol. (a, b, c, d) We find out the given relations that α and β are the roots of the equation

$x\cos\theta + y\sin\theta = 2a$

$\Rightarrow \quad (x\cos\theta - 2a)^2 = (-y\sin\theta)^2$

$\Rightarrow \quad x^2\cos^2\theta - 4ax\cos\theta + 4a^2$
$= y^2\sin^2\theta = y^2(1-\cos^2\theta)$

$\Rightarrow \quad (x^2+y^2)\cos^2\theta - 4ax\cos\theta + 4a^2 - y^2 = 0$

which, being quadratic in $\cos\theta$, has two roots $\cos\alpha$ and $\cos\beta$, such that

$$\cos\alpha + \cos\beta = \frac{4ax}{x^2+y^2}$$

and $$\cos\alpha\cos\beta = \frac{4a^2-y^2}{x^2+y^2}$$

Similarly, we can write (1) as a quadratic in $\sin\theta$, giving two values $\sin\alpha$ and $\sin\beta$, such that

$$\sin\alpha + \sin\beta = \frac{4ay}{x^2+y^2}$$

and $$\sin\alpha\sin\beta = \frac{4a^2-x^2}{x^2+y^2}.$$

Ex. 61. *Let* $y = \sin^2 x + \cos^4 x$. *Then, for all real x*

(a) the maximum value of y is 2
(b) the minimum value of y is $\dfrac{3}{4}$
(c) $y \le 1$
(d) $y \ge \dfrac{1}{4}$

Sol. (b, c) $y = \cos^4 x - \cos^2 x + 1$

$$= \left(\cos^2 x - \frac{1}{2}\right)^2 + \frac{3}{4}$$

$\therefore y_{min} = \frac{3}{4}$ and y is maximum when $\left(\cos^2 x - \frac{1}{2}\right)^2$ is then maximum

$$\therefore \quad y_{max} = \frac{1}{4} + \frac{3}{4} = 1$$

Ex. 62. *If in ΔABC, $\tan A + \tan B + \tan C = 6$ and $\tan A \tan B = 2$, then $\sin^2 A : \sin^2 B : \sin^2 C$ is*

(a) 8 : 9 : 5 (b) 8 : 5 : 9
(c) 5 : 9 : 5 (d) 5 : 8 : 5

Sol. (*b, c*) $\tan A + \tan B + \tan C = 6$ ...(i)

$\Rightarrow \tan A \tan B \tan C = 6$

$$2 \tan C = 6$$

$$\therefore \quad \tan C = 3$$

$$\therefore \quad \sin^2 C = \frac{\tan^2 C}{1 + \tan^2 C} = \frac{9}{1+9} = \frac{9}{10}$$

From Eq. (i), $\tan A + \tan B = 3$ and $\tan A \tan B = 2$

$\tan A - \tan B$

$$= \pm \sqrt{\{(\tan A + \tan B)^2 - 4 \tan A \tan B)\}}$$

$$= \pm 1$$

we get, $\tan A = 2, 1$ and $\tan B = 1, 2$

$$\therefore \quad \sin^2 A = \frac{4}{1+4}, \frac{1}{1+1} \text{ and } \sin^2 B = \frac{1}{1+1}, \frac{4}{1+4}$$

$$\Rightarrow \quad \sin^2 A = \frac{8}{10}, \frac{5}{10} \text{ and } \sin^2 B = \frac{5}{10}, \frac{8}{10}$$

$$\therefore \quad \sin^2 A : \sin^2 B : \sin^2 C = 8:5:9 \text{ or } 5:8:9$$

Ex. 63. *If $0 \le x, y \le 180°$ and $\sin(x - y) = \cos(x + y) = \frac{1}{2}$, then the values of x and y are given by*

(a) $x = 45°, y = 15°$ (b) $x = 45°, y = 135°$
(c) $x = 165°, y = 15°$ (d) $x = 165°, y = 135°$

Sol. (*a, d*) $\sin(x - y) = \frac{1}{2} \Rightarrow x - y = 30°$ or $150°$ (1)

and $\cos(x + 1) = \frac{1}{2} \Rightarrow x + y = 60°$ or $300°$ (2)

Since x and y lie between $0°$ and $180°$, (1) and (2) are simultaneously true when $x = 45°$, $y = 15°$, or $x = 165°$, $y = 135°$. But, for the values given by (b) or (c), (1) and (2) do not hold simultaneously.

Ex. 64. *If $\sin\alpha + \sin\beta = l$, $\cos\alpha \cos\beta = m$ and $\tan\left(\frac{\alpha}{2}\right)\tan\left(\frac{\beta}{2}\right) = n (\neq 1)$, then*

(a) $\cos(\alpha - \beta) = \frac{l^2 + m^2 - 2}{2}$

(b) $\cos(\alpha + \beta) = \frac{m^2 - l^2}{m^2 + l^2}$

(c) $\frac{1+n}{1-n} = \frac{l^2 + m^2}{2n}$

(d) $\alpha + \beta = \frac{\pi}{2}$ if $l = m$

Sol. (*a, b, c, d*) Now, $l^2 = \sin^2\alpha + \sin^2\beta + 2\sin\alpha\sin\beta$ and $m^2 = \cos^2\alpha + \cos^2\beta + 2\cos\alpha\cos\beta$

$$2\cos(\alpha - \beta) = l^2 + m^2 - 2 \quad \text{(by adding)}$$

$$\Rightarrow \quad 2\cos 2\alpha + \cos 2\beta) = m^2 - l^2 \quad \text{(by subtracting)}$$

$$\Rightarrow \quad 2\cos(\alpha + \beta)\cos(\alpha - \beta) + 2\cos(\alpha + \beta) = m^2 - l^2$$

$$\Rightarrow \quad \cos(\alpha + \beta) = \frac{m^2 - l^2}{m^2 + l^2}.$$

Ex. 65. *Let $f(x) = ab \sin x + b\sqrt{1 - a^2} \cos x + c$, where $|a| < 1, b > 0$ then*

(a) maximum value of $f(x)$ if b is $c = 0$

(b) difference of maximum and minimum values of $f(x)$ is $2b$

(c) $f(x) = c$ if $x = -\cos^{-1} a$

(d) $f(x) = c$ if $x = \cos^{-1} a$

Sol. (*a, b, c*) $f(x) = ab \sin x + b\sqrt{1 - a^2} \cos x + c$, where $|a| < 1, b < 0$

$$f(x) = \sqrt{a^2b^2 + b^2 - b^2a^2} \sin(x + \alpha) + c$$

$$= b\sin(x + \alpha) + c, \text{ where } \tan\alpha = \frac{b\sqrt{1 - a^2}}{ab} = \frac{\sqrt{1 - a^2}}{a}$$

$$= b\cos(x - \alpha) + c, \text{ where } \tan\alpha = \frac{ab}{b\sqrt{1 - a^2}}$$

$$= \frac{a}{\sqrt{1 - a^2}}$$

$$f(x)_{max} - f(x)_{min} = c + b - (c - b) = 2b$$

$$f(x) = c \text{ if } x + \alpha = 0$$

or $x = -\alpha$ or $x = -\cos^{-1} a$

Ex. 66. *If $(x - a)\cos\theta + y\sin\theta = (x - a)\cos\phi + y\sin\phi = a$ and $\tan\left(\frac{\theta}{2}\right) - \tan\left(\frac{\phi}{2}\right) = 2b$, then*

(a) $y^2 = 2ax - (1 - b^2)x^2$ (b) $\tan\frac{\theta}{2} = \frac{1}{x}(y + bx)$

(c) $y^2 = 2bx - (1 - a^2)x^2$ (d) $\tan\frac{\phi}{2} = \frac{1}{x}(y - bx)$

Sol. (*a, b*) Let, $\tan\left(\frac{\theta}{2}\right) = \alpha$ and $\tan\left(\frac{\phi}{2}\right) = \beta$, so that $\alpha - \beta = 2b$.

$$\text{Also,} \quad \cos\theta = \frac{1 - \tan^2\left(\frac{\theta}{2}\right)}{1 + \tan^2\left(\frac{\theta}{2}\right)} = \frac{1 - \alpha^2}{1 + \alpha^2}$$

And $\sin\theta = \dfrac{2\tan\left(\dfrac{\theta}{2}\right)}{1+\tan^2\left(\dfrac{\theta}{2}\right)} = \dfrac{2\alpha}{1+\alpha^2}$

Similarly, $\cos\phi = \dfrac{1-\beta^2}{1+\beta^2}$ and $\sin\phi = \dfrac{2\beta}{1+\beta^2}$

Therefore, we have from the given relations

$$(x-a)\frac{1-\alpha^2}{1+\alpha^2} + y\left(\frac{2\alpha}{1+\alpha^2}\right) = a$$

$\Rightarrow$ $x\alpha^2 - 2y\alpha + 2a - x = 0$

Similarly $x\beta^2 - 2y\beta + 2a = 0$

We see that α and β are roots of the equation

$$xz^2 - 2yz + 2a - x = 0,$$

So that $\alpha + \beta = \dfrac{2y}{x}$ and $\alpha\beta = \dfrac{(2a-x)}{x}$.

Now, from $(\alpha+\beta)^2 = (\alpha-\beta)^2 + 4\alpha\beta$, we get

$\Rightarrow$ $\left(\dfrac{2y}{x}\right)^2 = (2b)^2 + \dfrac{4(2a-x)}{x}$

$\Rightarrow$ $y^2 = 2ax - (1-b^2)x^2$

Also, from $\alpha + \beta = \dfrac{2y}{x}$ and $\alpha - \beta = 2b$, we get

$$\alpha = \frac{y}{x} + b \text{ and } \beta = \frac{y}{x} - b$$

$\Rightarrow$ $\tan\dfrac{\theta}{2} = \dfrac{1}{x}(y + bx)$

and $\tan\dfrac{\phi}{2} = \dfrac{1}{x}(y - bx)$

Ex. 67. *If* $\cos(\beta-\gamma) + \cos(\gamma-\alpha) + \cos(\alpha-\beta) = -\dfrac{3}{2}$ *then*

(a) $\Sigma\cos\alpha = 0$ (b) $\Sigma\sin\alpha = 0$
(c) $\Sigma\cos\alpha\sin\alpha = 0$ (d) $\Sigma(\cos\alpha + \sin\alpha) = 0$

Sol. (*a, b, d*) The given expression can be written as

$$2[\cos\beta\cos\gamma + \cos\gamma\cos\alpha + \cos\alpha\cos\beta] + 2[\sin\beta\sin\gamma + \sin\gamma\sin\alpha + \sin\alpha\sin\beta] + (\sin^2\alpha + \cos^2\alpha) + (\sin^2\beta + \cos^2\beta) + (\sin^2\gamma + \cos^2\gamma) = 0$$

$\Rightarrow (\cos\alpha + \cos\beta + \cos\gamma)^2 + (\sin\alpha + \sin\beta + \sin\gamma)^2 = 0$

$\Rightarrow$ $\Sigma\cos\alpha = 0$ and $\Sigma\sin\alpha = 0$

$\Rightarrow$ $\Sigma(\cos\alpha + \sin\alpha) = 0$

JEE Type Solved Examples : Statement I and II Type Questions

Ex. 68. Statement I $\tan 5\theta - \tan 3\theta - \tan 2\theta = \tan 5\theta\tan 3\theta\tan 2\theta$.

Statement II $x = y + z$
$\Rightarrow \tan x - \tan y - \tan z = \tan x\tan y\tan z$

(a) A (b) B
(c) C (d) D

Sol. (*a*) $\because 5\theta = 3\theta + 2\theta$

$\Rightarrow$ $\tan 5\theta = \tan(3\theta + 2\theta) = \dfrac{\tan 3\theta + \tan 2\theta}{1 - \tan 3\theta\tan 2\theta}$

$\Rightarrow \tan 5\theta - \tan 5\theta\tan 3\theta\tan 2\theta = \tan 3\theta + \tan 2\theta$

$\Rightarrow$ $\tan 5\theta - \tan 3\theta - \tan 2\theta = \tan 5\theta\tan 3\theta\tan 2\theta$

Ex. 69. Statement I *The maximum value of* $\sin\theta + \cos\theta$ *is 2.*

Statement II *The maximum value of* $\sin\theta$ *is 1 and that of* $\cos\theta$ *is also 1.*

(a) A (b) B
(c) C (d) D

Sol. (*d*) $\because -\sqrt{2} \le \sin\theta + \cos\theta \le \sqrt{2}$

$\therefore$ Maximum value of $\sin\theta + \cos\theta$ is $\sqrt{2}$

But maximum value of $\sin\theta$ is 1 and that of $\cos\theta$ is also 1 which is always true.

Ex. 70. Statement I *If* $a, b, c \in R$ *and not all equal, then*

$$\sec\theta = \frac{(bc + ca + ab)}{(a^2 + b^2 + c^2)},$$

Statement II $\sec\theta \le -1$ *and* $\sec\theta \ge 1$

(a) A (b) B
(c) C (d) D

Sol. (*d*) $\because a^2 + b^2 + c^2 - ab - bc - ca$

$$= \frac{1}{2}\{(a-b)^2 + (b-c)^2 + (c-a)^2\} > 0$$

$\Rightarrow$ $a^2 + b^2 + c^2 > ab + bc + ca$

or $\dfrac{ab + bc + ca}{a^2 + b^2 + c^2} < 1$

$\Rightarrow$ $\sec\theta < 1$, which is false.

Ex. 71. Statement I $\prod_{r=1}^{n}(1 + \sec 2^r\theta) = \tan 2^n\theta\cot\theta$

Statement II $\prod_{r=1}^{n}\cos(2^{r-1}\theta) = \dfrac{\sin(2^n\theta)}{2^n\sin\theta}$

(a) A (b) B
(c) C (d) D

Sol. (*a*) $\because \prod_{r=1}^{n}(1+\sec 2^r\theta)=\dfrac{\prod_{r=1}^{n}(1+\cos 2^r\theta)}{\prod_{r=1}^{n}\cos 2^r\theta}$

$$=\frac{\prod_{r=1}^{n}2\cos^2(2^{r-1}\theta)}{\prod_{r=1}^{n}\cos(2^r\theta)}$$

$$=\frac{2^n\cdot\prod_{r=1}^{n}\cos(2^{r-1}\theta)\prod_{r=1}^{n}\cos(2^{r-1}\theta)}{\dfrac{\cos(2^n\theta)}{\cos\theta}\prod_{r=1}^{n}\cos(2^{r-1}\theta)}$$

$$=\frac{\dfrac{2^n\cdot\sin(2^n\theta)}{2^n\sin\theta}\cdot\cos\theta}{\cos(2^n\theta)}$$

$$=\tan(2^n\theta)\cdot\cot\theta$$

Ex. 72. **Statement I** $\cos 36^\circ>\sin 36^\circ$

Statement II $\cos 36^\circ>\tan 36^\circ$

(a) A (b) B
(c) C (d) D

Sol. (*b*) Since, $\cos\theta>\sin\theta$ for $0\le\theta\le\dfrac{\pi}{4}$

So, Statement I is true.

Now, $\cos 36^\circ>\tan 36^\circ$

$\Rightarrow\quad \cos 36^\circ>\dfrac{\sin 36^\circ}{\cos 36^\circ}$

$\Rightarrow\quad \cos^2 36^\circ>\sin 36^\circ$

$\Rightarrow\quad 1+\cos 72^\circ>2\sin 36^\circ=2\sin(30^\circ+6^\circ)$

$\Rightarrow\quad 1+2\sin 9^\circ\cos 9^\circ>\cos 6^\circ+2\cos 30^\circ\sin 6^\circ$

which is true

Ex. 73. **Statement I** $\cos^3\alpha+\cos^3\left(\alpha+\dfrac{2\pi}{3}\right)+\cos^3\left(\alpha+\dfrac{4\pi}{3}\right)$

$$=3\cos\alpha\cos\left(\alpha+\frac{2\pi}{3}\right)\cos\left(\alpha+\frac{4\pi}{3}\right)$$

Statement II *If* $a+b+c=0\Leftrightarrow a^3+b^3+c^3=3abc$

(a) A (b) B
(c) C (d) D

Sol. (*a*) $\because\ \cos\alpha+\cos\left(\alpha+\dfrac{2\pi}{3}\right)+\cos\left(\alpha+\dfrac{4\pi}{3}\right)$

$$=\cos\alpha+2\cos(\alpha+\pi)\cos\frac{\pi}{3}$$

$$=\cos\alpha+(-2\cos\alpha)\left(\frac{1}{2}\right)=0$$

$$\therefore \cos^3\alpha+\cos^3\left(\alpha+\frac{2\pi}{3}\right)+\cos^3\left(\alpha+\frac{4\pi}{3}\right)$$

$$=3\cos\alpha\cos\left(\alpha+\frac{2\pi}{3}\right)\cos\left(\alpha+\frac{4\pi}{3}\right)$$

Ex. 74. **Statement I** $\sin 2>\sin 3$

Statement II *If* $x, y\in\left(\dfrac{\pi}{2},\pi\right)$, $x<y$, *then* $\sin x>\sin y$

Sol. (*a*)

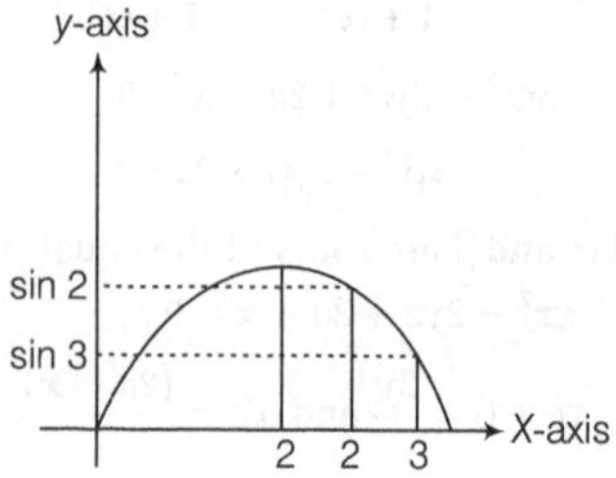

Ex. 75. *Let* $\alpha,\beta,\gamma>0$ *and* $\alpha+\beta+\gamma=\dfrac{\pi}{2}$

Statement I $\left|\tan\alpha\tan\beta-\dfrac{a!}{6}\right|+\left|\tan\beta\tan\gamma-\dfrac{b!}{2}\right|$ $+\left|\tan\gamma\tan\alpha-\dfrac{c!}{3}\right|\le 0$, *where* $n!=1.2\,....\,n$, *then* $\tan\alpha\tan\beta$, $\tan\beta\tan\gamma$, $\tan\gamma\tan\alpha$ *are in AP.*

Statement II $\tan\alpha\tan\beta+\tan\beta\tan\gamma+\tan\gamma\tan\alpha=1$

Sol. (*d*) Statement II $\alpha+\beta=\dfrac{\pi}{2}-\gamma$

$$\frac{\tan\alpha+\tan\beta}{1-\tan\alpha\tan\beta}=\frac{1}{\tan\gamma}$$

$\Rightarrow\quad \Sigma\tan\alpha\tan\beta=1$

$\therefore$ Statement II is true.

Statement I $\tan\alpha\tan\beta=\dfrac{a!}{6}$,

$\tan\beta\tan\gamma=\dfrac{b!}{2}$

and $\tan\alpha\tan\beta=\dfrac{c!}{3}$

$$\frac{a!}{6}+\frac{b!}{2}+\frac{c!}{3}=1$$

$\Rightarrow\quad a!=1\,b!=1\,c!=1$

$\Rightarrow\tan\alpha\tan\beta$, $\tan\gamma\tan\alpha$ and $\tan\beta\tan\gamma$ are not in AP.

$\therefore$Statement I is false.

Hence, (d) is the correct answer.

Ex. 76. **Statement I** *The triangle so obtained is an equilateral triangle.*

Statement II *If roots of the equation be* $\tan A$, $\tan B$ *and* $\tan C$, *then* $\tan A+\tan B+\tan C=3\sqrt{3}$

Sol. (*b*) $\tan A + \tan B + \tan C = 3\sqrt{3}$

and $\quad \tan A \tan B \tan C = 3$

$\therefore \quad \tan A + \tan B + \tan C$

$\neq \tan A \tan B \tan C$

$\Rightarrow$ triangle does not exists.

Ex. 77. *Let us define the function* $f(x) = x^2 + x + 1$

Statement I *The equation* $\sin x = f(x)$ *has no solution.*

Statement II *The curve* $y = \sin x$ *and* $y = f(x)$ *do not intersect each other when graph is observed.*

Sol. (*a*) Let $y = \sin x$ and $y = x^2 + x + 1$

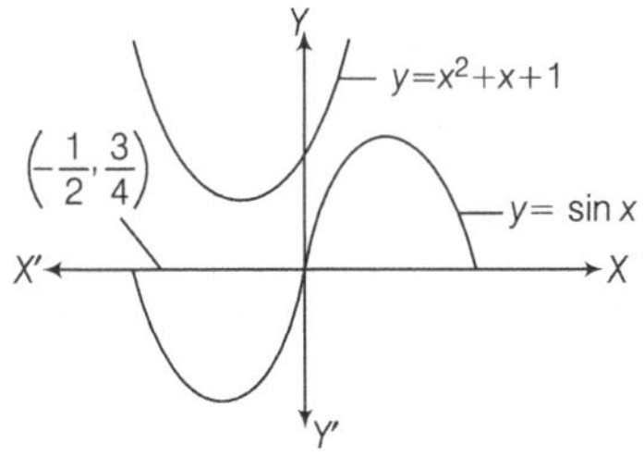

Since, $-1 \leq \sin x \leq 1$ and $y = \left(x + \frac{1}{2}\right)^2 + \frac{3}{4}$

It is clear from the graph that no two curves intersect each other.

JEE Type Solved Examples : Passage Based Questions

Passage I

(Ex. Nos. 78 to 80)

Consider, $f(x) = (x + 2a)(x + a - 4) \, (a \in R)$,

$g(x) - k(x^2 + x) + 3k + x (k \in R)$ and

$h(x) = (1 - \sin\theta)x^2 + 2(1 - \sin\theta)x - 3\sin\theta$

$\left(\theta \in R - (4n + 1)\frac{\pi}{2}, n \in I\right)$

Ex. 78. *If* $f(x) < 0$ *for* $-1 \leq x \leq 1$, *then 'a' satisfies*

(a) $\frac{1}{2} < a < 3$ (b) $-\frac{1}{2} < a < \frac{1}{2}$

(c) $-3 < a < -\frac{1}{2}$ (d) $-3 < a < \frac{1}{2}$

Sol. (*a*) Given, $f(x) = (x + 2a)(x + a - 4)$

$= x^2 + (3a - 4)x + 2a(a - 4)$.

$\left.\begin{matrix} f(-1) < 0 \\ f(1) < 0 \end{matrix}\right\} n$

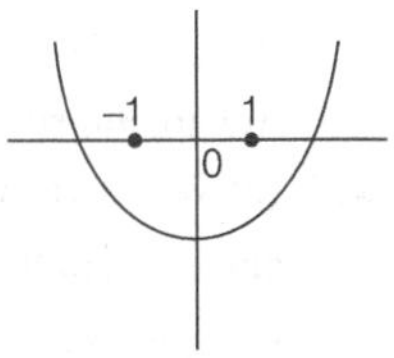

On solving, we get $\frac{1}{2} < a < 3$

Ex. 79. *If* $g(x) > -3$ *for all real x, then the values of k are given by*

(a) $-1 < k < \frac{1}{11}$ (b) $-1 < k < 0$

(c) $0 < k < \frac{1}{11}$ (d) $k < \frac{1}{11}$

Sol. (*d*) $g(x) = k(x^2 + x) + 3k + x > -3 \, \forall \, x$

$\Rightarrow \quad k(x^2 + x) + 3k + x + 3 > 0 \, \forall \, x$

or $\quad kx^2 + (k + 1)x + (3k + 3) > 0 \, \forall \, x$

$\left.\begin{matrix} k > 0 \\ D < 0 \end{matrix}\right\} n$

Here, $D - (k + 1)^2 - 4k.3(k + 1) < 0$

$\Rightarrow \quad k^2 + 2k + 1 - 12k^2 - 12 < 0$

$\Rightarrow \quad 11k^2 + 10k - 1 > 0$

$\Rightarrow \quad (k + 1)(11k - 1) > 0$

$\Rightarrow \quad k < -1 \quad$ or $\quad k > \frac{1}{11}$

$\Rightarrow \quad k > \frac{1}{11} \qquad (\because k > 0)$

Ex. 80. *If the quadratic equation* $h(x) = 0$ *has both roots complex, then* θ *belongs to*

(a) $\left(-\frac{\pi}{2}, \frac{\pi}{2}\right)$ (b) $\left(0, \frac{3\pi}{2}\right)$

(c) $\left(\frac{\pi}{6}, \frac{7\pi}{6}\right)$ (d) $\left(\frac{7\pi}{6}, \frac{11\pi}{6}\right)$

Sol. (*d*) Given, $(1 - \sin\theta)x^2 + 2(1 - \sin\theta)x - 3\sin\theta = 0$ has both roots complex, then $D < 0$

$(1 - \sin\theta)(1 + 2\sin\theta) < 0$

$\underbrace{(\sin\theta - 1)}_{(-)\text{ve number}}(2\sin\theta + 1) > 0$

$\Rightarrow \quad 2\sin\theta + 1 < 0$

$\sin\theta < -\frac{1}{2}$

$\Rightarrow \quad \theta \in \left(\frac{7\pi}{6}, \frac{11\pi}{6}\right)$

Passage II

(Ex. Nos. 81 to 83)

Let $f(\theta) = \sin\theta - \cos^2\theta - 1$, where $\theta \in R$ and $m \le f(\theta) \le M$.

Ex. 81. *Let N denotes the number of solution of the equation $f(\theta) = 0$ in $[0, 4\pi]$ then the value of*

$\log_{\sqrt{m^2}}(N) + \log_{\sqrt{m^2}}\left(\frac{1}{N+1}\right)$ *is equal to*

(a) $\frac{1}{2}$ (b) 1

(c) $\frac{-1}{2}$ (d) -1

Sol. (c) $f(\theta) = \sin\theta - (1 - \sin^2\theta) - 1$

$$= \sin^2\theta + \sin\theta - 2$$

$$= \left(\sin\theta + \frac{1}{2}\right)^2 - 2 - \frac{1}{4}$$

$$= \left(\sin\theta + \frac{1}{2}\right)^2 - \frac{9}{4}$$

$\therefore \quad f(\theta)_{\min} = 0 - \frac{9}{4} = \frac{-9}{4}$

$\Rightarrow \quad m = \frac{-9}{4}$

$$f(\theta)_{\max} = \left(1 + \frac{1}{2}\right)^2 - \frac{9}{4} = \frac{9}{4} - \frac{9}{4} = 0$$

$\Rightarrow \quad M = 0$

Hence, $m = \frac{-9}{4}$, $M = 0$

Now, $f(\theta) = 0$

$\Rightarrow \quad (\sin\theta + 2)(\sin\theta - 1) = 0$

$\Rightarrow \quad \sin\theta = 1$

$\Rightarrow \quad \theta = \frac{\pi}{2}$ and $\frac{5\pi}{2}$

Hence, $N = 2$, i.e. number of solution s of $\sin\theta = 1$ in $[0, 4\pi]$.

$\therefore \quad \log_{\sqrt{m^2}}(N) + \log_{\sqrt{m^2}}\left(\frac{1}{N+1}\right)$

$$= \log_{|m|}\left(\frac{N}{N+1}\right) v = \log_{|m|}\left(\frac{2}{3}\right)$$

$$= \log_{\frac{9}{4}}\left(\frac{2}{3}\right) = \frac{-1}{2}.$$

Ex. 82. *The value of $(4m + 13)$ is equal to*

(a) 0 (b) 4

(c) 5 (d) 6

Sol. (b) As $m = \frac{-9}{4}$, so $(4m + 13) = 4$

Ex. 83. *Sum of all values of x satisfying the equation*

$$x = \sqrt{\frac{1}{|m|} + \sqrt{\frac{1}{|m|} + \sqrt{\frac{1}{|m|} + \dots\dots\dots\dots \infty}}}, \text{ is}$$

(a) $\frac{1}{3}$ (b) $\frac{2}{3}$

(c) $\frac{3}{3}$ (d) $\frac{4}{3}$

Sol. (d) $x = \sqrt{\frac{4}{9} + \sqrt{\frac{4}{9} + \sqrt{\frac{4}{9} + \dots\dots\dots\dots \infty}}}$

$\Rightarrow \quad x = \sqrt{\frac{4}{9} + x}$

$\Rightarrow \quad x^2 = \frac{4}{9} + x$

$$9x^2 = 4 + 9x$$

$\Rightarrow \quad 9x^2 - 9x - 4 = 0$

$\Rightarrow \quad 9x^2 - 12x + 3x - 4 = 0$

$\Rightarrow \quad (3x - 4)(3x + 1) = 0$

$\therefore \quad x = \frac{4}{3}$ and $x = \frac{-1}{3}$ (rejected)

Passage III

(Ex. Nos. 84 to 88)

The method of eliminating 'θ' from two given equations involving trigonometrical functions of 'θ'. By using given equations involving 'θ' and trigonometrical identities, we shall obtain an equation not involving 'θ'.

On the basis of above information answer the following questions.

Ex. 84. *If $x\sin^3\theta + y\cos^3\theta = \sin\theta\cos\theta$ and $x\sin\theta - y\cos\theta = 0$ then (x, y) lie one*

(a) a circle (b) a parabola

(c) an ellipse (d) a hyperbola

Sol. (a) We have, $x\sin^3\theta + y\cos^3\theta = \sin\theta\cos\theta$ ...(i)

and $x\sin\theta - y\cos\theta = 0$ ...(ii)

From Eq. (ii), $\tan\theta = \frac{y}{x}$

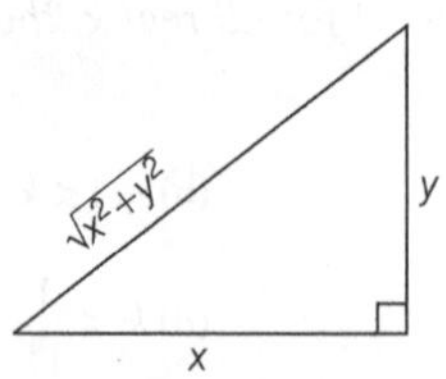

$\therefore \quad \sin\theta = \frac{y}{\sqrt{(x^2 + y^2)}}$ and $\cos\theta = \frac{x}{\sqrt{(x^2 + y^2)}}$

From Eq., (i) $x \times \dfrac{y^3}{(x^2+y^2)^{\frac{3}{2}}} + y \times \dfrac{x^3}{(x^2+y^2)^{\frac{3}{2}}}$

$$= \frac{xy}{(x^2+y^2)}$$

or $\quad \dfrac{(x^2+y^2)}{(x^2+y^2)^{\frac{3}{2}}} = \dfrac{1}{(x^2+y^2)} \Rightarrow (x^2+y^2)^{\frac{1}{2}} = 1$

or $\quad x^2 + y^2 = 1$ which is a circle

Ex. 85. *If* $\dfrac{x}{a\cos\theta} = \dfrac{y}{b\sin\theta}$ ...(i)

and $\dfrac{ax}{\cos\theta} - \dfrac{by}{\sin\theta} = a^2 - b^2$, *then* (x, y) *lie on*

(a) a circle (b) a parabola

(c) an ellipse (d) a hyperbola

Sol. (c) $\because \quad \dfrac{x}{a\cos\theta} = \dfrac{y}{b\sin\theta}$ ...(i)

and $\quad \dfrac{ax}{\cos\theta} - \dfrac{by}{\sin\theta} = a^2 - b^2$ ...(ii)

From Eq. (i), $\tan\theta = \dfrac{ay}{bx}$

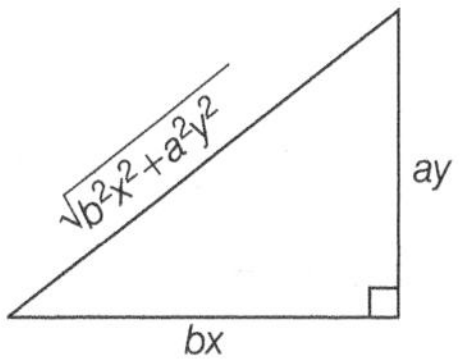

From Eq. (ii),

$$\frac{ax}{\dfrac{bx}{\sqrt{(b^2x^2+a^2y^2)}}} - \frac{by}{\dfrac{ay}{\sqrt{(b^2x^2+a^2y^2)}}} = (a^2 - b^2)$$

$\Rightarrow \quad (a^2 - b^2)\sqrt{(b^2x^2+a^2y^2)} = ab(a^2 - b^2)$

$\Rightarrow \quad b^2x^2 + a^2y^2 = a^2b^2$

$\therefore \quad \dfrac{x^2}{a^2} + \dfrac{y^2}{b^2} = 1$ which is an ellipse.

Ex. 86. *If* $\tan\theta + \sin\theta = m$ *and* $\tan\theta - \sin\theta = n$, *then* $(m^2 - n^2)^2$ *is*

(a) $4\sqrt{mn}$ (b) $4mn$

(c) $16\sqrt{mn}$ (d) $16mn$

Sol. (d) $\because m + n = 2\tan\theta,\ m - n = 2\sin\theta$ ...(i)

and $\quad mn = \tan^2\theta - \sin^2\theta = \sin^2\theta(\sec^2\theta - 1)$

$$= \sin^2\theta \tan^2\theta$$

$$= \left(\frac{m-n}{2}\right)^2 \left(\frac{m+n}{2}\right)^2 \quad \text{[from Eq. (i)]}$$

$\therefore \quad (m^2 - n^2)^2 = 16\,mn$

Ex. 87. *If* $\sin\theta + \cos\theta = a$ *and* $\sin^3\theta + \cos^2\theta = b$, *then we get* $\lambda a^3 + \mu b + va = 0$ *when* λ, μ, v *are independent of* θ, *then the value of* $\lambda^3 + \mu^3 + v^3$ *is*

(a) – 6 (b) – 18 (c) – 36 (d) – 98

Sol. (b) $\quad \sin\theta + \cos\theta = a$ ...(i)

$\sin^3\theta + \cos^3\theta = b$ ...(ii)

From Eq. (i),

$$\sin^2\theta + \cos^2\theta + 2\sin\theta\cos\theta = a^2$$

or $\quad \sin\theta\cos\theta = \dfrac{a^2-1}{2}$ ...(iii)

From Eq. (ii),

$$(\sin\theta + \cos\theta)^3 + 3\sin\theta\cos\theta(\sin\theta + \cos\theta) = b$$

$\Rightarrow \quad a^3 - \dfrac{3(a^2-1)}{2}a = b \quad$ [from Eqs. (i) and (iii)]

$\Rightarrow \quad 2a^3 - 3a^3 + 3a = 2b \Rightarrow a^3 + 2b - 3a = 0$

On comparing, we get

$$\lambda = 1, \mu = 2, v = -3$$

$\therefore \quad \lambda + \mu + v = 0$

$\therefore \quad \lambda^3 + \mu^3 + v^3 = 3\lambda\mu v = 3(1)(2)(-3) = -18$

Ex. 88. *After eliminating 'θ' from equations* $\dfrac{x\cos\theta}{a} + \dfrac{y\sin\theta}{b} = 1$ *and* $x\sin\theta - y\cos\theta = \sqrt{(a^2\sin^2\theta + b^2\cos^2\theta}$, *we get*

(a) $x^2 + y^2 = a^2 + b^2$ (b) $\dfrac{x^2}{a^2} + \dfrac{y^2}{b^2} = 1$

(c) $\dfrac{x^2}{a(a+b)} + \dfrac{y^2}{b(a+b)} = 1$ (d) $x^2 + y^2 = (a+b)^2$

Sol. (c) $\because \dfrac{x\cos\theta}{a} + \dfrac{y\sin\theta}{b} = 1$ ...(i)

and $\quad x\sin\theta - y\cos\theta = \sqrt{(a^2\sin^2\theta + b^2\cos^2\theta}$ ...(ii)

Squaring Eq. (i), we get

$$\frac{x^2}{a^2}\cos^2\theta + \frac{y^2}{b^2}\sin^2\theta + \frac{2xy}{ab}\sin\theta\cos\theta$$

$$= 1 = \sin^2\theta + \cos^2\theta$$

or $\quad \left(\dfrac{x^2}{a^2} - 1\right)\cos^2\theta + \left(\dfrac{y^2}{b^2} - 1\right)\sin^2\theta$

$$+ \frac{2xy}{ab}\sin\theta\cos\theta = 0 \quad \text{...(iii)}$$

and squaring Eq. (ii), we get

$$x^2\sin^2\theta + y^2\cos^2\theta - 2xy\sin\theta\cos\theta$$

$$= a^2\sin^2\theta + b^2\cos^2\theta$$

$$(x^2 - a^2)\sin^2\theta + (y^2 - b^2)\cos^2\theta - 2xy\sin\theta\cos\theta = 0$$

$\Rightarrow \quad \left(\dfrac{x^2-a^2}{ab}\right)\sin^2\theta + \left(\dfrac{y^2-b^2}{ab}\right)\cos^2\theta$

$$-\frac{2xy}{ab}\sin\theta\cos\theta = 0 \quad \text{...(iv)}$$

Adding Eqs. (iii) and (iv), we get

$$\left(\frac{x^2-a^2}{a}\right)\left(\frac{\sin^2\theta}{b}+\frac{\cos^2\theta}{a}\right)+\left(\frac{y^2-b^2}{b}\right)\left(\frac{\sin^2\theta}{b}+\frac{\cos^2\theta}{a}\right)=0$$

or $$\frac{x^2-a^2}{a}+\frac{y^2-b^2}{b}=0 \text{ or } \frac{x^2}{a}+\frac{y^2}{b}=(a+b)$$

JEE Type Solved Examples :

Matching Type Questions

Ex. 89. *Match the statement of Column I with values of Column II.*

Column-I	Column-II
(A) The number of real roots of the equation $\cos^7 x + \sin^4 x = 1$ in $(-\pi, \pi)$ is	(p) 1
(B) The value of $\sqrt{3}$ cosec $20° -$ sec $20°$ is	(q) 4
(C) $4\cos 36° - 4\cos 72° + 4\sin 18°\cos 36°$ equals	(r) 3
(D) The number of values of x where $x \in [-2\pi, 2\pi]$, which satisfy cosec $x = 1 + \cot x$	(s) 2

Sol. (A) → (r), (B) → (q), (C) → (r), (D) → (s)

(A) $\cos^7 x + \sin^4 x = 1$

$\cos^7 x = (1+\sin^2 x)\cos^2 x$

$\Rightarrow \quad \cos x = 0$ or $\cos^5 x = 1 + \sin^2 x$

$\cos x = 0 \Rightarrow x - \frac{\pi}{2}, \frac{\pi}{2}$; $\cos^5 x = 1 + \sin^2 x$

$\Rightarrow \quad x = 0 \quad$ [$\because$ LHS ≤ 1 and RHS ≥ 1]

$\therefore \quad x = -\frac{\pi}{2}, 0, \frac{\pi}{2}$.

(B) $\sqrt{3}$ cosec $20° -$ sec $20°$

$$= \frac{\sqrt{3}}{\sin 20°} - \frac{1}{\cos 20°} = \frac{\sqrt{3}\cos 20° - \sin 20°}{\sin 20°\cos 20°}$$

$$= \frac{2\left(\frac{\sqrt{3}}{2}\cos 20° - \frac{1}{2}\sin 20°\right)}{\sin 20°\cos 20°} = \frac{4\sin 40°}{\sin 40°} = 4$$

(C) $4\cos 36° - 4\cos 72° + 4\sin 18°\cdot\cos 36°$

$$= 4\left(\frac{\sqrt{5}+1}{4}\right) - 4\left(\frac{\sqrt{5}-1}{4}\right) + 4\left(\frac{\sqrt{5}-1}{4}\right)\left(\frac{\sqrt{5}+1}{4}\right)$$

$$= \sqrt{5}+1-\sqrt{5}+1+1 = 3$$

(D) cosec $x = 1 + \cot x$; $\frac{1}{\sin x} = \frac{\sin x + \cos x}{\sin x} \Rightarrow$

$\sin x + \cos x = 1$ and $\sin x \ne 0$

$$\cos\left(x - \frac{\pi}{4}\right) = \frac{1}{\sqrt{2}}$$

$\Rightarrow \quad x - \frac{\pi}{4} = -2\pi + \frac{\pi}{4}, \frac{\pi}{4}$

$$\left(\because x - \frac{\pi}{4} \in \left[-2\pi - \frac{\pi}{4}, 2\pi - \frac{\pi}{4}\right]\right)$$

$\Rightarrow \quad x = -\frac{3\pi}{2}, \frac{\pi}{2}$

Ex. 90. *Match the statement of Column I with values of Column II.*

Column-I	Column-II
(A) The number of solutions of the equation $\lvert\cot x\rvert = \cot x + \frac{1}{\sin x}$ $(0 < x < \pi)$	(p) No solution
(B) If $\sin\theta + \sin\phi = \frac{1}{2}$ and $\cos\theta + \cos\phi = 2$, then $\cot\left(\frac{\theta+\phi}{2}\right)$	(q) $\frac{1}{3}$
(C) $\sin^2\alpha + \sin\left(\frac{\pi}{3}-\alpha\right)\sin\left(\frac{\pi}{3}-\alpha\right)$	(r) 1
(D) If $\tan\theta = 3\tan\phi$, then maximum value of $\tan^2(\theta-\phi)$ is	(s) 4

Sol. (A) → (r), (B) → (s), (C) → (p), (D) → (q)

(A) $\lvert\cot x\rvert = \cot x + \frac{1}{\sin x}$

If $0 < x < \frac{\pi}{2} \Rightarrow \cot x > 0$

So, $\cot x = \cot x + \frac{1}{\sin x} \Rightarrow \frac{1}{\sin x} = 0$ no solution

If $\frac{\pi}{2} < \cot x < \pi$, $-\cot x = \cot x + \frac{1}{\sin x}$

$$\frac{2\cos x}{\sin x} + \frac{1}{\sin x} = 0$$

$1 + 2\cos x = 0$ and $x \ne 0 \Rightarrow x = \frac{2\pi}{3}$

(B) Since, $\sin\phi + \sin\theta = \frac{1}{2}$ and $\cos\theta + \cos\phi = 2$ has no solution.

(C) $\sin^2\alpha + \sin\left(\frac{\pi}{3}-\alpha\right)\cdot\sin\left(\frac{\pi}{3}+\alpha\right)$

$$= \sin^2\alpha + \sin^2\frac{\pi}{3} - \sin^2\alpha = \frac{3}{4}$$

(D) $\tan\theta = 3\tan\phi$

$$\tan(\theta - \phi) = \frac{\tan\theta - \tan\phi}{1 + \tan\theta\tan\phi} = \frac{2\tan\phi}{1 + 3\tan^2\phi}$$

$$= \frac{2}{\cot\phi + 3\tan\phi} \cdot \text{Max of } \tan\phi > 0$$

$$\frac{\cot\phi + 3\tan\phi}{2} \geq \sqrt{3} \qquad \text{(using AM} \geq \text{GM)}$$

$$\Rightarrow \quad (\cot\phi + 3\tan\phi)^2 \geq 12 \Rightarrow \tan^2(\theta - \phi) \leq \frac{1}{3}$$

Ex. 91. *Match the statement of Column I with values of Column II.*

Column-I		Column-II	
(A)	The tangents of two acute angles are 3 and 2. The sine of twice their difference is	(p)	1
(B)	If $n = \frac{\pi}{4\alpha}$, then $\tan\alpha\tan 2\alpha\tan 3\alpha \ldots \tan(2n-1)\alpha$ is equal to	(q)	0
(C)	If $x = y\cos\frac{2\pi}{3} = z\cos\frac{4\pi}{3}$, then $xy + yz + zx =$	(r)	$\frac{1}{2}$
(D)	The ratio of the greatest value of $2 - \cos x + \sin^2 x$ to its least value is	(s)	$\frac{7}{25}$
		(t)	$\frac{13}{4}$

Sol. (A) → (s), (B) → (p), (C) → (q), (D) → (t)

(A) Given, $\tan\alpha = 3$ and $\tan\beta = 2$

$$\therefore \quad \tan(\alpha - \beta) = \frac{\tan\alpha - \tan\beta}{1 - \tan\alpha\tan\beta} = \frac{3-2}{1+3\times 2} = \frac{1}{7}$$

$$\Rightarrow \quad \sin(\alpha - \beta) = \frac{1}{\sqrt{50}} \text{ and } \cos(\alpha - \beta) = \frac{7}{\sqrt{50}}$$

$$\therefore \quad \sin 2(\alpha - \beta) = 2\sin(\alpha - \beta)\cos(\alpha - \beta)$$

$$= 2 \times \frac{1}{\sqrt{50}} \times \frac{7}{\sqrt{50}} = \frac{7}{25}$$

(B) We have, $\tan\alpha \cdot \tan(2n-1)\alpha = \tan\alpha \cdot \tan\left(\frac{\pi}{2\alpha} - 1\right)\alpha$

$$= \tan\alpha \cdot \tan\left(\frac{\pi}{2} - \alpha\right) = \tan\alpha\cot\alpha = 1$$

∴ The given expression = 1.

(C) We have, $x = y\cos\frac{2\pi}{2} = z\cos\frac{4\pi}{3} = k$ (say)

$$\Rightarrow \quad \frac{1}{x} = \frac{1}{k}, \frac{1}{y} = \frac{\cos\frac{2\pi}{3}}{k}, \frac{1}{z} = \frac{\cos\frac{4\pi}{3}}{k}$$

$$\therefore \quad \frac{1}{x} + \frac{1}{y} + \frac{1}{z} = \frac{1}{k}\left(1 + \cos\frac{2\pi}{3} + \cos\frac{4\pi}{3}\right)$$

$$= \frac{1}{k}\left(1 - \frac{1}{2} - \frac{1}{2}\right) = 0$$

$$\Rightarrow \quad xy + xz + yz = 0$$

(D) We have, $2 - \cos x + \sin^2 x = 2 - \cos x + 1 - \cos^2 x$

$$= -(\cos^2 x + \cos x) + 3 = -\left[\left(\cos x + \frac{1}{2}\right)^2 - \frac{1}{4}\right] + 3$$

$$= \frac{13}{4} - \left(\cos x + \frac{1}{2}\right)^2$$

∴ Maximum value occurs at $\cos x = -\frac{1}{2}$ and it is $\frac{13}{4}$ and minimum value occurs at $\cos x = 1$ and it is 1

∴ The required ratio is $\frac{13}{4}$.

Ex. 92. *Match the statement of Column I with values of Column II*

Column-I		Column-II	
(A)	If α, β, γ and δ are four solutions of the equation $\tan\left(\theta + \frac{\pi}{4}\right) = 3\tan 3\theta$, no two of which have equal tangents, then the value of $\tan\alpha + \tan\beta + \tan\gamma + \tan\delta$ is	(p)	$\sqrt{2}$
(B)	If $\frac{\cos(\theta_1 - \theta_2)}{\cos(\theta_1 + \theta_2)} + \frac{\cos(\theta_3 + \theta_4)}{\cos(\theta_3 - \theta_4)} = 0$ then $\tan\theta_1\tan\theta_2\tan\theta_3\tan\theta_4 =$	(q)	$\sqrt{3}$
(C)	If $\sec(\alpha - \beta)$, $\sec\alpha$ and $\sec(\alpha + \beta)$ are in A.P. (with $\beta \neq 0$), then $\cos\alpha\sec\frac{\beta}{2} =$	(r)	−1
(D)	If $\cos\alpha = \frac{2\cos\beta - 1}{2 - \cos\beta}$ $(0 < \alpha < \beta < \pi)$, then $\frac{\tan\frac{\alpha}{2}}{\tan\frac{\beta}{2}}$ is equal to	(s)	0

Sol. (A) → (s), (B) → (r), (C) → (p), (D) → (q)

(A) Using $\tan\left(\theta + \frac{\pi}{4}\right) = \frac{1 + \tan\theta}{1 - \tan\theta}$

and $\quad 3\tan 3\theta = \frac{3(3\tan\theta - \tan^3\theta)}{1 - 3\tan^2\theta}$

the given equation becomes

$$3\tan^4\theta - 6\tan^2\theta + 8\tan\theta - 1 = 0$$

If $\tan\alpha$, $\tan\beta$, $\tan\gamma$ and $\tan\delta$ are the roots of this equation, then the sum of these roots, $\tan\alpha + \tan\beta + \tan\gamma + \tan\delta$ equals zero, since the coefficient of $\tan^3\theta$ is zero.

(B) The given equation can be written as

$$\Rightarrow \quad \frac{\cos\theta_1\cos\theta_2 + \sin\theta_1\sin\theta_2}{\cos\theta_1\cos\theta_2 - \sin\theta_1\sin\theta_2} + \frac{\cos\theta_3\cos\theta_4 - \sin\theta_3\sin\theta_4}{\cos\theta_3\cos\theta_4 + \sin\theta_3\sin\theta_4}$$

$$\Rightarrow \quad \frac{1 + \tan\theta_1\tan\theta_2}{1 - \tan\theta_1\tan\theta_2} + \frac{1 - \tan\theta_3\tan\theta_4}{1 + \tan\theta_3\tan\theta_4} = 0$$

$$\Rightarrow \frac{2+2\tan\theta_1\tan\theta_2\tan\theta_3\tan\theta_4}{(1-\tan\theta_1\tan\theta_2)(1+\tan\theta_3\tan\theta_4)}=0$$

Showing that $\tan\theta_1\tan\theta_2\tan\theta_3\tan\theta_4=-1$.

(C) For the given A.P., we have

$2\sec\alpha=\sec(\alpha-\beta)+\sec(\alpha+\beta)$, which gives

$$\frac{2}{\cos\alpha}=\frac{1}{\cos(\alpha-\beta)}+\frac{1}{\cos(\alpha+\beta)}$$

$$=\frac{2\cos\alpha\cos\beta}{\cos^2\alpha-\sin^2\beta}$$

$$\Rightarrow \cos^2\alpha-\sin^2\beta=\cos^2\alpha\cos\beta$$

$$\Rightarrow \cos^2\alpha(1-\cos\beta)=\sin^2\beta$$

$$\Rightarrow \cos^2\alpha\left(2\sin^2\frac{\beta}{2}\right)=4\sin^2\frac{\beta}{2}\cos^2\frac{\beta}{2}$$

$$\Rightarrow \cos^2\alpha\sec^2\frac{\beta}{2}=2$$

(D) $1+\cos\alpha=1+\dfrac{2\cos\beta-1}{2-\cos\beta}$

$$=\frac{2-\cos\beta+2\cos\beta-1}{2-\cos\beta}=\frac{1+\cos\beta}{2-\cos\beta}$$

$$\Rightarrow \cos^2\frac{\alpha}{2}=\frac{\cos^2\frac{\beta}{2}}{1+2\sin^2\frac{\beta}{2}} \quad \text{...(i)}$$

$$\Rightarrow 1-\cos^2\frac{\alpha}{2}=1-\frac{\cos^2\frac{\beta}{2}}{1+2\sin^2\frac{\beta}{2}}$$

$$=\frac{1+2\sin^2\frac{\beta}{2}-\cos^2\frac{\beta}{2}}{1+2\sin^2\frac{\beta}{2}}$$

$$=\frac{1+2\sin^2\frac{\beta}{2}-\left[1-\sin^2\frac{\beta}{2}\right]}{1+2\sin^2\frac{\beta}{2}}$$

$$\Rightarrow \sin^2\frac{\alpha}{2}=\frac{3\sin^2\frac{\beta}{2}}{1+2\sin^2\frac{\beta}{2}} \quad \text{...(ii)}$$

Dividing equation (ii) by (i), we get

$$\tan^2\frac{\alpha}{2}=3\tan^2\frac{\beta}{2}$$

$$\Rightarrow \frac{\tan\frac{\alpha}{2}}{\tan\frac{\beta}{2}}=\sqrt{3}$$

JEE Type Solved Examples :

Single Integer Answer Type Questions

• ***Ex. 93.*** $\tan 46^\circ\tan 14^\circ-\tan 74^\circ\tan 14^\circ+\tan 74^\circ\tan 46^\circ$ *is equal to*

Sol. (3) $\dfrac{\tan 46^\circ+\tan 14^\circ}{1-\tan 46^\circ\tan 14^\circ}=\tan(46^\circ+14^\circ)=\sqrt{3}$ ...(i)

$$\frac{\tan 74^\circ-\tan 14^\circ}{1+\tan 74^\circ\tan 14^\circ}=\tan(74^\circ-14^\circ)$$

$$=\sqrt{3} \quad \text{...(ii)}$$

$$\frac{\tan 74^\circ+\tan 46^\circ}{1-\tan 74^\circ\tan 46^\circ}=\tan(74^\circ+46^\circ)$$

$$=-\sqrt{3} \quad \text{...(iii)}$$

From Eqs. (i), (ii) and (iii)

$$\tan 46^\circ\tan 14^\circ=1-\frac{\tan 46^\circ+\tan 14^\circ}{\sqrt{3}}$$

$$\tan 74^\circ\tan 14^\circ=\frac{\tan 74^\circ-\tan 14^\circ}{\sqrt{3}}-1$$

$$\tan 74^\circ\tan 46^\circ=1-\frac{\tan 74^\circ+\tan 46^\circ}{-\sqrt{3}}$$

$\therefore \tan 46^\circ\tan 14^\circ-\tan 74^\circ\tan 14^\circ+\tan 74^\circ\tan 46^\circ=3$

• ***Ex. 94.*** *Maximum value of the expression* $\log_3(9-2\cos^2\theta-4\sec^2\theta)$ *is equal to*

Sol. (1) For the expression $a\cos^2\theta+b\sec^2\theta$ if $b>a$, then minimum value attains at $\cos^2\theta=\sec^2\theta=1$

$\Rightarrow$ max of $\{9-(2\cos^2\theta+4\sec^2\theta)\}=3$

So, maximum of $\log_3(9-2\cos^2\theta+4\sec^2\theta))=1$

• ***Ex. 95.*** *Let* $x\in\left(0,\dfrac{\pi}{2}\right)$ *and* $\log_{24\sin x}(24\cos x)=\dfrac{3}{2}$, *then find the value of* $\operatorname{cosec}^2 x$.

Sol. (9) $(24\sin x)^{3/2}=24\cos x$

$$\Rightarrow \sqrt{24}\,(\sin x)^{3/2}=\cos x$$

$$\Rightarrow 24\sin^3 x=\cos^2 x=1-\sin^2 x$$

Put $\sin x=t$, we get

$$24t^3+t^2-1=0$$

$$\Rightarrow (3t-1)\underbrace{(8t^2+3t+1)}_{>0}=0$$

$$\Rightarrow \qquad t = \frac{1}{3}$$

$$\therefore \qquad t = \frac{1}{3} \text{ i.e. } \sin x = \frac{1}{3}$$

$$\Rightarrow \qquad \operatorname{cosec} x = 3$$

$$\Rightarrow \qquad \operatorname{cosec}^2 x = 9$$

Ex. 96. *If x and y are non-zero real numbers satisfying $xy(x^2 - y^2) = x^2 + y^2$, then find the minimum value of $x^2 + y^2$.*

Sol. (4) Put $x = r\cos\theta$ and $y = r\sin\theta$

Hence, we have to minimise r^2?

Now, $r^2 \cos\theta \sin\theta r^2(\cos^2\theta - \sin^2\theta) = r^2$

$$r^2 \sin 2\theta \cos 2\theta = 2$$

$$r^2 \frac{\sin 4\theta}{4} = 1$$

$$r^2 = \frac{4}{\sin 4\theta}$$

$$r^2 = 4\operatorname{cosec}^2 4\theta$$

$$\therefore \qquad r^2_{\min} = 4$$

Ex. 97. *Using the identity $\sin^4 x = \frac{3}{8} - \frac{1}{2}\cos 2x + \frac{1}{8}\cos 4x$ or otherwise, if the value of $\sin^4\left(\frac{\pi}{7}\right) + \sin^4\left(\frac{3\pi}{7}\right) + \sin^4\left(\frac{5\pi}{7}\right) = \frac{a}{b}$, where a and b are coprime, find the value of $(a - b)$.*

Sol. (5) $\sin^4\left(\frac{\pi}{7}\right) = \frac{3}{8} - \frac{1}{2}\cos\left(\frac{2\pi}{7}\right) + \frac{1}{8}\cos\left(\frac{4\pi}{7}\right)$ ...(i)

and $\sin^4\left(\frac{3\pi}{7}\right) = \frac{3}{8} - \frac{1}{2}\cos\left(\frac{6\pi}{7}\right) + \frac{1}{8}\cos\left(\frac{12\pi}{7}\right)$

or $\sin^4\left(\frac{3\pi}{7}\right) = \frac{3}{8} + \frac{1}{2}\cos\left(\frac{\pi}{7}\right) - \frac{1}{8}\cos\left(\frac{5\pi}{7}\right)$ ...(ii)

Similarly, $\sin^4\left(\frac{5\pi}{7}\right) = \frac{3}{8} - \frac{1}{2}\cos\left(\frac{10\pi}{7}\right) + \frac{1}{8}\cos\left(\frac{20\pi}{7}\right)$

or $\sin^4\left(\frac{5\pi}{7}\right) = \frac{3}{8} + \frac{1}{2}\cos\left(\frac{3\pi}{7}\right) - \frac{1}{8}\cos\left(\frac{\pi}{7}\right)$ ...(iii)

On adding Eqs. (i), (ii) and (iii), we get

$$\sin^4\left(\frac{\pi}{7}\right) + \sin^4\left(\frac{3\pi}{7}\right) + \sin^4\left(\frac{5\pi}{7}\right)$$

$$= \frac{9}{8} + \frac{1}{2}\cos\left(\frac{5\pi}{7}\right) - \frac{1}{8}\cos\left(\frac{3\pi}{7}\right) + \frac{1}{2}\cos\left(\frac{\pi}{7}\right) - \frac{1}{8}\cos\left(\frac{5\pi}{7}\right) + \frac{1}{2}\cos\left(\frac{3\pi}{7}\right) - \frac{1}{8}\cos\left(\frac{\pi}{7}\right)$$

$$= \frac{9}{8} + \frac{3}{8}\underbrace{\left[\cos\left(\frac{\pi}{7}\right) + \cos\left(\frac{3\pi}{7}\right) + \cos\left(\frac{5\pi}{7}\right)\right]}_{\frac{1}{2}\left(\text{use } S = \frac{\sin\frac{n\alpha}{2}}{\sin\frac{\alpha}{2}}\cos\left(\frac{\theta_1 + \theta_2}{2}\right)\right)}$$

$$= \frac{9}{8} + \frac{3}{16} = \frac{21}{16}$$

$$\therefore \qquad a - b = 21 - 16 = 5.$$

Ex. 98. *In any triangle, if $(\sin A + \sin B + \sin C)(\sin A + \sin B - \sin C) = 3\sin A \sin B$, then the angle $\frac{C}{10}$ (in degree).*

Sol. (6) We have, $(\sin A + \sin B)^2 - \sin^2 C = 3\sin A \sin B$

$$\sin^2 A - \sin^2 C + \sin^2 B = \sin A \sin B$$

$$\sin(A + C)\sin(A - C) + \sin^2 B = \sin A \sin B$$

$$\sin B[\sin(A - C) + \sin(A + C)] = \sin A \sin B$$

[using, $\sin(A + C) = \sin B$]

$$2\sin A \cos C = \sin A \qquad (\sin B \neq 0)$$

$$\cos C = \frac{1}{2} \Rightarrow \frac{C}{10} = \frac{60^\circ}{10} = 6$$

Ex. 99. *Find the exact value of the expression*

$$T = \frac{\sin 40^\circ}{\sin 80^\circ} + \frac{\sin 80^\circ}{\sin 20^\circ} - \frac{\sin 20^\circ}{\sin 40^\circ}$$

Sol. (3) We have, $\frac{1}{2\cos 40^\circ} + 4\cos 40^\circ \cdot \cos 20^\circ - \frac{1}{2\cos 20^\circ}$

$$= \frac{1}{2}\left[\frac{1}{\cos 40^\circ} - \frac{1}{\cos 20^\circ}\right] + 2(\cos 60^\circ + \cos 20^\circ)$$

$$= \frac{1}{2}\left[\frac{\cos 20^\circ - \cos 40^\circ}{\cos 40^\circ \cos 20^\circ}\right] + 1 + 2\cos 20^\circ$$

$$= \frac{2\sin 30^\circ \sin 10^\circ}{2\cos 40^\circ \cos 20^\circ} + 1 + 2\cos 20^\circ$$

$$= \frac{2\sin 10^\circ \sin 20^\circ}{\sin 80^\circ} + 1 + 2\cos 20^\circ$$

$$= \frac{2\sin 10^\circ\, 2\sin 10^\circ \cos 10^\circ}{\cos 10^\circ} + 2\cos 20^\circ + 1$$

$$= 2(1 - \cos 20^\circ) + 2\cos 20^\circ + 1 = 3$$

Alternatively $T_1 = \frac{2\sin 20^\circ \cos 20^\circ}{\cos 10^\circ} = 2 \cdot 2 \cdot \sin 10^\circ \cdot \cos 20^\circ$

$$= 2(\sin 30^\circ - \sin 10^\circ)$$

$$T_1 = 1 - 2\sin 10^\circ$$

$$T_2 = \frac{\sin 80^\circ}{\sin 20^\circ} = \frac{2\sin 40^\circ \cos 40^\circ}{\sin 20^\circ}$$

$$= 4\cos 20^\circ \cdot \cos 40^\circ$$

$$T_2 = 2[\cos 60^\circ + \cos 20^\circ] = 1 + 2\cos 20^\circ$$

$$T_2 = \frac{\sin 20^\circ}{\sin 40^\circ} = \frac{1}{2\cos 20^\circ}$$

$$\therefore\ T = T_1 + T_2 + T_3$$
$$= (1 - 2\sin 10^\circ) + (1 + 2\cos 20^\circ) - \frac{1}{2\cos 20^\circ}$$
$$= 2 + 2(\cos 20^\circ - \sin 10^\circ) - \frac{1}{2\cos 20^\circ}$$
$$= 2 + 2(\cos 20^\circ - \cos 80^\circ) - \frac{1}{2\cos 20^\circ}$$
$$= 2 + 2\cdot 2\sin 50^\circ \cdot \sin 30^\circ - \frac{1}{2\cos 20^\circ}$$
$$= 2 + 2\sin 50^\circ - \frac{1}{2\cos 20^\circ}$$
$$T = 2 + \frac{4\sin 50^\circ \cos 20^\circ - 1}{2\cos 20^\circ}$$
$$= 2 + \frac{2[\sin 70^\circ + \sin 30^\circ] - 1}{2\cos 20^\circ}$$
$$= 2 + \frac{2\sin 70^\circ}{2\cos 20^\circ} = 2 + 1 = 3$$

Ex. 100. *If* $\cot(\theta - \alpha), 3\cot\theta, \cot(\theta + \alpha)$ *are in AP (where,* $\theta \neq \frac{n\pi}{2}, \alpha \neq k\pi, n, k \in I$*), then* $\frac{2\sin^2\theta}{\sin^2\alpha}$ *is equal to*

Sol. (3) We have, $6\cot\theta = \cot(\theta - \alpha) + \cot(\theta + \alpha)$

$$\Rightarrow \frac{6\cos\theta}{\sin\theta} = \frac{\sin 2\theta}{\sin(\theta - \alpha)\sin(\theta + \alpha)}$$
$$\Rightarrow \frac{6\cos\theta}{\sin\theta} = \frac{2\sin\theta\cos\theta}{\sin^2\theta - \sin^2\alpha}$$
$$\Rightarrow 3(\sin^2\theta - \sin^2\alpha) = \sin^2\theta$$
or $$2\sin^2\theta = 3\sin^2\alpha$$
$$\Rightarrow \frac{2\sin^2\theta}{\sin^2\alpha} = 3$$

Ex. 101. *If* $4\sin^2 x + \text{cosec}^2 x, a, \sin^2 y + 4\text{cosec}^2 y$ *are in AP, then minimum value of* $(2a)$ *is*

Sol. (9) $2a = 4\sin^2 x + \text{cosec}^2 x + \sin^2 y + 4\,\text{cosec}^2 y$

$$= (2\sin x - \text{cosec}\, x)^2 + 4 + (\sin y - \text{cosec}\, y)^2 + 3\,\text{cosec}^2 y + 2$$
$$= 6 + (2\sin x - \text{cosec}\, x)^2 + (\sin y - \text{cosec}\, y)^2 + 3\,\text{cosec}^2 y$$

$\therefore$ Minimum value of

$$2a = 6 + 3 = 9,$$

when $2\sin x = \text{cosec}\, x$

and $\sin y = \text{cosec}\, y$

Ex. 102. *If* $\sin\alpha, \sin\beta, \sin\gamma$ *are in AP and* $\cos\alpha, \cos\beta, \cos\gamma$ *are in GP, then the value of* $\frac{\cos^2\alpha + \cos^2\gamma + 4\cos\alpha\cos\gamma - 2\sin\alpha\sin\gamma - 2}{1 - 2\sin^2\beta}$*, where* $\beta \neq \frac{\pi}{4}$*, is equal to*

Sol. (4) Now, $\sin\alpha + \sin\gamma = 2\sin\beta$ and $\cos^2\beta = \cos\alpha \cdot \cos\gamma$

$$= \frac{\cos^2\alpha + \cos^2\gamma + 4\cos\alpha\cos\gamma - 2 - 2\sin\alpha\sin\gamma}{1 - 2\sin^2\beta}$$
$$= \frac{-\sin^2\alpha - \sin^2\gamma - 2\sin\alpha\sin\gamma + 4\cos\alpha\cos\gamma}{1 - 2\sin^2\beta}$$
$$= \frac{-(\sin\alpha + \sin\gamma)^2 + 4\cos\alpha\cos\gamma}{1 - 2\sin^2\beta}$$
$$= \frac{-4\sin^2\beta + 4\cos^2\beta}{\cos 2\beta} = 4$$

Ex. 103. *Let*

$$\prod_{r=1}^{51} \tan\left(\frac{\pi}{3}\left(1 + \frac{3^r}{3^{50} - 1}\right)\right) = k\prod_{r=1}^{51} \cot\left[\frac{\pi}{3}\left(1 - \frac{3^r}{3^{50} - 1}\right)\right]$$

On solving equation, we get, $1 - 3\tan^2\left(\frac{\pi}{3^{50} - 1}\right) = \frac{a}{bk - 1}$*, (*$a, b \in I$*), then value of* $(a - b)$ *is equal to*

Sol. (5) We have,

$$\prod_{r=1}^{51} \tan\left(\frac{\pi}{3}\left(1 + \frac{3^r}{3^{50} - 1}\right)\right) = \prod_{r=1}^{51} \cot\left[\frac{\pi}{3}\left(1 - \frac{3^r}{3^{50} - 1}\right)\right]$$

Let $\frac{3^{r-1}\pi}{3^{50} - 1} = \theta_r$

$$\prod_{r-1}^{51} \tan\left(\frac{\pi}{3} + \theta_r\right)\tan\left(\frac{\pi}{3} + \theta_r\right) = k$$
$$\prod_{r-1}^{51} \frac{\tan 3\theta_r}{\tan\theta^r} = k$$
$$k = \frac{\tan\theta_2}{\tan\theta_1} \times \frac{\tan\theta_3}{\tan\theta_2} \times \ldots\ldots \frac{\tan\theta_{52}}{\tan\theta_{51}}$$
$$= \frac{\tan\theta_{52}}{\tan\theta} = \frac{\tan\left(\frac{3^{51}\pi}{3^{50} - 1}\right)}{\tan\left(\frac{\pi}{3^{50} - 1}\right)}$$
$$= \frac{\tan\left(3\pi + \frac{3\pi}{3^{50} - 1}\right)}{\tan\left(\frac{\pi}{3^{50} - 1}\right)}$$

$$= \frac{\tan\left(\frac{3\pi}{3^{50}-1}\right)}{\tan\left(\frac{\pi}{3^{50}-1}\right)}$$

Let, $\alpha = \frac{\pi}{3^{50}-1}$;

$k = \frac{\tan 3\alpha}{\tan\alpha}$;

$1 - 3\tan^2\alpha = \frac{8}{3k-1}$

So, $a = 8, b = 3$

$a - b = 5$

Ex. 104. *If* $\sec A \tan B + \tan A \sec B = 91$*, then the value of* $(\sec A \sec B + \tan A \tan B)^2$ *is equal to*

Sol. (8282) $(\sec A \sec B + \tan A \tan B)^2 - (\sec A \tan B + \tan A \sec B)^2$

$$= \left[\frac{1+\sin A\sin B}{\cos A\cos B}\right]^2 - \left[\frac{\sin B + \sin A}{\cos A\cos B}\right]^2$$

$$= \frac{1+\sin^2 A\sin^2 B - \sin^2 B - \sin^2 A}{\cos^2 A\cos^2 B}$$

$$= \frac{1-\sin^2 B\cos^2 A - \sin^2 A}{\cos^2 A\cos^2 B}$$

$$= \frac{\cos^2 A\cos^2 B}{\cos^2 A\cos^2 B} = 1$$

$\Rightarrow (\sec A \sec B + \tan A \tan B)^2 = (91)^2 + 1 = 8282.$

Ex. 105. *If* $(25)^2 + a^2 + 50a\cos\theta = (31)^2 + b^2 + 62\,b\cos\theta = 1$ *and* $775 + ab + (31a + 25b)\cos\theta = 0$*, then the value of* $\operatorname{cosec}^2\theta$ *is*

Sol. (1586) We can write $(a + 25\cos\theta)^2 + (25)^2 - (25\cos\theta)^2 = 1$

and

$\Rightarrow \quad (a + 25\cos\theta)^2 = 1 - (25\sin\theta)^2$

Similarly $\quad (b + 31\cos\theta)^2 = 1 - (31\sin\theta)^2$

Multiplying we get

$$[(a + 25\cos\theta)(b + 31\cos\theta)]^2 = [1 - (25\sin\theta)^2][1 - (31\sin\theta)^2]$$

$\Rightarrow \quad [ab + (31a + 25b)\cos\theta + 775\cos^2\theta]^2 = 1 - (625 + 961)\sin^2\theta + (775\sin^2\theta)^2$

$\Rightarrow \quad (-775 + 775\cos^2\theta)^2 = 1 - 1586\sin^2\theta + (775\sin^2\theta)^2$

$\Rightarrow \quad \operatorname{cosec}^2\theta = 1586$

Ex. 106. *If* $\sin x_1 + \sin x_2 + \sin x_3 + \ldots + \sin x_{2008} = 2008$*, then find the value of* $\sin^{2008} x_1 + \sin^{2008} x_2 + \sin^{2008} x_3 + \ldots + \sin^{2008} x_{2008}$.

Sol. (2008) We know that, $\sin x_i \le 1\ \forall\ i$

$\Rightarrow \sin x_1 + \sin x_2 + \sin x_3 + \ldots + \sin x_{2008} \le 2008$

Thus, equality holds only when each of the terms is 1

i.e. $\quad \sin x_i = 1\ \forall\ i = 1, 2, 3, \ldots, 2008.$

and consequently

$$\cos x_i = 0, \forall i = 1,2,3, \ldots, 2008$$

Now, $\sin^{2008} x_1 + \sin^{2008} x_2 + \sin^{2008} x_3 + \ldots \sin^{2008} x_{2008}$

$$= 1 + 1 + 1 + \ldots + 1 = 2008$$

Ex. 107. *If* $4\sin 27^\circ = \sqrt{\alpha} + \sqrt{\beta}$*, then the value of* $(\alpha + \beta - \alpha\beta + 2)^4$ *must be*

Sol. (400), We know $(\cos 27^\circ + \sin 27^\circ)^2$

$= 1 + \sin 54 = 1 + \cos 36^\circ$

$\Rightarrow \quad \cos 27^\circ + \sin 27^\circ = \sqrt{(1 + \cos 36^\circ)} \quad [\because \text{LHS} > 0]$

Also, $\quad \cos 27^\circ - \sin 27^\circ = \sqrt{(1 - \cos 36^\circ)}$

$[\because \cos 27^\circ > \sin 27^\circ]$

$\therefore \quad 2\sin 27^\circ = \sqrt{(1 + \cos 36^\circ)} - \sqrt{(1 - \cos 36^\circ)}$

$$= \sqrt{\left(1 + \left(\frac{\sqrt{5}+1}{4}\right)\right)} - \sqrt{\left(1 - \left(\frac{\sqrt{5}+1}{4}\right)\right)}$$

$\therefore \quad 4\sin 27^\circ = \sqrt{(5 + \sqrt{5})} - \sqrt{(3 - \sqrt{5})}$

On comparing, we get

$$\alpha = 5 + \sqrt{5}, \beta = 3 - \sqrt{5}$$

$\therefore \quad \alpha + \beta = 8, \alpha\beta = 10 - 2\sqrt{5}$

$\alpha + \beta - \alpha\beta + 2 = 2\sqrt{5}$

$\therefore (\alpha + \beta - \alpha\beta + 2)^4 = 400$

Ex. 108. *If* $0 < A < \frac{\pi}{2}$ *and* $\sin A + \cos A + \tan A + \cot A + \sec A + \operatorname{cosec} A = 7$ *and* $\sin A$ *and* $\cos A$ *are the roots of the equation* $4x^2 - 3x + a = 0$*, then the value of* $25a$ *must be*

Sol. (28) $\sin A$ and $\cos A$ are the roots of the equation $4x^2 - 3x + a = 0$, then

$$\sin A + \cos A = \frac{3}{4}, \sin A\cos A = \frac{a}{4} \quad \ldots\text{(i)}$$

Also, $\sin A + \cos A + \tan A + \cot A + \sec A + \operatorname{cosec} A = 7$

$$\Rightarrow \quad (\sin A + \cos A) + \left(\frac{\sin A}{\cos A} + \frac{\cos A}{\sin A}\right) + \left(\frac{1}{\cos A} + \frac{1}{\sin A}\right) = 7$$

$$\Rightarrow \quad (\sin A + \cos A) + \frac{1}{\sin A \cos A} + \frac{(\sin A + \cos A)}{\sin A \cos A} = 7$$

$$\Rightarrow \quad \frac{3}{4} + \frac{4}{a} + \frac{3}{a} = 7$$

$$\Rightarrow \quad \frac{3}{4} + \frac{7}{a} = 7$$

$$\Rightarrow \quad \frac{7}{a} = 7 - \frac{3}{4} = \frac{25}{4}$$

$$\therefore \quad 25a = 28$$

Ex. 109. *Given that,* $f(n\theta) = \dfrac{2\sin\theta}{\cos 2\theta - \cos 4n\theta}$, *and* $f(\theta) + f(2\theta) + f(3\theta) + \dots + f(n\theta) = \dfrac{\sin \lambda\theta}{\sin\theta \sin\mu\theta}$, *then the value of* $\mu - \lambda$ *is*

Sol. (1) $f(n\theta) = \dfrac{2\sin 2\theta}{\cos 2\theta - \cos 4n\theta}$

$$= \frac{2\sin 2\theta}{2\sin(2n+1)\theta \sin(2n-1)\theta}$$

$$= \frac{\sin((2n+1)\theta - (2n-1)\theta)}{\sin(2n+1)\theta \sin(2n-1)\theta}$$

$$= \frac{\sin(2n+1)\theta \cos(2n-1)\theta - \cos(2n+1)\theta \sin(2n-1)\theta}{\sin(2n+1)\theta \sin(2n-1)\theta}$$

$$= \cot(2n-1)\theta - \cot(2n+1)\theta$$

$\therefore f(\theta) + f(2\theta) + f(3\theta) + \dots + f(n\theta)$

$$= \cot\theta - \cot(2n+1)\theta$$

$$= \frac{\sin(2n+1)\theta\cos\theta - \cos(2n+1)\theta\sin\theta}{\sin(2n+1)\theta\sin\theta}$$

$$= \frac{\sin 2n\theta}{\sin(2n+1)\theta\sin\theta}$$

$\therefore \quad \pi = 2n$ and $\mu = 2n+1$

Hence, $\mu - \lambda = 2n + 1 - 2n = 1$

Ex. 110. *If* $\dfrac{1}{\cos 290^\circ} + \dfrac{1}{\sqrt{3}\sin 250^\circ} = \lambda$, *then the value of* $9\lambda^4 + 81\lambda^2 + 97$ *must be*

Sol. (785) Here, $\cos 290^\circ = \cos(270^\circ + 20^\circ) = \sin 20^\circ$ and $\sin 250^\circ = \sin(270^\circ - 20^\circ) = -\cos 20^\circ$

$\therefore$ The given expression $= \dfrac{1}{\sin 20^\circ} - \dfrac{1}{\sqrt{3}\cos 120^\circ} = \lambda$

$$\Rightarrow \quad \frac{1}{\sin 20^\circ} - \frac{\cos 60^\circ}{\sin 60^\circ \cos 20^\circ} = \lambda$$

$$\Rightarrow \quad \frac{\sin 60^\circ\cos 20^\circ - \cos 60^\circ \sin 20^\circ}{\sin 20^\circ \cos 20^\circ \sin 60^\circ} = \lambda$$

$$\Rightarrow \quad \frac{\sin(60^\circ - 20^\circ)}{\frac{\sin 40^\circ}{2} \times \frac{\sqrt{3}}{2}} = \lambda$$

$$\therefore \quad \lambda = \frac{4}{\sqrt{3}}$$

$$\Rightarrow \quad \lambda^2 = \frac{16}{3}$$

Then, $9\lambda^4 + 81\lambda^2 + 97 = 9 \times \dfrac{256}{9} + 81 \times \dfrac{16}{3} + 97$

$$= 256 + 432 + 97 = 785$$

Ex. 111. *If* $\log_{10}\sin x + \log_{10}\cos x = -1$ *and* $\log_{10}(\sin x + \cos x) = \dfrac{(\log_{10} n) - 1}{2}$, *then the value of 'n/3' is* ....

Sol. (4) Given, $\log_{10}\left(\dfrac{\sin 2x}{2}\right) = -1$

or $\quad \dfrac{\sin 2x}{2} = \dfrac{1}{10}$

or $\quad \sin 2x = \dfrac{1}{5}$

Also $\quad \log_{10}(\sin x + \cos x) = \dfrac{\log_{10}\left(\frac{n}{10}\right)}{2}$

or $\quad \log_{10}(\sin x + \cos x)^2 = \log_{10}\left(\dfrac{n}{10}\right)$

or $\quad 1 + \sin 2x = \dfrac{n}{10}$

or $\quad 1 + \dfrac{1}{5} = \dfrac{n}{10}$ or $\dfrac{6}{5} = \dfrac{n}{10}$

or $\quad \dfrac{n}{3} = 4$

Ex. 112. *If* $498[16\cos x + 12\sin x] = 2k + 60$, *then the maximum value of* k *is*

Sol. (4950) $16\cos x + 12\sin x = \sqrt{16^2 + 12^2}\cos(x - \alpha)$, $\alpha = \tan^{-1}\left(\dfrac{3}{4}\right)$.

$$\Rightarrow \quad |2k + 60| \le 498 \times 20 \text{ as } |\cos(x-\alpha)| \le 1$$

$$\Rightarrow \quad k \le 4950$$

Ex. 113. *If* $a \tan\alpha + \sqrt{(a^2-1)}\tan\beta + \sqrt{(a^2+1)}\tan\gamma = 2a$, *where a is constant and* α, β, γ *are variable angles. Then the least value of* $2727(\tan^2\alpha + \tan^2\beta + \tan^2\gamma)$ *must be*

Sol. (3636) We have,

$$(a\tan\beta - \sqrt{(a^2-1)}\tan\alpha)^2 + (\sqrt{(a^2-1)}\tan\gamma - \sqrt{(a^2+1)}\tan\beta)^2 + (\sqrt{(a^2+1)}\tan\alpha - a\tan\gamma)^2 \ge 0$$

$$\Rightarrow (a^2 + a^2 - 1 + a^2 + 1)(\tan^2\alpha + \tan^2\beta + \tan^2\gamma) - \{a\tan\alpha + \sqrt{(a^2-1)}\tan\beta + \sqrt{(a^2+1)}\tan\gamma\}^2 \ge 0$$

(using Lagrange's identity)

$$\Rightarrow \quad 3a^2(\tan^2\alpha + \tan^2\beta + \tan^2\gamma) - (2a)^2 \ge 0$$

$$\therefore \quad 3(\tan^2\alpha + \tan^2\beta + \tan^2\gamma) \ge 4$$

Hence, $2727 \quad (\tan^2\alpha + \tan^2\beta + \tan^2\gamma) \ge 3636$

$\therefore$ Least value is 3636.

Ex. 114. *If* $\dfrac{\tan x}{2} = \dfrac{\tan y}{3} = \dfrac{\tan z}{5}$, $x + y + z = \pi$ *and* $\tan^2 x + \tan^2 y + \tan^2 z = \dfrac{38}{K}$ *then* $K = \ldots$..

Sol. (3) $\tan x = 2t$, $\tan y = 3t$, $\tan z = 5t$

Also $\quad x + y + z = \pi$

$\therefore \tan x + \tan y + \tan z = \tan x \tan y \tan z$

$$\Rightarrow \quad t^2 = \frac{1}{3}$$

$$\Rightarrow \quad \tan^2 x + \tan^2 y + \tan^2 z = t^2(4 + 9 + 25) = 38t^2,$$

$$\therefore \quad K = 3$$

Ex. 115. *If* $\tan\left(\dfrac{3\pi}{11}\right) + 4\sin\left(\dfrac{2\pi}{11}\right) = \lambda$, *then the value of* $1 + \lambda^2 + \lambda^4 + \lambda^6$ *must be.*

Sol. (1464) Let $\lambda = \tan\left(\dfrac{3\pi}{11}\right) + 4\sin\left(\dfrac{2\pi}{11}\right)$

$$= \frac{1}{\cos\left(\frac{3\pi}{11}\right)}\left\{\sin\left(\frac{3\pi}{11}\right) + 4\sin\left(\frac{2\pi}{11}\right)\cos\left(\frac{3\pi}{11}\right)\right\}$$

$$\lambda\cos\left(\frac{3\pi}{11}\right) = \sin\left(\frac{3\pi}{11}\right) + 4\sin\left(\frac{2\pi}{11}\right)\cos\left(\frac{3\pi}{11}\right)$$

$$\lambda^2\cos^2\left(\frac{3\pi}{11}\right) = \sin^2\left(\frac{3\pi}{11}\right) + 16\sin^2\left(\frac{2\pi}{11}\right)\cos^2\left(\frac{3\pi}{11}\right) + 8\sin\left(\frac{2\pi}{11}\right)\sin\left(\frac{3\pi}{11}\right)\cos\left(\frac{3\pi}{11}\right)$$

$$\lambda^2\left(2\cos^2\left(\frac{3\pi}{11}\right)\right) = 2\sin^2\left(\frac{3\pi}{11}\right) + 32\sin^2\left(\frac{2\pi}{11}\right)\cos^2\left(\frac{3\pi}{11}\right) + 8\sin\left(\frac{2\pi}{11}\right)\sin\left(\frac{6\pi}{11}\right)$$

$$= \left(1 - \cos\left(\frac{6\pi}{11}\right)\right) + 8\left(1 - \cos\frac{4\pi}{11}\right)\left(1 + \cos\frac{6\pi}{11}\right) + 4\left(\cos\frac{4\pi}{11} - \cos\frac{8\pi}{11}\right)$$

$$= 9 + 7\cos\left(\frac{6\pi}{11}\right) - 4\cos\left(\frac{4\pi}{11}\right) - 8\cos\left(\frac{4\pi}{11}\right)\cos\left(\frac{6\pi}{11}\right) - 4\cos\left(\frac{8\pi}{11}\right)$$

$$= 9 + 7\cos\left(\frac{6\pi}{11}\right) - 4\cos\left(\frac{4\pi}{11}\right) - 4\left(\cos\left(\frac{10x}{11}\right) + \cos\left(\frac{2\pi}{11}\right)\right) - 4\cos\left(\frac{8\pi}{11}\right)$$

$$= 9 + 11\cos\left(\frac{6\pi}{11}\right) - 4\left\{\cos\left(\frac{2\pi}{11}\right) + \cos\left(\frac{4\pi}{11}\right) + \cos\left(\frac{6\pi}{11}\right) + \cos\left(\frac{8\pi}{11}\right) + \cos\left(\frac{10\pi}{11}\right)\right\}$$

$$= 9 + 11\cos\left(\frac{6\pi}{11}\right) - \frac{4\cdot\cos\left(\frac{6\pi}{11}\right)\sin\left(\frac{5\pi}{11}\right)}{\sin\left(\frac{\pi}{11}\right)}$$

$$= 9 + 11\cos\left(\frac{6\pi}{11}\right) - \frac{2\left\{\sin\pi - \sin\left(\frac{\pi}{11}\right)\right\}}{\sin\left(\frac{\pi}{11}\right)}$$

$$= 9 + 11\cos\left(\frac{6\pi}{11}\right) + 2$$

$$= 11\left(1 + \cos\left(\frac{6\pi}{11}\right)\right) = 11\left(2\cos^2\left(\frac{3\pi}{11}\right)\right)$$

$$\therefore \quad \lambda^2 = 11$$

Then, $1 + \lambda^2 + \lambda^4 + \lambda^6 = 1 + 11 + 121 + 1331 = 1464$

Subjective Type Examples

Ex. 116. *For all θ in $[0, \pi/2]$, show that* $\cos(\sin\theta) > \sin(\cos\theta)$.

Sol. We know,

$$\cos\theta + \sin\theta = \sqrt{2}\left[\frac{1}{\sqrt{2}}\cos\theta + \frac{1}{\sqrt{2}}\sin\theta\right]$$

$$= \sqrt{2}\left[\sin\frac{\pi}{4}\cos\theta + \cos\frac{\pi}{4}\sin\theta\right]$$

$$= \sqrt{2}\sin\left(\frac{\pi}{4} + \theta\right)$$

$$\Rightarrow \quad \cos\theta + \sin\theta \le \sqrt{2} < \frac{\pi}{2} \quad \{\text{as } \sqrt{2} = 1.414\}$$

$$\Rightarrow \quad \cos\theta + \sin\theta < \frac{\pi}{2} \quad [\pi/2 = 1.57 \text{ approx}]$$

$$\Rightarrow \quad \cos\theta < \frac{\pi}{2} - \sin\theta$$

On taking sine both sides;

$$\sin(\cos\theta) < \sin\left(\frac{\pi}{2} - \sin\theta\right)$$

$$\Rightarrow \quad \sin(\cos\theta) < \cos(\sin\theta)$$

$$\therefore \quad \cos(\sin\theta) > \sin(\cos\theta)$$

Alternate Method

For $0 \le x \le \dfrac{\pi}{2}$

$$x \ge \sin x \quad \text{...(i)}$$

Replace x by $\cos\theta$, we get

$$\cos\theta \ge \sin(\cos\theta) \quad \text{...(ii)}$$

Also, we know $\cos\theta$ is decreasing for $0 \le \theta \le \dfrac{\pi}{2}$.

As $\theta_1 < \theta_2 \Rightarrow \cos\theta_1 > \cos\theta_2$ when $\theta_1, \theta_2 \in [0, \pi/2]$

$\therefore$ Taking cos on both side of Eq. (i) and putting θ for x, we get

$$\cos\theta \le \cos(\sin\theta) \quad \text{...(iii)}$$

Using Eqs. (ii) and (iii),

$$\cos(\sin\theta) \ge \cos\theta \ge \sin(\cos\theta)$$

$$\Rightarrow \quad \cos(\sin\theta) > \sin(\cos\theta)$$

Ex. 117. *Show that* $2^{\sin x} + 2^{\cos x} \ge 2^{1 - \frac{1}{\sqrt{2}}}$ *for all real x.*

Sol. Clearly, $2^{\sin x}$ and $2^{\cos x}$ are positive, so their AM $\ge$ GM

$$\therefore \quad \frac{2^{\sin x} + 2^{\cos x}}{2} \ge \sqrt{2^{\sin x} . 2^{\cos x}} = \sqrt{2^{\sin x + \cos x}} \quad \text{...(i)}$$

As we know,

$$\sin x + \cos x \ge -\sqrt{2}$$

$$[\text{using } -\sqrt{a^2 + b^2} \le a\sin x + b\cos x \le \sqrt{a^2 + b^2}]$$

$$2^{\sin x + \cos x} \ge 2^{-\sqrt{2}}$$

$$\text{or} \quad \sqrt{2^{\sin x + \cos x}} \ge 2^{-\sqrt{2}/2}$$

$$\text{or} \quad \sqrt{2^{\sin x + \cos x}} \ge 2^{-1/\sqrt{2}} \quad \text{...(ii)}$$

From Eqs. (i) and (ii),

$$\frac{2^{\sin x} + 2^{\cos x}}{2} \ge \sqrt{2^{\sin x + \cos x}} \ge 2^{-1/\sqrt{2}}$$

$$\therefore \quad 2^{\sin x} + 2^{\cos x} \ge 2 \cdot 2^{-1/\sqrt{2}}$$

$$\text{or} \quad 2^{\sin x} + 2^{\cos x} \ge 2^{1 - 1/\sqrt{2}} \text{ for all values of } x.$$

Ex. 118. *Eliminate θ and ϕ if*

$$a\sin^2\theta + b\cos^2\theta = m$$

$$b\sin^2\phi + a\cos^2\phi = n$$

$$\textit{and} \quad a\tan\theta = b\tan\phi$$

Sol. Dividing $a\sin^2\theta + b\cos^2\theta = m$ by $\cos^2\theta$, we get

$$a\tan^2\theta + b = m\sec^2\theta$$

$$\text{or} \quad (a - m)\tan^2\theta = (m - b) \quad \text{...(i)}$$

Dividing $b\sin^2\phi + a\cos^2\phi = n$ by $\cos^2\phi$, we get

$$b\tan^2\phi + a = n\sec^2\phi$$

$$\text{or} \quad (b - n)\tan^2\phi = (n - a) \quad \text{...(ii)}$$

On dividing Eq. (i) by Eq. (ii), we get

$$\frac{(a - m)}{(b - n)} \cdot \left(\frac{\tan\theta}{\tan\phi}\right)^2 = \frac{m - b}{n - a}$$

$$\Rightarrow \quad \left(\frac{a - m}{b - n}\right) \cdot \frac{b^2}{a^2} = \frac{m - b}{n - a} \quad [\text{given, } a\tan\theta = b\tan\phi]$$

$$\text{or} \quad b^2(a - m)(n - a) = a^2(b - n)(m - b)$$

$$\text{or} \quad b^2\{(m + n)a - a^2 - mn\} = a^2\{(m + n)b - b^2 - mn\}$$

$$\text{or} \quad (m + n)(ab^2 - a^2b) + mn(a^2 - b^2) = 0$$

$$\text{or} \quad (m + n)ab(b - a) + mn(a - b)(a + b) = 0$$

$$\text{or} \quad (m + n)(ab) = mn(a + b)[a - b \ne 0]$$

Ex. 119. *Let* $\cos A + \cos B + \cos C = \dfrac{3}{2}$ *in a* ΔABC, *show that the triangle is equilateral.*

Sol. In a triangle, $A + B + C = \pi$

$$\Rightarrow \quad \cos A + \cos B + \cos C = 2\cos\left(\frac{A + B}{2}\right).$$

$$\cos\left(\frac{A - B}{2}\right) + \cos C = \frac{3}{2}$$

$$\Rightarrow \quad 2\cos\left(\frac{\pi}{2} - \frac{C}{2}\right) \cdot \cos\left(\frac{A - B}{2}\right) + 1 - 2\sin^2\frac{C}{2} = \frac{3}{2}$$

$$\Rightarrow \quad 2\sin^2\frac{C}{2} - 2\sin\left(\frac{C}{2}\right)\cos\left(\frac{A-B}{2}\right) - 1 + \frac{3}{2} = 0$$

$$\Rightarrow \quad 4\sin^2\frac{C}{2} - 4\sin\left(\frac{C}{2}\right)\cos\left(\frac{A-B}{2}\right) + 1 = 0 \quad \text{...(i)}$$

Now, Eq. (i) is quadratic in $(\sin C/2)$ and is real.

$$\therefore \quad D \geq 0$$

$$\Rightarrow \quad 16\cos^2\left(\frac{A-B}{2}\right) - 16 \geq 0 \Rightarrow \cos^2\left(\frac{A-B}{2}\right) - 1 \geq 0$$

$$\Rightarrow \quad \cos^2\left(\frac{A-B}{2}\right) \geq 1$$

which is only possible if $\cos^2\left(\frac{A-B}{2}\right) = 1$

$$\Rightarrow \quad \cos^2\left(\frac{A-B}{2}\right) = 1$$

$$\frac{A-B}{2} = 0$$

$$\Rightarrow \quad A = B \quad \text{...(ii)}$$

Similarly, we can show $B = C, C = A$. Hence, the triangle is equilateral.

Ex. 120. *If* $\frac{\tan 3A}{\tan A} = k$, *show that* $\frac{\sin 3A}{\sin A} = \frac{2k}{k-1}$ *and hence or otherwise prove that either* $k > 3$ *or* $k < \frac{1}{3}$.

Sol. $$\frac{\tan 3A}{\tan A} = \frac{3\tan A - \tan^3 A}{1 - 3\tan^2 A} \cdot \frac{1}{\tan A} = k$$

$$\Rightarrow \quad \frac{3 - \tan^2 A}{1 - 3\tan^2 A} = k$$

$$\Rightarrow \quad (3 - \tan^2 A) = k(1 - 3\tan^2 A)$$

$$\Rightarrow \quad (3k - 1)\tan^2 A = k - 3$$

$$\Rightarrow \quad \tan^2 A = \left(\frac{k-3}{3k-1}\right) \quad \text{...(i)}$$

Now, $$\frac{\sin 3A}{\sin A} = \frac{3\sin A - 4\sin^3 A}{\sin A} = 3 - 4\sin^2 A$$

$$\Rightarrow \quad 3 - 4\sin^2 A = 3 - \frac{4}{\operatorname{cosec}^2 A} = 3 - \frac{4}{1 + \cot^2 A}$$

$$= 3 - \frac{4}{1 + \frac{1}{\tan^2 A}}$$

$$\Rightarrow \quad 3 - \frac{4}{1 + \left(\frac{3k-1}{k-3}\right)} \quad \text{[using Eq. (i)]}$$

$$\Rightarrow \quad 3 - \frac{4(k-3)}{4(k-1)} = \frac{3k - 3 - k + 3}{k-1} = \frac{2k}{k-1} \quad \text{...(ii)}$$

Again from Eq. (i),

$$\tan^2 A = \frac{k-3}{3k-1} \quad [\tan A \neq 0 \text{ and } \tan^2 A > 0]$$

$$\Rightarrow \quad \frac{k-3}{3k-1} > 0, \text{ using number line rule.}$$

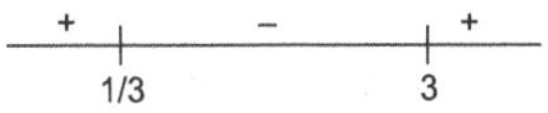

which shows $k < 1/3$ or $k > 3$

Ex. 121. *Let A, B, C be three angles such that* $A = \pi/4$ *and* $\tan B \tan C = p$. *Find all possible values of p such that A, B, C are the angles of triangles.*

Sol. Let us assume ΔABC.

$$\therefore \quad A + B + C = \pi$$

$$\Rightarrow \quad B + C = \pi - \frac{\pi}{4} = \frac{3\pi}{4} \quad \text{...(i) } [\because A = \pi/4, \text{ given}]$$

Also, $0 < B, C < 3\pi/4$

$$\Rightarrow \quad \tan B \tan C = p$$

$$\Rightarrow \quad \frac{\sin B \cdot \sin C}{\cos B \cdot \cos C} = \frac{p}{1}$$

$$\Rightarrow \quad \frac{\sin B \cdot \sin C + \cos B \cdot \cos C}{\cos B \cdot \cos C - \sin B \cdot \sin C} = \frac{1+p}{1-p}$$

$$\Rightarrow \quad \frac{\cos(B-C)}{\cos(B+C)} = \frac{1+p}{1-p}$$

$$\Rightarrow \quad \cos(B-C) = \left(\frac{1+p}{1-p}\right)\cos\left(\frac{3\pi}{4}\right)$$

[using Eq. (i), $B + C = 3\pi/4$]

$$\Rightarrow \quad \cos(B-C) = \frac{1+p}{\sqrt{2}(p-1)} \quad \text{...(ii)}$$

Since, B or C can vary from 0 to $3\pi/4$.

$$\therefore \quad 0 \leq (B - C) < 3\pi/4$$

$$\Rightarrow \quad -\frac{1}{\sqrt{2}} < \cos(B-C) \leq 1 \quad \text{...(iii)}$$

From Eqs. (ii) and (iii), we get

$$-\frac{1}{\sqrt{2}} < \frac{1+p}{\sqrt{2}(p-1)} \leq 1$$

$$\therefore \quad -\frac{1}{\sqrt{2}} < \frac{1+p}{\sqrt{2}(p-1)} \text{ and } \frac{1+p}{\sqrt{2}(p-1)} \leq 1$$

$$\Rightarrow \quad 0 < 1 + \frac{p+1}{p-1} \text{ and } \frac{(p+1) - \sqrt{2}(p-1)}{\sqrt{2}(p-1)} \leq 0$$

$$\Rightarrow \quad \frac{2p}{(p-1)} > 0 \text{ and } \frac{[p - (\sqrt{2}+1)^2]}{(p-1)} \geq 0$$

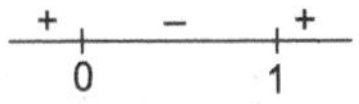

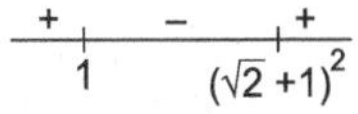

$$\Rightarrow \quad p < 0 \text{ or } p > 1 \text{ and } p < 1 \text{ or } p \geq (\sqrt{2}+1)^2$$

The combining above expressions;

$$p < 0 \text{ or } p \geq (\sqrt{2}+1)^2$$

i.e. $$p \in (-\infty, 0) \cup [(\sqrt{2}+1)^2, \infty).$$

Ex. 122. *If ABC is a triangle and* $\tan\frac{A}{2}, \tan\frac{B}{2}, \tan\frac{C}{2}$ *are in HP, then find minimum value of* $\cot\frac{B}{2}$.

Sol.
$$A+B+C=\pi$$
$$\Rightarrow \frac{A}{2}+\frac{B}{2}=\frac{\pi}{2}-\frac{C}{2}$$
$$\Rightarrow \cot\left(\frac{A}{2}+\frac{B}{2}\right)=\cot\left(\frac{\pi}{2}-\frac{C}{2}\right)$$
$$\Rightarrow \frac{\cot\frac{A}{2}\cdot\cot\frac{B}{2}-1}{\cot\frac{A}{2}+\cot\frac{B}{2}}=\tan\left(\frac{C}{2}\right)=\frac{1}{\cot\left(\frac{C}{2}\right)}$$
$$\Rightarrow \cot\frac{A}{2}\cdot\cot\frac{B}{2}\cdot\cot\frac{C}{2}=\cot\frac{A}{2}+\cot\frac{B}{2}+\cot\frac{C}{2} \quad \text{...(i)}$$

But $\tan\frac{A}{2}, \tan\frac{B}{2}, \tan\frac{C}{2}$ are in HP
$$\Rightarrow \cot\frac{A}{2}, \cot\frac{B}{2}, \cot\frac{C}{2} \text{ are in AP}$$
$$\therefore \cot\frac{A}{2}+\cot\frac{C}{2}=2\cot\frac{B}{2} \quad \text{...(ii)}$$

From Eqs. (i) and (ii), we get
$$\cot\frac{A}{2}\cdot\cot\frac{B}{2}\cot\frac{C}{2}=3\cot\frac{B}{2}$$
$$\Rightarrow \cot\frac{A}{2}\cdot\cot\frac{C}{2}=3 \quad \text{...(iii)}$$

As we know, AM ≥ GM
$$\Rightarrow \frac{\cot\frac{A}{2}+\cot\frac{C}{2}}{2}\geq\sqrt{\cot\frac{A}{2}\cdot\cot\frac{C}{2}}$$
$$\Rightarrow \frac{2\cot\frac{B}{2}}{2}\geq\sqrt{3} \quad \text{[\{using Eq. (iii)]}$$
$$\Rightarrow \cot\frac{B}{2}\geq\sqrt{3}$$

$\therefore$ Minimum value of $\cot\frac{B}{2}$ is $\sqrt{3}$.

Ex 123. *(i) If* $\tan A-\tan B=x$ *and* $\cot B-\cot A=y$.

Prove that $\cot(A-B)=\frac{1}{x}+\frac{1}{y}$.

(ii) If $2\cos A=x+\frac{1}{x}$, $2\cos B=y+\frac{1}{y}$, *then show that*

$2\cos(A-B)=\frac{x}{y}+\frac{y}{x}$.

Sol. (i) If $\cot B-\cot A=y \Rightarrow \frac{\tan A-\tan B}{\tan A\tan B}=y$
$$\therefore \frac{x}{y}=\tan A\tan B$$

Now,
$$\cot(A-B)=\frac{1}{\tan(A-B)}$$
$$=\frac{1+\tan A\tan B}{\tan A-\tan B}$$
$$=\frac{1+\frac{x}{y}}{x}=\frac{1}{x}+\frac{1}{y}=\text{RHS}$$

(ii) $2\cos A=x+\frac{1}{x}$, since $4\sin^2 A=4$
$$-4\cos^2 A=4-\left(x+\frac{1}{x}\right)^2$$
$$\therefore 4\sin^2 A=-\left[\left(x+\frac{1}{x}\right)^2-4\right]$$
or
$$4\sin^2 A=i^2\left[\left(x-\frac{1}{x}\right)^2\right]$$
$$\Rightarrow 2\sin A=i\left(x-\frac{1}{x}\right) \quad \text{...(i)}$$

Similarly,
$$2\cos B=y+\frac{1}{y}$$
$$\Rightarrow 2\sin B=i\left(y-\frac{1}{y}\right) \quad \text{...(ii)}$$

Now, $2\cos(A-B)=2[\cos A\cos B+\sin A\sin B]$
$$=\frac{2}{4}\left[\left(x+\frac{1}{x}\right)\left(y+\frac{1}{y}\right)+i^2\left(x-\frac{1}{x}\right)\left(y-\frac{1}{y}\right)\right]$$
$$=\frac{1}{2}\left[\left\{xy+\frac{1}{xy}+\frac{y}{x}+\frac{x}{y}\right\}-\left\{xy+\frac{1}{xy}-\frac{y}{x}-\frac{x}{y}\right\}\right]$$
$$=\frac{1}{2}\left[2\frac{y}{x}+2\frac{x}{y}\right]=\frac{x}{y}+\frac{y}{x}=\text{RHS}$$

Ex. 124. *If* $\tan\theta\tan\phi=\sqrt{\frac{a-b}{a+b}}$, *then prove that* $(a-b\cos2\theta)(a-b\cos2\phi)$ *is independent of* θ *and* ϕ.

Sol. Let us put,
$$\tan\theta=t_1 \text{ and } \tan\phi=t_2$$
$$\therefore t_1^2.t_2^2=\frac{a-b}{a+b} \quad \text{...(i)}$$
$$\left\{\text{given, } \tan\theta\tan\phi=\sqrt{\frac{a-b}{a+b}}\right\}$$

Also,
$$\cos2\theta=\frac{1-\tan^2\theta}{1+\tan^2\theta}=\frac{1-t_1^2}{1+t_1^2} \quad \text{...(ii)}$$
$$\cos2\phi=\frac{1-\tan^2\phi}{1+\tan^2\phi}=\frac{1-t_2^2}{1+t_2^2} \quad \text{...(iii)}$$

Now, $(a-b\cos2\theta).(a-b\cos2\phi)$

$$= \left\{a - b\left(\frac{1-t_1^2}{1+t_1^2}\right)\right\} \cdot \left\{a - b\left(\frac{1-t_2^2}{1+t_2^2}\right)\right\} \text{ [using Eqs. (ii) and (iii)]}$$

$$= \left\{\frac{(a-b)+(a+b)t_1^2}{1+t_1^2}\right\} \cdot \left\{\frac{(a-b)+(a+b)t_2^2}{1+t_2^2}\right\}$$

$$= \frac{(a+b)}{(1+t_1^2)}\left[\frac{a-b}{a+b} + t_1^2\right] \cdot \frac{(a+b)}{(1+t_2^2)}\left[\frac{a-b}{a+b} + t_2^2\right]$$

$$= \frac{(a+b)}{(1+t_1^2)}[t_1^2 t_2^2 + t_1^2] \cdot \frac{(a+b)}{(1+t_2^2)}[t_1^2 t_2^2 + t_2^2] \quad \text{[using Eq. (i)]}$$

$$= (a+b)^2 \cdot (t_1^2 . t_2^2) = (a+b)^2 \cdot \frac{(a-b)}{(a+b)} = (a^2 - b^2)$$

So, $(a - b\cos 2\theta)(a - b\cos 2\phi) = a^2 - b^2$, which is independent of θ and ϕ.

Ex. 125. *Find all possible real values of x and y satisfying.*

$$\sin^2 x + 4\sin^2 y - \sin x - 2\sin y - 2\sin x \sin y + 1 = 0,$$
$$\forall\ x, y \in [0, \pi/2]$$

Sol. Given, equation can be rewritten as,

$$\sin^2 x - \sin x(1 + 2\sin y) + (4\sin^2 y - 2\sin y + 1) = 0$$

$$\Rightarrow \sin x = \frac{(1+2\sin y) \pm \sqrt{(1+2\sin y)^2 - 4(4\sin^2 y - 2\sin y + 1)}}{2}$$

$$= \frac{(1+2\sin y) \pm \sqrt{-3 - 12\sin^2 y + 12\sin y}}{2}$$

$$= \frac{(1+2\sin y) \pm \sqrt{-3(2\sin y - 1)^2}}{2} \quad \text{...(i)}$$

Since, $\sin x$ is real.

$\therefore$ From equation (i) is real only if,

$$2\sin y - 1 = 0 \text{ or } \sin y = \frac{1}{2} \text{ and } \sin x = \frac{1+1}{2} = 1.$$

$$\Rightarrow \quad y = \frac{\pi}{6} \text{ and } x = \frac{\pi}{2} \text{ as } x, y \in [0, \pi/2].$$

Ex. 126. *Find the roots of the following cubic equations* $2x^3 - 3x^2\cos(A-B) - 2x\cos^2(A+B) + \sin 2A \sin 2B\cos(A-B) = 0$.

Sol. We know,

$$\sin 2A \sin 2B = \frac{1}{2}[\cos(2A - 2B) - \cos(2A - 2B)]$$

$$= \frac{1}{2}[2\cos^2(A-B) - 1 - 2\cos^2(A+B) + 1]$$

$$= \cos^2(A-B) - \cos^2(A+B)$$

$$\therefore \quad \sin 2A . \sin 2B = \cos^2(A-B) - \cos^2(A+B) \quad \text{...(i)}$$

Now, $2x^3 - 3x^2\cos(A-B) - 2x\cos^2(A+B) + \sin 2A . \sin 2B . \cos(A-B) = 0$

$$\Rightarrow 2x^3 - 3x^2\cos(A-B) - 2x\cos^2(A+B) + \cos^3(A-B) - \cos^2(A+B).\cos(A-B) = 0$$

By inspection, we find that $x = -\frac{1}{2}\cos(A-B)$ because

$$\left(-\frac{1}{4} - \frac{3}{4} + 1\right)\cos^3(A-B) + \cos^2(A+B)\cos(A-B) - \cos^2(A+B)\cos(A-B) = 0$$

Hence, $2x + \cos(A-B)$ is factor of the given equation which when divided by it, given the other factor as,

$$x^2 - 2x\cos(A-B) + \cos^2(A-B) - \cos^2(A+B) = 0$$

$$\text{So, } x = \frac{2\cos(A-B) \pm \sqrt{4\cos^2(A-B) - 4\cos^2(A-B) + 4\cos^2(A+B)}}{2}$$

$$x = \frac{2\cos(A-B) \pm 2\cos(A+B)}{2}$$

$x = \cos(A-B) + \cos(A+B)$ or $\cos(A-B) - \cos(A+B)$

Hence, the roots are,

$$2\cos A\cos B,\ 2\sin A\sin B \text{ and } -\frac{1}{2}\cos(A-B).$$

Ex. 127. *If* $m^2 + m'^2 + 2mm'\cos\theta = 1.$

$$n^2 + n'^2 + 2nn'\cos\theta = 1$$

and $mn + m'n' + (mn' + m'n)\cos\theta = 0$, *then prove that* $m^2 + n^2 = \operatorname{cosec}^2\theta$.

Sol. $m^2 + m'^2 + 2mm'\cos\theta = 1$

or $\quad (m^2\cos^2\theta + m^2\sin^2\theta) + m'^2 + 2mm'\cos\theta = 1$

or $\quad m^2\cos^2\theta + 2mm'\cos\theta + m'^2 = 1 - m^2\sin^2\theta$

or $\quad (m\cos\theta + m')^2 = 1 - m^2\sin^2\theta \quad$ ...(i)

Similarly, $\quad n^2 + n'^2 + 2nn'\cos\theta = 1$

$\Rightarrow \quad (n\cos\theta + n')^2 = 1 - n^2\sin^2\theta \quad$ ...(ii)

Finally, $mn + m'n' + (mn' + m'n)\cos\theta = 0$

$$\Rightarrow (mn\cos^2\theta + mn\sin^2\theta) + m'n' + mn'\cos\theta + m'n\cos\theta = 0$$

$$\Rightarrow mn\cos^2\theta + m'n\cos\theta + m'n' + mn'\cos\theta = -mn\sin^2\theta$$

$$\Rightarrow n\cos\theta(m\cos\theta + m') + n'(m' + m\cos\theta) = -mn\sin^2\theta$$

$$\Rightarrow (m\cos\theta + m')(n\cos\theta + n') = -mn\sin^2\theta$$

$$\text{or} \quad (m\cos\theta + m')^2(n\cos\theta + n')^2 = m^2n^2\sin^4\theta$$

$$\Rightarrow (1 - m^2\sin^2\theta)(1 - n^2\sin^2\theta) = m^2n^2\sin^4\theta.$$

[using Eqs. (i) and (ii)]

$$\Rightarrow 1 - (m^2 + n^2)\sin^2\theta + m^2n^2\sin^4\theta = m^2n^2\sin^4\theta$$

$$\Rightarrow (m^2 + n^2)\sin^2\theta = 1$$

$$\Rightarrow m^2 + n^2 = \operatorname{cosec}^2\theta$$

Ex 128. *Prove that from the equality* $\dfrac{\sin^4\alpha}{a}+\dfrac{\cos^4\alpha}{b}=\dfrac{1}{a+b}$ *follows the relation,*

$$\frac{\sin^8\alpha}{a^3}+\frac{\cos^8\alpha}{b^3}=\frac{1}{(a+b)^3} \text{ and}$$

$$\frac{\sin^{4n}x}{a^{2n-1}}+\frac{\cos^{4n}x}{b^{2n-1}}=\frac{1}{(a+b)^{2n-1}}.$$

Sol. Given condition can be rewritten as,

$$b(\sin^2\alpha)^2+a\cos^4\alpha=\frac{ab}{a+b}$$

$$\Rightarrow \quad b(1-\cos^2\alpha)^2+a\cos^4\alpha=\frac{ab}{a+b}$$

$$\Rightarrow \quad b(\cos^4\alpha-2\cos^2\alpha+1)+a\cos^4\alpha=\frac{ab}{a+b}$$

$$\Rightarrow \quad (a+b)^2\cos^4\alpha-2b(a+b)\cos^2\alpha+b(a+b)=ab$$

$$(a+b)^2\cos^4\alpha-2b(a+b)\cos^2\alpha+b^2=0$$

$$\Rightarrow \quad [(a+b)\cos^2\alpha-b]^2=0$$

$$\Rightarrow \quad \cos^2\alpha=\frac{b}{a+b} \Rightarrow \sin^2\alpha=\frac{a}{a+b} \quad \text{...(i)}$$

$$\therefore \quad \frac{\sin^8\alpha}{a^3}+\frac{\cos^8\alpha}{b^3}=\frac{a^4}{a^3(a+b)^4}+\frac{b^4}{b^3(a+b)^4}$$

$$=\frac{a}{(a+b)^4}+\frac{b}{(a+b)^4}$$

$$=\frac{a+b}{(a+b)^4}=\frac{1}{(a+b)^3}$$

$$\therefore \quad \frac{\sin^8\alpha}{a^3}+\frac{\cos^8\alpha}{b^3}=\frac{1}{(a+b)^3}$$

Now, $$\frac{\sin^{4n}x}{a^{2n-1}}+\frac{\cos^{4n}x}{b^{2n-1}}=\frac{a^{2n}}{a^{2n-1}(a+b)^{2n}}+\frac{b^{2n}}{b^{2n-1}(a+b)^{2n}}$$

$$=\frac{a+b}{(a+b)^{2n}}=\frac{1}{(a+b)^{2n-1}}.$$

Ex 129. *If* $a_{r+1}=\sqrt{\dfrac{1}{2}(1+a_r)}$, *the prove that*

$$\cos\left(\frac{\sqrt{1-a_0^2}}{a_1\cdot a_2\cdot a_3\ldots\text{to}\,\infty}\right)=a_0.$$

Sol. Let $a_0=\cos\theta$, then $a_{r+1}=\sqrt{\dfrac{1}{2}(1+a_r)}$ gives

$$a_1=\sqrt{\frac{1}{2}(1+a_0)}=\sqrt{\frac{1}{2}(1+\cos\theta)}=\cos\frac{\theta}{2}$$

$$a_2=\sqrt{\frac{1}{2}(1+a_1)}=\sqrt{\frac{1}{2}\left(1+\cos\frac{\theta}{2}\right)}=\cos\frac{\theta}{2^2}$$

$$a_3=\sqrt{\frac{1}{2}(1+a_2)}=\sqrt{\frac{1}{2}\left(1+\cos\frac{\theta}{2^2}\right)}=\cos\frac{\theta}{2^3}, \text{... etc.}$$

$$\therefore \quad a_1\cdot a_2\cdot a_3\ldots a_n$$

$$\Rightarrow \quad \cos\frac{\theta}{2}\cdot\cos\frac{\theta}{2^2}\cdot\cos\frac{\theta}{2^3}\ldots\cos\frac{\theta}{2^n}$$

$$\Rightarrow \quad \frac{\cos\frac{\theta}{2}\cdot\cos\frac{\theta}{2^2}\cdot\cos\frac{\theta}{2^3}\ldots\cos\frac{\theta}{2^n}\cdot 2\sin\frac{\theta}{2^n}}{2\sin\frac{\theta}{2^n}}$$

$$\Rightarrow \quad \frac{\cos\frac{\theta}{2}\cdot\cos\frac{\theta}{2^2}\ldots\cos\frac{\theta}{2^{n-1}}\cdot\sin\frac{\theta}{2^{n-1}}}{2\sin\frac{\theta}{2^n}}$$

$$[\because 2\sin\alpha.\cos\alpha=\sin 2\alpha]$$

$$\Rightarrow \quad \frac{\cos\frac{\theta}{2}\cdot\cos\frac{\theta}{2^2}\ldots\cos\frac{\theta}{2^{n-2}}\cdot 2\sin\frac{\theta}{2^{n-2}}}{2^2\sin\frac{\theta}{2^n}}$$

$$=\ldots\ldots\ldots\ldots=\frac{\cos\frac{\theta}{2^{n-(n-1)}}\cdot\sin\frac{\theta}{2^{n-(n-1)}}}{2^{n-1}\cdot\sin\frac{\theta}{2^n}}$$

$$=\frac{\cos\frac{\theta}{2}\cdot\sin\frac{\theta}{2}}{2^{n-1}\cdot\sin\frac{\theta}{2^n}}=\frac{\sin\theta}{2^n.\sin\frac{\theta}{2^n}}$$

$$\therefore \quad a_1.a_2\ldots\text{to}\,\infty=\lim_{n\to\infty}\frac{\sin\theta}{2^n.\sin\frac{\theta}{2^n}}=\lim_{n\to\infty}\frac{\sin\theta}{\theta\cdot\frac{\sin(\theta/2^n)}{(\theta/2^n)}}=\frac{\sin\theta}{\theta}$$

$$\therefore \quad \frac{\sqrt{1-a_0^2}}{a_1a_2a_3\ldots\text{to}\,\infty}=\frac{\sqrt{1-\cos^2\theta}}{\frac{\sin\theta}{\theta}}=\theta$$

$$\therefore \quad \cos\left(\frac{\sqrt{1-a_0^2}}{a_1a_2a_3\ldots\text{to}\,\infty}\right)=\cos\theta=a_0.$$

Ex. 130. *Evaluate* $\sum_{r=2}^{n}\sin r\alpha$, *where* $(n+2)\alpha=2\pi$ *(without using formula.)*

Sol. Let $S=\sum_{r=2}^{n}\sin r\alpha=\sin 2\alpha+\sin 3\alpha+\sin 4\alpha+\ldots+\sin n\alpha$

$$\therefore \quad 2\sin\frac{\alpha}{2}\cdot S=2\sin\alpha/2\sin 2\alpha+2\sin\frac{\alpha}{2}\cdot\sin 3\alpha+2\sin\frac{\alpha}{2}\cdot\sin 4\alpha+\ldots+2\sin\alpha/2.\sin n\alpha$$

$$=\left\{\cos\frac{3\alpha}{2}-\cos\frac{5\alpha}{2}\right\}+\left\{\cos\frac{5\alpha}{2}-\cos\frac{7\alpha}{2}\right\}+\ldots+\left\{\cos\left(n-\frac{1}{2}\right)\alpha-\cos\left(n+\frac{1}{2}\right)\alpha\right\}$$

$$=\cos\frac{3\alpha}{2}-\cos\left(n+\frac{1}{2}\right)\alpha$$

$$= 2\sin\left(\frac{\frac{3\alpha}{2}+\left(n+\frac{1}{2}\right)\alpha}{2}\right)\sin\left(\frac{\left(n+\frac{1}{2}\right)\alpha-\frac{3\alpha}{2}}{2}\right)$$

$$2\sin\frac{\alpha}{2}\cdot S = 2\sin\frac{(n+2)\alpha}{2}\cdot\sin\frac{(n-1)\alpha}{2}$$

$$\Rightarrow \quad S = \frac{\sin\frac{(n-1)\alpha}{2}}{\sin\frac{\alpha}{2}}\cdot\sin\frac{(n+2)\alpha}{2}$$

$$= \frac{\sin\frac{(n-1)\alpha}{2}}{\sin\frac{\alpha}{2}}\cdot\sin\frac{2\pi}{2} = 0$$

Ex. 131. *Sum the series* $\sqrt{1+\cos\alpha}+\sqrt{1+\cos 2\alpha}+\sqrt{1+\cos 3\alpha}+\dots$ *to n terms.*

Sol. We have,

$$\sqrt{1+\cos\alpha}+\sqrt{1+\cos 2\alpha}+\sqrt{1+\cos 3\alpha}+\dots+\sqrt{1+\cos n\alpha}$$

$$= \sqrt{2\cos^2\alpha/2}+\sqrt{2\cos^2\alpha}+\sqrt{2\cos^2\frac{3\alpha}{2}}+\dots \text{ to } n \text{ terms}$$

$$= \sqrt{2}\left\{\cos\frac{\alpha}{2}+\cos\frac{2\alpha}{2}+\cos\frac{3\alpha}{2}+\dots+\text{ to } n \text{ terms}\right\}$$

$$= \sqrt{2}\cdot\frac{\sin\frac{n\alpha}{4}}{\sin\frac{\alpha}{4}}\cdot\cos\left\{\frac{\alpha}{2}+(n-1)\frac{\alpha}{4}\right\} \quad \{\text{using formula}\}$$

$$= \sqrt{2}\frac{\sin\frac{n\alpha}{4}}{\sin\frac{\alpha}{4}}.\cos\left\{(n+1)\frac{\alpha}{4}\right\}$$

Ex. 132. *If* $A+B+C=\pi$, *show that*

$$\cot A+\cot B+\cot C-\operatorname{cosec} A\operatorname{cosec} B\cdot\operatorname{cosec} C = \cot A\cdot\cot B\cdot\cot C$$

Sol. $$\text{LHS} = \frac{\cos A}{\sin A}+\frac{\cos A}{\sin B}+\frac{\cos C}{\sin C}-\frac{1}{\sin A\cdot\sin B\cdot\sin C}$$

$$= \frac{\cos A\cdot\sin B\sin C+\cos B\sin A\sin C+\cos C\sin A\sin B-1}{\sin A\sin B\sin C}$$

$$= \frac{\sin C(\cos A\sin B+\cos B\sin A)+\cos C\sin A\sin B-1}{\sin A\sin B\sin C}$$

$$= \frac{\sin C\sin(A+B)+\cos C\sin A\sin B-1}{\sin A\sin B\sin C}$$

$$= \frac{\sin^2 C+\cos C\sin A\sin B-1}{\sin A\sin B\sin C}$$

[using $\sin(A+B)=\sin(\pi-C)=\sin C$]

$$= \frac{\cos C.\sin A\sin B-\cos^2 C}{\sin A\sin B\sin C}$$

$$= \frac{\cos C\{\sin A\sin B-\cos C\}}{\sin A\sin B\sin C}$$

[$\because \cos C=\cos(\pi-(A+B))=-\cos(A+B)$]

$$= \frac{\cos C\{\sin A\sin B+\cos(A+B)\}}{\sin A\sin B\sin C}$$

$$= \frac{\cos C\{\sin A\sin B+\cos A\cos B-\sin A\sin B\}}{\sin A\sin B\sin C}$$

$$= \frac{\cos A\cos B\cos C}{\sin A\sin B\sin C} = \cot A\cdot\cot B\cdot\cot C = \text{RHS}$$

Ex. 133. *In* ΔABC, *if* $\cot\theta=\cot A+\cot B+\cot C$, *prove that* $\sin^3\theta=\sin(A-\theta)\sin(B-\theta)\sin(C-\theta)$.

Sol. We have, $\cot\theta=\cot A+\cot B+\cot C$

$$\Rightarrow \quad \cot(\theta)-\cot(A)=\cot B+\cot C$$

$$\Rightarrow \quad \frac{\cos\theta}{\sin\theta}-\frac{\cos A}{\sin A}=\frac{\cos B}{\sin B}+\frac{\cos C}{\sin C}$$

$$\Rightarrow \quad \frac{\cos\theta\sin A-\cos A\sin\theta}{\sin\theta\sin A}=\frac{\cos B\sin C+\sin B\cos C}{\sin B\sin C}$$

$$\Rightarrow \quad \frac{\sin(A-\theta)}{\sin A\sin\theta}=\frac{\sin(B+C)}{\sin B\sin C}$$

$$\Rightarrow \quad \sin(A-\theta)=\frac{\sin^2 A\sin\theta}{\sin B\sin C} \quad \dots\text{(i)}$$

Similarly, $$\sin(B-\theta)=\frac{\sin^2 B\sin\theta}{\sin A\sin C} \quad \dots\text{(ii)}$$

and $$\sin(C-\theta)=\frac{\sin^2 C\sin\theta}{\sin A\sin B} \quad \dots\text{(iii)}$$

Multiplying Eqs. (i), (ii) and (iii), we get

$$\sin(A-\theta)\sin(B-\theta)\sin(C-\theta)=\sin^3\theta.$$

Ex. 134. *If A, B, C and D are angles of a quadrilateral and* $\sin\frac{A}{2}\sin\frac{B}{2}\sin\frac{C}{2}\sin\frac{D}{2}=\frac{1}{4}$, *then prove that* $A=B=C=D=\pi/2$.

Sol. Now, $$\left(2\sin\frac{A}{2}\cdot\sin\frac{B}{2}\right)\cdot\left(2\sin\frac{C}{2}\cdot\sin\frac{D}{2}\right)=1$$

$$\Rightarrow \left\{\cos\left(\frac{A-B}{2}\right)-\cos\left(\frac{A+B}{2}\right)\right\}\left\{\cos\left(\frac{C-D}{2}\right)-\cos\left(\frac{C+D}{2}\right)\right\}=1$$

Since, $A+B=2\pi-(C+D)$, the above equation becomes,

$$\left(\cos\frac{A-B}{2}-\cos\frac{A+B}{2}\right)\left(\cos\frac{C-D}{2}+\cos\frac{A+B}{2}\right)=1$$

$$\Rightarrow \cos^2\left(\frac{A+B}{2}\right)-\cos\left(\frac{A+B}{2}\right)\left\{\cos\left(\frac{A-B}{2}\right)-\cos\left(\frac{C-D}{2}\right)\right\}+1-\cos\left(\frac{A-B}{2}\right)\cos\left(\frac{C-D}{2}\right)=0.$$

This is a quadratic equation in $\cos\left(\frac{A+B}{2}\right)$ which has real roots.

$$\Rightarrow \left\{\cos\left(\frac{A-B}{2}\right)-\cos\left(\frac{C-D}{2}\right)\right\}^2 -$$

$$4\left\{1-\cos\left(\frac{A-B}{2}\right)\cdot\cos\left(\frac{C-D}{2}\right)\right\}\geq 0$$

$$\Rightarrow \left(\cos\frac{A-B}{2}+\cos\frac{C-D}{2}\right)^2 \geq 4$$

$$\Rightarrow \cos\frac{A-B}{2}+\cos\frac{C-D}{2}\geq 2$$

Now, both $\cos\frac{A-B}{2}$ and $\cos\frac{C-D}{2}\leq 1$

$$\Rightarrow \cos\frac{A-B}{2}=1=\cos\frac{C-D}{2}$$

$$\Rightarrow \frac{A-B}{2}=0=\frac{C-D}{2}$$

$$\Rightarrow A=B, C=D$$

Similarly, $A=C, B=D$

$$\Rightarrow A=B=C=D=\pi/2$$

Ex. 135. *If α, β are two different values of θ which satisfy is $bc\cos\theta\cos\phi+ac\sin\theta\sin\phi=ab$, then prove that $(b^2+c^2-a^2)\cos\alpha\cos\beta+ac\sin\alpha\sin\beta=a^2+b^2-c^2$.*

Sol. We have, $bc\cos\theta\cos\phi=ab-ac\sin\theta\sin\phi$.

$$\Rightarrow b^2c^2\cos^2\theta\cos^2\phi=a^2b^2+a^2c^2\sin^2\theta\sin^2\phi-2a^2bc\sin\theta\sin\phi$$

$$\Rightarrow (a^2c^2\sin^2\phi+b^2c^2\cos^2\phi)\sin^2\theta-2a^2bc\sin\theta\sin\phi+a^2b^2-b^2c^2\cos^2\phi=0$$

$$\Rightarrow \sin\alpha\sin\beta=\frac{a^2b^2-b^2c^2\cos^2\phi}{a^2c^2\sin^2\phi+b^2c^2\cos^2\phi} \quad \text{...(i)}$$

Similarly, $ac\sin\theta\sin\phi=ab-bc\cos\phi\cos\theta$

$$\Rightarrow a^2c^2\sin^2\theta\sin^2\phi=a^2b^2+b^2c^2\cos^2\theta\cos^2\phi-2ab^2c\cos\theta\cos\phi$$

$$\therefore \cos\alpha\cos\beta=\frac{a^2b^2-a^2c^2\sin^2\phi}{a^2c^2\sin^2\phi+b^2c^2\cos^2\phi} \quad \text{...(ii)}$$

On substituting the value from Eqs. (i) and (ii) in

$(b^2+c^2-a^2)\cos\alpha\cos\beta+ac\sin\alpha\sin\beta$, we get

$$\Rightarrow \frac{(b^2+c^2-a^2)(a^2b^2-a^2c^2\sin^2\phi)+ac(a^2b^2-b^2c^2\cos^2\phi)}{a^2c^2\sin^2\phi+b^2c^2\cos^2\phi}$$

$$\Rightarrow (a^2+b^2-c^2)=\text{RHS}$$

Ex. 136. *Find all number pairs x, y that satisfy the equaion;*

$$\tan^4 x+\tan^4 y+2\cot^2 x\cot^2 y=3+\sin^2(x+y);\ \forall\ x, y\in\left[0, \frac{\pi}{2}\right]$$

Sol. We know, $a^4+b^4\geq 2a^2b^2$ {AM ≥ GM}

$$\therefore \tan^4 x+\tan^4 y\geq 2\tan^2 x\tan^2 y \quad \text{...(i)}$$

Equality occurring only when $\tan^2 x=\tan^2 y=1$.

Also, $\tan^2 x\tan^2 y+\cot^2 x\cot^2 y\geq 2$ ...(ii)

Since, $a+\frac{1}{a}\geq 2$ and equality occurring only when

$a=1$, i.e. $\tan^2 x\tan^2 y=1$

From Eqs. (i) and (ii);

$$\tan^4 x+\tan^4 y+2\cot^2 x\cot^2 y\geq 4 \quad \text{...(iii)}$$

Also, $\text{RHS}=3+\sin^2(x+y)\leq 4$ ...(iv)

From Eqs. (iii) and (iv),

$$\text{LHS}=\text{RHS}=4$$

$$\Rightarrow \tan^2 x=\tan^2 y=\tan^2 x\tan^2 y=1$$

$$\Rightarrow \tan x=\tan y=\pm 1$$

$$\Rightarrow \tan x=\tan y=1 \quad \{\text{as } x, y\in[0, \pi/2]\}$$

$$\therefore x=y=\pi/4$$

Only one solution i.e. $(x=\pi/4, y=\pi/4)$.

Ex. 137. *Prove that $\tan\frac{3\pi}{11}+4\sin\frac{2\pi}{11}=\sqrt{11}$.*

Sol. Let $y=\tan\frac{3\pi}{11}+4\sin\frac{2\pi}{11}=\frac{1}{\cos\frac{3\pi}{11}}\left(\sin\frac{3\pi}{11}+4\sin\frac{2\pi}{11}\cos\frac{3\pi}{11}\right)$

$$y^2.\cos^2\frac{3\pi}{11}=\sin^2\frac{3\pi}{11}+16\sin^2\frac{2\pi}{11}\cdot\cos^2\frac{3\pi}{11}+8\sin\frac{2\pi}{11}\cdot\cos\frac{3\pi}{11}\cdot\sin\frac{3\pi}{11}$$

$$\Rightarrow 2\cos^2\frac{3\pi}{11}y^2=2\sin^2\frac{3\pi}{11}+32\sin^2\frac{2\pi}{11}\cdot\cos^2\frac{3\pi}{11}+8\sin\frac{2\pi}{11}\cdot\sin\frac{6\pi}{11}$$

$$=\left(1-\cos\frac{6\pi}{11}\right)+8\left(1-\cos\frac{4\pi}{11}\right)\cdot\left(1+\cos\frac{6\pi}{11}\right)+4\left(\cos\frac{4\pi}{11}-\cos\frac{8\pi}{11}\right)$$

$$=9+7\cos\frac{6\pi}{11}-4\cos\frac{4\pi}{11}-8\cos\frac{4\pi}{11}\cdot\cos\frac{6\pi}{11}-4\cos\frac{8\pi}{11}$$

$$=9+7\cos\frac{6\pi}{11}-4\cos\frac{4\pi}{11}-4\left(\cos\frac{10\pi}{11}+\cos\frac{2\pi}{11}\right)-4\cos\frac{8\pi}{11}$$

$$=9+11\cos\frac{6\pi}{11}-4\left(\cos\frac{2\pi}{11}+\cos\frac{4\pi}{11}+\cos\frac{6\pi}{11}+\cos\frac{8\pi}{11}+\cos\frac{10\pi}{11}\right)$$

$$=9+11\cos\frac{6\pi}{11}-4\left(\frac{\cos\left(\frac{2\pi}{11}+2\cdot\frac{2\pi}{11}\right)\cdot\sin\left(\frac{5\pi}{11}\right)}{\sin\pi/11}\right)$$

$$=9+11\cos\frac{6\pi}{11}-\frac{4\cos\frac{6\pi}{11}\cdot\sin\frac{5\pi}{11}}{\sin\frac{\pi}{11}}$$

$$= 9 + 11\cos\frac{6\pi}{11} - \frac{2\sin\frac{12\pi}{11}}{\sin\frac{\pi}{11}}$$

$$= 9 + 11\cos\frac{6\pi}{11} + 2 = 11\left(1 + \cos\frac{6\pi}{11}\right)$$

$$2y^2\cdot\left(\cos^2\frac{3\pi}{11}\right) = 22\cos^2\frac{3\pi}{11} \Rightarrow y^2 = 11$$

$$\Rightarrow \quad y = \sqrt{11} \qquad [\text{as } y > 0]$$

Ex. 138. *Prove that* $\sin\frac{2\pi}{7} + \sin\frac{4\pi}{7} + \sin\frac{8\pi}{7} = \frac{\sqrt{7}}{2}$.

Sol. Put, $7\theta = 2n\pi$, where n is any integer, then

$$4\theta = 2n\pi - 3\theta$$

$$\Rightarrow \quad \sin(4\theta) = \sin(2n\pi - 3\theta) = -\sin 3\theta \qquad \text{...(i)}$$

This means $\sin\theta$ takes the values; $0, \pm\sin\frac{2\pi}{7}, \pm\sin\frac{4\pi}{7}$ and $\pm\sin\frac{8\pi}{7}$.

Since, $\sin\frac{6\pi}{7} = -\sin\left(\frac{8\pi}{7}\right)$

From Eq. (i), we now get $2\sin 2\theta.\cos 2\theta = 4\sin^3\theta - 3\sin\theta$

$$\Rightarrow \quad 4\sin\theta\cos\theta(1 - 2\sin^2\theta) = \sin\theta(4\sin^2\theta - 3)$$

$$\Rightarrow \quad 4\cos\theta(1 - 2\sin^2\theta) = 4\sin^2\theta - 3$$

$$\Rightarrow \quad 16\cos^2\theta(1 - 2\sin^2\theta)^2 = (4\sin^2\theta - 3)^2$$

$$\Rightarrow 16(1 - \sin^2\theta)(1 - 4\sin^2\theta + 4\sin^4\theta) = 16\sin^4\theta - 24\sin^2\theta + 9$$

$$\Rightarrow \quad 64\sin^6\theta - 112\sin^4\theta + 56\sin^2\theta - 7 = 0$$

This is a cubic in $\sin^2\theta$ with the roots,

$$\sin^2\left(\frac{2\pi}{7}\right), \sin^2\left(\frac{4\pi}{7}\right), \sin^2\left(\frac{8\pi}{7}\right)$$

$\therefore$ Sum of the roots is

$$\sin^2\left(\frac{2\pi}{7}\right) + \sin^2\left(\frac{4\pi}{7}\right) + \sin^2\left(\frac{8\pi}{7}\right) = \frac{112}{64} = \frac{7}{4}$$

We already proved

$$\sin\frac{2\pi}{7}\cdot\sin\frac{4\pi}{7} + \sin\frac{4\pi}{7}\cdot\sin\frac{8\pi}{7} + \sin\frac{8\pi}{7}\sin\frac{\pi}{7} = 0$$

So, $$\left(\sin\frac{2\pi}{7} + \sin\frac{4\pi}{7} + \sin\frac{8\pi}{7}\right)^2 = \frac{7}{4}$$

$$\Rightarrow \quad \sin\frac{2\pi}{7} + \sin\frac{4\pi}{7} + \sin\frac{8\pi}{7} = \frac{\sqrt{7}}{2}$$

Alternate Method

$$x^7 - 1 = 0$$

[assuming x as the seventh root of unity]

$$x^7 = 1 + 0.i = \cos(2k\pi) + i\sin(2k\pi)$$

$$x = \left(\cos\frac{2k\pi}{7} + i\sin\frac{2k\pi}{7}\right)$$

$$\Rightarrow \quad x = e^{i2k\pi/7} \qquad [\text{where, } k = 0, 1, 2, 3, 4, 5, 6]$$

$$\Rightarrow \quad \sum_{k=0}^{6} e^{i2k\pi/7} = 0$$

$$\Rightarrow \quad 1 + \sum_{k=1}^{6} e^{i2k\pi/7} = 0$$

$$\Rightarrow \quad 1 + \sum_{k=1}^{3}(e^{i2k\pi/7} + e^{-2k\pi/7}) = 0$$

$$\Rightarrow \quad 1 + \sum_{k=1}^{3} 2.\cos\frac{2k\pi}{7} = 0$$

$$\Rightarrow \quad 1 + 2\sum_{k=1}^{3}\left(1 - 2\sin^2\frac{k\pi}{7}\right) = 0$$

$$\Rightarrow \quad 1 + 2\left[3 - 2\left(\sin^2\frac{\pi}{7} + \sin^2\frac{2\pi}{7} + \sin^2\frac{3\pi}{7}\right)\right] = 0$$

$$\Rightarrow \quad \sin^2\frac{\pi}{7} + \sin^2\frac{2\pi}{7} + \sin^2\frac{3\pi}{7} = \frac{7}{4}$$

$$\Rightarrow \quad \sin^2\frac{8\pi}{7} + \sin^2\frac{2\pi}{7} + \sin^2\frac{4\pi}{7} = \frac{7}{4}$$

$$\Rightarrow \quad \sin^2\frac{2\pi}{7} + \sin^2\frac{4\pi}{7} + \sin^2\frac{8\pi}{7} = \frac{7}{4} \qquad \text{...(i)}$$

and $\sin\frac{2\pi}{7}\cdot\sin\frac{4\pi}{7} + \sin\frac{4\pi}{7}\cdot\sin\frac{8\pi}{7} + \sin\frac{8\pi}{7}\cdot\sin\frac{2\pi}{7}$

$$= \frac{1}{2}\left[\cos\frac{2\pi}{7} - \cos\frac{6\pi}{7} + \cos\frac{4\pi}{7} - \cos\frac{12\pi}{7} + \cos\frac{6\pi}{7} - \cos\frac{10\pi}{7}\right]$$

$$= \frac{1}{2}\left[\cos\frac{2\pi}{7} + \cos\frac{4\pi}{7} - \cos\left(2\pi - \frac{2\pi}{7}\right) - \cos\left(2\pi - \frac{4\pi}{7}\right)\right]$$

$$= \frac{1}{2}\left[\cos\frac{2\pi}{7} + \cos\frac{4\pi}{7} - \cos\frac{2\pi}{7} - \cos\frac{4\pi}{7}\right] = 0 \qquad \text{...(ii)}$$

From Eqs. (i) and (ii), we get

$$\left(\sin\frac{2\pi}{7} + \sin\frac{4\pi}{7} + \sin\frac{8\pi}{7}\right)^2 = \sin^2\frac{2\pi}{7} + \sin^2\frac{4\pi}{7} + \sin^2\frac{8\pi}{7} = \frac{7}{4}$$

$$\Rightarrow \quad \sin\frac{2\pi}{7} + \sin\frac{4\pi}{7} + \sin\frac{8\pi}{7} = \frac{\sqrt{7}}{2}$$

Ex. 139. *In a ΔABC, $\tan A + \tan B + \tan C = k$, then find the interval in which k should lie so that*

(A) there exists exactly one isosceles triangle ABC

(B) there exists exactly two isosceles triangle ABC

(C) can there exist three non-similar isosceles triangles for any real value of k.

Sol. Let $A = B$, then $2A + C = 180^\circ$

and $\quad 2\tan A + \tan C = k \qquad \text{...(i)}$

Now, $\quad 2A + C = 180^\circ$

$\Rightarrow \quad \tan 2A = -\tan C$

Also, $\quad 2\tan A + \tan C = k$

$\Rightarrow \quad 2\tan A + \tan(180 - 2A) = k$

$\Rightarrow \quad 2\tan A - \tan 2A = k$

$\Rightarrow \quad 2\tan A - \dfrac{2\tan A}{1 - \tan^2 A} = k$

$\Rightarrow \quad 2\tan A(1 - \tan^2 A - 1) = k - k\tan^2 A$

$\Rightarrow \quad 2\tan^3 A - k\tan^2 A + k = 0$

Let, $\quad \tan A = x, x > 0 \text{ (as } A < 90^\circ)$

Then let, $\quad f(x) = 2x^3 - kx^2 + k \qquad \text{...(ii)}$

$f'(x) = 6x^2 - 2kx = 0$

$\Rightarrow \quad x = k/3, 0$

Following cases arises

(i) $k < 0$, three graphs of cubic equation (ii) are possible. Clearly, in all these case, only one triangle is possible and the condition for that triangle to be possible is $f(0) < 0 \Rightarrow k < 0$ so for $k < 0$ only one isosceles triangle is possible.

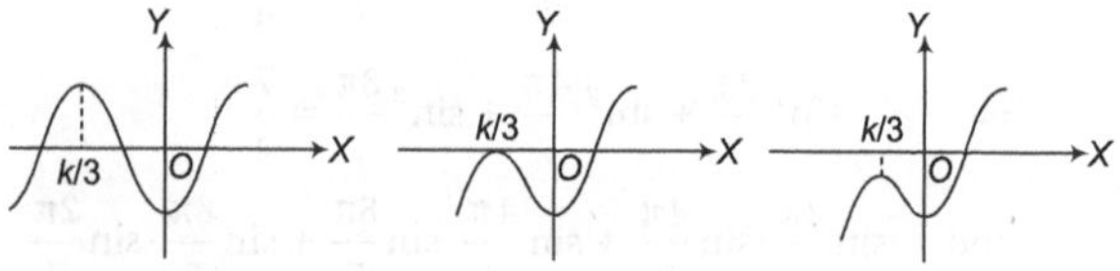

(ii) $k > 0$, three graphs of the cubic equation (ii) are possible. In fig. (i), two such triangle are possible. The condition is $f(k/3) < 0$.

$\Rightarrow \quad k\left(1 - \dfrac{k^2}{27}\right) < 0 \Rightarrow k > 3\sqrt{3}$

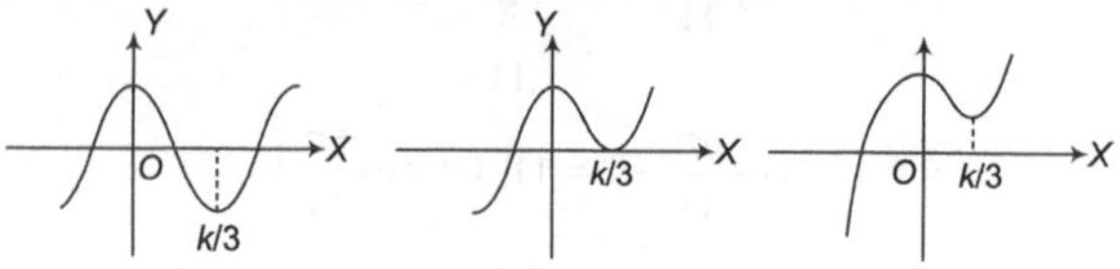

In figure (ii), one such triangle is possible. The condition is $f(k/3) = 0$

$\Rightarrow \quad k = 3\sqrt{3}.$

In figure (iii), no such triangle is possible. The condition is $f(k/3) > 0$

$\Rightarrow \quad k\left(1 - \dfrac{k^2}{27}\right) > 0$

$\Rightarrow \quad k < 3\sqrt{3}.$

(iii) $k = 0$, graph will be shown as, so no such triangle is possible. Hence, the solution for mentioned conditions;

$\therefore$ (A) either $k < 0$ or $k = 3\sqrt{3}$

(B) $k > 3\sqrt{3}$

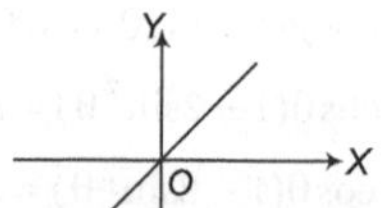

(C) Clearly, there will never exists three or more than three non-similar isosceles triangle for any value of k.

Trigonometric Functions and Identities Exercise 1 : Single Option Correct Type Questions

1. The value of $\sum_{n=1}^{10}\left(\sin\frac{2n\pi}{11}-\cos\frac{2n\pi}{11}\right)$ is equal to

(a) 2 (b) 1
(c) 0 (d) -1

2. Given, $a^2+2a+\operatorname{cosec}^2\left(\frac{\pi}{2}(a+x)\right)=0$, then which of the following holds good?

(a) $a=1; \frac{x}{2}\in 1$
(b) $a=-1; \frac{x}{2}\in I$
(c) $a\in R; x\in\phi$
(d) a, x are finite but not possible to find

3. The minimum value of the function $f(x)=(3\sin x-4\cos x-10)(3\sin x+4\cos x-10)$, is

(a) 49 (b) $\frac{195-60\sqrt{2}}{2}$
(c) 84 (d) 48

4. The value of expression $\sum_{\theta=0}^{8}\frac{1}{1+\tan^3(10\theta)^\circ}$ equal is to

(a) 5 (b) $\frac{21}{4}$
(c) $\frac{14}{3}$ (d) $\frac{9}{2}$

5. The value of $\sqrt{1-\sin^2 110^\circ}\cdot\sec 110^\circ$ is equal to

(a) 2 (b) -1
(c) -2 (d) 1

6. If $\tan\alpha$, $\tan\beta$ are the roots of the equation $x^2+px+q=0$, then the value of $\sin^2(\alpha+\beta)+p\sin(\alpha+\beta)\cos(\alpha+\beta)+q\cos^2(\alpha+\beta)$ is

(a) independent of p but dependent on q
(b) independent of q but dependent on p
(c) independent of both p and q
(d) dependent on both p and q

7. The value of the product

$$\sin\left(\frac{\pi}{2^{2009}}\right)\cos\left(\frac{\pi}{2^{2009}}\right)\cos\left(\frac{\pi}{2^{2008}}\right)\cos\left(\frac{\pi}{2^{2007}}\right)\cos\left(\frac{\pi}{2^{2006}}\right)\ldots\cos\left(\frac{\pi}{2^3}\right)\cos\left(\frac{\pi}{2^2}\right),$$ is

(a) $\frac{1}{2^{2007}}$ (b) $\frac{1}{2^{2008}}$
(c) $\frac{1}{2^{2009}}$ (d) $\frac{1}{2^{2010}}$

8. If $\tan B=\frac{n\sin A\cos A}{1-n\cos^2 A}$, then $\tan(A+B)$ equals to

(a) $\frac{\sin A}{(1-n)\cos A}$ (b) $\frac{(n-1)\cos A}{\sin A}$
(c) $\frac{\sin A}{(n-1)\cos A}$ (d) $\frac{\sin A}{(n+1)\cos A}$

9. If $P=(\tan(3^{n+1}\theta)-\tan\theta)$ and $Q=\sum_{r=0}^{n}\frac{\sin(3^r\theta)}{\cos(3^{r+1}\theta)}$, then

(a) $P=2Q$ (b) $P=3Q$
(c) $2P=Q$ (d) $3P=Q$

10. The value of $(\cos^4 1^\circ+\cos^4 2^\circ+\cos^4 3^\circ+\ldots+\cos^4 179^\circ)-(\sin^4 1^\circ+\sin^4 2^\circ+\sin^4 3^\circ+\ldots+\sin^4 179^\circ)$ equals to

(a) $2\cos 1^\circ$ (b) -1
(c) $2\sin 1^\circ$ (d) 0

11. Suppose that 'a' is a non-zero real number for which $\sin x+\sin y=a$ and $\cos x+\cos y=2a$. The value of $\cos(x-y)$, is

(a) $\frac{3a^2-2}{2}$ (b) $\frac{7a^2-2}{2}$
(c) $\frac{9a^2-2}{2}$ (d) $\frac{5a^2-2}{2}$

12. Let $P(x)=\sqrt{(\cos x+\cos 2x+\cos 3x)^2+(\sin x+\sin 2x+\sin 3x)^2}$, then $P(x)$ is equal to

(a) $1+2\cos x$ (b) $1+\sin 2x$
(c) $1-2\cos x$ (d) None of these

13. If the maximum value of the expression $\frac{1}{5\sec^2\theta-\tan^2\theta+4\operatorname{cosec}^2\theta}$ is equal to $\frac{p}{q}$ (where, p and q are coprime), then the value of $(p+q)$ is

(a) 14 (b) 15
(c) 16 (d) 18

14. Let $f_n(a)=\frac{\sin\alpha+\sin 3\alpha+\sin 5\alpha+\ldots+\sin(2n-1)\alpha}{\cos\alpha+\cos 3\alpha+\cos 5\alpha+\ldots+\cos(2n-1)\alpha}$.

Then, the value of $f_4\left(\frac{\pi}{32}\right)$ is equal to

(a) $\sqrt{2}+1$ (b) $\sqrt{2}-1$
(c) $2+\sqrt{3}$ (d) $2-\sqrt{3}$

15. The minimum value of $\left|\sin x+\cos x+\frac{\cos x+\sin x}{\cos^4 x-\sin^4 x}\right|$ is

(a) 2 (b) $\frac{3}{2}$ (c) $\sqrt{2}$ (d) 1

16. If $a = \cos(2012\pi)$, $b = \sec(2013\pi)$ and $c = \tan(2014\pi)$, then
(a) $a < b < c$ (b) $b < c < a$
(c) $c < b < a$ (d) $a = b < c$

17. In a ΔABC, the minimum value of $\sec^2\frac{A}{2} + \sec^2\frac{B}{2} + \sec^2\frac{C}{2}$ is equal to
(a) 3 (b) 4
(c) 5 (d) 6

18. The number of ordered pairs (x, y) of real numbers satisfying $4x^2 - 4x + 2 = \sin^2 y$ and $x^2 + y^2 \le 3$, is equal to
(a) 0 (b) 2
(c) 4 (d) 8

19. In a ΔABC, $3\sin A + 4\cos B = 6$ and $3\cos A + 4\sin B = 1$, then $\angle C$ can be
(a) $30°$ (b) $60°$
(c) $90°$ (d) $150°$

20. An equilateral triangle has side length 8. The area of the region containing all points outside the triangle but not more than 3 units from a point on the triangle is :
(a) $9(8 + \pi)$
(b) $8(9 + \pi)$
(c) $9\left(8 + \frac{\pi}{2}\right)$
(d) $8\left(9 + \frac{\pi}{2}\right)$

21. If $a\cos^3\alpha + 3a\cos\alpha\sin^2\alpha = m$ and $a\sin^3\alpha + 3a\cos^2\alpha\sin\alpha = n$. Then, $(m+n)^{2/3} + (m-n)^{2/3}$ is equal to
(a) $2a^2$ (b) $2a^{1/3}$
(c) $2a^{2/3}$ (d) $2a^3$

22. As shown in the figure, AD is the altitude on BC and AD produced meets the circumcircle of ΔABC at P where $DP = x$. Similarly, $EQ = y$ and $FR = z$. If a, b, c respectively denotes the sides BC, CA and AB, then $\frac{a}{2x} + \frac{b}{2y} + \frac{c}{2z}$ has the value equal to

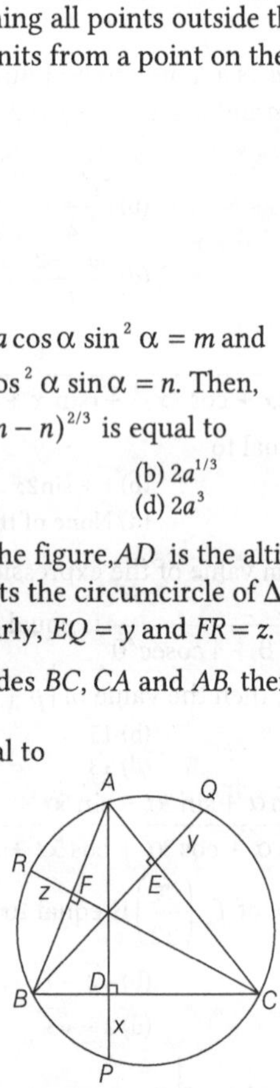

(a) $\tan A + \tan B + \tan C$
(b) $\cot A + \cos B + \cot C$
(c) $\cos A + \cos B + \cos C$
(d) $\text{cosec}\, A + \text{cosec}\, B + \text{cosec}\, C$

23. One side of a rectangular piece of paper is 6 cm, the adjacent sides being longer than 6 cm. One corner of the paper is folded so that is sets on the opposite longer side. If the length of the crease is l cm and it makes an angle θ with the long side as shown, then l is

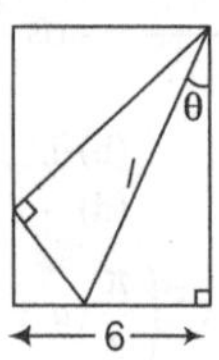

(a) $\frac{3}{\sin\theta\cos^2\theta}$ (b) $\frac{6}{\sin^2\theta\cos\theta}$
(c) $\frac{3}{\sin\theta\cos\theta}$ (d) $\frac{3}{\sin^2\theta}$

24. The average of the numbers $n\sin n°$ for $n = 2, 4, 6, \ldots 180$
(a) 1 (b) $\cot 1°$
(c) $\tan 1°$ (d) $\frac{1}{2}$

25. A circle is inscribed inside a regular pentagon and another circle is circumscribed about this pentagon. Similarly, a circle is inscribed in a regular heptagon and another circumscribed about the heptagon. The area of the regions between the two circles in two cases are A_1 and A_2, respectively. If each polygon has a side length of 2 units, then which one of the following is true ?
(a) $A_1 = \frac{5}{7}A_2$ (b) $A_1 = \frac{25}{49}A_2$
(c) $A_1 = \frac{49}{25}A_2$ (d) $A_1 = A_2$

26. The value of $\sum_{r=1}^{18}\cos^2(5r)°$, where $x°$ denotes the x degrees, is equal to
(a) 0 (b) $\frac{7}{2}$
(c) $\frac{17}{2}$ (d) $\frac{25}{2}$

27. Minimum value of $4x^2 - 4x|\sin x| - \cos^2\theta$ is equal to
(a) -2 (b) -1
(c) $-\frac{1}{2}$ (d) 0

28. If in a triangle ABC, $\cos 3A + \cos 3B + \cos 3C = 1$, then one angle must be exactly equal to
(a) $\frac{\pi}{3}$ (b) $\frac{2\pi}{3}$ (c) π (d) $\frac{4\pi}{3}$

29. If $|\tan A| < 1$ and $|A|$ is acute, then $\frac{\sqrt{(1+\sin 2A)} + \sqrt{(1-\sin 2A)}}{\sqrt{(1+\sin 2A)} - \sqrt{(1-\sin 2A)}}$ is equal to
(a) $\tan A$ (b) $-\tan A$
(c) $\cot A$ (d) $-\cot A$

30. For any real θ, the maximum value of $\cos^2(\cos\theta)+\sin^2(\sin\theta)$ is

(a) 1 (b) $1+\sin^2 1$
(c) $1+\cos^2 1$ (d) does not exist

31. Minimum value of $27^{\cos 2x}\cdot 81^{\sin 2x}$ is

(a) -5 (b) $\frac{1}{5}$
(c) $\frac{1}{243}$ (d) $\frac{1}{27}$

32. $ABCD$ is a trapezium, such that AB and CD are parallel $BC\perp CD$. If $\angle ADB=\theta$, $BC=p$ and $CD=q$, then AB is equal to

(a) $\frac{(p^2+q^2)\sin\theta}{p\cos\theta+q\sin\theta}$ (b) $\frac{p^2+q^2\cos\theta}{p\cos\theta+q\sin\theta}$
(c) $\frac{p^2+q^2}{p^2\cos\theta+q^2\sin\theta}$ (d) $\frac{(p^2+q^2)\sin\theta}{(p\cos\theta+q\sin\theta)^2}$

33. If $4n\alpha=\pi$, then $\cot\alpha\cot 2\alpha\cot 3\alpha\ldots\cot(2n-1)\alpha$ is equal to

(a) 0 (b) 1
(c) n (d) None of these

34. If in a triangle $ABC(\sin A+\sin B+\sin C)$ $(\sin A+\sin B-\sin C)=3\sin A\sin B$, then angle C is equal to

(a) 30° (b) 45°
(c) 60° (d) 75°

35. If α,β,γ are acute angles and $\cos\theta=\frac{\sin\beta}{\sin\alpha}$, $\cos\phi=\frac{\sin\gamma}{\sin\alpha}$ and $\cos(\theta-\phi)=\sin\beta\sin\gamma$, then $\tan^2\alpha-\tan^2\beta-\tan^2\gamma$ is equal to

(a) -1 (b) 0
(c) 1 (d) None of these

36. If $\tan\beta=\frac{n\sin\alpha\cos\alpha}{1-n\sin^2\alpha}$, then $\tan(\alpha-\beta)$ is equal to

(a) $n\tan\alpha$ (b) $(1-n)\tan\alpha$
(c) $(1+n)\tan\alpha$ (d) None of these

37. If $\frac{\cos\theta}{a}=\frac{\sin\theta}{b}$, then $\frac{a}{\sec 2\theta}+\frac{b}{\operatorname{cosec} 2\theta}$ is equal to

(a) a (b) b
(c) $\frac{a}{b}$ (d) $a+b$

38. The graph of the function $\cos x\cos(x+2)-\cos^2(x+1)$ is

(a) a straight line passing through $(0,-\sin^2\theta)$ with slope 2
(b) a straight line passing through (0, 0)
(c) a parabola with vertex $(1,-\sin^2 1)$
(d) a straight line passing through the point $\left(\frac{\pi}{2},-\sin^2 1\right)$ and parallel to the X-axis

39. $f(\theta)=|\sin\theta|+|\cos\theta|$, $\theta\in R$, then

(a) $f(\theta)\in[0,2]$ (b) $f(\theta)\in[0,\sqrt{2}]$
(c) $f(\theta)\in[0,1]$ (d) $f(\theta)\in[1,\sqrt{2}]$

40. If $A=\cos(\cos x)+\sin(\cos x)$ the least and greatest value of A are

(a) 0 and 2 (b) -1 and 1
(c) $-\sqrt{2}$ and $\sqrt{2}$ (d) 0 and $\sqrt{2}$

41. If $U_n=\sin n\theta\sec^n\theta$, $V_n=\cos n\theta\sec^n\theta\neq 1$, then $\frac{V_n-V_{n-1}}{U_{n-1}}+\frac{1}{n}\frac{U_n}{V_n}$ is equal to

(a) 0 (b) $\tan\theta$
(c) $-\tan\theta+\frac{\tan n\theta}{n}$ (d) $\tan\theta+\frac{\tan n\theta}{n}$

42. If $0\le x\le\frac{\pi}{3}$ then range of $f(x)=\sec\left(\frac{\pi}{6}-x\right)+\sec\left(\frac{\pi}{6}+x\right)$ is

(a) $\left(\frac{4}{\sqrt{3}},\infty\right)$ (b) $\left[\frac{4}{\sqrt{3}},\infty\right)$
(c) $\left(0,\frac{4}{\sqrt{3}}\right]$ (d) $\left(0,\frac{4}{\sqrt{3}}\right)$

43. If $A=\sin^8\theta+\cos^{14}\theta$, then for all values of θ,

(a) $A\ge 1$ (b) $0<A\le 1$
(c) $1<2a\le 3$ (d) None of these

44. The expression $3\left\{\sin^4\left(\frac{3\pi}{2}-\alpha\right)+\sin^4(3\pi+\alpha)\right\}-2\left\{\sin^6\left(\frac{\pi}{2}+\alpha\right)+\sin^6(5\pi-\alpha)\right\}$ is equal to

(a) 0 (b) -1
(c) 1 (d) 3

45. The maximum value of $\sin\left(x+\frac{\pi}{6}\right)+\cos\left(x+\frac{\pi}{6}\right)$ in the interval $\left(0,\frac{\pi}{2}\right)$ is attained at

(a) $\frac{\pi}{12}$ (b) $\frac{\pi}{6}$ (c) $\frac{\pi}{3}$ (d) $\frac{\pi}{2}$

46. If $\cot^2 x=\cot(x-y)\cdot\cot(x-z)$, then $\cot 2x$ is equal to $\left(x\neq\pm\frac{\pi}{4}\right)$

(a) $\frac{1}{2}(\tan y+\tan z)$ (b) $\frac{1}{2}(\cot y+\cot z)$
(c) $\frac{1}{2}(\sin y+\sin z)$ (d) None of these

47. The minimum value of the expression $\sin\alpha+\sin\beta+\sin\gamma$, where α,β,γ are real numbers satisfying $\alpha+\beta+\gamma=\pi$, is

(a) positive (b) zero
(c) negative (d) None of these

48. If $\cos x - \sin\alpha \cot\beta \sin x = \cos\alpha$, then $\tan\frac{x}{2}$ is equal to

(a) $\cot\frac{\alpha}{2}\tan\frac{\beta}{2}$ (b) $\cot\frac{\beta}{2}\tan\frac{\alpha}{2}$

(c) $\tan\frac{\alpha}{2}\tan\frac{\beta}{2}$ (d) None of these

49. If $\cos^4\theta\sec^2\alpha, \frac{1}{2}$ and $\sin^4\theta\,\text{cosec}^2\alpha$ are in AP, then $\cos^8\theta\sec^6\alpha, \frac{1}{2}$ and $\sin^8\theta\cdot\text{cosec}^6\alpha$ are in

(a) AP (b) GP

(c) HP (d) None of these

50. The maximum value of $\cos\alpha_1\cdot\cos\alpha_2\cdot\cos\alpha_3\cdot\ldots\cdot\cos\alpha_n$ under the restriction $0\le\alpha_1,\alpha_2,\ldots,\alpha_n\le\frac{\pi}{2}$ and $\cot\alpha_1\cdot\cot\alpha_2\cdot\ldots\cdot\cot\alpha_n=1$ is

(a) $\frac{1}{2^{\frac{n}{2}}}$ (b) $\frac{1}{2^n}$

(c) $\frac{-1}{2^n}$ (d) 1

51. If $x\in(0,\pi)$ and $\sin x\cdot\cos^3 x > \cos x\cdot\sin^3 x$, then complete set of values of x is

(a) $x\in\left(0,\frac{\pi}{4}\right)\cup\left(\frac{\pi}{2},\frac{3\pi}{4}\right)$

(b) $x\in\left(\frac{\pi}{4},\frac{\pi}{2}\right)\cup\left(\frac{3\pi}{4},\pi\right)$

(c) $x\in\left(\frac{\pi}{4},\frac{\pi}{2}\right)$

(d) None of the above

52. If $u=\sqrt{a^2\cos^2\theta+b^2\sin^2\theta}+\sqrt{a^2\sin^2\theta+b^2\cos^2\theta}$ then the difference between the maximum and minimum values of u^2 is given by

(a) $2(a^2+b^2)$ (b) $2\sqrt{a^2+b^2}$

(c) $(a+b)^2$ (d) $(a-b)^2$

53. For a positive integer n, let $f_n(\theta)=\frac{\tan\theta}{2}(1+\sec\theta)(1+\sec 2\theta)\ldots(1+\sec 2^n\theta)$, then

(a) $f_2\left(\frac{\pi}{16}\right)=0$ (b) $f_3\left(\frac{\pi}{32}\right)=-1$

(c) $f_4\left(\frac{\pi}{64}\right)=-1$ (d) $f_5\left(\frac{\pi}{128}\right)=1$

Trigonometric Functions and Identities Exercise 2 : More than One Option Correct Type Questions

54. Suppose $\cos x=0$ and $\cos(x+z)=\frac{1}{2}$. Then, the possible value(s) of z is (are).

(a) $\frac{\pi}{6}$ (b) $\frac{5\pi}{6}$

(c) $\frac{7\pi}{6}$ (d) $\frac{11\pi}{6}$

55. Let $f_n(\theta)=2\sin\frac{\theta}{2}\sin\frac{3\theta}{2}+2\sin\frac{\theta}{2}\sin\frac{5\theta}{2}+2\sin\frac{\theta}{2}\sin\frac{7\theta}{2}+\ldots+2\sin\frac{\theta}{2}\sin(2n+1)\frac{\theta}{2}, n\in N$, then which of the following is/are correct?

(a) $f_9\left(\frac{\pi}{4}\right)=\frac{1}{\sqrt{2}}$ (b) $f_n\left(\frac{2\pi}{n}\right)=0, n\in N$

(c) $f_5\left(\frac{2\pi}{7}\right)=0$ (d) $f_9\left(\frac{\pi}{4}\right)=-\frac{1}{\sqrt{2}}$

56. Let $P=\sin 25°\sin 35°\sin 60°\sin 85°$ and $Q=\sin 20°\sin 40°\sin 75°\sin 80°$. Which of the following relation(s) is (are) correct ?

(a) $P+Q=0$ (b) $P-Q=0$

(c) $P^2+Q^2=1$ (d) $P^2-Q^2=0$

57. For $0<\theta<\frac{\pi}{2}$, if $x=\sum_{n=0}^{\infty}\cos^{2n}\theta, y=\sum_{n=0}^{\infty}\sin^{2n}\theta,$

$$z=\sum_{n=0}^{\infty}\cos^{2n}\theta\sin^{2n}\theta, \text{ then}$$

(a) $xyz=xz+y$

(b) $xyz=xy+z$

(c) $xyz=x+y+z$

(d) $xyz=yz+x$

58. Let $P(x)=\cot^2 x\left(\frac{1+\tan x+\tan^2 x}{1+\cot x+\cot^2 x}\right)+\left(\frac{\cos x-\cos 3x+\sin 3x-\sin x}{2(\sin 2x+\cos 2x)}\right)^2$. Then, which of the following is (are) correct?

(a) The value of $P(18°)+P(72°)$ is 2.

(b) The value of $P(18°)+P(72°)$ is 3.

(c) The value of $P\left(\frac{4\pi}{3}\right)+P\left(\frac{7\pi}{6}\right)$ is 3.

(d) The value of $P\left(\frac{4\pi}{3}\right)+P\left(\frac{7\pi}{6}\right)$ is 2.

59. It is known that $\sin\beta = \frac{4}{5}$ and $0 < \beta < \pi$, then thes value of $\dfrac{\sqrt{3}\sin(\alpha+\beta) - \dfrac{2}{\cos\frac{11\pi}{6}}\cos(\alpha+\beta)}{\sin\alpha}$ is

(a) independent of α for all β in $(0, \pi)$

(b) $\frac{5}{\sqrt{3}}$ for $\tan\beta > 0$

(c) $\frac{\sqrt{3}(7 + 24\cot\alpha)}{15}$ for $\tan\beta < 0$

(d) zero for $\tan\beta > 0$

60. In cyclic quadrilateral $ABCD$, if $\cot A = \frac{3}{4}$ and $\tan B = \frac{-12}{5}$, then which of the following is (are) correct?

(a) $\sin D = \frac{12}{13}$ (b) $\sin(A + B) = \frac{16}{65}$

(c) $\cos D = \frac{-15}{13}$ (d) $\sin(C + D) = \frac{-16}{65}$

61. If the equation $2\cos^2 x + \cos x - a = 0$ has solutions, then a can be

(a) $\frac{-1}{4}$ (b) $\frac{-1}{8}$

(c) 2 (d) 5

62. If $A = \sin 44° + \cos 44°$, $B = \sin 45° + \cos 45°$ and $C = \sin 46° + \cos 46°$. Then, correct option(s) is/are

(a) $A < B < C$ (b) $C < B < A$

(c) $B > A$ (d) $A = C$

63. If $\tan(2\alpha + \beta) = x$ & $\tan(\alpha + 2\beta) = y$, then $[\tan 3(\alpha + \beta)]$. $[\tan(\alpha - \beta)]$ is equal to (wherever defined)

(a) $\frac{x^2 + y^2}{1 - x^2y^2}$ (b) $\frac{x^2 - y^2}{1 + x^2y^2}$

(c) $\frac{x^2 + y^2}{1 + x^2y^2}$ (d) $\frac{x^2 - y^2}{1 - x^2y^2}$

64. If $x = \sec\phi - \tan\phi$ and $y = \operatorname{cosec}\phi + \cot\phi$, then

(a) $x = \frac{y + 1}{y - 1}$ (b) $x = \frac{y - 1}{y + 1}$

(c) $y = \frac{1 + x}{1 - x}$ (d) $xy + x - y + 1 = 0$

65. If $\tan\left(\frac{x}{2}\right) = \operatorname{cosec} x - \sin x$, then $\tan^2\left(\frac{x}{2}\right)$ is equal to

(a) $2 - \sqrt{5}$ (b) $\sqrt{5} - 2$

(c) $(9 - 4\sqrt{5})(2 + \sqrt{5})$ (d) $(9 + 4\sqrt{5})(2 - \sqrt{5})$

66 If $y = \frac{\sqrt{1 - \sin 4A} + 1}{\sqrt{1 + \sin 4A} - 1}$, then one of the value of y is

(a) $-\tan A$ (b) $\cot A$

(c) $\tan\left(\frac{\pi}{4} + A\right)$ (d) $-\cot\left(\frac{\pi}{4} + A\right)$

67. If $3\sin\beta = \sin(2\alpha + \beta)$, then

(a) $[\cot\alpha + \cot(\alpha + \beta)][\cot\beta - 3\cot(2\alpha + \beta)] = 6$

(b) $\sin\beta = \cos(\alpha + \beta)\sin\alpha$

(c) $2\sin\beta = \sin(\alpha + \beta)\cos\alpha$

(d) $\tan(\alpha + \beta) = 2\tan\alpha$

68. Let $P_n(u)$ be a polynomial is u of degree n. Then, for every positive integer n, $\sin 2nx$ is expressible is

(a) $P_{2n}(\sin x)$ (b) $P_{2n}(\cos x)$

(c) $\cos x\, P_{2n-1}(\sin x)$ (d) $\sin x\, P_{2n-1}(\cos x)$

69. If $\tan\theta = \frac{\sin\alpha - \cos\alpha}{\sin\alpha + \cos\alpha}$, then

(a) $\sin\alpha - \cos\alpha = \pm\sqrt{2}\sin\theta$

(b) $\sin\alpha + \cos\alpha = \pm\sqrt{2}\cos\theta$

(c) $\cos 2\theta = \sin 2\alpha$

(d) $\sin 2\theta + \cos 2\alpha = 0$

70. If $\cos 5\theta = a\cos\theta + b\cos^3\theta + c\cos^5\theta + d$, then

(a) $a = 20$

(b) $b = -20$

(c) $c = 16$

(d) $d = 5$

71. $x = \sqrt{a^2\cos^2\alpha + b^2\sin^2\alpha} = \sqrt{a^2\sin^2\alpha + b^2\cos^2\alpha}$ then $x^2 = a^2 + b^2 + 2\sqrt{p(a^2 + b^2) - p^2}$, where p is equal to

(a) $a^2\cos^2\alpha + b^2\sin^2\alpha$

(b) $a^2\sin^2\alpha + b^2\cos^2\alpha$

(c) $\frac{1}{2}[a^2 + b^2 + (a^2 - b^2)\cos 2\alpha]$

(d) $\frac{1}{2}[a^2 + b^2 - (a^2 - b^2)\cos 2\alpha]$

72. $\left(\frac{\cos A + \cos B}{\sin A - \sin B}\right)^n + \left(\frac{\sin A + \sin B}{\cos A - \cos B}\right)^n$ (n, even or odd) is equal to

(a) $2\tan^n\left(\frac{A - B}{2}\right)$ (b) $2\cot^n\left(\frac{A - B}{2}\right)$

(c) 0 (d) None of these

73. Let $P(k) = \left(1 + \cos\frac{\pi}{4k}\right)\left(1 + \cos\frac{(2k-1)\pi}{4k}\right)\left(1 + \cos\frac{(2k+1)\pi}{4k}\right)\left(1 + \cos\frac{(4k-1)\pi}{4k}\right)$. Then

(a) $P(3) = \frac{1}{16}$ (b) $P(4) = \frac{2 - \sqrt{2}}{16}$

(c) $P(5) = \frac{3 - \sqrt{5}}{32}$ (d) $P(6) = \frac{2 - \sqrt{3}}{16}$

74. If $x = a\cos^3\theta\sin^2\theta$, $y = a\sin^3\theta\cos^2\theta$ and $\frac{(x^2 + y^2)^p}{(xy)^q}$ $(p, q \in N)$ is independent of θ, then

(a) $p = 4$ (b) $p = 5$

(c) $q = 4$ (d) $q = 5$

Trigonometric Functions and Identities Exercise 3: Statement I and II Type Questions

- This section contains 11 questions. Each question contains **Statement I** (Assertion) and **Statement II** (Reason). Each question has 4 choices (a), (b), (c) and (d) out of which only one is correct. The choices are
  - (a) Both Statement I and Statement II are individually true and R is the correct explanation of Statement I.
  - (b) Both Statement I and Statement II are individually true but Statement II is not the correct explanation of Statement I.
  - (c) Statement I is true but Statement II is false.
  - (d) Statement I is false but Statement II is true.

75. Statement I $\tan\alpha + 2\tan 2\alpha + 4\tan 4\alpha + 8\tan 8\alpha + 16\cot 16\alpha = \cot\alpha$

Statement II $\cot\alpha - \tan\alpha = 2\cot 2\alpha$

76. Statement I If $xy + yz + zx = 1$, then $\Sigma \dfrac{x}{(1+x^2)} = \dfrac{2}{\sqrt{\Pi(1+x^2)}}$.

Statement II In a ΔABC $\sin 2A + \sin 2B - \sin 2C = 4\cos A\cos B\sin C$

77. Statement I If α and β are two distinct solutions of the equation $a\cos x + b\sin x = c$, then $\tan\left(\dfrac{\alpha+\beta}{2}\right)$ is independent of c.

Statement II Solution of $a\cos x + b\sin x = c$ is possible, if $-\sqrt{(a^2+b^2)} \le c \le \sqrt{(a^2+b^2)}$

78. Statement I If A is obtuse angle in ΔABC, then $\tan B\tan C > 1$.

Statement II In ΔABC, $\tan A = \dfrac{\tan B + \tan C}{\tan B\tan C - 1}$

79. Statement I $\sin\left(\dfrac{2\pi}{7}\right) + \sin\left(\dfrac{4\pi}{7}\right) + \sin\left(\dfrac{8\pi}{7}\right) = -\dfrac{1}{2}$.

Statement II $\cos\dfrac{2\pi}{7} + i\sin\dfrac{2\pi}{7}$ is complex 7th root of unity.

80. Statement I The curve $y = 81^{\sin^2 x} + 81^{\cos^2 x} - 30$ intersects X-axis at eight points in the region $-\pi \le x \le \pi$.

Statement II The curve $y = \sin x$ or $y = \cos x$ intersects the X-axis at infinitely many points.

81. Statement I The numbers $\sin 18°$ and $-\sin 54°$ are the roots of a quadratic equation with integer coefficients.

Statement II If $x = 18°$, $\cos 3x = \sin 2x$ and if $y = -54°$ $\sin 2y = \cos 3y$.

82. Statement I The minimum value of the expression $\sin\alpha + \sin\beta + \sin\gamma$ where α, β, γ are real numbers such that $\alpha + \beta + \gamma = \pi$ is negative.

Statement II If $\alpha + \beta + \gamma = \pi$, then α, β, γ are the angles of a triangle.

83. Statement I If $2\sin\left(\dfrac{\theta}{2}\right) = \sqrt{1+\sin\theta} + \sqrt{1-\sin\theta}$ then $\dfrac{\theta}{2}$ lies between $2n\pi + \dfrac{\pi}{4}$ and $2n\pi + \dfrac{3\pi}{4}$.

Statement II If $\dfrac{\pi}{4} \le \theta \le \dfrac{3\pi}{4}$ then $\sin\dfrac{\theta}{2} > 0$.

84. Statement I If $2\cos\theta + \sin\theta = 1\left(\theta \ne \dfrac{\pi}{2}\right)$ then the value of $7\cos\theta + 6\sin\theta$ is 2.

Statement II If $\cos 2\theta - \sin\theta = \dfrac{1}{2}, 0 < \theta < \dfrac{\pi}{2}$, then $\sin\theta + \cos 6\theta = 0$.

85. Statement I If $A > 0$, $B > 0$ and $A + B = \dfrac{\pi}{3}$, then the maximum value of $\tan A\tan B$ is $\dfrac{1}{3}$.

Statement II If $a_1 + a_2 + a_3 + \ldots + a_n = k$ (constant), then the value $a_1 a_2 a_3 \ldots a_n$ is greatest when $a_1 = a_2 = a_3 = \ldots = a_n$

Trigonometric Functions and Identities Exercise 4 : Passage Based Questions

Passage I

(Q. Nos. 86 and 87)

If a, b, c are the sides of ΔABC such that $3^{2a^2} - 2\cdot 3^{a^2+b^2+c^2} + 3^{2b^2+2c^2} = 0$, then

86. Triangle ABC is

(a) equilateral (b) right angled
(c) isosceles right angled (d) obtuse angled

87. If sides of ΔPQR are a, $b\sec C$, $c\operatorname{cosec} C$. Then, area of ΔPQR is

(a) $\frac{\sqrt{3}}{4}a^2$ (b) $\frac{\sqrt{3}}{4}b^2$ (c) $\frac{\sqrt{3}}{4}c^2$ (d) $\frac{1}{2}abc$

Passage II

(Q. Nos. 88 to 90)

For $0 < x < \frac{\pi}{2}$, let $P_{mn}(x) = m\log_{\cos x}(\sin x) + n\log_{\cos x}(\cot x)$; where $m, n \in \{1, 2, \dots, 9\}$

[For example :

$$P_{29}(x) = 2\log_{\cos x}(\sin x) + 9\log_{\cos x}(\cot x) \text{ and}$$
$$P_{77}(x) = 7\log_{\cos x}(\sin x) + 7\log_{\cos x}(\cot x)]$$

On the basis of above information , answer the following questions :

88. Which of the following is always correct?

(a) $P_{mn}(x) \ge m \ \forall\ m \ge n$ (b) $P_{mn}(x) \ge n \ \forall\ m \ge n$
(c) $2P_{mn}(x) \le n \ \forall\ m \le n$ (d) $2P_{mn}(x) \le m \ \forall\ m \le n$

89. The mean proportional of numbers $P_{49}\left(\frac{\pi}{4}\right)$ and $P_{94}\left(\frac{\pi}{4}\right)$ is equal to

(a) 4 (b) 6
(c) 9 (d) 10

90. If $P_{34}(x) = P_{22}(x)$, then the value of $\sin x$ is expressed as $\left(\frac{\sqrt{q}-1}{p}\right)$, then $(p+q)$ equals

(a) 3 (b) 4
(c) 7 (d) 9

Note Mean proportional of a and b $(a > 0, b > 0)$ is $\sqrt{ab}$]

Passage III

(Q. Nos. 91 to 93)

If $7\theta = (2n+1)\pi$, where $n = 0, 1, 2, 3, 4, 5, 6$, then answer the following questions.

91. The equations whose roots are $\cos\frac{\pi}{7}, \cos\frac{3\pi}{7}, \cos\frac{5\pi}{7}$ is

(a) $8x^3 + 4x^2 + 4x + 1 = 0$
(b) $8x^3 - 4x^2 - 4x - 1 = 0$
(c) $8x^3 - 4x^2 - 4x - 1 = 0$
(d) $8x^3 + 4x^2 + 4x - 1 = 0$

92. The value of $\sec\frac{\pi}{7} + \sec\frac{3\pi}{7} + \sec\frac{5\pi}{7}$ is

(a) 4 (b) – 4
(c) 3 (d) – 3

93. The value of $\sec^2\frac{\pi}{7} + \sec^2\frac{3\pi}{7} + \sec^2\frac{5\pi}{7}$ is

(a) – 24 (b) 80 (c) 24 (d) – 80

Passage IV

(Q. Nos. 94 to 96)

If $1 + 2\sin x + 3\sin^2 x + 4\sin^3 x + \dots$ upto infinite terms = 4 and number of solutions of the equation in $\left[\frac{-3\pi}{2}, 4\pi\right]$ is k.

94. The value of k is equal to

(a) 4 (b) 5 (c) 6 (d) 7

95. The value of $\left|\frac{\cos 2x - 1}{\sin 2x}\right|$ is equal to

(a) 1 (b) $\sqrt{3}$
(c) $2-\sqrt{3}$ (d) $\frac{1}{\sqrt{3}}$

96. Sum of all internal angles of a k-sided regular polygon is

(a) 5π (b) 4π
(c) 3π (d) 2π

Passage V

(Q. Nos. 97 to 98)

Let α is a root of the equation $(2\sin x - \cos x)(1+\cos x) = \sin^2 x$, β is a root of the equation $3\cos^2 x - 10\cos x + 3 = 0$ and γ is a root of the equation $1 - \sin 2x = \cos x - \sin x$, $0 \le \alpha, \beta, \gamma \le \frac{\pi}{2}$.

97. $\cos\alpha + \cos\beta + \cos\gamma$ can be equal to

(a) $\frac{3\sqrt{6}+2\sqrt{2}+6}{6\sqrt{2}}$ (b) $\frac{3\sqrt{3}+8}{6}$
(c) $\frac{3\sqrt{3}+2}{6}$ (d) None of these

98. $\sin(\alpha - \beta)$ is equal to

(a) 1 (b) 0
(c) $\frac{1-2\sqrt{6}}{6}$ (d) $\frac{\sqrt{3}-2\sqrt{2}}{6}$

Trigonometric Functions and Identities Exercise 5: Matching Type Questions

99. Match the statement of Column I with values of Column II.

Column I		Column II
(A) If $\theta + \phi = \frac{\pi}{2}$, where θ and ϕ are positive, then $(\sin\theta + \sin\phi)\sin\left(\frac{\pi}{4}\right)$ is always less than	(p)	1
(B) If $\sin\theta - \sin\phi = a$ and $\cos\theta + \cos\phi = b$, then $a^2 + b^2$ cannot exceed	(q)	2
(C) If $3\sin\theta + 5\cos\theta = 5$, $(\theta \neq 0)$ then the value of $5\sin\theta - 3\cos\theta$ is	(r)	3
	(s)	4
	(t)	5

100. Match the statement of Column I with values of Column II.

Column I		Column II
(A) If maximum and minimum values of $\frac{7 + 6\tan\theta - \tan^2\theta}{(1 + \tan^2\theta)}$ for all real values of $\theta \sim \frac{\pi}{2}$ are λ and μ respectively, then	(p)	$\lambda + \mu = 2$
(B) If maximum and minimum values of $5\cos\theta + 3\cos\left(\theta + \frac{\pi}{3}\right) + 3$ for all real values of θ are λ and μ respectively, then	(q)	$\lambda - \mu = 6$
(C) If maximum and minimum values of $1 + \sin\left(\frac{\pi}{4} + \theta\right) + 2\cos\left(\frac{\pi}{4} - \theta\right)$ for all real values of θ and λ and μ respectively, then	(r)	$\lambda + \mu = 6$
	(s)	$\lambda - \mu = 10$
	(t)	$\lambda - \mu = 14$

101. Match the statement of Column I with values of Column II.

Column I		Column II
(A) The number of solutions of the equation $\lvert\cot x\rvert = \cot x + \frac{1}{\sin x}$ $(0 < x < \pi)$ is	(p)	no solution
(B) If $\sin\theta + \sin\phi = \frac{1}{2}$ and $\cos\theta + \cos\phi = 2$, then value of $\cot\left(\frac{\theta + \phi}{2}\right)$ is	(q)	$\frac{1}{3}$
(C) The value of $\sin^2\alpha + \sin\left(\frac{\pi}{3} - \alpha\right)\sin\left(\frac{\pi}{3} + \alpha\right)$ is	(r)	1
(D) If $\tan\theta = 3\tan\phi$, then maximum value of $\tan^2(\theta - \phi)$ is	(s)	2
	(t)	4

102. Match the statement of Column I with values of Column II.

Column I		Column II
(A) In a ΔABC, $\sin\left(\frac{A}{2}\right) + \sin\left(\frac{B}{2}\right) + \sin\left(\frac{C}{2}\right) =$	(p)	$-1 + 4\sin\left(\frac{\pi + A}{4}\right)\sin\left(\frac{\pi + B}{4}\right)\cos\left(\frac{\pi + C}{4}\right)$
(B) In a ΔABC, $\sin\left(\frac{A}{2}\right) + \sin\left(\frac{B}{2}\right) - \sin\left(\frac{C}{2}\right) =$	(q)	$4\cos\left(\frac{\pi + A}{4}\right)\cos\left(\frac{\pi + B}{4}\right)\cos\left(\frac{\pi - C}{4}\right)$
(C) In a ΔABC, $\cos\left(\frac{A}{2}\right) + \cos\left(\frac{B}{2}\right) - \cos\left(\frac{C}{2}\right)$	(r)	$1 + 4\sin\left(\frac{\pi - A}{4}\right)\sin\left(\frac{\pi - B}{4}\right)\sin\left(\frac{\pi - C}{4}\right)$
	(s)	$-1 + 4\cos\left(\frac{\pi - A}{4}\right)\cos\left(\frac{\pi - B}{4}\right)\sin\left(\frac{\pi - C}{4}\right)$
	(t)	$1 + 4\cos\left(\frac{\pi + A}{4}\right)\cos\left(\frac{\pi + B}{4}\right)\sin\left(\frac{\pi - C}{4}\right)$

Trigonometric Functions and Identities Exercise 6: Single Integer Answer Type Questions

103. In a ΔABC, $\dfrac{1}{1+\tan^2\dfrac{A}{2}}+\dfrac{1}{1+\tan^2\dfrac{B}{2}}+\dfrac{1}{1+\tan^2\dfrac{C}{2}}=k\left[1+\sin\dfrac{A}{2}\sin\dfrac{B}{2}\sin\dfrac{C}{2}\right]$, then the value of k is

104. If $\dfrac{\sin\alpha}{\sin\beta}=\dfrac{\cos\gamma}{\cos\delta}$, then $\dfrac{\sin\left(\dfrac{\alpha-\beta}{2}\right)\cdot\cos\left(\dfrac{\alpha+\beta}{2}\right)\cdot\cos\delta}{\sin\left(\dfrac{\delta-\gamma}{2}\right)\cdot\sin\left(\dfrac{\delta+\gamma}{2}\right)\cdot\sin\beta}$ is equal to

105. Find the exact value of the expression
$\tan\dfrac{\pi}{20}-\tan\dfrac{3\pi}{20}+\tan\dfrac{5\pi}{20}-\tan\dfrac{7\pi}{20}+\tan\dfrac{9\pi}{20}$.

106. Let $x=\dfrac{\sum_{n=1}^{44}\cos n^\circ}{\sum_{n=1}^{44}\sin n^\circ}$, find the greatest integer that does not exceed.

107. Find θ (in degree) satisfying the equation,
$\tan 15^\circ\cdot\tan 25^\circ\cdot\tan 35^\circ=\tan\theta$, where $\theta\in(0, 45^\circ)$

108. Find the exact value of $\operatorname{cosec}10^\circ+\operatorname{cosec}50^\circ-\operatorname{cosec}70^\circ$.

109. If $\cos 5\alpha=\cos^5\alpha$, where $\alpha\in\left(0,\dfrac{\pi}{2}\right)$, then find the possible values of $(\sec^2\alpha+\operatorname{cosec}^2\alpha+\cot^2\alpha)$.

110. Compute the value of the expression
$\tan^2\dfrac{\pi}{16}+\tan^2\dfrac{2\pi}{16}+\tan^2\dfrac{3\pi}{16}+\ldots+\tan^2\dfrac{7\pi}{16}$.

111. Compute the square of the value of the expression
$\dfrac{4+\sec 20^\circ}{\operatorname{cosec}20^\circ}$

112. In ΔABC, if $\dfrac{\sin A}{3}=\dfrac{\cos B}{3}=\dfrac{\tan C}{2}$, then the value of $\left(\dfrac{\sin A}{\cot 2A}+\dfrac{\cos B}{\cot 2B}+\dfrac{\tan C}{\cot 2C}\right)$ is

113. Let f and g be function defined by $f(\theta)=\cos^2\theta$ and $g(\theta)=\tan^2\theta$, suppose α and β satisfy $2f(\alpha)-g(\beta)=1$, then value of $2f(\beta)-g(\alpha)$ is

114. If sum of the series $1+x\log_{\left|\frac{1-\sin x}{\cos x}\right|}\left(\dfrac{1+\sin x}{\cos x}\right)^{1/2}+x^2\log_{\left|\frac{1-\sin x}{\cos x}\right|}\left(\dfrac{1+\sin x}{\cos x}\right)^{1/4}+\ldots\infty$ (wherever defined) is equal to $\dfrac{k(1-x)}{(2-x)}$, then k is equal to

115. If $\dfrac{9x}{\cos\theta}+\dfrac{5y}{\sin\theta}=56$ and $\dfrac{9x\sin\theta}{\cos^2\theta}-\dfrac{5y\cos\theta}{\sin^2\theta}=0$ then the value of $\left[(9x)^{\frac{2}{3}}+(5y)^{\frac{2}{3}}\right]^3$ is

116. The angle A of the ΔABC is obtuse.
$x=2635-\tan B\tan C$, if $[x]$ denotes the greatest integer function, the value of $[x]$ is

117. If $4\cos 36^\circ+\cot\left(7\dfrac{1}{2}^\circ\right)=\sqrt{n_1}+\sqrt{n_2}+\sqrt{n_3}+\sqrt{n_4}+\sqrt{n_5}+\sqrt{n_6}$, then the value of $\sum_{i=1}^{6}n_i^2$ must be

118. If $\sin^2 A=x$ and $\prod_{r=1}^{4}\sin(rA)=ax^2+bx^3+cx^4+dx^5$, then the value of $10a-7b+15c-5d$ must be

119. If $x, y\in R$ satisfies $(x+5)^2+(y-12)^2=(14)^2$, then the minimum value of $\sqrt{x^2+y^2}$ is ....

120. The least degree of a polynomial with integer coefficient whose one of the roots may be $\cos 12^\circ$ is

121. If $A+B+C=180^\circ$, $\dfrac{\sin 2A+\sin 2B+\sin 2C}{\sin A+\sin B+\sin C}=k\sin\dfrac{A}{2}\sin\dfrac{B}{2}\sin\dfrac{C}{2}$ then the value of $3k^3+2k^2+k+1$ is equal to

122. The value of $f(x)=x^4+4x^3+2x^2-4x+7$, when $x=\cot\dfrac{11\pi}{8}$ is ....

123. In any ΔABC, then minimum value of $2020\sum\dfrac{\sqrt{(\sin A)}}{(\sqrt{(\sin B)}+\sqrt{(\sin C)}-\sqrt{(\sin A)})}$ must be

124. If $\sin\theta+\sin^2\theta+\sin^3\theta=1$, then the value of $\cos^6\theta-4\cos^4\theta+8\cos^2\theta$ must be

125. $16\left(\cos\theta-\cos\frac{\pi}{8}\right)\left(\cos\theta-\cos\frac{3\pi}{8}\right)\left(\cos\theta-\cos\frac{5\pi}{8}\right)\left(\cos\theta-\cos\frac{7\pi}{8}\right)=\lambda\cos 4\theta$, then the value of λ is

....

126. If $\frac{1}{\sin 20^\circ}+\frac{1}{\sqrt{3}\cos 20^\circ}=2k\cos 40^\circ$, then $18k^4+162k^2+369$ is equal to

Trigonometric Functions and Identities Exercise 7 : Subjective Type Questions

127. Prove that $\tan 82\frac{1}{2}^\circ=(\sqrt{3}+\sqrt{2})(\sqrt{2}+1)$

or $\cot 7\frac{1}{2}^\circ=\sqrt{2}+\sqrt{3}+\sqrt{4}+\sqrt{6}$.

128. If $m\sin(\alpha+\beta)=\cos(\alpha-\beta)$, prove that

$$\frac{1}{1-m\sin 2\alpha}+\frac{1}{1-m\sin 2\beta}=\frac{2}{1-m^2}.$$

129. If $\alpha+\beta+\gamma=\pi$ and

$\tan\frac{1}{4}(\beta+\gamma-\alpha)\tan\frac{1}{4}(\gamma+\alpha-\beta)\tan\frac{1}{4}(\alpha+\beta-\gamma)=1$,

then prove that $1+\cos\alpha+\cos\beta+\cos\gamma=0$.

130. Find the value of a for which the equation $\sin^4 x+\cos^4 x=a$ has real solutions.

131. If a and b are positive quantities and $a\geq b$, then find the minimum positive values of $a\sec\theta-b\tan\theta$.

132. If a, b, c and k are constant quantities and α, β, γ are variable subjects to the relation $a\tan\alpha+b\tan\beta+c\tan\gamma=k$, then find the minimum value of $\tan^2\alpha+\tan^2\beta+\tan^2\gamma$.

133. If $\frac{x}{\tan(\theta+\alpha)}=\frac{y}{\tan(\theta+\beta)}=\frac{z}{\tan(\theta+\gamma)}$, prove that :

$$\Sigma\frac{x+y}{x-y}\sin^2(\alpha-\beta)=0.$$

134. Let $a_1, a_2, \ldots, a_n$ be real constants, x be a real variable and $f(x)=\cos(a_1+x)+\frac{1}{2}\cos(a_2+x)+\frac{1}{4}\cos(a_3+x)+\ldots+\frac{1}{2^{n-1}}\cos(a_n+x)$

Given that $f(x_1)=f(x_2)=0$, prove that $x_2-x_1=m\pi$ for some integer m.

135. Eliminate θ from the equations $\tan(n\theta+\alpha)-\tan(n\theta+\beta)=x$ and $\cot(n\theta+\alpha)-\cot(n\theta+\beta)=y$.

136. If $\{\sin(\alpha-\beta)+\cos(\alpha+2\beta)\sin\beta\}^2=4\cos\alpha\sin\beta\sin(\alpha+\beta)$. Then, prove that $\tan\alpha+\tan\beta=\frac{\tan\beta}{(\sqrt{2}\cos\beta-1)^2}$; $\alpha,\beta\in(0,\pi/4)$.

137. If A, B, C are the angle of a triangle and

$$\begin{vmatrix}\sin A & \sin B & \sin C\\ \cos A & \cos B & \cos C\\ \cos^3 A & \cos^3 B & \cos^3 C\end{vmatrix}=0,$$

then show that ΔABC is an isosceles.

138. In any ΔABC, prove that

$$\Sigma\frac{\sqrt{\sin A}}{\sqrt{\sin B}+\sqrt{\sin C}-\sqrt{\sin A}}\geq 3$$

and the equality holds if and only if triangle is equilateral.

139. If $2(\cos(\alpha-\beta)+\cos(\beta-\gamma)+\cos(\gamma-\alpha)+3=0$, prove that

$$\frac{d\alpha}{\sin(\beta+\theta)\sin(\gamma+\theta)}+\frac{d\beta}{\sin(\alpha+\theta)\sin(\gamma+\theta)}+\frac{d\gamma}{\sin(\alpha+\theta)\sin(\beta+\theta)}=0,$$

where, 'θ' is any real angle such that $\alpha+\theta, \beta+\theta, \gamma+\theta$ are not the multiple of π.

140. If the quadratic equation $4^{\sec^2\alpha}x^2+2x+\left(\beta^2-\beta+\frac{1}{2}\right)=0$ have real roots, then find all the possible values of $\cos\alpha+\cos^{-1}\beta$.

141. Four real constants a, b, A, B are given and $f(\theta)=1-a\cos\theta-b\sin\theta-A\cos 2\theta-B\sin 2\theta$. Prove that if $f(\theta)\geq 0, \forall\,\theta\in R$, then $a^2+b^2\leq 2$ and $A^2+B^2\leq 1$.

142. If $\frac{\cos\theta_1}{\cos\theta_2}+\frac{\sin\theta_1}{\sin\theta_2}=\frac{\cos\theta_0}{\cos\theta_2}+\frac{\sin\theta_0}{\sin\theta_2}=1$, where θ_1 and θ_0 do not differ by an even multiple of π, prove that

$$\frac{\cos\theta_1\cdot\cos\theta_0}{\cos^2\theta_2}+\frac{\sin\theta_1\cdot\sin\theta_0}{\sin^2\theta_2}=-1$$

143. Prove that

$$\sum_{k=1}^{n-1} {}^nC_k [\cos kx . \cos(n+k)x + \sin(n-k)x . \sin(2n-k)x] = (2^n - 2)\cos nx.$$

144. Determine all the values of x in the interval $x \in [0, 2\pi]$ which satisfy the inequality
$2\cos x \le |\sqrt{1+\sin 2x} - \sqrt{1-\sin 2x}| \le \sqrt{2}$.

145. Find all the solutions of this equation
$x^2 - 3\left[\sin\left(x - \frac{\pi}{6}\right)\right] = 3$, where [.] represents the greatest integer function.

146. In a ΔABC, prove that

$$\sum_{r=0}^{n} {}^nC_r a^r b^{n-r} \cos(rB - (n-r)A) = c^n.$$

147. Resolve $z^5 + 1$ into linear and quadratic factors with real coefficients. Hence, or otherwise deduce that,
$4\sin\left(\frac{\pi}{10}\right) \cdot \cos\left(\frac{\pi}{5}\right) = 1.$

148. Prove that the roots of the equation
$8x^3 - 4x^2 - 4x + 1 = 0$ are $\cos\frac{\pi}{7}, \cos\frac{3\pi}{7}, \cos\frac{5\pi}{7}$ and hence, show that $\sec\frac{\pi}{7} + \sec\frac{3\pi}{7} + \sec\frac{5\pi}{7} = 4$ and deduce the equation whose roots are $\tan^2\frac{\pi}{7}, \tan^2\frac{3\pi}{7}, \tan^2\frac{5\pi}{7}$.

Trigonometric Functions and Identities Exercise 8 :
Questions Asked in Previous 10 Years Exam

149. Let α and β be non-zero real numbers such that $2(\cos\beta - \cos\alpha) + \cos\alpha\cos\beta = 1$. Then which of the following is/are true?

[More than one correct option 2017 Adv.]

(a) $\sqrt{3}\tan\left(\frac{\alpha}{2}\right) - \tan\left(\frac{\beta}{2}\right) = 0$

(b) $\tan\left(\frac{\alpha}{2}\right) - \sqrt{3}\tan\left(\frac{\beta}{2}\right) = 0$

(c) $\tan\left(\frac{\alpha}{2}\right) + \sqrt{3}\tan\left(\frac{\beta}{2}\right) = 0$

(d) $\sqrt{3}\tan\left(\frac{\alpha}{2}\right) + \tan\left(\frac{\beta}{2}\right) = 0$

150. Let $-\frac{\pi}{6} < \theta < -\frac{\pi}{12}$. Suppose α_1 and β_1 are the roots of the equation $x^2 - 2x\sec\theta + 1 = 0$, and α_2 and β_2 are the roots of the equation $x^2 + 2x\tan\theta - 1 = 0$. If $\alpha_1 > \beta_1$ and $\alpha_2 > \beta_2$, then $\alpha_1 + \beta_2$ equals to

[Single correct option 2016 Adv.]

(a) $2(\sec\theta - \tan\theta)$ (b) $2\sec\theta$
(c) $-2\tan\theta$ (d) 0

151. The value of $\sum_{k=1}^{13} \frac{1}{\sin\left(\frac{\pi}{4} + \frac{(k-1)\pi}{6}\right)\sin\left(\frac{\pi}{4} + \frac{k\pi}{6}\right)}$ is equal to

[Single correct option 2016 Adv.]

(a) $3 - \sqrt{3}$ (b) $2(3 - \sqrt{3})$ (c) $2(\sqrt{3} - 1)$ (d) $2(2 + \sqrt{3})$

152. Let $f : (-1, 1) \to R$ be such that $f(\cos 4\theta) = \frac{2}{2 - \sec^2\theta}$ for $\theta \in \left(0, \frac{\pi}{4}\right) \cup \left(\frac{\pi}{4}, \frac{\pi}{2}\right)$. Then, the value(s) of $f\left(\frac{1}{3}\right)$ is/are

[More than one correct option 2012]

(a) $1 - \sqrt{\frac{3}{2}}$ (b) $1 + \sqrt{\frac{3}{2}}$ (c) $1 - \sqrt{\frac{2}{3}}$ (d) $1 + \sqrt{\frac{2}{3}}$

153. The number of all possible values of θ, where $0 < \theta < \pi$, for which the system of equations
$(y+z)\cos 3\theta = (xyz)\sin 3\theta$
$x\sin 3\theta = \frac{2\cos 3\theta}{y} + \frac{2\sin 3\theta}{z}$
and $(xyz)\sin 3\theta = (y+2z)\cos 3\theta + y\sin 3\theta$ have a solution (x_0, y_0, z_0) with $y_0 z_0 \ne 0$, is ......

[Integer Answer Type 2010]

154. For $0 < \theta < \frac{\pi}{2}$, the solution(s) of
$\sum_{m=1}^{6} \operatorname{cosec}\left(\theta + \frac{(m-1)\pi}{4}\right)\operatorname{cosec}\left(\theta + \frac{m\pi}{4}\right) = 4\sqrt{2}$ is/are

[More than one correct option 2009]

(a) $\frac{\pi}{4}$ (b) $\frac{\pi}{6}$ (c) $\frac{\pi}{12}$ (d) $\frac{5\pi}{12}$

155. If $\frac{\sin^4 x}{2} + \frac{\cos^4 x}{3} = \frac{1}{5}$, then

[Single correct option 2009]

(a) $\tan^2 x = \frac{2}{3}$ (b) $\frac{\sin^8 x}{8} + \frac{\cos^8 x}{27} = \frac{1}{125}$
(c) $\tan^2 x = \frac{1}{3}$ (d) $\frac{\sin^8 x}{8} + \frac{\cos^8 x}{27} = \frac{2}{125}$

156. Let $\theta \in \left(0, \frac{\pi}{4}\right)$ and $t_1 = (\tan\theta)^{\tan\theta}$, $t_2 = (\tan\theta)^{\cot\theta}$, $t_3 = (\cot\theta)^{\tan\theta}$ and $t_4 = (\cot\theta)^{\cot\theta}$, then

[Single correct option 2006]

(a) $t_1 > t_2 > t_3 > t_4$ (b) $t_4 > t_3 > t_1 > t_2$
(c) $t_3 > t_1 > t_2 > t_4$ (d) $t_2 > t_3 > t_1 > t_4$

157. The number of ordered pairs (α, β), where $\alpha, \beta \in (-\pi, \pi)$ satisfying $\cos(\alpha - \beta) = 1$ and $\cos(\alpha + \beta) = \frac{1}{e}$ is **[Single correct option 2005]**

(a) 0 (b) 1
(c) 2 (d) 4

II. JEE Mains and AIEEE

158. $5(\tan^2 x - \cos^2 x) = 2\cos 2x + 9$, then the value of $\cos 4x$ is **[2017 JEE Main]**

(a) $-\frac{3}{5}$ (b) $\frac{1}{3}$ (c) $\frac{2}{9}$ (d) $-\frac{7}{9}$

159. If $f_k(x) = \frac{1}{k}(\sin^k x + \cos^k x)$, where $x \in R, k \geq 1$, then $f_4(x) - f_6(x)$ is equal to **[2014 JEE Main]**

(a) $\frac{1}{6}$ (b) $\frac{1}{3}$ (c) $\frac{1}{4}$ (d) $\frac{1}{12}$

160. The expression $\frac{\tan A}{1 - \cot A} + \frac{\cot A}{1 - \tan A}$ can be written as **[2013 JEE Main]**

(a) $\sin A \cos A + 1$ (b) $\sec A \operatorname{cosec} A + 1$
(c) $\tan A + \cot A$ (d) $\sec A + \operatorname{cosec} A$

161. In a ΔPQR, if $3\sin P + 4\cos Q = 6$ and $4\sin Q + 3\cos P = 1$, then the angle R is equal to **[2012 AIEEE]**

(a) $\frac{5\pi}{6}$ (b) $\frac{\pi}{6}$
(c) $\frac{\pi}{4}$ (d) $\frac{3\pi}{4}$

162. If $A = \sin^2 x + \cos^4 x$, then for all real x **[2011 AIEEE]**

(a) $\frac{13}{16} \leq A \leq 1$ (b) $1 \leq A \leq 2$
(c) $\frac{3}{4} \leq A \leq \frac{13}{16}$ (d) $\frac{3}{4} \leq A \leq 1$

163. Let $\cos(\alpha + \beta) = \frac{4}{5}$ and $\sin(\alpha - \beta) = \frac{5}{13}$, where $0 \leq \alpha, \beta \leq \frac{\pi}{4}$. Then, $\tan 2\alpha$ is equal to **[2010 AIEEE]**

(a) $\frac{25}{16}$ (b) $\frac{56}{33}$ (c) $\frac{19}{12}$ (d) $\frac{20}{7}$

164. Let A and B denote the statements

$A: \cos\alpha + \cos\beta + \cos\gamma = 0$

$B: \sin\alpha + \sin\beta + \sin\gamma = 0$

If $\cos(\beta - \gamma) + \cos(\gamma - \alpha) + \cos(\alpha - \beta) = -\frac{3}{2}$, then **[2009 AIEEE]**

(a) A is true and B is false
(b) A is false and B is true
(c) Both A and B are true
(d) Both A and B are false

165. A triangular park is enclosed on two sides by a fence and on the third side by a straight river bank. The two sides having fence are of same length x. The maximum area enclosed by the park is **[2006 AIEEE]**

(a) $\sqrt{\frac{x^3}{8}}$ (b) $\frac{1}{2}x^2$ (c) πx^2 (d) $\frac{3}{2}x^2$

166. If $0 < x < \pi$ and $\cos x + \sin x = \frac{1}{2}$, then $\tan x$ is **[2006 AIEEE]**

(a) $\frac{(4 - \sqrt{7})}{3}$ (b) $-\frac{(4 + \sqrt{7})}{3}$
(c) $\frac{(1 + \sqrt{7})}{4}$ (d) $\frac{(1 - \sqrt{7})}{4}$

167. In a ΔPQR, $\angle R = \frac{\pi}{2}$. If $\tan\left(\frac{P}{2}\right)$ and $\tan\left(\frac{Q}{2}\right)$ are the roots of $ax^2 + bx + c = 0, a \neq 0$, then **[2005 AIEEE]**

(a) $b = a + c$ (b) $b = c$
(c) $c = a + b$ (d) $a = b + c$

Answers

Exercise for Session 1

1. 72°, 18° 2. $\frac{\pi}{2}$ cm 3. 70 m 4. 45° 5. 8π
6. $\left(\frac{11}{90}\right)^c$ 7. 252 cm 8. 880 cm/s 9. 1.7 cm 10. 7

Exercise for Session 2

3. -3 4. $\pm\sqrt{a^2 + b^2 - c^2}$ 5. 13 6. 2 7. $\frac{1}{12}$
10. 4

Exercise for Session 3

1. $\frac{2k}{k^2 + 1}$ 2. 1 4. $x = 1, y = 0$ 6. 0
7. $\frac{x^2}{a^2} + \frac{y^2}{b^2} = 2$ 9. 10 10. $\frac{7}{5}$

Exercise for Session 4

1. 28 3. -2
4.

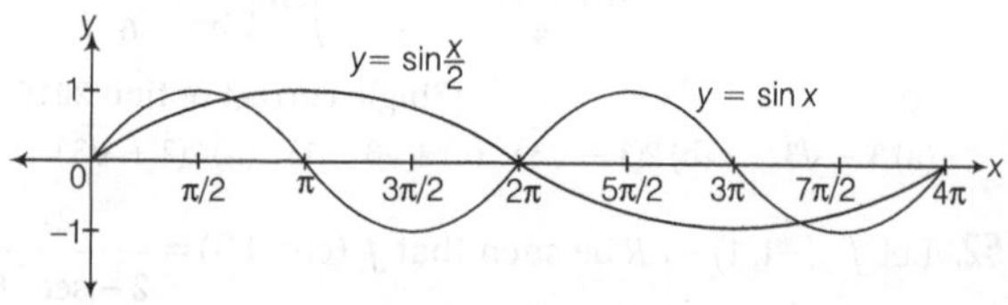

5.

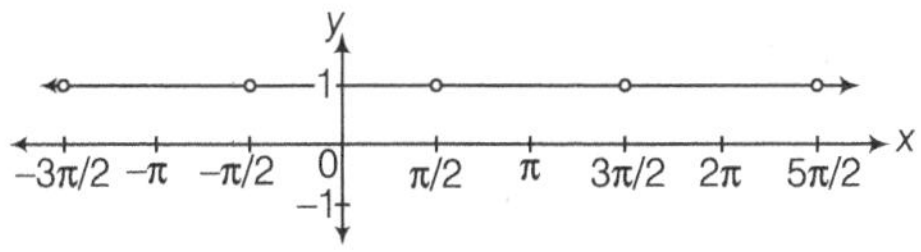

From the graph, the period of the function is π.

7. $\left[\frac{\pi}{4}, \frac{5\pi}{4}\right]$

8.

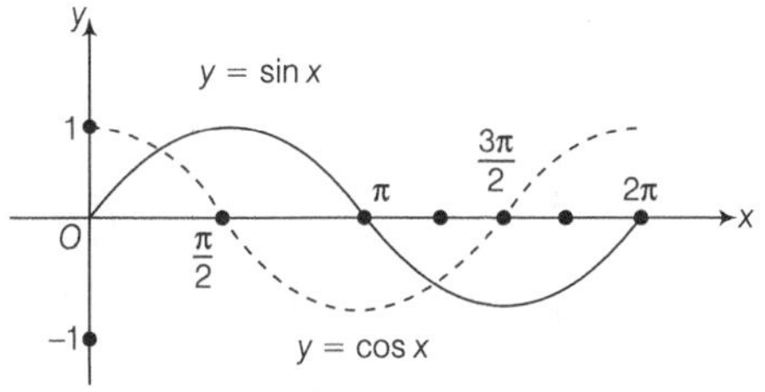

9.

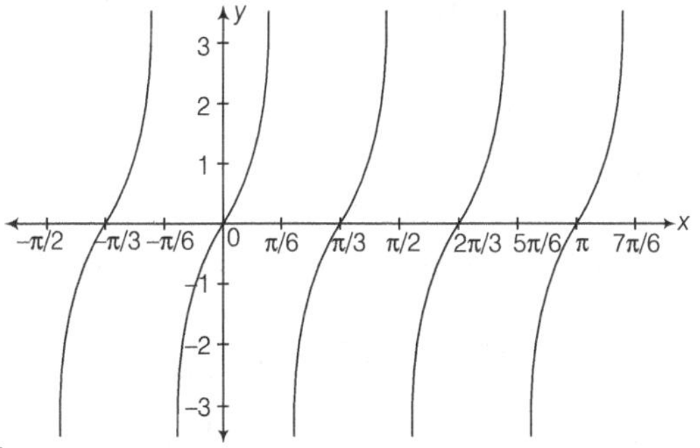

10. $\frac{1}{4}$

Exercise for Session 5

1. $\sqrt{3}$ 2. Negative 3. − 1 4. 1 5. 1 6. 1
7. 2 8. does not exists any real solutions 9. 2
10. 0

Exercise for Session 6

1. I and IV 3. $\frac{\pi}{4}$ 4. $\frac{65}{33}$ 5. $\tan\beta + 2\tan\gamma$
6. $a^2 + b^2$ 8. 27

Exercise for Session 7

4. $\frac{4}{5}$ 5. 0 6. $2\cot^n\left(\frac{A-B}{2}\right)$ 7. $\frac{1}{8}$ 8. $\frac{2\pi}{8}$

Exercise for Session 8

1. False 2. $\tan\theta$ 3. $\frac{2}{\sqrt{5}}$ 5. $\alpha = -1, \beta = 3$ 6. $\frac{24}{7}$
7. − 1 8. 1 9. 15 10. 1

Exercise for Session 9

1. $-2 \pm \sqrt{5}$ 2. $\frac{3+2\sqrt{2}}{8}$ 3. $\frac{4}{\sqrt{3}}$ 4. $2\cos n\theta$, 10. $\cot\frac{\pi}{5}$

Exercise for Session 10

4. $\frac{1}{2}$ 5. 2 7. − 1 8. 1

Exercise for Session 11

2. 0 4. 2 5. positive 6. $\frac{3}{4} \le A \le 1$ 7. 4.
8. $-\sqrt{2}$ and $\sqrt{2}$ 10. 13 : 4 or 1 : 4

Chapter Exercises

1. (b) 2. (b) 3. (a) 4. (a) 5. (b) 6. (a)
7. (b) 8. (a) 9. (a) 10. (b) 11. (d) 12. (d)
13. (d) 14. (b) 15. (a) 16. (b) 17. (b) 18. (b)
19. (a) 20. (a) 21. (c) 22. (a) 23. (a) 24. (b)
25. (d) 26. (c) 27. (b) 28. (b) 29. (c) 30. (b)
31. (c) 32. (a) 33. (b) 34. (c) 35. (b) 36. (b)
37. (a) 38. (d) 39. (d) 40. (c) 41. (c)
42. (b) 43. (b) 44. (c) 45. (a) 46. (b) 47. (c)
48. (b) 49. (a) 50. (a) 51. (a) 52. (d) 53. (a)
54. (a,b,c,d) 55. (a,b,c) 56. (b,d) 57. (b,c) 58. (b,c)
59. (b,c) 60. (a,b,d) 61. (b,c) 62. (c,d) 63. (d) 64. (b, c, d)
65. (b, c) 66. (a, b, c, d) 67. (a, b, c, d) 68. (c, d)
69. (a, b, c, d) 70. (b, c) 71. (a, b, c, d) 72. (b, c)
73. (a, b, c, d) 74. (b, c) 75. (a) 76. (b) 77. (b)
78. (d) 79. (d) 80. (a) 81. (a) 82. (c) 83. (b)
84. (b) 85. (b) 86. (b) 87. (a) 88. (b) 89. (b)
90. (c) 91. (b) 92. (a) 93. (c) 94. (b) 95. (d)
96. (c) 97. (b) 98. (c)
99. A—(p, q, r, s, t); B—(s, t); C—(r)
100. A—(r, s); B—(r, t); C—(p, q)
101. A—(r); B—(p); C—(p); D—(q)
102. A—(r, t); B—(p, s); C—q
103. (2) 104. (1) 105. (5) 106. (2) 107. (5) 108. (6)
109. (5) 110. (35) 111. (3) 112. (2) 113. (1) 114. (2)
115. (3136) 116. (2634) 117. (91) 118. (3448)
119. (1) 120. (4) 121. (1673) 122. (6) 123. (6060)
124. (4) 125. (2) 126. (1745)
130. $\frac{1}{2} \le a \le 1$ 131. Minimum value is $\sqrt{a^2 - b^2}$
132. Minimum value is $\left(\frac{k^2}{a^2 + b^2 + c^2}\right)$
135. $\cot(\alpha + \beta) = \frac{1}{x} - \frac{1}{y}$
140. $\cos\alpha - \cos^{-1}\beta = \begin{cases} \frac{\pi}{3} - 1, \text{ when } n \text{ is an even integer} \\ \frac{\pi}{3} + 1, \text{ when } n \text{ is an even integer} \end{cases}$
144. $x \in \left[\frac{\pi}{4}, \frac{7\pi}{4}\right]$
145. Only two solutions, $x = 0, \sqrt{3}$
148. Required equation is, $z^3 - 21z^2 + 35z - 7 = 0$ whose roots are $\tan^2\frac{\pi}{7}, \tan^2\frac{3\pi}{7}, \tan^2\frac{5\pi}{7}$
149. (b, c) 150. (c) 151. (c) 152. (a,b) 153. (3)
154. (c,d) 155. (b) 156. (b) 157. (d) 158. (d) 159. (d)
160. (b) 161. (b) 162. (d) 163. (b) 164. (c) 165. (b)
166. (b) 167. (c)

Solutions

1. Given series

$$= \left(\sin\frac{2\pi}{11} - \cos\frac{2\pi}{11}\right) + \left(\sin\frac{4\pi}{11} - \cos\frac{4\pi}{11}\right)$$

$$+ \left(\sin\frac{6\pi}{11} - \cos\frac{6\pi}{11}\right) + \ldots + \left(\sin\frac{20\pi}{11} - \cos\frac{20\pi}{11}\right)$$

$$= \left(\sin\frac{2\pi}{11} + \sin\frac{4\pi}{11} + \ldots + \sin\frac{20\pi}{11}\right)$$

$$-\left(\cos\frac{2\pi}{11} + \cos\frac{4\pi}{11} + \ldots + \cos\frac{20\pi}{11}\right)$$

$$= \frac{\sin\pi \cdot \sin\frac{10\pi}{11}}{\sin\frac{\pi}{11}} - \frac{\cos\pi \cdot \sin\frac{10\pi}{11}}{\sin\frac{\pi}{11}}$$

$$= 0 + \frac{\sin\left(\pi - \frac{\pi}{11}\right)}{\sin\frac{\pi}{11}} = 1$$

2. $(a+1)^2 + \text{cosec}^2\left(\frac{\pi a}{2} + \frac{\pi x}{2}\right) - 1 = 0$

or $\quad (a+1)^2 + \cot^2\left(\frac{\pi a}{2} + \frac{\pi x}{2}\right) = 0$

From option [b], if $a = -1$ and $\cot^2\left(\frac{-\pi}{2} + \frac{\pi x}{2}\right) = 0$

$\Rightarrow \quad \tan^2\left(\frac{\pi x}{2}\right) = 0$

$\Rightarrow \quad \frac{\pi}{2} = 1$

3. $f(x) = 9\sin^2 x - 16\cos^2 x - 10(3\sin x - 4\cos x)$

$$-10(3\sin x + 4\cos x) + 100$$

$$= 25\sin^2 x - 60\sin x + 84$$

$$= (5\sin x - 6)^2 + 48$$

$\therefore f(x)_{\min}$ occurs when $\sin x = 1$

Minimum value = 49

4. $S = \frac{1}{1+\tan^3 0°} + \frac{1}{1+\tan^3 10°} + \ldots + \frac{1}{1+\tan^3 80°}$

Now, $\frac{1}{1+\tan^3\theta} + \frac{1}{1+\tan^3(90-\theta)}$

$$= \frac{1}{1+\tan^3\theta} + \frac{1}{1+\cot^3\theta}$$

$$= \frac{1}{1+\tan^3\theta} + \frac{\tan^3\theta}{1+\tan^3\theta}$$

$$= \frac{1+\tan^3\theta}{1+\tan^3\theta} = 1$$

Hence, $S = 1 + (1+1+1+1) = 5$

5. Clearly, $\sqrt{1-\sin^2 110°} \cdot \sec 110°$

$$= |\cos 110°| \sec 110°$$

$$= -\cos 110° \sec 110° = -1$$

6. $\tan\alpha + \tan\beta = -p$

$\tan\alpha \tan\beta = q$

$\tan(\alpha+\beta) = \frac{-p}{1-q} = \frac{p}{q-1}$

$$\frac{1}{1+\tan^2(\alpha+\beta)}[\tan^2(\alpha+\beta) + p\tan(\alpha+\beta) + q]$$

$$= \frac{1}{1+\frac{p^2}{(q-1)^2}}\left[\frac{p^2}{(q-1)^2} + \frac{p^2}{(q-1)} + q\right]$$

$$= \frac{1}{(q-1)^2+p^2}[p^2 + p^2(q-1) + q(q-1)^2]$$

$$= \frac{1}{p^2+(q-1)^2}[p^2 q + q(q-1)^2]$$

$$= q\left[\frac{p^2+(q-1)^2}{p^2+(q-1)^2}\right] = q$$

7. Let A be the expression. Multiplying A by 2^{2008} and using $2\sin\theta\cos\theta = \sin 2\theta$,

we have $\quad 2^{2008}A = \sin\frac{\pi}{2} = 1$. $A = \frac{1}{2^{2008}}$

Alternatively $\sin\left(\frac{\pi}{2^{2009}}\right)\cos\left(\frac{\pi}{2^{2009}}\right) = \frac{1}{2}\sin\left(\frac{\pi}{2^{2008}}\right)$

$$= \frac{1}{2^2}\cdot 2\sin\left(\frac{\pi}{2^{2008}}\right)\cos\left(\frac{\pi}{2^{2008}}\right)$$

$$= \frac{1}{2^2}\sin\left(\frac{\pi}{2^{2007}}\right)$$

Similarly, continued product upto,

$$\cos\left(\frac{\pi}{2^2}\right) = \frac{1}{2^{2008}}\sin\left(\frac{\pi}{2}\right) = \frac{1}{2^{2008}}$$

8. $\tan(A+B) = \frac{\tan A + \tan B}{1 - \tan A \tan B}$

$$= \frac{\tan A + \frac{n\sin A\cos A}{1-n\cos^2 A}}{1 - \tan A \cdot \frac{n\sin A\cos A}{1-n\cos^2 A}}$$

$$= \frac{\sin A(1-n\cos^2 A) + n\sin A\cos^2 A}{\cos A(1-n\cos^2 A) - n\sin^2 A\cos A}$$

$$= \frac{\sin A - 0}{\cos A(1-n\cos^2 A - n\sin^2 A)}$$

$$= \frac{\sin A}{(1-n)\cos A}$$

9. We have, $\quad Q = \sum_{r=0}^{n} \frac{\sin(3^r\theta)}{\cos(3^{r+1}\theta)}$

$$= \frac{\sin\theta}{\cos 3\theta} + \frac{\sin 3\theta}{\cos 9\theta} + \frac{\sin 9\theta}{\cos 27\theta} + \ldots\ldots + \frac{\sin(3^n\theta)}{\cos(3^{n+1}\theta)}$$

As, $\dfrac{\sin\theta}{\cos 3\theta} = \dfrac{2\sin\theta\cos\theta}{2\cos\theta\cos 3\theta} = \dfrac{\sin 2\theta}{2\cos\theta\cos 3\theta}$

$$= \frac{1}{2}\left[\frac{\sin(3\theta - \theta)}{\cos\theta\cos 3\theta}\right]$$

$$= \frac{1}{2}(\tan 3\theta - \tan\theta)$$

$\therefore$ $Q = \dfrac{1}{2}[(\tan 3\theta - \tan\theta) + (\tan 9\theta - \tan 3\theta) + \ldots + \tan 3^{n+1}\theta - \tan 3^n\theta)]$

$\Rightarrow$ $Q = \dfrac{P}{2} \Rightarrow P = 2Q$

10. Expression
$(\cos^4 1° + \cos^4 2° + \cos^4 3° + \ldots + \cos^4 179°)$

$$-(\sin^4 1° + \sin^4 2° + \sin^4 3° + \ldots + \sin^4 179°)$$

$$= \cos 2° + \cos 4° + \cos 6° + \ldots + \cos(358°)$$

$$= \frac{\cos\left(\frac{2° + 358°}{2}\right)\cdot \sin(179 \times 1°)}{\sin 1°}$$

$$= \cos(180°) = -1.$$

11. $\sin x + \sin y = a$ ...(i)

$\cos x + \cos y = 2a$ ...(ii)

On squaring and adding Eqs. (i) and (ii), we get

$$2 + 2\cos(x - y) = 5a^2$$

$$\cos(x - y) = \frac{5a^2 - 2}{2}$$

12. $P(x) = \sqrt{3 + 2(\cos x + \cos x + \cos 2x)}$

$$= \sqrt{3 + 2(2\cos x + 2\cos x^2 x - 1}$$

$$= \sqrt{4\cos^2 x + 4\cos^2 x + 1}$$

$$= |2\cos x + 1|$$

13. Consider $y = 5\sec c^2\theta - \tan^2\theta + 4\text{cosec}^2\theta$

$\therefore$ $y = 5 + 5\tan^2\theta - \tan^2\theta + 4 + \cot^2\theta$

$$y = 9 + 4(\tan^2 + \cot^2)$$

$$= 9 + 4[(\tan\theta - \cot\theta)^2 + 2]$$

$\therefore$ $y_{\min} = 9 + 8 = 17$

$\Rightarrow$ Maximum value of the expression is $\dfrac{1}{17} = \dfrac{p}{q}$

$\Rightarrow$ $p + q = 1 + 17 = 18$

14. $f_n(\alpha) = \tan n\alpha$ and $f_n\left(\dfrac{\pi}{32}\right) = \tan\dfrac{\pi}{8} = \sqrt{2} - 1$

15. Let $\sin x + \cos x = t$

$\therefore$ $y = \left|t + \dfrac{1}{t}\right|$

Hence, minimum value of y is 2.

16. $a = \cos(2012\,\pi) = 1$

$b = \sec(2013\,\pi) = -1$

$c = \tan(2014\,\pi) = 0$

$\therefore b < c < a$

17. InΔABC, $\sum \tan\dfrac{A}{2}\cdot\tan\dfrac{B}{2} = 1$

$\therefore$ $\sum \tan^2\dfrac{A}{2} \geq \sum \tan\dfrac{A}{2}\tan\dfrac{B}{2} = 1$

$[\because a^2 + b^2 + c^2 - ab - bc - ca \geq 0, \forall\, a, b, c \in R]$

$\therefore$ $3 + \sum \tan^2\dfrac{A}{2} \geq 4$

$\Rightarrow$ $3 + \tan^2\dfrac{A}{2} + \tan^2\dfrac{B}{2} + \tan^2\dfrac{C}{2} \geq 1 + 3$

$\Rightarrow$ $\sec^2\dfrac{A}{2} + \sec^2\dfrac{B}{2} + \sec^2\dfrac{C}{2} \geq 4$

18. The given equation can be rewritten as

$(2x - 1)^2 + 1 = \sin^2 y$, which is possible only when $x = \dfrac{1}{2}$,

$\sin^2 y = 1$

$\Rightarrow y = \dfrac{-\pi}{2}, \dfrac{\pi}{2}$ [as $x^2 + y^2 \leq 3$]

Thus, there are only two pairs (x, y) satisfying the given equation. They are $\left(\dfrac{1}{2}, \dfrac{-\pi}{2}\right)$ and $\left(\dfrac{1}{2}, \dfrac{\pi}{2}\right)$.]

19. Given,

$3\sin A + 4\cos B = 6$ ...(i)

$3\cos A + 4\sin B = 1$ ...(ii)

On squaring and adding Eqs. (i) and (ii), we get

$$9 + 16 + 24\sin(A + B) = 37$$

$$24\sin(A + B) = 12$$

$$\sin(A + B) = \frac{1}{2}$$

$\Rightarrow$ $\sin C = \dfrac{1}{2}$

$$C = 30° \text{ or } 150°$$

If $C = 150°$, then even of $B = 0$ and $A = 30°$.

The quantity $3\sin A + 4\cos B$

$$3\cdot\frac{1}{2} + 4 = 5\frac{1}{2} < 6$$

Hence, $C = 150°$ is not possible

$\Rightarrow$ $\angle C = 30°$ only

20. Area $= 3\cdot(8\cdot 3) + 3\cdot\dfrac{1}{2}r^2\theta$

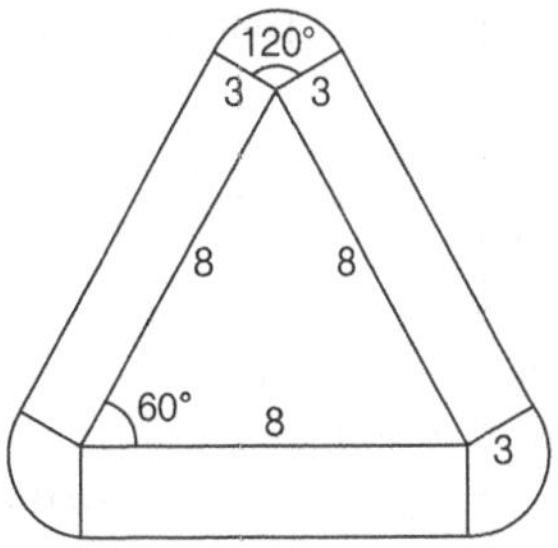

$$= 72 + \frac{3}{2}\cdot 9\cdot\frac{2\pi}{3}$$

$$= 72 + 9\pi$$

$$= 9(8 + \pi)$$

21. $m+n = a\{(\cos^3\alpha + \sin^3\alpha) + 3\cos\alpha\sin\alpha(\cos\alpha + \sin\alpha)\}$

$m+n = a\{\cos\alpha + \sin\alpha\}^3$

Similarly, $\quad m - n = a\{\cos\alpha - \sin\alpha\}^3$

$$(m+n)^{2/3} = a^{2/3}(\cos\alpha + \sin\alpha)^2 \quad \text{...(i)}$$

Similarly, $(m-n)^{2/3} = a^{2/3}(\cos\alpha - \sin\alpha)^2 \quad$ ...(ii)

On adding Eqs. (i) and (ii), we get

$(m+n)^{2/3} + (m-n)^{2/3} = a^{2/3}\,(2)$

$\Rightarrow \qquad = 2a^{2/3}$

22. $BD = x\tan C$ in ΔPDB

and $DC = x\tan B$ for ΔPDC

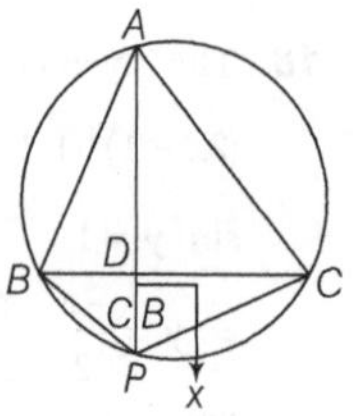

$\therefore \quad BD + DC = a = x(\tan B + \tan C)$

$$\frac{a}{x} = \tan B + \tan C$$

Similarly, $\quad \dfrac{b}{y} = \tan A + \tan C$

$$\frac{c}{z} = \tan A + \tan B$$

$$\therefore \quad \frac{a}{2x} + \frac{b}{2y} + \frac{c}{2z} = \frac{1}{2}\left(\frac{a}{x} + \frac{b}{y} + \frac{c}{z}\right) \tan A + \tan B + \tan C$$

23. $\sin\theta = \dfrac{x}{l} \quad$ ...(i)

Also, $\quad \cos 2\theta = \dfrac{6-x}{x}$

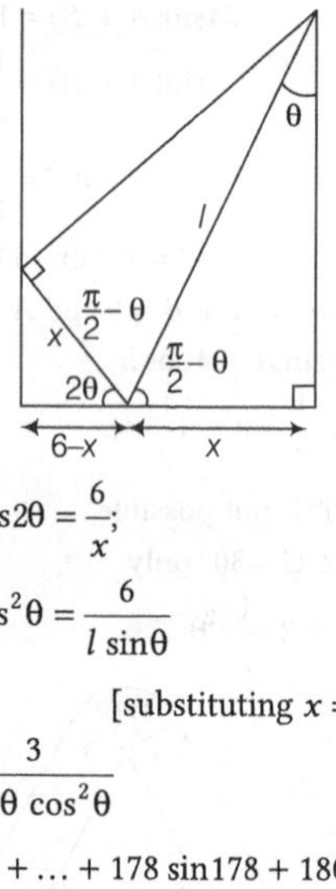

$$1 + \cos 2\theta = \frac{6}{x};$$

$$2\cos^2\theta = \frac{6}{l\sin\theta}$$

[substituting $x = l\sin\theta$ from Eq. (i)]

$$l = \frac{3}{\sin\theta\cos^2\theta}$$

24. $S = 2\sin 2^\circ + 4\sin 4^\circ + \dots + 178\sin 178 + 180^\circ\sin 180^\circ$

$S = 2[\sin 2^\circ + 2\sin 4^\circ + 3\sin 6^\circ + \dots + 89\sin 178^\circ] \quad$ ...(i)

$S = 2[89\sin 178^\circ + 88\sin 176^\circ + \dots + 1\cdot\sin 2^\circ] \quad$ ...(ii)

On adding Eqs. (i) and (ii), we get

[converting in reverse order]

$2S = 2[90(\sin 2^\circ + \sin 4^\circ + \sin 6^\circ + \dots + \sin 178^\circ)]$

$$S = 90\cdot\frac{\sin\left(\frac{n\theta}{2}\right)}{\sin\frac{\theta}{2}}\sin\left(\frac{(n+1)\theta}{2}\right)$$

$$= \frac{90\sin(89^\circ)}{\sin 1^\circ}\cdot\sin 90^\circ \qquad [\theta = 2^\circ]$$

$$S = 90\cot 1^\circ$$

$$\text{Average value} = \frac{90\cot 1^\circ}{90} = \cot 1^\circ$$

25. In 1st case, $r = \cot\dfrac{\pi}{5}$; $R = \operatorname{cosec}\dfrac{\pi}{5}$

2nd case, $r_1 = \cot\dfrac{\pi}{7}$; $R_1 = \operatorname{cosec}\dfrac{\pi}{7}$

$$\therefore \quad A_1 = \pi(R^2 - r^2) = \pi\left(\operatorname{cosec}^2\frac{\pi}{5} - \cot^2\frac{\pi}{5}\right) = \pi$$

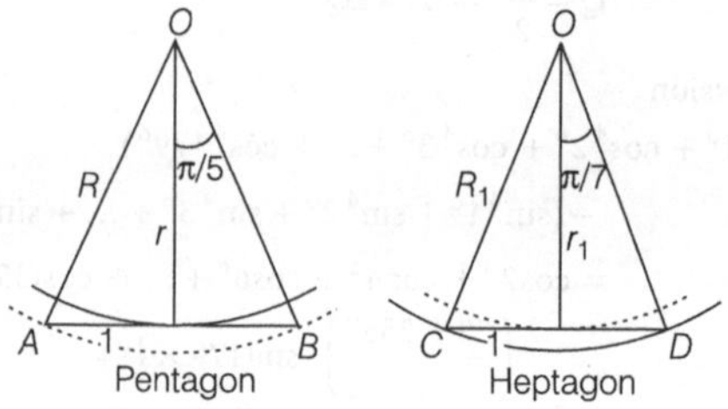

M $\quad A_2 = \pi(R_1^2 - r_1^2)$

$$= \pi\left(\operatorname{cosec}^2\frac{\pi}{7} - \cot^2\frac{\pi}{7}\right) = \pi$$

$\Rightarrow \quad A_1 = A_2$

26. $\displaystyle\sum_{r=1}^{18}\cos^2(5r)^\circ = \cos^2 5^\circ + \cos^2 10^\circ$

$$+ \cos^2 15^\circ + \dots + \cos^2 85^\circ + \cos^2 90^\circ$$

$$= (\cos^2 5^\circ + \cos^2 85^\circ) + (\cos^2 10^\circ + \cos^2 80^\circ)$$

$$+ (\cos^2 15^\circ + \cos^2 75^\circ) + \dots + (\cos^2 40^\circ + \cos^2 50^\circ) + \cos^2 45^\circ$$

$$= (\cos^2 5^\circ + \sin^2 5^\circ) + (\cos^2 10^\circ + \sin^2 10^\circ)$$

$$+ (\cos^2 15^\circ + \sin^2 15^\circ) + \dots + (\cos^2 40^\circ + \sin^2 40^\circ) + \cos^2 45^\circ$$

$$= 1 + 1 + 1 + \dots + 1 + \frac{1}{2} = 8 + \frac{1}{2} = \frac{17}{2}$$

27. $4x^2 - 4x|\sin\theta| - (1 - \sin^2\theta)$

$$= -1 + (2x - |\sin\theta|)^2$$

$\therefore$ Minimum value $= -1$

28. $\because \cos 3A + \cos 3B + \cos 3C = 1$

$\Rightarrow \quad \cos 3A + \cos 3B + \cos 3C - 1 = 0$

$\Rightarrow \quad \cos 3A + \cos 3B + \cos 3C + \cos 3\pi = 0$

$$\Rightarrow 2\cos\left(\frac{3A+3B}{2}\right)\cos\left(\frac{3A-3B}{2}\right) + 2\cos\left(\frac{3\pi+3C}{2}\right)\cos\left(\frac{3\pi-3C}{2}\right) = 0$$

$$\Rightarrow 2\cos\left(\frac{3\pi-3C}{2}\right)\left\{\cos\left(\frac{3A-3B}{2}\right) + \cos\left(\frac{3\pi+3C}{2}\right)\right\} = 0$$

$$\Rightarrow 2\cos\left(\frac{3\pi}{2} - \frac{3C}{2}\right)\cdot 2\cos\left(\frac{3\pi+3C+3A-3B}{4}\right)\cdot\cos\left(\frac{3\pi+3C-3A+3B}{4}\right) = 0$$

$$\Rightarrow \quad 2\cos\left(\frac{3\pi}{2}-\frac{3C}{2}\right)\cdot 2\cos\left(\frac{3\pi}{2}-\frac{3B}{2}\right)\cdot\cos\left(\frac{3\pi}{2}-\frac{3A}{2}\right)=0$$

$$\Rightarrow \quad -4\sin\left(\frac{3A}{2}\right)\sin\left(\frac{3B}{2}\right)\sin\left(\frac{3C}{2}\right)=0$$

$$\Rightarrow \quad \sin\left(\frac{3A}{2}\right)\sin\left(\frac{3B}{2}\right)\sin\left(\frac{3C}{2}\right)=0$$

$$\therefore \quad \frac{3A}{2}=\pi \text{ or } \frac{3B}{2}=\pi \text{ or } \frac{3C}{2}=\pi$$

$$\therefore \quad A=\frac{2\pi}{3} \text{ or } B=\frac{2\pi}{3} \text{ or } C=\frac{2\pi}{3}$$

29. $\because |\tan A|<1$

$$\Rightarrow \quad -1<\tan A<1 \text{ and } 0\le|A|<\frac{\pi}{2}$$

$$\Rightarrow \quad -\frac{\pi}{2}<A<\frac{\pi}{2}$$

$$\because \quad \sqrt{(1+\sin 2A)}=\sqrt{1+\frac{2\tan A}{1+\tan^2 A}}$$

$$=\frac{|1+\tan A|}{\sqrt{(1+\tan^2 A)}}=\frac{(1+\tan A)}{\sqrt{(1+\tan^2 A)}}$$

$$\text{and } \sqrt{(1-\sin 2A)}=\sqrt{1-\left(\frac{2\tan A}{1+\tan^2 A}\right)}$$

$$=\frac{|1-\tan A|}{\sqrt{(1+\tan^2 A)}}$$

$$=\frac{(1-\tan A)}{\sqrt{(1-\tan^2 A)}}$$

$$\therefore \quad \frac{\sqrt{(1+\sin 2A)}+\sqrt{(1-\sin 2A)}}{\sqrt{(1+\sin 2A)}-\sqrt{(1-\sin 2A)}}=\frac{2}{2\tan A}=\cot A$$

30. Let $f(\theta)=\cos^2(\cos\theta)+\sin^2(\sin\theta)$

$\because \quad -1\le\cos\theta\le 1$ and $-1\le\sin\theta\le 1$

$\therefore \quad \cos 1\le\cos(\cos\theta)\le 1$ and $-\sin 1\le\sin(\sin\theta)\le\sin 1$

$\therefore \quad \cos^2 1\le\cos^2(\cos\theta)\le 1$ and $0\le\sin^2(\sin\theta)\le\sin^2 1$

$\therefore$ Maximum value of $f(\theta)=1+\sin^2 1$

31. Let $f(x)=27^{\cos 2x}\,81^{\sin 2x}=3^{3\cos 2x+4\sin 2x}$

$$=3^{5\left(\frac{3}{5}\cos 2x+\frac{4}{5}\sin 2x\right)}$$

Let $\frac{3}{5}=\sin\phi$ and $\frac{4}{5}=\cos\phi$

Thus, $\quad f(x)=3^{5(\sin\phi\cos 2x+\cos\phi\sin 2x)}=3^{5[(\sin(\phi+2x)]}$

For minimum value of given function, $\sin(\phi+2x)$ will be minimum.

i.e. $\quad \sin(\phi+2x)=-1$

$$\therefore \quad f(x)=3^{5(-1)}=\frac{1}{243}$$

Alternate Method

Let $\quad f(x)=27^{\cos 2x}\,81^{\sin 2x}=3^{4\sin 2x}=3^{3\cos 2x+4\sin 2x}$

For minimum value of given function, $3\cos 2x+4\sin 2x$ will be minimum.

$$\therefore \quad -\sqrt{3^2+4^2}\le 3\cos 2x+4\sin 2x\le\sqrt{3^2+4^2}$$

$$\Rightarrow \quad -5\le 3\cos 2x+4\sin 2x\le 5$$

$\therefore$ Minimum of $3\cos 2x+4\sin 2x=-5$

So, $\quad \min f(x)=3^{-5}=\frac{1}{243}$

32. Let $AB=x$ and $\angle BDC=\alpha$

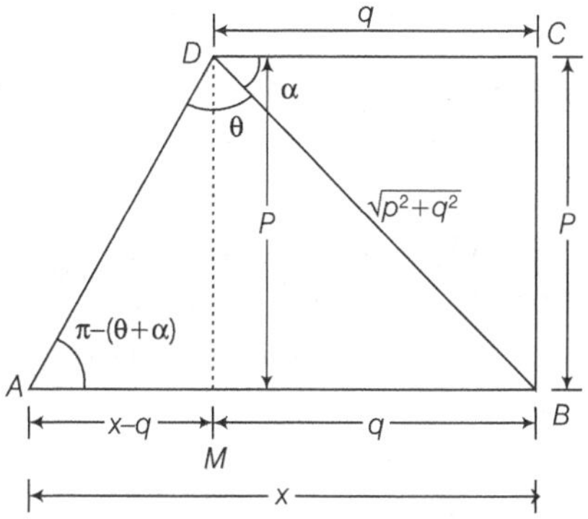

In ΔDAM, $\tan(\pi-\theta-\alpha)=\frac{p}{x-q}$

$$\Rightarrow \quad \tan(\theta+\alpha)=\frac{p}{q-x}\Rightarrow q-x=p\cos(\theta+\alpha)$$

$$\Rightarrow \quad x=q-p\cot(\theta+\alpha)=q-p\left(\frac{\cot\theta\cot\alpha-1}{\cot\alpha+\cot\theta}\right)$$

$$=\frac{q(\cot\alpha+\cot\theta)+p(\cot\theta\cot\alpha)+p}{\cot\alpha+\cot\theta}$$

$$=\frac{q\left(\frac{q}{p}+\frac{\cos\theta}{\sin\theta}\right)-p\left(\frac{q}{p}\frac{\cos\theta}{\sin\theta}\right)+p}{\frac{q}{p}+\frac{\cos\theta}{\sin\theta}} \qquad \left[\because \cot\alpha=\frac{q}{p}\right]$$

$$=\frac{\frac{q^2}{p}+\frac{q\cos\theta}{\sin\theta}-q\frac{\cos\theta}{\sin\theta}+p}{\frac{q\sin\theta+p\cos\theta}{p\sin\theta}}=\frac{(q^2+p^2)\sin\theta}{q\sin\theta+p\sin\theta}$$

33. Given $4n\alpha=\pi \Rightarrow 2n\alpha=\frac{\pi}{2}$

Now $\cot\alpha\cdot\cot(2n-1)\alpha=\cot\alpha\cot\left(\frac{\pi}{2}-\alpha\right)$

$$=\cot\alpha\cdot\tan\alpha=1$$

Similarly, $\cot 2\alpha\cos(2n-2)\alpha=1$,

$\cot 3\alpha\cdot\cot(2n-3)\alpha=1, \ldots, \cot(n-1)\alpha\cot(n+1)\alpha=1$

Thus $\cot\alpha\cot 2\alpha\cot 3\alpha\ldots\cot(2n-1)\alpha$

$=\{\cot\alpha\cot(2n-1)\alpha\}\{\cot 2\alpha\cot(2n-2)\alpha\}$

$\ldots\{\cot(n-1)\alpha\cot(n+1)\alpha\}\cdot\cot n\alpha$

$=1\cdot 1\cdot 1\ldots 1\cdot 1 \qquad \left[\because \cot n\alpha=\cot\frac{\pi}{4}=1\right]$

$=1$

34. We have $(\sin A+\sin B+\sin C)(\sin A+\sin B-\sin C)$

$$=3\sin A\sin B$$

$$\Rightarrow \quad (\sin A+\sin B)^2-\sin^2 C=3\sin A\sin B$$

$$\Rightarrow \quad \sin^2 A+\sin^2 B-\sin^2 C=\sin A\sin B$$

$$\Rightarrow \quad \sin^2 A + \sin(B+C)\sin(B-C) = \sin A \sin B$$

$$\Rightarrow \quad \sin A\,[\sin(B+C) + \sin(B-C)] = \sin A \sin B \qquad [\because A+B+C=\pi]$$

$$\Rightarrow \quad \sin A(2\sin B\cos C) = \sin A \sin B$$

$$\therefore \quad \cos C = \frac{1}{2} \Rightarrow C = 60°$$

35. From the third relation we get

$$\cos\theta\cos\phi + \sin\theta\sin\phi = \sin\beta\sin\gamma$$

$$\Rightarrow \sin^2\theta\sin^2\phi = (\cos\theta\cos\phi - \sin\beta\sin\gamma)^2$$

$$\Rightarrow \left(1 - \frac{\sin^2\beta}{\sin^2\alpha}\right)\left(1 - \frac{\sin^2\gamma}{\sin^2\alpha}\right) = \left(\frac{\sin\beta\sin\gamma}{\sin^2\alpha} - \sin\beta\sin\gamma\right)^2$$

[from the first and second relations]

$$\Rightarrow (\sin^2\alpha - \sin^2\beta)(\sin^2\alpha - \sin^2\gamma)$$
$$= \sin^2\beta\sin^2\gamma(1 - \sin^2\alpha)^2$$

$$\Rightarrow \sin^4\alpha(1 - \sin^2\beta\sin^2\gamma)$$
$$- \sin^2\alpha(\sin^2\beta + \sin^2\gamma - 2\sin^2\beta\sin^2\gamma) = 0$$

$$\therefore \quad \sin^2\alpha = \frac{\sin^2\beta - \sin^2\gamma - 2\sin^2\beta\sin^2\gamma}{1 - \sin^2\beta\sin^2\gamma} \qquad [\because \sin\alpha \ne 0]$$

$$\text{and } \cos^2\alpha = \frac{1 - \sin^2\beta - \sin^2\gamma + \sin^2\beta\sin^2\gamma}{1 - \sin^2\beta\sin^2\gamma}$$

$$\Rightarrow \quad \tan^2\alpha = \frac{\sin^2\beta - \sin^2\beta - \sin^2\gamma + \sin^2\gamma - \sin^2\beta\sin^2\gamma}{\cos^2\beta - \sin^2\gamma(1 - \sin^2\beta)}$$

$$= \frac{\sin^2\beta\cos^2\gamma + \cos^2\beta\sin^2\gamma}{\cos^2\beta\cos^2\gamma}$$

$$= \tan^2\beta + \tan^2\gamma$$

$$\Rightarrow \tan^2\alpha - \tan^2\beta - \tan^2\gamma = 0$$

36. $\tan\beta = \dfrac{n\tan\alpha}{1 + \tan^2\alpha - n\tan^2\alpha}$

$$= \frac{n\tan\alpha}{1 + (1-n)\tan^2\alpha}$$

$$\Rightarrow \tan(\alpha - \beta) = \frac{\tan\alpha - \dfrac{n\tan\alpha}{1+(1-n)\tan^2\alpha}}{1 + \dfrac{\tan\alpha\cdot n\tan\alpha}{1+(1-n)\tan^2\alpha}}$$

$$= \frac{\tan\alpha + (1-n)\tan^3\alpha - n\tan\alpha}{1 + (1-n)\tan^2\alpha + n\tan^2\alpha}$$

$$= \frac{(1-n)\tan\alpha(1+\tan^2\alpha)}{1+\tan^2\alpha} = (1-n)\tan\alpha$$

37. Let $\dfrac{(\cos\theta)}{a} = \dfrac{(\sin\theta)}{b} = k$, so that $\cos\theta = ak$ and $\sin\theta = bk$. Then

$$a\cos 2\theta + b\sin 2\theta = a(1 - 2\sin^2\theta) + 2b\sin\theta\cos\theta$$
$$= a - 2ab^2k^2 + 2b\cdot bk\cdot ak$$
$$= a - 2ab^2k^2 + 2ab^2k^2 = a$$

38. Let $y = \cos x\cos(x+2) - \cos^2(x+1)$

$$= \frac{1}{2}[\cos(2x+2) + \cos 2] - \cos^2(x+1)$$

$$= \frac{1}{2}[2\cos^2(x+1) - 1 + \cos 2] - \cos^2(x+1)$$

$$= -\frac{1}{2}(1 - \cos 2) = -\frac{1}{2}(2\sin^2 1) = -\sin^2 1$$

This shows that $y = -\sin^2 1$ is a straight line which is parallel to X-axis and clearly passes through the point $\left(\dfrac{\pi}{2}, -\sin^2 1\right)$.

39. $f(\theta) = |\sin\theta| + |\cos\theta|, \forall\, \theta \in R$ Clearly, $f(\theta) > 0$.

Also, $\quad f^2(\theta) = \sin^2\theta + \cos^2\theta + |2\sin\theta\cdot\cos\theta|$

$$= 1 + |\sin 2\theta|$$
$$0 \le |\sin 2\theta| \le 1$$
$$\Rightarrow \quad 1 \le f^2(\theta) \le 2 \Rightarrow 1 \le f(\theta) \le \sqrt{2}$$

40. $A = \cos(\cos x) + \sin(\cos x)$

$$= \sqrt{2}\left\{\cos(\cos x)\cos\frac{\pi}{4} + \sin(\cos x)\sin\frac{\pi}{4}\right\}$$

$$= \sqrt{2}\left\{\cos\left(\cos x - \frac{\pi}{4}\right)\right\}$$

$$\because \quad -1 \le \cos\left(\cos x - \frac{\pi}{4}\right) \le 1$$

$$\therefore \quad -\sqrt{2} \le A \le \sqrt{2}$$

41. We have, $\dfrac{U_n}{V_n} = \tan n\theta$

$$\text{and } \frac{V_n - V_{n-1}}{U_{n-1}} = \frac{\cos n\theta\sec^n\theta - \cos(n-1)\theta\sec^{n-1}\theta}{\sin(n-1)\theta\sec^{n-1}\theta}$$

$$= \frac{\cos n\theta\sec\theta - \cos(n-1)\theta}{\sin(n-1)\theta}$$

$$= \frac{\cos n\theta - \cos(n-1)\theta\cos\theta}{\cos\theta\sin(n-1)\theta}$$

$$= \frac{\cos(n-1)\theta\cos\theta - \sin(n-1)\theta\sin\theta - \cos(n-1)\theta\cos\theta}{\cos\theta\sin(n-1)\theta}$$

$$= -\tan\theta$$

So, that $\dfrac{V_n - V_{n-1}}{U_{n-1}} + \dfrac{1}{n}\dfrac{U_n}{V_n} = -\tan\theta + \dfrac{\tan n\theta}{n} \ne 0$

42. If $a, b > 0$

Using A.M. ≥ G.M., we get

$$\frac{1}{a} + \frac{1}{b} \ge \frac{2}{\sqrt{ab}}$$

$$\Rightarrow \quad f(x) \ge \frac{2}{\sqrt{\cos\left(\dfrac{\pi}{6} - x\right)\cos\left(\dfrac{\pi}{6} + x\right)}}$$

$$= \frac{2}{\sqrt{\cos^2\dfrac{\pi}{6} - \sin^2 x}} = \frac{2}{\sqrt{\dfrac{3}{4} - \dfrac{1 - \cos 2x}{2}}}$$

$$= \frac{2}{\sqrt{\dfrac{1}{4} + \dfrac{\cos 2x}{2}}}$$

Now for $0 \le x \le \frac{\pi}{3}, \frac{-1}{2} \le \cos 2x \le 1$

$$\Rightarrow \quad 0 \le \sqrt{\frac{1}{4} + \frac{\cos 2x}{2}} \le \frac{\sqrt{3}}{2}$$

$$\Rightarrow \quad f(x) \ge \frac{4}{\sqrt{3}}$$

Since 'f' is continuous range of 'f' is $\left[\frac{4}{\sqrt{3}}, \infty\right)$.

43. $\because 0 \le \sin^2\theta \le 1$ and $0 \le \cos^2\theta \le 1$

$$\Rightarrow \quad 0 \le \sin^8\theta \le \sin^2\theta \text{ and } 0 \le \cos^{14}\theta \le \cos^2\theta$$

$$\therefore \quad 0 < \sin^8\theta + \cos^{14}\theta \le \sin^2\theta + \cos^2\theta$$

Hence, $0 < A \le 1$

44. $\sin\left(\frac{3\pi}{2} - \alpha\right) = -\cos\alpha, \sin\left(\frac{\pi}{2} + \alpha\right) = \cos\alpha$

$$\sin(3\pi + \alpha) = -\sin\alpha$$

$$\sin(5\pi - \alpha) = -\sin\alpha$$

$$\therefore \quad 3\left\{\sin^4\left(\frac{3\pi}{2} - \alpha\right) + \sin^4(3\pi + \alpha)\right\} - 2\left[\sin^6\left(\frac{\pi}{2} + \alpha\right) + \sin^6(5\pi - \alpha)\right]$$

$$= 3\{\cos^4\alpha + \sin^4\alpha\} - 2\{\cos^6\alpha + \sin^6\alpha\}$$

$$= 3\{1 - 2\sin^2\alpha\cos^2\alpha\} - 2\{1 - 3\sin^2\alpha\cos^2\alpha\} = 1$$

45. $\sin\left(x + \frac{\pi}{6}\right) + \cos\left(x + \frac{\pi}{6}\right)$

$$= \sqrt{2}\left[\sin\left(x + \frac{\pi}{6} + \frac{\pi}{4}\right)\right]$$

$$= \sqrt{2}\sin\left(x + \frac{5\pi}{12}\right) \le \sqrt{2}$$

Equality holds when $x + \frac{5\pi}{12} = \frac{\pi}{2}$ ie, $x = \frac{\pi}{12}$

Therefore, maximum value of given expression is attained at

$$x = \frac{\pi}{12}$$

46. $\cot^2 x = \cot(x - y)\cdot\cot(x - z)$

$$\Rightarrow \cot^2 x = \left(\frac{\cot x \cot y + 1}{\cot y - \cot x}\right)\left(\frac{\cot x \cdot \cot z + 1}{\cot z - \cot x}\right)$$

$$\Rightarrow \cot^2 x\cdot\cot y\cdot\cot z - \cot^3 x\cdot\cot y - \cot^3 x\cot z + \cot^4 x$$
$$= \cot^2 x\cdot\cot y\cdot\cot z + \cot x\cdot\cot y + \cot x\cdot\cot z + 1$$

$$\Rightarrow \cot x\cot y(1 + \cot^2 x) + \cot x\cot z(1 + \cot^2 x) + 1 - \cot^4 x = 0$$

$$\Rightarrow \cot x(\cot y + \cot z)(1 + \cot^2 x) + (1 - \cot^2 x)(1 + \cot^2 x) = 0$$

$$\Rightarrow \cot x(\cot y + \cot z) + (1 - \cot^2 x) = 0$$

$$\Rightarrow \quad \frac{\cot^2 x - 1}{2\cot x} = \frac{1}{2}(\cot y + \cot z)$$

$$\Rightarrow \quad \frac{1}{2}(\cot y + \cot z) = \cot 2x$$

47. Given that, $\alpha + \beta + \gamma = \pi$

Taking $\alpha = -\frac{\pi}{2}; \beta = -\frac{\pi}{2}$ and $\gamma = 2\pi$

$$\therefore \quad \sin\alpha + \sin\beta + \sin\gamma = -1 - 1 + 0 = -2$$

but $\sin\alpha + \sin\beta + \sin\gamma \ge -3$ for any α, β, γ

Hence, minimum value of $\sin\alpha + \sin\beta + \sin\gamma$ is negative.

48. $\cos x - \sin\alpha\cot\beta\sin x = \cos\alpha$

$$\Rightarrow \quad \sin\beta\cos x - \sin\alpha\cos\beta\sin x = \cos\alpha\sin\beta$$

$$\Rightarrow \sin\beta\left(1 - \tan^2\frac{x}{2}\right) - \sin\alpha\cos\beta\cdot 2\tan\frac{x}{2} = \cos\alpha\sin\beta\left(1 + \tan^2\frac{x}{2}\right)$$

$$\Rightarrow \tan^2\frac{x}{2}(-\sin\beta - \cos\alpha\sin\beta) - \sin\alpha\cos\beta\cdot 2\tan\frac{x}{2} + \sin\beta(1 - \cos\alpha) = 0$$

$$\Rightarrow \quad \tan\frac{x}{2} = \frac{-2\sin\alpha\cos\beta \pm \sqrt{4[\sin^2\alpha\cos^2\beta + \sin^2\beta(1 + \cos\alpha)(1 - \cos\alpha)]}}{2\sin\beta(1 + \cos\alpha)}$$

$$= \frac{-\sin\alpha\cos\beta \pm \sqrt{\sin^2\alpha(\sin^2\beta + \cos^2\beta)}}{\sin\beta(1 + \cos\alpha)}$$

$$= \frac{-\sin\alpha\cos\beta \pm \sin\alpha}{\sin\beta(1 + \cos\alpha)} = \frac{\sin\alpha(1 - \cos\beta \pm 1)}{\sin\beta(1 + \cos\alpha)}$$

$$= \tan\frac{\beta}{2}\tan\frac{\alpha}{2} \text{ or } -\tan\frac{\alpha}{2}\cot\frac{\beta}{2}$$

49. $\because \cos^4\theta\sec^4\alpha, \frac{1}{2}$ and $\sin^4\theta\operatorname{cosec}^2\alpha$ are in AP

$$1 = \cos^4\theta\sec^2\alpha + \sin^4\theta\operatorname{cosec}^2\alpha$$

$$\Rightarrow \quad 1 = \frac{\cos^4\theta}{\cos^2\alpha} + \frac{\sin^4\theta}{\sin^2\alpha}$$

$$\Rightarrow \quad (\sin^2\theta + \cos^2\theta)^2 = \frac{\cos^4\theta}{\cos^2\alpha} + \frac{\sin^4\theta}{\sin^2\alpha}$$

$$\Rightarrow \quad \cos^4\theta\left(\frac{1}{\cos^2\alpha} - 1\right) + \sin^4\theta\left(\frac{1}{\sin^2\alpha} - 1\right) - 2\sin^2\theta\cos^2\theta = 0$$

$$\Rightarrow \quad \sin^4\alpha\cos^4\theta + \sin^4\theta\cos^4\alpha - 2\sin^2\theta\cos^2\theta\sin^2\alpha\cos^2\alpha = 0$$

$$\Rightarrow \quad (\sin^2\alpha\cos^2\theta - \cos^2\alpha\sin^2\theta)^2 = 0$$

$$\Rightarrow \quad \tan^2\theta = \tan^2\alpha$$

$$\therefore \quad \theta = n\pi \pm \alpha, n \in I$$

Now, $\cos^8\theta\sec^6\alpha = \cos^8\alpha\sec^6\alpha = \cos^2\alpha$

and $\sin^8\theta\operatorname{cosec}^6\alpha = \sin^8\alpha\cdot\operatorname{cosec}^6\alpha = \sin^2\alpha$

Hence, $\cos^8\theta\sec^6\alpha, \frac{1}{2}, \sin^8\theta\operatorname{cosec}^6\alpha$

ie, $\cos^2\alpha, \frac{1}{2}, \sin^2\alpha$ are in AP.

50. Given, $(\cot \alpha_1)\cdot(\cot \alpha_2)\dots(\cot \alpha_n)=1$

$$\Rightarrow \prod_{i=1}^{n} \cos \alpha_1 = \prod_{i=1}^{n} \sin \alpha_1$$

$$\Rightarrow \prod_{i=1}^{n} \cos^2 \alpha_1 = \prod_{i=1}^{n} \sin \alpha_1 \cos \alpha_1 = \prod_{i=1}^{n} \frac{\sin 2\alpha_1}{2} \le \frac{1}{2^n}$$

$$\Rightarrow \prod_{i=1}^{n} \cos \alpha_1 \le \frac{1}{2^{\frac{n}{2}}}$$

Hence, maximum value of $\prod_{i=1}^{n} \cos \alpha_1$ is $\frac{1}{2^{\frac{n}{2}}}$.

51. $\sin x \cdot \cos^3 x > \cos x \cdot \sin^3 x$

$$\Rightarrow \sin x \cdot \cos x(\cos^2 x - \sin^2 x) > 0$$

$$\Rightarrow \sin x \cdot \cos x \cdot \cos 2x > 0$$

$$\Rightarrow \cos x \cdot \cos 2x > 0$$

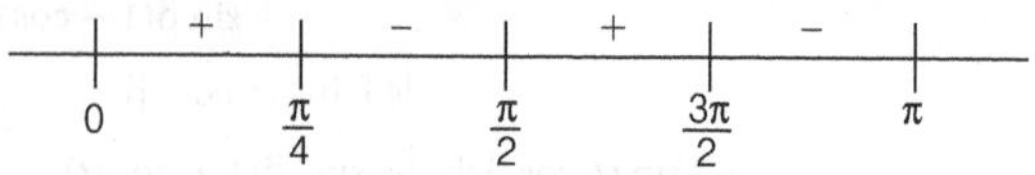

$$\therefore \quad x \in \left(0, \frac{\pi}{4}\right) \cup \left(\frac{\pi}{2}, \frac{3\pi}{4}\right)$$

52. $u^2 = a^2 + b^2$

$$+ 2\sqrt{a^2 \cos^2 \theta + b^2 \sin^2 \theta} \times \sqrt{a^2 \sin^2 \theta + b^2 \cos^2 \theta}$$

$$= a^2 + b^2 + 2\sqrt{\sin^2 \theta \cos^2 \theta(a^4 + b^4) + a^2b^2(\sin^4 \theta + \cos^4 \theta)}$$

$$= a^2 + b^2 + 2\sqrt{a^2b^2(1 - 2\sin^2\theta \cos^2\theta) + (a^4 + b^4)\sin^2\theta \cos^2\theta}$$

$$= (a^2 + b^2) + 2\sqrt{a^2b^2 + (a^2 - b^2)^2 \sin^2 \theta \cos^2 \theta}$$

$$= (a^2 + b^2) + 2\sqrt{a^2b^2 + \frac{(a^2 - b^2)^2}{4} \sin^2 2\theta}$$

Max. $u^2 = (a^2 + b^2) + 2\sqrt{a^2b^2 + \dfrac{(a^2 - b^2)^2}{4}}$

Min. $u^2 = (a^2 + b^2) + 2ab$

$\Rightarrow$ Difference $2\sqrt{a^2b^2 + \dfrac{(a^2 - b^2)^2}{4}} - 2ab$

$$= \sqrt{4a^2b^2 + a^4 + b^4 - 2a^2b^2} - 2ab$$

$$= \sqrt{(a^2 + b^2)^2} - 2ab$$

$$= a^2 + b^2 - 2ab = (a - b)^2$$

53. $\because \left(\tan \dfrac{\theta}{2}\right)(1 + \sec \theta) = \tan\left(\dfrac{\theta}{2}\right)\left(\dfrac{1 + \cos \theta}{\cos \theta}\right)$

$$= \frac{\sin \frac{\theta}{2}}{\cos \frac{\theta}{2}} \cdot \frac{2\cos^2 \frac{\theta}{2}}{\cos \theta} = \frac{2 \sin \frac{\theta}{2} \cdot \cos \frac{\theta}{2}}{\cos \theta}$$

$$= \frac{\sin \theta}{\cos \theta} = \tan \theta \quad \text{...(i)}$$

$\therefore$ By repeated use of Eq. (i), we have

$$f_n(\theta) = \tan \theta\,(1 + \sec 2\theta)(1 + \sec 4\theta)\dots(1 + \sec 2^n \theta)$$

$$= \tan 2\theta\,(1 + \sec 4\theta)\dots(1 + \sec 2^n \theta)$$

$$= \tan 4\theta\,(1 + \sec 8\theta)\dots(1 + \sec 2^n \theta)$$

$$= \dots = \tan 2^n \theta$$

Now, $f_2\left(\dfrac{\pi}{16}\right) = \tan\left(2^2 \dfrac{\pi}{16}\right) = \tan \dfrac{\pi}{4} = 1$

$$f_3\left(\frac{\pi}{32}\right) = \tan\left(2^3 \frac{\pi}{32}\right) = \tan \frac{\pi}{4} = 1$$

$$f_4\left(\frac{\pi}{64}\right) = \tan\left(2^4 \frac{\pi}{64}\right) = \tan \frac{\pi}{4} = 1$$

and $f_5\left(\dfrac{\pi}{128}\right) = \tan \dfrac{\pi}{4} = 1$

54. $\cos x = 0 \Rightarrow x = (2n + 1)\dfrac{\pi}{2}$

$$\Rightarrow \cos\left((2n + 1)\frac{\pi}{2} + z\right) = \frac{1}{2}$$

$$\Rightarrow \sin z = \frac{1}{2} \text{ or } \sin z = -\frac{1}{2}$$

$$\Rightarrow z = \frac{\pi}{6}, \frac{5\pi}{6}, \frac{7\pi}{6}, \frac{11\pi}{6}$$

55. $f_n(\theta) = \cos\theta - \cos 2\theta + \cos 2\theta - \cos 3\theta + \dots + \cos(n)\theta - \cos(n + 1)\theta$

$$f_n(\theta) = \cos\theta - \cos(n + 1)\theta$$

Now, check options.

56. $P = \sin 25° \sin 35° \sin 60° \sin 85°$

$$= \sin 25° \sin(60° - 25°)\ \sin 60° \sin(60° + 25°)$$

$$= \sin 60° \sin 25° \sin(60° - 25°) \sin(60° + 25°)$$

$$\therefore \quad P = \sin 60° \times \frac{1}{4} \sin 75° \quad \text{...(i)}$$

$$Q = \sin 20° \sin 40° \sin 75° \sin 80°$$

$$= \sin 20° \sin(60° - 20°) \sin 75° \sin(60° + 20°)$$

$$= \sin 75° \sin 20° \sin(60° - 20°) \sin(60° + 20°)$$

$$\therefore \quad Q = \sin 75° \times \frac{1}{4} \times \sin 60° \quad \text{...(ii)}$$

Hence, $P = Q$

57. $x = \dfrac{1}{1 - \cos^2\theta} = \dfrac{1}{\sin^2\theta}, y = \dfrac{1}{\cos^2\theta},$

$$z = \frac{1}{1 - \sin^2\theta \cos^2\theta}$$

$$\Rightarrow \frac{1}{x} + \frac{1}{y} = 1$$

$$\Rightarrow xy = x + y \Rightarrow \frac{1}{z} = 1 - \frac{1}{xy}$$

$$\Rightarrow xyz = xy + z = x + y + z$$

58. Given $P(x) = \cot^2 x\left(\dfrac{1 + \tan x + \tan^2 x}{1 + \cot x + \cot^2 x}\right) + \left(\dfrac{\cos x - \cos 3x + \sin 3x - \sin x}{2(\sin 2x + \cos 2x)}\right)^2$

$$= \frac{\cot^2 + \cot x + 1}{1 + \cot x\ \ \cot^2 x} + \left(\frac{2 \sin x\,(\sin 2x + \cos 2x)}{2(\sin 2x + \cos 2x)}\right)^2$$

$= 1 + \sin^2 x$

$\therefore \quad P(18°) = P(72°) = (1 + \sin^2 18°) + (1 + \sin^2 72°)$

$= 1 + 1 + (\sin^2 18° + \cos^2 18°) = 3$

59. $E = \dfrac{3\sin(\alpha+\beta) - 4\cos(\alpha+\beta)}{\sqrt{3}\sin\alpha}$

$= \dfrac{3(\sin\alpha\cos\beta + \cos\alpha\sin\beta) - 4(\cos\alpha\cos\beta + \sin\alpha\sin\beta)}{\sqrt{3}\sin\alpha}$

$= \dfrac{5}{\sqrt{3}}$ for $0 < \beta < \dfrac{\pi}{2}$

and $E = \dfrac{\sqrt{3}\,(7 + 24\cot\alpha)}{15}$ for $\dfrac{\pi}{2} < \beta < \pi$.

60. $\cot A = \dfrac{3}{4}$

$\Rightarrow \quad \cot C = \dfrac{-3}{4}$

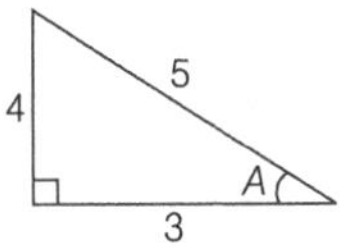

$\Rightarrow$ C is obtuse angle.

$\therefore \quad \sin C = \dfrac{4}{5}, \cos C = -\dfrac{3}{5}$

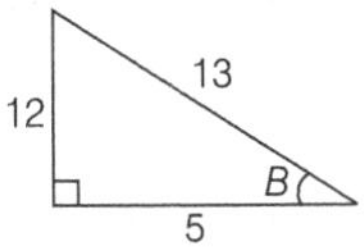

$\tan B = \dfrac{-12}{5}$

$\Rightarrow \quad \tan D = \dfrac{12}{5}$

$\Rightarrow$ D is an acute angle

$\therefore \quad \sin D = \dfrac{12}{13} \quad \cos D = \dfrac{5}{13}$

Hence, $\sin(C + D) = \sin C \cdot \cos D + \cos C \cdot \sin D$

$= \left(\dfrac{4}{5}\right)\left(\dfrac{5}{13}\right) + \left(\dfrac{-3}{5}\right)\left(\dfrac{12}{13}\right)$

$= \dfrac{20 - 36}{65} = \dfrac{-16}{65}$

Also, $\sin(A + B) = \sin(2\pi - (C + D))$

$= -\sin(C + D) = \dfrac{16}{65}$.

61. $2\left(\cos^2 x + \dfrac{1}{2}\cos x\right) = a$

$2\left(\cos x + \dfrac{1}{4}\right)^2 = a + \dfrac{1}{8}$

$\therefore \quad \left(\cos x + \dfrac{1}{4}\right)^2 = \dfrac{a}{2} + \dfrac{1}{16}$

$\left(\cos x + \dfrac{1}{4}\right)^2 \in \left[0, \dfrac{25}{16}\right]$

$\therefore \quad \dfrac{8a + 1}{16} \in \left[0, \dfrac{25}{16}\right]$

$\Rightarrow \quad 8a + 1 \in [0, 25]$

$\Rightarrow \quad a \in \left[\dfrac{-1}{8}, 3\right]$

62. $A = \sin 44° + \cos 44°$

$= \cos 46° + \sin 46° = C$

$B = \sin 45° + \cos 45° = \sqrt{2}[\sin 90°]$

$A = \sqrt{2}\left[\dfrac{1}{\sqrt{2}}\sin 44° + \dfrac{1}{\sqrt{2}}\cos 44°\right]$

$= \sqrt{2}\,[\sin 44° \cdot \cos 45° + \cos 44° \cdot \sin 45°]$

$= \sqrt{2}\sin 89°$

$\Rightarrow \quad B > A$

63. $\tan(2\alpha + \beta) = x$

$\tan(\alpha + 2\beta) = y$

$\Rightarrow \quad \tan(3(\alpha + \beta)).\tan(\alpha - \beta)$

$= \tan[(2\alpha + \beta) + (\alpha + 2\beta)].$

$\tan[2(\alpha + \beta) - (\alpha + 2\beta)]$

$= \dfrac{\tan(2\alpha + \beta) + \tan(\alpha + 2\beta)}{1 - \tan(2\alpha + \beta).\tan(\alpha + 2\beta)} \cdot \dfrac{\tan(2\alpha + \beta) - \tan(\alpha + 2\beta)}{1 + \tan(2\alpha + \beta).\tan(\alpha + 2\beta)}$

$= \dfrac{x + y}{1 - xy} \cdot \dfrac{x - y}{1 + xy} = \dfrac{x^2 - y^2}{1 - x^2y^2}$

64. We have $x = \dfrac{1 - \sin\phi}{\cos\phi}, y = \dfrac{1 + \cos\phi}{\sin\phi}$

Multiplying, we get $xy = \dfrac{(1 - \sin\phi)(1 + \cos\phi)}{\cos\phi\sin\phi}$

$\Rightarrow \quad xy + 1 = \dfrac{1 - \sin\phi + \cos\phi - \sin\phi\cos\phi + \sin\phi\cos\phi}{\cos\phi\sin\phi}$

$= \dfrac{1 - \sin\phi + \cos\phi}{\cos\phi\sin\phi}$

and $\quad x - y = \dfrac{(1 - \sin\phi)\sin\phi - \cos\phi(1 + \cos\phi)}{\cos\phi\sin\phi}$

$= \dfrac{\sin\phi - \sin^2\phi - \cos\phi - \cos^2\phi}{\cos\phi\sin\phi}$

$= \dfrac{\sin\phi - \cos\phi - 1}{\cos\phi\sin\phi} = -(xy + 1)$

Thus, $xy + x - y + 1 = 0$.

$\Rightarrow \quad x = \dfrac{y - 1}{y + 1}$ and $y = \dfrac{1 + x}{1 - x}$.

65. The given relation can be written as

$\tan\left(\dfrac{x}{2}\right) = \dfrac{1 - \sin^2 x}{\sin x} = \dfrac{\cos^2 x}{\sin x}$

$$\Rightarrow \quad 2\sin^2\left(\frac{x}{2}\right) = \left[\cos^2\left(\frac{x}{2}\right) - \sin^2\left(\frac{x}{2}\right)\right]^2$$

$$\Rightarrow \quad 2\tan^2\left(\frac{x}{2}\right) = \left[1 - \tan^2\left(\frac{x}{2}\right)\right]^2 / \left[1 + \tan^2\left(\frac{x}{2}\right)\right]$$

$$\Rightarrow \quad 2y(1+y) = (1-y)^2 \qquad \left[\text{where } y = \tan^2\frac{x}{2}\right]$$

$$\Rightarrow \quad y^2 + 4y - 1 = 0$$

$$\Rightarrow \quad y = \frac{-4 \pm \sqrt{16+4}}{2} = -2 \pm \sqrt{5}$$

Since $y > 0$, we get

$$y = \sqrt{5} - 2 = \frac{(\sqrt{5}-2)^2}{\sqrt{5}+2} \cdot \frac{2+\sqrt{5}}{2+\sqrt{5}}$$

$$= (9 - 4\sqrt{5})(2+\sqrt{5})$$

66. $y = \dfrac{\sqrt{(\cos 2A - \sin 2A)^2 + 1}}{\sqrt{(\cos 2A + \sin 2A)^2 - 1}}$

$$\Rightarrow \quad y = \frac{\pm(\cos 2A - \sin 2A) + 1}{\pm(\cos 2A + \sin 2A) - 1}$$

which gives us four values of y, say y_1, y_2, y_3 and y_4 We have,

$$y_1 = \frac{\cos 2A - \sin 2A + 1}{\cos 2A + \sin 2A - 1} = \frac{(1+\cos 2A) - \sin 2A}{(\cos 2A - 1) + \sin 2A}$$

$$= \frac{2\cos^2 A - 2\sin A\cos A}{-2\sin^2 A + 2\sin A\cos A}$$

$$= \frac{\cos A(\cos A - \sin A)}{\sin A(\cos A - \sin A)} = \cot A$$

$$y_2 = \frac{-(\cos 2A - \sin 2A) + 1}{-(\cos 2A + \sin 2A) - 1}$$

$$= \frac{(1 - \cos 2A) + \sin 2A}{-(1+\cos 2A) - \sin 2A}$$

$$= \frac{2\sin^2 A + 2\sin A\cos A}{-2\cos^2 A - 2\sin A\cos A} = -\tan A$$

$$y_3 = \frac{(\cos 2A - \sin 2A) + 1}{-(\cos 2A + \sin 2A) - 1}$$

$$= \frac{(1+\cos 2A) - \sin 2A}{-(1+\cos 2A) - \sin 2A}$$

$$= \frac{2\cos^2 A - 2\sin A\cos A}{-2\cos^2 A - 2\sin A\cos A}$$

$$= -\frac{\cos A - \sin A}{\cos A + \sin A}$$

$$= -\frac{1 - \tan A}{1 + \tan A} = -\tan\left(\frac{\pi}{4} - A\right) = -\cot\left(\frac{\pi}{4} + A\right)$$

$$y_4 = \frac{-(\cos 2A - \sin 2A) + 1}{(\cos 2A + \sin 2A) - 1}$$

$$= \frac{(1 - \cos 2A) + \sin 2A}{-(1 - \cos 2A) + \sin 2A}$$

$$= \frac{2\sin^2 A - 2\sin A\cos A}{-2\sin A + 2\sin A\cos A}$$

67. $\because$ $3\sin\beta = \sin(2\alpha + \beta)$

$$\Rightarrow \quad 2\sin\beta = \sin(2\alpha+\beta) - \sin\beta$$

$$= 2\cos(\alpha+\beta)\sin\alpha$$

$$\therefore \quad \sin\beta = \cos(\alpha+\beta)\sin\alpha \qquad \text{...(i)}$$

$\therefore$ **Alternate** (b) is correct

Also, $\sin\beta = \sin(\alpha+\beta) - \alpha)$

$$= \sin(\alpha+\beta)\cos\alpha - \cos(\alpha+\beta)\sin\alpha \qquad \text{...(ii)}$$

From Eqs. (i) and (ii)

$$\sin\beta = \sin(\alpha+\beta)\cos\alpha - \sin\beta$$

$$\therefore \quad 2\sin\beta = \sin(\alpha+\beta)\cos\alpha$$

($\therefore$ Alternate (c) is correct)

Alternate (a)

LHS $= (\cot\alpha + \cos(\alpha+\beta))(\cot\beta - 3\cot(2\alpha+\beta))$

$$= \left(\frac{\sin(2\alpha+\beta)}{\sin\alpha \cdot \sin(\alpha+\beta)}\right)\left(\frac{\cos\beta}{\sin\beta} - \frac{3\cos(2\alpha+\beta)}{\sin(2\alpha+\beta)}\right)$$

$$= \left(\frac{3\sin\beta}{\sin\alpha \cdot \sin(\alpha+\beta)}\right)\left(\frac{\cos\beta}{\sin\beta} - \frac{3\cos(2\alpha+\beta)}{3\sin\beta}\right)$$

($\because 3\sin\beta = \sin(2\alpha+\beta)$)

$$= \left(\frac{3\sin\beta}{\sin\alpha \cdot \sin(\alpha+\beta)}\right)\left(\frac{\cos\beta - \cos(2\alpha+\beta)}{\sin\beta}\right)$$

$$= \left(\frac{3\sin\beta}{\sin\alpha \cdot \sin(\alpha+\beta)}\right)\left(\frac{2\sin(\alpha+\beta)\sin\alpha}{\sin\beta}\right)$$

$$= 6$$

Alternate (d)

$$\because \quad \tan(\alpha+\beta) = 2\tan\alpha$$

$$\Rightarrow \quad \frac{\sin(\alpha+\beta)}{\cos(\alpha+\beta)} = \frac{2\sin\alpha}{\cos\alpha}$$

$$\Rightarrow \quad \sin(\alpha+\beta)\cos\alpha = 2\cos(\alpha+\beta)\sin\alpha$$

$$\Rightarrow \quad \sin(\alpha+\beta)\cos\alpha - \cos(\alpha+\beta)\sin\alpha = \cos(\alpha+\beta)\sin\alpha$$

$$\sin\beta = \cos(\alpha+\beta)\sin\alpha$$

[Alternate (b)]

68. $P_n(u)$ be a polynomial in u of degree n.

$$\therefore \quad \sin 2nx = 2\sin nx\cos nx$$

$$= \sin x\, P_{2n-1}(\cos x) \text{ or } \cos x\, P_{2n-1}(\sin x)$$

69. $\tan\theta = \dfrac{\sin\alpha - \cos\alpha}{\sin\alpha + \cos\alpha}$

$$= \frac{\tan\alpha - 1}{\tan\alpha + 1}$$

$$\tan\theta = \tan\left(\alpha - \frac{\pi}{4}\right)$$

$$\Rightarrow \quad \theta = n\pi + \alpha - \frac{\pi}{4}, n \in I$$

or $\quad 2\theta = 2n\pi + 2\alpha - \dfrac{\alpha}{2}$

$$\sin 2\theta = \sin\left(2\alpha - \frac{\pi}{2}\right) = -\cos 2\alpha$$

and $\quad \cos 2\theta = \cos\left(2\alpha - \dfrac{\pi}{2}\right) = \sin 2\alpha$

and $\sin\alpha - \cos\alpha = \sqrt{2}\sin\left(\alpha - \frac{\pi}{4}\right)$

$=\sqrt{2}\sin\{\theta - n\pi\} = \pm\sqrt{2}\sin\theta$

and $\sin\alpha + \cos\alpha = \sqrt{2}\sin\left(\alpha + \frac{\pi}{4}\right)$

$= \sqrt{2}\sin\left\{\frac{\pi}{2} + \theta - n\pi\right\}$

$= \sqrt{2}\cos(\theta - n\pi) = \pm\sqrt{2}\cos\theta$

70. $\cos 5\theta = \cos(4\theta + \theta) = \cos 4\theta\cos\theta - \sin 4\theta\sin\theta$

$= (2\cos^2 2\theta - 1)\cos\theta - 2\sin 2\theta\cos 2\theta\sin\theta$

$= [2(2\cos^2\theta - 1)^2 - 1]\cos\theta - 2\cdot 2\cos\theta \sin^2\theta(2\cos^2\theta - 1)$

$= [2(4\cos^4\theta - 4\cos^2\theta + 1) - 1]\cos\theta - 4\cos\theta(2\cos^2\theta - 1)(1 - \cos^2\theta)$

$= \cos\theta(8\cos^4\theta - 8\cos^2\theta + 1) - 4\cos\theta(3\cos^2\theta - 2\cos^4\theta - 1)$

$= \cos\theta(16\cos^4\theta - 20\cos^2\theta + 5)$

71. $x = \sqrt{a^2\cos^2\alpha + b^2\sin^2\alpha} + \sqrt{a^2\sin^2\alpha + b^2\cos^2\alpha}$

$\Rightarrow \quad x^2 = a^2 + b^2 + \sqrt{\begin{matrix}2(a^2\cos^2\alpha + b^2\sin^2\alpha)\\(a^2\sin^2\alpha + b^2\cos^2\alpha)\end{matrix}}$

$a^2 + b^2 + 2k,$

where $k = \sqrt{\begin{matrix}[(a^2 + b^2) - (a^2\sin^2\alpha + b^2\cos^2\alpha)]\\ \times(a^2\sin^2\alpha + b^2\cos^2\alpha)\end{matrix}}$

$\therefore \quad x = a^2 + b^2 + 2\sqrt{(a^2 + b^2)p - p^2}$

where $p = a^2\sin^2\alpha + b^2\cos^2\alpha$

$= \frac{a^2}{2}(1 - \cos 2\alpha) + \frac{b^2}{2}(1 + \cos 2\alpha)$

72. $\left(\frac{\cos A + \cos B}{\sin A - \sin B}\right)^n + \left(\frac{\sin A + \sin B}{\cos A - \cos B}\right)^n$

$= \cos^n\left(\frac{A - B}{2}\right) + \cot^n\left(\frac{B - A}{2}\right)$

If n even, $2\cot^n\left(\frac{A - B}{2}\right)$, if n odd, O

73. $P(k) = \left(1 + \cos\frac{\pi}{4k}\right)\left(1 + \cos\left(\frac{\pi}{2} - \frac{\pi}{4k}\right)\right)\left(1 + \cos\left(\frac{\pi}{2} + \frac{\pi}{4k}\right)\right)\left(1 + \cos\left(\pi - \frac{\pi}{4k}\right)\right)$

$= \left(1 + \cos\frac{\pi}{4k}\right)\left(1 + \sin\frac{\pi}{4k}\right)\left(1 - \sin\frac{\pi}{4k}\right)\left(1 - \cos\frac{\pi}{4k}\right)$

$= \left(1 - \cos^2\frac{\pi}{4k}\right)\left(1 - \sin^2\frac{\pi}{4k}\right)$

$= \frac{4\sin^2\frac{\pi}{4k}\cdot\cos^2\frac{\pi}{4k}}{4}$

$P(k) = \frac{1}{4}\sin^2\left(\frac{\pi}{2k}\right)$

$\Rightarrow \quad P(3) = \frac{1}{4}\cdot\frac{1}{4} = \frac{1}{16}$

$\Rightarrow \quad P(4) = \frac{\pi}{4}\sin^2\frac{\pi}{2k} = \frac{1}{4}\sin^2\frac{\pi}{8}$

$= \frac{1}{8}\left(1 - \cos\frac{\pi}{4}\right) = \frac{2 - \sqrt{2}}{16}$

$\Rightarrow \quad P(5) = \frac{1}{4}\sin^2\frac{\pi}{10} = \frac{1}{8}\left(2\sin^2\frac{\pi}{10}\right) = \frac{1}{8}(1 - \cos 36°)$

$= \frac{1}{8}\left(1 - \frac{\sqrt{5} + 1}{4}\right) = \frac{3 - \sqrt{5}}{32}$

$\Rightarrow \quad P(6) = \frac{1}{4}\sin^2\frac{\pi}{12} = \frac{1}{8}\left(2\sin^2\frac{\pi}{12}\right) = \frac{1}{8}\left(1 - \cos\frac{\pi}{6}\right)$

$= \frac{1}{8}\left(1 - \frac{\sqrt{3}}{2}\right) = \frac{2 - \sqrt{3}}{16}$

74. $x^2 + y^2 = a^2\sin^4\theta\cos^4\theta$

$xy = a^2\sin^5\theta\cos^5\theta$

$\therefore \quad \frac{(x^2 + y^2)^p}{(xy)^q} = \frac{a^{2p}(\sin\theta\cos\theta)^{4p}}{a^{2q}(\sin\theta\cos\theta)^{5q}}$

which is independent of θ if $4p = 5q$

i.e. $\quad p = 5, q = 4.$

75. LHS

$= \tan\alpha + 2\tan 2\alpha + 4\tan 4\alpha + 8\tan 8\alpha + 16\cot 16\alpha$

$= \cot\alpha - (\cot\alpha - \tan\alpha) + 2\tan 2\alpha + 4\tan 4\alpha + 8\tan 8\alpha + 16\cot 16\alpha$

$= \cot\alpha - 2(\cot 2\alpha - \tan 2\alpha) + 4\tan 4\alpha + 8\tan 8\alpha + 16\cot 16\alpha$

$(\because \cot\alpha - \tan\alpha = 2\cot 2\alpha)$

$= \cot\alpha - 4(\cot 4\alpha - \tan 4\alpha) + 8\tan 8\alpha + 16\cot 16\alpha$

$(\because \cot 2\alpha - \tan 2\alpha = 2\cot 4\alpha)$

$= \cot\alpha - 8(\cot 8\alpha - \tan 8\alpha) + 16\cot 16\alpha$

$= \cot\alpha - 16\cot 16\alpha + 16\cot 16\alpha$

$(\because \cot 8\alpha - \tan 8\alpha = 2\cot 16\alpha)$

$= \cot\alpha = \text{RHS}$

76. Let $x = \cot A, y = \cot B, z = \cot C$

$\therefore \quad \cot A\cot B + \cot B\cot C + \cot C\cot A = 1$

$\because \quad A + B + C = 180°$

$\therefore \quad \Sigma\frac{x}{(1 + x^2)} = \Sigma\frac{\cot A}{(1 + \cot^2 A)}$

$= \frac{1}{2}\Sigma\frac{2\tan A}{(1 + \tan^2 A)} = \frac{1}{2}\Sigma\sin 2A$

$= \frac{1}{2}(\sin 2A + \sin 2B + \sin 2C)$

$= \frac{1}{2}(4\sin A\sin B\sin C) = 2\sin A\sin B\sin C$

$= \frac{2}{\sqrt{(1 + \cot^2 A)(1 + \cot^2 B)(1 + \cot^2 C)}}$

$= \frac{2}{\sqrt{(1 + x^2)(1 + y^2)(1 + z^2)}} = \frac{2}{\sqrt{\Pi(1 + x^2)}}$

and $\sin 2A + \sin 2B - \sin 2C$

$$= 2\sin(A+B)\cos(A-B) - 2\sin C\cos C$$
$$= 2\sin C\{\cos(A-B) - \cos C\}$$
$$= 2\sin C\{\cos(A-C) + \cos(A+B)\}$$
$$= 2\sin C(2\cos A\cos B)$$
$$= 4\cos A\cos B\sin C$$

77. $\because a\cos x + b\sin x = c$

$$\Rightarrow \quad a\left\{\frac{1-\tan^2\left(\frac{x}{2}\right)}{1+\tan^2\left(\frac{x}{2}\right)}\right\} + b\left\{\frac{2\tan\left(\frac{x}{2}\right)}{1+\tan^2\left(\frac{x}{2}\right)}\right\} = c$$

$$\Rightarrow \quad (a+c)\tan^2\left(\frac{x}{2}\right) - 2b\tan\left(\frac{x}{2}\right) + (c-a) = 0$$

$$\therefore \quad \tan\left(\frac{\alpha}{2}\right) + \tan\left(\frac{\beta}{2}\right) = \frac{2b}{(a+c)}$$

and
$$\tan\left(\frac{\alpha}{2}\right)\tan\left(\frac{\beta}{2}\right) = \frac{c-a}{a+c}$$

Now,
$$\tan\left(\frac{\alpha+\beta}{2}\right) = \frac{\tan\left(\frac{\alpha}{2}\right) + \tan\left(\frac{\beta}{2}\right)}{1 - \tan\left(\frac{\alpha}{2}\right)\tan\left(\frac{\beta}{2}\right)}$$

$$= \frac{\frac{2b}{a+c}}{1-\left(\frac{c-a}{a+c}\right)} = \frac{b}{a} = \text{independent of } c$$

Also,
$$-\sqrt{(a^2+b^2)} \le a\cos x + b\sin x \le \sqrt{(a^2+b^2)}$$

$$\therefore \quad -\sqrt{(a^2+b^2)} \le c \le \sqrt{(a^2+b^2)}$$

78. $\because A + B + C = 180°$

$$\Rightarrow \quad A = 180° - (B+C)$$

$$\therefore \quad \tan A = \tan(180° - (B+C))$$
$$= -\tan(B+C) = -\left\{\frac{\tan B + \tan C}{1 - \tan B\tan C}\right\}$$
$$= \left(\frac{\tan B + \tan C}{\tan B\tan C - 1}\right)$$

Now, $\because A$ is obtuse

$\therefore \quad \tan A < 0,$

then $\quad \tan B + \tan C > 0$

$\therefore \quad \tan B\tan C - 1 < 0$

$\Rightarrow \quad \tan B\tan C < 1$

79. Let $S = \sin\left(\frac{2\pi}{7}\right) + \sin\left(\frac{4\pi}{7}\right) + \sin\left(\frac{8\pi}{7}\right)$

and $\quad C = \cos\left(\frac{2\pi}{7}\right) + \cos\left(\frac{4\pi}{7}\right) + \cos\left(\frac{8\pi}{7}\right)$

$$\therefore \quad C + iS = \alpha + \alpha^2 + \alpha^4 \qquad \text{...(i)}$$

Where $\alpha = \cos\left(\frac{2\pi}{7}\right) + i\sin\left(\frac{2\pi}{7}\right)$ is complex 7th root of unity.

Then, $\quad C - iS = \bar{\alpha} + \bar{\alpha}^2 + \bar{\alpha}^4$

$$= \alpha^6 + \alpha^5 + \alpha^3 \qquad \text{...(ii)}$$

Adding Eqs. (i) and (ii), then

$$2C = \alpha + \alpha^2 + \alpha^4 + \alpha^6 + \alpha^5 + \alpha^3 = -1$$

($\because$ sum of 7, 7th roots of unity is zero)

$$\therefore \quad C = -\frac{1}{2}$$

Also, multiplying Eqs. (i) and (ii), then $C^2 + S^2 = 2$

($\because \alpha^7 = 1$ and sum of 7, 7th roots of unity)

$$\Rightarrow \quad S^2 = 2 - \left(\frac{1}{2}\right)^2 = \frac{7}{4}$$

$$\therefore \quad S = \frac{\sqrt{7}}{2}$$

80. We observe that $y = 81^{\sin^2 x} + 81^{\cos^2 x} - 30 = 0$

$$\Rightarrow \quad 81^{\sin^2 x} + 81^{1-\sin^2 x} - 30 = 0$$

$$\Rightarrow \quad 81^{2\sin^2 x} - 30.81^{\sin^2 x} + 81 = 0$$

$$\Rightarrow \quad (81^{\sin^2 x} - 3)(81^{\sin^2 x} - 27) = 0$$

$$\Rightarrow \quad \sin^2 x = \frac{1}{4} \text{ or } \frac{3}{4}$$

$$\Rightarrow \quad \sin x = \pm\frac{1}{2} \text{ or } \sin x = \pm\frac{\sqrt{3}}{2}$$

$$\Rightarrow \quad x = \pm\frac{\pi}{6}, \pm\frac{5\pi}{6}, \text{ or } x = \pm\frac{\pi}{3}, \pm\frac{2\pi}{3}$$

$\Rightarrow$ The graph $y = 81^{\sin^2 x} + 81^{\cos^2 x} - 30$

Intersects the X-axis at eight points in $(-\pi \le x \le \pi)$.

$\Rightarrow$ Statement-1 is true.

81. Statement-2 is correct, using it we have $\cos 3x = \sin 2x$

$$\Rightarrow \quad 4\cos^3 x - 3\cos x = 2\sin x\cos x$$

Similarly $4\cos^3 y - 3\cos y = 2\sin y\cos y$

So, $\quad 4(1 - \sin^2 x) - 3 = 2\sin x$

$$\Rightarrow \quad 4\sin^2 x + 2\sin x - 1 = 0$$

and $\quad 4\sin^2 y + 2\sin y - 1 = 0$

Hence, $\sin x = \sin 18°$ and $\sin y = \sin(-54°) = -\sin 54°$ are the roots of a quadratic equation with integer coefficients.

82. The minimum value of the sum can be -3 provided $\sin\alpha = \sin\beta = \sin\gamma = -1$

$$\Rightarrow \quad \alpha = (4l-1)\frac{\pi}{2}, \beta = (4m-1)\frac{\pi}{2}, \gamma = (4n-1)\frac{\pi}{2}$$

Now $\quad \alpha + \beta + \gamma = \pi \Rightarrow [4(l+m+n) - 3]\frac{\pi}{2} = \pi$

$\Rightarrow 4(l+m+n) = 5$ which is not possible as l, m, n are integers.

1. minimum value can not be -3.

But for $\alpha = \frac{3\pi}{2}, \beta = \frac{3\pi}{2}, \gamma = -2\pi, \alpha + \beta + \gamma = \pi$

and $\quad \sin\alpha + \sin\beta + \sin\gamma = 2$

So, $\sin\alpha + \sin\beta + \sin\gamma$ can have negative values and thus the minimum value of the sum is negative proving that statement-1 is correct. But the statement-2 is false as $\alpha + \beta + \gamma = \pi$ for $\alpha = \beta = \frac{3\pi}{2}, \gamma = -2\pi$ which are not the angles of a triangle.

83. We have $2\sin\left(\frac{\theta}{2}\right)$

$$= \sqrt{\left(\cos\left(\frac{\theta}{2}\right) + \sin\left(\frac{\theta}{2}\right)\right)^2} + \sqrt{\left(\cos\left(\frac{\theta}{2}\right) - \sin\left(\frac{\theta}{2}\right)\right)^2}$$

$$= \left|\cos\left(\frac{\theta}{2}\right) + \sin\left(\frac{\theta}{2}\right)\right| + \left|\cos\left(\frac{\theta}{2}\right) - \sin\left(\frac{\theta}{2}\right)\right|$$

$$\Rightarrow \cos\left(\frac{\theta}{2}\right) + \sin\left(\frac{\theta}{2}\right) > 0 \text{ and } \cos\left(\frac{\theta}{2}\right) - \sin\left(\frac{\theta}{2}\right) < 0$$

$$\Rightarrow \sin\left(\frac{\theta}{2} + \frac{\pi}{4}\right) > 0 \text{ and } \cos\left(\frac{\theta}{2} + \frac{\pi}{4}\right) < 0$$

$$\Rightarrow 2n\pi + \frac{\pi}{2} < \frac{\theta}{2} + \frac{\pi}{4} > 2n\pi + \pi$$

$$\Rightarrow 2n\pi + \frac{\pi}{4} < \frac{\theta}{2} < 2n\pi + \frac{3\pi}{4}$$

So statement-1 is true but does not follow from statement-2 which is also true.

84. $2\cos\theta + \sin\theta = 1$

$$\Rightarrow \frac{2\left(1 - \tan^2\left(\frac{\theta}{2}\right)\right)}{1 + \tan^2\left(\frac{\theta}{2}\right)} + \frac{2\tan\left(\frac{\theta}{2}\right)}{1 + \tan^2\left(\frac{\theta}{2}\right)} = 1$$

$$\Rightarrow 3\tan^2\left(\frac{\theta}{2}\right) - 2\tan\left(\frac{\theta}{2}\right) - 1 = 0$$

$$\Rightarrow \tan\left(\frac{\theta}{2}\right) = -\frac{1}{3} \text{ as } \theta \neq \frac{\pi}{2}$$

Now $7\cos\theta + 6\sin\theta$

$$= \frac{7\left(1 - \tan^2\left(\frac{\theta}{2}\right)\right)}{1 + \tan^2\left(\frac{\theta}{2}\right)} + \frac{6 \times 2\tan\left(\frac{\theta}{2}\right)}{1 + \tan^2\left(\frac{\theta}{2}\right)}$$

$$= \frac{7 - 7\tan^2\left(\frac{\theta}{2}\right) + 12\tan\left(\frac{\theta}{2}\right)}{1 + \tan^2\left(\frac{\theta}{2}\right)} = \frac{7 - 7 \times \frac{1}{9} + 12\left(-\frac{1}{3}\right)}{1 + \frac{1}{9}} = 2$$

Showing that statement-1 is true.

In statement-2

$$\cos 2\theta - \sin\theta = \frac{1}{2}$$

$$\Rightarrow 2(1 - 2\sin^2\theta) - \sin\theta = 1$$

$$\Rightarrow 4\sin^2\theta + 2\sin\theta - 1 = 0$$

$$\Rightarrow \sin\theta = \frac{-2 \pm \sqrt{4 + 16}}{8} = \frac{-1 \pm \sqrt{5}}{4}$$

$$\Rightarrow \sin\theta = \frac{\sqrt{5} - 1}{4} \Rightarrow \theta = 18^\circ$$

$$\Rightarrow \cos 6\theta = \cos 108^\circ = \cos(90^\circ + 18^\circ) = -\sin 18^\circ$$

$$\Rightarrow \sin\theta + \cos 6\theta = 0$$

So statement-2 is also true but does not lead to statement-1.

85. $\because A + B = \frac{\pi}{3}$

$$\therefore \tan(A + B) = \tan\left(\frac{\pi}{3}\right) = \sqrt{3}$$

$$\Rightarrow \frac{\tan A + \tan B}{1 - \tan A \tan B} = \sqrt{3}$$

$$\Rightarrow \tan A \tan B = 1 - \frac{1}{\sqrt{3}}(\tan A + \tan B)$$

$\therefore \tan A \tan B$ will be maximum if $\tan A + \tan B$ is minimum.

But the minimum value of $\tan A + \tan B$ is obtained when $\tan A = \tan B$

$$\Rightarrow A = B = \frac{\pi}{6}$$

Hence, the maximum value of $\tan A \tan B$

$$= \tan\frac{\pi}{6} \cdot \tan\frac{\pi}{6} = \frac{1}{\sqrt{3}} \cdot \frac{1}{\sqrt{3}} = \frac{1}{3}$$

86. Let $3^{a^2} = A$ and $3^{b^2 + c^2} = B$

$$\Rightarrow A^2 - 2A.B + B^2 = 0$$

$$\Rightarrow A = B$$

$$\Rightarrow a^2 = b^2 + c^2$$

87.

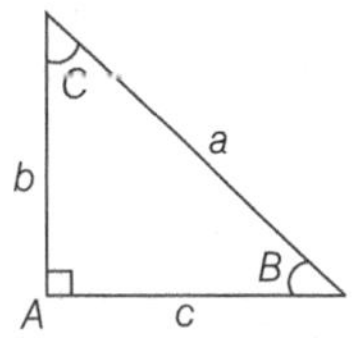

From figure, it is clear that $a = b\sec C = c\,\text{cosec}\,C$

$\Rightarrow$ Equilateral triangle $\Rightarrow$ Area $= \frac{\sqrt{3}}{4}a^2$

Sol. (Q. Nos. 88 to 90)

$$7\theta = (2n + 1)\pi,\ n = 0 \text{ to } 6$$

$$4\theta = (2n + 1)\pi - 3\theta$$

$$\cos 4\theta = -\cos 3\theta$$

$$\Rightarrow 2\cos^2 2\theta - 1 = -(4\cos^3\theta - 3\cos\theta)$$

$$\Rightarrow 2(2x^2 - 1) - 1 = -(4x^3 - 3x), \text{ where } x = \cos\theta$$

$$\Rightarrow 8x^4 + 4x^3 - 8x^2 - 3x + 1 = 0$$

$$(x + 1)(8x^3 - 4x^2 - 4x - 1) = 0$$

88. $P_{mn} = m\log_{\cos x}(\sin x) + n\log_{\cos x}(\cot x)$

$$\geq n(\log_{\cos x}(\sin x) + \log_{\cos x}(\cot x))\ \forall\ m \geq n$$

$$= n(\log_{\cos x}(\sin x \cdot \cot x))$$

$$= n\log_{\cos x}\cos x = n$$

Thus, $P_{mn} \geq n\ \forall\ m \geq n$

89. Clearly, $P_{49}\left(\frac{\pi}{4}\right) = 4\log_{\frac{1}{\sqrt{2}}}\left(\frac{1}{\sqrt{2}}\right) + 9\log_{\frac{1}{\sqrt{2}}}(1) = 4$

Similarly $P_{94} = \left(\frac{\pi}{4}\right) = 9$

Mean proportional of $P_{49}\left(\frac{\pi}{4}\right)$ and $P_{94}\left(\frac{\pi}{4}\right)$ is $\sqrt{9 \times 4} = 6$

90. $P_{34}(x) = P_{22}(x)$

$\Rightarrow 3\log_{\cos x}(\sin x) + 4\log_{\cos x}(\cot x)$
$= 2(\log_{\cos x}(\sin x) + \log_{\cos x}(\cot x))$

$\Rightarrow 3(\log_{\cos x}(\sin x) + \log_{\cos x}(\cot x)) + \log_{\cos x}(\cot x) = 2$

$\Rightarrow 3 + \log_{\cos x}(\cot x) = 2$

$\Rightarrow \log_{\cos x}(\cot x) = -1$

$\Rightarrow \cot x = (\cos x)^{-1} \Rightarrow \dfrac{\cos x}{\sin x} = \dfrac{1}{\cos x}$

$\Rightarrow \cos^2 x = \sin x$

$1 - \sin^2 x = \sin x \Rightarrow \sin^2 x + \sin x - 1 = 0$

$\Rightarrow \sin x = \dfrac{-1 \pm \sqrt{5}}{2}$

$\Rightarrow \sin x = \dfrac{\sqrt{5} - 1}{2}$ $(\because \sin x \nless -1)$

Thus, $p + q = 7$

Sol. (Q. Nos. 91 to 93)

$7\theta = (2n+1)\pi, n = 0$ to 6

$4\theta = (2n+1)\pi - 3\theta$

$\cos 4\theta = -\cos 3\theta$

$\Rightarrow 2\cos^2 2\theta - 1 = -(4\cos^3\theta - 3\cos\theta)$

$\Rightarrow 2(2x^2 - 1) - 1 = -(4x^3 - 3x)$, where $x = \cos\theta$

$\Rightarrow 8x^4 + 4x^3 - 8x^2 - 3x + 1 = 0$

$(x+1)(8x^3 - 4x^2 - 4x - 1) = 0$

91. The roots are $\cos\dfrac{\pi}{7}, \cos\dfrac{2\pi}{7}, \ldots, \cos\dfrac{13\pi}{7}$

where $\cos\dfrac{\pi}{7} = \cos\dfrac{13\pi}{7}, \cos\dfrac{3\pi}{7} = \cos\dfrac{11\pi}{7}, \cos\dfrac{5\pi}{7} = \cos\dfrac{9\pi}{7}$

$\therefore$ The roots of $8x^3 - 4x^2 - 4x + 1 = 0$ are $\cos\dfrac{\pi}{7}, \cos\dfrac{3\pi}{7}, \cos\dfrac{5\pi}{7}$.

92. $\sec\dfrac{\pi}{7}, \sec\dfrac{3\pi}{7}, \sec\dfrac{5\pi}{7}$ are roots of $\dfrac{8}{x^3} - \dfrac{4}{x^2} - \dfrac{4}{x} + 1 = 0$

$\Rightarrow x^3 - 4x^2 - 4x + 8 = 0$

$\therefore \sec\dfrac{\pi}{7} + \sec\dfrac{3\pi}{7} + \sec\dfrac{5\pi}{7} = 4$

93. $\sec^2\dfrac{\pi}{7}, \sec^2\dfrac{3\pi}{7}, \sec^2\dfrac{5\pi}{7}$ are roots of $f(\sqrt{x}) = 0$

$\Rightarrow (\sqrt{x})^3 - 4(\sqrt{x})^2 - 4\sqrt{x} + 8 = 0$

$\Rightarrow x^3 - 24x^2 + 80x - 64 = 0$

$\therefore \sec^2\dfrac{\pi}{7} + \sec^2\dfrac{3\pi}{7} + \sec^2\dfrac{5\pi}{7} = 24$

Sol. (Q. Nos. 94 to 96)

Let $S = 1 + 2\sin x + 3\sin^2 x + 4\sin^3 x + \ldots$

$\Rightarrow \sin x.\, S = \sin x + 2\sin^2 x + 3\sin^3 x + \ldots$

$\therefore (1 - \sin x) S = 1 \ \sin x + \sin^2 x + \ldots$

$(1 - \sin x) S = \dfrac{1}{1 - \sin x}$

$\therefore S = \dfrac{1}{(1 - \sin x)^2}$

Given $S = 4 \Rightarrow \dfrac{1}{(1 - \sin x)^2} = 4$

$\Rightarrow \sin x = \dfrac{1}{2}$ or $\dfrac{3}{2}$ (rejected)

Number of solutions in $\left[\dfrac{-3\pi}{2}, 4\pi\right]$ is $k = 5$.

94. $k = 5$

95. $\left|\dfrac{\cos 2x - 1}{\sin 2x}\right| = \left|\dfrac{2\sin^2 x}{2\cos x \sin x}\right| = |\tan x| = \dfrac{1}{\sqrt{3}}$

96. Sum of interior angles $= (k-2)\pi = 3\pi$

97. Now, $(2\sin x - \cos x)(1 + \cos x) = \sin^2 x$

$\Rightarrow (1 + \cos x)[2\sin x - \cos x - 1 + \cos x] = 0$

$\Rightarrow (1 + \cos x)(2\sin x - 1) = 0$

$\Rightarrow \cos x = -1$ or $\sin x = \dfrac{1}{2}$

So, $\sin\alpha = \dfrac{1}{2}$ $\left[\text{as } 0 \le \alpha \le \dfrac{\pi}{2}\right]$

$\therefore \cos\alpha = \dfrac{\sqrt{3}}{2}$

$3\cos^2 x - 10\cos x + 3 = 0$

$\Rightarrow \cos x = \dfrac{1}{3}, \cos x \ne 3$

$\Rightarrow \cos\beta = \dfrac{1}{3}, \sin\beta = \dfrac{2\sqrt{2}}{3}$

and $1 - \sin 2x = \cos x - \sin x$

$\Rightarrow \sin^2 x + \cos^2 x - 2\sin x\cos x = \cos x - \sin x$

$\Rightarrow (\cos x - \sin x)(\cos x - \sin x - 1) = 0$

$\Rightarrow \sin x = \cos x = \dfrac{1}{\sqrt{2}}$

or $\cos x - \sin x = 1$

$\Rightarrow \cos x = 1, \sin x = 0$

$\Rightarrow \cos\gamma = 1, \sin\gamma = 0$

$\therefore \cos\alpha + \cos\beta + \cos\gamma = \dfrac{\sqrt{3}}{2} + \dfrac{1}{3} + 1 = \dfrac{3\sqrt{3} + 8}{6}$

98. $\sin(\alpha - \beta) = \sin\alpha\cos\beta - \cos\alpha\sin\beta$

$= \dfrac{1}{2} \times \dfrac{1}{3} - \dfrac{\sqrt{3}}{2} \times \dfrac{2\sqrt{2}}{3} = \dfrac{1 - 2\sqrt{6}}{6}$

99. (A) If M is mid point of PQ, then $M = \left(\dfrac{\theta + \phi}{2}, \dfrac{\sin\theta + \sin\phi}{2}\right)$

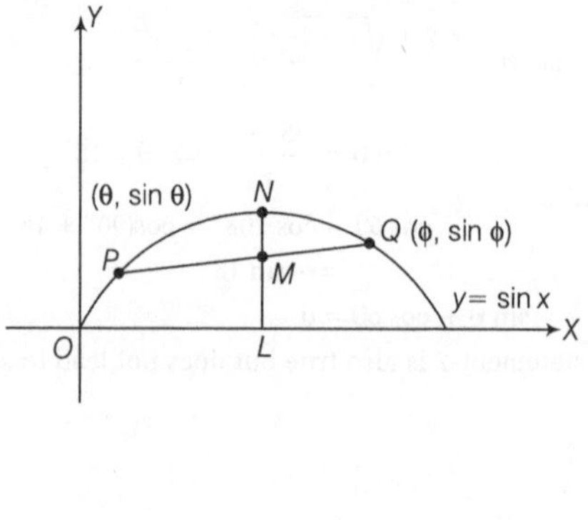

Also, $N \equiv \left(\frac{\theta+\phi}{2}, \sin\left(\frac{\theta+\phi}{2}\right)\right)$

It is clear from the figure.

$$ML \le NL$$

$$\Rightarrow \quad \frac{\sin\theta+\sin\phi}{2} \le \sin\left(\frac{\theta+\phi}{2}\right)$$

$$\Rightarrow \quad \sin\theta+\sin\phi \le 2\sin\left(\frac{\theta+\phi}{2}\right)$$

$$= 2\sin\left(\frac{\pi}{4}\right) = \sqrt{2}$$

$$\therefore \quad \sin\theta+\sin\phi \le \sqrt{2}$$

and $(\sin\theta+\sin\phi)\sin\frac{\pi}{4} \le 1$ (p, q, r, s, t)

(B) $\because a^2+b^2 = (\sin\theta-\sin\phi)^2+(\cos\theta+\cos\phi)^2$

$$= 2+2\cos(\theta+\phi)$$

$$= 4\cos^2\left(\frac{\theta+\phi}{2}\right) \le 4 \text{ (s, t)}$$

(C) $\because 3\sin\theta+5\cos\theta=5$

$$\Rightarrow \quad 3\sin\theta = 5(1-\cos\theta)$$

Squaring both sides, then

$$9\sin^2\theta = 25(1-\cos\theta)^2$$

$$\Rightarrow 9(1-\cos\theta)(1+\cos\theta) = 25(1-\cos\theta)^2$$

$$\Rightarrow \quad 9(1+\cos\theta) = 25(1-\cos\theta)(1-\cos\theta \ne 0)$$

$$\therefore \quad 34\cos\theta = 16$$

$$\cos\theta = \frac{8}{17}, \text{ then } \sin\theta = \frac{15}{17}$$

$$\therefore \quad 5\sin\theta - 3\cos\theta = \frac{75}{17}-\frac{24}{17} = 3 \text{ (r)}$$

Hence, $5\sin\theta-3\cos\theta=3$

100. (A) Let $y = \frac{7+6\tan\theta-\tan^2\theta}{(1+\tan^2\theta)}$

$$= 7\cos^2\theta+6\sin\theta\cos\theta-\sin^2\theta$$

$$= 7\left(\frac{1+\cos\theta}{2}\right)+3\sin 2\theta-\left(\frac{1-\cos 2\theta}{2}\right)$$

$$= 3\sin 2\theta+4\cos 2\theta+3$$

$$-\sqrt{(3^2+4^2)}+3 \le 3\sin 2\theta+4\cos 2\theta+3$$

$$\le \sqrt{(3^2+4^2)}+3$$

$$\therefore \quad -2 \le y \le 8 \Rightarrow \lambda = 8, \mu = -2$$

$$\Rightarrow \quad \lambda+\mu = 6, \lambda-\mu = 10 \text{ (R, S)}$$

(B) Let $y = 5\cos\theta+3\cos\left(\theta+\frac{\pi}{3}\right)+3$

$$= 5\cos\theta+3\left(\frac{1}{2}\cos\theta-\frac{\sqrt{3}}{2}\sin\theta\right)+3$$

$$= \frac{13}{2}\cos\theta-\frac{3\sqrt{3}}{2}\sin\theta+3$$

$$\therefore \quad 3-\sqrt{\left(\frac{13}{2}\right)^2+\left(\frac{-3\sqrt{3}}{2}\right)^2} \le \frac{13}{2}\cos\theta-\frac{3\sqrt{3}}{2}\sin\theta+3$$

$$\le 3+\sqrt{\left(\frac{13}{2}\right)^2+\left(\frac{-3\sqrt{3}}{2}\right)^2}$$

$$\Rightarrow \quad 3-7 \le y \le 3+7$$

$$\Rightarrow \quad -4 \le y \le 10$$

$$\therefore \quad \lambda = 10, \mu = -4$$

$$\Rightarrow \quad \lambda+\mu = 6, \lambda-\mu = 14 \text{ (R, T)}$$

(C) Let $y = 1+\sin\left(\frac{\pi}{4}+\theta\right)+2\cos\left(\frac{\pi}{4}-\theta\right)$

$$= 1+\cos\left(\frac{\pi}{2}-\left(\frac{\pi}{2}+\theta\right)\right)+2\cos\left(\frac{\pi}{4}-\theta\right)$$

$$= 1+\cos\left(\frac{\pi}{4}-\theta\right)+2\cos\left(\frac{\pi}{4}-\theta\right)$$

$$= 1+3\cos\left(\frac{\pi}{4}-\theta\right)$$

$$\therefore \quad -1 \le \cos\left(\frac{\pi}{4}-\theta\right) \le 1$$

$$\Rightarrow \quad -3 \le 3\cos\left(\frac{\pi}{4}-\theta\right) \le 3$$

$$\Rightarrow \quad 1-3 \le 1+3\cos\left(\frac{\pi}{4}-\theta\right) \le 1+3$$

$$\therefore \quad -2 \le y \le 4$$

$$\Rightarrow \quad \lambda = 4, \mu = -2$$

$$\therefore \quad \lambda+\mu = 2, \lambda-\mu = 6 \text{ (P, Q)}$$

101. (A) $|\cot x| = \cot x + \frac{1}{\sin x}$

If $0 < x < \frac{\pi}{2} \Rightarrow \cot x > 0$

So $\cot x = \cot x + \frac{1}{\sin x} \Rightarrow \frac{1}{\sin x} = 0$, no solution

If $\frac{\pi}{2} < \cot x < \pi$, $-\cot x = \cot x + \frac{1}{\sin x}$

$$\Rightarrow \quad \frac{2\cos x}{\sin x}+\frac{1}{\sin x} = 0$$

$$\Rightarrow \quad 1+2\cos x = 0 \text{ and } \sin x \ne 0 \Rightarrow x = \frac{2\pi}{3}.$$

(B) since $\sin\phi+\sin\theta = \frac{1}{2}$ ...(i)

and $\cos\theta+\cos\phi = 2$ ...(ii)

(ii) is true only if $\theta=\phi=0$ or 2π but $\theta=\phi=0$ or 2π do not satisfy (i)

Hence given system of equation has no solution.

(C) $\sin^2\alpha+\sin\left(\frac{\pi}{3}-\alpha\right)\cdot\sin\left(\frac{\pi}{3}+\alpha\right)$

$$= \sin^2\alpha+\sin^2\frac{\pi}{3}-\sin^2\alpha = \frac{3}{4}.$$

(D) $\tan\theta = 3\tan\phi$

$$\tan(\theta-\phi) = \frac{\tan\theta-\tan\phi}{1+\tan\theta\tan\phi} = \frac{2\tan\phi}{1+3\tan\phi}$$

$$= \frac{2}{\cot\phi+3\tan\phi} \cdot \text{Max if } \tan\phi > 0$$

$$\frac{\cos\phi + 3\tan\phi}{2} \ge \sqrt{3} \qquad \text{(Using AM} \ge \text{GM)}$$

$$\Rightarrow \quad (\cot\phi + 3\tan\phi)^2 \ge 12 \Rightarrow \left[\frac{2}{\tan(\theta - \phi)}\right]^2 \ge 12$$

$$\Rightarrow \quad \tan^2(\theta - \phi) \le \frac{1}{3}.$$

102. (A) $\sin\left(\frac{A}{2}\right) + \sin\left(\frac{B}{2}\right) + \sin\left(\frac{C}{2}\right)$

$$= 1 + \left(\sin\left(\frac{A}{2}\right) + \sin\left(\frac{B}{2}\right)\right) - \left(\sin\frac{\pi}{2} - \sin\frac{C}{2}\right)$$

$$= 1 + 2\sin\left(\frac{A+B}{4}\right)\cos\left(\frac{A-B}{4}\right) - 2\cos\left(\frac{\pi+C}{4}\right)\sin\left(\frac{\pi-C}{4}\right)$$

$$= 1 + 2\sin\left(\frac{\pi-C}{4}\right)\left\{\cos\left(\frac{A-B}{4}\right) - \cos\left(\frac{\pi+C}{4}\right)\right\}$$

$(\because A+B+C=\pi)$

$$= 1 + 2\sin\left(\frac{\pi-C}{4}\right)\left\{2\sin\left(\frac{\pi+C+A-B}{8}\right)\sin\left(\frac{\pi+C+B-A}{8}\right)\right\}$$

$$= 1 + 4\sin\left(\frac{\pi-C}{4}\right)\sin\left(\frac{\pi-B}{4}\right)\sin\left(\frac{\pi-A}{4}\right)$$

$$= 1 + 4\sin\left(\frac{\pi-A}{4}\right)\sin\left(\frac{\pi-B}{4}\right)\sin\left(\frac{\pi-C}{4}\right)$$

$$= 1 + 4\cos\left(\frac{\pi}{2} - \frac{\pi-A}{4}\right)\cos\left(\frac{\pi}{2} - \frac{\pi-B}{4}\right)\sin\left(\frac{\pi-C}{4}\right)$$

$$= 1 + 4\cos\left(\frac{\pi+A}{4}\right)\cos\left(\frac{\pi+B}{4}\right)\sin\left(\frac{\pi-C}{4}\right)$$

(B) $\sin\left(\frac{A}{2}\right) + \sin\left(\frac{B}{2}\right) - \sin\left(\frac{C}{2}\right)$

$$= -1 + \left(\sin\left(\frac{A}{2}\right) + \sin\left(\frac{B}{2}\right)\right) + \left(\sin\left(\frac{\pi}{2}\right) - \sin\left(\frac{C}{2}\right)\right)$$

$$= -1 + 2\sin\left(\frac{A+B}{4}\right)\cos\left(\frac{A-B}{4}\right) + 2\cos\left(\frac{\pi+C}{4}\right)\sin\left(\frac{\pi-C}{4}\right)$$

$$= -1 + 2\sin\left(\frac{\pi-C}{4}\right)\left\{\cos\left(\frac{A-B}{4}\right) + \cos\left(\frac{\pi+C}{4}\right)\right\}$$

$(\because A+B+C=\pi)$

$$= -1 + 2\sin\left(\frac{\pi-C}{4}\right)\left\{2\cos\left(\frac{\pi+C+A-B}{8}\right)\cos\left(\frac{\pi+C+B-A}{8}\right)\right\}$$

$$= -1 + 4\sin\left(\frac{\pi-C}{4}\right)\cos\left(\frac{\pi-B}{4}\right)\cos\left(\frac{\pi-A}{4}\right)$$

$$= -1 + 4\cos\left(\frac{\pi-A}{4}\right)\cos\left(\frac{\pi-B}{4}\right)\sin\left(\frac{\pi-C}{4}\right)$$

$$= -1 + 4\sin\left(\frac{\pi}{2} - \frac{\pi-A}{4}\right)\sin\left(\frac{\pi}{2} - \frac{\pi-B}{4}\right)\cos\left(\frac{\pi}{2} - \frac{\pi-C}{4}\right)$$

$$= -1 + 4\sin\left(\frac{\pi+A}{4}\right)\sin\left(\frac{\pi+B}{4}\right)\cos\left(\frac{\pi+C}{4}\right)$$

(C) $\cos\left(\frac{A}{2}\right) + \cos\left(\frac{B}{2}\right) - \cos\left(\frac{C}{2}\right)$

$$= \left(\cos\left(\frac{A}{2}\right) + \cos\left(\frac{B}{2}\right)\right) - \left(\cos\left(\frac{C}{2}\right) + \cos\left(\frac{\pi}{2}\right)\right)$$

$$= 2\cos\left(\frac{A+B}{4}\right)\cos\left(\frac{A-B}{4}\right) - \cos\left(\frac{\pi+C}{4}\right)\cos\left(\frac{\pi-C}{4}\right)$$

$$= 2\cos\left(\frac{\pi-C}{4}\right)\left\{\cos\left(\frac{A-B}{4}\right) - \cos\left(\frac{\pi+C}{4}\right)\right\}$$

$(\because A+B+C=\pi)$

$$= 2\cos\left(\frac{\pi-C}{4}\right)\left\{2\sin\left(\frac{\pi+C+A-B}{8}\right)\sin\left(\frac{\pi+C+B-A}{8}\right)\right\}$$

$$= 4\cos\left(\frac{\pi-C}{4}\right)\sin\left(\frac{\pi-B}{4}\right)\sin\left(\frac{\pi-A}{4}\right)$$

$$= 4\cos\left(\frac{\pi-C}{4}\right)\cos\left(\frac{\pi}{2} - \frac{\pi-B}{4}\right)\cos\left(\frac{\pi}{2} - \frac{\pi-A}{4}\right)$$

$$= 4\cos\left(\frac{\pi+A}{4}\right)\cos\left(\frac{\pi+B}{4}\right)\cos\left(\frac{\pi-C}{4}\right)$$

103. $\dfrac{1}{1+\tan^2\frac{A}{2}} + \dfrac{1}{1+\tan^2\frac{B}{2}} + \dfrac{1}{1+\tan^2\frac{C}{2}}$

$$= k\left[1 + \sin\frac{A}{2}\sin\frac{B}{2}\sin\frac{C}{2}\right]$$

$$\Rightarrow \cos^2\frac{A}{2} + \cos^2\frac{B}{2} + \cos^2\frac{C}{2}$$

$$= 2\left[1 + \sin\frac{A}{2}\sin\frac{B}{2}\sin\frac{C}{2}\right] \qquad \text{[by using identity]}$$

104. $\dfrac{\sin\alpha}{\sin\beta} = \dfrac{\cos\gamma}{\cos\delta}$

$$\Rightarrow \frac{\sin\alpha - \sin\beta}{\sin\beta} = \frac{\cos\gamma - \cos\delta}{\cos\delta} \qquad \text{(using dividendo)}$$

$$\Rightarrow \frac{2\cos\left(\frac{\alpha+\beta}{2}\right)\sin\left(\frac{\alpha-\beta}{2}\right)}{\sin\beta} = \frac{2\sin\left(\frac{\gamma+\delta}{2}\right)\sin\left(\frac{\delta-\gamma}{2}\right)}{\cos\delta}$$

105. Let $\frac{\pi}{20} = \theta \Rightarrow 10\theta = \frac{\pi}{2}$

$\Rightarrow \quad 2\theta = 18°$ or $\theta = 9°$

Now, $\tan\theta - \tan3\theta + \tan5\theta - \tan7\theta + \tan9\theta$

$\tan\theta - \tan3\theta + \tan5\theta - \cot3\theta + \cot\theta$

$(\tan\theta + \cot\theta) - (\tan3\theta + \cot3\theta) + \tan45°$

[using $\tan5\theta = \tan45°$]

$$E = \frac{2}{\sin2\theta} - \frac{2}{\sin6\theta} + 1$$

$$E = 2\left(\frac{1}{\sin 2\theta} - \frac{1}{\cos 4\theta}\right) + 1$$

$$E = 2\left(\frac{1}{\sin 18^\circ} - \frac{1}{\cos 36^\circ}\right) + 1$$

$$E = 2\left(\frac{4}{\sqrt{5}-1} - \frac{4}{\sqrt{5}+1}\right) + 1 = 5$$

Aliter $E = 1 + 2\left(\dfrac{\sin 6\theta - \sin 2\theta}{\sin 2\theta \cdot \sin 6\theta}\right)$

$$= 1 + 2\left(\frac{2\cos 4\theta \cdot \sin 2\theta}{\sin 2\theta \cdot \cos 4\theta}\right)$$

$$= 1 + 4 = 5$$

106. $x = \dfrac{\cos 1^\circ + \cos 2^\circ + \ldots + \cos 44^\circ}{\sin 1^\circ + \sin 2^\circ + \ldots + \sin 44^\circ}$

$$= \frac{\dfrac{\sin 22^\circ}{\sin(1/2)^\circ} \cdot \cos 22.5^\circ}{\dfrac{\sin 22^\circ}{\sin(1/2)^\circ} \cdot \sin 22.5^\circ} = \cot 22.5^\circ$$

[using the formula of sum of cos series]

$$S = \frac{\sin(n\theta/2)}{\sin(\theta/2)} \cos\frac{(n+1)\theta}{2},$$

for sine series, $S = \dfrac{\sin(n\theta/2)}{\sin(\theta/2)} \sin\dfrac{(n+1)\theta}{2}$

$$\cot\left(\frac{\pi}{8}\right) = \sqrt{2} + 1 = 2.414\ldots$$

$\therefore$ $x = 2.414\ldots$

Greatest integer = 2.

107. LHS = $\tan 15^\circ \cdot \tan(30^\circ - 5^\circ) \cdot \tan(30^\circ + 5^\circ)$

Let $t = \tan 30^\circ$ and $m = \tan 5^\circ$

$$\therefore \text{LHS} = \tan 15^\circ \cdot \frac{t-m}{1+tm} \cdot \frac{t+m}{1-tm} = \tan(3(5^\circ)) \cdot \frac{t^2 - m^2}{1 - t^2m^2}$$

$$= \frac{3m - m^3}{1 - 3m^2} \frac{1 - 3m^2}{3 - m^2}$$

$$= \frac{m(3 - m^2)}{(1 - 3m^2)} \cdot \frac{(1 - 3m^2)}{3 - m^2} = m = \tan 5^\circ$$

Hence, $\tan\theta = \tan 5^\circ$

$\Rightarrow$ $\theta = 5^\circ$.

108. We have, $\dfrac{1}{\sin 10^\circ} + \dfrac{1}{\sin 50^\circ} - \dfrac{1}{\sin 70^\circ}$

$$= \frac{1}{\cos 80^\circ} + \frac{1}{\cos 40^\circ} - \frac{1}{\cos 20^\circ}$$

$$= \frac{\cos 40^\circ \cos 20^\circ + \cos 80^\circ \cos 20^\circ - \cos 40^\circ \cos 80^\circ}{\cos 20^\circ . \cos 40^\circ . \cos 80^\circ}$$

$$= 8\,[\cos 20^\circ (\cos 40^\circ + \cos 80^\circ) - \cos 40^\circ \cos 80^\circ]$$

$$= 8[2\cos 20^\circ \cos 60^\circ \cos 20^\circ - \cos 40^\circ \cos 80^\circ]$$

$$= 4[2\cos^2 20^\circ - 2\cos 40^\circ \cos 80^\circ]$$

$$= 4[1 + \cos 40^\circ - (\cos 120^\circ + \cos 40^\circ)]$$

$$= 4 \cdot \frac{3}{2} = 6$$

109. We have, $\cos 5\alpha = \cos^5\alpha$

$$\cos 5\alpha = \cos(3\alpha + 2\alpha) = \cos 3\alpha \cos 2\alpha - \sin 3\alpha \sin 2\alpha$$

$$= (4\cos^3\alpha - 3\cos\alpha)(2\cos^2\alpha - 1) - (3\sin\alpha - 4\sin^3\alpha) 2\sin\alpha\cos\alpha$$

$$= (4\cos^3\alpha - 3\cos\alpha)(2\cos^2\alpha - 1) - (1 \cos^2\alpha)(3 - 4 + 4\cos^2\alpha) 2\cos\alpha$$

$$= (4\cos^3\alpha - 3\cos\alpha)(2\cos^2\alpha - 1) - (2\cos^2\alpha - 2\cos^3\alpha)(4\cos^2\alpha - 1)$$

$$= 8\cos^5\alpha - 4\cos^3\alpha - 6\cos^3\alpha + 3\cos\alpha - [8\cos^3\alpha - 2\cos\alpha - 8\cos^5\alpha + 2\cos^3\alpha]$$

$$= 16\cos^5\alpha - 20\cos^3\alpha + 5\cos\alpha$$

$\therefore$ $16\cos^5\alpha - 20\cos^3\alpha + 5\cos\alpha = \cos^5\alpha$

$$15\cos^5\alpha - 20\cos^3\alpha + 5\cos\alpha = 0$$

$$5\cos\alpha\,[3\cos^4\alpha - 4\cos^2\alpha + 1] = 0$$

Also $\cos\alpha = 0$

$$3\cos^4\alpha - 3\cos^2\alpha - \cos^2\alpha + 1 = 0$$

$$3\cos^2\alpha(\cos^2\alpha - 1) - (\cos^2\alpha - 1) = 0$$

$$(3\cos^2\alpha - 1)(1 - \cos^2\alpha) = 0$$

$$\cos^2\alpha = 1$$

$$\cos^2\alpha = \frac{1}{3}$$

$\Rightarrow$ $\sec^2\alpha = 3;\ \text{cosec}^2\alpha = \dfrac{3}{2};\ \cot^2\alpha = \dfrac{1}{2}$

$\therefore$ $(\sec^2\alpha + \text{cosec}^2\alpha + \cot^2\alpha) = 3 + \dfrac{3}{2} + \dfrac{1}{2} = 5$

110. We have,

$$\tan^2\frac{\pi}{16} + \tan^2\frac{2\pi}{16} + \tan^2\frac{3\pi}{16} + \ldots + \tan^2\frac{7\pi}{16}$$

$$= \left(\tan^2\frac{\pi}{16} + \cot^2\frac{\pi}{16} + 2\right) + \left(\tan^2\frac{2\pi}{16} + \cot^2\frac{2\pi}{16} + 2\right) + \left(\tan^2\frac{3\pi}{16} + \cot^2\frac{3\pi}{16} + 2\right) + \tan^2\frac{4\pi}{16} - 6$$

$$\left[\begin{array}{l}\text{If } A + B = \dfrac{\pi}{2}, \text{ then } \tan B = \tan\left(\dfrac{\pi}{2} - A\right) = \cot A \\ \text{So, } \tan^2\dfrac{7\pi}{16} = \tan^2\left(\dfrac{8\pi}{16} - \dfrac{\pi}{16}\right) = \cot^2\dfrac{\pi}{16} \text{ etc.}\end{array}\right]$$

$$= \left(\tan\frac{\pi}{16} + \cot\frac{\pi}{16}\right)^2 + \left(\tan\frac{2\pi}{16} + \cot\frac{2\pi}{16}\right)^2 + \left(\tan\frac{3\pi}{16} + \cot\frac{3\pi}{16}\right)^2 - 5$$

$$= \frac{4}{\sin^2\frac{\pi}{8}} + \frac{4}{\sin^2\frac{3\pi}{8}} + \left(\frac{4}{\sin^2\frac{\pi}{4}} - 5\right)$$

$$= \frac{4\left(\sin^2\frac{3\pi}{8} + \sin^2\frac{\pi}{8}\right)}{\sin^2\frac{\pi}{8}\sin^2\frac{3\pi}{8}} + 3 = \frac{16}{\sin^2\frac{\pi}{4}} + 3 = 32 + 3 = 35$$

111. We have, $\dfrac{4+\sec 20^\circ}{\text{cosec } 20^\circ}$

$$= \frac{\sin 20^\circ}{\cos 20^\circ}(4\cos 20^\circ + 1)$$
$$= \frac{2\sin 40^\circ + \sin 20^\circ}{\cos 20^\circ}$$
$$= \frac{\sin 40^\circ + (\sin 40^\circ + \sin 20^\circ)}{\cos 20^\circ}$$
$$= \frac{\sin 40^\circ + 2\sin 30^\circ \cos 10^\circ}{\cos 20^\circ}$$
$$= \frac{\sin 40^\circ + \sin 80^\circ}{\cos 20^\circ}$$
$$= \frac{2\sin 60^\circ \cos 20^\circ}{\cos 20^\circ} = 2 \times \frac{\sqrt{3}}{2} = \sqrt{3}$$

Hence, square of the value of expression = 3

112. $A + B + C = \pi$ ...(i)

$$\frac{\sin A}{3} = \frac{\cos B}{3} = \frac{\tan C}{2} \quad \text{...(ii)}$$

$\Rightarrow \quad \sin A = \cos B \Rightarrow A + B = \dfrac{\pi}{2}$ (rejected)

Or $\quad A - B = \dfrac{\pi}{2}$ ...(iii)

$\Rightarrow \quad 2A + C = \dfrac{3\pi}{2}$ [from Eqs. (i) and (ii)]

Now $\quad \dfrac{\sin A}{\tan C} = \dfrac{3}{2}$ [from Eq. (ii)]

$\Rightarrow \quad \dfrac{\sin A}{\tan\left(\dfrac{3\pi}{2} - 2A\right)} = \dfrac{3}{2}$ [from Eq. (iii)]

$\Rightarrow \quad 2\sin A = 3\cot 2A$

$\Rightarrow \quad 2\sin A = \dfrac{3\cdot(2\cos^2 A - 1)}{2\sin A\cos A}$

$\Rightarrow \quad 4\cos A(1-\cos^2 A) = 3(2\cos^2 A - 1)$

$\Rightarrow \quad 4\cos^3 A + 6\cos^2 A - 4\cos A - 3 = 0$

Put $\cos A = -\dfrac{1}{2}$

$\Rightarrow \quad (2\cos A + 1)(2\cos^2 A + 2\cos A - 3) = 0$

$\Rightarrow \quad \cos A = -\dfrac{1}{2},$

$$\cos A = \frac{-2 \pm \sqrt{4+24}}{4} = -1 \pm \sqrt{7} \text{ (rejected)}$$

$\Rightarrow \quad A = \dfrac{2\pi}{3}, B = \dfrac{\pi}{6}, C = \dfrac{\pi}{6}$

$\therefore \quad \dfrac{\sin A}{\cos 2A} + \dfrac{\cos B}{\cot 2B} + \dfrac{\tan C}{\cot 2C} = \dfrac{1}{2} + \dfrac{1}{2} + 1 = 2$

113. $f(\theta) = \dfrac{1}{1+g(\theta)}$ ...(i)

Given, $2f(\alpha) - g(\beta) = 1$

$2f(\alpha) = 1 + g(\beta) = \dfrac{1}{f(\beta)}$ [from Eq. (i)]

$2f(\alpha)f(\beta) = 1$ ...(ii)

Now, $2f(\beta) - g(\alpha) = 2f(\beta) + 1 - (1 + g(\alpha))$

$$= 2f(\beta) + 1 - \frac{1}{f(\alpha)}$$
$$= \frac{2f(\alpha)f(\beta) - 1}{f(\alpha)} + 1 = 1 \quad \text{[from Eq. (ii)]}$$

114. As we know that $\dfrac{1-\sin x}{\cos x} = \dfrac{\cos x}{1+\sin x}$

$$\log_{\left|\frac{1-\sin x}{\cos x}\right|}\left|\frac{1+\sin x}{\cos x}\right| = -1$$

Now, series is

$$\text{Let } S = 1 - \frac{x}{2} - \frac{x^2}{4} - \ldots = 1 - \frac{\frac{x}{2}}{1-\frac{x}{2}}$$
$$= 1 - \frac{-x}{2-x} = \frac{2-x-x}{2-x} = \frac{2(1-x)}{(2-x)} = \frac{k(1-x)}{(2-x)}$$

Thus, $k = 2$

115. From the second relation $9x\sin^3\theta = 5y\cos^3\theta$.

$\Rightarrow \quad \dfrac{\cos^3\theta}{9x} = \dfrac{\sin^3\theta}{5y} = k^3$ (say)

$\Rightarrow \quad \cos\theta = k(9x)^{\frac{1}{3}}$ and $\sin\theta = k(5y)^{\frac{1}{3}}$

Squaring and adding, we get

$$1 = \cos^2\theta + \sin^2\theta = k^2\left[(9x)^{\frac{2}{3}} + (5y)^{\frac{2}{3}}\right]$$

and $\quad \dfrac{9x}{k(9x)^{\frac{1}{3}}} + \dfrac{5y}{k(5y)^{\frac{1}{3}}} = 56$ (From Ist relation)

$\Rightarrow \quad (9x)^{\frac{2}{3}} + (5y)^{\frac{2}{3}} = 56k$

$\Rightarrow \quad \left[(9x)^{\frac{2}{3}} + (5y)^{\frac{2}{3}}\right]^2 = (56)^2k^2 = \dfrac{(56)^2}{(9x)^{\frac{2}{3}} + (5y)^{\frac{2}{3}}}$

$\Rightarrow \quad \left[(9x)^{\frac{2}{3}} + (5y)^{\frac{2}{3}}\right]^3 = (56)^2 = 3136.$

116. $A > \dfrac{\pi}{2} \Rightarrow B + C < \dfrac{\pi}{2}$

$\Rightarrow \quad \tan(B+C) > 0 \Rightarrow \dfrac{\tan B + \tan C}{1 - \tan B\tan C} > 0$

$\Rightarrow \quad \tan B\tan C < 1$ as $\tan B > 0, \tan C > 0$

$\Rightarrow \quad [x] = 2635 - 1 = 2634$

117. $\because \cot\left(7\dfrac{1^\circ}{2}\right) = \dfrac{1+\cos 15^\circ}{\sin 15^\circ}$

$$= \frac{1 + \frac{\sqrt{3}+1}{2\sqrt{2}}}{\frac{\sqrt{3}-1}{2\sqrt{2}}} = \frac{2\sqrt{2}+\sqrt{3}+1}{\sqrt{3}-1}$$
$$= \frac{(2\sqrt{2}+\sqrt{3}+1)(\sqrt{3}+1)}{2}$$

$$= \frac{2\sqrt{6} + 2\sqrt{2} + 3 + \sqrt{3} + \sqrt{3} + 1}{2}$$

$$= \sqrt{6} + \sqrt{2} + 2 + \sqrt{3}$$

$$= \sqrt{2} + \sqrt{3} + \sqrt{4} + \sqrt{6}$$

and $4 \cos 36° = 4\left(\frac{\sqrt{5}+1}{4}\right) = \sqrt{5} + 1 = \sqrt{5} + \sqrt{1}$

Hence, $4 \cos 36° + \cot\left(7\frac{1°}{2}\right)$

$$= \sqrt{1} + \sqrt{2} + \sqrt{3} + \sqrt{4} + \sqrt{5} + \sqrt{6}$$

$\therefore \quad n_1 = 1, n_2 = 2, n_3 = 3, n_4 = 4, n_5 = 5$ and $n_6 = 6$

$\therefore \quad \sum_{i=1}^{6} n_i^2 = n_1^2 + n_2^2 + n_3^2 + n_4^2 + n_5^3 + n_6^2$

$$= 1^2 + 2^2 + 3^2 + 4^2 + 5^2 + 6^2$$

$$= 91$$

118. $\because \prod_{r=1}^{4} \sin(rA) = \sin A \sin 2A \sin 3A \sin 4A$

$$= \sin A \cdot 2 \sin A \cos A \cdot (3 \sin A - 4 \sin^3 A \cdot 2 \sin 2A \cos 2A)$$

$$= 2 \sin^2 A \cos A \cdot \sin A(3 - 4 \sin^2 A)$$

$\cdot\, 4 \sin A \cos A \cdot (1 - 2 \sin^2 A)$

$$= 8x^2(1 - x)(3 - 4x)(1 - 2x)$$

$$= 24x^2 - 104x^3 + 144x^4 - 64x^5$$

On comparing, we get $a = 24, b = -104, c = 144, d = -64$

$10a - 7b + 15c - 5d$

$$= 10 \times 24 - 7 \times -104 + 15 \times 144 - 5 \times -64$$

$$= 240 + 728 + 2160 + 320 = 3448$$

119. Let $x + 5 = 14 \cos \theta$ and $y - 12 = 14 \sin \theta$

$\therefore x^2 + y^2 = (14 \cos \theta - 5)^2 + (14 \sin \theta + 12)^2$

$$= 196 + 25 + 144 + 28(12 \sin \theta - 5 \cos \theta)$$

$$= 365 + 28(12 \sin \theta - 5 \cos \theta)$$

$\therefore \quad \sqrt{x^2 + y^2}\Big|_{\min} = \sqrt{365 - 28 \times 13} = \sqrt{365 - 364} = 1$

120. $\because 12° \times 5 = 60°$

Let $\quad 12° = \theta$

$\therefore \quad 5\theta = 60°$

$\Rightarrow \quad 3\theta + 2\theta = 60°$

$\therefore \quad \cos(3\theta + 2\theta) = \cos 60°$

$\Rightarrow \cos 3\theta \cos 2\theta - \sin 3\theta \sin 2\theta = \frac{1}{2}$

$\Rightarrow (4 \cos^3 \theta - 3 \cos \theta)(2 \cos^2 \theta - 1) - (3 \sin \theta - 4 \sin^3 \theta)$

$$2 \sin \theta \cos \theta = \frac{1}{2}$$

Let $\quad \cos \theta = x$

$\therefore \quad (4x^3 - 3x)(2x^2 - 1) - 2x(3 - 4(1 - x^2))(1 - x^2) = \frac{1}{2}$

$\Rightarrow \quad (8x^5 - 10x^3 + 3x) - (2x - 2x^3)(4x^2 - 1) = \frac{1}{2}$

$\Rightarrow (16x^5 - 20x^3 + 6x) - (4x - 4x^3)(4x^2 - 1) - 1 = 0$

$\Rightarrow \quad 32x^5 - 40x^3 + 10x - 1 = 0$

$\Rightarrow \quad \left(x - \frac{1}{2}\right)(32x^4 + 16x^3 - 32x^2 - 16x + 2) = 0$

but $x \neq \frac{1}{2}$,

$\therefore \quad 16x^4 + 8x^3 - 16x^2 - 8x + 1 = 0$

$\therefore$ Degree is 4.

121. From conditional identities, we have

$$\frac{\sin 2A + \sin 2B + \sin 2C}{\sin A + \sin B + \sin C}$$

$$= \frac{4 \sin A \sin B \sin C}{4 \cos\left(\frac{A}{2}\right) \cos\left(\frac{B}{2}\right) \cos\left(\frac{C}{2}\right)}$$

$$= 8 \sin\left(\frac{A}{2}\right) \sin\left(\frac{B}{2}\right) \sin\left(\frac{C}{2}\right) \Rightarrow k = 8$$

and $3k^3 + 2k^2 + k + 1 = 1536 + 128 + 8 + 1 = 1673$

122. $x = \cot \frac{11\pi}{8} = \cot\left(\pi + \frac{3\pi}{8}\right) = \cot \frac{3\pi}{8} = \sqrt{2} - 1$

$\Rightarrow \quad (x + 1)^2 = 2$

$\therefore \quad x^2 + 2x - 1 = 0$

Now, $\quad f(x) = x^4 + 4x^3 + 2x^2 - 4x + 7$

$$= x^2(x^2 + 2x - 1) + 2x^3 + 3x^2 - 4x + 7$$

$$= 0 + 2x^3 + 3x^2 - 4x + 7$$

$$= 2x(x^2 + 2x - 1) - x^2 - 2x + 7 = -x^2 - 2x + 7$$

$$= -(x^2 + 2x - 1) + 6 = 0 + 6 = 6$$

123. $\frac{\sqrt{(\sin A)}}{\sqrt{(\sin B)} + \sqrt{(\sin C)} - \sqrt{(\sin A)}} = \frac{\sqrt{a}}{\sqrt{b} + \sqrt{c} - \sqrt{a}}$

Now, $\sqrt{b} + \sqrt{c} - \sqrt{a} = \frac{(\sqrt{b} + \sqrt{c} - \sqrt{a})(\sqrt{b} + \sqrt{c} + \sqrt{a})}{(\sqrt{b} + \sqrt{c} + \sqrt{a})}$

$$= \frac{(\sqrt{b} + \sqrt{c})^2 - a}{(\sqrt{b} + \sqrt{c} + \sqrt{a})} = \frac{(b + c - a) + 2\sqrt{bc}}{(\sqrt{b} + \sqrt{c} + \sqrt{a})} > 0$$

Hence, $\quad \sqrt{b} + \sqrt{c} - \sqrt{a} > 0$

Now, let $\sqrt{b} + \sqrt{c} - \sqrt{a} = x, \sqrt{c} + \sqrt{a} - \sqrt{b} = y$

and $\quad \sqrt{a} + \sqrt{b} - \sqrt{c} = z$

$\therefore \quad \frac{\sqrt{\sin A}}{\sqrt{\sin B} + \sqrt{\sin C} - \sqrt{\sin A}} = \frac{y + z}{2x}$

$\Rightarrow \sum \frac{\sqrt{(\sin A)}}{\sqrt{(\sin B)} + \sqrt{(\sin C)} - \sqrt{(\sin A)}}$

$$= \frac{1}{2}\left\{\frac{y}{x} + \frac{z}{x}\right\} + \frac{1}{2}\left\{\frac{z}{y} + \frac{x}{y}\right\} + \frac{1}{2}\left\{\frac{x}{z} + \frac{y}{z}\right\}$$

$$= \frac{1}{2}\left(\frac{x}{y} + \frac{y}{x}\right) + \frac{1}{2}\left(\frac{y}{z} + \frac{z}{y}\right) + \frac{1}{2}\left(\frac{z}{x} + \frac{x}{z}\right)$$

$$\geq 1 + 1 + 1 \quad (\because \text{AM} \geq \text{GM})$$

$2020 \sum \frac{\sqrt{(\sin A)}}{\sqrt{(\sin B)} + \sqrt{(\sin C)} - \sqrt{(\sin A)}} \leq 6060$

$\therefore$ Minimum value is 6060.

124. We have, $\sin\theta(1+\sin^2\theta)=1-\sin^2\theta$

$\Rightarrow \quad \sin\theta(2-\cos^2\theta)=\cos^2\theta$

Squaring both sides, we get

$$\sin^2\theta(2-\cos^2\theta)^2=\cos^4\theta$$

$\Rightarrow \quad (1-\cos^2\theta)(4-4\cos^2\theta+\cos^4\theta)=\cos^4\theta$

$\Rightarrow \quad -\cos^6\theta+5\cos^4\theta-8\cos^2\theta+4=\cos^4\theta$

$\therefore \quad \cos^6\theta-4\cos^4\theta+8\cos^2\theta=4$

125. $16\left(\cos\theta-\cos\frac{\pi}{8}\right)\left(\cos\theta-\cos\frac{3\pi}{8}\right)\left(\cos\theta-\cos\frac{5\pi}{8}\right)\left(\cos\theta-\cos\frac{7\pi}{8}\right)$

$$=16\left(\cos\theta-\cos\frac{\pi}{8}\right)\left(\cos\theta-\cos\frac{7\pi}{8}\right)\times\left(\cos\theta-\cos\frac{3\pi}{8}\right)\left(\cos\theta-\cos\frac{5\pi}{8}\right)$$

$$=16\left(\cos\theta-\cos\frac{\pi}{8}\right)\left(\cos\theta+\cos\frac{\pi}{8}\right)\times\left(\cos\theta-\cos\frac{3\pi}{8}\right)\left(\cos\theta+\cos\frac{3\pi}{8}\right)$$

$$=16\left(\cos^2\theta-\cos^2\frac{\pi}{8}\right)\left(\cos^2\theta-\cos^2\frac{3\pi}{8}\right)$$

$$=16\left(\cos^2\theta-\cos^2\frac{\pi}{8}\right)\left(\cos^2\theta-\sin^2\frac{\pi}{8}\right)$$

$$=16\left(\cos^4\theta-\cos^2\theta+\sin^2\frac{\pi}{8}\cos^2\frac{\pi}{8}\right)$$

$$=16\left(\cos^4\theta-\cos^2\theta+\frac{1}{8}\right)$$

$$=16\left(-\cos^2\theta\sin^2\theta+\frac{1}{8}\right)=16\left(\frac{-\sin^2 2\theta}{4}+\frac{1}{8}\right)$$

$$=16\left(\frac{1-2\sin^2 2\theta}{8}\right)=\frac{16\cos 4\theta}{8}=2\cos 4\theta$$

$\therefore \quad \lambda=2$

126. $2k\cos\cos 40°=\frac{1}{\sin 20°}+\frac{1}{\sqrt{3}\cos 20°}$

$$=\frac{\sqrt{3}\cos 20°+\sin 20°}{\sqrt{3}\sin 20°\cos 20°}$$

$$=\frac{\frac{\sqrt{3}}{2}\cos 20°+\frac{1}{2}\sin 20°}{\frac{\sqrt{3}}{4}\sin 40°}$$

$$=\frac{\sin 60°\cos 20°+\cos 60°\sin 20°}{\left(\frac{\sqrt{3}}{4}\right)\sin 40°}$$

$$=\left(\frac{4}{\sqrt{3}}\right)2\cos 40°$$

$\Rightarrow \quad 2k^2=16$

so $18k^4+162k^2+369=1745$

127. $\tan 82\frac{1}{2}°=\cot 7\frac{1}{2}°=\dfrac{\cos 7\frac{1}{2}°}{\sin 7\frac{1}{2}°}$

On multiplying numerator and denominator by $2\cos 7\frac{1}{2}°$, we get

$$\tan 82\frac{1}{2}°=\frac{2\cos^2 7\frac{1}{2}°}{2\sin 7\frac{1}{2}°\cos 7\frac{1}{2}°}=\frac{1+\cos 15°}{\sin 15°}$$

$$=\frac{1+\cos(45°-30°)}{\sin(45°-30°)}=\frac{1+\frac{\sqrt{3}+1}{2\sqrt{2}}}{\frac{\sqrt{3}-1}{2\sqrt{2}}}$$

$$=\frac{2\sqrt{2}+\sqrt{3}+1}{\sqrt{3}-1}\times\frac{\sqrt{3}+1}{\sqrt{3}+1}$$

$$=\frac{2\sqrt{2}(\sqrt{3}+1)+(\sqrt{3}+1)^2}{2}=\frac{2\sqrt{2}(\sqrt{3}+1)+(4+2\sqrt{3})}{2}$$

$$=\sqrt{2}(\sqrt{3}+1)+(2+\sqrt{3})=\sqrt{6}+\sqrt{2}+\sqrt{4}+\sqrt{3}$$

$$=\sqrt{2}+\sqrt{3}+\sqrt{4}+\sqrt{6}=(\sqrt{3}+\sqrt{2})(\sqrt{2}+1)$$

128. LHS $=\dfrac{2-m(\sin 2\alpha+\sin 2\beta)}{1-m(\sin 2\alpha+\sin 2\beta)+m^2\sin 2\alpha\sin 2\beta}$

$$=\frac{2-2m\sin(\alpha+\beta)\cos(\alpha-\beta)}{1-2m\sin(\alpha+\beta).\cos(\alpha-\beta)+4m^2\sin\alpha\cos\alpha\sin\beta\cos\beta}$$

$$=\frac{2\{1-\cos^2(\alpha-\beta)\}}{1-2m\sin(\alpha+\beta).\cos(\alpha-\beta)+4m^2\sin\alpha\cos\alpha\sin\beta\cos\beta}$$

[using $m\sin(\alpha+\beta)=\cos(\alpha-\beta)$]

$$=\frac{2\sin^2(\alpha-\beta)}{1-2\cos^2(\alpha-\beta)+m^2[\sin(\alpha+\beta)+\sin(\alpha-\beta)][\sin(\alpha+\beta)-\sin(\alpha-\beta)]}$$

$$=\frac{2\sin^2(\alpha-\beta)}{1-2\cos^2(\alpha-\beta)+m^2\sin^2(\alpha+\beta)-m^2\sin^2(\alpha-\beta)}$$

$$=\frac{2\sin^2(\alpha-\beta)}{1-\cos^2(\alpha-\beta)-m^2\sin^2(\alpha-\beta)}$$

$$=\frac{2\sin^2(\alpha-\beta)}{\sin^2(\alpha-\beta)-m^2\sin^2(\alpha-\beta)}=\frac{2}{1-m^2}$$

129. Given $\tan\frac{1}{4}(\beta+\gamma-2\alpha).\tan\frac{1}{4}(\gamma+\alpha-\beta)\tan\frac{1}{4}(\alpha+\beta-\gamma)=1$,

where $\alpha+\beta+\gamma=\pi$

$\Rightarrow \quad \tan\left(\frac{\pi-2\alpha}{4}\right)\tan\left(\frac{\pi-2\beta}{4}\right)\tan\left(\frac{\pi-2\gamma}{4}\right)=1$

$\Rightarrow \quad \left(1-\tan\frac{\alpha}{2}\right)\left(1-\tan\frac{\beta}{2}\right)\left(1-\tan\frac{\gamma}{2}\right)$

$$=\left(1+\tan\frac{\alpha}{2}\right)\left(1+\tan\frac{\beta}{2}\right)\left(1+\tan\frac{\gamma}{2}\right)$$

$\Rightarrow \quad \tan\frac{\alpha}{2}+\tan\frac{\beta}{2}+\tan\frac{\gamma}{2}+\tan\frac{\alpha}{2}\tan\frac{\beta}{2}\tan\frac{\gamma}{2}=0$

$\Rightarrow \quad \tan\frac{\alpha}{2}+\tan\frac{\beta}{2}+\tan\frac{\gamma}{2}=-\tan\frac{\alpha}{2}\tan\frac{\beta}{2}\tan\frac{\gamma}{2} \quad$ ...(i)

Also, $\tan\frac{\alpha}{2}\tan\frac{\beta}{2}+\tan\frac{\beta}{2}\tan\frac{\gamma}{2}+\tan\frac{\gamma}{2}\tan\frac{\alpha}{2}=1$ ...(ii)

On squaring Eq. (i) and using Eq. (ii); $\left\{\because \frac{\alpha}{2}+\frac{\beta}{2}+\frac{\gamma}{2}=\frac{\pi}{2}\right\}$

$$\tan^2\frac{\alpha}{2}+\tan^2\frac{\beta}{2}+\tan^2\frac{\gamma}{2}+2=\tan^2\frac{\alpha}{2}\tan^2\frac{\beta}{2}\tan^2\frac{\gamma}{2} \quad ...(iii)$$

The equation to be prove is

$$1+\cos\alpha+\cos\beta+\cos\gamma=0$$

$$\Rightarrow \quad 1+\frac{1-\tan^2\frac{\alpha}{2}}{1+\tan^2\frac{\alpha}{2}}+\frac{1-\tan^2\frac{\beta}{2}}{1+\tan^2\frac{\beta}{2}}+\frac{1-\tan^2\frac{\gamma}{2}}{1+\tan^2\frac{\gamma}{2}}=0$$

$$\Rightarrow \quad \frac{2}{1+\tan^2\frac{\alpha}{2}}+\frac{2\left(1-\tan^2\frac{\beta}{2}\tan^2\frac{\gamma}{2}\right)}{\left(1+\tan^2\frac{\beta}{2}\right)\left(1+\tan^2\frac{\gamma}{2}\right)}=0$$

$$\Rightarrow \left(1+\tan^2\frac{\alpha}{2}\right)\left(1-\tan^2\frac{\beta}{2}\tan^2\frac{\gamma}{2}\right)+\left(1+\tan^2\frac{\beta}{2}\right)\left(1+\tan^2\frac{\gamma}{2}\right)=0$$

$$\Rightarrow \left(1+\tan^2\frac{\alpha}{2}-\tan^2\frac{\beta}{2}\tan^2\frac{\gamma}{2}-\tan^2\frac{\alpha}{2}\tan^2\frac{\beta}{2}\tan^2\frac{\gamma}{2}\right)$$

$$+\left(1+\tan^2\frac{\beta}{2}+\tan^2\frac{\gamma}{2}+\tan^2\frac{\beta}{2}\tan^2\frac{\gamma}{2}\right)=0$$

$$\Rightarrow 2+\tan^2\frac{\alpha}{2}+\tan^2\frac{\beta}{2}+\tan^2\frac{\gamma}{2}=\tan^2\frac{\alpha}{2}\tan^2\frac{\beta}{2}\tan^2\frac{\gamma}{2} \quad ...(iv)$$

From Eq. (iii) and (iv)

$$1+\cos\alpha+\cos\beta+\cos\gamma=0$$

130. We have, $\sin^4 x+\cos^4 x\le\sin^2 x+\cos^2 x$,

as $\quad |\sin x|\le 1$ and $|\cos x|\le 1$

$\Rightarrow \quad a\le 1$ ...(i)

Next, $\sin^4 x+\cos^4 x=a$

$\Rightarrow \quad (\sin^2 x+\cos^2 x)^2-2\sin^2 x\cos^2 x=a$

$\Rightarrow \quad 1-\frac{1}{2}\sin^2 2x=a$

$\Rightarrow \quad \frac{1}{2}\sin^2 2x=1-a$

$\Rightarrow \quad 1-a\le\frac{1}{2}$ $\quad\left[\because \frac{1}{2}\sin^2 x\le\frac{1}{2}\right]$

$\Rightarrow \quad a\ge\frac{1}{2}$ ...(ii)

From Eqs. (i) and (ii),

$$\frac{1}{2}\le a\le 1$$

131. Let $a\sec\theta-b\tan\theta=x$

So, $\quad a^2\sec^2\theta=(x+b\tan\theta)^2$

$\Rightarrow \quad a^2(1+\tan^2\theta)=x^2+2bx\tan\theta+b^2\tan^2\theta$

$\Rightarrow \quad \tan^2\theta(a^2-b^2)-2bx\tan\theta+(a^2-x^2)=0$

$$\Rightarrow \quad \left(\tan\theta-\frac{bx}{a^2-b^2}\right)^2=\frac{a^2(x^2+b^2-a^2)}{(a^2-b^2)^2}$$

Thus, $\quad x^2+(b^2-a^2)\ge 0$

$$\Rightarrow \quad x^2\ge a^2-b^2$$

Thus, the minimum value of x is $\sqrt{a^2-b^2}$, which is attained at $\theta=\sin^{-1}\left(\frac{b}{a}\right)$.

132. We can write

$(b\tan\gamma-c\tan\beta)^2+(c\tan\alpha-a\tan\gamma)^2+(a\tan\beta-b\tan\alpha)^2$

$=(a^2+b^2+c^2)(\tan^2\alpha+\tan^2\beta+\tan^2\gamma)-(a\tan\alpha+b\tan\beta+c\tan\gamma)^2$

The minimum value of the LHS being zero, that of

$$(a^2+b^2+c^2)(\tan^2\alpha+\tan^2\beta+\tan^2\gamma)-k^2\ge 0$$

$$\Rightarrow \quad \tan^2\alpha+\tan^2\beta+\tan^2\gamma\ge\frac{k^2}{a^2+b^2+c^2}$$

Hence, minimum value of $\tan^2\alpha+\tan^2\beta+\tan\gamma$ is $\left(\frac{k^2}{a^2+b^2+c^2}\right)$.

133. Here, $\frac{x}{y}=\frac{\tan(\theta+\alpha)}{\tan(\theta+\beta)}$. By componendo and dividendo

$$\frac{x+y}{x-y}=\frac{\tan(\theta+\alpha)+\tan(\theta+\beta)}{\tan(\theta+\alpha)-\tan(\theta+\beta)}=\frac{\sin(2\theta+\alpha+\beta)}{\sin(\alpha-\beta)}$$

$$\therefore \quad \frac{x+y}{x-y}\sin^2(\alpha-\beta)=\sin(2\theta+\alpha+\beta)\cdot\sin(\alpha-\beta)$$

$$\frac{x+y}{x-y}\cdot\sin^2(\alpha-\beta)=\frac{1}{2}\{\cos 2(\theta+\beta)-\cos 2(\theta+\alpha)\} \quad ...(i)$$

Similarly,

$$\frac{y+z}{y-z}\cdot\sin^2(\beta-\gamma)=\frac{1}{2}\{\cos 2(\theta+\gamma)-\cos 2(\theta+\beta)\} \quad ...(ii)$$

and $\frac{z+x}{z-x}\cdot\sin^2(\gamma-\alpha)=\frac{1}{2}\{\cos 2(\theta+\alpha)-\cos 2(\theta+\gamma)\}$ ...(iii)

From Eqs. (i), (ii) and (iii), we get

$$\Sigma\frac{x+y}{x-y}\cdot\sin^2(\alpha-\beta)=0$$

134. $f(x)$ may be written as $f(x)=\sum_{k=1}^{n}\frac{1}{2^{k-1}}\cos(a_k+x)$

$$=\sum_{k=1}^{n}\frac{1}{2^{k-1}}(\cos a_k\cos x-\sin a_k\sin x)$$

$$=\left(\sum_{k=1}^{n}\frac{1}{2^{k-1}}\cdot\cos a_k\right)\cdot\cos x\left(-\sum_{k=1}^{n}\frac{1}{2^{k-1}}\cdot\sin a_k\right)\cdot\sin x$$

where, $A=\sum_{k=1}^{n}\frac{1}{2^{k-1}}\cos a_k$, $B=\sum_{k=1}^{n}\frac{1}{2^{k-1}}\sin a_k$. Now, A and B both cannot be zero, for if they were then $f(x)$ would vanish identically.

Now,

$$f(x_1)=A\cos x_1-B\sin x_1=0$$

$$f(x_2)=A\cos x_2-B\sin x_2=0$$

$$\Rightarrow \quad \tan x_1=\frac{A}{B} \text{ and } \tan x_2=\frac{A}{B}$$

$$\Rightarrow \quad \tan x_1=\tan x_2\Rightarrow x_2-x_1=m\pi.$$

135. $x = \tan(n\theta + \alpha) - \tan(n\theta + \beta)$

$$= \frac{\sin(n\theta + \alpha)}{\cos(n\theta + \alpha)} - \frac{\sin(n\theta + \beta)}{\cos(n\theta + \beta)}$$

$$= \frac{\sin(n\theta + \alpha - n\theta - \beta)}{\cos(n\theta + \alpha)\cos(n\theta + \beta)} = \frac{2\sin(\alpha - \beta)}{\cos(2n\theta + \alpha + \beta) + \cos(\alpha - \beta)}$$

$$\Rightarrow \cos(2n\theta + \alpha + \beta) + \cos(\alpha - \beta) = \frac{2\sin(\alpha - \beta)}{x} \quad \text{...(i)}$$

Again $y = \cot(n\theta + \alpha) - \cot(n\theta + \beta)$

$$= \frac{\cos(n\theta + \alpha)}{\sin(n\theta + \alpha)} - \frac{\cos(n\theta + \beta)}{\sin(n\theta + \beta)} = \frac{\sin(n\theta + \beta - n\theta - \alpha)}{\sin(n\theta + \alpha)\sin(n\theta + \beta)}$$

$$\Rightarrow \quad y = \frac{\sin(\beta - \alpha)}{\cos(\alpha - \beta) - \cos(2n\theta + \alpha + \beta)}$$

$$\Rightarrow \cos(\alpha - \beta) - \cos(2n\theta + \alpha + \beta) = \frac{2\sin(\beta - \alpha)}{y} \quad \text{...(ii)}$$

On adding Eqs. (i) and (ii), we get

$$2\cos(\alpha - \beta) = \frac{2\sin(\alpha - \beta)}{x} + \frac{2\sin(\beta - \alpha)}{y}$$

$$\Rightarrow \quad \cot(\alpha - \beta) = \frac{1}{x} - \frac{1}{y}$$

136. $\{\sin(\alpha - \beta) + \cos(\alpha + 2\beta)\cdot\sin\beta\}^2 = 4\cos\alpha\cdot\sin\beta\cdot\sin(\alpha + \beta)$

$$\Rightarrow \{\sin\alpha\cos\beta - \sin\beta\cos\alpha + (\cos\alpha\cos 2\beta - \sin\alpha\sin 2\beta)\sin\beta\}^2$$

$$= 4\cos\alpha\sin\beta\sin(\alpha + \beta)$$

$$\Rightarrow \{\tan\alpha - \tan\beta + \cos 2\beta\cdot\tan\beta - \sin 2\beta\cdot\tan\alpha\tan\beta\}^2$$

$$= 4\tan\beta(\tan\alpha + \tan\beta) \qquad \{\because \text{dividing by } \cos^2\alpha.\cos^2\beta\}$$

$\Rightarrow$

$$\{\tan\alpha\cdot\cos 2\beta - \tan\beta + \cos 2\beta\cdot\tan\beta\}^2 = 4\tan\beta\{\tan\alpha + \tan\beta\}$$

$$\Rightarrow \{(\tan\alpha + \tan\beta)\cdot\cos 2\beta - \tan\beta\}^2 = 4\tan\beta(\tan\alpha + \tan\beta) \quad \text{...(i)}$$

If
$$(\tan\alpha + \tan\beta) = \frac{\tan\beta}{x} \quad \text{...(ii)}$$

Eq. (i) becomes;

$$\left\{\frac{\tan\beta}{x}\cdot\cos 2\beta - \tan\beta\right\}^2 = 4\tan\beta\cdot\frac{\tan\beta}{x}$$

$$\Rightarrow \quad (\cos 2\beta - x)^2 = 4x$$

$$\Rightarrow \quad \cos^2 2\beta + x^2 - 2x\cos 2\beta = 4x$$

$$\Rightarrow \quad x^2 - 2x(\cos 2\beta + 2) + \cos^2 2\beta = 0$$

$$\Rightarrow \quad x = (\cos 2\beta + 2) \pm 2\sqrt{1 + \cos 2\beta}$$

$$\Rightarrow \quad x = \cos 2\beta + 2 \pm 2\sqrt{2\cos^2\beta}$$

$$\Rightarrow \quad x = 2\cos^2\beta - 1 + 2 \pm 2\sqrt{2}\cos\beta$$

$$= (\sqrt{2}\cos\beta \pm 1)^2$$

$$\Rightarrow \quad \tan\alpha + \tan\beta = \frac{\tan\beta}{x} = \frac{\tan\beta}{(\sqrt{2}\cos\beta - 1)^2} \qquad [\text{since, } x < 1]$$

137. Let
$$\Delta = \begin{vmatrix} \sin A & \sin B & \sin C \\ \cos A & \cos B & \cos C \\ \cos^3 A & \cos^3 B & \cos^3 C \end{vmatrix}$$

$$= \cos A\cos B\cos C \begin{vmatrix} \tan A & \tan B & \tan C \\ 1 & 1 & 1 \\ \cos^2 A & \cos^2 B & \cos^2 C \end{vmatrix}$$

$$= \cos A\cos B\cos C \begin{vmatrix} \tan A & \tan B - \tan A & \tan C - \tan A \\ 1 & 0 & 0 \\ \cos^2 A & \cos^2 B - \cos^2 A & \cos^2 A - \cos^2 A \end{vmatrix}$$

$$\left[\text{since, } \tan B - \tan A = -\frac{\sin(A - B)}{\cos A\cos B},\right.$$

$$\left.\cos^2 B - \cos^2 A = \sin(A - B)\sin(A + B)\right]$$

$$\therefore \quad \Delta = -\cos A\cos B\cos C \begin{vmatrix} -\dfrac{\sin(A - B)}{\cos A\cos B} & -\dfrac{\sin(A - C)}{\cos A\cos C} \\ \sin(A - B)\sin(A + B) & \sin(A - C)\sin(A + C) \end{vmatrix}$$

$$= \cos A\cos B\cos C\cdot \frac{\sin(A - B).\sin(A - C)}{\cos A\cos B\cos C}\begin{vmatrix} \cos C & \cos B \\ \sin(A + B) & \sin(A + C) \end{vmatrix}$$

$$= -\sin(B - C)\sin(C - A)\sin(A - B) = 0$$

If $B = C$ or $C = A$ or $A = B$

Hence, ΔABC is an isosceles.

138. Here,
$$\frac{\sqrt{\sin A}}{\sqrt{\sin B} + \sqrt{\sin C} - \sqrt{\sin A}} = \frac{\sqrt{a}}{\sqrt{b} + \sqrt{c} - \sqrt{a}}$$

Now,
$$\sqrt{b} + \sqrt{c} - \sqrt{a} = \frac{(\sqrt{b} + \sqrt{c} - \sqrt{a})(\sqrt{b} + \sqrt{c} + \sqrt{a})}{(\sqrt{b} + \sqrt{c} + \sqrt{a})}$$

$$= \frac{b + c - a + 2\sqrt{bc}}{\sqrt{b} + \sqrt{c} + \sqrt{a}} > 0$$

Hence, $\sqrt{b} + \sqrt{c} - \sqrt{a} = 0$

Let $\sqrt{b} + \sqrt{c} - \sqrt{a} = x$, $\sqrt{c} + \sqrt{a} - \sqrt{b} = y$, $\sqrt{a} + \sqrt{b} - \sqrt{c} = z$

$$\therefore \quad \frac{\sqrt{\sin A}}{\sqrt{\sin B} + \sqrt{\sin C} - \sqrt{\sin A}} = \frac{y + z}{2x}$$

$$\Rightarrow \Sigma = \frac{\sqrt{\sin A}}{\sqrt{\sin B} + \sqrt{\sin C} - \sqrt{\sin A}}$$

$$= \frac{1}{2}\left\{\frac{y}{x} + \frac{z}{x}\right\} + \frac{1}{2}\left\{\frac{y}{z} + \frac{z}{y}\right\} + \frac{1}{2}\left\{\frac{z}{x} + \frac{x}{y}\right\}$$

which is greater than or equal to 3, as each term $\left(\frac{1}{2}\left(\frac{x}{y} + \frac{y}{x}\right)\text{etc.}\right)$ is greater than or equal to 1.

(using AM $\geq$ GM)

Now, equality hold if and only if,

$$\frac{x}{y} = \frac{y}{x}, \frac{y}{z} = \frac{z}{y}$$

and
$$\frac{z}{x} = \frac{x}{y} \quad \text{i.e. } x = y = z$$

$\Rightarrow$ $a = b = c$ i.e. triangle is equilateral.

139. $2(\cos(\alpha - \beta) + \cos(\beta - \gamma) + \cos(\gamma - \alpha)) + 3 = 0$

$$\Rightarrow 2(\cos(\alpha + \theta - (\beta + \theta)) + \cos(\beta + \theta - (\gamma + \theta))$$
$$+ \cos(\gamma + \theta - (\alpha + \theta))) + 3 = 0$$

$$\Rightarrow 2(\cos(\alpha + \theta).\cos(\beta + \theta) + \sin(\alpha + \theta).\sin(\beta + \theta) + \ldots + \ldots))$$
$$+ \{(\sin^2(\alpha + \theta) + \cos^2(\alpha + \theta)) + (\sin^2(\beta + \theta) + \cos^2(\beta + \theta))$$
$$+ (\sin^2(\gamma + \theta) + \cos^2(\gamma + \theta))\} = 0$$

$\Rightarrow \quad (\sin(\gamma+\theta)+\sin(\beta+\theta)+\sin(\alpha+\theta))^2+(\cos(\alpha+\theta)$
$+\cos(\beta+\theta)+\cos(\gamma+\theta))^2=0$

which is only possible if;

$$\sin(\alpha+\theta)+\cos(\beta+\theta)+\sin(\gamma+\theta)=0 \quad \text{...(i)}$$
$$\cos(\alpha+\theta)+\cos(\beta+\theta)+\cos(\gamma+\theta)=0 \quad \text{...(ii)}$$

From Eq. (ii), we get

$$d(\cos(\alpha+\theta)+\cos(\beta+\theta)+\cos(\gamma+\theta))=0$$

$\Rightarrow \sin(\alpha+\theta)\cdot d\alpha+\sin(\beta+\theta)\cdot d\beta+\sin(\gamma+\theta)\cdot d\gamma=0$

$$\Rightarrow \frac{d\alpha}{\sin(\beta+\theta)\cdot\sin(\gamma+\theta)}+\frac{d\beta}{\sin(\alpha+\theta)\cdot\sin(\gamma+\theta)}+\frac{d\gamma}{\sin(\alpha+\theta)\cdot\sin(\beta+\theta)}=0$$

140. The quadratic equation,

$4^{\sec^2\alpha}x^2+2x+\left(\beta^2-\beta+\frac{1}{2}\right)=0$ have real roots

$\Rightarrow$ Discriminant $=4-4\cdot4^{\sec^2\alpha}\left(\beta^2-\beta+\frac{1}{2}\right)\geq 0$

$$\Rightarrow 4^{\sec^2\alpha}\left(\beta^2-\beta+\frac{1}{2}\right)\leq 1$$

$$\left[\text{but } 4^{\sec^2\alpha}\geq 4,\ \beta^2-\beta+\frac{1}{2}=\left(\beta+\frac{1}{2}\right)^2+\frac{1}{4}\geq\frac{1}{4}\right]$$

i.e. the equation will be satisfied only when $4^{\sec^2\alpha}=4$ and

$$\beta^2-\beta+\frac{1}{2}=\frac{1}{4}$$

$$\Rightarrow \quad \sec^2\alpha=1 \text{ and } \left(\beta-\frac{1}{2}\right)^2=0$$

$$\Rightarrow \quad \cos^2\alpha=1 \text{ and } \beta=\frac{1}{2}$$

$$\Rightarrow \quad \alpha=n\pi \text{ and } \beta=\frac{1}{2}$$

$$\cos\alpha+\cos^{-1}\beta=\cos n\pi+\cos^{-1}\left(\frac{1}{2}\right)$$

$=1+\frac{\pi}{3}$, when n is an even integer.

$=-1+\frac{\pi}{3}$, when n is an odd integer.

i.e. values of $\cos\alpha+\cos^{-1}\beta$ is $\frac{\pi}{3}-1, \frac{\pi}{3}+1$.

141. $f(\theta)=1-(a\cos\theta+b\sin\theta)-(A\cos2\theta+B\sin2\theta)$

$\Rightarrow \quad f(\theta)=1-\sqrt{a^2+b^2}\cos(\theta-\alpha)-\sqrt{A^2+B^2}\cos(2\theta-\beta)$

Now, $f\left(\alpha+\frac{\pi}{4}\right)=1-\frac{\sqrt{a^2+b^2}}{\sqrt2}-\sqrt{A^2+B^2}\cos\left(\frac{\pi}{2}+2\alpha-\beta\right)$

$$=1-\frac{\sqrt{a^2+b^2}}{\sqrt2}+\sqrt{A^2+B^2}\sin(2\alpha-\beta) \quad \text{...(i)}$$

and $f\left(\alpha-\frac{\pi}{4}\right)=1-\frac{\sqrt{a^2+b^2}}{\sqrt2}-\sqrt{A^2+B^2}\cos\left(2\alpha-\beta-\frac{\pi}{2}\right)$

$$=1-\frac{\sqrt{a^2+b^2}}{\sqrt2}-\sqrt{A^2+B^2}\sin(2\alpha-\beta) \quad \text{...(ii)}$$

On adding Eqs. (i) and (ii),

$$f\left(\alpha+\frac{\pi}{4}\right)+f\left(\alpha-\frac{\pi}{4}\right)=2-2\frac{\sqrt{a^2+b^2}}{\sqrt2}\geq 0$$

$$\Rightarrow \quad \sqrt{a^2+b^2}\leq\sqrt2$$

$$\Rightarrow \quad a^2+b^2\leq 2$$

Similarly putting $\theta=\beta$ and $\beta+\pi$. We have,

$$f(\beta)+f(\beta+\pi)=2-2\sqrt{A^2+B^2}\geq 0$$

$$\Rightarrow \quad \sqrt{A^2+B^2}\leq 1 \Rightarrow A^2+B^2\leq 1$$

142. Clearly θ_1, θ_0 are roots of ; $\frac{\cos\theta}{\cos\theta_2}+\frac{\sin\theta}{\sin\theta_2}=1$

$$\Rightarrow \quad \frac{\cos\theta}{\cos\theta_2}=1-\frac{\sin\theta}{\sin\theta_2}\Rightarrow\frac{\cos^2\theta}{\cos^2\theta_2}=1+\frac{\sin^2\theta}{\sin^2\theta_2}-\frac{2\sin\theta}{\sin\theta_2}$$

$$\Rightarrow \sin^2\theta\left(\frac{1}{\sin^2\theta_2}+\frac{1}{\cos^2\theta_2}\right)-\frac{2\sin\theta}{\sin\theta_2}+\left(1-\frac{1}{\cos^2\theta_2}\right)=0$$

The roots of the equation are θ_0 and θ_1.

$$\Rightarrow \quad \sin\theta_0\cdot\sin\theta_1=\frac{1-\frac{1}{\cos^2\theta_2}}{\frac{1}{\sin^2\theta_2}+\frac{1}{\cos^2\theta_2}}$$

$$=(\cos^2\theta_2-1)\cdot\sin^2\theta_2=-\sin^4\theta_2$$

$$\Rightarrow \quad \frac{\sin\theta_0\cdot\sin\theta_1}{\sin^2\theta_2}=-\sin^2\theta_2 \quad \text{...(i)}$$

Similarly, taking a quadratic in $\cos\theta$, we get

$$\Rightarrow \quad \frac{\cos\theta_0\cdot\cos\theta_1}{\sin^2\theta_2}=-\cos^2\theta_2 \quad \text{...(ii)}$$

On adding Eqs. (i) and (ii), we get

$$\frac{\sin\theta_0\sin\theta_1}{\sin^2\theta_2}+\frac{\cos\theta_0\cos\theta_1}{\cos^2\theta_2}=-1$$

143. Let the given expression be E, then E can be written as,

$$E=\sum_{k=1}^{n-1}{}^nC_k$$
$$\cos kx\cdot\cos(n+k)x+\sum_{k=1}^{n-1}{}^nC_k\sin(n-k)x.\sin(2n-k)x$$

or $\quad E=\sum_{k=1}^{n-1}{}^nC_k\cos kx$

$$\cos(n+k)x+\sum_{k=1}^{n-1}{}^nC_k\sin(k)x\cdot\sin(n+k)x$$

[replacing k by $(n-k)$ in the second]

Sum and using ${}^nC_k={}^nC_{n-k}$

$$E=\sum_{k=1}^{n-1}{}^nC_k(\cos kx\cos(n+k)x+\sin kx\cdot\sin(n+k)x)$$

$$=\sum_{k=1}^{n-1}{}^nC_k\cos nx$$

$$=\cos nx\{({}^nC_0+{}^nC_1+\ldots+{}^nC_n)-{}^nC_0-{}^nC_n\}$$

$$=\cos nx\{2^n-2\}$$

$\therefore \quad E=(2^n-2)\cos nx$

144. It is evident from the inequality that,

$$|\sqrt{1+\sin 2x}-\sqrt{1-\sin 2x}| \le \sqrt{2}\ \forall\ x \in [0, 2\pi]$$

as $\quad |\sqrt{1+\sin 2x}-\sqrt{1-\sin 2x}| \le \sqrt{1+\sin 2x} \le \sqrt{2}$

Now,

$2\cos x \le |\sqrt{1+\sin 2x}-\sqrt{1-\sin 2x}|$ holds for all x for which $\cos x \le 0$.

$$\Rightarrow \quad x \le \left[\frac{\pi}{2}, \frac{3\pi}{2}\right] \quad \text{...(i)}$$

Now, if $\cos x > 0$

Then, $\quad 4\cos^2 x \le 1+\sin 2x + 1 - \sin 2x - \sqrt{1-\sin^2 2x}$

$\Rightarrow \quad 2 + 2\cos 2x \le 2 - 2|\cos x|$

$\Rightarrow \quad |\cos 2x| \le -\cos 2x$

$\Rightarrow \quad \cos 2x \le 0$

$$\Rightarrow \quad x \in \left[\frac{\pi}{4}, \frac{3\pi}{4}\right] \cup \left[\frac{5\pi}{4}, \frac{7\pi}{4}\right] \quad \text{...(ii)}$$

Hence, from Eqs. (i) and (ii)

$$x \in \left[\frac{\pi}{4}, \frac{7\pi}{4}\right]$$

145. The given equation can be rewritten as, $x^2 - 3 = 3\left[\sin\left(x - \frac{\pi}{6}\right)\right]$

Here, right hand side can take only the values −3, 0, 3.

Case I When $x^2 - 3 = -3 \Rightarrow x = 0$

At $x = 0, \left[\sin\left(x - \frac{\pi}{6}\right)\right] = -1$, so $x = 0$ is a solution.

Case II When $x^2 - 3 = 0 \Rightarrow x = \pm\sqrt{3}$

Now at $x = \sqrt{3}, \left[\sin\left(x - \frac{\pi}{6}\right)\right] = 0 \Rightarrow x = \sqrt{3}$

But at $x = -\sqrt{3}, \left[\sin\left(x - \frac{\pi}{6}\right)\right] = -1$, hence $x = -\sqrt{3}$ is not a solution.

Case III When $x^2 - 3 = 3 \Rightarrow x = \pm\sqrt{6}$

But $\left[\sin\left(\pm\sqrt{6} - \frac{\pi}{6}\right)\right] \ne 1 \Rightarrow x = \pm\sqrt{6}$ is not a solution.

Hence, the given equation has only two solutions $x = 0$ and $\sqrt{3}$.

146. $\sum_{r=0}^{n} {}^nC_r a^r b^{n-r} \cos(rB - (n-r)A)$

$= \text{real part of } \sum_{r=0}^{n} {}^nC_r a^r b^{n-r} e^{i\{rB-(n-r)A\}}$

Now, $\sum_{r=0}^{n} {}^nC_r a^r b^{n-r} e^{i\{rB-(n-r)A\}}$

$= \sum_{r=0}^{n} {}^nC_r (ae^{iB})^r (be^{-iA})^{n-r} = (ae^{iB} + be^{-iA})^n$

$= (a\cos B + ia\sin B + b\cos A - bi\sin A)^n$

$= \{(a\cos B + b\cos A) + i(a\sin B - b\sin A)\}^n$

$= \{C + i\cdot 0\}^n = C^n$

147. Let $z^5 + 1 = 0 \Rightarrow z^5 = -1 = (\cos(2r+1) + i\sin(2r+1)\pi)$

$\Rightarrow \quad z = e^{i\left(\frac{2r+1}{5}\right)\pi}, r = 0, 1, 2, 3, 4$

$\Rightarrow$ Roots of $z^k + 1 = 0$ are $e^{i\pi/5}, e^{i3\pi/5}, e^{i\pi}, e^{i7\pi/5}, e^{i9\pi/5}$. Clearly $e^{i7\pi/5}, e^{i9\pi/5}$ and $e^{i3\pi/5}, e^{i7\pi/5}$ are pairwise conjugate.

$\Rightarrow z^5 + 1 = (z - e^{i\pi})(z - e^{i\pi/5})(z - e^{-i\pi/5})(z - e^{-i3\pi/5})(z - e^{i7\pi/5})$

$$\Rightarrow z^5 + 1 = (z+1)\left(z^2 - 2z\cos\frac{\pi}{5} + 1\right)\left(z^2 - 2z\cos\frac{3\pi}{5} + 1\right) \quad \text{...(i)}$$

It is required factorisation of $z^5 + 1$.

$$\text{Now,} \quad \frac{z^5+1}{z+1} = 1 - z + z^2 - z^2 + z^4 \quad \text{...(ii)}$$

$\Rightarrow \quad 1 - z + z^2 - z^3 + z^4 =$

$$\left(z^2 - 2z\cos\frac{\pi}{5} + 1\right)\left(z^2 - 2z\cos\frac{3\pi}{5} + 1\right)$$

[using Eqs. (i) and (ii)}

On dividing both side by z^2

$$\left(z^2 + \frac{1}{z^2}\right) - \left(z + \frac{1}{z}\right) + 1$$

$$= \left(z + \frac{1}{z} - 2\cos\frac{\pi}{5}\right)\left(z + \frac{1}{z} - 2\cos\frac{3\pi}{5}\right)$$

Let $z = e^{i\theta}$

$$\Rightarrow \quad z^2 + \frac{1}{z^2} = 2\cos 2\theta,\ z + \frac{1}{z} = 2\cos\theta$$

$$\Rightarrow 2\cos 2\theta - 2\cos\theta + 1 = 4\left(\cos\theta - \cos\frac{\pi}{5}\right)\left(\cos\theta - \cos\frac{3\pi}{5}\right)$$

Putting $\theta = 0$, we get

$$\frac{1}{4} = \left(1 - \cos\frac{\pi}{5}\right)\left(1 - \cos\frac{3\pi}{5}\right)$$

$$\Rightarrow \quad \frac{1}{4} = 2\sin^2\frac{\pi}{10}\cdot 2\sin^2\frac{3\pi}{10}$$

$$\Rightarrow \quad \sin\frac{\pi}{10}\cdot\sin\frac{3\pi}{10} = \frac{1}{4}$$

$$\Rightarrow \quad 4\sin\frac{\pi}{10}\cdot\cos\left(\frac{\pi}{2} - \frac{3\pi}{10}\right) = 1$$

$$\Rightarrow \quad 4\sin\frac{\pi}{10}\cdot\cos\frac{\pi}{5} = 1$$

148. Let $\theta = \frac{(2n+1)\pi}{7}$, where $n = 0, 1, 2, 3, 4, 5, 6$.

Then, $4\theta = (2n+1)\pi - 3\theta$

$\Rightarrow \quad \cos 4\theta = -\cos 3\theta$

$\Rightarrow \quad 2\cos^2 2\theta - 1 = -(4\cos^3\theta - 3\cos\theta)$

$\Rightarrow \quad 2(2\cos^2\theta - 1)^2 - 1 = -4\cos^3\theta + 3\cos\theta$

$\Rightarrow \quad 2(2x^2 - 1)^2 - 1 = -4x^3 + 3x \quad$ [put $x = \cos\theta$]

$\Rightarrow \quad 8x^4 + 4x^3 - 8x^2 - 3x + 1 = 0$

$\Rightarrow \quad (x+1)(8x^3 - 4x^2 - 4x + 1) = 0$

The roots of this equation are,

$$\cos\frac{\pi}{7}, \cos\frac{3\pi}{7}, \cos\frac{5\pi}{7}, \cos\frac{7\pi}{7}, \cos\frac{9\pi}{7}, \cos\frac{11\pi}{7}, \cos\frac{13\pi}{7}$$

$\therefore$ The roots of $8x^3 - 4x^2 - 4x + 1 = 0$

are $\quad \cos\frac{\pi}{7}, \cos\frac{3\pi}{7}, \cos\frac{5\pi}{7} \quad$ ...(i)

Put $x = \frac{1}{y}$ in Eq. (i) (i.e. $y = \sec\theta$), then

$\sec\frac{\pi}{7}, \sec\frac{3\pi}{7}, \sec\frac{5\pi}{7}$ are the roots of the equation.

$$\frac{8}{y^3} - \frac{4}{y^2} - \frac{4}{y} + 1 = 0$$

$$\Rightarrow \quad y^3 - 4y^2 - 4y + 8 = 0$$

$$\Rightarrow \quad \sec\frac{\pi}{7} + \sec\frac{3\pi}{7} + \sec\frac{5\pi}{7} = 4$$

Again putting $\frac{1}{x^2} = y$ in Eq. (i)

(i.e. $y = \sec^2\theta$)

$$\frac{8}{y^{3/2}} - \frac{4}{y} - \frac{4}{y^{1/2}} + 1 = 0$$

$$\Rightarrow \quad 8 - 4y^{1/2} - 4y + y^{3/2} = 0$$

$$\Rightarrow \quad y^{1/2}(y - 4) = 4(y - 2)$$

$$\Rightarrow \quad y(y - 4)^2 = 16(y - 2)^2$$

$$\Rightarrow \quad y^3 - 24y^2 + 80y - 64 = 0 \quad \text{...(ii)}$$

Hence, the roots are

$$\sec^2\frac{\pi}{7}, \sec^2\frac{3\pi}{7}, \sec^2\frac{5\pi}{7}$$

Now, putting $y = 1 + z$, (i.e. $z = \tan^2\theta$)

We have,

$$(1 + z)^3 - 24(1 + z)^2 + 80(1 + z) - 64 = 0$$

$$\Rightarrow \quad z^3 - 21z^2 + 35z - 7 = 0$$

whose roots are $\tan^2\frac{\pi}{7}, \tan^2\frac{3\pi}{7}, \tan^2\frac{5\pi}{7}$.

149. We have, $2(\cos\beta - \cos\alpha) + \cos\alpha\cos\beta = 1$

or $4(\cos\beta - \cos\alpha) + 2\cos\alpha\cos\beta = 2$

$$\Rightarrow \quad 1 - \cos\alpha + \cos\beta - \cos\alpha\cos\beta = 3 + 3\cos\alpha - 3\cos\beta - 3\cos\alpha\cos\beta$$

$$\Rightarrow \quad (1 - \cos\alpha)(1 + \cos\beta) = 3(1 + \cos\alpha)(1 - \cos\beta)$$

$$\Rightarrow \quad \frac{(1 - \cos\alpha)}{(1 + \cos\alpha)} = \frac{3(1 - \cos\beta)}{1 + \cos\beta}$$

$$\Rightarrow \quad \tan^2\frac{\alpha}{2} = 3\tan^2\frac{\beta}{2}$$

$$\therefore \quad \tan\frac{\alpha}{2} \pm \sqrt{3}\tan\frac{\beta}{2} = 0$$

150. Here, $x^2 - 2x\sec\theta + 1 = 0$ has roots α_1 and β_1.

$$\therefore \quad \alpha_1, \beta_1 = \frac{2\sec\theta \pm \sqrt{4\sec^2\theta - 4}}{2 \times 1} = \frac{2\sec\theta \pm 2|\tan\theta|}{2}$$

Since, $\theta \in \left(-\frac{\pi}{6}, -\frac{\pi}{12}\right)$,

i.e. $\theta \in$ IV quadrant $= \frac{2\sec\theta \mp 2\tan\theta}{2}$

$\therefore$ $\alpha_1 = \sec\theta - \tan\theta$ and $\beta_1 = \sec\theta + \tan\theta$ [as $\alpha_1 > \beta_1$]

and $x^2 + 2x\tan\theta - 1 = 0$ has roots α_2 and β_2.

i.e. $\alpha_2, \beta_2 = \frac{-2\tan\theta \pm \sqrt{4\tan^2\theta + 4}}{2}$

$\therefore$ $\alpha_2 = -\tan\theta + \sec\theta$

and $\beta_2 = -\tan\theta - \sec\theta$ [as $\alpha_2 > \beta_2$]

Thus, $\alpha_1 + \beta_2 = -2\tan\theta$

151. Here, $\displaystyle\sum_{k=1}^{13} \frac{1}{\sin\left\{\frac{\pi}{4} + \frac{(k-1)\pi}{6}\right\}\sin\left(\frac{\pi}{4} + \frac{k\pi}{6}\right)}$

Converting into differences, by multiplying and dividing by $\sin\left[\left(\frac{\pi}{4} + \frac{k\pi}{6}\right) - \left\{\frac{\pi}{4} + \frac{(k-1)\pi}{6}\right\}\right]$, i.e. $\sin\left(\frac{\pi}{6}\right)$.

$$\therefore \sum_{k=1}^{13} \frac{\sin\left[\left(\frac{\pi}{4} + k\frac{\pi}{6}\right) - \left\{\frac{\pi}{4} + (k-1)\frac{\pi}{6}\right\}\right]}{\sin\frac{\pi}{6}\left\{\sin\left\{\frac{\pi}{4} + (k-1)\frac{\pi}{6}\right\}\sin\left(\frac{\pi}{4} + k\frac{\pi}{6}\right)\right\}}$$

$$= 2\sum_{k=1}^{13} \frac{\left[\begin{array}{l}\sin\left(\frac{\pi}{4} + \frac{k\pi}{6}\right)\cos\left\{\frac{\pi}{4} + (k-1)\frac{\pi}{6}\right\} \\ - \sin\left\{\frac{\pi}{4} + (k-1)\frac{\pi}{6}\right\}\cos\left(\frac{\pi}{4} + \frac{k\pi}{6}\right)\end{array}\right]}{\sin\left\{\frac{\pi}{4} + (k-1)\frac{\pi}{6}\right\}\sin\left(\frac{\pi}{4} + k\frac{\pi}{6}\right)}$$

$$= 2\sum_{k=1}^{13}\left[\cot\left\{\frac{\pi}{4} + (k-1)\frac{\pi}{6}\right\} - \cot\left(\frac{\pi}{4} + k\frac{\pi}{6}\right)\right]$$

$$= 2\left[\left\{\cot\left(\frac{\pi}{4}\right) - \cot\left(\frac{\pi}{4} + \frac{\pi}{6}\right)\right\} + \left\{\cot\left(\frac{\pi}{4} + \frac{\pi}{6}\right) - \cot\left(\frac{\pi}{4} + \frac{2\pi}{6}\right)\right\} + \ldots + \left\{\cot\left(\frac{\pi}{4} + 12\frac{\pi}{6}\right) - \cot\left(\frac{\pi}{4} + 13\frac{\pi}{6}\right)\right\}\right]$$

$$= 2\left\{\cot\frac{\pi}{4} - \cot\left(\frac{\pi}{4} + 13\frac{\pi}{6}\right)\right\}$$

$$= 2\left[1 - \cot\left(\frac{29\pi}{12}\right)\right] = 2\left[1 - \cot\left(2\pi + \frac{5\pi}{12}\right)\right]$$

$$= 2\left[1 - \cot\frac{5\pi}{12}\right] \quad \left[\because \cot\frac{5\pi}{12} = (2 - \sqrt{3})\right]$$

$$= 2(1 - 2 + \sqrt{3})$$

$$= 2(\sqrt{3} - 1)$$

152. $f(\cos 4\theta) = \frac{2}{2 - \sec^2\theta}$ ...(i)

At $\cos 4\theta = \frac{1}{3} \Rightarrow 2\cos^2 2\theta - 1 = \frac{1}{3}$

$$\Rightarrow \quad \cos^2 2\theta = \frac{2}{3} \Rightarrow \cos 2\theta = \pm\sqrt{\frac{2}{3}} \quad \text{...(ii)}$$

$$\therefore \quad f(\cos 4\theta) = \frac{2\cdot\cos^2\theta}{2\cos^2\theta - 1} = \frac{1 + \cos 2\theta}{\cos 2\theta}$$

$$\Rightarrow \quad f\left(\frac{1}{3}\right) = 1 \pm \sqrt{\frac{3}{2}} \quad \text{[from Eq. (ii)]}$$

153. Given equations can be written as

$$x\sin 3\theta - \frac{\cos 3\theta}{y} - \frac{\cos 3\theta}{z} = 0 \quad \text{...(i)}$$

$$x\sin 3\theta - \frac{2\cos 3\theta}{y} - \frac{2\sin 3\theta}{z} = 0 \quad \text{...(ii)}$$

and $x\sin 3\theta - \frac{2}{y}\cos 3\theta - \frac{1}{z}(\cos 3\theta + \sin 3\theta) = 0$ ...(iii)

Eqs. (ii) and (iii), implies

$2\sin 3\theta = \cos 3\theta + \sin 3\theta \Rightarrow \sin 3\theta = \cos 3\theta$

$\therefore \quad \tan 3\theta = 1$

$\Rightarrow \quad 3\theta = \frac{\pi}{4}, \frac{5\pi}{4}, \frac{9\pi}{4}$ or $\theta = \frac{\pi}{12}, \frac{5\pi}{12}, \frac{9\pi}{12}$

154. For $0 < \theta < \frac{\pi}{2}$

$$\sum_{m=1}^{6} \operatorname{cosec}\left(\theta + \frac{(m-1)\pi}{4}\right)\operatorname{cosec}\left(\theta + \frac{m\pi}{4}\right) = 4\sqrt{2}$$

$$\Rightarrow \sum_{m=1}^{6} \frac{1}{\sin\left(\theta + \frac{(m-1)\pi}{4}\right)\sin\left(\theta + \frac{m\pi}{4}\right)} = 4\sqrt{2}$$

$$\Rightarrow \sum_{m=1}^{6} \frac{\sin\left[\theta + \frac{m\pi}{4} - \left(\theta + \frac{(m-1)\pi}{4}\right)\right]}{\sin\frac{\pi}{4}\left\{\sin\left(\theta + \frac{(m-1)\pi}{4}\right)\sin\left(\theta + \frac{m\pi}{4}\right)\right\}} = 4\sqrt{2}$$

$$\Rightarrow \sum_{m=1}^{6} \frac{\cot\left(\theta + \frac{(m-1)\pi}{4}\right) - \cot\left(\theta + \frac{m\pi}{4}\right)}{1/\sqrt{2}} = 4\sqrt{2}$$

$$\Rightarrow \sum_{m=1}^{6} \left[\cot\left(\theta + \frac{(m-1)\pi}{4}\right) - \cot\left(\theta + \frac{m\pi}{4}\right)\right] = 4$$

$$\Rightarrow \cot(\theta) - \cot\left(\theta + \frac{\pi}{4}\right) + \cot\left(\theta + \frac{\pi}{4}\right) - \cot\left(\theta + \frac{2\pi}{4}\right)$$

$$+ \dots + \cot\left(\theta + \frac{5\pi}{4}\right) - \cot\left(\theta + \frac{6\pi}{4}\right) = 4$$

$$\Rightarrow \cot\theta - \cot\left(\frac{3\pi}{2} + \theta\right) = 4$$

$$\Rightarrow \cot\theta + \tan\theta = 4$$

$$\Rightarrow \tan^2\theta - 4\tan\theta + 1 = 0$$

$$\Rightarrow (\tan\theta - 2)^2 - 3 = 0$$

$$\Rightarrow (\tan\theta - 2 + \sqrt{3})(\tan\theta - 2 - \sqrt{3}) = 0$$

$$\Rightarrow \tan\theta = 2 - \sqrt{3}$$

or $$\tan\theta = 2 + \sqrt{3}$$

$$\Rightarrow \theta = \frac{\pi}{12};\ \theta = \frac{5\pi}{12} \quad \left[\because \theta \in \left(0, \frac{\pi}{2}\right)\right]$$

155. $\frac{\sin^4 x}{2} + \frac{\cos^4 x}{3} = \frac{1}{5}$

$$\Rightarrow \frac{\sin^4 x}{2} + \frac{(1 - \sin^2 x)^2}{3} = \frac{1}{5}$$

$$\Rightarrow \frac{\sin^4 x}{2} + \frac{1 + \sin^4 x - 2\sin^2 x}{3} = \frac{1}{5}$$

$$\Rightarrow 5\sin^4 x - 4\sin^2 x + 2 = \frac{6}{5}$$

$$\Rightarrow 25\sin^4 x - 20\sin^2 x + 4 = 0$$

$$\Rightarrow (5\sin^2 x - 2)^2 = 0$$

$$\Rightarrow \sin^2 x = \frac{2}{5}$$

$$\cos^2 x = \frac{3}{5},\ \tan^2 x = \frac{2}{3}$$

$$\therefore \frac{\sin^8 x}{8} + \frac{\cos^8 x}{27} = \frac{1}{125}$$

156. As when $\theta \in \left(0, \frac{\pi}{4}\right)$, $\tan\theta < \cot\theta$

Since, $\tan\theta < 1$ and $\cot\theta > 1$

$\therefore \quad (\tan\theta)^{\cot\theta} < 1$ and $(\cot\theta)^{\tan\theta} > 1$

$\therefore$ $t_4 > t_1$ which only holds in (b).

Therefore, (b) is the answer.

157. Since, $\cos(\alpha - \beta) = 1$

$\Rightarrow \quad \alpha - \beta = 2n\pi$

But $\quad -2\pi < \alpha - \beta < 2\pi \quad$ [as $\alpha, \beta \in (-\pi, \pi)$]

$\therefore \quad \alpha - \beta = 0$ ...(i)

Given, $\quad \cos(\alpha + \beta) = \frac{1}{e}$

$\Rightarrow \cos 2\alpha = \frac{1}{e} < 1$, which is true for four values of α.

[as $-2\pi < 2\alpha < 2\pi$]

158. Given, $5(\tan^2 x - \cos^2 x) = 2\cos 2x + 9$

$$\Rightarrow 5\left(\frac{1 - \cos 2x}{1 + \cos 2x} - \frac{1 + \cos 2x}{2}\right) = 2\cos 2x + 9$$

Put $\cos 2x = y$, we have

$$5\left(\frac{1 - y}{1 + y} - \frac{1 + y}{2}\right) = 2y + 9$$

$$\Rightarrow 5(2 - 2y - 1 - y^2 - 2y) = 2(1 + y)(2y + 9)$$

$$\Rightarrow 5(1 - 4y - y^2) = 2(2y + 9 + 2y^2 + 9y)$$

$$\Rightarrow 5 - 20y - 5y^2 = 22y + 18 + 4y^2$$

$$\Rightarrow 9y^2 + 42y + 13 = 0$$

$$\Rightarrow 9y^2 + 3y + 39y + 13 = 0$$

$$\Rightarrow 3y(3y + 1) + 13(3y + 1) = 0$$

$$\Rightarrow (3y + 1)(3y + 13) = 0$$

$$\Rightarrow y = -\frac{1}{3}, -\frac{13}{3}$$

$$\therefore \cos 2x = -\frac{1}{3}, -\frac{13}{3}$$

$$\Rightarrow \cos 2x = -\frac{1}{3} \quad \left[\because \cos 2x \neq -\frac{13}{3}\right]$$

Now, $\cos 4x = 2\cos^2 2x - 1 = 2\left(-\frac{1}{3}\right)^2 - 1$

$$= \frac{2}{9} - 1 = -\frac{7}{9}$$

159. $f_k(x) = \frac{1}{k}(\sin^k x + \cos^k x)$, where $x \in R$ and $k \geq 1$

Now, $f_4(x) - f_6(x)$

$$= \frac{1}{4}(\sin^4 x + \cos^4 x) - \frac{1}{6}(\sin^6 x + \cos^6 x)$$

$$= \frac{1}{4}(1 - 2\sin^2 x \cdot \cos^2 x) - \frac{1}{6}(1 - 3\sin^2 x \cdot \cos^2 x)$$

$$= \frac{1}{4} - \frac{1}{6} = \frac{1}{12}$$

160. Given expression is

$$\frac{\tan A}{1 - \cot A} + \frac{\cot A}{1 - \tan A} = \frac{\sin A}{\cos A} \times \frac{\sin A}{\sin A - \cos A} + \frac{\cos A}{\sin A} \times \frac{\cos A}{\cos A - \sin A}$$

$$= \frac{1}{\sin A - \cos A}\left\{\frac{\sin^3 A - \cos^3 A}{\cos A \sin A}\right\}$$

$$= \frac{\sin^2 A + \sin A \cos A + \cos^2 A}{\sin A \cos A}$$

$$= \frac{1 + \sin A \cos A}{\sin A \cos A}$$

$$= 1 + \sec A \operatorname{cosec} A$$

161. Given A ΔPQR such that

$$3 \sin P + 4 \cos Q = 6 \quad \text{...(i)}$$

$$4 \sin Q + 3 \cos P = 1 \quad \text{...(ii)}$$

On squaring and adding Eqs. (i) and (ii) both sides we get

$$(3 \sin P + 4 \cos Q)^2 + (4 \sin Q + 3 \cos P)^2 = 36 + 1$$

$$\Rightarrow 9(\sin^2 P + \cos^2 P) + 16(\sin^2 Q + \cos^2 Q) + 2 \times 3 \times 4(\sin P \cos Q + \sin Q \cos P) = 37$$

$$\Rightarrow 24[\sin(P + Q)] = 37 - 25$$

$$\Rightarrow \sin(P + Q) = \frac{1}{2}$$

Since, P and Q are angles of ΔPQR, hence $0° < P, Q < 180°$.

$$\Rightarrow P + Q = 30° \text{ or } 150°$$

$$\Rightarrow R = 150° \text{ or } 30°$$

Hence, two cases arise here.

***Case* I** $\quad R = 150°$

$$R = 150° \Rightarrow P + Q = 30°$$

$$\Rightarrow 0 < P, Q < 30°$$

$$\Rightarrow \sin P < \frac{1}{2}, \cos Q < 1$$

$$\Rightarrow 3 \sin P + 4 \cos Q < \frac{3}{2} + 4$$

$$\Rightarrow 3 \sin P + 4 \cos Q < \frac{11}{2} < 6$$

$\Rightarrow 3 \sin P + 4 \cos Q \Rightarrow 6$ is not possible.

***Case* II** $\quad R = 30°$

Hence, $R = 30°$ is the only possibility.

162.

$$A = \sin^2 x + \cos^4 x$$

$$\Rightarrow A = 1 - \cos^2 x + \cos^4 x$$

$$= \cos^4 x - \cos^2 x + \frac{1}{4} + \frac{3}{4}$$

$$= \left(\cos^2 x - \frac{1}{2}\right)^2 + \frac{3}{4} \quad \text{...(i)}$$

where, $\quad 0 \leq \left(\cos^2 x - \frac{1}{2}\right)^2 \leq \frac{1}{4} \quad$ ...(ii)

$$\therefore \quad \frac{3}{4} \leq A \leq 1$$

163. $\cos(\alpha + \beta) = \frac{4}{5} \Rightarrow \alpha + \beta \in$ Ist quadrant

and $\quad \sin(\alpha - \beta) = \frac{5}{13} \Rightarrow \alpha - \beta \in$ Ist quadrant

Now, $\quad 2\alpha = (\alpha + \beta) + (\alpha - \beta)$

$$\therefore \quad \tan 2\alpha = \frac{\tan(\alpha + \beta) + \tan(\alpha - \beta)}{1 - \tan(\alpha + \beta)\tan(\alpha - \beta)} = \frac{\frac{3}{4} + \frac{5}{12}}{1 - \frac{3}{4} \cdot \frac{5}{12}} = \frac{56}{33}$$

164. $\cos(\beta - \gamma) + \cos(\gamma - \alpha) + \cos(\alpha - \beta) = -\frac{3}{2}$

$$\Rightarrow 2[\cos(\beta - \gamma) + \cos(\gamma - \alpha) + \cos(\alpha - \beta)] + 3 = 0$$

$$\Rightarrow 2[\cos(\beta - \gamma) + \cos(\gamma - \alpha) + \cos(\alpha - \beta)] + \sin^2\alpha + \cos^2\alpha + \sin^2\beta + \cos^2\beta + \sin^2\gamma + \cos^2\gamma = 0$$

$$\Rightarrow (\sin\alpha + \sin\beta + \sin\gamma)^2 + (\cos\alpha + \cos\beta + \cos\gamma)^2 = 0$$

It is possible when,

$$\sin\alpha + \sin\beta + \sin\gamma = 0$$

and $\quad \cos\alpha + \cos\beta + \cos\gamma = 0$

Hence, both statements A and B are true.

165. Area $= \frac{1}{2} \times$ Base $\times$ Altitude

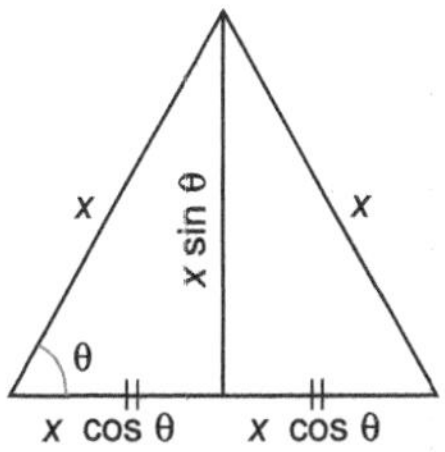

$$= \frac{1}{2} \times (2x \cos\theta) \times (x \sin\theta) = \frac{1}{2}x^2 \sin 2\theta$$

[since, maximum value of $\sin 2\theta$ is 1]

$\therefore$ Maximum area $= \frac{1}{2}x^2$

166. Given, $\quad \cos x + \sin x = \frac{1}{2}$

$$\therefore \quad \frac{1 - \tan^2 \frac{x}{2}}{1 + \tan^2 \frac{x}{2}} + \frac{2\tan\frac{x}{2}}{1 + \tan^2\frac{x}{2}} = \frac{1}{2}$$

Let $\quad \tan\frac{x}{2} = t \Rightarrow \frac{1 - t^2}{1 + t^2} + \frac{2t}{1 + t^2} = \frac{1}{2}$

$$\Rightarrow \quad 2(1-t^2+2t)=1+t^2 \Rightarrow 3t^2-4t-1=0$$

$$\Rightarrow \quad t=\frac{2\pm\sqrt{7}}{3}$$

As $0<x<\pi \Rightarrow 0<\frac{x}{2}<\frac{\pi}{2}$

So, $\tan\frac{x}{2}$ is positive.

$$\therefore \quad t=\tan\frac{x}{2}=\frac{2+\sqrt{7}}{3}$$

Now, $$\tan x=\frac{2\tan\frac{x}{2}}{1-\tan^2\frac{x}{2}}=\frac{2t}{1-t^2}$$

$$\Rightarrow \quad \tan x=\frac{2\left(\frac{2+\sqrt{7}}{3}\right)}{1-\left(\frac{2+\sqrt{7}}{3}\right)^2}$$

$$\Rightarrow \quad \tan x=\frac{-3(2+\sqrt{7})}{1+2\sqrt{7}}\times\frac{1-2\sqrt{7}}{1-2\sqrt{7}}$$

$$\Rightarrow \quad \tan x=-\left(\frac{4+\sqrt{7}}{3}\right)$$

167. Since, $\tan\frac{P}{2}$ and $\tan\frac{Q}{2}$ are the roots of equation

$$ax^2+bx+c=0$$

$$\therefore \quad \tan\frac{P}{2}+\tan\frac{Q}{2}=-\frac{b}{a} \quad \text{...(i)}$$

and $$\tan\frac{P}{2}\tan\frac{Q}{2}=\frac{c}{a}$$

Also, $$\frac{P}{2}+\frac{Q}{2}+\frac{R}{2}=\frac{\pi}{2} \quad [\because P+Q+R=\pi]$$

$$\Rightarrow \quad \frac{P+Q}{2}=\frac{\pi}{2}-\frac{R}{2}$$

$$\Rightarrow \quad \frac{P+Q}{2}=\frac{\pi}{4} \quad \left[\because \angle R=\frac{\pi}{2}\text{ (given)}\right]$$

$$\Rightarrow \quad \tan\left(\frac{P}{2}+\frac{Q}{2}\right)=1$$

$$\Rightarrow \quad \frac{\tan\frac{P}{2}+\tan\frac{Q}{2}}{1-\tan\frac{P}{2}\tan\frac{Q}{2}}=1$$

$$\Rightarrow \quad \frac{-\frac{b}{a}}{1-\frac{c}{a}}=1$$

$$\Rightarrow \quad -\frac{b}{a}=1-\frac{c}{a} \quad \text{[from Eq. (i)]}$$

$$\Rightarrow \quad -b=a-c$$

$$\Rightarrow \quad c=a+b$$

Alternate Solution

Since, $$\angle R=\frac{\pi}{2}$$

$$\Rightarrow \quad \angle P+\angle Q=\frac{\pi}{2}$$

$$\Rightarrow \quad \frac{\angle P}{2}=\frac{\pi}{4}-\frac{\angle Q}{2}$$

$$\therefore \quad \tan\frac{P}{2}=\tan\left(\frac{\pi}{4}-\frac{Q}{2}\right)$$

$$=\frac{\tan\frac{\pi}{4}-\tan\frac{Q}{2}}{1+\tan\frac{\pi}{4}\tan\frac{Q}{2}}$$

$$\Rightarrow \quad \tan\frac{P}{2}+\tan\frac{P}{2}\tan\frac{Q}{2}=1-\tan\frac{Q}{2}$$

$$\Rightarrow \quad \tan\frac{P}{2}+\tan\frac{Q}{2}=1-\tan\frac{P}{2}\tan\frac{Q}{2}$$

$$\Rightarrow \quad -\frac{b}{a}=1-\frac{c}{a}$$

$$\Rightarrow \quad -b=a-c$$

$$\Rightarrow \quad c=a+b$$

CHAPTER

02

Trigonometric Equations and Inequations

Learning Part

Session 1

- Trigonometric Equations
- General Solution
- Principal Solution

Session 2

- Equation of the Form $a \cos \theta + b \sin \theta = c$
- Some Particular Equations

Session 3

- Solution of Simultaneous Trigonometric Equations
- Problems Based on Extreme Values of $\sin x$ and $\cos x$

Session 4

- Trigonometric Inequality

Practice Part

- JEE Type Examples
- Chapter Exercises

Arihant on Your Mobile !

Exercises with the symbol can be practised on your mobile. See inside cover page to activate for free.

Session 1

Trigonometric Equations, Principal Solution and General Solution

Trigonometric Equations

The equations involving trigonometric functions of unknown angles are known as Trigonometric equations.

e.g. $\cos\theta = 0, \cos^2\theta - 4\cos\theta = 1,$

$\sin^2\theta + \sin\theta = 2, \cos^2\theta - 4\sin\theta = 1.$

A solution of a trigonometric equation is the value of the unknown angle that satisfies the equation.

e.g. $\sin\theta = \frac{1}{\sqrt{2}} \Rightarrow \theta = \frac{\pi}{4}$

or $\theta = \frac{\pi}{4}, \frac{3\pi}{4}, \frac{9\pi}{4}, \frac{11\pi}{4}, \dots$

Thus, the trigonometric equation may have infinite number of solutions and can be classified as

(i) Principal solution (ii) General solution

Principal Solution

The solutions lying in the interval $[0, 2\pi]$ are called principal solutions.

General Solution

Since, trigonometric functions are periodic, a solution generalised by means of periodicity of the trigonometrical functions. The solution consisting of all possible solutions of a trigonometric equation is called its general solution.

We use following results for solving the trigonometric equations.

Result 1 $\sin\theta = 0 \Leftrightarrow \theta = n\pi, n \in I.$

We know that $\sin\theta = 0$ for all integral multiples of π.

$\therefore \quad \sin\theta = 0 \Leftrightarrow \theta = 0, \pm\pi, \pm 2\pi, \pm 3\pi, \dots$

$\Rightarrow \quad \theta = n\pi, n \in I$

Thus, $\sin\theta = 0$

$\Leftrightarrow \quad \theta = n\pi, n \in I.$

Result 2 $\cos\theta = 0 \Leftrightarrow \theta = (2n+1)\frac{\pi}{2}, n \in I.$

We know that $\cos\theta = 0$ for all odd multiples of $\frac{\pi}{2}$.

$\therefore \quad \cos\theta = 0 \Leftrightarrow \theta = \pm\frac{\pi}{2}, \pm\frac{3\pi}{2}, \pm\frac{5\pi}{2}, \dots$

$\Rightarrow \quad \theta = (2n+1)\frac{\pi}{2}, n \in I$

Thus, $\cos\theta = 0$

$\Leftrightarrow \quad \theta = (2n+1)\frac{\pi}{2}, n \in I$

Result 3 $\tan\theta = 0 \Leftrightarrow \theta = n\pi, n \in I.$

We know that $\tan\theta = 0$ for all integral multiple of π.

$\therefore \quad \tan\theta = 0 \Leftrightarrow \theta = 0, \pm\pi, \pm 2\pi, \pm 3\pi, \dots$

$\Rightarrow \quad \theta = n\pi, n \in I$

Thus, $\tan\theta = 0 \quad \Leftrightarrow \quad \theta = n\pi, n \in I$

Result 4 $\sin\theta = \sin\alpha \Leftrightarrow \theta = n\pi + (-1)^n\alpha$, where $n \in I$ and $\alpha \in \left[-\frac{\pi}{2}, \frac{\pi}{2}\right]$

We have, $\sin\theta = \sin\alpha$, where $\alpha \in \left[-\frac{\pi}{2}, \frac{\pi}{2}\right]$

Now, $\sin\theta - \sin\alpha = 0$

$\Leftrightarrow \quad 2\cos\left(\frac{\theta+\alpha}{2}\right)\sin\left(\frac{\theta-\alpha}{2}\right) = 0$

$\Leftrightarrow \quad \cos\left(\frac{\theta+\alpha}{2}\right) = 0 \text{ or } \sin\left(\frac{\theta-\alpha}{2}\right) = 0$

$\Leftrightarrow \quad \left(\frac{\theta+\alpha}{2}\right) = (2m+1)\frac{\pi}{2}, m \in I$

or $\left(\frac{\theta-\alpha}{2}\right) = m\pi, m \in I$

$\Leftrightarrow \quad (\theta+\alpha) = (2m+1)\pi, m \in I$

or $(\theta-\alpha) = 2m\pi, m \in I$

$\Leftrightarrow \quad \theta = (2m+1)\pi - \alpha, m \in I$

or $\theta = (2m\pi) + \alpha, m \in I$

$\Leftrightarrow \quad \theta = (\text{any odd multiple of } \pi) - \alpha$

or $\theta = (\text{any even multiple of } \pi) + \alpha$

$\Leftrightarrow \quad \theta = n\pi + (-1)^n\alpha$, where $n \in I$

Thus, $\sin\theta = \sin\alpha$

$\Leftrightarrow \quad \theta = n\pi + (-1)^n\alpha$, where $n \in I$ and $\alpha \in \left[-\frac{\pi}{2}, \frac{\pi}{2}\right]$

Note $\sin\theta = 1 \Leftrightarrow \sin\theta = \sin\frac{\pi}{2} \Leftrightarrow \theta = n\pi + (-1)^n \frac{\pi}{2}$

$\Rightarrow \quad \theta = (4n+1)\frac{\pi}{2}, n \in N.$

Result 5 $\cos\theta = \cos\alpha \Leftrightarrow \theta = 2n\pi \pm \alpha, n \in I$ and $\alpha \in [0, \pi]$.

We have,

$$\cos\theta = \cos\alpha, \quad \text{where } \alpha \in [0, \pi]$$

Now, $\cos\theta - \cos\alpha = 0$

$\Leftrightarrow \quad -2\sin\left(\frac{\theta+\alpha}{2}\right)\cdot\sin\left(\frac{\theta-\alpha}{2}\right) = 0$

$\Leftrightarrow \quad \sin\left(\frac{\theta+\alpha}{2}\right) = 0 \text{ or } \sin\left(\frac{\theta-\alpha}{2}\right) = 0$

$\Leftrightarrow \quad \frac{\theta+\alpha}{2} = n\pi \text{ or } \frac{\theta-\alpha}{2} = n\pi, n \in I$

$\Leftrightarrow \quad \theta+\alpha = 2n\pi \text{ or } \theta-\alpha = 2n\pi, n \in I$

$\Leftrightarrow \quad \theta = 2n\pi - \alpha \text{ or } \theta = 2n\pi + \alpha, n \in I$

$\Leftrightarrow \quad \theta = 2n\pi \pm \alpha, n \in I$

Thus, $\cos\theta = \cos\alpha$

$\Leftrightarrow \quad \theta = 2n\pi \pm \alpha, n \in I, \text{ where } \alpha \in [0, \pi]$

Note

(i) $\cos\theta = 1 \Leftrightarrow \cos\theta = \cos 0 \Leftrightarrow \theta = 2n\pi \pm 0 \Rightarrow \theta = 2n\pi$

(ii) $\cos\theta = -1 \Leftrightarrow \cos\theta = \cos\pi \Leftrightarrow \theta = 2n\pi \pm \pi$

$\Leftrightarrow \theta = (2n \pm 1)\pi \Rightarrow \theta = (2n+1)\pi$

(iii) $\sin\theta = \sin\alpha$ and $\cos\theta = \cos\alpha \Leftrightarrow \sin\left(\frac{\theta-\alpha}{2}\right) = 0$

$\Rightarrow \frac{\theta-\alpha}{2} = n\pi \Rightarrow \theta = 2n\pi + \alpha$

Result 6 $\tan\theta = \tan\alpha \Leftrightarrow \theta = n\pi + \alpha, n \in I$ where $\alpha \in \left(-\frac{\pi}{2}, \frac{\pi}{2}\right)$.

We have, $\tan\theta = \tan\alpha$, where $\alpha \in \left(-\frac{\pi}{2}, \frac{\pi}{2}\right)$

$\Leftrightarrow \quad \frac{\sin\theta}{\cos\theta} = \frac{\sin\alpha}{\cos\alpha}$

$\Leftrightarrow \sin\theta\cos\alpha - \cos\theta\sin\alpha = 0$

$\Leftrightarrow \quad \sin(\theta-\alpha) = 0$

$\Leftrightarrow \quad \theta - \alpha = n\pi, n \in I$

$\Leftrightarrow \quad \theta = n\pi + \alpha, n \in I$

Thus, $\tan\theta = \tan\alpha$

$\Leftrightarrow \quad \theta = n\pi + \alpha \text{ where } \alpha \in \left(-\frac{\pi}{2}, \frac{\pi}{2}\right)$

Result 7 $\sin^2\theta = \sin^2\alpha, \cos^2\theta = \cos^2\alpha, \tan^2\theta = \tan^2\alpha \Rightarrow \theta = n\pi \pm \alpha$

(i) $\sin^2\theta = \sin^2\alpha$

$\Leftrightarrow \quad \frac{1-\cos 2\theta}{2} = \frac{1-\cos 2\alpha}{2}$

$\Leftrightarrow \quad \cos 2\theta = \cos 2\alpha$

$\Leftrightarrow \quad 2\theta = 2n\pi \pm 2\alpha, n \in I$

$\Leftrightarrow \quad \theta = n\pi \pm \alpha, n \in I$

(ii) $\cos^2\theta = \cos^2\alpha$

$\Leftrightarrow \quad \frac{1+\cos 2\theta}{2} = \frac{1+\cos 2\alpha}{2}$

$\Leftrightarrow \quad \cos 2\theta = \cos 2\alpha$

$\Leftrightarrow \quad 2\theta = 2n\pi \pm 2\alpha, n \in I$

$\Leftrightarrow \quad \theta = n\pi \pm \alpha, n \in I$

(iii) $\tan^2\theta = \tan^2\alpha$

$\Leftrightarrow \quad \frac{1-\tan^2\theta}{1+\tan^2\theta} = \frac{1-\tan^2\alpha}{1+\tan^2\alpha}$

[applying componendo and dividendo]

$\Leftrightarrow \quad \cos 2\theta = \cos 2\alpha$

$\Leftrightarrow \quad 2\theta = 2n\pi \pm 2\alpha, n \in I$

$\Leftrightarrow \quad \theta = n\pi \pm \alpha, n \in I$

Summary of Above Results

1. $\sin\theta = 0 \Leftrightarrow \theta = n\pi$
2. $\cos\theta = 0 \Leftrightarrow \theta = (2n+1)\frac{\pi}{2}$
3. $\tan\theta = 0 \Leftrightarrow \theta = n\pi$
4. $\sin\theta = \sin\alpha \Leftrightarrow \theta = n\pi + (-1)^n\alpha$, where $\alpha \in \left[-\frac{\pi}{2}, \frac{\pi}{2}\right]$
5. $\cos\theta = \cos\alpha \Leftrightarrow \theta = 2n\pi \pm \alpha$, where $\alpha \in [0, \pi]$
6. $\tan\theta = \tan\alpha \Leftrightarrow \theta = n\pi + \alpha$, where $\alpha \in \left(-\frac{\pi}{2}, \frac{\pi}{2}\right)$
7. $\sin^2\theta = \sin^2\alpha, \cos^2\theta = \cos^2\alpha, \tan^2\theta = \tan^2\alpha \Leftrightarrow \theta = n\pi \pm \alpha$
8. $\sin\theta = 1 \Leftrightarrow \theta = (4n+1)\frac{\pi}{2}$
9. $\cos\theta = 1 \Leftrightarrow \theta = 2n\pi$
10. $\cos\theta = -1 \Leftrightarrow \theta = (2n+1)\pi$
11. $\sin\theta = \sin\alpha$ and $\cos\theta = \cos\alpha \Leftrightarrow \theta = 2n\pi + \alpha$

Note

(i) In this chapter 'n' is taken as an integer, if not stated otherwise.

(ii) The general solution should be given unless the solution is required in a specified interval or range.

Example 1. If $\sin\alpha$, 1, $\cos 2\alpha$ are in GP, then find the general solution for α.

Sol. Since, $\sin\alpha$, 1, $\cos 2\alpha$ are in GP.

$$\therefore \quad 1 = \sin\alpha\cos 2\alpha$$
$$\Rightarrow \quad 1 = \sin\alpha(1 - 2\sin^2\alpha)$$
$$\Rightarrow \quad 2\sin^3\alpha - \sin\alpha + 1 = 0$$
$$\Rightarrow \quad (\sin\alpha + 1)(2\sin^2\alpha - 2\sin\alpha + 1) = 0$$
$$\Rightarrow \quad \sin\alpha + 1 = 0 \quad (\text{as } 2\sin^2\alpha - 2\sin\alpha + 1 \neq 0)$$
$$\therefore \quad \sin\alpha = -1$$
$$\Rightarrow \quad \sin\alpha = \sin\left(-\frac{\pi}{2}\right)$$
$$\Rightarrow \quad \alpha = n\pi + (-1)^n\left(-\frac{\pi}{2}\right), n \in Z$$
$$\Rightarrow \quad \alpha = n\pi + (-1)^{n+1}\left(\frac{\pi}{2}\right), n \in Z$$

Example 2. If $\frac{1}{6}\sin\theta$, $\cos\theta$ and $\tan\theta$ are in GP, then find the general solution for θ.

Sol. Since, $\frac{1}{6}\sin\theta$, $\cos\theta$, $\tan\theta$ are in GP.

$$\therefore \quad \cos^2\theta = \frac{1}{6}\sin\theta\cdot\tan\theta$$
$$\Rightarrow \quad 6\cos^3\theta + \cos^2\theta - 1 = 0$$

Note that $\cos\theta = \frac{1}{2}$ satisfies the equation (by trial),

$$\therefore \quad (2\cos\theta - 1)(3\cos^2\theta + 2\cos\theta + 1) = 0$$
$$\Rightarrow \quad \cos\theta = \frac{1}{2} \quad (\text{other values of } \cos\theta \text{ are imaginary})$$
$$\Rightarrow \quad \cos\theta = \cos\frac{\pi}{3}$$
$$\Rightarrow \quad \theta = 2n\pi \pm \frac{\pi}{3}, n \in Z$$

Example 3. Solve $\sin^2\theta - \cos\theta = \frac{1}{4}$ for θ and write the values of θ in the interval $0 \le \theta \le 2\pi$.

Sol. The given equation can be written as;

$$1 - \cos^2\theta - \cos\theta = \frac{1}{4}$$
$$\Rightarrow \quad \cos^2\theta + \cos\theta - \frac{3}{4} = 0$$
$$\Rightarrow \quad 4\cos^2\theta + 4\cos\theta - 3 = 0$$
$$\Rightarrow \quad (2\cos\theta - 1)(2\cos\theta + 3) = 0$$
$$\Rightarrow \quad \cos\theta = \frac{1}{2}, -\frac{3}{2}$$

Since, $\cos\theta = -\frac{3}{2}$ is not possible as; $-1 \le \cos\theta \le 1$

$$\therefore \quad \cos\theta = \frac{1}{2}$$
$$\Rightarrow \quad \cos\theta = \cos\frac{\pi}{3}$$
$$\Rightarrow \quad \theta = 2n\pi \pm \frac{\pi}{3}$$

For the given interval, $n = 0$ and $n = 1$.

$$\Rightarrow \quad \theta = \frac{\pi}{3}, \frac{5\pi}{3}$$

Example 4. Solve $\cos\theta + \cos 3\theta + \cos 5\theta + \cos 7\theta = 0$.

Sol. We have,

$$\cos\theta + \cos 3\theta + \cos 5\theta + \cos 7\theta = 0$$
$$\Rightarrow \quad (\cos\theta + \cos 7\theta) + (\cos 3\theta + \cos 5\theta) = 0$$
$$\Rightarrow \quad 2\cos 4\theta\cdot\cos 3\theta + 2\cos 4\theta\cdot\cos\theta = 0$$
$$\Rightarrow \quad \cos 4\theta(\cos 3\theta + \cos\theta) = 0$$
$$\Rightarrow \quad \cos 4\theta(2\cos 2\theta\cos\theta) = 0$$

Either $\cos\theta = 0$

$$\Rightarrow \quad \theta = (2n+1)\frac{\pi}{2}$$

or $\cos 2\theta = 0 \Rightarrow \theta = (2n+1)\frac{\pi}{4}$

or $\cos 4\theta = 0 \Rightarrow \theta = (2n+1)\frac{\pi}{8}$

$$\therefore \quad \theta = (2n+1)\frac{\pi}{2}, (2n+1)\frac{\pi}{4}, (2n+1)\frac{\pi}{8}$$

Example 5. Find the number of solutions for,

$$\sin 5\theta\cdot\cos 3\theta = \sin 9\theta\cdot\cos 7\theta \text{ in } \left[0, \frac{\pi}{2}\right].$$

Sol. Here, $2\sin 5\theta\cdot\cos 3\theta = 2\sin 9\theta\cdot\cos 7\theta$

$$\Rightarrow \quad \sin 8\theta + \sin 2\theta = \sin 16\theta + \sin 2\theta$$
$$\Rightarrow \quad \sin 8\theta = \sin 16\theta \text{ or } \sin 16\theta = \sin 8\theta$$
$$\therefore \quad 16\theta = n\pi + (-1)^n 8\theta \quad \ldots\text{(i)}$$

when n is even Eq. (i) becomes;

$$8\theta = n\pi \Rightarrow \theta = \frac{n\pi}{8} \quad \ldots\text{(ii)}$$

when n is odd Eq. (i) becomes;

$$24\theta = n\pi \Rightarrow \theta = \frac{n\pi}{24} \quad \ldots\text{(iii)}$$

$\therefore$ For the given interval $\left[0, \frac{\pi}{2}\right]$ Eq. (ii) and (iii) gives the solution as,

$$\theta = 0, \frac{\pi}{4}, \frac{\pi}{2} \text{ and } \frac{\pi}{24}, \frac{\pi}{8}, \frac{5\pi}{24}, \frac{7\pi}{24}, \frac{3\pi}{8}, \frac{11\pi}{24}$$

$\therefore$ Number of solutions is 9.

Example 6. Solve $\dfrac{\sin x + i\cos x}{1+i}$, $i=\sqrt{-1}$ when it is purely imaginary.

Sol. Here, $\dfrac{\sin x + i\cos x}{1+i} = \dfrac{(1-i)(\sin x + i\cos x)}{(1-i)(1+i)}$

$$= \frac{\sin x + \cos x + i(\cos x - \sin x)}{2}$$

which will be purely imaginary, if

$$\sin x + \cos x = 0$$

$$\Rightarrow \quad \tan x = -1 = \tan\left(-\frac{\pi}{4}\right)$$

$$\Rightarrow \quad x = n\pi - \frac{\pi}{4}, \text{ is the general solution.}$$

Example 7. Find the general solutions of

$$2^{1+|\cos x|+|\cos x|^2+|\cos x|^3+\ldots \text{ to } \infty} = 4$$

Sol. $2^{1+|\cos x|+|\cos x|^2+|\cos x|^3+\ldots \text{ to } \infty}$

$$\Rightarrow \quad \frac{1}{2^{1-|\cos x|}} = 2^2 \quad \Rightarrow \quad \frac{1}{1-|\cos x|} = 2$$

$$\Rightarrow \quad |\cos x| = \frac{1}{2} \quad \Rightarrow \quad \cos x = \pm\frac{1}{2}$$

$$\therefore \quad x = 2n\pi \pm \frac{\pi}{3}, 2n\pi \pm \left(\pi - \frac{\pi}{3}\right)$$

$$= 2n\pi \pm \frac{\pi}{3}, (2n \pm 1)\pi \mp \frac{\pi}{3}$$

$$\Rightarrow \quad x = n\pi \pm \frac{\pi}{3}$$

Example 8. If $x \neq \dfrac{n\pi}{2}$ and $(\cos x)^{\sin^2 x - 3\sin x + 2} = 1$, then find the general solutions of x.

Sol. As $x \neq \dfrac{n\pi}{2} \Rightarrow \cos x \neq 0, 1, -1$

So, $(\cos x)^{\sin^2 x - 3\sin x + 2} = 1$

$$\Rightarrow \quad \sin^2 x - 3\sin x + 2 = 0$$

$$\therefore \quad (\sin x - 2)(\sin x - 1) = 0$$

$$\Rightarrow \quad \sin x = 1, 2$$

where, $\sin x = 2$ is not possible and $\sin x = 1$ does not satisfy the equation.

$\therefore$ No general solution is possible.

Some Important Results

1. While solving a trigonometric equation, squaring the equation at any step should be avoided as far as possible. If squaring is necessary, check the solution for extraneous values.
2. Never cancel terms containing unknown terms on the two sides, which are in product. It may cause loss of the genuine solution.
3. The answer should not contain such values of angles which make any of the terms undefined or infinite.
4. Domain should not change. If it changes, necessary corrections must be made.
5. Check that denominator is not zero at any stage while solving equations.

Example 9. Find the set of values of x for which $\dfrac{\tan 3x - \tan 2x}{1 + \tan 3x \cdot \tan 2x} = 1$.

Sol. We have, $\dfrac{\tan 3x - \tan 2x}{1 + \tan 3x \cdot \tan 2x} = 1 \Rightarrow \tan(3x - 2x) = 1$

$$\Rightarrow \quad \tan x = 1 \quad \Rightarrow \quad \tan x = \tan\frac{\pi}{4}$$

$$\Rightarrow \quad x = n\pi + \frac{\pi}{4} \quad [\text{using } \tan\theta = \tan\alpha \Leftrightarrow \theta = n\pi + \alpha]$$

But for this value of x,

$\tan 2x = \tan\left(2n\pi + \dfrac{\pi}{2}\right) = \infty$, which does not satisfy the given equation as it reduces to indeterminate form.

Hence, the solution set for x is ϕ.

Example 10. Solve $\sin x = 0$ and $\dfrac{\sin x}{\cos\frac{x}{2}\cos\frac{3x}{2}} = 0$ and show their solutions are different.

Sol. We have, $\sin x = 0 \Rightarrow x = n\pi$,

i.e. $\quad x = 0, \pi, 2\pi, 3\pi, \ldots$ ...(i)

Where as,

$$\frac{\sin x}{\cos\frac{x}{2}\cos\frac{3x}{2}} = 0, \text{ where } \cos\frac{x}{2} \neq 0 \text{ and } \cos\frac{3x}{2} \neq 0$$

i.e. $\quad \dfrac{x}{2} \neq \dfrac{\pi}{2}, \dfrac{3\pi}{2}, \dfrac{5\pi}{2}, \ldots$ and $\dfrac{3x}{2} \neq \dfrac{\pi}{2}, \dfrac{3\pi}{2}, \dfrac{5\pi}{2}, \ldots$

i.e. $\quad x \neq \pi, 3\pi, 5\pi, \ldots$ and $\dfrac{\pi}{3}, \pi, \dfrac{5\pi}{3}, \ldots$ ...(ii)

But $\quad \dfrac{\sin x}{\cos\frac{x}{2}\cos\frac{3x}{2}} = 0 \Rightarrow \sin x = 0$

$\Rightarrow \quad x = \pi, 2\pi, 3\pi, 4\pi, \ldots$ ...(iii)

From Eqs. (ii) and (iii),

$\quad x = 2\pi, 4\pi, 6\pi, \ldots$ ...(iv)

From Eqs. (i) and (iv);

The two equations are not equivalent. Since, some solutions of the first do not satisfy the second equation.

Example 11. Find number of solutions of $\tan x + \sec x = 2\cos x$ in $[0, 2\pi]$.

Sol. Here, $\tan x + \sec x = 2\cos x \Rightarrow \sin x + 1 = 2\cos^2 x$

$$\Rightarrow \quad 2\sin^2 x + \sin x - 1 = 0 \Rightarrow \sin x = \frac{1}{2}, -1$$

But $\sin x = -1 \Rightarrow x = \frac{3\pi}{2}$ which does not satisfy $\tan x + \sec x = 2\cos x$.

Thus, $\sin x = \frac{1}{2} \Rightarrow x = \frac{\pi}{6}, \frac{5\pi}{6}$

Number of solutions of $\tan x + \sec x = 2\cos x$ is 2.

Example 12. Solve $\sec x - 1 = (\sqrt{2} - 1)\tan x$.

Sol. We have, $\sec x - 1 = (\sqrt{2} - 1)\tan x$

$$\Rightarrow \quad 1 - \cos x = (\sqrt{2} - 1)\sin x$$

$$\Rightarrow \quad 2\sin^2 \frac{x}{2} - (\sqrt{2} - 1)\cdot 2\sin\frac{x}{2}\cos\frac{x}{2} = 0$$

$$\Rightarrow \quad 2\sin\frac{x}{2}\left\{\sin\frac{x}{2} - (\sqrt{2} - 1)\cos\frac{x}{2}\right\} = 0$$

$$\Rightarrow \quad \sin\frac{x}{2}\left[\sin\frac{x}{2} - (\sqrt{2} - 1)\cos\frac{x}{2}\right] = 0$$

$$\Rightarrow \quad \sin\frac{x}{2} = 0 \text{ or } \tan\frac{x}{2} = (\sqrt{2} - 1) = \tan\frac{\pi}{8}$$

$$\Rightarrow \quad \frac{x}{2} = n\pi \text{ or } \frac{x}{2} = n\pi + \frac{\pi}{8}$$

$$\therefore \quad x = 2n\pi, 2n\pi + \frac{\pi}{4}$$

Note

$\theta \neq (2n+1)\frac{\pi}{2}$, otherwise the equation will be meaningless.

Example 13. Solve $\tan\theta + \tan 2\theta + \tan\theta \cdot \tan 2\theta = 1$.

Sol. We can re-write the given equation as;

$$\tan\theta + \tan 2\theta = 1 - \tan\theta \cdot \tan 2\theta$$

$$\Rightarrow \quad \frac{\tan\theta + \tan 2\theta}{1 - \tan\theta\tan 2\theta} = 1 \Rightarrow \tan 3\theta = 1 = \tan\frac{\pi}{4}$$

$$\Rightarrow \quad 3\theta = n\pi + \frac{\pi}{4} \Rightarrow \theta = \frac{n\pi}{3} + \frac{\pi}{12}$$

Example 14. Find the number of solutions of $|\cos x| = \sin x$, $0 \leq x \leq 4\pi$.

Sol. Case I If $\cos x \geq 0$, i.e. $x \in \left[0, \frac{\pi}{2}\right] \cup \left[\frac{3\pi}{2}, \frac{5\pi}{2}\right] \cup \left[\frac{7\pi}{2}, 4\pi\right]$

then, $\cos x = \sin x$.

$$\Rightarrow \quad \tan x = 1 \Rightarrow x = n\pi + \frac{\pi}{4} = \frac{\pi}{4}, \frac{5\pi}{4}, \frac{9\pi}{4}, \frac{13\pi}{4}$$

$\therefore$ If $\cos x \geq 0$, the possible values of x are $\frac{\pi}{4}, \frac{9\pi}{4}$.

Case II If $\cos x < 0$, i.e. $x \in \left(\frac{\pi}{2}, \frac{3\pi}{2}\right) \cup \left(\frac{5\pi}{2}, \frac{7\pi}{2}\right)$, then

$-\cos x = \sin x$

$$\Rightarrow \quad \tan x = -1 \Rightarrow x = n\pi - \frac{\pi}{4} = \frac{3\pi}{4}, \frac{7\pi}{4}, \frac{11\pi}{4}, \frac{15\pi}{4}$$

$\therefore$ If $\cos x < 0$, the possible values of x are $\frac{3\pi}{4}, \frac{11\pi}{4}$.

Thus, the possible number of solutions are 4.

Example 15. Solve $\cot\theta = \sin 2\theta$ by substituting $\sin 2\theta = \frac{2\tan\theta}{1 + \tan^2\theta}$ and again by substituting $\sin 2\theta = 2\sin\theta \cdot \cos\theta$ and check whether the two answer are same or not.

Sol. Method I

Put $\sin 2\theta = \frac{2\tan\theta}{1 + \tan^2\theta}$

$$\therefore \quad \cot\theta = \sin 2\theta \Rightarrow \cot\theta = \frac{2\tan\theta}{1 + \tan^2\theta}$$

$$\Rightarrow \quad \frac{1}{\tan\theta} = \frac{2\tan\theta}{1 + \tan^2\theta}$$

$$\Rightarrow \quad 2\tan^2\theta = 1 + \tan^2\theta$$

$$\Rightarrow \quad \tan^2\theta = 1 \Rightarrow \tan^2\theta = (1)^2 = \tan^2\frac{\pi}{4}$$

$$\Rightarrow \quad \theta = n\pi \pm \frac{\pi}{4} \quad \text{...(i)}$$

Method II Put $\sin 2\theta = 2\sin\theta\cos\theta$

$$\therefore \quad \cot\theta = \sin 2\theta$$

$$\Rightarrow \quad \frac{\cos\theta}{\sin\theta} = 2\sin\theta\cos\theta$$

$$\Rightarrow \quad \cos\theta = 2\sin^2\theta\cos\theta$$

$$\Rightarrow \quad \cos\theta(1 - 2\sin^2\theta) = 0$$

$$\Rightarrow \quad \cos\theta = 0 \text{ or } \sin^2\theta = \frac{1}{2} = \left(\frac{1}{\sqrt{2}}\right)^2 = \sin^2\frac{\pi}{4}$$

$$\Rightarrow \quad \theta = (2n+1)\frac{\pi}{2} \text{ or } \theta = n\pi \pm \frac{\pi}{4} \quad \text{...(ii)}$$

From Eqs. (i) and (ii), it is clear, first method gives less number of roots then the second method.

Note As far as possible, avoid the use of following formulae

$$\sin 2x = \frac{2\tan x}{1 + \tan^2 x}, \cos 2x = \frac{1 - \tan^2 x}{1 + \tan^2 x},$$

$$\tan 2x = \frac{2\tan x}{1 - \tan^2 x} \text{ and } \tan 3x = \frac{3\tan x - \tan^3 x}{1 - 3\tan^2 x}$$

As these formulae are not defined for some real values of x. Hence, in many cases the solution obtained with use of these formulae may not be the complete solutions set of the given equation.

Exercise for Session 1

1. Solve $\sin 5x = \cos 2x$.
2. Find the number of values of x in $[0, 5\pi]$ satisfying the equation $3\cos^2 x - 10\cos x + 7 = 0$.
3. If $2\tan^2 x - 5\sec x$ is equal to 1 for exactly 7 distinct values of $x \in \left[0, \frac{n\pi}{2}\right], n \in N$, then find the greatest value of n.
4. Find the general solution of equation $\sec^2 x = \sqrt{2}(1 - \tan^2 x)$.
5. Solve $7\cos^2\theta + 3\sin^2\theta = 4$.
6. Find the general solution of the equation $\tan^2\alpha + 2\sqrt{3}\tan\alpha = 1$.
7. Find the number of solutions of $\sin^2 x - \sin x - 1 = 0$.
8. Find the general values of θ satisfying $\tan\theta + \tan\left(\frac{3\pi}{4} + \theta\right) = 2$.
9. Find the general solution of $\sin x + \sin 5x = \sin 2x + \sin 4x$.
10. Solve $\cos\theta\cos 2\theta\cos 3\theta = \frac{1}{4}$.
11. Solve $2\cot 2x - 3\cot 3x = \tan 2x$.
12. Find the roots of the equation $\cot x - \cos x = 1 - \cot x \cos x$.
13. If the equation $x^2 + 4x\sin\theta + \tan\theta = 0 \left(0 < \theta < \frac{\pi}{2}\right)$ has repeated roots, then find the value of θ.
14. Find the number of solutions of the equation $2\sin^3 x + 6\sin^2 x - \sin x - 3 = 0$ in $(0, 2\pi)$.
15. Find the number of roots of the equation $16\sec^3\theta - 12\tan^2\theta - 4\sec\theta = 9$ in interval $(-\pi, \pi)$.

Session 2

Equation of the Form *a* cos θ + *b* sin θ = *c* and Some Particular Equations

Equation of the Form $a\cos\theta + b\sin\theta = c$

To solve the equation $a\cos\theta + b\cos\theta = c$, put $a = r\cos\phi$ and $b = r\sin\phi$, where

$$r = \sqrt{a^2 + b^2} \text{ and } \phi = \tan^{-1}\frac{b}{a}$$

Substituting these values in the equation, we get,
$r\cos\phi\cos\theta + r\sin\phi\sin\theta = c$

$$\cos(\theta - \phi) = \frac{c}{r} \Rightarrow \cos(\theta - \phi) = \frac{c}{\sqrt{a^2 + b^2}}$$

If $|c| > \sqrt{a^2 + b^2}$, then the equation $a\cos\theta + b\sin\theta = c$ has no solution.

If $|c| \le \sqrt{a^2 + b^2}$, then put $\frac{|c|}{\sqrt{a^2 + b^2}} = \cos\alpha$, so that

$\cos(\theta - \phi) = \cos\alpha$

$\Rightarrow \quad (\theta - \phi) = 2n\pi \pm \alpha$

$\Rightarrow \quad \theta = 2n\pi \pm \alpha + \phi$, where $n \in I$

Working Rule

Step I Check whether $|c| \le \sqrt{a^2 + b^2}$ or not. If it is satisfied, no real solution exists.

Step II If the above condition is satisfied, divide both sides of the equation by $\sqrt{a^2 + b^2}$.

Example 16. Prove that the equation $p \cos x - q \sin x = r$ admits solution for x only if $-\sqrt{p^2 + q^2} < r < \sqrt{p^2 + q^2}$.

Sol. Here, $\quad p \cos x - q \sin x = r$

On dividing both sides by $\sqrt{p^2 + q^2}$, we get

$$\frac{p}{\sqrt{p^2 + q^2}} \cos x - \frac{q}{\sqrt{p^2 + q^2}} \sin x = \frac{r}{\sqrt{p^2 + q^2}} \quad \text{...(i)}$$

Put, $\dfrac{p}{\sqrt{p^2 + q^2}} = \cos \phi$, $\dfrac{q}{\sqrt{p^2 + q^2}} = \sin \phi$ in Eq. (i), we get

$$\cos \phi \cos x - \sin \phi \sin x = \frac{r}{\sqrt{p^2 + q^2}}$$

$$\Rightarrow \quad \cos (x + \phi) = \frac{r}{\sqrt{p^2 + q^2}}$$

As we know, $\quad -1 \le \cos(x + \phi) \le 1$

$\therefore$ The above equation posses solution only if,

$$-1 \le \frac{r}{\sqrt{p^2 + q^2}} \le 1 \quad \text{or} \quad -\sqrt{p^2 + q^2} \le r \le \sqrt{p^2 + q^2}$$

Example 17. Solve $\sin x + \sqrt{3} \cos x = \sqrt{2}$.

Sol. Given, $\sqrt{3} \cos x + \sin x = \sqrt{2}$, dividing both sides by $\sqrt{a^2 + b^2} = \sqrt{(\sqrt{3})^2 + 1^2} = 2$, we get

$$\Rightarrow \quad \frac{\sqrt{3}}{2} \cos x + \frac{1}{2} \sin x = \frac{\sqrt{2}}{2} = \frac{1}{\sqrt{2}}$$

$$\Rightarrow \quad \cos\left(x - \frac{\pi}{6}\right) = \cos\left(\frac{\pi}{4}\right)$$

$$\Rightarrow \quad x - \frac{\pi}{6} = 2n\pi \pm \frac{\pi}{4}$$

$$\Rightarrow \quad x = 2n\pi \pm \frac{\pi}{4} + \frac{\pi}{6}$$

$$\Rightarrow \quad x = 2n\pi + \frac{5\pi}{12}, 2n\pi - \frac{\pi}{12}, \text{ where } n \in I$$

Example 18. Find the number of distinct solutions of $\sec x + \tan x = \sqrt{3}$, where $0 \le x \le 3\pi$.

Sol. Here, we have $\sec x + \tan x = \sqrt{3}$

$$\Rightarrow \quad 1 + \sin x = \sqrt{3} \cos x$$

$$\text{or} \quad \sqrt{3} \cos x - \sin x = 1$$

On dividing both sides by $\sqrt{a^2 + b^2}$ i.e. $\sqrt{4} = 2$, we get

$$\Rightarrow \quad \frac{\sqrt{3}}{2} \cos x - \frac{1}{2} \sin x = \frac{1}{2}$$

$$\Rightarrow \quad \cos \frac{\pi}{6} \cos x - \sin \frac{\pi}{6} \sin x = \frac{1}{2}$$

$$\Rightarrow \quad \cos\left(x + \frac{\pi}{6}\right) = \cos\left(\frac{\pi}{3}\right)$$

$$\Rightarrow \quad x + \frac{\pi}{6} = 2n\pi \pm \frac{\pi}{3}$$

$$\text{or} \quad x = 2n\pi + \frac{\pi}{6}, 2n\pi - \frac{\pi}{2}$$

Now, when $x = 2n\pi + \dfrac{\pi}{6}$, there are solution for $n = 0, 1$ and

when $x = 2n\pi - \dfrac{\pi}{2}$, there are solution for $n = 1$.

Thus, total number of solutions are 3.

Example 19. Prove that the equation $k \cos x - 3 \sin x = k + 1$ is solvable only if $k \in (-\infty, 4]$.

Sol. Here,

$k \cos x - 3 \sin x = k + 1$, could be re-written as;

$$\frac{k}{\sqrt{k^2 + 9}} \cos x - \frac{3}{\sqrt{k^2 + 9}} \sin x = \frac{k + 1}{\sqrt{k^2 + 9}}$$

$$\text{or} \quad \cos(x + \phi) = \frac{k + 1}{\sqrt{k^2 + 9}}$$

which posses solution only if, $-1 \le \dfrac{k + 1}{\sqrt{k^2 + 9}} \le 1$

$$\text{i.e.} \quad \left|\frac{k + 1}{\sqrt{k^2 + 9}}\right| \le 1$$

$$\text{i.e.} \quad (k + 1)^2 \le k^2 + 9$$

$$\text{i.e.} \quad k^2 + 2k + 1 \le k^2 + 9$$

$$\text{or} \quad k \le 4$$

Thus, the interval in which, $k \cos x - 3 \sin x = k + 1$ admits solution for k is $(-\infty, 4]$.

Example 20. Let [.] denotes the greatest integer less than or equal to x and $f(x) = \sin x + \cos x$. Then, find the most general solution of $f(x) = \left[f\left(\dfrac{\pi}{10}\right)\right]$.

Sol. Here,

$f\left(\dfrac{\pi}{10}\right) = \sin 18° + \cos 18° = \sqrt{2} [\sin (45° + 18°)] = \sqrt{2} \sin 63°$.

As $\sin 63° > \sin 45° = \dfrac{1}{\sqrt{2}}$ and $\sin 63° < 1$

$\therefore \quad 1 < f\left(\frac{\pi}{10}\right) < \sqrt{2} \Rightarrow \left[f\left(\frac{\pi}{10}\right)\right] = 1$

So, the equation is sin x + cos x = 1.

$$\therefore \quad \cos\left(x - \frac{\pi}{4}\right) = \frac{1}{\sqrt{2}} \text{ or } x - \frac{\pi}{4} = 2n\pi \pm \frac{\pi}{4}$$

$$\Rightarrow \quad x = 2n\pi + \frac{\pi}{2}, 2n\pi$$

Example 21. Find the number of solutions of

$$\cos x = |1 + \sin x|, 0 \le x \le 3\pi$$

Sol. As we know, $1 + \sin x \ge 0$, for all x

So, $\cos x = 1 + \sin x$, for all $x \Rightarrow \cos x - \sin x = 1$

On dividing both sides by $\sqrt{a^2 + b^2}$ i.e. by $\sqrt{2}$, we get

$$\frac{1}{\sqrt{2}} \cos x - \frac{1}{\sqrt{2}} \sin x = \frac{1}{\sqrt{2}}$$

$$\Rightarrow \quad \cos\frac{\pi}{4} \cos x - \sin\frac{\pi}{4} \sin x = \cos\frac{\pi}{4}$$

$$\Rightarrow \quad \cos\left(x + \frac{\pi}{4}\right) = \cos\frac{\pi}{4} \Rightarrow x + \frac{\pi}{4} = 2n\pi \pm \frac{\pi}{4}$$

$$\Rightarrow \quad x = 2n\pi, 2n\pi - \frac{\pi}{2}, \quad \text{where } 0 \le x \le 3\pi$$

$$\Rightarrow \quad x = 0, \frac{3\pi}{2}, 2\pi \text{ are the only solution.}$$

Thus, number of solutions are 3.

Some Particular Equations

Equation of the Form

$$a_0 \sin^n x + a_1 \sin^{n-1} x \cos x + a_2 \sin^{n-2} x \cos^2 x + \ldots + a_n \cos^n x = 0,$$

where $a_0, a_1, \ldots, a_n$ are real number and the sum of the exponents of sin x and cos x in each term is equal to n, are said to be homogeneous with respect to sin x and cos x. For $\cos x \ne 0$, above equation can be written as,

$$a_0 \tan^n x + a_1 \tan^{n-1} x + \ldots + a_n = 0$$

Example 22. Solve $3\cos^2\theta - 2\sqrt{3} \sin\theta\cos\theta - 3\sin^2\theta = 0$.

Sol. The given equation can be written as:

$3\tan^2\theta + 2\sqrt{3}\tan\theta - 3 = 0$

$$\Rightarrow \quad \tan\theta = \frac{-2\sqrt{3} \pm \sqrt{12+36}}{6} = \frac{1}{\sqrt{3}}, -\sqrt{3}$$

Either, $\tan\theta = \frac{1}{\sqrt{3}} = \tan\frac{\pi}{6}$

$$\Rightarrow \quad \theta = n\pi + \frac{\pi}{6} \quad \ldots\text{(i)}$$

or $\tan\theta = -\sqrt{3} = \tan\left(\frac{-\pi}{3}\right)$

$$\Rightarrow \quad \theta = n\pi - \frac{\pi}{3} \quad \ldots\text{(ii)}$$

$$\therefore \quad \theta = n\pi + \frac{\pi}{6} \text{ or } \theta = n\pi - \frac{\pi}{3}$$

Example 23. Solve the equation $5\sin^2 x - 7\sin x \cos x + 16\cos^2 x = 4$.

Sol. To solve this equation, we use the fundamental trigonometric identity, $\sin^2 x + \cos^2 x = 1$

Given equation can be written as

$4(\sin^2 x + \cos^2 x) + \sin^2 x - 7\sin x \cos x + 12\cos^2 x = 4$

$\Rightarrow \quad \sin^2 x - 7 \sin x \cos x + 12 \cos^2 x = 0$

On dividing by $\cos^2 x$ both sides, we get

$$\tan^2 x - 7 \tan x + 12 = 0$$

Now, it can be factorised as;

$$(\tan x - 3)(\tan x - 4) = 0$$

$\Rightarrow \quad \tan x = 3, 4$

i.e. $\quad \tan x = \tan(\tan^{-1} 3)$

or $\quad \tan x = \tan(\tan^{-1} 4)$

$\Rightarrow \quad x = n\pi + \tan^{-1} 3$

or $\quad x = n\pi + \tan^{-1} 4$

Equation of the Form

$R(\sin kx, \cos nx, \tan mx, \cot lx) = 0$,

where R is a rational function of the indicated arguments and k, l, m, n are natural numbers, can be reduced to a rational equation with respect to the arguments sin x, cos x, tan x and cot x by means of the formulae for trigonometric functions of the sum of angles (in particular, the formulas for double and triple angle) and then reduce the obtained equation to a rational equation with respect to the unknown, $t = \tan\frac{x}{2}$, by means of following formulae;

$$\sin x = \frac{2\tan\frac{x}{2}}{1 + \tan^2\frac{x}{2}}, \cos x = \frac{1 - \tan^2\frac{x}{2}}{1 + \tan^2\frac{x}{2}},$$

$$\tan x = \frac{2\tan\frac{x}{2}}{1 - \tan^2\frac{x}{2}}, \cot x = \frac{1 - \tan^2\frac{x}{2}}{2\tan\frac{x}{2}}$$

Example 24. Solve the equation

$$(\cos x-\sin x)\left(2\tan x+\frac{1}{\cos x}\right)+2=0$$

Sol. Using above formulae, we get

$$\left\{\frac{1-\tan^2\frac{x}{2}}{1+\tan^2\frac{x}{2}}-\frac{2\tan\frac{x}{2}}{1+\tan^2\frac{x}{2}}\right\}\left\{\frac{4\tan\frac{x}{2}}{1-\tan^2\frac{x}{2}}+\frac{1+\tan^2\frac{x}{2}}{1-\tan^2\frac{x}{2}}\right\}+2=0$$

$$\left(\frac{1-t^2}{1+t^2}-\frac{2t}{1+t^2}\right)\left(\frac{4t}{1-t^2}+\frac{1+t^2}{1-t^2}\right)+2=0$$

$$\left(\text{Taking } \tan\frac{x}{2}=t\right)$$

$$\Rightarrow \quad \frac{3t^4+6t^3+8t^2-2t-3}{(t^2+1)(1-t^2)}=0$$

Its roots are $t_1=\frac{1}{\sqrt{3}}$ and $t_2=-\frac{1}{\sqrt{3}}$.

Thus, the solution of the equation reduces to that of two elementary euqations,

$$\tan\frac{x}{2}=\frac{1}{\sqrt{3}} \text{ and } \tan\frac{x}{2}=-\frac{1}{\sqrt{3}} \Rightarrow \frac{x}{2}=n\pi\pm\frac{\pi}{6}$$

$$\Rightarrow \quad x=2n\pi\pm\frac{\pi}{3}, \text{ is required solution.}$$

Example 25. Solve the equation

$$\sin^{10}x+\cos^{10}x=\frac{29}{16}\cos^4 2x$$

Sol. Using half-angle formulae, we can represent the given equation in the form;

$$\left(\frac{1-\cos 2x}{2}\right)^5+\left(\frac{1+\cos 2x}{2}\right)^5=\frac{29}{16}\cos^4 2x$$

Put $\cos 2x=t$,

$$\left(\frac{1-t}{2}\right)^5+\left(\frac{1+t}{2}\right)^5=\frac{29}{16}t^4$$

$$\Rightarrow \quad 24t^4-10t^2-1=0$$

whose only real root is, $t^2=\frac{1}{2}$.

$$\therefore \quad \cos^2 2x=\frac{1}{2}\Rightarrow 1+\cos 4x=1$$

$$\Rightarrow \quad \cos 4x=0\Rightarrow 4x=(2n+1)\frac{\pi}{2}$$

$$\Rightarrow \quad x=\frac{n\pi}{4}+\frac{\pi}{8};\ n\in I$$

Note

Some trigonometric equations can sometimes be simplified by lowering their degrees. If the exponent of the sines and cosines occurring into an equation are even, the lowering of the degree can be done by half angle formulas as in above example.

Equation of the Form

$R(\sin x+\cos x, \sin x\cdot\cos x)=0$,

where R is a rational function of the arguments in brackets. To solve such equations, put $\sin x+\cos x=t$ ...(i) and use the following identity

$$(\sin x+\cos x)^2=\sin^2 x+\cos^2 x+2\sin x\cos x$$

$$=1+2\sin x\cos x$$

$$\Rightarrow \quad \sin x\cos x=\frac{t^2-1}{2} \quad \text{...(ii)}$$

Taking (i) and (ii) into account, we can reduce given equation into;

$$R\left(t,\frac{t^2-1}{2}\right)=0$$

Similarly, by the substitution $(\sin x-\cos x)=t$, we can reduce the equation of the form;

$$R(\sin x-\cos x, \sin x\cos x)=0$$

to an equation; $\quad R\left(t,\frac{1-t^2}{2}\right)=0$

Example 26. Solve the equation

$$\sin x+\cos x-2\sqrt{2}\sin x\cos x=0$$

Sol. Let $(\sin x+\cos x)=t$ and using the equation

$$\sin x\cdot\cos x=\frac{t^2-1}{2}, \text{ we get}$$

$$t-2\sqrt{2}\left(\frac{t^2-1}{2}\right)=0$$

$$\Rightarrow \quad \sqrt{2}t^2-t-\sqrt{2}=0$$

The numbers $t_1=\sqrt{2}, t_2=-\frac{1}{\sqrt{2}}$ are roots of this quadratic equation.

Thus, the solution of the given equation reduces to the solution of two trigonometric equations

$$\sin x+\cos x=\sqrt{2} \quad \text{and} \quad \sin x+\cos x=-\frac{1}{\sqrt{2}}$$

$$\text{or} \quad \frac{1}{\sqrt{2}}\sin x+\frac{1}{\sqrt{2}}\cos x=1 \quad \text{and} \quad \frac{1}{\sqrt{2}}\sin x+\frac{1}{\sqrt{2}}\cos x=-\frac{1}{2}$$

$$\text{or} \quad \sin x\cdot\cos\frac{\pi}{4}+\sin\frac{\pi}{4}\cos x=1 \quad \text{and} \quad \sin x\cos\frac{\pi}{4}+\sin\frac{\pi}{4}\cos x=-\frac{1}{2}$$

$$\Rightarrow \quad \sin\left(x+\frac{\pi}{4}\right)=1 \quad \text{and} \quad \sin\left(x+\frac{\pi}{4}\right)=-\frac{1}{2}$$

$$\Rightarrow \quad x+\frac{\pi}{4}=(4n+1)\frac{\pi}{2} \quad \text{and} \quad x+\frac{\pi}{4}=n\pi+(-1)^n\cdot\left(-\frac{\pi}{6}\right)$$

$$\Rightarrow \quad x=2n\pi+\frac{\pi}{4} \quad \text{and} \quad x=n\pi+(-1)^{n+1}\frac{\pi}{6}-\left(\frac{\pi}{4}\right)$$

Exercise for Session 2

1. Solve the equation $\sin x + \cos x = 1$.
2. Solve $\sqrt{3}\cos\theta - 3\sin\theta = 4\sin 2\theta \cos 3\theta$.
3. Solve $\cot\theta + \operatorname{cosec}\theta = \sqrt{3}$.
4. Solve $\sqrt{2}\sec\theta + \tan\theta = 1$.
5. Find the general solution of $(\sqrt{3}-1)\sin\theta + (\sqrt{3}+1)\cos\theta = 2$.
6. Find the number of integral values of k for which equation $7\cos x + 5\sin x = 2k + 1$ has at least one solution.

   [Hint : $a\cos\theta + b\sin\theta = c$ has solution only when $|c| \le \sqrt{a^2+b^2}$].
7. Solve $2\sin^2 x - 5\sin x \cos x - 8\cos^2 x = -2$.
8. Solve the equation $(1-\tan\theta)(1+\sin 2\theta) = 1+\tan\theta$.

Session 3

Solution of Simultaneous Trigonometric Equations and Problems Based on Extreme Values of sin *x* and cos *x*

Solution of Simultaneous Trigonometric Equations

Here, we discuss problems related to the solution of two equations satisfied simultaneously.

We may divide the problem into two categories :

(i) Two equations in one 'unknown' satisfied simultaneously.

(ii) Two equations in two 'unknowns' satisfied simultaneously.

Example 27. Find the most general values of θ which satisfies the equations $\sin\theta = -\frac{1}{2}$ and $\tan\theta = \frac{1}{\sqrt{3}}$.

Sol. First, find the values of θ lying between 0 and 2π and satisfying the two given equations separately. Select the value of θ which satisfies both the equations, then generalise it.

$$\sin\theta = -\frac{1}{2} \Rightarrow \theta = \frac{7\pi}{6} \text{ or } \frac{11\pi}{6}$$

$$\tan\theta = \frac{1}{\sqrt{3}} \Rightarrow \theta = \frac{\pi}{6} \text{ or } \frac{7\pi}{6} \text{ [value between 0 and } 2\pi]$$

Common value of $\theta = \frac{7\pi}{6}$

The required solution is, $\theta = 2n\pi + \frac{7\pi}{6}$

Example 28. If $\tan(A-B) = 1$ and $\sec(A+B) = \frac{2}{\sqrt{3}}$, then find the smallest positive values of A and B and their most general values.

Sol. For the smallest positive values, find $A + B$ and $A - B$ between 0 and 2π from the given equations.

Since, A and B are positive angles, $A + B > A - B$. Solve the two to get A and B.

For the most general values, find the general values of $A - B$ and $A + B$ by solving the given equations separately. Solve two to get A and B

$$\tan(A-B) = 1 \Rightarrow A - B = \frac{\pi}{4} \text{ or } \frac{5\pi}{4} \quad \text{...(i)}$$

Also, $\sec(A+B) = \frac{2}{\sqrt{3}} \Rightarrow A + B = \frac{\pi}{6}$ or $\frac{11\pi}{6}$

Since, $A + B > A - B$,

$\therefore$ $$A + B = \frac{11\pi}{6} \quad \text{...(ii)}$$

On solving Eqs. (i) and (ii), we get

$$A = \frac{25\pi}{24}, B = \frac{19\pi}{24}$$

or $$A = \frac{37\pi}{24}, B = \frac{7\pi}{24}$$

Solve for the most general values

$$\tan(A - B) = 1 \Rightarrow A - B = n\pi + \frac{\pi}{4} \quad \text{...(iii)}$$

$$\sec(A+B) = \frac{2}{\sqrt{3}} \Rightarrow \cos(A+B) = \frac{\sqrt{3}}{2}$$

$$\Rightarrow \quad A + B = 2m\pi \pm \frac{\pi}{6} \quad \text{...(iv)}$$

On solving Eqs. (iii) and (iv), we get,

$$A = \frac{1}{2}\left[(2m+n)\pi + \frac{\pi}{4} \pm \frac{\pi}{6}\right]$$

$$B = \frac{1}{2}\left[(2m-n)\pi - \frac{\pi}{4} \pm \frac{\pi}{6}\right] \text{ where } m, n \in I$$

Example 29. Solve the system of equations

$$x + y = \frac{2\pi}{3} \text{ and } \frac{\sin x}{\sin y} = 2$$

Sol. Let us reduce the second equation of the system to the form,

$$\sin x = 2\sin y \quad \text{...(i)}$$

Using $x + y = \frac{2\pi}{3}$ we get, $\sin x = 2\sin\left(\frac{2\pi}{3} - x\right)$

$$\Rightarrow \quad \sin x = 2\left(\sin\frac{2\pi}{3}\cdot\cos x - \cos\frac{2\pi}{3}\cdot\sin x\right)$$

$$= 2\left(\frac{\sqrt{3}}{2}\cdot\cos x + \frac{1}{2}\cdot\sin x\right)$$

$$\Rightarrow \quad \sin x = \sqrt{3}\cos x + \sin x$$

$$\Rightarrow \quad \cos x = 0$$

$$\Rightarrow \quad x = (2n+1)\frac{\pi}{2}$$

$$\Rightarrow \quad x = \frac{\pi}{2} + n\pi$$

Substituting in $x + y = \frac{2\pi}{3}$, we get

$$y = -n\pi + \frac{\pi}{6}$$

$$\therefore \quad x = \frac{\pi}{2} + n\pi, y = \frac{\pi}{6} - n\pi, \text{ where } n \in I.$$

Example 30. If $r > 0$, $-\pi \le \theta \le \pi$ and r, θ satisfy $r\sin\theta = 3$ and $r = 4(1+\sin\theta)$, then find the possible solutions of the pair (r, θ).

Sol. Here, $r = 4(1+\sin\theta)$ and $r\sin\theta = 3$

On eliminating θ from above equations; $r = 4\left(1 + \frac{3}{r}\right)$

$$\Rightarrow \quad r^2 - 4r - 12 = 0$$

$$\Rightarrow \quad (r-6)(r+2) = 0$$

$$\Rightarrow \quad r = 6 \text{ or } r = -2$$

$$\therefore \quad r\sin\theta = 3 \Rightarrow \sin\theta = \frac{1}{2} \text{ or } \sin\theta = -\frac{3}{2}$$

Neglecting $\sin\theta = -\frac{3}{2}$, we get $\theta = \frac{\pi}{6}, \frac{5\pi}{6}$

$$\therefore \quad (r, \theta) = \left(6, \frac{\pi}{6}\right) \text{ and } \left(6, \frac{5\pi}{6}\right) \text{ are the required pairs.}$$

Problems Based on Extreme Values of sin *x* and cos *x*

Example 31. Solve $2\cos^2\frac{x}{2}\cdot\sin^2 x = x^2 + \frac{1}{x^2}$, $0 < x \le \frac{\pi}{2}$.

Sol. In this problem, terms on the two sides of the equation are different in nature.

LHS is in trigonometric form, whereas RHS is in algebraic form. Hence, we will used inequality method.

Here, LHS $= 2\cos^2\frac{x}{2}\cdot\sin^2 x$

$$= (1+\cos x)\sin^2 x < 2$$

$$[\because 1 + \cos x < 2, \sin^2 x \le 1]$$

and $\text{RHS} = x^2 + \frac{1}{x^2} \ge 2$ $\quad$ [$\because$ A.M. $\ge$ GM]

Hence, LHS is never equal to RHS

$\therefore$ The given equation has no solution.

Example 32. Solve $\sin^6 x = 1 + \cos^4 3x$.

Sol. LHS $= \sin^6 x \le 1$

RHS $= 1 + \cos^4 3x \ge 1$

Hence, $\sin^6 x = 1 + \cos^4 3x$ is possible only when.

LHS = RHS = 1

$$\Rightarrow \quad \sin^6 x = 1 \quad \text{and} \quad 1 + \cos^4 3x = 1$$

$$\Rightarrow \quad \sin^2 x = 1 \quad \text{and} \quad \cos^4 3x = 0$$

$$\Rightarrow \quad \cos^2 x = 0 \quad \text{and} \quad \cos 3x = 0$$

$$\Rightarrow \quad \cos x = 0 \quad \text{and} \quad \cos 3x = 0$$

$\Rightarrow \quad x = (2m+1)\frac{\pi}{2}$

and $\quad 3x = (2n+1)\frac{\pi}{2}$, where $m, n \in I$

$\Rightarrow \quad x = (2m+1)\frac{\pi}{2}$ and $x = (2n+1)\frac{\pi}{6}$,

where $m, n \in I$

$\Rightarrow \quad x = (2m+1)\frac{\pi}{2}$

and $\quad x = (2n+1)\frac{\pi}{6}$

Common values of x is $(2n+1)\frac{\pi}{2}$, where $n \in I$. The required solution,

$$x = (2n+1)\frac{\pi}{2}, n \in I$$

Example 33. Solve $\sin^4 x = 1 + \tan^8 x$.

Sol. LHS $= \sin^4 x \le 1$

RHS $= 1 + \tan^8 x \ge 1$

$\Rightarrow$ LHS = RHS only when

$\sin^4 x = 1$ and $1 + \tan^8 x = 1$

$\Rightarrow \quad \sin^2 x = 1$ and $\tan^8 x = 0$

which in never possible, since $\sin x$ and $\tan x$ vanish simultaneously.

Therefore, the given equation has no solution.

Example 34. Solve $\sin^2 x + \cos^2 y = 2\sec^2 z$.

Sol. LHS $= \sin^2 x + \cos^2 y \le 2$

RHS $= 2\sec^2 z \ge 2$

Hence LHS = RHS only when,

$\sin^2 x = 1, \cos^2 y = 1$ and $\sec^2 z = 1$

$\Rightarrow \quad \cos^2 x = 0, \sin^2 y = 0$ and $\cos^2 z = 1$

$\Rightarrow \quad \cos x = 0, \sin y = 0, \sin z = 0$

$\Rightarrow x = (2m+1)\frac{\pi}{2}, y = n\pi, z = t\pi$,

where m, n, t are integer.

Example 35. Solve the equation $(\sin x + \cos x)^{1+\sin 2x} = 2$, when $-\pi \le x \le \pi$.

Sol. We know, $-\sqrt{a^2+b^2} \le a\sin\theta + b\cos\theta \le \sqrt{a^2+b^2}$ and $-1 \le \sin\theta \le 1$.

$\therefore (\sin x + \cos x)$ admits the maximum value as $\sqrt{2}$ and minimum value as $-\sqrt{2}$,

$(1 + \sin 2x)$ admits the maximum value as 2. Also, $(\pm\sqrt{2})^2 = 2$.

$\therefore$ The equation could hold only when,

$\sin x + \cos x = \pm\sqrt{2}$ and $1 + \sin 2x = 2$

Now, $\sin x + \cos x = \sqrt{2}$

$\Rightarrow \quad \cos\left(x - \frac{\pi}{4}\right) = 1$

$\Rightarrow \quad x = 2n\pi + \frac{\pi}{4}$ ...(i)

and $\quad \sin x + \cos x = -\sqrt{2}$

$\cos\left(x - \frac{\pi}{4}\right) = -1$

$x = (2n \pm 1)\pi + \frac{\pi}{4}$ ...(ii)

and $\quad 1 + \sin 2x = 2$

$\Rightarrow \quad \sin 2x = 1 = \sin\frac{\pi}{2}$

$\Rightarrow \quad 2x = n\pi + (-1)^n \frac{\pi}{2}$

$\Rightarrow \quad x = \frac{n\pi}{2} + (-1)^n \frac{\pi}{4}$ ...(iii)

The value in $[-\pi, \pi]$ satisfying Eqs. (i), (ii) and (iii) is $x = \frac{\pi}{4}, \frac{-3\pi}{4}$ (when $n = 0, -1$).

Example 36. Find the most general solutions for $2^{\sin x} + 2^{\cos x} = 2^{1-1/\sqrt{2}}$.

Sol. As we know, AM $\ge$ GM

$\therefore \quad \frac{2^{\sin x} + 2^{\cos x}}{2} \ge \sqrt{2^{\sin x} \cdot 2^{\cos x}}$ ...(i)

Now, Eq. (i) admits minimum value when

$\sin x + \cos x$ is $(-\sqrt{2})$

{using $-\sqrt{a^2+b^2} \le a\cos x + b\sin x \le \sqrt{a^2+b^2}$}

$\therefore \quad 2^{\sin x} + 2^{\cos x} \ge 2 \cdot \sqrt{2^{-\sqrt{2}}}$

or $\quad 2^{\sin x} + 2^{\cos x} \ge 2 \cdot 2^{-\frac{\sqrt{2}}{2}}$

or $\quad 2^{\sin x} + 2^{\cos x} \ge 2^{1-\frac{1}{\sqrt{2}}}$

Thus, the equation holds only when,

$\sin x + \cos x = -\sqrt{2}$

$\Rightarrow \quad \cos\left(x - \frac{\pi}{4}\right) = -1$

$\Rightarrow \quad x - \frac{\pi}{4} = 2n\pi \pm \pi$

$\Rightarrow \quad x = (2n \pm 1)\pi + \frac{\pi}{4}$

Example 37. Solve $\left|\sqrt{3}\cos x - \sin x\right| \ge 2$ for $x \in [0, 4\pi]$.

Sol. We know, $\left|\sqrt{3}\cos x - \sin x\right| \le \sqrt{3+1} = 2$ ...(i)

and $\quad \left|\sqrt{3}\cos x - \sin x\right| \ge 2$ (given) ...(ii)

Thus, from Eqs. (i) and (ii), we must have

$$\left|\sqrt{3}\cos x - \sin x\right| = 2$$

$$\Rightarrow \quad \left|\frac{\sqrt{3}}{2}\cos x - \frac{1}{2}\sin x\right| = 1$$

$$\Rightarrow \quad \left|\cos\frac{\pi}{6}\cos x - \sin\frac{\pi}{6}\cdot\sin x\right| = 1$$

$$\Rightarrow \quad \left|\cos\left(x+\frac{\pi}{6}\right)\right| = 1$$

$$\Rightarrow \quad \cos\left(x+\frac{\pi}{6}\right) = 1 \text{ or } \cos\left(x+\frac{\pi}{6}\right) = -1$$

$$\Rightarrow \quad x+\frac{\pi}{6} = 0, 2\pi, 4\pi, \ldots \text{ or } \pi, 3\pi, 5\pi, \ldots$$

$$\Rightarrow \quad x = \frac{11\pi}{6}, \frac{23\pi}{6}, \frac{5\pi}{6}, \frac{17\pi}{6}, \text{ for } x \in [0, 4\pi]$$

Hence, $x \in \left\{\frac{11\pi}{6}, \frac{23\pi}{6}, \frac{5\pi}{6}, \frac{17\pi}{6}\right\}$

Example 38. Show that the equation, $\sin x = [1+\sin x] + [1-\cos x]$ has no solution for $x \in R$. (where [.] represents greatest integers function).

Sol As we know that the period of $\sin x$ is 2π, we need to check the solution for $x \in [0, 2\pi]$.

Let us first check at those points on which $\sin x$ and $\cos x$ are integer and then for the values lying between them.

Case I (a) At $x = 0$

$[1+\sin x] = 1$ and $[1-\cos x] = 0$

$\therefore \quad \sin x = 1+0 = 1$ or $\sin 0 = 1$ (as $x = 0$)

$\Rightarrow 0 = 1$ (absurd)

(b) At $x = \frac{\pi}{2}$;

$[1+\sin x] = 2$ and $[1-\cos x] = 1$

$\therefore \quad \sin x = 2+1 = 3$ (absurd)

(c) At $x = \pi$

$[1+\sin x] = 1$ and $[1-\cos x] = 2$

$\therefore \quad \sin x = 1+2 = 3$ (absurd)

(d) At $x = \frac{3\pi}{2}$,

$[1+\sin x] = 0$ and $[1-\cos x] = 1$

$\therefore \quad \sin x = 0+1$ or $\sin\frac{3\pi}{2} = 1 \left(\text{as } x = \frac{3\pi}{2}\right)$

$\Rightarrow \quad -1 = 1$ (absurd)

(e) At $x = 2\pi$;

$[1+\sin x] = 1$ and $[1-\cos x] = 0$

$\therefore \quad \sin x = 1+0$

or $\quad \sin 0 = 1$ (as $x = 0$)

$\Rightarrow \quad 0 = 1$ (absurd)

$\therefore \quad$ At $x = \left\{0, \frac{\pi}{2}, \pi, \frac{3\pi}{2}, 2\pi\right\}$

we do not have any solution.

Now, to check for the values lying between them.

Case II (a) When $x \in \left(0, \frac{\pi}{2}\right)$

$[1+\sin x] = 1, [1-\cos x] = 0$

$\Rightarrow \quad \sin x = 1+0$

But since $x \in \left(0, \frac{\pi}{2}\right)$, $\sin x \neq 1$

$\therefore \quad \sin x = 1$ is absurd when $x \in \left(0, \frac{\pi}{2}\right)$

(b) When $x \in \left(\frac{\pi}{2}, \pi\right)$

$[1+\sin x] = 1, [1-\cos x] = 1$

$\Rightarrow \quad \sin x = 1+1 = 2$ (absurd)

(c) When $x \in \left(\pi, \frac{3\pi}{2}\right)$

$[1+\sin x] = 0, [1-\cos x] = 1 \Rightarrow$
$\sin x = 0+1 = 1$

But $x \in \left(\pi, \frac{3\pi}{2}\right)$ in which $\sin x \neq 1$

$\therefore \sin x = 1$ is absurd, when $x \in \left(\pi, \frac{3\pi}{2}\right)$

(d) When $x \in \left(\frac{3\pi}{2}, 2\pi\right)$

$[1+\sin x] = 0, [1-\cos x] = 0$

$\therefore \quad \sin x = 0+0 = 0$

But $\sin x \neq 0$ when $x \in \left(\frac{3\pi}{2}, 2\pi\right)$

Thus, the given equation does not posses any solution for $x \in [0, 2\pi]$ or in general, $\sin x = [1+\sin x] + [1-\cos x]$ does not posses any solution for $x \in R$.

Exercise for Session 3

1. Find the general values of θ which satisfies the equations $\tan\theta = -1$ and $\cos\theta = \frac{1}{\sqrt{2}}$.
2. Find the most general solution of $\text{cosec}\, x = -2$ and $\cot x = \sqrt{3}$.
3. Find the common roots of the equations $2\sin^2 x + \sin^2 2x = 2$ and $\sin 2x + \cos 2x = \tan x$.
4. Solve the equations, $\sqrt{3}\sin 2A = \sin 2B$ and $\sqrt{3}\sin^2 A + \sin^2 B = \frac{1}{2}(\sqrt{3}-1)$.
5. Find the number of solutions of $\sin^2 x \cos^2 x = 1 + \cos^2 x \sin^4 x$ in the interval $[0, \pi]$.
6. Solve: $1 + \sin x \sin^2 \frac{x}{2} = 0$.
7. Solve: $\cos^{50} x - \sin^{50} x = 1$.
8. Find the number of real solutions of the equation $(\cos x)^5 + (\sin x)^3 = 1$ in the interval $[0, 2\pi]$.
9. Find the number of solutions of the equation
$$1 + e^{\cot^2 x} = \sqrt{2|\sin x| - 1} + \frac{1 - \cos 2x}{1 + \sin^4 x} \text{ for } x \in (0, 5\pi).$$
10. Find the number of solutions as ordered pair (x, y) of the equation $2^{\sec^2 x} + 2^{\text{cosec}^2 y} = 2\cos^2 x(1 - \cos^2 2y)$ in $[0, 2\pi]$.

Session 4

Trigonometric Inequality

Trigonometric Inequality

An inequality involving trigonometric function of an unknown angle is called a trigonometric inequality.

Solution of Trigonometric Inequality

To solve the trigonometric inequation of type $f(x) \le a$, or $f(x) \ge a$ where $f(x)$ is some trigonometric ratio, the following steps should be taken:

1. Draw the graph of $f(x)$ is an interval length equal to the fundamental period of $f(x)$.
2. Draw the line $y = a$.
3. Take the portion of the graph for which the inequalities satisfied.
4. To generalize, add $nT (n \in I)$ and take union over the set of integers, where T, is the fundamental period of $f(x)$.

Example 39. Find the solution set of inequality $\sin x > \frac{1}{2}$.

Sol. When $\sin x = \frac{1}{2}$, the two values of x between 0 and 2π are $\frac{\pi}{6}$ and $\frac{5\pi}{6}$. From the graph of $y = \sin x$, it is obvious that

$$\sin x > \frac{1}{2} \text{ for } \frac{\pi}{6} < x < \frac{5\pi}{6}$$

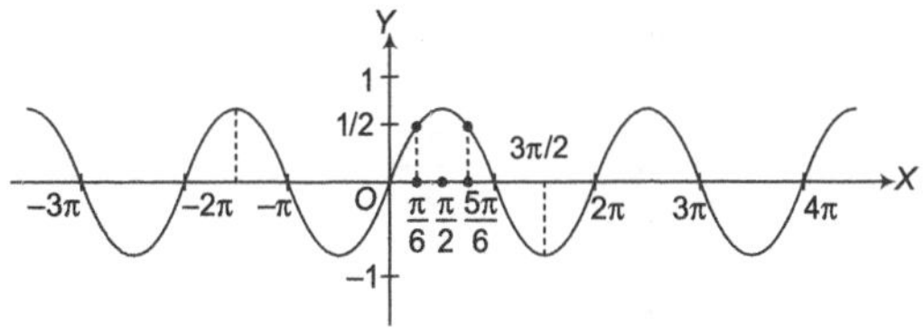

Hence, $\quad \sin x > \frac{1}{2}$

$$\Rightarrow \quad 2n\pi + \frac{\pi}{6} < x < 2n\pi + \frac{5\pi}{6}$$

The required solution set is $\bigcup_{n \in I} \left(2n\pi + \frac{\pi}{6}, 2n\pi + \frac{5\pi}{6}\right)$

Note

We added $2n\pi$ to find the general solution as period of $\sin x$ is 2π.

Example 40. Find the solution set of inequality $\cos x \geq -\frac{1}{2}$.

Sol. From the graph of $y = \cos x$, it is obvious that $\cos x \geq -\frac{1}{2}$.

For $\quad -\frac{2\pi}{3} \leq x \leq \frac{2\pi}{3}$

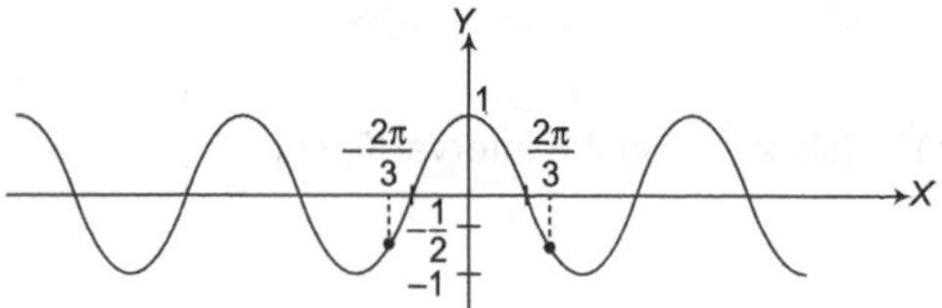

Hence, $\quad \cos x \geq -\frac{1}{2}$

$$\Rightarrow \quad 2n\pi - \frac{2\pi}{3} \leq x \leq 2n\pi + \frac{2\pi}{3}$$

The required solution set is

$$\bigcup_{n \in I} \left(2n\pi - \frac{2\pi}{3}, 2n\pi + \frac{2\pi}{3}\right)$$

Example 41. Find the solution set for, $4\sin^2 x - 8\sin x + 3 \leq 0$ where $x \in [0, 2\pi]$.

Sol. Here, $\quad 4\sin^2 x - 8\sin x + 3 \leq 0$

$\Rightarrow \quad (2\sin x - 1)(2\sin x - 3) \leq 0$

Here, $2\sin x - 3$ is always negative.

$\therefore \quad 2\sin x - 1 \geq 0$ i.e. $\sin x \geq \frac{1}{2}$

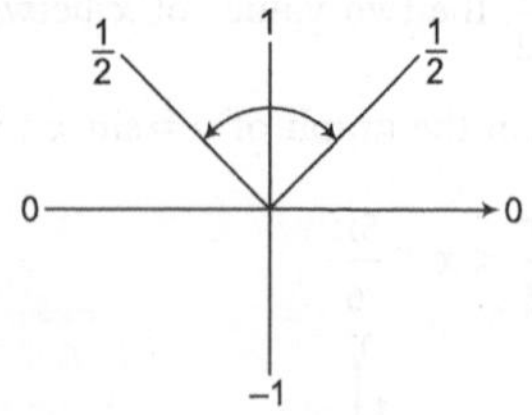

$\therefore$ From the figure, $\frac{\pi}{6} \leq x \leq \frac{5\pi}{6}$

Example 42. Solve $2\cos^2\theta + \sin\theta \leq 2$, where $\pi/2 \leq \theta \leq \frac{3\pi}{2}$

Sol. $2\cos^2\theta + \sin\theta \leq 2$

or $\quad 2(1 - \sin^2\theta) + \sin\theta \leq 2$

or $\quad -2\sin^2\theta + \sin\theta \leq 0$

or $\quad 2\sin^2\theta - \sin\theta \geq 0$

or $\quad \sin\theta(2\sin\theta - 1) \geq 0$

or $\quad \sin\theta(\sin\theta - 1/2) \geq 0$

which is possible if $\sin\theta \leq 0$ or $\sin\theta \geq 1/2$

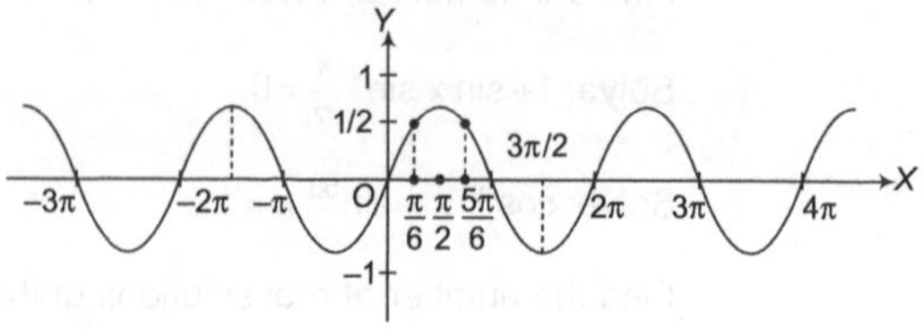

From the graph

$\sin\theta \geq 1/2 \Rightarrow \pi/2 \leq \theta \leq 5\pi/6$

$\sin\theta \leq 0 \Rightarrow \pi \leq \theta \leq 3\pi/2$

Hence, the required values of θ are given by

$$\theta \in [\pi/2, 5\pi/6] \cup [\pi, 3\pi/2]$$

Example 43. Solve $\sin^2\theta > \cos^2\theta$

Sol. We have, $\quad \sin^2\theta > \cos^2\theta \Rightarrow \cos 2\theta < 0$

$\Rightarrow \quad (\pi/2) < 2\theta < (3\pi/2)$ or $(\pi/4) < \theta < (3\pi/4)$

Taking general values i.e., adding $2n\pi$, we get

$$2n\pi + \pi/2 < 2\theta < 2n\pi + 3\pi/2, n \in Z$$

or $\quad n\pi + \pi/4 < \theta < n\pi + 3\pi/4$

Example 44. Find the solution set for, $|\tan x| \leq 1$ when $x \in [-\pi, \pi]$.

Sol. Here, $|\tan x| \leq 1$

$\Rightarrow -1 \leq \tan x \leq 1$, the value scheme for this is shown below.

From the figure,

$$-\frac{\pi}{4} \leq x \leq \frac{\pi}{4}$$

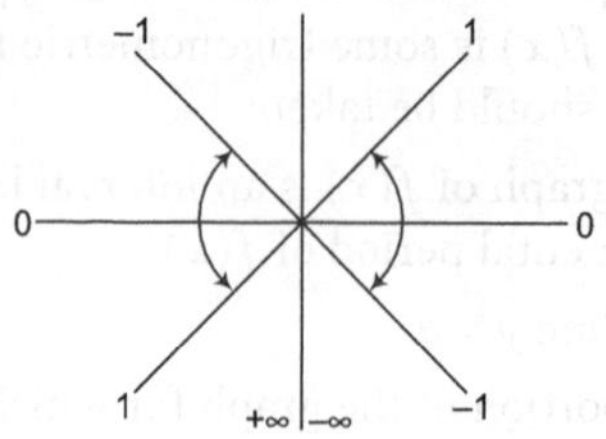

or $$-\pi \le x \le -\frac{3\pi}{4}$$

or $$\frac{3\pi}{4} \le x \le \pi$$

$\therefore$ $$x \in \left[-\pi, -\frac{3\pi}{4}\right] \cup \left[-\frac{\pi}{4}, \frac{\pi}{4}\right] \cup \left[\frac{3\pi}{4}, \pi\right]$$

Example 45. Solve $\sin 2x > \sqrt{2}\sin^2 x + (2-\sqrt{2})\cos^2 x$

Sol. $\sin 2x > \sqrt{2}\sin^2 x + (2-\sqrt{2})\cos^2 x$

$\Rightarrow$ $2\sin x \cos x > \sqrt{2}\sin^2 x + (2-\sqrt{2})\cos^2 x$

$\Rightarrow$ $\tan^2 x - \sqrt{2}\tan x + (\sqrt{2}-1) < 0$

$\Rightarrow$ $(\tan x - 1)(\tan x - (\sqrt{2}-1)) < 0$

$\Rightarrow$ $(\sqrt{2}-1) < \tan x < 1$

$\Rightarrow$ $\pi/8 < x < \pi/4$

$\therefore$ $x \in \left(n\pi + \frac{\pi}{8}, n\pi + \frac{\pi}{4}\right)$ where $n \in Z$.

Example 46. Solve $\tan^3 x + 3 > 3\tan x + \tan^2 x$

Sol. $\tan^3 x - \tan^2 x + 3 - 3\tan x > 0$

$\Rightarrow$ $\tan^2 x(\tan x - 1) - 3(\tan x - 1) > 0$

$\Rightarrow$ $(\tan x - 1)(\tan^2 x - 3) > 0$

$(\tan x - 1)(\tan x + \sqrt{3})(\tan x - \sqrt{3}) > 0$

$\Rightarrow$ $(y-1)(y+\sqrt{3})(y-\sqrt{3}) > 0$, where $\tan x = y$

Sign-scheme of above inequality is as follows:

(–) $-\sqrt{3}$ (+) 1 (–) $\sqrt{3}$ (+)

$\therefore$ $-\sqrt{3} < y < 1$ or $y > \sqrt{3}$

$\Rightarrow$ $-\sqrt{3} < \tan x < 1$ or $\tan x > \sqrt{3}$

For $-\pi/2 < x < \pi/2$

$$-\pi/3 < x < x/4 \text{ or } \pi/3 < x < \pi/2$$

$\therefore$ General solution is

$$x \in \left(n\pi + \frac{\pi}{3}, n\pi + \frac{\pi}{2}\right) \cup \left(n\pi - \frac{\pi}{3}, n\pi + \frac{\pi}{4}\right)$$

where $n \in Z$.

Exercise for Session 4

1. If $2\cos x < \sqrt{3}$ and $x \in [-\pi, \pi]$, then find the solution set for x.
2. Find the set of all x in the interval $[0, \pi]$ for which $2\sin^2 x - 3\sin x + 1 \ge 0$.
3. If $\cos x - \sin x \ge 1$ and $0 \le x \le 2\pi$, then find the solution set for x.
4. Solve $\sin\theta + \sqrt{3}\cos\theta \ge 1$, $-\pi < \theta \le \pi$.
5. Find the set of values of x, which satisfy $\sin x \cdot \cos^3 x > \cos x \cdot \sin^3 x$, $0 \le x \le 2\pi$.
6. Find the set of all x in $\left(\frac{-\pi}{2}, \frac{\pi}{2}\right)$ which satisfy $|4\sin x - 1| < \sqrt{5}$.
7. Solve $\sin^4\left(\frac{x}{3}\right) + \cos^4\left(\frac{x}{3}\right) > \frac{1}{2}$.
8. Solve $\tan x - \tan^2 x > 0$ and $|2\sin x| < 1$.

JEE Type Solved Examples : Single Option Correct Type Questions

Ex. 1. *The number of solutions of equation* $8[x^2 - x] + 4[x] = 13 + 12[\sin x]$, $[.]$ *denotes GIF is*

(a) 0 (b) 2
(c) 4 (d) 6

Sol. (a) $8[x^2 - x] + 4[x] = 13 + 12[\sin x]$

LHS is always even and RHS is always odd. Hence, no solution.

Ex. 2. *The number of ordered pairs* (x, y) *satisfying* $|x| + |y| = 2$ *and* $\sin\left(\frac{\pi x^2}{3}\right) = 1$ *is/are*

(a) 1 (b) 2
(c) 3 (d) 4

Sol. (d) $|x| + |y| = 2$

$$\Rightarrow \quad |x|, |y| \in [0, 2]$$

Also $$\sin\left(\frac{\pi x^2}{3}\right) = 1$$

$$\Rightarrow \quad \frac{\pi x^2}{3} = (4n+1)\frac{\pi}{2}$$

$$\Rightarrow \quad x^2 = (4n+1)\frac{3}{2}$$

$\because$ $|x| \in [0, 2]$, then only possible value of x^2 is $\frac{3}{2}$

$$\therefore \quad |x| = \sqrt{\frac{3}{2}}, |y| = 2 - \sqrt{\frac{3}{2}}$$

Hence, total number of ordered pairs is 4.

Ex. 3. *Number of solutions of*

$$\cos^2\left[\frac{\pi}{4}(\sin x + \sqrt{2}\cos^2 x)\right] - \tan^2\left[x + \frac{\pi}{4}\tan^2 x\right] = 1, x \in [-2\pi, 2\pi] \text{ is}$$

(a) 1 (b) 2
(c) 4 (d) 8

Sol. (b) The solution is only possible, when

$$\cos^2\left(\frac{\pi}{4}(\sin x + \sqrt{2}\cos^2 x)\right) = 1 \quad \text{...(i)}$$

and $$\tan^2\left(x + \frac{\pi}{4}\tan^2 x\right) = 0 \quad \text{...(ii)}$$

Let's solved Eq. (i)

$$\frac{\pi}{4}(\sin x + \sqrt{2}\cos^2 x) = k\pi$$

$$\Rightarrow \quad \sin x + \sqrt{2}\cos^2 x = 4k$$

Now as $|\sin x + \sqrt{2}\cos^2 x| \le |\sin x| + \sqrt{2}\,|\cos^2 x| \le 1 + \sqrt{2} < 4$

$\therefore$ $k = 0$ is only possible value.

$$\Rightarrow \quad \sin x + \sqrt{2}\cos^2 x = 0$$

$$\Rightarrow \quad \sqrt{2}\sin^2 x - \sin x - \sqrt{2} = 0$$

$$\Rightarrow \quad \sin x = \frac{-1}{\sqrt{2}}, \sqrt{2}$$

$$\Rightarrow \quad \sin x = -\frac{1}{\sqrt{2}} \quad (\because \sin x \ne \sqrt{2})$$

$$\Rightarrow \quad x = 2n\pi - \frac{\pi}{4} \text{ or } 2n\pi + \frac{5\pi}{4}$$

From Eq. (ii) we can say that the only solution possible is $x = 2n\pi - \frac{\pi}{4}$

Hence, for $x \in [-2\pi, 2\pi]$ we have 2 solutions.

Ex. 4. *The general solution of*

$$\sin^2\theta\sec\theta + \sqrt{3}\tan\theta = 0 \text{ is}$$

(a) $\theta = n\pi + (-1)^{n+1}\frac{\pi}{3}, \theta = n\pi; n \in I$

(b) $\theta = n\pi, n \in I$

(c) $\theta = \frac{n\pi}{2}, n \in I$

(d) $\theta = n\pi + (-1)^{n+1}\frac{\pi}{3}, n \in I$

Sol. (b) $\sin^2\theta\sec\theta + \sqrt{3}\tan\theta = 0$

$$\Rightarrow \quad \sin\theta(\sin\theta + \sqrt{3})\sec\theta = 0$$

$$\Rightarrow \quad \sin\theta = 0$$

$$\Rightarrow \quad \theta = n\pi, n \in I \quad (\because \sin\theta \ne -\sqrt{3}, \sec\theta \ne 0)$$

Ex. 5. *The number of solutions of the equation* $\sin\left(\frac{\pi x}{2\sqrt{3}}\right) = x^2 - 2\sqrt{3}x + 4$ *is*

(a) 0 (b) 2
(c) more than 2 (d) 1

Sol. (d) We know that $-1 \le \sin\frac{\pi x}{2\sqrt{3}} \le 1$,

therefore, we must have

$$-1 \le x^2 - 2\sqrt{3}x + 4 \le 1$$

$$\Rightarrow \quad -1 \le (x - \sqrt{3})^2 + 1 \le 1$$

$$\Rightarrow \quad -2 \le (x - \sqrt{3})^2 \le 0$$

But, square of a real number cannot be negative, therefore, we must have $(x-\sqrt{3})^2=0$

$\Rightarrow \qquad x=\sqrt{3}$

Note that $x=\sqrt{3}$ satisfies the given equation.

Ex. 6. *x_1 and x_2 are two solutions of the equation $e^x \cos x = 1$. The minimum number of the solutions of the equation $e^x \sin x = 1$, lying between x_1 and x_2 can be*

(a) 0 (b) 1

(c) 3 (d) None of these

Sol. (b) We have $e^x \cos x = 1$ or $\cos x = e^{-x}$

Consider the function, $f(x) = \cos x - e^{-x}$

$\Rightarrow \qquad f(x_1) = 0 = f(x_2).$

Clearly, $f(x)$ is continuous in $[x_1, x_2]$ and differentiable in (x_1, x_2).

Hence, by Rolle's theorem, there is atleast one $x \in (x_1, x_2)$ such that $f'(x) = -\sin x + e^{-x} = 0$

$\Rightarrow \qquad \sin xe^x = 1$ has atleast one solution $\in (x_1, x_2)$.

Ex. 7. *The product of common differences of all possible AP which are made from values of 'x' satisfying*

$$\cos^2\left(\frac{1}{2}\lambda x\right) + \cos^2\left(\frac{1}{2}\mu x\right) = 1$$

(a) $\dfrac{4\pi^2}{\lambda^2 - \mu^2}$ (b) $\dfrac{4\pi}{\lambda - \mu}$

(c) $\dfrac{2\pi^2}{\lambda^2 - \mu^2}$ (d) None of these

Sol. (a) $\dfrac{1+\cos(\lambda x)}{2} + \dfrac{1+\cos(\mu x)}{2} = 1$

$$\Rightarrow \qquad \cos(\lambda x) + \cos(\mu x) = 0$$

$$\Rightarrow \qquad 2\cos\left(\frac{(\lambda+\mu)x}{2}\right)\cdot\cos\left(\frac{(\lambda-\mu)x}{2}\right) = 0$$

$$\Rightarrow \qquad \frac{(\lambda+\mu)x}{2} = (2n+1)\frac{\pi}{2}$$

$$\text{or} \qquad \frac{(\lambda-\mu)x}{2} = (2n+1)\frac{\pi}{2}$$

$$x = \frac{(2n+1)\pi}{\lambda+\mu} \text{ or } x = \frac{(2n+1)\pi}{\lambda-\mu}$$

Thus, common difference can be

$$\frac{2\pi}{\lambda+\mu} \text{ or } \frac{2\pi}{\lambda-\mu}$$

Now, product $= \dfrac{4\pi^2}{\lambda^2-\mu^2}$

Ex. 8. *Number of solutions of the equation $\cos^4 2x + 2\sin^2 2x = 17(\cos x + \sin x)^8, 0 < x < 2\pi$ is*

(a) 4 (b) 8

(c) 10 (d) 16

Sol. (a) Let $\sin 2x = y$, then $1 + y^4 = 17(1+y)^4$

Clearly, $\qquad y = \sin 2x = -\dfrac{1}{2}$ is the only possibility

$\therefore \qquad x = 105°, 165°, 285°, 345°.$

Ex. 9. *The number of values of θ in the interval $\left(-\dfrac{\pi}{2}, \dfrac{\pi}{2}\right)$ satisfying the equation $(\sqrt{3})^{\sec^2\theta} = \tan^4\theta + 2\tan^2\theta$ is*

(a) 2 (b) 4

(c) 0 (d) 1

Sol. (a) $\tan^4\theta + 2\tan^2\theta = (\tan^2\theta + 1)^2 - 1$

$$= (\sec^2\theta)^2 - 1 = \sec^4\theta - 1$$

Put $\qquad \sec^2\theta = t$

$\Rightarrow \qquad (\sqrt{3})^t = t^2 - 1$

$\Rightarrow t = 2$ is only solution as $t > 1$

$\therefore$ There will be 2 values of θ in given interval.

Ex. 10. *Number of solutions of the equation $\cot(\theta) + \cot\left(\theta + \dfrac{\pi}{3}\right) + \cot\left(\theta - \dfrac{\pi}{3}\right) + \cot(3\theta) = 0$, where $\theta \in \left(0, \dfrac{\pi}{2}\right)$*

(a) Infinite (b) 0

(c) 1 (d) None of these

Sol. (c) $\cot(\theta) + \cot\left(\theta + \dfrac{\pi}{3}\right) + \cot\left(\theta - \dfrac{\pi}{3}\right) + \cot(3\theta) = 0$

Put $\theta = \dfrac{\pi}{2} - \alpha$

$$\tan\alpha + \tan\left(\alpha - \frac{\pi}{3}\right) + \tan\left(\alpha + \frac{\pi}{3}\right) + \tan 3\alpha = 0$$

$$\tan\alpha + \frac{\tan\alpha - \tan\frac{\pi}{3}}{1 + \tan\alpha\tan\frac{\pi}{3}} + \frac{\tan\alpha + \tan\frac{\pi}{3}}{1 - \tan\alpha\tan\frac{\pi}{3}} + \tan 3\alpha = 0$$

$$3\left(\frac{3\tan\alpha - \tan^3\alpha}{1 - 3\tan^2\alpha}\right) + \tan 3\alpha = 0$$

$$4\tan 3\alpha = 0$$

$$\Rightarrow \qquad \tan 3\alpha = 0$$

$$\Rightarrow \qquad 3\alpha = n\pi$$

$$\Rightarrow \qquad \alpha = \frac{\pi}{3}$$

JEE Type Solved Examples : More than One Correct Option Type Questions

Ex. 11. $0 < a < 2\pi$, $\sin^{-1}(\sin a) < x^2 - 2x$ *for all* $x \in I$ *then* $a \in$

(a) $(0, \pi + 1)$ (b) $\left(\pi + 1, \frac{3\pi}{2}\right)$

(c) $\left(\frac{3\pi}{2}, 2\pi - 1\right)$ (d) $(2\pi - 1, 2\pi)$

Sol. (b,c) $\sin^{-1}(\sin a) < x^2 - 2x$ for all x

$\therefore \quad \sin^{-1}(\sin a) < \min(x^2 - 2x) = -1$

For $a \in \left(0, \frac{\pi}{2}\right)$, $a < -1 \Rightarrow a \in \phi$

For $a \in \left(\frac{\pi}{2}, \frac{3\pi}{2}\right)$, $\sin^{-1}(\sin a) = \pi - a$

$$\pi - a < -1$$

$$a > \pi + 1 \Rightarrow a \in \left(\pi + 1, \frac{3\pi}{2}\right)$$

For $a \in \left(\frac{3\pi}{2}, 2\pi\right)$, $\sin^{-1}(\sin a) = a - 2\pi$

$$a - 2\pi < -1$$

$$a < 2\pi - 1$$

$$a < 2\pi - 1 \Rightarrow a \in \left(\frac{3\pi}{2}, 2\pi - 1\right).$$

Ex. 12. *If* $\left(\cos^2 x + \frac{1}{\cos^2 x}\right)(1 + \tan^2 2y)(3 + \sin 3z) = 4$, *then*

(a) x may be a multiple of π

(b) x cannot be an even multiple of π

(c) z can be a multiple of π

(d) y can be a multiple of $\frac{\pi}{2}$

Sol. (a,d) $\left(\cos^2 x + \frac{1}{\cos^2 x}\right)(1 + \tan^2 2y)(3 + \sin 3z) = 4$

Since, $\cos^2 x + \frac{1}{\cos^2 x} > 2, 1 + \tan^2 2y \geq 1, 2 \leq 3 + \sin 3z \leq 4$

so, the only possibility is

$$\cos^2 x + \frac{1}{\cos^2 x} = 2, 1 + \tan^2 2y = 1, 3 + \sin 3z = 2$$

$\Rightarrow \quad \cos x = \pm 1 \Rightarrow x = n\pi$

$\tan 2y = 0 \Rightarrow y = \frac{m\pi}{2}$

$\sin 3z = -1$

$\Rightarrow \quad z = (4k - 1)\frac{\pi}{6}; m, n, k \in I$

Ex. 13. *The value of* x *in* $\left(0, \frac{\pi}{2}\right)$ *satisfying equation* $\frac{\sqrt{5} - 1}{\sin x} + \frac{\sqrt{10 + 2\sqrt{5}}}{\cos x} = 8$ *is*

(a) $\frac{\pi}{10}$ (b) $\frac{3\pi}{10}$

(c) $\frac{9\pi}{10}$ (d) $\frac{7\pi}{10}$

Sol. (a,b) $\frac{\sqrt{5} - 1}{4} \times \frac{1}{\sin x} + \frac{\sqrt{10 + 2\sqrt{5}}}{4} \times \frac{1}{\cos x} = 2$

$$\frac{\sqrt{5} - 1}{4}\cos x + \frac{\sqrt{10 + 2\sqrt{5}}}{4} \times \sin x = 2\sin x \cdot \cos x$$

$$\sin\left[x + \frac{\pi}{10}\right] = \sin 2x$$

$$x + \frac{\pi}{10} = 2x, \ x = \frac{\pi}{10}$$

or $\quad \sin\left(x + \frac{\pi}{10}\right) = \sin(\pi - 2x)$

$\Rightarrow \quad x + \frac{\pi}{10} = \pi - 2x$

$\Rightarrow \quad 3x = \pi - \frac{\pi}{10} = \frac{9\pi}{10}$

$\Rightarrow \quad x = \frac{3\pi}{10}$

Ex. 14. *Given* $2\sin t = \frac{1 - 2x + 5x^2}{3x^2 - 2x - 1}$ *the real values of* $t \in \left[-\frac{\pi}{2}, \frac{\pi}{2}\right]$ *satisfying this lie in the interval*

(a) $\left[-\frac{\pi}{2}, \frac{\pi}{2}\right]$ (b) $\left[\frac{3\pi}{10}, \frac{\pi}{2}\right]$

(c) $\left[-\frac{\pi}{2}, \frac{\pi}{10}\right] \cup \left[\frac{3\pi}{10}, \frac{\pi}{2}\right]$ (d) All of these

Sol. (a,b,c,d) Given $x^2(6\sin t - 5) + x(2 - 4\sin t) - (1 + 2\sin t) = 0$, since $x \in R$, $\Delta > 0$

$\Rightarrow \quad \sin t < \frac{1 - \sqrt{5}}{4}$

or $\quad \sin t > \frac{1 + \sqrt{5}}{4}$

$\therefore \quad t \in \left[-\frac{\pi}{2}, \frac{\pi}{10}\right] \cup \left[\frac{3\pi}{10}, \frac{\pi}{2}\right]$

Ex. 15. *The system of equations*

$$\tan x = a\cot x,\ \tan 2x = b\cos y$$

(a) Cannot have a solution if $a = 0$

(b) Cannot have a solution if $a = 1$

(c) Cannot have a solution if $2\sqrt{a} > |b(1-a)|$

(d) has a solution for all a and b

Sol. (b,c) If $a = 0$, then $\tan x = 0 \Rightarrow x = n\pi$ and then for any value of y such that $\cos y = 0$ the second equation satisfies option (a) is false.

If $a = 1$ then $\tan x = \cot x \Rightarrow \tan^2 x = 1$

$\Rightarrow$ $\tan 2x$ is not defined.

$\Rightarrow$ option (b) is true

Now from the first equation $\tan x = \sqrt{a}$

$\Rightarrow$ a must be positive

$$|\cos y| = \left|\frac{2\tan x}{b(1-\tan^2 x)}\right| = \left|\frac{2\sqrt{a}}{b(1-a)}\right| \le 1$$

$$\Rightarrow \quad 2\sqrt{a} \le |b(1-a)|$$

Ex. 16. *If* $\dfrac{y+3}{2y+5} = \sin^2 x + 2\cos x + 1$, *then the value of* y *lie in the interval*

(a) $\left(-\infty, -\frac{8}{3}\right]$ (b) $\left[-\frac{12}{5}, \infty\right)$

(c) $\left[-\frac{8}{3}, -\frac{12}{5}\right]$ (d) $\left(-\infty, -\frac{8}{3}\right] \cup \left[-\frac{12}{5}, \infty\right)$

Sol. (a,b,d) $\cos^2 x - 2\cos x + 1 = 3 - \dfrac{y+3}{2y+5}$

$$(\cos x - 1)^2 = \frac{5y+12}{2y+5}$$

Also, $-1 \le \cos x \le 1 \Rightarrow -2 \le \cos x - 1 \le 0$

$$\Rightarrow \quad 0 \le (\cos x - 1)^2 \le 4$$

$$\Rightarrow \quad 0 \le \frac{5y+12}{2y+5} \le 4$$

Now, $\dfrac{5y+12}{2y+5} - 4 \le 0 \Rightarrow \dfrac{-3y-8}{2y+5} \le 0$

$$\Rightarrow \quad \frac{3y+8}{2y+5} \ge 0 \Rightarrow (3y+8)(2y+5) \ge 0$$

and $\dfrac{5y+12}{2y+5} \ge 0 \Rightarrow (5y+12)(2y+5) \ge 0$

$$\Rightarrow \quad \left(-\infty, -\frac{8}{3}\right] \cup \left[-\frac{5}{2}, \infty\right)$$

and $\left(-\infty, -\frac{5}{2}\right) \cup \left[-\frac{12}{5}, \infty\right)$

$$\therefore \quad y \in \left(-\infty, -\frac{8}{3}\right] \cup \left[-\frac{12}{5}, \infty\right)$$

Ex. 17. *Which of the following set of values of 'x' satisfies the equation*

$$2^{\left(2\sin^2 x - 3\sin x + 1\right)} + 2^{\left(2 - 2\sin^2 x + 3\sin x\right)} = 9$$

(a) $x = n\pi \pm \frac{\pi}{6}, n \in I$ (b) $x = n\pi \pm \frac{\pi}{3}, n \in I$

(c) $x = n\pi, n \in I$ (d) $x = 2n\pi + \frac{\pi}{2}, n \in I$

Sol. (a,d) $2^{\left(2\sin^2 x - 3\sin x + 1\right)} + 2^{3-\left(2\sin^2 x - 3\sin x + 1\right)} = 9$

Let $2^{\left(2\sin^2 x - 3\sin x + 1\right)} = t$

$$\Rightarrow \quad t + \frac{8}{t} = 9 \Rightarrow t^2 - 9t + 8 = 0 \Rightarrow t = 1, 8$$

$$\Rightarrow \quad 2\sin^2 x - 3\sin x + 1 = 3$$

or $\quad 2\sin^2 x - 3\sin x + 1 = 0$

$$\Rightarrow \quad \sin x = -\frac{1}{2},$$

$$\sin x = \frac{1}{2},\ \sin x = 1$$

Ex. 18. *For* $0 < \theta < \frac{\pi}{2}$, *the solution(s) of*

$$\sum_{m=1}^{6} \operatorname{cosec}\left(\theta + \frac{(m-1)\pi}{4}\right)\operatorname{cosec}\left(\theta + \frac{m\pi}{4}\right) = 4\sqrt{2} \text{ is (are)}$$

(a) $\frac{\pi}{4}$ (b) $\frac{\pi}{6}$

(c) $\frac{\pi}{12}$ (d) $\frac{5\pi}{12}$

Sol. (c, d) We have,

$$\sum_{m=1}^{6} \operatorname{cosec}\left[\theta + \frac{(m-1)\pi}{4}\right]\operatorname{cosec}\left[\theta + \frac{m\pi}{4}\right] = 4\sqrt{2}$$

$$\Rightarrow \quad \sum_{m=1}^{6} \frac{\sin\frac{\pi}{4}}{\sin\left[\theta + \frac{(m-1)\pi}{4}\right]\sin\left[\theta + \frac{m\pi}{4}\right]} = 4$$

$$\Rightarrow \quad \sum_{m=1}^{6} \frac{\sin\left[\left(\theta + \frac{m\pi}{4}\right) - \left(\theta + \frac{(m-1)\pi}{4}\right)\right]}{\sin\left(\theta + \frac{(m-1)\pi}{4}\right)\sin\left(\theta + \frac{m\pi}{4}\right)} = 4$$

$$\Rightarrow \sum_{m=1}^{6}\left[\frac{\sin\left(\theta + \frac{m\pi}{4}\right)\cos\left(\theta + \frac{(m-1)\pi}{4}\right) - \cos\left(\theta + \frac{m\pi}{4}\right)\sin\left(\theta + \frac{(m-1)\pi}{4}\right)}{\sin\left(\theta + \frac{(m-1)\pi}{4}\right)\sin\left(\theta + \frac{m\pi}{4}\right)}\right] = 4$$

$$\Rightarrow \quad \sum_{m=1}^{6}\left[\cot\left(\theta + \frac{(m-1)\pi}{4}\right) - \cot\left(\theta + \frac{m\pi}{4}\right)\right] = 4$$

$$\Rightarrow \left[\cot\theta - \cot\left(\theta + \frac{\pi}{4}\right)\right] + \left[\cot\left(\theta + \frac{\pi}{4}\right) - \cot\left(\theta + \frac{2\pi}{4}\right)\right]$$

$$+ \ldots + \left[\cot\left(\theta + \frac{5\pi}{4}\right) - \cot\left(\theta + \frac{6\pi}{4}\right)\right] = 4$$

$$\Rightarrow \quad \cot\theta - \cot\left(\theta + \frac{3\pi}{2}\right) = 4$$

$$\Rightarrow \quad \cot\theta + \tan\theta = 4$$

$$\Rightarrow \quad \cos^2\theta + \sin^2\theta = 4\sin\theta\cos\theta$$

$$\Rightarrow \quad \sin 2\theta = \frac{1}{2} \Rightarrow 2\theta = \frac{\pi}{6} \text{ or } \frac{5\pi}{6}$$

$$\Rightarrow \quad \theta = \frac{\pi}{12} \text{ or } \frac{5\pi}{12}$$

Ex. 19. *If* $\dfrac{1+\sin 6^\circ}{\cos 6^\circ} = \tan A = \sqrt{\dfrac{1+\sin B}{1-\sin B}}$*; where A and* $B \in (0.90^\circ)$*, then*

(a) $A = 8B$ (b) $8A = B$
(c) $A - 7B = 6^\circ$ (d) $A + B = 54^\circ$

Sol. (a,c,d) $\dfrac{(\sin 3^\circ + \cos 3^\circ)^2}{(\cos^2 3^\circ - \sin^2 3^\circ)} = \tan A = \sqrt{\left(\dfrac{\sin\frac{B}{2} + \cos\frac{B}{2}}{\cos\frac{B}{2} - \sin\frac{B}{2}}\right)^2}$

$$\frac{\sin 3^\circ + \cos 3^\circ}{\cos 3^\circ - \sin 3^\circ} = \tan A = \frac{\sin\frac{B}{2} + \cos\frac{B}{2}}{\cos\frac{B}{2} - \sin\frac{B}{2}}$$

$$\frac{(\tan 3^\circ + 1)}{(-\tan 3^\circ + 1)} = \tan A = \frac{1 + \tan\frac{B}{2}}{1 - \tan\frac{B}{2}}$$

$$\tan(45^\circ + 3^\circ) = \tan A = \tan\left(45^\circ + \frac{B}{2}\right)$$

$$\Rightarrow \quad A = 48^\circ \text{ and } B = 6^\circ$$

Ex. 20. *If* $\sqrt{1+\sin A} - \sqrt{1-\sin A} = 2\cos\frac{A}{2}$*, then value of A can be*

(a) 110° (b) 260°
(c) 300° (d) 190°

Sol. (a,b,d) $\sqrt{1+\sin A} - \sqrt{1-\sin A} = 2\cos\frac{A}{2}$

$$\left|\sin\frac{A}{2} + \cos\frac{A}{2}\right| - \left|\sin\frac{A}{2} - \cos\frac{A}{2}\right| = 2\cos\frac{A}{2}$$

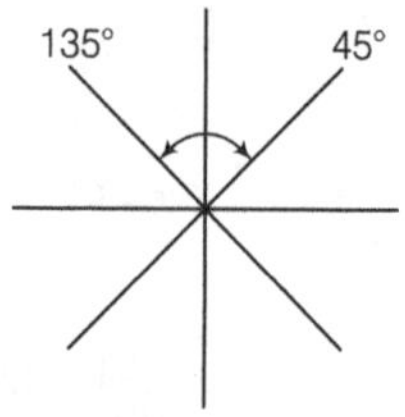

So, $\sin\frac{A}{2} + \cos\frac{A}{2} > 0$ and $\cos\frac{A}{2} < \sin\frac{A}{2}$

$$\Rightarrow \quad 45^\circ < \frac{A}{2} < 135^\circ \Rightarrow 90^\circ < A < 270^\circ$$

JEE Type Solved Examples : Passage Based Questions

Passage I

(Ex. Nos. 21 to 23)

Consider a circle, in which a point P is lying inside the circle such that $(PA)(PB) = (PC)(PD)$ (as shown in figure).

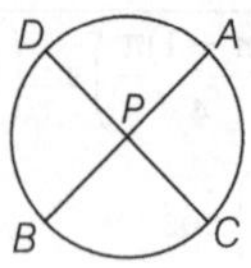

On the basis of above information, answer the following questions :

21. Let $PA = 4$, $PB = 3$ and CD is diameter of the circle having the length 8. If $PC > PD$, then $\dfrac{PC}{PD}$ is equal to

(a) 3 (b) 4
(c) 5 (d) 6

Sol. (a) (PA) (PB) = (PC) (PD)

$$4 \times 3 = x(2r - x)$$

$$\Rightarrow \quad 12 = x(8 - x)$$

$$x^2 - 8x + 12 = 0 \Rightarrow x = 6, 2$$

$$\frac{PC}{PD} = \frac{6}{2} = 3$$

22. If $PA = |\cos\theta + \sin\theta|$ and $PB = |\cos\theta - \sin\theta|$, then maximum value of $(PC)(PD)$, is equal to

(a) 1 (b) $2\sqrt{2}$
(c) $\sqrt{2}$ (d) 2

Sol. (a) $PC \cdot PD = (PA)(PB)$

$$= |\cos\theta + \sin\theta||\cos\theta - \sin\theta|$$

$$= |\cos^2\theta - \sin^2\theta| = \cos 2\theta$$

Maximum value = 1

23. If $\log_{PA} x = 2$, $\log_{PB} x = 3$, $\log_x PC = 4$, then $\log_{PD} x$ is equal to

(a) $\frac{7}{12}$ (b) $\frac{12}{7}$

(c) $-\frac{7}{12}$ (d) $-\frac{6}{19}$

Sol. (d) $(PA)(PB) = (PC)(PD)$

$$\log_x((PA)(PB)) = \log_x((PC)(PD))$$

$$\log_x PA + \log_x PB = \log_x PC + \log_x PD$$

$$\frac{1}{2} + \frac{1}{3} = 4 + \log_x PD$$

$$\log_x PD = \frac{5}{6} - 4 = -\frac{19}{6}$$

Passage II

(Ex. Nos. 24 to 26)

PA and *PB* are two tangents drawn from point *P* to circle of radius 5. A line is drawn from point *P* which cuts circle at *C* and *D* such that $PC = 5$ and $PD = 15$ and $\angle APB = \theta$.

On the basis of above information, answer the following questions :

24. Area of ΔAPB is

(a) $\frac{25\sqrt{3}}{2}$ (b) $25\sqrt{3}$

(c) $\frac{75\sqrt{3}}{2}$ (d) $\frac{75\sqrt{3}}{4}$

25. Value of $\sin 2\theta + \cos 4\theta + \sin 5\theta + \tan 7\theta + \cot 8\theta$ is equal to

(a) $\frac{4\sqrt{3}-1}{2}$ (b) $\frac{4-\sqrt{3}}{2\sqrt{3}}$

(c) $\frac{4+\sqrt{3}}{2\sqrt{3}}$ (d) $\frac{4\sqrt{3}+1}{2}$

26. Number of solution(s) of the equation $\log_{\cos\theta}(x+2) = 2 + 3\log_{(x+2)} \sin\left(\frac{5\theta}{2}\right)$ is

(a) 0 (b) 1

(c) 2 (d) 3

Sol. (Ex. Nos. 24 to 26)

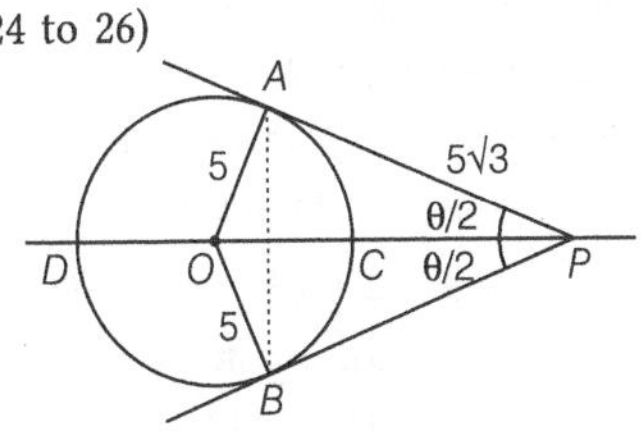

$$OD = OC = 5,\ PD = DC + CP$$

$$\Rightarrow \quad CP = 15 - 10 = 5,\ OP = 10$$

$$\Rightarrow \quad \sin\frac{\theta}{2} = \frac{OA}{OP} = \frac{5}{10} = \frac{1}{2} \Rightarrow \frac{\theta}{2} = 30^\circ$$

$$\Rightarrow \quad \theta = 60^\circ$$

24. (d) $\therefore$ *APB* is an equilateral triangle.

$$\text{Area} = \frac{\sqrt{3}}{4}(5\sqrt{3})^2 = \frac{75\sqrt{3}}{4}$$

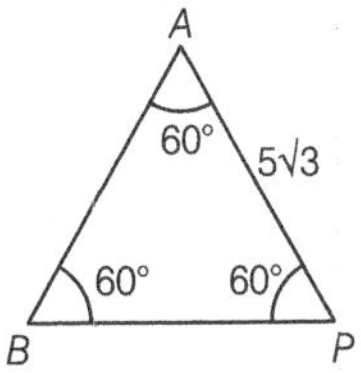

25. (b) $\sin 2\theta + \cos 4\theta + \sin 5\theta + \tan 7\theta + \cot 8\theta$

$$= \sin 120^\circ + \cos 240^\circ + \sin 300^\circ + \tan 420^\circ + \cot 480^\circ$$

$$= \frac{\sqrt{3}}{2} - \frac{1}{2} - \frac{\sqrt{3}}{2} + \sqrt{3} - \frac{1}{\sqrt{3}} = \frac{4-\sqrt{3}}{2\sqrt{3}}$$

26. (c) Given $\log_{1/2}(x+2) = 2 + 3\log_{(x+2)}\left(\frac{1}{2}\right)$

Let $\log_{1/2}(x+2) = t$

$$t = 2 + \frac{3}{t}$$

$$\Rightarrow \quad t^2 - 2t - 3 = 0$$

$$\Rightarrow \quad t = 3 \text{ or } -1$$

$$\Rightarrow \quad \log_{1/2}(x+2) = 3$$

$$\text{or} \quad \log_{1/2}(x+2) = -1$$

$$\Rightarrow \quad x = -\frac{15}{8} \text{ or } x = 0$$

$\Rightarrow$ two solutions.

Passage III

(Ex. Nos. 27 and 28)

If $3\sin^2 x - 7\sin x + 2 = 0$, $x \in \left[0, \frac{\pi}{2}\right]$ and $f_n(\theta) = \sin^n\theta + \cos^n\theta$.

On the basis of above information, answer the following questions :

27. The value of $f_4(x)$ is

(a) $\frac{97}{81}$ (b) $\frac{57}{81}$ (c) $\frac{65}{81}$ (d) $\frac{73}{81}$

28. The value of $\frac{\sin 5x + \sin 4x}{1 + 2\cos 3x}$ is

(a) $\frac{3+2\sqrt{2}}{9}$ (b) $\frac{3+4\sqrt{2}}{9}$

(c) $\frac{4\sqrt{2}-2}{9}$ (d) $\frac{4\sqrt{2}-3}{9}$

Sol. (Ex. Nos. 27 to 28) $3\sin^2 x - 7\sin x + 2 = 0$

$$\Rightarrow \quad \sin x = \frac{1}{3} \text{ or } \sin x = 2 \text{ (reject)}$$

27. (c) $f_4(x) = \sin^4 x + \cos^4 x = 1 - 2(\sin x \cos x)^2$

$$= 1 - 2\left(\frac{1}{3} \cdot \frac{2\sqrt{2}}{3}\right)^2 = \frac{65}{81}$$

28. (b) $\dfrac{\sin 5x + \sin 4x}{1 + 2\cos 3x}$

$$= \frac{(\sin 5x + \sin 4x)\sin 3x}{(1 + 2\cos 3x)\sin 3x} = \frac{(\sin 5x + \sin 4x)\sin 3x}{\sin 3x + \sin 6x}$$

$$= \frac{2\sin\frac{9x}{2}\cos\frac{x}{2}\sin 3x}{2\sin\frac{9x}{2}\cos\frac{3x}{2}} = \frac{\cos\frac{x}{2} \cdot 2\sin\frac{3x}{2}\cos\frac{3x}{2}}{\cos\frac{3x}{2}}$$

$$= \sin 2x + \sin x$$

$$= \frac{4\sqrt{2}}{9} + \frac{1}{3} = \frac{4\sqrt{2} + 3}{9}$$

JEE Type Solved Examples : Single Integer Answer Type Questions

● ***Ex. 29.*** *Number of integral solutions of the equation* $\log_{\sin x}\sqrt{\sin^2 x} + \log_{\cos x}\sqrt{\cos^2 x} = 2$, *where* $x \in [0, 6\pi]$ *is*

Sol. (4) $\log_{\sin x}\sqrt{\sin^2 x} + \log_{\cos x}\sqrt{\cos^2 x} = 2$

$\therefore$ $\sin x > 0$ and $\sin x \neq 1$

$\cos x > 0$ and $\cos x \neq 1$

Domain $x \in \left(0, \frac{\pi}{2}\right) \cup \left(2\pi, \frac{5\pi}{2}\right) \cup \left(4\pi, \frac{9\pi}{2}\right)$

$\Rightarrow$ Number of integers $= 1 + 1 + 2 = 4$

● ***Ex. 30.*** *If* $x_k = (\sec\theta)^{\frac{1}{2^k}} + (\tan\theta)^{\frac{1}{2^k}}$ *and* $y_k = (\sec\theta)^{\frac{1}{2^k}} - (\tan\theta)^{\frac{1}{2^k}}$, *then value of* $3y_n \prod_{k=0}^{n}(x_k)$ *is equal to*

Sol. (3) $x_k \cdot y_k = (\sec\theta)^{\frac{1}{2^{k-1}}} - (\tan\theta)^{\frac{1}{2^{k-1}}}$

$\Rightarrow$ $x_k \cdot y_k = y_{k-1}$

Now, $y_n \cdot \prod_{k=0}^{n} x_k = y_n \cdot \prod_{k=0}^{n} \frac{y_{k-1}}{y_k}$

$$= y_n \times \frac{y_{-1}}{y_0} \times \frac{y_0}{y_1} \times \ldots\ldots \times \frac{y_{n-1}}{y_n} = y_{-1}$$

$$= (\sec\theta)^2 - (\tan\theta)^2 = 1$$

● ***Ex. 31.*** *The number of ordered pairs* (α, β), *where* $\alpha, \beta \in [0, 2\pi]$ *satisfying* $\log_{2\sec x}(\beta^2 - 6\beta + 10) = \log_3|\cos\alpha|$ *is*

Sol. (2) $\log_{2\sec x}(\beta^2 - 6\beta + 10) = \log_3|\cos\alpha|$

it is only possible when

$\beta^2 - 6\beta + 10 = 1$ and $\cos\alpha = 1$

$\Rightarrow$ $\beta = 3$ and $\alpha = 0, 2\pi$

$\therefore$ Two ordered pairs (0, 3) and $(2\pi, 3)$.

● ***Ex. 32.*** *If* $\dfrac{\cos^3\theta}{(1 - \sin\theta)} + \dfrac{\sin^3\theta}{(1 + \cos\theta)} = 1 + \cos\theta$, *then number of possible values of* θ *is (where* $\theta \in [0, 2\pi]$*).*

Sol. (0) $\dfrac{\cos\theta\cos^2\theta}{(1 - \sin\theta)} + \dfrac{\sin\theta\sin^2\theta}{(1 + \cos\theta)} = \cos\theta + 1$

$$\cos\theta\frac{(1 - \sin^2\theta)}{(1 - \sin\theta)} + \frac{(1 - \cos^2\theta)\sin\theta}{(1 + \cos\theta)} = \cos\theta + 1$$

$$\cos\theta(1 + \sin\theta) + \sin\theta(1 - \cos\theta) = \cos\theta + 1$$

$$\sin\theta + \cos\theta = \cos\theta + 1 \Rightarrow \sin\theta = 1$$

$\theta = \frac{\pi}{2}$ which is not possible.

● ***Ex. 33.*** *If the sum of all values of x satisfying the system of equations*

$$\tan x + \tan y + \tan x \cdot \tan y = 5$$
$$\sin(x + y) = 4\cos x \cdot \cos y$$

is $\frac{k\pi}{2}$, *where* $x \in \left(0, \frac{\pi}{2}\right)$ *then find the values of k.*

Sol. (1) Given,

$\tan x + \tan y + \tan x \cdot \tan y = 5$ ...(i)

and $\sin(x + y) = 4\cos x \cdot \cos y$ ...(ii)

Now, from Eq. (ii), we get

$$\sin x\cos y + \cos x\sin y = 4\cos x\cos y$$

On dividing by $\cos x \cdot \cos y$, we get

$\tan x + \tan y = 4$ ...(iii)

$\therefore$ Eqs. (i) and (iii), $\tan x(4 - \tan x) = 1$

$\Rightarrow$ $\tan^2 x - 4\tan x + 1 = 0$

$\Rightarrow$ $\tan x = 2 + \sqrt{3}$ or $2 - \sqrt{3}$

$\Rightarrow$ $x = \frac{5\pi}{12}$ or $\frac{\pi}{12}$

$\therefore$ Sum $= \frac{6\pi}{12} = \frac{\pi}{2} = \frac{k\pi}{2} \Rightarrow k = 1$

JEE Type Solved Examples : Statement I and II Type Questions

- This section contains **2** questions. Each question contains **Statement I (Assertion)** and **Statement II (Reason)**. Each question has 4 choices (a), (b), (c) and (d) out of which only one is correct. The choices are
  - (a) Statement I is true, Statement II is true; Statement II is a correct explanation for Statement I.
  - (b) Statement I is true, Statement II is true; Statement II is not a correct explanation for Statement II.
  - (c) Statement I is true, Statement II is false.
  - (d) Statement I is false, Statement II true.

Ex. 34. Statement I *$x=\frac{k\pi}{13}, k\in I$ does not represent the general solution of trigonometric equation.*

$\sin 13x - \sin 13x \cos 2x = 0$

Statement II *Both $x=r\pi, r\in I$ and $x=\frac{k\pi}{13}, k\in I$ satisfies the trigonometric equation.*

$$\sin 13x - \sin 13x \cos 2x = 0$$

Sol. (d) $\sin 13x(1-\cos 2x)=0$

$\Rightarrow \quad \sin 13x=0 \text{ or } \cos 2x=1$

$\therefore \quad 13x=k\pi \text{ or } \cos 2x=1$

$x=\frac{k\pi}{13}, \; k\in \text{Integer.}$

$\Rightarrow \quad 2x=2n\pi \Rightarrow x=n\pi, n\in \text{Integer}$

$\Rightarrow$ Statement I is false and Statement II is true.

Ex. 35. Statement I *Common value(s) of 'x' satisfying the equations.*

$\log_{\sin x}(\sec x+8)>0$ and $\log_{\sin x}\cos x+\log_{\cos x}\sin x=2$ in $(0,4\pi)$ does not exist.

Statement II *On solving above trigonometric equations we have to take intersection of trigonometric chains given by*

$$\sec x>1 \text{ and } x=n\pi+\frac{\pi}{4}, n\in I$$

Sol. (c) $\log_{\sin x}\cos x+\log_{\cos x}\sin x=2$ only

When $\sin x=\cos x$

$$\Rightarrow \quad x=n\pi+\frac{\pi}{4} \quad \text{...(i)}$$

Also, $\log_{\sin x}(\sec x+8)>0$

$$\Rightarrow \quad \sec x+8<1 \Rightarrow \sec x<7 \quad \text{...(ii)}$$

Clearly, $x=\frac{\pi}{4}+n\pi$ satisfy Eq.(ii)

$\therefore$ Statement I is true and Statement II is false.

JEE Type Solved Examples : Matching Type Questions

Ex. 36. *Match the statement of Column I with the value of Column II.*

	Column I		Column II
A	The number of solutions of the equation $\lvert\tan 2x\rvert=\sin x; x\in[0,\pi]$	p	1
B	The value of $4\tan\frac{\pi}{16}-4\tan^3\frac{\pi}{16}+6\tan^2\frac{\pi}{16}-\tan^4\frac{\pi}{16}+1$	q	4
C	If the equation $\tan(p\cot x)=\cot p(\tan x)$ has a solution in $(0,x)-\left\{\frac{\pi}{2}\right\}$, then $\frac{4}{\pi}P_{max}$ is	r	3
D	The value of $\frac{2x}{\pi}$ in $[0,2\pi]$, if $5^{\cos^2 2x+2\sin^2 x}+5^{2\cos^2 x+\sin^2 2x}=126$ has a solution	s	2

Sol. (A) $\rightarrow$(q); (B) $\rightarrow$ (s); (C) $\rightarrow$ (p); (D) $\rightarrow$ (p,r)

(A) Clearly, number of solutions of $\lvert\tan 2x\rvert=\sin x$ in $[0,\pi]$ are 4.

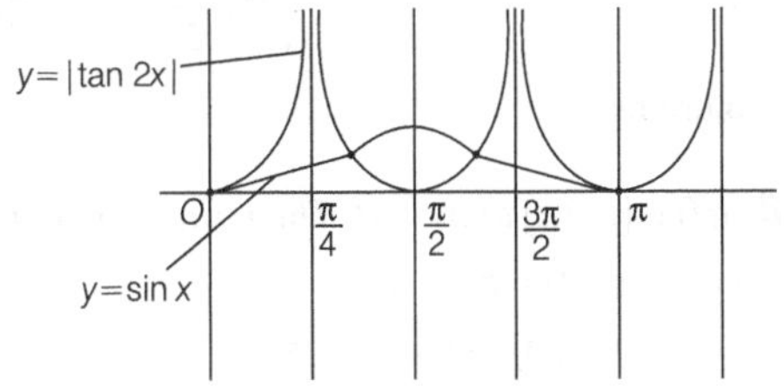

(B) $\tan 4A=\frac{2\tan 2A}{1-\tan^2 2A}=\frac{\frac{4\tan A}{1-\tan^2 A}}{1-\left(\frac{2\tan A}{1-\tan^2 A}\right)^2}$

$$\Rightarrow \quad \tan 4A=\frac{4\tan A(1-\tan^2 A)}{1+\tan^4 A-6\tan^2 A}$$

$= 4\tan A - 4\tan^3 A + (6\tan^2 A - \tan^2 A - 1)\tan 4A = 0$

If $A = \dfrac{\pi}{16}$

$\Rightarrow \quad 4\tan\dfrac{\pi}{16} - 4\tan^3\dfrac{\pi}{6} + 6\tan^2\dfrac{\pi}{16} - \tan^4\dfrac{\pi}{16} - 1 = 0$

$\therefore$ Required value is 2.

(C) $\tan(p\cot x) = \cot(p\tan x)$

$$\tan(p\cot x) = \tan\left(\frac{\pi}{2} - p\tan x\right)$$

$$p\cot x = n\pi + \frac{\pi}{2} - p\tan x$$

$$p = \frac{n\pi + \frac{\pi}{2}}{\tan x + \cot x} = \frac{\pi}{2}\sin x\cos x \qquad (\because x \in [0, \pi])$$

$$P_{max} = \frac{\pi}{4};\ \frac{4P_{max}}{\pi} = 1$$

(D) $5^{\cos^2 2x + 2\sin^2 x} + 5^{2\cos^2 x + \sin^2 2x} = 126$

$$5^{\cos^2 2x + 2\sin^2 x} + 5^{2 - 2\sin^2 x + 1 - \cos^2 2x}$$

$$5^{\cos^2 2x + 2\sin^2 x} + 5^{3 - \cos^2 2x + 2\sin^2 x}$$

Put $\cos^2 2x + 2\sin^2 x = y$, we get

$$5^y + 5^{3-y} = 126$$

$$5^y + \frac{125}{5^y} = 126 \Rightarrow 5^y = 125, 1 \Rightarrow y = 3, 0$$

$$\cos^2 2x + 2\sin^2 x = 3 \Rightarrow x = \frac{\pi}{2}, \frac{3\pi}{2}$$

$$\cos^2 2x + 2\sin^2 x \neq 0\ ;\ \frac{2x}{\pi} = 1, 3$$

Subjective Type Examples

Ex. 37. *If $0 \le x \le 3\pi$, $0 \le y \le 3\pi$ and $\cos x \cdot \sin y = 1$, then find the possible number of values of the ordered pair (x, y).*

Sol. Clearly, $\cos x \cdot \sin y = 1$

$\Rightarrow \quad \cos x = 1, \sin y = 1$

or $\quad \cos x = -1, \sin y = -1$

Now, $\cos x = 1;\ \sin y = 1$

$\Rightarrow \quad x = 0,\ 2\pi$ and $y = \dfrac{\pi}{2}, \dfrac{5\pi}{2}$ ...(i)

and, $\cos x = -1, \sin y = -1$

$\Rightarrow \quad x = \pi, 3\pi$ and $y = \dfrac{3\pi}{2}$ ...(ii)

$\therefore$ From Eqs. (i) and (ii), the required of ordered pairs are

$$\left(0, \frac{\pi}{2}\right)\left(0, \frac{5\pi}{2}\right)\left(2\pi, \frac{\pi}{2}\right)$$

$$\left(2\pi, \frac{5\pi}{2}\right)\left(\pi, \frac{3\pi}{2}\right)\left(3\pi, \frac{3\pi}{2}\right)$$

i.e. 6 solutions.

Ex. 38. *If $\theta \in [0, 3\pi]$ and $r \in R$. Then, find the pairs of (r, θ) satisfying $2\sin\theta = r^4 - 2r^2 + 3$.*

Sol. Here, $\quad 2\sin\theta = r^4 - 2r^2 + 3$

$\Rightarrow \quad 2\sin\theta = (r^2 - 1)^2 + 2$

$\Rightarrow \quad 2\sin\theta = (r^2 - 1)^2 + 2 \ge 2$ ...(i)

But $\max(\sin\theta) = 1$

$\Rightarrow \quad \max(2\sin\theta) = 2$ ...(ii)

$\therefore$ From Eqs. (i) and (ii) ;

$\quad \sin\theta = 1$ and $r^2 - 1 = 0$

$\Rightarrow \quad \theta = \dfrac{\pi}{2}, \dfrac{5\pi}{2}$ and $r = \pm 1$

$\therefore$ Required ordered pairs are

$$\left(\frac{\pi}{2}, 1\right)\left(\frac{\pi}{2}, -1\right)\left(\frac{5\pi}{2}, 1\right)\left(\frac{5\pi}{2}, -1\right)$$

i.e. 4 ordered pairs.

Ex. 39. *Find all the values of θ satisfying the equation, $\sin 7\theta = \sin\theta + \sin 3\theta$ such that $0 \le \theta \le \pi$.*

Sol. Given, $\sin 7\theta = \sin\theta + \sin 3\theta$

$\Rightarrow \quad \sin 7\theta - \sin\theta - \sin 3\theta = 0$

$\therefore \quad 2\sin 3\theta \cdot \cos 4\theta - \sin 3\theta = 0$

$\therefore \quad \sin 3\theta(2\cos 4\theta - 1) = 0$

$\Rightarrow \quad \sin 3\theta = 0$ or $\cos 4\theta = \dfrac{1}{2}$

i.e. $\quad 3\theta = n\pi, n \in I$ or $4\theta = 2n\pi \pm \dfrac{\pi}{3}, n \in I$

$\Rightarrow \quad \theta = \dfrac{n\pi}{3}$ or $\theta = \dfrac{n\pi}{2} \pm \dfrac{\pi}{12}$

put $n = 0, \theta = 0$ or $\theta = \pm\dfrac{\pi}{12}, \left(-\dfrac{\pi}{12} \text{ rejected}\right)$

put $\quad n = 1, \theta = \dfrac{\pi}{3}$

or $\quad \theta = \dfrac{\pi}{2} + \dfrac{\pi}{12}, \dfrac{\pi}{2} - \dfrac{\pi}{12}$

$\quad \theta = \dfrac{7\pi}{12}, \dfrac{5\pi}{12}$

put $\quad n = 2, \theta = \dfrac{2\pi}{3}$

or $\theta = \pi \pm \frac{\pi}{12}\left(\pi + \frac{\pi}{12} \text{ rejected}\right)$

put $n = 3, \theta = \pi$

$\therefore$ Solutions are $0, \frac{\pi}{12}, \frac{\pi}{3}, \frac{5\pi}{12}, \frac{7\pi}{12}, \frac{2\pi}{3}, \frac{11\pi}{12}$ and π.

Ex. 40. *Solve* $\sin 3x + \cos 2x = -2$.

Sol. Since, $\sin 3x \geq -1$ and $\cos 2x \geq -1$. We have, $\sin 3x + \cos 2x \geq -2$.

Thus, the equality holds true if and only if;

$\sin 3x = -1$ and $\cos 2x = -1$

$\Rightarrow \quad 3x = n\pi + (-1)^n\left(-\frac{\pi}{2}\right)$ and $2x = 2n\pi \pm \pi$.

i.e. $\quad x = \frac{n\pi}{3} + (-1)^n\left(-\frac{\pi}{6}\right)$ and $x = n\pi \pm \frac{\pi}{2}, n \in I$

$\therefore$ Solution set is,

$$\left\{x \mid x = \frac{n\pi}{3} + (-1)^n\left(-\frac{\pi}{6}\right)\right\} \cap \left\{x \mid x = n\pi \pm \frac{\pi}{2}\right\}$$

Note

Here unlike all other problems the solution set consists of the intersection of two solution sets and not the union of the solution sets.

Ex. 41. *Find all values of* θ *which satisfy,*

$\sin(3\theta + \alpha) + \sin(3\theta - \alpha) + \sin(\alpha - \theta) - \sin(\alpha + \theta) = \cos\alpha$ *given* $\cos\alpha \neq 0$.

Sol. Given, $\sin(3\theta + \alpha) + \sin(3\theta - \alpha) + \sin(\alpha - \theta) - \sin(\alpha + \theta) = \cos\alpha$

$\Rightarrow \quad 2\sin(3\theta)\cdot\cos(\alpha) + 2\sin(-\theta)\cos(\alpha) = \cos\alpha$

$\Rightarrow \quad 2(\sin 3\theta - \sin\theta)\cos\alpha = \cos\alpha$

$\Rightarrow \quad 2(\sin 3\theta - \sin\theta) = 1$ (as $\cos\alpha \neq 0$ given)

$\Rightarrow \quad 2\cdot 2\sin\theta\cdot\cos 2\theta = 1$

$\Rightarrow \quad 4\sin\theta(1 - 2\sin^2\theta) = 1$

$\Rightarrow \quad 4\sin\theta - 8\sin^3\theta = 1$

$\Rightarrow \quad 8\sin^3\theta - 4\sin\theta + 1 = 0$

$\Rightarrow \quad (2\sin\theta - 1)(4\sin^2\theta + 2\sin\theta - 1) = 0$

$\Rightarrow \quad \sin\theta = \frac{1}{2}$ or $\sin\theta = \frac{-2 \pm \sqrt{4 + 16}}{8} = \frac{-1 \pm \sqrt{5}}{4}$

$\Rightarrow \quad \theta = n\pi + (-1)^n\frac{\pi}{6}, n\pi + (-1)^n\sin^{-1}\left(\frac{\sqrt{5} - 1}{4}\right)$

or $\quad n\pi + (-1)^n\sin^{-1}\left(\frac{-1 - \sqrt{5}}{4}\right)$

$\Rightarrow \quad \theta = n\pi + (-1)^n\frac{\pi}{6}, n\pi + (-1)^n\frac{\pi}{10}$

or $\quad n\pi + (-1)^n\left(\frac{-3\pi}{10}\right), n \in I$

$$\left[\because \sin\frac{\pi}{10} = \sin 18^\circ = \frac{\sqrt{5} - 1}{4}, \sin\left(-\frac{3\pi}{10}\right) = \sin(-54^\circ) = \frac{-1 - \sqrt{5}}{4}\right]$$

$\therefore$ General solution set

$$= \left\{\theta \mid \theta = n\pi + (-1)^n\frac{\pi}{6}\right\} \cup \left\{\theta \mid \theta = n\pi + (-1)^n\frac{\pi}{10}\right\} \cup \left\{\theta \mid \theta = n\pi - (-1)^n\frac{3\pi}{10}\right\}$$

Ex. 42. *Solve the equation*

$$\sin^4 x + \cos^4 x = \frac{7}{2}\sin x \cdot \cos x$$

Sol. Given, $\sin^4 x + \cos^4 x = \frac{7}{2}\sin x \cdot \cos x$

$\Rightarrow \quad (\sin^2 x + \cos^2 x)^2 - 2\sin^2 x\cos^2 x = \frac{7}{2}\sin x\cos x$

$\Rightarrow \quad 1 - \frac{1}{2}(\sin 2x)^2 = \frac{7}{4}(\sin 2x)$

$\Rightarrow \quad 2\sin^2 2x + 7\sin 2x - 4 = 0$

$\Rightarrow \quad (2\sin 2x - 1)(\sin 2x + 4) = 0$

$\Rightarrow \quad \sin 2x = \frac{1}{2}$

or $\quad \sin 2x = -4 < -1$ (Rejected)

$\Rightarrow \quad 2x = n\pi + (-1)^n\frac{\pi}{6}$

i.e. $\quad x = \frac{n\pi}{2} + (-1)^n\frac{\pi}{12}$

Ex. 43. *Find all the solutions of*

$$4\cos^2 x\sin x - 2\sin^2 x = 3\sin x$$

Sol. Given, $4\cos^2 x\sin x - 2\sin^2 x = 3\sin x$

$\Rightarrow \quad 4(1 - \sin^2 x)\sin x - 2\sin^2 x - 3\sin x = 0$

$\Rightarrow \quad 4\sin x - 4\sin^3 x - 2\sin^2 x - 3\sin x = 0$

$\Rightarrow \quad -4\sin^3 x - 2\sin^2 x + \sin x = 0$

$\Rightarrow \quad -\sin x(4\sin^2 x + 2\sin x - 1) = 0$

$\Rightarrow \quad \sin x = 0$ or $4\sin^2 x + 2\sin x - 1 = 0$

$\Rightarrow \quad \sin x = \sin 0$ or $\sin x = \frac{-2 \pm \sqrt{4 + 16}}{2(4)}$

$\Rightarrow \quad x = n\pi$ or $\sin x = \frac{-1 \pm \sqrt{5}}{4} = \sin\frac{\pi}{10}, \sin\left(-\frac{3\pi}{10}\right)$

$\Rightarrow \quad x = n\pi, x = n\pi + (-1)^n\frac{\pi}{10}, x = n\pi + (-1)^n\left(-\frac{3\pi}{10}\right)$

$\therefore$ General solution set

$$= \{x \mid x = n\pi\} \cup \left\{x \mid x = n\pi + (-1)^n\frac{\pi}{10}\right\} \cup \left\{x \mid x = n\pi + (-1)^n\left(-\frac{3\pi}{10}\right)\right\}$$

Ex. 44. *Solve the equation* $1+2\operatorname{cosec} x=\dfrac{-\sec^2\frac{x}{2}}{2}$.

Sol. Here, $1+2\operatorname{cosec} x=\dfrac{-\sec^2\frac{x}{2}}{2}$

$\Rightarrow \quad 1+\dfrac{2}{\sin x}=-\dfrac{\left[1+\tan^2\frac{x}{2}\right]}{2}$

i.e $\quad 2[2+\sin x]=-\left[1+\tan^2\frac{x}{2}\right]\sin x$

$\therefore \quad 2\left[\dfrac{2\tan\left(\frac{x}{2}\right)}{1+\tan^2\left(\frac{x}{2}\right)}+2\right]=-\left[1+\tan^2\frac{x}{2}\right]\cdot\left[\dfrac{2\tan\frac{x}{2}}{1+\tan^2\frac{x}{2}}\right]$

Put $\tan\frac{x}{2}=t$,

$$2\left[\frac{2t}{1+t^2}+2\right]=-(1+t^2)\left(\frac{2t}{1+t^2}\right)$$

$\therefore \quad 4t^2+4t+4+2t(1+t^2)=0$

$\therefore \quad 2t^3+4t^2+6t+4=0$

i.e. $\quad t^3+2t^2+3t+2=0$

$\therefore \quad (t+1)(t^2+t+2)=0$

$\Rightarrow \quad t=-1 \quad [\because t^2+t+2\neq 0]$

or $\quad \tan\frac{x}{2}=t=-1$

$\Rightarrow \quad \frac{x}{2}=n\pi-\frac{\pi}{4}, n\in I$

Thus, $x=2n\pi-\frac{\pi}{2}$, $n\in I$ is the required solution.

Ex. 45. *Find all values of* θ *lying between 0 and* 2π, *satisfying the equations*

$$r\sin\theta=\sqrt{3} \quad \text{...(i)}$$

$$r+4\sin\theta=2(\sqrt{3}+1) \quad \text{...(ii)}$$

Sol. We have to solve for θ,

$\therefore$ We shall eliminate r from Eqs. (i) and (ii),

From Eq. (i),

$$r=\frac{\sqrt{3}}{\sin\theta}$$

$\therefore$ From Eq. (ii), $\quad \dfrac{\sqrt{3}}{\sin\theta}+4\sin\theta=2(\sqrt{3}+1)$

$\Rightarrow \quad \sqrt{3}+4\sin^2\theta=2(\sqrt{3}+1)\cdot\sin\theta$

$\Rightarrow \quad 4\sin^2\theta-2\sqrt{3}\sin\theta-2\sin\theta+\sqrt{3}=0$

$\Rightarrow \quad 2\sin\theta(2\sin\theta-\sqrt{3})-1(2\sin\theta-\sqrt{3})=0$

$\Rightarrow \quad (2\sin\theta-1)(2\sin\theta-\sqrt{3})=0$

$\Rightarrow \quad \sin\theta=\frac{1}{2}$ or $\sin\theta=\frac{\sqrt{3}}{2}$

$\Rightarrow \quad \theta=n\pi+(-1)^n\frac{\pi}{6}$

or $\quad \theta=n\pi+(-1)^n\frac{\pi}{3}$

$\Rightarrow$ General solution set is,

$$\left\{\theta\,|\,\theta=n\pi+(-1)^n\frac{\pi}{6}\right\}\cup\left\{\theta\,|\,\theta=n\pi+(-1)^n\frac{\pi}{3}\right\}$$

put $n=0$, $\theta=\frac{\pi}{6}$ or $\frac{\pi}{3}$

put $n=1$, $\theta=\pi-\frac{\pi}{6}=\frac{5\pi}{6}$ or $\pi-\frac{\pi}{3}=\frac{2\pi}{3}$

$\therefore$ Solutions are $\frac{\pi}{6},\frac{\pi}{3},\frac{2\pi}{3}$ and $\frac{5\pi}{6}$.

Ex. 46. *Solve the following system of equations.*

$\sin x+\cos y=1$, $\cos 2x-\cos 2y=1$

Sol. Given, $\quad \sin x+\cos y=1 \quad$ ...(i)

and $\quad (1-2\sin^2 x)-(2\cos^2 y-1)=1$

i.e. $\quad \sin^2 x+\cos^2 y=\frac{1}{2} \quad$ ...(ii)

Put $\sin x=u$ and $\cos y=v$ in Eqs. (i) and (ii),

$$u+v=1 \text{ and } u^2+v^2=\frac{1}{2}$$

Solving above equations; $u=\frac{1}{2}$ and $v=\frac{1}{2}$ $\Rightarrow \sin x=\frac{1}{2}$

$\Rightarrow \quad x=n\pi+(-1)^n\frac{\pi}{6}, n\in I$ and $\cos y=\frac{1}{2}$

$\Rightarrow \quad y=2m\pi\pm\frac{\pi}{3}, m\in I$

$\therefore$ The given equations have solutions,

$x=n\pi+(-1)^n\frac{\pi}{6}, n\in I$ and $y=2m\pi\pm\frac{\pi}{3}, m\in I$

Ex. 47. *Find the coordinates of the points of intersection of the curves* $y=\cos x$, $y=\sin 3x$ *if* $-\frac{\pi}{2}\le x\le\frac{\pi}{2}$.

Sol. The point of intersection is given by

$$\sin 3x=\cos x=\sin\left(\frac{\pi}{2}-x\right)$$

$\Rightarrow \quad 3x=n\pi+(-1)^n\left(\frac{\pi}{2}-x\right)$

(i) Let n be even i.e. $n=2m$

$\Rightarrow \quad 3x=2m\pi+\left(\frac{\pi}{2}-x\right)$

$\Rightarrow \quad x=\frac{m\pi}{2}+\frac{\pi}{8} \quad$ ...(i)

(ii) Let n be odd i.e. $n = 2m + 1$

$$\Rightarrow \quad 3x = (2m+1)\pi - \left(\frac{\pi}{2} - x\right)$$

$$\Rightarrow \quad 3x = 2m\pi + \frac{\pi}{2} + x \Rightarrow 2x = 2m\pi + \frac{\pi}{2}$$

$$\Rightarrow \quad x = m\pi + \frac{\pi}{4} \qquad \text{...(ii)}$$

Now, as $-\frac{\pi}{2} \le x \le \frac{\pi}{2}$

$$\therefore \quad x = \frac{\pi}{8}, \frac{\pi}{4}, -\frac{3\pi}{8} \qquad \text{[From Eqs. (i) and (ii)]}$$

Thus, point of intersection are

$$\left(\frac{\pi}{8}, \cos\frac{\pi}{8}\right), \left(\frac{\pi}{4}, \cos\frac{\pi}{4}\right), \left(-\frac{3\pi}{8}, \cos\frac{3\pi}{8}\right)$$

Ex. 48. *Find the range of y such that the equation in x, $y + \cos x = \sin x$ has a real solutions. For $y = 1$, find x such that $0 < x < 2\pi$.*

Sol. We have, $y = \sin x - \cos x = \sqrt{2}\left(\frac{1}{\sqrt{2}}\sin x - \frac{1}{\sqrt{2}}\cos x\right)$

$$-\sqrt{2}\sin\left(x - \frac{\pi}{4}\right)$$

$$\Rightarrow \quad -\sqrt{2} \le y \le \sqrt{2} \qquad \text{...(i)}$$

Now, for $y = 1$

$$\sin x - \cos x = 1$$

$$\Rightarrow \quad \frac{1}{\sqrt{2}}\sin x - \frac{1}{\sqrt{2}}\cos x = \frac{1}{\sqrt{2}}$$

$$\Rightarrow \quad \sin\left(x - \frac{\pi}{4}\right) = \sin\left(\frac{\pi}{4}\right)$$

$$\Rightarrow \quad x - \frac{\pi}{4} = n\pi + (-1)^n \frac{\pi}{4}$$

$$\Rightarrow \quad x = n\pi + (-1)^n \frac{\pi}{4} + \frac{\pi}{4}$$

$$\therefore \quad x = \frac{\pi}{2}, \pi$$

Ex. 49. *A triangle ABC is such that $\sin(2A + B) = \frac{1}{2}$. If A, B, C are in AP, then find the value of A, B and C.*

Sol. We have, $\sin(2A + B) = \frac{1}{2} = \sin\left(\frac{\pi}{6}\right)$

$$\Rightarrow \quad 2A + B = n\pi + (-1)^n \frac{\pi}{6} \qquad \text{...(i)}$$

Also, we have, $A + B + C = \pi$ and $2B = A + C$

$$\Rightarrow \quad 3B = \pi \Rightarrow B = \frac{\pi}{3} \qquad \text{...(ii)}$$

From Eq. (i), for $n = 1$

$$2A + B = \frac{5\pi}{6}$$

$$\Rightarrow \quad 2A = \frac{5\pi}{6} - \frac{\pi}{3} \qquad \text{[from Eq. (ii)]}$$

$$\Rightarrow \quad A = \frac{\pi}{4} \Rightarrow C = \frac{5\pi}{12}$$

Ex. 50. *Find the number of values of θ in the interval $\left(-\frac{\pi}{2}, \frac{\pi}{2}\right)$, satisfying the equation,*

$(1 - \tan\theta)(1 + \tan\theta)\cdot\sec^2\theta + 2^{\tan^2\theta} = 0$. *Also, find $\theta \in \left[-\frac{\pi}{3}, \frac{\pi}{3}\right]$, satisfying the given equation.*

Sol. Given, $(1 - \tan\theta)(1 + \tan\theta)\sec^2\theta + 2^{\tan^2\theta} = 0$

$$\Rightarrow \quad (1 - \tan^2\theta)(1 + \tan^2\theta) + 2^{\tan^2\theta} = 0$$

$$\Rightarrow \quad 1 - \tan^4\theta + 2^{\tan^2\theta} = 0$$

$$\Rightarrow \quad 1 + 2^{\tan^2\theta} = \tan^4\theta$$

By observation, we have $\tan^2\theta = 3$.

$$\Rightarrow \quad \theta = n\pi \pm \left(\frac{\pi}{3}\right)$$

Moreover there will be values of θ, satisfying, $3 < \tan^2\theta < 4$ and satisfying the given equation as if $f(x) = x^2 - 2^x - 1$, then $f(3^+)\,f(4^-) < 0$.

So the number of values of θ is 4.

And θ, lying in the interval $\left[-\frac{\pi}{3}, \frac{\pi}{3}\right]$ is $\pm\frac{\pi}{3}$.

Ex. 51. *Find the general solutions of the equation*

$$\left(\cos\frac{x}{4} - 2\sin x\right)\sin x + \left(1 + \sin\frac{x}{4} - 2\cos x\right)\cos x = 0.$$

Sol. Here,

$$\left(\cos\frac{x}{4} - 2\sin x\right)\sin x + \left(1 + \sin\frac{x}{4} - 2\cos x\right)\cos x = 0$$

$$\Rightarrow \cos\frac{x}{4}\cdot\sin x - 2\sin^2 x + \cos x + \sin\frac{x}{4}\cdot\cos x - 2\cos^2 x = 0$$

$$\Rightarrow \left(\sin x\cdot\cos\left(\frac{x}{4}\right) + \sin\left(\frac{x}{4}\right)\cdot\cos x\right) + \cos x - 2(\sin^2 x + \cos^2 x) = 0$$

$$\Rightarrow \quad \sin\left(x + \frac{x}{4}\right) + \cos x = 2$$

Since, the greatest value of $\sin\left(\frac{5x}{4}\right)$ and $\cos(x)$ is 1.

Therefore their sum is equal to 2 only if,

$\sin\left(\frac{5x}{4}\right) = 1$ and $\cos(x) = 1$.

$$\Rightarrow \quad \frac{5x}{4} = 2n\pi + \frac{\pi}{2} \text{ and } x = 2k\pi \ (n, k \in I)$$

Since, we have two choose those values of x which satisfy both of these equations, we have

$$2k\pi = \frac{8n\pi}{5} + \frac{2\pi}{5}$$

$\Rightarrow k = \frac{4n+1}{5}$, where both k and n are integers. We write,

$$k = n - \frac{(n-1)}{5}$$

for $\frac{n-1}{5} = m$, we have $n = 1 + 5m$ and $k = 1 + 4m$ $(m \in I)$

$\therefore \quad x = 2\pi + 8m\pi, m \in I.$

Ex. 52. *Find all possible triplets (x, y, z) such that $(x+y)+(y+2z)\cos 2\theta + (z-x)\sin^2\theta = 0$, for all θ.*

Sol. We have,

$$(x+y)+(y+2z)\cos 2\theta + (z-x)\left(\frac{1-\cos 2\theta}{2}\right) = 0$$

$$\Rightarrow (2x+2y)+(2y+4z)\cos 2\theta + (z-x) - (z-x)\cos 2\theta = 0$$

$$\Rightarrow (2x+2y+z-x)+(2y+4z-z+x)\cos 2\theta = 0$$

$$\Rightarrow (x+2y+z)+(x+2y+3z)\cos 2\theta = 0$$

Which is zero for all values of θ, if

$x + 2y + z = 0$ and $x + 2y + 3z = 0$

$\Rightarrow \quad x + 2y + z = 0$

and $\quad x + 2y + 3z = 0$

$$\frac{x}{6-2} = \frac{y}{1-3} = \frac{z}{2-2} = k$$

$\Rightarrow \quad x = 4k, y = -2k, z = 0$

or $\quad x = 2k, y = -k, z = 0$

i.e. $(2k, -k, 0)$ for any $k \in R$.

Hence, there are infinite number of triplets.

Ex. 53. *For every real number find all the real solutions to equation $\sin x + \cos(a+x) + \cos(a-x) = 2$.*

Sol. Given, $\sin x + \cos(a+x) + \cos(a-x) = 2$

$\Rightarrow \quad 1 \cdot \sin x + 2 \cdot \cos a \cdot \cos x = 2$

Let $\quad 1 = r\sin\phi, 2\cos a = r\cos\phi$

$\Rightarrow \quad r(\cos x\cos\phi + \sin x \cdot \sin\phi) = 2$

$\Rightarrow \quad r\cos(x-\phi) = 2, \begin{cases} \text{where } r^2 = 1 + 4\cos^2 a \\ \Rightarrow r = \sqrt{1+4\cos^2 a} \end{cases}$

$$\Rightarrow \quad \cos(x-\phi) = \frac{2}{r}$$

$$\Rightarrow \quad \cos(x-\phi) = \frac{2}{\sqrt{1+4\cos^2 a}} \quad \text{...(i)}$$

This equation has real solution if,

$$\frac{2}{\sqrt{1+4\cos^2 a}} \le 1$$

or $\quad 2 \le \sqrt{1+4\cos^2 a}$ or $4\cos^2 a \ge 3$

or $\quad \cos^2 a \ge \frac{3}{4}$

$$-\cos^2 a \le -\frac{3}{4} \Rightarrow 1 - \cos^2 a \le 1 - \frac{3}{4}$$

$\Rightarrow \quad \sin^2 a \le \frac{1}{4}$

$\Rightarrow \quad (\sin a + 1/2)(\sin a - 1/2) \le 0$

$\Rightarrow \quad -\frac{1}{2} \le \sin a \le \frac{1}{2}$

$\therefore \quad m\pi - \frac{\pi}{6} \le a \le m\pi + \frac{\pi}{6}$, where $m \in I$

Clearly, $x - \phi = 2n\pi \pm \cos^{-1}\left(\frac{2}{\sqrt{1+4\cos^2 a}}\right)$ [From Eq. (i)]

$$\Rightarrow \quad x = \phi + 2n\pi \pm \cos^{-1}\left(\frac{2}{\sqrt{1+4\cos^2 a}}\right)$$

where $\cos\phi = \frac{2\cos a}{\sqrt{1+4\cos^2 a}}$ or $\phi = \cos^{-1}\left(\frac{2\cos a}{\sqrt{1+4\cos^2 a}}\right)$

Ex. 54. *Find the solutions of the equation $(\sin x + \cos x)\sin 2x = a(\sin^3 x + \cos^3 x)$ located between $\frac{\pi}{2}$ and π and for which values of 'a' does this equation have at most one solution satisfying the condition $\frac{\pi}{2} \le x \le \pi$.*

Sol. $(\sin x + \cos x)\sin 2x = a(\sin x + \cos x)[\sin^2 x + \cos^2 x - \sin x\cos x]$

$$\Rightarrow (\sin x + \cos x)\left[\sin 2x - a\left(1 - \frac{1}{2}\sin 2x\right)\right] = 0 \quad \text{...(i)}$$

Now, $\quad \sin x + \cos x = 0 \Rightarrow \tan x = -1 = \tan\frac{3\pi}{4}$

$\therefore \quad x = \frac{3\pi}{4} \quad \left[\because \frac{\pi}{2} \le x \le \pi\right]$

Hence, there is always at least one root lying in $\frac{\pi}{2}$ and π for any value of the parameter a.

Now, $\quad 2\sin 2x - 2a + a\sin 2x = 0$ [from Eq. (i)]

$$\Rightarrow \quad \sin 2x = \frac{2a}{2+a} \quad \text{...(ii)}$$

Since, $\quad \frac{\pi}{2} < x < \pi$ or $\pi < 2x < 2\pi$

$$\Rightarrow \quad -1 \le \sin 2x < 0 \quad \text{...(iii)}$$

Now, from Eqs. (ii) and (iii), we have

$$-1 \le \frac{2a}{2+a} < 0, \text{ where } a \ne -2$$

$$\Rightarrow \quad -1 \le \frac{2a}{2+a} \text{ and } \frac{2a}{2+a} < 0$$

$$\Rightarrow \qquad 0 \le \frac{2a}{2+a} + 1 \text{ and } a(2+a) < 0$$

Using number line rules, we get

$$a < -2 \qquad \text{or} \qquad a \ge -\frac{2}{3} \qquad \text{...(A)}$$

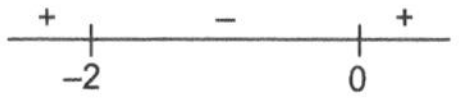

$$\text{and} \qquad -2 < a \le 0 \qquad \text{...(B)}$$

$$\text{From (A) and (B)} \qquad -\frac{2}{3} \le a \le 0 \qquad \text{...(iv)}$$

[i.e. common to both (A) and (B)]

Hence, for every value of 'a' satisfying the condition $-\frac{2}{3} \le a \le 0$ the equation, $\sin 2x = \frac{2a}{2+a}$ has the roots lying between $\frac{\pi}{2}$ and π.

Now, we have to find the solution of the equation

$\sin 2x = \frac{2a}{2+a}$, where $-\frac{2}{3} \le a \le 0$ and $\frac{\pi}{2} < x < \pi$

Clearly, $\pi < 2x < 2\pi$

Case I $\pi < 2x < \frac{3\pi}{2}$

$$0 < 2x - \pi < \frac{\pi}{2}$$

$$\because \qquad \sin(2x) = \frac{2a}{2+a}$$

$$\therefore \qquad \sin(2x - \pi) = -\frac{2a}{2+a}; \quad x = \frac{\pi}{2} - \frac{1}{2}\sin^{-1}\left(\frac{2a}{2+a}\right)$$

Case II $\frac{3\pi}{2} < 2x < 2\pi$

$$-\frac{\pi}{2} < 2x - 2\pi < 0$$

$$\text{Since,} \qquad \sin 2x = \frac{2a}{2+a}$$

$$\therefore \qquad \sin(2x - 2\pi) = \frac{2a}{2+a}$$

$$2x - 2\pi = \sin^{-1}\left(\frac{2a}{2+a}\right)$$

$$x = \pi + \frac{1}{2}\sin^{-1}\left(\frac{2a}{2+a}\right)$$

$$\text{Thus,} \quad x = \begin{cases} \frac{3\pi}{4}, \text{ for } a \leftarrow (-\infty, \infty) \\ \frac{\pi}{2} - \frac{1}{2}\sin^{-1}\left(\frac{2a}{2+a}\right), & \text{for } \frac{-2}{3} \le a \le 0 \\ \pi + \frac{1}{2}\sin^{-1}\left(\frac{2a}{2+a}\right), & \text{for } \frac{-2}{3} \le a \le 0 \end{cases}$$

Ex. 55. *Solve the equation*

$$(\tan x)^{\cos^2 x} = (\cot x)^{\sin x}.$$

Sol. $(\tan x)^{\cos^2 x} = \left(\frac{1}{\tan x}\right)^{\sin x}$

$$\Rightarrow \qquad (\tan x)^{\cos^2 x + \sin x} = 1$$

Now, here two case arises,

Case I When $\tan x = 1$, the power $(\cos^2 x + \sin x)$ can take any value.

$$\therefore \qquad \tan x = 1$$

$$\Rightarrow \qquad x = n\pi + \frac{\pi}{4}$$

Case II When $\tan x \ne 1, 0$ and $\cos^2 x + \sin x = 0$

$$\text{or} \qquad x \ne n\pi, n\pi + \frac{\pi}{4}$$

$$\text{and} \qquad 1 - \sin^2 x + \sin x = 0.$$

$$\Rightarrow \qquad \sin^2 x - \sin x - 1 = 0$$

$$\therefore \qquad \sin x = \frac{1 \pm \sqrt{1+4}}{2} = \frac{\sqrt{5}+1}{2}, \frac{1-\sqrt{5}}{2}$$

Here, $\frac{\sqrt{5}+1}{2} > 1$ which is not possible.

$$\sin x = \frac{1-\sqrt{5}}{2} = \sin\alpha$$

$$\therefore \; x = n\pi + (-1)^n\alpha, \text{ where } \alpha = \sin^{-1}\left(\frac{1-\sqrt{5}}{2}\right)$$

Thus, $x = n\pi + \frac{\pi}{4}$ or $x = n\pi + (-1)^n\alpha$,

where $\alpha = \sin^{-1}\left(\frac{1-\sqrt{5}}{2}\right)$

Ex. 56. *Solve the equation*

$$a\cos x + \cot x + 1 = \operatorname{cosec} x.$$

Sol. $a\cos x + \frac{\cos x}{\sin x} + 1 = \frac{1}{\sin x}$

$$\Rightarrow \qquad a\sin x\cos x + \cos x + \sin x = 1 \; (\sin x \ne 0)$$

$$\Rightarrow \qquad \sin x + \cos x = 1 - a\sin x\cos x$$

On squaring both sides, we get

$$1 + 2\sin x\cos x = 1 + a^2\sin^2 x\cos^2 x - 2a\sin x \cdot \cos x$$

$$a^2\sin^2 x\cos^2 x - 2(a+1)\sin x\cos x = 0$$

$$\sin x\cos x[a^2\sin x\cos x - 2(a+1)] = 0$$

$$\sin 2x[a^2\sin 2x - 4(a+1)] = 0$$

$$\therefore \qquad \sin 2x = 0 \text{ for any value of } a.$$

$$2x = n\pi$$

$$x = \frac{n\pi}{2} \text{ for } a \in (-\infty, \infty)$$

$$\text{or} \qquad a^2\sin 2x - 4(a+1) = 0$$

$$\Rightarrow \quad \sin 2x = \frac{4(a+1)}{a^2}$$

$\because -1 \le \sin 2x \le 1$ for all values of x.

$$\therefore \quad -1 \le \frac{4(a+1)}{a^2} \le 1$$

$$-a^2 \le 4(a+1) \le a^2$$

i.e. $\quad -a^2 \le 4a + 4$ or $4a + 4 \le a^2$

$$\Rightarrow \quad a^2 + 4a + 4 \ge 0$$

or $\quad a^2 - 4a - 4 \ge 0$

$$\Rightarrow \quad (a+2)^2 \ge 0$$

or $\quad (a-2)^2 - 8 \ge 0$

which is true for all 'a' or $(a-2)^2 \ge 8$

$$\Rightarrow \quad a - 2 \ge 2\sqrt{2}$$

or $\quad a - 2 \le -2\sqrt{2}$

$$\Rightarrow \quad a \le 2 - 2\sqrt{2} \quad \text{or} \quad a \ge 2 + 2\sqrt{2}$$

or $\quad a \in (-\infty, 2 - 2\sqrt{2}] \cup [2 + 2\sqrt{2}, \infty)$

$$\Rightarrow \quad 2x = n\pi + (-1)^n \sin^{-1}\left(\frac{4(a+1)}{a^2}\right)$$

$$x = \frac{n\pi}{2} + \frac{(-1)^n}{2}\sin^{-1}\left(\frac{4(a+1)}{a^2}\right)$$

$$\Rightarrow \quad x = \begin{cases} \dfrac{n\pi}{2}; \text{ for } a \in (-\infty, \infty) \\ \dfrac{n\pi}{2} + \dfrac{(-1)^n}{2}\sin^{-1}\left(\dfrac{4(a+1)}{2}\right); \end{cases}$$

$$\text{for } a \in (-\infty, 2 - 2\sqrt{2}] \cup [2 + 2\sqrt{2}, \infty)$$

Ex. 57. *Find the values of 'a' which the system of equations $\sin x \cdot \cos y = a^2$ and $\sin y \cdot \cos x = a$ have a solution.*

Sol. We have,

$$\sin x \cdot \cos y = a^2 \text{ and } \cos x \cdot \sin y = a$$

adding above equations,

$$\sin x \cos y + \cos x \sin y = a^2 + a$$

$$\Rightarrow \quad \sin(x+y) = a^2 + a \quad \text{...(i)}$$

As we know,

$$-1 \le \sin(x+y) \le 1$$

$$\therefore \quad -1 \le a^2 + a \le 1$$

i.e. $\quad -1 \le a^2 + a$ and $a^2 + a \le 1$

$$\Rightarrow \quad a^2 + a + 1 \ge 0 \text{ and } a^2 + a - 1 \le 0$$

$$\Rightarrow \quad \left(a + \frac{1}{2}\right)^2 + \frac{3}{4} \ge 0 \text{ and } \left(a + \frac{1}{2}\right)^2 - \frac{5}{4} \le 0$$

$$\Rightarrow \quad \text{true for all real '}a\text{' and } -\frac{\sqrt{5}}{2} \le a + \frac{1}{2} \le \frac{\sqrt{5}}{2}$$

$$\Rightarrow \quad -\left(\frac{\sqrt{5}+1}{2}\right) \le a \le \frac{\sqrt{5}-1}{2} \quad \text{...(ii)}$$

Again, on subtracting the two given equations,

$$\sin x \cos y - \cos x \sin y = a^2 - a$$

$$\Rightarrow \quad \sin(x-y) = a^2 - a \quad \text{...(iii)}$$

As we know,

$$-1 \le \sin(x-y) \le 1$$

$$\Rightarrow \quad -1 \le a^2 - a \le 1$$

$$\Rightarrow \quad -1 \le a^2 - a \quad \text{and} \quad a^2 - a \le 1$$

$$\Rightarrow \quad a^2 - a + 1 \ge 0 \text{ and } a^2 - a - 1 \le 0$$

$$\Rightarrow \quad \left(a - \frac{1}{2}\right)^2 + \frac{3}{4} \ge 0 \text{ and } \left(a - \frac{1}{2}\right)^2 \le \frac{5}{4}$$

$$\Rightarrow \text{True for all real '}a\text{' and } -\frac{\sqrt{5}}{2} \le a - \frac{1}{2} \le \frac{\sqrt{5}}{2}$$

$$\Rightarrow \quad \left(\frac{1-\sqrt{5}}{2}\right) \le a \le \frac{1+\sqrt{5}}{2} \quad \text{...(iv)}$$

From Eqs. (ii) and (iv) common solution is

$$\left(\frac{1-\sqrt{5}}{2}\right) \le a \le \left(\frac{\sqrt{5}-1}{2}\right)$$

or $\quad a \in \left[\dfrac{1-\sqrt{5}}{2}, \dfrac{\sqrt{5}-1}{2}\right]$

Ex. 58. *Find all the numbers 'a' for which any root of the equation* $\sin 3x = a \sin x + (4 - 2|a|)\sin^2 x$ *is a root of the equation* $\sin 3x + \cos 2x = 1 + 2\sin x \cos 2x$ *and any root of the latter equation is a root of the former.*

Sol. The first equation of the system can be written as;

$$3\sin x - 4\sin^3 x = a\sin x + (4 - 2|a|)\sin^2 x$$

$$\sin x\{4\sin^2 x + (4 - 2|a|)\sin x + (a-3)\} = 0 \quad \text{...(i)}$$

The second equation is,

$$\sin 3x + \cos 2x = 1 + 2\sin x \cos 2x$$

$$\sin 3x + \cos 2x = 1 + \sin 3x - \sin x$$

$$\cos 2x = 1 - \sin x$$

$$1 - 2\sin^2 x = 1 - \sin x$$

$$\sin x(2\sin x - 1) = 0 \quad \text{...(ii)}$$

$\because$ Both equations have a common solution, therefore

$$\sin x = 0$$

Also, second root of Eq. (ii) i.e.

$$\sin x = \frac{1}{2}, \text{ satisfy Eq. (i).}$$

From Eq. (i), $4\left(\dfrac{1}{2}\right)^2 + (4 - 2|a|)\dfrac{1}{2} + (a-3) = 0$

$$\Rightarrow \quad 1 + 2 - |a| + (a-3) = 0$$

$$\Rightarrow \quad |a| = a \text{ or } a \ge 0 \text{ for } \sin x = \frac{1}{2} \quad \text{...(iii)}$$

Again from Eq. (i);

$$4\sin^2 x + (4 - 2|a|)\sin x + (a - 3) = 0$$

For real x, $(4 - 2|a|)^2 - 4 \cdot 4 \cdot (a - 3) \geq 0$

$\Rightarrow$ $(2 - a)^2 - 4(a - 3) \geq 0$

$\Rightarrow$ $4 - 4a + a^2 - 4a + 12 \geq 0$

$\Rightarrow$ $a^2 - 8a + 16 \geq 0$

$\Rightarrow$ $(a - 4)^2 \geq 0$

$\therefore$ $\sin x = \dfrac{-(4 - 2|a|) \pm 2\sqrt{(a-4)^2}}{2 \cdot 4}$

$\Rightarrow$ $\sin x = \dfrac{|a| - 2 \pm \sqrt{(a-4)^2}}{4}$

$\Rightarrow$ $\sin x = \dfrac{|a| - 2 + a - 4}{4}$

or $\sin x = \dfrac{|a| - 2 - a + 4}{4}$

$\Rightarrow$ $\sin x = \dfrac{2a - 6}{4}$ or $\sin x = \dfrac{1}{2}$ [$\because |a| = a$ from Eq. (iii)]

$\Rightarrow$ $\sin x = \dfrac{a - 3}{2}$ or $\sin x = \dfrac{1}{2}$

For real x the values of 'a' will be suitable in the following three cases (also $a \geq 0$).

(i) $\dfrac{a-3}{2} = 0 \Rightarrow a = 3$

(ii) $\dfrac{a-3}{2} = \dfrac{1}{2} \Rightarrow a = 4$

(iii) $\left|\dfrac{a-3}{2}\right| \leq 1$ or $|a - 3| \leq 2$ or $-2 \leq a - 3 \leq 2$

$\Rightarrow$ $1 \leq a \leq 5$

$\therefore$ Hence, $a \in [1, 5]$.

Ex. 59. *Solve the inequality*

$$\frac{5}{4}\sin^2 x + \sin^2 x \cdot \cos^2 x > \cos 2x.$$

Sol. Re-writing the inequality in the form,

$$5(1 - \cos 2x) + 2(1 - \cos^2 2x) > 8\cos 2x$$

or $2\cos^2 2x + 13\cos 2x - 7 < 0$

Putting $y = \cos 2x$, we get

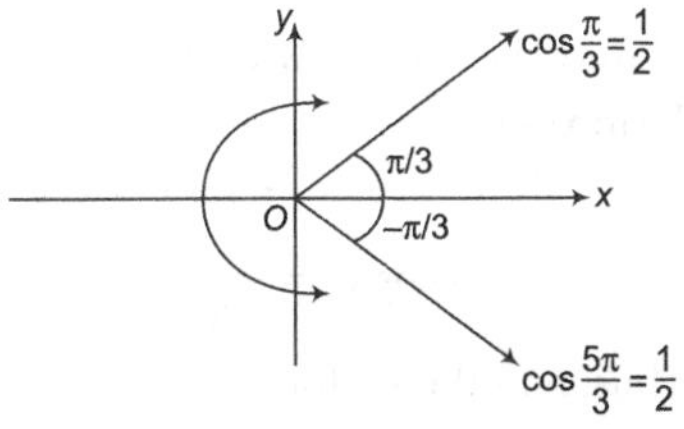

$$2y^2 + 13y - 7 < 0 \text{ or } (2y - 1)(y + 7) < 0$$

or y lies between -7 and $\dfrac{1}{2}$ or $-7 < \cos 2x < \dfrac{1}{2}$

Since, $-1 \leq \cos 2x$ for all x.

$\therefore$ $-1 \leq \cos 2x < \dfrac{1}{2}$,

Using figure, we get $2n\pi + \dfrac{\pi}{3} < 2x < 2n\pi + \dfrac{5\pi}{3}$

$\Rightarrow$ $n\pi + \dfrac{\pi}{6} < x < n\pi + \dfrac{5\pi}{6}$

Ex. 60. *Solve the inequality,*

$$\sin x \cos x + \frac{1}{2}\tan x \geq 1$$

Sol. Left hand side is defined for all x except,

$$x = n\pi + \frac{\pi}{2}, \text{ where } n \in I$$

Now, we have $\dfrac{2\tan x}{1 + \tan^2 x} + \tan x \geq 2$ {from given equation}

Putting $y = \tan x$, we get

$$\frac{2y}{1 + y^2} + y \geq 2 \text{ or } \frac{2y + y(1 + y^2) - 2(1 + y^2)}{(1 + y^2)} \geq 0$$

$\because$ $1 + y^2 > 0$ for all y.

$\therefore$ $2y + y(1 + y^2) - 2(1 + y^2) \geq 0$

$\Rightarrow$ $y^3 - 2y^2 + 3y - 2 \geq 0$

or $y^2(y - 1) - y(y - 1) + 2(y - 1) \geq 0$

or $(y - 1)(y^2 - y + 2) \geq 0$

where $y^2 - y + 2 = \left(y - \dfrac{1}{2}\right)^2 + \dfrac{7}{4} > 0$ for all y.

$\therefore$ $(y - 1) \geq 0$ or $\tan x \geq 1$

From figure, we get

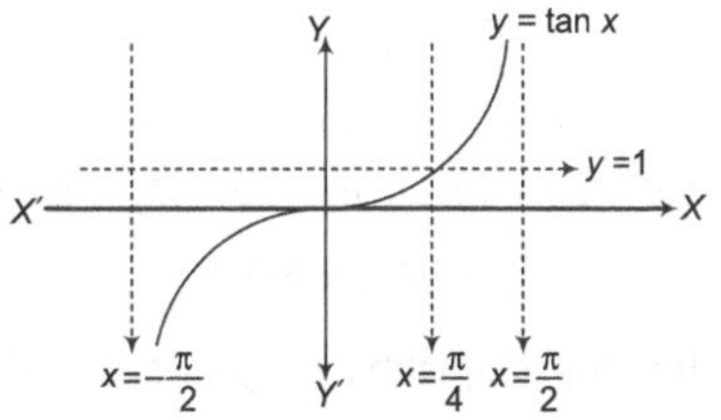

Hence, the solution of the inequality lies in the interval,

$$n\pi + \frac{\pi}{4} \leq x < n\pi + \frac{\pi}{2}, n \in I$$

i.e. $x \in \left[n\pi + \dfrac{\pi}{4}, n\pi + \dfrac{\pi}{2}\right)$

Ex. 61. *If* $0 \leq x \leq 2\pi$, *then solve the inequality,*

$$2^{\text{cosec}^2 x}\sqrt{\frac{1}{2}y^2 - y + 1} \leq \sqrt{2}.$$

Sol. The given inequality can be written as;

$$2^{\text{cosec}^2 x}\sqrt{(y-1)^2 + 1} \leq 2 \quad \text{...(i)}$$

Since, $\text{cosec}^2 x \geq 1$ for all real x.

$$\therefore \quad 2^{\text{cosec}^2 x} \geq 2 \quad \text{...(ii)}$$

Also, $(y-1)^2 + 1 \geq 1$

$$\Rightarrow \quad \sqrt{(y-1)^2 + 1} \geq 1 \quad \text{...(iii)}$$

From Eqs. (ii) and (iii),

$$2^{\text{cosec}^2 x} \sqrt{(y+1)^2 + 1} \geq 2 \quad \text{...(iv)}$$

$\therefore$ From Eqs. (i) and (iv), equality holds only when,

$$2^{\text{cosec}^2 x} = 2 \text{ and } \sqrt{(y-1)^2 + 1} = 1$$

or $\quad \text{cosec}^2 x = 1$

and $\quad (y-1)^2 + 1 = 1$

or $\quad \sin x = \pm 1$ and $y = 1$

or $\quad x = \dfrac{\pi}{2}, \dfrac{3\pi}{2}$ and $y = 1$

Hence, the solution of the given inequality is $x = \dfrac{\pi}{2}, \dfrac{3\pi}{2}$ and $y = 1$.

Ex. 62. *Solve the inequality,*

$$5 + 2\cos 2x \leq 3\,|\,2\sin x - 1\,|.$$

Sol. The given inequality can be written as;

$$5 + 2(1 - 2\sin^2 x) \leq 3\,|\,2\sin x - 1\,|$$

or $\quad 7 - 4\sin^2 x \leq 3\,|\,2\sin x - 1\,|$

Putting $y = \sin x$,

$$7 - 4y^2 \leq 3\,|\,2y - 1\,| \quad \text{...(i)}$$

Now, consider the two cases

(i) $2y - 1 \geq 0$ or $y \geq \dfrac{1}{2}$

then, $\quad 7 - 4y^2 \leq 3(2y - 1)$

$[\because$ for $y \geq a, |\,y - a\,| = (y - a)]$

or $\quad 4y^2 + 6y - 10 \geq 0$ or $2y^2 + 3y - 5 \geq 0$

or $\quad (y-1)(2y+5) \geq 0$

Solving this inequality, we get $y \leq -\dfrac{5}{2}$ or $y \geq 1$.

But from the condition $y \geq \dfrac{1}{2}$, we have $y \geq 1$.

i.e. $\quad \sin x \geq 1$.

The inequality holds true only for x satisfying the equation $\sin x = 1$ that is

$$x = 2n\pi + \frac{\pi}{2} \text{ when } n \in I.$$

(ii) Let $2y - 1 < 0$ or $y < \dfrac{1}{2}$

then, $\quad 7 - 4y^2 \leq -3(2y - 1)$

$[\because$ for $y < a, |\,y - a\,| = -(y - a)]$

or $\quad 2y^2 - 3y - 2 \geq 0$

or $\quad (y-2)(2y+1) \geq 0$

On solving the inequality, we get

$$y \leq -\frac{1}{2} \text{ or } y \geq 2.$$

But from condition $y < \dfrac{1}{2}$, we have $y \leq -\dfrac{1}{2}$.

or $\quad \sin x \leq -\dfrac{1}{2}$

On solving the inequality (from graph we have)

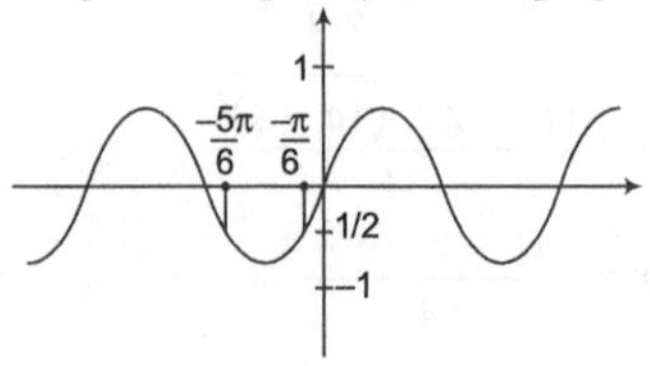

we get, $\quad 2n\pi - \dfrac{5\pi}{6} \leq x \leq 2n\pi - \dfrac{\pi}{6}$

Thus, $\quad x = 2n\pi + \dfrac{\pi}{2}$ or $x \in \left[2n\pi - \dfrac{5\pi}{6}, 2n\pi - \dfrac{\pi}{6}\right]$

Ex. 63. *Solve*

$$|\cos x - 2\sin 2x - \cos 3x| = 1 - 2\sin x - \cos 2x.$$

Sol. Here, LHS $= |\cos x - 2\sin 2x - \cos 3x|$

$= |(\cos x - \cos 3x) - 2\sin 2x|$

$= |2\sin 2x \cdot \sin x - 2\sin 2x|$

$= |2\sin 2x(\sin x - 1)|$

and RHS $= 1 - 2\sin x - \cos 2x$

$= 2\sin^2 x - 2\sin x = 2\sin x(\sin x - 1)$

Thus, $|\cos x - 2\sin 2x - \cos 3x| = 1 - 2\sin x - \cos 2x$ could be rewritten as,

$|\,2\sin 2x\,(\sin x - 1)| = 2\sin x(\sin x - 1)$

where $1 - \sin x \geq 0$, for all real x

$\therefore \quad |\,2\sin 2x\,|(1 - \sin x) = 2\sin x(\sin x - 1)$

$\Rightarrow \quad (|\,2\sin 2x\,| + 2\sin x)(1 - \sin x) = 0$

Either $\sin x = 1$

or $4|\sin x||\cos x| + 2\sin x = 0$. So, two cases arises

Case I $\sin x = 1$

$$\Rightarrow \quad x = 2n\pi + \frac{\pi}{2}, n \in I$$

Case II $2|\sin x||\cos x| + \sin x = 0$

If $\sin x > 0$

$\Rightarrow \quad 2\sin x\,|\cos x\,| + \sin x = 0$

or $\quad |\cos x\,| = -\dfrac{1}{2}$ $\quad$ (not possible)

Thus, no solution for

$\sin x > 0$.

Consider $\sin x \leq 0$

$\Rightarrow \quad -2\sin x\,|\cos x\,| + \sin x = 0$

$\Rightarrow \quad \sin x = 0$ i.e. $x = n\pi, n \in I$.

or $$|\cos x| = \frac{1}{2}$$

$\Rightarrow$ $$\cos x = \pm\frac{1}{2} \text{ and } \sin x < 0$$

$\Rightarrow$ Those values which lie in III or IV quadrant

$$\therefore \quad x = \begin{cases} k\pi + \dfrac{\pi}{3}, & \text{for } k = (2m+1), m \in I \\ k\pi - \dfrac{\pi}{3}, & \text{for } k = (2m), m \in I \end{cases}$$

$$\text{Hence, } \quad x = \begin{cases} 2n\pi + \dfrac{\pi}{2}, n \in I \\ (2n+1)\pi + \dfrac{\pi}{3}, n \in I \\ 2n\pi - \dfrac{\pi}{3}, n \in I \end{cases}$$

Ex. 64. *Prove that the equation $2\sin x = |x| + a$ has no solution for $a \in \left(\dfrac{3\sqrt{3}-\pi}{3}, \infty\right)$.*

Sol. We have, $\dfrac{3\sqrt{3}-\pi}{3} > 0$ and hence their arises three cases

Case I When $|x| \geq 2$, we have

$|x| + a > 2$, whereas $2\sin x \leq 2$

Hence the equation,

$2\sin x = |x| + a$, possesses no solution for $|x| \geq 2$

Case II When, $-2 \leq x \leq 0$, we have $0 \leq |x| \leq 2$

$\Rightarrow$ $2\sin x \leq 0$ and $|x| + a > 0$

$\therefore$ The equation,

$2\sin x = |x| + a$ has no solution.

Case III When, $0 < x < 2$, we have

In this case the given equation reduces to

$$2\sin x = x + a$$

Let $f(x) = 2\sin x - x$

$\Rightarrow$ $f'(x) = 2\cos x - 1 = 0$

$\Rightarrow$ $x = \dfrac{\pi}{3} \in (0, 2)$ is a critical point.

$$f''(x) = -2\sin x < 0 \text{ for } x = \frac{\pi}{3}$$

$\Rightarrow$ $x = \dfrac{\pi}{3}$ is a point of maxima.

$$(f(x))_{\max} = 2\sin\frac{\pi}{3} - \frac{\pi}{3} = \frac{2\sqrt{3}}{2} - \frac{\pi}{3} = \frac{3\sqrt{3}-\pi}{3}$$

$\therefore$ $a = 2\sin x - x$, cannot be greater $\dfrac{3\sqrt{3}-\pi}{3}$ for the equation to have a solution. Hence, the result.

Aliter We have, $\sin x = \dfrac{1}{2}|x| + \dfrac{a}{2}$.

Now, consider the graphs of $y_1 = \sin x$ and $y_2 = \dfrac{1}{2}|x|$.

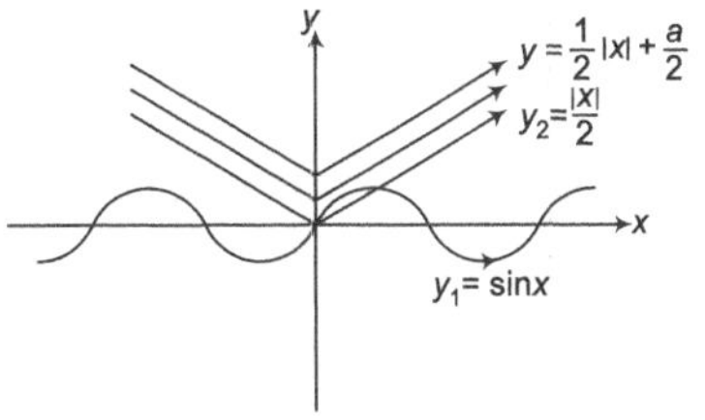

The equation $\sin x = \dfrac{1}{2}|x| + \dfrac{a}{2}$ will have solution, if the line $y = \dfrac{1}{2}|x| + \dfrac{a}{2}$ (parallel to y_2) intersects or touches the curve $y_1 = \sin x$ at least one point. In this case we must have $\dfrac{dy_1}{dx} = \cos x = \dfrac{1}{2}$ (i.e. the slope of the line) $\Rightarrow x = \dfrac{\pi}{3}$.

Hence, the solution exists if,

$$\frac{1}{2}\cdot\frac{\pi}{3} + \frac{1}{2}a \leq \sin\frac{\pi}{3} \Rightarrow a \leq \frac{3\sqrt{3}-\pi}{3}$$

Ex. 65. *In ΔABC, prove that*

$$\text{cosec}\frac{A}{2} + \text{cosec}\frac{B}{2} + \text{cosec}\frac{C}{2} \geq 6.$$

Sol. Since $\dfrac{A}{2}, \dfrac{B}{2}, \dfrac{C}{2}$ all are acute angles, we can use AM $\geq$ GM.

i.e. $$\frac{\text{cosec}\frac{A}{2} + \text{cosec}\frac{B}{2} + \text{cosec}\frac{C}{2}}{3}$$

$$\geq \left(\text{cosec}\frac{A}{2}\cdot\text{cosec}\frac{B}{2}\cdot\text{cosec}\frac{C}{2}\right)^{1/3} \quad \text{...(i)}$$

Consider, $\sin\dfrac{A}{2}\cdot\sin\dfrac{B}{2}\cdot\sin\dfrac{C}{2} = \dfrac{1}{2}\sin\dfrac{A}{2}\cdot\left(2\sin\dfrac{B}{2}\cdot\sin\dfrac{C}{2}\right)$

$$= \frac{1}{2}\sin\frac{A}{2}\left(\cos\left(\frac{B-C}{2}\right) - \cos\left(\frac{B+C}{2}\right)\right)$$

$$= \frac{1}{2}\sin\frac{A}{2}\left(\cos\left(\frac{B-C}{2}\right) - \sin\left(\frac{A}{2}\right)\right)$$

$$\leq \frac{1}{2}\sin\frac{A}{2}\left(1 - \sin\left(\frac{A}{2}\right)\right), \text{ as } \cos\left(\frac{B-C}{2}\right) \leq 1$$

$$\leq \frac{1}{2}\left(\sin\frac{A}{2} - \sin^2\frac{A}{2}\right)$$

$$\leq \frac{1}{2}\left(\frac{1}{4} - \left(\frac{1}{2} - \sin\frac{A}{2}\right)^2\right) \leq \frac{1}{8}$$

$$\therefore \quad \text{cosec}\frac{A}{2}\cdot\text{cosec}\frac{B}{2}\cdot\text{cosec}\frac{C}{2} \geq 8 \quad \text{...(ii)}$$

From Eqs. (i) and (ii),

$$\frac{\text{cosec}\frac{A}{2} + \text{cosec}\frac{B}{2} + \text{cosec}\frac{C}{2}}{3} \geq (8)^{1/3}$$

$$\Rightarrow \quad \text{cosec}\frac{A}{2} + \text{cosec}\frac{B}{2} + \text{cosec}\frac{C}{2} \geq 6$$

Ex. 66. *If* $\dfrac{1}{\cos\alpha\cdot\cos\beta}+\tan\alpha\cdot\tan\beta=\tan\gamma$, *where* $0<\gamma<\dfrac{\pi}{2}$ *and* α,β *are positive acute angles, show that* $\dfrac{\pi}{4}<\gamma<\dfrac{\pi}{2}$.

Sol. Since, $\tan\gamma=\dfrac{1}{\cos\alpha\cos\beta}+\tan\alpha\cdot\tan\beta$, where $0<\gamma<\dfrac{\pi}{2}$

Now, $$1-\tan^2\gamma=1-\left(\frac{1}{\cos\alpha\cos\beta}+\tan\alpha\cdot\tan\beta\right)^2$$

$$=1-\left(\frac{1+\sin\alpha\sin\beta}{\cos\alpha\cos\beta}\right)^2$$

$$=\frac{\cos^2\alpha\cos^2\beta-\sin^2\alpha\sin^2\beta-1-2\sin\alpha\sin\beta}{\cos^2\alpha.\cos^2\beta}$$

$$=\frac{(1-\sin^2\alpha)(1-\sin^2\beta)-\sin^2\alpha\sin^2\beta-1-2\sin\alpha\sin\beta}{\cos^2\alpha\cos^2\beta}$$

$$=\frac{1-\sin^2\beta-\sin^2\alpha+\sin^2\alpha\sin^2\beta-\sin^2\alpha\sin^2\beta-1-2\sin\alpha\sin\beta}{\cos^2\alpha\cos^2\beta}$$

$$=-\frac{(\sin^2\alpha+\sin^2\beta+2\sin\alpha\sin\beta)}{\cos^2\alpha\cdot\cos^2\beta}=-\frac{(\sin\alpha+\sin\beta)^2}{\cos^2\alpha\cdot\cos^2\beta}$$

which is <0. ...(i)

Because, if it is equal to zero then $\sin\alpha+\sin\beta=0$

or $$2\sin\left(\frac{\alpha+\beta}{2}\right)\cos\left(\frac{\alpha-\beta}{2}\right)=0$$

$\Rightarrow$ either $\dfrac{\alpha-\beta}{2}=\dfrac{\pi}{2}$ or $\dfrac{\alpha+\beta}{2}=0$ which are impossible.

Thus, from Eq. (i);

$$1-\tan^2\gamma<0\Rightarrow\tan^2\gamma>1$$

$$\Rightarrow\quad\tan\gamma>1\Rightarrow\gamma>\frac{\pi}{4}$$

$$\therefore\quad\frac{\pi}{4}<\gamma<\frac{\pi}{2}.$$

Ex. 67. *Find the quadrants of the coordinate planes such that for each point* (x,y) *on these quadrants (where* $x\neq0,y\neq0$*), the equation;*

$$\frac{\sin^4\theta}{x}+\frac{\cos^4\theta}{y}=\frac{1}{x+y}\text{ is solvable for }\theta.$$

Sol. Here, $$\frac{\sin^4\theta}{x}+\frac{\cos^4\theta}{y}=\frac{1}{x+y}$$

$$\Rightarrow\quad(x+y)y\sin^4\theta+(x+y)x\cos^4\theta=xy$$

$$\Rightarrow\quad xy(\sin^4\theta+\cos^4\theta)+(x^2\cos^4\theta+y^2\sin^4\theta)=xy$$

$$\Rightarrow\quad xy\{(\sin^2\theta+\cos^2\theta)^2-2\sin^2\theta\cos^2\theta)+(x\cos^2\theta-y\sin^2\theta)^2+2xy\cos^2\theta\sin^2\theta\}=xy$$

$$\Rightarrow\quad xy(1-2\sin^2\theta\cos^2\theta)+\{(x\cos^2\theta-y\sin^2\theta)^2+2xy\cos^2\theta\sin^2\theta\}=xy$$

$$\Rightarrow\quad xy-2xy\sin^2\theta\cos^2\theta+(x\cos^2\theta-y\sin^2\theta)^2+2xy\cos^2\theta\sin^2\theta=xy$$

$$\Rightarrow\quad(x\cos^2\theta-y\sin^2\theta)^2=0\Rightarrow\tan^2\theta=\frac{x}{y}$$

$\Rightarrow$ x and y must be of same sign, which is true in 1st and 3rd quadrant only.

Ex. 68. *For what values of 'b' does the equation* $\dfrac{b\cos x}{2\cos 2x-1}=\dfrac{b+\sin x}{(\cos^2x-3\sin^2x)\tan x}$ *possess solutions.*

Sol. The condition for the existence of solution are,

1. $2\cos 2x-1\neq0$ i.e. $\cos 2x\neq\dfrac{1}{2}$ i.e. $x\neq\dfrac{\pi}{6}$
2. $\tan x\neq0$ i.e. $x\neq0,\pm\dfrac{\pi}{2}$
3. $\cos^2x-3\sin^2x\neq0$ i.e. $\tan^2x\neq\dfrac{1}{3}$ i.e. $x\neq\pm\dfrac{\pi}{6}$

...(i)

Note $2\cos 2x-1=2(\cos^2x-\sin^2x)-(\cos^2x+\sin^2x)$
$$=\cos^2x-3\sin^2x$$

Subject to the above condition, the equation reduces to

$$b\sin x=b+\sin x\Rightarrow\sin x=\frac{b}{b-1}\qquad\text{...(ii)}$$

which is only possible if ; $-1\le\dfrac{b}{b-1}\le1$

i.e. $$\frac{b}{b-1}+1\ge0\text{ and }\frac{b}{b-1}-1\le0$$

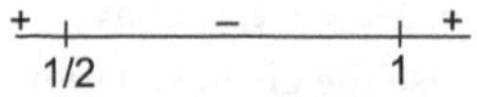

and $\dfrac{1}{b-1}\le0$

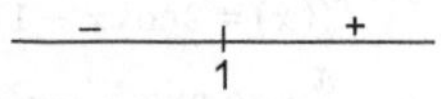

i.e. $$b\le\frac{1}{2}\text{ or }b>1\text{ and }b\le1\qquad\text{...(iii)}$$

From Eqs. (i), (ii) and (iii)

$$b\le\frac{1}{2}\qquad\text{...(iv)}$$

When $b=\dfrac{1}{2}$, $x=\dfrac{-\pi}{2}$ (from Eq. (ii))

But from Eq. (i) $\quad x\neq\dfrac{-\pi}{2}$

$$\Rightarrow\quad b\neq\frac{1}{2}\qquad\text{...(v)}$$

From (iv) and (v) the equation possess solutions only when $b<\dfrac{1}{2}$.

Trigonometric Equations and Inequations Exercise 1 : Single Option Correct Type Questions

1. The equation $2\sin\frac{x}{2}\cos^2 x - 2\sin\frac{x}{2}\sin^2 x$ $= \cos^2 x - \sin^2 x$ has a root for which the false statement is

   (a) $\sin 2x = 1$ (b) $\cos x = \frac{1}{2}$

   (c) $\cos 2x = -\frac{1}{2}$ (d) $\cos x = 1$

2. Let the smallest positive value of x for which the function $f(x) = \sin\frac{x}{3} + \sin\frac{x}{11}, (x \in R)$ achieves its maximum value be x_0. Express x_0 in degree i.e. $x_0 = \alpha^0$. Then, the sum of the digits in α is

   (a) 15 (b) 17

   (c) 16 (d) 18

3. The number of solutions of the equation $16(\sin^5 x + \cos^5 x) = 11(\sin x + \cos x)$ in the interval $[0, 2\pi]$ is

   (a) 6 (b) 7

   (c) 8 (d) 9

4. $x\sin a + y\sin 2a + z\sin 3a = \sin 4a$

   $x\sin b + y\sin 2b + z\sin 3b = \sin 4b$

   $x\sin c + y\sin 2c + z\sin 3c = \sin 4c$.

   Then, the roots of the equation.

   $$t^3 - \frac{z}{2}t^2 - \frac{y+2}{4}t + \frac{z-x}{8} = 0 (a, b, c \neq n\pi) \text{ are}$$

   (a) $\sin a, \sin b, \sin c$

   (b) $\cos a, \cos b, \cos c$

   (c) $\sin 2a, \sin 2b, \sin 2c$

   (d) $\cos 2a, \cos 2b, \cos 2c$

5. The least positive value of x satisfying

   $$\frac{\sin^2 2x + 4\sin^4 x - 4\sin^2 x\cos^2 x}{4 - \sin^2 2x - 4\sin^2 x} = \frac{1}{9} \text{ is}$$

   (a) $\pi/3$ (b) $\pi/6$

   (c) $2\pi/3$ (d) $5\pi/6$

6. The maximum value of the expression $\left|\sqrt{\sin^2 x + 2a^2} - \sqrt{2a^2 - 1 - \cos^2 x}\right|$, where a and x are real number, is

   (a) 1 (b) 2

   (c) $\sqrt{2}$ (d) $\sqrt{3}$

7. The general solution of $8\tan^2\frac{x}{2} = 1 + \sec x$ is

   (a) $x = 2n\pi \pm \cos^{-1}\left(\frac{-1}{3}\right)$ (b) $x = 2n\pi \pm \frac{\pi}{6}$

   (c) $x = 2n\pi \pm \cos^{-1}\left(\frac{1}{3}\right)$ (d) None of these

8. The general solution of $\tan\theta + \tan 4\theta + \tan 7\theta$ $= \tan\theta \cdot \tan 4\theta \cdot \tan 7\theta$

   (a) $\theta = \frac{n\pi}{4}$ (b) $\theta = \frac{n\pi}{12}$

   (c) $\theta = \frac{n\pi}{12}$ (d) None of these

9. The solution of the equation $e^{\sin x} - e^{-\sin x} - 4 = 0$ is

   (a) $x = 0$

   (b) $x = \sin^{-1}[\log(2-\sqrt{5})]$

   (c) no real solution

   (d) None of the above

10. The number of the solution of the equation $\cos(\pi\sqrt{x-4}) \cdot \cos\pi\sqrt{x} = 1$ is

    (a) >2 (b) 2

    (c) 1 (d) 0

11. The number of real solution of the equation. $\sin(e^x) = 5^x + 5^{-x}$ is

    (a) 0 (b) 1

    (c) 2 (d) Infinitely many

12. ABC is a triangle such that $\sin(2A+B) = \sin(C-A)$ $= -\sin(B+2C) = 1/2$. If A, B and C are in AP, then the value of A, B and C are

    (a) 45°, 60°, 75° (b) 30°, 60°, 90°

    (c) 20°, 60°, 100° (d) None of these

13. Let $2\sin^2 x + 3\sin x - 2 > 0$ and $x^2 - x - 2 < 0$ (x is measured in radian). Then 'x' lies in the internal.

    (a) $\left(\frac{\pi}{6}, \frac{5\pi}{6}\right)$ (b) $\left(-1, \frac{5\pi}{6}\right)$

    (c) $(-1, 2)$ (d) $\left(\frac{\pi}{6}, 2\right)$

14. The number of points of intersection of the two curves $y = 2\sin x$ and $y = 5x^2 + 2x + 3$ is

    (a) 0 (b) 1

    (c) 2 (d) ∞

15. The number of all possible triplets (a_1, a_2, a_3) such that $a_1 + a_2 \cos 2x + a_3 \sin^2 x = 0$ for all x is

(a) 0 (b) 1
(c) 3 (d) Infinite

16. The equation $\sin^4 x - (k+2)\sin^2 x - (k+3) = 0$ possesses a solution if

(a) $k > -3$ (b) $k < -2$
(c) $-3 \le k \le -2$ (d) k is any (+ ve) value

17. In interval $\left[\frac{-\pi}{2}, \frac{\pi}{2}\right]$, the equation $\log_{\sin\theta}(\cos^2\theta) = 2$ has

(a) no solution (b) a unique solution
(c) two solution (d) infinitely many solution

18. If $\sum_{i=1}^{n} \cos\theta_i = n$, then $\sum_{i=1}^{n} \sin\theta_i$ is

(a) $n-1$ (b) 0
(c) n (d) $n+1$

19. If $0 < x < \pi/2$ and $\sin^n x + \cos^n x \ge 1$, then

(a) $n \in [2, \infty)$ (b) $(-\infty, 2]$
(c) $n \in [-1, 1]$ (d) None of these

20. The most general values of 'x' for which $\sin x + \cos x = \min_{a \in R} [1, a^2 - 4a + 6]$ are given by

(a) $2n\pi$ (b) $2n\pi + \frac{\pi}{2}$
(c) $n\pi + (-1)^n \frac{\pi}{4} - \frac{\pi}{4}$ (d) None of these

21. Value of 'x' and 'y' satisfying the equation $\sin^7 y = |x^3 - x^2 - 9x + 9| + |x^3 - 4x - x^2 + 4| + \sec^2 2y + \cos^4 y$ are

(a) $x=1, y=n\pi$ (b) $x=1, y=2n\pi + \frac{\pi}{2}$
(c) $x=1, y=2n\pi$ (d) None of these

22. If $\max\{5\sin\theta + 3\sin(\theta - \alpha)\} = 7$, then the set of possible value of α is $\theta \in R$

(a) $\left\{x : x = 2n\pi \pm \frac{\pi}{3}, n \in I\right\}$ (b) $\left\{x : x = 2n\pi \pm \frac{2\pi}{3}, n \in I\right\}$
(c) $\left[\frac{\pi}{3}, \frac{2\pi}{3}\right]$ (d) None of these

23. The number of integral values of 'n' so that $\sin x(\sin x + \cos x) = n$ has atleast one solution, is

(a) 2 (b) 1
(c) 3 (d) zero

24. Total number of solution of $\sin\{x\} = \cos\{x\}$, where $\{\cdot\}$ denotes the fractional part, in $[0, 2\pi]$ is equal to

(a) 5 (b) 6
(c) 7 (d) None of these

25. If $a, b \in [0, \pi]$ and the equation $x^2 + 4 + 3\sin(ax + b) - 2x = 0$ has atleast one solution, then the value of $(a+b)$ can be:

(a) $\frac{7\pi}{2}$ (b) $\frac{3\pi}{2}$
(c) $\frac{9\pi}{2}$ (d) None of these

26. The value of a for which the equation $4\operatorname{cosec}^2(\pi(a+x)) + a^2 - 4a = 0$ has a real solution, is

(a) $a=1$ (b) $a=2$
(c) $a=10$ (d) None of these

27. If the equation $2\cos x + \cos 2\lambda x = 3$ has only one solution, then λ is

(a) 1 (b) A rational number
(c) An irrational number (d) None of these

28. Let n be positive integer such that $\sin\frac{\pi}{2n} + \cos\frac{\pi}{2n} = \frac{\sqrt{n}}{2}$. Then

(a) $6 \le n \le 8$ (b) $4 < n \le 8$
(c) $6 < n < 8$ (d) $4 < n < 8$

29. The number of solutions of the equation $5\sec\theta - 13 = 12\tan\theta$ in $[0, 2\pi]$ is

(a) 2 (b) 1
(c) 4 (d) 0

30. The number of solution of equation $x^3 + x^2 + 4x + 2\sin x = 0$ in $0 \le x \le 2\pi$ is

(a) Zero (b) One
(c) Two (d) Four

31. If $\tan\left(\frac{\pi}{2}\sin\theta\right) = \cot\left(\frac{\pi}{2}\cos\theta\right)$, then $\sin\theta + \cos\theta$ is equal to

(a) 0 (b) 1
(c) -1 (d) 1 or -1

32. The equation $\sin x + \sin y + \sin z = -3$ for $0 \le x \le 2\pi$, $0 \le y \le 2\pi$, $0 \le z \le 2\pi$ has

(a) one solution (b) two sets of solutions
(c) four sets of solutions (d) no solution

33. If $x = n\pi + (-1)^n\alpha, n \in I$ and $x = n\pi + (-1)^n\beta$ are the roots of $4\cos x - 3\sec x = \tan x$, then $4(\sin\alpha + \sin\beta)$ is

(a) -1 (b) 1
(c) 2 (d) None of these

34. If $\tan m\theta = \tan n\theta$ and general value of θ are in AP, then common difference is

(a) $\frac{1}{m-n}$ (b) $\frac{\pi}{m+n}$
(c) $\frac{\pi}{m-n}$ (d) None of these

35. If $\sin 3\alpha = 4\sin\alpha \sin(x+\alpha)\sin(x-\alpha)$, then

(a) $x=n\pi \pm \frac{\pi}{3}, n\in I$

(b) $x=n\pi \pm \frac{\pi}{6}, n\in I$

(c) $x=n\pi \pm \frac{\pi}{2}, n\in I$

(d) None of the above

36. $\lambda\cos x - 3\sin x = \lambda + 1$ is solvable only, if

(a) $\lambda\in[0,5]$ (b) $\lambda\in[4,5]$

(c) $\lambda\in(-\infty,4]$ (d) None of these

37. $\cos 2x - 3\cos x + 1 = \dfrac{1}{(\cot 2x - \cot x)\sin(x-\pi)}$ holds, if

(a) $\cos x=0$ (b) $\cos x=1$

(c) $\cos x=\frac{5}{2}$ (d) for no value of x

38. If $\sec x \cos 5x = -1$ and $0 < x < \frac{\pi}{4}$, then x is equal to

(a) $\frac{\pi}{6}$ (b) $\frac{\pi}{3}$

(c) $\frac{\pi}{4}$ (d) None of these

39. If $\sin^{100}\theta - \cos^{100}\theta = 1$, then θ is

(a) $2n\pi + \frac{\pi}{3}, n\in I$ (b) $n\pi + \frac{\pi}{2}, n\in I$

(c) $n\pi + \frac{\pi}{4}, n\in I$ (d) $2n\pi - \frac{\pi}{3}, n\in I$

40. If $\sqrt{3}\sin x - \cos x = \min\limits_{\alpha\in R}\{2, e^2, \pi, \alpha^2 - 4\alpha + 7\}$, then

(a) $x=2n\pi, n\in I$ (b) $x=2n\pi + \frac{2\pi}{3}, n\in I$

(c) $x=n\pi + (-1)^n\frac{\pi}{4} + \frac{\pi}{6}, n\in I$ (d) $x=n\pi + (-1)^{n+1}\frac{\pi}{4} - \frac{\pi}{3}, n\in I$

41. The number of solutions of the equation $\cos 4x + 6 = 7\cos 2x$, when $x\in[315°, 317°]$ is

(a) 0 (b) 1

(c) 2 (d) 4

42. The number of solutions of $\cot(5\pi\sin\theta) = \tan(5\pi\cos\theta)$, $\forall\,\theta\in(0,2\pi)$ is

(a) 7 (b) 14

(c) 21 (d) 28

43. If exp $[(\sin^2 x + \sin^4 x + \sin^6 x + \dots\infty)$ In 2] satisfies the equation $y^2 - 9y + 8 = 0$, then the value of $\dfrac{\cos x}{\cos x + \sin x}, 0 < x < \frac{\pi}{2}$, is

(a) $\sqrt{3}+1$ (b) $\frac{\sqrt{3}-1}{2}$

(c) $\sqrt{3}-1$ (d) None of these

44. Total number of solutions of $\cos x = \sqrt{1-\sin 2x}$ in $[0, 2\pi]$, is equal to

(a) 2 (b) 3

(c) 5 (d) None of these

45. If the equation $\cos 3x\cos^3 x + \sin 3x\sin^3 x = 0$, then x is equal to

(a) $(2n+1)\frac{\pi}{4}$ (b) $(2n-1)\frac{\pi}{4}$

(c) $\frac{n\pi}{4}$ (d) None of these

46. Total number of solutions of $\sin x = \frac{|x|}{10}$ is equal to

(a) 4 (b) 6

(c) 7 (d) None of these

47. The number of all possible 5-tuples $(a_1, a_2, a_3, a_4, a_5)$ such that $a_1 + a_2\sin x + a_3\cos x + a_4\sin 2x + a_5\cos 2x = 0$ holds for all x is

(a) zero (b) 1

(c) 2 (d) infinite

48. x_1 and x_2 are two positive values of x for which $2\cos x, |\cos x|$ and $3\sin^2 x - 2$ are in GP. The minimum value of $|x_1 - x_2|$ is equal to

(a) $\frac{4\pi}{3}$ (b) $\frac{\pi}{3}$

(c) $2\cos^{-1}\left(\frac{2}{3}\right)$ (d) $\cos^{-1}\left(\frac{2}{3}\right)$

49. If $\cos x - \dfrac{\cot\beta\sin x}{2} = \dfrac{\sqrt{3}}{2}$, then the value of $\tan\frac{x}{2}$ is

(a) $\tan\frac{\beta}{2}\tan 15°$

(b) $\tan\frac{\beta}{2}$

(c) $\tan 15°$

(d) None of the above

50. The expression $n\sin^2\theta + 2n\cos(\theta+\alpha)\sin\alpha\sin\theta + \cos 2(\alpha+\theta)$ is independent of 'θ', the value of n is

(a) 1 (b) 2

(c) 3 (d) 4

51. The value of the determinant $\begin{vmatrix} 1 & a & a^2 \\ \cos(n-1)x & \cos nx & \cos(n+1)x \\ \sin(n-1)x & \sin nx & \sin(n+1)x \end{vmatrix}$ is zero if

(a) $x=n\pi$ (b) $x=n\pi/2$

(c) $x=(2n+1)\pi/2$ (d) $x=\frac{1+a^2}{2a} n\in I$

52. $\dfrac{\sin 3\alpha}{\cos 2\alpha} < 0$ if α lies in
(a) $(13\pi/48, 14\pi/48)$ (b) $(14\pi/48, 18\pi/48)$
(c) $(18\pi/48, 23\pi/48)$ (d) any of these intervals

53. If $f(x) = \begin{vmatrix} \sin^2\theta & \cos^2\theta & x \\ \cos^2\theta & x & \sin^2\theta \\ x & \sin^2\theta & \cos^2\theta \end{vmatrix}$ $\theta \in (0, \pi/2)$, then roots of $f(x) = 0$ are
(a) $1/2, -1$ (b) $1/2, -1, 0$
(c) $-1/2, 1, 0$ (d) $-1/2, -1, 0$

54. The equation $\sin x + \sin y + \sin z = -3$ for $0 \le x \le 2\pi$, $0 \le y \le 2\pi$, $0 \le z \le 2\pi$, has
(a) One solution (b) Two sets of solutions
(c) Four sets of solutions (d) No solutions

55. If $\sec x \cos 5x + 1 = 0$, where $0 < x < 2\pi$, then $x =$
(a) $\dfrac{\pi}{5}, \dfrac{\pi}{4}$ (b) $\dfrac{\pi}{5}$
(c) $\dfrac{\pi}{4}$ (d) None of these

56. If $|k| = 5$ and $0° \le \theta \le 360°$, then the number of different solutions of $3\cos\theta + 4\sin\theta = k$ is
(a) Zero (b) Two
(c) One (d) Infinite

57. If $\cot(\alpha + \beta) = 0$, then $\sin(\alpha + 2\beta) =$
(a) $\sin\alpha$ (b) $\cos\alpha$
(c) $\sin\beta$ (d) $\cos 2\beta$

58. If $\cot\theta + \cot\left(\dfrac{\pi}{4} + \theta\right) = 2$, then the general value of θ is
(a) $2n\pi \pm \dfrac{\pi}{6}$ (b) $2n\pi \pm \dfrac{\pi}{3}$
(c) $n\pi \pm \dfrac{\pi}{3}$ (d) $n\pi \pm \dfrac{\pi}{6}$

59. If $\cos 2\theta = (\sqrt{2} + 1)\left(\cos\theta - \dfrac{1}{\sqrt{2}}\right)$, then the value of θ is
(a) $2n\pi + \dfrac{\pi}{4}$ (b) $2n\pi \pm \dfrac{\pi}{4}$
(c) $2n\pi - \dfrac{\pi}{4}$ (d) None of these

60. If $\left|1 - \dfrac{|\sin x|}{1 + \sin x}\right| \ge \dfrac{2}{3}$, then $\sin x$ lies in
(a) $\left(-\infty, \dfrac{-1}{2}\right] \cup \left[\dfrac{1}{2}, \infty\right)$
(b) $\left(-\dfrac{1}{2}, \dfrac{1}{2}\right)$
(c) $\left[-\dfrac{1}{2}, \dfrac{1}{2}\right]$
(d) None of the above

Trigonometric Equations and Inequations Exercise 2 : More than One Option Correct Type Questions

61. The value of 't' which satisfies $(t - [|\sin x|])! = 3!5!7!$ is/are .... where [.] is GIF
(a) 9 (b) 10
(c) 11 (d) 12

62. Let $f(x) = \cos(a_1 + x) + \dfrac{1}{2}\cos(a_2 + x) + \dfrac{1}{2^2}\cos(a_3 + x)$
$$+ \ldots + \frac{1}{2^{n-1}}\cos(a_n + x)$$
where $a_1, a_2, \ldots, a_n \in R$. If $f(x_1) = f(x_2) = 0$, then $|x_2 - x_1|$ may be equal to
(a) π (b) 2π
(c) 3π (d) $\dfrac{\pi}{2}$

63. Let α, β, γ be parametric angles of 3 points P, Q and R respectively lying on $x^2 + y^2 = 1$. If the lengths of chords AP, AQ and AR are in GP where A is $(1, 0)$, then [Given $\alpha, \beta, \gamma \in (0, 2\pi)$].
(a) $\sin\dfrac{\alpha + \gamma}{4}\cos\dfrac{\alpha - \gamma}{4} \ge \sin\dfrac{\beta}{2}$
(b) $\sin\left(\dfrac{\alpha + \gamma}{4}\right)\cos\left(\dfrac{\alpha - \gamma}{4}\right) \le \sin\dfrac{\beta}{2}$
(c) $\sin\dfrac{\alpha}{2}\sin\dfrac{\gamma}{2} \ge \sin\dfrac{\beta}{2}$
(d) $\sin\dfrac{\alpha}{2}\sin\dfrac{\gamma}{2} \le \sin\dfrac{\beta}{2}$

64. Let x, y, z be elements from interval $[0, 2\pi]$ satisfying the inequality $(4 + \sin 4x)(2 + \cot^2 y)(1 + \sin^4 z) \le 12\sin^2 z$, then
(a) the number of ordered pairs (x, y) is 5
(b) the number of ordered pairs (y, z) is 8
(c) the number of ordered pairs (z, x) is 8
(d) the number of pairs (y, z) such that $z = y$ is 2

65. The number of integral values of a for which the system of linear equations $x \sin\theta - 2y \cos\theta - az = 0$, $x + 2y + z = 0, - x + y + z = 0$ may have non- trivial solutions, then

(a) at $a = 2$ the given system will have finite solutions for $\theta \in R$
(b) number of possible integral values of a is 3
(c) for $a = 1$ the system will have infinite solutions
(d) for $a = 3$ the system will have unique solution

66. The equation $2\sin^3\theta + (2\lambda - 3)\sin^2\theta - (3\lambda + 2)\sin\theta - 2\lambda = 0$ has exactly three roots in $(0, 2\pi)$, then λ can be equal to

(a) 0 (b) $\frac{1}{2}$
(c) 1 (d) -1

67. If $x + y = 2\pi/3$ and $\sin x/\sin y = 2$, then the

(a) number of values of $x \in [0, 4\pi]$ are 4
(b) number of values of $x \in [0, 4\pi]$ are 2
(c) number of values of $y \in [0, 4\pi]$ are 2
(d) number of values of $y \in [0, 4\pi]$ are 8

68. If $0 \le x \le 2\pi$ and $|\cos x| \le \sin x$, then

(a) the set of all values of x is $\left[\frac{\pi}{4}, \frac{3\pi}{4}\right]$
(b) the number of solutions that are integral multiple of $\frac{\pi}{4}$ is four
(c) the sum of the largest and the smallest solution is π
(d) the set of all values of x is $x \in \left[\frac{\pi}{4}, \frac{\pi}{2}\right) \cup \left(\frac{\pi}{2}, \frac{3\pi}{4}\right]$

69. If x and y are positive acute angles such that $(x + y)$ and $(x - y)$ satisfy the equation $\tan^2\theta - 4\tan\theta + 1 = 0$, then

(a) $x = \frac{\pi}{6}$ (b) $y = \frac{\pi}{4}$
(c) $y = \frac{\pi}{6}$ (d) $x = \frac{\pi}{4}$

70. If $x + y = \frac{4\pi}{3}$ and $\sin x = 2\sin y$, then

(a) $x = n\pi + \frac{\pi}{2}, n \in I$
(b) $y = \frac{5\pi}{6} - n\pi, n \in I$
(c) Both (a) and (b)
(d) None of the above

71. The number of solutions of the equations $y = \frac{1}{3}[\sin\theta + [\sin\theta + [\sin\theta]]]$ and $[y + [y]] = 2\cos\theta$ [where, [·] denotes the greatest integer function] is /are

(a) 0 (b) 1
(c) 2 (d) infinite

72. If $[\sin x] + [\sqrt{2}\cos x] = -3$, $x \in [0, 2\pi]$, (where, [·] denotes the greatest integer function), then

(a) $x \in \left(\pi, \frac{5\pi}{4}\right)$ (b) $x \in \left(\pi, \frac{7\pi}{6}\right)$
(c) $x \in \left[\pi, \frac{5\pi}{4}\right]$ (d) None of these

73. If $\alpha \in [-2\pi, 2\pi]$ and $\cos\frac{\alpha}{2} + \sin\frac{\alpha}{2} = \sqrt{2}(\cos 36° - \sin 18°)$ then a values of α is

(a) $\frac{7\pi}{6}$ (b) $\frac{\pi}{6}$
(c) $-\frac{5\pi}{6}$ (d) $-\frac{\pi}{6}$

74. The number of values of α in the interval $[-\pi, 0]$ satisfying $\sin\alpha + \int_{\alpha}^{2\alpha} \cos 2x \, dx = 0$, then

(a) $\alpha = 0$
(b) $\alpha = 0, -\pi, -\frac{\pi}{3}$
(c) $\alpha = \frac{\pi}{6}, \frac{5\pi}{6}$
(d) None of the above

75. The solutions of $\theta \in [0, 2\pi]$ satisfying the equation $\log_{\sqrt{3}} \tan\theta \left(\sqrt{\log_{\tan\theta} 3 + \log_{\sqrt{3}} 3\sqrt{3}}\right) = -1$, then

(a) $\theta = \frac{\pi}{6}$ (b) $\frac{\pi}{3}, \frac{5\pi}{3}$
(c) has sum $\frac{4\pi}{3}$ (d) >2

76. If α and β are the solutions of $a\cos\theta + b\sin\theta = c$, then

(a) $\sin\alpha + \sin\beta = \frac{2bc}{a^2 + b^2}$
(b) $\sin\alpha \sin\beta = \frac{c^2 - a^2}{a^2 + b^2}$
(c) $\sin\alpha + \sin\beta = \frac{2ac}{b^2 + c^2}$
(d) $\sin\alpha + \sin\beta = \frac{c}{a^2 + b^2}$

77. The solution of the equation $\sin 2x + \sin 4x = 2\sin 3x$ is

(a) $x = \frac{n\pi}{3}$
(b) $x = n\pi$
(c) $x = 2n\pi$
(d) None of the above

78. The general solution of $4\sin^4 x + \cos^4 x = 1$ is

(a) $(2n+1)\frac{\pi}{2}$ (b) $n\pi$
(c) $n\pi \pm \sin^{-1}\sqrt{\frac{2}{5}}$ (d) None of these

79. The values of $x, 0 \le x \le \frac{\pi}{2}$ whcih satisfy the equation

$81^{\sin^2 x} + 81^{\cos^2 x} = 30$ are

(a) $\frac{\pi}{3}$ (b) $\frac{\pi}{4}$

(c) $\frac{\pi}{6}$ (d) $\frac{7\pi}{18}$

80. All values of $x \in \left(0, \frac{\pi}{2}\right)$ such that $\frac{\sqrt{3}-1}{\sin x} + \frac{\sqrt{3}+1}{\cos x} = 4\sqrt{2}$ are

(a) $\frac{\pi}{15}$ (b) $\frac{\pi}{12}$

(c) $\frac{11\pi}{36}$ (d) $\frac{3\pi}{10}$

81. $\frac{\alpha^2}{1-\tan^2 x} = \frac{\sin^2 x + \alpha^2 - 2}{\cos 2x}$ has a solution if

(a) $\alpha \le -1$ (b) $\alpha \ge 1$

(c) $\alpha = 1/2$ (d) α is any real number

82. The equation $4\sin(x+\pi/3)\cos(x-\pi/6) = a^2 + \sqrt{3}\sin 2x - \cos 2x$ has a solution if the value of

(a) -2 (b) 0

(c) 2 (d) $a, a \in]-2, 2[$

83. Which of the following is/are correct.

(a) $(\tan x)^{\ln(\sin x)} > (\cot x)^{\ln(\sin x)}, \forall x \in (0, \pi/4)$

(b) $4^{\ln \operatorname{cosec} x} < 5^{\ln \operatorname{cosec} x}, \forall x \in (0, \pi/2)$

(c) $(1/2)^{\ln(\cos x)} < (1/3)^{\ln(\cos x)}, \forall x \in (0, \pi/2)$

(d) $2^{\ln(\tan x)} > 2^{\ln(\tan x)}, \forall x \in (0, \pi/2)$

84. The value of θ, lying between $\theta = 0$ and $\theta = \frac{\pi}{2}$ and satisfying the equation.

$$\begin{vmatrix} 1+\cos^2\theta & \sin^2\theta & 4\sin 4\theta \\ \cos^2\theta & 1+\sin^2\theta & 4\sin 4\theta \\ \cos^2\theta & \sin^2\theta & 1+4\sin 4\theta \end{vmatrix} = 0, \text{ is}$$

(a) $\frac{11\pi}{24}$ (b) $\frac{7\pi}{24}$

(c) $\frac{5\pi}{24}$ (d) None of these

85. If $[x]$ denotes the greatest integer less than or equal to x then the equation $\sin x = [1+\sin x] + [1-\cos x]$ has no solution in

(a) $\left[-\frac{\pi}{2}, \frac{\pi}{2}\right]$ (b) $\left[\frac{\pi}{2}, \pi\right]$

(c) $\left[\pi, \frac{3\pi}{2}\right]$ (d) R

Trigonometric Equations and Inequations Exercise 3 : Passage Based Questions

Passage I

(Q. Nos. 86 to 88)

If number of solutions and sum of solutions of the equation $3\sin^2 x - 7\sin x + 2 = 0, x \in [0, 2\pi]$ are respectively N and S and $f_n(\theta) = \sin^n\theta + \cos^n\theta$. On the basis of above information, answer the following questions.

86. Value of N is

(a) 1 (b) 2 (c) 3 (d) 4

87. Value of S is

(a) $\frac{5\pi}{6}$ (b) $\frac{7\pi}{6}$ (c) 2π (d) π

88. If α is solution of equation $3\sin^2 x - 7\sin x + 2 = 0$, $x \in [0, 2\pi]$, then the value of $f_4(\alpha)$ is.

(a) $\frac{97}{81}$ (b) $\frac{57}{81}$ (c) $\frac{65}{81}$ (d) 0

Passage II

(Q. Nos. 89 to 90)

Let $\log_a N = \alpha + \beta$ where α is integer and $\beta = [0, 1)$. Then, On the basis of above information, answer the following questions.

89. The difference of largest and smallest integral value of N satisfying $\alpha = 3$ and $a = 5$, is

(a) 499 (b) 500 (c) 501 (d) 502

90. If N_1 is number of integers when $a = 2$ and $\alpha = 2$ and N_2 is number of integers when $\alpha = 1$ and $a = 3$, then the minimum value of $(N_1 \sec^2\theta + N_2 \operatorname{cosec}^2\theta)$

(a) $10 + 4\sqrt{6}$ (b) $10 + \sqrt{6}$ (c) 10 (d) 100

Passage III

(Q. Nos. 91 to 93)

If an angle and a side of a right angle triangle is known, then rest of the sides and angles can be found as follows

In ΔABC (Figure 1), if $\angle B = 90°$, $\angle C = \theta$ and $BC = x$, then $AB = x\tan\theta$ and $AC = x\sec\theta$.

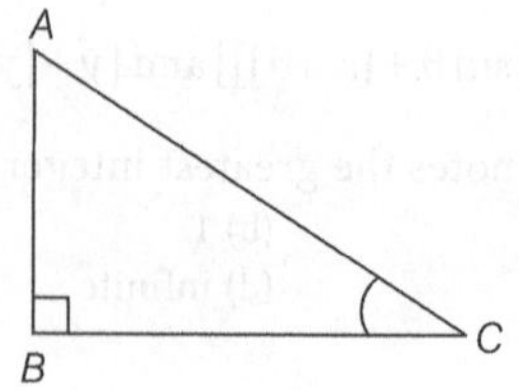

Now, consider an isosceles triangle PQR (Figure 2),

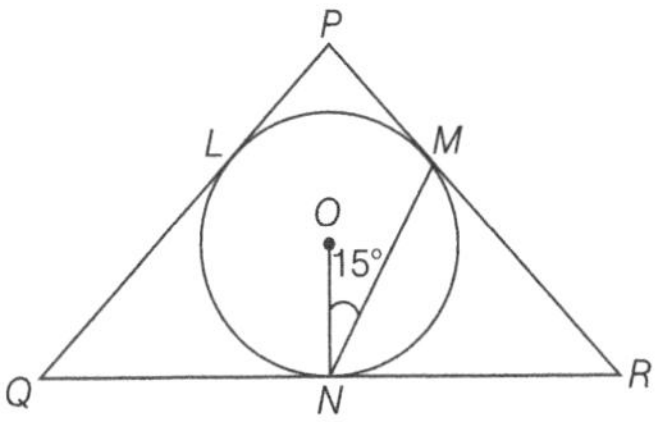

where $PQ = PR$ and $20N = \sqrt{3}$

On the basis of above information answer the following

91. The angle of triangle PQR are

(a) 150°, 15° 15° (b) 60°, 60°, 60°
(c) 120°, 30°, 30° (d) 75°, 52.5°, 52.5°

92. Area of circumcircle of quadrilateral $PLOM$ is

(a) π (b) $\frac{\pi}{4}$ (c) $\frac{3\pi}{4}$ (d) 3π

93. Length of the side QR is

(a) $\tan 15°$ (b) $\sqrt{3}\tan 15°$
(c) $\cot 15°$ (d) $\sqrt{3}\cot 15°$

Passage IV

(Q. Nos. 94 to 96)

α is a root of equation $(2\sin x - \cos x)(1+\cos x) = \sin^2 x$, β is a root of the equation $3\cos 2x - 10\cos x + 3 = 0$ and γ is a root of the equation $1 - \sin 2x = \cos x - \sin x : 0 \le \alpha, \beta, \gamma \le \pi/2$

94. $\cos\alpha + \cos\beta + \cos\gamma$ can be equal to

(a) $\frac{3\sqrt{6}+2\sqrt{2}+6}{6\sqrt{2}}$ (b) $\frac{3\sqrt{3}-8}{6}$
(c) $\frac{3\sqrt{3}+2}{6}$ (d) None of these

95. $\sin\alpha + \sin\beta + \sin\gamma$ can be equal to

(a) $\frac{14-3\sqrt{2}}{6\sqrt{2}}$ (b) $5/6$
(c) $\frac{3+4\sqrt{2}}{6}$ (d) $\frac{1+\sqrt{2}}{2}$

96. $\sin(\alpha - \beta)$ is equal to

(a) 1 (b) 0
(c) $\frac{1-2\sqrt{6}}{6}$ (d) $\frac{\sqrt{3}-2\sqrt{2}}{6}$

Passage V

(Q. Nos. 97 to 99)

Consider the equations

$$5\sin^2 x + 3\sin x\cos x - 3\cos^2 x = 2 \quad \text{...(i)}$$
$$\sin^2 x - \cos 2x = 2 - \sin 2x \quad \text{...(ii)}$$

97. If α is a root of (i) and β is a root of (ii), then $\tan\alpha + \tan\beta$ can be equal to

(a) $1+\sqrt{69}/6$ (b) $-1-\sqrt{69}/6$
(c) $\frac{-3+\sqrt{69}}{6}$ (d) $\frac{3-\sqrt{69}}{3}$

98. If $\tan\alpha$, $\tan\beta$ satisfy (i) and $\cos\gamma$, $\cos\delta$ satisfy (ii), then $\tan\alpha\cdot\tan\beta + \cos\gamma + \cos\delta$ can be equal to

(a) -1 (b) $-\frac{5}{3}+\frac{2}{\sqrt{13}}$
(c) $\frac{5}{3}-\frac{2}{\sqrt{13}}$ (d) $\frac{5}{3}+\frac{2}{\sqrt{13}}$

99. The number of solutions common to (i) and (ii) is

(a) 0 (b) 1
(c) finite (d) infinite

Trigonometric Equations and Inequations Exercise 4 : Single Integer Answer Type Questions

100. Let Δ_k be the area of triangle AP_kB which is inscribed in a circle of radius 2 units. If AB diameter of circle, $\angle ABP_k = \frac{k\pi}{2n}$ and $\sum_{k=1}^{n+1} \Delta_k = 4\cot\frac{\pi}{32}$, then $\frac{n}{2}$ is equal to

101. If the sum of the roots of the equation $\cos 4x + 6 = 7\cos 2x$ in the interval $[0, 314]$ is $k\pi$, $k \in R$ Find $(k - 4948)$.

102. If equation $x^2\tan^2\theta - (2\tan\theta)x + 1 = 0$ and
$$\left(\frac{1}{1+\log_b ac}\right)x^2 + \left(\frac{1}{1+\log_c ab}\right)x + \left(\frac{1}{1+\log_a bc} - 1\right) = 0$$
(where $a, b, c, > 1$) have a common root and then 2nd equation has equal roots, then number of possible value of θ in $(0, \pi)$ is

103. Number of ordered pairs (x, y) which satisfies the relation $\frac{x^4+1}{8x^2} = \sin^2 y.\cos^2 y$, where $y \in [0, 2\pi]$.

104. The number of solutions for,
$$\left.\begin{array}{l}\sin\left(x-\frac{\pi}{4}\right)-\cos\left(x+\frac{3\pi}{4}\right)=1\\ \frac{2\cos 7x}{\cos 3+\sin 3}>2^{\cos 2x}\end{array}\right\}\text{ in } (0, 2\pi), \text{ is}$$

105. If $\cos A \sin\left(A-\frac{\pi}{6}\right)$ is maximum, when the values of A is equal to $\frac{\pi}{\lambda}$, then the value of λ is

106. Let $p, q \in N$ and $q > p$, the number of solutions of the equation $q|\sin\theta| = p|\cos\theta|$ in the interval $[0, 2\pi]$ is

107. If $\theta_1, \theta_2, \theta_3$ are three values lying in $[0, 2\pi]$ for which $\tan\theta = \lambda$, then $\left|\tan\frac{\theta_1}{3}\cdot\tan\frac{\theta_2}{3}+\tan\frac{\theta_2}{3}\cdot\tan\frac{\theta_3}{3}+\tan\frac{\theta_1}{3}\cdot\tan\frac{\theta_3}{3}\right|$ is equal to

108. If α be the smallest positive root of the equation $\sqrt{\sin(1-x)} = \sqrt{\cos x}$, then the approximate integral value of α must be

109. If x and y are the solutions of the equation $12\sin x + 5\cos x = 2y^2 - 8y + 21$, the value of $12\cot\left(\frac{xy}{2}\right)$ must be

110. If $\tan(\pi\cos\theta) = \cot(\pi\sin\theta)$, then $\cos^2(\theta - \pi/4)$ is equal to

111. If $3\sin x + 4\cos x = 5$, then the value of $90\tan^2(x/2) - 60\tan(x/2) + 10$ is equal to

Trigonometric Equations and Inequations Exercise 5 : Statement I and II Type Questions

- This section contains 6 questions. Each question contains **Statement I (Assertion)** and **Statement II (Reason)**. Each question has 4 choices (a), (b), (c) and (d) out of which only one is correct. The choices are
  (a) Statement I is true, Statement II is true; Statement II is a correct explanation for Statement I.
  (b) Statement I is true, Statement II is true, Statement II is not a correct explanation for Statement II.
  (c) Statement I is true, Statement II is false.
  (d) Statement I is false, Statement II is true.

112. Statement I $\sin x = a$, where $-1 < a < 0$, then for $x \in [0, n\pi]$ has $2(n-1)$ solutions $\forall\, n \in N$.

Statement II $\sin x$ takes value a exactly two times when we take one complete rotation covering all the quadrants starting from $x = 0$.

113. Statement I The number of solutions of the equation $|\sin x| = |x|$ is only one.

Statement II $|\sin x| \ge 0\, \forall x \in R$.

114. Statement I If $2\sin 2x - \cos 2x = 1$, $x \ne (2n+1)\pi/2$, n is the integer, then $\sin 2x + \cos 2x$ is equal to $1/5$.

Statement II $\sin 2x + \cos 2x = \frac{1 + 2\tan x - \tan^2 x}{1 + \tan^2 x}$

115. Statement I The system of linear equations

$$x + (\sin\alpha)y + (\cos\alpha)z = 0$$
$$x + (\cos\alpha)y + (\sin\alpha)z = 0$$
$$-x + (\sin\alpha)y - (\cos\alpha)z = 0$$

has a non trivial solution for only one value of α lying between 0 and π.

Statement II $\begin{vmatrix} \sin x & \cos x & \cos x \\ \cos x & \sin x & \cos x \\ \cos x & \cos x & \sin x \end{vmatrix} = 0$

has no solution in the interval $-\pi/4 < x < \pi/4$.

116. Let $\theta \in (\pi/4, \pi/2)$, then

Statement I $(\cos\theta)^{\sin\theta} < (\cos\theta)^{\cos\theta} < (\sin\theta)^{\cos\theta}$

Statement II The equation $e^{\sin\theta} - e^{-\sin\theta} = 4$ has a unique solution.

117. Statement I If $\exp\{(\sin^2 x + \sin^4 x + \sin^6 x + \ldots \text{inf}) \log_e 2\}$ satisfying the equation $x^2 - 9x + 8 = 0$, then the value of $\frac{\cos x}{\cos x + \sin x}$ is $\frac{\sqrt{3}-1}{2} \cdot (0 < x < \pi/2)$

Statement II $\sin^2 x + \sin^4 x + \sin^6 x + \ldots \text{inf} = \sec^2 x$

Trigonometric Equations and Inequations Exercise 6 : Matching Type Questions

- Math the statement of Column I with the value of Column II.

118.

	Column I		Column II
(A)	If α, β are the solutions of $\sin x = -\frac{1}{2}$ in $[0, 2\pi]$ and α, γ are the solutions of $\cos x = \frac{-\sqrt{3}}{2}$ in $[0, 2\pi]$, then	(p)	$\alpha - \beta = \pi$
(B)	If α, β are the solutions of $\cot x = -\sqrt{3}$ in $[0, 2\pi]$ and α, γ are the solutions of $\operatorname{cosec} x = -2$ in $[0, 2\pi]$, then	(q)	$\beta - \gamma = \pi$
(C)	If α, β are the solutions of $\sin x = -\frac{1}{2}$ in $[0, 2\pi]$ and α, γ are the solutions of $\tan x = \frac{1}{\sqrt{3}}$ in $[0, 2\pi]$, then	(r)	$\alpha - \gamma = \pi$
		(s)	$\alpha + \beta = 3\pi$
		(t)	$\beta + \gamma = 2\pi$

(a) A → (q, s) ; B → (p, t); C → (r, s, t)
(b) A → (q) ; B → (t); C → (r)
(c) A → (r, t) ; B → (t); C → (p, q)
(d) A → (p, q) ; B → (q, r) C → (r, s, t)

119.

	Column I		Column II
(A)	$2\sin\theta\|\cos\theta\| = \frac{1}{\sqrt{2}}$	(p)	$\theta = 3\pi/8$
(B)	$2\cos 2\theta \cos 4\theta + 2\cos^2 2\theta - 1 = 0$	(q)	$\theta = 7\pi/8$
(C)	$8\cos^2\theta \sin\theta - 4\cos^2\theta - 2\sin\theta + 1 = 0$	(r)	$\theta = 2\pi/3$
(D)	$\sin 4\theta = \pm 1$	(s)	$\theta = \pi/6$

(a) A → (p, q); B → (p, q, r); C → (r, s); D → (p, q)
(b) A → (r, s); B → (q, r); C → (r); D → (p, s)
(c) A → (p); B → (q); C → (r); D → (s)
(d) A → (s); B → (q, r); C → (r, s); D → (p, q)

120. If $f_n(\theta) = \dfrac{\sin\theta + \sin 3\theta + \sin 5\theta + \ldots + \sin((2n-1)\theta)}{\sin\theta + \cos\theta + \cos 5\theta + \ldots + \cos((2n-1)\theta)}$

	Column I		Column II
(A)	$f_3\left(\frac{\pi}{12}\right)$	(p)	$\sqrt{2} - 1$
(B)	$f_4\left(\frac{\pi}{32}\right)$	(q)	$2 - \sqrt{3}$
(C)	$f_6\left(\frac{\pi}{16}\right)$	(r)	$\sqrt{2} + 1$
(D)	$f_7\left(\frac{\pi}{84}\right)$	(s)	$2 + \sqrt{3}$
		(t)	1

(a) A → (p); B → (q); C → (r); D → (s,t)
(b) A → (t); B → (p); C → (r); D → (q)
(c) A → (q); B → (r); C → (s); D → (t)
(d) A → (r,t); B → (s); C → (p); D → (q)

Trigonometric Equations and Inequations Exercise 7 : Subjective Type Questions

121. Find the number of solutions of the equations;

(i) $|\cot x| = \cot x + \frac{1}{\sin x}$, when $x \in [0, 2\pi]$

(ii) $\sin^3 x \cos x + \sin^2 x \cdot \cos^2 x + \sin x \cdot \cos^3 x = 1$, when $x \in [0, 2\pi]$

(iii) $2^{\cos x} = |\sin x|$, when $x \in [-2\pi, 2\pi]$

(iv) $|\cos x| = [x]$, (where [.] denotes the greatest integer function).

(v) $x + 2\tan x = \frac{\pi}{2}$, when $x \in [0, 2\pi]$.

122. Find all value of α for which the equation $\sin^4 x + \cos^4 x + \sin 2x + \alpha = 0$ is valid. Also, find the general solution of the equation.

123. If $32\tan^8\theta = 2\cos^2\alpha - 3\cos\alpha$ and $3\cos 2\theta = 1$, then find the general value of α.

124. Solve for x and y

$$4^{\sin x} + 3^{\frac{1}{\cos y}} = 11$$

$$5.16^{\sin x} - 2.3^{\frac{1}{\cos y}} = 2$$

125. Find all numbers x, y that satisfy the equation

$$\left(\sin^2 x + \frac{1}{\sin^2 x}\right)^2 + \left(\cos^2 x + \frac{1}{\cos^2 x}\right)^2 = 12 + \frac{1}{2}\sin y.$$

126. Find all the solutions of x, y in the equation

$$4\left(3\sqrt{4x - x^2}\sin^2\left(\frac{x+y}{2}\right) + 2\cos(x+y)\right)$$
$$= 13 + 4\cos^2(x+y)$$

127. Solve for x and y, $1 - 2x - x^2 = \tan^2(x+y) + \cot^2(x+y)$.

128. Solve the system of equations

$$\tan^2 x + \cot^2 x = 2\cos^2 y$$
$$\cos^2 y + \sin^2 z = 1$$

129. Find all pairs of x, y that satisfy the equation

$$\cos x + \cos y + \cos(x+y) = -\frac{3}{2}$$

130. Solve the equation $\cot\left(\frac{\theta}{2}\right) - \text{cosecc}\left(\frac{\theta}{2}\right) = \cot\theta$.

131. Find the general solution of $1 + \sin^3 x + \cos^3 x = \frac{3}{2}\sin 2x$.

132. Solve $\log_{(\sin x)} 2\log_{(\sin^2 x)} a = -1$ stating any condition on 'a' that may be required for the existence of the solution.

133. Find all the values of 'a' $(a \neq 0)$ for which the equation $\int_0^x (t^2 - 8t + 13)dt = x\sin\frac{a}{x}$ has a solution. Find the solution.

134. Find all values between 0 and π which satisfies the equation

$$\sin^8 x + \cos^8 x = \frac{17}{16}\cos^2 2x$$

135. Find all number pairs x, y that satisfy the equation

$$\tan^4 x + \tan^4 y + 2\cot^2 x . \cot^2 y = 3 + \sin^2(x+y).$$

136. Determine all values of 'a' for which the equation $\cos^4 x - (a+2)\cos^2 x - (a+3) = 0$, possesses solution. Find the solutions.

137. For $x \in (-\pi, \pi)$ find the value of x for which the given equation

$$(\sqrt{3}\sin x + \cos x)^{\sqrt{\sqrt{3}\sin 2x - \cos 2x + 2}} = 4 \text{ is satisfied.}$$

138. Show that the equation

$\sec\theta + \text{cosec}\,\theta = c$ has two roots between 0 and 2π, if $c^2 < 8$ and four root if $c^2 > 8$.

139. Solve the equation for x and y,

$|\sin x + \cos x|^{\sin^2 x - 1/4} = 1 + |\sin y|$ and

$\cos^2 y = 1 + \sin^2 y$.

Trigonometric Equations and Inequations Exercise 8 : Questions Asked in Previous 10 Years' Exam

(i) JEE Advanced & IIT-JEE

140. Let $S = \left\{x \in (-\pi, \pi) : x \neq 0, \pm\frac{\pi}{2}\right\}$. The sum of all distinct solutions of the equation $\sqrt{3}\sec x + \text{cosec}\,x + 2(\tan x - \cot x) = 0$ in the set S is equal to

[Single Correct Option 2016 Adv.]

(a) $-\frac{7\pi}{9}$ (b) $-\frac{2\pi}{9}$

(c) 0 (d) $\frac{5\pi}{9}$

141. The number of distinct solutions of the equation $\frac{5}{4}\cos^2 2x + \cos^4 x + \sin^4 x + \cos^6 x + \sin^6 x = 2$ in the interval $[0, 2\pi]$ is **[Integer Answer Type 2015 Adv.]**

142. For $x \in (0, \pi)$, the equation $\sin x + 2\sin 2x - \sin 3x = 3$ has

(a) infinitely many solutions

[Single Correct Option 2014 Adv.]

(b) three solutions

(c) one solution

(d) no solution

143. Let $\theta, \phi \in [0, 2\pi]$ be such that $2\cos\theta(1 - \sin\phi) = \sin^2\theta$

$\left(\tan\frac{\theta}{2} + \cot\frac{\theta}{2}\right)\cos\phi - 1$, $\tan(2\pi - \theta) > 0$

and $-1 < \sin\theta < -\frac{\sqrt{3}}{2}$. Then, ϕ cannot satisfy

[More than One Correct Option 2012]

(a) $0 < \phi < \frac{\pi}{2}$ (b) $\frac{\pi}{2} < \phi < \frac{4\pi}{3}$

(c) $\frac{4\pi}{3} < \phi < \frac{3\pi}{2}$ (d) $\frac{3\pi}{2} < \phi < 2\pi$

144. If $P = \{\theta : \sin\theta - \cos\theta = \sqrt{2}\cos\theta\}$ and $Q = \{\theta : \sin\theta + \cos\theta = \sqrt{2}\sin\theta\}$ be two sets. Then, **[Single Correct Option 2011]**

(a) $P \subset Q$ and $Q - P \neq \phi$ (b) $Q \not\subset P$
(c) $P \not\subset Q$ (d) $P = Q$

145. The positive integer value of $n > 3$ satisfying the equation $\dfrac{1}{\sin\left(\dfrac{\pi}{n}\right)} = \dfrac{1}{\sin\left(\dfrac{2\pi}{n}\right)} + \dfrac{1}{\sin\left(\dfrac{3\pi}{n}\right)}$ is ... **[Integer Answer Type 2011]**

146. The number of values of θ in the interval $\left(-\dfrac{\pi}{2}, \dfrac{\pi}{2}\right)$ such that $\theta \neq \dfrac{n\pi}{5}$ for $n = 0, \pm 1, \pm 2$ and $\tan\theta = \cot 5\theta$ as well as $\sin 2\theta = \cos 4\theta$ is...... **[Integer Answer Type 2010]**

147. The number of solutions of the pair of equations $2\sin^2\theta - \cos 2\theta = 0$ and $2\cos^2\theta - 3\sin\theta = 0$ in the interval $[0, 2\pi]$ is **[Single Correct Option 2007]**

(a) 0
(b) 1
(c) 2
(d) 4

148. The set of values of θ satisfying the inequation $2\sin^2\theta - 5\sin\theta + 2 > 0$, where $0 < \theta < 2\pi$, is **[Single Correct Option 2006]**

(a) $\left(0, \dfrac{\pi}{6}\right) \cup \left(\dfrac{5\pi}{6}, 2\pi\right)$
(b) $\left[0, \dfrac{\pi}{6}\right] \cup \left[\dfrac{5\pi}{6}, 2\pi\right]$
(c) $\left[0, \dfrac{\pi}{3}\right] \cup \left[\dfrac{2\pi}{3}, 2\pi\right]$
(d) None of the above

(ii) JEE Main & AIEEE

149. If $0 \leq x < 2\pi$, then the number of real values of x, which satisfy the equation

$$\cos x + \cos 2x + \cos 3x + \cos 4x = 0, \text{ is}$$

[2016 JEE Main]

(a) 3 (b) 5
(c) 7 (d) 9

150. The possible values of $\theta \in (0, \pi)$ such that $\sin(\theta) + \sin(4\theta) + \sin(7\theta) = 0$ are **[2011 AIEEE]**

(a) $\dfrac{2\pi}{9}, \dfrac{\pi}{4}, \dfrac{4\pi}{9}, \dfrac{\pi}{2}, \dfrac{3\pi}{4}, \dfrac{8\pi}{9}$ (b) $\dfrac{\pi}{4}, \dfrac{5\pi}{12}, \dfrac{\pi}{2}, \dfrac{2\pi}{3}, \dfrac{3\pi}{4}, \dfrac{8\pi}{9}$
(c) $\dfrac{2\pi}{9}, \dfrac{\pi}{4}, \dfrac{\pi}{2}, \dfrac{2\pi}{3}, \dfrac{3\pi}{4}, \dfrac{35\pi}{36}$ (d) $\dfrac{2\pi}{9}, \dfrac{\pi}{4}, \dfrac{\pi}{2}, \dfrac{2\pi}{3}, \dfrac{3\pi}{4}, \dfrac{8\pi}{9}$

151. The number of values of x in the interval $[0, 3\pi]$ satisfying the equation $2\sin^2 x + 5\sin x - 3 = 0$, is **[2006 AIEEE]**

(a) 6 (b) 1
(c) 2 (d) 4

Answers

Exercise for Session 1

1. $x=(4n+1)\frac{\pi}{14}, -(4n-1)\frac{\pi}{6}$ 2. 3 3. 15 4. $n\pi \pm \frac{\pi}{8}$

5. $n\pi \pm \frac{\pi}{8}$ 6. $(6n+1)\frac{\pi}{12}$ 7. 4 8. $n\pi \pm \frac{\pi}{3}$

9. $\frac{n\pi}{3}$ 10. $(2n+1)\frac{\pi}{8}, n\pi \pm \frac{\pi}{3}$ 11. No solution

12. $n\pi + \frac{\pi}{4}$ 13. $\frac{\pi}{12}$ or $\frac{5\pi}{12}$ 14. 4 15. 0

Exercise for Session 2

1. $2n\pi$ or $(4n+1)\frac{\pi}{2}$ 2. $\frac{n\pi}{3} + \frac{\pi}{18}$ or $\frac{-n\pi}{2} + \frac{\pi}{6}$ 3. $2n\pi + \frac{\pi}{3}$

4. $2n\pi - \frac{\pi}{4}$ 5. $2n\pi + \frac{\pi}{4} + \frac{\pi}{12}$ 6. 8

7. $n\pi + \tan^{-1} 2$ or $n\pi + \tan^{-1}\left(-\frac{3}{4}\right)$ 8. $n\pi, n\pi - \frac{\pi}{4}$

Exercise for Session 3

1. $2n\pi + \frac{7\pi}{4}$ 2. $2n\pi + \frac{7\pi}{6}$ 3. $(2n+1)\frac{\pi}{4}$

4. $A = n\pi \pm \frac{\pi}{12}; B = n\pi \pm \frac{\pi}{6}$ 5. 0 6. No solution

7. $n\pi$ 8. 3 9. 5 10. 0

Exercise for Session 4

1. $\left[-\pi, -\frac{\pi}{6}\right) \cup \left(\frac{\pi}{6}, \pi\right]$ 2. $\left\{\frac{\pi}{2}\right\} \cup \left[0, \frac{\pi}{6}\right] \cup \left[\frac{5\pi}{6}, \pi\right]$

3. $\left[\frac{3\pi}{2}, 2\pi\right] \cup \{0\}$ 4. $\left[-\frac{\pi}{6}, \frac{\pi}{2}\right]$

5. $\left(0, \frac{\pi}{4}\right)$ 6. $\left(-\frac{\pi}{10}, \frac{3\pi}{10}\right)$

7. $R - \left\{x : x = \frac{3n\pi}{2} \pm \frac{3\pi}{4}, n \in Z\right\}$ 8. $\left(n\pi, n\pi + \frac{\pi}{6}\right)$

Chapter Exercises

1. (d) 2. (d) 3. (a) 4. (b) 5. (b) 6. (c)
7. (c) 8. (b) 9. (c) 10. (c) 11. (a) 12. (a)
13. (d) 14. (a) 15. (d) 16. (c) 17. (b) 18. (b)
19. (b) 20. (c) 21. (b) 22. (a) 23. (a) 24. (b)
25. (b) 26. (b) 27. (c) 28. (d) 29. (a) 30. (b)
31. (d) 32. (a) 33. (a) 34. (c) 35. (a) 36. (c)
37. (a) 38. (a) 39. (b) 40. (b) 41. (a) 42. (b)
43. (b) 44. (a) 45. (a) 46. (b) 47. (b) 48. (c)
49. (a) 50. (b) 51. (a) 52. (a) 53. (a) 54. (a)
55. (c) 56. (b) 57. (a) 58. (d) 59. (b) 60. (c)
61. (c,b) 62. (a,b,c) 63. (a,d) 64. (c,d) 65. (b,c,d)
66. (a,c,d) 67. (a,c) 68. (a,c) 69. (c,d) 70. (c)
71. (a) 72. (a) 73. (a,d) 74. (a,b) 75. (a,c) 76. (a,b)
77. (a,c) 78. (b,c) 79. (a,c) 80. (b,c) 81. (a,b) 82. (a,b,c,d)
83. (a,b,c,d) 84. (a,b) 85. (a,b,c,d)
86. (b) 87. (d) 88. (c) 89. (a) 90. (a) 91. (c)
92. (b) 93. (d) 94. (a) 95. (c) 96. (c) 97. (a)
98. (b) 99. (a) 100. (8) 101. (2) 102. (1) 103. (8)
104. (1) 105. (3) 106. (4) 107. (3) 108. (2) 109. (5)
110. (2) 111. (0) 112. (d) 113. (b) 114. (d) 115. (b)
116. (c) 117. (c) 118. (a) 119. (a) 120. (b)

121. (i) $\rightarrow$ (2), (ii) No solution, (iii) 4, (iv) 0, (v) 3

122. $\alpha \in \left[\frac{-3}{2}, \frac{1}{2}\right]$ and $x = \frac{nx}{2} + \frac{(-1)^n}{2}\sin^{-1}(1 - \sqrt{2\alpha + 3})$

123. $\alpha = 2n\pi \pm \frac{2\pi}{3}, n \in Z$

124. $x = n\pi + (-1)^n \frac{\pi}{6}$ and $y = 2m\pi \pm \frac{\pi}{3}, m, n \in Z$

125. $x = (2m+1)\frac{\pi}{4}$ and $y = 2n\pi + \frac{\pi}{2}, n \in I$

126. $\left(2, 2n\pi \pm \frac{2\pi}{3} - 2\right)$

127. $x = -1$ and $y = n\pi \pm \frac{\pi}{4} + 1, n \in I$

128. $x = k\pi + \frac{\pi}{4}$, $y = m\pi$ and $z = n\pi$ where $k, m, n \in Z$

129. $x = 2m\pi \pm \frac{2\pi}{3}$ and $y = 2(m-n)\pi \pm \frac{2\pi}{3}, m, n \in I$

130. $\theta = 4n\pi \pm \frac{2\pi}{3}$

131. $x = 2n\pi \pm \frac{3\pi}{4} + \frac{\pi}{4}$

132. $x = n\pi + (-1)^n \sin^{-1}\{2^{-\sqrt{(-\log_2 a)/2}}\}$ and the condition is $0 < a < 1$

133. $a = 3\pi(4n+1)$

134. $x = \frac{n\pi}{2} \pm (-1)^n \frac{\pi}{8}$

135. $x = y = m\pi \pm \frac{\pi}{4}, n \in I$

136. $x = n\pi \pm \cos^{-1}\sqrt{a+3}$, where $n \in z$ and $a \in [-3, -2]$

137. $x = \frac{\pi}{3}$

139. $x = 2m\pi + \frac{\pi}{2}, 2m\pi, n\pi \pm \frac{\pi}{6}$ and $y = k\pi$; $m, n, k, \in I$

140. (c) 141. (8) 142. (d) 143. (a,c,d) 144. (d)
145. (7) 146. (3) 147. (c) 148. (a) 149. (c)
150. (a) 151. (d)

Solutions

1. $(\cos^2 x - \sin^2 x)\left(2\sin\frac{x}{2} - 1\right) = 0$

$$\cos 2x = 0 \text{ or } \sin\frac{x}{2} = \frac{1}{2}$$

Hence, option (d) is false.

2. The maximum possible value is 2.

$\sin\left(\frac{x}{3}\right)$ takes the value 1 when

$$\frac{x}{3} = 2m\pi + \frac{\pi}{2}$$

i.e. $$\frac{x}{3} = 90 + 360m$$

$\sin\left(\frac{x}{11}\right)$ takes the value 1

when $$\frac{x}{11} = 2n\pi + \frac{\pi}{2}$$

i.e. $$\frac{x}{11} = 90 + 360n$$

We are looking for a common solution,

we have $3m - 11n = 2$.

Clearly, the smallest positive solution to this is $m = 8$, $n = 2$, thus $x_0 = 8910°$, giving $\alpha = 8910$.

3. $16(\sin^5 x + \cos^5 x) - 11(\sin x + \cos x) = 0$

$\Rightarrow (\sin x + \cos x)\{16(\sin^4 x - \sin^3 x \cos x + \sin^2 x \cos^2 x - \sin x \cos^3 x + \cos^4 x) - 11\} = 0$

$\Rightarrow (\sin x + \cos x)\{16(1 - \sin^2 x \cos^2 x - \sin x \cos x) - 11\} = 0$

$\Rightarrow (\sin x + \cos x)(4\sin x \cos x - 1)(4\sin x \cos x + 5) = 0$

As $4\sin x \cos x + 5 \neq 0$, we have

$\sin x + \cos x = 0$, $4\sin x \cos x - 1 = 0$

The required values are $\frac{\pi}{12}, \frac{5\pi}{12}, \frac{9\pi}{12}, \frac{13\pi}{12}, \frac{17\pi}{12}, \frac{21\pi}{12}$.

There are 6 solutions in $[0, 2\pi]$.

4. a, b, c are roots of equation

$x\sin\theta + y\sin 2\theta + + z\sin 3\theta = \sin 4\theta$

$\Rightarrow x\sin\theta + y(2\sin\theta\cos\theta) + z(3\sin\theta - 4\sin^3\theta) = 4\sin\theta\cos\theta\cos 2\theta$

$\Rightarrow \cos^3\theta - \frac{z}{2}\cos^2\theta - \frac{y+2}{4}\cos\theta + \frac{z-x}{8} = 0$

5. $\frac{\sin^2 2x + 4\sin^4 x - 4\sin^2 x\cos^2 x}{4 - \sin^2 2x - 4\sin^2 x} = \frac{1}{9}$

$$\Rightarrow \frac{4\sin^4 x}{4\cos^2 x - 4\sin^2 x\cos^2 x} = \frac{1}{9}$$

$$\Rightarrow \frac{\sin^4 x}{\cos^4 x} = \frac{1}{9} \Rightarrow \tan x = \pm\frac{1}{\sqrt{3}}$$

$$\Rightarrow x = \frac{\pi}{6}$$

6. $\left|\sqrt{m} - \sqrt{n}\right| \leq \sqrt{|m-n|}$

$\Rightarrow \left|\sqrt{\sin^2 x + 2a^2} - \sqrt{2a^2 - 1 - \cos^2 x}\right|$

$\leq \sqrt{\sin^2 x + 2a^2 - 2a^2 + 1 + \cos^2 x} \leq \sqrt{2}$

7. Consider, $\tan^2\frac{x}{2} = 1 + \sec x$

$$\Rightarrow 8\left(\frac{1-\cos x}{1+\cos x}\right) = 1 + \frac{1}{\cos x}$$

$$\Rightarrow 8\cos x - 8\cos^2 x = (1+\cos x)^2$$

$$\Rightarrow 8\cos x - 8\cos^2 x = 1 + \cos^2 x + 2\cos x$$

$$\Rightarrow 9\cos^2 x - 6\cos x + 1 = 0$$

$$\Rightarrow (3\cos x - 1)^2 = 0$$

$$\Rightarrow \cos x = \frac{1}{3} = \cos\alpha$$

$$\Rightarrow x = 2n\pi \pm \alpha \text{ where } `\alpha\text{'} = \cos^{-1}\frac{1}{3}$$

[By using $\cos\theta = \cos\alpha \Rightarrow \theta = 2n\pi \pm \alpha$]

$$\Rightarrow x = 2n\pi \pm \cos^{-1}\left(\frac{1}{3}\right)$$

8. We have $\tan\theta + \tan 4\theta + \tan 7\theta = \tan\theta\cdot\tan 4\theta\cdot\tan 7\theta$

$\Rightarrow \tan\theta + \tan 4\theta = -\tan 7\theta + \tan\theta\cdot\tan 4\theta\cdot\tan 7\theta$

$\Rightarrow \tan\theta + \tan 4\theta = -\tan 7\theta(1 - \tan\theta\cdot\tan 4\theta)$

$\Rightarrow \frac{\tan\theta + \tan 4\theta}{1 - \tan\theta\cdot\tan 4\theta} = -\tan 7\theta$

$$\Rightarrow \tan(\theta + 4\theta) = -\tan 7\theta$$

$$\Rightarrow \tan 5\theta = \tan(-7\theta)$$

$$\Rightarrow 5\theta = n\pi + (-7\theta)$$

$$\Rightarrow 12\theta = n\pi \Rightarrow \theta = \frac{n\pi}{12} \text{ (where } n \in I)$$

9. Put $e^{\sin x} = t$

Given equation becomes $t^2 - 4t - 1 = 0$

$$\Rightarrow t = 2 \pm \sqrt{5} \Rightarrow e^{\sin x} = 2 \pm \sqrt{5}$$

Either $\sin x = \log_e(2+\sqrt{5})$

or $\sin x = \log_e(2-\sqrt{5})$

as $2+\sqrt{5} > e$ or as $(2-\sqrt{5})$ is $(-\text{ve})$ and log is not defined for $(-\text{ve})$ values.

$\Rightarrow \sin x > 1 \Rightarrow$ no solution.

10. For $\sqrt{x-4}$ to be real $x \geq 4$, for which $\sqrt{x}$ is also real.

Now, if $\cos(\pi\sqrt{x}) < 1$, then $\cos(\pi\sqrt{x-4}) > 1$

and, if $\cos(\pi\sqrt{x}) > 1$, then $\cos(\pi\sqrt{x-4}) < 1$

(since their product = 1)

But both of these are not possible as $\cos\theta$ cannot be greater than 1.

$$\Rightarrow \cos(\pi\sqrt{x-4}) = 1 \text{ and } \cos(\pi\sqrt{x}) = 1$$

$$\Rightarrow x - 4 = 0 \text{ and } x = 0$$

$$\Rightarrow x = 4 \text{ or } x = 0$$

But $x = 0$ is not possible (as $x \geq 4$)

$\Rightarrow x = 4$ is only solution.

11. Consider R.H.S i.e., $5^x + 5^{-x}$

$\Rightarrow \quad \dfrac{5^x + 5^{-x}}{2} \geq (5^x \cdot 5^{-x})^{1/2}$

$\Rightarrow \quad 5^x + 5^{-x} \geq 2$ [By using A.M $\geq$ G.M]

$\Rightarrow$ From the given equation $\sin(e^x) \geq 2$

which is not possible for any real values of 'x'. Thus, the given equation has no solution.

12. We know that in a triangle $A+B+C=180°$

and $2B = A+C$ (As A, B, C are in A.P)

$\Rightarrow \quad 3B = 180$ or $B = 60°$

Now 'θ' be the common difference between A, B and C,

then $\quad C - A = 2\theta$ ...(i)

$\Rightarrow \quad \sin(C-A) = 1/2$ (given)

$\Rightarrow \quad \sin(2\theta) = 1/2$ [Using Eq. (i)]

$\Rightarrow \quad 2\theta = \dfrac{\pi}{6}$ or $\dfrac{5\pi}{6} \Rightarrow \theta = \dfrac{\pi}{12}$ or $\dfrac{5\pi}{12}$

Now, $\quad A = B - \theta \Rightarrow \dfrac{\pi}{3} - \dfrac{\pi}{12}$ or $\dfrac{\pi}{3} - \dfrac{5\pi}{12}$

$\Rightarrow A = \dfrac{\pi}{4}$ as a cannot be less than '0' and $C = \dfrac{\pi}{3} + \dfrac{\pi}{12} = \dfrac{5\pi}{12}$

$\Rightarrow \quad A = 45°, B = 60°, C = 75°$

13. Consider : Ist equation i.e. $2\sin^2 x + 3\sin x - 2 > 0$

$\Rightarrow \quad (2\sin x - 1)(\sin x + 2) > 0$

$\Rightarrow \quad (2\sin x - 1) > 0$ [As $\sin x + 2 > 0 \,\forall\, x \in R$]

$\Rightarrow \quad \sin x > \dfrac{1}{2} \Rightarrow x \in \left(\dfrac{\pi}{6}, \dfrac{5\pi}{6}\right)$ ...(i)

Consider $x^2 - x - 2 < 0$

$\Rightarrow \quad (x-2)(x+1) < 0 \Rightarrow -1 < x < 2$ ...(ii)

From Eqs. (i) and (ii) ,

Now, as, $2 < \dfrac{5\pi}{6}$, we obtain that 'x' must lie in $\left(\dfrac{\pi}{6}, 2\right)$.

14. Consider : $y = 5x^2 + 2x + 3 = 5\left[x^2 + \dfrac{2}{5}x + \dfrac{3}{5}\right]$

$= 5\left[\left(x + \dfrac{1}{5}\right)^2 + \dfrac{3}{5} - \dfrac{1}{25}\right] = 5\left(x + \dfrac{1}{5}\right)^2 + \dfrac{14}{5} > 2$

As $y = 2\sin x \leq 2$, so there cannot be any point of intersection.

15. We have $a_1 + a_2 \cos 2x + a_3 \sin^2 x = 0$

$\Rightarrow a_1 + a_2 \cos 2x + a_3(1 - \cos 2x)/2 = 0$

$\Rightarrow$ which is zero for all value of 'x'.

If $a_1 = -\dfrac{a_3}{2} = -a_2$ or $a_1 = \dfrac{-k}{2}, a_2 = \dfrac{k}{2}, a_3 = k$

For any $k \in R$

Hence, the required number of triplets is infinite.

16. We have, $\sin^4 x - (k+2)\sin^2 x - (k+3) = 0$

$\Rightarrow \quad \sin^2 x = \dfrac{(k+2) \pm \sqrt{(k+2)^2 + 4(k+3)}}{2} = \dfrac{(k+2) \pm (k+4)}{2}$

$\Rightarrow$ Either $\sin^2 x = k+3$ or $\sin^2 x = -1$

$\Rightarrow \quad 0 \leq \sin^2 x \leq 1$ or not possible

$\Rightarrow \quad 0 \leq k+3 \leq 1$

$\Rightarrow \quad -3 \leq k \leq -2$

17. $\theta \in \left[\dfrac{-\pi}{2}, \dfrac{\pi}{2}\right] \Rightarrow -1 \leq \sin\theta \leq 1$

But here $0 < \sin\theta < 1$ [As $\log_a x$ is define for $a > 0$ or $0 < a < 1$]

Now, $\log_{\sin\theta} \cos^2\theta = 2$ [By using $\log_a b = c \Rightarrow b = a^c$]

$\Rightarrow \quad \cos^2\theta = \sin^2\theta \Rightarrow \tan^2\theta = 1$

$\Rightarrow \quad \theta = n\pi + \dfrac{\pi}{4} \,\forall\, \theta \in \left[\dfrac{-\pi}{2}, \dfrac{\pi}{2}\right]$

$\Rightarrow \quad \theta = \dfrac{\pi}{4}, \theta = \dfrac{-\pi}{4}$

(Reject as $\sin\left(\dfrac{-\pi}{4}\right) = \dfrac{-1}{\sqrt{2}} < 0$)

$\Rightarrow$ The given equation has unique solution.

18. Consider $\sum_{i=1}^{n} \cos\theta_i = n$

$\cos\theta_1 + \cos\theta_2 + \cos\theta_3 + \ldots + \cos\theta_n = \underbrace{1 + 1 + 1 \ldots + n}_{n \text{ times}}$ is valid only when

$\cos\theta_1 = 1, \cos\theta_2 = 1, \cos\theta_3 = 1, \ldots, \cos\theta_n = 1$

$\Rightarrow \quad \theta_1 = \theta_2 = \theta_3 = \ldots = \theta_n = 0$

$\Rightarrow \quad \sum_{i=1}^{n} \sin\theta_i = 0$

19. **Case I** : For $n = 2$, $\sin^2 x + \cos^2 x = 1$.

Case II : If $n > 2$, $\sin^n x$ and $\cos^n x$ both decrease then

$\sin^n x + \cos^n x < 1$ (as $0 < x < \pi/2$)

Case III : If $n < 2$, $\sin^n x$ and $\cos^n x$ both increases then

$\sin^n x + \cos^n x > 1$ (as $0 < x < \pi/2$)

Then, $\sin^n x + \cos^n x \geq 1$ for $n \leq 2$

$\Rightarrow n \in (-\infty, 2]$

20. $\sin x + \cos x = \min_{a \in R}\{1, a^2 - 4a + 6\}$ ...(i)

As, $a^2 - 4a + 6 = (a-2)^2 + 2 > 2$ for all a

$\Rightarrow$ (i) becomes $\sin x + \cos x = 1$

$\Rightarrow \quad \sin\left(x + \dfrac{\pi}{4}\right) = \dfrac{1}{\sqrt{2}}$

$\Rightarrow \quad x + \dfrac{\pi}{4} = n\pi + (-1)^n \dfrac{\pi}{4}$

$\Rightarrow \quad x = n\pi + (-1)^n \dfrac{\pi}{4} - \dfrac{\pi}{4}$

21. $\sin^7 y = |x^3 - x^2 - 9x + 9| + |x^3 - 4x - x^2 + 4| + \sec^2 2y + \cos^4 y$

Now, for $x = 1$ (According to the choices)

$\sin^7 y = \sec^2 2y + \cos^4 y$

$\Rightarrow \quad \sin^7 y \cdot \cos^2 2y = 1 + \cos^4 y \cdot \cos^2 2y$

Now, R.H.S ≥ 1 and L.H.S ≤ 1

$\Rightarrow \quad$ L.H.S $= 1$

$\Rightarrow \quad \sin^7 y \cdot \cos^2 2y = 1$

$\Rightarrow \quad \sin^7 y = 1$ and $\cos^2 2y = 1$

$\Rightarrow \quad y = \dfrac{\pi}{2}$

General values of 'y' is $2n\pi + \dfrac{\pi}{2}$

Hence, $\quad x = 1$ and $y = 2n\pi + \dfrac{\pi}{2}$

22. As : $5\sin\theta+3\sin(\theta-\alpha)=5\sin\theta+3$
$(\sin\theta\cos\alpha-\cos\theta\sin\alpha)=(5+3\cos\alpha)$
$\sin\theta-3\sin\alpha\cos\theta$

Now, $-\sqrt{(5+3\cos\alpha)^2+9\sin^2\alpha}\le 5\sin\theta+3\sin(\theta-\alpha)\le\sqrt{(5+3\cos\alpha)^2+9\sin^2\alpha}$

$\Rightarrow \max_{\theta\in R}\{5\sin\theta+3\sin(\theta-\alpha)\}=\sqrt{34+30\cos\alpha}$

$\Rightarrow \sqrt{34+30\cos\alpha}=7\Rightarrow\cos\alpha=\frac{49-34}{30}$

$\Rightarrow \cos\alpha=\frac{1}{2}=\cos\frac{\pi}{3}\Rightarrow\alpha=2n\pi\pm\frac{\pi}{3}$

23. Consider : $\sin x(\sin x+\cos x)=n$

$\Rightarrow \sin^2x+\sin x\cdot\cos x=n\Rightarrow\frac{1-\cos 2x}{2}+\frac{\sin 2x}{2}=n$

$\Rightarrow \sin 2x-\cos 2x=2n-1$

$\Rightarrow$ As $-\sqrt{2}\le\sin 2x-\cos 2x\le\sqrt{2}$

$\Rightarrow -\sqrt{2}\le 2n-1\le\sqrt{2}$

$\Rightarrow \frac{1-\sqrt{2}}{2}\le n\le\frac{1+\sqrt{2}}{2}\Rightarrow n=0,\ 1$

24. $\sin\{x\}=\cos\{x\}$ graph of $y=\sin\{x\}$ and $y\cos\{x\}$ meet exactly 6 times in $[0, 2\pi]$.

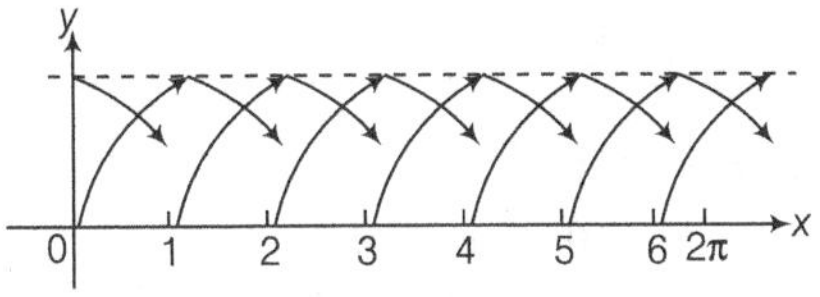

Points of intersection are at

$$x=\frac{\pi}{4},1+\frac{\pi}{4},2+\frac{\pi}{4},3+\frac{\pi}{4},5+\frac{\pi}{4}$$

25. $x^2+4-2x+3\sin(ax+b)=0$
$(x-1)^2+3+3\sin(ax+b)=0$
$\Rightarrow x=1$ and $\sin(ax+b)=-1$

$\Rightarrow \sin(a+b)=-1\Rightarrow a+b=\frac{3\pi}{2}$

26. Here, $4[1+\cot^2\pi(a+x)]+a^2-4a=0$

$\Rightarrow 4\cot^2\pi(a+x)+(a-2)^2=0$

$\Rightarrow a-2=0$ and $\cot^2\pi(a+x)=0\Rightarrow a=2$

27. As max $\cos\theta=1$, $2\cos x+\cos 2\lambda x=3$ is possible only when $\cos x=1$ and $\cos 2\lambda x=1$,
i.e. $\cos x=1$ and $\sin\lambda x=0$
Clearly, if λ is rational, say p/q, then $x=2q\pi$, $q\in I$, satisfies both the equations. Therefore, for exactly one solution, $x=0$, λ should be irrational.

28. $\sin\frac{\pi}{2n}+\cos\frac{\pi}{2n}=\sqrt{2}\sin\left(\frac{\pi}{2n}+\frac{\pi}{4}\right)$ or $\sin\left(\frac{\pi}{2n}+\frac{\pi}{4}\right)=\frac{\sqrt{n}}{2\sqrt{2}}$

. Since $\frac{\pi}{4}<\frac{\pi}{2n}+\frac{\pi}{4}<\frac{3\pi}{4}$ for $n>1$

or $\frac{1}{\sqrt{2}}<\frac{\sqrt{n}}{2\sqrt{2}}\le 1$ or $2<\sqrt{n}\le 2\sqrt{2}$ or $4<n\le 8$.

If $n=1$, L.H.S $=1$, R.H.S $=1/2$

Similarly for $n=8$, $\sin\left(\frac{\pi}{16}+\frac{\pi}{4}\right)\ne 1$

$\therefore\quad 4<n<8$

29. $5\sec\theta-13=12\tan\theta$

or, $13\cos\theta+12\sin\theta=5$

or, $\frac{13}{\sqrt{13^2+12^2}}\cos\theta+\frac{12}{\sqrt{13^2+12^2}}\sin\theta=\frac{5}{\sqrt{13^2+12^2}}$

or, $\cos(\theta-\alpha)=\frac{5}{\sqrt{313}}$, where $\cos\alpha=\frac{13}{\sqrt{313}}$

$\therefore\quad \theta=2n\pi\pm\cos^{-1}\frac{5}{\sqrt{313}}+\alpha$

$=2n\pi\pm\cos^{-1}\frac{5}{\sqrt{313}}+\cos^{-1}\frac{13}{\sqrt{313}}$

As $\cos^{-1}\frac{5}{\sqrt{313}}>\cos^{-1}\frac{13}{\sqrt{313}}$, then

$\theta\in[0,2\pi]$, when $n=0$ (One value, taking positive sign) and when $n=1$ (One value, taking negative sign).

30. Here, $x^3+(x+2)^2+2\sin x=4$.
Clearly, $x=0$ satisfies the equation.
If $0<x\le\pi$, $x^3+(x+2)^2+2\sin x>4$
If $\pi<x\le 2\pi$,
$x^3+(x+2)^2+2\sin x>27+25-2$
So, $x=0$ is the only solution.

31. If $\tan\left(\frac{\pi}{2}\sin\theta\right)=\cot\left(\frac{\pi}{2}\cos\theta\right)$

$\Rightarrow \tan\left(\frac{\pi}{2}\sin\theta\right)=\tan\left(\frac{\pi}{2}-\frac{\pi}{2}\cos\theta\right)$

$\Rightarrow \frac{\pi}{2}\sin\theta=n\pi+\frac{\pi}{2}-\frac{\pi}{2}\cos\theta, n\in I$

$\Rightarrow \sin\theta+\cos\theta=2n+1, n\in I$

$\Rightarrow \sin\theta+\cos\theta=2n+1,$

$n\in I$; but $-\sqrt{2}\le\sin\theta+\cos\theta\le\sqrt{2}$, therefore, $\sin\theta+\cos\theta=1$ or -1

32. We have, $\sin x+\sin y+\sin z=-3$
$0\le x\le 2\pi$
$0\le y\le 2\pi$
$0\le z\le 2\pi$
It is possible only, when $\sin x=\sin y=\sin z=-1$

$\therefore\quad x=y=z=\frac{3\pi}{2}$ for $x,y,z\in[0,2\pi]$

33. We have, $4\cos x-3\sec x=\tan x$

$\left(\cos x\ne 0\text{ i.e., } x \text{ odd multiplied of } \frac{\pi}{2}\right)$

Then, $4\cos x-\frac{3}{\cos x}=\frac{\sin x}{\cos x}$

$4\cos^2x-3=\sin x$
$4-4\sin^2x-3=\sin x$
$4\sin^2x+\sin x-1=0$

$\therefore \quad \sin x = \dfrac{-1+\sqrt{17}}{8}$

Either $\quad \sin x = \dfrac{-1+\sqrt{17}}{8} = \sin\alpha$ (let)

or $\quad \sin x = \dfrac{-1-\sqrt{17}}{8} = \sin\beta$ (let)

$\therefore \quad x = n\pi + (-1)^n \alpha$ or $x = n\pi + (-1)^n \beta$

$\therefore \quad \sin\alpha = \dfrac{-1+\sqrt{17}}{8}, \sin\beta = \dfrac{-1-\sqrt{17}}{8}$

So, $4(\sin\alpha + \sin\beta) = 4\left[\dfrac{-1+\sqrt{17}}{8} + \dfrac{-1-\sqrt{17}}{8}\right] = 4\left[\dfrac{-2}{8}\right] = -1$

34. We have, $\tan m\theta = \tan n\theta$

$m\theta = k\pi + n\theta, \forall k \in I$ (Using formula)

$m\theta - n\theta = k\pi, \forall k \in I$

$\theta = k \cdot \dfrac{\pi}{m-n}, \forall k \in I$ ($\because$ m and n are constants)

Then, if we put $k = 1, 2, 3, \ldots$ we get

$\theta = \dfrac{\pi}{m-n}, \dfrac{2\pi}{m-n}, \dfrac{3\pi}{m-n}, \ldots$, which is AP.

Thus, common difference $= \dfrac{\pi}{m-n}$

35. $\because \sin 3\alpha = 4\sin\alpha \sin(x-\alpha)\sin(x+\alpha)$

$3\sin\alpha - 4\sin^3\alpha = 4\sin\alpha(\sin^2 x - \sin^2\alpha)$

On dividing both sides by $\sin\alpha$, we get

$3 - 4\sin^2\alpha = 4\sin^2 x - 4\sin^2\alpha$

$\Rightarrow \quad 4\sin^2 x = 3$

$\Rightarrow \quad \sin^2 x = \dfrac{3}{4}$

$\Rightarrow \quad \sin^2 x = \left(\dfrac{\sqrt{3}}{2}\right)^2 = \sin^2\dfrac{\pi}{3}$

$\therefore \quad x = n\pi \pm \dfrac{\pi}{3}, \forall n \in I$ (using formula)

36. We know that,

$a\sin\theta \pm b\cos\theta = c$ is solvable, if $|c| \le \sqrt{a^2+b^2}$.

Now, $\lambda\cos x - 3\sin x = \lambda + 1$ is solvable, if

$|\lambda+1| \le \sqrt{\lambda^2+9}$

$(\lambda+1)^2 \le \lambda^2 + 9$

$\lambda^2 + 1 + 2\lambda \le \lambda^2 + 9$

$2\lambda \le 9 - 1 \Rightarrow \lambda \le 4$

$\therefore \quad \lambda \in (-\infty, 4]$

37. We have, $\cos 2x - 3\cos x + 1 = \dfrac{1}{(\cot 2x - \cot x)\sin(x-\pi)}$

$2\cos^2 x - 1 - 3\cos x + 1 = \dfrac{1}{\dfrac{\sin(x-2x)}{\sin 2x \cdot \sin x} \cdot (-\sin x)}$

$\left[\text{using formula}, \cot A - \cot B = \dfrac{\sin(B-A)}{\sin A \cdot \sin B}\right]$

$2\cos^2 x - 3\cos x = \dfrac{2\sin x \cdot \cos x \cdot \sin x}{\sin^2 x}$

$2\cos^2 x - 3\cos x - 2\cos x = 0$

$2\cos^2 x - 5\cos x = 0$

$\cos x(2\cos x - 5) = 0$

Either $2\cos x - 5 = 0$

$\Rightarrow \quad \cos x = \dfrac{5}{2}$ (which is not possible)

Then, $\quad \cos x = 0$

38. $\because \sec x \cdot \cos 5x = -1$ and $0 < x < \dfrac{\pi}{4}$

$\dfrac{\cos 5x}{\cos x} = -1$

$\Rightarrow \quad \cos 5x = -\cos x$

$\cos 5x + \cos x = 0$

$\Rightarrow \quad 2\cos 3x \cdot \cos 2x = 0$

Either $2\cos 3x = 0$ or $\cos 2x = 0$

$\cos 3x = 0$ or $2x = (2n+1)\dfrac{\pi}{2}$

$3x = (2n+1)\dfrac{\pi}{2}$ or $x = (2n+1)\dfrac{\pi}{4}, \forall n \in I$ (not possible)

$\therefore \quad x = (2n+1)\dfrac{\pi}{6}$

Put $n = 0$, then $x = \dfrac{\pi}{6}$

$\therefore \quad x = \dfrac{\pi}{6}$

39. $\because \sin^{100}\theta - \cos^{100}\theta = 1$

$\sin^{100}\theta = 1 + \cos^{100}\theta$

This equation is valid, if $\cos^{100}\theta = 0$ and $\sin^{100}\theta = 1$

($\because 0 \le \cos^2\theta \le 1, 0 \le \sin^2\theta \le 1$)

$\therefore \quad \cos\theta = 0$, then $\sin\theta = 1$

$\theta = (2n+1)\dfrac{\pi}{2}, \forall n \in I$

$\theta = n\pi + \dfrac{\pi}{2}, \forall n \in I$

40. $\because \alpha^2 - 4\alpha + 7 = \alpha^2 - 4\alpha + 4 + 3 = (\alpha-2)^2 + 3 \ge 3$

and $\quad 2 < e < 3$ and $\pi = 3.14$

$4 < e^2 < 9$

Now, $\sqrt{3}\sin x - \cos x = \min\limits_{\alpha \in R}\{2, e^2, \pi, \alpha^2 - 4\alpha + 7\}$

$\sqrt{3}\sin x - \cos x = 2$

$\dfrac{\sqrt{3}}{2}\sin x - \dfrac{1}{2}\cos x = 1$

$\sin x \cdot \cos\dfrac{\pi}{6} - \cos x \cdot \sin\dfrac{\pi}{6} = 1$

$\sin\left(x - \dfrac{\pi}{6}\right) = 1$

$\therefore \quad x - \dfrac{\pi}{6} = 2n\pi + \dfrac{\pi}{2}, \forall n \in I$

$x = 2n\pi + \dfrac{\pi}{2} + \dfrac{\pi}{6}, \forall n \in I$

$x = 2n\pi + \dfrac{2\pi}{3}, \forall n \in I$

41. We have, $\cos 4x + 6 = 7\cos 2x$

$$2\cos^2 2x - 1 + 6 - 7\cos 2x = 0$$
$$2\cos^2 2x - 7\cos 2x + 5 = 0$$
$$(2\cos x - 5)(\cos 2x - 1) = 0$$

Thus, $\cos 2x = 1$

and $\cos 2x = \frac{5}{2}$ (which is not possible, $-1 \le \cos\theta \le 1$)

$\therefore$ $\cos 2x = 1$

$2x = 2n\pi, \forall n \in I$ (using formula)

$x = n\pi, \forall n \in I$

$x = \pi, 2\pi, 3\pi, \ldots\ldots$

i.e., $x = 180°, 360°, 540°, \ldots\ldots.$

$x \notin [315°, 317°]$

So, $x = n\pi \notin [315°, 317°], \forall n \in I$

Hence, number of solutions is 0.

42. Let $A = 5\pi\sin\theta$ and $B = 5\pi\cos\theta$

Then, $\cot A - \tan B = 0$

$$\frac{\cos A}{\sin A} - \frac{\sin B}{\cos B} = 0 \Rightarrow \cos A \cdot \cos B - \sin A \cdot \sin B = 0$$

$\therefore$ $\cos(A + B) = 0$

$\Rightarrow$ $A + B = (2n + 1)\frac{\pi}{2}, \forall n \in I$ ...(i)

Now, $5\pi\sin\theta + 5\pi\cos\theta = (2n + 1)\frac{\pi}{2}, \forall n \in I$

$$\sin\theta + \cos\theta = \frac{2n + 1}{10}$$

$$\sin\theta \cdot \frac{1}{\sqrt{2}} + \cos\theta \cdot \frac{1}{\sqrt{2}} = \frac{2n + 1}{10\sqrt{2}}$$

$$\sin\left(\theta + \frac{\pi}{4}\right) = \frac{2n + 1}{10\sqrt{2}} \quad \text{...(ii)}$$

$\because$ $-1 \le \sin\left(\theta + \frac{\pi}{4}\right) \le 1$

$\Rightarrow$ $-1 \le \frac{2n + 1}{10\sqrt{2}} \le 1$

$\Rightarrow$ $\frac{-10\sqrt{2} - 1}{2} \le n \le \frac{10\sqrt{2} - 1}{2}$

$-7.5 \le n \le 6.5$ $[\because \sqrt{2} = 1.4 \text{ (let)}]$

$\therefore n = -7, -6, -5, -4, -3, -2, -1, 0, 1, 2, 3, 4, 5, 6$

Hence, number of solutions is 14.

43. $\sin^2 x + \sin^4 x + \sin^6 x + \ldots\infty = \frac{\sin^2 x}{1 - \sin^2 x} = \tan^2 x$

$\Rightarrow$ $\exp^{[(\sin^2 x + \sin^4 x + \ldots)\ln 2]} = e^{\tan^2 x \ln} = e^{\ln 2^{\tan^2 x}}$

The given equation is $y^2 - 9y + 8 \Rightarrow (y - 1)(y - 8) = 0$

Either $y = 1 \Rightarrow 2^{\tan x} = 1 = 2^0 \Rightarrow \tan^2 x = 0$, but $x \in \left(0, \frac{\pi}{2}\right)$

$\therefore$ Neglecting $x = 0$ or $y = 2^3 \Rightarrow \tan^2 x = 3$

$\Rightarrow$ $\tan x = \pm\sqrt{3} \Rightarrow x = \frac{\pi}{3}$, as $0 < x < \frac{\pi}{2}$

$\Rightarrow$ $\frac{\cos x}{\cos x + \sin x} = \frac{1/2}{1/2 + \sqrt{3}/2} = \frac{1}{\sqrt{3} + 1} = \frac{\sqrt{3} - 1}{2}$

44. $\cos x = \sqrt{1 - \sin 2x} = |\sin x - \cos x|$

(i) $\sin x \le \cos x$

$\Rightarrow$ $\cos x = \cos x - \sin x \Rightarrow \sin x = 0$

where, $x \in \left[0, \frac{\pi}{4}\right) \cup \left(\frac{5\pi}{4}, 2\pi\right]$,

$\therefore$ $\sin x = 0$

$\Rightarrow$ $x = 2\pi$, neglecting $x = \pi$

(ii) $\sin x > \cos x \Rightarrow \tan x = 2$

where $x \in \left(\frac{\pi}{4}, \frac{5\pi}{4}\right)$, $\therefore \tan x = 2 \Rightarrow x = \tan^{-1}(2)$

Thus, the given equation has two solutions.

45. We have, $\sin 3x = 3\sin x - 4\sin^3 x \Rightarrow \sin^3 x = \frac{1}{4}(3\sin x - \sin 3x)$

and $\cos 3x = 4\cos^3 x - 3\cos x$

$\Rightarrow$ $\cos^3 x = \frac{1}{4}(\cos 3x + 3\cos x)$

$\therefore$ $\cos 3x \cos^3 x + \sin 3x \sin^3 x$

$$= \frac{1}{4}[\cos^2 3x + 3\cos x\cos 3x + 3\sin x\sin 3x - \sin^2 3x]$$

$$= \frac{1}{4}[3\cos 2x + \cos 6x] = \cos^3 2x$$

$\Rightarrow$ $\cos 2x = 0 \Rightarrow 2x = (2n + 1)\frac{\pi}{2}$

$\Rightarrow$ $x = (2n + 1)\frac{\pi}{4}$

46. Graphs of $y = \sin x$ and $y = \frac{|x|}{10}$ meet exactly six times. Hence, there are six solutions.

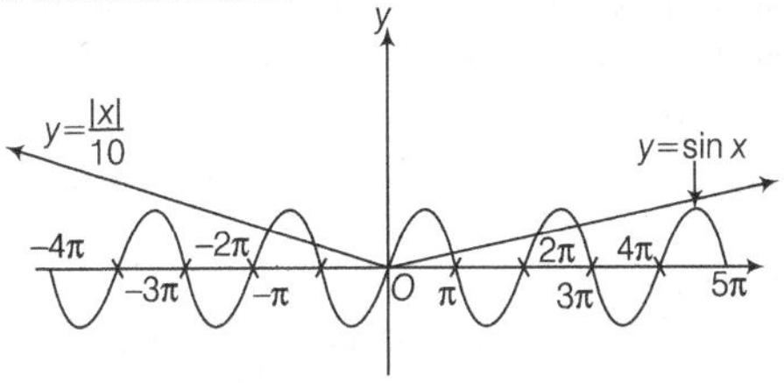

47. Since, the equation $a_1 + a_2\sin x + a_3\cos x + a_4\sin 2x + a_5\cos 2x = 0$ holds for all values of x,

$a_1 + a_3 + a_5 = 0$ (on putting $x = 0$)

$a_1 - a_3 + a_5 = 0$ (on putting $x = \pi$)

$\Rightarrow$ $a_3 = 0$ and $a_1 + a_5 = 0$ ...(i)

Putting $x = \frac{\pi}{2}$ and $\frac{3\pi}{2}$, we get

$a_1 + a_2 - a_5 = 0$ and $a_1 - a_2 - a_5 = 0$

$\Rightarrow$ $a_2 = 0$ and $a_1 - a_5 = 0$ ...(ii)

Eqs. (i) and (ii) give $a_1 = a_2 = a_3 = a_5 = 0$

The given equation reduces to $a_4 \sin 2x = 0$. This is true for all values of x, therefore $a_4 = 0$

Hence, $a_1 = a_2 = a_3 = a_4 = a_5 = 0$

Thus, the number of 5-tuples is one.

48. $\cos^2 x = 2\cos x(3\sin^2 x - 2)$

$\Rightarrow \cos x[\cos x - 2\{3(1 - \cos^2 x) - 2\}] = 0$

$\Rightarrow \cos x(6\cos^2 x - 2 + \cos x) \Rightarrow \cos x = 0$, which is not possible.

or $\qquad 6\cos^2 x+\cos x-2=0 \Rightarrow \cos x=\frac{1}{2},-\frac{2}{3}$

$\Rightarrow \qquad x=\frac{\pi}{3},\frac{5\pi}{3}$ or $x=\pi-\cos^{-1}\left(\frac{2}{3}\right),\ \pi+\cos^{-1}\left(\frac{2}{3}\right)$

$\Rightarrow \qquad |x_1-x_3|=\frac{4\pi}{3}$ or $|x_1-x_2|=2\cos^{-1}\left(\frac{2}{3}\right)$

$\Rightarrow \qquad |x_1-x_2|_{\min}=2\cos^{-1}\left(\frac{2}{3}\right)$

49. $\cos x-\frac{\cot\beta\sin x}{2}=\frac{\sqrt3}{2} \Rightarrow \frac{1-\tan^2\frac{x}{2}}{1+\tan^2\frac{x}{2}}-\frac{\cot\beta\tan\frac{x}{2}}{1+\tan^2\frac{x}{2}}=\frac{\sqrt3}{2}$

$\Rightarrow \qquad 1-\tan^2\frac{x}{2}-\cot\beta\tan\frac{x}{2}=\frac{\sqrt3}{2}\left(1+\tan^2\frac{x}{2}\right)$

$\Rightarrow \qquad (2+\sqrt3)\tan^2\frac{x}{2}+2\cot\beta\tan\frac{x}{2}+(\sqrt3-2)=0$

$\Rightarrow \qquad \tan\frac{x}{2}=\frac{-2\cot\beta\pm\sqrt{4\cot^2\beta+4}}{2(2+\sqrt3)}=\frac{-2\cot\beta\pm2\text{cosec}\beta}{2(2+\sqrt3)}$

$\Rightarrow \qquad \tan\frac{x}{2}=\frac{-\cot\beta+\text{cosec}\beta}{(2+\sqrt3)}$

or $\qquad \tan\frac{x}{2}=\frac{-\cot\beta-\text{cosec}\beta}{(2+\sqrt3)}$

$\Rightarrow \qquad \tan\frac{x}{2}=\tan\frac{\beta}{2}\tan15^\circ$

50. $n\sin^2\theta+2n\cos(\theta+\alpha)\sin\alpha\sin\theta+\cos2(\alpha+\theta)$

$=n\sin^2\theta+n\cos(\theta+\alpha)\{\cos(\theta-\alpha)-\cos(\theta+\alpha)\}+2\cos^2(\theta+\alpha)-1$

$=n\sin^2\theta+n(\cos^2\theta-\sin^2\alpha)-n\cos^2(\theta+\alpha)+2\cos^2(\alpha+\theta)-1$

$=n\sin^2\theta+n\cos^2\theta-n\sin^2\alpha+(2-n)\cos^2(\theta+\alpha)-1$

$=(n-1)-n\sin^2\alpha+(2-n)\cos^2(\theta+\alpha)\Rightarrow n=2$

51. Applying $C_1\to C_1-2\cos x\,C_2+C_3$ to the given determinant, we get

$$\begin{vmatrix}1-2a\cos x+a^2 & a & a^2\\ 0&\cos nx&\cos(n+1)x\\ 0&\sin nx&\sin(n+1)x\end{vmatrix}=(1-2a\cos x+a^2)\sin x=0$$

if $\sin x=0$ or $\cos x=(1+a^2)/2a$ i.e., if $n=n\pi,\ n\in I$

52. $\frac{\sin3\alpha}{\cos2\alpha}<0$ if $\sin3\alpha>0$ and $\cos2\alpha<0$ or $\sin3\alpha<0$ and $\cos2\alpha>0$

i.e., if $3\alpha\in(0,\pi)$ and $2\alpha\in(\pi/2,3\pi/2)$

or $3\alpha\in(\pi,2\pi)$ and $2\alpha\in(-\pi/2,\pi/2)$

i.e., if $\alpha\in(0,\pi/3)$ and $\alpha\in(\pi/4,3\pi/4)$

or $\alpha\in(\pi/3,2\pi/3)$ and $\alpha\in(-\pi/4,\pi/4)$ i.e., if $\alpha\in(\pi/4,\pi/3)$

since $(13\pi/48,14\pi/48)\subset(\pi/4,\pi/3)$,

option (a) is correct.

53. $f(x)=(\sin^2\theta+\cos^2\theta+x)\begin{vmatrix}1&\cos^2\theta&x\\ 1&x&\sin^2\theta\\ 1&\sin^2\theta&\cos^2\theta\end{vmatrix}$

$$=(x+1)\begin{vmatrix}1&\cos^2\theta&x\\ 0&x-\cos^2\theta&\sin^2\theta-x\\ 0&\sin^2\theta-\cos^2\theta&\cos^2\theta-x\end{vmatrix}$$

$=(x+1)[(x-\cos^2\theta)(\cos^2\theta-x)-(\sin^2\theta-\cos^2\theta)(\sin^2\theta-x)]$

$=(x+1)[-x^2-\cos^4\theta+2x\cos^2\theta$

$\qquad -x\cos^2\theta+x\sin^2\theta-\sin^4\theta+\sin^2\theta\cos^2\theta]$

$=(-1/2)(x+1)[(x-\sin^2\theta)^2+(x-\cos^2\theta)^2+(\sin^2\theta-\cos^2\theta)]^2$

So, $f(x)=0$, if $x=-1$ or $x=\sin^2\theta=\cos^2\theta$

$\sin^2\theta=\cos^2\theta$

$\Rightarrow \qquad \theta=\pi/4\Rightarrow x=1/2$

Hence, $x=-1,\ 1/2$

54. Given, $\sin x+\sin y+\sin z=-3$ is satisfied only when $x=y=z=\frac{3\pi}{2}$, for $x,y,z\in[0,2\pi]$.

55. $\sec x\cos5x=-1\Rightarrow\cos5x=-\cos x\Rightarrow5x=2n\pi\pm(\pi-x)$

$\Rightarrow x=\frac{(2n+1)\pi}{6}$ or $\frac{(2n-1)\pi}{4}$

Hence, $x=\frac{\pi}{4},\frac{\pi}{2},\frac{3\pi}{4},\frac{5\pi}{6},\frac{5\pi}{6},\frac{7\pi}{6},\frac{7\pi}{4},\frac{9\pi}{6},\frac{11\pi}{6}$

56. $3\cos\theta+4\sin\theta=5\left[\frac{3}{5}\cos\theta+\frac{4}{5}\sin\theta\right]=5\cos(\theta-\alpha)$

where $\cos\alpha=3/5,\ \sin\alpha=4/5$

Now, $3\cos\theta+4\sin\theta=k$

$\therefore \qquad 5\cos(\theta-\alpha)=\pm5\Rightarrow\cos(\theta-\alpha)=\pm1$

$\Rightarrow \qquad \theta-\alpha=0^\circ,180^\circ\Rightarrow\theta=\alpha,\ 180^\circ+\alpha$

57. Given, $\cot(\alpha+\beta)=0$

$\Rightarrow \cos(\alpha+\beta)=0\Rightarrow\alpha+\beta=(2n+1)\frac{\pi}{2},n\in I$

$\therefore \sin(a+2\beta)=\sin(2\alpha+2\beta-\alpha)=\sin[(2n+1)\pi-\alpha]$

$=\sin(2n\pi+\pi-\alpha)=\sin(\pi-\alpha)=\sin\alpha$

58. $\cot\theta+\cot\left(\frac{\pi}{4}+\theta\right)=2$

$\Rightarrow \qquad \frac{\cos\theta}{\sin\theta}+\frac{\cos\{(\pi/4)+\theta\}}{\sin\{(\pi/4)+\theta\}}=2$

$\Rightarrow \qquad \sin\left(\frac{\pi}{2}+2\theta\right)=2\sin\theta\sin\left(\frac{\pi}{4}+\theta\right)$

$\Rightarrow \qquad \sin\left(\frac{\pi}{4}+2\theta\right)+\cos\left(\frac{\pi}{4}+2\theta\right)=\frac{1}{\sqrt2}$

$\Rightarrow \qquad \cos2\theta=\frac{1}{2}\Rightarrow 2\theta=2n\pi\pm\frac{\pi}{3}\Rightarrow\theta=n\pi\pm\frac{\pi}{6}.$

59. $2\cos^2\theta-(\sqrt2+1)\cos\theta-1+\frac{(\sqrt2+1)}{\sqrt2}=0$

$\Rightarrow \qquad \cos\theta=\frac{(\sqrt2+1)\pm\sqrt{(\sqrt2+1)^2-\frac{8}{\sqrt2}}}{4}$

$\Rightarrow \qquad \cos\theta=\cos\left(\frac{\pi}{4}\right)\Rightarrow\theta=2n\pi\pm\frac{\pi}{4}$

Trick : Since $\theta=\frac{\pi}{4}$ satisfies the equation and therefore the general value should be $2n\pi\pm\frac{\pi}{4}$.

60. $|\sin x|\ge0\Rightarrow\frac{|\sin x|}{1+|\sin x|}<1\Rightarrow 1-\frac{|\sin x|}{1+|\sin x|}>0$

So, the given in equation becomes

$$1-\frac{|\sin x|}{1+|\sin x|}\geq\frac{2}{3}$$

$\Rightarrow \quad \frac{1}{3}\geq\frac{|\sin x|}{1+|\sin x|} \Rightarrow |\sin x|\leq\frac{1}{2}$

$\Rightarrow \quad -\frac{1}{2}\leq\sin x\leq\frac{1}{2} \Rightarrow \sin x\in\left[-\frac{1}{2},\frac{1}{2}\right]$

61. $(t-[|\sin x|])! = 3!5!7!$

if $x = n\pi + \frac{\pi}{2} (n \in I)$

then $(t-1)! = 3!\,5!\,7!$

$\Rightarrow \quad (t-1)! = 10!$

$\Rightarrow \quad t-1 = 10$

$\Rightarrow \quad t = 11$

If $x \neq n\pi + \frac{\pi}{2} (n \in I)$, then

$(t-0)! = 10!$

$\Rightarrow \quad t = 10$

62. $f(x) = \left(\cos a_1 + \frac{\cos a_2}{2} + \dots + \frac{\cos a_n}{2^{n-1}}\right)$

$\cos x - \left(\frac{\sin a_1}{1} + \frac{\sin a_2}{2} + \dots + \frac{\sin a_n}{2^{n-1}}\right)\sin x$

$\Rightarrow \quad f(x) = A\cos x - B\sin x$

Now, $\quad f(x_1) = f(x_2) = 0$

$\Rightarrow \quad \left.\begin{aligned}A\cos x_1 - B\sin x_1 = 0\\ A\cos x_2 - B\sin x_2 = 0\end{aligned}\right\}$

$\Rightarrow \quad \tan x_1 = \tan x_2$

$\Rightarrow \quad x_1 = n\pi + x_2$

$\Rightarrow \quad x_1 - x_2 = n\pi$

63. $AP = \sqrt{(1-\cos\alpha)^2 + \sin^2\alpha}$

$= 2|\sin\alpha/2| = 2\sin\frac{\alpha}{2}$

Similarly $AQ = 2\sin\frac{\beta}{2}$ and $AR = 2\sin\frac{\gamma}{2}$

Now as AP, AQ, AR are in GP.

$\therefore \sin\frac{\alpha}{2}, \sin\frac{\beta}{2}, \sin\frac{\gamma}{2}$ are in GP.

$\Rightarrow \quad \frac{\sin\frac{\alpha}{2}+\sin\frac{\gamma}{2}}{2} \geq \sin\frac{\beta}{2}$

$\Rightarrow \quad \sin\frac{\alpha+\gamma}{2}\cos\frac{\alpha-\gamma}{2} \geq \sin\frac{\beta}{2}$

Also, $\sin\frac{\alpha}{2}\sin\frac{\gamma}{2} \leq \sin\frac{\beta}{2}$

64. $\sin^2 z + \operatorname{cosec}^2 z \geq 2,\ 2+\cot^2 y \geq 2,\ 4+\sin 4x \geq 3$

$\Rightarrow \quad \sin^2 z = 1, \cot^2 y = 0, \sin 4x = -1$

$\Rightarrow \quad z \in \left\{\frac{\pi}{2},\frac{3\pi}{2}\right\},$

$y \in \left\{\frac{\pi}{2},\frac{3\pi}{2}\right\},\ x \in \left\{\frac{3\pi}{8},\frac{7\pi}{8},\frac{11\pi}{8},\frac{15\pi}{8}\right\}.$

65. For non-trivial solution

$$\begin{vmatrix}\sin\theta & -2\cos\theta & -a\\ -1 & 1 & 1\\ 1 & 2 & 1\end{vmatrix} = 0$$

$\Rightarrow \quad \sin\theta + 4\cos\theta = 3a$

$-\frac{\sqrt{17}}{3} \leq a \leq \frac{\sqrt{17}}{3}$

So, 3 integral values.

66. The equation becomes $(\sin\theta - 2)(\sin\theta + \lambda)(2\sin\theta + 1) = 0$

$\Rightarrow \quad \lambda = \pm 1, 0$

67. $x+y=2\pi/3$ or $y=(2\pi/3)-x$

$\therefore \quad \sin x = 2\sin\left(\frac{2\pi}{3}-x\right)$

$= 2\left[\left(\frac{\sqrt{3}}{2}\right)\cos x + \left(\frac{1}{2}\right)\sin x\right]$

$= \sqrt{3}\cos x + \sin x \Rightarrow \cos x = 0$

$\Rightarrow \quad x = n\pi + \frac{\pi}{2}, n\in Z$

$\Rightarrow \quad y = \frac{2\pi}{3} - n\pi - \frac{\pi}{2} = \frac{\pi}{6} - n\pi$

Hence, for $x\in[0,4\pi]$, $x=\pi/2, 3\pi/2, 5\pi/2, 7\pi/2$ and for $y\in[0,4\pi]$, $y=\pi/6, 7\pi/6, 13\pi/6, 19\pi/6$

68. It is easier to solve the inequality using graphical method. The graphs of $y=|\cos x|$ and $y=\sin x$ are shown in the following figure.

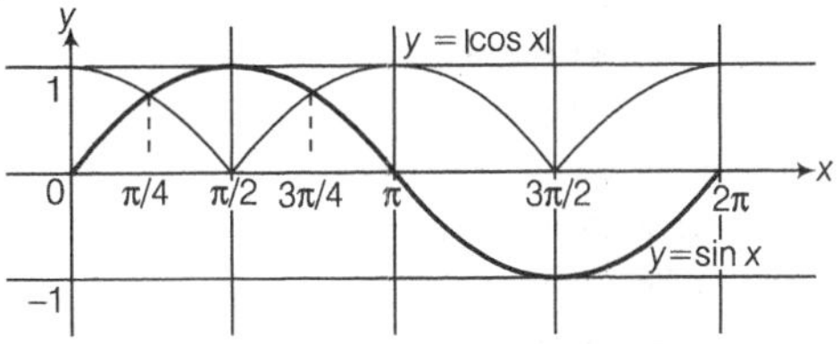

From the figure, $|\cos x|\leq\sin x$ for $x\in\left[\frac{\pi}{4},\frac{3\pi}{4}\right]$

69. $(x+y)$ and $(x-y)$ satisfy the equation $\tan^2\theta - 4\tan\theta + 1 = 0$. Thus,

$\tan(x+y) + \tan(x-y) = 4$

and $\quad \tan(x+y)\tan(x-y) = 1$

or $\quad \tan 2x = \tan((x+y)+(x-y))$

or $\quad \tan 2x = \frac{\tan(x+y)+\tan(x-y)}{1-\tan(x+y)\tan(x-y)}$

$\therefore \quad \tan 2x = \infty$ or $2x = 90^\circ$ or $x = 45^\circ = \frac{\pi}{4}$

$\therefore \quad y = \frac{\pi}{6}$

70. $\because x+y = \frac{4\pi}{3}$ ...(i)

and $\quad \sin x = 2\sin y$ (given)

$\sin x = 2\sin\left(\frac{4\pi}{3}-x\right)$ [From Eq. (i)]

$$\sin x=2\left[\sin\frac{4\pi}{3}\cdot\cos x-\cos\frac{4\pi}{3}\cdot\sin x\right]$$

$$\sin x=2\left[\frac{-\sqrt{3}}{2}\cos x+\frac{1}{2}\sin x\right]$$

$$\sin x=-\sqrt{3}\cos x+\sin x$$

$$\cos x=0$$

$$x=(2n+1)\frac{\pi}{2},\forall n\in I$$

$$x=n\pi+\frac{\pi}{2}$$

Putting the value of x in Eq. (i) we get

$$y=\frac{4\pi}{3}-n\pi-\frac{\pi}{2}$$

$$y=\frac{5\pi}{6}-n\pi,\ \forall n\in I$$

$\therefore$ $$(x,y)=\left(n\pi+\frac{\pi}{2},\frac{5\pi}{6}-n\pi\right),\forall n\in I$$

71. We know that,

$[x+I]=[x]+I$, if I is an integer.

then, $$y=\frac{1}{3}[\sin\theta+[\sin\theta+[\sin\theta]]]$$

$$=\frac{1}{3}[\sin\theta+[\sin\theta]+[\sin\theta]]$$

$$=\frac{1}{3}([\sin\theta]+[\sin\theta]+[\sin\theta])$$

$$y=[\sin\theta] \quad \text{...(i)}$$

$$[y+[y]]=2\cos\theta$$

$$[y]+[y]=2\cos\theta$$

$$[y]=\cos\theta \quad \text{...(ii)}$$

$\because$ $-1\le\sin\theta\le1$, then three cases arise:

Case I If $-1\le\sin\theta<0$, then $[\sin\theta]=-1$

$y=-1$ put in Eq. (ii), then

$$\cos\theta=[-1]=-1$$

$\Rightarrow$ $\sin\theta=0$

But $-1\le\sin\theta<0$

Hence, this case is not possible.

Case II If $0\le\sin\theta<1$, then $[\sin\theta]=0$

and $y=0$ [from Eq.(i)]

Put in Eq.(ii), $\cos\theta=0$

$\Rightarrow$ $\sin\theta=1$

But we have $0\le\sin\theta<1$, so this case is not possible.

Case III If $\sin\theta=1$, then $[\sin\theta]=1$

$y=1$ put in Eq. (ii), $\cos\theta=1 \Rightarrow \sin\theta=0$

But, we have $\sin\theta=1$, so this case is not possible.

$\therefore$ Number of solutions is 0.

72. $\because [\sin x]+[\sqrt{2}\cos x]=-3,\ x\in[0,2\pi]$

It is possible only when $[\sin x]=-1$ and $[\sqrt{2}\cos x]=-2$

If $[\sin x]=-1 \Rightarrow -1\le\sin x<0$,

$\therefore$ $x\in(\pi,2\pi)$

If $[\sqrt{2}\cos x]=-2$

$\Rightarrow$ $$-2\le\sqrt{2}\cos x<-1 \Rightarrow -\sqrt{2}\le\cos x<-\frac{1}{\sqrt{2}}$$

$\Rightarrow$ $$-1\le\cos x<-\frac{1}{\sqrt{2}} \Rightarrow x\in\left[\pi,\frac{5\pi}{4}\right)$$

From Eqs. (i) and (ii), we get $x\in\left(\pi,\frac{5\pi}{4}\right)$

73. Here, $\cos\left(\frac{\alpha}{2}-\frac{\pi}{4}\right)=\frac{\sqrt{5}+1}{4}-\frac{\sqrt{5}-1}{4}=\frac{1}{2}\Rightarrow\frac{\alpha}{2}-\frac{\pi}{4}=2n\pi\pm\frac{\pi}{3}$

$\therefore$ $$\alpha=4n\pi+\frac{\pi}{2}\pm\frac{2\pi}{3}=4n\pi+\frac{7\pi}{6},\ 4n\pi-\frac{\pi}{6}$$

For, $n=0,\ \alpha=\frac{7\pi}{6},-\frac{\pi}{6}$

74. $\sin\alpha+\left[\frac{\sin 2x}{2}\right]_{\alpha}^{2\alpha}=0\Rightarrow\sin\alpha(1+\cos3\alpha)=0\Rightarrow\alpha=0,\ \cos3\alpha=-1$

$3\alpha=-\pi,-3\pi\ \alpha=-\frac{\pi}{3},-\pi$

75. $\log_{\sqrt{3}}\tan\theta\left[\sqrt{\frac{\log_{\sqrt{3}}3}{\log_{\sqrt{3}}\tan\theta}+\frac{\log(\sqrt{3})^3}{\log_{\sqrt{3}}}}\right]=-1$

$\Rightarrow$ $$\log_{\sqrt{3}}\tan\theta\left[\sqrt{\frac{2}{\log_{\sqrt{3}}\tan\theta}+3}\right]=-1$$

Let $\log_{\sqrt{3}}\tan\theta=y\Rightarrow y\sqrt{\frac{2}{y}+3}=-1\Rightarrow\sqrt{\frac{2}{y}+3}=-\frac{1}{y}$

$\Rightarrow$ $$\frac{2}{y}+3=\frac{1}{y^2}\text{ or }y^2(2+3y)=y$$

$\Rightarrow$ $$y[3y^2+2y-1]=0$$

$\because$ $$y<0$$

$$y(3y-1)(y+1)=0$$

$$y=-1 \quad (\because y \text{ cannot be positive})$$

$\Rightarrow$ $$\log_{\sqrt{3}}\tan\theta=-1$$

$$\tan\theta=\frac{1}{\sqrt{3}}\ \therefore\ \theta=\frac{\pi}{6}\text{ and }\frac{7\pi}{6}$$

$\therefore$ There are two values of θ in $[0,2\pi]$

76. Consider $a\cos\theta+b\sin\theta=c$

$\Rightarrow$ $$a\cos\theta=c-b\sin\theta$$

$\Rightarrow$ $$a^2\cos^2\theta=(c-b\sin\theta)^2 \quad \text{(Squaring both sides)}$$

$\Rightarrow$ $$a^2(1-\sin^2\theta)=c^2-2bc\sin\theta+b^2\sin^2\theta$$

$\Rightarrow$ $$(a^2+b^2)\sin^2\theta-2bc\sin\theta+c^2-a^2=0$$

$\Rightarrow$ As α and β are values of 'θ' as given:

$\therefore$ roots of above equation are $\sin\alpha$ and $\sin\beta$.

$\Rightarrow$ $$\sin\alpha+\sin\beta=\text{sum of roots}=\frac{2bc}{a^2+b^2}$$

$$\sin\alpha\cdot\sin\beta=\text{products of roots}=\frac{c^2-a^2}{a^2+b^2}.$$

77. Consider $\sin2x+\sin4x=2\sin3x$

$\Rightarrow$ $$(\sin2x+\sin4x)-2\sin3x=0$$

$\Rightarrow$ $$2\sin3x\cdot\cos x-2\sin3x=0$$

$$\left[\text{By using }\sin A-\sin B=2\sin\frac{A+B}{2}\cdot\cos\frac{A-B}{2}\right]$$

$\Rightarrow \quad 2\sin 3x(\cos x-1)=0$

$\Rightarrow$ Either $\sin 3x=0$ or $\cos x=1$

$\Rightarrow \quad 3x=n\pi \Rightarrow x=2n\pi$

$\Rightarrow \quad x=\dfrac{n\pi}{3}$ or $x=2n\pi$ [where $n\in I$]

78. We have, $4\sin^4 x+\cos^4 x=1$

$\Rightarrow \quad 4\sin^4 x=1-\cos^4 x=1\ (1-\cos^2 x)(1+\cos^2 x)$

$\Rightarrow \quad 4\sin^4 x-[(1-\cos^2 x)(1+\cos^2 x)]=0$

$\Rightarrow \quad 4\sin^4 x-[\sin^2 x(1+\cos^2 x)]=0$

$\Rightarrow \quad \sin^2 x[4\sin^2 x-(1+\cos^2 x)]=0$

$\Rightarrow \quad \sin^2 x[4\sin^2 x-(1+1-\sin^2 x)]=0$

$\Rightarrow \quad \sin^2 x[4\sin^2 x-1-(1-\sin^2 x)]=0$

$\Rightarrow \quad \sin^2 x(5\sin^2 x-2)=0$

Either $\sin x=0$ or $\sin x=\pm\sqrt{\dfrac{2}{5}}$

$\Rightarrow \quad x=n\pi$ or $x=n\pi\pm\alpha$

where $\alpha=\sin^{-1}\sqrt{\dfrac{2}{5}}$ and $n\in I$

79. Let $81^{\sin^2 x}=y$ ...(i)

then $\quad 81^{\cos^2 x}=81^{(1-\sin^2 x)}$

$=81.81^{-\sin^2 x}=81\cdot y^{-1}$ ...(ii)

So, the given can be written as

$y^2-30y+81=0 \Rightarrow y=3$ or $y=27$

By using Eqs. (i) and (ii)

$\Rightarrow$ Either $81^{\sin^2 x}=3$ or $81^{\sin^2 x}=27$

$\Rightarrow \quad 4\sin^2 x=1$ or $4\sin^2 x=3$

$\Rightarrow \quad \sin x=\dfrac{1}{2}$ or $\sin x=\dfrac{\sqrt{3}}{2}\left[\text{as } 0\le x\le\dfrac{\pi}{2}\right]$

$\Rightarrow \quad x=\dfrac{\pi}{6}$ or $x=\dfrac{\pi}{3}$ are only the solution.

80. The given equation can be written as

$$\frac{\left(\dfrac{\sqrt{3}-1}{2\sqrt{2}}\right)}{\sin x}+\frac{\left(\dfrac{\sqrt{3}+1}{2\sqrt{2}}\right)}{\cos x}=2$$

$$\Rightarrow \quad \frac{\sin\dfrac{\pi}{12}}{\sin x}+\frac{\cos\dfrac{\pi}{12}}{\cos x}=2$$

$$\Rightarrow \quad \sin\left(\frac{\pi}{12}+x\right)=\sin 2x$$

$$\therefore \quad \frac{\pi}{12}+x=2x \text{ or } \frac{\pi}{12}+x=\pi-2x$$

$$\Rightarrow \quad x=\frac{\pi}{12} \text{ or } \frac{11\pi}{36}$$

81. $\dfrac{\alpha^2}{1-\tan^2 x}=\dfrac{(\sin^2 x+\alpha^2-2)(1+\tan^2 x)}{1-\tan^2 x}$

$\Rightarrow \quad \alpha^2\cos^2 x=\sin^2 x+\alpha^2-2$

$\Rightarrow \quad 2=\sin^2 x(1+\alpha^2)$

$\Rightarrow \quad \sin^2 x=\dfrac{2}{1+\alpha^2}$

$\Rightarrow \quad 0\le\dfrac{2}{1+\alpha^2}\le 1$

$\Rightarrow \quad \alpha^2\ge 1\Rightarrow\alpha\le -1$ or $\alpha\ge 1$

82. $4[\sin x\cos\pi/3+\cos x\sin\pi/3]\times[\cos x\cos\pi/6+\sin x\sin\pi/6]$

$=a^2+\sqrt{3}\sin 2x-\cos 2x$

$$\Rightarrow 4\left[\frac{1}{2}\sin x+\frac{\sqrt{3}}{2}\cos x\right]\left[\frac{\sqrt{3}}{2}\cos x+\frac{1}{2}\sin x\right]$$

$$=a^2+\sqrt{3}\sin 2x-\cos 2x$$

$\Rightarrow \sqrt{3}\sin 2x+3\cos^2 x+\sin^2 x=a^2+\sqrt{3}\sin 2x-\cos 2x$

$\Rightarrow \cos 2x+2=a^2-\cos 2x$

$\Rightarrow \quad \cos 2x=\dfrac{a^2-2}{2} \Rightarrow -1\le\dfrac{a^2-2}{2}\le 1$

$\Rightarrow \quad 0\le a^2\le 4 \Rightarrow -2\le a\le 2$

All values of a given in (a), (b), (c), (d) satisfy this relation.

83. (a) For $x\in\left(0,\dfrac{\pi}{4}\right)$, $\tan x<\cot x$

Also $\ln(\sin x)<0$

$\Rightarrow \quad (\tan x)^{\ln(\sin x)}>(\cot x)^{\ln(\sin x)}$

(b) For $x\in\left(0,\dfrac{\pi}{2}\right)$, $\text{cosec}\, x\ge 1$

$\Rightarrow \quad \ln(\text{cosec}\, x)\ge 0\Rightarrow 4^{\ln \text{cosec}\, x}<5^{\ln(\text{cosec}\, x)}$

(c) $x\in\left(0,\dfrac{\pi}{2}\right) \quad \Rightarrow \cos x\in(0,1)$

$\Rightarrow \ln(\cos x)<0$ Also, $\dfrac{1}{2}>\dfrac{1}{3}$

$\Rightarrow \quad \left(\dfrac{1}{2}\right)^{\ln(\cos x)}<\left(\dfrac{1}{3}\right)^{\ln(\cos x)}$

(d) For $x\in\left(0,\dfrac{\pi}{2}\right)$

Since, $\sin x<\tan x$, we get

$\ln(\sin x)<\ln(\tan x)$

$\Rightarrow \quad 2^{\ln(\sin x)}<2^{\ln(\tan x)}$

84. The given equation can be written as

$$\begin{vmatrix} 1+\cos^2\theta & \sin^2\theta & 4\sin 4\theta \\ -1 & 1 & 0 \\ 0 & -1 & 1 \end{vmatrix}=0$$

[Applying $R_3\to R_3-R_2$ and $R_2\to R_2-R_1$]

$$\Rightarrow \begin{vmatrix} 2 & \sin^2\theta & 4\sin 4\theta \\ 0 & 1 & 0 \\ -1 & -1 & 1 \end{vmatrix}=0 \quad [\text{Applying } C_1\to C_1+C_2]$$

$\Rightarrow \quad 2+4\sin 4\theta=0\Rightarrow\sin 4\theta=-\dfrac{1}{2}$

$\Rightarrow \quad 4\theta=n\pi+(-1)^n\left(-\dfrac{\pi}{6}\right) \Rightarrow \theta=\dfrac{n\pi}{4}+(-1)^{n+1}\dfrac{\pi}{24}$

We have to choose values of θ s.t $0<\theta<\dfrac{\pi}{2}$

$\therefore \quad \theta=\dfrac{7\pi}{24},\dfrac{11\pi}{24}$

85. At $x = -\frac{\pi}{2}, \frac{3\pi}{2}$

$\Rightarrow$ $[1+\sin x]=0, [1-\cos x]=1$

$\therefore$ $\sin x = 0+1 \Rightarrow -1=1$ (Absurd)

At $x=0$

$(1+\sin x)=1, (1-\cos x)=0$

$\therefore$ $\sin x = 1+0 \Rightarrow 0=1$ (Absurd)

At $x=\frac{\pi}{2}$

$[1+\sin x]=2 \cdot [1-\cos x]=1$

$\sin x = 2+1=3$ (Absurd)

At $x=\pi$

$\sin x = 1+2=3$ (Absurd)

In $\left(-\frac{\pi}{2}, 0\right)$, $[1+\sin x]=0, [1-\cos x]=0$

$\therefore$ $\sin x = 0+0=0$ (Absurd)

In $\left(0, \frac{\pi}{2}\right)$, $[1+\sin x]=1, [1-\cos x]=0$

$\therefore$ $\sin x = 1+0=1$ (Absurd)

In $\left(\frac{\pi}{2}, \pi\right)$, $[1+\sin x]=1, [1-\cos x]=1$

$\therefore$ $\sin x = 1+1=2$ (Absurd)

In $\left(\pi, \frac{3\pi}{2}\right)$, $[1+\sin x]=0, [1-\cos x]=1$

$\therefore$ $\sin x = 0+1=1$ (Absurd)

$\therefore$ All the four results hold.

Sol. (Q. Nos. 86 to 88)

$$3\sin^2 x - 7\sin x + 2 = 0$$

$$\sin x = \frac{1}{3} \text{ or } 2 \text{(Reject)}$$

86. $N = 2$

One value in first quadrant and other lies in second quadrant.

87. Let $x + \alpha$, then two values α and $\pi - \alpha$.

$\Rightarrow$ sum is π

88. $\because$ $\sin\alpha = \frac{1}{3} \Rightarrow \cos\alpha = \pm\frac{2\sqrt{2}}{3}$

$$f_4(\alpha) = \sin^4 x + \cos^4 \alpha = 1 - 2\sin^2\alpha\cos^2\alpha = \frac{65}{81}$$

Sol. (Q. Nos. 89 to 90)

89. If $\alpha = 3$ and $a = 5$, then $N \in [5^3, 5^4]$

largest value of N is $5^4 - 1 = 624$

smallest value of N is $5^3 = 125$

Difference of largest and smallest integral value of N $= 624 - 125 = 499$

90. If $\alpha = 2$ and $a = 2$, then

N_1 is $(2^3) - (2^2) = 4$

If $\alpha = 1$ and $a = 3$ then N_2 is $3^2 - 3^1 = 6$

$$y = (N_1 \sec^2\theta + N_2 \operatorname{cosec}^2\theta)$$
$$= (4\sec^2\theta + 6\operatorname{cosec}^2\theta)$$

$\therefore$ The minimum values of y is $(2+\sqrt{6})^2 = 10 + 4\sqrt{6}$

Sol. (Q. Nos. 91 to 93)

91. $\because$ $\angle OMN = 15° = \angle ONM$

$\therefore$ $\angle MON = 180° - 15° - 15° = 150°$

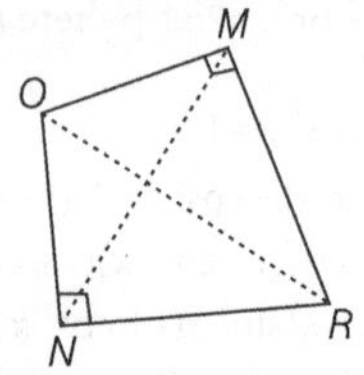

Now, quadrilateral $ONRM$ is cyclic

$[\because \angle OMR = \angle ONR = 90°]$

$\therefore$ $\angle R = 180° - 150° = 30° = \angle Q$ $[\because PR = PQ]$

$\Rightarrow$ $\angle P = 120°$

92. $\angle PLO = \angle PMO = 90°$

$\therefore$ Quadrilateral $PLOM$ is cyclic and OP is diameter of circumcircle

$\Rightarrow$ $\angle LOM = 180° - \angle P = 60°$

$\Rightarrow$ $\angle POM = 30°$

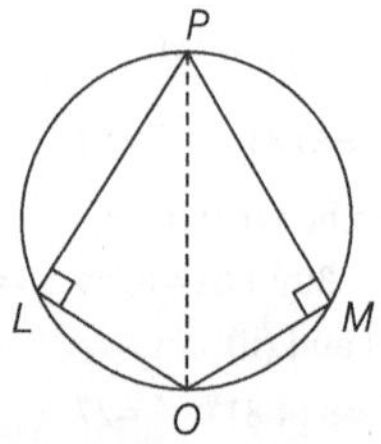

Now, is right angled $\Delta\, POM$,

$$OP = \frac{\sqrt{3}}{2}\sec 30° \quad [\because AC = x\sec\theta]$$

$$= \frac{\sqrt{3}}{2} \cdot \frac{2}{\sqrt{3}} = 1$$

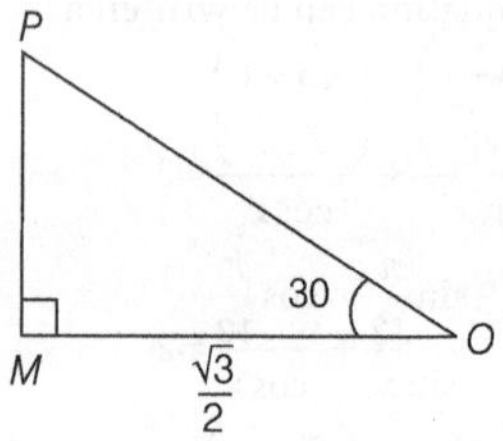

Radius of circumcircle $= \frac{1}{2}$

$\therefore$ Area $= \pi\left(\frac{1}{2}\right)^2 = \frac{\pi}{4}$

93. $\because$ $\angle OQN = 15°$

$$ON = \frac{\sqrt{3}}{2}$$

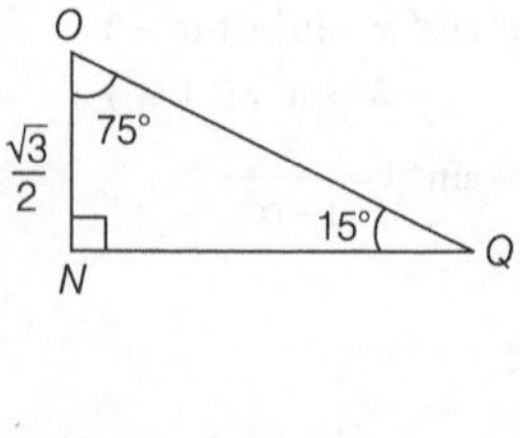

$\therefore$ In right angled ΔQON

$$NQ = ON \tan 75° = \frac{\sqrt{3}}{2} \tan 75°$$

$$\Rightarrow \quad QR = \sqrt{3} \tan 75° = \sqrt{3} \cot 15°$$

Sol. (Q, Nos. 94 to 96)

$$(2\sin x - \cos x)(1 + \cos x) = \sin^2 x$$

$$\Rightarrow \quad (1+\cos x)[2\sin x - \cos x - 1 + \cos x] = 0$$

$$\Rightarrow \quad (1+\cos x)(2\sin x - 1) = 0$$

$$\Rightarrow \quad \cos x = -1 \text{ or } \sin x = 1/2$$

So $\quad \sin\alpha = 1/2$ [as $0 \le \alpha \le \pi/2$]

$$\Rightarrow \quad \cos\alpha = \sqrt{3}/2$$

Next, $3\cos^2 x - 10\cos x + 3 = 0$ ...(i)

$$\Rightarrow \quad (3\cos x - 1)(\cos x - 3) = 0$$

$$\Rightarrow \quad \cos x = 1/3 \text{ as } \cos x \ne 3$$

So, $\quad \cos\beta = 1/3, \sin\beta = \frac{2\sqrt{2}}{3}$ ...(ii)

and $\quad 1 - \sin 2x = \cos x - \sin x$

$$\Rightarrow \sin^2 x + \cos^2 x - 2\sin x \cos x = \cos x - \sin x$$

$$\Rightarrow \quad (\cos x - \sin x)(\cos x - \sin x - 1) = 0$$

$\Rightarrow$ Either $\sin x = \cos x \Rightarrow \sin\gamma = \cos\gamma = 1/\sqrt{2}$ ...(iii)

or $\quad \cos x - \sin x = 1 \Rightarrow \cos x = 1, \sin x = 0$

$$\Rightarrow \quad \cos\gamma = 1, \sin\gamma = 0$$

94. $\cos\alpha + \cos\beta + \cos\gamma$ can be equal to ...(iv)

$$\frac{\sqrt{3}}{2} + \frac{1}{3} + \frac{1}{\sqrt{2}} \text{ or } \frac{\sqrt{3}}{2} + \frac{1}{3} + 1$$

i.e., $\quad \dfrac{3\sqrt{6} + 2\sqrt{2} + 6}{6\sqrt{2}}$ or $\dfrac{3\sqrt{3} + 8}{6}$

95. $\sin\alpha + \sin\beta + \sin\gamma$ can be equal to

$$\frac{1}{2} + \frac{2\sqrt{2}}{3} + \frac{1}{\sqrt{2}} \text{ or } \frac{1}{2} + \frac{2\sqrt{2}}{3} + 0$$

i.e., $\quad \dfrac{3\sqrt{2} + 14}{6\sqrt{2}}$ or $\dfrac{3 + 4\sqrt{2}}{6}$

96. $\sin(\alpha - \beta)$ is equal to

$$\sin\alpha\cos\beta - \cos\alpha\sin\beta = \frac{1}{2} \times \frac{1}{3} - \frac{\sqrt{3}}{2} \times \frac{2\sqrt{2}}{3}$$

$$= \frac{1 - 2\sqrt{6}}{6}$$

97. $5\sin^2 x + 3\sin x \cos x - 3\cos^2 x = 2(\sin^2 x + \cos^2 x)$

$$\Rightarrow 3\tan^2 x + 3\tan x - 5 = 0 \Rightarrow \tan x = \frac{-3 \pm \sqrt{69}}{6}$$

and $\quad \sin^2 x - \cos 2x = 2 - \sin 2x$

$$\Rightarrow \quad 3\sin^2 x + 2\sin x \cos x = 3(\sin^2 x + \cos^2 x)$$

$$\Rightarrow \quad \cos x(2\sin x - 3\cos x) = 0$$

Either $\cos x = 0$ or $\tan x = \frac{3}{2} \Rightarrow \cos x = \pm\frac{2}{\sqrt{13}}$

$\Rightarrow \quad$ Taking $\tan\alpha = \dfrac{-3 \pm \sqrt{69}}{6}$, $\tan\beta = \dfrac{3}{2}$

we get $\tan\alpha + \tan\beta = 1 \pm \sqrt{69}/6$

98. Taking $\tan\alpha = \dfrac{-3 + \sqrt{69}}{6}$,

$$\tan\beta = \frac{-3 - \sqrt{69}}{6}$$

$$\cos\gamma = 0, \cos\delta = \pm\frac{2}{\sqrt{13}}$$

we get $\tan\alpha\tan\beta + \cos\gamma + \cos\delta = -\frac{5}{3} \pm \frac{2}{\sqrt{13}}$

99. (1) and (2) have no solution common.

100. Let $\theta_k = \dfrac{k\pi}{2n}$

$\because$ AB is diameter of circle.

$\therefore \Delta AP_kB$ is right angled triangle.

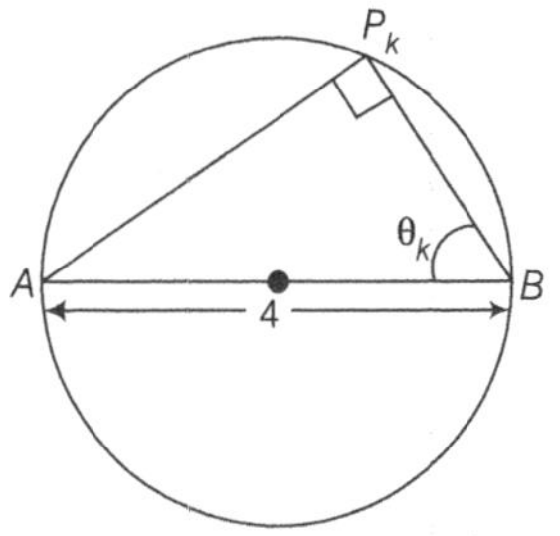

$$\Rightarrow \quad \Delta_k = \frac{1}{2} \cdot AB.\cos\theta_k \cdot AB\sin\theta_k$$

$$= 4\sin 2\theta_k \quad \text{...(i)}$$

Now, $\displaystyle\sum_{k=1}^{n+1} \sin 2\theta_k = \sin 2\theta_1 + \sin 2\theta_2 + \ldots + \sin 2\theta_{n+1}$

$$= \sin\frac{\pi}{n} + \sin\frac{2\pi}{n} + \ldots + \sin\frac{(n+1)\pi}{n}$$

$$= \frac{\sin\left(n.\frac{\pi}{2n}\right)\sin\left(\frac{\pi}{n} + (n-1)\frac{\pi}{2n}\right)}{\sin\frac{\pi}{2n}}$$

$$= \frac{\sin\left(\frac{\pi(n+1)}{2n}\right)}{\sin\frac{\pi}{2n}} = \cot\frac{\pi}{2n} = \cot\frac{\pi}{32} \Rightarrow n = 16$$

$$\therefore \quad \frac{n}{2} = 8$$

101. $2\cos^2 2x - 7\cos 2x + 5 = 0$

$$\Rightarrow \quad \cos 2x = 1$$

$$\Rightarrow \quad x = n\pi$$

$$k = 1 + 2 + \ldots + 99$$

$$= \frac{99 \times 100}{2} = 4950$$

Now, $k - 4948 = 4950 - 4948 = 2$

102. In 2nd equation sum of coefficients is zero, hence its one root is 1 and second root is also 1 as it has equal roots.

Common root of first equation is 1.

$$\Rightarrow \quad \tan^2\theta - 2\tan\theta + 1 = 0$$

$$\Rightarrow \quad (\tan\theta - 1)^2 = 0$$

$\Rightarrow \tan\theta = 1$, hence one solution in $(0, \pi)$.

103. $\dfrac{x^4+1}{8x^2} = \dfrac{(2\sin y\cos y)^2}{4}$

$\Rightarrow \quad x^2 + \dfrac{1}{x^2} = 2\sin^2 2y$

LHS ≥ 2 and RHS ≤ 2

LHS = RHS will hold

only if LHS = 2 = RHS

i.e. $\quad x^2 = 1$ and $\sin^2 2y = 1$

$\Rightarrow \quad x = \pm 1$ and $2y = \dfrac{\pi}{2}, \dfrac{3\pi}{2}, \dfrac{5\pi}{2}, \dfrac{7\pi}{2}$

So, total number of ordered pairs are 8.

104. We have,

$$\sin\left(x-\frac{\pi}{4}\right)-\cos\left(x+\frac{3\pi}{4}\right)=1$$

$$\sin\left(x-\frac{\pi}{4}\right)+\sin\left(x+\frac{\pi}{4}\right)=1$$

$\Rightarrow \quad 2\sin x\cos\dfrac{\pi}{4}=1 \qquad [\because \sin\theta=\sin(\pi-\theta)]$

$\Rightarrow \quad \sin x=\dfrac{1}{\sqrt{2}}$

$\Rightarrow \quad x=\dfrac{\pi}{4}, \dfrac{3\pi}{4}$

We know that 3 radians $\cong 171^\circ 22'$.

Therefore, $\sin 3>0$, $\cos 3<0$ and $|\cos 3|>\sin 3$

$\therefore \quad \cos 3+\sin 3<0$

Now, $\quad \dfrac{2\cos 7x}{\cos 3+\sin 3}>2^{\cos 2x}$

$\Rightarrow \quad 2\cos 7x<2^{\cos 2x}(\cos 3+\sin 3) \qquad (\because \cos 3+\sin 3<0)$

$\Rightarrow \quad 2\cos 7x<0 \Rightarrow \cos 7x<0 \qquad (\therefore 2^{\cos x}>0)$

Clearly, $x=\dfrac{3\pi}{4}$ satisfies this equation.

Hence, $x=\dfrac{3\pi}{4}$ is the required solution.

Hence, only one solution.

105. $\because \cos A\sin\left(A-\dfrac{\pi}{6}\right)=\dfrac{1}{2}\left[2\cos A\sin\left(A-\dfrac{\pi}{6}\right)\right]$

$$=\frac{1}{2}\left[\sin\left(A+A-\frac{\pi}{6}\right)-\sin\left(A-A+\frac{\pi}{6}\right)\right]$$

$$=\frac{1}{2}\left[\sin\left(2A-\frac{\pi}{6}\right)-\sin\frac{\pi}{6}\right]$$

So, it is maximum, when $\sin\left(2A-\dfrac{\pi}{6}\right)$ is maximum.

i.e., $\quad 2A-\dfrac{\pi}{6}=\dfrac{\pi}{2}$

$$2A=\frac{\pi}{2}+\frac{\pi}{6}$$

$$A=\frac{\pi}{3}$$

but $\quad A=\dfrac{\pi}{\lambda}$

Hence, $\lambda=3$

106. Draw the curves $p|\cos\theta|$ and $q|\sin\theta|$ and find the number of intersection points.

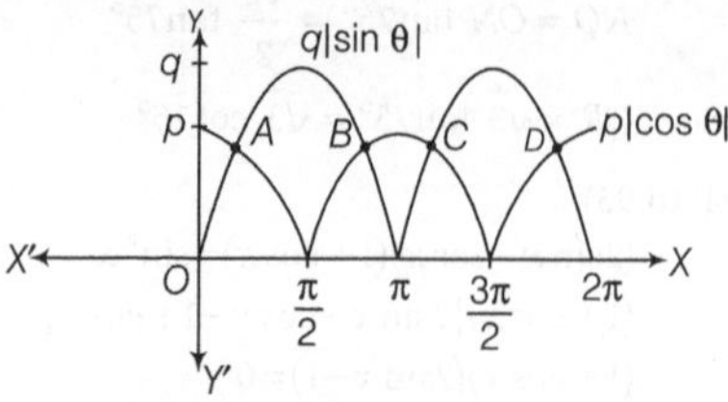

Hence, intersection points are A, B, C, D

$\therefore$ Number of solutions is 4.

107. $\tan\theta=\dfrac{3\tan\dfrac{\theta}{3}-\tan^3\dfrac{\theta}{3}}{1-3\tan^2\dfrac{\theta}{3}}$

$\Rightarrow \quad \tan^3\dfrac{\theta}{3}-3\lambda\tan\dfrac{\theta}{3}-3\tan\dfrac{\theta}{3}+\lambda=0$

$\therefore \quad \tan\dfrac{\theta_1}{3}\cdot\tan\dfrac{\theta_2}{3}+\tan\dfrac{\theta_2}{3}\cdot\tan\dfrac{\theta_3}{3}+\tan\dfrac{\theta_3}{3}\cdot\tan\dfrac{\theta_1}{3}=-3$

$\Rightarrow \quad \left|\tan\dfrac{\theta_1}{3}\cdot\tan\dfrac{\theta_2}{3}+\tan\dfrac{\theta_2}{3}\cdot\tan\dfrac{\theta_3}{2}+\tan\dfrac{\theta_1}{3}\cdot\tan\dfrac{\theta_3}{3}\right|=3$

108. $\sin(1-x)\geq 0$, $\cos x\geq 0$

$$\cos x=\cos\left\{\frac{\pi}{2}-(1-x)\right\}$$

$\therefore \quad x=2n\pi+\left(\dfrac{\pi}{2}-(1-x)\right)$

$\Rightarrow \quad x=n\pi-\dfrac{\pi}{4}+\dfrac{1}{2}$

Put $n=1$, $x=\dfrac{3\pi}{4}+\dfrac{1}{2}\Rightarrow[x]=2$

109. L.H.S ≤ 13 and R.H.S $= 2(y-2)^2+13\geq 13$

Roots of eqn. exist if $y=2$ and $\sin\left\{x+\tan^2\dfrac{5}{12}\right\}=1$

Now, consider $\sin\left\{x+\tan^{-1}\dfrac{5}{12}\right\}=1$

$\Rightarrow \quad x+\tan^{-1}\dfrac{5}{12}=\sin^{-1}(1)$

$\Rightarrow \quad x+\tan^{-1}\dfrac{5}{12}=\dfrac{\pi}{2}$

$\Rightarrow \quad x=\dfrac{\pi}{2}-\tan^{-1}\dfrac{5}{12}$

$\Rightarrow \quad x=\cot^{-1}\dfrac{5}{12}$

$\Rightarrow \quad \cot x=\dfrac{5}{12}$

Hence, $12\cot\left(\dfrac{xy}{2}\right)=12\cot x=12\times\dfrac{5}{12}=5$

110. $\tan(\pi\cos\theta)=\tan(\pi/2-\pi\sin\theta)$

$\Rightarrow \pi\cos\theta=n\pi+\pi/2-\pi\sin\theta\,(n\in I)$

$\Rightarrow \pi(\sin\theta+\cos\theta)=(2n+1)\dfrac{\pi}{2}$

$\Rightarrow \quad \sin\theta+\cos\theta=\dfrac{2n+1}{2}$

$\Rightarrow \quad \cos(\pi/4-\theta)=\dfrac{2n+1}{2\sqrt{2}}$

Since $\quad -1\le\cos(\pi/4-\theta)\le 1$

$\Rightarrow \quad -1\le\dfrac{2n+1}{2\sqrt{2}}\le 1$

$\Rightarrow n=0$ or -1 as n is an integer

$\Rightarrow \quad \cos(\pi/4-\theta)=\pm(1/2\sqrt{2})$

$\Rightarrow \quad 8\cos^2(\pi/4-\theta)=1$

$\Rightarrow \quad 16\cos^2\left(\dfrac{\pi}{4}-\theta\right)=2$

111. From the given equation we have

$$3\frac{2\tan(x/2)}{1+\tan^2(x/2)}+4\frac{1-\tan^2(x/2)}{1+\tan^2(x/2)}=5$$

$\Rightarrow \quad 6\tan(x/2)+4-4\tan^2(x/2)=5+5\tan^2(x/2)$

$\Rightarrow \quad 9\tan^2(x/2)-6\tan(x/2)+1=0$

$\Rightarrow \quad 90\tan^2(x/2)-60\tan(x/2)=-10$

$\because \quad 90\tan^2\dfrac{x}{2}-60\tan\dfrac{x}{2}+10=-10+10=0$

112. When $n=1$,. we have interval $[0,\pi]$, which covers only the first and second quadrants in which.

$\sin x=-1/2$ is not possible. Hence, the number of solutions zero. Also, from $2(n-1)$, we have zero solution when $n=1$.

For $n=2$, we have interval $[0,2\pi]$ which covers all the quadrants only once. Hence, the number of solutions is two.

Also, from $2(n-1)$, we have two solutions, when $n=2$.

Fro $n=3$, we have interval $[0,3\pi]$, which covers the third and fourth quadrants only once. Hence, the number of solutions is two. But from $2(n-1)$, we have four solutions which contradict.

Hence, Statement I is false, and Statement II is true.

113. The graphs of $y=|\sin x|$ and $y=|x|$ is

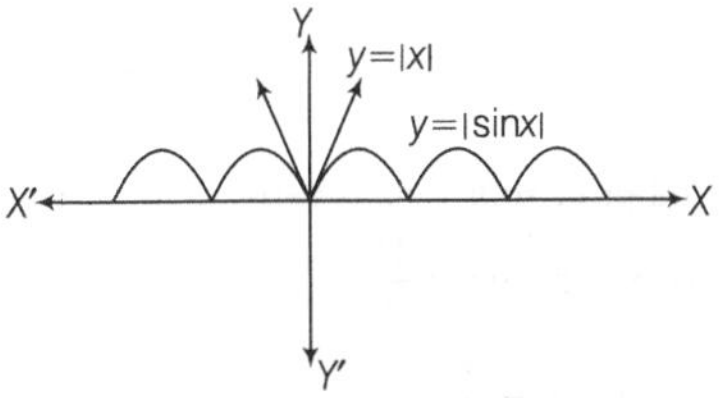

$|\sin x|=|x|$ has only one solution $x=0$. But Statement II is not the only explanation of Statement I.

114. $\sin 2x=\dfrac{2\tan x}{1+\tan^2 x}$, $\cos 2x=\dfrac{1-\tan^2 x}{1+\tan^2 x}$

$\Rightarrow$ Statement II is correct.

In Statement I, we have

$$\cos^2 x=\sin 2x$$

$\Rightarrow \quad \cos x(\cos x-2\sin x)=0$

$\Rightarrow \quad \tan x=1/2$ as $\cos x\ne 0$

From Statement II we get

$$\sin 2x+\cos 2x=\frac{1+1-1/4}{1+1/4}=\frac{7}{5}$$

Statement I is false

115. In statement, $\begin{vmatrix} 1 & \sin\alpha & \cos\alpha \\ 1 & \cos\alpha & \sin\alpha \\ -1 & \sin\alpha & -\cos\alpha \end{vmatrix}=0$

$\Rightarrow$ either $\sin\alpha=0$ or $\tan\alpha=1$

$\Rightarrow \quad \alpha=\pi/4$ (as $0<\alpha<\pi$)

and Statement II, $(\sin x+2\cos x)$

$$\begin{vmatrix} 1 & \cos x & \cos\alpha \\ 1 & \sin x & \cos x \\ 1 & \cos x & \sin x \end{vmatrix}=0$$

$\Rightarrow \ (\sin x+2\cos x)(\cos x-\sin x)^2=0$

$\Rightarrow \ (\sin x+2\cos x)(\cos x-\sin x)^2=0$

which does not hold for any value of x as $-\pi/4<x<\pi/4$

116. For $(\pi/4,\pi/2)$, $0<\cos\theta<\dfrac{1}{\sqrt{2}}<\sin\theta<1$

So, $\quad (\cos\theta)^{\cos\theta}<(\sin\theta)^{\cos\theta}$

and $\quad (\cos\theta)^{\sin\theta}<(\sin\theta)^{\cos\theta}$

$\Rightarrow \quad (\cos\theta)^{\sin\theta}+(\cos\theta)^{\cos\theta}<(\sin\theta)^{\cos\theta}$

Showing that Statement I is true.

In Statement II let, $e^{\sin\theta}=t$

Then, $\quad t^2-4t-1=0$

$\Rightarrow \quad t=\dfrac{4\pm\sqrt{16+4}}{2}=2\pm\sqrt{5}$

$\Rightarrow \quad e^{\sin\theta}=2\pm\sqrt{5} \Rightarrow \sin\theta=\log(2\pm\sqrt{5})$

Since, $2-\sqrt{5}<0$, $\sin\theta=\log(2+\sqrt{5})>\log_e$

$\Rightarrow \ \sin\theta>1$ which is not possible.

So, the give equation has no solution and the statement II is false.

117. $\sin^2 x+\sin^4 x+\sin^6 x+\ldots\text{inf}$

$$=\frac{\sin^2 x}{1-\sin^2 x}=\tan^2 x$$

$\Rightarrow$ Statement II is false.

Now, $\exp\{\tan^2 x\log_e 2\}=2^{\tan^2 x}$

So $2^{\tan^2 x}=1$ or $2^{\tan^2 x}=8$

$\Rightarrow \quad 2^{\tan^2 x}=8 \qquad [\because \tan x>0\Rightarrow 2^{\tan^2 x}>1]$

$\Rightarrow \quad \tan^2 x=3\Rightarrow\tan x=\sqrt{3}$

Now, $\quad \dfrac{\cos x}{\cos x+\sin x}=\dfrac{1}{1+\tan x}=\dfrac{1}{\sqrt{3}+1}=\dfrac{\sqrt{3}-1}{2}$

118. (A) $\sin x=-\dfrac{1}{2}$ and $\cos x=\dfrac{-\sqrt{3}}{2}$

$$\alpha=\frac{7\pi}{6},\beta=\frac{11\pi}{6}\ \alpha=\frac{7\pi}{6},\gamma=\frac{5\pi}{6}$$

$$\beta-\gamma-\frac{11\pi}{6}-\frac{5\pi}{6}=\pi$$

$$\alpha+\beta=\frac{7\pi}{6}+\frac{11\pi}{6}=3\pi$$

$\therefore \quad A\to q,s$

(B) $\cot x=-\sqrt{3}$ and $\operatorname{cosec} x=-2$

$$\alpha=\frac{11\pi}{6},\beta=\frac{5\pi}{6},\gamma=\frac{7\pi}{6}$$

$\therefore \quad \alpha-\beta=\pi$ and $\beta+\gamma=2\pi$

$\therefore \quad \beta \to p, t$

(C) $\sin x=-\dfrac{1}{2}$ and $\tan x=\dfrac{1}{\sqrt{3}}$

$$\alpha=\frac{7\pi}{6}, \beta=\frac{11\pi}{6}, \gamma=\frac{\pi}{6}$$

$$\alpha-\gamma=\pi,\ \alpha+\beta=3\pi,\ \beta+\gamma=2\pi$$

$\therefore \quad C \to r, s, t$

119. (A) $2\sin\cos\theta=1/\sqrt{2}$ if $\cos\theta>0$

$\Rightarrow \quad \sin 2\theta = 1/\sqrt{2} \Rightarrow \theta = \pi/8$ or $3\pi/8$

$\sin 2\theta = -1/\sqrt{2}$ if $\cos\theta<0$

$\Rightarrow \quad \theta = 5\pi/8$ or $7\pi/8$

$\therefore A \to p, q$

(B) $2\cos 2\theta \cos 4\theta + \cos 4\theta = 0$

$\Rightarrow \quad (2\cos 2\theta + 1)\cos 4\theta = 0$

Either $\cos 4\theta = 0$

$\Rightarrow \quad \theta = \dfrac{\pi}{8}, \dfrac{3\pi}{8}, \dfrac{5\pi}{8}, \dfrac{7\pi}{8}$ or $\cos 2\theta = -1/2$

$\Rightarrow \quad \theta = \pi/3$ or $2\pi/3$

$\therefore B \to p, q, r$

(C) $(4\cos^2\theta - 1)(2\sin\theta - 1) = 0$

$\Rightarrow \quad \cos^2\theta = 1/4 \Rightarrow \cos\theta = \pm 1/2$

$\Rightarrow \quad \theta = \pi/3, 2\pi/3$ or $\sin\theta = 1/2$

$\Rightarrow \quad \theta = \pi/6$

$\therefore C \to r, s$

(D) If $\sin 4\theta = \pm 1$, $\theta = \dfrac{\pi}{8}, \dfrac{3\pi}{8}, \dfrac{5\pi}{8}, \dfrac{7\pi}{8}$

$\therefore D \to p, q$

120. $f_n(\theta) = \dfrac{\sin\theta + \sin 3\theta + \sin 5\theta + \ldots + \sin((2n-1)\theta)}{\cos\theta + \cos 3\theta + \cos 5\theta + \cdots + \cos[(2n-1)\theta]}$

$$= \frac{\dfrac{\sin\left[\theta + \dfrac{(n-1)2\theta}{2}\right]\sin\dfrac{n}{2}(2\theta)}{\sin\dfrac{2\theta}{2}}}{\dfrac{\cos\left[\theta + \dfrac{(n-1)2\theta}{2}\right]\sin\dfrac{n}{2}(2\theta)}{\sin\dfrac{2\theta}{2}}} = \frac{\sin(n\theta)}{\cos(n\theta)} = \tan n\theta$$

(A) $f_3\left(\dfrac{\pi}{12}\right) = \tan\left(\dfrac{\pi}{4}\right) = 1$

(B) $f_4\left(\dfrac{\pi}{32}\right) = \tan\left(\dfrac{\pi}{8}\right) = \sqrt{2} - 1$

(C) $f_6\left(\dfrac{\pi}{16}\right) = \tan\left(\dfrac{3\pi}{8}\right) = \sqrt{2} + 1$

(D) $f_7\left(\dfrac{\pi}{84}\right) = \tan\left(\dfrac{\pi}{12}\right) = 2 - \sqrt{3}$

121. (i) If $\cot x > 0$, then $\dfrac{1}{\sin x} = 0$ which is not possible.

Now if $\cot x < 0$,

then, $\quad -\cot x = \cot x + \dfrac{1}{\sin x}$

$\Rightarrow \quad \dfrac{2\cos x + 1}{\sin x} = 0 \Rightarrow \cos x = -\dfrac{1}{2}$

$\Rightarrow \quad x = 2n\pi \pm \dfrac{2\pi}{3}, n \in I$ and $0 \le x \le 2\pi$.

$\therefore \quad x = \dfrac{2\pi}{3}, \dfrac{4\pi}{3}$

(ii) We have, $\sin^3 x \cos x + \sin^2 x . \cos^2 x + \sin x . \cos^3 x = 1$

$\Rightarrow \quad \sin x \cos x(\sin^2 x + \sin x \cos x + \cos^2 x) = 1$

$\Rightarrow \quad \dfrac{\sin 2x}{2}\left(1 + \dfrac{\sin 2x}{2}\right) = 1$

$\Rightarrow \quad \sin 2x(2 + \sin 2x) = 4$

$\Rightarrow \quad \sin^2 2x + 2\sin 2x - 4 = 0$

$\Rightarrow \quad \sin 2x = \dfrac{-2 \pm \sqrt{4 + 16}}{2} = -1 \pm \sqrt{5}$

This is not possible, as $-1 \le \sin 2x \le 1$.

Hence, the given equation has no solution.

(iii) We have, $2^{\cos x} = |\sin x|$

It is true only for $\cos x = 0$ and $|\sin x| = 1$

$\Rightarrow \quad \cos x = \cos\dfrac{\pi}{2}$ and $\sin x = \pm 1 = \sin\left(\pm\dfrac{\pi}{2}\right)$

$\Rightarrow \quad x = 2n\pi \pm \dfrac{\pi}{2}$

But, $x \in [-2\pi, 2\pi]$

$x = \dfrac{\pi}{2}, -\dfrac{\pi}{2}, \dfrac{3\pi}{2}, -\dfrac{3\pi}{2}$ Hence, number of solutions = 4.

(iv) We have, $\quad |\cos x| = [x] = y$ (say)

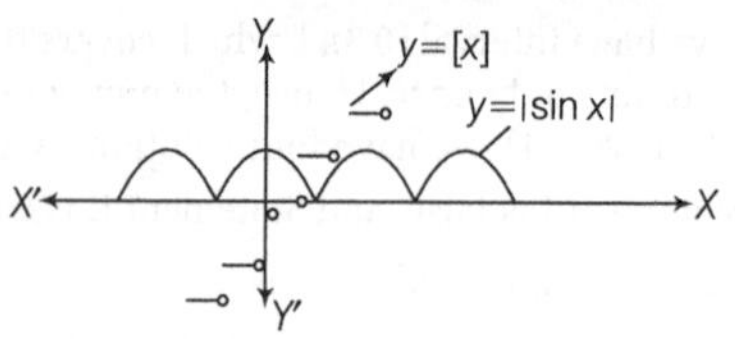

Graph of $|\cos x|$ and $[x]$ don't cut each other for any real value of x.

Hence, number of solutions is zero.

(v) We have $x + 2\tan x = \dfrac{\pi}{2}$

or $\quad \tan x = \dfrac{\pi}{4} - \dfrac{x}{2}$

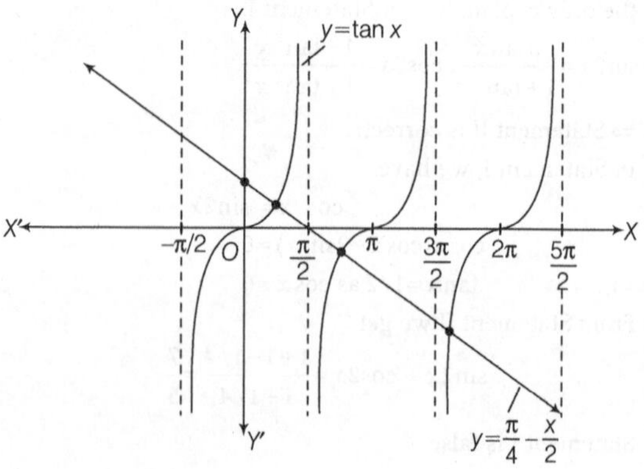

Now the graph of the curve $y = \tan x$ and $y = \frac{\pi}{4} - \frac{x}{2}$, in the interval $[0, 2\pi]$ intersect at three points.

Hence, number of solutions is three.

122. Here, $(\sin^2 x + \cos^2 x)^2 - 2\sin^2 x.\cos^2 x + \sin 2x + \alpha = 0$

$$\Rightarrow \quad 1 - \frac{1}{2}\sin^2 2x + \sin 2x + \alpha = 0$$

$$\Rightarrow \quad \sin^2 2x - 2\sin 2x - 2(1+\alpha) = 0$$

$$\Rightarrow \quad \sin 2x = \frac{2 \pm \sqrt{4 + 8(1+\alpha)}}{2} = 1 \pm \sqrt{2\alpha + 3} \quad \text{...(i)}$$

$$\Rightarrow \quad \sin 2x = 1 - \sqrt{2\alpha+3} \quad [\because 1 + \sqrt{2\alpha+3} \geq 1]$$

and $\sin 2x = 1$ is already included in the solution of $\sin 2x = 1 - \sqrt{2\alpha+3}$

But $\sin 2x$ is real, so $2\alpha + 3 \geq 0$ i.e., $\alpha \geq -\frac{3}{2}$

Also, $-1 \leq \sin x \leq 1$

$$\therefore \quad -1 < 1 - 1\sqrt{2\alpha+3} \leq 1$$

$$\Rightarrow \quad -2 \leq -\sqrt{2\alpha+3} \leq 0$$

Squaring both sides,

$$0 \leq 2\alpha + 3 \leq 4$$

$$\Rightarrow \quad -3 \leq 2\alpha \leq 1$$

$$-\frac{3}{2} \leq \alpha \leq \frac{1}{2} \text{ or } \alpha \in \left[-\frac{3}{2}, \frac{1}{2}\right]_n$$

Also the general solution is, $x = \frac{n\pi}{2} + \frac{(-1)^n}{2}\sin^{-1}\{1 - \sqrt{2\alpha+3}\}$

123. Here, $3\cos 2\theta = 1 \Rightarrow 3(2\cos^2\theta - 1) = 1$

$$\Rightarrow \quad 6\cos^2\theta = 4; \Rightarrow \cos^2\theta = \frac{2}{3}$$

$$\text{Now,} \quad \tan^2\theta = \frac{\sin^2\theta}{\cos^2\theta} = \frac{1-\cos^2\theta}{\frac{2}{3}} = \frac{1-\frac{2}{3}}{\frac{2}{3}} = \frac{1}{2}$$

$$\therefore \quad 32\tan^8\theta = 2\cos^2\alpha - 3\cos\alpha$$

$$\Rightarrow \quad 32\left(\frac{1}{2}\right)^4 = 2\cos^2\alpha - 3\cos\alpha$$

$$\Rightarrow \quad 2\cos^2\alpha - 3\cos\alpha - 2 = 0$$

$$\Rightarrow \quad (2\cos\alpha + 1)(\cos\alpha - 2) = 0$$

$$\Rightarrow \quad 2\cos\alpha + 1 = 0 \text{ or } \cos\alpha - 2 = 0$$

$$\Rightarrow \quad \cos\alpha = -\frac{1}{2} \text{ or } \cos\alpha = 2 \text{ (impossible)}$$

$$\Rightarrow \quad \alpha = 2n\pi \pm \left(\pi - \frac{\pi}{3}\right) = 2n\pi \pm \frac{2\pi}{3}, n \in Z$$

124. Let $4^{\sin x} = \lambda$ and $3^{1/\cos y} = \mu$

Then, the equation becomes, $\lambda + \mu = 11$

$5\lambda^2 - 2\mu = 2$, solving these equations we get,

$$2\lambda + 5\lambda^2 = 24$$

$$\text{or} \quad 5\lambda^2 + 2\lambda - 24 = 0 \text{ or } (5\lambda + 12)(\lambda - 2) = 0$$

$$\text{So} \quad \lambda = 2, -\frac{12}{5}$$

$$\text{If} \quad \lambda = 2,\ 4^{\sin x} = 2,\ \therefore\ 2^{2\sin x} = 2^1.$$

$$\therefore \quad \sin x = \frac{1}{2}$$

If $\lambda = -\frac{12}{5}$ then $4^{\sin x} = -\frac{12}{5}$, which is impossible as $4^{\sin x} > 0$

when $\lambda = 2$, we get $\mu = 11 - 2 = 9$

$$\Rightarrow \quad \frac{1}{3^{\cos y}} = 9 = 3^2$$

$$\Rightarrow \quad y = 2m\pi + \frac{\pi}{3}, m \in z \Rightarrow \cos y = \frac{1}{2}$$

Thus, $x = n\pi + (-1)^n\frac{\pi}{6}$ and $y = 2m\pi \pm \frac{\pi}{3}$, where $m, n \in Z$.

125. The given equation is,

$$\sin y = 2\left(\sin^4 x + \frac{1}{\sin^4 x} + 2\right) + 2\left(\cos^4 x + \frac{1}{\cos^4 x} + 2\right) - 24$$

$$= 2(\sin^4 x + \cos^4 x) + 2\left(\frac{1}{\sin^4 x} + \frac{1}{\cos^4 x}\right) + 8 - 24$$

$$= 2(\sin^4 x + \cos^4 x)\left\{1 + \frac{1}{\sin^4 x \cdot \cos^4 x}\right\} - 16$$

$$\Rightarrow \quad \sin y = 2\left(1 - \frac{1}{2}\sin^2 2x\right)\left(1 + \frac{16}{\sin^4 2x}\right) - 16$$

$$\Rightarrow \quad 16 + \sin y = (2 - \sin^2 2x)\left(1 + \frac{16}{\sin^4 2x}\right) \quad \text{...(i)}$$

Since, $\sin y \leq 1$

$$\Rightarrow \quad 16 + \sin y \leq 17$$

$\therefore$ L.H.S. is not greater than 17, on the other hand $0 \leq \sin^2 2x \leq 1$

$$\Rightarrow \quad 2 - \sin^2 2x \geq 1 \text{ and } \sin^4 2x \leq 1$$

$$\Rightarrow \quad \frac{16}{\sin^4 2x} \geq 16$$

$$\therefore \quad 1 + \frac{16}{\sin^4 2x} \geq 17$$

$$\Rightarrow \quad (2 - \sin^2 2x)\left(1 + \frac{16}{\sin^4 2x}\right) \geq 17$$

This shows that right member of equation (i) is not less than 17. Thus the inequality holds only when

$$\sin y = 1 \text{ and } \sin^2 2x = 1$$

$$\Rightarrow \quad y = (4n+1)\frac{\pi}{2}, n \in I$$

from second equation $\sin^2 2x = 1 = \sin^2\frac{\pi}{2}$

$$\Rightarrow \quad 2x = m\pi \pm \frac{\pi}{2}, m \in I$$

$$x = (2m+1)\frac{\pi}{4}$$

Thus, the solution are $\left((2m \pm 1)\frac{\pi}{y}, (4n+1)\frac{\pi}{2}\right)$

126. We have,

$$4\left[3\sqrt{4x - x^2}\left(\frac{1 - \cos(x+y)}{2}\right) + 2\cos(x+y)\right] = 13 + 4\cos^2(x+y)$$

$$\Rightarrow 6\sqrt{4x - x^2} - 6\sqrt{4x - x^2}\cos(x+y) + 8\cos(x+y)$$

$$= 13 + 4\cos^2(x+y)$$

$$\Rightarrow \quad 4\cos^2(x+y) + (6\sqrt{4x-x^2} - 8)\cos(x+y) + (13 - 6\sqrt{4x-x^2}) = 0 \quad ...(i)$$

Let, $\cos(x+y) = t$ and $6\sqrt{4x-x^2} - 8 = a \Rightarrow 6\sqrt{4x-x^2} = 8 + a$

Clearly, $\quad 4x - x^2 \geq 0$ or $x(x-4) \leq 0$ or $0 \leq x \leq 4 \quad ...(ii)$

Now, equation (i) reduces to;

$$4t^2 + at + (5-a) = 0$$

which is quadratic in t.

$$\therefore \quad D \geq 0$$

$$a^2 - 4.4(5-a) \geq 0$$

$$a^2 + 16a - 80 \geq 0$$

$$(a+20)(a-4) \geq 0$$

$$\Rightarrow \quad a \leq -20 \text{ or } a \geq 4 \quad ...(iii)$$

However, according to substitution;

$$a = 6\sqrt{4x-x^2} - 8$$

$$= 6\sqrt{4-(x-2)^2} - 8$$

$$-8 \leq a \leq 4 \quad ...(iv)$$

$$\Rightarrow \quad a = 4 \quad \{\text{from (iii) and (iv)}\}$$

Hence, $\quad 6\sqrt{4x-x^2} - 8 = 4$

$$\sqrt{4x-x^2} = 2$$

$$4x - x^2 = 4$$

$$x^2 - 4x + 4 = 0$$

$$\Rightarrow \quad (x-2)^2 = 0 \text{ or } x = 2$$

Now, equation (i) becomes;

$$\Rightarrow \quad 4\cos^2(2+y) + 4\cos(2+y) + 1 = 0$$

$$\Rightarrow \quad (2\cos(2+y)+1)^2 = 0$$

$$\Rightarrow \quad 2\cos(2+y) + 1 = 0$$

$$\Rightarrow \quad \cos(2+y) = -\frac{1}{2} = \cos\left(\pi - \frac{\pi}{3}\right)$$

$$\Rightarrow \quad 2 + y = 2n\pi \pm \frac{2\pi}{3}$$

$$\Rightarrow \quad y = 2n\pi \pm \frac{2\pi}{3} - 2, n \in I$$

Thus, the solutions are $\left(2, 2n\pi \pm \frac{2\pi}{3} - 2\right)$

127. Rewriting the given equation as,

$\{\tan(x+y) - \cot(x+y)\}^2 = -(1 + 2x + x^2) = -(1+x)^2$

or $\{\tan(x+y) - \cot(x+y)\}^2 + (1+x)^2 = 0$

which is possible only when,

$\tan(x+y) - \cot(x+y) = 0$ and $1 + x = 0$

$$\tan^2(x+y) = 1 = \tan^2\frac{\pi}{4} \text{ and } x = -1$$

Now, $\tan^2(x+y) = \tan^2\frac{\pi}{4} \Rightarrow x + y = n\pi \pm \frac{\pi}{4}$

$$\Rightarrow \quad y = n\pi \pm \frac{\pi}{4} + 1, n \in I$$

$\therefore$ The required solutions are $\left(-1, n\pi \pm \frac{\pi}{4} + 1\right), n \in I.$

128. As we know, A.M. $\geq$ G.M.

$$\therefore \quad \frac{\tan^2 x + \cot^2 x}{2} \geq \sqrt{\tan^2 x \cdot \cot^2 x}$$

$$\Rightarrow \quad \tan^2 x + \cot^2 x \geq 2$$

Now, from the first equation,

$$2\cos^2 y \geq 2$$

$\cos^2 y \geq 1$, which is only possible when,

$$\cos^2 y = 1$$

Putting $\cos^2 y = 1$ in second equation we get,

$$\sin^2 z = 0 \Rightarrow z = n\pi \quad ...(ii)$$

Similarly $\quad \cos^2 y = 1 \Rightarrow y = m\pi \quad ...(iii)$

Alos, $\quad \tan^2 x + \cot^2 x = 2\cos^2 y$

$$\Rightarrow \quad \tan^2 x + \cot^2 x = 2 \Rightarrow \tan^2 x = \cos^2 x = 1$$

$$\Rightarrow \quad x = k\pi \pm \frac{\pi}{4} \quad ...(iv)$$

Hence, the solutions are

$$x = k\pi \pm \frac{\pi}{4}, y = m\pi, z = n\pi \text{ where } k, m, n \in Z$$

129. The given equation can be rewrite as

$$2\cos\left(\frac{x+y}{2}\right)\cos\left(\frac{x-y}{2}\right) + 2\cos^2\left(\frac{x+y}{2}\right) - 1 = -\frac{3}{2}$$

$$\text{or} \quad 4\cos^2\left(\frac{x+y}{2}\right) + 4\cos\left(\frac{x+y}{2}\right)\cos\left(\frac{x-y}{2}\right) + 1 = 0$$

$$\text{or} \quad \left(2\cos\left(\frac{x+y}{2}\right) + \cos\left(\frac{x-y}{2}\right)\right)^2 + \sin^2\left(\frac{x-y}{2}\right) = 0$$

$$\Rightarrow 2\cos\left(\frac{x+y}{2}\right) = -\cos\left(\frac{x-y}{2}\right) \text{ and } \sin^2\left(\frac{x-y}{2}\right) = 0$$

from second equation, we get,

$$x - y = 2n\pi, \text{ where } n \in I$$

$$\text{or} \quad y = x - 2n\pi$$

Substituting the values of y in the first equation, we get

$$2\cos(x - n\pi) = -\cos n\pi$$

$$\Rightarrow \quad 2\cos x \cdot \cos n\pi = -\cos n\pi$$

$$\Rightarrow \quad \cos x = -\frac{1}{2} = \cos\left(\pi - \frac{\pi}{3}\right) \Rightarrow x = 2m\pi \pm \frac{2\pi}{3}, m \in I$$

$\therefore$ Solutions are,

$$\left\{\left(2m\pi \pm \frac{2\pi}{3}\right), \left(2(m-n)\pi \pm \frac{2\pi}{3}\right)\right\}$$

130. Here $\cot\frac{\theta}{2} - \cot\theta = \operatorname{cosec}\frac{\theta}{2}$

$$\Rightarrow \quad \frac{\cos\frac{\theta}{2}}{\sin\frac{\theta}{2}} - \frac{\cos\theta}{\sin\theta} = \operatorname{cosec}\left(\frac{\theta}{2}\right)$$

$$\Rightarrow \quad \frac{2\cos^2\frac{\theta}{2} - \cos\theta}{\sin\theta} = \operatorname{cosec}\frac{\theta}{2}$$

$$\Rightarrow \quad 2\cos^2\frac{\theta}{2} - \left(\cos^2\frac{\theta}{2} - \sin^2\frac{\theta}{2}\right) = \sin\theta \cdot \operatorname{cosec}\frac{\theta}{2}$$

$$\Rightarrow \quad 1 = \operatorname{cosec}\frac{\theta}{2} \cdot \sin\theta$$

$\Rightarrow \quad \sin\theta = \sin\frac{\theta}{2} \Rightarrow \sin\frac{\theta}{2}\left(2\cos\frac{\theta}{2} - 1\right) = 0$

$\Rightarrow \quad \sin\frac{\theta}{2} = 0$ or $\cos\frac{\theta}{2} = \frac{1}{2}$

$\Rightarrow \quad \theta = 2n\pi$ or $\theta = 2\left(2n\pi \pm \frac{\pi}{3}\right)$

for $\theta = 2n\pi$, $\cot\frac{\theta}{2}$ and $\cot\theta$ are undefined,

Hence do not satisfy the given equation.

$\therefore$ The only solution is $\theta = 4n\pi \pm \frac{2\pi}{3}$

131. We know, $a + b + c = 0, \Leftrightarrow a^3 + b^3 + c^3 = 3abc$

Put $a = \sin x$, $b = \cos x$, $c = 1$

$$1 + \sin^3 x + \cos^3 x = 3\sin x \,.\, \cos x = \frac{3}{2}\sin 2x$$

$\Rightarrow a + b + c = 0 \Rightarrow 1 + \sin x + \cos x = 0$

$\Rightarrow \quad \sqrt{2}\cos\left(x - \frac{\pi}{4}\right) = -1$

$\Rightarrow \quad \cos\left(x - \frac{\pi}{4}\right) = \cos\frac{3\pi}{4}$

$\Rightarrow \quad x = 2n\pi \pm \frac{3\pi}{4} + \frac{\pi}{4}$

132. We have, $\log_{\sin x} 2 \,.\, \log_{\sin^2 x} a = -1$

$\Rightarrow \quad \log_{\sin x} 2 \,.\, \log_{\sin x} a = -2 \qquad \begin{pmatrix} \sin x > 0 \\ a > 0 \end{pmatrix}$ ...(i)

$\Rightarrow \quad \log_{\sin x} 2 \,.\, \frac{\log_2 a}{\log_2 \sin x} = -2$

$\Rightarrow \quad \log_2 a = -2(\log_2 \sin x)^2$

$\Rightarrow \quad \log_2 \sin x = \pm\sqrt{\frac{(-\log_2 a)}{2}}$

$\left\{\text{as } \sin x < 1 \Rightarrow \log_2 \sin x < 0, \text{ so, reject } \log_2 \sin x = \sqrt{\frac{(-\log_2 a)}{2}}\right\}$

$\Rightarrow \quad \log_2 \sin x = -\sqrt{\frac{(-\log_2 a)}{2}}$

$\Rightarrow \quad \sin x = 2^{-\sqrt{(-\log_2 a)/2}}$

$\Rightarrow x = n\pi + (-1)^n \sin^{-1}\{2^{-\sqrt{(-\log_2 a)/2}}\}$ and the condition is $0 < a < 1$ so that $\log_2 a$ is defined and $\log_2 a < 0$.

133. We have, $\int_0^x (t^2 - 8t + 13)\,dt = x\sin\frac{a}{x}$

$\Rightarrow \quad \left(\frac{t^3}{3} - \frac{8t^2}{2} + 13t\right)_0^x = x\sin\frac{a}{x}$

$\Rightarrow \quad \frac{x^3}{3} - 4x^2 + 13x = x\sin\frac{a}{x}$

$\Rightarrow \quad x^2 - 12x + 39 = 3\sin\frac{a}{x} \qquad \{\because x \neq 0\}$

$\Rightarrow \quad (x - 6)^2 + 3 = 3\sin\left(\frac{a}{x}\right)$

min. L.H.S. = 3 and max. R.H.S. = 3

$\Rightarrow \quad (x - 6)^2 + 3 = 3$ and $3\sin\left(\frac{a}{x}\right) = 3$

$\Rightarrow \quad x = 6$ and $\sin\frac{a}{x} = 1$

$\Rightarrow \quad x = 6$ and $\frac{a}{x} = (4n + 1)\frac{\pi}{2} \Rightarrow a = 3\pi(4n + 1)$

134. The given equation can be rewritten as

$$(\sin^4 x + \cos^4 x)^2 - 2\sin^4 x\cos^4 x = \frac{17}{16}(1 - \sin^2 2x)$$

Put $t = \sin x \,.\, \cos x$,

$$(1 - 2t^2)^2 - 2t^4 = \frac{17}{16}(1 - 4t^2)$$

$\Rightarrow \quad 32t^4 + 4t^2 - 1 = 0$

$\Rightarrow \quad 32t^4 + 8t^2 - 4t^2 - 1 = 0$

$\Rightarrow \quad (4t^2 + 1)(8t^2 - 1) = 0$

$\Rightarrow \quad t = \pm\frac{1}{2\sqrt{2}} \qquad [\because 4t^2 + 1 \neq 0]$

$\Rightarrow \quad \sin x \,.\, \cos x = \pm\frac{1}{2\sqrt{2}} \Rightarrow \sin 2x = \pm\frac{1}{\sqrt{2}}$

$\Rightarrow \quad 2x = n\pi \pm (-1)^n\left(\frac{\pi}{4}\right) \Rightarrow x = \frac{n\pi}{2} \pm (-1)^n\frac{\pi}{8}$

135. We have, $\tan^4 x + \tan^4 y + 2\cot^2 x\cot^2 y$

$$= 3 + \sin^2(x + y)$$

$\Rightarrow \tan^4 x + \tan^4 y + 2\cot^2 x\cot^2 y - 2 = 1 + \sin^2(x + y)$

$\Rightarrow (\tan^2 x - \tan^2 y) + 2(\tan x\tan y - \cot x\cot y)^2 = -1 + \sin^2(x + y)$

Clearly, L.H.S ≥ 0 and R.H.S ≤ 0

$\therefore \quad$ L.H.S = R.H.S = 0

$\Rightarrow \tan^2 x = \tan^2 y$, $\tan x\tan y = \cot x\cot y$ and $\sin^2(x + y) = 1$

$\Rightarrow \tan^2 x = \tan^2 y$, $\tan^2 x\tan^2 y = 1$ and $\sin^2(x + y) = 1$

$\Rightarrow \tan^2 x = 1$ and $x + y = n\pi \pm \frac{\pi}{2}$, $n, m, \in z$

$\Rightarrow \quad x = m\pi \pm \frac{\pi}{4}$ and $y = (n - m)\pi \mp \frac{\pi}{4}$, $n, m, \in z$

$\Rightarrow$ Thus, $x = y = p\pi \pm \frac{\pi}{4}, p \in z$

136. $\cos^2 x = \dfrac{(a + 2) \pm \sqrt{(a + 2)^2 + 4(a + 3)}}{2}$

$$\cos^2 x = \frac{(a + 2) \pm (a + 4)}{2}$$

either, $\quad \cos^2 x = -1$ (not possible)

or $\quad \cos^2 x = a + 3$

Since, $\quad 0 \leq \cos^2 x \leq 1$

$\therefore \quad 0 \leq a + 3 \leq 1 \Rightarrow -3 \leq a \leq -2$

Also, $\quad \cos^2 x = (a + 3)$

$\Rightarrow \quad x = n\pi \pm \cos^{-1}\sqrt{a + 3}$, where $n \in I$ and $a \in [-3, -2]$

137. Here $\quad \left[2\sin\left(x + \frac{\pi}{6}\right)\right]^{\sqrt{2\sin^2 x + 2\sqrt{3}\sin x \,.\, \cos x + 1}} = 4$

$\Rightarrow \quad \left[2\sin\left(x + \frac{\pi}{6}\right)\right]^{\sqrt{3\sin^2 x + \cos^2 x + 2\sqrt{3}\sin x \,.\, \cos x}} = 4$

$\Rightarrow \quad \left[2\sin\left(x + \frac{\pi}{6}\right)\right]^{|\sqrt{3}\sin x + \cos x|} = 4$

$$\Rightarrow \quad \left[2\sin\left(x+\frac{\pi}{6}\right)\right]^{|2\sin(x+\pi/6)|} = 4$$

$$\Rightarrow \quad \left\{\sin\left(x+\frac{\pi}{6}\right)\right\}^{|\sin(x+\pi/6)|} = 1$$

To make the R.H.S. well defined it is necessary that,

$$\sin\left(x+\frac{\pi}{6}\right) > 0 \Rightarrow x \in \left(-\frac{\pi}{6}, \frac{5\pi}{6}\right)$$

Now, we have $\left\{\sin\left(x+\frac{\pi}{6}\right)\right\}^{\sin(x+\pi/6)} = 1$

$$\Rightarrow \quad \sin\left(x+\frac{\pi}{6}\right) = 1 \Rightarrow x = \frac{\pi}{3}$$

138. We have,

$$\sec\theta + \operatorname{cosec}\theta = c$$

$$\sin\theta + \cos\theta = c\sin\theta \cdot \cos\theta$$

Squaring both sides, we get

$$1 + 2\sin\theta\cos\theta = c^2\sin^2\theta \cdot \cos^2\theta$$

$$1 + \sin 2\theta = \frac{c^2}{4}\sin^2 2\theta$$

Put $\sin 2\theta = t$

$$c^2t^2 - 4t - 4 = 0, t \in [-1, 1]$$

Thus, the equation must have atleast one root $\in [-1, 1]$

Case I : $D \ge 0$, $f(-1) > 0$, $f(1) > 0$

$\Rightarrow 16 + 16c^2 \ge 0, c^2 + 4 - 4 > 0, c^2 - 4 - 4 > 0 \Rightarrow c^2 > 8$

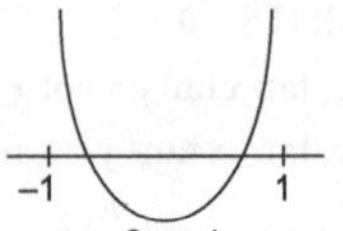

Case I

In this case we will get two distinct values of $\sin z\theta$, resulting in 4 distinct values of $\theta \in (0, 2\pi)$ two of them would be repeated.

Case II : $f(1)f(-1) < 0$

$$\Rightarrow \quad (c^2 - 8)c^2 < 0 \Rightarrow c^2 < 8$$

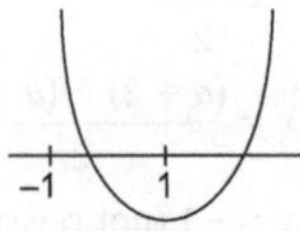

Case II

In this case we will get one value of $\sin 2\theta$, resulting in 2 distinct values of $\theta \in (0, 2\pi)$.

139. $|\sin x + \cos x|^{\sin^2 x - 1/4} = 1 + |\sin y|$ ...(i)

and $\cos^2 y = 1 + \sin^2 y$ ...(ii)

$\Rightarrow 1 - \sin^2 y = 1 + \sin^2 y \Rightarrow \sin^2 y = 0, y = k\pi, k \in I$

$\Rightarrow |\sin x + \cos x|^{\sin^2 x - 1/4} = 1$

Then, either $\sin^2 x = \frac{1}{4}$ or $\sin x + \cos x = 1$

$\Rightarrow x = n\pi \pm \frac{\pi}{6}$ or $\cos\left(x - \frac{\pi}{4}\right) = \cos\frac{\pi}{4}$

Now, $\cos\left(x - \frac{\pi}{4}\right) = \cos\frac{\pi}{4} \Rightarrow x - \frac{\pi}{4} = 2m\pi \pm \frac{\pi}{4}$

$$\Rightarrow \quad x = 2m\pi + \frac{\pi}{2} \text{ or } x = 2m\pi$$

$$\Rightarrow \quad x = 2m\pi \text{ or } 2m\pi + \frac{\pi}{2}$$

$$\Rightarrow x = 2m\pi + \frac{\pi}{2}, 2m\pi, n\pi \pm \frac{\pi}{6} \text{ and } y = k\pi; m, n, k \in I$$

140. Given, $\sqrt{3}\sec x + \operatorname{cosec} x + 2(\tan x - \cot x) = 0$,

$$(-\pi < x < \pi) - \{0, \pm \pi/2\}$$

$$\Rightarrow \quad \sqrt{3}\sin x + \cos x + 2(\sin^2 x - \cos^2 x) = 0$$

$$\Rightarrow \quad \sqrt{3}\sin x + \cos x - 2\cos 2x = 0$$

Multiplying and dividing by $\sqrt{a^2 + b^2}$, i.e. $\sqrt{3+1} = 2$, we get

$$2\left(\frac{\sqrt{3}}{2}\sin x + \frac{1}{2}\cos x\right) - 2\cos 2x = 0$$

$$\Rightarrow \quad \left(\cos x \cdot \cos\frac{\pi}{3} + \sin x \cdot \sin\frac{\pi}{3}\right) - \cos 2x = 0$$

$$\Rightarrow \quad \cos\left(x - \frac{\pi}{3}\right) = \cos 2x$$

$$\therefore \quad 2x = 2n\pi \pm \left(x - \frac{\pi}{3}\right)$$

$$[\because \cos\theta = \cos\alpha \Rightarrow \theta = 2n\pi \pm \alpha]$$

$$\Rightarrow \quad 2x = 2n\pi + x - \frac{\pi}{3} \text{ or } 2x = 2n\pi - x + \frac{\pi}{3}$$

$$\Rightarrow \quad x = 2n\pi - \frac{\pi}{3} \text{ or } 3x = 2n\pi + \frac{\pi}{3}$$

$$\Rightarrow \quad x = 2n\pi - \frac{\pi}{3} \text{ or } x = \frac{2n\pi}{3} + \frac{\pi}{9}$$

$$\therefore \quad x = \frac{-\pi}{3} \text{ or } x = \frac{\pi}{9}, \frac{-5\pi}{9}, \frac{7\pi}{9}$$

Now, sum of all distinct solutions $= \frac{-\pi}{3} + \frac{\pi}{9} - \frac{5\pi}{9} + \frac{7\pi}{9} = 0$

141. Here, $\frac{5}{4}\cos^2 2x + (\cos^4 x + \sin^4 x) + (\cos^6 x + \sin^6 x) = 2$

$$\Rightarrow \quad \frac{5}{4}\cot 2x + [(\cos^2 x + \sin^2 x)^2 - 2\sin^2 x\cos^2 x]$$
$$+ (\cos^2 x + \sin^2 x)[(\cos^2 x + \sin^2 x)^2 - 3\sin^2 x\cos^2 x] = 2$$

$$\Rightarrow \frac{5}{4}\cos^2 2x + (1 - 2\sin^2 x\cos^2 x) + (1 - 3\cos^2 x\sin^2 x) = 2$$

$$\Rightarrow \quad \frac{5}{4}\cos^2 2x - 5\sin^2 x\cos^2 x = 0$$

$$\Rightarrow \quad \frac{5}{4}\cos^2 2x - \frac{5}{4}\sin^2 2x = 0$$

$$\Rightarrow \quad \frac{5}{4}\cos^2 2x - \frac{5}{4} + \frac{5}{4}\cos^2 2x = 0$$

$$\Rightarrow \quad \frac{5}{2}\cos^2 2x = \frac{5}{4}$$

$$\Rightarrow \quad \cos^2 2x = \frac{1}{2} \Rightarrow 2\cos^2 2x = 1$$

$$\Rightarrow \quad 1 + \cos 4x = 1$$

$$\Rightarrow \quad \cos 4x = 0, \text{ as } 0 \le x \le 2\pi$$

$$\therefore \quad 4x = \left\{\frac{\pi}{2}, \frac{3\pi}{2}, \frac{5\pi}{2}, \frac{7\pi}{2}, \frac{9\pi}{2}, \frac{11\pi}{2}, \frac{13\pi}{2}, \frac{15\pi}{2}\right\}$$

as $\quad 0 \le 4x \le 8\pi$

$$\therefore \quad x = \left\{\frac{\pi}{8}, \frac{3\pi}{8}, \frac{5\pi}{8}, \frac{7\pi}{8}, \frac{9\pi}{8}, \frac{11\pi}{8}, \frac{13\pi}{8}, \frac{15\pi}{8}\right\}$$

Hence, the total number of solutions are 8.

142. **Plan** For solving this type of questions, obtain the LHS and RHS in equation and examine, the two are equal or not for a given interval.

Given, trigonometrical equation

$$(\sin x - \sin 3x) + 2\sin 2x = 3$$

$$\Rightarrow -2\cos 2x \sin x + 4\sin x\cos x = 3$$

$$\left[\because \sin C - \sin D = 2\cos\left(\frac{C+D}{2}\right)\sin\left(\frac{C-D}{2}\right) \text{ and } \sin 2\theta = 2\sin\theta\cos\theta\right]$$

$$\Rightarrow \quad 2\sin x\,(2\cos x - \cos 2x) = 3$$

$$\Rightarrow \quad 2\sin x\,(2\cos x - 2\cos^2 x + 1) = 3$$

$$\Rightarrow \quad 2\sin x\left[\frac{3}{2} - 2\left(\cos x - \frac{1}{2}\right)^2\right] = 3$$

$$\Rightarrow \quad 3\sin x - 3 = 4\left(\cos x - \frac{1}{2}\right)^2 \sin x$$

As $x \in (0, \pi)$ LHS ≤ 0 and RHS ≥ 0

For solution to exist, LHS = RHS = 0

Now, LHS = 0 $\Rightarrow$ $3\sin x - 3 = 0$

$$\Rightarrow \quad \sin x = 1 \Rightarrow x = \frac{\pi}{2}$$

For $\quad x = \frac{\pi}{2}$,

$$\text{RHS} = 4\left(\cos\frac{\pi}{2} - \frac{1}{2}\right)^2 \sin\frac{\pi}{2} = 4\left(\frac{1}{4}\right)(1) = 1 \ne 0$$

$\therefore$ No solution of the equation exists.

143. **Plan** It is based on range of $\sin x$, i.e. $[-1,1]$ and the internal for $a<x<b$.

Description of Situation As $\theta, \phi \in [0, 2\pi]$ and

$$\tan(2\pi - \theta) > 0, -1 < \sin\theta < -\frac{\sqrt{3}}{2}$$

$$\tan(2\pi - \theta) > 0 \Rightarrow -\tan\theta > 0$$

$\therefore$ $\theta \in$ II or IV quadrant.

Also, $\quad -1 < \sin\theta < -\dfrac{\sqrt{3}}{2}$

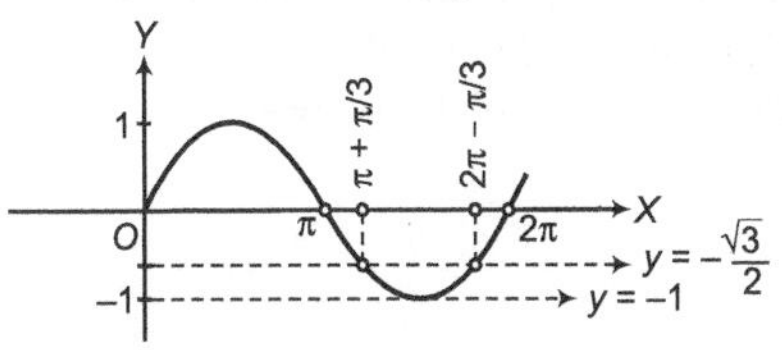

$$\Rightarrow \quad \frac{4\pi}{3} < \theta < \frac{5\pi}{3} \text{ but } \theta \in \text{II or IV quadrant}$$

$$\Rightarrow \quad \frac{3\pi}{2} < \theta < \frac{5\pi}{3} \qquad \text{...(i)}$$

Here, $\quad 2\cos\theta(1 - \sin\phi) = \sin^2\theta\left(\tan\dfrac{\theta}{2} + \cot\dfrac{\theta}{2}\right)\cos\phi - 1$

$$\Rightarrow \quad 2\cos\theta - 2\cos\theta\sin\phi = \sin^2\theta\left(\frac{\sin^2\frac{\theta}{2} + \cos^2\frac{\theta}{2}}{\sin\frac{\theta}{2}\cos\frac{\theta}{2}}\right)\cos\phi - 1$$

$$\Rightarrow \quad 2\cos\theta - 2\cos\theta\sin\phi = 2\sin^2\theta\left(\frac{1}{\sin\theta}\right)\cos\phi - 1$$

$$\Rightarrow \quad 2\cos\theta + 1 = 2\sin\phi\cos\theta + 2\sin\theta\cos\phi$$

$$\Rightarrow \quad 2\cos\theta + 1 = 2\sin(\theta + \phi) \qquad \text{...(ii)}$$

From Eq. (i),

$$\frac{3\pi}{2} < \theta < \frac{5\pi}{3} \Rightarrow 2\cos\theta + 1 \in (1, 2)$$

$$\therefore \quad 1 < 2\sin(\theta + \phi) < 2$$

$$\Rightarrow \quad \frac{1}{2} < \sin(\theta + \phi) < 1 \qquad \text{...(iii)}$$

$$\Rightarrow \quad \frac{\pi}{6} < \theta + \phi < \frac{5\pi}{6} \text{ or } \frac{13\pi}{6} < \theta + \phi < \frac{17\pi}{6}$$

$$\therefore \quad \frac{\pi}{6} - \theta < \phi < \frac{5\pi}{6} - \theta \text{ or } \frac{13\pi}{6} - \theta < \phi < \left(\frac{17\pi}{6}\right) - \theta$$

$$\Rightarrow \phi \in \left(-\frac{3\pi}{2}, -\frac{2\pi}{3}\right) \text{ or } \left(\frac{2\pi}{3}, \frac{7\pi}{6}\right), \text{ as } \theta \in \left(\frac{3\pi}{2}, \frac{5\pi}{3}\right)$$

144. $P = \{\theta : \sin\theta - \cos\theta = \sqrt{2}\cos\theta\}$

$$\Rightarrow \quad \cos\theta(\sqrt{2} + 1) = \sin\theta$$

$$\Rightarrow \quad \tan\theta = \sqrt{2} + 1$$

$$\Rightarrow \quad Q = \{\theta : \sin\theta + \cos\theta\} = \sqrt{2}\sin\theta$$

$$\Rightarrow \quad \sin\theta(\sqrt{2} - 1) = \cos\theta$$

$$\Rightarrow \quad \tan\theta = \frac{1}{\sqrt{2} - 1} \times \frac{\sqrt{2} + 1}{\sqrt{2} + 1} = (\sqrt{2} + 1)$$

$$\therefore \quad P = Q$$

145. Given, $n > 3 \in$ Integer

and $\quad \dfrac{1}{\sin\left(\frac{\pi}{n}\right)} = \dfrac{1}{\sin\left(\frac{2\pi}{n}\right)} + \dfrac{1}{\sin\left(\frac{3\pi}{n}\right)}$

$$\Rightarrow \quad \frac{1}{\sin\frac{\pi}{n}} - \frac{1}{\sin\frac{3\pi}{n}} = \frac{1}{\sin\frac{2\pi}{n}}$$

$$\Rightarrow \quad \frac{\sin\frac{3\pi}{n} - \sin\frac{\pi}{n}}{\sin\frac{\pi}{n}\cdot\sin\frac{3\pi}{n}} = \frac{1}{\sin\frac{2\pi}{n}}$$

$$\Rightarrow \quad 2\cos\left(\frac{2\pi}{n}\right)\cdot\sin\frac{\pi}{n} = \frac{\sin\frac{\pi}{n}\cdot\sin\frac{3\pi}{n}}{\sin\frac{2\pi}{n}}$$

$$\Rightarrow \quad 2\sin\frac{2\pi}{n}\cdot\cos\frac{2\pi}{n} = \sin\frac{3\pi}{n}$$

$$\Rightarrow \quad \sin\frac{4\pi}{n} = \sin\frac{3\pi}{n} \Rightarrow \frac{4\pi}{n} = \pi - \frac{3\pi}{n}$$

$$\Rightarrow \quad \frac{7\pi}{n} = \pi \Rightarrow n = 7$$

146. Given, $\tan\theta = \cot 5\theta$

$\Rightarrow \quad \tan\theta = \tan\left(\frac{\pi}{2} - 5\theta\right) \Rightarrow \frac{\pi}{2} - 5\theta = n\pi + \theta$

$\Rightarrow \quad 6\theta = \frac{\pi}{2} - n\pi \Rightarrow \theta = \frac{\pi}{12} - \frac{n\pi}{6}$

Also, $\cos 4\theta = \sin 2\theta = \cos\left(\frac{\pi}{2} - 2\theta\right)$

$\Rightarrow \quad 4\theta = 2n\pi \pm \left(\frac{\pi}{2} - 2\theta\right)$

Taking positive sign,

$$6\theta = 2n\pi + \frac{\pi}{2} \Rightarrow \theta = \frac{n\pi}{3} + \frac{\pi}{12}$$

Taking negative sign,

$$2\theta = 2n\pi - \frac{\pi}{2} \Rightarrow \theta = n\pi - \frac{\pi}{4}$$

Above values of θ suggest that there are only 3 common solutions.

147. $2\sin^2\theta - \cos 2\theta = 0 \Rightarrow \sin^2\theta = \frac{1}{4}$

Also, $2\cos^2\theta = 3\sin\theta \Rightarrow \sin\theta = \frac{1}{2}$

$\Rightarrow$ Two solutions exist in the interval $[0, 2\pi]$.

148. Since, $2\sin^2\theta - 5\sin\theta + 2 > 0$

$\Rightarrow \quad (2\sin\theta - 1)(\sin\theta - 2) > 0$

[where, $(\sin\theta - 2) < 0, \forall\, \theta \in R$]

$\therefore \quad (2\sin\theta - 1) < 0$

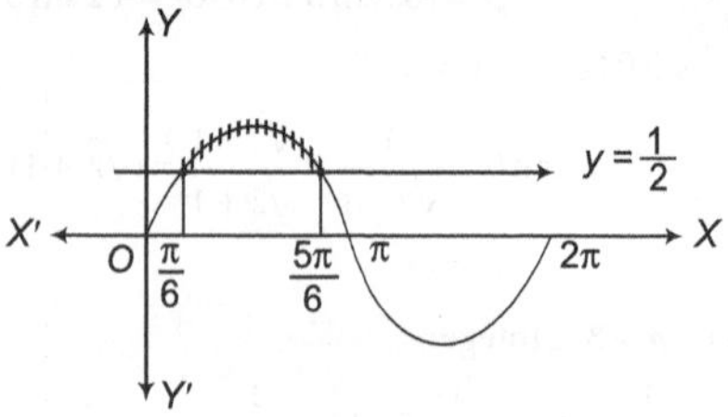

$\Rightarrow \quad \sin\theta < \frac{1}{2}$

$\therefore$ From the graph, $\theta \in \left(0, \frac{\pi}{6}\right) \cup \left(\frac{5\pi}{6}, 2\pi\right)$

149. Given equation is

$$\cos x + \cos 2x + \cos 3x + \cos 4x = 0$$

$\Rightarrow \quad (\cos x + \cos 3x) + (\cos 2x + \cos 4x) = 0$

$\Rightarrow \quad 2\cos 2x\cos x + 2\cos 3x\cos x = 0$

$\Rightarrow \quad 2\cos x(\cos 2x + \cos 3x) = 0$

$\Rightarrow \quad 2\cos x\left(2\cos\frac{5x}{2}\cos\frac{x}{2}\right) = 0$

$\Rightarrow \quad \cos x \cdot \cos\frac{5x}{2} \cdot \cos\frac{x}{2} = 0$

$\Rightarrow \quad \cos x = 0$ or $\cos\frac{5x}{2} = 0$ or $\cos\frac{x}{2} = 0$

Now, $\cos x = 0$

$\Rightarrow \quad x = \frac{\pi}{2}, \frac{3\pi}{2} \qquad [\because 0 \le x < 2\pi]$

$\cos\frac{5x}{2} = 0$

$\Rightarrow \quad \frac{5x}{2} = \frac{\pi}{2}, \frac{3\pi}{2}, \frac{5\pi}{2}, \frac{7\pi}{2}, \frac{9\pi}{2}, \frac{11\pi}{2}, \ldots$

$\Rightarrow \quad x = \frac{\pi}{5}, \frac{3\pi}{5}, \pi, \frac{7\pi}{5}, \frac{9\pi}{5} \qquad [\because 0 \le x < 2\pi]$

and $\cos\frac{x}{2} = 0 \Rightarrow \frac{x}{2} = \frac{\pi}{2}, \frac{3\pi}{2}, \frac{5\pi}{2}, \ldots$

$\Rightarrow \quad x = \pi \qquad [\because 0 \le x < 2\pi]$

Hence, $x = \frac{\pi}{2}, \frac{3\pi}{2}, \pi, \frac{\pi}{5}, \frac{3\pi}{5}, \frac{7\pi}{5}, \frac{9\pi}{5}$

150. $\sin\theta + \sin 4\theta + \sin 7\theta = 0$

$\Rightarrow \quad \sin 4\theta + (\sin\theta + \sin 7\theta) = 0$

$\Rightarrow \quad \sin 4\theta + 2\sin 4\theta \cdot \cos 3\theta = 0$

$\Rightarrow \quad \sin 4\theta\{1 + 2\cos 3\theta\} = 0 \Rightarrow \sin 4\theta = 0, \cos 3\theta = -\frac{1}{2}$

As, $\quad 0 < \theta < \pi$

$\therefore \quad 0 < 4\theta < 4\pi$

$\therefore \quad 4\theta = \pi, 2\pi, 3\pi$

$\cos 3\theta = -\frac{1}{2}$

$0 < 3\theta < 3\pi \Rightarrow 3\theta = \frac{2\pi}{3}, \frac{4\pi}{3}, \frac{8\pi}{3}$

$\Rightarrow \quad \theta = \frac{\pi}{4}, \frac{\pi}{2}, \frac{3\pi}{4}, \frac{2\pi}{9}, \frac{4\pi}{9}, \frac{8\pi}{9}$

151. Given equation is $2\sin^2 x + 5\sin x - 3 = 0$.

$\Rightarrow \quad (2\sin x - 1)(\sin x + 3) = 0$

$\Rightarrow \quad \sin x = \frac{1}{2} \qquad [\because \sin x \ne -3]$

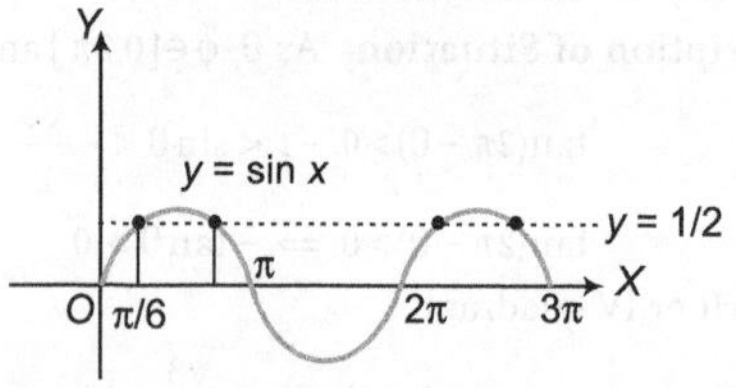

It is clear from figure that the curve intersect the line at four points in the given interval.

Hence, number of solutions are 4.

CHAPTER

03

Properties and Solutions of Triangles

Learning Part

Session 1

- Basic Relation between the Sides and Angles of Triangle

Session 2

- Auxiliary Formulae

Session 3

- Circles Connected with Triangle

Session 4

- Orthocentre and its Distance from the Angular Points of a Triangle and Pedal Triangle
- Centroid of Triangle

Session 5

- Regular Polygons and Radii of the Inscribed and Circumscribing Circle a Regular Polygon

Session 6

- Quadrilaterals and Cyclic Quadrilaterals

Session 7

- Solution of Triangles

Session 8

- Height and Distance

Practice Part

- JEE Type Examples
- Chapter Exercises

Session 1

Basic Relation between the Sides and Angles of Triangle

Basic Relation between the Sides and Angles of Triangle

In a ΔABC, the angles are denoted by capital letters A, B and C and the lengths of the sides opposite to these angles are denoted by small letters a, b and c, respectively.

Semi-perimeter of the triangle is written as :

$$s = \frac{a+b+c}{2},$$

and its area is denoted by Δ.

Some geometrical properties of A,B,C and a,b,c. In ΔABC

(i) $A + B + C = 180°$

(ii) $a + b > c, b + c > a, c + a > b$

(iii) $a > 0, b > 0, c > 0$

Sine Formula or Sine Rule

In any ΔABC, the sides are proportional to sines of the opposite angles,

i.e. $$\frac{a}{\sin A} = \frac{b}{\sin B} = \frac{c}{\sin C}$$

Proof : *Case* I. When $\angle C$ is acute :

From A draw $AD \perp BC$

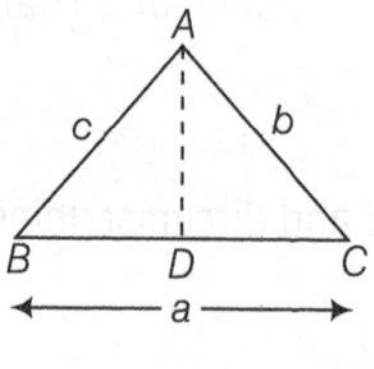

In ΔABD,

$$\sin B = \frac{AD}{AB} = \frac{AD}{c}$$

or $$AD = c \sin B \quad \text{...(i)}$$

In ΔACD, $$\sin C = \frac{AD}{AC} = \frac{AD}{b}$$

or $$AD = b \sin C \quad \text{...(ii)}$$

From Eqs. (i) and (ii), we get $c \sin B = b \sin C$ ...(1)

***Case* II. When $\angle C$ is obtuse :**

From A draw $AD \perp BC$

In ΔABD,

$$\sin B = \frac{AD}{AB} = \frac{AD}{c}$$

$\therefore$ $$AD = c \sin B \quad \text{...(iii)}$$

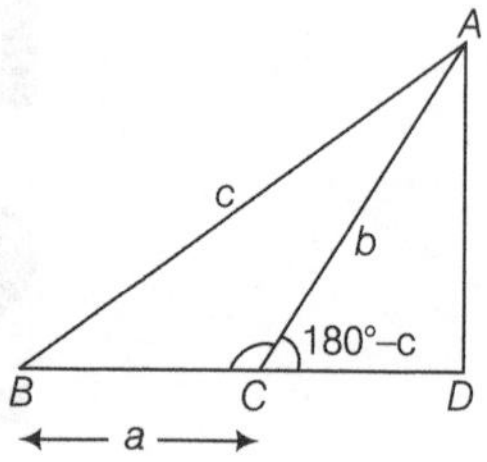

From ΔACD,

$$\sin(180° - C) = \frac{AD}{AC} = \frac{AD}{b}$$

$\therefore AD = b \sin(180° - C) = b \sin C$ ...(iv)

From Eqs. (iii) and (iv), we get

$$c \sin B = b \sin C \quad \text{...(2)}$$

***Case* III. When $\angle C = 90°$**

Draw $AD \perp BC$

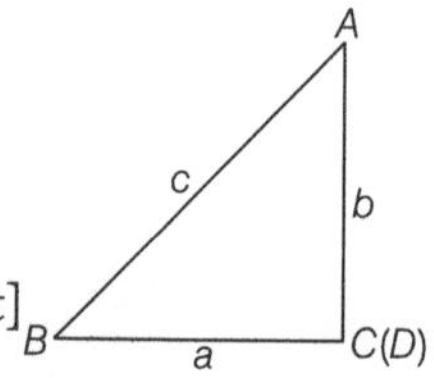

In ΔABC, $\sin B = \frac{AD}{AB} = \frac{AD}{c}$

$\therefore$ $AD = c \sin B$

or $AC = c \sin B$

[$\because$ D and C are same point]

or $b = c \sin B$

or $b \sin C = c \sin B$ [$\because C = 90°$] ...(3)

Thus from (1), (2) and (3), it follows that in all cases

$$b \sin C = c \sin B \text{ or } \frac{b}{\sin B} = \frac{c}{\sin C} \quad \text{...(4)}$$

Similarly by drawing perpendicular from from C to AB, we can prove that

$$\frac{a}{\sin A} = \frac{b}{\sin C} \quad \text{...(5)}$$

From (4) and (5), we get

$$\frac{a}{\sin A} = \frac{b}{\sin B} = \frac{c}{\sin C}$$

Cosine Formula or Cosine Rule

(i) $\cos A = \dfrac{b^2 + c^2 - a^2}{2bc}$ or $a^2 = b^2 + c^2 - 2bc \cos A$

(ii) $\cos B = \dfrac{c^2 + a^2 - b^2}{2ac}$

or $b^2 = c^2 + a^2 - 2ac \cos B$

(iii) $\cos C = \dfrac{a^2 + b^2 - c^2}{2ab}$

or $c^2 = a^2 + b^2 - 2ab \cos C$

Proof :

***Case* I. When $\angle A$ is acute**

Draw $BD \perp AC$

In ΔADB, $\quad \sin A = \dfrac{BD}{AB} = \dfrac{BD}{c}$

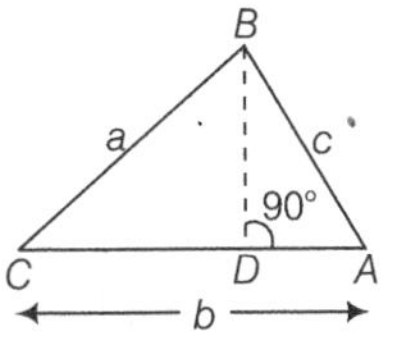

$\therefore \quad BD = c \sin A \quad$ ...(i)

and $\quad \cos A = \dfrac{AD}{AB} = \dfrac{AD}{c}$

$\therefore \quad AD = c \cos A \quad$ ...(ii)

Now $\quad CD = AC - AD$

$= b - c \cos A \quad$ ...(iii)

***Case* II. When $\angle A$ is obtuse**

In ΔADB,

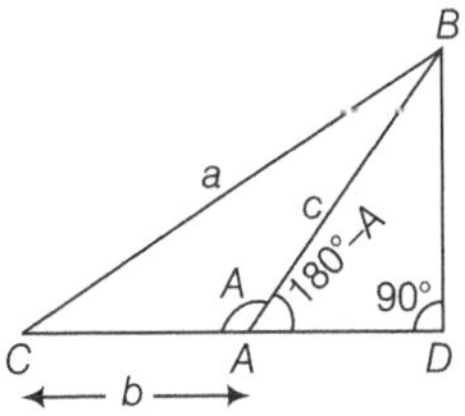

$$\sin(180° - A) = \frac{BD}{AB}$$

$\therefore \quad \sin A + \dfrac{BD}{c}$

or $\quad BD = c \sin A$ and $\cos(180° - A) = \dfrac{AD}{AB}$

$\therefore \quad -\cos A = \dfrac{AD}{c}$ or $AD = -c \cos A$

Now, $\quad CD = AC + AD = b - c \cos A$

***Case* III. When $\angle A = 90°$**

Here D and A are same points

In ΔACB, $\quad BD = BA = c = c \sin A$

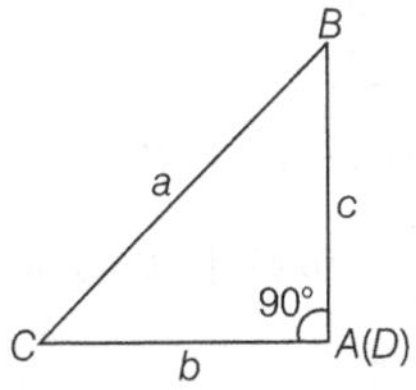

$[\because \angle A - 90° \therefore \sin A = \sin 90° = 1]$ ...(i)

and $\quad AD = 0 = c \cos A \quad [\because \cos A = \cos 90° = 0]$ ...(ii)

$\therefore \quad CD = AC - AD = b - c \cos A \quad$ ...(iii)

Thus in all cases $BD = c \sin A$ and $CD = b - c \cos A$

Now, in ΔBCD, $BC^2 = CD^2 + BD^2$

$$\therefore \quad a^2 = (b - c\cos A)^2 + c^2 \sin^2 A$$

$$= b^2 + c^2 \cos^2 A - 2bc \cos A + c^2 \sin^2 A$$

$$= b^2 + c^2(\cos^2 A + \sin^2 A) - 2bc \cos A$$

$$= b^2 + c^2 - 2bc \cos A \quad ...(1)$$

or $\quad 2bc \cos A = b^2 + c^2 - a^2$

$$\therefore \quad \cos A = \frac{b^2 + c^2 - a^2}{2bc}$$

Also, from (1), $\quad a^2 = b^2 + c^2 - 2bc \cos A$

Similarly, we can prove that $\cos B = \dfrac{c^2 + a^2 - b^2}{2ca}$

or $\quad b^2 = a^2 + c^2 - 2ca \cos B$

and $\quad \cos C = \dfrac{a^2 + b^2 - c^2}{2ab}$

or $\quad c^2 = a^2 + b^2 - 2ab \cos C$

Projection Formulae

(i) $c = a \cos B + b \cos A$

(ii) $b = a \cos C + c \cos A$

(iii) $a = b \cos C + c \cos B$

Proof :

***Case* I. When $\angle B$ is acute**

From ΔCBD,

$$\cos B = \frac{BD}{BC} = \frac{BD}{a}$$

$\therefore \quad BD = a \cos B \quad$ ...(i)

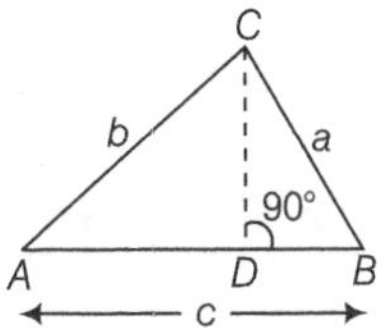

In ΔADC,

$$\cos A = \frac{AD}{AC} = \frac{AD}{b}$$

$\therefore \quad AD = b \cos A \quad$ ...(ii)

Now, $\quad c = AB = AD + DB = b \cos A + a \cos B$

Thus, $\quad c = b \cos A + a \cos B$

***Case* II. When $\angle B$ is obtuse**

From ΔCBD,

$$\cos(180° - B) = \frac{BD}{BC}$$

$\therefore \qquad -\cos B = \dfrac{BD}{a}$

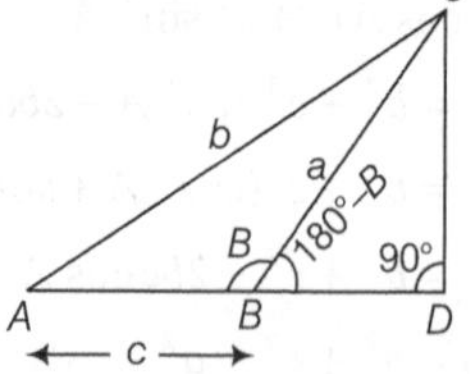

or $\qquad BD = -a\cos B$

In ΔCAD

$$\cos A = \frac{AD}{AC} = \frac{AD}{b}$$

or $\qquad AD = b\cos A$

Now, $c = AB = AD - BD = b\cos A + a\cos B \qquad$ ...(ii)

Case III. When $\angle B = 90°$

In ΔCAB,

$$\cos A = \frac{AB}{AC} = \frac{c}{b}$$

$\therefore \qquad c = b\cos A$

$$= b\cos A + c\cos B$$

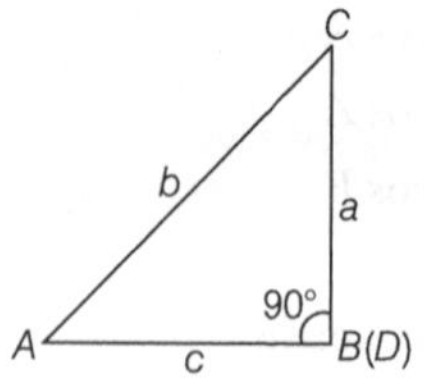

$[\because c\cos B = c\cos 90° = 0, \text{ as } \cos 90° = 0]$

Thus in all cases, $c = b\cos A + a\cos B$

Similarly, we can prove that

$$b = c\cos A + a\cos C$$

and $\qquad a = b\cos C + c\cos B$

Tangent Rule or Napier's Analogy

In any ΔABC, $\tan\left(\dfrac{A-B}{2}\right) = \dfrac{a-b}{a+b}\cot\dfrac{C}{2}$

Proof: In ΔABC, we know

$$\frac{a}{\sin A} = \frac{b}{\sin B} = \frac{c}{\sin C} = k \text{ (say)} \qquad \text{[sine law]}$$

$\therefore \qquad a = k\sin A, b = k\sin B, c = k\sin C \qquad$ ...(i)

Now, RHS

$$= \left(\frac{a-b}{a+b}\right)\cdot\cot\frac{C}{2}$$

$$= \left(\frac{k\sin A - k\sin B}{k\sin A + k\sin B}\right)\cdot\cot\frac{C}{2} \qquad \text{[using Eq(i)]}$$

$$= \frac{2\cos\left(\frac{A+B}{2}\right)\cdot\sin\left(\frac{A-B}{2}\right)}{2\sin\left(\frac{A+B}{2}\right)\cdot\cos\left(\frac{A-B}{2}\right)}\cdot\frac{\cos\frac{C}{2}}{\sin\frac{C}{2}}$$

$$= \frac{\sin\left(\frac{A-B}{2}\right)}{\cos\left(\frac{A-B}{2}\right)} \qquad \left[\begin{array}{l}\text{as, } \cos\left(\frac{A+B}{2}\right) = \sin\left(\frac{C}{2}\right)\\ \text{and } \sin\left(\frac{A+B}{2}\right) = \cos\left(\frac{C}{2}\right)\end{array}\right]$$

$$= \tan\left(\frac{A-B}{2}\right) = \text{LHS}$$

$\Rightarrow \qquad \tan\left(\dfrac{A-B}{2}\right) = \left(\dfrac{a-b}{a+b}\right)\cot\dfrac{C}{2}$

Similarly it could be shown,

$$\tan\left(\frac{B-C}{2}\right) = \frac{b-c}{b+c}\cot\frac{A}{2}, \tan\frac{C-A}{2} = \frac{c-a}{c+a}\cot\frac{B}{2}$$

Example 1. Find the angles of the triangles whose sides are $3+\sqrt{3}, 2\sqrt{3}$ and $\sqrt{6}$.

Sol. Let $a = 3+\sqrt{3}$, $b = 2\sqrt{3}$, $c = \sqrt{6}$

$\Rightarrow \qquad \cos A = \dfrac{b^2 + c^2 - a^2}{2bc} = \dfrac{12 + 6 - (9 + 3 + 6\sqrt{3})}{12\sqrt{2}}$

$$= \frac{6 - 6\sqrt{3}}{12\sqrt{2}} = \frac{1-\sqrt{3}}{2\sqrt{2}}$$

$$\cos A = \cos(60° + 45°)$$

$$\left[\text{as } \cos(60° + 45°) = \frac{1-\sqrt{3}}{2\sqrt{2}}\right]$$

$\therefore \qquad A = 105°$

Applying Sine formula $\qquad \dfrac{a}{\sin A} = \dfrac{b}{\sin B}$

$\Rightarrow \qquad \sin B = \dfrac{b}{a}\sin A = \dfrac{2\sqrt{3}}{3+\sqrt{3}}\sin(105°)$

$$= \frac{2\sqrt{3}}{3+\sqrt{3}}\{\sin 60°.\cos 45° + \cos 60°\cdot\sin 45°\}$$

$$= \frac{2\sqrt{3}}{\sqrt{3}(\sqrt{3}+1)}\left\{\frac{\sqrt{3}+1}{2\sqrt{2}}\right\}$$

$$\sin B = \frac{1}{\sqrt{2}} = \sin 45° \; [\because B \neq 180 - 45° \text{ as } B + A < 180°]$$

$\Rightarrow \qquad B = 45°$

Here, $\qquad A = 105°, B = 45°$

$\Rightarrow \qquad C = 180° - (A + B) = 180° - (150°) = 30°$

$\therefore \qquad \angle A = 105°, \angle B = 45°$ and $\angle C = 30°$

Example 2. The sides of a triangle are 8 cm, 10 cm and 12 cm. Prove that the greatest angle is double of the smallest angle.

Sol. Let $a = 8$ cm, $b = 10$ cm and $c = 12$ cm. Hence, greatest angle is C and the smallest angle is A, {as we know greatest angle is opposite to greatest side and smallest angle is opposite to smallest side.} Here, we have to prove $C = 2A$, applying cosine law, we get

$$\cos C = \frac{a^2 + b^2 - c^2}{2ab} = \frac{64 + 100 - 144}{2 \cdot 8 \cdot 10} = \frac{1}{8} \quad \text{...(i)}$$

and $$\cos A = \frac{b^2 + c^2 - a^2}{2bc}$$

$$= \frac{100 + 144 - 64}{2 \cdot 10 \cdot 12} = \frac{3}{4} \quad \text{...(ii)}$$

$$\cos 2A = 2\cos^2 A - 1 = 2 \cdot \frac{9}{16} - 1 \quad \text{[using Eq. (ii)]}$$

$$\therefore \quad \cos 2A = \frac{1}{8} \quad \text{...(iii)}$$

From Eqs. (i) and (iii), we get

$$\cos 2A = \cos C$$

$$\Rightarrow \quad C = 2A.$$

Example 3. With usual notations, if in a ΔABC, $\frac{b+c}{11} = \frac{c+a}{12} = \frac{a+b}{13}$, then prove that

$$\frac{\cos A}{7} = \frac{\cos B}{19} = \frac{\cos C}{25}.$$

Sol. Let, $\frac{b+c}{11} = \frac{c+a}{12} = \frac{a+b}{13} = k$

$$\Rightarrow \quad 2(a + b + c) = 36k \quad \text{...(i)}$$

$$b + c = 11k, c + a = 12k, a + b = 13k \quad \text{...(ii)}$$

On solving Eqs. (i) and (ii), we get

$$a = 7k, b = 6k, c = 5k$$

Hence, $$\cos A = \frac{b^2 + c^2 - a^2}{2bc} = \frac{36k^2 + 25k^2 - 49k^2}{60k^2}$$

$$= \frac{12}{60} = \frac{1}{5} = \frac{7}{35}$$

$$\cos B = \frac{a^2 + c^2 - b^2}{2ac} = \frac{49k^2 + 25k^2 - 36k^2}{70k^2}$$

$$= \frac{38}{70} = \frac{19}{35}$$

$$\cos C = \frac{a^2 + b^2 - c^2}{2ab} = \frac{49k^2 + 36k^2 - 25k^2}{84k^2}$$

$$= \frac{60}{84} = \frac{5}{7}$$

$$= \frac{25}{35}$$

$$\Rightarrow \quad \frac{\cos A}{7} = \frac{\cos B}{19} = \frac{\cos C}{25}$$

Example 4. The sides of a triangle are three consecutive natural numbers and its largest angle is twice the smallest one. Determine the sides of the triangle.

Sol. Let the sides be $n, n + 1, n + 2$

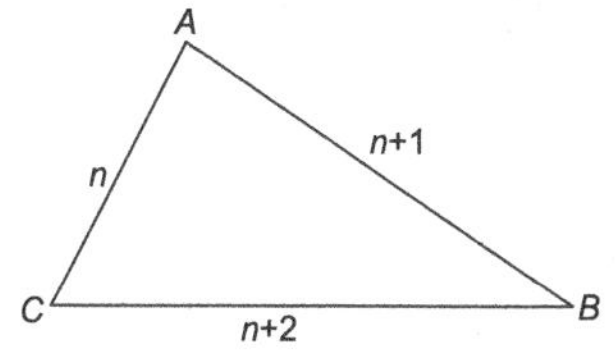

i.e. $\quad AC = n, AB = n + 1, BC = n + 2$

Smallest angle is B and largest one is A.

Here, $\quad A = 2B$

Also, $\quad A + B + C = 180°$

$$\Rightarrow \quad 3B + C = 180° \Rightarrow C = 180° - 3B$$

Using, sine rule, $\frac{\sin A}{n+2} = \frac{\sin B}{n} = \frac{\sin C}{n+1}$

$$\Rightarrow \quad \frac{\sin 2B}{n+2} = \frac{\sin B}{n} = \frac{\sin(180° - 3B)}{n+1}$$

$$\Rightarrow \quad \underset{\text{(i)}}{\frac{\sin 2B}{n+2}} = \underset{\text{(ii)}}{\frac{\sin B}{n}} = \underset{\text{(iii)}}{\frac{\sin 3B}{n+1}}$$

From Eqs. (i) and (ii); $\frac{2\sin B \cos B}{n+2} = \frac{\sin B}{n}$

$$\cos B = \frac{n+2}{2n} \quad \text{...(iv)}$$

and from Eqs. (ii) and (iii);

$$\frac{\sin B}{n} = \frac{3\sin B - 4\sin^3 B}{n+1}$$

$$\Rightarrow \quad \frac{\sin B}{n} = \frac{\sin B\{3 - 4\sin^2 B\}}{n+1}$$

$$\Rightarrow \quad \frac{n+1}{n} = 3 - 4(1 - \cos^2 B)$$

$$\frac{n+1}{n} = -1 + 4\cos^2 B \quad \text{...(v)}$$

From Eqs. (iv) and (v), we get

$$\frac{n+1}{n} = -1 + 4\left(\frac{n+2}{2n}\right)^2$$

$$\frac{n+1}{n} + 1 = \left(\frac{n^2 + 4n + 4}{n^2}\right)$$

$$\frac{2n+1}{n} = \frac{n^2 + 4n + 4}{n^2}$$

$$2n^2 + n = n^2 + 4n + 4$$

$$\Rightarrow \quad n^2 - 3n - 4 = 0 \Rightarrow (n - 4)(n + 1) = 0$$

$$n = 4 \text{ or } -1 \qquad \text{where } n \neq -1$$

$\therefore \quad n = 4$. Hence, the sides are 4, 5, 6.

Example 5. Let O be a point inside a ΔABC, such that $\angle OAB = \angle OBC = \angle OCA = w$. Then, show that
$$\cot w = \cot A + \cot B + \cot C.$$

Sol. In ΔOAC, Using sine rule,
$$\frac{\sin(A-w)}{OC} = \frac{\sin(180-A)}{b}$$
$$\Rightarrow \quad \frac{\sin(A-w)}{OC} = \frac{\sin A}{b} \quad \text{...(i)}$$

Also, in ΔOBC

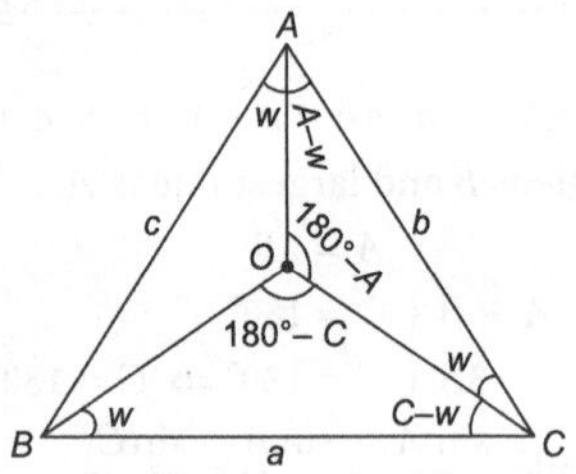

$$\frac{\sin w}{OC} = \frac{\sin(180-C)}{a}$$
$$\Rightarrow \quad \frac{\sin w}{OC} = \frac{\sin C}{a} \quad \text{...(ii)}$$

On dividing Eqs. (i) and (ii); $\dfrac{\sin(A-w)}{\sin w} = \dfrac{a\sin A}{b\sin C}$
$$\left\{\text{as, we know, } \frac{a}{\sin A} = \frac{b}{\sin B} = \frac{c}{\sin C} = k\right\}$$
$$\Rightarrow \quad \frac{\sin(A-w)}{\sin w} = \frac{k\sin A \cdot \sin A}{k\sin B \cdot \sin C}$$
$$\Rightarrow \quad \frac{\sin A\cos w - \cos A\sin w}{\sin w} = \sin A\frac{\{\sin(\pi-(B+C))\}}{\sin B\sin C}$$
$$\Rightarrow \quad \sin A(\cot w) - \cos A = \sin A\left(\frac{\sin B\cos C + \cos B\sin C}{\sin B\sin C}\right)$$
$$\Rightarrow \quad \sin A(\cot w) - \cos A = \sin A(\cot C + \cot B)$$
$$\Rightarrow \quad \cot w - \cot A = \cot B + \cot C$$
$$\Rightarrow \quad \cot w = \cot A + \cot B + \cot C.$$

Example 6. Solve
$b\cos^2\dfrac{C}{2} + c\cos^2\dfrac{B}{2}$ in terms of k, where k is perimeter of the ΔABC.

Sol. We have, $b\cos^2\dfrac{C}{2} + c\cos^2\dfrac{B}{2}$
$$\Rightarrow \quad \frac{b}{2}(1+\cos C) + \frac{c}{2}(1+\cos B)$$
$$\Rightarrow \quad \frac{b+c}{2} + \frac{1}{2}(b\cos C + c\cos B)$$
$$\Rightarrow \quad \frac{b+c}{2} + \frac{1}{2}a \quad \text{[using projection formula]}$$
$$\Rightarrow \quad \frac{a+b+c}{2}$$
$$\therefore \quad b\cos^2\frac{C}{2} + c\cos^2\frac{B}{2} = \frac{k}{2} \quad \text{[where } k = a+b+c\text{, given]}$$

Example 7. In a ΔABC, $c\cos^2\dfrac{A}{2} + a\cos^2\dfrac{C}{2} = \dfrac{3b}{2}$, then show a, b, c are in AP.

Sol. We have, $c\cos^2\dfrac{A}{2} + a\cos^2\dfrac{C}{2} = \dfrac{3b}{2}$
$$\Rightarrow \quad \frac{c}{2}(1+\cos A) + \frac{a}{2}(1+\cos C) = \frac{3b}{2}$$
$$\Rightarrow \quad a + c + (c\cos A + a\cos C) = 3b$$
$$\Rightarrow \quad a + c + b = 3b \quad \text{[using projection formula]}$$
$$\Rightarrow \quad a + c = 2b$$
which shows a, b, c are in AP.

Example 8. In a ΔABC, $a = 2b$ and $|A-B| = \dfrac{\pi}{3}$.

Determine the $\angle C$.

Sol. Given, $\quad a = 2b \quad$ ...(i)
$$\Rightarrow \quad \angle A > \angle B \quad \text{[as } a > b \text{ and we know greater angle is opposite to greater side]}$$
$$\Rightarrow \quad |A-B| = \frac{\pi}{3}$$
$$\text{or} \quad A - B = \frac{\pi}{3} \quad \text{...(ii) [as } A > B\text{]}$$

Using Napier's analogy, we have
$$\tan\left(\frac{A-B}{2}\right) = \frac{a-b}{a+b}\cot\left(\frac{C}{2}\right)$$
$$\Rightarrow \quad \tan\left(\frac{\pi}{6}\right) = \frac{2b-b}{2b+b}\cot\left(\frac{C}{2}\right)$$
[using Eqs. (i) and (ii)]
$$\Rightarrow \quad \frac{1}{\sqrt{3}} = \frac{1}{3}\cot\left(\frac{C}{2}\right)$$
$$\Rightarrow \quad \cot\left(\frac{C}{2}\right) = \sqrt{3}$$
$$\text{i.e.} \quad \frac{C}{2} = \frac{\pi}{6} \Rightarrow C = \frac{\pi}{3}$$

Example 9. In a ΔABC, the tangent of half the difference of two angles is one-third the tangent of half the sum of the angles. Determine the ratio of the sides opposite to the angles.

Sol. We have, $\quad \tan\left(\dfrac{A-B}{2}\right) = \dfrac{1}{3}\tan\left(\dfrac{A+B}{2}\right) \quad$ ...(i)

Using Napier's analogy,
$$\tan\left(\frac{A-B}{2}\right) = \frac{a-b}{a+b}\cdot\cot\left(\frac{C}{2}\right) \quad \text{...(ii)}$$

From Eqs. (i) and (ii);

$$\frac{1}{3}\tan\left(\frac{A+B}{2}\right) = \frac{a-b}{a+b}\cdot\cot\left(\frac{C}{2}\right)$$

$$\Rightarrow \quad \frac{1}{3}\cot\left(\frac{C}{2}\right) = \frac{a-b}{a+b}\cdot\cot\left(\frac{C}{2}\right)$$

$$\left[\text{as } A+B+C=\pi \ \therefore\ \tan\left(\frac{B+C}{2}\right) = \tan\left(\frac{\pi}{2}-\frac{C}{2}\right) = \cot\frac{C}{2}\right]$$

$$\Rightarrow \quad \frac{a-b}{a+b} = \frac{1}{3} \quad \text{or} \quad 3a-3b = a+b$$

$$2a = 4b$$

or

$$\frac{b}{a} = \frac{1}{2}$$

Thus, the ratio of the sides opposite to the angles is $b : a = 1 : 2$.

Exercise for Session 1

1. In the given figure, if $AB = AC$, $\angle BAD = 30°$ and $AE = AD$, then x is equal to

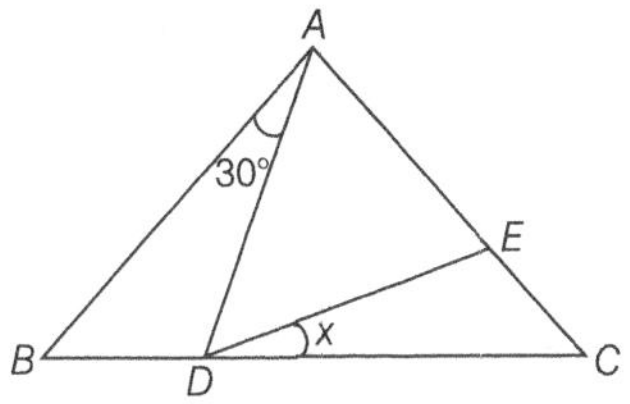

(a) 15° (b) 10° (c) $12\frac{1}{2}$ (d) $7\frac{1}{2}$

2. In ΔABC, $a = 4$, $b = 12$ and $B = 60°$, then the value of $\sin A$ is

(a) $\frac{1}{2\sqrt{3}}$ (b) $\frac{1}{3\sqrt{2}}$ (c) $\frac{2}{\sqrt{3}}$ (d) $\frac{\sqrt{3}}{2}$

3. Let ABC be a triangle such that $\angle A = 45°$, $\angle B = 75°$, then $a + c\sqrt{2}$ is equal to

(a) 0 (b) b (c) $2b$ (d) $-b$

4. Angles A, B and C of a ΔABC are in AP. If $\frac{b}{c} = \frac{\sqrt{3}}{\sqrt{2}}$, then $\angle A$ is equal to

(a) $\frac{\pi}{6}$ (b) $\frac{\pi}{4}$ (c) $\frac{5\pi}{12}$ (d) $\frac{\pi}{2}$

5. If $\cot\frac{A}{2} = \frac{b+c}{a}$, then ΔABC is

(a) Isosceles (b) Equilateral (c) Right angled (d) None of these

6. If in a ΔABC, $\frac{a^2-b^2}{a^2+b^2} = \frac{\sin(A-B)}{\sin(A+B)}$, then the triangle is

(a) Right angled or isosceles (b) Right angled and isosceles
(c) Equilateral (d) None of these

7. In any triangle ABC, $\frac{a^2\sin(B-C)}{\sin B+\sin C} + \frac{b^2\sin(C-A)}{\sin C+\sin A} + \frac{c^2\sin(A-B)}{\sin A+\sin B} =$

(a) $a+b+c$ (b) $a+b-c$ (c) $a-b+c$ (d) 0

8. In any ΔABC, if $2\cos B = \frac{a}{2}$, then the triangle is

(a) right angled (b) equilateral (c) isosceles (d) None of these

9. The expression $\dfrac{(a+b+c)(b+c-a)(c+a-b)(a+b-c)}{4b^2c^2}$ is equal to

(a) $\cos^2 A$ (b) $\sin^2 A$ (c) $\cos A \cos B \cos C$ (d) None of these

10. InΔABC, if $a\cos A = b\cos B$, then the triangle is

(a) Isosceles (b) Right angled
(c) Isosceles or right angled (d) Right angled isosceles

11. In a ΔABC, $(a+b+c)(b+c-a) = \lambda bc$, if

(a) $\lambda < 0$ (b) $\lambda > 0$
(c) $0 < \lambda < 4$ (d) $\lambda < 4$

12. If $a = 9, b = 8$ and $c = x$ satisfies $3\cos C = 2$, then

(a) $x = 5$ (b) $x = 6$ (c) $x = 4$ (d) $x = 7$

13. In ΔABC, if $\sin^2 A + \sin^2 B = \sin^2 C$, then the triangle is

(a) equilateral (b) isosceles (c) right angled (d) None of these

14. The sides of a triangle are $\alpha - \beta$, $\alpha + \beta$ and $\sqrt{3\alpha^2 + \beta^2}$, $(\alpha > \beta > 0)$. Its largest angle is

(a) $\dfrac{2\pi}{3}$ (b) $\dfrac{\pi}{2}$ (c) $\dfrac{3\pi}{4}$ (d) $\dfrac{5\pi}{6}$

15. In any triangle, $\dfrac{1+\cos(A-B)\cos C}{1+\cos(A-C)\cos B} =$

(a) $\dfrac{a^2+b^2}{a^2+c^2}$ (b) $\dfrac{b^2+c^2}{b^2-c^2}$ (c) $\dfrac{c^2-a^2}{a^2+b^2}$ (d) None of these

16. If the sides of a ΔABC are in AP and a is the smallest side, then $\cos A$ equals

(a) $\dfrac{3c-4b}{2c}$ (b) $\dfrac{3c-4b}{2b}$ (c) $\dfrac{4c-3b}{2c}$ (d) None of these

17. In a ΔABC, $a^2\cos 2B + b^2\cos 2A + 2ab\cos(A-B) =$

(a) a^2 (b) c^2 (c) b^2 (d) $a^2 + b^2$

18. In any ΔABC, $2[bc\cos A + ca\cos B + ab\cos C] =$

(a) $a^2+b^2+c^2$ (b) $a^2+b^2-c^2$ (c) $a^2-b^2+c^2$ (d) None of these

19. In a ΔABC, $\tan\frac{1}{2}(A+B)\cdot\cot\frac{1}{2}(A-B)$ is equal to

(a) $\dfrac{a-b}{a+b}$ (b) $\dfrac{a+b}{c}$ (c) $\dfrac{a+b}{a-b}$ (d) $\dfrac{a-b}{2(a+b)}$

20. If in a ΔABC, $b = \sqrt{3}$, $c = 1$ and $B - C = 90°$, then $\angle A$ is

(a) 30° (b) 45° (c) 75° (d) 15°

Session 2

Auxiliary Formulae

Trigonometric Ratios of Half-angles

In any ΔABC, we have

$$\cos A = \frac{b^2 + c^2 - a^2}{2bc} \quad \text{...(i) [using cosine law]}$$

$$2\sin^2 \frac{A}{2} = 1 - \cos A \quad \text{...(ii)}$$

$$2\sin^2 \frac{A}{2} = 1 - \left(\frac{b^2 + c^2 - a^2}{2bc}\right)$$

[using Eqs. (i) and (ii)]

$$= \frac{2bc - b^2 - c^2 + a^2}{2bc}$$

$$= \frac{a^2 - (b^2 + c^2 - 2bc)}{2bc} = \frac{a^2 - (b-c)^2}{2bc}$$

$$= \frac{(a+b-c)(a-b+c)}{2bc}$$

[we know $a + b + c = 2s$
$\Rightarrow a + b = 2s - c$ and $a + c = 2s - b$]

$$\therefore \quad 2\sin^2 \frac{A}{2} = \frac{(2s - c - c)(2s - b - b)}{2bc}$$

$$2\sin^2 \frac{A}{2} = \frac{4(s-c)(s-b)}{2bc}$$

$$\Rightarrow \quad \sin^2 \frac{A}{2} = \frac{(s-b)(s-c)}{bc}$$

[since in a triangle, A is always less than 180°,
$\therefore \sin A/2$ is (+ve)]

$$\text{or} \quad \sin \frac{A}{2} = \sqrt{\frac{(s-b)(s-c)}{bc}} \quad \text{...(A)}$$

Similarly, it may be proved,

$$\sin \frac{B}{2} = \sqrt{\frac{(s-a)(s-c)}{ac}} \quad \text{...(B)}$$

$$\sin \frac{C}{2} = \sqrt{\frac{(s-a)(s-b)}{ab}} \quad \text{...(C)}$$

Again, $2\cos^2 \frac{A}{2} = 1 + \cos A = 1 + \frac{b^2 + c^2 - a^2}{2bc}$

$$= \frac{2bc + b^2 + c^2 - a^2}{2bc}$$

$$= \frac{(b+c)^2 - a^2}{2bc}$$

$$= \frac{(b+c-a)(b+c+a)}{2bc}$$

[where $a + b + c = 2s$, $b + c = 2s - a$]

$$2\cos^2 \frac{A}{2} = \frac{2s \times 2(s-a)}{2bc}; \quad \cos^2 \frac{A}{2} = \frac{s(s-a)}{bc}$$

Since, $A/2$ is less than 90°,

$\therefore \quad \cos A/2 > 0.$

$$\therefore \quad \cos \frac{A}{2} = \sqrt{\frac{s(s-a)}{bc}}$$

Similarly, $\cos \frac{B}{2} = \sqrt{\frac{s(s-b)}{ac}}$ and $\cos \frac{C}{2} = \sqrt{\frac{s(s-c)}{ab}}$

Also, $\tan \frac{A}{2} = \frac{\sin A/2}{\cos A/2} = \sqrt{\frac{(s-b)(s-c)}{bc}} \times \sqrt{\frac{bc}{s(s-a)}}$

$$\therefore \quad \tan \frac{A}{2} = \sqrt{\frac{(s-b)(s-c)}{s(s-a)}}$$

Similarly, $\tan \frac{B}{2} = \sqrt{\frac{(s-a)(s-c)}{s(s-b)}}$

and $\tan \frac{C}{2} = \sqrt{\frac{(s-a)(s-b)}{s(s-c)}}$

Area of Triangle

If Δ represents the area of a triangle ABC, then

area of $\Delta = \frac{1}{2} BC \cdot AD$, $\quad \left[\because \Delta = \frac{1}{2}\text{(base) (height)}\right]$

$= \frac{1}{2} a \cdot (c \sin B) \quad \left[\text{as } \sin B = \frac{AD}{c}\right]$

$$\Delta = \frac{1}{2} ac \cdot \sin B$$

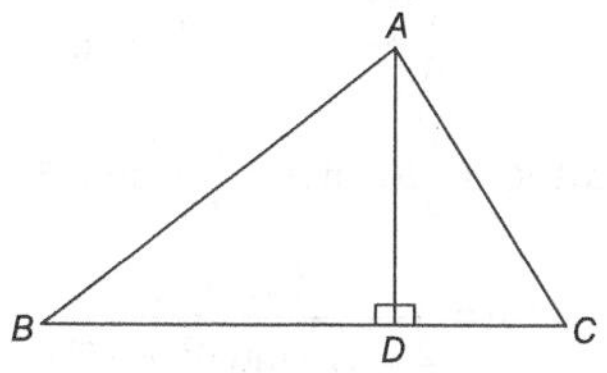

Also, $\sin C = \frac{AD}{b} \Rightarrow AD = b \sin C$

$$\therefore \quad \Delta = \frac{1}{2} a \cdot b \cdot \sin C$$

Similarly; $\quad \Delta = \frac{1}{2} bc \cdot \sin A$

$$\therefore \quad \boldsymbol{\Delta = \frac{1}{2} ab \sin C = \frac{1}{2} bc \sin A = \frac{1}{2} ca \sin B}$$

Note

(i) Area of a triangle in terms of sides (Heron's formula) :

$$\Delta = \frac{1}{2} bc \sin A = \frac{1}{2} bc \cdot 2\sin\frac{A}{2} \cdot \cos\frac{A}{2}$$

$$= bc\sqrt{\frac{(s-b)(s-c)}{bc}}\sqrt{\frac{s(s-a)}{bc}}$$

$$\therefore \quad \Delta = \sqrt{s(s-a)(s-b)(s-c)}.$$

(ii) Area of triangle in terms of one side and sine of three angles:

$$\Delta = \frac{1}{2} bc \sin A = \frac{1}{2}(k \sin B)(k \sin C) \cdot \sin A$$

$$= \frac{1}{2} k^2 \sin A \sin B \sin C = \frac{1}{2}\left(\frac{a}{\sin A}\right)^2 \cdot \sin A \cdot \sin B \cdot \sin C$$

$$\therefore \quad \Delta = \frac{a^2}{2} \cdot \frac{\sin B \sin C}{\sin A} = \frac{b^2}{2} \cdot \frac{\sin A \sin C}{\sin B} = \frac{c^2}{2} \cdot \frac{\sin A \sin B}{\sin C}$$

Example 10. If the angles of a triangle are 30° and 45°, and the included side is $(\sqrt{3}+1)$ cm, then prove that the area of the triangle is $\frac{1}{2}(\sqrt{3}+1)$.

Sol. We have, $\quad \frac{\sin A}{a} = \frac{\sin B}{b} = \frac{\sin C}{c}$

$$\Rightarrow \quad \frac{\sin 105^\circ}{\sqrt{3}+1} = \frac{\sin 30^\circ}{b} = \frac{\sin 45^\circ}{c}$$

$$\Rightarrow \quad b = \frac{\sqrt{3}+1}{2\sin 105^\circ}, \; c = \frac{\sqrt{3}+1}{\sqrt{2}\sin 105^\circ}$$

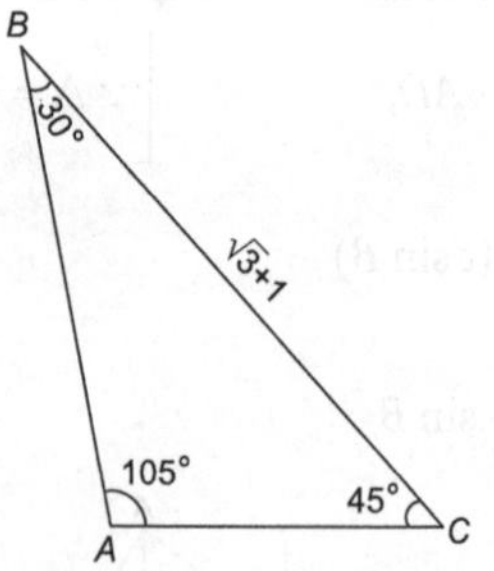

So, area of $\Delta ABC = \frac{1}{2} bc \sin A = \frac{1}{2} bc \sin 105^\circ$

$$= \frac{1}{2} \cdot \frac{(\sqrt{3}+1)^2}{2\sqrt{2}.\sin(60^\circ + 45^\circ)}$$

$$= \frac{(\sqrt{3}+1)^2}{4\sqrt{2}\left\{\frac{\sqrt{3}}{2} \cdot \frac{1}{\sqrt{2}} + \frac{1}{\sqrt{2}} \cdot \frac{1}{2}\right\}} = \frac{(\sqrt{3}+1)^2}{2(\sqrt{3}+1)} = \frac{1}{2}(\sqrt{3}+1)$$

Thus, area of Δ is $\frac{1}{2}(\sqrt{3}+1)$.

Aliter In above example we have $\angle A = 105^\circ$, $\angle B = 30^\circ$, $\angle C = 45^\circ$ and $a = \sqrt{3}+1$.

Thus, area of $\quad \Delta = \frac{a^2}{2} \frac{\sin B \sin C}{\sin A}$ $\quad$ [using note (ii)]

$$= \frac{(\sqrt{3}+1)^2}{2} \cdot \frac{\sin 30^\circ . \sin 45^\circ}{\sin(105^\circ)} = \frac{1}{2}(\sqrt{3}+1)$$

Example 11. Consider the following statements concerning in ΔABC

(i) The sides a, b, c and area Δ are rational.

(ii) $a, \tan\frac{B}{2}, \tan\frac{C}{2}$ are rational.

(iii) $a, \sin A, \sin B, \sin C$ are rational.

Prove that (i) ⇒ (ii) ⇒ (iii) ⇒ (i).

Sol. a, b, c, Δ are rational (given).

$$\Rightarrow \quad s, s-a, s-b, s-c \text{ are rational.}$$

Now, $\quad \tan\frac{B}{2} = \sqrt{\frac{(s-c)(s-a)}{s(s-b)}}$

$$= \sqrt{\frac{s(s-a)(s-b)(s-c)}{s^2(s-b)^2}} = \frac{\Delta}{s(s-b)}$$

$\therefore \quad \tan\frac{B}{2}$ = rational $\quad$ [as $\Delta, s, (s-a)$ are rational]

Similarly, $\tan\frac{C}{2}$ is rational. Hence (i) ⇒ (ii)

Now, $\sin B = \dfrac{2\tan\frac{B}{2}}{1+\tan^2\frac{B}{2}}$ is rational by (ii).

Similarly, $\sin C$ is also rational.

$$\tan\frac{A}{2} = \cot\frac{B+C}{2} = \frac{1 - \tan\frac{B}{2} \cdot \tan\frac{C}{2}}{\tan\frac{B}{2} + \tan\frac{C}{2}} = \text{rational by (ii)}$$

⇒ $\sin A$ is rational

Hence, (ii) ⇒ (iii).

Now, $\frac{a}{\sin A} = \frac{b}{\sin B} = \frac{c}{\sin C} = k$, which is rational since 'a' and '$\sin A$' are rational.

⇒ $\frac{b}{\sin B}$ and $\frac{c}{\sin C}$ are rational. But $\sin B$ and $\sin C$ are rational by (iii)

⇒ $\quad b$ and c are rational.

⇒ $\quad \Delta = \frac{1}{2} bc \sin A$ is also rational.

Hence (iii) ⇒ (i).

Exercise for Session 2

1. If in a ΔABC, $(s-a)(s-b)=s(s-c)$, then angle C is equal to
(a) 90° (b) 45° (c) 30° (d) 60°

2. In any ΔABC, if $\cot\frac{A}{2}, \cot\frac{B}{2}, \cot\frac{C}{2}$ are in AP, then a, b, c are in
(a) AP (b) GP (c) HP (d) None of these

3. In any ΔABC, $\dfrac{\tan\frac{A}{2}-\tan\frac{B}{2}}{\tan\frac{A}{2}+\tan\frac{B}{2}}=$
(a) $\frac{a-b}{a+b}$ (b) $\frac{a-b}{c}$ (c) $\frac{a-b}{a+b+c}$ (d) $\frac{c}{a+b}$

4. In a ΔABC, $bc\cos^2\frac{A}{2}+ca\cos^2\frac{B}{2}+ab\cos^2\frac{C}{2}$ is equal to
(a) $(s-a)^2$ (b) $(s-b)^2$ (c) $(s-c)^2$ (d) s^2

5. In a ΔABC, if $\cos A+\cos C+4\sin^2\left(\frac{B}{2}\right)$, then a, b, c are in
(a) AP (b) GP (c) HP (d) None of these

6. If in a ΔABC, $3a=b+c$, then the value of $\cos\frac{B}{2}\cot\frac{C}{2}$ is
(a) 1 (b) $\sqrt{3}$ (c) 2 (d) None of these

7. In any ΔABC, $\left(\frac{b-c}{a}\right)\cos^2\left(\frac{A}{2}\right)+\left(\frac{c-a}{b}\right)\cos^2\left(\frac{b}{2}\right)+\left(\frac{a-b}{2}\right)\cos^2\left(\frac{C}{2}\right)$ is equal to
(a) $\frac{b^2-c^2}{a^2}$ (b) $\frac{c^2-a^2}{b^2}$ (c) $\frac{a^2-b^2}{c^2}$ (d) 0

8. In a ΔABC, the tangent of half difference of two angles is one-third the tangent of half the sum of the two angles. The ratio of the sides opposite the angles is
(a) 2 : 3 (b) 1 : 3 (c) 2 : 1 (d) 3 : 4

9. If in a triangle, $a\cos^2\frac{C}{2}+c\cos^2\frac{A}{2}=\frac{3b}{2}$, then its sides will be in -
(a) AP (b) GP (c) HP (d) AGP

10. In the adjacent figure 'P' is any interior point of the equilateral triangle ABC of side length 2 unit

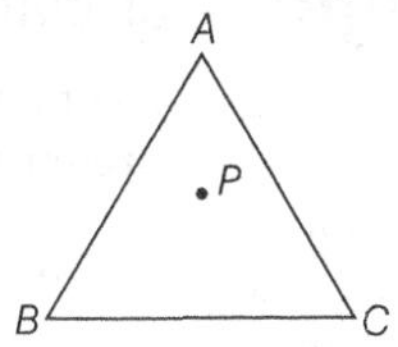

If x_a, x_b and x_c represent the distance of P from the sides BC, CA and AB respectively then $x_a+x_b+x_c$ is equal to
(a) 6 (b) $\sqrt{3}$ (c) $\frac{\sqrt{3}}{2}$ (d) $2\sqrt{3}x$

11. If $c^2=a^2+b^2$, then $4s(s-a)(s-b)(s-c)$ is equal to
(a) s^4 (b) b^2c^2 (c) c^2a^2 (d) a^2b^2

12. The number of possible $\angle ABC$ in which $BC=\sqrt{11}$ cm, $CA=\sqrt{13}$ cm and $A=60°$ is
(a) 0 (b) 1 (c) 2 (d) None of these

13. If two sides a, b and the $\angle A$ be such that two triangles are formed, then the sum of the two values of the third side is

(a) $b^2 - a^2$ (b) $2b \cos A$ (c) $2b \sin A$ (d) $\frac{b-c}{b+c}$

14. If in a ΔABC, $\sin A = \sin^2 B$ and $2\cos^2 A = 3\cos^2 B$, then the ΔABC is

(a) right angled (b) obtuse angled (c) isosceles (d) equilateral

15. If $a \cos A = b \cos B$, then the triangle is

(a) equilateral (b) right angled (c) isosceles (d) isosceles or right angled

16. Points D, E are taken on the side BC of a triangle ABC such that $BD = DE = EC$. If $\angle BAD = x$, $\angle DAE = y$, $\angle EAC = z$, then the value of $\frac{\sin(x+y)\sin(y+z)}{\sin x \sin z}$ is equal to

(a) 1 (b) 2 (c) 4 (d) None of these

17. If the base angles of a triangle are $22\frac{1°}{2}$ and $112\frac{1°}{2}$, then the height of the triangle is equal to

(a) half the base (b) the base (c) twice the base (d) four times the base

18. In a ΔABC, $a = 1$ and the perimeter is six times the AM of the sines of the angles. The measure of $\angle A$ is

(a) $\frac{\pi}{3}$ (b) $\frac{\pi}{2}$ (c) $\frac{\pi}{6}$ (d) $\frac{\pi}{4}$

19. In a ΔABC, if median AD is perpendicular to AB, then $\tan A + 2\tan B$ is equal to

(a) 1 (b) 3 (c) 0 (d) $\frac{1}{2}$

20. If p is the product of the sines of angles of a triangle, and q the product of their cosines, then tangents of the angles are roots of the equation

(a) $qx^3 - px^2 + (1+q)x - p = 0$ (b) $px^3 - qx^2 + (1+p)x - q = 0$

(c) $(1+q)x^3 - px^2 + qx - p = 0$ (d) None of these

Session 3

Circles Connected with Triangle

Circles Connected with Triangle

Circumcircle of a Triangle

The circle which passes through the angular points of a ΔABC is called its Circumcircle. The centre of this circle is the point of intersection of perpendicular bisectors of the sides and called the Circumcentre. Its radius is always denoted by R.

Note

1. Circumcentre of an a cute-angled triangle lies inside the triangle.
2. Circumcentre of an obtuse-angled triangle lies outside the triangle.
3. In a right angled triangle the circumcentre is the mid-point of hypotenuse.

Circum-radius

The radius of the circumcircle of a ΔABC is called the circum-radius given by;

(i) $R = \frac{a}{2\sin A} = \frac{b}{2\sin B} = \frac{c}{2\sin C}$ (ii) $R = \frac{abc}{4\Delta}$

Proof

(i) Here, the perpendicular bisectors of the sides BC, CA and AB intersect at O.

$\therefore$ O is the circumcentre such that,

$$OA = OB = OC = R$$

We have, $\angle BOC = 2\angle A$

$\therefore$ $\angle BOD = \angle COD = \angle A$

In ΔOBD, $\quad \sin A = \dfrac{BD}{OB} = \dfrac{a/2}{R}$

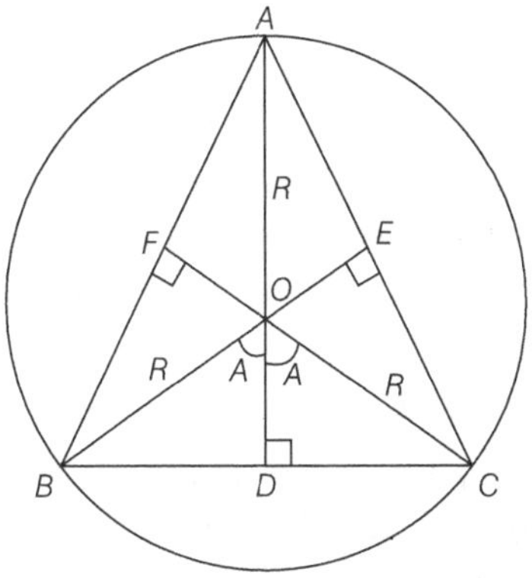

$\Rightarrow \quad R = \dfrac{a}{2\sin A}$

Similarly, $\quad R = \dfrac{b}{2\sin B}$ and $R = \dfrac{c}{2\sin C}$

Hence, $\boxed{R = \dfrac{a}{2\sin A} = \dfrac{b}{2\sin B} = \dfrac{c}{2\sin C}}$

(ii) As discussed, Area of $\Delta = \dfrac{1}{2} bc \sin A$

$\Rightarrow \quad \sin A = \dfrac{2\Delta}{bc} \quad$ ...(i)

Also, $\quad R = \dfrac{a}{2\sin A} \quad$ ...(ii)

$\therefore$ From Eqs. (i) and (ii);

$$R = \frac{a}{2\left(\frac{2\Delta}{bc}\right)} = \frac{abc}{4\Delta} \Rightarrow R = \frac{abc}{4\Delta}.$$

In-circle or Inscribed Circle of a Triangle

The circle that can be inscribed with in a triangle so as to touch each of its sides is called its inscribed circle or In-circle. The centre of this circle is the points of intersection of bisectors of the angles of the triangle. The radius of the circle is always denoted by 'r' and is equal to the length of perpendicular from its centre to any one of the sides of triangle.

In-radius The radius of the inscribed circle of a triangle is called the in-radius. It is denoted by 'r' and is given by

(i) $r = \dfrac{\Delta}{s}$

(ii) $r = (s-a)\tan\dfrac{A}{2} = (s-b)\tan\dfrac{B}{2} = (s-c)\tan\dfrac{C}{2}$.

(iii) $r = \dfrac{a\sin B/2 \cdot \sin C/2}{\cos A/2}$

$r = \dfrac{b\sin C/2 \cdot \sin A/2}{\cos B/2} \Rightarrow r = \dfrac{c\sin A/2 . \sin B/2}{\cos C/2}$

(iv) $r = 4R\sin A/2 \sin B/2 \sin C/2$

Proof Let the internal bisectors of the angles of the triangle ABC meet at I. Suppose the circle touches the sides BC, CA and AB at D, E and F, respectively.

Then, ID, IE, IF are perpendicular to these sides and $ID = IE = IF = r$.

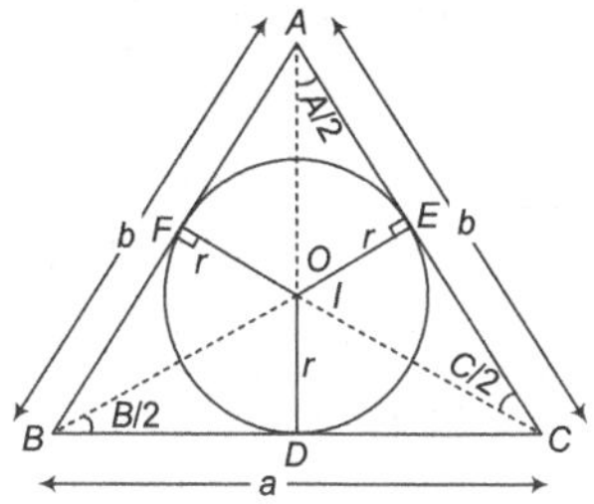

(i) We have, area of ΔABC = area of ΔIBC + area of ΔIAB + area of ΔICA

$$\Delta = \frac{1}{2}ar + \frac{1}{2}cr + \frac{1}{2}br$$

$$\Delta = \frac{1}{2}(a+b+c)r = sr \qquad \left[\text{as; } s = \frac{a+b+c}{2}\right]$$

$\Rightarrow \quad \Delta = sr$

or $\quad r = \dfrac{\Delta}{s}$.

(ii) Since, the lengths of the tangents to a circle from a given points are equal, therefore

$AE = AF, BD = BF$ and $CD = CE$. ...(I)

Now, $\quad 2s = a + b + c = BC + CA + AB$

$= (BD + DC) + (CE + EA) + (AF + FB)$

$= (BD + BF) + (AE + AF) + (CD + CE)$

$= 2(BD + AE + CD) = 2(BC + AE) = 2(a + AE)$

$\Rightarrow \quad s = a + AE$

$\Rightarrow \quad AE = (s-a)$

Now, in ΔIAE,

$$\tan\frac{A}{2} = \frac{r}{AE}$$

$\Rightarrow \quad r = AE\tan(A/2) = (s-a)\tan A/2$

$\therefore \quad r = (s-a)\tan A/2$

Similarly, $r = (s-b)\tan B/2$ and $r = (s-c)\tan C/2$

Hence, $\quad r = (s-a)\tan A/2$

$= (s-b)\tan B/2 = (s-c)\tan C/2$

(iii) In ΔIBD and ΔICD, we have,

$$\tan B/2 = \frac{r}{BD} \text{ and } \tan\frac{C}{2} = \frac{r}{CD}$$

$$\therefore \quad BD = \frac{r}{\tan B/2} \text{ and } CD = \frac{r}{\tan C/2}$$

Now, $a = BD + CD$

$$\Rightarrow \quad a = \frac{r}{\tan\left(\frac{B}{2}\right)} + \frac{r}{\tan\left(\frac{C}{2}\right)}$$

$$\Rightarrow \quad a = r\left[\frac{\cos B/2}{\sin B/2} + \frac{\cos C/2}{\sin C/2}\right]$$

$$\Rightarrow \quad a = r\left[\frac{\cos B/2 \cdot \sin C/2 + \sin B/2 \cdot \cos C/2}{\sin B/2 \cdot \sin C/2}\right]$$

$$a = \frac{r\sin(B/2 + C/2)}{\sin B/2 \cdot \sin C/2} \quad [\because A + B + C = \pi]$$

$$\therefore \quad \sin(B/2 + C/2) = \sin\left(\frac{\pi}{2} - \frac{A}{2}\right) = \cos A/2.$$

$$\Rightarrow \quad a = \frac{r\cos A/2}{\sin B/2 \cdot \sin C/2}$$

$$\therefore \quad r = \frac{a\sin B/2 \cdot \sin C/2}{\cos A/2}, r = \frac{b\sin A/2 \cdot \sin C/2}{\cos B/2} \text{ and}$$

$$r = \frac{c\sin A/2 \cdot \sin B/2}{\cos C/2}$$

(iv) We have, $r = \dfrac{a\sin B/2 \cdot \sin C/2}{\cos A/2}$ and $R = \dfrac{a}{2\sin A}$

$$\Rightarrow \quad r = \frac{2R\sin A \cdot \sin B/2 \cdot \sin C/2}{\cos A/2}$$

$$\Rightarrow \quad r = \frac{2R \cdot (2\sin A/2 \cdot \cos A/2) \cdot \sin B/2 \cdot \sin C/2}{\cos A/2}$$

$$\Rightarrow \quad r = 4R\sin A/2 \cdot \sin B/2 \cdot \sin C/2$$

Escribed Circles of a Triangle

The circle which touches the sides BC and two sides AB and AC produced of a triangle ABC is called the **Escribed circle opposite to the angle A.**

Its radius is denoted by r_1. Similarly, r_2 and r_3 denote the radii of the escribed circles opposite to the angles B and C, respectively. The centres of the escribed circles are called the ex-centres.

The centre of the escribed circles opposite to the angle A is the point of Intersection of external bisector of angles B and C. The internal bisector also passes through the same point. This centre is generally denoted by I_1.

Formulae for r_1, r_2, r_3

In any ΔABC, we have

(i) $r_1 = \dfrac{\Delta}{s-a}, r_2 = \dfrac{\Delta}{s-b}, r_3 = \dfrac{\Delta}{s-c}$

(ii) $r_1 = s\tan A/2, r_2 = s\tan B/2, r_3 = s\tan C/2$

(iii) $r_1 = \dfrac{a\cos B/2 \cdot \cos C/2}{\cos A/2}, r_2 = \dfrac{b\cos A/2 \cdot \cos C/2}{\cos B/2},$

$r_3 = \dfrac{c\cos A/2 \cdot \cos B/2}{\cos C/2}$

(iv) $r_1 = 4R\sin A/2\cos B/2.\cos C/2,$

$r_2 = 4R\cos A/2\sin B/2.\cos C/2,$

$r_3 = 4R\cos A/2\cos B/2.\sin C/2$

Proof (i) Let the ΔABC be as;

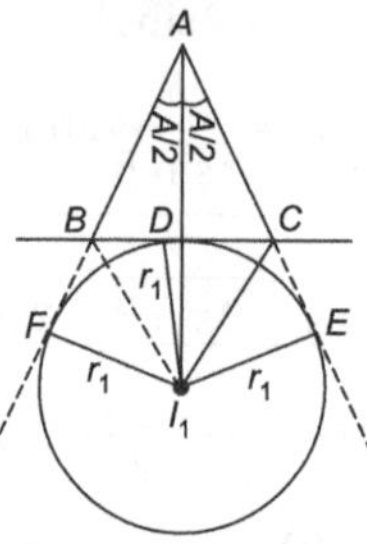

We have,

$$I_1D = I_1E = I_1F = r_1$$

Now, area of ΔABC = area of ΔI_1AC + area of ΔI_1AB − area of ΔI_1BC

$$\Rightarrow \quad \Delta = \frac{1}{2}I_1E \cdot AC + \frac{1}{2}I_1F \cdot AB - \frac{1}{2}I_1D \cdot BC$$

$$\Delta = \frac{1}{2}r_1b + \frac{1}{2}r_1c - \frac{1}{2}r_1a$$

$$\Delta = \frac{r_1}{2}(b + c - a)$$

$$\Delta = \frac{r_1}{2}(2s - 2a) \quad [\text{using } a + b + c = 2s]$$

$$\Rightarrow \quad r_1 = \frac{\Delta}{s-a}$$

Similarly, $r_2 = \dfrac{\Delta}{s-b}$ and $r_3 = \dfrac{\Delta}{s-c}$

(ii) Since, the lengths of tangents to a circle from an external points are equal,

$\therefore \quad AE = AF, BD = BF$ and $CD = CE$

Now, $AE + AF = (AC + CE) + (AB + BF)$

$= (AC + CD) + (AB + BD)$

$= AC + AB + CD + BD$

$$= AC + AB + BC$$
$$= a + b + c = 2s.$$
$$\Rightarrow \quad 2AF = 2s$$
$$\Rightarrow \quad AE = AF = s$$

In ΔI_1AF, $\tan A/2 = \dfrac{I_1F}{AF} = \dfrac{r_1}{AF}$

$$\Rightarrow \quad \tan A/2 = \frac{r_1}{s}$$
$$\Rightarrow \quad r_1 = s\tan A/2$$

Similarly, $r_2 = s\tan B/2$ and $r_3 = s\tan C/2$.

(iii) In ΔI_1BD, we have

$$\tan\left(\frac{B}{2}\right) = \frac{I_1D}{BD} = \frac{r_1}{BD}$$
$$\Rightarrow \quad BD = r_1\tan\frac{B}{2}$$

Similarly, in ΔI_1CD, we have

$$CD = r_1\tan\frac{C}{2}$$

Now, $a = BC = BD + CD = r_1\tan\dfrac{B}{2} + r_1\tan\dfrac{C}{2}$

$$= r_1\left(\tan\frac{B}{2} + \tan\frac{C}{2}\right) = r_1\frac{\cos A/2}{\cos B/2\cos C/2}$$
$$\Rightarrow \quad r_1 = \frac{a\cos B/2\cos C/2}{\cos A/2}$$

Similarly,

$$r_2 = \frac{b\cos A/2\cdot\cos C/2}{\cos B/2} \text{ and } r_3 = \frac{c\cos A/2\cdot\cos B/2}{\cos C/2}$$

(iv) We have, $r_1 = \dfrac{a\cos B/2\cdot\cos C/2}{\cos A/2}$ and $R = \dfrac{a}{2\sin A}$

$$\Rightarrow \quad r_1 = \frac{2R\sin A\cdot\cos B/2\cdot\cos C/2}{\cos A/2}$$
$$\Rightarrow \quad r_1 = \frac{4R\sin A/2\cdot\cos A/2\cdot\cos B/2\cdot\cos C/2}{\cos A/2}$$
$$\therefore \quad r_1 = 4R\sin A/2\cdot\cos B/2\cdot\cos C/2$$

Similarly, $r_2 = 4R\cos A/2\cdot\sin B/2\cdot\cos C/2$

$$r_3 = 4R\cos A/2\cdot\cos B/2\cdot\sin C/2$$

Example 12. Show that $\dfrac{b-c}{r_1} + \dfrac{c-a}{r_2} + \dfrac{a-b}{r_3} = 0$.

Sol. LHS $\dfrac{(b-c)}{r_1} + \dfrac{(c-a)}{r_2} + \dfrac{(a-b)}{r_3}$

$$\Rightarrow \quad (b-c)\cdot\left(\frac{s-a}{\Delta}\right) + (c-a)\cdot\left(\frac{s-b}{\Delta}\right) + (a-b)\cdot\left(\frac{s-c}{\Delta}\right)$$
$$\Rightarrow \quad \frac{(s-a)(b-c) + (s-b)(c-a) + (s-c)(a-b)}{\Delta}$$
$$= \frac{s(b-c+c-a+a-b) - [ab-ac+bc-ba+ac-bc]}{\Delta}$$
$$= \frac{0}{\Delta} = 0 = \text{RHS}$$

Thus, $\dfrac{b-c}{r_1} + \dfrac{c-a}{r_2} + \dfrac{a-b}{r_3} = 0$

Example 13. If $r_1 = r_2 + r_3 + r$, then prove that the triangle is right angled.

Sol. We have, $r_1 - r = r_2 + r_3$

$$\Rightarrow \quad \frac{\Delta}{s-a} - \frac{\Delta}{s} = \frac{\Delta}{s-b} + \frac{\Delta}{s-c}$$
$$\Rightarrow \quad \frac{s-s+a}{s(s-a)} = \frac{s-c+s-b}{(s-b)(s-c)}$$
$$\Rightarrow \quad \frac{a}{s(s-a)} = \frac{2s-(b+c)}{(s-b)(s-c)} \quad [\text{as, } 2s = a+b+c]$$
$$\Rightarrow \quad \frac{a}{s(s-a)} = \frac{a}{(s-b)(s-c)}$$
$$\Rightarrow \quad s^2 - (b+c)s + bc = s^2 - as$$
$$\Rightarrow \quad s(-a+b+c) = bc$$
$$\Rightarrow \quad \frac{(b+c-a)(a+b+c)}{2} = bc$$
$$\Rightarrow \quad (b+c)^2 - (a)^2 = 2bc$$
$$\Rightarrow \quad b^2 + c^2 + 2bc - a^2 = 2bc$$
$$\Rightarrow \quad b^2 + c^2 = a^2$$
$$\therefore \quad \angle A = 90^\circ$$

Example 14. Prove that $r\cot\dfrac{B}{2}\cdot\cot\dfrac{C}{2} = r_1$.

Sol. LHS $\quad r\cot B/2\cdot\cot C/2$

$$\Rightarrow \quad 4R\sin A/2\cdot\sin B/2\cdot\sin C/2\cdot\frac{\cos B/2}{\sin B/2}\cdot\frac{\cos C/2}{\sin C/2}$$

[as, $r = 4R\sin A/2\cdot\sin B/2\cdot\sin C/2$]

$$\Rightarrow \quad 4R\cdot\sin A/2\cdot\cos B/2\cdot\cos C/2$$
$$\Rightarrow \quad r_1 = \text{RHS } [\text{as, } r_1 = 4R\sin A/2\cdot\cos B/2\cdot\cos C/2]$$
$$\therefore r\cot B/2\cdot\cot C/2 = r_1$$

Example 15. In a right angled triangle, prove that $r + 2R = s$.

Sol. In a right angled triangle, the circum centre lies on the hypotenuse.

$$\Rightarrow \quad R = \frac{a}{2} \quad \text{...(i) } [\because \angle A = 90^\circ]$$

Also, $\quad r = (s-a)\tan A/2 = (s-a)\tan 45^\circ$

$$r = (s-a) \quad \text{...(ii)}$$

From Eqs. (i) and (ii), we get $r = s - 2R$

$$\Rightarrow \quad r + 2R = s.$$

Example 16. The ex-radii r_1, r_2, r_3 of a ΔABC are in HP, show that its sides a, b, c are in AP.

Solution. r_1, r_2, r_3 are in HP.

$$\Rightarrow \quad \frac{2}{r_2} = \frac{1}{r_1} + \frac{1}{r_3}$$

$$\Rightarrow \quad \frac{2(s-b)}{\Delta} = \frac{(s-a)}{\Delta} + \frac{(s-c)}{\Delta}$$

$$\Rightarrow \quad 2s - 2b = 2s - (a+c)$$

$$\Rightarrow \quad 2b = a + c$$

Hence, a, b, c are in AP.

Example 17. If A, B, C are the angles of a triangle, then prove that

$$\cos A + \cos B + \cos C = 1 + \frac{r}{R}.$$

Sol. $\cos A + \cos B + \cos C$

$$= 2\cos\left(\frac{A+B}{2}\right)\cdot\cos\left(\frac{A-B}{2}\right) + \cos C$$

$$= 2\sin\frac{C}{2}\cdot\cos\left(\frac{A-B}{2}\right) + 1 - 2\sin^2\frac{C}{2}$$

$$= 1 + 2\sin\frac{C}{2}\left[\cos\left(\frac{A-B}{2}\right) - \sin\left(\frac{C}{2}\right)\right]$$

$$= 1 + 2\sin\frac{C}{2}\left[\cos\left(\frac{A-B}{2}\right) - \cos\left(\frac{A+B}{2}\right)\right]$$

$$\left[\because \frac{C}{2} = 90^\circ - \left(\frac{A+B}{2}\right)\right]$$

$$= 1 + 2\sin\frac{C}{2}\cdot 2\sin\frac{A}{2}\cdot\sin\frac{B}{2}$$

$$= 1 + 4\sin\frac{A}{2}\cdot\sin\frac{B}{2}\cdot\sin\frac{C}{2} = 1 + \frac{r}{R}$$

[as, $r = 4R\sin A/2\cdot\sin B/2\cdot\sin C/2$]

$$\Rightarrow \quad \cos A + \cos B + \cos C = 1 + \frac{r}{R}$$

Example 18. Find the ratio of the circum-radius and the inradius of ΔABC, whose sides are in the ratio $4:5:6$.

Sol. Here, $a = 4k,\ b = 5k,\ c = 6k$

$$\therefore \quad s = \frac{15k}{2} \qquad \text{...(i)}$$

$$\therefore \quad \Delta = \sqrt{s(s-a)(s-b)(s-c)}$$

$$= \sqrt{\frac{15k}{2}\left(\frac{15k}{2} - 4k\right)\left(\frac{15k}{2} - 5k\right)\left(\frac{15k}{2} - 6k\right)}$$

$$= \frac{15\sqrt{7}}{4}k^2 \qquad \text{...(ii)}$$

and $$R = \frac{abc}{4\Delta} = \frac{4k\cdot 5k\cdot 6k}{4\cdot\frac{15}{4}\sqrt{7}k^2} \qquad \text{[using Eq. (ii)]}$$

$$\therefore \quad R = \frac{8}{\sqrt{7}}k \qquad \text{...(iii)}$$

and $$r = \frac{\Delta}{s} = \frac{15\sqrt{7}}{4}k^2\cdot\frac{2}{15k} \qquad \text{[using Eqs. (i) and (ii)]}$$

$$r = \frac{\sqrt{7}}{2}k \qquad \text{...(iv)}$$

$$\therefore \quad \frac{R}{r} = \frac{8k/\sqrt{7}}{\sqrt{7}k/2} = \frac{16}{7} \qquad \text{[using Eqs. (iii) and (iv)]}$$

$$\Rightarrow \quad R : r = 16 : 7$$

Example 19. Find the ratio of $IA : IB : IC$, where I is the incentre of ΔABC.

Sol. Here, $BD : DC = c : b$

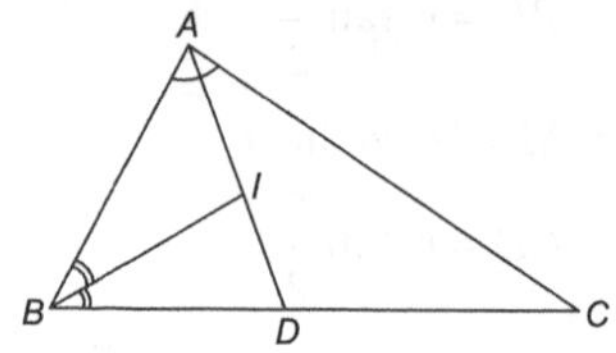

But $BD + DC = a$;

$$\therefore \quad BD = \frac{c}{b+c}\cdot a \qquad \text{...(i)}$$

In ΔABD, $$\frac{BD}{\sin A/2} = \frac{AD}{\sin B}$$

$$\therefore \quad AD = \frac{ac}{b+c}\cdot\frac{\sin B}{\sin A/2} = \frac{2\Delta}{b+c}\cdot\text{cosec } A/2 \qquad \text{...(ii)}$$

Also, $$\frac{AI}{ID} = \frac{AB}{BD} = \frac{c}{\frac{ac}{b+c}} = \frac{b+c}{a} \qquad \text{[using Eq. (i)]}$$

or $$\frac{ID}{AI} = \frac{a}{b+c}$$

On adding '1', we get

$$\frac{ID}{AI} + 1 = \frac{a}{b+c} + 1 \Rightarrow \frac{ID + AI}{AI} = \frac{a+b+c}{b+c}$$

$$\Rightarrow \quad AI = \frac{b+c}{a+b+c}\cdot AD$$

$$\therefore \quad AI = \frac{b+c}{a+b+c}\cdot\frac{2\Delta}{b+c}\cdot\text{cosec } A/2 = \frac{\Delta}{s}\cdot\text{cosec} A/2$$

Similarly, $$BI = \frac{\Delta}{s}\cdot\text{cosec } B/2$$

$$CI = \frac{\Delta}{s}\text{cosec } C/2$$

$$\Rightarrow \quad IA : IB : IC = \frac{\Delta}{s}\cdot\text{cosec } A/2 : \frac{\Delta}{s}\text{cosec } B/2 : \frac{\Delta}{s}\text{cosec } C/2$$

$$\therefore \quad IA : IB : IC = \text{cosec } A/2 : \text{cosec } B/2 : \text{cosec } C/2$$

Note

Student are advised to remember the above result i.e.
$IA = r \operatorname{cosec} A/2$, $IB = r \operatorname{cosec} B/2$, $IC = r \operatorname{cosec} C/2$.

Example 20. If the sides of a triangle are in GP and the largest angle is twice the smallest angle, then find the relation for r.

Sol. Let the sides of Δ be $a, b = ar, c = ar^2$, where $r > 1$

Here, $c = 2A$ (given)

So, $B = \pi - A - C = \pi - 3A$

$$\therefore \quad \frac{a}{\sin A} = \frac{b}{\sin B} = \frac{c}{\sin C} \Rightarrow \frac{1}{\sin A} = \frac{r}{\sin B} = \frac{r^2}{\sin C}$$

$$\Rightarrow \quad \frac{1}{\sin A} = \frac{r}{\sin 3A} = \frac{r^2}{\sin 2A}$$

$$\therefore \quad r^2 = 2\cos A \text{ and } r = \frac{\sin 3A}{\sin A} = 3 - 4\sin^2 A$$

$$r = 4\cos^2 A - 1$$

$$\therefore \quad r = r^4 - 1$$

Thus, the required relation is $r^4 - r - 1 = 0$.

Example 21. The equation $ax^2 + bx + c = 0$, where a, b, c are the sides of a ΔABC, and the equation $x^2 + \sqrt{2}x + 1 = 0$ have a common root. Find measure for $\angle C$.

Sol. Clearly, the roots of $x^2 + \sqrt{2}x + 1 = 0$ are non-real complex.

So, the one root common implies both roots are common.

So, $$\frac{a}{1} = \frac{b}{\sqrt{2}} = \frac{c}{1} = k$$

$$\therefore \quad \cos C = \frac{a^2 + b^2 - c^2}{2ab}$$

$$= \frac{k^2 + 2k^2 - k^2}{2k.\sqrt{2}k} = \frac{1}{\sqrt{2}}$$

$$\Rightarrow \quad \angle C = 45^\circ$$

Example 22. If in a ΔABC, the value of $\cot A, \cot B, \cot C$ are in AP show a^2, b^2, c^2 are in A.P.

Sol. Here; $\cot A$, $\cot B$, $\cot C$ are in AP.

$$\Rightarrow \quad \frac{b^2 + c^2 - a^2}{2bc \cdot \frac{a}{2R}}, \frac{c^2 + a^2 - b^2}{2ac \cdot \frac{b}{2R}}, \frac{a^2 + b^2 - c^2}{2ab \cdot \frac{c}{2R}} \text{ are in AP}$$

$$\Rightarrow \quad b^2 + c^2 - a^2, c^2 + a^2 - b^2, a^2 + b^2 - c^2 \text{ are in AP}$$

$$\left[\text{multiplying by } \frac{abc}{R}\right]$$

$$\Rightarrow \quad -2a^2, -2b^2, -2c^2 \text{ are in AP}$$

[subtracting $a^2 + b^2 + c^2$ from each]

$$\Rightarrow \quad a^2, b^2, c^2 \text{ are in AP.}$$

Aliter $2\cot B = \cot A + \cot C$

$$\Rightarrow \quad \frac{2(a^2 + c^2 - b^2)}{2ac \cdot kb} = \frac{b^2 + c^2 - a^2}{2bc \cdot ka} + \frac{a^2 + b^2 - c^2}{2ab \cdot kc}$$

[using sine and cosine law]

$$\Rightarrow \quad 2(a^2 + c^2 - b^2) = b^2 + c^2 - a^2 + a^2 + b^2 - c^2$$

$$\Rightarrow \quad 2(a^2 + c^2 - b^2) = 2b^2$$

$$\Rightarrow \quad a^2 + c^2 - b^2 = b^2 \quad \text{or} \quad a^2 + c^2 = 2b^2$$

i.e. a^2, b^2, c^2 are in AP.

Exercise for Session 3

1. The side of a triangle are 22 cm, 28 cm and 36 cm. So, find the area of the circumscribed circle.
2. If the lengths of the side of a triangle are 3, 4 and 5 units, then find the circum radius R.
3. In an equilateral triangle of side $2\sqrt{3}$ cm. The find circum-radius.
4. If $8R^2 = a^2 + b^2 + c^2$, then prove that the Δ is right angled.
5. In a ΔABC, show that $2R^2 \sin A \sin B \sin C = \Delta$.
6. In a ΔABC, show that $\dfrac{a\cos A + b\cos B + c\cos C}{a + b + c} = \dfrac{r}{R}$
7. If the sides of a triangles are 3 : 7 : 8, then find ratio $R : r$.
8. In an equilateral triangle show that the in-radius and the circum-radius are connected by $r = \dfrac{R}{2}$.
9. In any ΔABC, find $\sin A + \sin B + \sin C$.

10. In any ΔABC, show that $\cos A + \cos B + \cos C = \left(1 + \frac{r}{R}\right)$.

11. If the side be a, b and c, then show that $\frac{r_1 + r}{a} + \frac{r_2 - r}{b} = \frac{c}{r_3}$.

12. Show that $r_2 r_3 + r_3 r_1 + r_1 r_2 = s^2$

13. Show that $(r_1 + r_2)(r_2 + r_3)(r_3 + r_1) = 4Rs^2$

14. If $r_1 = r_2 + r_3 + r$, then show that Δ is right angled.

15. In an equilateral triangle, show that the in-radius, circumradius and one of the ex-radii are in the ratio 1 : 2 : 3.

16. Show that $\left(\frac{1}{r} - \frac{1}{r_1}\right)\left(\frac{1}{r} - \frac{1}{r_2}\right)\left(\frac{1}{r} - \frac{1}{r_3}\right) = \frac{16R}{r^2(\Sigma a)^2}$

17. If r_1, r_2, r_3 in a triangle be in HP, then show that the sides are in AP.

18. In a ΔABC, show that $r_1 r_2 r_3 = \Delta^2$.

19. If l_1, l_2, l_3 are respectively the perpendicular from the vertices of a triangle on the opposite side, then show that $l_1 l_2 l_3 = \frac{a^2 b^2 c^2}{8R}$.

20. If the angle of a triangle are in the ratio 1 : 2 : 3, then show that the sides opposite to the respective angle are in the ratio $1 : \sqrt{3} : 2$.

21. Show that, $4Rr \cos\frac{A}{2} \cos\frac{B}{2} \cos\frac{C}{2} = S$

22. If $(a - b)(s - c) = (b - c)(s - a)$, then show that r_1, r_2, r_3 are in HP.

23. To show that $\frac{1}{r_1^2} + \frac{1}{r_2^2} + \frac{1}{r_3^2} + \frac{1}{r^2} = \frac{\Sigma a^2}{S^2}$

24. Show that $(r_1 - r)(r_2 - r)(r_3 - r) = 4Rr^2$

25. Show that $\left(\frac{1}{r_1} + \frac{1}{r_2}\right)\left(\frac{1}{r_2} + \frac{1}{r_3}\right)\left(\frac{1}{r_3} + \frac{1}{r_1}\right) = \frac{64R^3}{a^2 b^2 c^2}$

26. If the sides be a, b and c, then find the value of $(r + r_1)\tan\frac{B - C}{2} + (r + r_2)\tan\frac{C - A}{2} + (r + r_3)\tan\frac{A - B}{2}$

27. If the sides be a, b, c, then find value of $\frac{b - c}{r_1} + \frac{c - a}{r_2} + \frac{a - b}{r_3}$.

28. If the sides be a, b, c, then find $(r_1 - r)(r_2 + r_3)$.

29. If a, b, c are in AP, then show that r_1, r_2, r_3 are in HP.

30. Show that $\frac{r_1}{bc} + \frac{r_2}{ca} + \frac{r_3}{ab} = 2R - r$

31. Show that $r_1 + r_2 = c \cot\left(\frac{C}{2}\right)$

32. Show that $Rr(\sin A + \sin B + \sin C) = \Delta$

33. Show that $16R^2 r\, r_1 r_2 r_3 = a^2 b^2 c^2$

34. If $\frac{r}{r_1} = \frac{r_2}{r_3}$, then show that $c = 90°$.

Session 4

Orthocentre and Its Distance from the Angular Points of a Triangle and Pedal Triangle and Centroid of Triangle

Orthocentre and Its Distance from the Angular Points of a Triangle and Pedal Triangle

Let ABC be any triangle, and let AK, BL and CM be the perpendicular from A, B and C upon the opposite sides of the triangle. These three perpendiculars meet at a point O' which is called the orthocentre of the triangle ABC. **The triangle KLM, formed by joining the feet of these perpendiculars is called the pedal triangle of ABC.**

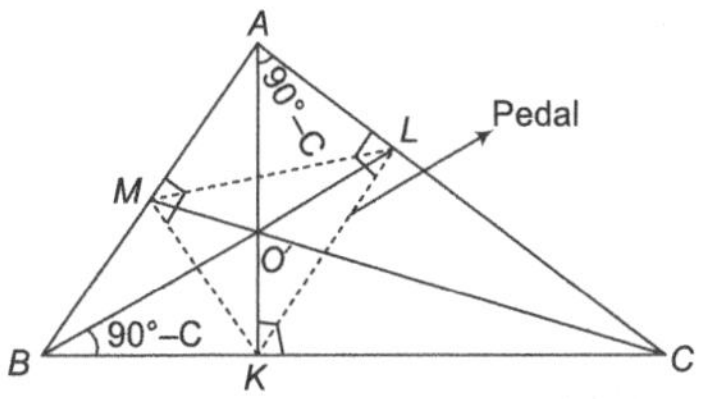

In $\Delta O'BK$, $\tan(90° - C) = \dfrac{O'K}{KB}$

$$\Rightarrow \quad O'K = KB.\tan(90° - C) = KB \cot C$$

$$= AB \cos B \cot C \left[\because \text{from } \Delta ABK, \cos B = \frac{BK}{AB}\right]$$

$$= c \cdot \cos B \cdot \frac{\cos C}{\sin C}$$

$$\Rightarrow O'K = 2R \cos B \cos C \left[\because R = \frac{a}{2\sin A} = \frac{b}{2\sin B} = \frac{c}{2\sin C}\right]$$

Similarly,

$$O'L = 2R \cos A \cos C \text{ and } O'M = 2R \cos A \cos B$$

In $\Delta AO'L$, $\cos(90° - C) = \dfrac{AL}{AO'}$

$$\Rightarrow \quad AO' = AL \cdot \operatorname{cosec} C \Rightarrow AO' = c \cos A \cdot \operatorname{cosec} C$$

$$\left[\because \text{from } \Delta ALB, \cos A = \frac{AL}{AB}\right]$$

$$\Rightarrow \quad AO' = c \cdot \cos A \cdot \frac{1}{\sin C}$$

$$\Rightarrow \quad AO' = 2R \cos A$$

Similarly, $BO' = 2R \cos B$ and $CO' = 2R \cos C$

Thus, the distance of the orthocentre of the triangle from the angular points are,

$$AO' = 2R \cos A$$
$$BO' = 2R \cos B$$
$$CO' = 2R \cos C,$$

and its distance from the sides are,

$$O'K = 2R \cos B \cos C.$$
$$O'L = 2R \cos C \cos A.$$
$$O'M = 2R \cos A \cos B.$$

Some Relations between Orthocentre, Incentre, Escribed Circles, Centroid, Circum-centre and Pedal Triangle

(i) Orthocentre of the triangle is the incentre of the pedal triangle.

(ii) If I_1, I_2 and I_3 be the centres of escribed circles which are opposite to A, B and C respectively and I is the centre of incircle then ΔABC is the pedal triangle of the $\Delta I_1 I_2 I_3$ and I is the orthocentre of the $\Delta I_1 I_2 I_3$.

(iii) The centroid of the triangle lies on the line joining the circumcentre to the orthocentre and divides it in the ratio 1 : 2.

(iv) Circle circumscribing the pedal triangle of a given triangle bisects the sides of the given triangle and also the lines joining the vertices of the given triangle to the orthocentre of the given triangle. This circle is known as **nine point circle.**

(v) Circum-centre of the pedal triangle of a given triangle bisects the line joining the circumcentre of the triangle to the **orthocentre.**

Example 23. In ΔABC, a, b and c represents the sides, thus find the sides and angles of the pedal triangle.

Sol. Let ΔABC be any triangle and let D, E, F be the feet of perpendicular from the angular points on the opposite sides of the ΔABC, then the ΔDEF is known as **Pedal triangle of ABC.**

Here, $\angle HDC$ and $\angle HEC$ are 90° each. Thus, points H, D, C and E are concyclic,

$$\therefore \quad \angle HDE = \angle HCE = 90^\circ - A \quad \ldots(i)$$

Similarly H, D, B, F are concyclic.

$$\therefore \quad \angle HDF = \angle HBF = 90^\circ - A \quad \ldots(ii)$$

Hence, $\angle FDE = 180^\circ - 2A$ [using Eqs. (i) and (ii)]

So, $\angle DEF = 180^\circ - 2B$

and $\angle EFD = 180^\circ - 2C$

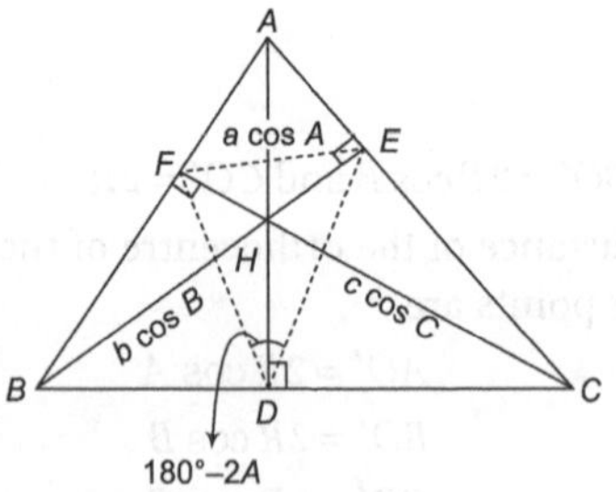

Thus, angles of pedal triangle FDE are

$$180^\circ - 2A,\ 180^\circ - 2B,\ 180^\circ - 2C.$$

Again in ΔBFD,

$$\angle FDB = 90^\circ - \angle HDF = 90^\circ - (90^\circ - A) = A$$

$$\therefore \quad \frac{FD}{\sin B} = \frac{BF}{\sin A}$$

$$\therefore \quad FD = \frac{\sin B}{\sin A} \cdot BF$$

$$= \frac{\sin B}{\sin A} \cdot BC \cdot \cos B \quad \left[\because \text{In } \Delta BFC, \frac{BF}{BC} = \cos B\right]$$

$$= \frac{a \sin B \cdot \cos B}{\sin A} = 2R \sin B \cdot \cos B$$

$$FD = b \cos B$$

Similarly, $EF = a \cos A$ and $DE = c \cos C$

$\Rightarrow$ Sides of pedal triangle :

$a \cos A$, $b \cos B$ and $c \cos C$ or $R \sin 2A$, $R \sin 2B$ and $R \sin 2C$

Note

If given ΔABC is obtuse, then angles are represented by $2A$, $2B$, $2C - 180^\circ$ and the sides are $a\cos A$, $b\cos B$, $-c\cos C$.

Example 24. Find the area, circum-radius and in-radius of the pedal triangle.

Sol. We know, area of $\Delta = \frac{1}{2}$ (product of the sides) $\times$ (sine of the included angle)

$$= \frac{1}{2}(R \sin 2B)(R \sin 2C) \cdot \sin(180^\circ - 2A)$$

$$= \frac{1}{2} R^2 \cdot \sin 2A \cdot \sin 2B \cdot \sin 2C$$

$$\text{The circum-radius} = \frac{EF}{2 \sin FDE} = \frac{R \sin 2A}{2 \sin(180^\circ - 2A)} = \frac{R}{2}$$

The in-radius of the pedal ΔDEF

$$= \frac{\text{ar}(\Delta DEF)}{\text{Semi-perimeter of } \Delta DEF}$$

$$= \frac{1}{2} \cdot \frac{R^2 \sin 2A \cdot \sin 2B \cdot \sin 2C}{2R \sin A \cdot \sin B \cdot \sin C}$$

$$= 2R \cos A . \cos B . \cos C$$

Thus, area of pedal $\Delta = \frac{1}{2} R^2 \sin 2A \cdot \sin 2B \sin 2C$

Circum-radius $= \frac{R}{2}$

In-radius $= 2R \cos A \cdot \cos B \cdot \cos C$

Centroid of Triangle

In ΔABC, the mid-points of the sides BC, CA and AB are D, E and F, respectively. The lines, AD, BE and CF are called medians of the triangle ABC, the points of concurrency of three medians is called centroid. Generally, it is represented by G.

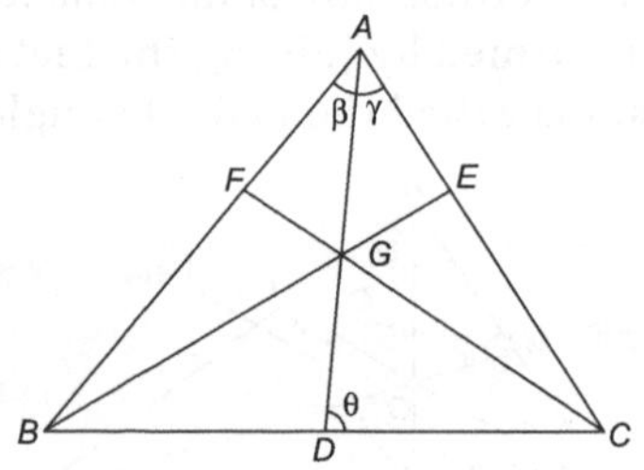

By analytical geometry :

$$AG = \frac{2}{3} AD \ ; \ BG = \frac{2}{3} BE \text{ and } CG = \frac{2}{3} CF$$

Length of Medians and the Angles that the Median Makes with Sides

In above figure, $AD^2 = AC^2 + CD^2 - 2AC \cdot CD \cdot \cos C$

$$\therefore \quad AD^2 = b^2 + \frac{a^2}{4} - ab \cos C$$

$$\therefore \quad AD^2 = b^2 + \frac{a^2}{4} - ab \cdot \left(\frac{b^2 + a^2 - c^2}{2ab}\right)$$

$$AD^2 = \frac{2b^2 + 2c^2 - a^2}{4}$$

$$\Rightarrow \quad \mathbf{AD = \frac{1}{2}\sqrt{2b^2 + 2c^2 - a^2}}$$

$$\text{or} \quad AD = \frac{1}{2}\sqrt{b^2 + c^2 + 2bc \cos A}$$

$$\text{Similarly,} \quad \mathbf{BE = \frac{1}{2}\sqrt{2c^2 + 2a^2 - b^2}}$$

and $$CF = \frac{1}{2}\sqrt{2a^2 + 2b^2 - c^2}$$

Let $\angle BAD = \beta$ and $\angle CAD = \gamma$, we have

$$\frac{\sin\gamma}{\sin C} = \frac{DC}{AD} = \frac{a}{2 \cdot \frac{1}{2}\sqrt{2b^2 + 2c^2 - a^2}}$$

$$\therefore \quad \sin\gamma = \frac{a\sin C}{\sqrt{2b^2 + 2c^2 - a^2}}$$

Similarly, $\sin\beta = \dfrac{a\sin B}{\sqrt{2b^2 + 2c^2 - a^2}}$

again $$\frac{\sin\theta}{\sin C} = \frac{b}{\frac{1}{2}\sqrt{2b^2 + 2c^2 - a^2}}$$

$$\therefore \quad \sin\theta = \frac{2b\sin C}{\sqrt{2b^2 + 2c^2 - a^2}}$$

Ex-central Triangle

Let ABC be a triangle and I be the centre of incircle. Let I_1, I_2 and I_3 be the centres of the escribed circles which are opposite to A, B, C respectively, then I_1, I_2, I_3 is called the Ex-central triangle of ΔABC.

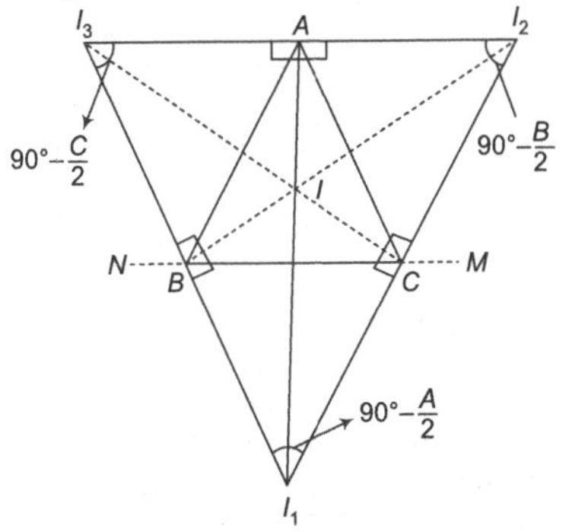

By geometry IC bisects the $\angle ACB$ and I_2C bisects the $\angle ACM$.

$$\therefore \quad \angle ICI_2 = \angle ACI + \angle ACI_2$$
$$= \frac{1}{2}\angle ACB + \frac{1}{2}\angle ACM$$
$$= \frac{1}{2}(180°) = 90°$$

Similarly, $\angle ICI_1 = 90°$

Hence, I_1I_2 is perpendicular to IC.

Similarly, AI is perpendicular to I_2I_3 and BI is perpendicular to I_1I_3.

Hence, $I_1I_2I_3$ is a triangle, thus the triangle ABC is the pedal triangle of its ex-central triangle $I_1I_2I_3$.

Sides and Angles of the Ex-central Triangle

In above figure,

$$\angle BI_1C = \angle BI_1I + \angle CI_1I$$
$$= \angle BCI + \angle CBI$$
$$= \frac{C}{2} + \frac{B}{2} = 90° - \frac{A}{2}$$

or $$\angle I_2I_1I_3 = 90° - \frac{A}{2}$$

Thus the angles are,

$$90° - \frac{A}{2}, 90° - \frac{B}{2}, 90° - \frac{C}{2}$$

Again in right angled ΔI_1I_3C,

$$\frac{I_1C}{I_1I_3} = \cos\left(90° - \frac{A}{2}\right)$$

$$I_1C = I_1I_3 \sin\frac{A}{2} \quad \text{...(i)}$$

In ΔBI_1C, $$\frac{I_1C}{\sin\angle I_1BC} = \frac{BC}{\sin\angle BI_1C}$$

$$\Rightarrow \quad \frac{I_1I_3 \sin A/2}{\sin\left(\frac{180° - B}{2}\right)} = \frac{a}{\sin(90° - A/2)}$$

$$\Rightarrow \quad I_1I_3 = \frac{a\cos(B/2)}{\sin A/2 \cdot \cos A/2} = \frac{2R \cdot \sin A \cdot \cos(B/2)}{\frac{1}{2}\left[2\sin\left(\frac{A}{2}\right) \cdot \cos\left(\frac{A}{2}\right)\right]}$$

$$\therefore \quad \mathbf{I_1I_3 = 4R\cos\frac{B}{2}}$$

Similarly, $$\mathbf{I_1I_2 = 4R\cos\frac{C}{2}}$$

$$\mathbf{I_2I_3 = 4R\cos\frac{A}{2}}$$

Area and Circum-radius of the Ex-central Triangle

$\because$ Area of triangle $= \frac{1}{2}$ (Product of two sides) $\times$ (Sine of include angles)

$$= \frac{1}{2}(4R\cos B/2) \cdot (4R\cos C/2) \times \sin(90° - A/2)$$

$$\Delta = 8R^2 \cos A/2 \cdot \cos B/2 \cdot \cos C/2$$

The circum-radius $\dfrac{I_2I_3}{2\sin I_2I_1I_3} = \dfrac{4R\cos A/2}{2\sin(90° - A/2)} = 2R.$

$\Rightarrow$ **Circum-radius $= 2R$.**

Distance between the In-centre and Ex-centres

Here, $\angle BI I_1 = \angle CI I_1 = 90°$

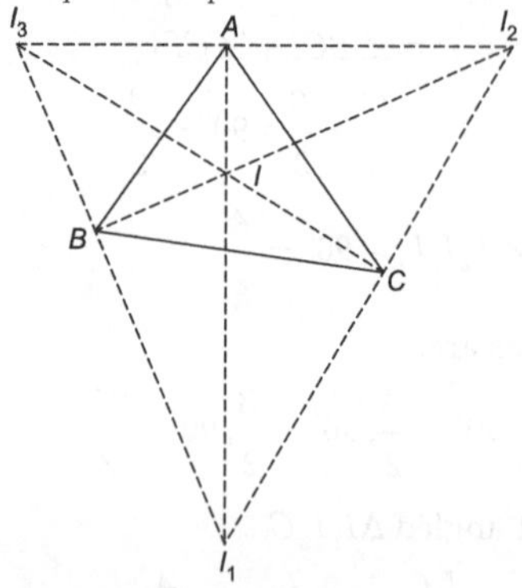

$\therefore$ II_1 is the diameter of the circumcircle of ΔBCI_1

$$\therefore \quad II_1 = \frac{BC}{\sin BI_1C} = \frac{a}{\sin(90° - A/2)} = \frac{2R\sin A}{\cos A/2}$$

$$II_1 = 4R \cdot \sin\left(\frac{A}{2}\right)$$

Similarly, $\quad II_2 = 4R\sin\left(\frac{B}{2}\right)$

and $\quad II_3 = 4R\sin\left(\frac{C}{2}\right)$

Further, $\quad \angle BI_1I = \angle BCI = \frac{C}{2}$

$$\therefore \quad BI = II_1 \sin\frac{C}{2}$$

$\Rightarrow$ $\quad \mathbf{BI = 4R\sin A/2 \cdot \sin C/2}$

Similarly $\quad \mathbf{AI = 4R\sin B/2 \cdot \sin C/2}$

$\quad \mathbf{CI = 4R\sin A/2 \cdot \sin B/2}$

Distance between an Ex-centre and Circum Centre

Let O be the circum centre and I be the in-centre, then AI produced passes through the ex-centre I_1.

Let AI, meet the circum-circle in D, join CI, BI, CD, BD, CI_1, BI_1.

Draw I_1E_1 perpendicular to AC. Produce I_1O to meet the circle in L join CL.

The angle $\angle BI_1$ and $\angle CI_1$ are right angles, hence the circle on I_1I as diameter passes through B and C.

The chord BD and CD of the circum-centre subtend equal angles at A and are therefore, equal.

$$\therefore \quad DB = DC = DI$$

Hence D is the centre of the circle IBI_1C.

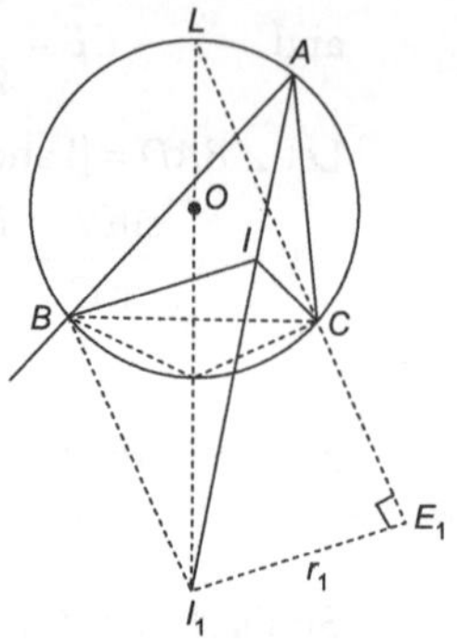

$$\therefore \quad DI_1 = DC = 4R\sin\frac{A}{2}$$

Now, $OI_1^2 - R^2 =$ Square of tangent from $I_1 = I_1D \cdot I_1A$

$$= 2R\sin\frac{A}{2} \cdot r_1 \text{cosec}\,\frac{A}{2}$$

$$\therefore \quad OI_1^2 = R^2 + 2Rr_1$$

or

$$\mathbf{OI_1 = R\sqrt{1 + 8\sin A/2 \cdot \cos B/2 \cdot \cos C/2}}$$

Similarly, $\quad OI_2 = \sqrt{R^2 + 2Rr_2}$

$$\mathbf{OI_2 = R\sqrt{1 + 8\cos A/2 \cdot \sin B/2 \cdot \cos C/2}}$$

and $\quad OI_3 = \sqrt{R^2 + 2Rr_3}$

$$\mathbf{OI_3 = R\sqrt{1 + 8\cos A/2 \cdot \cos B/2 \cdot \sin C/2}}$$

Example 25. Show that $II_1 \cdot II_2 \cdot II_3 = 16R^2r$.

Sol. Since, $\quad II_1 = 4R\sin A/2$

$II_2 = 4R\sin B/2$

and $\quad II_3 = 4R\sin C/2$

$$\therefore \quad II_1 \cdot II_2 \cdot II_3 = 64R^3 \cdot \sin\frac{A}{2} \cdot \sin\frac{B}{2} \cdot \sin\frac{C}{2}$$

Since $\quad r = 4R\sin A/2 \cdot \sin B/2 \cdot \sin C/2$

$$\therefore \quad II_1 \cdot II_2 \cdot II_3 = 64R^3 \cdot \frac{r}{4R} = 16R^2r$$

Example 26. Prove that

$$\frac{II_1 \cdot I_2I_3}{\sin A} = \frac{II_2 \cdot I_3I_1}{\sin B}$$

Sol. LHS $\quad \frac{II_1 \cdot I_2I_3}{\sin A} = \frac{4R\sin A/2 \cdot 4R\cos A/2}{\sin A}$

$$= \frac{16R^2\sin A/2 \cdot \cos A/2}{2\sin A/2 \cdot \cos A/2} = 8R^2$$

RHS $\quad \frac{II_2 \cdot I_3I_1}{\sin B} = \frac{4R\sin B/2 \cdot 4R\cos B/2}{\sin B}$

$$= \frac{16R^2\sin B/2 \cdot \cos B/2}{2\sin B/2 \cdot \cos B/2}$$

$$= 8R^2$$

$\Rightarrow \quad$ LHS = RHS

Example 27. If g, h, k denotes the side of a pedal triangle, then prove that

$$\frac{g}{a^2} + \frac{h}{b^2} + \frac{k}{c^2} = \frac{a^2 + b^2 + c^2}{2abc}$$

Sol. We have,

$$g = a\cos A, h = b\cos B, k = c\cos C \quad [\because \text{ sides of pedal } \Delta]$$

$$\therefore \quad \frac{g}{a^2} + \frac{h}{b^2} + \frac{k}{c^2} = \frac{\cos A}{a} + \frac{\cos B}{b} + \frac{\cos C}{c}$$

$$= \frac{b^2 + c^2 - a^2}{2abc} + \frac{a^2 + c^2 - b^2}{2abc} + \frac{a^2 + b^2 - c^2}{2abc}$$

$$= \frac{a^2 + b^2 + c^2}{2abc}$$

$$\therefore \quad \frac{g}{a^2} + \frac{h}{b^2} + \frac{k}{c^2} = \frac{a^2 + b^2 + c^2}{2abc}$$

Example 28. If x, y, z are perpendicular from the circum centre of the sides of the ΔABC respectively.

Prove that $\frac{a}{x} + \frac{b}{y} + \frac{c}{z} = \frac{abc}{4xyz}$

Sol. In ΔOBM, $\tan A = \frac{\frac{a}{2}}{x} = \frac{a}{2x}$

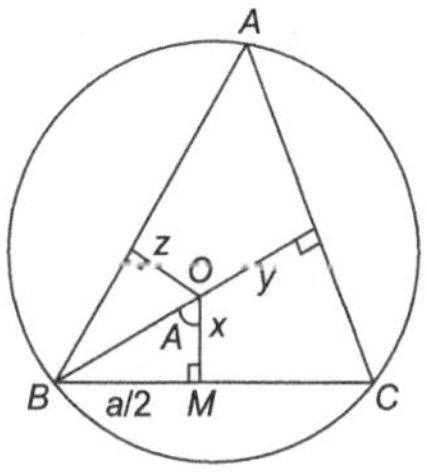

Similarly, $\tan B = \frac{b}{2y}$

and $\tan C = \frac{c}{2z}$

Since, $A + B = \pi - C$

$\Rightarrow \quad \tan(A + B) = \tan(\pi - C)$

$\Rightarrow \quad \tan A + \tan B + \tan C = \tan A \cdot \tan B \cdot \tan C$

$$\Rightarrow \quad \frac{a}{2x} + \frac{b}{2y} + \frac{c}{2z} = \frac{abc}{8xyz}$$

$$\Rightarrow \quad \frac{a}{x} + \frac{b}{y} + \frac{c}{z} = \frac{abc}{4xyz}$$

Example 29. If O, H and G represents circum centre, orthocentre and centroid respectively, then show $HG : GO = 2 : 1$. We have,

Sol. We have, two Δ's AGH and GMO are equiangular.

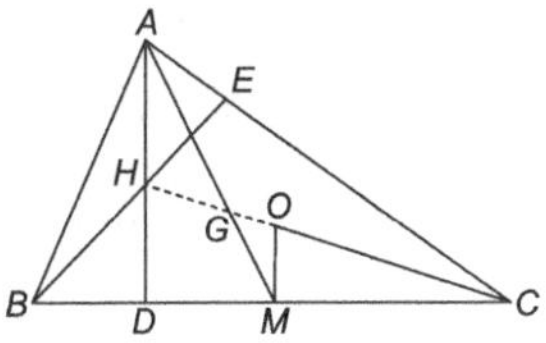

Also, $AH = 2R\cos A$

$OM = R\cos A$

$$\therefore \quad \frac{AH}{OM} = \frac{2R\cos A}{R\cos A} = \frac{2}{1}$$

Hence by similar Δ's,

$$\frac{AH}{OM} = \frac{AG}{GM} = \frac{HG}{GO} = 2$$

$\therefore \Rightarrow$ G divides HO in the ratio of $2 : 1$

or $\quad HG : GO = 2 : 1$

Example 30. Prove that the distance between the circumcentre and the orthocentre of a triangle ABC is $R\sqrt{1 - 8\cos A\cos B\cos C}$.

Sol. Let O and P be the circumcentre and the orthocentre, respectively.

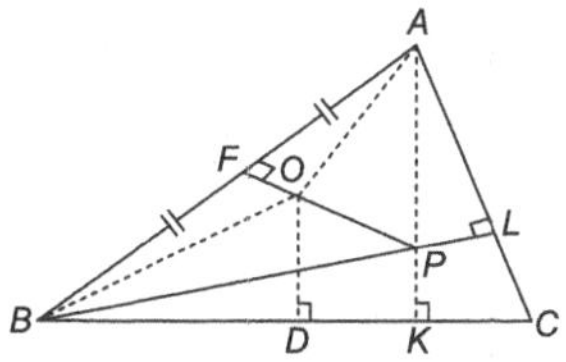

If OF is the perpendicular to AB, we have

$$\angle OAF = 90^\circ - \angle AOF = 90^\circ - C$$

Also, $\angle OAP = A - \angle OAF - \angle PAL$

$$= A - 2(90^\circ - C)$$

$$= A + 2C - 180^\circ$$

$$= A + 2C - (A + B + C)$$

$$= C - B$$

Also, $OA = R$ and $PA = 2R\cos A$

Now in ΔAOP, $OP^2 = OA^2 + PA^2 - 2OA.PA.\cos\angle OAP$

$$= R^2 + 4R^2\cos^2 A - 4R^2\cos A\cos(C - B)$$

$$= R^2 + 4R^2\cos A\{\cos A - \cos(C - B)\}$$

$$= R^2 - 4R^2\cos A\{\cos(B + C) + \cos(C - B)\}$$

$$= R^2 - 8R^2\cos A\cos B\cos C$$

Hence, $\quad OP = R\sqrt{1 - 8\cos A\cos B\cos C}$

Example 31. Find the distance between the circumcentre and the incentre of the ΔABC.

Sol. Let O be circumcentre and OF be the perpendicular to AB. Let I be the incentre and IE be the perpendicular to AC.

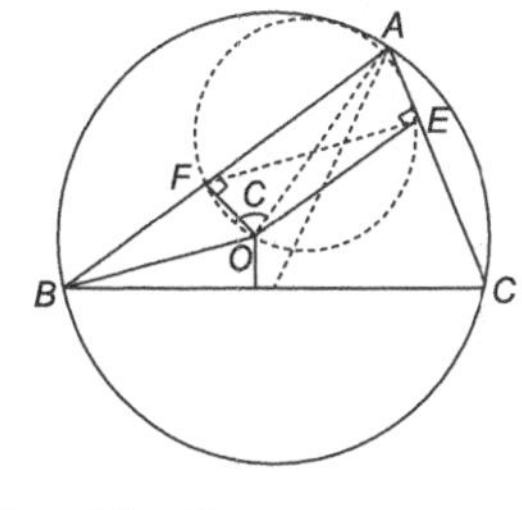

Then, $\angle OAF = 90^\circ - C$

$\Rightarrow \quad \angle OAI = \angle IAF - \angle OAF$

$$= \frac{A}{2} - (90^\circ - C)$$

$$= \frac{A}{2} + C - \left(\frac{A + B + C}{2}\right) = \frac{C - B}{2}$$

Also, $\quad AI = \frac{IE}{\sin A/2} = \frac{r}{\sin A/2}$

$$= 4R\sin B/2\sin C/2$$

Hence,

$$OI^2 = OA^2 + AI^2 - 2OA \cdot AI \cdot \cos\angle OAI$$

$$= R^2 + 16R^2 \sin^2 B/2 \cdot \sin^2 C/2 - 8R^2 \sin B/2 \sin C/2 \cos\left(\frac{C-B}{2}\right)$$

$$\Rightarrow \frac{OI^2}{R^2} = 1 + 16\sin^2\frac{B}{2}\cdot\sin^2\frac{C}{2} - 8\sin\frac{B}{2}\sin\frac{C}{2}\left(\cos\frac{B}{2}\cos\frac{C}{2} + \sin\frac{B}{2}\cdot\sin\frac{C}{2}\right)$$

$$= 1 - 8\sin\frac{B}{2}\cdot\sin\frac{C}{2}\left(\cos\frac{B}{2}\cdot\cos\frac{C}{2} - \sin\frac{B}{2}\cdot\sin\frac{C}{2}\right)$$

$$= 1 - 8\sin\frac{B}{2}\cdot\sin\frac{C}{2}\cdot\cos\left(\frac{B+C}{2}\right)$$

$$= 1 - 8\sin\frac{B}{2}\cdot\sin\frac{C}{2}\cdot\sin\frac{A}{2}$$

$$\frac{OI}{R} = \sqrt{1 - 8\sin B/2 \cdot \sin C/2 \cdot \sin A/2}$$

or $$OI = R\sqrt{1 - \frac{2r}{R}}$$

$$OI = \sqrt{R^2 - 2Rr}.$$

Exercise for Session 4

1. If H is the orthocentre of the ΔABC, then find AH.
2. A circle touches two of the smaller sides of a ΔABC $(a < b < c)$ and has its centre on the greatest side. Then, find the radius of the circles.
3. If the sides be a, b, c, then show that $a\cos A + b\cos B + c\cos C = 4R\sin A\sin B\sin C$
4. If the altitudes of a triangle be 3, 4, 6, then find its in-radius.
5. In a ΔABC, if $a = 3, b = 4, c = 5$, then find the distance between its incentre and circumcentre.
6. If p_1, p_2, p_3 are respectively the perpendicular from the vertices of a triangle to the opposite sides, then find the value of $p_1 p_2 p_3$.
7. Show that the distance between the circumcentre and the incentre of the triangle ABC is $\sqrt{R^2 - 2Rr}$.
8. Show that the distance between the circumcentre and the orthocentre of a triangle ABC is $R\sqrt{1 - 8\cos A\cos B\cos C}$.
9. If in a ΔABC, AD, BE and CF are the altitudes and R is the circumradius, then find the radius of the DEF.
10. If I, I_1, I_2 and I_3 be respectively the centre of the in-circle and the three escribed circles of a ΔABC, then find $I_2 I_3$.

Session 5

Regular Polygons and Radii of the Inscribed and Circumscribing Circle a Regular Polygon

A regular polygon is a polygon which has all its sides as well as all its angle equal.

If the polygon has 'n' sides, Sum of the internal angles is $(n-2)\pi$ and each angle is $\frac{(n-2)\pi}{n}$.

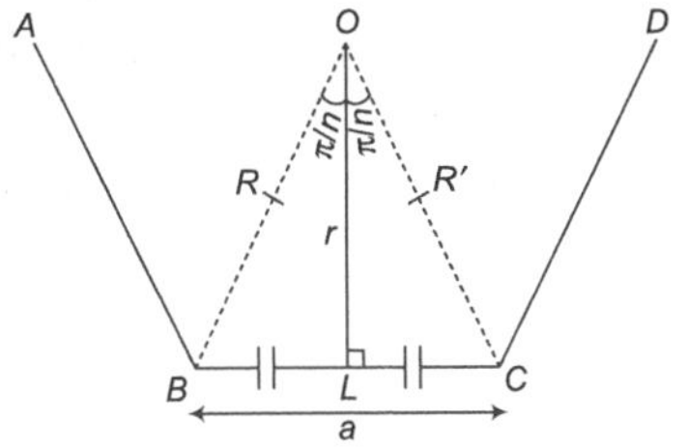

Let AB, BC and CD be three consecutive sides of the regular polygon and n be the number of its sides. Let O be the point of intersection of the bisector of the angles $\angle ABC$ and $\angle BCD$.

The point O is both the incentre and circumcentre of polygon and so $BL = LC$. Hence we have,

$OB = OC = R$, the radius of the circumcircle and $OL = r$, the radius of the incircle.

It can be seen,

$$\boldsymbol{R = \frac{a}{2}\operatorname{cosec}\left(\frac{\pi}{n}\right)} \text{ and } \boldsymbol{r = \frac{a}{2}\cot\left(\frac{\pi}{n}\right)}$$

where 'a' is length of a side of the polygon.

The area of the polygon $= n$ area of ΔOBC

$$= n(OL)(BL) = n \cdot \frac{a}{2} \cdot \cot\left(\frac{\pi}{n}\right) \cdot \frac{a}{2}$$

$$= \boldsymbol{\frac{1}{4} na^2 \cdot \cot\left(\frac{\pi}{n}\right)}$$

Also, the area of the polygon.

$$= n(OL)(BL) = n(OL)(OL \tan \angle BOL)$$

$$= \boldsymbol{nr^2 \tan\left(\frac{\pi}{n}\right)}$$

Again, the area $= n\,(OL)(BL)$.

$$= n\, OB \cdot \cos\frac{\pi}{n} \cdot OB \cdot \sin\frac{\pi}{n}$$

$$= nR^2 \cdot \cos\frac{\pi}{n} \cdot \sin\frac{\pi}{n}$$

$$= \boldsymbol{\frac{n}{2} R^2 \sin\frac{2\pi}{n}}$$

Example 32. If A_0, A_1, A_2, A_3, A_4 and A_5 be the consecutive vertices of a regular hexagon inscribed in a unit circle. Then, find the product of length of A_0A_1, A_0A_2 and A_0A_4.

Sol. We know, in hexagon central angle is $\frac{360°}{6} = 60°$ and each

$$\text{angle} = \frac{(2n-4)\pi}{2n}$$

$$= \frac{(6-2)\times 180°}{6} = 120°$$

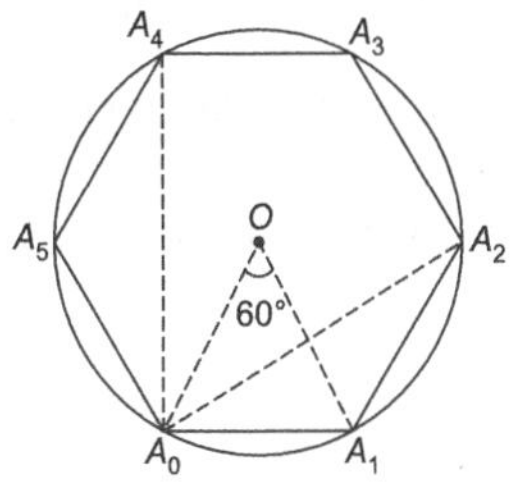

As the unit circle,

$\therefore$ radius $OA_0 = 1 = r$

In $\Delta A_0A_1A_2$,

$$\Rightarrow \quad \cos 120° = \frac{A_0A_1^2 + A_1A_2^2 - A_0A_2^2}{2A_0A_1 \cdot A_1A_2}$$

$$= \frac{1 + 1 - A_0A_2^2}{2 \cdot 1 \cdot 1}$$

$$\Rightarrow \quad A_0A_2 = \sqrt{3} \quad \text{...(i)}$$

Similarly in $\Delta A_0A_5A_4$, we have

$$A_0A_4 = \sqrt{3} \quad \text{...(ii)}$$

Thus, the value of,

$$(A_0A_1)\cdot(A_0A_2)\cdot(A_0A_4) = 1 \cdot \sqrt{3} \cdot \sqrt{3}$$

$$= 3 \text{ square units}$$

Example 33. If the area of circle is A_1 and area of regular pentagon inscribed in the circle is A_2, then find the ratio of area of two.

Sol. In ΔOAB, $OA = OB = r$ and $\angle AOB = \dfrac{360^\circ}{5} = 72^\circ$

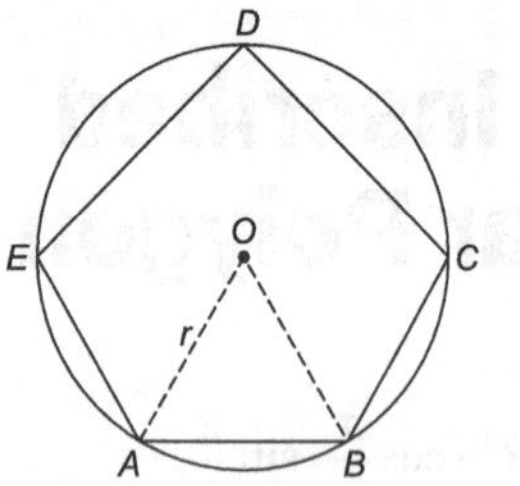

$$\therefore \quad \text{Area of } \Delta AOB = \frac{1}{2}\cdot r.r.\sin 72^\circ$$

$$\therefore \quad \text{Area of } \Delta AOB = \frac{1}{2} r^2 \cos 18^\circ \qquad \text{...(i)}$$

Area of pentagon = 5 (area of ΔAOB)

$$\Rightarrow \quad A_2 = 5\left\{\frac{1}{2} r^2 \cos 18^\circ\right\} \qquad \text{...(ii)}$$

Also we know, Area of circle $= \pi r^2$

$$\Rightarrow \quad A_1 = \pi r^2 \qquad \text{...(iii)}$$

Thus,
$$\frac{A_1}{A_2} = \frac{\pi r^2}{\frac{5}{2} r^2 \cos 18^\circ} = \frac{2\pi}{5} \sec\left(\frac{\pi}{10}\right)$$

| Example 34. If the area of cyclic quadrilateral $ABCD$ is $\left(\dfrac{3\sqrt{3}}{4}\right)$. The radius of the circle circumscribing it is 1. If $AB = 1$, $BD = \sqrt{3}$, then evaluate $BC \cdot CD$.

Sol. Here, In ΔBOD

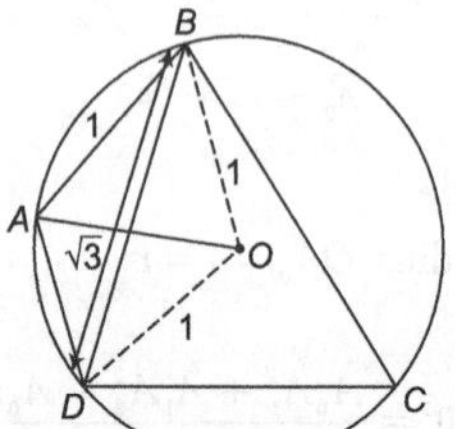

$$\Rightarrow \quad \angle BOD = 2C$$

$$\therefore \quad \cos 2C = \frac{1^2 + 1^2 - (\sqrt{3})^2}{2\cdot 1\cdot 1}$$

or
$$\cos 2C = -\frac{1}{2}$$

$$\therefore \quad \angle C = 60^\circ \qquad \text{...(i)}$$

Also,
$$\angle A + \angle C = 180^\circ$$

[since $ABCD$ is cyclic quadrilateral]

$$\Rightarrow \quad \angle A = 120^\circ \qquad \text{...(ii)}$$

$\therefore$ In ΔABD,

$$\cos 120^\circ = \frac{1^2 + AD^2 - (\sqrt{3})^2}{2\cdot AD\cdot 1}$$

$$\Rightarrow \quad -\frac{1}{2} = \frac{AD^2 - 2}{2AD}$$

$$AD^2 + AD - 2 = 0 \text{ or } AD = 1 \qquad \text{...(iii)}$$

Now, $\quad \text{ar}(ABCD) = \text{ar}(\Delta ABD) + \text{ar}(\Delta BCD)$

$$\Rightarrow \quad \frac{3\sqrt{3}}{4} = \frac{1}{2}\cdot 1\cdot 1\cdot \sin 120^\circ + \frac{1}{2}\cdot BC\cdot DC\cdot \sin 60^\circ$$

$$\Rightarrow \quad \frac{3\sqrt{3}}{4} = \frac{\sqrt{3}}{4} + \frac{1}{2}\cdot\frac{\sqrt{3}}{2} BC\cdot CD$$

$$\Rightarrow \quad \frac{2\sqrt{3}}{4} = \frac{\sqrt{3}}{4} BC\cdot CD$$

or
$$BC\cdot CD = 2$$

| Example 35. A regular pentagon and a regular decagon have the same perimeter, prove that their areas are as $2 : \sqrt{5}$.

Sol. Let AB be one of the n sides of a regular polygon. If $AB = a$,

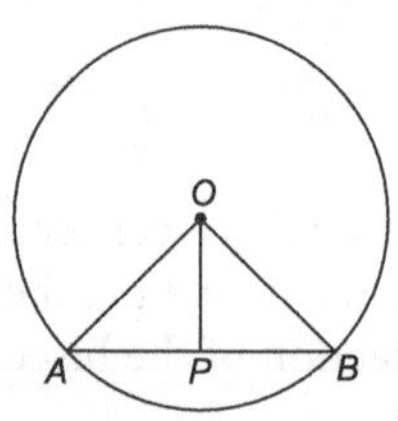

$$\text{area of polygon} = \frac{n}{2}\cdot AB\cdot OD$$

$$= \frac{na^2}{4}\cot\frac{\pi}{n} \qquad \text{...(i)}$$

Let the perimeter of the pentagon and decagon be $10x$. Then, each side of the pentagon is $2x$ and its area is

$$5x^2 \cot\frac{\pi}{5} \qquad \text{...(ii)}$$

[using Eq. (i) where $n = 5$ and $a = 2x$] again, each side of the decagon is x and its area is

$$\frac{5}{2} x^2 \cot\frac{\pi}{10} \qquad \text{...(iii)}$$

[using Eq. (i) where $n = 10$ and $a = x$]

$$\frac{\text{Area of petagon}}{\text{Area of decagon}} = \frac{2\cot 36^\circ}{\cot 18^\circ} = \frac{2\cos 36^\circ\cdot \sin 18^\circ}{\sin 36^\circ\cdot \cos 18^\circ}$$

$$= \frac{2\cos 36^\circ\cdot\sin 18^\circ}{2\sin 18^\circ\cdot\cos 18^\circ\cdot\cos 18^\circ} = \frac{\cos 36^\circ}{\cos^2 18^\circ}$$

$$= \frac{2\cos 36^\circ}{2\cos^2 18^\circ} = \frac{2\cos 36^\circ}{1 + \cos 36^\circ}$$

$$= \frac{2(\sqrt{5}+1)}{4\left\{1 + \dfrac{\sqrt{5}+1}{4}\right\}}$$

$$= \frac{2(\sqrt{5}+1)}{5+\sqrt{5}} = \frac{2}{\sqrt{5}}$$

Exercise for Session 5

1. Find the sum of the radii of the circles, which are respectively inscribed and circumscribed about a regular polygon of n sides.
2. Find the radius of the circumscribing circle of a regular polygon of n sides each of length is a.
3. If A, A_1, A_2, A_3 be the area of the in-circle and ex-circles, the show that $\frac{1}{\sqrt{A_1}}+\frac{1}{\sqrt{A_2}}+\frac{1}{\sqrt{A_3}}=\frac{1}{\sqrt{A}}$.
4. A regular polygon of nine sides, each of length 2, is inscribed in a circle, then find the radius of the circle.
5. Show that the area of the circle and the regular polygon of n-sides and of equal perimeter are in the ratio of $\tan\left(\frac{\pi}{n}\right):\frac{\pi}{n}$.
6. Let $A_1, A_2, A_3, \ldots, A_n$ be the vertices of an n-sided regular polygon such that $\frac{1}{A_1A_2}=\frac{1}{A_1A_3}+\frac{1}{A_1A_4}$. Find the value of n. Prove or disprove the converse of this result.
7. Let I_n is the area of n-sided regular polygon inscribed in a circle of unit radius and O_n be the area of the polygon circumscribing the given circle. Then, prove that $I_n=\frac{O_n}{2}\left(1+\sqrt{1-\left(\frac{2I_n}{n}\right)^2}\right)$

Session 6

Quadrilaterals and Cyclic Quadrilaterals

Area of Quadrilateral

$ABCD$ is any quadrilateral where $AB=a, BC=b, CD=c, AD=d$ and $\angle DPA=\alpha$.

Let s denotes the area of quadrilateral, then

area of ΔDAC = area of ΔAPD + area of ΔDPC

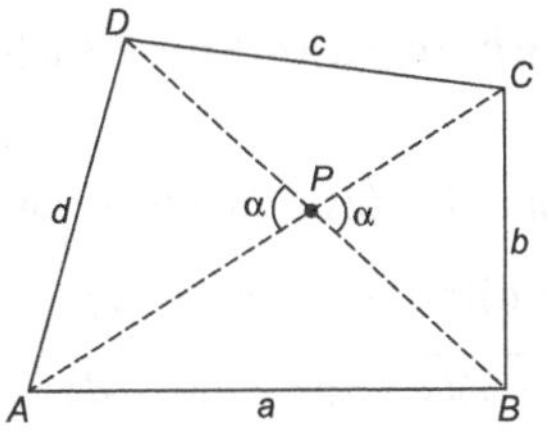

$$=\frac{1}{2}DP\cdot AP\cdot\sin\alpha+\frac{1}{2}\cdot DP\cdot PC\cdot\sin(\pi-\alpha)$$

$$=\frac{1}{2}DP.(AP+PC)\sin\alpha$$

$$\text{Area of }\Delta DAC=\frac{1}{2}DP\cdot AC\cdot\sin\alpha \quad\ldots(\text{i})$$

Similarly,

$$\text{area of }\Delta ABC=\frac{1}{2}BP\cdot AC\cdot\sin\alpha \quad\ldots(\text{ii})$$

$\therefore$ s = area of ΔDAC + area of ΔABC

$$=\frac{1}{2}DP\cdot AC\cdot\sin\alpha+\frac{1}{2}BP\cdot AC\cdot\sin\alpha$$

[using Eqs. (i) and (ii)]

$$=\frac{1}{2}(DP+BP)\cdot AC\cdot\sin\alpha$$

$$\Rightarrow\quad s=\frac{1}{2}BD\cdot AC\cdot\sin\alpha$$

$\therefore$ **Area of quadrilateral** $=\frac{1}{2}$ **(product of the diagonals) $\times$ (sine of included angle).**

Again,

We can express the area of Δ in terms of sides and the sum of two opposite angles :

In ΔABD, $\quad BD^2 = a^2 + d^2 - 2ad\cos A \quad$ ...(i)

In ΔBCD, $\quad BD^2 = b^2 + c^2 - 2bc\cos C \quad$ ...(ii)

From Eqs. (i) and (ii),

$$a^2 + d^2 - 2ad\cos A = b^2 + c^2 - 2bc\cos C$$

$$\therefore \quad a^2 + d^2 - b^2 - c^2 = 2ad\cos A - 2bc\cos C \quad \text{...(iii)}$$

Also, $\quad s = \Delta BAD + \Delta BCD$

$$= \frac{1}{2}ad\sin A + \frac{1}{2}bc\sin C$$

$$\Rightarrow \quad 4s = 2ad\sin A + 2bc\sin C \quad \text{...(iv)}$$

On squaring and adding Eqs. (iii) and (iv) both sides, we get

$$16s^2 + (a^2 + d^2 - b^2 - c^2)^2$$
$$= 4a^2d^2 + 4b^2c^2 - 8abcd\cos(A + C) \quad \text{...(v)}$$

Let $A + C = 2\alpha$, then

$$\cos(A + C) = \cos 2\alpha = 2\cos^2\alpha - 1$$

From Eq. (v), we get

$$16s^2 = 4a^2d^2 + 4b^2c^2 - 8abcd(2\cos^2\alpha - 1) - (a^2 + d^2 - b^2 - c^2)^2$$

$$= \{2(ad + bc)\}^2 - (a^2 + b^2 - c^2 - d^2)^2 - 16abcd\cos^2\alpha$$

$$= \{(a + d)^2 - (b + c)^2\}\{(b + c)^2 - (a - d)^2\} - 16abcd\cos^2\alpha$$

$$= (a + d + b - c)(a + d - b + c)(b + c + a - d)(b + c - a + d) - 16abcd\cos^2\alpha$$

Let, $2s = a + b + c + d$

$\therefore$

$$16s^2 = (2s - 2a)(2s - 2b)(2s - 2c)(2s - 2d) - 16abcd\cos^2\alpha$$

$$s^2 = (s - a)(s - b)(s - c)(s - d) - abcd\cos^2\alpha$$

$$\Rightarrow \quad s = \sqrt{(s - a)(s - b)(s - c)(s - d) - abcd\cos^2\alpha}$$

where $2\alpha = A + C$ and $2s = a + b + c + d$

Thus, area of quadrilateral;

$$\frac{1}{2}BD \cdot AC \cdot \sin\alpha, \text{ where } \angle DPA = \alpha$$

or $\quad \boldsymbol{A = \sqrt{(s - a)(s - b)(s - c)(s - d) - abcd\cos^2\alpha}}$,

where $2\alpha = A + C$

Area of Cyclic Quadrilateral

A quadrilateral is cyclic quadrilateral if its vertices lie on a circle.

Let $ABCD$ be a cyclic quadrilateral such that $AB = a$, $BC = b$, $CD = c$ and $DA = d$.

Then, $\angle B + \angle D = 180°$ and $\angle A + \angle C = 180°$.

Let $2s = a + b + c + d$ be the perimeter of the quadrilateral.

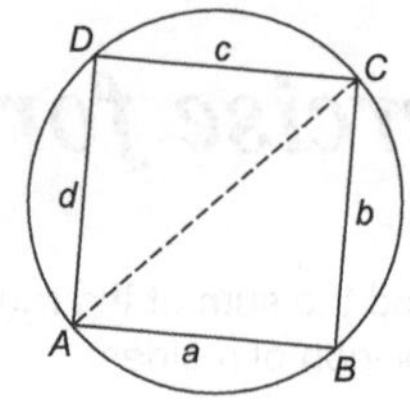

Now, $\quad \Delta$ = area of cyclic quadrilateral $ABCD$

= area of ΔABC + area of ΔACD

$$= \frac{1}{2}ab\sin B + \frac{1}{2}cd\sin D$$

$$= \frac{1}{2}ab\sin B + \frac{1}{2}cd\sin(\pi - B) = \frac{1}{2}ab\sin B + \frac{1}{2}cd\sin B$$

$$\Rightarrow \quad \Delta = \frac{1}{2}(ab + cd)\sin B \quad \text{...(i)}$$

Using cosine formula in a ΔABC and ΔACD, we have

$$AC^2 = AB^2 + BC^2 - 2AB \cdot BC \cdot \cos B$$

$$\Rightarrow \quad AC^2 = a^2 + b^2 - 2ab\cos B \quad \text{...(ii)}$$

and $\quad AC^2 = AD^2 + CD^2 - 2AD \cdot CD \cdot \cos D$

$$\Rightarrow \quad AC^2 = d^2 + c^2 - 2cd \cdot \cos(\pi - B) \quad \text{...(iii)}$$

From Eqs. (ii) and (iii), we have

$$a^2 + b^2 - 2ab\cos B = d^2 + c^2 - 2cd\cos B$$

$$\Rightarrow \quad 2(ab + cd)\cos B = a^2 + b^2 - c^2 - d^2 \quad \text{...(iv)}$$

$$\Rightarrow \quad 4(ab + cd)^2\cos^2 B = (a^2 + b^2 - c^2 - d^2)^2$$

$$\Rightarrow \quad 4(ab + cd)^2 \cdot (1 - \sin^2 B) = (a^2 + b^2 - c^2 - d^2)^2$$

$$\Rightarrow \quad 4(ab + cd)^2\sin^2 B = 4(ab + cd)^2 - (a^2 + b^2 - c^2 - d^2)^2$$

$$\Rightarrow \quad 4(ab + cd)^2 \cdot \sin^2 B = \{2(ab + cd) + (a^2 + b^2 - c^2 - d^2)\}\{2(ab + cd) - (a^2 + b^2 - c^2 - d^2)\}$$

$$\Rightarrow \quad 4(ab + cd)^2 \cdot \sin^2 B = \{(a + b)^2 - (c - d)^2\} \cdot \{(c + d)^2 - (a - b)^2\}$$

$$\Rightarrow \quad 4(ab + cd)^2 \cdot \sin^2 B = \{(a + b + c - d)(a + b - c + d)(c + d - a + b)(c + d + a - b)$$

$$\Rightarrow \quad 4(ab + cd)^2 . \sin^2 B = (2s - 2d)(2s - 2c)(2s - 2b)(2s - 2a)$$

$$\Rightarrow \quad 16\Delta^2 = 16(s - a)(s - b)(s - c)(s - d) \quad \text{[using Eq. (i)]}$$

$$\Rightarrow \quad \Delta = \sqrt{(s - a)(s - b)(s - c)(s - d)} \quad \text{...(v)}$$

$\therefore$ From Eqs. (i), (iv) and (v),

Area of cyclic quadrilateral;

$$\boldsymbol{\Delta = \frac{1}{2}(ab + cd) \cdot \sin B, \ \Delta = \sqrt{(s - a)(s - b)(s - c)(s - d)}}$$

and $\quad \cos B = \dfrac{a^2 + b^2 - c^2 - d^2}{2(ab + cd)}$

Ptolemy's Theorem

In a cyclic quadrilateral *ABCD*,

$$AC \cdot BD = AB \cdot CD + BC \cdot AD$$

i.e. in a cyclic quadrilateral the product of diagonals is equal to the sum of the products of the lengths of the opposite sides.

Proof: Let *ABCD* be a cyclic quadrilateral, where

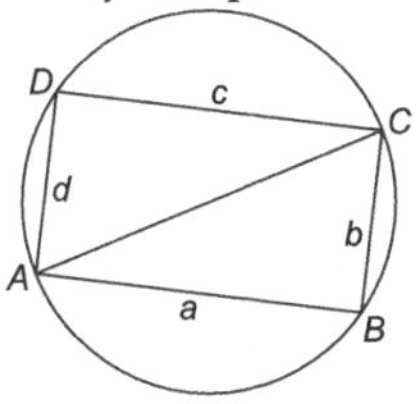

$$AC^2 = a^2 + b^2 - 2ab\cos B$$

and $$\cos B = \frac{a^2 + b^2 - c^2 - d^2}{2(ab + cd)}$$

$$\Rightarrow \quad AC^2 = a^2 + b^2 - 2ab \cdot \left\{\frac{a^2 + b^2 - c^2 - d^2}{2(ab + cd)}\right\}$$

$$\Rightarrow \quad AC^2 = \frac{(a^2 + b^2)(ab + cd) - ab(a^2 + b^2 - c^2 - d^2)}{(ab + cd)}$$

$$\Rightarrow \quad AC^2 = \frac{(a^2 + b^2) \cdot cd + ab(c^2 + d^2)}{ab + cd}$$

$$\Rightarrow \quad AC^2 = \frac{(ac + bd) \cdot (ad + bc)}{ab + cd} \quad \text{...(i)}$$

Similarly,

$$BD^2 = \frac{(ab + bd) \cdot (ac + bd)}{ad + bc} \quad \text{...(ii)}$$

$$\Rightarrow \quad AC^2 \cdot BD^2 = (ac + bd)^2$$

$$\Rightarrow \quad AC \cdot BD = (ac + bd)$$

$$\Rightarrow \quad AC \cdot BD = AB \cdot CD + BC \cdot AD \text{[Ptolemy's theorem]}$$

Circum-radius of a Cyclic Quadrilateral

Let *ABCD* be a cyclic quadrilateral. Then the circum-circle of the quadrilateral *ABCD* is also the circum circle of ΔABC.

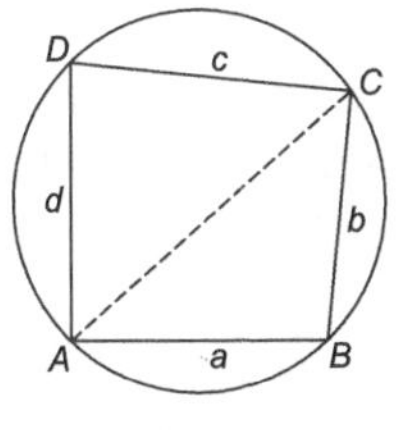

Hence, the circum-radius of the cyclic quadrilateral

$$ABCD = R = \text{Circum-radius of } \Delta ABC = \frac{AC}{2\sin B}$$

$$= \frac{AC(ab + cd)}{4\Delta} \qquad \left[\text{using } \Delta = \frac{1}{2}(ab + cd)\sin B\right]$$

But, $$AC = \sqrt{\frac{(ac + bd)(ad + bc)}{(ab + cd)}} \qquad \text{[using Eq. (i)]}$$

Hence, $$\boldsymbol{R = \frac{1}{4\Delta}\sqrt{(ac + bd)(ad + bc)(ab + cd)}}$$

$$\therefore \quad \boldsymbol{R = \frac{1}{4}\sqrt{\frac{(ac + bd)(ad + bc)(ab + cd)}{(s - a)(s - b)(s - c)(s - d)}}}$$

Example 36. If the sides of a cyclic quadrilateral are 3, 3, 4, 4. Show that a circle can be inscribed in it.

Sol. By geometry,

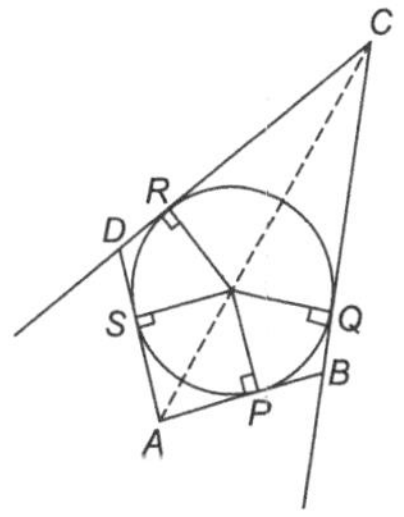

$$AP = AS \quad \text{...(i)}$$
$$BP = BQ \quad \text{...(ii)}$$
$$DR = DS \quad \text{...(iii)}$$
$$CR = CQ \quad \text{...(iv)}$$

Adding all four equations, we get

$$AB + CD = AD + CB \quad \text{...(v)}$$

Now, the sides of cyclic quadrilateral 3, 3, 4, 4 inscribe the circle in it, if it satisfy (v).

i.e. let $AB + CD = 3 + 4 = 7$, $AD + CB = 3 + 4 = 7$

i.e. 3, 3, 4, 4 i.e. $3 = AB$, $3 = BC$, $4 = CD$, $4 = CB$

satisfy condition (v) or a circle can be inscribed.

Note

If sum of opposite side of a quadrilateral is equal, then and only then a circle can be inscribed in the quadrilateral.

Example 37. The two adjacent sides of a cyclic quadrilateral are 2 and 5 the angle between them is 60°. If the area of the quadrilateral is $4\sqrt{3}$, then find the remaining two sides.

Sol. Let $AD = 2$, $AB = 5$, $\angle DAB = 60^\circ$

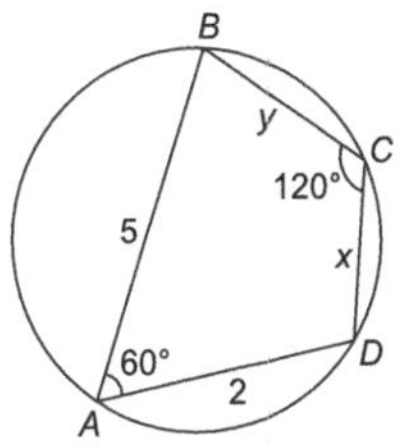

Since, the quadrilateral is cyclic, $\angle BCD = 120^\circ$

Area of

$$\Delta ABD = \frac{1}{2} \cdot 2 \cdot 5 \sin 60^\circ = 5 \cdot \frac{\sqrt{3}}{2} \quad \text{...(i)}$$

Area of ΔBCD = Area of quadrilateral *ABCD* – area of ΔABD

$$= 4\sqrt{3} - \frac{5\sqrt{3}}{2} = \frac{3\sqrt{3}}{2}$$

...(ii) [using Eq. (i)]

Let $CD = x$ and $BC = y$

Now, area of $\Delta BCD = \frac{1}{2}xy\sin 120^\circ$

or $\dfrac{3\sqrt{3}}{2} = \dfrac{1}{2}xy\dfrac{\sqrt{3}}{2}$ ...(iii) [using Eq. (ii)]

$\Rightarrow$ $xy = 6$ ...(iv)

Applying cosine rule in ΔBAD, we get

$$\cos 60^\circ = \frac{AD^2 + AB^2 - BD^2}{2AD \cdot AB}$$

or $\dfrac{1}{2} = \dfrac{2^2 + 5^2 - BD^2}{2 \cdot 2 \cdot 5}$

$\Rightarrow$ $BD^2 = 19$

Applying cosine rule in ΔBCD, we get

$$\cos 120^\circ = \frac{x^2 + y^2 - 19}{2xy}$$

or $-\dfrac{1}{2} = \dfrac{x^2 + y^2 - 19}{2xy}$

or $x^2 + y^2 + xy = 19$ [using $xy = 6$]

or $x^2 + y^2 = 13$

Now, $x^2 + y^2 + 2xy = 13 + 12 = 25$

$\Rightarrow$ $(x + y)^2 = 25$

$\Rightarrow$ $x + y = 5$

and $x^2 + y^2 - 2xy = 13 - 12$

$\Rightarrow$ $x - y = \pm 1$

Solving we get, $(x = 3, y = 2)$ or $(x = 2, y = 3)$

Example 38. If a, b, c, d are the sides of a quadrilateral, then find the minimum value of $\dfrac{a^2 + b^2 + c^2}{d^2}$.

Sol. Here, $AB = a$, $BC = b$, $CD = c$ and $AD = d$ are the sides of quadrilateral $ABCD$.

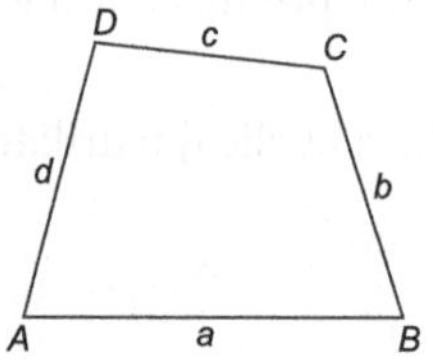

And we know, $(a - b)^2 + (b - c)^2 + (c - a)^2 \geq 0$

$\Rightarrow$ $2(a^2 + b^2 + c^2) \geq 2(ab + bc + ca)$

$\Rightarrow$ $3(a^2 + b^2 + c^2) \geq (a^2 + b^2 + c^2) + 2(ab + bc + a)$

[i.e., adding $(a^2 + b^2 + c^2)$ to both sides]

$\Rightarrow$ $3(a^2 + b^2 + c^2) \geq (a + b + c)^2$

[$\because$ sum of any three sides of quadrilateral is greater than fourth]

$\Rightarrow$ $3(a^2 + b^2 + c^2) \geq (a + b + c)^2 > d^2$ [$\because a + b + c > d$]

$\Rightarrow$ $\dfrac{a^2 + b^2 + c^2}{d^2} > \dfrac{1}{3}$

$\therefore$ Minimum value of $\dfrac{a^2 + b^2 + c^2}{d^2}$ is $\dfrac{1}{3}$.

Exercise for Session 6

1. The area of a cyclic quadrilateral $ABCD$ is $\dfrac{(3\sqrt{3})}{4}$. The radius of the circle circumscribing cyclic quadrilateral is 1. If $AB = 1$, $BD = \sqrt{3}$, then find $BC \cdot CD$.
2. If two adjacent sides of a cyclic quadrilateral are 2 and 5 and the angle between them is 60°. If the third side is 3, then find the remaining fourth side.
3. The ratio of the area of a regular polygon of n sides inscribed in a circle to that of the polygon of same number of sides circumscribing the same circle is 3 : 4. Then, the value of n is
4. A right angled trapezium is circumscribed about a circle. Find the radius of the circle. If the lengths of the bases (i.e. parallel sides) are equal to a and b.
5. If A, B, C, D are the angles of quadrilateral, then find $\dfrac{\Sigma \tan A}{\Sigma \cot A}$.

Session 7

Solution of Triangles

In a triangle, there are six variables. *viz.* three sides a, b, c and three angles A, B, C when any three of these six variables (except all the three angles) of a triangles are given, the triangle is known completely; that is the other three variables can be expressed in terms of the given variables and can be evaluated. This process is called the solution of triangles.

Solution of a Right Angled Triangle

***Case* I** When two sides are given.

Let the triangle be right angled at C, then we can determine the remaining variables as given in the following table :

Given	Required
(i) a, b	$\tan A = \frac{a}{b}$, $B = 90^\circ - A$, $C = \frac{a}{\sin A}$
(ii) a, c	$\sin A = \frac{a}{c}$, $b = c\cos A$, $B = 90^\circ - A$ where $\angle C = 90^\circ$

***Case* II** When a side and an acute angle is given. In this case we can determine the remaining variables as given in the following table :

Given	Required
(i) a, A	$B = 90^\circ - A$, $b = a\cot A$, $c = \frac{a}{\sin A}$
(ii) c, A	$B = 90^\circ - A$, $a = c\sin A$, $b = c\cos A$

Solution of a Triangle in General

***Case* I** When three sides a, b and c are given.

In this case the remaining variables are determined by using the following table :

Given	Required
a, b, c	(i) Area of $\Delta = \sqrt{s(s-a)(s-b)(s-c)}$, $2s = a + b + c$
	(ii) $\sin A = \frac{2\Delta}{bc}$, $\sin B = \frac{2\Delta}{ac}$, $\sin C = \frac{2\Delta}{ab}$
	(iii) $\tan\frac{A}{2} = \frac{\Delta}{s(s-a)}$, $\tan\frac{B}{2} = \frac{\Delta}{s(s-b)}$, $\tan\frac{C}{2} = \frac{\Delta}{s(s-c)}$

***Case* II** When two sides a, b and included $\angle C$ are given. In this case we use the following table :

Given	Required
a, b and $\angle C$	(i) Area of $\Delta = \frac{1}{2}ab\sin C$
	(ii) $\tan\left(\frac{A-B}{2}\right) = \frac{a-b}{a+b}\cdot\cot\left(\frac{C}{2}\right)$
	(iii) $c = \frac{a\sin C}{\sin A}$

***Case* III** When one side a and two angles A and B are given. In this case we use the following table :

Given	Required
a and $\angle A$, $\angle B$	(i) $\angle C = 180^\circ - (\angle A + \angle B)$
	(ii) $b = \frac{a\sin B}{\sin A}$ and $c = \frac{a\sin C}{\sin A}$
	(iii) $\Delta = \frac{1}{2}ac\sin B$

***Case* IV** When two sides a, b and $\angle A$ opposite to one side is given then,

$$\sin B = \frac{b}{a}\sin A \qquad \text{...(i)}$$

$$\angle C = 180^\circ - (\angle A + \angle B)\,;\; c = \frac{a\sin C}{\sin A}$$

From Eq. (i), the following possibilities will arise :

(a) **When A is an acute angle and $a < b\sin A$** In this relation $\sin B = \frac{b}{a}\sin A$ gives that $\sin B > 1$ which is impossible.

Hence, no triangle is possible.

(b) **When A is an acute angle and $a = b\sin A$** In this case only one triangle is possible which is right angled at B.

$\therefore \quad a = b\sin A \Rightarrow \angle B = 90^\circ$

(c) **When A is an acute angle and $a > b\sin A$** In this case there are two values of B given by $\sin B = \frac{b\sin A}{a}$, say B_1 and B_2 such that $B_1 + B_2 = 180^\circ$ side 'c' can be obtained by using $c = \frac{a\sin C}{\sin A}$.

$\therefore \quad a > b\sin A \Rightarrow$ two triangle are possible.

Note

(i) In any right angled triangle, the orthocentre coincides with the vertex containing the right angled.

(ii) The mid-point of the hypotenuse of a right angled triangle is equidistant from the three vertices of the triangle.

(iii) The mid-point of the hypotenuse of the right angled triangle is the circum-centre of the triangle.

Example 39. In any ΔABC, the sides are 6 cm, 10 cm and 14 cm. Show that the triangle is obtuse-angled with the obtuse angle equal to 120°.

Sol. Let $a = 14$, $b = 10$ and $c = 6$ cm.

$$\Rightarrow \quad s = \frac{14 + 10 + 6}{2} = 15 \text{ cm.}$$

As we know largest angle is opposite to the largest side,

$$\therefore \quad \cos A = \frac{b^2 + c^2 - a^2}{2bc}$$

$$= \frac{100 + 36 - 196}{2(10)(6)} = -\frac{1}{2}$$

$$\Rightarrow \quad A = 120^\circ$$

Example 40. If a, b and A are given in a triangle and C_1, C_2 are the possible values of the third side, prove that : $C_1^2 + C_2^2 - 2C_1C_2 \cos 2A = 4a^2 \cos^2 A$

Sol. $\cos A = \dfrac{b^2 + c^2 - a^2}{2bc}$

$\Rightarrow c^2 - 2bc \cos A + b^2 - a^2 = 0$, which is quadratic is 'c'.

$$\left.\begin{aligned} \therefore \quad C_1 + C_2 &= 2b \cos A \\ \text{and} \quad C_1 C_2 &= b^2 - a^2 \end{aligned}\right\} \quad \text{...(i)}$$

$$\therefore \quad C_1^2 + C_2^{\,2} - 2C_1C_2 \cos 2A$$

$$\Rightarrow \quad (C_1 + C_2)^2 - 2C_1C_2 - 2C_1C_2 \cos 2A$$

[using Eq. (i)]

$$\Rightarrow \quad (C_1 + C_2)^2 - 2C_1C_2(1 + \cos 2A)$$

$$\Rightarrow \quad 4b^2 \cos^2 A - 2(b^2 - a^2) \cdot 2 \cos^2 A$$

$$\Rightarrow \quad 4a^2 \cos^2 A.$$

$$\therefore \quad C_1^2 + C_2^2 - 2C_1C_2 \cos A = 4a^2 \cos A$$

Example 41. In a ΔABC, the median to the side BC is of length $\dfrac{1}{\sqrt{11 - 6\sqrt{3}}}$ and it divides the $\angle A$ into angles of 30° and 45°. Find the length of the side BC.

Sol. $\angle C = 180^\circ - (75^\circ + B)$

$$\angle C = 105^\circ - B$$

In ΔABD,

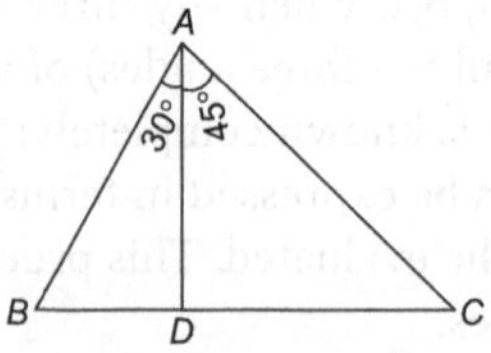

$$\frac{BD}{\sin 30^\circ} = \frac{AD}{\sin B} \Rightarrow BD = \frac{AD}{2 \sin B} \quad \text{...(i)}$$

In ΔADC,

$$\frac{CD}{\sin 45^\circ} = \frac{AD}{\sin C}$$

$$\Rightarrow \quad CD = \frac{AD}{\sqrt{2} \sin(105^\circ - B)} \quad \text{...(ii)}$$

Now, $\quad BD = CD$

$$\Rightarrow \quad \frac{AD}{2 \sin B} = \frac{AD}{\sqrt{2} \sin(105^\circ - B)}$$

$$\Rightarrow \quad \sqrt{2} \sin B = \sin(105^\circ - B)$$

$$\Rightarrow \quad \sqrt{2} \sin B = \sin 105^\circ \cdot \cos B - \cos 105^\circ \cdot \sin B$$

$$\Rightarrow \quad \sqrt{2} \sin B = \frac{(\sqrt{3} + 1)}{2\sqrt{2}} \cos B + \frac{(\sqrt{3} - 1)}{2\sqrt{2}} \sin B$$

$$\Rightarrow \quad 4 \sin B = (\sqrt{3} + 1) \cos B + (\sqrt{3} - 1) \sin B$$

$$\Rightarrow \quad \cot B = 3\sqrt{3} - 4$$

$$\Rightarrow \quad \sin B = \frac{1}{2\sqrt{11 - 6\sqrt{3}}}$$

Hence, $\quad BC = 2BD = \dfrac{AD}{\sin B} = \dfrac{2\sqrt{11 - 6\sqrt{3}}}{\sqrt{11 - 6\sqrt{3}}} = 2$

Exercise for Session 7

1. In ΔABC, $a : b : c = (1 + x) : 1 : (1 - x)$, where $x \in (0, 1)$. If $\angle A = \dfrac{\pi}{2} + \angle C$, then find the value of x.

2. In a ΔABC, $2s$ = perimeter and R = circumradius. Then, find $\dfrac{s}{R}$.

3. If in a ΔABC, $\angle C = 90^\circ$, then find the maximum value of $\sin A \sin B$.

4. If the area of a triangle is 81 square cm and its perimeter is 27 cm, then find its in-radius in centi-metres.

5. In a ΔABC, if $r_1 = 2r_2 = 3r_3$, then show that $\dfrac{a}{b} = \dfrac{5}{4}$.

6. Find in-radius of the triangle formed by the axes and the line $4x + 3y - 12 = 0$.

7. In a ΔPQR as shown in figure given that $x : y : z :: 2 : 3 : 6$, then find value of $\angle QPR$.

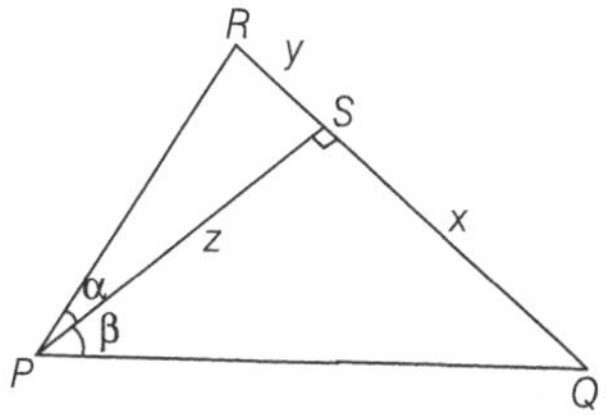

8. In a ΔABC, if $\dfrac{R}{r} \le 2$, then show that the triangle is equilateral.

9. The angle of a right-angled triangle are in AP. Then, find the ratio of the in-radius and the perimeter.

10. If in a triangle $\left(1 - \dfrac{r_1}{r_2}\right)\left(1 - \dfrac{r_1}{r_3}\right) = 2$, then show that the triangle is right angled.

Session 8

Height and Distance

Angle of Elevation

If 'O' be the observer's eye and OX be the horizontal line through O. If object P is at a higher level than eye, then $\angle POX$ is called the angle of elevation.

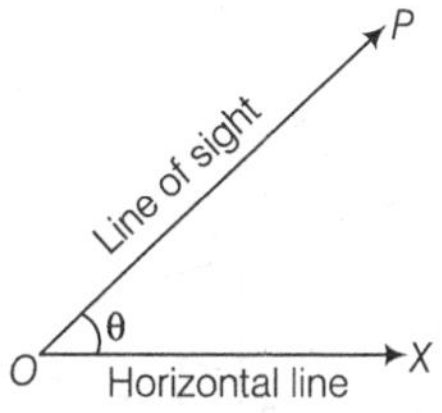

Angle of Depression

If 'O' be the observer's eye and OX is a horizontal line object P is at a lower level than O, then the $\angle POX$ is called the angle of depression.

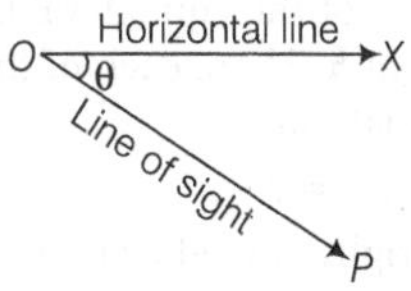

Note

(i) Angle of elevation and depression are always acute angle.

(ii) Angle of elevation of an object from an observer is same as angle O depression of an observer from the object.

Bearing

If the observer and the object are O and P be on the same level respectively, then bearings is defined. To measure the bearing the four standard direction East, West, North and South are taken as the cardinal directions. Angle between the line of observation. i.e., OP and any one standard direction is measured.

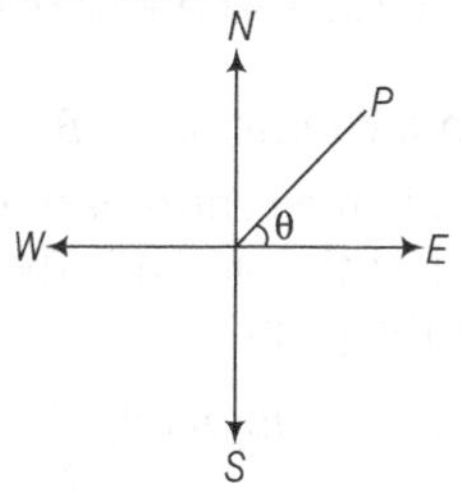

Thus, $\angle POE$ is called the bearing of the point P with respect to O measured from East to North.

In other words the bearing of P as seen from O is the direction in which P is seen from O.

Note

North-East means equally inclined to North and East. ENE means equally inclined to East and North-East.

Example 42. Two flagstaffs stand on a horizontal plane. A and B are two points on the line joining their feet and between them. The angles of elevation of the tops of the flagstaffs as seen from A are 30° and 60° and as seen from B are 60° and 45°. If AB is 30 m, the distance between the flagstaffs in metres is

(a) $30 + 15\sqrt{3}$ (b) $45 + 15\sqrt{3}$

(c) $60 - 15\sqrt{3}$ (d) $60 + 15\sqrt{3}$

Sol. (d) Let x and y be the heights of the flagstaffs at P and Q respectively.

Then, $AP = x\cot 60^\circ = \dfrac{x}{\sqrt{3}}$, $AQ = y\cot 30^\circ = y\sqrt{3}$

$$BP = x\cot 45^\circ = x,\ BQ = y\cot 60^\circ = \frac{y}{\sqrt{3}}$$

$$\Rightarrow \quad BP - AP = x - \frac{x}{\sqrt{3}} = AB$$

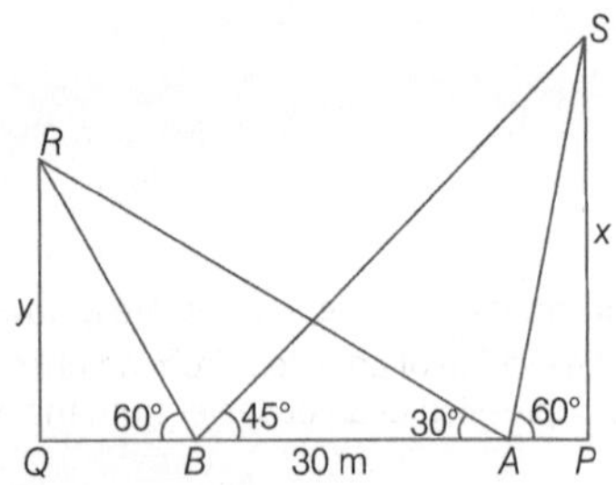

$$\Rightarrow \quad 30\sqrt{3} = (\sqrt{3} - 1)x$$

$$\Rightarrow \quad x = 15(3 + \sqrt{3})$$

Similarly, $30 = y\left(\sqrt{3} - \dfrac{1}{\sqrt{3}}\right) \Rightarrow y = 15\sqrt{3}$

So that, $PQ = BP + BQ = x + \dfrac{y}{\sqrt{3}}$

$$= 15(3 + \sqrt{13}) + 15 = (60 + 15\sqrt{3})\text{ m}$$

Example 43. In a cubical hall $ABCD$, $PQRS$ with each side 10 m, G is the centre of the wall $BCRQ$ and T is the mid-point of the side AB. The angle of elevation of G at the point T is

(a) $\sin^{-1}\left(\dfrac{1}{\sqrt{3}}\right)$ (b) $\cos^{-1}\left(\dfrac{1}{\sqrt{3}}\right)$

(c) $\tan^{-1}\left(\dfrac{1}{\sqrt{3}}\right)$ (d) $\cot^{-1}\left(\dfrac{1}{\sqrt{3}}\right)$

Sol. (a)

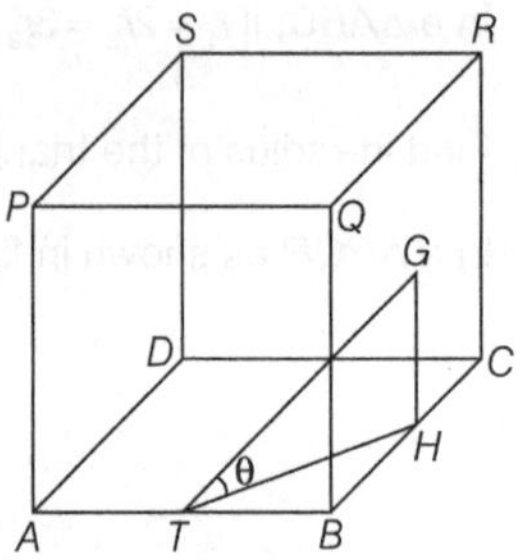

Let H be the mid-point of BC since

$$\angle TBH = 90^\circ,$$
$$(TH)^2 = (BT)^2 + (BH)^2$$
$$= 5^2 + 5^2 = 50$$

Also, $\angle THG = 90^\circ$,

$$(TG)^2 = (TH)^2 + (GH)^2$$
$$= 50 + 25 = 75$$

Let θ be the required angle of elevation of G at T.

Then, $\sin\theta = \dfrac{GH}{TG}$

$$= \frac{5}{5\sqrt{3}} = \frac{1}{\sqrt{3}}$$

$$\Rightarrow \quad \theta = \sin^{-1}\left(\frac{1}{\sqrt{3}}\right)$$

Example 44. Each side of an equilateral triangle subtends an angle of 60° at the top of a tower h m high located at the centre of the triangle. If a is the length of each side of the triangle, then

(a) $3a^2 = 2h^2$ (b) $2a^2 = 3h^2$

(b) $a^2 = 3h^2$ (d) $3a^2 = h^2$

Sol. (b)

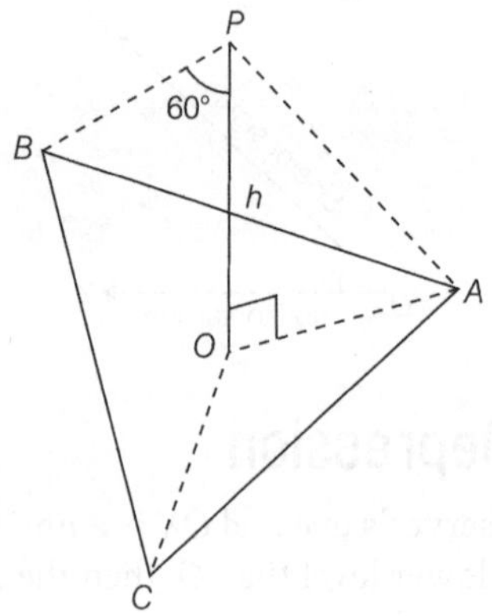

Let O be the centre of the equilateral triangle ABC and OP the tower of height h. Then, each of the triangles PAB, PBC and PCA are equilateral.

Thus, $PA = PB = PC = a$.

Therefore, from right-angled triangle POA, we have $PA^2 = PO^2 + OA^2$.

$$\Rightarrow \quad a^2 = h^2 + \left(\frac{a}{2}\sec 30^\circ\right)^2$$
$$= h^2 + \frac{a^2}{4}\cdot\frac{4}{3}$$
$$= h^2 + \frac{a^2}{3}$$
$$\Rightarrow \quad \frac{2}{3}a^2 = h^2$$
$$\Rightarrow \quad 2a^2 = 3h^2$$

Example 45. A vertical tower PQ subtends the same angle 30° at each of two places A and B, 60 m apart on the ground, AB subtends an angle 120° at the foot of the tower. If h is the height of the tower, then $9h^2 + h + 1$ is equal to

(a) 3121 (b) 2136
(c) 3600 (d) None of these

Sol. (d) $AP = PB = h\cot 30^\circ = \sqrt{3}h$

$$-\frac{1}{2} = \cos 120^\circ = \frac{3h^2 + 3h^2 - 60^2}{2\cdot 3h^2}$$
$$-3h^2 = 3h^2 + 3h^2 - 3600$$
$$9h^2 = 3600$$
$$h = 20$$
$$\therefore \quad 9h^2 + h + 1 = 3600 + 20 + 1$$
$$= 3681$$

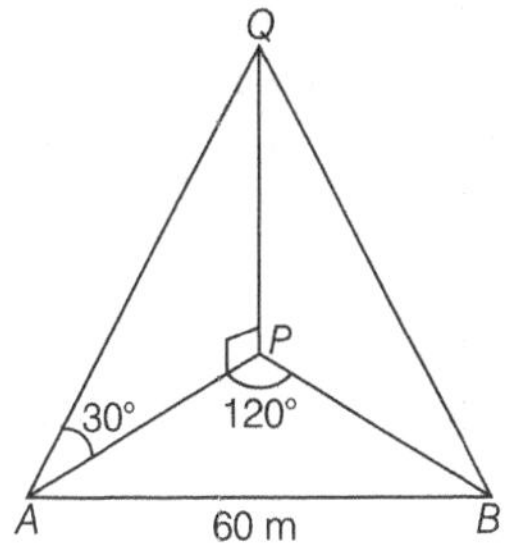

Exercise for Session 8

1. If a tower subtends angles $\theta, 2\theta$ and 3θ at three points A, B and C respectively, lying on the same side of a horizontal line through the foot of the tower, show that $\frac{AB}{BC}\frac{\cot\theta - \cot 2\theta}{\cot 2\theta - \cot 3\theta}$.

2. A person stands at a point A due south of a tower of height h and observes that its elevation is 60°. He then walks westwards towards B, where the elevation is 45°. At a point C on AB produced, show that if he find it to be 30°. OA, OB, OC are in GP.

3. A train travelling on one of two intersecting railway lines, subtends at a certain station on the other line, an angle α when the front of the carriage reaches the junction and an angle β when the end of the carriage reaches it. Then, the two lines are inclined to each other at an angle θ, show that $2\cot\theta = \cot\alpha - \cot\beta$, $\cot\alpha + \cot\beta$

4. The angle of elevation of the top of the tower observed from each of the three points A, B, C on the ground, forming a triangle is the same angle α. If R is the circum-radius of the triangle ABC, then find the height of the tower $R\tan\alpha$.

5. The length of the shadow of a pole inclined at 10° to the vertical towards the sun is 2.05 metres, when the elevation of the sun is 38°. Then, find the length of the pole.

JEE Type Solved Examples :
Single Option Correct Type Questions

Ex. 1. *If* $\tan^2\frac{\pi - A}{4} + \tan^2\frac{\pi - B}{4} + \tan^2\frac{\pi - C}{4} = 1$, *then* ΔABC *is*

(a) equilateral (b) isosceles
(c) scalene (d) None of these

Sol. (a) Let $\alpha = \frac{\pi - A}{4}, \beta = \frac{\pi - B}{4}, \gamma = \frac{\pi - C}{4}$ ...(i)

$$\Rightarrow \quad \alpha + \beta + \gamma = \frac{\pi}{2}$$

$$\Rightarrow \quad \Sigma \tan\alpha \cdot \tan\beta = 1$$

$$\Sigma \tan^2\alpha = 1 = \Sigma \tan\alpha \tan\beta$$

$$\Rightarrow \quad \tan\alpha = \tan\beta = \tan\gamma$$

$$\therefore \quad \alpha = \beta = \gamma$$

$$\frac{\pi - A}{4} = \frac{\pi - B}{4} = \frac{\pi - C}{4} \quad \text{[From Eq. (i)]}$$

$$\therefore \quad A = B = C$$

Ex. 2. *In* $\Delta ABC, a^2 + c^2 = 2002b^2$, *then* $\frac{\cot A + \cot C}{\cot B}$ *equals to*

(a) $\frac{1}{2001}$ (b) $\frac{2}{2001}$
(c) $\frac{3}{2001}$ (d) $\frac{4}{2001}$

Sol. (b) $\frac{\cot A + \cot C}{\cot B} = \frac{\sin(A + C)\sin B}{\sin A \sin C \sin B}$

$$= \frac{\sin^2 B}{\sin A \cos B \sin C} = \frac{4R^2b^2}{4R^2\, ac \cos B}$$

$$= \frac{2b^2}{2ac \cos B} = \frac{2b^2}{a^2 + c^2 - b^2}$$

$$= \frac{2b^2}{2002\, b^2 - b^2} = \frac{2}{2001}$$

Ex. 3. *A triangle has vertices A, B and C and the respective opposite sides have lengths a, b and c. This triangle is inscribed in a circle of radius R. If b = c = 1 and the altitude from A to side BC has length* $\sqrt{\frac{2}{3}}$, *then R equals*

(a) $\frac{1}{\sqrt{3}}$ (b) $\frac{2}{\sqrt{3}}$
(c) $\frac{\sqrt{3}}{2}$ (d) $\frac{\sqrt{3}}{2\sqrt{2}}$

Sol. (d) $a = \frac{2}{\sqrt{3}}$ and $\Delta = \frac{1}{2}\ ah$

$$= \frac{1}{2} \cdot \frac{2}{\sqrt{3}} \cdot \frac{\sqrt{2}}{\sqrt{3}} = \frac{\sqrt{2}}{3}$$

$$\therefore \quad R = \frac{abc}{4\Delta}$$

$$= \frac{(2)(1)(3)}{\sqrt{3}\,(4)\sqrt{2}}$$

$$= \frac{\sqrt{3}}{2\sqrt{2}}$$

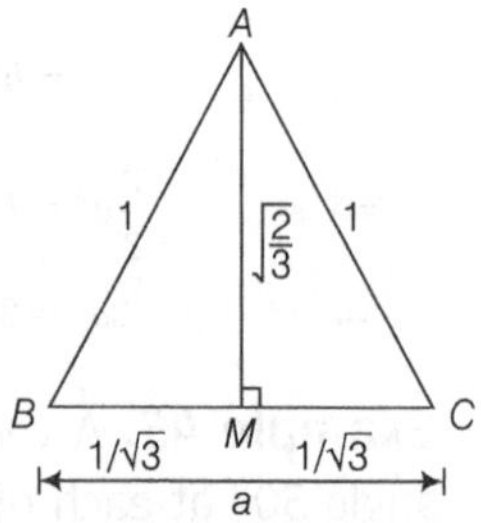

Ex. 4. *In* ΔABC, *if* $AC = 8, BC = 7$ *and D lies between A and B such that* $AD = 2, BD = 4$, *then the length CD equals*

(a) $\sqrt{46}$ (b) $\sqrt{48}$ (c) $\sqrt{51}$ (d) $\sqrt{75}$

Sol. (c) $l^2 = 2^2 + 8^2 - 2 \cdot 2 \cdot 8 \cos A$

$$= 4 + 64 - 32 \cos A$$

and $\cos A = \frac{6^2 + 8^2 - 7^2}{2 \cdot 6 \cdot 8} = \frac{51}{16 \times 6} = \frac{17}{32}$

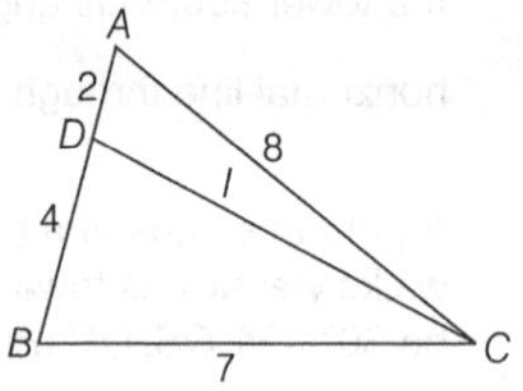

$$l^2 = 68 - 32 \times \frac{17}{32} = 51$$

$$l = \sqrt{51}$$

Ex. 5. *In a triangle, if* $(a + b + c)(a + b - c)(b + c - a)(c + a - b) = \frac{8a^2b^2c^2}{a^2 + b^2 + c^2}$, *then the triangle is*

(a) isosceles (b) right angled
(c) equilateral (d) obtuse angled

Sol. (b) We have, $s(s - a)(s - b)(s - c) = \frac{a^2b^2c^2}{2(a^2 + b^2 + c^2)}$

$$\Rightarrow \quad a^2 + b^2 + c^2 = \frac{a^2b^2c^2}{2\Delta^2} \quad \text{...(i)}$$

As, $\Delta = \frac{1}{2}bc \sin A = \frac{abc}{R}$

So, Eq. (i) becomes

$$a^2 + b^2 + c^2 = \frac{(a^2b^2c^2)}{2a^2 \cdot b^2 \cdot c^2} = 8R^2$$

$$\Rightarrow \quad 4R^2(\sin^2 A + \sin^2 B + \sin^2 C) = 8R^2$$

$$\Rightarrow \quad \sin^2 A + \sin^2 B + \sin^2 C = 2$$

or $$2 + 2\cos A \cdot \cos B \cdot \cos C = 2$$

$$\Rightarrow \quad \cos A \cdot \cos B \cdot \cos C = 0$$

$\therefore \Delta ABC$ must be right angled.

Ex. 6. *Consider a ΔABC and let a, b and c denote the lengths of the sides opposite to vertices A, B and C, respectively. If $a = 1$, $b = 3$ and $C = 60°$, then $\sin^2 B$ is equal to*

(a) $\frac{27}{28}$ (b) $\frac{3}{28}$

(c) $\frac{81}{28}$ (d) $\frac{1}{3}$

Sol. (a) By Cosine law, $\cos 60° = \dfrac{1^2 + 3^2 - c^2}{2 \cdot 1 \cdot 3}$

$$\Rightarrow \quad c = \sqrt{7}$$

Now, $\dfrac{b}{\sin B} = \dfrac{c}{\sin C}$ [By Sine law]

$$\Rightarrow \quad \sin B = \frac{b\sin C}{c} = \frac{3 \times \frac{\sqrt{3}}{2}}{\sqrt{7}} = \frac{3\sqrt{3}}{2\sqrt{7}}$$

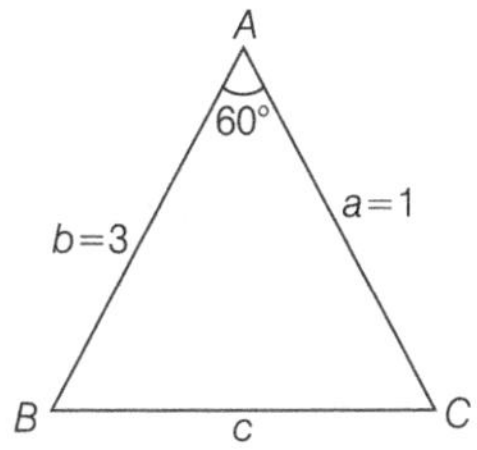

Hence, $\sin^2 B = \dfrac{27}{28}$.

Ex. 7. *In ΔABC, if $\cos A + \sin A - \dfrac{2}{\cos B + \sin B} = 0$, then $\dfrac{a+b}{c}$ is equal to*

(a) $\sqrt{2}$ (b) 1 (c) $\frac{1}{\sqrt{2}}$ (d) $2\sqrt{2}$

Sol. (a) We have,

$\Rightarrow \cos A \cos B + \sin A \sin B + \cos A \sin B + \sin A \cos B = 2$

$$\Rightarrow \quad \cos(A - B) + \sin(A + B) = 2$$

$\therefore$ $$\cos(A - B) = 1$$

and $$\sin(A + B) = 1$$

$\Rightarrow$ $A = B$, so $a = b$

and $$\sin 2A = 1$$

$$\Rightarrow \quad A = 45°$$

or $A = 135°$ (Not possible)

Hence, $\dfrac{a+b}{c} = \dfrac{2a}{a\sqrt{2}} = \sqrt{2}$

Ex. 8. *In ΔABC, if $\angle A = 30°$, $b = 10$ and $a = x$, then the values of x for which there are 2 possible triangles is given by (All symbol used have usual meaning in a triangle)*

(a) $5 < x < 10$ (b) $x < \frac{5}{2}$

(c) $\frac{5}{3} < x < 10$ (d) $\frac{5}{2} < x < 10$

Sol. (a) If c is the third side, then the altitude to c has length $10\sin 30° = 5$,

there are two triangles if x is greater than this value and less than the length of b.

If $x > 10$, point B will come to the left of A and $\angle A$ would be obtuse in the case.

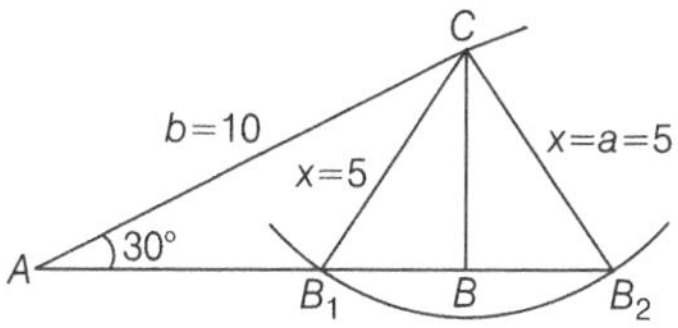

Aliter We have, $\dfrac{\sqrt{3}}{2} = \cos 30°$

$$= \frac{100 + c^2 - x^2}{2(10)(c)} \quad \text{[by using cosine rule]}$$

$$\Rightarrow \quad c^2 - 10\sqrt{3}c + (100 - x^2) = 0$$

$$c = \frac{10\sqrt{3} \pm \sqrt{4x^2 - 100}}{2}$$

Now, for 2 distinct positive values of c, we must have

$$10\sqrt{3} > \sqrt{4x^2 - 100}$$

$$\Rightarrow \quad 300 > 4x^2 - 100$$

$$\Rightarrow \quad x^2 < 100 \Rightarrow 5 < x < 10$$

Ex. 9. *In a ΔABC, $AB = AC$, P and Q are points on AC and AB respectively such that $CB = BP = PQ = QA$. If $\angle AQP = \theta$, then $\tan^2 \theta$ is a root of the equation*

(a) $y^3 + 21y^2 - 35y - 12 = 0$

(b) $y^3 - 21y^2 + 35y - 12 = 0$

(c) $y^3 - 21y^2 + 35y - 7 = 0$

(d) $12y^3 - 35y^2 + 35y - 12 = 0$

Sol. (c) $\angle QAP = \angle QPA = 90 - \dfrac{\theta}{2}$

$$\angle PQB = \angle PBQ = 180 - \theta$$

$$\angle BCA = \angle ABC = \angle BPC = 45 + \frac{\theta}{4}$$

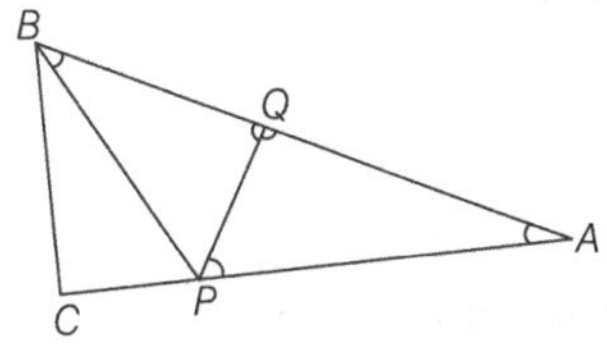

Now, $\left(90° - \frac{\theta}{2}\right) + (2\theta - 180°) + \left(45° + \frac{\theta}{4}\right) = 180°$

$\Rightarrow \quad \theta = \frac{5\pi}{7}$

$\Rightarrow \quad 7\theta = 5\pi$

$\Rightarrow \quad 4\theta = 5\pi - 3\theta$

$\Rightarrow \quad \tan 4\theta = -\tan 3\theta$

$\Rightarrow \quad \frac{2\tan 2\theta}{1 - \tan^2 2\theta} = -\frac{3\tan\theta - \tan^3\theta}{1 - 3\tan^2\theta}$

$\Rightarrow \quad \frac{2\left[\frac{2t}{1-t^2}\right]}{1 - \left(\frac{2t}{1-t^2}\right)^2} = \frac{3t - t^3}{1 - 3t^2}$, where $t = \tan\theta$

$\Rightarrow \quad \frac{4(1-t^2)}{(1-t^2)^2 - 4t^2} = \frac{t^2 - 3}{1 - 3t^2}$

$\Rightarrow \quad 4(1 - 4t^2 + 3t^4) = (t^2 - 3)(1 - 6t^2 + t^4)$

$\Rightarrow \quad t^6 - 21t^4 + 35t^2 - 7 = 0$

$\therefore \tan^2\theta$ is the root of the equation $y^3 - 21y^2 + 35y - 7 = 0$.

Ex. 10. *The angle of elevation of tower from a point A due south of it is 30° and from a point B due west of it is 45°. If the height of the tower be 100 m, then AB=*

(a) 150 m (b) 200 m

(c) 173.2 m (d) 141.4 m

Sol. (b) $OB = 100 \cot 45°$

$OA = 100 \cot 30°$

$AB = \sqrt{(OA^2 + OB^2)}$

$AB = 200$ m

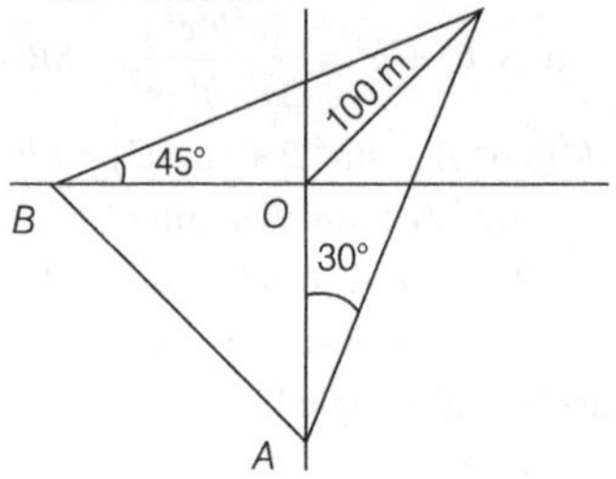

Ex. 11. *An aeroplane flying horizontally 1 km above the ground is observed at an elevation of 60° and after 10 seconds the elevation is observed to be 30°. The uniform speed of the aeroplane in km/h is*

(a) 240 (b) $240\sqrt{3}$

(c) $60\sqrt{3}$ (d) None of these

Sol. (b) $d = H \cot 30° - H \cot 60°$

Time taken = 10 s

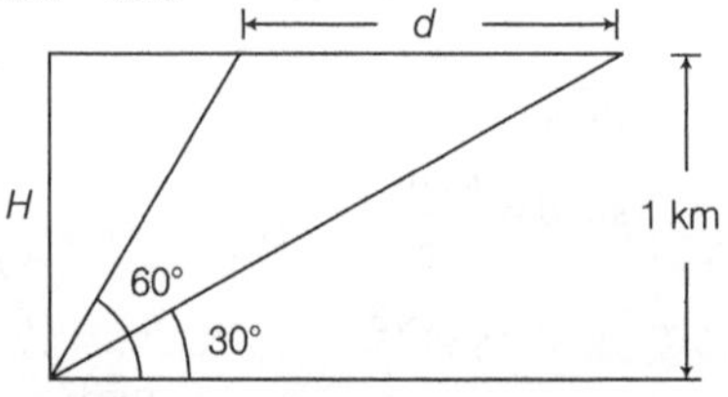

$$\text{Speed} = \frac{\cot 30° - \cot 60°}{10} \times 60 \times 60 = 240\sqrt{3}$$

JEE Type Solved Examples : More than One Correct Option Type Questions

Ex. 12 *In ΔABC, the ratio $\frac{a}{\sin A} = \frac{b}{\sin B} = \frac{c}{\sin C}$ is always equal to (All symbols used have usual meaning in a triangle.)*

(a) $2R$, where R is the circumradius

(b) $\frac{abc}{2\Delta}$, where Δ is the area of the triangle

(c) $\frac{2}{3}(a^2 + b^2 + c^2)^{\frac{1}{2}}$

(d) $\frac{(abc)^{\frac{2}{3}}}{(h_1h_2h_3)^{\frac{1}{3}}}$

Sol. (a,b,d) We know that

$$\frac{a}{\sin A} = \frac{b}{\sin B} = \frac{c}{\sin C} = 2R$$

and $R = \frac{abc}{4\Delta} \Rightarrow 2R = \frac{abc}{2\Delta}$

$\therefore$ a, b are true

Now, $\frac{2}{3}(a^2 + b^2 + c^2)^{1/2}$

If B true iff $a = b = c$

$\therefore$ 'c' is incorrect

Now, for option 'd'

We have, $\frac{(abc)^{2/3}}{(h_1h_2h_3)^{1/3}}$

We know, $\frac{1}{2}ah_1 = \frac{\pi}{2}bh_2 = \frac{1}{2}ch_3 = \Delta$

$$\Rightarrow \quad h_1 = \frac{2\Delta}{a}, h_2 = \frac{2\Delta}{b}, h_3 = \frac{2\Delta}{c}$$

$$\because \quad h_1 h_2 h_3 = \frac{8\Delta^3}{abc}$$

$$\therefore \quad \frac{(abc)^{2/3}}{(h_1 h_2 h_3)^{1/3}} = \frac{(abc)^{2/3}}{\left(\dfrac{8\Delta^3}{abc}\right)^{1/3}} = \frac{(abc)^{2/3}(abc)^{1/3}}{2\Delta}$$

$$= \frac{abc}{2\Delta} = 2R$$

Hence option, a, b and d are correct.

Ex. 13. *Let ABCD be a cyclic quadrilateral such that $AB = 2$, $BC = 3$, $\angle B = 120°$ and area of quadrilateral $= 4\sqrt{3}$. Which of the following is/are correct?*

(a) The value of $(AC)^2$ is equal to 19

(b) The sum of all possible values of product $AC \cdot BD$ is equal to 35

(c) The sum of all possible values of $(AD)^2$ is equal to 29

(d) The value of $(CD)^2$ can be 4

Sol. (a,c,d) Area of quadrilateral

$$= 4\sqrt{3} = \frac{1}{2} \times 2 \times 3 \sin 120° + \frac{1}{2} xy \sin 60°$$

$$4\sqrt{3} = \frac{\sqrt{3}}{2}\left[3 + \frac{xy}{2}\right]$$

$$16 = 6 + xy \Rightarrow xy = 10$$

$$AC^2 = 2^2 + 3^2 - 2 \cdot 2 \cdot 3 \times \cos 120°$$

$$= 4 + 9 + 6 = 19$$

$$= x^2 + y^2 - 2xy \cos 60°$$

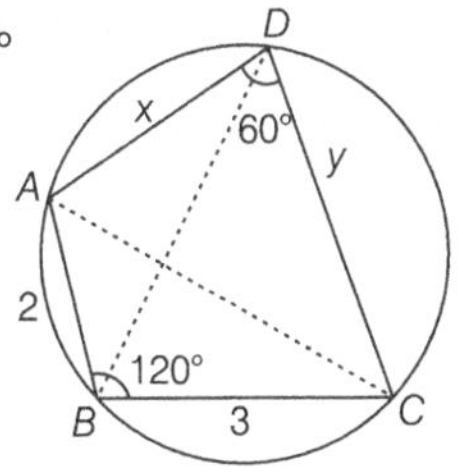

$$x^2 + y^2 - xy = 19$$

or $\quad x^2 + y^2 = 29$

$\Rightarrow \quad x = 5, y = 2$

or $\quad x = 2, y = 5$

Ex. 14. *In a ΔABC, which of the following quantities denote the area of the triangle?*

(a) $\dfrac{a^2 - b^2}{2}\left(\dfrac{\sin A \sin B}{\sin(A - B)}\right)$

(b) $\dfrac{r_1 r_2 r_3}{\sqrt{\Sigma r_1 r_2}}$

(c) $\dfrac{a^2 + b^2 + c^2}{\cot A + \cot B + \cot C}$

(d) $r^2 \cot\dfrac{A}{2} \cdot \cot\dfrac{B}{2} \cdot \cot\dfrac{C}{2}$

Sol. (a,b,d) (a) $\dfrac{4R^2}{2}\left(\dfrac{\sin^2 A - \sin^2 B}{\sin(A - B)}\right) \cdot \dfrac{ab}{4R^2} = \dfrac{1}{2}\sin(A + B)\, ab$

$$= \frac{1}{2}(ab)\sin C = \frac{abc}{4R} = \Delta$$

(b) $r_1 r_2 r_3 = \dfrac{\Delta^3 . s}{s(s - a)(s - b)(s - c)} = \Delta \cdot s$

$$\sqrt{\Sigma r_1 r_2} = \sqrt{s^2\left(\Sigma \tan\frac{A}{2}\tan\frac{B}{2}\right)} = s \quad \left[\because \Sigma \tan\frac{A}{2}\tan\frac{B}{2} = 1\right]$$

$$\therefore \quad \frac{r_1 r_2 r_3}{\sqrt{\Sigma r_1 r_2}} = \Delta$$

(c) Denominator $= \dfrac{(b^2 + c^2 - a^2)2R}{2bca} + \ldots + = \dfrac{R}{abc}$

$$[b^2 + c^2 - a^2 + c^2 + a^2 - b^2 + a^2 + b^2 - c^2]$$

$$= \frac{(a^2 + b^2 + c^2)R}{abc}$$

$$\therefore \quad \frac{N^r}{D^r} = \frac{\left(\Sigma a^2\right) abc}{\left(\Sigma a^2\right) R} = \frac{abc}{R} = 4\Delta$$

$\Rightarrow \quad C$ is not correct.

(d) $\dfrac{\Delta^2}{s^2} \cdot \dfrac{s(s - a)}{\Delta} \cdot \dfrac{s(s - b)}{\Delta} \cdot \dfrac{s(s - c)}{\Delta}$

$$= \frac{s(s - a)(s - b)(s - c)}{\Delta} = \frac{\Delta^2}{\Delta} = \Delta$$

Hence, (a), (b) and (d) are correct.

Ex. 15. *Consider the system of equations $\sin x \cos 2y = (a^2 - 1)^2 + 1$ and $\cos x \sin 2y = a + 1$. Which of the following ordered pairs (x, y) of real numbers can satisfy the given system of equations for permissible real values of a?*

(a) $\left(\dfrac{-\pi}{2}, \dfrac{-\pi}{2}\right)$ (b) $\left(\dfrac{\pi}{2}, \dfrac{3\pi}{2}\right)$

(c) $\left(\dfrac{3\pi}{2}, \dfrac{-\pi}{2}\right)$ (d) $\left(\dfrac{-\pi}{2}, \dfrac{3\pi}{2}\right)$

Sol. (a,c,d) *For permissible values of 'a', we must have*

$(a^2 - 1)^2 + 1 \le 1$ and $|a + 1| \le 1$

$\Rightarrow (a^2 - 1)^2 \le 0$ and $-1 \le a + 1 \le 1$

$\Rightarrow \quad -2 \le a \le 0 \Rightarrow a^2 - 1 = 0$

$\Rightarrow \quad a = 1$ or -1

$\therefore$ Permissible value of $a = -1$

Hence, the system of equations becomes $\sin x \cos 2y = 1$ and $\cos x \sin 2y = 0$.

Now, verify (a), (c), (d) alternatives.

Ex. 16. *In a ΔABC, let $2a^2 + 4b^2 + c^2 = 2a(2b + c)$, then which of the following holds good?*

[**Note** All symbols used have usual meaning in a triangle.]

(a) $\cos B = \dfrac{-7}{8}$ (b) $\sin(A - C) = 0$

(c) $\dfrac{r}{r_1} = \dfrac{1}{5}$ (d) $\sin A : \sin B : \sin C = 1 : 2 : 1$

Sol. (b,c) We have, $2a^2 + 4b^2 + c^2 - 4ab - 2ac = 0$

$$\Rightarrow \quad (a - 2b)^2 + (a - c)^2 = 0$$
$$\Rightarrow \quad a = 2b, a = c$$
$$\Rightarrow \quad b = \frac{a}{2}, c = a$$
$$\therefore \quad \cos A = \frac{b^2 + c^2 - a^2}{2bc}$$
$$= \frac{\frac{a^2}{4} + a^2 - a^2}{2\left(\frac{a}{2}\right)a} = \frac{1}{4}$$

Similarly, $\cos B = \frac{7}{8}$ and $\cos C = \frac{1}{4}$

Hence, option (a) is not correct.

As $\quad a = c \Rightarrow A = C$

$\Rightarrow \quad A - C = 0 \Rightarrow \sin(A - C) = 0$

Hence, option (b) is correct.

$$\text{As} \quad \frac{r}{r_1} = \frac{(s-a)\tan\frac{A}{2}}{s\tan\frac{A}{2}} = \frac{(s-a)}{s}$$
$$= 1 - \frac{a}{s} = 1 - \frac{a}{\frac{5a}{4}} = 1 - \frac{4}{5} = \frac{1}{5}$$

Hence, option (c) is correct.

Also, $a : b : c = a : \frac{a}{2} : a = 2 : 1 : 2 = \sin A : \sin B : \sin C$

Hence, option (d) is not correct.

Ex. 17. *In ΔABC, angles A, B and C are in the ratio 1 : 2 : 3, then which of the following is(are) correct? (All symbol used have usual meaning in a triangle).*

(a) Circum-radius of $\Delta ABC = c$

(b) $a : b : c = 1 : \sqrt{3} : 2$

(c) Perimeter of $\Delta ABC = 3 + \sqrt{3}$

(d) Area of $\Delta ABC = \frac{\sqrt{3}}{8}c^2$

Sol. (b, d) Given, $A + 2A + 3A = 180°$,

$B = 60°$ and $C = 90°$

Now, $\quad R = \frac{c}{2}$

$$\frac{a}{\sin A} = \frac{b}{\sin B} = \frac{c}{\sin C} = 2\left(\frac{c}{2}\right)$$

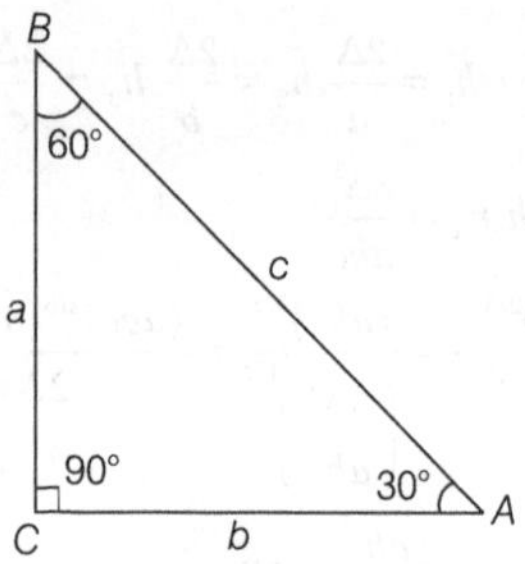

$$a = \frac{c}{2}, b = \frac{\sqrt{3}c}{2}$$

So, $\quad a : b : c = \frac{c}{2} : \frac{\sqrt{3}c}{2} : c = \frac{1}{2} : \frac{\sqrt{3}}{2} : 1 = 1 : \sqrt{3} : 2$

Perimeter $= (3 + \sqrt{3})k, (k \in R)$

Area of $\Delta ABC = \frac{1}{2}ab$

$$= \frac{1}{2}\left(\frac{c}{2}\right)\left(\frac{\sqrt{3}c}{2}\right) = \frac{\sqrt{3}}{8}c^2$$

Ex. 18. *If the length of tangents from A, B, C to the incircle of ΔABC are 4, 6, 8, then which of the following is(are) correct? (All symbols used have usual meaning in a triangle.)*

(a) Area of ΔABC is $12\sqrt{6}$ (b) r_1, r_2, r_3 are in HP

(c) a, b, c are in AP (d) $r = \frac{4\sqrt{6}}{3}$

Sol. (b, c, d) $s - a = 4, s - b = 6, s - c = 8$

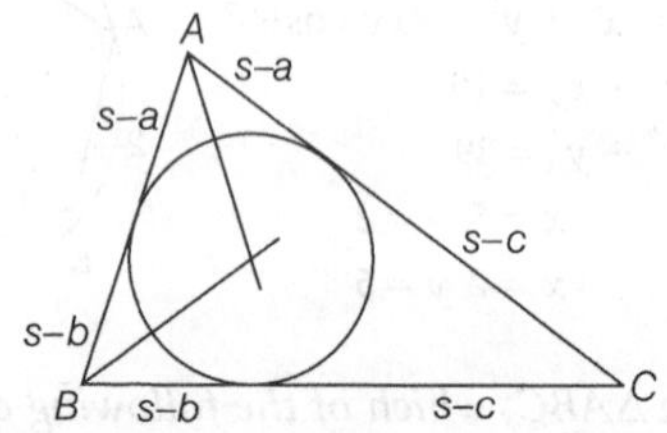

$$\therefore \quad s = 18$$
$$\therefore \quad \Delta = \sqrt{18 \times 4 \times 6 \times 8} = 24\sqrt{6}$$
$$a = 14, b = 12 \text{ and } c = 10$$

$s - a, s - b, s - c$ are in AP.

$\therefore$ a, b, c are in AP.

$\frac{\Delta}{s-a}, \frac{\Delta}{s-b}, \frac{\Delta}{s-c}$ are in HP.

$\therefore r_1, r_2, r_3$ are in HP $r = \frac{\Delta}{s} = \frac{4\sqrt{6}}{3}$

Ex. 19. *In ΔABC, let $b=10$, $c=10\sqrt{2}$ and $R=5\sqrt{2}$, then which of the following statement (s) is (are) correct?*

(a) Area of triangle ABC is 50.

(b) Distance between orthocentre and circumcentre is $5\sqrt{2}$.

(c) Sum of circum-radius and in-radius of ΔABC is equal to 10.

(d) Length of internal angle bisector of $\angle ACB$ of ΔABC is $\frac{5}{2\sqrt{2}}$.

Sol. (a,b,c) We know, $\frac{b}{\sin B} = 2R$

$$\Rightarrow \quad \sin B = \frac{10}{10\sqrt{2}} = \frac{1}{\sqrt{2}}$$

$$\Rightarrow \quad B = 45°$$

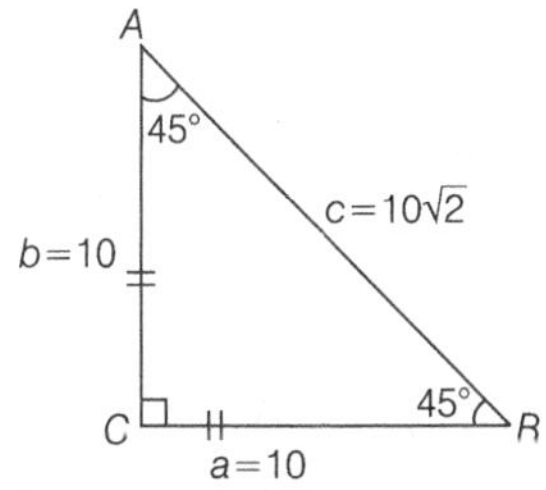

Also, $\quad \frac{c}{\sin C} = 2R$

$$\Rightarrow \quad \sin C = \frac{10\sqrt{2}}{10\sqrt{2}} = 1$$

$\therefore \quad A = 45° \Rightarrow C = 90°$

$\therefore \Delta ABC$ is isosceles right angled at C.

$\because \quad a = 10, b = 10, c = 10\sqrt{2}$

Area of $\Delta ABC = \frac{1}{2} \times 10 \times 10 = 50$

Distance between orthocentre and circumcentre

$$BD = DC = 5\sqrt{2}$$

$$R + r = 5\sqrt{2} + 10 - 5\sqrt{2} = 10 \qquad [\because r = 10 - 5\sqrt{2}]$$

Ex. 20. *Let 'ℓ' is the length of medians from the vertex A to the side BC of a ΔABC, then*

(a) $4\ell^2 = 2b^2 + 2c^2 - a^2$

(b) $4\ell^2 = b^2 + c^2 + 2bc \cos A$

(c) $4\ell^2 = a^2 + 4bc \cos A$

(d) $4\ell^2 = (2s - a)^2 - 4bc \sin^2 \frac{A}{2}$

Sol. (a, b, c, d) D is the mid-point of line BC.

$$AB^2 + AC^2 = 2(AD^2 + BD^2)$$

$$c^2 + b^2 = 2\left[l^2 + \left(\frac{a}{2}\right)^2\right]$$

$$\Rightarrow \quad c^2 + b^2 = 2l^2 + \frac{a^2}{2}$$

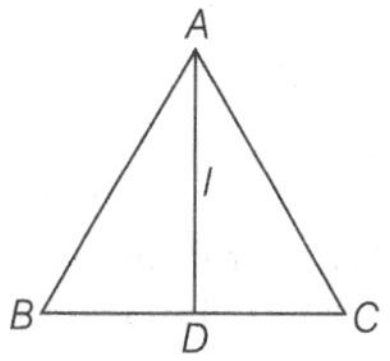

$$\begin{aligned} 4l^2 &= 2b^2 + 2c^2 - a^2 \\ &= b^2 + c^2 + (b^2 + c^2 - a^2) \\ &= b^2 + c^2 + 2bc \cos A \\ &= (b^2 + c^2 - a^2) + a^2 + 2bc \cos A \\ &= 2bc \cos A + a^2 + 2bc \cos A \\ &= 4bc \cos A + a^2 \end{aligned}$$

Ex. 21. *If a right angled ΔABC of maximum area is inscribed within a circle of radius R, then (Δ represents area of ΔABC and r, r_1, r_2, r_3 represent in-radius and ex-radii, and s is the semi-perimeter of ΔABC, then*

(a) $\Delta = R^2$ (b) $\frac{1}{r_1} + \frac{1}{r_2} + \frac{1}{r_3} = \frac{\sqrt{2}+1}{R}$

(c) $r = (\sqrt{2} - 1)R$ (d) $s = (1 + \sqrt{2})R$

Sol. (a, b, c, d) For a right angled triangle inscribed in a circle of radius R the length of the hypotenuse is $2R$.

Then, area is maximum when its is isosceles triangle

With each side $= \sqrt{2}\, R$

$$\therefore \quad S = \frac{1}{2}(2\sqrt{2} + 2)R = (\sqrt{2} + 1)R$$

$$\Delta = \frac{1}{2}\sqrt{2}R \cdot \sqrt{2}R = R^2$$

$$r = \frac{\Delta}{S} = \frac{R^2}{(\sqrt{2}+1)R}$$

$$\Rightarrow \quad r = (\sqrt{2} - 1)R$$

$$\Rightarrow \quad \frac{1}{r_1} + \frac{1}{r_2} + \frac{1}{r_3} = \frac{1}{r} = \frac{1}{(\sqrt{2}-1)R} = \frac{\sqrt{2}+1}{R}$$

JEE Type Solved Examples : Statement I and II Answer Type Questions

- This section contains 3 questions. Each question contains **Statement I (Assertion)** and **Statement II (Reason)**. Each question has 4 choices (a), (b), (c) and (d) out of which only one is correct. The choices are
  - (a) Statement I is True, Statement II is True; Statement II is a correct explanation for Statement I.
  - (b) Statement I is True, Statement II is True; Statement II is NOT a correct explanation for Statement I.
  - (c) Statement I is True, Statement II is False.
  - (d) Statement I is False, Statement II is True.

Ex. 22. Statement I *In a* ΔABC, *if* $a < b < c$ *and* r *is inradius and* r_1, r_2, r_3 *are the exradii opposite to angle A, B, C respectively, then* $r < r_1 < r_2 < r_3$.

Statement II *For,* ΔABC $r_1r_2 + r_2r_3 + r_3r_1 = \frac{r_1r_2r_3}{r}$

Sol. (b) **Statement I** $a < b < c$

$$s-a > s-b > s-c$$
$$s > s-a > s-b > s-c$$
$$\frac{\Delta}{s} < \frac{\Delta}{s-a} < \frac{\Delta}{s-b} < \frac{\Delta}{s-c}$$
$$r < r_1 < r_2 < r_3$$

Statement II $r_1 = \frac{\Delta}{s-a}, r_2 = \frac{\Delta}{s-b}, r_3 = \frac{\Delta}{s-c}, r = \frac{\Delta}{s}$

$$\frac{1}{r_1} + \frac{1}{r_2} + \frac{1}{r_3} = \frac{s}{\Delta} = \frac{1}{r}$$

Ex. 23. Statement I *If the sides of a triangle are 13, 14, 15 then the radius of in circle = 4*

Statement II *In a* ΔABC, $\Delta = \sqrt{s(s-a)(s-b)(s-c)}$ *where* $s = \frac{a+b+c}{2}$ *and* $r = \frac{\Delta}{s}$

Sol. (a) $s = 21$

$$\Delta = \sqrt{21\cdot 8\cdot 7\cdot 6} = \sqrt{3\cdot 7\cdot 2^4\cdot 7\cdot 3} = 3\cdot 7\cdot 4 = 84$$
$$r = \frac{\Delta}{s} = \frac{84}{21} = 4$$

Ex. 24. Statement I *In a* ΔABC, $\Sigma\frac{\cos^2\frac{A}{2}}{a}$ *has the value equal to* $\frac{s^2}{abc}$.

Statement II *In a* ΔABC, $\cos\frac{A}{2} = \sqrt{\frac{(s-b)(s-c)}{bc}}$

$$\cos\frac{\beta}{2} = \sqrt{\frac{(s-a)(s-c)}{ac}}, \cos\frac{c}{2} = \sqrt{\frac{(s-a)(s-b)}{ab}}$$

Sol. (c) $\frac{\cos^2\frac{A}{2}}{a} = \frac{s(s-a)}{abc}$

$$\therefore \quad \Sigma\frac{\cos^2\frac{A}{2}}{a}\ \frac{s^2}{abc}$$

JEE Type Solved Examples : Passage Based Questions

Passage I

(Ex. Nos. 25 to 27)

In a ΔABC, let $\tan A = 1$, $\tan B = 2$, $\tan C = 3$ and $c = 3$.

25. Area of the ΔABC is equal to

(a) $\frac{3\sqrt{2}}{2}$ (b) 3
(c) $2\sqrt{3}$ (d) $3\sqrt{2}$

26. The radius of the circle circumscribing the triangle ABC, is equal to

(a) $\frac{\sqrt{10}}{2}$ (b) $\sqrt{5}$ (c) $\sqrt{10}$ (d) $\frac{\sqrt{5}}{2}$

27. Let Δ denote the area of the ΔABC and Δp be the area of its pedal triangle. If $\Delta = k\,\Delta p$, then k is equal to

(a) $\sqrt{10}$ (b) $2\sqrt{5}$
(c) 5 (d) $2\sqrt{10}$

Sol. (Ex. Nos. 25 to 27) We have,

$$\tan A = 1 \Rightarrow \sin A = \frac{1}{\sqrt{2}};$$
$$\tan B = 2 \Rightarrow \sin B = \frac{2}{\sqrt{5}};$$
$$\tan C = 3 \Rightarrow \sin C = \frac{3}{\sqrt{10}}$$

Using Sine law,

$$a\sqrt{2} = \frac{b\sqrt{5}}{2} = \frac{c\sqrt{10}}{3}$$

$\therefore$ $$a\sqrt{2} = \frac{b\sqrt{5}}{2} = \sqrt{10} \quad [\text{As } c = 3]$$

Now, $a = \sqrt{5}; b = 2\sqrt{2}; c = 3$

25. (b) $\therefore \Delta = \frac{1}{2}\cdot\sqrt{5}\cdot 2\sqrt{2}\sin C = \sqrt{10}\cdot\frac{3}{\sqrt{10}} = 3$

26. (a) $R = \frac{abc}{4\Delta} = \frac{\sqrt{5}\cdot 2\sqrt{2}\cdot 3}{4\cdot 3} = \frac{\sqrt{10}}{2}$

27. (c) We know that, $\Delta_p = 2\Delta \cos A \cos B \cos C$

$$= 2\times\frac{1}{\sqrt{2}}\times\frac{1}{\sqrt{5}}\times\frac{1}{\sqrt{10}}\times\Delta$$

$$= \frac{\Delta}{5}$$

$\therefore$ $\Delta = 5\Delta_p$

$\Rightarrow$ $k = 5$

Passage II

(Ex. Nos. 28 to 30)

Let ABC be any triangle and P be a point inside it such that $\angle PAB = \frac{\pi}{18}, \angle PBA = \frac{\pi}{9}, \angle PCA = \frac{\pi}{6}, \angle PAC = \frac{2\pi}{9}$. Let $\angle PCB = x$

28. x is equal to

(a) $\frac{\pi}{9}$ (b) $\frac{2\pi}{9}$

(c) $\frac{\pi}{3}$ (d) None of these

29. ΔABC is

(a) Equilateral (b) Isosceless

(c) Scalene (d) Right angled

30. Which of the following is true

(a) $BC > AC$ (b) $AC = AB$

(c) $AC > AB$ (d) $BC = AC$

Sol. (Ex. Nos 28 to 30)

28. (a) In ΔPAB, $\frac{PA}{\sin 20} = \frac{PB}{\sin 10}$

$$\Rightarrow \frac{PA}{PB} = \frac{\sin 20^\circ}{\sin 10^\circ}$$

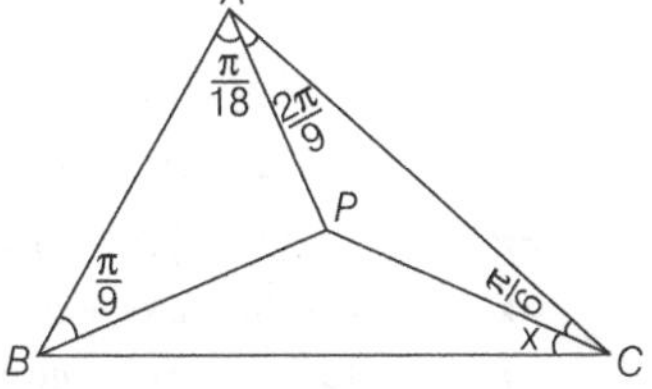

Similarly, In ΔPBC and ΔPAC, $\frac{PB}{PC} = \frac{\sin x}{\sin(80^\circ - x)}$,

$$\frac{PC}{PA} = \frac{\sin 40^\circ}{\sin 30^\circ}.$$

$$\frac{PA}{PB}\times\frac{PB}{PC}\times\frac{PC}{PA} = 1$$

$$\Rightarrow \frac{\sin 20^\circ}{\sin 10^\circ}\times\frac{\sin x}{\sin(80^\circ - x)}\times\frac{\sin 40^\circ}{\sin 30^\circ} = 1$$

$$\sin(80^\circ - x) = 4\cos 10^\circ \sin 40^\circ \sin x = 2\sin 50^\circ \sin x + \sin x$$

$$\Rightarrow \sin(80^\circ - x)\sin x = 2\sin 50^\circ \sin x$$

$$\Rightarrow \sin(80^\circ - x)\cos 40^\circ = 2\sin 50^\circ \sin x$$

$$\Rightarrow \sin(40^\circ - x) = \sin x$$

$$\Rightarrow x = 20^\circ$$

29. (b) $\angle A = \angle C = 50^\circ$

30. (c) $\angle ABC = 80^\circ$

$\therefore$ AC is longest side.

Passage III

(Ex. Nos. 31 to 33)

Let ΔABC be any triangle and D, E, F feet of perpendicular from vertices A, B, C on opposite side BC, CA, AB, respectively. Then, the ΔDEF is know as pedal Δ of ABC. H is orthocentre of the ΔABC. We note that $\angle HDC = \angle HEC = 90^\circ$, so the points H, D, C and E are concyclic.

In question ΔABC is assumed an acute angled triangle. In case ΔABC be obtuse angled with A as obtuse angle. The angle of pedal Δ will be $2A - 180^\circ, 2B, 2C$ and side will be represented by $-a\cos A, b\cos B, c\cos C$.

31. If l, m, n denote the side of a pedal triangle, then $\frac{l}{a^2} + \frac{m}{b^2} + \frac{n}{c^2}$ is equal to

(a) $\frac{a^2+b^2+c^2}{a^3+b^3+c^3}$ (b) $\frac{a^2+b^2+c^2}{2abc}$

(c) $\frac{a^3+b^3+c^3}{abc(a+b+c)}$ (d) $\frac{1}{a}+\frac{1}{b}+\frac{1}{c}$

32. If R be circum-radius of a Δ, then circum-radius of a pedal Δ is

(a) R (b) $\frac{2R}{3}$

(c) $\frac{R}{3}$ (d) $\frac{R}{2}$

33. The in-radius of pedal Δ of a ΔABC

(a) $\frac{R}{2}$ (b) $R\sin A\sin B\sin C$

(c) $2R\cos A\cos B\cos C$ (d) $4R\sin\frac{A}{2}\sin\frac{B}{2}\sin\frac{C}{2}$

Sol. (Ex. Nos 31 to 33)

31. (b) $l = -a\cos A, m = b\cos B, n = c\cos C$

$$\frac{l}{a^2}+\frac{m}{b^2}+\frac{n}{c^2} = \frac{\cos A}{a}+\frac{\cos B}{b}+\frac{\cos C}{c}$$

$$= \frac{a^2+b^2+c^2}{2abc}$$

32. (d) Circum-radius $= \dfrac{EF}{2\sin(\angle FDE)}$

$$= \frac{R\sin 2A}{2\sin(180^\circ - 2A)} = \frac{R}{2}$$

33. (c) Area of pedal $\Delta = \frac{1}{2}(DE)\, DF \sin(\angle EDF) = \frac{1}{2}(R\sin 2B)$

$$(R\sin 2C)\sin(180^\circ - 2A)$$

$$= \frac{1}{2}R^2 \sin 2A \sin 2B \sin 2C$$

Semi-perimeter of pedal Δ,

$$S = \frac{1}{2}R(\sin 2A + \sin 2B + \sin 2C)$$

In-radius $= \dfrac{\Delta}{S} = 2R\cos A \cos B \cos C.$

JEE Type Solved Examples : Matching Type Questions

Ex. 34. *Match the statement of Column I with value of Column II.*

Column I	Column II
(A) In a triangle ABC if $a^4 - 2(b^2 + c^2)a^2 + b^4 + b^2c^2 + c^4 = 0$, then $\angle A$	p. 30°
(B) In a triangle ABC, If $a^4 + b^4 + c^4 = a^2b^2 + 2b^2c^2 + 2c^2a^2$, then $\angle C$ is	q. 60°
(C) In a triangle ABC, If $a^4 + b^4 + c^4 + 2a^2c^2 = 2a^2b^2 + 2b^2c^2$, then $\angle B$ is	r. 90°
	s. 120°
	t. 150°

Sol. (A) → (q, s); (B) → (p, t); C → (r)

(A) $\because a^4 - 2(b^2 + c^2)a^2 + b^4 + b^2c^2 + c^4 = 0$

$$\Rightarrow (b^2 + c^2 - a^2)^2 - (b^2 + c^2)^2 + b^4 + b^2c^2 + c^4 = 0$$

$$\Rightarrow (b^2 + c^2 - a^2)^2 = b^2c^2$$

$$\Rightarrow \left(\frac{b^2 + c^2 - a^2}{2bc}\right)^2 = \frac{1}{4}$$

$$\Rightarrow \cos^2 A = \frac{1}{4}$$

$$\therefore \cos A = \pm\frac{1}{2}$$

$$\therefore A = 60^\circ \text{ or } 120^\circ\ (q, s)$$

(B) $\because \cos C = \dfrac{a^2 + b^2 - c^2}{2ab}$

or $a^2 + b^2 - c^2 + 2ab\cos C$

Squaring both sides, then

$$a^4 + b^4 + c^4 + 2a^2b^2 - 2a^2c^2 - 2b^2c^2 = 4a^2b^2\cos^2 C$$

$$\Rightarrow 3a^2b^2 = 4a^2b^2\cos^2 C$$

$$(\because a^4 + b^4 + c^4 = a^2b^2 + 2b^2c^2 + 2c^2a^2)$$

or $\cos^2 C = \dfrac{3}{4}$

$$\cos C = \pm\frac{\sqrt{3}}{2}$$

$$\therefore C = 30^\circ \text{ or } 150^\circ\ (p, t)$$

(C) $\because a^4 + b^4 + c^4 + 2a^2c^2 = 2a^2b^2 + 2b^2c^2$

$$\Rightarrow b^4 - 2(a^2 + c^2)b^2 + (a^2 + c^2)^2 = 0$$

$$\Rightarrow \{b^2 - (a^2 + c^2)\}^2 = 0$$

$$\Rightarrow \therefore a^2 + c^2 = b^2$$

$$\Rightarrow \cos B = 0$$

$$\therefore B = 90^\circ\ (r)$$

JEE Type Solved Examples : Single Integer Answer Type Questions

Ex. 35. *In a ΔABC, if $r_1 + r_3 + r = r_2$, then find the value of $(\sec^2 A + \cos^2 B - \cot^2 C)$.*

[**Note** All symbols used have usual meaning in a triangle.]

Sol. (1) We have, $r_1 + r_3 + r = r_2$

$$\Rightarrow \left(\frac{\Delta}{s-a} + \frac{\Delta}{s-c}\right) = \left(\frac{\Delta}{s-b} - \frac{\Delta}{s}\right)$$

$$\Rightarrow \frac{2s - (a + c)}{(s-a)(s-c)} = \frac{b}{s(s-b)}$$

$$\Rightarrow (s-a)(s-c) = s(s-b)$$

$$\Rightarrow s^2 - (a + c)s + ac = s^2 - bs$$

$$\Rightarrow (a + c - b)s - ac = 0$$

$$\Rightarrow (a + c - b)\frac{(a + b + c)}{2} = ac$$

$$\Rightarrow \quad (a+c)^2 - b^2 = 2ac$$
$$\Rightarrow \quad a^2 + c^2 + 2ac - b^2 = 2ac$$
$$\Rightarrow \quad a^2 + c^2 = b^2$$
So, $\angle B = \frac{\pi}{2}$ and $\angle A + \angle C = 90°$

Now,
$$\sec^2 A + \cos^2 B - \cot^2 C = \sec^2(90° - C) + \cos^2 90° - \cot^2 C$$
$$= \text{cosec}^2 C + 0 - \cot^2 C$$
$$= 1$$
Hence, $(\sec^2 A + \cos^2 B - \cot^2 C) = 1$

Alternatively We have,
$$(r_1 + r_3) - (r_2 - r) = 0 \quad \text{...(i)}$$
As $r_1 + r_3 = 4R\sin\frac{A}{2}\cos\frac{B}{2}\cos\frac{C}{2} + 4R\cos\frac{A}{2}\cos\frac{B}{2}\sin\frac{C}{2}$
$$= 4R\cos\frac{B}{2}\sin\left(\frac{A+C}{2}\right) = 4R\cos^2\frac{B}{2} \quad \text{...(ii)}$$
Also, $r_2 - r = 4R\cos\frac{A}{2}\sin\frac{B}{2}\cos\frac{C}{2} - 4R\sin\frac{A}{2}\sin\frac{B}{2}\sin\frac{C}{2}$
$$= 4R\sin\frac{B}{2}\cos\left(\frac{A+C}{2}\right) = 4R\sin^2\frac{B}{2} \quad \text{...(iii)}$$
$\therefore$ Using Eqs. (ii) and (iii) in Eq. (i), we get
$$4R\left(\cos^2\frac{B}{2} - \sin^2\frac{B}{2}\right) = 0$$
$$\Rightarrow \quad 4R\cos B = 0$$
$$\Rightarrow \quad \angle B = \frac{\pi}{2} \text{ and } \angle A + \angle C = 90°$$
Hence, $(\sec^2 A + \cos^2 B - \cot^2 C) = 1$

Ex. 36. *In ΔABC, let $b = 6, c = 10$ and $r_1 = r_2 + r_3 + r$ then find area of ΔABC.*

[**Note** All symbols used have usual meaning in a triangle.]

Sol. (30) We have, $r_1 = r_2 + r_3 + r$ [given]
$$\Rightarrow \quad (r_1 - r) = (r_2 + r_3)$$
$$\Rightarrow \quad \frac{s-(s-a)}{s(s-a)} = \frac{2s-(b+c)}{(s-b)(s-c)}$$
$$\Rightarrow \quad \frac{(s-b)(s-c)}{s(s-a)} = 1$$
$$\Rightarrow \quad \tan^2\frac{A}{2} = 1 \Rightarrow \frac{A}{2} = 45°$$
Hence, $A = 90°$

Now, area of $\Delta ABC = \frac{1}{2}bc\sin A$
$$= \frac{1}{2}(6)(10)\sin 90°$$
$$= 30 \text{ sq units.}$$

Ex. 37. *Consider an obtuse angled triangles with side 8 cm, 15 cm and x cm (largest side being 15 cm). If x is an integer, then find the number of possible triangles.*

Sol. (5) Since, 15 be the largest side
$$\cos\theta = \frac{x^2 + 8^2 - 15^2}{16x}$$
for obtuse angled $-1 < \cos\theta < 0$
$$-16x < x^2 - 161 < 0$$

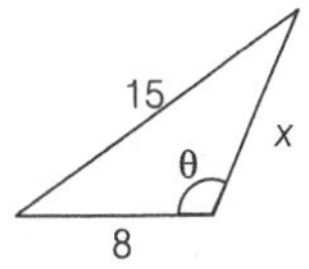

$$\Rightarrow \quad x^2 < 161$$
$$\Rightarrow \quad x < 13 \quad \text{...(i)}$$
and $\quad x^2 + 16x - 161 > 0$
$$\Rightarrow \quad (x+23)(x-7) > 0$$
or $\quad x > 7 \quad$ ...(ii)
$$\therefore \quad 7 < x < 13$$
$$\Rightarrow \quad x \in \{8, 9, 10, 11, 12\}$$
Hence, number of possible values of x is 5.

Ex. 38. *Let ABC be a right angled triangle at C. If the inscribed circle touches the side AB to D and $(AD)(BD) = 11$, then find the area of ΔABC.*

Sol. (11) We have,

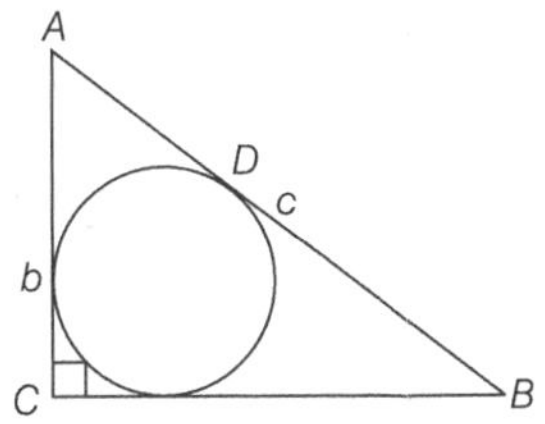

$$(AD)(BD) = 11$$
$$\Rightarrow \quad (s-a)(s-b) = 11$$
$$\Rightarrow \quad (2s-2a)(2s-2b) = 44$$
$$\Rightarrow \quad (b+c-a)(a+c-b) = 44$$
$$\Rightarrow \quad c^2 - (b-a)^2 = 44$$
$$\Rightarrow \quad 2ab = 44$$
$$[\text{As } c^2 = a^2 + b^2]$$
$$\Rightarrow \quad ab = 22$$
Now, area $(\Delta ABC) = \frac{1}{2}ab = \frac{1}{2}(22) = 11$

Ex. 39. *Consider a ΔABC and let a, b and c denote the lengths of the sides opposite to vertices A, B and C, respecively. Suppose $a = 2, b = 3, c = 4$ and H be the orthocentre. Find $15(HA)^2$.*

Sol. (196) We know that, $HA = 2R\cos A$, where
$$\cos A = \frac{b^2 + c^2 - a^2}{2bc} = \frac{9 + 16 - 4}{24}$$
$$= \frac{21}{24} = \frac{7}{8}$$
$$\therefore \quad \sin A = \sqrt{1 - \frac{49}{64}} = \frac{\sqrt{15}}{8}$$

Now, $\dfrac{a}{\sin A} = 2R$

$$\Rightarrow \quad 2R = \frac{2 \times 8}{\sqrt{15}} = \frac{16}{\sqrt{15}}$$

$$\therefore \quad HA = 2R \cos A = \frac{16}{\sqrt{15}} \times \frac{7}{8}$$

$$= \frac{14}{\sqrt{15}}$$

Hence, $15(AH)^2 = 196$.

Ex. 40. *In a ΔABC, the internal angle bisector of $\angle ABC$ meets AC at K. If $BC = 2$, $CK = 1$ and $BK = \dfrac{3\sqrt{2}}{2}$, then find the length of side AB.*

Sol. (3) Using cosine law in ΔBKC,

$$\cos \frac{B}{2} = \frac{4 + \frac{18}{4} - 1}{2 \cdot \frac{3\sqrt{2}}{2} \cdot 2}$$

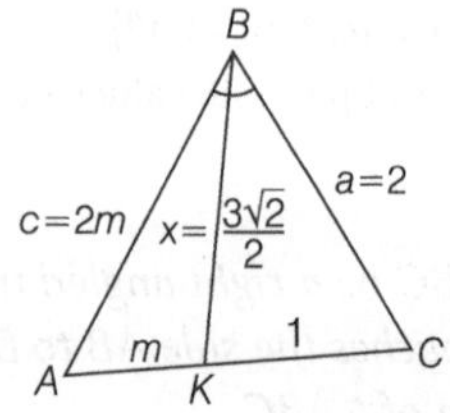

$$= \frac{3 + \frac{9}{2}}{6\sqrt{2}} = \frac{15}{12\sqrt{2}} = \frac{5}{4\sqrt{2}}$$

Now, $$x = \frac{2ac}{a+c} \cos \frac{B}{2} = \frac{(2 \cdot 2 \cdot 2\text{m})}{2m + 2} \cdot \frac{5}{4\sqrt{2}}$$

$$\Rightarrow \quad \frac{3\sqrt{2}}{2} = \frac{4m}{m+1} \cdot \frac{5}{4\sqrt{2}}$$

$$\Rightarrow \quad 3 = \frac{5m}{m+1}$$

$$\Rightarrow \quad 3m + 3 = 5m$$

$$m = \frac{3}{2}$$

$$\therefore \quad AB = 2m$$

$$\therefore \quad c = 3$$

Ex. 41. *In ΔABC has $AC = 13$, $AB = 15$ and $BC = 14$. Let O be the circumcentre of the ΔABC. If the length of perpendicular from the point 'O' on BC can be expressed as a rational $\dfrac{m}{n}$ in the lowest form, then find $(m + n)$.*

Sol. $(41) \Delta = \sqrt{s(s-a)(s-b)(s-c)}$

$$= \sqrt{21 \cdot 8 \cdot 6 \cdot 7} = \sqrt{84}$$

$$R = \frac{abc}{4\Delta} = \frac{(14)(13)(15)}{4 \cdot 84} = \frac{2 \cdot 13 \cdot 15}{4 \cdot 12} = \frac{65}{8}$$

$$\therefore \quad p = \sqrt{R^2 - 7^2}$$

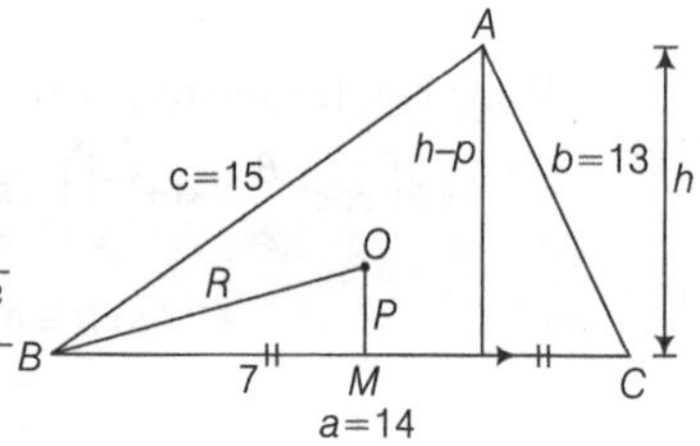

$$= \sqrt{\left(\frac{65}{8}\right)^2 - 7^2}$$

$$= \frac{\sqrt{(65)^2 - (56)^2}}{8}$$

$$= \frac{\sqrt{121 \cdot 9}}{8}$$

$$= \frac{33}{8} = \frac{m}{n}$$

$$\Rightarrow \quad m + n = 33 + 8 = 41.$$

Subjective Type Examples

Ex. 42. *Two sides of a triangle are given by the roots of the equation $x^2 - 2\sqrt{3}x + 2 = 0$. The angle between the sides is $\dfrac{\pi}{3}$. Find the perimeter of Δ.*

Sol. We have, two sides of a triangle are given by roots of the equation $x^2 - 2\sqrt{3}x + 2 = 0$.

$$\Rightarrow \quad a + b = 2\sqrt{3}$$

and $\quad ab = 2$ with $\angle C = \dfrac{\pi}{3}$ ...(i)

using cosine law, we have

$$\cos C = \frac{a^2 + b^2 - c^2}{2ab}$$

$$\Rightarrow \quad \cos \frac{\pi}{3} = \frac{a^2 + b^2 - c^2}{2ab} \quad \text{[using Eq. (i)]}$$

$$\Rightarrow \quad \frac{1}{2} = \frac{a^2 + b^2 - c^2}{2ab}$$

$$\Rightarrow \quad a^2 + b^2 - c^2 = ab$$

$$\Rightarrow \quad (a+b)^2 - 2ab - c^2 = ab \quad \text{[using Eq. (i)]}$$

$$\Rightarrow \quad 12 - 4 - c^2 = 2$$

$$c^2 = 6; \; c = \sqrt{6}$$

$\therefore$ The perimeter of Δ

$$= a + b + c = 2\sqrt{3} + \sqrt{6}$$

Ex. 43. *If in ΔABC, $\angle A = 90^\circ$ and c, $\sin B$, $\cos B$ are rational numbers, then show a and b are rational.*

Sol. Let AD be perpendicular from A to BC.

Then, $\cos B = \dfrac{BD}{c}$

$\Rightarrow$ $BD = c\cos B$

$\Rightarrow$ BD is rational.

Similarly, $AD = c\sin B$

$\Rightarrow$ AD is rational.

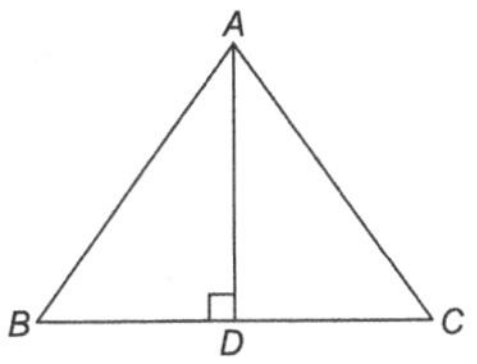

Now, $\sin C = \cos B = \dfrac{AD}{b}$

$\Rightarrow$ b is rational.

Since, $\cos C = \sin B = \dfrac{DC}{b}$

$\Rightarrow$ DC is rational.

Hence, $a = BD + DC$ is rational.

Thus, a is rational and b is rational.

Ex. 44. *If the sides of a triangle ABC are in AP and 'a' is the smallest side, then express $\cos A$ in terms of b and c.*

Sol. Since sides of the triangle are in AP.

i.e. a, b, c are in AP and let $a < b < c$.

$\therefore$ $2b = a + c$ ...(i)

Now, $\cos A = \dfrac{b^2 + c^2 - a^2}{2bc}$

$= \dfrac{b^2 + c^2 - (2b - c)^2}{2bc}$ [using Eq. (i)]

$= \dfrac{b^2 + c^2 - 4b^2 - c^2 + 4bc}{2bc}$

$= \dfrac{4bc - 3b^2}{2bc}$

$\therefore$ $\cos A = \dfrac{4c - 3b}{2c}$

Ex. 45. *If A, B and C are angles of a triangle such that $\angle A$ is obtuse, then show $\tan B \tan C < 1$.*

Sol. Since, A is obtuse angle,

then $90^\circ < A < 180^\circ$

$\Rightarrow$ $90^\circ < 180 - (B + C) < 180^\circ$

$\Rightarrow$ $-90^\circ < -(B + C) < 0$

$\Rightarrow$ $90^\circ > B + C > 0$

$\Rightarrow$ $B + C < 90^\circ$

$\Rightarrow$ $B < 90^\circ - C$

$\therefore$ $\tan B < \tan(90^\circ - C)$

$\tan B < \cot C$

or $\tan B \cdot \tan C < 1$

If A is obtuse, then $\tan A \tan C < 1$.

Ex. 46. *If A is the area and $2s$ the sum of the sides of a triangle, then show $A \le \dfrac{s^2}{3\sqrt{3}}$.*

Sol. We have, $2s = a + b + c$,

$A^2 = s(s - a)(s - b)(s - c)$

Now, AM $\ge$ GM

$\Rightarrow$ $\dfrac{s + (s - a) + (s - b) + (s - c)}{4} \ge \{(s - a)(s - b)(s - c)\}^{1/4}$

$\Rightarrow$ $\dfrac{4s - 2s}{4} \ge (A^2)^{1/4} \Rightarrow \dfrac{s}{2} \ge A^{1/2}$

$\Rightarrow$ $A \le \dfrac{s^2}{4}$ ...(i)

Also, $\dfrac{(s - a) + (s - b) + (s - c)}{3} \ge \{(s - a)(s - b)(s - c)\}^{1/3}$

or $\dfrac{3s - 2s}{3} \ge \left(\dfrac{A^2}{s}\right)^{1/3}$

or $\dfrac{s}{3} \ge \left(\dfrac{A^2}{s}\right)^{1/3}$ or $\dfrac{A^2}{s} \le \dfrac{s^3}{27}$

$\Rightarrow$ $A \le \dfrac{s^2}{3\sqrt{3}}$ ...(ii)

Thus, from Eqs. (i) and (ii);

$A \le \dfrac{s^2}{3\sqrt{3}}$

Ex. 47. *In a triangle, if $r_1 > r_2 > r_3$, then show $a > b > c$.*

Sol. We have, $r_1 > r_2 > r_3$

$\Rightarrow$ $\dfrac{\Delta}{s - a} > \dfrac{\Delta}{s - b} > \dfrac{\Delta}{s - c}$

$\Rightarrow$ $\dfrac{s - a}{\Delta} < \dfrac{s - b}{\Delta} < \dfrac{s - c}{\Delta}$

$\Rightarrow$ $s - a < s - b < s - c$

$\Rightarrow$ $-a < -b < -c$ or $a > b > c$

Ex. 48. *ABC is a triangle and D is the middle point of BC. If AD is perpendicular to AC, then prove that*

$$\cos A \cdot \cos C = \frac{2(c^2 - a^2)}{3ac}.$$

Sol. D is the mid-point of BC

$\Rightarrow$ $BD = DC = \dfrac{a}{2}$

Draw BE perpendicular to CA (produced).

$$\angle DAC = \angle BEC = 90^\circ$$

In Δ's ADC and EBC,

$$\angle BEC = \angle DAC = 90^\circ$$

and $\angle C$ is common.

Hence, $\angle CDA = \angle CBE = 90^\circ - C$

In Δ's ADC and EBC are similar

$$\Rightarrow \quad \frac{DC}{BC} = \frac{DA}{BE} = \frac{CA}{CE} \Rightarrow EA = b$$

and if $DA = y$, then $BE = 2y$

Now, $\cos C = \dfrac{b}{a/2} = \dfrac{2b}{a}$

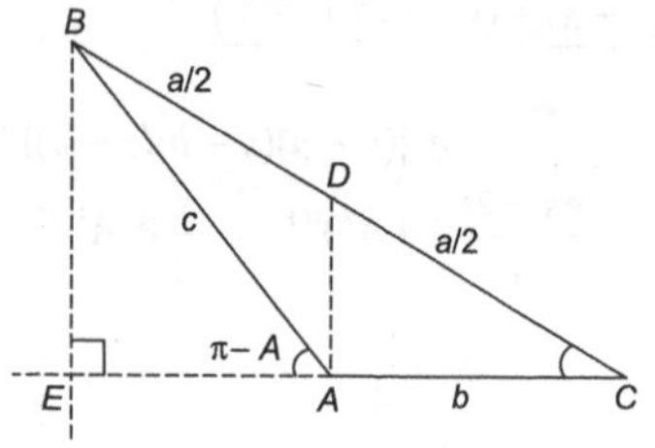

and $\cos(\pi - A) = \dfrac{b}{c}$

$$\cos A = -\frac{b}{c}$$

$$\therefore \quad \cos A \cdot \cos C = -\frac{2b^2}{ac} \quad \text{...(i)}$$

In ΔADC, $y^2 + b^2 = \dfrac{a^2}{4}$

$$\Rightarrow \quad 4y^2 + 4b^2 = a^2$$

In ΔBAE, $4y^2 + b^2 = c^2 \Rightarrow c^2 + 3b^2 = a^2$

$$\Rightarrow \quad b^2 = \frac{a^2 - c^2}{3} \quad \text{...(ii)}$$

From Eqs. (i) and (ii), $\cos A \cdot \cos C = \dfrac{2(c^2 - a^2)}{3ac}$

Ex. 49. *In a ΔABC, prove that*

$$a^3\cos(B - C) + b^3\cos(C - A) + c^3\cos(A - B) = 3abc$$

Sol. First term of LHS $= a^3\cos(B - C)$

$$= a^2[a\cos(B - C)], \text{where } a = 2R\sin A$$
$$= a^2 \cdot 2R\sin A \cdot \cos(B - C)$$
$$= a^2 2R\sin(B + C)\cos(B - C) \quad [\text{as } A + B + C = \pi$$
$$\therefore \sin A = \sin(\pi - (B + C)) = \sin(B + C)]$$
$$= a^2 \cdot R[2\sin(B + C)\cdot\cos(B - C)]$$
$$= a^2 \cdot R[\sin 2B + \sin 2C]$$
$$= a^2 \cdot R[2\sin B\cos B + 2\sin C\cos C]$$
$$= a^2[b\cos B + c\cos C] \quad \text{...(i)}$$

[as, $2R\sin B = b$ and $2R\sin C = c$]

Similarly,

$$b^3\cos(C - A) = b^2[a\cos A + c\cos C] \quad \text{...(ii)}$$

and $c^3\cos(A - B) = c^2[a\cos A + b\cos B]$ ...(iii)

$\therefore$ From Eqs. (i), (ii) and (iii), we get,

$$a^3\cos(B - C) + b^3\cos(C - A) + c^3\cos(A - B)$$
$$= a^2b\cos B + a^2c\cos C + b^2a\cos A + b^2c\cos C + c^2a\cos A + c^2b\cos B$$
$$= ab(a\cos B + b\cos A) + ac(a\cos C + c\cos A) + bc(b\cos C + c\cos B)$$
$$= ab\cdot c + ac\cdot b + bc\cdot a \quad [\text{using projection formula}]$$
$$= 3abc$$

$$\therefore \quad a^3\cos(B - C) + b^3(C - A) + c^3\cos(A - B) = 3abc$$

Ex. 50. *In a triangle of base 'a', the ratio of the other sides is r (<1). Show that the altitude of the triangle is less than or equal to $\dfrac{ar}{1 - r^2}$.*

Sol. Let D be the foot of the altitude from A,

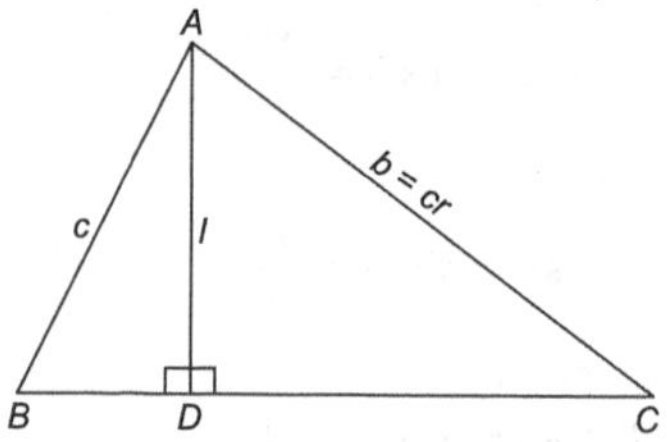

and $BC = a$, $AB = c$, $AC = cr$, $AD = l = C\sin B$

Now, $\dfrac{ar}{1 - r^2} = \dfrac{ac^2r}{c^2 - r^2c^2} = \dfrac{abc}{c^2 - b^2}$ [using Sine law]

$$= \frac{a\sin B\sin C}{\sin^2 C - \sin^2 B}$$
$$= \frac{a\sin B\sin C}{\sin(C - B).\sin(C + B)}$$
$$= \frac{a\sin B.\sin C}{\sin A\sin(C - B)}$$
$$= \frac{c\sin B}{\sin(C - B)} = \frac{l}{\sin(C - B)}$$

$$\Rightarrow \quad l \le \frac{ar}{1 - r^2}.$$

Ex. 51. *Three circles touch one-another externally. The tangents at their points of contact meet at a point whose distance from a point of contact is 4. Find the ratio of the product of the radii to the sum of the radii of circles.*

Sol. Let r_1, r_2 and r_3 be the radii of the three circles with centres at C_1, C_2 and C_3. Let the circles touch at P, Q and R.

Also,

$$C_1C_2 = r_1 + r_2$$
$$C_2C_3 = r_2 + r_3$$
$$C_3C_1 = r_3 + r_1$$

Let O be the point whose distance from the points of contact is 4.

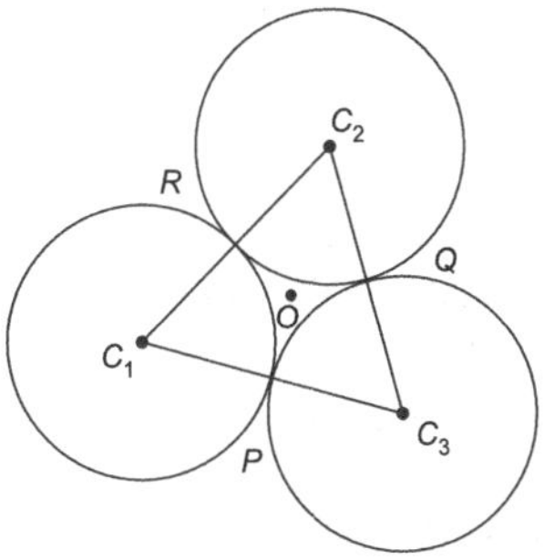

Then, O is the in-centre of $\Delta C_1C_2C_3$ with $OP = OQ = OR = 4$ being the radius of the in-circle.

Hence, $$4 = \frac{\Delta C_1C_2C_3}{\frac{1}{2}[C_1C_2 + C_2C_3 + C_3C_1]} = \frac{S}{s} \quad \text{...(i)}$$

where, $s = r_1 + r_2 + r_3$,

$$S^2 = s(s - C_1C_2)(s - C_2C_3)(s - C_3C_1)$$
$$= s r_1 r_2 r_3.$$

$\therefore$ Eq. (i) gives, $$16 = \frac{S^2}{s^2} = \frac{s\, r_1r_2r_3}{s^2} = \frac{r_1r_2r_3}{r_1 + r_2 + r_3}$$

Hence, the ratio of the product of the radii to the sum of the radii = 16 : 1.

Ex. 52. *The internal bisectors of the angles of a ΔABC meet the sides BC, CA, AB in D, E and F, respectively. Show that the area of the ΔDEF is equal to,*

$$\frac{2\Delta abc}{(b+c)(c+a)(a+b)}$$

Sol. AD is the internal bisector of $\angle A$.

$$\Rightarrow \quad \frac{BD}{DC} = \frac{BA}{AC} = \frac{c}{b}$$

$$\Rightarrow \quad \frac{BD}{c} = \frac{DC}{b} = \frac{BD + DC}{c + b} = \frac{a}{b + c}$$

$$\Rightarrow \quad BD = \frac{ac}{b+c}, \; DC = \frac{ab}{b+c}$$

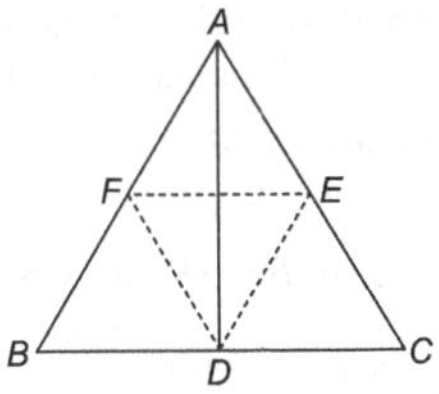

Similarly, $$BF = \frac{ac}{a+b}$$

$$\Rightarrow \quad \text{Area of } \Delta BFD = \frac{1}{2}(BF \cdot BD) \cdot \sin B$$

$$\Rightarrow \quad \frac{\text{Area of } \Delta BFD}{\text{Area of } \Delta ABC} = \frac{(BF \cdot BD) \cdot \sin B}{ac \sin B}$$

$$= \frac{ac}{(a+b)(b+c)}$$

Now, area of $\Delta DEF = \Delta ABC - (\Delta BFD + \Delta DEC + \Delta AFE)$

$$\Rightarrow \quad \frac{\text{Area of } \Delta DEF}{\text{Area of } \Delta ABC} = 1 - \left(\frac{\Delta BFD + \Delta DEC + \Delta AFE}{\Delta ABC}\right)$$

$$= 1 - \left\{\frac{ac}{(a+b)(b+c)} + \frac{cb}{(c+a)(b+a)} + \frac{ab}{(a+c)(b+c)}\right\}$$

$$= \frac{2abc\,\Delta}{(a+b)(b+c)(c+a)}$$

$$\Rightarrow \quad \text{Area of } \Delta DEF = \frac{2abc\,\Delta}{(a+b)(b+c)(c+a)}$$

Ex. 53. *Let D be a point on the side BC of a ΔABC such that $BD : DC = m : n$ and $\angle ADC = \theta$, $\angle BAD = \alpha$ and $\angle DAC = \beta$. Prove that*

(i) $(m+n)\cot\theta = m\cot\alpha - n\cot\beta$

(ii) $(m+n)\cot\theta = n\cot B - m\cot C$

Sol. (i) Given, $\dfrac{BD}{DC} = \dfrac{m}{n}$ and $\angle ADC = \theta$

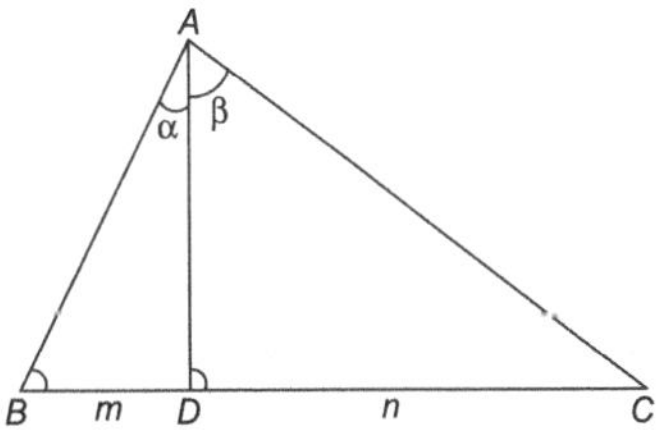

Since, $\angle ADB = (180° - \theta)$, $\angle BAD = \alpha$

and $\angle DAC = \beta$

$\angle ABD = 180° - (\alpha + 180° - \theta) = \theta - \alpha$

and $\angle ACD = 180° - (\theta + \beta)$

From ΔABD,

$$\frac{BD}{\sin\alpha} = \frac{AD}{\sin(\theta - \alpha)} \quad \text{...(i)}$$

From ΔADC,

$$\frac{DC}{\sin\beta} = \frac{AB}{\sin\{180° - (\theta + \beta)\}}$$

or $$\frac{DC}{\sin\beta} = \frac{AD}{\sin(\theta + \beta)} \quad \text{...(ii)}$$

On dividing Eq. (i) by Eq. (ii), we get

$$\frac{BD\sin\beta}{DC\sin\alpha} = \frac{\sin(\theta + \beta)}{\sin(\theta - \alpha)}$$

or $$\frac{m}{n} \cdot \frac{\sin\beta}{\sin\alpha} = \frac{\sin\theta\cos\beta + \cos\theta\sin\beta}{\sin\theta\cos\alpha - \cos\theta\sin\alpha}$$

or $$m\sin\beta(\sin\theta\cos\alpha - \cos\theta\sin\alpha) = n\sin\alpha(\sin\theta\cos\beta + \cos\theta\sin\beta)$$

or $$(m+n)\cot\theta = m\cot\alpha - n\cot\beta$$

(ii) Substitute $\alpha = \theta - B$ and $\beta = 180° - (\theta + C)$ in the above result,

Simplify and obtain $(m+n)\cot\theta = n\cot B - m\cot C$.

Note This is known as $m - n$ theorem.

Ex. 54. *The base of a triangle is divided into three equal parts. If t_1, t_2, t_3 be the tangents of the angles subtended by these parts at the opposite vertex, prove that :*

$$\left(\frac{1}{t_1}+\frac{1}{t_2}\right)\left(\frac{1}{t_2}+\frac{1}{t_3}\right)=4\left(1+\frac{1}{t_2^2}\right)$$

Sol. Let the points P and Q divide the side BC in three equal parts

Such that $BP = PQ = QC = x$

Also let, $\angle BAP = \alpha, \angle PAQ = \beta, \angle QAC = \gamma$

and $\angle AQC = \theta$

From question, $\tan\alpha = t_1, \tan\beta = t_2, \tan\gamma = t_3$.

Applying, $m:n$ rule in triangle ABC, we get

$$(2x + x)\cot\theta = 2x\cot(\alpha+\beta) - x\cot\gamma \quad \text{...(i)}$$

From ΔAPC, we get

$$(x + x)\cot\theta = x\cot\beta - x\cot\gamma \quad \text{...(ii)}$$

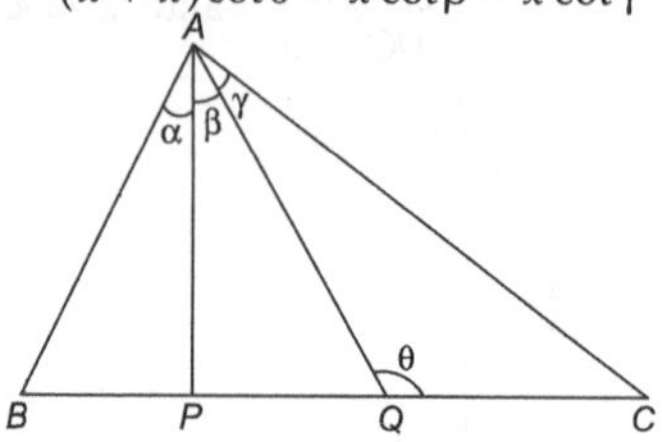

On dividing Eq. (i) and Eq. (ii), we get

$$\frac{3}{2}=\frac{2\cot(\alpha+\beta)-\cot\gamma}{\cot\beta-\cot\gamma}$$

or $$3\cot\beta-\cot\gamma=\frac{4(\cot\alpha\cdot\cot\beta-1)}{\cot\beta+\cot\alpha}$$

or $3\cot^2\beta - \cot\beta\cot\gamma + 3\cot\alpha\cdot\cot\beta - \cot\alpha\cdot\cot\gamma$
$= 4\cot\alpha\cdot\cot\beta - 4$

or $4 + 4\cot^2\beta = \cot^2\beta + \cot\alpha\cdot\cot\beta +$
$\cot\beta\cdot\cot\gamma + \cot\gamma\cdot\cot\alpha$

or $4(1+\cot^2\beta) = (\cot\beta + \cot\alpha)(\cot\beta + \cot\gamma)$

or $$4\left(1+\frac{1}{t_2^2}\right)=\left(\frac{1}{t_1}+\frac{1}{t_2}\right)\left(\frac{1}{t_2}+\frac{t}{t_3}\right)$$

Hence, the result.

Ex. 55. *In a ΔABC, if $\cot A+\cot B+\cot C=\sqrt{3}$. Prove that the triangle is equilateral.*

Sol. $\cot A + \cot B + \cot C = \sqrt{3}$

On squaring both sides, we get

$\cot^2 A + \cot^2 B + \cot^2 C + 2\cot A\cdot\cot B + 2\cot B\cdot\cot C +$
$2\cot C\cdot\cot A = 3$ ...(i)

Also, $\cot A\cot B + \cot B\cot C + \cot C\cot A = 1$ ...(ii)

From Eqs. (i) and Eqs. (ii), we have

$\cot^2 A + \cot^2 B + \cot^2 C - \cot A\cot B - \cot C\cot B -$
$\cot C\cot A = 0$

or $\frac{1}{2}\{(\cot A - \cot B)^2 + (\cot B - \cot C)^2$
$+ (\cot C - \cot A)^2\} = 0$

$\Rightarrow \quad \cot A = \cot B = \cot C$

[Since RHS is zero, thus each square must be zero]

$\Rightarrow \quad A = B = C$

i.e. the triangle is equilateral.

Ex. 56. *In the ΔABC, if $(a^2+b^2)\sin(A-B)=(a^2-b^2)$ $\sin(A+B)$. Prove that the triangle is either isosceles or right angled.*

Sol. Let $$\frac{a}{\sin A}=\frac{b}{\sin B}=\frac{c}{\sin C}=K \text{ (say)}.$$

$\Rightarrow \quad a = K\sin A, \; b = K\sin B$

and $\quad c = K\sin C$.

Now, the given relation is,

$(a^2+b^2)\sin(A-B) = (a^2-b^2)\sin(A-B)$

$\Rightarrow \quad K^2(\sin^2 A + \sin^2 B)\sin(A-B)$
$= K^2(\sin^2 A - \sin^2 B)\sin(A+B)$

$\Rightarrow \quad (\sin^2 A + \sin^2 B)\sin(A-B) = \sin^2(A+B)\cdot\sin(A-B)$

or $\quad \sin(A-B)\,[\sin^2 A + \sin^2 B - \sin^2 C] = 0$

Hence, either the first factor = 0;

or the second factor = 0

If $\sin(A-B) = 0 \Rightarrow A - B = 0 \Rightarrow A = B$

$\Rightarrow$ triangle is isosceles.

If $\sin^2 A + \sin^2 B - \sin^2 C = 0$

$$\Rightarrow \quad \frac{a^2}{K^2}+\frac{b^2}{K^2}-\frac{c^2}{K^2}=0 \quad \text{or} \quad a^2+b^2=c^2$$

$\Rightarrow$ the triangle is right angled.

Hence, the result.

Ex. 57. *The sides of a ΔABC are in AP. If the $\angle A$ and $\angle C$ are the greatest and smallest angles respectively, prove that $4(1-\cos A)(1-\cos C)=\cos A+\cos C$*

Sol. Since A is the greatest and C is the smallest angle, 'a' is the greatest and 'c' is the smallest side.

$\Rightarrow \quad a, b, c$ are in AP.

$\Rightarrow \quad 2b = a + c.$

or $\quad 4R\sin B = 2R[\sin A + \sin C]$

or $$2\cdot 2\sin\frac{B}{2}\cdot\cos\frac{B}{2}=2\sin\left(\frac{A+C}{2}\right)\cdot\cos\left(\frac{A-C}{2}\right)$$

or $$2\sin\frac{B}{2}=\cos\left(\frac{A-C}{2}\right) \quad \left[\because \frac{A+C}{2}=90^\circ-\frac{B}{2}\right]$$

or $$2\cos\left(\frac{A+C}{2}\right)=\cos\left(\frac{A-C}{2}\right) \quad \text{...(i)}$$

Now, LHS $= 4(1-\cos A)(1-\cos C)$

$= 4\cdot 2\sin^2 A/2\cdot 2\sin^2 C/2$

$= 4(2\sin A/2\cdot\sin C/2)^2$

$$= 4\left[\cos\frac{A-C}{2} - \cos\frac{A+C}{2}\right]^2 \quad \text{[using Eq. (i)]}$$

$$= \left[2\cos\frac{A-C}{2} - \cos\frac{A+C}{2}\right]^2 = 4\cos^2\left(\frac{A+C}{2}\right)$$

$$\text{RHS} = \cos A + \cos C = 2\cos\frac{A+C}{2}\cdot\cos\frac{A-C}{2}$$

$$= 2\cos\frac{A+C}{2}\cdot\cos\frac{A+C}{2} = 4\cos^2\frac{A+C}{2}$$

[using Eq. (i)]

Hence Proved.

Ex. 58. *Perpendiculars are drawn from the angles A, B, C of an acute angled Δ on the opposite sides and produced to meet the circumscribing circle. If these produced parts be α, β, γ respectively, then show that*

$$\frac{a}{\alpha} + \frac{b}{\beta} + \frac{c}{\gamma} = 2(\tan A + \tan B + \tan C)$$

Sol. Let AD be perpendicular from A on BC when AD is produced, it meets the circumscribing circle at E.

From question, $DE = \alpha$.

Since, angles in the same segment are equal,

$$\angle AEB = \angle ACB = \angle C$$

and $\angle AEC = \angle ABC = \angle B$

From the right angled triangle BDE,

$$\tan C = \frac{BD}{DE} \quad \text{...(i)}$$

From the right angled triangle CDE,

$$\tan B = \frac{CD}{DE} \quad \text{...(ii)}$$

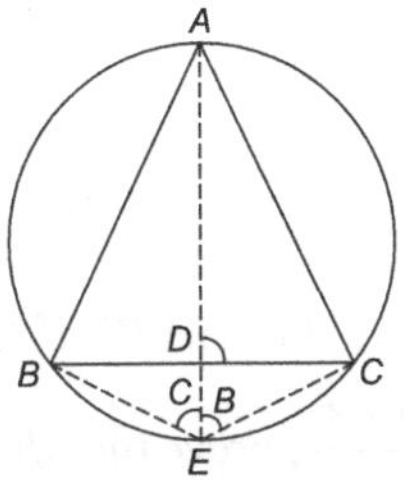

On adding Eqs. (i) and (ii), we get

$$\tan B + \tan C = \frac{BD + CD}{DE}$$

$$= \frac{BC}{DE} = \frac{a}{\alpha} \quad \text{...(iii)}$$

Similarly, $\tan C + \tan A = \dfrac{b}{\beta}$ ...(iv)

and $\tan A + \tan B = \dfrac{c}{\gamma}$ ...(v)

On adding Eqs. (iii), (iv) and (v), we get

$$\frac{a}{\alpha} + \frac{b}{\beta} + \frac{c}{\gamma} = 2(\tan A + \tan B + \tan C)$$

Ex. 59. *In any ΔABC, if*

$$\cos\theta = \frac{a}{b+c}, \cos\phi = \frac{b}{a+c}, \cos\psi = \frac{c}{a+b}$$

where θ, φ and ψ lie between 0 and π, prove that

$$\tan^2\frac{\theta}{2} + \tan^2\frac{\phi}{2} + \tan^2\frac{\psi}{2} = 1$$

Sol. Given, $\cos\theta = \dfrac{a}{b+c}$

$$\Rightarrow \frac{1-\tan^2\frac{\theta}{2}}{1+\tan^2\frac{\phi}{2}} = \frac{a}{b+c}$$

$$\Rightarrow \tan^2\frac{\theta}{2} = \frac{b+c-a}{a+b+c} \quad \text{...(i)}$$

Similarly, $\tan^2\dfrac{\phi}{2} = \dfrac{a+c-b}{a+b+c}$ ...(ii)

and $\tan^2\dfrac{\psi}{2} = \dfrac{a+b-c}{a+b+c}$ ...(iii)

On adding Eqs. (i), (ii) and (iii), we get

$$\tan^2\frac{\theta}{2} + \tan^2\frac{\phi}{2} + \tan^2\frac{\psi}{2} = \frac{a+b+c}{a+b+c}$$

$$\Rightarrow \tan^2\frac{\theta}{2} + \tan^2\frac{\phi}{2} + \tan^2\frac{\psi}{2} = 1$$

Ex. 60. *The product of the sines of the angles of a triangle is p and the product of their cosines is q. Show that the tangents of the angles are the roots of the equation;*

$$qx^2 - px^2 + (1+q)x - p = 0$$

Sol. From the question, $\sin A \cdot \sin B \cdot \cos C = p$

and $\cos A \cdot \cos B \cdot \cos C = p$

$$\Rightarrow \tan A \cdot \tan B \cdot \tan C = \frac{p}{q} \quad \text{...(i)}$$

Also, $\tan A + \tan B + \tan C = \tan A \cdot \tan B \cdot \tan C$

$$\Rightarrow \tan A + \tan B + \tan C = \frac{p}{q} \quad \text{...(ii)}$$

Now, $\tan A \tan B + \tan B \tan C + \tan C \tan A$

$$= \frac{\sin A \cdot \sin B \cdot \cos C + \sin B \cdot \sin C \cdot \cos A + \sin C \cdot \sin A \cdot \cos B}{\cos A \cdot \cos B \cdot \cos C}$$

$$= \frac{1}{q}[\sin A \sin B \cos C + \sin C(\sin B \cdot \cos A + \cos B \cdot \sin A \sin C)]$$

$$= \frac{1}{q}[\sin A \sin B \cos C + \sin C \sin(A+B)]$$

$$= \frac{1}{q}[\sin A \cdot \sin B \cdot \cos C + \sin^2 C]$$

$$= \frac{1}{q}[1 - \cos^2 C + \sin A \sin B \cos C]$$

$$= \frac{1}{q}[1 + \cos C\{- \cos C + \sin A \sin B\}]$$

$$= \frac{1}{q}[1 + \cos C(\cos(A + B) + \sin A \sin B)]$$

$$= \frac{1}{q}[1 + \cos A \cos B \cos C] = \frac{1}{q}[1 + q] \quad \text{...(iii)}$$

The equation whose roots are $\tan A$, $\tan B$, $\tan C$ will be given by

$$x^3 - (\tan A + \tan B + \tan C)x^2 + (\tan A \tan B + \tan B \tan C + \tan C \tan A)x - \tan A \cdot \tan B \cdot \tan C = 0$$

or $\quad x^3 - \frac{p}{q}x^2 + \left(\frac{1+a}{q}\right)x - \frac{p}{q} = 0$

[using Eqs. (i), (ii) and (iii)]

or $\quad qx^3 - px^2 + (1 + q)x - p = 0$ **Hence Proved.**

Ex. 61. *Given the base 'a' of a triangle, the opposite angle A, and the product k^2 of the other two sides, show that it is not possible for 'a' to be less than $2k \sin \frac{A}{2}$.*

Sol. Given, $b \cdot c = k^2$

Now, $\quad \cos A = \frac{b^2 + c^2 - a^2}{2bc}$

or $\quad 2k^2 \cos A = b^2 + \left(\frac{k^2}{b}\right)^2 - a^2$

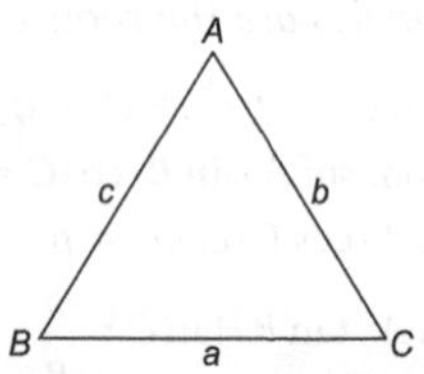

Since b^2 is real,

$$(a^2 + 2k^2 \cos A)^2 - 4k^4 \geq 0$$

$\Rightarrow \quad (a^2 + 2k^2 \cos A + 2k^2)(a^2 + 2k^2 \cos A - 2k^2) \geq 0$

$\Rightarrow \quad (a^2 + 2k^2 \cdot 2\cos^2 A/2)(a^2 - 2k^2 \cdot 2\sin^2 A/2) \geq 0$

$\Rightarrow (a^2 + 4k^2 \cdot \cos^2 A/2)(a^2 - 4k^2 \sin^2 A/2) \geq 0$

$\Rightarrow a^2 - 4k^2 \sin^2 A/2 \geq 0$

[as, $a^2 + 4k^2 \cos^2 A/2$ is always positive]

$\Rightarrow (a + 2k \sin A/2)(a - 2k \sin A/2) \geq 0$

$\Rightarrow \quad a \leq -2k \sin A/2$

or $\quad a \geq 2k \sin A/2 \quad$ [since, $2k \sin(A/2)$ is real]

But 'a' must be positive.

$\Rightarrow \quad a \leq -2k \sin A/2$ is rejected

Hence, $a \geq 2k \sin A/2$

i.e. for the triangle to exist,

$$a \geq 2k \sin A/2$$

Ex. 62. *In a ΔABC, having sides a, b, c if $C = 60°$, prove that*

$$\frac{1}{a+c} + \frac{1}{b+c} = \frac{3}{a+b+c}$$

Sol. $\quad \cos C = \frac{a^2 + b^2 - c^2}{2ab}$ but $C = 60°$

$\therefore \quad \frac{1}{2} = \frac{a^2 + b^2 - c^2}{2ab}$

$\therefore \quad ab = a^2 + b^2 - c^2$

On adding $2ab$ both sides,

$\therefore \quad 3ab = a^2 + b^2 + 2ab - c^2$

$\Rightarrow \quad 3ab = (a + b)^2 - c^2$

$\Rightarrow \quad 3ab = (a + b + c)(a + b - c) \quad \text{...(i)}$

Now, $\quad \frac{1}{a+c} + \frac{1}{b+c} = \frac{a + b + 2c}{ab + c(a + b + c)}$

On multiplying numerator and denominator by 3,

$$= \frac{3(a + b + 2c)}{3ab + 3c(a + b + c)}$$

$$= \frac{3(a + b + 2c)}{(a + b + c)(a + b - c) + 3c(a + b + c)}$$

[using Eq. (i) for $3ab$]

$$= \frac{3(a + b + 2c)}{(a + b + c)(a + b + 2c)}$$

$$= \frac{3}{(a + b + c)}$$

$\therefore \quad \frac{1}{a+c} + \frac{1}{b+c} = \frac{3}{(a + b + c)}$

Ex. 63. *Let $1 < m < 3$. In a ΔABC, if $2b = (m + 1)a$ and $\cos A = \frac{1}{2}\sqrt{\frac{(m-1)(m+3)}{m}}$, prove that there are two values to the third side, one of which is m times the other.*

Sol. From the formula for $\cos A$,

we can write $2bc \cos A = b^2 + c^2 - a^2$

but $\quad b = \left(\frac{m + 1}{2}\right)a$

$\therefore \quad (m + 1)ac \cos A = \left\{\frac{(m + 1)^2}{4} - 1\right\}a^2 + c^2$

$$= \left\{\frac{(m + 1)^2 - 4}{4}\right\}a^2 + c^2$$

$$= \left\{\frac{(m-1)(m+3)}{4}\right\}a^2 + c^2 \quad \text{...(i)}$$

But from the given values of $\cos A$,
we can write

$$\frac{(m-1)(m+3)}{4} = m\cos^2 A,$$

$\therefore$ Eq. (i) gives as

$$(m+1)ac\cos A = ma^2\cos^2 A + c^2$$

or $$mac\cos A - c^2 = ma^2\cos^2 A - ac\cos A$$

$\therefore$ $$(ma\cos A - c) = a\cos A(ma\cos A - c)$$

$\therefore$ $$(c - a\cos A)(c - ma\cos A) = 0$$

This implies 'c' has two values;
$a\cos A$ and $ma\cos A$, and the latter is m times the former.

Hence proved

Ex. 64. *In any ΔABC, if D be any points of the base BC, such that $\frac{BD}{DC} = \frac{m}{n}$ and $\angle BAD = \alpha$, $\angle DAC = \beta$, $\angle CDA = \theta$ and $AD = x$ then prove that*

$$(m+n)^2 \cdot x^2 = (m+n)(mb^2 + nc^2) - mna^2$$

Sol. In ΔADC,

$$AD^2 + DC^2 - 2AD \cdot DC \cdot \cos\theta = AC^2$$

i.e. $$x^2 + \left(\frac{na}{m+n}\right)^2 - 2x\left(\frac{na}{m+n}\right)\cdot\cos\theta = b^2 \quad \text{...(i)}$$

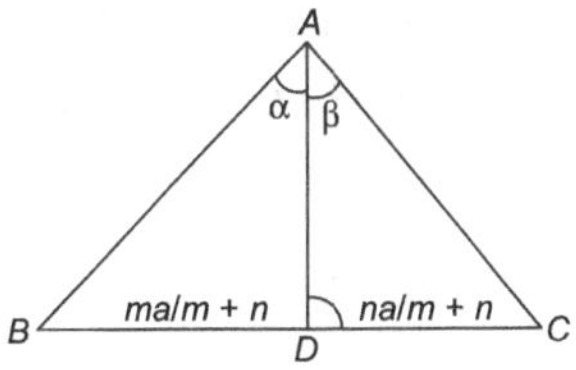

In ΔABD, $AD^2 + BD^2 - 2AD \cdot BD \cdot \cos(\pi - \theta) = AB^2$

i.e. $$x^2 + \left(\frac{ma}{m+n}\right)^2 - 2x\left(\frac{ma}{m+n}\right)\cdot\cos\theta = c^2 \quad \text{...(ii)}$$

On adding m times Eq. (i) and n times Eq. (ii), we get

$$(m+n)x^2 + \frac{mn\,a^2}{m+n} = mb^2 + nc^2$$

$\Rightarrow$ $$(m+n)^2 . x^2 = (m+n)(mb^2 + nc^2) - mna^2$$

Ex. 65. *ABCD is a trapezium such that AB, DC are parallel and BC is perpendicular to them. If $\angle ADB = \theta$, $BC = p$ and $CD = q$, show that*

$$AB = \frac{(p^2 + q^2)\sin\theta}{p\cos\theta + q\sin\theta}$$

Sol. Let $\angle BDC = \alpha$, then $\angle DAB = 180° - (\theta + \alpha)$ applying sine formula to ΔABD, we get

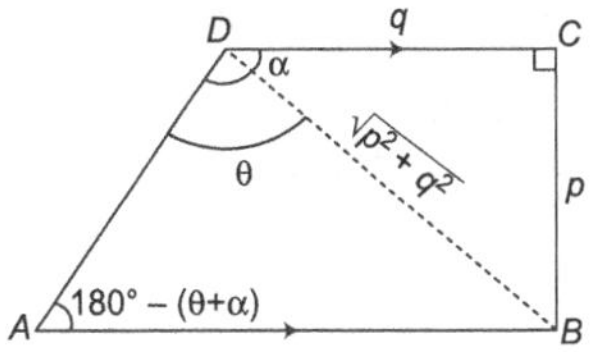

$$\frac{AB}{\sin\theta} = \frac{BD}{\sin(\theta + \alpha)}$$

$\therefore$ $$\frac{AB}{\sin\theta} = \frac{\sqrt{p^2+q^2}}{\sin\theta\cos\alpha + \cos\theta\sin\alpha}$$

$$\left[\begin{array}{l}\text{but in } \Delta BCD, \\ \sin\alpha = \dfrac{p}{\sqrt{p^2+q^2}} \text{ and } \cos\alpha = \dfrac{q}{\sqrt{p^2+q^2}}\end{array}\right]$$

$\therefore$ $$AB = \frac{(p^2+q^2)\sin\theta}{q\sin\theta + p\cos\theta}$$ **Hence proved.**

Ex. 66. *Prove that in ΔABC,*

$$\frac{(a+b+c)^2}{a^2+b^2+c^2} \Rightarrow \frac{\cot\frac{A}{2} + \cot\frac{B}{2} + \cot\frac{C}{2}}{\cot A + \cot B + \cot C}$$

Sol. LHS (denominator) $= a^2 + b^2 + c^2$

$$= [(b^2 + c^2 - a^2) + (c^2 + a^2 - b^2) + (a^2 + b^2 - c^2)]$$

$$= [2bc\cos A + 2ca\cos B + 2ab\cos C]$$

$$= 2abc\left[\frac{\cos A}{a} + \frac{\cos B}{b} + \frac{\cos C}{c}\right]$$

but, $$\frac{a}{\sin A} = \frac{b}{\sin B} = \frac{c}{\sin C} = K$$

$\therefore$ Denominator (on LHS)

$$= \frac{2abc}{K}[\cot A + \cot B + \cot C] \quad \text{...(i)}$$

LHS (numerator) $= (a+b+c)^2$

$$= K^2[\sin A + \sin B + \sin C]^2$$

$$= 16K^2\cos^2\frac{A}{2}\cos^2\frac{B}{2}\cos^2\frac{C}{2} \quad \left[\text{using } \Sigma\sin A = 4\Pi\cos\frac{A}{2}\right]$$

On multiplying and dividing by $2K^3\sin A\sin B\sin C$, we get

$$= (2K^3\sin A\sin B\sin C)\frac{16K^2\cos^2\frac{A}{2}\cdot\cos^2\frac{B}{2}\cdot\cos^2\frac{C}{2}}{2K^3\sin A\cdot\sin B\cdot\sin C}$$

$$= (2abc)\frac{8}{K}\cdot\frac{\cos^2\frac{A}{2}\cdot\cos^2\frac{B}{2}\cdot\cos^2\frac{C}{2}}{8\left(\sin\frac{A}{2}\cos\frac{A}{2}\right)\left(\sin\frac{B}{2}\cos\frac{B}{2}\right)\left(\sin\frac{C}{2}\cos\frac{C}{2}\right)}$$

$$= \frac{2abc}{K}\left\{\cot\frac{A}{2}\cdot\cot\frac{B}{2}\cdot\cot\frac{C}{2}\right\} \quad \text{...(ii)}$$

Since A, B, C are angles of a triangle,

$$\cot\frac{A}{2}\cdot\cot\frac{B}{2}\cdot\cot\frac{C}{2}=\cot\frac{A}{2}+\cot\frac{B}{2}+\cot\frac{C}{2} \qquad ...(iii)$$

From Eqs. (ii) and (iii)

$$\text{Numerator (on LHS)} = \frac{2abc}{K}[\Sigma\cot A/2] \qquad ...(iv)$$

On dividing Eq. (iv) by Eq. (i), we get

$$\frac{\cot A/2+\cot B/2+\cot C/2}{\cot A+\cot B+\cot C} \Rightarrow \frac{(a+b+c)^2}{(a^2+b^2+c^2)}$$

Ex. 67. *If the sides of a triangle are in AP, and its greatest angle exceeds the least angle by α, show that the sides are in the ratio $1+x:1:1-x$, where $x=\sqrt{\frac{1-\cos\alpha}{7-\cos\alpha}}$.*

Sol. Let ΔABC be the given Δ in which A is the greatest and C is the least angle.

Then, according to the hypothesis

$$A-C=\alpha \qquad ...(i)$$

But $$A+C=\pi-B \qquad ...(ii)$$

$\therefore$ From Eq. (i) and Eq. (ii), we get

$$A=\frac{\pi}{2}+\frac{\alpha-B}{2}$$

and $$C=\frac{\pi}{2}-\frac{\alpha+B}{2} \qquad ...(iii)$$

Again by hypothesis,

$$a+c=2b$$

$$\therefore \quad \sin A+\sin C=2\sin B$$

$$\Rightarrow 2\sin\left(\frac{A+C}{2}\right)\cos\left(\frac{A-C}{2}\right)=4\sin\frac{B}{2}\cdot\cos\frac{B}{2}$$

$$\left[\frac{A+C}{2}=\frac{\pi}{2}-\frac{B}{2}\right]$$

$$\Rightarrow 2\cos\frac{B}{2}\cos\left(\frac{A-C}{2}\right)=4\sin\frac{B}{2}\cdot\cos\frac{B}{2}$$

$$\Rightarrow \cos\left(\frac{\alpha}{2}\right)=2\sin\frac{B}{2} \qquad \text{[using Eq. (i)]}$$

$$\therefore \quad \sin\frac{B}{2}=\frac{\cos\frac{\alpha}{2}}{2}$$

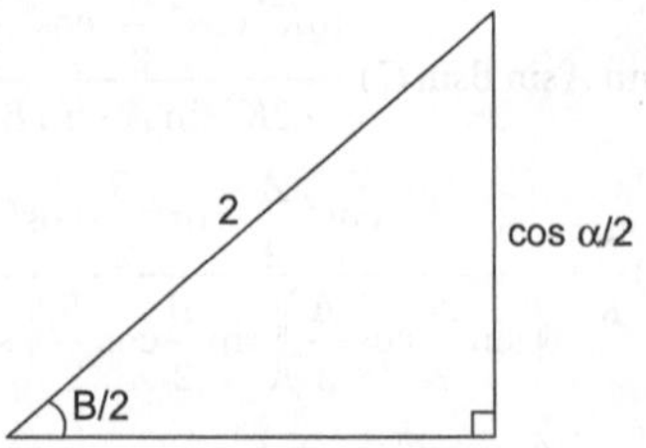

$$\sqrt{4-\cos^2\alpha/2}=\sqrt{4-\left(\frac{1+\cos\alpha}{2}\right)}=\sqrt{\frac{7-\cos\alpha}{2}}$$

But $$\frac{a}{c}=\frac{\sin A}{\sin C}=\frac{\cos\left(\frac{\alpha-B}{2}\right)}{\cos\left(\frac{\alpha+B}{2}\right)} \qquad \text{[using Eq. (iii)]}$$

$$=\frac{\cos\alpha/2\cdot\cos B/2+\sin\alpha/2\cdot\sin B/2}{\cos\alpha/2\cdot\cos B/2-\sin\alpha/2\cdot\sin B/2}$$

On dividing numerator and denominator by $\cos\alpha/2.\cos B/2$

$$=\frac{1+\tan\alpha/2\,.\,\tan B/2}{1-\tan\alpha/2\,.\,\tan B/2}$$

Write $\tan\alpha/2\cdot\tan B/2$ from figure;

$$=\frac{1+\sqrt{\frac{2\sin^2\alpha/2}{7-\cos\alpha}}}{1-\sqrt{\frac{2\sin^2\alpha/2}{7-\cos\alpha}}}=\frac{1+\sqrt{\frac{1-\cos\alpha}{7-\cos\alpha}}}{1-\sqrt{\frac{1-\cos\alpha}{7-\cos\alpha}}}=\frac{1+x}{1-x}$$

$$\left[\text{given } x=\sqrt{\frac{1-\cos\alpha}{7-\cos\alpha}}\right]$$

$$\therefore \quad \frac{a}{1+x}=\frac{c}{1-x}=\frac{a+c}{2}=b$$

$$\therefore \quad \frac{a}{1+x}=\frac{b}{1}=\frac{c}{1-x}$$

Thus, the sides are in the ratio $1+x:1:1-x$.

Ex. 68. *In a ΔABC, if $\tan A/2$, $\tan B/2$, $\tan C/2$ are in AP. Show that $\cos A, \cos B, \cos C$ are in AP.*

Sol. Given $\tan A/2$, $\tan B/2$, $\tan C/2$ are in AP.

$$\therefore \tan A/2-\tan B/2=\tan B/2-\tan C/2$$

$$\therefore \quad \frac{\sin A/2}{\cos A/2}-\frac{\sin B/2}{\cos B/2}=\frac{\sin B/2}{\cos B/2}-\frac{\sin C/2}{\cos C/2}$$

$$\Rightarrow \frac{\sin A/2\cdot\cos B/2-\sin B/2\cdot\cos A/2}{\cos A/2\cdot\cos B/2}=\frac{\sin B/2\cdot\cos C/2-\sin C/2\cdot\cos B/2}{\cos B/2\cdot\cos C/2}$$

$$\Rightarrow \frac{\sin\left(\frac{A-B}{2}\right)}{\cos\frac{A}{2}}=\frac{\sin\left(\frac{B-C}{2}\right)}{\cos\frac{C}{2}} \qquad ...(i)$$

but $$\cos\frac{A}{2}=\sin\frac{B+C}{2}$$

and $$\cos\frac{C}{2}=\sin\left(\frac{A+B}{2}\right) \qquad ...(ii)$$

$$\Rightarrow \sin\left(\frac{A-B}{2}\right)\cdot\sin\left(\frac{A+B}{2}\right)$$

$$=\sin\left(\frac{B-C}{2}\right)\cdot\sin\left(\frac{B+C}{2}\right) \qquad \text{[from Eqs. (i) and (ii)]}$$

$$\Rightarrow \cos B-\cos A=\cos C-\cos B$$

Hence, $\cos A, \cos B, \cos C$ are in AP.

Ex. 69. *If a, b, c are in HP, then prove that $\sin^2\frac{A}{2}$, $\sin^2\frac{B}{2}$, $\sin^2\frac{C}{2}$ are also in HP.*

Sol. Given that $\frac{2}{b}=\frac{1}{a}+\frac{1}{c}$

i.e. $\quad 2ac = b(a+c) \quad$ ...(i)

we want to prove

$$\frac{2}{\sin^2\frac{B}{2}}=\frac{1}{\sin^2\frac{A}{2}}+\frac{1}{\sin^2\frac{C}{2}}$$

i.e.,
$$\frac{2ac}{(s-a)(s-c)}=\frac{bc}{(s-b)(s-c)}+\frac{ab}{(s-a)(s-b)}$$

$$\left[\text{using } \sin\frac{A}{2}=\sqrt{\frac{(s-b)(s-c)}{bc}},\ \sin\frac{B}{2}=\sqrt{\frac{(s-a)(s-c)}{ac}}\ \text{and } \sin\frac{C}{2}=\sqrt{\frac{(s-a)(s-b)}{ab}}\right]$$

Consider LHS

$$\frac{2ac}{(s-a)(s-c)}=\frac{b(a+c)}{(s-a)(s-c)} \quad \text{[using Eq. (i)]}$$

$$=\frac{b(a+c)(s-b)}{(s-a)(s-b)(s-c)}=\frac{b\,[s(a+c)-b(a+c)]}{(s-a)(s-b)(s-c)}$$

$$=\frac{b\,(sa+sc-2ac)}{(s-a)(s-b)(s-c)} \quad \text{[using Eq. (i)]}$$

$$=\frac{b\,(sa-ac+sc-ac)}{(s-a)(s-b)(s-c)}$$

$$=\frac{b\,[a(s-c)+c\,(s-a)]}{(s-a)(s-b)(s-c)}$$

$$=\frac{ab}{(s-a)(s-b)}+\frac{bc}{(s-b)(s-c)}=\text{RHS}$$

Ex. 70. *If r_1, r_2, r_3 are the ex-radii of ΔABC, then prove that*

$$\frac{bc}{r_1}+\frac{ca}{r_2}+\frac{ab}{r_3}=2R\left[\left(\frac{a}{b}+\frac{b}{a}\right)+\left(\frac{b}{c}+\frac{c}{b}\right)+\left(\frac{c}{a}+\frac{a}{c}\right)-3\right]$$

Sol. We know, $\quad r_1=\frac{\Delta}{s-a}, r_2=\frac{\Delta}{s-b}, r_3=\frac{\Delta}{s-c}$

$$\therefore \quad \text{LHS} =\Sigma\frac{bc}{r_1}=\Sigma\frac{bc(s-a)}{\Delta}$$

$$=\frac{1}{\Delta}\Sigma bc(s-a)=\frac{1}{\Delta}[\Sigma bcs-\Sigma(bc)a]$$

$$=\frac{1}{\Delta}\left[\Sigma\frac{abc}{a}\cdot s-3abc\right]=\frac{1}{\Delta}\left[abc\,\Sigma\frac{s}{a}-3abc\right]$$

$$=\frac{abc}{\Delta}\left[\Sigma\frac{s}{a}-3\right] \quad \left[\text{but } \frac{abc}{\Delta}=4R\right]$$

$$=4R\left[\Sigma\frac{s}{a}-3\right]=2R\left[\Sigma\frac{2s}{a}-6\right]$$

$$=2R\left[\Sigma\frac{a+b+c}{a}-6\right]$$

$$=2R\left[\Sigma\left(1+\frac{b+c}{a}\right)-6\right]$$

$$=2R\left[3+\Sigma\left(\frac{b+c}{a}\right)-6\right]$$

$$=2R\left[\left(\frac{b+c}{a}+\frac{c+a}{b}+\frac{a+b}{c}\right)-3\right]$$

$$=2R\left[\frac{b}{a}+\frac{c}{a}+\frac{c}{b}+\frac{a}{b}+\frac{a}{c}+\frac{b}{c}-3\right]$$

$$=2R\left[\left(\frac{a}{b}+\frac{b}{a}\right)+\left(\frac{b}{c}+\frac{c}{b}\right)+\left(\frac{c}{a}+\frac{a}{c}\right)-3\right]$$

Hence proved.

Ex. 71. *If r and R are radii of the incircle and circum-circle of ΔABC, then prove that :*

$$8rR\{\cos^2 A/2+\cos^2 B/2+\cos^2 C/2\}=2bc+2ca+2ab-a^2-b^2-c^2.$$

Sol. LHS $=8\left(\frac{\Delta}{s}\right)\cdot\left(\frac{abc}{4\Delta}\right)\{\Sigma\cos^2 A/2\}$

$$=\frac{abc}{s}\Sigma(2\cos^2 A/2)=\frac{abc}{s}\Sigma(1+\cos A)$$

$$=\frac{abc}{s}\Sigma\left(1+\frac{b^2+c^2-a^2}{2bc}\right)$$

$$=\frac{abc}{s}\Sigma\left(\frac{2bc+b^2+c^2-a^2}{2bc}\right)=\frac{abc}{s}\Sigma\left\{\frac{(b+c)^2-a^2}{2bc}\right\}$$

$$=\frac{(abc)}{s}\Sigma\frac{(a+b+c)(b+c-a)}{2bc}$$

$$=\frac{(abc)}{s}2s\Sigma\left(\frac{b+c-a}{2bc}\right), \quad \text{where } a+b+c=2s$$

$$=\Sigma a(b+c-a)=\Sigma(ab+bc-a^2)$$

$$=2bc+2ca+2ab-a^2-b^2-c^2.$$

$$\therefore\ 8rR\{\cos^2 A/2+\cos^2 B/2+\cos^2 C/2\}=2bc+2ab+2ca-a^2-b^2-c^2$$

Ex. 72. Prove that

$$r_1^2+r_2^2+r_3^3+r^2=16R^2-a^2-b^2-c^2.$$

where r = in-radius, R = circumradius, r_1, r_2, r_3 are ex-radii.

Sol. LHS $=r^2+r_1^2+r_2^2+r_3^2$

$$=16R^2\sin^2 A/2\cdot\sin^2 B/2\sin^2 C/2+16R^2\sin^2 A/2\cdot\cos^2 B/2\cdot\cos^2 C/2$$
$$+16R^2\cos^2 A/2\sin^2 B/2.\cos^2 C/2+16R^2\cos^2 A/2\cdot\cos^2 B/2\cdot\sin^2 C/2$$

$= 16R^2 \sin^2 A/2\,[\sin^2 B/2 \cdot \sin^2 C/2 + \cos^2 B/2 \cdot \cos^2 C/2]$
$\quad + 16R^2 \cos^2 A/2\,[\sin^2 B/2 \cdot \cos^2 C/2 + \cos^2 B/2 \cdot \sin^2 C/2]$
$= 4R^2 \sin^2 A/2\,[(2\sin^2 B/2)(2\sin^2 C/2) + (2\cos^2 B/2)(2\cos^2 C/2)]$
$\quad + 4R^2 \cos^2 A/2\,[(2\sin^2 B/2)(2\cos^2 C/2) + (2\cos^2 B/2)(2\cos^2 C/2)]$
$= 4R^2 \sin^2 A/2\,[(1-\cos B)(1-\cos C) + (1+\cos B)(1+\cos C)] + 4R^2 \cos^2 A/2\,[(1-\cos B)(1+\cos C) + (1+\cos B)(1-\cos C)]$
$= 4R^2 \sin^2 A/2\,[2 + 2\cos B \cos C] + 4R^2 \cos^2 A/2\,[2 - 2\cos B \cos C]$
$= 8R^2 + 8R^2 \cos B \cos C(\sin^2 A/2 - \cos^2 A/2)$
$= 8R^2 - 8R^2 \cos A . \cos B \cos C$
$= 8R^2 + 4R^2\,[\cos(A+B) + \cos(A-B)] \cdot \cos(A+B)$
$= 8R^2 + 4R^2\,[\cos^2 C + \cos^2 A - \sin^2 B]$
$= 8R^2 + 4R^2\,[1 - \sin^2 C + 1 - \sin^2 A - \sin^2 B]$
$= 16R^2 - 4R^2 \sin^2 A - 4R^2 \sin^2 B - 4R^2 \sin^2 C$
$= 16R^2 - a^2 - b^2 - c^2 = \text{RHS}$

Ex. 73. *Tangents are parallel to the three sides are drawn to the in-circle. If x, y, z are the lengths of the parts of the tangents with in the triangle, then prove that* $\frac{x}{a} + \frac{y}{b} + \frac{z}{c} = 1$.

Sol. Let $PQ = x$, PQ parallel to BC.
$SR = y$, SR parallel to AC.
$TU = z$, TU parallel to AB.

Let I be the in-centre of the ΔABC.

Consider ΔAPQ,

By sine rule, $\frac{x}{\sin A} = \frac{AQ}{\sin B} = \frac{AP}{\sin C}$

$\therefore \quad AQ = \frac{\sin B}{\sin A} x = \frac{b}{a} x$

$AP = \frac{\sin C}{\sin A} x = \frac{c}{a} x$

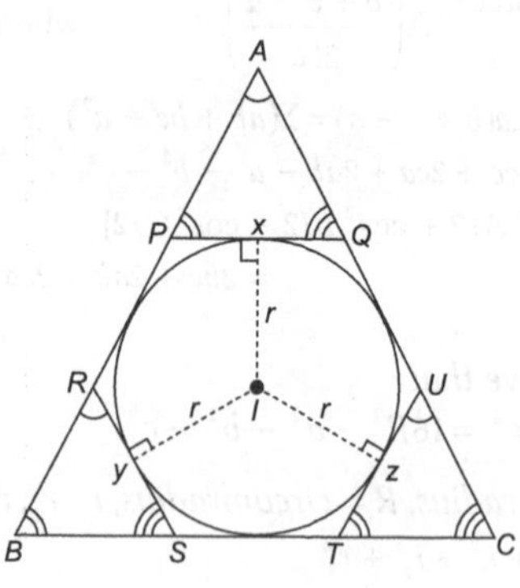

Also 'r' will be ex-radius of ΔAPQ.

$$\therefore \quad r = \left(\frac{x + PA + AQ}{2}\right)\tan A/2$$

$$= \left[\frac{x + \frac{bx}{a} + \frac{cx}{a}}{2}\right]\tan A/2$$

$$= \frac{(a+b+c)}{2a} x \tan A/2 = \frac{sx}{a}\tan A/2.$$

Similarly in ΔBRS,

$$BR = \frac{c}{b}y;\ BS = \frac{a}{b}y \text{ and } r = \frac{sy}{b}\tan B/2$$

$$\left[\because BR + BS + y = \frac{c}{b}y + \frac{c}{b}y + y = \frac{sy}{b}\right]$$

and in ΔCTU, $CT = \frac{a}{c}z$, $CU = \frac{b}{c}z$ and $r = \frac{sz}{c}\tan\frac{C}{2}$

We also know that,

$$r = (s-a)\tan A/2 = (s-b)\tan B/2 = (s-c)\tan C/2$$

$$\therefore \quad \frac{sx}{a}\tan A/2 = (s-a)\tan A/2 \Rightarrow \frac{sx}{a} = s - a$$

$$\frac{sx}{b}\tan B/2 = (s-b)\tan B/2 \Rightarrow \frac{sx}{b} = s - b$$

$$\frac{sx}{c}\tan C/2 = (s-c)\tan C/2 \Rightarrow \frac{sx}{c} = s - c$$

On adding $\quad s\left(\frac{x}{a} + \frac{y}{b} + \frac{z}{c}\right) = 3s - (a+b+c) = s$

$$\therefore \quad \left(\frac{x}{a} + \frac{y}{b} + \frac{z}{c}\right) = 1$$

Ex. 74. *In a* ΔABC, *if* $\cos A + 2\cos B + \cos C = 2$. *Prove that the sides of the triangle are in AP.*

Sol. $\cos A + 2\cos B + \cos C = 2$

or $\quad \cos A + \cos C = 2(1 - \cos B)$

$$\Rightarrow \quad 2\cos\left(\frac{A+C}{2}\right) \cdot \cos\left(\frac{A-C}{2}\right) = 4\sin^2 B/2$$

$$\Rightarrow \quad \cos\left(\frac{A-C}{2}\right) = 2\sin\frac{B}{2}$$

$$\left[\text{as } \cos\left(\frac{A+C}{2}\right) = \cos\left(\frac{\pi}{2} - \frac{B}{2}\right) = \sin\frac{B}{2}\right]$$

$$\Rightarrow \quad \cos\left(\frac{A-C}{2}\right) = 2\cos\left(\frac{A+C}{2}\right)$$

$$\Rightarrow \quad \cos\frac{A}{2} \cdot \cos\frac{C}{2} + \sin\frac{A}{2} \cdot \sin\frac{C}{2} = 2\cos\frac{A}{2} \cdot \cos\frac{C}{2} - 2\sin\frac{A}{2} \cdot \sin\frac{C}{2}$$

$$\Rightarrow \quad \cot\frac{A}{2} \cdot \cot\frac{C}{2} = 3$$

$$\Rightarrow \quad \sqrt{\frac{s(s-a)}{(s-b)(s-c)}} \cdot \sqrt{\frac{s(s-c)}{(s-a)(s-b)}} = 3$$

$$\Rightarrow \quad \frac{s}{(s-b)} = 3 \Rightarrow s = 3s - 3b \Rightarrow 2s = 3b$$

$$\Rightarrow \quad a + c = 2b$$

$\therefore$ a, b, c are in AP.

Ex. 75. *In a cyclic quadrilateral ABCD, prove that* $\tan^2 \frac{B}{2} = \frac{(s-a)(s-b)}{(s-c)(s-d)}$, *a, b, c and d being the lengths of sides AB, BC, CD and DA respectively and s is semi-perimeter of quadrilateral.*

Sol. From ΔABC, we get

$$AC^2 = a^2 + b^2 + 2ab\cos B \quad \text{...(i)}$$

and from ΔADC,

We have, $AC^2 = c^2 + d^2 + 2cd\cos D$

$$= c^2 + d^2 + 2cd\cos(180^\circ - B)$$

$$\Rightarrow \quad AC^2 = c^2 + d^2 - 2cd\cos B \quad \text{...(ii)}$$

From Eq. (i) and Eq. (ii)

$$\cos B = \frac{a^2 + b^2 - c^2 - d^2}{2(ab + cd)}$$

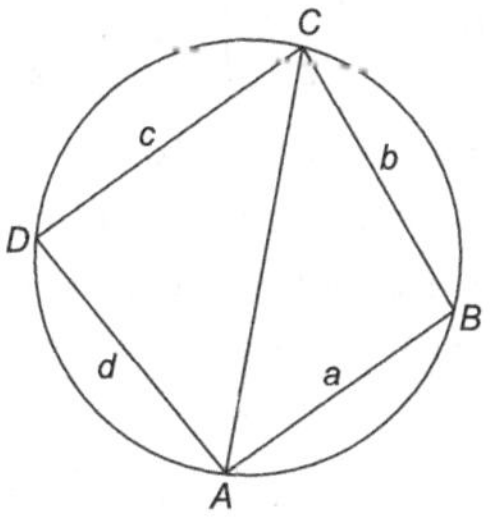

Now since, $\tan^2 \frac{B}{2} = \frac{1 - \cos B}{1 + \cos B}$

$$\tan^2 \frac{B}{2} = \frac{2(ab + cd) - (a^2 + b^2 - c^2 - d^2)}{2(ab + cd) + (a^2 + b^2 - c^2 - d^2)}$$

$$= \frac{(c+d)^2 - (a-b)^2}{(a+b)^2 - (c-d)^2}$$

$$= \frac{(c+d+a-b)(c+d-a+b)}{(a+b-c+d)(a+b+c-d)}$$

$$= \frac{(2s-2b)(2s-2a)}{(2s-2c)(2s-2d)} = \frac{(s-a)(s-b)}{(s-d)(s-c)}$$

[since $a + b + c + d = 2s$]

$$\therefore \quad \tan^2 \frac{B}{2} = \frac{(s-a)(s-b)}{(s-d)(s-c)}$$

Ex. 76. *If x, y, z are the distances of the vertices of the* ΔABC *respectively from the orthocentre, then show that* $\frac{a}{x} + \frac{b}{y} + \frac{c}{z} = \frac{abc}{xyz}$.

Sol. Let H be the orthocentre. Then,

$$\angle BHC = 180^\circ - \angle HBC - \angle HCB$$

$$= 180^\circ - (90^\circ - C) - (90^\circ - B)$$

$$= B + C = \pi - A$$

So, $\text{ar}(\Delta HBC) = \frac{1}{2} BH \cdot CH \cdot \sin \angle BHC$

$$= \frac{1}{2} yz \cdot \sin(\pi - A) = \frac{1}{2} yz \sin A \quad \text{...(i)}$$

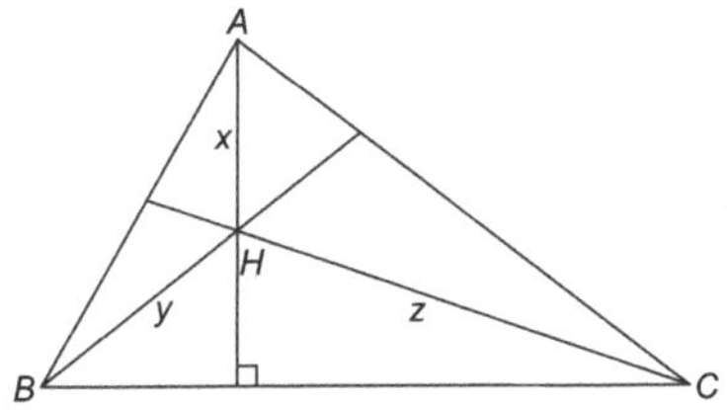

Similarly,

$$\text{ar}(\Delta CHA) = \frac{1}{2} zx \sin B \quad \text{...(ii)}$$

$$\text{ar}(\Delta AHB) = \frac{1}{2} xy \sin C \quad \text{...(iii)}$$

$$\therefore \ (\Delta ABC) = \frac{1}{2} yz\sin A + \frac{1}{2} zx \sin B + \frac{1}{2} xy \sin C$$

[from Eqs. (i), (ii) and (iii)]

$$= \frac{1}{2} xyz \left\{ \frac{\sin A}{x} + \frac{\sin B}{y} + \frac{\sin C}{z} \right\}$$

$$= \frac{1}{2} xyz \cdot \frac{1}{2R} \left\{ \frac{a}{x} + \frac{b}{y} + \frac{c}{z} \right\} \quad \text{...(iv)}$$

$$\left[\text{as we know, } R = \frac{abc}{4\Delta}, \text{i.e. } \Delta = \frac{abc}{4R} \right]$$

$\therefore$ Eq. (iv) reduces to;

$$\frac{abc}{4} R = \frac{xyz}{4R} \left(\frac{a}{x} + \frac{b}{y} + \frac{c}{z} \right)$$

$$\Rightarrow \quad \frac{a}{x} + \frac{b}{y} + \frac{c}{z} = \frac{abc}{xyz}$$

Ex. 77. *In the* ΔABC, *a similar* $\Delta A'B'C'$ *is inscribed so that* $B'C' = \lambda BC$. *If* $B'C'$ *is inclined at an angle* θ *with BC, then prove that* $\lambda \cos\theta = \frac{1}{2}$.

Sol. ΔABC and $\Delta A'B'C'$ are similar,

where $\angle B'A'C' = \angle BAC = A$

$\angle A'B'C' = \angle ABC = B$

$\angle B'C'A' = \Delta BCA = C$

In $\Delta AC'B', \angle AB'C' = \theta + C$

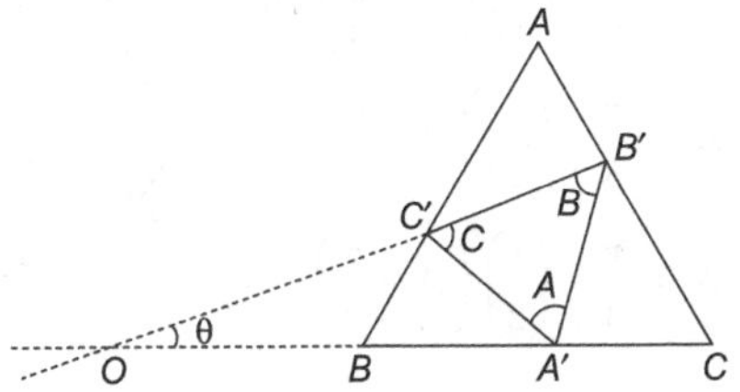

So, using sine law on $\Delta AC'B'$,

$$\frac{AC'}{\sin(\theta + C)} = \frac{B'C'}{\sin A}$$

or $$\frac{AC'}{\sin(\theta + C)} = \frac{\lambda BC}{\sin A} = \frac{\lambda a}{\sin A} = 2\lambda R$$

In $\Delta BA'C'$, $\angle BA'C' = \angle A'C'B' - \angle A'OC' = C - \theta$

So, $$\frac{BC'}{\sin(C - \theta)} = \frac{A'C'}{\sin B}$$

$$\Rightarrow \quad \frac{BC'}{\sin(C - \theta)} = \frac{\lambda AC}{\sin B} = \frac{\lambda b}{\sin B} = 2\lambda R$$

$$\left[\frac{B'C'}{BC} = \frac{A'C'}{AC} = \lambda \text{ from similar } \Delta\right]$$

Thus we get, $AC' = 2\lambda R\sin(\theta + C)$

and $BC' = 2\lambda R\sin(C - \theta)$

$\therefore \quad C = AB = AC' + BC' = 2\lambda R\ \{\sin(C + \theta) + \sin(C - \theta)\}$

$= 2\lambda R \cdot 2\sin C \cdot \cos\theta$

$$\therefore \quad \cos\theta = \frac{C}{4\lambda R\sin C} = \frac{1}{4\lambda R} \cdot \frac{C}{\sin C} = \frac{1}{4\lambda R} \cdot 2R = \frac{1}{2\lambda}$$

$$\therefore \quad \lambda\cos\theta = \frac{1}{2}.$$

Ex. 78. *The circle inscribed in the triangle ABC touches the side BC, CA and AB in the point A_1, B_1 and C_1 respectively. Similarly the circle inscribed in the $\Delta A_1B_1C_1$ touches the sides in A_2, B_2, C_2 respectively and so on. If $A_nB_nC_n$ be the nth Δ so formed, prove that its angle are $\frac{\pi}{3} - (2)^{-n}\left(A - \frac{\pi}{3}\right), \frac{\pi}{3} - (2)^{-n}\left(B - \frac{\pi}{3}\right)$ and $\frac{\pi}{3} - (2)^{-n}\left(C - \frac{\pi}{3}\right)$. Hence prove that the triangle so formed is ultimately equilateral.*

Sol. Let I be incentre of the in-circle, $\angle B_1A_1C_1 = \frac{1}{2}(\pi - A)$

$$A_1 = \frac{\pi}{3} - \frac{1}{2}\left(A - \frac{\pi}{3}\right)$$

Similarly, $$A_2 = \frac{\pi}{3} - \frac{1}{2}\left(A_1 - \frac{\pi}{3}\right)$$

$$= \frac{\pi}{3} - \frac{1}{2}\left[\frac{\pi}{3} - \frac{1}{2}\left(A - \frac{\pi}{3}\right) - \frac{\pi}{3}\right]$$

$$= \frac{\pi}{3} + \left(\frac{1}{2}\right)^2\left(A - \frac{\pi}{3}\right)$$

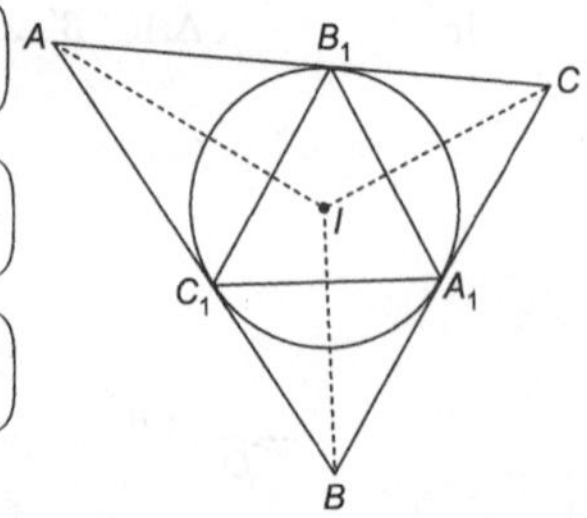

$$A_3 = \frac{\pi}{3} - \left(\frac{1}{2}\right)^3\left(A - \frac{\pi}{3}\right)$$

$$A_n = \frac{\pi}{3} - \left(\frac{1}{2}\right)^n\left(A - \frac{\pi}{3}\right)$$

So on,

$$B_n = \frac{\pi}{3} - \left(\frac{1}{2}\right)^n\left(B - \frac{\pi}{3}\right)$$

$$C_n = \frac{\pi}{3} - \left(\frac{1}{2}\right)^n\left(C - \frac{\pi}{3}\right)$$

When $n \to \infty$ then $\left(\frac{1}{2}\right)^n$ becomes zeros or

$$A_n = B_n = C_n = \frac{\pi}{3}.$$

Hence, the triangle will be equilateral.

Ex. 79. *In a ΔABC, prove that :*

$$2r \le \frac{a\cot A + b\cot B + c\cot C}{3} \le R$$

Sol. $$r = 4R\sin\frac{A}{2} \cdot \sin\frac{B}{2} \cdot \sin\frac{C}{2}$$

$$= 2R\left[\cos\left(\frac{A-B}{2}\right) - \cos\left(\frac{A+B}{2}\right)\right] \cdot \sin\frac{C}{2}$$

$$= R\left[2\cos\left(\frac{A+B}{2}\right) \cdot \cos\left(\frac{A-B}{2}\right) - 2\sin^2\frac{C}{2}\right]$$

$$= R[\cos A + \cos B + \cos C - 1]$$

$\Rightarrow \quad 2r + 2R = 2R\,[\cos A + \cos B + \cos C]$

$\Rightarrow \quad a\cot A + b\cot B + c\cot C = 2r + 2R \ge 2r + 4r$

$$\Rightarrow \quad 2r \le \frac{a\cot A + b\cot B + c\cot C}{3}$$

$$= \frac{2R}{3}(\cos A + \cos B + \cos C)$$

$$\le \frac{2}{3}R \cdot \frac{3}{2}$$

$$\Rightarrow \quad 2r \le \frac{a\cot A + b\cot B + \cot C}{3} \le R$$

Ex. 80. *If A, B and C are angles of a triangle, then prove that* $$E = \frac{\cos\left(\frac{B-C}{2}\right)}{\cos\left(\frac{B+C}{2}\right)} + \frac{\cos\left(\frac{C-A}{2}\right)}{\cos\left(\frac{C+A}{2}\right)} + \frac{\cos\left(\frac{A-B}{2}\right)}{\cos\left(\frac{A+B}{2}\right)} \ge 6$$

Sol. Here, $A + B + C = \pi$

$$\Rightarrow E = \frac{\cos\left(\frac{B-C}{2}\right)}{\cos\left(\frac{B+C}{2}\right)} + \frac{\cos\left(\frac{C-A}{2}\right)}{\cos\left(\frac{C+A}{2}\right)} + \frac{\cos\left(\frac{A-B}{2}\right)}{\cos\left(\frac{A+B}{2}\right)}$$

$$= \frac{\cos\left(\frac{B-C}{2}\right)}{\sin\left(\frac{A}{2}\right)} + \frac{\cos\left(\frac{C-A}{2}\right)}{\sin\left(\frac{B}{2}\right)} + \frac{\cos\left(\frac{A-B}{2}\right)}{\sin\left(\frac{C}{2}\right)}$$

$$= \frac{2\cos\frac{A}{2}\cos\left(\frac{B-C}{2}\right)}{\sin A} + \frac{2\cos\frac{B}{2}\cos\left(\frac{C-A}{2}\right)}{\sin B} + \frac{2\sin\frac{C}{2}\cos\left(\frac{A-B}{2}\right)}{\sin C}$$

$$= \frac{2\sin\left(\frac{B+C}{2}\right)\cos\left(\frac{B-C}{2}\right)}{\sin A} + \frac{2\sin\left(\frac{A+C}{2}\right)\cos\left(\frac{C-A}{2}\right)}{\sin B} + \frac{2\sin\left(\frac{A+B}{2}\right)\cos\left(\frac{A-B}{2}\right)}{\sin C}$$

$$= \left(\frac{\sin B + \sin C}{\sin A}\right) + \left(\frac{\sin C + \sin A}{\sin B}\right) + \left(\frac{\sin A + \sin B}{\sin C}\right)$$

$$= \left(\frac{\sin B}{\sin A} + \frac{\sin A}{\sin B}\right) + \left(\frac{\sin C}{\sin B} + \frac{\sin B}{\sin C}\right) + \left(\frac{\sin A}{\sin C} + \frac{\sin C}{\sin A}\right)$$

as A, B, C are angles of triangle

$\Rightarrow \quad 0 < A, B, C < \pi$

$\Rightarrow \quad \sin A, \sin B, \sin C > 0$

$\Rightarrow \quad E \geq 2 + 2 + 2 \quad \left[\text{as } x + \frac{1}{x} \geq 2, \text{if } x > 0\right]$

$\Rightarrow \quad E \geq 6$

Ex. 81. *If Δ is the area of a Δ with side lengths a, b, c then show that $\Delta \leq \frac{1}{4}\sqrt{(a+b+c)\,abc}$. Also, show that equality occurs in the above inequality if and only if $a = b = c$.*

Sol. We know, $\Delta = \sqrt{s(s-a)(s-b)(s-c)}$

$$= \sqrt{\frac{s}{8}(b+c-a)(c+a-b)(a+b-c)}$$

Since, sum of two sides is always greater than third side.

$\therefore \quad b+c-a, c+a-b,$

$a+b-c > 0.$

$\Rightarrow \quad (s-a)(s-b)(s-c) > 0$

let $(s-a) = x, (s-b) = y, (s-c) = z$

Now, $\quad x + y = 2s - a - b = c, y + z = a$

and $\quad z + x = b$

Since, AM $\geq$ GM, then $2\sqrt{xy} \leq x + y = c;$

$2\sqrt{yz} \leq y + z = a; 2\sqrt{zx} \leq z + x = b$

$\therefore \quad 8xyz \leq abc$

$\Rightarrow \quad (s-a)(s-b)(s-c) \leq \frac{1}{8}(abc)$

$\Rightarrow \quad s(s-a)(s-b)(s-c) \leq \frac{1}{8}s(abc) \leq \frac{1}{16}(a+b+c)(abc)$

$\Rightarrow \quad \Delta \leq \frac{1}{4}\sqrt{abc(a+b+c)}$ and equality holds if

$x = y = z \Rightarrow a = b = c$

Aliter RHS $= \frac{1}{4}\sqrt{abc(a+b+c)}$

$$= \frac{1}{4}\sqrt{2s \cdot 4R\,\Delta} = \frac{1}{4}\sqrt{8.\frac{\Delta}{r}\cdot R.\Delta}$$

$$= \Delta\sqrt{\frac{R}{2r}} = \frac{\Delta}{\sqrt{8\sin\frac{A}{2}.\sin\frac{B}{2}.\sin\frac{C}{2}}} \quad \text{...(i)}$$

Consider, $\sin\frac{A}{2}\cdot\sin\frac{B}{2}\cdot\sin\frac{C}{2}$

$$= \frac{1}{2}\left[\cos\left(\frac{A-B}{2}\right) - \cos\left(\frac{A+B}{2}\right)\right]\cdot\cos\left(\frac{A+B}{2}\right)$$

$$= \frac{1}{2}\left[\frac{1}{4}\cos^2\left(\frac{A-B}{2}\right) - \left\{x - \frac{1}{2}\cos\left(\frac{A-B}{2}\right)\right\}^2\right]$$

$$\leq \frac{1}{8}\cos^2\left(\frac{A-B}{2}\right) \leq \frac{1}{8}$$

$\Rightarrow \quad \dfrac{1}{8\sin\frac{A}{2}.\sin\frac{B}{2}.\sin\frac{C}{2}} \geq 1 \quad \text{...(ii)}$

Thus, from Eqs. (i) and (ii); $\Delta \leq \frac{1}{4}\sqrt{(a+b+c)abc}$

and equality holds, $x = \frac{1}{2}\cos\left(\frac{A-B}{2}\right)$

and $\quad \cos\left(\frac{A-B}{2}\right) = 1$

$\Rightarrow \quad A = B$ and $\cos\left(\frac{A+B}{2}\right) = \frac{1}{2}$

$\Rightarrow \quad \sin\frac{C}{2} = \frac{1}{2} \Rightarrow C = 60°$

$\therefore \quad A = B = 60°$

Thus, the equality holds of triangle is equilateral.

Properties and Solutions of Triangles Exercise 1 :
Single Option Correct Type Questions

1. In the adjoining figure, the circle meets the sides of an equilateral triangle at six points.

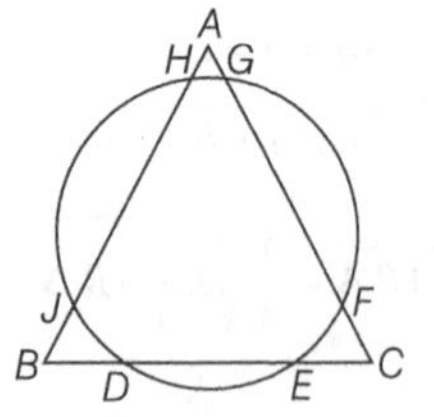

If $AG = 2$, $GF = 13$, $FC = 1$ and $HJ = 7$, then DE equals to
(a) $2\sqrt{22}$ (b) $7\sqrt{3}$
(c) 9 (d) 10

2. In a ΔABC, if $a = 13$, $b = 14$ and $c = 15$, then $\angle A$ is equal to (All symbols used have their usual meaning in a triangle.)
(a) $\sin^{-1}\frac{4}{5}$ (b) $\sin^{-1}\frac{3}{5}$ (c) $\sin^{-1}\frac{3}{4}$ (d) $\sin^{-1}\frac{2}{3}$

3. In a ΔABC, if $b = (\sqrt{3} - 1)a$ and $\angle C = 30°$, then the value of $(A - B)$ is equal to (All symbols used have usual meaning in the triangle.)
(a) 30° (b) 45°
(c) 60° (d) 75°

4. In a ΔABC, if $\angle C = 105°$, $\angle B = 45°$ and length of side $AC = 2$ units, then the length of the side AB is equal to
(a) $\sqrt{2}$ (b) $\sqrt{3}$
(c) $\sqrt{2} + 1$ (d) $\sqrt{3} + 1$

5. If P is a point on the altitude AD of the ΔABC such that $\angle CBP = \frac{B}{3}$, then AP is equal to
(a) $2a\sin\frac{C}{3}$ (b) $2b\sin\frac{A}{3}$
(c) $2c\sin\frac{B}{3}$ (d) $2c\sin\frac{C}{3}$

6. In ΔABC, if $2b = a + c$ and $A - C = 90°$, then $\sin B$ equals
[**Note** All symbols used have usual meaning in ΔABC.]
(a) $\frac{\sqrt{7}}{5}$ (b) $\frac{\sqrt{5}}{8}$ (c) $\frac{\sqrt{7}}{4}$ (d) $\frac{\sqrt{5}}{3}$

7. Let ABC be a right triangle with length of side $AB = 3$ and hyotenuse $AC = 5$. If D is a point on BC such that $\frac{BD}{DC} = \frac{AB}{AC}$, then AD is equal to
(a) $\frac{4\sqrt{3}}{3}$ (b) $\frac{3\sqrt{5}}{2}$ (c) $\frac{4\sqrt{5}}{3}$ (d) $\frac{5\sqrt{3}}{4}$

8. Two medians drawn from the acute angles of a right angled triangle intersect at an angle $\frac{\pi}{6}$. If the length of the hypotenuse of the triangle is 3 units, then the area of the triangle (in sq units) is $\sqrt{K}$, then K is
(a) 3 (b) $\frac{3\sqrt{5}}{2}$ (c) $\sqrt{3}$ (d) None of these

9. If in a right angle ΔABC, $4\sin A\cos B - 1 = 0$ and $\tan A$ is finite, then
(a) angles are in AP (b) angles are in GP
(c) angles are in HP (d) None of these

10. Let $A = \begin{bmatrix} a & b & c \\ p & q & r \\ 1 & 1 & 1 \end{bmatrix}$ and $B = A^2$

If $(a - b)^2 + (p - q)^2 = 25$, $(b - c)^2 + (q - r)^2 = 36$ and $(c - a)^2 + (r - p)^2 = 49$, then det B is
(a) 192 (b) 864
(c) 3456 (d) $25 \times 36 \times 47$

11. If in a ΔABC, the incircle passing through the point of intersection of perpendicular bisector of sides BC, AB, then $4\sin\frac{A}{2}\sin\frac{B}{2}\sin\frac{C}{2}$ equals to
(a) $\sqrt{2}$ (b) $\sqrt{2} - 1$
(c) $\sqrt{2} + 1$ (d) $\frac{1}{2}$

12. If two sides of a triangle are roots of the equation $x^2 - 7x + 8 = 0$ and the angle between these sides is 60°, then the product of in-radius and circum-radius of the triangle is
(a) $\frac{8}{7}$ (b) $\frac{5}{3}$
(c) $\frac{5\sqrt{2}}{3}$ (d) 8

13. If median AD of a triangle ABC makes angle $\frac{\pi}{6}$ with side BC, then the value of $(\cot B - \cot C)^2$ is equal to
(a) 6 (b) 9
(c) 12 (d) 15

14. If the perimeter of the triangle formed by foot of altitudes of the ΔABC is equal to four times the circumradius of ΔABC, then ΔABC is
(a) isosceles triangle (b) equilateral triangle
(c) right angled triangle (d) None of these

15. In a triangle with one angle $\frac{2\pi}{3}$, the lengths of the sides form an A.P. If the length of the greatest side is 7 cm, the radius of the circumcircle of the triangle is

(a) $\frac{7\sqrt{3}}{3}$ cm (b) $\frac{5\sqrt{3}}{3}$ cm (c) $\frac{2\sqrt{3}}{3}$cm (d) $\sqrt{3}$ cm

16. Sides of a triangle ABC are in AP. If $a < \min\{b, c\}$, then $\cos A$ may be equal to

(a) $\frac{3c-4b}{2b}$ (b) $\frac{3c-4b}{2c}$

(c) $\frac{4c-3b}{2b}$ (d) $\frac{4c-3b}{2c}$

17. The product of the sines of the angles of a triangle is p and the product of their cosines is q. Then, the tangents of the angles are the roots of the equation

(a) $qx^3 - px^2 + (1+q)x - p = 0$

(b) $qx^3 - px^2 - (1-q)x - p = 0$

(c) $qx^3 - px^2 + (1+q)x + p = 0$

(d) None of the above

18. Let C be incircle of ΔABC. If the tangents of lengths t_1, t_2 and t_3 are drawn inside the given triangle parallel to sides a, b and c, respectively, then $\frac{t_1}{a} + \frac{t_2}{b} + \frac{t_3}{c}$ is equal to

(a) 0 (b) 1 (c) 2 (d) 3

19. If the sine of the angles of ΔABC satisfy the equation

$c^3x^3 - c^2(a+b+c)x^2 + lx + m = 0$

(where a, b, c are the sides of ΔABC), then ΔABC is

(a) always right angled for any l, m

(b) right angled only when $l = c(ab+bc+ca) = c\,\Sigma ab$, $m = -abc$

(c) right angled only when $l = \frac{c\Sigma ab}{4}$, $m = -\frac{abc}{8}$

(d) never right angled

20. In a triangle ABC, medians AD and CE are drawn. If $AD = 5$, $\angle DAC = \frac{\pi}{8}$ and $\angle ACE = \frac{\pi}{4}$, the area of the triangle is

(a) $\frac{50}{9}$ (b) $\frac{25}{9}$ (c) $\frac{25}{3}$ (d) $\frac{25}{7}$

21. In a triangle ABC, $a \ge b \ge c$. If

$\frac{a^3+b^3+c^3}{\sin^3 A + \sin^3 B + \sin^3 C} = 8$, then the maximum value of a is

(a) $\frac{1}{2}$ (b) 2 (c) 8 (d) 64

22. A triangle ABC exists such that

(a) $(b+c+a)(b+c-a) = 5bc$

(b) the sides are of lengths $\sqrt{19}, \sqrt{38}, \sqrt{116}$

(c) $\left(\frac{b^2-c^2}{a^2}\right) + \left(\frac{c^2-a^2}{b^2}\right) + \left(\frac{a^2-b^2}{c^2}\right) = 0$

(d) $\cos\left(\frac{B-C}{2}\right) = (\sin B + \sin C)\cos\left(\frac{B+C}{2}\right)$

23. If a ΔABC, a, b, A are given and b_1, b_2 are two values of the third sides b such that $b_2 = 2b_1$. Then, $\sin A$ is equal to

(a) $\sqrt{\frac{9a^2-c^2}{8a^2}}$ (b) $\sqrt{\frac{9a^2-c^2}{8c^2}}$

(c) $\sqrt{\frac{9a^2-c^2}{8b^2}}$ (d) None of these

24. In a triangle ABC, if $\cot A = (x^3+x^2+x)^{\frac{1}{2}}$, $\cot B = (x + x^{-1} + 1)^{\frac{1}{2}}$ and $\cot C = (x^{-3} + x^{-2} + x^{-1})^{\frac{-1}{2}}$, then the triangle is

(a) equilateral (b) isosceles

(c) right angled (d) obtuse angled

25. In a ΔABC, a, b, A are given and c_1, c_2 are two values of the third side c. The sum of the areas two triangles with sides a, b, c_1 and a, b, c_2 is

(a) $\frac{1}{2}b^2\sin 2A$

(b) $\frac{1}{2}a^2\sin 2A$

(c) $b^2\sin 2A$

(d) None of these

26. In ΔABC, if $a = 10$ and $b\cot B + c\cot C = 2(r+R)$, then the maximum area of ΔABC will be

(a) 50 (b) $\sqrt{50}$

(c) 25 (d) 5

27. Three circles touch one-another externally. The tangents at their point of contact meet at a point whose distance from a point contact is 4. Then, the ratio of the product of the radii to the sum of the radii of circles is

(a) 16 : 1 (b) 1 : 16

(c) 8 : 1 (d) None of these

28. Let a, b, c be the sides of a triangle. No two of them are equal and $\lambda \in R$. If the roots of the equation $x^2 + 2(a+b+c)x + 3\lambda(ab+bc+ca) = 0$ are real distinct, then

(a) $\lambda < \frac{4}{3}$ (b) $\lambda > \frac{5}{3}$

(c) $\lambda \in \left(\frac{1}{3}, \frac{5}{3}\right)$ (d) $\lambda \in \left(\frac{4}{3}, \frac{5}{3}\right)$

29. In triangle ABC, if P, Q, R divides sides BC, AC and AB, respectively, in the ratio $k:1$ (in order). If the ratio $\left(\frac{\text{area } \Delta PQR}{\text{area } \Delta ABC}\right)$ is $\frac{1}{3}$, then k is equal to

(a) $\frac{1}{3}$ (b) 2

(c) 3 (d) None of these

30. Let $f(x+y)=f(x)\cdot f(y)$ for all x and y and $f(1)=2$. If in a triangle ABC, $a=f(3)$, $b=f(1)+f(3)$, $c=f(2)+f(3)$,

(a) C (b) $2C$

(c) $3C$ (d) $4C$

31. Let a, b, c be given positive numbers, then values of x, y and $z \in R^+$ which satisfies equations $x+y+z=a+b+c$ and $4xyz=-(a^2x+b^2y+c^2z)=abc$ are respectively.

(a) $\frac{b+c}{2}, \frac{a+c}{2}, \frac{a+b}{2}$ (b) $\frac{a}{2}, \frac{b}{2}, \frac{c}{2}$

(c) $\frac{a+b}{2}, \frac{a+c}{2}, \frac{b+c}{2}$ (d) None of these

32. If 't_1', 't_2' and 't_3' are the lengths of the tangents drawn from centre of x-circle to the circumcircle of the ΔABC, then $\frac{1}{t_1^2}+\frac{1}{t_2^2}+\frac{1}{t_3^2}$ is equal to

(a) $\frac{abc}{a+b+c}$ (b) $\frac{a+b+c}{abc}$

(c) $\frac{a+b+c}{2abc}$ (d) $\frac{2abc}{a+b+c}$

33. In triangle ABC, $\angle A > \frac{\pi}{2}$. AA_1 and AA_2 are the median and altitude, espectively. If $\angle BAA_1 = \angle A_1AA_2 = \angle A_2AC$, then $\sin^3\frac{A}{3}\cdot\cos\frac{A}{3}$ is equal to

(a) $\frac{3a^3}{16b^2c}$ (b) $\frac{3a^3}{64b^2c}$

(c) $\frac{3a^2}{4b^2c}$ (d) $\frac{3a^3}{12b^2c}$

34. In an ambiguous case of solving a triangle when $a=\sqrt{5}$, $b=2$, $\angle A=\frac{\pi}{6}$ and the two possible values of third side are c_1 and c_2, then

(a) $|c_1-c_2|=2\sqrt{6}$ (b) $|c_1-c_2|=4\sqrt{6}$

(c) $|c_1-c_2|=4$ (d) $|c_1-c_2|=6$

35. If R_1 is the circumradius of the pedal triangle of a given triangle and R_2 is the circumradius of the pedal triangle of the pedal triangle formed, and so on $R_3, R_4 \ldots$, then the value of $\sum_{i=1}^{\infty} R_i$, where R (circumradius) of ΔABC is 5 is

(a) 8 (b) 10 (c) 12 (d) 15

36. If in a triangle $\left(1-\frac{r_1}{r_2}\right)\left(1-\frac{r_1}{r_3}\right)=2$, then the triangle is

(a) right angled (b) isosceles

(c) equilateral (d) None of these

37. If the median AD of a triangle ABC makes an angle θ with side AB, then $\sin(A-\theta)$ is equal to

(a) $\frac{b}{c}\sin\theta$ (b) $\frac{c}{b}\sin\theta$

(c) $\frac{c}{b}\cos\theta$ (d) None of these

38. In a ΔABC, angles A, B, C are in AP, then $\lim_{A\to C}\frac{\sqrt{3-4\sin A\sin C}}{|A-C|}$ is

(a) 1 (b) 2

(c) 3 (d) 4

39. In a triangle ABC, $(a+b+c)(b+c-a)=\lambda bc$ if

(a) $\lambda<0$ (b) $\lambda>6$

(c) $0<\lambda<4$ (d) $\lambda>4$

40. In the triangle ABC, if $(a^2+b^2)\sin(A-B)$ $=(a^2-b^2)\sin(A+B)$, then the triangle is

(a) either isosceles or right angled

(b) only right angled

(c) only isosceles triangle

(d) None of the above

41. In a ΔABC, sides a, b, c are in AP and $\frac{2}{1!9!}+\frac{2}{3!7!}+\frac{1}{5!5!}=\frac{8^a}{(2b)!}$, then the maximum value of $\tan A\tan B$ is equal to

(a) $\frac{1}{2}$ (b) $\frac{1}{3}$ (c) $\frac{1}{4}$ (d) $\frac{1}{5}$

42. If a, b, c be the sides of a triangle ABC and if roots of the equation $a(b-c)x^2+b(c-a)x+c(a-b)=0$ are equal, then $\sin^2\left(\frac{A}{2}\right), \sin^2\left(\frac{B}{2}\right), \sin^2\left(\frac{C}{2}\right)$ are in

(a) AP (b) GP

(c) HP (d) AGP

43. The ratio of the area of a regular polygon of n sides inscribed in a circle to that of the polygon of same number of sides circumscribing the same circle is 3 : 4. Then, the value of n is

(a) 6 (b) 4

(c) 8 (d) 12

44. In any ΔABC, $\Pi\left(\frac{\sin^2 A+\sin A+1}{\sin A}\right)$ is always greater than

(a) 9 (b) 3

(c) 27 (d) None of these

45. If the incircle of the triangle ABC, passes through it's circumcentre, then the $\cos A + \cos B + \cos C$ is

(a) -2 (b) $\sqrt{2}$

(c) $-\sqrt{2}$ (d) None of these

46. The perimeter of a triangle is 6 times the arithmetic mean of the sines of its angles. If the side a is 1, then A is equal to

(a) 30° (b) 60° (c) 90° (d) 120°

47. If there are only two linear functions f and g which map [1, 2] on [4, 6] and in a ΔABC, $c = f(1) + g(1)$ and a is the maximum value of r^2, where r is the distance of a variable point on the curve $x^2 + y^2 - xy = 10$ from the origin, then $\sin A : \sin C$ is

(a) 1 : 2 (b) 2 : 1

(c) 1 : 1 (d) None of these

48. A circle is inscribed in an equilateral triangle of side a. The area of any square inscribed in this circle is

(a) a^2 (b) $\frac{a^2}{4}$ (c) $\frac{a^2}{3}$ (d) $\frac{a^2}{6}$

49. In any triangle ABC, if $\sin A$, $\sin B$, $\sin C$ are in AP, then the maximum value of $\tan \frac{B}{2}$ is

(a) $-\frac{1}{\sqrt{3}}$ (b) $\frac{1}{\sqrt{3}}$

(c) $\frac{1}{3}$ (d) None of these

50. In a ΔABC, $2\cos A = \frac{\sin B}{\sin C}$ and $2^{\tan^2 B}$ is a solution of equation $x^2 - 9x + 8 = 0$, then ΔABC is

(a) equilateral (b) isosceles

(c) scalene (d) right angled

51. A triangle is inscribed in a circle. The vertices of the triangle divide the circle into three arcs of length 3, 4 and 5 unit, then the area of the triangle is equal to

(a) $\frac{9\sqrt{3}(1+\sqrt{3})}{\pi^2}$ sq unit (b) $\frac{9\sqrt{3}(\sqrt{3}-1)}{\pi^2}$ sq unit

(c) $\frac{9\sqrt{3}(1+\sqrt{3})}{2\pi^2}$ sq unit (d) $\frac{9\sqrt{3}(\sqrt{3}-1)}{2\pi^2}$ sq unit

52. If a, b and c are the sides of a triangle such that $b \cdot c = \lambda^2$, then the relation in a, λ and A is

(a) $c \geq 2\lambda \sin\left(\frac{C}{2}\right)$ (b) $b \geq 2\lambda \sin\left(\frac{A}{2}\right)$

(c) $a \geq 2\lambda \sin\left(\frac{A}{2}\right)$ (d) None of these

53. In ΔABC, AD is an altitude from A on side BC. If $b > c$, $\angle c = 23°$ and $AD = \frac{abc}{b^2 - c^2}$, then $\angle B$ is

(a) 110° (b) 113° (c) 120° (d) 130°

54. If in a triangle ABC, $a = 5$, $b = 4$ and $\cos(A - B) = \frac{31}{32}$, then the third side c is equal to

(a) 3 (b) 6

(c) 7 (d) 9

55. In ΔABC, if $AB = x$, $BC = x + 1$, $\angle C = \frac{\pi}{3}$, then the least integer value of x is

(a) 6 (b) 7

(c) 8 (d) None of these

56. Three circular coins each of radii 1 cm are kept in an equilateral triangle so that all the three coins touch each other and also the sides of the triangle. Area of the triangle is

(a) $(4 + 2\sqrt{3})$ cm^2 (b) $\frac{1}{4}(12 + 7\sqrt{3})$ cm^2

(c) $\frac{1}{4}(48 + 7\sqrt{3})$ cm^2 (d) $(6 + 4\sqrt{3})$ cm^2

57. The sides of a triangle are in AP. If the angles A and C are the greatest and smallest angle respectively, then $4(1 - \cos A)(1 - \cos C)$ is equal to

(a) $\cos A - \cos C$ (b) $\cos A \cos C$

(c) $\cos A + \cos C$ (d) $\cos C - \cos A$

58. If in ΔABC, $c(a + b)\cos \frac{1}{2} B = b(a + c)\cos \frac{1}{2} C$, the triangle is

(a) isosceles

(b) equilateral

(c) right angled but not isosceles

(d) right angled and isosceles

59. In a triangle, the line joining the circumcentre to the incentre is parallel to BC, then $\cos B + \cos C$ is equal to

(a) $\frac{3}{2}$ (b) 1

(c) $\frac{3}{4}$ (d) $\frac{1}{2}$

60. In the given figure, AB is the diameter of the circle, centered at O. If $\angle COA = 60°$, $AB = 2r$, $AC = d$ and $CD = l$, then l is equal to

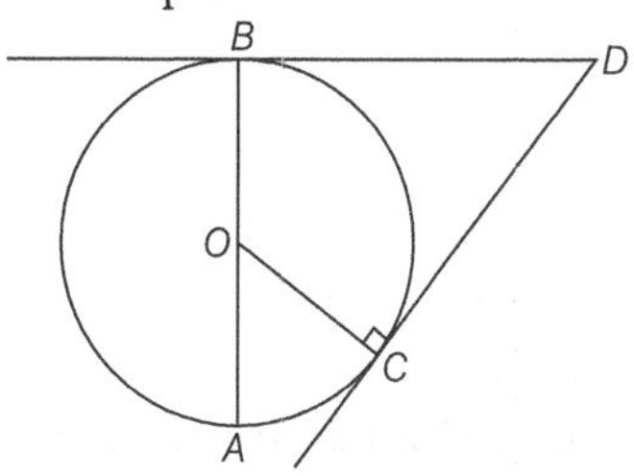

(a) $d\sqrt{3}$ (b) $\frac{d}{\sqrt{3}}$

(c) $3d$ (d) $\frac{\sqrt{3}d}{2}$

61. In a triangle ABC; AD, BE and CF are the altitudes and R is the circum radius, then the radius of the circle DEF is
(a) $2R$ (b) R
(c) $\frac{R}{2}$ (d) None of these

62. In a right angled triangle ABC, the bisector of the right angle C divides AB into segment x and y and $\tan\left(\frac{A-B}{2}\right) = t$, then $x : y$ is equal to
(a) $(1+t) = (1-t)$ (b) $(1-t) : (1+t)$
(c) $1 : (1+t)$ (d) $(1-t) : 1$

63. A variable triangle ABC is circumscribed about a fixed circle of unit radius. Side BC always touches the circle at D and has fixed direction. If B and C vary in such a way that $(BD)\cdot(CD) = 2$, then locus of vertex A will be a straight line
(a) parallel to side BC
(b) right angle to side BC
(c) making and angle $\frac{\pi}{6}$ with BC
(d) making an angle $\sin^{-1}\left(\frac{2}{3}\right)$ with BC

65. A tower of height b subtends an angle at a point O on the level of the foot of the tower and at a distance a from the foot of the tower. If a pole mounted on the tower also subtends an equal angle at O, the height of the pole is
(a) $b\left(\frac{a^2-b^2}{a^2+b^2}\right)$ (b) $b\left(\frac{a^2+b^2}{a^2-b^2}\right)$
(c) $a\left(\frac{a^2-b^2}{a^2+b^2}\right)$ (d) $a\left(\frac{a^2+b^2}{a^2-b^2}\right)$

66. A balloon is observed simultaneously from three points A, B and C on a straight road directly under it. The angular elevation at B is twice and at C is thrice that of A. If the distance between A and B is 200 m and the distance between B and C is 100 m, then the height of balloon is given by
(a) 50 m (b) $50\sqrt{3}$ m
(c) $50\sqrt{2}$ m (d) None of these

67. A vertical pole (more than 100 m high) consists of two portions, the lower being one third of the whole. If the upper portion subtends an angle $\tan^{-1}\left(\frac{1}{2}\right)$ at a point in a horizontal plane through the foot of the pole and distance 40 ft from it, then the height of the pole is
(a) 100 ft (b) 120 ft
(c) 150 ft (d) None of these

Properties and Solutions of Triangles Exercise 2 : More than One Correct Type Questions

68. If area of ΔABC, Δ and $\angle C$, are given and if the side c opposite to given angles is minimum, then
(a) $a = \sqrt{\frac{2\Delta}{\sin C}}$ (b) $b = \sqrt{\frac{2\Delta}{\sin C}}$
(c) $a = \frac{4\Delta}{\sin C}$ (d) $b = \frac{4\Delta}{\sin^2 C}$

69. If Δ represents area of acute angled ΔABC, then $\sqrt{a^2b^2 - 4\Delta^2} + \sqrt{b^2c^2 - 4\Delta^2} + \sqrt{c^2a^2 - 4\Delta^2}$ equals to
(a) $a^2 + b^2 + c^2$
(b) $\frac{a^2+b^2+c^2}{2}$
(c) $ab\cos C + bc\cos A + ca\cos B$
(d) $ab\sin C + bc\sin A + ca\sin B$

70. In ΔABC, the value of $c\cos(A-\theta) + a\cos(C+\theta)$ cannot exceed $(\theta \in (0, 2\pi))$ [Letters have usual meanings]
(a) a (b) b
(c) c (d) s

71. In ΔABC, if $ac = 3$, $bc = 4$ and $\cos(A-B) = \frac{3}{4}$, then
(a) measure of $\angle A$ is $\frac{\pi}{2}$
(b) measure of $\angle B$ is $\frac{\pi}{2}$
(c) $\cot\frac{C}{2} = \sqrt{7}$
(d) circumradius of ΔABC is $\frac{2}{7^{1/4}}$

72. In ΔABC, let $a = 5$, $b = 4$ and $\cos(A-B) = \frac{31}{32}$, then which of the following statement(s) is (are) correct?
[**Note** All symbols used have usual meaning in a triangle]
(a) The perimeter of ΔABC equals $\frac{15}{2}$
(b) The radius of circle inscribed in ΔABC equals $\frac{\sqrt{7}}{2}$
(c) The measure of $\angle C$ equals $\cos^{-1}\frac{1}{8}$
(d) The value of $R(b^2\sin 2C + c^2\sin 2B)$ equals 120

73. In which of the following situations, it is possible to have a ΔABC?

(All symbols used have usual meaning in a triangle)

(a) $(a + c - b)(a - c + b) = 4bc$
(b) $b^2 \sin 2C + \cos^2 \sin 2B = ab$
(c) $a = 3$, $b = 5$, $c = 7$ and $C = \frac{2\pi}{3}$
(d) $\cos\left(\frac{A-C}{2}\right) = \cos\left(\frac{A+C}{2}\right)$

74. In a ΔABC, let $BC = 1$, $AC = 2$ and measure of $\angle C$ is 30°. Which of the following statement(s) is(are) correct?

(a) $2\sin A = \sin B$
(b) Length of side AB equals $5 - 2\sqrt{3}$
(c) measure of $\angle A$ is less than 30°
(d) Circumradius of $\angle ABC$ is equal to length of side AB

75. Let one angle of a triangle be 60°, the area of triangle is $10\sqrt{3}$ and perimeter is 20 cm. If $a > b > c$ where a, b and c denote lengths of sides opposi9te to vertices A, B and C respectively, then which of the following is(are) correct?

(a) Inradius of triangle is $\sqrt{3}$
(b) Length of longest side of triangle is 7
(c) Circum-radius of triangles is $\frac{7}{\sqrt{3}}$
(d) Radius of largest escribed circle is $\frac{1}{12}$

76. In a ΔABC, if $a = 4$, $b = 8$ and $\angle C = 60^\circ$, then which of the following relations is(are) correct?

[**Note** All symbols used have usual meaning in triangles ABC.]

(a) The area of ΔABC is $8\sqrt{3}$
(b) The value of $\Sigma \sin^2 A = 2$
(c) Inradius of triangle ABC is $\frac{2\sqrt{3}}{3+\sqrt{3}}$
(d) The length of internal angle bisector of $\angle C$ is $\frac{4}{\sqrt{3}}$.

77. Given an isosceles triangle with equal sides of length b, base angle $\alpha < \frac{\pi}{4}$ and R, r the radii and O, I the centres of the circumcircle and incircle, respectively. Then

(a) $R = \frac{1}{2} b \operatorname{cosec} \alpha$
(b) $\Delta = 2b^2 \sin 2\alpha$
(c) $r = \frac{b \sin 2\alpha}{2(1+\cos\alpha)}$
(d) $OI = \left|\frac{b\cos\left(\frac{3\alpha}{2}\right)}{2\sin\alpha\cos\left(\frac{\alpha}{2}\right)}\right|$

78. There exist a triangle ABC satisfying

(a) $\tan A + \tan B + \tan C = 0$
(b) $\frac{\sin A}{2} = \frac{\sin B}{3} = \frac{\sin C}{7}$
(c) $(a+b)^2 = c^2 + ab$ and $\sqrt{2}(\sin A + \cos A) = \sqrt{3}$
(d) $\sin A + \sin B = \left(\frac{\sqrt{3}+1}{2}\right) \cos A \cos B = \frac{\sqrt{3}}{4} = \sin A \sin B$

79. Let a, b, c be the sides of triangle whose perimeter is P and area is A, then

(a) $P^3 \le 27(b+c-a)(c+a-b)(a+b-c)$
(b) $P^2 \le 3(a^2+b^2+c^2)$
(c) $a^2+b^2+c^2 \ge 4\sqrt{3}A$
(d) $P^4 \le 25 < A$

80. If in ΔABC, $A = 90^\circ$ and c, $\sin B$ and $\cos B$ are rational number, then

(a) a is rational
(b) a is irrational
(c) b is rational
(d) b is irrational

81. In ΔABC, which of the following statements are true

(a) maximum value of $\sin 2A + \sin 2B + \sin 2C$ is same as the maximum value of $\sin A + \sin B + \sin C$
(b) $R \ge 2r$, where R is circumradius and r is inradius
(c) $R^2 \ge \frac{abc}{(a+b+c)}$
(d) ΔABC is right angled if $r + 2R = s$, where s is semiperimeter

82. If l is the length of median from the the vertex A to the side BC of a ΔABC, then

(a) $4l^2 = 2b^2 + 2c^2 - a^2$
(b) $4l^2 = b^2 + c^2 + 2bc\cos A$
(c) $4l^2 = a^2 + 4bc\cos A$
(d) $4l^2 - (2s-a)^2 - 4bc\sin^2\left(\frac{A}{2}\right)$

83. If A, A_1, A_2, A_3 are the areas of the inscribed and escribed of a ΔABC, then

(a) $\sqrt{A_1} + \sqrt{A_2} + \sqrt{A_3} = \sqrt{\pi}(r_1 + r_2 + r_3)$
(b) $\frac{1}{\sqrt{A_1}} + \frac{1}{\sqrt{A_2}} + \frac{1}{\sqrt{A_3}} = \frac{1}{\sqrt{A}}$
(c) $\frac{1}{\sqrt{A_1}} + \frac{1}{\sqrt{A_2}} + \frac{1}{\sqrt{A_3}} = \frac{s^2}{\sqrt{\pi r_1 r_2 r_3}}$
(d) $\sqrt{A_1} + \sqrt{A_2} + \sqrt{A_3} = \sqrt{\pi}(4R + r)$

84. If a, b, A be given in a triangle and c_1 and c_2 two possible values of third side such that $c_1^2 + c_1c_2 + c_2^2 = a^2$, then A is equal to

(a) 30° (b) 60°
(c) 90° (d) 120°

85. D, E and F are the middle points of the sides of the triangle ABC, then

(a) centroid of the triangle DEF is the same as that of ABC
(b) orthocentre of the triangle DEF is the circumcentre of ABC
(c) orthocentre of the triangle DEF is the incentre of ABC
(d) centroid of the triangle DEF is not the same as that of ABC

86. The sides of ΔABC satisfy the equation $2a^2 + 4b^2 + c^2 = 4ab + 2ac$. Then

(a) the triangle is isosceles
(b) the triangle is obtuse
(c) $B = \cos^{-1}\left(\frac{7}{8}\right)$ (d) $A = \cos^{-1}\left(\frac{1}{4}\right)$

87. If Δ represents the area of acute angled triangle ABC, then $\sqrt{a^2b^2 - 4\Delta^2} + \sqrt{b^2c^2 - 4\Delta^2} + \sqrt{c^2a^2 - 4\Delta^2}$ is equal to

(a) $a^2 + b^2 + c^2$ (b) $\frac{a^2 + b^2 + c^2}{2}$
(c) $ab\cos C + bc\cos A + ca\cos B$
(d) $ab\sin C + bc\sin A + ca\sin B$

88. In triangle, ABC, if $2a^2b^2 + 2b^2c^2 = a^4 + b^4 + c^4$, then angle B is equal to

(a) 45° (b) 135°
(c) 120° (d) 60°

89. If H is the orthocentre of triangle ABC, R = circumradius and $P = AH + BH + CH$, then

(a) $P = 2(R + r)$
(b) max. of P is $3R$
(c) min. of P is $3R$
(d) $P = 2(R - r)$

90. If inside a big circle exactly $n(n \geq 3)$ small circles, each of radius r, can be drawn in such a way that each small circle touches the big circle and also touches both its adjacent small circles, then the radius of big circle is

(a) $r\left(1 + \operatorname{cosec}\frac{\pi}{n}\right)$ (b) $\left(\frac{1 + \tan\frac{\pi}{n}}{\cos\frac{\pi}{n}}\right)$
(c) $r\left[1 + \operatorname{cosec}\frac{2\pi}{n}\right]$ (d) $\frac{r\left[\sin\frac{\pi}{2n} + \cos\frac{2\pi}{n}\right]^2}{\sin\frac{\pi}{n}}$

91. If in triangle ABC, a, b, c and angle A are given and $c\sin A < a < c$, then (b_1 and b_2 are values of b)

(a) $b_1 + b_2 = 2c\cos A$ (b) $b_1 + b_2 = c\cos A$
(c) $b_1b_2 = c^2 - a^2$ (d) $b_1b_2 = c^2 + a^2$

Properties and Solutions of Triangles Exercise 3 : Statement I and II Answer Type Questions

- This section contains 15 questions. Each question contains **Statement I (Assertion)** and **Statement II (Reason).** Each question has 4 choices (a), (b) and (d) out of which only one is correct. Choices are
  (a) Both Statement I and Statement II are correct and Statement II is the correct explanation of Statement I
  (b) Both Statement I and Statement II are correct and Statement II is not the correct explanation of Statement I
  (c) Statement I is correct but Statement II is incorrect
  (d) Statement II is correct but Statement I is incorrect

92. In a ΔABC,

$$a^3 + b^3 + c^3 = c^2(a + b + c)$$

(All symbol used have usual meaning in a triangle.)

Statement I The value of $\angle C = 60°$.

Statement II ΔABC must be equilateral.

93. In a ΔABC, let $a = 6$, $b = 3$ and $\cos(A - B) = \frac{4}{5}$.

[**Note** All symbols used have usual meaning in a triangle.]

Statement I $\angle B = \frac{\pi}{2}$

Statement II $\sin A = \frac{2}{\sqrt{5}}$

94. Statement I If in a triangle ABC $\sin^2 A + \sin^2 B + \sin^2 C = 2$, then one of the angles must be 90°.

Statement II In any triangles ABC

$\cos 2A + \cos 2B + \cos 2C = -1 - 4\cos A\cos b\cos C$

95. Statement I If A, B, C, D are angles of a cyclic quadrilateral then $\Sigma\sin A = 0$.

Statement II If A, B, C, D are angles of cyclic quadrilateral then, $\Sigma\cos A = 0$.

96. Statement I In any triangle ABC, the square of the length of the bisector AD is $bc\left(1 - \frac{a^2}{(b + c)^2}\right)$.

Statement II In any triangle ABC length of bisector AD is $\frac{2bc}{(b + c)}\cos\left(\frac{A}{2}\right)$.

97. **Statement I** If I is incentre of ΔABC and I_1 excentre opposite to A and P is intersection of II_1 and BC, then $IP \cdot I_1P = BP \cdot PC$

Statement II In a ΔABC, I is incentre and I_1 is excentre opposite to A, then IBI, I_1, C must be square.

98. All the notations used in statement I and statement II are usual.

Statement I In triangle ABC, if $\frac{\cos A}{a} = \frac{\cos B}{b} = \frac{\cos C}{c}$, then value of $\frac{r_1 + r_2 + r_3}{r}$ is equal to 9.

Statement II If $\Delta ABC : \frac{a}{\sin A} = \frac{b}{\sin B} = \frac{c}{\sin C} - 2R$, where R is circumradius.

99. **Statement I** In a triangle ABC if $\tan A : \tan B : \tan C = 1 : 2 : 3$, then $A = 45°$

Statement II If $p : q : r = 1 : 2 : 3$, then $p = 1$

100. **Statement I** In any right angled triangle $\frac{a^2 + b^2 + c^2}{R^2}$ is always equal to 8.

Statement II $a^2 = b^2 + c^2$

101. **Statement I** perimeter of a regular pentagon inscribed in a circle with centre O and radius a cm equals $10a \sin 36°$ cm.

Statement II Perimeter of a regular polygon inscribed in a circle with centre O and radius a cm equals $(3n - 5)\sin\left(\frac{360°}{2n}\right)$ cm, then it is n sided, where $n \geq 3$.

102. **Statement I** In any triangle ABC
$a \cos A + b \cos B + c \cos C \leq s$.

Statement II In any triangle ABC
$\sin\left(\frac{A}{2}\right)\sin\left(\frac{B}{2}\right)\sin\left(\frac{C}{2}\right) \leq \frac{1}{8}$

103. **Statement I** In a ΔABC, if $\cos^2 \frac{A}{2} + \cos^2 \frac{B}{2} + \cos^2 \frac{C}{2} = y\left(x^2 + \frac{1}{x^2}\right)$, then the maximum value of y is $\frac{9}{8}$.

Statement II In a ΔABC, $\sin \frac{A}{2} \cdot \sin \frac{B}{2} \cdot \sin \frac{C}{2} \leq \frac{1}{8}$

104. **Statement I** In any triangle
$a \cos A + b \cos B + c \cos C \leq s$

Statement II In any triangle $\sin\left(\frac{A}{2}\right)\sin\left(\frac{B}{2}\right)\sin\left(\frac{C}{2}\right) \leq \frac{1}{8}$

105. **Statement I** In triangle ABC, $\frac{a^2 + b^2 + c^2}{\Delta} \geq 4\sqrt{3}$

Statement II If $a_i > 0$, $i = 1, 2, 3, \ldots, n$ which are not identical, then

$\frac{a_1^m + a_2^m + \ldots + a_n^m}{n} > \left(\frac{a_1 + a_2 + \ldots + a_n}{n}\right)^m$, if $m < 0$ or $m > 1$.

106. **Statement I** AA_1, BB_1, CC_1 are the medians of triangle ABC whose centroid is G. If the points A, C_1, G and B are concyclic, then c^2, a^2, b^2 are in AP.

Statement II $BG \cdot CC_1 = BC_1 \cdot BA$

Properties and Solutions of Triangles Exercise 4 : Passage Based Questions

Passage I

(Q. Nos. 107 to 109)

R is circumradii of ΔABC, H is orthocentre. R_1, R_2, R_3 are circumradii of ΔAHB, ΔAHC, ΔBHC. If AH produced meet the circumradii of ABC at M and intersect BC at L.

$$\angle AHB = 180° - C$$

$$\frac{c}{\sin(180° - C)} = 2R_1$$

$$\frac{c}{\sin C} = 2R_1$$

$$R_1 = R$$

107. $R_1R_2 + R_2R_3 + R_1 R_3$ is equal to

(a) $2R^2$ (b) $3R^2$
(c) $5R^2$ (d) R^2

108. Area of ΔAHB

(a) $2R \cos A \cos B \cos C$
(b) $R^2 \cos A \cos B \cos C$
(c) $2R^2 \cos A \cos B \sin C$
(d) None of the above

109. Ratio of area of ΔAHB to ΔBML, is

(a) $\cos B : 2\cos A$ (b) $2 : 1$
(c) $\cos A : \cos B \ \cos C$ (d) None of these

Passage II

(Q. Nos. 110 to 112)

Let ABC be an acute triangle with $BC = a$, $CA = b$ and $AB = c$, where $a \neq b \neq c$. From any point 'p' inside ΔABC let D, E, F denote foot of perpendiculars from 'P' onto the sides BC, CA and AB, respectively. Now, answer the following questions.

110. All positions of point 'P' for which ΔDEF is isosceles lie on

(a) the incircle of ΔABC
(b) line of internal angle bisectors from A, B and C
(c) arcs of 3 circles
(d) None of the above

111. Let $A(7, 0)$, $B(4, 4)$ and $C(0, 0)$ and ΔDEF is isosceles with $DE = DF$. Then, the curve on which 'P' may lie

(a) $x = 4$ or $x + y = 7$ or $4x = 3y$
(b) $x = 4$ or $x^2 + y^2 = 4x + 4y$
(c) $3(x^2 + y^2) + 196 = 49(x + y)$
(d) None of the above

112. If ΔDEF is equilateral, then 'P'

(a) coincides with incentre of ΔABC
(b) coincides with orthocentre of ΔABC
(c) lies on pedal Δ of ABC
(d) None of the above

Passage III

(Q. Nos. 113 to 115)

In an acute angled ΔABC, let AD, BE and CF be the perpendicular from A, B and C upon the opposite sides of the triangle. (All symbols used have usual meaning in a triangle.)

113. The ratio of the product of the side lengths of the ΔDEF and ΔABC, is equal to

(a) $\frac{3(abc)^{\frac{1}{3}}}{4(a+b+c)}$
(b) $\frac{1}{4}$
(c) $\cos A \cos B \cos C$
(d) $\sin\left(\frac{A}{2}\right) \sin\left(\frac{B}{2}\right) \sin\left(\frac{C}{2}\right)$

114. The orthocentre of the ΔABC, is the

(a) centroid of the ΔDEF
(b) circum-centre of the ΔDEF
(c) incentre of the ΔDEF
(d) orthocentre of the ΔDEF

115. The circum-radius of the ΔDEF can be equal to

(a) $\frac{abc}{8\Delta}$ (b) $\frac{a}{4 \sin A}$
(c) $\frac{R}{2}$ (d) $\frac{r}{8} \operatorname{cosec}\frac{A}{2} \operatorname{cosec}\frac{B}{2} \operatorname{cosec}\frac{C}{2}$

Passage IV

(Q.Nos. 116 to 118)

Let a, b, c are the sides opposite to angles A, B, C respectively in a ΔABC $\tan\frac{A-B}{2} = \frac{a-b}{a+b}\cot\frac{C}{2}$ and $\frac{a}{\sin A} = \frac{b}{\sin B} = \frac{c}{\sin C}$.

If $a = 6$, $b = 3$ and $\cos(A - B) = \frac{4}{5}$

116. Angle C is equal to

(a) $\frac{\pi}{4}$ (b) $\frac{\pi}{2}$ (c) $\frac{3\pi}{4}$ (d) $\frac{2\pi}{3}$

117. Area of the triangle is equal to

(a) 8 (b) 9
(c) 10 (d) 11

118. Value of $\sin A$ is equal to

(a) $\frac{1}{\sqrt{5}}$ (b) $\frac{2}{\sqrt{5}}$
(c) $\frac{1}{2\sqrt{5}}$ (d) $\frac{1}{\sqrt{3}}$

Passage V

(Q. Nos. 119 to 123)

When any two sides and one of the opposite acute angle are given, under certain additional conditions two triangles are possible. The case when two triangles are possible is called the ambiguous case.

In fact when any two sides and the angle opposite to one of them are given either no triangle is possible or only one triangle is possible or two triangles are possible.

In the ambiguous case, let a, b and $\angle A$ are given and c_1, c_2 are two values of the third side c.

On the basis of above information, answer the following questions

119. Two different triangles are possible when

(a) $b \sin A < a$ (b) $b \sin A < a$ and $b > a$
(c) $b \sin A < a$ and $b < a$ (d) $b \sin A < a$ and $a = b$

120. The difference between two values of c is

(a) $2\sqrt{(a^2 - b^2)}$ (b) $\sqrt{(a^2 - b^2)}$
(c) $2\sqrt{(a^2 - b^2 \sin^2 A)}$ (d) $\sqrt{(a^2 - b^2 \sin^2 A)}$

121. The value of $c_1^2 - 2c_1c_2 \cos 2A + c_2^2$ is

(a) $4a \cos A$ (b) $4a^2 \cos A$
(c) $4a \cos^2 A$ (d) $4a^2 \cos^2 A$

122. If $\angle A = 45°$ and in ambiguous case (a, b, A are given) c_1, c_2 are two values of c and if θ be the angle between the two positions of the ambiguous side c then $\cos\theta$ is

(a) $\frac{c_1c_2}{c_1^2 + c_2^2}$ (b) $\frac{2c_1c_2}{c_1^2 + c_2^2}$
(c) $\frac{\sqrt{c_1c_2}}{(c_1 + c_2)}$ (d) $\frac{2\sqrt{c_1c_2}}{(c_1 + c_2)}$

123. If $2b=(m+1)a$ and $\cos A=\frac{1}{2}\sqrt{\left(\frac{(m-1)(m+3)}{m}\right)}$, where $1<m<3$, then $\frac{c_1}{c_2}$ is

(a) m or $\frac{1}{m}$ (b) $(m-1)$ or $\frac{1}{(m-1)}$

(c) $(m+1)$ or $\frac{1}{(m+1)}$ (d) $(m+3)$ or $\frac{1}{(m+3)}$

Passage VI

(Q.Nos. 124 to 126)

Consider a triangle ABC, where x, y, z are the length of perpendicular drawn from the vertices of the triangle to the opposite sides a, b, c respectively. Let the letters R, r, S, Δ denote the circumradius, inradius semi-perimeter and area of the triangle respectively.

124. If $\frac{bx}{c}+\frac{cy}{a}+\frac{az}{b}=\frac{a^2+b^2+c^2}{k}$, then the value of k is

(a) R (b) S (c) $2R$ (d) $\frac{3}{2}R$

125. If $\cot A+\cot B+\cot C=k\left(\frac{1}{x^2}+\frac{1}{y^2}+\frac{1}{z^2}\right)$, then the value of k is

(a) R^2 (b) rR

(c) Δ (d) $a^2+b^2+c^2$

126. The value of $\frac{c\sin B+b\sin C}{x}+\frac{a\sin C+c\sin A}{y}+\frac{b\sin A+a\sin B}{z}$ is equal to

(a) $\frac{R}{r}$ (b) $\frac{S}{R}$ (c) 2 (d) 6

Passage VII

(Q.Nos. 127 to 131)

AL, BM and CN are perpendicular from angular points of a triangle ABC on the opposite sides BC, CA and AB respectively. Δ is the area of triangle ABC, (r) and R are the inradius and circumradius.

127. If perimeters of ΔLMN and ΔABC and λ and μ, then the value of $\frac{\lambda}{\mu}$ is

(a) $\frac{r}{R}$ (b) $\frac{R}{r}$

(c) $\frac{rR}{\Delta}$ (d) $\frac{\Delta}{rR}$

128. If areas of Δ's AMN, BNL and CLM are Δ_1, Δ_2 and Δ_3 respectively, then the value of $\Delta_1+\Delta_2+\Delta_3$ is

(a) $\Delta(2+2\cos A\cos B\cos C)$

(b) $\Delta(2+2\sin A\sin B\sin C)$

(c) $\Delta(1-2\cos A\cos B\cos C)$

(d) $\Delta(1-2\sin A\sin B\sin C)$

129. If area of ΔLMN is Δ', then the value of $\frac{\Delta'}{\Delta}$ is

(a) $2\sin A\sin B\sin C$ (b) $2\cos A\cos B\cos C$

(c) $\sin A\sin B\sin C$ (d) $\cos A\cos B\cos C$

130. Radius of the circum circle of ΔLMN is

(a) $2R$ (b) R

(c) $\frac{R}{2}$ (d) $\frac{R}{4}$

131. If radius of the incircle of ΔLMN is r', then the value of $r'\sec A\sec B\sec C$ is

(a) $4R$ (b) $3R$

(c) $2R$ (d) R

Properties and Solutions of Triangles Exercise 5 : Matching Type Questions

132. Match the statement of **Column I** with values of **Column II.**

	Column I		Column II
(A)	In a ΔABC, let $\angle C=\frac{\pi}{2}$, $r=$ inradius, $R=$ circumradius then $2(r+R)$	(p)	$a+b+c$
(B)	If l, m, n are perpendicular drawn from the vertices of triangle having sides a, b and c then $\sqrt{2R\left(\frac{bl}{c}+\frac{cm}{a}+\frac{an}{b}\right)+2ab+2bc+2ca}$	(q)	$a-b$
(C)	In a ΔABC, $R(b^2\sin 2C+c^2\sin 2B)$ equals	(r)	$a+b$
(D)	In a right angle triangle ABC, $\angle C=\frac{\pi}{2}$, then $4R\sin\frac{A+B}{2}\sin\frac{(A-B)}{2}$	(s)	abc

133. Match the statement of **Column I** with the values of **Column II.**

Column I	Column II
(A) In a triangle ABC $(c-a)^2 = b^2 - ac$ and $\cos B + \sin C = \dfrac{3}{2}$	(p) $A = 30°$
(B) A, B, C are in A.P. and $C = 3A$	(q) $B = 60°$
(C) The length of the bisector of angle $B = \dfrac{\sqrt{3}ca}{(c+a)}$ and $a = b$	(r) $C = 90°$
(D) $\dfrac{1}{a+b} + \dfrac{1}{b+c} = \dfrac{3}{a+b+c}$	(s) $A = B = C = 60°$

134. Match the statement of **Column I** with values of **Column II.**

Column I	Column II
(A) In a ΔABC, $(a+b+c)(b+c-a) = \lambda bc$, where $\lambda \in I$, then greatest value of λ is	(p) 3
(B) In a ΔABC, $\tan A + \tan B + \tan C = 9$. If $\tan^2 A + \tan^2 B + \tan^2 C = k$, then least value of k satisfying is	(q) $9(3)^{1/3}$
(C) In a triangle ABC, then line joining the circumcentre to the incentre is parallel to BC, then value of $\cos B + \cos C$ is	(r) 1
(D) If in a ΔABC, $a = 5$, $b = 4$ and $\cos(A-B) = \dfrac{31}{32}$, then the third side c is equal to	(s) 6

135. Match the statement of Column I with values of Column II

Column I	Column II
(A) In a ΔABC, if $2a^2 + b^2 + c^2 = 2ac + 2ab$, then	(p) ΔABC is equilateral triangle
(B) In a ΔABC, if $a^2 + b^2 + c^2 = \sqrt{2}b(c+a)$, then	(q) ΔABC is right angled triangle
(C) In a ΔABC, if $a^2 + b^2 + c^2 = bc + ca\sqrt{3}$, then	(r) ΔABC is scalane triangle
	(s) ΔABC is scalane right angled triangle
	(t) Angles B, C, A are in AP

Properties and Solutions of Triangles Exercise 6 : Single Integer Answer Type Questions

136. If in ΔABC, $\angle C = \dfrac{\pi}{8}$, $a = \sqrt{2}$ and $b = \sqrt{2+\sqrt{2}}$, then find the sum of digits in the measure of angle A (in degree).

137. In the figure as shown, find the number of digits in the length of AB.

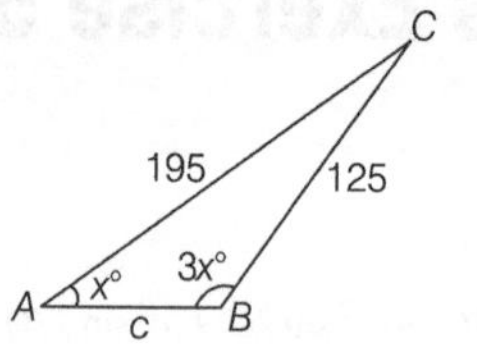

138. If $A = \dfrac{\pi}{7}$, $B = \dfrac{2\pi}{7}$ and $C = \dfrac{4\pi}{7}$ then in ΔABC, $\dfrac{a^2+b^2+c^2}{R^2}$ equals to

139. If A, B, C the angles of an acute angled ΔABC and

$$D = \begin{vmatrix} (\tan B + \tan C)^2 & \tan^2 A & \tan^2 A \\ \tan^2 B & (\tan A + \tan C)^2 & \tan^2 A \\ \tan^2 C & \tan^2 C & (\tan + \tan B)^2 \end{vmatrix},$$

then the least integral values of $\dfrac{D}{1000}$ is

140. In a ΔABC, P and Q are the mid-point or AB and AC, respectively. If O is the circum-centre of the ΔABC, then the value of $\left(\dfrac{\text{Area of } \Delta ABC}{\text{Area of } \Delta OPQ}\right) \cot B \cot C$ equals to

141. With usual notation in ΔABC, the numerical value of $\left(\dfrac{a+b+c}{r_1+r_2+r_3}\right)\left(\dfrac{a}{r_1}+\dfrac{b}{r_2}+\dfrac{c}{r_3}\right)$ is

142. In a ΔABC, $\cos A \cdot \cos C + \dfrac{\lambda(c^2-a^2)}{3ca}$, where AD is the median through A and $AD \perp AC$, then the value of λ is ....

143. In a triangle ABC, medians AD and CE are drawn. If $AD = 5$, $\angle DAC = \dfrac{\pi}{8}$ and $\angle ACE = \dfrac{\pi}{4}$, then the area of the triangle ABC is equal to $\dfrac{5a}{b}$, then $a + b$ is equal to ....

144. In ΔABC, $\dfrac{r}{r_1} = \dfrac{1}{2}$, then the value of $16\left(\Sigma \tan\left(\dfrac{A}{2}\right)\right)$ must be.

145. In a ΔABC, the maximum value of $120\left(\frac{\Sigma a\cos^2\left(\frac{A}{2}\right)}{a+b+c}\right)$ must be

146. The sides of triangle are three consecutive natural numbers and its largest angle is twice the smaller one. The largest side of the triangle must be

147. In ΔABC, $\angle C = 2\angle A$, and $AC = 2BC$, then the value of $\frac{a^2+b^2+c^2}{R^2}$ (where R is circumradius of triangle) is ....

148. If a, b and A are given in a triangle and c_1, c_2 are the possible values of the third side, and $c_1^2 + c_2^2 - 2c_1c_2\cos A = \lambda a^2\cos^2 A$, then the value of λ is ....

149. In triangle ABC, $a = 5$, $b = 4$, $c = 3$. G is the centroid of triangle. If R_1 be the circumradius of triangle GAB then the value of $\frac{a}{65}R_1^2$ must be

150. A triangle ABC is inscribed in a circle of radius 1 and centre at O. The lines AO, BO, CO meet the opposite sides at D, E, F. Then $\frac{1}{AD}+\frac{1}{BE}+\frac{1}{CF}$ is equal to ....

151. In ΔABC, $a \ge b \ge c$ and if $\frac{a^3+b^3+c^3}{\sin^3 A+\sin^3 B+\sin^3 C} = 8$, than the maximum value of a is ....

152. In a cyclic quadrilateral $PQRS$, $PQ = 2$ units, $QR = 5$ units, $RS = 3$ units and $PQR = 60°$, then SP is ....

Properties and Solutions of Triangles Exercise 7 :
Subjective Type Questions

153. In a ΔABC, the angles A and B are two values of θ satisfying $\sqrt{3}\cos\theta + \sin\theta = K$; where $|K| < 2$, then show triangles is obtuse angled.

154. If in an obtuse angled triangle the obtuse angle is $\frac{3\pi}{4}$ and the other two angles are equal to two values of θ satisfying $a\tan\theta + b\sec\theta = c$, where $|b| \le \sqrt{a^2+c^2}$, then find the value of $a^2 - c^2$.

155. In a ΔABC, a, c, A are given and b_1, b_2 are two values of third side b such that $b_2 = 2b_1$. Then, the value of $\sin A$.

156. If P is a point on the altitude AD of the ΔABC, such that $\angle CBP = \frac{B}{3}$, then find the value of AP.

157. If R denotes circum-radius of ΔABC, evaluate $\frac{b^2-c^2}{2aR}$.

158. In ΔABC, $A = \frac{2\pi}{3}$, $b - c = 3\sqrt{3}$ cm and ar (ΔABC) $= \frac{9\sqrt{3}}{2}\text{ cm}^2$. Solve for side a.

159. Find the value of $\tan A$, if area of ΔABC is $a^2 - (b-c)^2$.

160. In a ΔABC, $B = 90°$, $AC = h$ and the length of perpendicular from B to AC is p such that $h = 4p$. If $AB < BC$, then measure $\angle C$.

161. If in a ΔABC, $\sin^3 A + \sin^3 B + \sin^3 C$ $= 3\sin A \cdot \sin B \cdot \sin C$, then find the value of determinant $\begin{vmatrix} a & b & c \\ b & c & a \\ c & a & b \end{vmatrix}$.

162. In a ΔABC, the side a, b and c are such that they are the roots of $x^3 - 11x^2 + 38x - 40 = 0$. Then, the value of $\frac{\cos A}{a}+\frac{\cos B}{b}+\frac{\cos C}{c}$.

163. In a ΔABC the sides a, b and c are in AP. Evaluate $\left(\tan\frac{A}{2}+\tan\frac{C}{2}\right) : \cot\frac{B}{2}$.

164. The sides of a Δ are in AP. and its area is $\frac{3}{5}\times$ (area of an equilateral triangle of the same perimeter). Find the ratio of its sides.

165. If AD, BE and CF are the medians of a ΔABC, then evaluate $(AD^2 + BE^2 + CF^2) : (BC^2 + CA^2 + AB^2)$.

166. AD is a median of the ΔABC. If AE and AF are medians of the ΔABD and ΔADC respectively, and $AD = m_1$, $AE = m_2$, $AF = m_3$, then find the value of $\frac{a^2}{8}$.

167. If $X = \tan\frac{B-C}{2}\cdot\tan\frac{A}{2}; Y = \tan\frac{A-B}{2}\cdot\tan\frac{C}{2};$ $Z = \tan\frac{C-A}{2}\cdot\tan\frac{B}{2}$, then find the value of $X+Y+Z+XYZ$.

168. Let ΔABC be equilateral on side BA produced, we choose a point P such that A lies between P and B. We now denote 'a' as the length of a side of ΔABC; r_1 as the radius of incircle of ΔPAC; and r_2 as the radius of the excircle of ΔPBC with respect to side BC. Determine the sum (r_1+r_2) as a function of 'a' alone.

169. A hexagon is inscribed in a circle of radius r. Two of its sides have length 1, two have length 2 and the last two have length 3. Prove that r is a root of the equation $2r^3-7r-3=0$.

170. The base of a triangle is divided into three equal parts. If $\theta_1, \theta_2, \theta_3$ be the angles subtended by these parts at the vertex, then prove that

$$(\cot\theta_1+\cot\theta_2)(\cot\theta_2+\cot\theta_3) = 4\operatorname{cosec}^2\theta_2$$

171. If the circum-radius of a Δ is $\frac{54}{\sqrt{1463}}$, and its sides are in GP with ratio $\frac{3}{2}$, then find the sides of the triangle.

172. Prove that $a^2+b^2+c^2+2abc<2$, where a, b, c are the sides of triangle ABC such that $a+b+c=2$.

173. Let points $P_1, P_2, P_3, \ldots, P_{n-1}$ divides the side BC of a ΔABC into n parts. Let $r_1, r_2, r_3, \ldots, r_n$ be the radii of inscribed circles and let $p_1, p_2, \ldots, p_n$ be the radii of excribed circles corresponding to vertex Δ for triangle $ABP_1, AP_1P_2, \ldots, AP_{n-1}C$ and let r and P be the corresponding radii for the triangle ABC. Show that $\frac{r_1}{P_1}\cdot\frac{r_2}{P_2}\cdots\frac{r_n}{P_n} = \frac{r}{P}$.

174. A polygon of n sides, inscribed in a circle, is such that its sides subtend angles $2\alpha, 4\alpha, \ldots, 2n\alpha$ at the centre of the circles. Prove that its area A_1, is to the area A_2 of the regular polygon of n sides inscribed in the same circle, as $\sin n\alpha : n\sin\alpha$.

175. $A_1, A_2, A_3, \ldots, A_n$ is a regular polygon of n sides circumscribed about a circle of centre O and radius 'a'. P is any point distant 'c' from O. Show that the sum of the squares of the perpendiculars from P on the sides of the polygon is $n\left(a^2+\frac{c^2}{2}\right)$.

176. Show that in any ΔABC, $a^3\cos 3B + 3a^2b\cos(2B-A) + 3ab^2\cos(B-2A) + b^3\cos 3A = c^3$

177. Let ABC be a Δ with altitudes h_1, h_2, h_3 and inradius r. Prove that $\frac{h_1+r}{h_1-r}+\frac{h_2+r}{h_2-r}+\frac{h_3+r}{h_3-r}\geq 6$.

178. If in a ΔABC, $\frac{a\cos A+b\cos B+c\cos C}{a\sin B+b\sin C+c\sin A} = \frac{a+b+c}{9R}$, then prove that Δ is equilateral.

179. In ΔABC, 'h' is the length of altitude drawn from vertex A on the side BC. Prove that : $2(b^2+c^2)\geq 4h^2+a^2$. Also, discuss the case when equality holds true.

180. Consider a ΔABC. A directly similar $\Delta A_1B_1C_1$ is inscribed in the ΔABC such that A_1, B_1 and C_1 are the interior points of the sides AC, AB and BC, respectively. Prove that $\frac{\text{Area}(\Delta A_1B_1C_1)}{\text{Area}(\Delta ABC)} \geq \frac{1}{\operatorname{cosec}^2 A+\operatorname{cosec}^2 B+\operatorname{cosec}^2 C}$

181. Find the angle at the vertex of an isosceles triangle having the maximum area for the given length 'l' of the median to one of its equal sides.

182. Consider a ΔABC and points A_1 and B_1 on side BC such that $\angle BAA_1 = \angle B_1AC$. If incircle of ΔBAA_1 and B_1AC touch the sides BA_1 and B_1C at M and N respectively, prove that $\frac{1}{BM}+\frac{1}{MA_1} = \frac{1}{B_1N}+\frac{1}{NC}$.

183. An equilateral triangle PQR is circumscribed about a given ΔABC. Prove that the maximum area of ΔPQR is $2\Delta + \frac{a^2+b^2+c^2}{2\sqrt{3}}$. Where a, b, c are the sides of ΔABC and Δ is its area.

184. In a ΔABC, r_A, r_B, r_C are the radii of the circles which touch the incircle and the sides emanting from the vertices A, B, C respectively. Prove that, $\sqrt{r_Ar_B}+\sqrt{r_Br_C}+\sqrt{r_Cr_A} = r$

185. Find the points inside a Δ from which the sum of the squares of distance to the three sides is minimum. Also, find the minimum values of the sum of squares of distances.

186. In a scalene acute ΔABC, it is known that line joining circumcentre and orthocentre is parallel to BC. Prove that the angle $A\in\left(\frac{\pi}{3}, \frac{\pi}{2}\right)$.

187. Consider an acute angled ΔABC. Let AD, BE and CF be the altitudes drawn from the vertices to the opposite sides. Prove that : $\frac{EF}{a}+\frac{FD}{b}+\frac{DE}{c} = \frac{R+r}{R}$.

188. Two circle, the sum of whose radii is 'a' are placed in the same plane with their distance '$2a$' apart. An endless string is fully stretched so as partly to surround the circle and to cross between them. Prove that length of string is $\left(\frac{4\pi}{3}+2\sqrt{3}\right)a$.

189. If Δ_0 is the area of Δ formed by joining the points of contact of incircle with the sides of the given triangle whose area is Δ. Similarly Δ_1, Δ_2 and Δ_3 are the corresponding area of the Δ formed by joining the points of contact of excircles with the sides. Prove that $\frac{\Delta_1}{\Delta}+\frac{\Delta_2}{\Delta}+\frac{\Delta_3}{\Delta}-\frac{\Delta_0}{\Delta}=2$.

190. Let P be the point inside the ΔABC. Such that $\angle APB = \angle BPC = \angle CPA$. Prove that

$PA + PB + PC = \sqrt{\frac{a^2+b^2+c^2}{2}+2\sqrt{3}\Delta}$, where a, b, c, Δ are the sides and the area of ΔABC.

191. In an acute angled ΔABC, the points A', B' and C' are located such that A' is the point where altitude from A on BC meets the outward facing semi-circle drawn on BC as diameter, points B', C' are located similarly. Prove that

$\{\text{ar}(BCA')\}^2 + \{\text{ar}(CAB')\}^2 + \{\text{ar}(ABC')\}^2 = \{\text{ar}(ABC)^2\}$

Properties and Solutions of Triangles Exercise 8 : Questions Asked in Previous 10 Years' Exam

(i) JEE Advanced & IIT-JEE

192. In a ΔXYZ, let x, y, z be the lengths of sides opposite to the angles X, Y, Z respectively and $2s = x + y + z$. If $\frac{s-x}{4}=\frac{s-y}{3}=\frac{s-z}{2}$ and area of incircle of the ΔXYZ is $\frac{8\pi}{3}$, then **[More than one correct option 2016 Adv.]**

(a) area of the ΔXYZ is $6\sqrt{6}$

(b) the radius of circum-circle of the ΔXYZ is $\frac{35}{6}\sqrt{6}$

(c) $\sin\frac{X}{2}\sin\frac{Y}{2}\sin\frac{Z}{2}=\frac{4}{35}$

(d) $\sin^2\left(\frac{X+Y}{2}\right)=\frac{3}{5}$

193. In a triangle, the sum of two sides is x and the product of the same two sides is y. If $x^2 - c^2 = y$, where c is the third side of the triangle, then the ratio of the in-radius to the circum-radius of the triangle is **[Single correct option 2014 Adv.]**

(a) $\frac{3y}{2x(x+c)}$ (b) $\frac{3y}{2c(x+c)}$

(c) $\frac{3y}{4x(x+c)}$ (d) $\frac{3y}{4c(x+c)}$

194. Consider a ΔABC and let a, b and c denote the lengths of the sides opposite to vertices A, B and C, respectively. $a = 6$, $b = 10$ and the area of the triangle is $15\sqrt{3}$. If $\angle ACB$ is obtuse and if r denotes the radius of the incircle of the triangle, then r^2 is equal to...... **[Intger Answer Type 2013 Adv.]**

195. In a ΔPQR, P is the largest angle and $\cos P = \frac{1}{3}$. Further in circle of the triangle touches the sides PQ, QR and RP at N, L and M respectively, such that the lengths of PN, QL and RM are consecutive even integers. Then, possible length(s) of the side(s) of the triangle is (are) **[More than one correct option 2012]**

(a) 16 (b) 18 (c) 20 (d) 22

196. If ΔPQR is a triangle of area Δ with $a = 2$, $b = \frac{7}{2}$ and $c = \frac{5}{2}$, where a, b and c are the lengths of the sides of the triangle opposite to the angles at P, Q and R, respectively. Then, $\frac{2\sin P - \sin 2P}{2\sin P + \sin 2P}$ equals **[More than one correct option 2012]**

(a) $\frac{3}{4\Delta}$ (b) $\frac{45}{4\Delta}$

(c) $\left(\frac{3}{4\Delta}\right)^2$ (d) $\left(\frac{45}{4\Delta}\right)^5$

197. If the angles A, B and C of a triangle are in an arithmetic progression and if a, b and c denote the lengths of the sides opposite to A, B and C respectively, then the value of the expression $\frac{a}{c}\sin 2C+\frac{c}{a}\sin 2A$ is **[Single correct option 2010]**

(a) $\frac{1}{2}$ (b) $\frac{\sqrt{3}}{2}$ (c) 1 (d) $\sqrt{3}$

198. Let ABC be a triangle such that $\angle ACB = \frac{\pi}{6}$. If a, b and c denote the lengths of the sides opposite to A, B and C, respectively. Then, the value(s) of x for which $a = x^2 + x + 1$, $b = x^2 - 1$ and $c = 2x + 1$ is (are) **[Single correct option 2010]**

(a) $-(2+\sqrt{3})$ (b) $1+\sqrt{3}$

(c) $2+\sqrt{3}$ (d) $4\sqrt{3}$

199. In a ΔABC with fixed base BC, the vertex A moves such that $\cos B + \cos C = 4\sin^2 \frac{A}{2}$. If a, b and c denote the lengths of the sides of the triangle opposite to the angles A, B and C respectively, then **[More than one correct option 2009]**

(a) $b + c = 4a$
(b) $b + c = 2a$
(c) locus of point A is an ellipse
(d) locus of point A is a pair of straight line

200. Let ABC and ABC' be two non-congruent triangles with sides $AB = 4$, $AC = AC' = 2\sqrt{2}$ and $\angle B = 30°$. The absolute value of the difference between the areas of these triangles is **[Integer Answer Type 2009]**

201. A straight line through the vertex P of a ΔPQR intersects the side QR at the point S and the circum-circle of the ΔPQR at the point T. If S is not the centre of the circumcircle, then **[More than one correct option 2008]**

(a) $\frac{1}{PS} + \frac{1}{ST} < \frac{2}{\sqrt{QS \times SR}}$
(b) $\frac{1}{PS} + \frac{1}{ST} > \frac{2}{\sqrt{QS \times SR}}$
(c) $\frac{1}{PS} + \frac{1}{ST} < \frac{4}{QR}$
(d) $\frac{1}{PS} + \frac{1}{ST} > \frac{4}{QR}$

Passage

(Q.Nos. 202 to 204)

Consider the circle $x^2 + y^2 = 9$ and the parabola $y^2 = 8x$. They intersect at P and Q in the first and the fourth quadrants, respectively. Tangents to the circle at P and Q intersect the X-axis at R and tangents to the parabola at P and Q intersect the X-axis at S.

[One correct option 2007]

202. The radius of the in-circle of ΔPQR is

(a) 4 (b) 3
(c) $\frac{8}{3}$ (d) 2

203. The radius of the circum-circle of the ΔPRS is

(a) 5 (b) $3\sqrt{3}$
(c) $3\sqrt{2}$ (d) $2\sqrt{3}$

204. The ratio of the areas of ΔPQS and ΔPQR is

(a) $1 : \sqrt{2}$ (b) $1 : 2$
(c) $1 : 4$ (d) $1 : 8$

205. Internal bisector of $\angle A$ of ΔABC meets side BC at D. A line drawn through D perpendicular to AD intersects the side AC at E and side AB at F. If a, b, c represent sides of ΔABC, then **[More than one correct option 2006]**

(a) AE is HM of b and c (b) $AD = \frac{2bc}{b+c}\cos\frac{A}{2}$
(c) $EF = \frac{4bc}{b+c}\sin\frac{A}{2}$ (d) ΔAEF is isosceles

206. In-radius of a circle which is inscribed in a isosceles triangle one of whose angle is $2\pi/3$ is $\sqrt{3}$, then area of triangle (in sq units) is **[Single correct option 2006]**

(a) $4\sqrt{3}$
(b) $12 - 7\sqrt{3}$
(c) $12 + 7\sqrt{3}$
(d) None of the above

207. In a ΔABC, among the following which one is true? **[Single correct option 2005]**

(a) $(b + c)\cos\frac{A}{2} = a\sin\left(\frac{B+C}{2}\right)$
(b) $(b + c)\cos\left(\frac{B+C}{2}\right) = a\sin\frac{A}{2}$
(c) $(b - c)\cos\left(\frac{B-C}{2}\right) = a\cos\left(\frac{A}{2}\right)$
(d) $(b - c)\cos\frac{A}{2} = a\sin\left(\frac{B-C}{2}\right)$

(ii) JEE Main & AIEEE

208. Let a vertical tower AB have its end A on the level ground. Let C be the mid-point of AB and P be a point on the ground such that $AP = 2AB$. If $\angle BPC = \beta$, then $\tan\beta$ is equal to **[2017 JEE Main]**

(a) $\frac{6}{7}$ (b) $\frac{1}{4}$
(c) $\frac{2}{9}$ (d) $\frac{4}{9}$

209. $ABCD$ is a trapezium such that AB and CD are parallel and $BC \perp CD$, if $\angle ADB = \theta$, $BC = p$ and $CD = q$, then AB is equal to **[2013 JEE Main]**

(a) $\frac{(p^2 + q^2)\sin\theta}{p\cos\theta + q\sin\theta}$ (b) $\frac{p^2 + q^2\cos\theta}{p\cos\theta + q\sin\theta}$
(c) $\frac{p^2 + q^2}{p^2\cos\theta + q^2\sin\theta}$ (d) $\frac{(p^2 + q^2)\sin\theta}{(p\cos\theta + q\sin\theta)^2}$

210. For a regular polygon, let r and R be the radii of the inscribed and the circumscribed circles. A false statement among the following is **[2010 AIEEE]**

(a) there is a regular polygon with $\frac{r}{R} = \frac{1}{2}$
(b) there is a regular polygon with $\frac{r}{R} = \frac{1}{\sqrt{2}}$
(c) there is a regular polygon with $\frac{r}{R} = \frac{2}{3}$
(d) there is a regular polygon with $\frac{r}{R} = \frac{\sqrt{3}}{2}$

211. In a ΔABC, let $\angle C = \frac{\pi}{2}$, if r is the in-radius and R is the circum-radius of the ΔABC, then $2(r + R)$ is equal to **[2005 AIEEE]**

(a) $c + a$ (b) $a + b + c$
(c) $a + b$ (d) $b + c$

212 If in ΔABC, the altitudes from the vertices A, B and C on opposite sides are in HP, then $\sin A \sin B$ and $\sin C$ are in **[2005 AIEEE]**

(a) HP (b) AGP
(c) AP (d) GP

Answers

Exercise for Session 1

1. (a) 2. (a) 3. (c) 4. (c) 5. (c) 6. (a) 7. (d) 8. (c) 9. (b) 10. (c) 11. (c) 12. (d) 13. (c) 14. (a) 15. (a) 16. (c) 17. (b) 18. (a) 19. (c) 20. (a)

Exercise for Session 2

1. (a) 2. (a) 3. (b) 4. (d) 5. (a) 6. (c) 7. (d) 8. (c) 9. (a) 10. (b) 11. (d) 12. (c) 13. (b) 14. (b) 15. (d) 16. (c) 17. (a) 18. (c) 19. (c) 20. (a)

Exercise for Session 3

1. 1018.81 sq. cm 2. 2.5 3. 2 cm 7. 7:2 9. $\frac{S}{R}$
26. (1) 27. 0 28. a^2b^2

Exercise for Session 4

1. $a \cot A$ 2. $\frac{2A}{a+b}$ 4. 4/3 5. $\frac{\sqrt{5}}{2}$ 6. $2a^2b^2c^2$
9. $\frac{R}{2}$ 10. $a\,\text{cosec}(A/2)$

Exercise for Session 5

1. $\frac{a}{2}\cot\frac{\pi}{2n}$ 2. $\frac{a}{2}\text{cosec}\frac{\pi}{n}$ 4. $\text{cosec}\,\frac{\pi}{9}$
6. 7, 3

Exercise for Session 6

1. 2 2. 2 3. 6 4. $\frac{ab}{a+b}$ 5. $\pi \tan A$

Exercise for Session 7

1. $\frac{1}{\sqrt{7}}$ 2. $\sin A + \sin B + \sin C$ 3. $\frac{1}{2}$ 4. 6 6. $r = 1$
7. $\frac{\pi}{4}$ 9. $(2-\sqrt{3}):2\sqrt{3}$

Exercise for Session 8

4. $R \tan \alpha$ 5. $\frac{2.05 \sin 38^\circ}{\sin 42^\circ}$

Chapter Exercises

1. (a) 2. (a) 3. (c) 4. (c) 5. (c) 6. (c) 7. (b) 8. (c 9. (a) 10. (b) 11. (b) 12. (b) 13. (c) 14. (d) 15. (a) 16. (d) 17. (a) 18. (b) 19. (b) 20. (c) 21. (b) 22. (d) 23. (b) 24. (c) 25. (a) 26. (c) 27. (a) 28. (a) 29. (b) 30. (a) 31. (a) 32. (b) 33. (d) 34. (c) 35. (b) 36. (a) 37. (b) 38. (a) 39. (c) 40. (a) 41. (b) 42. (c) 43. (a) 44. (c) 45. (b) 46. (a) 47. (c) 48. (d) 49. (b) 50. (a) 51. (a) 52. (c) 53. (b) 54. (b) 55. (b) 56. (d) 57. (c) 58. (a) 59. (b) 60. (a) 61. (c) 62. (b) 63. (a) 64. (a) 65. (b) 66. (d) 67. (b) 68. (a) 69. (c)
70. (b,d) 71. (b,c,d) 72. (b,c,d) 73. (b,c) 74. (a,c,d) 75. (a,c) 76. (a,b) 77. (a,c,d) 78. (c,d) 79. (b,c) 80. (a,c) 81. (a,b,c,d) 82. (a,b,c,d) 83. (a,b,c,d) 84. (b,c) 85. (a,b) 86. (a,c,d) 87. (b,c) 88. (a,b) 89. (a,b) 90. (a,d) 91. (a,c)
92. (c) 93. (d) 94. (a) 95. (d) 96. (a) 97. (c) 98. (a) 99. (c) 100. (a) 101. (c) 102. (a) 103. (a) 104. (a) 105. (a) 106. (b) 107. (b) 108. (c) 109. (c) 110. (c) 111. (c) 112. (d) 113. (c) 114. (c) 115. (a,b,c,d) 116. (b) 117. (b) 118. (b) 119. (b) 120. (b) 121. (d) 122. (b) 123. (a) 124. (c) 125. (c) 126. (d) 127. (b) 128. (c) 129. (d) 130. (b) 131. (a)
132. (A) $\rightarrow$ (r), (B) $\rightarrow$ (p), C $\rightarrow$ (s) , (D) $\rightarrow$ (q)
133. (A) $\rightarrow$ (p, q, r), (B) $\rightarrow$ (p, q, r), C $\rightarrow$ (q, s) , (D) $\rightarrow$ q
134. (A) $\rightarrow$ (p), (B) $\rightarrow$ (q), C $\rightarrow$ (r) , (D) $\rightarrow$ (s)
135. (A) $\rightarrow$ (p, t), (B) $\rightarrow$ (q, s), C $\rightarrow$ (q, r, t)
136. (9) 137. (3) 138. (7) 139. (2) 140. (4) 141. (4) 142. (2) 143. (8) 144. (8) 145. (9) 146. (6) 147. (8) 148. (4) 149. (5) 150. (2) 151. (2) 152. (2) 154. $2ac$
155. $\sqrt{\frac{9a^2 - c^2}{8c^2}}$ 156. $2c \sin\left(\frac{B}{3}\right)$ 157. $(\sin(B - C))$
158. $(\sqrt{63} - 2\sqrt{3})$ 159. $\left(\frac{8}{15}\right)$ 160. (15°)
161. (0) 162. $\left(\frac{9}{16}\right)$ 163. $\left(\frac{2}{13}\right)$
164. (3 : 5 : 7) 165. (3 : 4) 166. $m_2^2 + m_3^2 - 2m_1^2$
167. (0) 168. $\left(\frac{a\sqrt{3}}{2}\right)$ 171. $\left(1, \frac{3}{2}, \frac{9}{4}\right)$
181. $\left(\frac{4}{5}\right)$ 185. $\Delta_{\min} = \frac{4(s-a)(s-b)(s-c)s}{a^2+b^2+c^2}$
192. (a,c,d) 193. (b) 194. (3) 195. (a, b) 196. (b,c,d) 197. (d) 198. (b) 199. (b, c) 200. 4 sq units 201. (d) 202. (d) 203. (b) 204. (c) 205. (a,b,c,d) 206. (c) 207. (d) 208. (c) 209. (a) 210. (a,b,d) 211. (c) 212. (c)

Solutions

1.

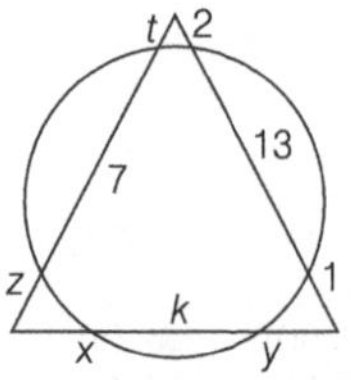

$$x + k + y = 16$$

Also, $(t + 7) \times t = 2 \times (2 + 13)$

$\Rightarrow$ $t = 3$

Now, $x(x + k) = 6 \times (6 + 7)$

and $y(y + k) = 1 \times 14$

Solving, we get $x = 10 - \sqrt{22}$, $y = 6 - \sqrt{22}$ and $k = 2\sqrt{22}$.

2. We have, $s = \dfrac{a + b + c}{2} = 21$

No2, $\sin A = \dfrac{2\Delta}{bc}$

$$= \frac{2}{14 \cdot 15} \sqrt{21 \cdot 8 \cdot 7 \cdot 6} = \frac{4}{5}$$

$\Rightarrow$ $A = \sin^{-1} \dfrac{4}{5}$

Alternatively By using cosine rule in ΔABC, we get

$$\cos A = \frac{b^2 + c^2 - a^2}{2bc}$$

$$= \frac{(14)^2 + (15)^2 - (13)^2}{2(14)(15)}$$

$$= \frac{196 + 225 - 169}{420}$$

$$= \frac{252}{420} = \frac{3}{5}\left(0 < A < \frac{\pi}{2}\right)$$

Hence, $A = \sin^{-1} \dfrac{4}{5}$

3. Using Napier's Analogy

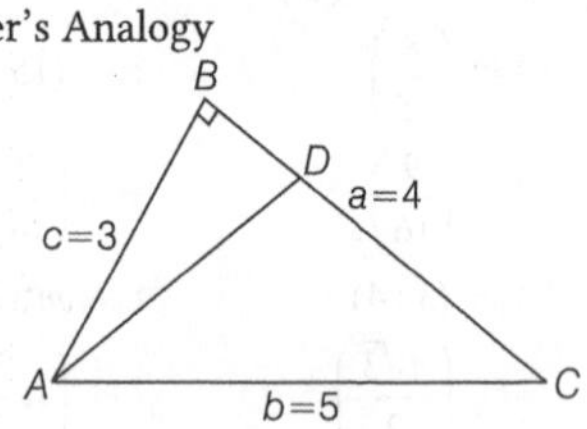

$$\tan\left(\frac{A - B}{2}\right) = \frac{a - b}{a + b} \cos\frac{C}{2}$$

$$= \frac{1 - (\sqrt{3} - 1)}{1 + (\sqrt{3} - 1)}(2 + \sqrt{3}) = \frac{1}{\sqrt{3}}$$

$\therefore$ $\dfrac{A - B}{2} = 30°$

$\Rightarrow$ $A - B = 60°$

4. $\dfrac{x}{\sin 105°} = \dfrac{2}{\sin 45°}$

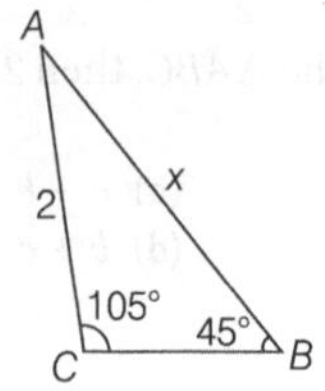

$$x = \frac{2\cos 15°}{\sin 45°} = \frac{2\sqrt{2}(\sqrt{3} + 1)}{2\sqrt{2}}$$

$$= \sqrt{3} + 1$$

5. $\dfrac{AP}{\sin\dfrac{2B}{3}} = \dfrac{c}{\sin\left(90° + \dfrac{B}{3}\right)}$

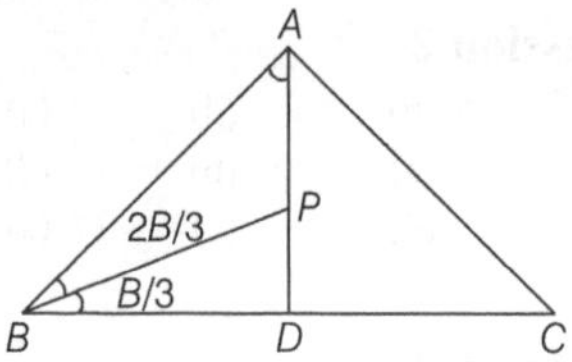

$\Rightarrow$ $AP = 2c \sin\dfrac{B}{3}$

6. Given, $2b = a + c$

$\Rightarrow$ $2\sin B = \sin A + \sin C$

$\Rightarrow$ $2\left(2\sin\dfrac{B}{2}\cos\dfrac{B}{2}\right) = 2\sin\left(\dfrac{A + C}{2}\right)\cos\left(\dfrac{A - C}{2}\right)$

$\Rightarrow$ $2\sin\dfrac{B}{2} = \cos\left(\dfrac{90°}{2}\right) = \dfrac{1}{\sqrt{2}}$

$\Rightarrow$ $\sin\dfrac{B}{2} = \dfrac{1}{2\sqrt{2}}$

$\Rightarrow$ $\cos\dfrac{B}{2} = \sqrt{1 - \dfrac{1}{8}} = \dfrac{\sqrt{7}}{2\sqrt{2}}$

Hence, $\sin B = 2\sin\dfrac{B}{2}\cos\dfrac{B}{2} = \dfrac{\sqrt{7}}{4}$

7. $\dfrac{BD}{DC} = \dfrac{AB}{AC}$

$\Rightarrow$ AD is the angle bisector

$\therefore$ $AD = \dfrac{2bc}{b + c}\cos\dfrac{A}{2}$

$$= \frac{2 \times 5 \times 3}{8}\sqrt{\frac{1 + \cos A}{2}}$$

$$= \frac{15}{4}\sqrt{\frac{1 + \frac{3}{5}}{2}}$$

$$= \frac{15}{4} \times \frac{2}{\sqrt{5}} = \frac{3 \times \sqrt{5} \times \sqrt{5}}{2 \times \sqrt{5}}$$

$$= \frac{3\sqrt{5}}{2}$$

8. Slope of $GC = \frac{-2b}{a}$, slope of $AG = \frac{-b}{2a}$

$$\tan 30^\circ = \frac{\frac{3b}{2a}}{1 + \frac{b^2}{a^2}} \text{ and } a^2 + b^2 = 9$$

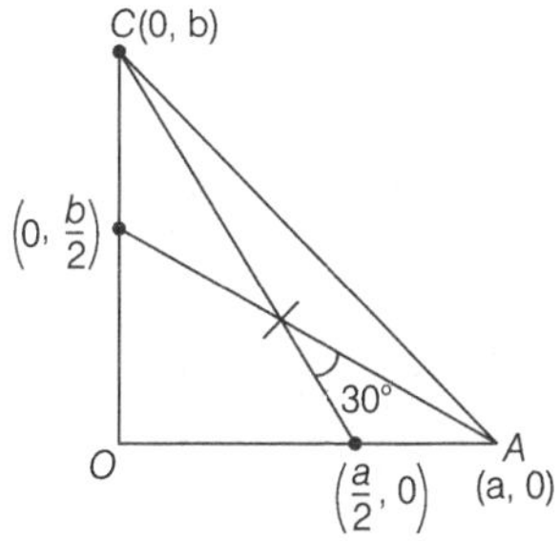

$$\Rightarrow \quad \frac{1}{\sqrt{3}} = \frac{3ab}{2(a^2 + b^2)}$$

$$\Rightarrow \quad \frac{1}{2}ab = \left(\frac{a^2 + b^2}{3\sqrt{3}}\right)$$

$$\Rightarrow \quad \sqrt{k} = \frac{9}{3\sqrt{3}} = \sqrt{3}$$

$$\Rightarrow \quad k = 3$$

9. $4\sin A \cos B = 1$, so A and B cannot be $\frac{\pi}{2}$

[as if $B = \frac{\pi}{2}$, then $\cos B = 0$ and if $A = \frac{\pi}{2}$, $\tan A$ is not defined)

$$\Rightarrow \quad C = \frac{\pi}{2}, B = \frac{\pi}{2} - A$$

$$\Rightarrow \quad 4\sin A \cos\left(\frac{\pi}{2} - A\right) = 1$$

$$\Rightarrow \quad \sin^2 A = \frac{1}{4} \quad \Rightarrow \quad \sin A = \frac{1}{2}$$

$$\Rightarrow \quad A = \frac{\pi}{6} \quad \Rightarrow \quad B = \frac{\pi}{3}$$

So angles are in AP.

10. det A is twice the area of the triangle with vertices (a, p) (b, q) (c, r) with sides 5, 6,7.

$$\Delta^2 = s(s - a)(s - b)(s - c)$$

$$\Rightarrow \quad 16\Delta^2 = 18 \cdot 8 \cdot 6 \cdot 4$$

$$\det B = (\det A)^2 = 4\Delta^2 = 18 \cdot 8 \cdot 6 = 864$$

11. Given that the circle passes through the circumcentre of ΔABC. Therefore, the distance between circumcenter, incentre $= \sqrt{R^2 - 2rR} = r$

$$\Rightarrow \quad r^2 + 2rR - R^2 = 0$$

$$\Rightarrow \quad \left(\frac{r}{R}\right)^2 + 2\frac{r}{R} - 1 = 0$$

$$\Rightarrow \quad \frac{r}{R} = \sqrt{2} - 1 \text{ and}$$

$$\cos A + \cos B + \cos C = 1 + \frac{r}{R} = 1 + 4\sin\frac{A}{2}\sin\frac{B}{2}\sin\frac{C}{2}$$

12. Let a and b the roots of $x^2 - 7x + 8 = 0$. Then, $a + b = 7$, $ab = 8$

Also, $\quad C = 60^\circ$

Now $\quad \frac{1}{2} = \frac{a^2 + b^2 - c^2}{2ab}$

$$\Rightarrow \quad ab = (a + b)^2 - 2ab - c^2$$

$$\therefore \quad c^2 = (a + b)^2 - 3ab = 49 - 24 = 25$$

$$\Rightarrow \quad c = 5$$

Thus, $\quad r . R = \frac{abc}{2(a + b + c)} = \frac{8 \times 5}{2(7 + 5)} = \frac{5}{3}$

13. Applying $m-n$ theorem

$$(BD + DC)\cot\frac{\pi}{6} = DC \cos B - BD \cot C$$

$$\Rightarrow \quad (\cot B - \cot C)^2 = 12$$

14. $A'C = b \cos C$, $B'C = a \cos C$

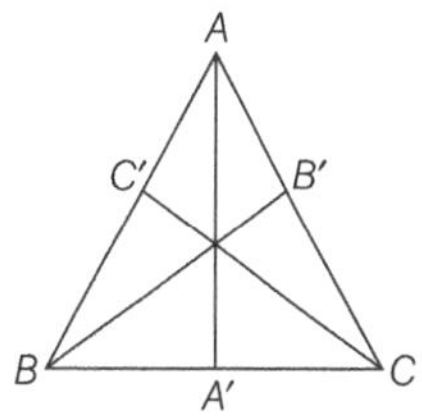

$$A'B' = c \cos C$$

Similarly, $A'C' = b \cos B$ and $B'C' = a \cos A$

Now, $\quad 4R = a \cos A + b \cos B + c \cos C$

$$\Rightarrow \quad \sin A \sin B \sin C = 1$$

$\Rightarrow$ This is only possible when $\angle A = \angle B = \angle C = \frac{\pi}{2}$, so triangle is not possible.

15. Let the sides of the triangle be $7, 7 - d, 7 - 2d$.

Since, the given angle is the greatest (being obtuse) angle of the triangle, it is opposite to the greatest side of the triangle and we have,

$$7^2 = (7 - d)^2 + (7 - 2d)^2 - 2(7 - d)(7 - 2d)\cos\frac{2\pi}{3}$$

$$\Rightarrow \quad 7^2 = 2 \times 7^2 - 42d + 5d^2 - 2(7^2 - 21d + 2d^2)\left(\frac{-1}{2}\right)$$

$$\Rightarrow \quad d^2 - 9d + 14 = 0$$

$$\Rightarrow \quad (d - 7)(d - 2) = 0$$

$$\Rightarrow \quad d = 2 \qquad (d = 7 \text{ is not possible})$$

Therefore, the sides of the triangle are 7 cm, 5 cm, 3 cm.

Area of the triangle $\Delta = \frac{1}{2} \times 5 \times 3 \times \sin\frac{2\pi}{3}$

$= \frac{15\sqrt{3}}{4}$ cm^2 and the radius of the circumcircle

$$R = \frac{7 \times 5 \times 3}{4\left(\frac{15\sqrt{3}}{4}\right)}$$

$$= \frac{7 \times 5 \times 3}{15\sqrt{3}} = \frac{7\sqrt{3}}{3} \text{ cm}$$

16. $\because$ Sides are in AP and $a < \min\{b, c\}$

***Case* I** If $\min(b, c) = b$

Then, a, b, c are in AP.

i.e, $2b = a + c$

$$\therefore \quad \cos A = \frac{b^2 + c^2 - a^2}{2bc} = \frac{b^2 + c^2 - (2b - c)^2}{2bc} = \frac{4bc - 3b^2}{2bc} = \frac{4c - 3b}{2c}$$

***Case* II** If $\min\{b, c\} = c$

Then a, c, b are in AP.

i.e, $2c = a + b$

$$\therefore \quad \cos A = \frac{b^2 + c^2 - a^2}{2bc} = \frac{b^2 + c^2 - (2c - b)^2}{2bc} = \frac{4bc - 3c^2}{2bc} = \frac{4b - 3c}{2b}$$

17. Given, $\sin A \sin B \sin C = p$

and $\cos A \cos B \cos C = q$

$$\Rightarrow \quad \tan A \tan B \tan C = \frac{p}{q}$$

Also, $\tan A + \tan b + \tan C = \tan A \tan B \tan C$

$$\Rightarrow \quad \tan A + \tan B + \tan C = \frac{p}{q}$$

Now, $\tan A \tan B + \tan B \tan C + \tan C \tan A$

$$= \frac{\sin A \sin B \cos C + \sin B \sin C \cos A + \sin C \sin A \cos B}{\cos A \cos B \cos C}$$

$$= \frac{1}{q}[\sin A \sin B \cos C + \sin C(\sin B \cos A + \cos B \sin A)]$$

$$= \frac{1}{q}[\sin A \sin B \cos C + \sin C \sin(A + B)]$$

$$= \frac{1}{q}[\sin A \sin B \cos C + \sin^2 C]$$

$$= \frac{1}{q}[1 - \cos^2 C + \sin A \sin B \cos C]$$

$$= \frac{1}{q}[1 + \cos C(-\cos C + \sin A \sin B)]$$

$$= \frac{1}{q}[1 + \cos C(\cos(A + B) + \sin A \sin B)]$$

$$= \frac{1}{q}[1 + \cos A \cos B \cos C] = \frac{1}{q}[1 + q]$$

The equation whose roots are $\tan A, \tan B, \tan C$ will be give by

$$x^3 - (\tan A + \tan B + \tan C)x^2 + (\tan A \tan B + \tan B \tan C + \tan C \tan A)x - \tan A \tan B \tan C = 0$$

or $$x^3 - \frac{p}{q}x^2 + \frac{1 + q}{q}x - \frac{p}{q} = 0$$

or $$qx^3 + px^2 + (1 + q)x - p = 0$$

Hence, (a) is the correct answer.

18.

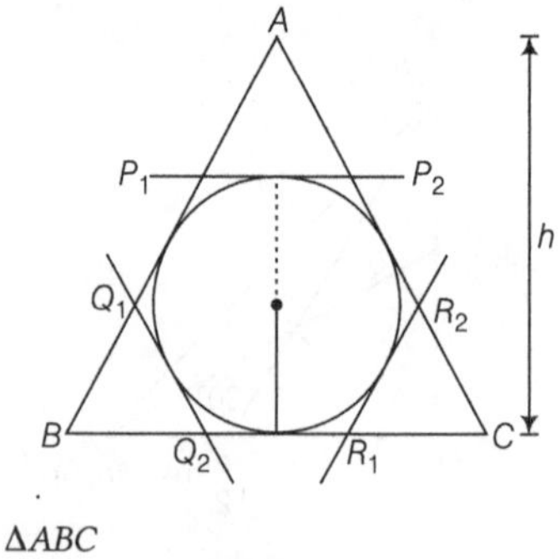

$\Delta AP_1P_2 \sim \Delta ABC$

$$\Rightarrow \quad \frac{t_1}{a} = \frac{h - 2r}{h} = 1 - \frac{2r}{h}$$

$$\Rightarrow \quad \frac{t_1}{a} = 1 - \frac{2\Delta}{sh} \qquad \text{(where } \Delta \equiv \text{ar}(\Delta ABC))$$

$$\Rightarrow \quad \frac{t_1}{a} = 1 - \frac{2\frac{1}{2}ah}{sh} \Rightarrow \frac{t_1}{a} = 1 - \frac{a}{s}$$

Similarly, $\frac{t_2}{b} = 1 - \frac{b}{s}$ and $\frac{t_3}{c} = 1 - \frac{c}{s}$

$$\therefore \quad \frac{t_1}{a} + \frac{t_2}{b} + \frac{t_3}{c} = 3 - \frac{(a + b + c)}{s}$$

$$= 3 - 2 = 1$$

19. $\sin A, \sin B, \sin C$ are the roots of the equation $c^3x^3 - c^2(a + b + c)x^2 + lx + m = 0$

$$\therefore \sin A + \sin B + \sin C = \frac{c^2(a + b + c)}{c^3} = \frac{a + b + c}{c}$$

or $$\frac{a}{2R} + \frac{b}{2R} + \frac{c}{2R} = \frac{a + b + c}{c}$$

Then, $c = 2R, a + b + c \neq 0$

ie, $2R \sin C = 2R$.

20. Let O be the point of intersection of the medians of triangle ABC (Fig. 12.10). Then, the area of ΔABC is three times that of ΔAOC, O being the centroid of ΔABC, divides the median through B in the ratio 2 : 1.

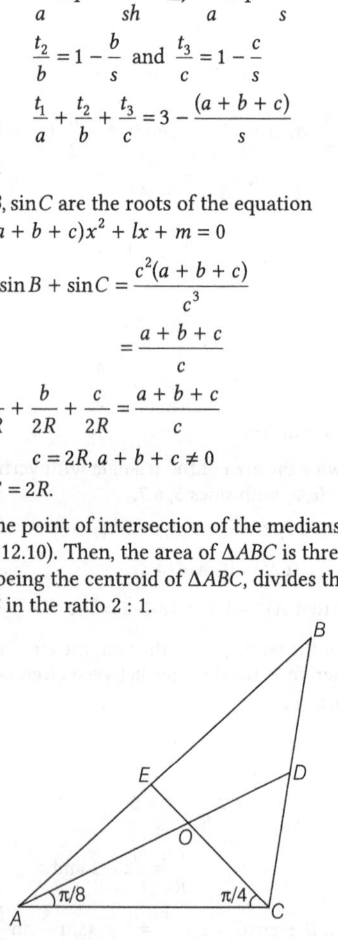

And the height of ΔAOC is one-third that of ΔABC. Now, in ΔAOC, $AO = \left(\frac{2}{3}\right) AD = \frac{10}{3}$. Therefore, applying the sine rule to ΔAOC, we get

$$\frac{OC}{\sin\left(\frac{\pi}{8}\right)} = \frac{AO}{\sin\left(\frac{\pi}{4}\right)} \Rightarrow OC = \frac{10}{3} \cdot \frac{\sin\left(\frac{\pi}{8}\right)}{\sin\left(\frac{\pi}{4}\right)}$$

Area of $\Delta AOC = \frac{1}{2} \cdot AO \cdot OC \cdot \sin \angle AOC$

$$= \frac{1}{2} \cdot \frac{10}{3} \cdot \frac{10}{3} \cdot \frac{\sin\left(\frac{\pi}{8}\right)}{\sin\left(\frac{\pi}{4}\right)} \cdot \sin\left(\frac{\pi}{2} + \frac{\pi}{8}\right)$$

$$= \frac{50}{9} \cdot \frac{\sin\left(\frac{\pi}{8}\right)\cos\left(\frac{\pi}{8}\right)}{\sin\left(\frac{\pi}{4}\right)} + \frac{50}{18} = \frac{25}{9}$$

$\Rightarrow$ Area of $\Delta ABC = 3 \cdot \frac{25}{9} = \frac{25}{3}$

21. We have, $\frac{(2R)^2(\sin^3 A + \sin^3 B + \sin^3 C)}{\sin^3 A + \sin^3 + \sin^3 C} = 8$

$\Rightarrow R = 1$, the radius of the circumcircle is 1.

Greatest length of a side of a triangle inscribed in a circle can be equal to the diameter of the circle and hence the maximum value of the greatest side a is equal to 2.

22. (a) : $(b + c + a)(b + c - a) = 5bc$

$\Rightarrow \quad (b + c)^2 - a^2 = 5bc$

$\Rightarrow \quad b^2 + c^2 + 2bc - a^2 = 5bc$

$\Rightarrow \quad 2bc \cos A = 3bc$

$\cos A = \frac{3}{2}$ impossible.

(b) : Let $a = \sqrt{19}, b = \sqrt{38}, c = \sqrt{116}$

$$\cos A = \frac{19 + 38 - 116}{2\sqrt{19}\sqrt{38}}$$

$$= -\frac{59}{2\sqrt{722}} = -\frac{29.5}{\sqrt{722}} < -1 \text{ impossible.}$$

(c) : $\because \frac{b^2 - c^2}{a^2} = \frac{\sin^2 B - \sin^2 C}{\sin^2 A}$

$$= \frac{\sin(B + C)\sin(B - C)}{\sin^2 A}$$

$$= \frac{\sin(B - C)}{\sin A}$$

$$= \frac{\sin(B - C)}{\sin(B + C)} \neq 0 \qquad (\because B \neq C)$$

$\because$ If $\quad B > C$

Then, $\quad B + C > B - C$

$$\frac{\sin(B + C)}{\sin(B - C)} > 1$$

and if $\quad B < C$

$$C - B < C + B$$

or $\quad \frac{\sin(C - B)}{\sin(C + B)} < 1$

or $\quad \frac{\sin(B - C)}{\sin(B + C)} > -1$

Hence, $\frac{b^2 - c^2}{a^2} + \frac{c^2 - a^2}{b^2} + \frac{a^2 - b^2}{c^2} \neq 0$

(d) :

$$\cos\left(\frac{B - C}{2}\right) = (\sin B + \sin C)\cos\left(\frac{B + C}{2}\right)$$

$$\cos\left(\frac{B - C}{2}\right) = 2\sin\left(\frac{B + C}{2}\right)\cos\left(\frac{B - c}{2}\right)\cos\left(\frac{B + C}{2}\right)$$

$$\cos\left(\frac{B - C}{2}\right) \neq 0$$

$\therefore \quad \sin(B + C) = 1$

$\therefore \quad B + C = \frac{\pi}{2}$

Then, $\quad \angle A = \frac{\pi}{2}$

23. We have, $\cos A = \frac{b^2 + c^2 - a^2}{2bc}$

$\Rightarrow b^2 - 2bc \cos A + (c^2 - a^2) = 0$

It is given that b_1 and b_2 are roots of this equation. Therefore, $b_1 + b_2 = 2c \cos A$ and $b_1 b_2 = c^2 - a^2$

$\Rightarrow \quad 3b_1 = 2c \cos A$ and $2b_1^2 = c^2 - a^2$

(since $b_2 = 2b_1$ given)

$\Rightarrow \quad 2\left(\frac{2c}{3}\cos A\right)^2 = c^2 - a^2$

$\Rightarrow \quad 8c^2(1 - \sin^2 A) = 9c^2 - 9a^2$

$\Rightarrow \quad \sin A = \sqrt{\frac{9a^2 - c^2}{8c^2}}$

Hence, (b) is the correct answer.

24. We have, $\cot(A + B) = \frac{\cot A \cot B - 1}{\cot B + \cot A}$

$$= \frac{x^{\frac{1}{2}}(x^2 + x + 1)^{\frac{1}{2}} \cdot x^{\frac{-1}{2}}(x^2 + x + 1)^{\frac{1}{2}} - 1}{x^{\frac{-1}{2}}(x^2 + x + 1)^{\frac{1}{2}} + x^{\frac{1}{2}}(x^2 + x + 1)^{\frac{1}{2}}}$$

$$= \frac{x^2 + x + 1 - 1}{(x^2 + x + 1)^{\frac{1}{2}} + (x^2 + x^{\frac{-1}{2}})}$$

$$= \frac{x(x + 1)x^{\frac{1}{2}}}{(x^2 + x + 1)^{\frac{1}{2}}(x + 1)} = x^{\frac{3}{2}}(x^2 + x + 1)^{\frac{-1}{2}}$$

$$= x^{-1} + x^{-2} + x^{-3})^{\frac{1}{2}} = \cot C$$

$\Rightarrow \quad A + B = C$ and $A + B + C = \pi$

$\therefore \quad C = \frac{\pi}{2}$

25. We have, $\cos A = \dfrac{c^2 + b^2 - a^2}{2bc}$

$\Rightarrow c^2 - 2bc\cos A + b^2 - a^2 = 0$

It is given that c_1 and c_2 are roots of this equation.

Therefore, $c_1 + c_2 = 2b\cos A$ and $c_1c_2 = b^2 - a^2$

$\Rightarrow \quad k(\sin C_1 + \sin C_2) = 2k\sin B\cos A$

$\Rightarrow \quad \sin C_1 + \sin C_2 = 2\sin B\cos A$

Now, sum of the area of two triangles

$$= \frac{1}{2}ab\sin C_1 + \frac{1}{2}ab\sin C_2$$

$$= \frac{1}{2}ab(\sin C_1 + \sin C_2)$$

$$= \frac{1}{2}ab(2\sin B\cos A)$$

$$= b\cdot b\sin A\cos A = \frac{1}{2}b^2\sin 2A$$

Hence, (a) is correct answer

26. $b\cot B + c\cot C = 2(r + R)$

$\Rightarrow \quad 2R\sin B\cdot\dfrac{\cos B}{\sin B} + 2R\sin C\cdot\dfrac{\cos C}{\sin C} = 2(r + R)$

$\Rightarrow \quad \cos B + \cos C = 1 + \dfrac{r}{R}$

$\Rightarrow \quad \cos B + \cos C = 1 + 4\sin\dfrac{A}{2}\sin\dfrac{B}{2}\sin\dfrac{C}{2}$

$\Rightarrow \quad \cos B + \cos C = \cos A + \cos B + \cos C$

$\therefore \quad \cos A = 0$

$\Rightarrow \quad A = \dfrac{\pi}{2}$

$\Rightarrow \quad a^2 = b^2 + c^2$

$\Rightarrow \quad b^2 + c^2 = 100$

Using A.M $\geq$ G.M., we get

$$\frac{b^2 + c^2}{2} \geq \sqrt{b^2c^2} \Rightarrow bc \leq 50$$

Hence, area of $\Delta ABC = \dfrac{1}{2}bc \leq 25$

27. Let r_1, r_2 and r_3 be the radii of the three circles with centres at C_1, C_2 and C_3. Let the circles touch at P, Q and R.

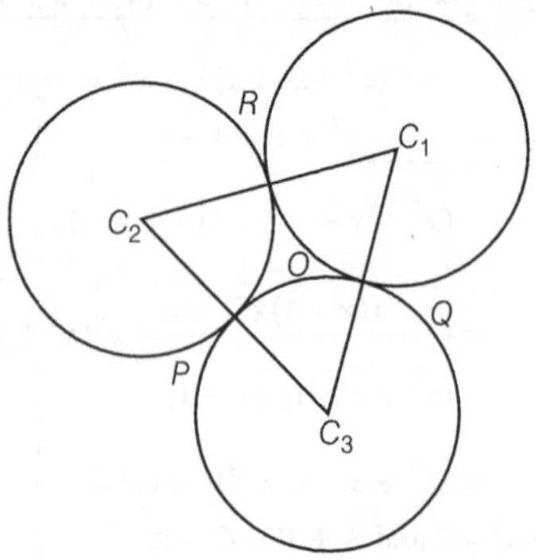

Also, $C_1C_2 = r_1 + r_2$, $C_2C_3 = r_2 + r_3$, $C_3C_1 = r_3 + r_1$

Let O be the point whose distance from the points of contact is 4.

Then, O is the incentre of the $\Delta C_1C_2C_3$ with $OP = OQ = OR = 4$, being the radius of the incircle

Hence, $4 = \dfrac{\Delta C_1C_2C_3}{\frac{1}{2}[C_1C_2 + C_2C_3 + C_3C_1]} = \dfrac{\Delta}{s}$ ...(i)

Where $s = r_1 + r_2 + r_3$,

$$\Delta^2 = s(s - C_1C_2)(s - C_2C_3)(s - C_3C_1) = s(r_1)(r_2)(r_3)$$

(i) gives $16 = \dfrac{\Delta^2}{s^2} = \dfrac{s(r_3)(r_1)(r_2)}{s^2} = \dfrac{r_1r_2r_3}{r_1 + r_2 + r_3}$

Hence, the ratio of the product of the radii to the sum of the radii = 16 : 1.

Hence, (a) is the correct answer.

28. $\because$ Roots are real and distinct.

$\therefore \quad \Delta > 0$

$\Rightarrow \quad 4(a + b + c)^2 - 12\lambda(ab + bc + ca) > 0$

$\Rightarrow \quad (a^2 + b^2 + c^2 + 2(ab + bc + ca)) - 3\lambda(ab + bc + ca) > 0$

$\Rightarrow \quad \dfrac{\Sigma a^2}{\Sigma ab} > (3\lambda - 2)$ ...(i)

Now, in a triangle

Difference of two sides < third side

ie, $\quad |a - b| < c, |b - c| < a$ and $|c - a| < b$

$\Rightarrow \quad (a - b)^2 + (b - c)^2 + (c - a)^2 < a^2 + b^2 + c^2$

$\Rightarrow \quad a^2 + b^2 + c^2 < 2(ab + bc + ca)$

or $\quad \dfrac{\Sigma a^2}{\Sigma ab} < 2$ ...(ii)

From Eqs. (i) and (ii), we get

$$3\lambda - 2 < \frac{\Sigma a^2}{\Sigma ab} < 2 \Rightarrow 3\lambda - 2 < 2 \Rightarrow \lambda < \frac{4}{3}$$

29. Area of triangle

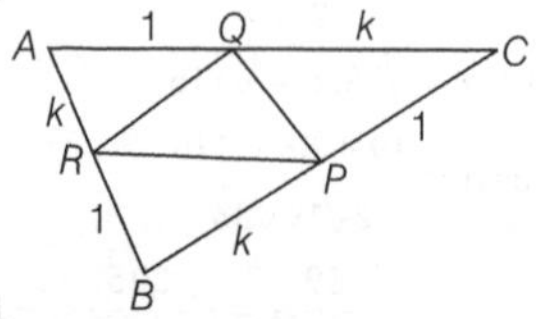

Area of ΔARQ,

$$= \frac{b}{(k+1)} \times \frac{kc}{k+1} \times \frac{1}{2}\sin A = \frac{k\Delta}{(k+1)^2}$$

Similarly, area of ΔBPR = Area of $\Delta PCQ = \dfrac{k\Delta}{(k+1)^2}$

Now, $\quad \dfrac{\Delta PQR}{\Delta ABC} = \dfrac{1}{3}$

or $\quad \dfrac{\Delta - \dfrac{3k}{(k+1)^2}\Delta}{\Delta} = \dfrac{1}{3}$

or $\quad 2k^2 - 5k + 2 = 0$

or $\quad k = \dfrac{1}{2}, 2$

30. $\therefore f(x+y)=f(x)\cdot f(y)$

for $\quad x=y=1$

$$f(2)=2^2$$

for $\quad x=1, y=2$

$$f(3)=f(1)f(2)=2^3$$

$$\therefore \quad f(n)=2^n$$

$$\therefore \quad a=f(3)=2^3=8\quad,$$

$$b=f(1)+f(3)=2+8=10$$

and $c=f(2)+f(3)=2^2+2^3=12$

$$\cos A=\frac{3}{4} \text{ and } \cos C=\frac{1}{8}$$

$$\cos 2A=2\cos^2 A-1$$

$$=2\times\frac{9}{16}-1$$

$$=\frac{1}{8}=\cos C$$

$$\therefore \quad 2A=C$$

31. Given, $x+y+z=a+b+c$ ...(i)

and $a^2x+b^2y+c^2z+abc=4xyz$ divide by $4xyz$, we get

$$\frac{a^2}{4yz}+\frac{b^2}{4xz}+\frac{c^2}{4xyz}=1$$

$$\left(\frac{a}{2\sqrt{yz}}\right)^2+\left(\frac{b}{2\sqrt{xz}}\right)^2+\left(\frac{c}{2\sqrt{xy}}\right)^2$$

$$+2\left(\frac{a}{2\sqrt{yz}}\right)\left(\frac{b}{2\sqrt{xz}}\right)\left(\frac{c}{2\sqrt{xy}}\right)=1$$

For using the trigonometric identify

$$\cos^2 A+\cos^2 B+\cos^2 C+2\cos A\cos b\cos C$$

$$=1,\ A+B+C=\pi$$

Let $\dfrac{a}{2\sqrt{yz}}=\cos A, \dfrac{b}{2\sqrt{xz}}=\cos B, \dfrac{c}{2\sqrt{xy}}=\cos C$

Where A, B, C are acute angle's

From Eq. (i)

$\because\ x+y+z=a+b+c$

$\Rightarrow x+y+z=2\sqrt{yz}\cos A+2\sqrt{xz}\cos B+2\sqrt{xy}\cos C$

$\Rightarrow x+y+z-2\sqrt{yz}\cos A-2\sqrt{xz}$

$$\cos B-2\sqrt{xy}\cos(\pi-(A+B))=0$$

$\Rightarrow x+y+z-2\sqrt{yz}\cos A-2\sqrt{xz}$

$\cos B+2\sqrt{xy}\cos(A+B)=0$

$\Rightarrow x+y+z-2\sqrt{yz}\cos A-2\sqrt{xz}$

$$\cos B+2\sqrt{xy}(\cos A\cos B-\sin A\sin B)=0$$

$\Rightarrow x(\sin^2 B+\cos^2 B)+y(\sin^2 A+\cos^2 A)$

$$+z-2\sqrt{yz}\cos A-2\sqrt{xz}\cos B$$

$\Rightarrow 2\sqrt{xy}(\cos A\cos B-\sin A\sin B)=0$

$\Rightarrow (\sqrt{x}\sin B-\sqrt{y}\sin A)^2+(\sqrt{x}\cos B+\sqrt{y}\cos A-\sqrt{z})^2=0$

$\Rightarrow \sqrt{x}\sin B=\sqrt{y}\sin A$ and $\sqrt{x}\cos B+\sqrt{y}\cos A=\sqrt{z}$

$$\Rightarrow \quad \sqrt{x}\frac{b}{2\sqrt{xz}}+\sqrt{y}\frac{a}{2\sqrt{yz}}=\sqrt{z}$$

$$\Rightarrow \quad \frac{b}{2\sqrt{z}}+\frac{a}{2\sqrt{z}}+\sqrt{z} \Rightarrow \frac{a+b}{2}=z$$

$\therefore$ By symmetricity $x=\dfrac{b+c}{2}, y=\dfrac{a+c}{2}, z=\dfrac{a+b}{2}$

32. Let S and I_1 be respectively the centres of the circumcircle and the excircle touching BC.

It can be shown that

$$SI_1=\sqrt{R^2+2Rr_1}$$

In ΔSI_1P, $SI_1^2=R^2+t_1^2$

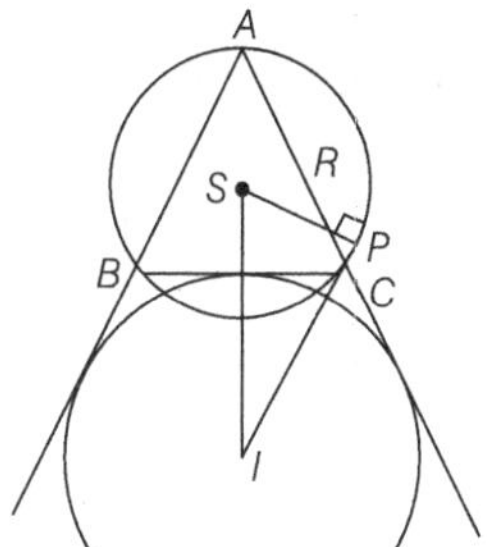

$$R^2=2Rr_1=R^2+t_1^2, \frac{1}{t_1^2}=\frac{1}{2Rr_1}$$

Similarly, $\dfrac{1}{t_2^2}=\dfrac{1}{2Rr_2}, \dfrac{1}{t_3^2}=\dfrac{1}{2Rr_3}$

$$\frac{1}{t_1^2}+\frac{1}{t_2^2}+\frac{1}{t_3^2}=\frac{1}{2R}\left[\frac{1}{r_1}+\frac{1}{r_2}+\frac{1}{r_3}\right]$$

$$=\frac{1}{2R}\left[\frac{s-a}{\Delta}+\frac{s-b}{\Delta}+\frac{s-c}{\Delta}\right]$$

$$=\frac{1}{2R}\frac{s}{\Delta}=\frac{s}{2R\Delta}=\frac{a+b+c}{abc}$$

33. $\angle BAA_1=\angle A_1AA_2=\angle A_2AC$

Clearly, triangle AA_1C is isosceles

$$\Rightarrow \quad AA_1=AC=b$$

$$A_1A_2=A_2C=\frac{a}{4}$$

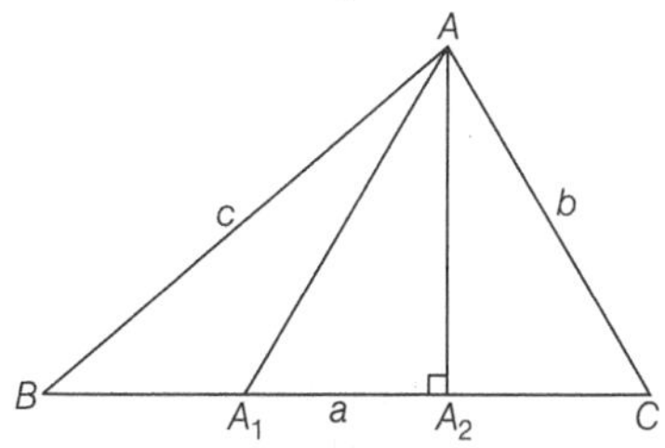

Now, $\sin^3\dfrac{A}{3}\cdot\cos\dfrac{A}{3}=\dfrac{1}{2}\sin^2\dfrac{A}{3}\cdot 2\sin\dfrac{A}{3}\cdot\dfrac{A}{3}$

$$=\frac{1}{2}\sin^2\frac{A}{3}\cdot\sin\frac{2A}{3}=\frac{1}{2}\left(\frac{A_2C}{b}\right)^2\cdot\frac{BA_2}{AB}$$

$$=\frac{1}{2}\cdot\frac{a^2}{16b^2}\cdot\frac{3a}{4}\cdot\frac{1}{c}=\frac{3a^3}{128b^2c}$$

34. $\because \cos A=\dfrac{b^2+c^2-a^2}{2bc}$

or $\quad c^2-2bc\cos A+b^2-a^2=0$

$$\therefore \quad c_1+c_2=2b\cos A=2\times 2\times\frac{\sqrt{3}}{2}=2\sqrt{3}$$

and $\quad c_1c_2=b^2-a^2=4-5=-1$

$$\therefore \quad |c_1-c_2|=\sqrt{(c_1+c_2)^2-4c_1c_2}$$

$$=\sqrt{12+4}=\sqrt{14}=4$$

35. Circumradius of triangle ABC, $R = 5$

$\therefore$ Circumradius of pedal triangle, $R_1 = \frac{5}{2}$, and so on.

Now, $\sum_{i=1}^{\infty} R_i = R_1 + R_2 + R_3 + \ldots\infty$

$$= 5 + \frac{5}{2} + \frac{5}{2^2} + \ldots\infty$$

$$= 5 \cdot \frac{5}{1 - \frac{1}{2}} = 10$$

36. $\left(1 - \frac{r_1}{r_2}\right)\left(1 - \frac{r_1}{r_3}\right) = 2$

$$\Rightarrow \left(1 - \frac{\frac{\Delta}{s-a}}{\frac{\Delta}{s-b}}\right)\left(1 - \frac{\frac{\Delta}{s-a}}{\frac{\Delta}{s-b}}\right) = 2$$

$$\Rightarrow \left(1 - \frac{s-b}{s-a}\right)\left(1 - \frac{s-c}{s-a}\right) = 2$$

$$\Rightarrow (b-a)(c-a) = 2(s-a)^2$$

$$\Rightarrow 2(bc - ab - ac + a^2) = (2s - 2a)^2$$

$$\Rightarrow 2bc - 2ab - 2ac + 2a^2 = (b + c - a)^2$$

$$\Rightarrow 2bc - 2ab - 2ac + 2a^2 = b^2 + c^2 + a^2 + 2bc - 2ac - 2ab$$

$$\Rightarrow b^2 + c^2 = a^2$$

$$\Rightarrow 2bc\cos A = 0$$

$$\therefore \angle A = 90^\circ$$

37. Prove that $\sin(A - \theta) = \frac{c}{b}\sin\theta$

In ΔABD,

$$\frac{BD}{\sin\theta} = \frac{AD}{\sin B}$$

$$\Rightarrow AD = BD\frac{\sin(B)}{\sin\theta} \quad \ldots(i)$$

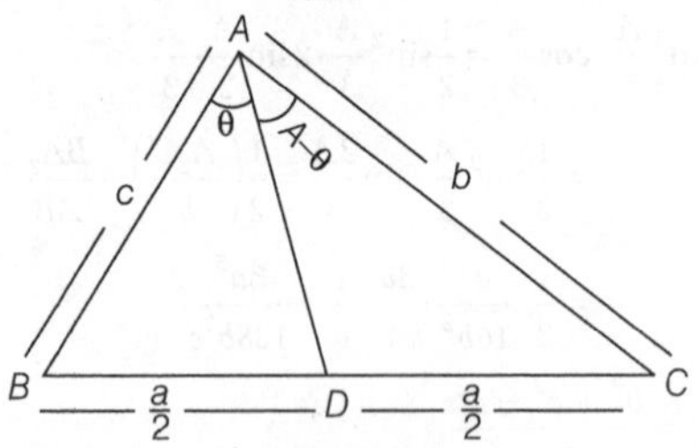

In ΔACD,

$$\frac{CD}{\sin(A-\theta)} = \frac{AD}{\sin C}$$

$$\Rightarrow AD + CD \cdot \frac{\sin C}{\sin(A-\theta)} \quad \ldots(ii)$$

From Eqs. (i) and (ii), we get

$$(BD)\frac{\sin B}{\sin\theta} = \frac{\sin C}{\sin(A-\theta)}(CD)$$

$$\Rightarrow \frac{\sin B}{\sin C} = \frac{\sin\theta}{\sin(A-\theta)}$$

$$\frac{b}{c} = \frac{\sin(\theta)}{\sin(A-\theta)}$$

$$\Rightarrow \sin(A-\theta) = \frac{c}{b}\sin\theta$$

38. A, B, C are in A.P. $\Rightarrow B = 60^\circ$

$$\Rightarrow \cos B = \cos 60^\circ = \frac{1}{2} = \frac{c^2 + a^2 - b^2}{2ca}$$

$$\Rightarrow a^2 + c^2 = b^2 + ac$$

$$\Rightarrow (a-c)^2 = b^2 - ac$$

$$\Rightarrow |a - c| = \sqrt{b^2 - ac}$$

$$\Rightarrow |\sin A - \sin C| = \sqrt{\sin^2 B - \sin A\sin C}$$

$$\Rightarrow 2\cos\frac{A+C}{2}\left|\sin\frac{A-C}{2}\right| = \sqrt{\frac{3}{4} - \sin A\sin C}$$

$$\Rightarrow 2\left|\sin\frac{A-C}{2}\right| = \sqrt{3 - 4\sin A\sin C}$$

So, that $\lim_{A\to C}\frac{\sqrt{3 - 4\sin A\sin C}}{|A-C|} = \lim_{A\to C}\frac{2\sin\left|\frac{A-C}{2}\right|}{|A-C|} = 1$

39. We are given that $(a + b + c)(b + c - a) = \lambda bc$, or $(b+c)^2 - a^2 = \lambda bc$. That is, $b^2 + c^2 + 2bc - a^2 = \lambda bc$, or $b^2 + c^2 - a^2 = (\lambda - 2)bc$.

Therefore, $\cos A = \frac{b^2 + c^2 - a^2}{2bc} = \frac{\lambda - 2}{2}$

As A is the angle of a triangle, $-1 < \cos A < 1$.

Therefore,

$$-1 < \frac{(\lambda - 2)}{2} < 1$$

$$\Rightarrow 0 < \lambda < 4$$

40. Let $\frac{a}{\sin A} = \frac{b}{\sin B} = \frac{c}{\sin C} = k$ (say)

$\Rightarrow a = k\sin A$, $b = k\sin B$, $c = k\sin C$

Now, the given relation is

$(a^2 + b^2)\sin(A - B) = (a^2 - b^2)\sin(A + B)$

or $k^2[\sin^2 A + \sin^2 B]\sin(A - B)$

$$= k^2[\sin^2 A - \sin^2 B]\sin(A + B)$$

$$\Rightarrow [\sin^2 A + \sin^2 B]\sin(A - B) = \sin^2(A + B)\sin(A - B)$$

or $\sin(A - B)[\sin^2 A + \sin^2 B - \sin^2 C] = 0$

Hence, either the first factor = 0, or the second factor = 0

If $\sin(A - B) = 0$

$\Rightarrow A - B = 0$

$\Rightarrow A = B$

$\Rightarrow$ Triangle is isosceles.

If $\sin^2 A + \sin^2 B - \sin^2 C = 0$

$\Rightarrow \quad \frac{a^2}{k^2} + \frac{b^2}{k^2} - \frac{c^2}{k^2} = 0$

or $\quad a^2 + b^2 - c^2 = 0$

or $\quad a^2 + b^2 = c^2$

$\Rightarrow$ The triangle is right angled.

41. $\because \frac{2}{1!9!} + \frac{2}{3!7!} + \frac{1}{5!5!} = \frac{8^a}{(2b)!}$

$\Rightarrow \quad \frac{1}{1!9!} + \frac{1}{3!7!} + \frac{1}{5!5!} + \frac{1}{3!7!} + \frac{1}{9!1!} = \frac{8^a}{(2b)!}$

$\Rightarrow \quad \frac{1}{10!}\left(\frac{10!}{1!9!} + \frac{10!}{3!7!} + \frac{10!}{5!5!} + \frac{10!}{7!3!} + \frac{10!}{9!1!}\right) = \frac{8^a}{(2b)!}$

$\frac{1}{10!}({}^{10}C_1 + {}^{10}C_3 + {}^{10}C_5 + {}^{10}C_7 + {}^{10}C_9) = \frac{8^a}{(2b)!}$

$\Rightarrow \quad \frac{2^9}{10!} = \frac{8^a}{(2b)!} = \frac{2^{3a}}{(2b)!}$

$\Rightarrow \quad a = 3, b = 5$

Also, $\quad 2b + a + c \Rightarrow 10 = 3 + c$

$\Rightarrow \quad c = 7$

$\therefore \quad a = 3, b = 5, c = 7$

$\because \quad \frac{\tan A + \tan B}{2} \ge \sqrt{\tan A \tan B}$ ...(i)

Also, $\quad \cos C = \frac{a^2 + b^2 - c^2}{2ab}$

$= \frac{9 + 25 - 49}{30} = -\frac{1}{2}$

$\therefore \quad C = 120^\circ$ and $A, B < 60^\circ$

$\tan A + \tan B + \tan C = \tan A \tan B \tan C$

$\Rightarrow \tan A + \tan B - \sqrt{3} = -\sqrt{3}\tan A \tan B$

$\Rightarrow \tan A + \tan B - \sqrt{3} = -\sqrt{3}\tan A \tan B$

$\therefore \tan A + \tan B = \sqrt{3}(1 - \tan A \tan B)$ ...(ii)

Also, $\quad \tan A + \tan B > 0$

$\Rightarrow \quad \sqrt{3}(1 - \tan A \tan B) > 0$

$\Rightarrow \quad \tan A \tan B < 1$ ...(iii)

From Eqs. (i) and (ii), we get

$\frac{\sqrt{3}(1 - \tan A \tan B)}{2} \ge \sqrt{(\tan A \tan B)}$

Let $\tan A \tan b = \lambda$

$\therefore \quad \sqrt{3}(1 - \lambda) \ge 2\sqrt{\lambda}$

$\Rightarrow \quad 3\lambda^2 - 10\lambda + 3 \ge 0$

$\Rightarrow \quad (3\lambda - 1)(\lambda - 3) \ge 0$

$\because \quad \lambda - 3 < 0$ [from Eq. (iii)]

$\therefore \quad 3\lambda - 1 \le 0$

$\Rightarrow \quad \lambda \le \frac{1}{3}$

$\tan A \tan b \le \frac{1}{3}$

42. $\because a(b - c) + b(c - a) + c(a - b) = 0$

$\therefore \quad x = 1$ is a root of the equation

$a(b - c)x^2 + b(c - a)x + c(a - b) = 0$

Then, other root = 1 ($\because$ roots are equal)

$\therefore \quad |X| = \frac{c(a - b)}{a(b - c)}$

$\Rightarrow \quad ab - ac = ca - bc$

$\therefore \quad b + \frac{2ac}{a + c}$

$\therefore a, b, c$ are in HP

Then, $\frac{1}{a}, \frac{1}{b}, \frac{1}{c}$ are in AP.

$\Rightarrow \quad \frac{s}{a}, \frac{s}{b}, \frac{s}{c}$ are in AP.

$\Rightarrow \quad \frac{s}{a} - 1, \frac{s}{b} - 1, \frac{s}{c} - 1$ are in AP

$\Rightarrow \quad \frac{(s - a)}{a}, \frac{(s - b)}{b}, \frac{(s - c)}{c}$ in AP

Multiplying in each by $\frac{abc}{(s - a)(s - b)(s - c)}$, then

$\frac{bc}{(s - b)(s - c)}, \frac{ca}{(s - c)(s - a)}, \frac{ab}{(s - a)(s - b)}$ are in AP.

$\Rightarrow \frac{(s - b)(s - c)}{bc}, \frac{(s - c)(s - a)}{ca}, \frac{(s - a)(s - b)}{ab}$ are in HP.

or $\sin^2\left(\frac{A}{2}\right), \sin^2\left(\frac{B}{2}\right), \sin^2\left(\frac{C}{2}\right)$ are in HP.

43. Let a be the radius of the circle, then the ratio of the area of regular polygons on n sides inscribed to circumscribing the same circle is given by

$$\frac{S_1}{S_2} = \frac{\frac{1}{2}na^2 \sin\left(\frac{2\pi}{n}\right)}{na^2 \tan\left(\frac{\pi}{n}\right)} = \frac{3}{4} \Rightarrow \cos^2\left(\frac{\pi}{n}\right) = \frac{3}{4}$$

or $\quad \cos\left(\frac{\pi}{n}\right) = \frac{\sqrt{3}}{2}$ or $\frac{\pi}{n} = \frac{\pi}{6}$

$\Rightarrow \quad n = 6$

44. $\Pi\left(\sin A + 1 + \frac{1}{\sin A}\right)$

$\because \sin A + \frac{1}{\sin A} \ge 2$ (if $A = 90^\circ$)

Then, $\sin B + \frac{1}{\sin B} > 2$

and $\sin C + \frac{1}{\sin C} > 2$

$\therefore \left(\sin A + \frac{1}{\sin A} + 1\right)\left(\sin B + \frac{1}{\sin B} + 1\right)\left(\sin C + \frac{1}{\sin C} + 1\right) > 27$

$\therefore \quad \Pi\left(\sin A + \frac{1}{\sin A} + 1\right) > 27$

or $\quad \Pi\left(\frac{\sin A + \sin A + 1}{\sin A}\right) > 27$

45. From the given information it is evident that distance between I (incentre) and O (circumcentre) should be equal to inradius of triangle.

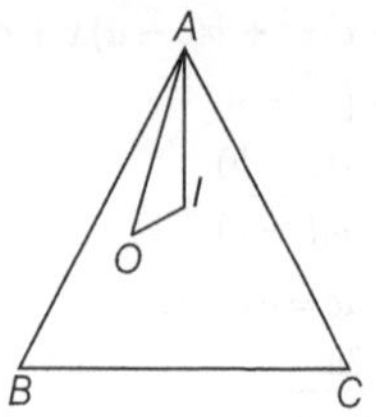

If 'r' and 'R' be the inradius and circumradius respectively, then

$$AI = r\,\text{cosec}\,\frac{A}{2} = 4R\sin\frac{B}{2}\sin\frac{C}{2}$$

and $$AO = R,\ \angle OAI = \frac{A}{2} - \left(\frac{\pi}{2} - C\right) = \frac{C-B}{2}$$

Now, $IO^2 = OA^2 + AI^2 - 2(OA)(AI)\cos\frac{C-B}{2}$

$$\Rightarrow \quad r^2 = R^2 + 16R^2\sin^2\frac{B}{2}\sin^2\frac{C}{2} - 8R^2\sin\frac{B}{2}\sin\frac{C}{2}\cos\left(\frac{C-B}{2}\right)$$

$$\Rightarrow r^2 = R^2\left(1 + 8\sin\frac{B}{2}\sin\frac{C}{2}\left(2\sin\frac{B}{2}\sin\frac{C}{2} - \cos\left(\frac{C-B}{2}\right)\right)\right)$$

$$= R^2\left(1 + 8\sin\frac{B}{2}\sin\frac{C}{2}\left(\sin\frac{B}{2}\sin\frac{C}{2} - \cos\frac{B}{2}\cos\frac{C}{2}\right)\right)$$

$$= R^2(1 + 8\sin\frac{B}{2}\sin\frac{C}{2}\left(\sin\frac{B}{2}\sin\frac{C}{2} - \cos\frac{B}{2}\cos\frac{C}{2}\right)$$

$$= R^2\left(1 - 8\sin\frac{B}{2}\sin\frac{C}{2}\cos\left(\frac{B+C}{2}\right)\right)$$

$$= R^2\left(1 - 8\sin\frac{A}{2}\sin\frac{B}{2}\sin\frac{C}{2}\right)$$

$$= R^2\left(1 - 8\cdot\frac{r}{4R}\right) = R^2 - 2rR$$

$$\Rightarrow \quad \left(\frac{r}{R}\right)^2 + 2\left(\frac{r}{R}\right) - 1 = 0$$

$$\Rightarrow \quad \frac{r}{R} = \frac{-2 \pm \sqrt{4+4}}{2} = \sqrt{2} - 1, \text{ as } \frac{r}{R} > 0$$

$$\Rightarrow \quad 1 + \frac{r}{R} = \sqrt{2}$$

$$\Rightarrow \quad \cos A + \cos B + \cos C = \sqrt{2}$$

Hence, (b) is the correct answer

46. $a + b + c = 6\dfrac{(\sin A + \sin B + \sin C)}{3}$

$$\Rightarrow \quad a + b + c = 2(\sin A + \sin B + \sin C)$$

$$\Rightarrow \quad 1 + \frac{\sin B}{\sin A} + \frac{\sin C}{\sin A} = 2(\sin A + \sin B + \sin C)$$

$$\left(\text{using } \frac{a}{\sin A} = \frac{b}{\sin B} = \frac{c}{\sin C}\right)$$

$$\Rightarrow \quad \sin A + \sin B + \sin C = 2\sin A(\sin A + \sin B + \sin C)$$

$$\Rightarrow \quad \sin A = \frac{1}{2}$$

$$\Rightarrow \quad A = 30^\circ$$

47. Let linear function is $F(x) = Ax + B$

$\because \quad [1, 2] \to [4, 6]$

$\Rightarrow \quad F(1) = 4$

$\Rightarrow \quad A + B = 4$

and $\quad F(2) = 6$

$\therefore \quad A = 2, B = 2$

Then, one function is $F(x) = 2x + 2 = f(x)$ (say)

$$\left.\begin{array}{l} F(1) = 6 \Rightarrow A + B = 6 \\ F(2) = 4 \Rightarrow 2A + B = 4 \end{array}\right\} A = -2,\ B = 8$$

Then, other function is $F(x) = -2x + 8 = g(x)$ (say)

$\therefore \quad c = f(1) + g(1) = 4 + 6 = 10$

Now, $x^2 + y^2 - xy = 10$

$$\Rightarrow \quad \frac{\left(x - \frac{y}{2}\right)^2}{\sqrt{(10)^2}} + \frac{y^2}{\left(\frac{\sqrt{10}}{\sqrt{3}}\right)^2} = 1$$

is an ellipse whose cente (0, 0).

Maximum distance from origin on any point on ellipse = Semi major axis = $\sqrt{10}$

$\therefore \quad r = \sqrt{10}$

Then, $\quad a = r^2 = 10$

$\because \quad a = c = 10$

$\therefore \quad \sin A : \sin C = 1 : 1$

48. Let a be the length of each side of the equilateral triangle ABC.

Then r, the radius of the in-circle $= \left(\frac{1}{3}\right)AP$ (the altitude, median and the angle bisector of angle A)

$$\Rightarrow \quad r = \frac{1}{3}\sqrt{a^2 - \frac{a^2}{4}} = \frac{a}{2\sqrt{3}}$$

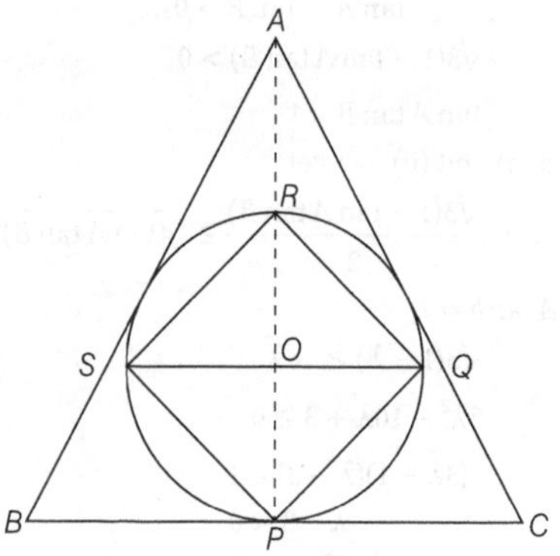

Area of the square $PQRS$ inscribed in this circle

$$= PQ^2 = OP^2 + OQ^2$$

$$= 2r^2 = 2 \times \frac{a^2}{4 \times 3} = \frac{a^2}{6}$$

49. $2\sin B = \sin A + \sin C$

$= \sin A + \sin(A + B)$

$= \sin A + \sin A \cos B + \cos A \sin B$

$\Rightarrow \quad \sin B(2 - \cos A) = \sin A(1 + \cos B)$

$\Rightarrow \quad \dfrac{\sin B}{1 + \cos B} = \dfrac{\sin A}{2 - \cos A}$

$\Rightarrow \quad \tan\dfrac{B}{2} = \dfrac{\sin A}{2 - \cos A} = k$ (suppose)

$\Rightarrow \quad \sin A + 2k - k\cos A$

$\Rightarrow \quad \sin A + k\cos A = 2k$

$\Rightarrow \quad \sin(A + \alpha) = \dfrac{2k}{\sqrt{1 + k^2}}$, where $\tan\alpha = \dfrac{k}{1}$

$\Rightarrow \quad \dfrac{2k}{\sqrt{1 + k^2}} \le 1 \Rightarrow 2k \le \sqrt{1 + k^2}$

$\Rightarrow \quad 4k^2 \le 1 + k^2 \Rightarrow 3k^2 - 1 \le 0$

$\Rightarrow \quad |k| \le \dfrac{1}{\sqrt{3}}$

$\Rightarrow$ Maximum value of $\tan\dfrac{B}{2}$ is $\dfrac{1}{\sqrt{3}}$.

50. $\because x^2 - 9x + 8 = 0$

$\therefore \quad x = 1, 8$

$\Rightarrow \quad 2^{\tan^2 B} = 1, 8 \Rightarrow 2^{\tan^2 B} = 2^0, 2^3$

$\tan^2 \ne 0$

$\therefore \quad \tan^2 B = 3 \Rightarrow \tan B = \sqrt{3}$

$\therefore \quad \angle B = 60°$

Also given, $2\cos A = \dfrac{\sin B}{\sin C}$

$\Rightarrow \quad 2\cos A \sin C = \sin B$

$\Rightarrow \quad \sin(A + C) - \sin(A - C) = \sin B$

$\Rightarrow \quad \sin(A - B) - \sin(A - C) = \sin B$

Then, $\sin(A - C) = 0$

$\Rightarrow \quad A = C$

$\Rightarrow \quad A + B + C = 180°$

$A + 60° + A = 180°$

$A = 60° = C, B = 60°$

51. We have,

$\text{arc}(BC) = 3$

$\text{arc}(CA) = 4$

$\text{arc}(AB) = 5$

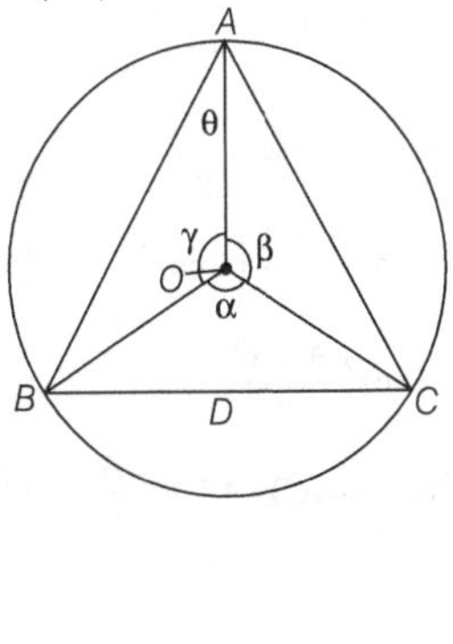

Let r be the radius of the circle, then

$$3 = r\alpha,\ 4 = r\beta,\ 5 = r\gamma \qquad \left(\because \text{Angle} = \frac{\text{arc}}{\text{radius}}\right)$$

Now, $3 + 4 + 5 = r(\alpha + \beta + \gamma) = r \cdot 2\pi$

$\therefore \quad r = \dfrac{6}{\pi}$

$\Delta ABC = \Delta OBC + \Delta OCA + \Delta OAB$

$= \dfrac{1}{2}r^2\sin\alpha + \dfrac{1}{2}r^2\sin\beta + \dfrac{1}{2}r^2\sin\gamma$

$= \dfrac{1}{2}r^2\left\{\sin\left(\dfrac{3}{r}\right) + \sin\left(\dfrac{4}{r}\right) + \sin\left(\dfrac{5}{r}\right)\right\}$

$= \dfrac{1}{2} \times \dfrac{36}{\pi^2}\left[\sin\dfrac{\pi}{2} + \sin\dfrac{2\pi}{3} + \sin\dfrac{5\pi}{6}\right]$

$= \dfrac{18}{\pi^2}\left\{1 + \dfrac{\sqrt{3}}{2} + \dfrac{1}{2}\right\}$

$= \dfrac{9\sqrt{3}(1 + \sqrt{3})}{\pi^2}$ sq unit

52. $\because \dfrac{b + c}{2} \ge \sqrt{bc}$

$\Rightarrow \quad \dfrac{b + c}{2} \ge \lambda$ ...(i)

and $\dfrac{a}{\sin A} = \dfrac{b}{\sin B} = \dfrac{c}{\sin C} = \dfrac{b + c}{\sin B + \sin C}$

$= \dfrac{(b + c)}{2\sin\left(\dfrac{B + C}{2}\right)\cos\left(\dfrac{B - C}{2}\right)}$

$= \dfrac{(b + c)}{2\cos\dfrac{A}{2}\cos\left(\dfrac{B - C}{2}\right)}$

$\therefore \quad a = \left(\dfrac{b + c}{2}\right) \cdot \dfrac{\sin A}{\cos\dfrac{A}{2}\cos\left(\dfrac{B - C}{2}\right)}$

$= \left(\dfrac{b + c}{2}\right) \cdot \dfrac{2\sin\dfrac{A}{2}}{\cos\left(\dfrac{B - C}{2}\right)}$

$\ge \left(\dfrac{b + c}{2}\right) \cdot 2\sin\dfrac{A}{2} \qquad \left(\because \sec\left(\dfrac{B - C}{2}\right) \ge 1\right)$

$\ge 2\lambda\sin\dfrac{A}{2}$ [from Eq. (i)]

Hence, $a \ge 2\lambda \lim\dfrac{A}{2}$

53. $\cos B = \dfrac{a^2 + c^2 - b^2}{2ac}$

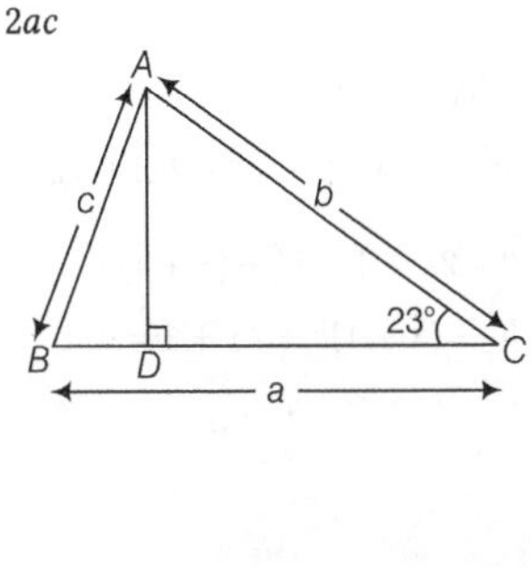

$$= \frac{a}{2c} - \frac{b^2 - c^2}{2ac} = \frac{a}{2c} - \frac{b^2 - c^2}{abc} \cdot \frac{b}{2}$$

$$= \frac{a}{2c} - \frac{b}{2(AD)}$$

$$= \frac{\sin A}{2\sin C} - \frac{1}{2\sin C}$$

In ΔACD, $\cos B = \dfrac{\sin A - 1}{2\sin C}$

$$\frac{AD}{b} = \sin C \Rightarrow \frac{b}{AD} = \frac{1}{\sin C}$$

$\Rightarrow \quad 2\cos B \sin C = \sin A - 1$

$\Rightarrow \quad \sin(B + C) - \sin(B - C) = \sin A - 1$

$\Rightarrow \quad \sin A - \sin(B - C) = \sin A - 1$

$\Rightarrow \quad \sin(B - C) = 1$

$$B - C = \frac{\pi}{2}$$

$$B = C + \frac{\pi}{2} = 23 + 90 = 113^\circ$$

$$\angle B = 113^\circ$$

54. We have, $\dfrac{31}{32} = \cos(A - B) = \dfrac{1 - \tan^2 \frac{A - B}{2}}{1 + \tan^2 \frac{A - B}{2}}$

$\Rightarrow \quad 63\tan^2 \dfrac{A - B}{2} = 1$

$\Rightarrow \quad \tan \dfrac{A - B}{2} = \dfrac{1}{\sqrt{63}}$

Now, $\quad \tan \dfrac{A - B}{2} = \dfrac{a - b}{a + b} \cot \dfrac{C}{2}$

$\Rightarrow \quad \dfrac{1}{\sqrt{63}} = \dfrac{5 - 4}{5 + 4} \cot \dfrac{C}{2}$

$\Rightarrow \quad \tan \dfrac{C}{2} = \dfrac{\sqrt{63}}{9}$

Also $\quad \cos C = \dfrac{1 - \tan^2\left(\frac{C}{2}\right)}{1 + \tan^2\left(\frac{C}{2}\right)}$

$$= \frac{1 - \frac{63}{81}}{1 + \frac{63}{81}} = \frac{18}{144} = \frac{1}{8}$$

$\therefore \quad c^2 = a^2 + b^2 - 2ab\cos C$

$$= 25 + 16 - 2 \cdot 5 \cdot 4 \cdot \left(\frac{1}{8}\right) = 36$$

Hence $c = 6$

55. Using cosine rule, we get

$$x^2 = (x + 1)^2 + b^2 - 2(x + 1)b\cos\frac{\pi}{3}$$

$\Rightarrow \quad 0 = 2x + 1 + b^2 - (x + 1)b$

$\Rightarrow \quad b^2 - (x + 1)b + 2x + 1 = 0$

Since b is real, we have

$\Rightarrow \quad (x + 1)^2 - 4(2x + 1) \geq 0$

$\Rightarrow \quad x^2 - 6x - 3 \geq 0$

$\Rightarrow \quad x \geq 3 + \sqrt{12}$

The least integral value of x is 7.

56. Let $r = 1$ cm

$$a = BC = BD + DE + EC$$
$$= BD + C_1C_2 + BD$$
$$= 2BD + C_1C_2$$
$$= 2r\cot 30^\circ + 2r$$
$$= 2r(\sqrt{3} + 1)$$
$$= 2(\sqrt{3} + 1) \text{ cm}$$

$\therefore$ Area of triangle

$$ABC = \frac{\sqrt{3}}{4}a^2 = \frac{\sqrt{3}}{4} \cdot 4(\sqrt{3} + 1)^2 \text{ cm}^2$$
$$= \sqrt{3}(4 + 2\sqrt{3}) \text{ cm}^2$$
$$= (6 + 4\sqrt{3}) \text{ cm}^2$$

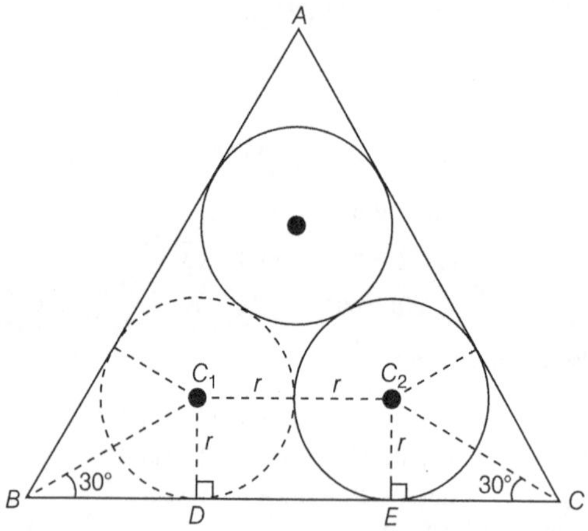

57. We have, $2b = a + c$

$\therefore \quad 2\sin B = \sin A + \sin C$

$$4\sin\frac{B}{2}\cos\frac{B}{2} = 2\sin\frac{A + C}{2}\cos\frac{A - C}{2}$$

$$= 2\cos\frac{B}{2}\cos\frac{A - C}{2}$$

$\therefore \quad 2\sin\dfrac{B}{2} = \cos\dfrac{A - C}{2}$

i.e,. $\quad 2\cos\dfrac{A + C}{2} = \cos\dfrac{A - C}{2}$ ...(i)

Now, $\cos A + \cos C = 2\cos\dfrac{A + C}{2} \cdot \cos\dfrac{A - C}{2}$

$$= 2\cos\frac{A + C}{2}\left(2\cos\frac{A + C}{2}\right) \quad \text{[using Eq. (i).]}$$

$$= 4\cos^2\frac{A + C}{2} \quad \text{...(ii)}$$

$$4(1 - \cos A)(1 - \cos C) = 4.2\sin^2\frac{A}{2} \cdot 2\sin^2\frac{C}{2}$$

$$= 4\left(2\sin\frac{A}{2}\sin\frac{C}{2}\right)^2 = 4\left\{\cos\frac{A-C}{2} - \cos\frac{A+C}{2}\right\}^2$$

$$= 4\left\{\cos\frac{A+C}{2} - \cos\frac{A+C}{2}\right\}^2 = 4\cos^2\frac{A+C}{2} \quad \text{...(iii)}$$

From Eqs. (ii) and (iii), we get

$2\cos A + \cos C = 4(1 - \cos A)(1 - \cos C)$

58. We have, $c(a+b)\cos\frac{1}{2}B = b(a+c)\cos\frac{1}{2}C$

$$\therefore \quad c(a+b)\sqrt{\left\{\frac{s(s-b)}{ca}\right\}} = b(a+c)\sqrt{\left\{\frac{s(s-c)}{ab}\right\}}$$

or $(a+b)\sqrt{c(s-b)} = (a+c)\sqrt{b(s-c)}$

Squaring,

$(a+b)^2 c(s-b) = (a+c)^2 b(s-c)$

or $s[c(a+b)^2 - b(a+c)^2] - bc[(a+b)^2 - (a+c)^2] = 0$

or $s\{ca^2 + 2abc + cb^2 - ba^2 - 2abc - bc^2\}$

$$- bc(b-c)(2a+b+c) = 0$$

or $s\{bc(b-c) - a^2(b-c)\} - bc(b-c)(2a+b+c) = 0$

or $(b-c)\{s(bc-a^2) - bc(2a+b+c)\} = 0$

or $\quad (b-c)\{s(bc-a^2) - bc(2s+a\} = 0$

or $\quad -(b-c)\{s(bc+a^2) + abc\} = 0$

Since, a, b, c are all positive and so $s(bc+a^2) + abc \neq 0$

It follows that $b - c = 0$.

Hence ΔABC is isoceles.

59. M is the mid point of BC

$$MC = \frac{a}{2}$$

$$OC = R$$

$$OM = ID = r$$

$\therefore \Delta OMC$

$$R^2 = r^2 + \frac{a^2}{4}$$

$$R^2 = r^2 + \frac{(2R\sin A)^2}{4}$$

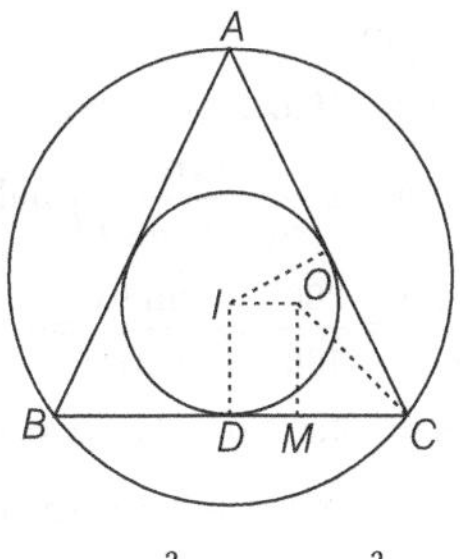

$\Rightarrow \quad R^2\cos 2A = r^2$

$\Rightarrow \quad \cos A = \frac{r}{R} \Rightarrow r = R\cos A$

$\because \quad \cos A + \cos B + \cos C$

$$= 1 + 4\sin\frac{A}{2}\sin\frac{B}{2}\sin\frac{C}{2} = 1 + \frac{r}{R} = 1 + \cos A$$

$\therefore \quad \cos B + \cos C = 1$

60.

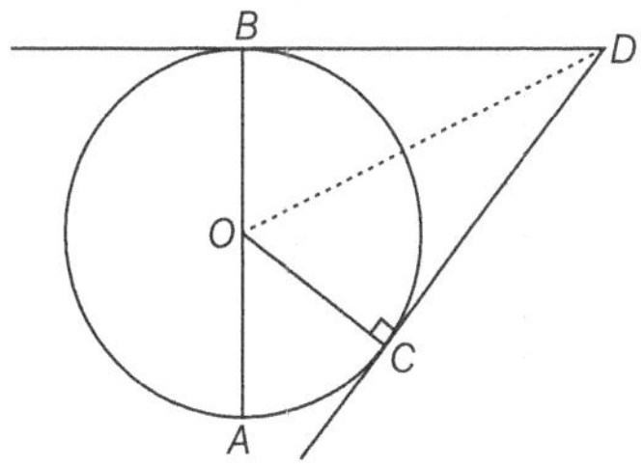

$$AC = d, OA = OB = r, CD = BD = l, \angle COA = \frac{\pi}{3}$$

$\therefore \quad AC^2 = OA^2 + OC^2 - 2AO\,OC\cos\frac{\pi}{3}$

or $\quad d^2 = 2r^2 - 2r^2\frac{1}{2} = r^2$

Also, $\quad \Delta BOD = \Delta = \frac{2\pi}{3\times 2} = \frac{\pi}{3}$

or $\quad \tan\frac{\pi}{3} = \frac{BD}{OB} = \frac{1}{r}$

$\Rightarrow \quad l = r\sqrt{3} = d\sqrt{3}$

61. In pedal ΔDEF,

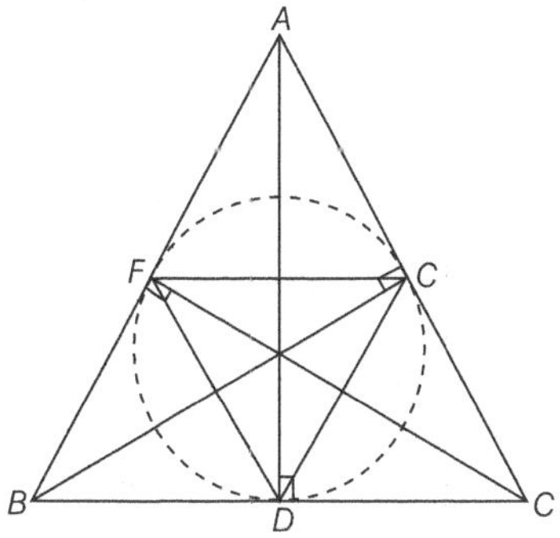

$$EF = a\cos A$$

$$DE = c\cos C$$

$$DF = b\cos B$$

If circum radius of ΔDEF is R_1

Then, $R_1 = \dfrac{(a\cos A)(b\cos B)(c\cos C)}{4\cdot\frac{1}{2}\cdot DF\cdot DE\sin(\angle EDF)}$

(Here, $\Delta EDF = 180° - 2A$)

$$= \frac{abc\cos A\cos B\cos C}{4\cdot\frac{1}{2}\cdot b\cos B\cdot c\cos C\cdot\sin(180° - 2A)}$$

$$= \frac{a\cos A}{2\sin 2A} = \frac{2R\sin A\cos A}{4\sin A\cos A}$$

$\therefore \quad R_1 = \frac{R}{2}$

62. Let CD be the bisector of C, so that $AD = x$ and $BD = y$

in ΔACD, $\dfrac{AD}{\sin 45°} = \dfrac{CD}{\sin A} \Rightarrow \sin A = \dfrac{CD}{\sqrt{2}x}$

Similarly in ΔBCD, $\dfrac{BD}{\sin 45°} = \dfrac{CD}{\sin B} \Rightarrow \sin B = \dfrac{CD}{\sqrt{2}y}$

$$\frac{\sin A}{\sin B} = \frac{y}{x} \Rightarrow \frac{\sin A - \sin B}{\sin A + \sin B} = \frac{y - x}{y + x}$$

[By Componendo and Dividendo]

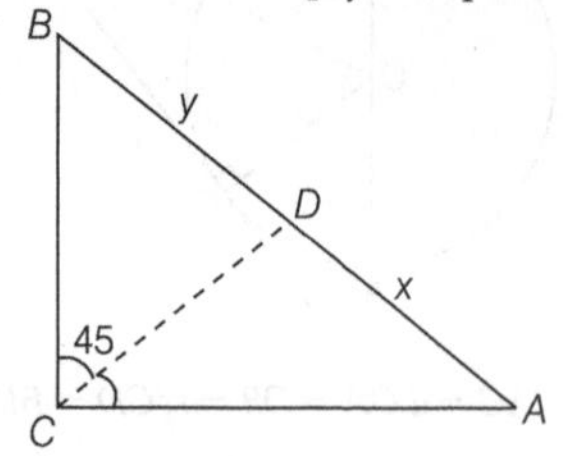

$$\Rightarrow \frac{2\cos\frac{A + B}{2}\sin\frac{A - B}{2}}{2\sin\frac{A + B}{2}\cos\frac{A - B}{2}} = \frac{y - x}{y + x}$$

$$\Rightarrow \tan\frac{A - B}{2} = \frac{y - x}{y + x} \quad [\text{As } A + B = 90°]$$

$$\Rightarrow \frac{y - x}{y + x} = \frac{t}{1} \quad \text{(given)}$$

$\Rightarrow$ again by Componendo and Dividendo, $\frac{y}{x} = \frac{1 + t}{1 - t}$

$$\Rightarrow x : y = (1 - t) : (1 + t)$$

63. $BD = (s - b), CD = (s - c) \Rightarrow (s - b)(s - c) = 2$

$$\Rightarrow s(s - a)(s - b)(s - c) = 2s(s - a)$$

$$\Rightarrow \Delta^2 = 2s(s - a)$$

$$\Rightarrow \frac{\Delta^2}{s^2} = \frac{2(s - a)}{s} = 1 \quad \text{(radius of incircle of triangle } ABC)$$

$$\Rightarrow \frac{a}{s} = \text{constant.}$$

Now, $\Delta = \frac{1}{2}aH_a$, where 'H_a' is the distance of 'A' from BC.

$$\Rightarrow \frac{\Delta}{s} = \frac{1}{2}\frac{aH_a}{s} = 1$$

$$\Rightarrow H_a = \frac{2s}{a} = \text{constant}$$

$\Rightarrow$ Locus of 'A' will be a straight line parallel to side BC.

64. Let AE is vertical lamp-post.

Given, $AE = 12$ m

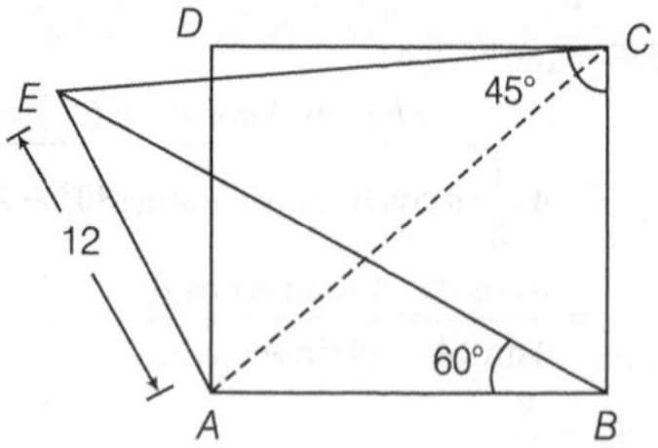

$$\tan 45° = \frac{AE}{AC}$$

$$AC = AE = 12 \text{ m}$$

$$\tan 60° = \frac{AE}{AB}$$

$$AB = \frac{AE}{\sqrt{3}} = 4\sqrt{3}$$

$$BC = \sqrt{AC^2 - AB^2} = \sqrt{144 - 48} = \sqrt{96} = 4\sqrt{6}$$

Area $= AB \times BC = 4\sqrt{3} \times 4\sqrt{6} = 48\sqrt{2}$ sq m.

65. $\tan\alpha = \frac{b}{a}$, $\tan 2\alpha = \frac{2(b/a)}{1-(b/a)^2} = \frac{p + b}{a}$

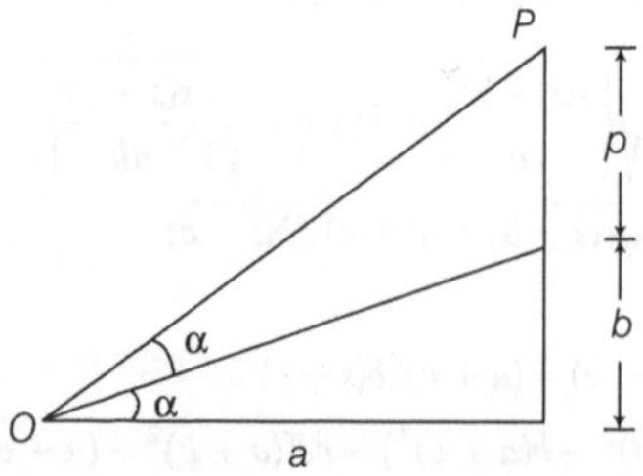

$$\Rightarrow \frac{2ba}{a^2 - b^2} = \frac{p + b}{a}$$

$$\Rightarrow \frac{2ba^2 - a^2b + b^3}{a^2 - b^2} = p \Rightarrow p = \frac{b(a^2 + b^2)}{(a^2 - b^2)}$$

66. $x = h\cot 3\alpha$ ...(i)

$(x + 100) = h\cot 2\alpha$ ... (ii)

$(x + 300) = h\cot\alpha$ ...(iii)

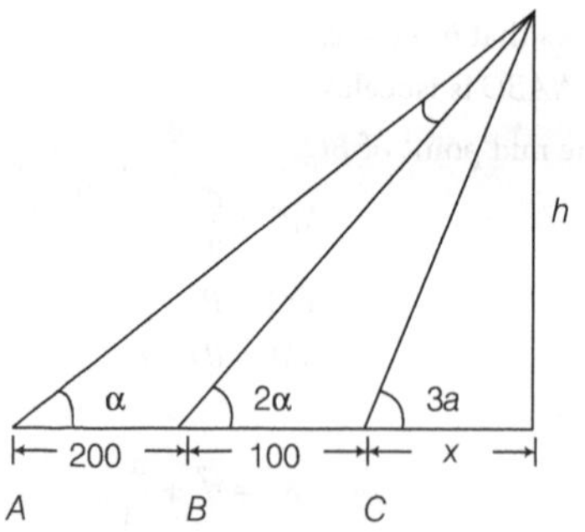

From Eqs. (i) and (ii)

$$-100 = h(\cot 3\alpha - \cot 2\alpha)$$

From Eqs. (ii) and (iii)

$$-200 = h(\cot 2\alpha - \cot\alpha)$$

$$100 = h\left(\frac{\sin\alpha}{\sin 3\alpha \sin 2\alpha}\right) \text{ and } 200 = h\left(\frac{\sin\alpha}{\sin 2\alpha \sin\alpha}\right)$$

or $$\frac{\sin 3\alpha}{\sin\alpha} = \frac{200}{100} \Rightarrow \frac{\sin 3\alpha}{\sin\alpha} = 2$$

$$\Rightarrow 3\sin\alpha - 4\sin^3\alpha - 2\sin\alpha = 0$$

$$\Rightarrow 4\sin^3\alpha - \sin\alpha = 0 \Rightarrow \sin\alpha = 0$$

or $$\sin^2\alpha = \frac{1}{4} = \sin^2\left(\frac{\pi}{6}\right)$$

$$\Rightarrow \alpha = \frac{\pi}{6}$$

Hence, $H = 200\sin\frac{\pi}{3} = 200\frac{\sqrt{3}}{2} = 100\sqrt{3}$ [form Eq. (i)]

67. Obviously, from figure

$$\tan\alpha = \frac{h/3}{40} = \frac{h}{120} \quad ...(i)$$

$$\tan\beta = \frac{h}{40} = \frac{3h}{120} \quad ...(ii)$$

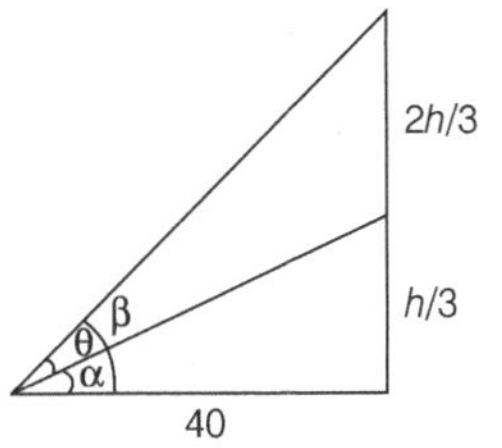

Therefore $\tan\theta = \tan(\beta - \alpha)$

$$\Rightarrow \quad \frac{1}{2} = \frac{\frac{3h}{120} - \frac{h}{120}}{1 + \frac{3h^2}{14400}} \Rightarrow h = 120, 40$$

But $h = 40$ cannot be taken according to the condition, therefore $h = 120$ ft.

68. $c^2 = a^2 + b^2 - 2ab\cos C = (a-b)^2 + 2ab(1 - \cos C)$

$$= (a-b)^2 + \frac{4\Delta}{\sin C}(1 - \cos C)$$

Hence, for c to be minimum $a = b$

Also $\quad \Delta = \frac{1}{2}ab\sin C \Rightarrow a^2 = \frac{2\Delta}{\sin C} = b^2$

69. $2\Delta = ab\sin C$

$$\Rightarrow \quad a^2b^2 - 4\Delta^2 = a^2b^2\cos^2 C$$

$$\therefore \quad \sqrt{a^2b^2 - 4\Delta^2} + \sqrt{b^2c^2 - 4\Delta^2} + \sqrt{c^2a^2 - 4\Delta^2}$$

$$= ab\cos C + bc\cos A + ca\cos B$$

70. $c\cos(A - \theta) + a\cos(C + \theta)$

$$= \cos\theta(c\cos A + a\cos C) + \sin\theta(c\sin A - a\sin C) = b\cos\theta \ge b$$

71. As $ac = 3$ and $bc = 4$

$$\therefore \quad \frac{b}{a} = \frac{4}{3}$$

$$\Rightarrow \quad \frac{b-a}{b+a} = \frac{4-3}{4+3} = \frac{1}{7} \quad [\angle B > \angle A]$$

$$\tan\left(\frac{B-A}{2}\right) = \frac{b-a}{b+a}\cot\frac{C}{2}$$

$$\Rightarrow \quad \sqrt{\frac{1 - \cos(B-A)}{1 + \cos(B-A)}} = \frac{4-3}{4+3}\cot\frac{C}{2}$$

$$\Rightarrow \quad \sqrt{\frac{1 - \frac{3}{4}}{1 + \frac{3}{4}}} = \frac{1}{7}\cot\frac{C}{2}$$

$$\sqrt{\frac{1}{7}} = \frac{1}{7}\cot\frac{C}{2} \Rightarrow \cot\frac{C}{2} = \sqrt{7}$$

or $\quad \cos C = \dfrac{1 - \tan^2\frac{C}{2}}{1 + \tan^2\frac{C}{2}} = \dfrac{1 - \frac{1}{7}}{1 + \frac{1}{7}} = \dfrac{6}{8} = \dfrac{3}{4}$

$\therefore \quad \cos C = \cos(A - B)$ or $\cos C = \cos(B - A)$

$\therefore \quad C = A - B \quad$ or $\quad C = B - A$

$\pi - (A + B) = A - B$ or $\pi - (A + B) = B - A$

$\therefore \quad 2A = \pi$ or $2B = \pi$, $A = \frac{\pi}{2}$ (Rejected) or $B = \frac{\pi}{2}$

But $\quad \angle B = \frac{\pi}{2} \quad [\text{As } \angle B > \angle A]$

But $\Rightarrow \frac{b}{a} = \frac{4b}{3a} = \frac{4}{3} = k$ (let)

$\Rightarrow \quad b = 4k, a = 3k$

Now, $ac = 3$

$$\Rightarrow \quad 3\sqrt{7}\,k^2 = 3$$

$$\Rightarrow \quad k^2 = \frac{1}{\sqrt{7}}$$

$$\Rightarrow \quad k = \frac{1}{7^{1/4}}$$

$$\therefore \quad R = 2k = \frac{2}{7^{1/4}}$$

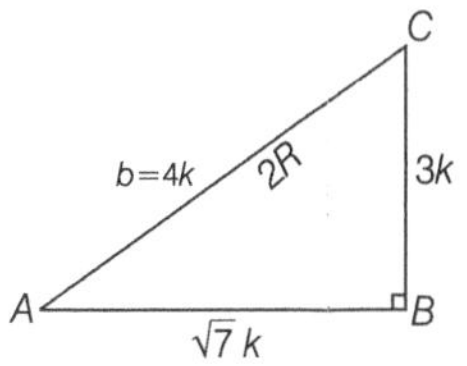

72. We have, $a = 5, b = 4$ and $\cos(A - B) = \frac{31}{32}$

As, $\tan\left(\frac{A-B}{2}\right) = \left(\frac{a-b}{a+b}\right)\cot\frac{C}{2} = \frac{1}{9}\cot\frac{C}{2}$

$$\Rightarrow \quad \cos(A - B) = \frac{1 - \tan^2\left(\frac{A-B}{2}\right)}{1 + \tan^2\left(\frac{A-B}{2}\right)}$$

$$\Rightarrow \quad \frac{31}{32} = \frac{1 - \frac{1}{81}\cot^2\frac{C}{2}}{1 + \frac{1}{81}\cot^2\frac{C}{2}}$$

$$\Rightarrow \quad 31 + \frac{31}{81}\cot^2\frac{C}{2} = 32 - \frac{32}{81}\cot^2\frac{C}{2}$$

$$\Rightarrow \quad \frac{7}{9}\cot^2\frac{C}{2} = 1 \Rightarrow \cot^2\frac{C}{2} = \frac{9}{7}$$

As, $\quad \cos C = \dfrac{1 - \tan^2\frac{C}{2}}{1 + \tan^2\frac{C}{2}} = \dfrac{1 - \frac{7}{9}}{1 + \frac{7}{9}} = \dfrac{2}{16} = \dfrac{1}{8}$

Also, $\quad \cos C = \dfrac{a^2 + b^2 - c^2}{2ab}$

$$\Rightarrow \quad \frac{25 + 16 - c^2}{2(5)(4)} = \frac{1}{8} \Rightarrow c = 6$$

Now, verify alternatives.

(d) LHS $= R(b^2\sin 2C + c^2\sin 2B)$

$= R(2b^2\sin C\cos C + 2c^2\sin B\cos B)$

$= 2R[4R^2\sin^2 B\sin C\cos C + 4R^2\sin^2 C\cos B]$

$= 2R(4R^2)\sin B\sin C[\sin B\cos C + \cos B\sin C]$

$= 8R^2\sin A\sin B\sin C$

$= (2R\sin A)(2R\sin B)(2R\sin C) = abc$

$(5)(4)(6) = 120$

73. (a) $a^2 - (c-b)^2 = 4bc$

$$\Rightarrow a^2 - (c^2 + b^2 - 2bc) = 4bc$$

$$a^2 = b^2 + c^2 + 2bc$$

$$\frac{-2bc}{2bc} = \frac{b^2 + c^2 - a^2}{2bc}$$

$\Rightarrow \cos B = -1$ which is not possible.

(b) $2b^2 \sin C \cos C + 2c^2 \sin B \cos B = ab$

$$\frac{(b^2 c \cos C + c^2 b \cos B)}{R} = ab$$

$$\Rightarrow \quad \frac{bc}{R}(b\cos C + c\cos B) = ab$$

$$\Rightarrow \quad \frac{abc}{R} = ab, \Rightarrow \frac{c}{2R} = \frac{1}{2}$$

$$\Rightarrow \quad \sin C = \frac{1}{2} \Rightarrow C = \frac{\pi}{6}$$

(c) For $a = 3$, $b = 5$, $c = 7$, we have

$$\cos C = \frac{a^2 + b^2 - c^2}{2ab} = \frac{9 + 25 - 49}{2 \times 3 \times 5}$$

$$= \frac{-15}{30} = \frac{-1}{2}$$

$$\therefore \quad C = \frac{2\pi}{3}$$

(d) As $\cos\left(\frac{A-C}{2}\right) = \cos\left(\frac{A+C}{2}\right)$

$\Rightarrow 2\sin\frac{A}{2}\sin\frac{C}{2} = 0$ which is not possible in triangle.

74. We have, $\frac{1}{\sin A} = \frac{2}{\sin B} = \frac{c}{\sin 30°}$

So, $\quad 2\sin A = \sin B$

Hence, option (a) is correct.

Also, $c^2 = 4 + 1 - 4\cos 30°$

$$= 5 - 2\sqrt{3}$$

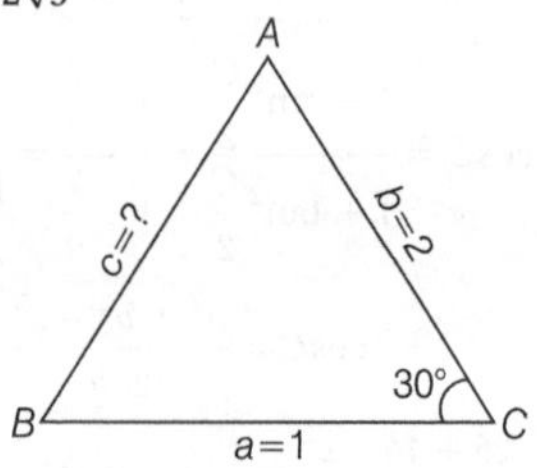

$$\Rightarrow \quad c = \sqrt{5 - 2\sqrt{3}}$$

As $c > a \Rightarrow \angle A < \angle C$ and $\angle B$ is obtuse.

Also, $\quad \Delta = \frac{1}{2} \times 1 \times 2 \times \frac{1}{2} = \frac{1}{2}$

So, $\quad R = \frac{abc}{4\Delta}$

$$= \frac{1 \times 2 \times c}{4\Delta} = c = AB$$

Hence, option (d) is correct.

75. $\Delta = 10\sqrt{3}$, $s = 10$

$$\Rightarrow \quad r = \frac{\Delta}{s} = \sqrt{3} \text{ and}$$

$$r = (s-b)\tan\frac{B}{2}$$

$$\Rightarrow \quad s - b = 3 \Rightarrow b = 7$$

$$\Rightarrow \quad a + c = 13 \quad \text{...(i)}$$

$$\Rightarrow \quad ac = 40 \quad \text{...(ii)}$$

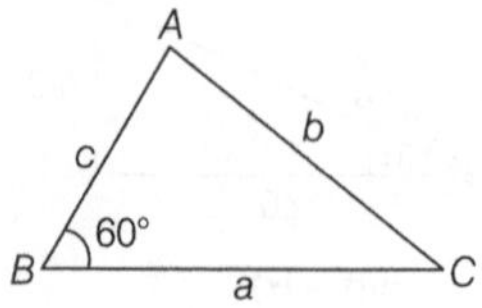

From Eq. (ii), $a^2 - 13a + 40 = 0$

$$\Rightarrow \quad a = 5, a = 8$$

$$\Rightarrow \quad c = 8 \text{ and } c = 5.$$

Hence, $a = 8$, $b = 7$ and $c = 5$ (As $a > b > c$)

$\Rightarrow r_1 = \frac{\Delta}{s-a}, r_2 = \frac{\Delta}{s-b}, r_3 = \frac{\Delta}{s-c}$ will give $\frac{1}{12} : \frac{1}{13} : \frac{1}{15}$

Also, $\quad \frac{b}{\sin B} = 2R$

$$\Rightarrow \quad 2R = \frac{7}{\frac{\sqrt{3}}{2}} \Rightarrow R = \frac{7}{\sqrt{3}}$$

76. $\cos C = \frac{1}{2} = \frac{16 + 64 - c^2}{64}$

$$\Rightarrow \quad c^2 = 48 \Rightarrow c = 4\sqrt{3}$$

Also, $\quad \frac{a}{\sin A} = \frac{c}{\sin C}$

$$\Rightarrow \quad \frac{4}{\sin A} = \frac{4\sqrt{3}}{\frac{\sqrt{3}}{2}}$$

$$\Rightarrow \quad A = 30°$$

As $\quad \angle C = 60°, \angle A = 30°$

$$\Rightarrow \quad \angle B = 90°$$

Now, $\quad r = \frac{\Delta}{s} = \frac{\frac{1}{2}ac}{\frac{a+b+c}{2}} = \frac{ac}{a+b+c}$

$$= \frac{4(4\sqrt{3})}{4 + 4\sqrt{3} + 8}$$

$$= \frac{16\sqrt{3}}{12 + 4\sqrt{3}} = \frac{4\sqrt{3}}{3 + \sqrt{3}}$$

Now, verify alternatives.

(c) $r = \frac{4\sqrt{3}}{3 + \sqrt{3}}$

(d) T the length of internal angle bisector of $\angle C$ is $\frac{8}{\sqrt{3}}$.

Hence, options a and b are correct.

77. $R = \dfrac{b}{2\sin\alpha} = \dfrac{b}{2}\operatorname{cosec}\alpha$

$$\Delta = \frac{1}{2}b^2\sin(180^\circ - 2\alpha) = \frac{1}{2}b^2\sin 2\alpha$$

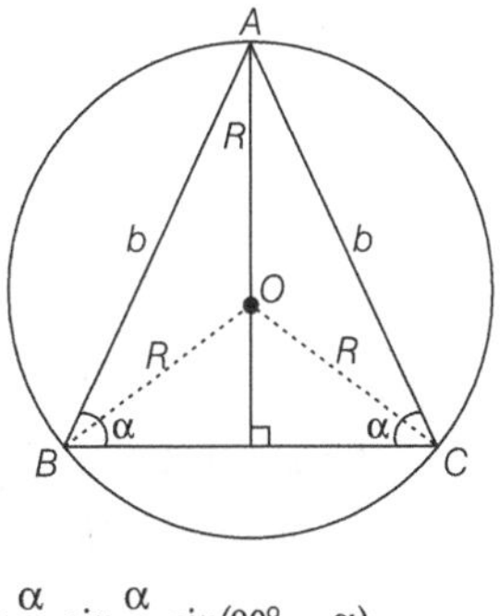

and $r = 4R\sin\dfrac{\alpha}{2}\cdot\sin\dfrac{\alpha}{2}\cdot\sin(90^\circ - \alpha)$

$$= \frac{2b}{\sin\alpha}\cdot\sin^2\frac{\alpha}{2}\cdot\cos\alpha\cdot\frac{\sin\alpha}{\sin\alpha}$$

$$= \frac{b(1-\cos\alpha)\sin 2\alpha}{2(1+\cos^2\alpha)} = \frac{b\sin 2\alpha}{2(1+\cos\alpha)}$$

and $OI = \sqrt{(R^2 - 2Rr)} = R\sqrt{\left(1 - \dfrac{2r}{R}\right)}$

$$= R\sqrt{1 - 4\cos\alpha + 4\cos^2\alpha} = R(2\cos\alpha - 1)$$

$$= R\left\{2\left(2\cos^2\frac{\alpha}{2} - 1\right) - 1\right\}$$

$$= R\left(4\cos^2\frac{\alpha}{2} - 3\right)$$

$$= \frac{R\left(4\cos^3\frac{\alpha}{2} - 3\cos\frac{\alpha}{2}\right)}{\cos\frac{\alpha}{2}}$$

$$= \frac{R\cos\left(\frac{3\alpha}{2}\right)}{\cos\frac{\alpha}{2}} = \left|\frac{b\cos\left(\frac{3\alpha}{2}\right)}{2\sin\alpha\cos\left(\frac{\alpha}{2}\right)}\right|$$

78. (a) Since, $\tan A + \tan B + \tan C = \tan A\tan B\tan C$

But here, $\tan A + \tan B + \tan C = 0$, impossible

(b)

$$\frac{\sin A}{2} = \frac{\sin B}{3} = \frac{\sin C}{7}$$

or $\dfrac{a}{2} = \dfrac{b}{3} = \dfrac{c}{7}$

$\Rightarrow$ $\dfrac{a+b}{5} = \dfrac{c}{7}$

or $\dfrac{a+b}{c} = \dfrac{5}{7} < 1$

$\therefore$ $a + b < c$ impossible.

(c) $(a+b)^2 = c^2 + ab$

$\Rightarrow$ $a^2 + b^2 - c^2 = -ab$

$\Rightarrow$ $\cos C = -\dfrac{1}{2}$

$$\angle C = 120^\circ$$

and $\sqrt{2}(\sin A + \cos A) = \sqrt{3}$

$$\sqrt{2}\cdot\sqrt{2}\sin\left(A + \frac{\pi}{4}\right) = \sqrt{3}$$

$$\sin\left(A + \frac{\pi}{4}\right) = \frac{\sqrt{3}}{2}$$

$$A + \frac{\pi}{4} = \frac{\pi}{3}$$

$\therefore$ $A = \dfrac{\pi}{12}$ possible

(d)

$\because$ $\sin A + \sin B = \dfrac{\sqrt{3}+1}{2}$ ...(i)

and $\cos A\cos B = \dfrac{\sqrt{3}}{4} = \sin A\sin B$

$\therefore$ $\cos A\cos B - \sin A\sin B = 0$

$$\cos(A + B) = 0$$

$$A + B = \frac{\pi}{2}$$

$\therefore$ $B = \dfrac{\pi}{2} - A$

From Eq. (i),

$$\sin A + \cos A = \frac{\sqrt{3}+1}{2}$$

$\Rightarrow$ $\sin\left(A + \dfrac{\pi}{4}\right) = \dfrac{\sqrt{3}+1}{2\sqrt{2}} = \sin\dfrac{5\pi}{12}$

$\Rightarrow$ $A + \dfrac{\pi}{4} = \dfrac{5\pi}{12}$

$\Rightarrow$ $A = \dfrac{\pi}{6}$

$\therefore$ $B = \dfrac{\pi}{3}$

Then, $C = \dfrac{\pi}{2}$ possible.

79. In ΔABC, $b + c - a > 0$, $c + a - b > 0$, $a + b - c > 0$, so

$$\frac{(b+c-a) + (c+a-b)(a+b-c)}{3} \geq \{(b+c-a)(c+a-b)(a+b-c)\}^{1/3}$$

$\Rightarrow$ $p^3 \geq 27\,(b+c-a)\,(c+a-b)\,(a+b-c)$

Also, $\dfrac{(a+b+c) + (b+c-a) + (c+a-b) + (a+b-c)}{4}$

$$> \{(a+b+c)(b+c-a)(c+a-b)(a+b-c)\}^{1/4}$$

$\Rightarrow$ $\dfrac{2P}{4} > (16A)^{1/4}$

$\Rightarrow$ $P > 4A^{1/4}$ or $P^4 = 256A$

For a given parameter, equilateral triangle has the largest area, so the area of triangle

$$A \leq \frac{\sqrt{3}}{4}\left(\frac{P}{3}\right)^2$$

(for equilateral triangle, $a = b = c = \frac{P}{3}$)

Now, $P^2 = (a + b + c)^2$

$= a^2 + b^2 + c^2 + 2(ab + bc + ca)$

$= 3(a^2 + b^2 + c^2) - 2(a^2 + b^2 + c^2 - ab - bc - ca \leq 3(a^2 + b^2 + c^2)$

(equality holds iff $a = b = c$)

Thus, $A \leq \frac{\sqrt{3}}{4}\frac{a^2 + b^2 + c^2}{3}$

$\Rightarrow$ $a^2 + b^2 + c^2 \geq 4\sqrt{3}A$

80. Let AD be a perpendicular from$A$ on BC.

Then, $\frac{BD}{c} = \cos B$

$\Rightarrow BD$ is rational, similarly AD is rational.

Now, $\sin C = \cos B = \frac{AD}{b} \Rightarrow b$ is rational

Since, $\cos C = \sin B = \frac{DC}{b}$

$\Rightarrow$ DC is rational

Hence, $a = BD + DC$ is rational

Hence, both a and b are rational numbers.

81. (a) : $\because$ Maximum value of

$\sin 2A + \sin 2B + \sin 2C$

and $\sin A + \sin B + \sin C$ is same that is 3.

(b) : $\because \sin\frac{A}{2}\sin\frac{B}{2}\sin\frac{C}{2} \leq \frac{1}{8}$

$\Rightarrow$ $\frac{r}{4R} \leq \frac{1}{8} \Rightarrow R \geq 2r$

(c) $\because \frac{abc}{(a + b + c)} = \frac{4R\Delta}{2s} = 2R \cdot r = R(2r) \leq R^2$

$\therefore$ $R^2 \geq \frac{abc}{(a + b + c)}$ $(\because \beta \geq 2r)$

(d)

$\angle B = 90°$

$\therefore$ $r = (s - b)\tan\frac{B}{2} = s - b$

$R = \frac{b}{2\sin B} = \frac{b}{2}$

$\Rightarrow$ $2R = b$

$\therefore$ $r + 2R = s$

82. $\because D$ is the mid point

$\therefore$ $AB^2 + AC^2 = 2[AD^2 + BD^2]$

$$c^2 + b^2 = 2\left\{l^2 + \left(\frac{a}{2}\right)^2\right\}$$

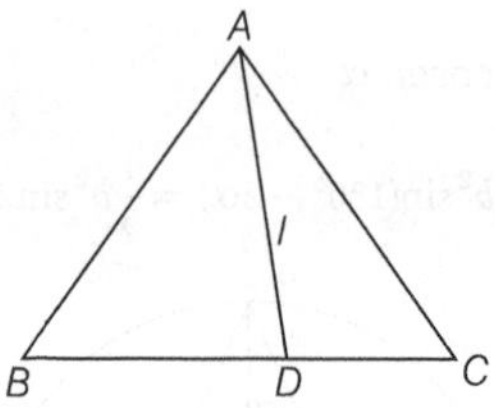

$\Rightarrow$ $c^2 + b^2 = 2l^2 + \frac{a^2}{2}$

$\Rightarrow$ $4l^2 + 2b^2 + 2c^2 - a^2$

$= b^2 + c^2 + (b^2 + c^2 - a^2)$

$= b^2 + c^2 + 2bc\cos A$

$= (b^2 + c^2 - a^2) + a^2 + 2bc\cos A$

$= 2bc\cos A + a^2 + 2bc\cos A$

$= 4bc\cos A + a^2$

83. $\because A = \pi r^2, A_1 = \pi r_1^2, A_2 = \pi r_2^2$ and $A_3 = \pi r_3^2$

(a) :

$$\sqrt{A_1} + \sqrt{A_2} + \sqrt{A_3} = \sqrt{\pi}\,(r_1 + r_2 + r_3)$$

(b) :

$$\frac{1}{\sqrt{A_1}} + \frac{1}{\sqrt{A_2}} + \frac{1}{\sqrt{A_3}}$$

$$= \frac{1}{\sqrt{\pi}}\left(\frac{1}{r_1} + \frac{1}{r_2} + \frac{1}{r_3}\right) = \frac{1}{\sqrt{\pi}}\left(\frac{1}{r}\right)$$

$$= \frac{1}{\sqrt{\pi r^2}} = \frac{1}{\sqrt{A}}$$

(c) :

$$\because \frac{s^2}{\sqrt{\pi r_1 r_2 r_3}} = \frac{s^2}{\sqrt{\pi}\frac{\Delta^3}{(s-a)(s-b)(s-c)}}$$

$$= \frac{s^2(s-a)(s-b)(s-c)}{\Delta^3\sqrt{\pi}} = \frac{s}{\Delta\sqrt{\pi}}$$

$$= \frac{1}{r\sqrt{\pi}} = \frac{1}{\sqrt{\pi r^2}} = \frac{1}{\sqrt{A}}$$

$$= \frac{1}{\sqrt{A_1}} + \frac{1}{\sqrt{A_2}} + \frac{1}{\sqrt{A_3}}$$

(d) :

$\because$ $\sqrt{A_1} + \sqrt{A_2} + \sqrt{A_3}$

$= \sqrt{\pi}(r_1 + r_2 + r_3) = \sqrt{\pi}(4R + r)$

84. $\cos A = \frac{b^2 + c^2 - a^2}{2bc}$

$\Rightarrow$ $c^2 - 2bc\cos A + b^2 - a^2 = 0$

$\therefore$ $c_1 + c_2 = 2b\cos A, c_1c_2 = b^2 - a^2$

Given, $c_1^2 + c_1c_2 + c_2^2 = a^2$

$\Rightarrow$ $(c_1 + c_2)^2 - c_1c_2 = a^2$

$\Rightarrow$ $4b^2\cos^2 A - (b^2 - a^2) = a^2$

$\Rightarrow \quad 4\cos^2 A = 1$

$\Rightarrow \quad 2(1 + \cos 2A) = 1$

or $\quad \cos 2A = -\frac{1}{2} = \cos\frac{2\pi}{3}$

$\therefore \quad 2A = 2n\pi \pm \frac{2\pi}{3}$

$\therefore \quad A = n\pi \pm \frac{\pi}{3}, n \in I$

$\therefore \quad A = \frac{\pi}{3}$ or $\frac{2\pi}{3}$

i.e. $\quad A = 60°$ or $120°$

85. Let D, E and F be the mid points of the sides of ΔABC E and F are mid-points of AC andAB, respectively.

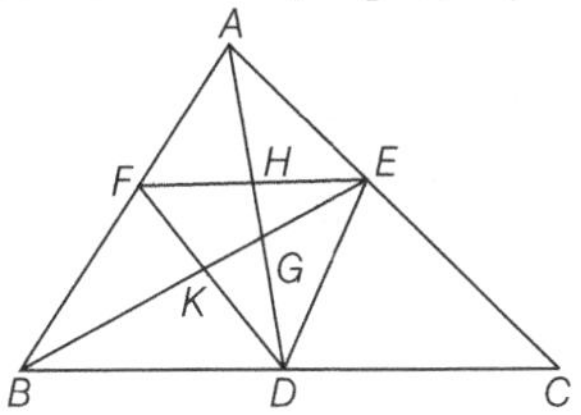

So, $EF \parallel BC$ and $EF = \frac{1}{2}BC$

Thus, AD will also divide EF into equal parts.

Hence, DH is also median of ΔDEF.

Similarly BE, median of ΔABC, is also median of ΔDEF.

These two lines meet at G. So, the centroids of both triangles are same.

The orthocenter, O of ΔDEF is the point of intersection of the perpendiculars DH and EK drawn from D and E, respectively.

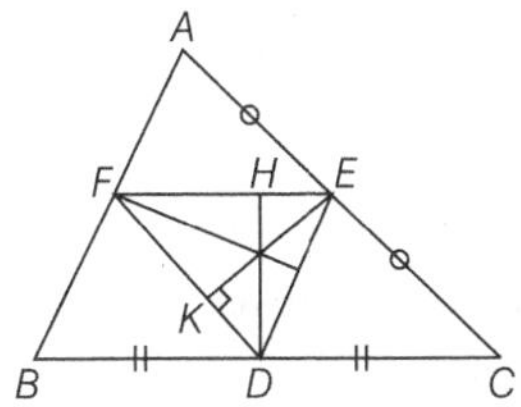

Since, $EF \parallel BC$, DH is perpendicular to BC also,

Similarly, EK is perpendicular to AC.

So, orthocenter of ΔDEF and circumentre of ΔABC is the same points.

86. $(a^2 - 2ac + c^2) + (a^2 - 4ab + 4b^2) = 0$

or $\quad (a - c)^2 + (a - 2b)^2 = 0$

$a = c$ and $a = 2b$

Therefore, the triangle is isoceles.

Also, $\cos B = \frac{a^2 + c^2 - b^2}{2ac} = \frac{7b^2}{8b^2} = \frac{7}{8}$

$\cos A = \frac{b^2 + c^2 - a^2}{2bc} = \frac{1}{4}$

87. $2\Delta = ab\sin C$

$\Rightarrow \quad a^2b^2 - 4\Delta^2 = a^2b^2\cos^2 C$

$\Rightarrow \quad \sqrt{a^2b^2 - 4\Delta^2} + \sqrt{b^2c^2 - 4\Delta^2} + \sqrt{c^2a^2 - 4\Delta^2}$

$= ab\cos C + bc\cos A + ca\cos B$

$= \frac{a^2 + b^2 + c^2}{2}$ (using cosine rule)

88. $2a^2b^2 + 2b^2c^2 = a^4 + b^4 + c^4$

Also, $\quad (a^2 - b^2 + c^2)^2 = a^4 + b^4 + c^4 - 2(a^2b^2 + b^2c^2 - c^2a^2)$

$\therefore \quad (a^2 - b^2 + c^2)^2 = 2c^2a^2$

$\therefore \quad \frac{a^2 - b^2 + c^2}{2ca} = \pm\frac{1}{\sqrt{2}} = \cos B$

or $\quad B = 45°$ or $135°$

89. $AH = 2R\cos A$, $BH = 2R\cos B$, $CH = 2R\cos C$

$\therefore P = 2R(\cos A + \cos b + \cos C) = 2R\left(1 + \frac{r}{R}\right) = 2(R + r)$

We know that in any triangle, $r \le \frac{R}{2}$

90.

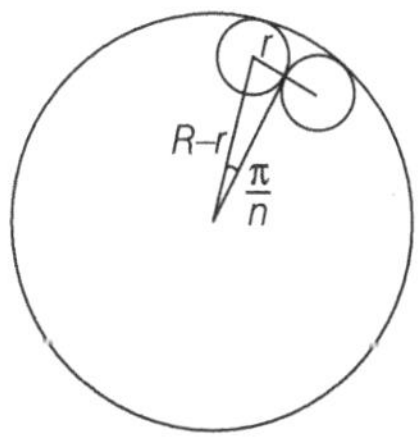

If R is radius of big circle, then $\sin\left(\frac{\pi}{n}\right) = \frac{r}{R - r}$

or $R = r\left(1 + \text{cosec}\left(\frac{\pi}{n}\right)\right) = \frac{r\left[\sin\frac{\pi}{2n} + \cos\frac{\pi}{2n}\right]^2}{\sin\frac{\pi}{n}}$

91. From the cosine formula, $\cos A = \frac{b^2 + c^2 - a^2}{2bc}$

or $\quad b^2 - (2c\cos A)b + (c^2 - a^2) = 0$

Which is s quadratic equation in b. Therefore,

$c\sin A < a < c$

Therefore, two triangles will be obtained. But this is possible when two values of the third side are also obtained. Clearly, two value of sides b will be b_1 and b_2. Let these be the roots of the above equation. Then,

Sum of roots $= b_1 + b_2 = 2c\cos A$ and $b_1b_2 = c^2 - a^2$

92. We have, $a^3 + b^3 + c^3 = c^2a + c^2b + c^3$

$\Rightarrow \quad a^3 + b^3 = c^2(a + b)$

$\Rightarrow \quad (a + b)(a^2 + b^2 - ab) = c^2(a + b)$

$\Rightarrow \quad \frac{a^2 + b^2 - c^2}{2ab} = \frac{1}{2} \Rightarrow \cos C = \frac{1}{2}$

$\angle C = 60°$

Hence, $A + B = 120°$

$\therefore$ ΔABC need not be equilateral.

Statement I is correct and statement II is false.

93. $S_1 : \cos(A-B) = \frac{4}{5}$

$$\Rightarrow \frac{1-\tan^2\left(\frac{A-B}{2}\right)}{1+\tan^2\left(\frac{A-B}{2}\right)} = \frac{4}{5}$$

$$\Rightarrow \frac{2\tan^2\left(\frac{A-B}{2}\right)}{2} = \frac{1}{9}$$

$$\Rightarrow \tan\left(\frac{A-B}{2}\right) = \frac{1}{3} \quad [\text{as } a > b \Rightarrow A > B]$$

Using, $\tan\left(\frac{A-B}{2}\right) = \left(\frac{a-b}{a+b}\right)\cot\frac{C}{2}$, we get

$$\frac{1}{3} = \frac{6-3}{6+3}\cot\frac{C}{2} \Rightarrow \cot\frac{C}{2} = 1$$

$$\Rightarrow \angle C = 90^\circ \Rightarrow \text{Statement I is false.}$$

S_2 : Using sine law in ΔABC, we get

$$\frac{a}{\sin A} = \frac{c}{\sin C}$$

$$\Rightarrow \frac{a}{\sin A} = \frac{\sqrt{a^2+b^2}}{\sin\frac{\pi}{2}}$$

$$\Rightarrow \frac{6}{\sin A} = \sqrt{45} \Rightarrow \sin A = \frac{2}{\sqrt{5}}$$

$\Rightarrow$ Statement II is true.

94. Statement II can be proved (by using $A + B + C = \pi$) to be true. (conditional identities) From which we get

$3 - 2(\sin^2 A + \sin^2 B + \sin^2 C) = -1 - 4\cos A\cos B\cos C$

$\Rightarrow 3 - 2(2) = -1 - 4\cos A\cos B\cos C$

$\Rightarrow \cos A\cos B\cos C = 0$

$\Rightarrow$ one of the angles A, B, C is equal to 90°.

95. $\because A + C = 180^\circ, B + D = 180^\circ$

$\therefore \cos A = -\cos C, \cos B = -\cos D$

$\Rightarrow \cos A + \cos B + \cos C + \cos D$

$= (\cos A + \cos C) + (\cos B + \cos D)$

$= 0 + 0$

$= 0$

$\therefore \quad \Sigma\cos A = 0$

and $\quad \sin A = \sin C$

and $\quad \sin B = \sin D$

Then, $\quad \Sigma\sin A = \sin A \ \sin B + \sin C + \sin D$

$= 2(\sin C + \sin D) \neq 0$

96. Area of ΔABC = Area of ΔABD + Area of ΔACD

$$\Rightarrow \frac{1}{2}bc\sin A = \frac{1}{2}c\cdot AD\sin\left(\frac{A}{2}\right) + \frac{1}{2}\cdot b\cdot AD\sin\left(\frac{A}{2}\right)$$

$$\Rightarrow AD = \frac{bc\sin A}{(b+c)\sin\left(\frac{A}{2}\right)} = \frac{2bc\cos\left(\frac{A}{2}\right)}{b+c}$$

Also, $$(AD)^2 = \frac{4b^2c^2}{(b+c)^2} \times \frac{s(s-a)}{bc}$$

$$= \frac{bc2s(2s-2a)}{(b+c)^2}$$

$$= bc\cdot\frac{(a+b+c)(b+c-a)}{(b+c)^2}$$

$$= \frac{bc\{(b+c)^2 - a^2\}}{(b+c)^2}$$

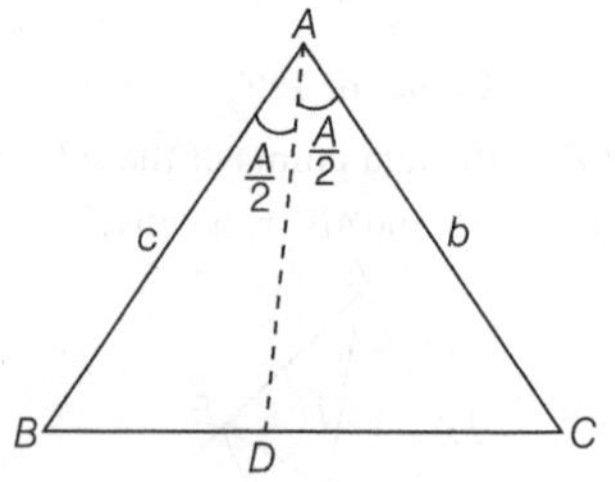

97. $ICI_1 = \frac{\pi}{2}$

$IBI_1 = \frac{\pi}{2}$,

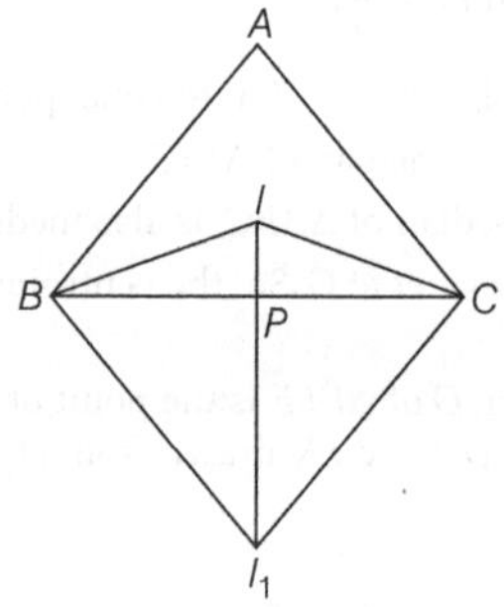

$\therefore BICI_1$, is cyclic

Quadrilateral $BP\cdot PC = IP\cdot I_1P$

Hence, (c) is the correct answer.

98. Statement II is true.

Statement I $\tan A = \tan B = \tan C$

$A = B = C$ ie, $a = b = c$

$r_1 = r_2 = r_3$

$$\therefore \quad \frac{r_1 + r_2 + r_3}{r} = 3\cdot\frac{r_1}{r}$$

$$= 3\cdot\frac{\frac{\Delta}{s-a}}{\frac{\Delta}{s}} = 3\frac{a+b+c}{b+c-a} = 9$$

Hence, (a) is the correct answer.

99. Statement II is False. Because if $p = 2, q = 4, r = 6$, then $p : q : r = 1 : 2 : 3$ but $p \neq 1$

For statement I, let $\tan A = k, \tan B = 2k, \tan C = 3k$, then from $\tan A + \tan B + \tan C = \tan A\tan B\tan C$ (in a triangle) we get $6k = 6k^3 \Rightarrow k = 0, 1, -1$ but $k = 0, -1$ is not possible. So $k = 1 \Rightarrow \tan A = 1 \Rightarrow A = 45^\circ$. So, statement I is correct.

100. $a^2 = b^2 + c^2$, then, $\angle A = \dfrac{\pi}{2}$

$\therefore \quad a = 2R\sin A = 2R$

$\Rightarrow \quad \dfrac{a^2 + b^2 + c^2}{R^2} = \dfrac{a^2 + a^2}{R^2} = \dfrac{2a^2}{R^2} = \dfrac{2(2R)^2}{R^2} = 8$

101. $\because \angle A_1OA_2 = \dfrac{2\pi}{n}$

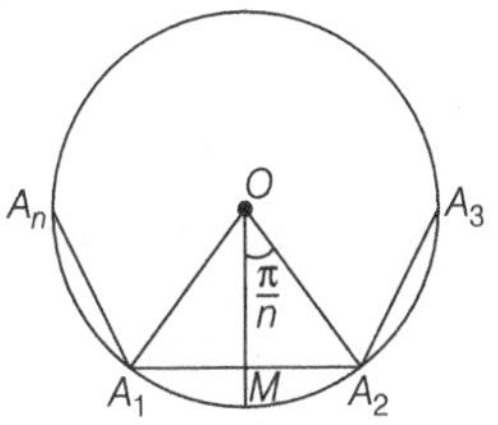

$\therefore \quad \angle A_1OM = \angle A_2OM = \dfrac{\pi}{n}$

$\therefore \quad A_1A_2 = 2MA_2$

$= 2a\sin\left(\dfrac{\pi}{n}\right)$ cm

$\therefore$ Perimeter $= 2an\ \text{in}\left(\dfrac{2\pi}{2n}\right)$ cm

$= 2an\sin\left(\dfrac{360^\circ}{2n}\right)$ cm

For, $\quad n = 5$

Perimeter is $10a\sin(36^\circ)$ cm

102. Since L.H.S of the inequality in statement II is a symmetric function of sines of the angles of the triangle, its maximum value is attained. When $A = B = C = 60^\circ$ and hence the statement II is true. Statement I is true if

$2R(\sin A\cos A + \sin B\cos B + \sin C\cos C)$

$\leq R(\sin A + \sin B + \sin C)$

or if $4\sin A\sin B\sin C \leq 4\cos\left(\dfrac{A}{2}\right)\cos\left(\dfrac{B}{2}\right)\cos\left(\dfrac{C}{2}\right)$

[From conditional identities]

or if $\sin\left(\dfrac{A}{2}\right)\sin\left(\dfrac{B}{2}\right)\sin\left(\dfrac{C}{2}\right) \leq \dfrac{1}{8}$

Which is true by statement II

So, statement I is correct

103 $\because \cos^2\dfrac{A}{2} + \cos^2\dfrac{B}{2} + \cos^2\dfrac{C}{2} = \dfrac{1}{2}(3 + \cos A + \cos B + \cos C)$

$\because \cos A + \cos B + \cos C = 1 + 4\sin\dfrac{A}{2}\sin\dfrac{B}{2}\sin\dfrac{C}{2}$ (from identity)

Then, $\cos^2\dfrac{A}{2} + \cos^2\dfrac{B}{2} + \cos^2\dfrac{C}{2}$

$= 2\left(1 + \sin\left(\dfrac{A}{2}\right)\sin\left(\dfrac{B}{2}\right)\sin\left(\dfrac{C}{2}\right)\right)$

$\because \quad 2\left(1 + \sin\left(\dfrac{A}{2}\right)\sin\left(\dfrac{B}{2}\right)\sin\left(\dfrac{C}{2}\right)\right)$

$= y\left(x^2 + \dfrac{1}{x^2}\right) \geq 2y$ $\quad (\because \text{AM} \geq \text{GM})$

$\therefore \quad y \leq 1 + \sin\left(\dfrac{A}{2}\right)\sin\left(\dfrac{B}{2}\right)\sin\left(\dfrac{C}{2}\right) \leq 1 + \dfrac{1}{8}$

i.e. $\quad y \leq \dfrac{9}{8}$

$\therefore$ Maximum value of y is $\dfrac{9}{8}$

104. $\because a\cos A + b\cos B + c\cos C \leq s$

$\Rightarrow 2R\sin A\cos A + 2R\sin B\cos B + 2R\sin C\cos C \leq \left(\dfrac{a + b + c}{2}\right)$

$\Rightarrow R(\sin 2A + \sin 2B + \sin 2C) \leq \dfrac{2}{2}(R\sin A + R\sin B + R\sin C)$

$\Rightarrow (\sin 2A + \sin 2B + \sin 2C) \leq (\sin A + \sin B + \sin C)$

$\Rightarrow (4\sin A\sin B\sin C) \leq \left\{4\cos\left(\dfrac{A}{2}\right)\cos\left(\dfrac{B}{2}\right)\cos\left(\dfrac{C}{2}\right)\right\}$

(from identities)

$\Rightarrow 4\left(2\sin\left(\dfrac{A}{2}\right)\cos\left(\dfrac{A}{2}\right)\right)\left(2\sin\left(\dfrac{B}{2}\right)\cos\left(\dfrac{B}{2}\right)\right)\left(2\sin\left(\dfrac{C}{2}\right)\cos\left(\dfrac{C}{2}\right)\right)$

$\leq 4\cos\left(\dfrac{A}{2}\right)\cos\left(\dfrac{B}{2}\right)\cos\left(\dfrac{C}{2}\right)$

$\Rightarrow \therefore \sin\left(\dfrac{A}{2}\right)\sin\left(\dfrac{B}{2}\right)\sin\left(\dfrac{C}{2}\right) \leq \dfrac{1}{8}$

105. We know that, in any triangle

$$\frac{s^2}{3\sqrt{3}} \geq \Delta \qquad \text{...(i)}$$

Now, $\dfrac{a^2 + b^2 + c^2}{3} \geq \left(\dfrac{a + b + c}{3}\right)^2 = \left(\dfrac{2s}{3}\right)^2$

$= \dfrac{4}{9}s^2 \geq \dfrac{4}{9} \times 3\sqrt{3}\Delta$ [from Eq. (i)]

$\therefore \quad \dfrac{a^2 + b^2 + c^2}{\Delta} \geq 4\sqrt{3}$

106. $\because A, C_1, G$ and B_1 are concyclic then

$BG \cdot BB_1 = BC_1 \cdot BA$

$\Rightarrow \quad \dfrac{2}{3}BB_1 \cdot BB_1 = \dfrac{C}{2} \cdot C$

$\Rightarrow \quad \dfrac{2}{3}(BB_1)^2 = \dfrac{c^2}{2}$

$\Rightarrow \quad \dfrac{2}{3}\left(\dfrac{2a^2 + 2c^2 - b^2}{4}\right) = \dfrac{c^2}{2}$

$\Rightarrow \quad 2a^2 + 2c^2 - b^2 = 3c^2$ or $2a^2 = b^2 + c^2$

$\therefore \quad c^2, a^2, b^2$ are in AP.

107. $R_1R_2 + R_2R_3 + R_1R_3 = 3R^2$ $\quad [\because R_1 = R_2 = R_3 = R]$

108. $AH = 2R\cos A$

$BH = 2R\cos B$

$\Delta = \dfrac{1}{2}(AH)(BH)\sin(180^\circ - C)$

$\Delta = 2R^2\cos A\cos B\sin C$

109. $2R^2 \cos A \cos B \sin C : 2R^2 \cos B`cosB \cos C \sin C$
$= \cos A : \cos B \cos C$

110. Let $DE = DF$ then

$BDPE$ is cyclic with BP as diameter

$\Rightarrow \quad BP = \dfrac{DE}{\sin B}$

$\Rightarrow \quad DF = BP \sin B$ and $DE = CP \sin C$

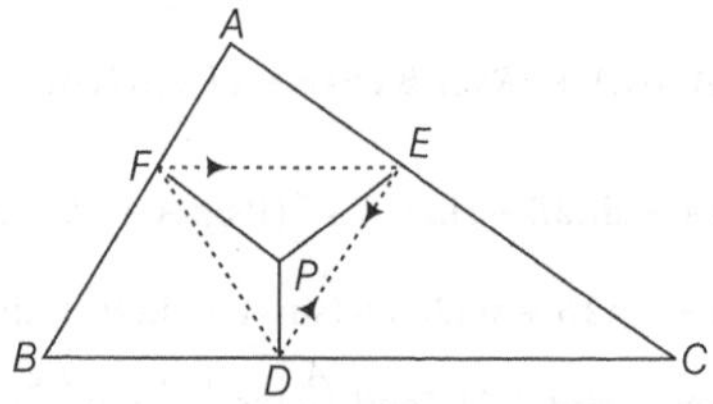

But $\quad DE = DF$

$\Rightarrow \quad \dfrac{BP}{CP} = \dfrac{\sin C}{\sin B} = \dfrac{c}{b}$

Hence, locus of P is arcs of 3 circles

111. $\dfrac{PB}{PC} = \dfrac{c}{b} = \dfrac{5}{7},$

Hence, P will lie on circle (C).

112. For $DE = DF = EF$, P is a point(s) where the three arcs intersect

113. ΔDEF is the pedal triangle of ΔABC.

Sides of the Pedal triangle are
$a \cos A, b \cos B, c \cos C$ i.e. $R \sin 2A$, $R \sin 2B$, $R \sin 2C$.

$\therefore$ Required ratio $= \dfrac{R^3 \sin 2A \sin 2B \sin 2C}{8R^3 \sin A \sin B \sin C} = \cos A \cos B \cos C$

114. Also, orthocentre of ΔABC is incentre of DEF.

115. Circum-radius of pedal triangle is $\dfrac{R}{2}$.

Sol. (Q.Nos. 116 to 118)

$$\cos(A - B) = \frac{4}{5} \Rightarrow \frac{1 - \tan^2 \frac{A - B}{2}}{1 + \tan^2 \frac{A - B}{2}} = \frac{4}{5}$$

$$\Rightarrow \quad \frac{2 \tan^2 \frac{A - B}{2}}{2} = \frac{1}{9}$$

$$\Rightarrow \quad \tan \frac{A - B}{2} = \frac{1}{3} \Rightarrow \frac{1}{3} = \frac{6 - 3}{6 + 3} \cot \frac{C}{2}$$

$$\Rightarrow \quad \cos \frac{C}{2} = 1 \Rightarrow C = 90^\circ$$

Area of triangle $= \dfrac{1}{2} ab \sin C \Rightarrow$ Area $= \dfrac{1}{2} \times 6 \times 3 \times 1 = 9$

$$\frac{a}{\sin A} = \frac{\sqrt{a^2 + b^2}}{1} \Rightarrow \frac{6}{\sin A} = \sqrt{45} \Rightarrow \sin A = \frac{2}{\sqrt{5}}$$

116. (b)

117. (b)

118. (b)

119. $\because \cos A = \dfrac{b^2 + c^2 - a^2}{2bc}$ or $c^2 - 2bc \cos A + b^2 - a^2 = 0$

For real roots, $D > 0$

$\Rightarrow \quad 4b^2 \cos^2 A - 4 \cdot 1 \cdot (b^2 - a^2) > 0$

$\Rightarrow \quad a^2 > b^2 \sin^2 A$

$\therefore \quad b \sin A < 0$

Also, $\quad c_1 + c_2 = 2b \cos A \quad$ ...(i)

and $\quad c_1 c_2 = b^2 - a^2 \quad$ ...(ii)

Then, $|c_1 - c_2| = \sqrt{(c_1 + c_2)^2 - 4c_1 c_2}$

$= \sqrt{4b^2 \cos^2 A - 4(b^2 - a^2)}$

$= 2\sqrt{(a^2 - b^2 \sin^2 A)} \quad$ ...(iii)

Consider smaller root, say c_2

Then, $\quad 2c_2 = (c_1 + c_2) - (c_1 - c_2)$

$= 2b \cos A - 2\sqrt{(a^2 - b^2 \sin^2 A)}$

$\Rightarrow \quad c_2 = b \cos A - \sqrt{(a^2 - b^2 \sin^2 A)} > 0$

$\Rightarrow \quad b \cos A > \sqrt{(a^2 - b^2 \sin^2 A)}$

$\Rightarrow \quad b^2 \cos^2 A > a^2 - b^2 \sin A$

$\Rightarrow \quad b^2 > a^2$ or $b > a$

Hence, two different triangle are possible if $b \sin A < a$ and $b > a$

120. From Eq. (iii), $|c_1 - c_2| = 2\sqrt{(a^2 - b^2 \sin^2 A)}$

121. $c_1^2 - 2c_1 c^2 \cos 2A + c_2^2 = (c_1 + c_2)^2 - 2c_1 c_2 (1 + \cos 2A)$

$= (c_1 + c_2)^2 - 4c_1 c_2 \cos^2 A$

$= (2b \cos A)^2 - 4(b^2 - a^2) \cos^2 A$

[from Eqs. (i) and (ii)]

$= 4 \cos^2 A \{b^2 - (b^2 - a^2)\}$

$= 4a^2 \cos^2 A$

122. $B_1 D = B_2 D = \left(\dfrac{c_1 - c_2}{2}\right)$

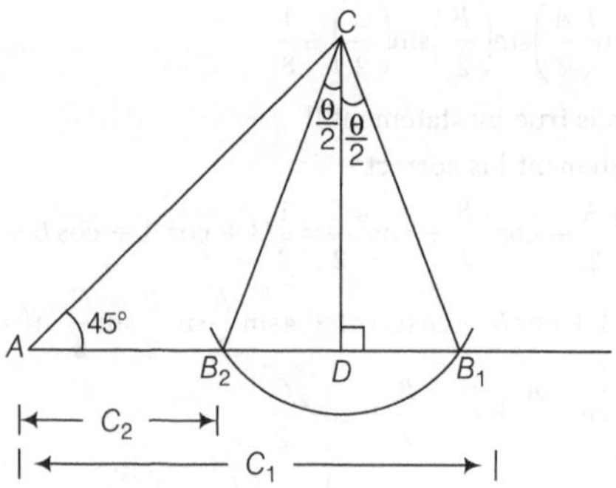

$\therefore \quad AD = AB_2 + B_2 D$

$= c_2 + \left(\dfrac{c_1 - c_2}{2}\right)$

$\therefore \quad AD = \left(\dfrac{c_1 + c_2}{2}\right)$

Now, in ΔACD, $\tan 45° = \dfrac{CD}{AD} = 1$

$$\therefore \quad CD = AD = \left(\frac{c_1 + c_2}{2}\right)$$

$$\therefore \text{In } \Delta B_2CD, \quad \tan\left(\frac{\theta}{2}\right) = \frac{B_2D}{CD} = \frac{\left(\frac{c_1 - c_2}{2}\right)}{\left(\frac{c_1 + c_2}{2}\right)} = \left(\frac{c_1 - c_2}{c_1 + c_2}\right)$$

$$\therefore \quad \cos\theta = \frac{1 - \tan^2\left(\frac{\theta}{2}\right)}{1 + \tan^2\left(\frac{\theta}{2}\right)} = \frac{1 - \left(\frac{c_1 - c_2}{c_1 + c_2}\right)^2}{1 + \left(\frac{c_1 - c_2}{c_1 + c_2}\right)^2}$$

$$= \frac{2c_1c_2}{(c_1^2 + c_2^2)}$$

123. $\because \cos A = \dfrac{b^2 + c^2 - a^2}{2bc}$

$$c^2 - 2bc\cos A + b^2 - a^2 = 0$$

$$\Rightarrow c^2 - 2(m + 1)\lambda c\cos A + (m^2 + 2m - 3)\lambda^2 = 0$$

$$\left(\because \frac{a}{2} = \frac{b}{m + 1} = \lambda \text{ (say)}\right)$$

$$\therefore \quad c_1 + c_2 = 2(m + 1)\lambda\cos A$$

or $$(c_1 + c_2)^2 = 4(m + 1)^2\lambda^2 \cdot \frac{(m - 1)(m + 3)}{4m}$$

$$= \frac{(m + 1)^2(m - 1)(m + 3)}{m} \quad \text{...(iv)}$$

and $(c_1 - c_2)^2 = (c_1 + c_2)^2 - 4c_1c_2$

$$= 4(m + 1)^2\lambda^2\cos^2 A - 4(m^2 + 2m - 3)\lambda^2$$

$$= 4\lambda^2\left\{(m + 1)^2 \cdot \frac{(m - 1)(m + 3)}{4} - (m + 3)(m - 1)\right\}$$

$$= 4\lambda^2\frac{(m - 1)(m + 3)}{4m}(m - 1)^2$$

$$= \frac{\lambda^2(m - 1)^2(m + 3)}{m} \quad \text{...(v)}$$

From Eqs. (iv) and (v),

$$\left(\frac{c_1 + c_2}{c_1 - c_2}\right)^2 = \left(\frac{m + 1}{m - 1}\right)^2$$

$$\Rightarrow \quad \frac{c_1 + c_2}{c_1 - c_2} = \pm\frac{m + 1}{m - 1}$$

$$= \frac{m + 1}{m - 1} \text{ or } \frac{m + 1}{1 - m}$$

At last using componendo and dividendo rule, we get

$$\frac{2c_1}{2c_2} = \frac{2m}{2} \text{ or } \frac{2}{m} = m \text{ or } \frac{1}{m}$$

Alternate method :

We have, $\cos A = \dfrac{b^2 + c^2 - a^2}{2bc}$

$$\Rightarrow \quad 2bc\cos A = b^2 + c^2 - a^2$$

$$\Rightarrow \quad (m + 1)ac\cos A = \frac{(m + 1)^2}{4}a^2 + c^2 - a^2$$

$$(\because 2b = (m + 1)a)$$

$$\Rightarrow \quad (m + 1)ac\cos A = \{(m + 1)^2 - 4\}\frac{a^2}{4} + c^2$$

$$mac\cos A + ac\cos A = (m - 1)(m + 3)\frac{a^2}{4} + c^2$$

$$= ma^2\cos^2 A + c^2$$

$$\left\{\because \frac{(m - 1)(m + 3)}{4} = m\cos^2 A\right\}$$

$$\Rightarrow \quad ma\cos A(c - a\cos A) - c(c - a\cos A) = 0$$

$$\Rightarrow \quad (c - a\cos A)(ma\cos A - c) = 0$$

$$\therefore \quad c = a\cos A \text{ and } c = ma\cos A$$

If $\quad c_2 = a\cos A$ and $c_1 = ma\cos A$

$$\therefore \quad \frac{c_1}{c_2} = m$$

and if $c_1 = a\cos A$ and $c_2 = ma\cos A$

Then, $\quad \dfrac{c_1}{c_2} = \dfrac{1}{m}$

$$\therefore \quad \frac{c_1}{c_2} = m \text{ or } \frac{1}{m}$$

124. $\dfrac{bx}{c} + \dfrac{cy}{a} + \dfrac{az}{b} = b\sin B + c\sin C + a\sin A$

$$= \frac{b^2 + c^2 + a^2}{2R}$$

$$\therefore \quad k = 2R$$

Hence, (c) is the correct answer.

125. $\dfrac{1}{x^2} + \dfrac{1}{y^2} + \dfrac{1}{z^2} = \dfrac{a^2}{4\Delta^2} + \dfrac{b^2}{4\Delta^2} + \dfrac{c^2}{4\Delta^2} = \dfrac{a^2 + b^2 + c^2}{4\Delta^2}$

$$\cot A + \cot B + \cot C = \frac{R}{abc}$$

$$(b^2 + c^2 - a^2 + c^2 + a^2 - b^2 + a^2 + b^2 - c^2)$$

$$= \frac{R}{abc}(b^2 + c^2 + a^2)$$

$$= \frac{R}{abc}\left(\frac{4\Delta^2}{x^2} + \frac{4\Delta^2}{y^2} + \frac{4\Delta^2}{z^2}\right)$$

$$= \frac{4\Delta^2R}{abc}\left(\frac{1}{x^2} + \frac{1}{y^2} + \frac{1}{z^2}\right)$$

$$= \frac{4\Delta R}{abc} \cdot \Delta\left(\frac{1}{x^2} + \frac{1}{y^2} + \frac{1}{z^2}\right)$$

$$= \Delta\left(\frac{1}{x_2} + \frac{1}{y^2} + \frac{1}{z^2}\right)$$

$$\therefore \quad k = \Delta$$

Hence, (c) is the correct answer.

126. $\Sigma\dfrac{c\sin B + b\sin C}{x}, \Sigma\dfrac{x + x}{x} = 6$

Hence, (d) is the correct answer.

Sol. (Q.Nos 127 to 131)

Here, ΔMNL is pedal triangle, then

$$MN = a\cos A$$

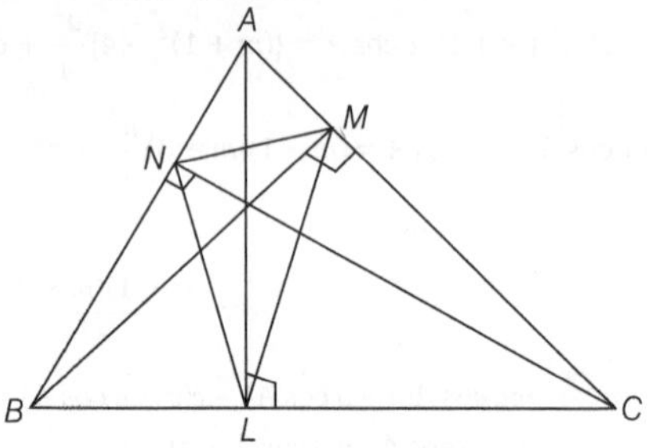

$NL = b\cos B$

and $ML = c\cos C$

and $\angle MLN = \pi - 2A$, $\angle LMN = \pi - 2B$

and $\angle MNL = \pi - 2C$

127. $\lambda = MN + NL + LM$

$$= a\cos A + b\cos B + c\cos C$$
$$= R(\sin 2A + \sin 2B + \sin 2C)$$
$$= 4R\sin A\sin B\sin C$$
$$= 4R\left(\frac{a}{2R}\right)\left(\frac{b}{2R}\right)\left(\frac{c}{2R}\right)$$
$$= \frac{abc}{2R^2} = \frac{4R\Delta}{2R^2} = \frac{2\Delta}{R}$$

and $$\mu = 2s = \frac{2\Delta}{r}$$

$\therefore$ $$\frac{\lambda}{\mu} = \frac{r}{R}$$

128. In ΔABM,

$$\cos A = \frac{AM}{AB} = \frac{AM}{c}$$

$\therefore$ $AM = c\cos A$ similarly $AN = b\cos A$

$\therefore$ $\Delta_1 =$ Area of ΔAMN

$$= \frac{1}{2}\cdot AM\cdot AN\cdot\sin A$$
$$= \frac{1}{2}(c\cos A)(b\cos A)(\sin A)$$
$$= \left(\frac{1}{2}bc\sin A\right)\cos^2 A = \Delta\cos^2 A$$
$$= \Delta\cos^2 A$$

Similarly, $\Delta_2 = \Delta\cos^2 B$ and $\Delta_3 = \Delta\cos^2 C$

$\therefore \Delta_1 + \Delta_2 + \Delta_3 = \Delta(\cos^2 A + \cos^2 B + \cos^2 C)$

$$= \Delta(1 - 2\cos A\cos B\cos C)$$

(from identity)

129. $\Delta' =$ Area of ΔLMN

$$= \frac{1}{2}\cdot ML\cdot NL\cdot\sin(\pi - 2A)$$
$$= \frac{1}{2}\cdot(c\cos C)(b\cos B)(\sin 2A)$$
$$= \frac{1}{2}bc\cos B\cos C(2\sin A\cos A)$$
$$= \left(\frac{1}{2}bc\sin A\right)(2\cos A\cos B\cos C)$$

$\therefore$ $$\Delta' = \Delta(2\cos A\cos B\cos C)$$

$\Rightarrow$ $$\frac{\Delta'}{\Delta} = 2\cos A\cos B\cos C$$

130. Let R' be the circumradius of ΔLMN, then $R' = \dfrac{MN}{2\sin(\angle MLN)}$

$$= \frac{a\cos A}{2\sin(\pi - 2A)}$$
$$= \frac{2R\sin A\cos A}{2\sin 2A} = \frac{R\sin 2A}{2\sin 2A} = \frac{R}{2}$$

131. $\because r' = 4R'\sin\left(\frac{L}{2}\right)\sin\left(\frac{M}{2}\right)\sin\left(\frac{N}{2}\right)$

Here, $L = \angle MLN = \pi - 2A$,

$M = \angle LMN = \pi - 2B$

and $N = \angle MNL = \pi - 2C$

and $R' = \dfrac{R}{2}$ (from Q. 4)

132. (A) $\to$ (r), (B) $\to$ (p), (C) $\to$ (s), (D) $\to$ (q)

(A) $2(r + R) = 2\left((s - c)\tan\frac{C}{2} + \frac{c}{2}\right)$

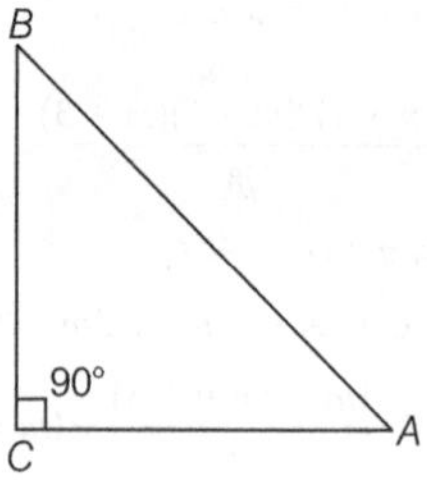

$$2\left(s - \frac{c}{2}\right) = a + b$$

(B) $\sin C = \dfrac{l}{b}$ ...(i)

$\sin B = \dfrac{n}{a}$ ...(ii)

$\sin A = \dfrac{m}{c}$ ...(iii)

$$2R\left(\frac{bl}{c}\right) + 2R\left(\frac{cm}{a}\right) + 2R\left(\frac{an}{b}\right)$$
$$\frac{c}{\sin c}\frac{bl}{c} + \frac{a}{\sin A}\frac{cm}{a} + \frac{b}{\sin B}\frac{an}{b}$$

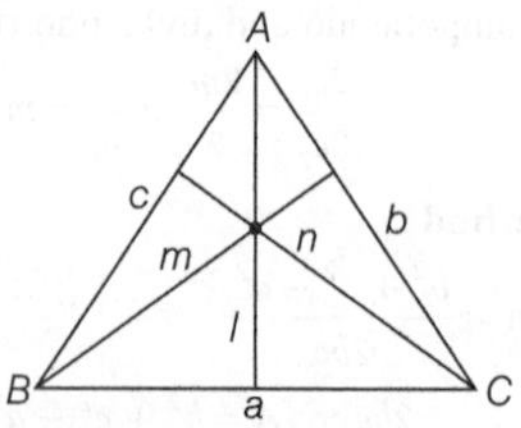

From Eqs. (i), (ii), (iii)

$$2R\left(\frac{bl}{c}+\frac{cm}{a}+\frac{an}{a}\right)=a^2+b^2+c^2$$

$$\therefore \quad \sqrt{2R\left(\frac{bl}{c}+\frac{cm}{a}+\frac{an}{b}\right)+2ab+2bc+2ac}=a+b+c$$

(C) $R=\dfrac{b}{2\sin B}=\dfrac{c}{2\sin C}$

Now, $\quad Rb^2\sin 2C+Rc^2\sin 2B$

$=b^2c\cos C+Rc^2b\cos B$

$=bc(b\cos C+c\cos B)$

abc

(D) We have,

$$4e\sin\left(\frac{A+B}{2}\right)\sin\frac{A-B}{2}$$

$$=2R\left(2\sin\frac{A+B}{2}\sin\frac{A-B}{2}\right)$$

$$=2R(\cos B-\cos A)=C\left(\frac{a}{c}-\frac{b}{c}\right) \qquad \left[\because R=\frac{C}{2}\right]$$

$$=a-b$$

133. (A) $\to$ (p, q, r); (B) $\to$ (p, q, r); (C) $\to$ (q, r); (D) $\to$ q

(A) $c^2+a^2-2ac=b^2-ac$

$$\Rightarrow \quad \frac{c^2+a^2-b^2}{2ac}$$

$$=\frac{1}{2}\Rightarrow \cos B=\frac{1}{2}\Rightarrow B=60^\circ \text{ and } \cos B+\sin C=\frac{3}{2}$$

$\Rightarrow \sin C=1\Rightarrow C=90^\circ$ and $A=30^\circ$

(B) $A+B+C=180^\circ,\ B=\dfrac{A+C}{2}$

$\Rightarrow\ B=60^\circ,\ C+A=120^\circ$

$C=3A\Rightarrow A=30^\circ,\ C=90^\circ$

(C) Length of the bisector of angle B is

$$\frac{2ca}{c+a}\cos(B/2)$$

$$=\frac{\sqrt{3}ca}{c+a}$$

$$\Rightarrow \quad \cos\left(\frac{B}{2}\right)=\frac{\sqrt{3}}{2}=\cos 30^\circ$$

$\Rightarrow\ B=60^\circ,\ a=b\Rightarrow A=B=60^\circ=C$

(D) We have, $(a+b+c)(a+2b+c)$

$=3(a+b)(b+c)$

$\Rightarrow \quad (a+c)^2+3b(a+c)+2b^2$

$=3(ab+bc+ca+b^2)$

$$\Rightarrow \quad \frac{a^2+c^2-b^2}{ac}=1$$

$$\Rightarrow \quad \cos B=\frac{1}{2}$$

$\Rightarrow \quad B=60^\circ$

134. (A) $\to$ (p), (B) $\to$ (q), (C) $\to$ (r), (D) $\to$ (s)

(A) $\quad (b+c)^2-a^2=\lambda bc$

or $\quad b^2+c^2-a^2=(\lambda-2)bc$

$$\frac{(b^2+c^2-a^2)}{2bc}=\frac{\lambda-2}{2}$$

$$\cos A=\frac{\lambda-2}{2}<1$$

or $\quad \lambda-2<2$

$\lambda<4$

or $\quad \lambda=3$

(B) $\tan A+\tan B+\tan C=C=9$ in any triangle

$\tan A+\tan B+\tan C=\tan A\tan B\tan C$

$$\frac{\tan^2 A+\tan^2 B+\tan^2 C}{3}\geq 2(\tan A\tan B\tan C)^{2/3}$$

$k\geq 3(9)^{2/3}$

$k\geq 9\cdot 3^{1/3}$

(C) Since, the line joining the circumcentre to the incentre is parallel to BC

$\therefore \quad r=R\cos A$

$\therefore \quad 4R\sin\dfrac{A}{2}\sin\dfrac{B}{2}\sin\dfrac{C}{2}=R\cos A$

$\therefore \quad -1+\cos A+\cos B+\cos C=\cos A$

$\therefore \quad \cos B+\cos C=1$

(D) $a=5,\ b=4$

$$\cos(A-B)=\frac{31}{32}$$

$$\tan\frac{A-B}{2}=\frac{a-b}{a+b}\cot\frac{C}{2}$$

$$=\frac{1}{9}\cot\frac{C}{2}$$

$$\cos(A-B)=\frac{1-\tan^2\frac{A-B}{2}}{1+\tan^2\frac{A-B}{2}}$$

$$\frac{31}{32}=\frac{1-\frac{1}{81}\cot^2\frac{C}{2}}{1+\frac{1}{81}\cot^2\frac{C}{2}}$$

$$31+\frac{31}{81}\cot^2\frac{C}{2}=32-\frac{32}{81}\cot^2\frac{C}{2}$$

$$\frac{7}{9}\cot^2\frac{C}{2}=1$$

$$\cot^2\frac{C}{2}=\frac{9}{7}=\tan^2\frac{C}{2}=\frac{7}{9}$$

$$\cos C=\frac{1-\tan^2\frac{C}{2}}{1+\tan^2\frac{C}{2}}=\frac{1-\frac{7}{9}}{1+\frac{7}{9}}=\frac{1}{8}$$

$$\because \quad C=\sqrt{a^2+b^2-2ab\cos C}$$

$$=\sqrt{25+16-5}=6$$

135. (A) $\to (p, t)$; (B) $\to (q, r)$; (C) $\to (q, r, t)$

(A) $\because 2a^2 + b^2 + c^2 = 2ac + 2ab$

$\Rightarrow \quad (a^2 + b^2 - 2ab) + (a^2 + c^2 - 2ac) = 0$

$(a - b)^2 + (a - c)^2 = 0$

Which is possible only when

$a - b = 0, a - c = 0$

$\therefore \quad a = b = c$

$\Rightarrow \Delta\, ABD$ is equilateral

$\therefore \quad \angle A = \angle B = \angle C = 60^\circ$

(B) $\because a^2 + b^2 + c^2 = b\sqrt{2}(c + a)$

$\Rightarrow \quad a^2 + b^2 + c^2 - bc\sqrt{2} - ab\sqrt{2} = 0$

$\Rightarrow \quad 2a^2 + 2b^2 + 2c^2 - 2bc\sqrt{2} - 2ab\sqrt{2} = 0$

$(b^2 - 2bc\sqrt{2} + 2c^2) + (b^2 - 2ab\sqrt{2} + 2a^2) = 0$

$\Rightarrow \quad (b - c\sqrt{2})^2 + (b - a\sqrt{2})^2 = 0$

Which is the possible only when

$b - c\sqrt{2} = 0, b - a\sqrt{2} = 0$

$\therefore \quad b = c\sqrt{2}, b = a\sqrt{2}$

i.e. $\quad b^2 + b^2 = (c\sqrt{2})^2 + (a\sqrt{2})^2$

$\Rightarrow \quad c^2 + a^2 = b^2$

$\therefore \quad \cos B = \dfrac{c^2 + a^2 - b^2}{2ac} = 0$

$\therefore \quad \angle B = 90^\circ$ also $c\sqrt{2} = a\sqrt{2}$

$\Rightarrow \quad c = a$

Then, $\quad \angle A = \angle C = 45^\circ$

Hence, $\quad \angle A = \angle = 45^\circ$

and $\quad \angle B = 90^\circ\, (Q, S)$

(C) $\because \quad a^2 + b^2 + c^2 = bc + ca\sqrt{3}$

$\Rightarrow \quad a^2 + b^2 + c^2 - bc - ca\sqrt{3} = 0$

$\Rightarrow \quad \left(\dfrac{c\sqrt{3}}{2} - a\right)^2 + \left(\dfrac{c}{2} - b\right)^2 = 0$

Which is possible only when

$\dfrac{c\sqrt{3}}{2} - a = 0$ and $\dfrac{c}{2} - b = 0$

$\therefore \quad a = \dfrac{c\sqrt{3}}{2}$ and $b = \dfrac{c}{2}$

Then, $\quad a^2 + b^2 = \dfrac{3c^2}{4} + \dfrac{c^2}{4} = c^2$

$\because \quad \cos C = \dfrac{a^2 + b^2 - c^2}{2ab} = 0$

$\because \quad \angle C = 90^\circ$

Also, $\quad \dfrac{a}{\sin A} = \dfrac{c}{\sin C} \Rightarrow \dfrac{\frac{c\sqrt{3}}{2}}{\sin A} = \dfrac{c}{1}$

$\therefore \quad \sin A = \dfrac{\sqrt{3}}{2} = \sin 60^\circ$

$\Rightarrow \quad \angle A = 60^\circ, \angle B = 30^\circ, \angle C = 90^\circ$

$\because \quad 2C = A + B$

136. Applying cosine rule, $2ab\cos C = a^2 + b^2 - c^2$

$\Rightarrow \quad 2\sqrt{2}\sqrt{2 + \sqrt{2}}\cos\dfrac{\pi}{8} = 2 + 2 + \sqrt{2} - c^2$

Using $\cos\dfrac{\pi}{8} = \dfrac{\sqrt{2 + \sqrt{2}}}{2}$

$\Rightarrow \quad \dfrac{2\sqrt{2}\,(2 + \sqrt{2})}{2} = 4 + \sqrt{2} - c^2$

$\Rightarrow \quad c^2 = 2 - \sqrt{2} \Rightarrow c = \sqrt{2 - \sqrt{2}}$

Applying sine rule, $\dfrac{\sin C}{c} = \dfrac{\sin A}{a}$

$\Rightarrow \quad \dfrac{\sqrt{2 - \sqrt{2}}}{2\sqrt{2 - \sqrt{2}}} = \dfrac{\sin A}{\sqrt{2}}$

$\Rightarrow \quad \sin A = \dfrac{1}{\sqrt{2}} \Rightarrow A = 45^\circ$

[$\because b$ is the longest side $\therefore \angle B$ will be the greatest $\therefore A \neq 135^\circ$)]

Sum of the digits is $= 4 + 5 = 9$

137. Using sine law, $\dfrac{125}{\sin x} = \dfrac{195}{\sin 3x} = \dfrac{c}{\sin 4x}$

$3 - 4\sin^2 x = \dfrac{195}{125} = \dfrac{39}{25}$

$\Rightarrow \quad 4\sin^2 x = 3 - \dfrac{39}{25} = \dfrac{75 - 39}{25} = \dfrac{36}{25}$

$\sin^2 x = \dfrac{9}{25} \Rightarrow \sin x = \dfrac{3}{5}$

$\therefore \quad c = \dfrac{125\sin 4x}{\sin x} = 125\dfrac{2\cos 2x\, 2\sin x\cos x}{\sin x}$

$= (125)(4)(\cos x)(1 - 2\sin^2 x)$

$= (500)\left(\dfrac{4}{5}\right)\left(1 - \dfrac{18}{25}\right) = 400\left(\dfrac{7}{25}\right)$

$= (16)(7) = 112$

$\therefore$ Number of digits in the length of side $= 3$

138. $a^2 + b^2 + c^2 = 2R^2(1 - \cos 2A + 1 - \cos 2B + 1 - \cos 2C)$

$= 2R^2\left[3 - (\cos\dfrac{2\pi}{7} + \cos\dfrac{4\pi}{7} + \cos\dfrac{6\pi}{7}\right]$

$= 2R^2\left(3 - \left(\dfrac{-1}{2}\right)\right) = 7R^2$

$\therefore \quad \dfrac{a^2 + b^2 + c^2}{R^2} = 7$

139. $D = 2\tan B\tan B\tan C\,(\tan A + \tan B + \tan C)^3$

$= 2(\tan A + \tan B + \tan C)^4$

$\geq 2(3\sqrt{3})^4 = 1458 = \dfrac{D}{1000} \geq 1.458$

$\therefore$ Least integer value is 2.

140. Clear $PQ = \dfrac{a}{2}, OA = R$

and $\quad \angle AOQ = B$

$OA\cos B = R\cos B$

now $\quad \angle AOQ = C$ since $PQ \parallel BC$

$\Rightarrow \quad |PQO = \frac{\pi}{2} - C$

$\Rightarrow$ Area of $\Delta OPQ = \frac{1}{2} PQ \cdot OQ \sin| PQO$

$$= \frac{aR}{4} \cos B \cos C$$

Area of $\Delta ABC = \frac{1}{2} ab \sin c = aR \sin B \sin C$

$$\therefore \quad \frac{\text{Area of } \Delta ABC}{\text{Area of } \Delta OPQ} = \frac{aR \sin B \sin C}{\frac{aR}{4} \cos B \cos C}$$

$$= 4 \tan B \tan C$$

$$\therefore \quad \left(\frac{\text{Area of } \Delta ABC}{\text{Area of } \Delta OPS} \right) \cot B \cdot \cot C = 4$$

141. $\sum \frac{a}{r_1} = \sum \frac{2R \sin \frac{A}{2} \cdot 2 \cos \frac{A}{2}}{4R \sin \frac{A}{2} \cos \frac{B}{2} \cos \frac{C}{2}} = \sum \frac{\sin\left(\frac{B}{2} + \frac{C}{2}\right)}{\cos \frac{B}{2} \cos \frac{C}{2}}$

$$\sum \left(\tan \frac{B}{2} + \tan \frac{C}{2} \right) 2 \sum (\tan \frac{A}{c}) = 2 \sum \frac{r_1}{s} = 2 \left[\frac{r_1 + r_2 + r_3}{s} \right]$$

$$= 2 \left[\frac{r_1 + r_2 + r_3}{\frac{a + b + c}{2}} \right]$$

$$= \frac{a}{r_1} + \frac{b}{r_2} + \frac{c}{r_3} = 4 \left(\frac{r_1 + r_2 + r_3}{a + b + c} \right)$$

$$\Rightarrow \left(\frac{a + b + c}{r_1 + r_2 + r_3} \right) \left(\frac{a}{r_1} + \frac{b}{r_2} + \frac{c}{r_3} \right) = 4$$

142. From the ΔABC,

$$\cos A = \frac{b^2 + c^2 - a^2}{2bc} \quad \text{...(i)}$$

From the ΔCAD,

$$\cos C = \frac{AC}{CD} = \frac{b}{a/2} = \frac{2b}{a} \quad \text{...(ii)}$$

From the ΔABD,

$$\frac{BD}{\sin(A - 90°)} = \frac{AB}{\sin \angle ADB}$$

or $\quad \frac{a/2}{-\cos A} = \frac{c}{\sin(90° + C)}$

or $\quad \frac{a}{-2 \cos A} = \frac{c}{\cos C}$

$\therefore \quad \cos A = \frac{a \cos C}{-2c}$

$$= \frac{a}{-2c} \cdot \frac{2b}{a} = -\frac{b}{c} \quad \text{[From Eq. (ii)]}$$

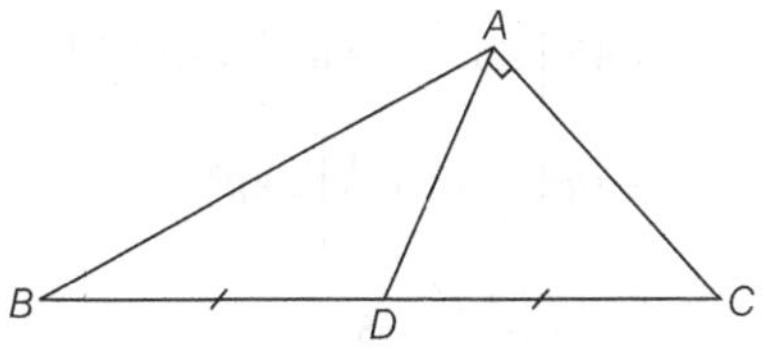

$\therefore$ From Eq. (i), $\frac{b^2 + c^2 - a^2}{2bc} = \frac{-b}{c}$

or $\quad b^2 + c^2 - a^2 = -2b^2$

or $\quad c^2 - a^2 = -3b^2 \quad \text{...(iii)}$

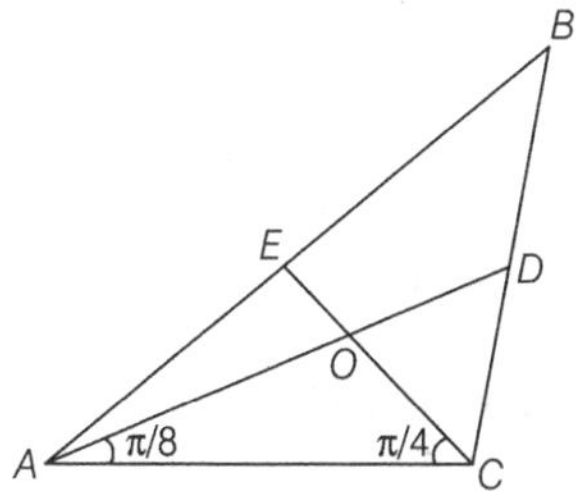

Now, $\cos A \cdot \cos C = \frac{b^2 + c^2 - a^2}{2bc} \cdot \frac{2b}{a}$

$$= \frac{b^2 + c^2 - a^2}{ca}$$

$$= \frac{3b^3 + 3(c^2 - a^2)}{3ca}$$

$$= \frac{a^2 - c^2 + 3(c^2 - a^2)}{3ca} \quad \text{[From Eq. (iii)]}$$

$$= \frac{2(c^2 - a^2)}{3ca}$$

$\Rightarrow \quad \lambda = 2$

143. Let O be the point of intersection of the medians of triangle ABC. Then, the area ofΔABC is three times that of ΔAOC. Now, in ΔAOC, $AO = \frac{2}{3} AD = \frac{10}{3}$. Therefore, applying the sine rule to ΔAOC, we get

$$\frac{OC}{\sin\left(\frac{\pi}{8}\right)} = \frac{AO}{\sin\left(\frac{\pi}{4}\right)}$$

$$\Rightarrow \quad OC = \frac{10}{3} \cdot \frac{\sin\left(\frac{\pi}{8}\right)}{\sin\left(\frac{\pi}{4}\right)}$$

Area of $\Delta AOC = \frac{1}{2} \cdot AO \cdot OC \cdot \sin \angle AOC$

$$= \frac{1}{2} \cdot \frac{10}{3} \cdot \frac{10}{3} \cdot \frac{\sin\left(\frac{\pi}{8}\right)}{\sin\left(\frac{\pi}{4}\right)} \cdot \sin\left(\frac{\pi}{2} + \frac{\pi}{8}\right)$$

$$= \frac{50}{9} \cdot \frac{\sin\left(\frac{\pi}{8}\right) \cos\left(\frac{\pi}{8}\right)}{\sin\left(\frac{\pi}{4}\right)} = \frac{50}{18} = \frac{25}{9}$$

$\therefore$ Area of $\Delta ABC = 3 \cdot \frac{25}{9} = \frac{25}{3} = \frac{5a}{b}$

$\Rightarrow \quad a = 5$ and $b = 3$

$\therefore \quad a + b = 5 + 3 = 8$

144. $\dfrac{r}{r_1} = \dfrac{4R\sin\dfrac{A}{2}\cdot\sin\dfrac{B}{2}\cdot\sin\dfrac{C}{2}}{4R\sin\dfrac{A}{2}\cdot\cos\dfrac{B}{2}\cdot\cos\dfrac{C}{2}} = \tan\dfrac{B}{2}\tan\dfrac{C}{2} = \dfrac{1}{2}$

Now, $\tan\dfrac{A}{2}\left(\tan\dfrac{B}{2} + \tan\dfrac{C}{2}\right)$

$$= \tan\frac{A}{2}\cdot\tan\frac{B}{2} + \tan\frac{A}{2}\cdot\tan\frac{C}{2}$$

$$= 1 - \tan\frac{B}{2}\cdot\tan\frac{C}{2} = 1 - \frac{1}{2} = \frac{1}{2}$$

$\therefore \quad 16\tan\dfrac{A}{2}\left(\tan\dfrac{B}{2} + \tan\dfrac{C}{2}\right) = \dfrac{1}{2}\times 16$

$$= 8$$

145. $\dfrac{a\cos^2\dfrac{A}{2} + b\cos^2\dfrac{B}{2} + c\cos^2\dfrac{C}{2}}{a+b+c}$

$$= \frac{a(1+\cos A) + b(1+\cos B) + c(1+\cos C)}{2(a+b+c)}$$

$$= \frac{(a+b+c) + (a\cos A + b\cos b + c\cos C)}{2(a+b+c)}$$

$$= \frac{(a+b+c) + (a\cos A + b\cos B + c\cos C)}{2(a+b+c)}$$

$$= \frac{1}{2} + \frac{a\cos A + b\cos B + c\cos C}{4s}$$

$$= \frac{1}{2} + \frac{R}{4s}(\sin 2A + \sin 2B + \sin 2C)$$

$$\left[\because \frac{a}{\sin A} = \frac{b}{\sin B} = \frac{c}{\sin C} = 2R\right]$$

$$= \frac{1}{2} + \frac{R}{4s}(4\sin A\sin B\sin C)$$

$$= \frac{1}{2} + \frac{R}{4s}\times\left(4\cdot\frac{a}{2R}\cdot\frac{b}{2R}\cdot\frac{c}{2R}\right)$$

$$= \frac{1}{2} + \frac{abc}{8R^2s} = \frac{1}{2} + \frac{4R\Delta}{8R^2s}\ \frac{1}{2} + \frac{r}{2R}$$

$$= \frac{1}{2}\left(1 + \frac{r}{R}\right) \le \frac{1}{2}\left(1 + \frac{1}{2}\right) \qquad \left[\because \frac{r}{R} \le \frac{1}{2}\right]$$

Hence, $\dfrac{a\cos^2\dfrac{A}{2} + b\cos^2\dfrac{B}{2} + c\cos^2\dfrac{C}{2}}{a+b+c} \le \dfrac{3}{4}$

$\therefore \quad 12\times\dfrac{\left(a\cos^2\dfrac{A}{2} + b\cos^2\dfrac{B}{2} + c\cos^2\dfrac{C}{2}\right)}{a+b+c} \le 12\times\dfrac{3}{4} = 9$

146. Let $AB = n$, $AC = n+1$, $BC = n+2$

Further,

Let $A = 2C$ (since AB is the smallest and BC is largest)

By sine rule, we have

$$\frac{\sin A}{n+2} = \frac{\sin B}{n+1} = \frac{\sin C}{n}$$

$\because \quad A = 2C$

and $\quad B = 180° - (A + 3C)$

$\Rightarrow \quad B = 180° - 3C$

$\therefore \quad \dfrac{\sin 2C}{n+2} = \dfrac{\sin 3C}{n+1} = \dfrac{\sin C}{n}$

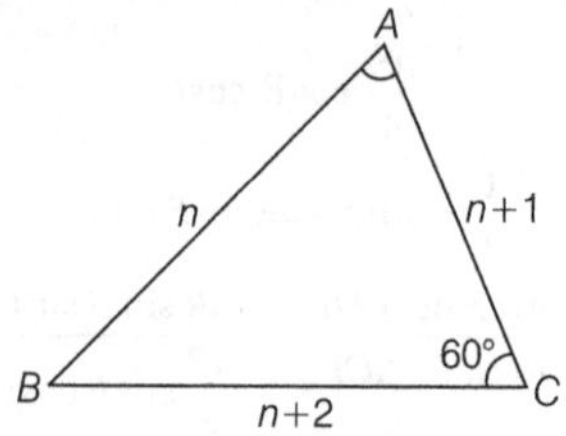

$\Rightarrow \quad \dfrac{2\cos C}{n+2} = \dfrac{3 - 4\sin^2 C}{n+1} = \dfrac{1}{n}$

$\therefore \quad \cos C = \dfrac{n+2}{2n}$ and $\sin^2 C = \dfrac{2n-1}{4n}$

$\therefore \quad \left(\dfrac{n+2}{2n}\right)^2 + \left(\dfrac{2n-1}{4n}\right) = 1$

$\Rightarrow \quad n^2 - 3n - 4 = 0$

$\therefore \quad n = 4, n = -1$

$\Rightarrow \quad n = 4, n \ne -1$

Then, sides are 4, 5, 6.

$\therefore$ Largest side is 6.

147. $A + B + C = \pi$

Given, $\quad C = 2A$

$\Rightarrow \quad B = \pi - 3A$

As $0 < C < \pi \Rightarrow 0 < 2A < \pi \Rightarrow 0\ A < \dfrac{\pi}{2}$

By sine rule, $\dfrac{a}{\sin A} = \dfrac{b}{\sin B}$

or $\quad \dfrac{a}{\sin A} = \dfrac{2a}{\sin(\pi - 3A)}$

or $\quad \dfrac{a}{1} = \dfrac{2a}{3 - 4\sin^2 A}$

or $\quad 3 - 4\sin^2 A = 2$

or $\quad \sin^2 A = \dfrac{1}{4}$ or $\sin A = \dfrac{1}{2}$

or $\quad A = \dfrac{\pi}{6}$ or $\dfrac{5\pi}{6}$

But $\quad 0 < A < \dfrac{\pi}{2}$

$\Rightarrow \quad A = \dfrac{\pi}{6}, \angle B = \dfrac{\pi}{2}$ and $\angle C = \dfrac{\pi}{3}$

$\Rightarrow \quad a^2 + b^2 + c^2$

$$= 4R^2[\sin^2 A + \sin^2 B + \sin^2 C]$$

$$= 4R^2\left[\sin^2\frac{\pi}{6} + \sin^2\frac{\pi}{2} + \sin^2\frac{\pi}{3}\right]$$

$$= 4R^2\left[\frac{1}{4} + 1 + \frac{3}{4}\right] = 8R^2$$

or $\quad \dfrac{a^2 + b^2 + c^2}{R^2} = 8$

148. We have, $\cos A = \dfrac{b^2 + c^2 - a^2}{2bc}$

$\Rightarrow c^2 - 2b\cos A + b^2 - a^2 = 0$, where is quadratic in 'c'

$$\left.\begin{aligned} \therefore c_1 + c_2 &= 2b\cos A \\ \text{and } c_1c_2 &= b^2 - a^2 \end{aligned}\right\} \quad ...(i)$$

$\therefore \quad c_1^2 + c_2^2 - 2c_1c_2\cos 2A$

$\Rightarrow \quad (c_1 + c_2)^2 - 2c_1c_2 - 2c_1c_2\cos 2A$ [Using Eq. (i)]

$\Rightarrow \quad (c_1 + c_2)^2 - 2c_1c_2(1 + \cos 2A)$

$\Rightarrow \quad 4b^2\cos^2 A - 2(b^2 - a^2)\cdot 2\cos^2 A$

$= 4a^2\cos^2 A$

$\therefore \quad c_1^2 + c_2^2 - 2c_1c_2\cos A$

$= 4a^2\cos^2 A$

Hence, the value of $\lambda = 4$

149. $AG = \dfrac{2}{3}AA_1,\ BG = \dfrac{2}{3}BB_1$

$\Rightarrow \quad AG = \dfrac{2}{3}\sqrt{2b^2 + 2c^2 - a^2}$

$BG = \dfrac{2}{3}\sqrt{2a^2 + 2c^2 - b^2}$

$\Rightarrow \quad AG = \dfrac{2}{3}a,\ BG = \dfrac{2}{3}\sqrt{b^2 + 4c^2}$ as $a^2 = b^2 + c^2$

$\Rightarrow \quad AG = \dfrac{10}{3},\ BG = \dfrac{2}{3}\sqrt{16 + 36} = \dfrac{4}{3}\sqrt{13}$

Also, $\quad AB = c = 3$ and $\Delta_{GAB} = \dfrac{1}{3}\Delta_{ABC} = 2$

If R_1 be the circumradius of ΔGAB, then

$R_1 = \dfrac{(AG)(BG)(AB)}{4\Delta_{GAB}}$

$= \dfrac{10}{3}\cdot\dfrac{4}{3}\sqrt{13}\cdot 3\cdot\dfrac{1}{4\cdot 2} = \dfrac{5\sqrt{13}}{3}$ unit

$\therefore \quad \dfrac{9}{65} \times \dfrac{25 \times 13}{9} = 5$

150. We have $\dfrac{1}{2}AD[c\cos C + b\cos B] = \Delta$

$\Rightarrow \quad \dfrac{1}{AD} + \dfrac{1}{BE} + \dfrac{1}{CF}$

$= \dfrac{a\cos A + b\cos B + c\cos C}{\Delta} = 2$

151. $\dfrac{a^3 + b^3 + c^3}{\sin^3 A + \sin^3 B + \sin^3 C} = 8$

$\Rightarrow \quad 8R^3 = 8 \Rightarrow R = 1$

Now, $\quad a = 2R\sin A$

$a = 2\sin A$

Hence, a is maximum when $\sin A$ is maximum.

152. From ΔPQR,

$PR^2 = 2^2 + 5^2 - 2\cdot 5\cdot 2\cdot\dfrac{1}{2} = 19$

In ΔPRS, $PR^2 = 3^2 + x^2 + 2\cdot 3x\cdot\dfrac{1}{2}$, where

$SP = x = x^2 + 3x + 9$

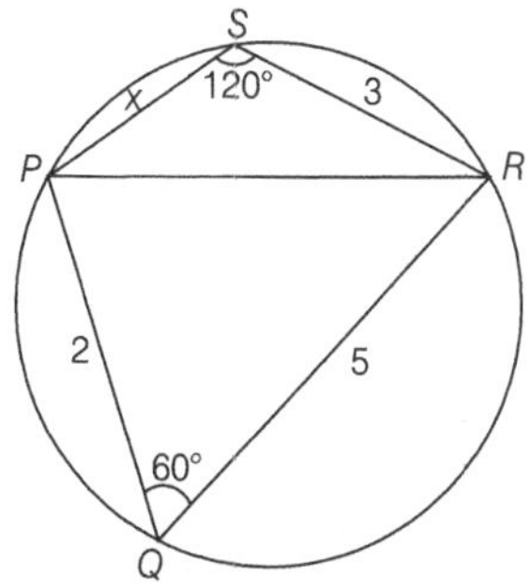

$\therefore \quad x^2 + 3x + 9 = 19$

$(x + 5)(x - 2) = 0$

$x + 5 = 0$ (not possible)

$\therefore \quad x = 2$

153. Since, A and B satisfying the given equation, therefore

$\sqrt{3}\cos A + \sin A = \sqrt{3}\cos B + \sin B$

$\Rightarrow \quad \dfrac{\sin A - \sin B}{\cos B - \cos A} = \sqrt{3}$

$\Rightarrow \quad \dfrac{2\cos\left(\dfrac{A + B}{2}\right)\sin\left(\dfrac{A - B}{2}\right)}{2\sin\left(\dfrac{A + B}{2}\right)\sin\left(\dfrac{A - B}{2}\right)} = \sqrt{3}$

$\Rightarrow \quad \tan\left(\dfrac{A + B}{2}\right) = \dfrac{1}{\sqrt{3}} = \tan\dfrac{\pi}{6}$

$\Rightarrow \quad \dfrac{A + B}{2} = \dfrac{\pi}{6}$

$\Rightarrow \quad A + B = \dfrac{\pi}{3}$

Now, $C = \pi - \dfrac{\pi}{3} = \dfrac{2\pi}{3}$

Hence, triangle is obtuse angled.

154. We have, $a\tan\theta + b\sec\theta = c$

$\Rightarrow \quad a\sin\theta + b = c\cos\theta$

$\Rightarrow \quad a\cos\theta - a\sin\theta = b \quad ...(i)$

Now, let α and β are the other two angles of the triangle, then according to given condition we get

$c\cos\alpha - a\sin\alpha = c\cos\beta - a\sin\beta$

$\Rightarrow \quad c(\cos\alpha - \cos\beta) = a(\sin\alpha - \sin\beta)$

$\Rightarrow \quad \dfrac{c}{a} = \dfrac{2\cos\left(\dfrac{\alpha + \beta}{2}\right)\sin\left(\dfrac{\alpha - \beta}{2}\right)}{-2\sin\left(\dfrac{\alpha + \beta}{2}\right)\sin\left(\dfrac{\alpha - \beta}{2}\right)}$

$\Rightarrow \quad \dfrac{c}{a} = -\cot\left(\dfrac{\alpha + \beta}{2}\right)$

$\Rightarrow \tan\left(\dfrac{\alpha + \beta}{2}\right) = -\dfrac{a}{c}$

$$\therefore \quad \tan(\alpha+\beta) = \frac{2\tan\left(\frac{\alpha+\beta}{2}\right)}{1-\tan^2\left(\frac{\alpha+\beta}{2}\right)} = \frac{2\left(-\frac{a}{c}\right)}{1-\frac{a^2}{c^2}} = \frac{2ac}{a^2-c^2}$$

$$\Rightarrow \quad \tan(\pi - C) = \frac{2ac}{a^2-c^2}$$

$$\Rightarrow \quad \tan\left(\pi - \frac{3\pi}{4}\right) = \frac{2ac}{a^2-c^2} \qquad \left[\because c = \frac{3\pi}{4}\right]$$

$$\Rightarrow \quad \tan\frac{\pi}{4} = \frac{2ac}{a^2-c^2}$$

$$\Rightarrow \quad a^2 - c^2 = 2ac$$

155. Here, the quadratic for third side b is given by

$$b^2 - 2bc\cos A + (c^2 - a^2) = 0$$

$\therefore \quad b_1 + b_2 = 2c\cos A \quad$ ...(i)

and $\quad b_1 b_2 = c^2 - a^2 \quad$ ...(ii)

also, $\quad b_2 = 2b_1 \quad$ ...(iii)

From Eqs. (i) and (iii), $3b_1 = 2c\cos A \quad$ ...(iv)

and from Eqs. (ii) and (iv),

$$2b_1^2 = c^2 - a^2 \quad \text{...(v)}$$

From Eqs. (iv) and (v), we have

$$\frac{2\cdot 4c^2\cos^2 A}{9} = c^2 - a^2$$

$$\Rightarrow \quad 8c^2(1-\sin^2 A) = 9c^2 - 9a^2$$

$$\Rightarrow \quad 1 - \sin^2 A = \frac{9(c^2-a^2)}{8c^2}$$

$$\Rightarrow \quad \sin^2 A = 1 - \frac{9(c^2-a^2)}{8c^2}$$

$$\Rightarrow \quad \sin A = \sqrt{\frac{8c^2 - 9(c^2-a^2)}{8c^2}}$$

$$\Rightarrow \quad \sin A = \sqrt{\frac{8c^2 - 9c^2 + 9a^2}{8c^2}}$$

$$\therefore \quad \sin A = \sqrt{\frac{9a^2 - c^2}{8c^2}}$$

156. $\angle BPA = 90° + \left(\frac{B}{3}\right), \angle ABP = \frac{2B}{3}$

In ΔABP, $\dfrac{AP}{\sin\left(\frac{2B}{3}\right)} = \dfrac{c}{\sin[90° + B/3)]} = \dfrac{c}{\cos(B/3)}$

[by sine rule]

$$\text{or} \quad AP = \frac{c\sin(2B/3)}{\cos(B/3)}$$

$$= \frac{2c\sin\left(\frac{B}{3}\right)\cos\left(\frac{B}{3}\right)}{\cos\left(\frac{B}{3}\right)}$$

$$= 2c\sin\left(\frac{B}{3}\right)$$

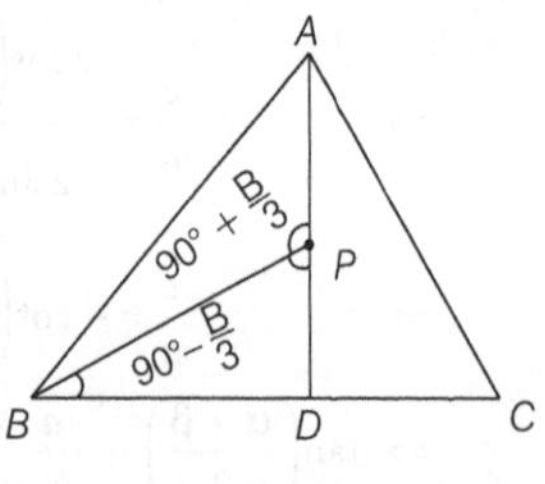

157. Using $\dfrac{a}{2\sin A} = \dfrac{b}{2\sin B} = \dfrac{c}{2\sin C} = R$ we get,

$$\frac{b^2-c^2}{2aR} = \frac{4R^2(\sin^2 B - \sin^2 C)}{4R^2\sin A}$$

$$= \frac{\sin(B+C)\sin(B-C)}{\sin A} = \sin(B-C)$$

158. We know that,

$$\text{or} \quad \text{ar}(\Delta ABC) = \frac{1}{2}bc\sin A$$

$$\Rightarrow \quad \frac{9\sqrt{3}}{2} = \frac{1}{2}bc\sin\frac{2\pi}{3}$$

$$\Rightarrow \quad 9\sqrt{3} = bc \times \frac{\sqrt{3}}{2}$$

$$\Rightarrow \quad bc = 18 \quad \text{...(i)}$$

$$\text{Also,} \quad b - c = 3\sqrt{3}$$

$$\Rightarrow \quad b = c + 3\sqrt{3} \quad \text{...(ii)}$$

From Eq. (i) and Eq. (ii), we get

$$(c + 3\sqrt{3})c = 18$$

$$\Rightarrow \quad c^2 + 3\sqrt{3}c - 18 = 0$$

$$\Rightarrow \quad c = \frac{-3\sqrt{3} \pm \sqrt{27+72}}{2}$$

$$= \frac{-3\sqrt{3} \pm \sqrt{99}}{2} = \frac{-3\sqrt{3} \pm 3\sqrt{11}}{2}$$

$$\therefore \quad c = \frac{3\sqrt{11} - 3\sqrt{3}}{2} \qquad [\because c > 0]$$

$$\Rightarrow \quad b = \frac{3\sqrt{11} - 3\sqrt{3}}{2} + 3\sqrt{3} = \frac{3\sqrt{11} + 3\sqrt{3}}{2}$$

$$\cos A = \frac{b^2 + c^2 - a^2}{2bc}$$

$$\Rightarrow \quad \frac{\sqrt{3}}{2} = \frac{b^2 + c^2 - a^2}{2bc}$$

$$\Rightarrow \quad \frac{\sqrt{3}}{2} = \frac{b^2 + c^2 - a^2}{2 \times \frac{9}{4}(11-3)}$$

$$\Rightarrow \quad 2\sqrt{3} = 63 - a^2 \qquad [\because b^2 + c^2 = 63]$$

$$\Rightarrow \quad a^2 = 63 - 2\sqrt{3}$$

$$\therefore \quad a = \sqrt{63 - 2\sqrt{3}}$$

159. We have, $\quad \Delta = a^2 - (b-c)^2$

$$= (a + b - c)(a - b + c)$$

$$\Rightarrow \quad \Delta^2 = (a + b - c)^2 (a - b + c)^2$$

$$= (2s - c - c)^2 (2s - b - b)^2$$

$$= (2s - 2c)^2 (2s - 2b)^2$$

$$= 16(s - c)^2 (s - b)^2$$

$$\therefore \quad s(s-a)(s-b)(s-c) = 16(s-c)^2(s-b)^2$$

$$\Rightarrow \quad \frac{(s-b)(s-c)}{s(s-a)} = \frac{1}{16}$$

$$\Rightarrow \quad \tan^2\frac{A}{2} = \frac{1}{16}$$

$$\Rightarrow \quad \tan\frac{A}{2} = \frac{1}{4}$$

$$\because \quad \tan A = \frac{2\tan\frac{A}{2}}{1-\tan^2\frac{A}{2}} = \frac{2\times\frac{1}{4}}{1-\left(\frac{1}{4}\right)^2} = \frac{\left(\frac{1}{2}\right)}{\left(\frac{15}{16}\right)} = \frac{8}{15}$$

160. We have,

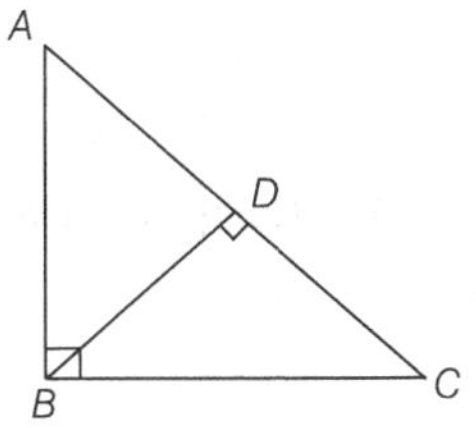

$\angle ABC = 90°$, $\angle BDC = 90°$, $AC = h$, $BD = p$ and $h = 4p$

Let $\angle C = \theta \Rightarrow \angle A = 90° - \theta$

Now, in ΔBDC

$$\frac{\sin C}{BD} = \frac{\sin D}{BC} \qquad \text{[using sine formula]}$$

$$\Rightarrow \quad \frac{\sin\theta}{p} = \frac{\sin 90°}{BC}$$

$$\Rightarrow \quad BC = \frac{p}{\sin\theta} \qquad \text{...(i)}$$

Again, in ΔABC

$$\frac{\sin B}{AC} = \frac{\sin A}{BC} \Rightarrow \frac{\sin 90°}{h} = \frac{\sin(90° - \theta)}{BC}$$

$$\Rightarrow \quad BC = h\cos\theta \qquad \text{...(ii)}$$

From Eqs. (i) and (ii), we have

$$\frac{p}{\sin\theta} = h\cos\theta \Rightarrow \frac{p}{h} = \sin\theta\cos\theta$$

$$\Rightarrow \quad \frac{p}{4p} = \sin\theta\cos\theta$$

$$\Rightarrow \quad \sin\theta\cos\theta = \frac{1}{4} \Rightarrow \sin 2\theta = \frac{1}{2}$$

$$\therefore \quad 2\theta = 30° \text{ or } 150°$$

$$\Rightarrow \quad \theta = 15° \text{ or } 75°$$

Hence, $\angle C = 15°$.

161. We have, $\sin^3 A + \sin^3 B + \sin^3 C = 3\sin A \cdot \sin B \cdot \sin C$

$$\Rightarrow \quad k^3a^3 + k^3b^3 + \; + k^3c^3 = 3k^3abc$$

$$\left[\therefore \frac{\sin A}{a} = \frac{\sin B}{b} = \frac{\sin C}{c} = k\right]$$

$$\Rightarrow \quad a^3 + b^3 + c^3 = 3abc$$

$$\Rightarrow \quad a + b + c = 0$$

$$[\because \text{if } a + b + c = 0, \text{ then } a^3 + b^3 + c^3 = 3abc]$$

Now, $\begin{vmatrix} a & b & c \\ b & c & a \\ c & a & b \end{vmatrix}$

on apply, $C_1 \to C_1 + C_2 + C_3$, we get-

$$= \begin{vmatrix} a+b+c & b & c \\ b+c+a & c & a \\ c+a+b & a & b \end{vmatrix}$$

$$= \begin{vmatrix} 0 & b & c \\ 0 & c & a \\ 0 & a & b \end{vmatrix} \qquad [\because a + b + c = 0]$$

$$= 0$$

162. We have,

$$a + b + c = 11, ab + bc + ca = 38 \text{ and } abc = 40$$

$$\therefore \quad \frac{\cos A}{a} + \frac{\cos B}{b} + \frac{\cos C}{c}$$

$$= \frac{\left(\frac{b^2+c^2-a^2}{2bc}\right)}{a} + \frac{\left(\frac{c^2+a^2-b^2}{2ca}\right)}{b} + \frac{\left(\frac{a^2+b^2-c^2}{2ab}\right)}{c}$$

$$= \frac{b^2+c^2-a^2}{2abc} + \frac{c^2+a^2-b^2}{2abc} + \frac{a^2+b^2-c^2}{2abc}$$

$$= \frac{a^2+b^2+c^2}{2abc} = \frac{(a+b+c)^2 - 2(ab+bc+ca)}{2abc}$$

$$= \frac{(11)^2 - 2(38)}{2\times 40} = \frac{121-76}{80}$$

$$= \frac{45}{80} = \frac{9}{16}$$

163. We know that,

$$\tan\frac{A}{2} = \sqrt{\frac{(s-b)(s-c)}{s(s-a)}}$$

and $$\tan\frac{C}{2} = \sqrt{\frac{(s-a)(s-b)}{s(s-c)}}$$

$$\therefore \quad \tan\frac{A}{2} + \tan\frac{C}{2} = \sqrt{\frac{s-b}{s}}\left[\sqrt{\frac{s-c}{s-a}} + \sqrt{\frac{s-a}{s-c}}\right]$$

$$= \sqrt{\frac{s-b}{s}}\left[\frac{s-c+s-a}{\sqrt{s-a}\sqrt{s-c}}\right]$$

$$= \sqrt{\frac{s-b}{s}}\left[\frac{2s-a-c}{\sqrt{s-a}\sqrt{s-c}}\right]$$

$$= \sqrt{\frac{s-b}{s}}\left[\frac{b}{\sqrt{s-a}\sqrt{s-c}}\right]$$

$$= \frac{b}{s}\sqrt{\frac{s(s-b)}{(s-a)(s-c)}}$$

$$= \frac{b}{s}\cot\frac{B}{2} \qquad \left[\because \cot\frac{B}{2} = \sqrt{\frac{s(s-b)}{(s-a)(s-c)}}\right]$$

$$\therefore \quad \frac{\tan\frac{A}{2} + \tan\frac{C}{2}}{\cot\frac{B}{2}} = \frac{b}{s} \qquad \text{...(i)}$$

Again, a, b, c are in AP

$$\therefore \quad a + c = 2b$$

or $$a + b + c = 3b \Rightarrow 2s = 3b$$

$$\Rightarrow \quad \frac{b}{s} = \frac{2}{3} \quad \text{...(ii)}$$

From Eqs. (i) and (ii), we have

$$\frac{\tan\frac{A}{2} + \tan\frac{C}{2}}{\cot\frac{B}{2}} = \frac{2}{3}$$

164. Let the sides of triangle be $a - d, a, a + d$.

$\therefore$ Perimeter $= a - d + a + a + d = 3a$

$\therefore$ Side of equilateral triangle having same perimeter will be a.

Semi-perimeter of given triangle

$$= \frac{a - d + a + a + d}{2} = \frac{3a}{2}$$

Now, according to question

$$\sqrt{\left(\frac{3a}{2}\right)\left(\frac{3a}{2} - a + d\right)\left(\frac{3a}{2} - a\right)\left(\frac{3a}{2} - a - d\right)} = \frac{3}{5} \times \frac{\sqrt{3}}{4} a^2$$

$$\Rightarrow \sqrt{\left(\frac{3a}{2}\right)\left(\frac{a}{2} + d\right)\left(\frac{a}{2}\right)\left(\frac{a}{2} - d\right)} = \frac{3\sqrt{3}}{20} a^2$$

$$\Rightarrow \quad = \frac{\sqrt{3}a}{2}\sqrt{\left(\frac{a^2}{4} - d^2\right)} = \frac{3\sqrt{3}}{20} a^2$$

$$\Rightarrow \quad \sqrt{\frac{a^2}{4} - d^2} = \frac{3a}{10}$$

$$\Rightarrow \quad \frac{a^2}{4} - d^2 = \frac{9a^2}{100}$$

$$\Rightarrow \quad d^2 = \frac{16}{100}a^2 \Rightarrow d = \frac{4}{10}a = \frac{2}{5}a$$

$\therefore$ Sides will be, $a - \frac{2}{5}a, a, a + \frac{2}{5}a$

or $\frac{3a}{5}, a, \frac{7a}{5}$

$\therefore$ Required ration $= \frac{3a}{5} : a : \frac{7a}{5} = 3 : 5 : 7$

165. Let the sides of ΔABC are a, b, c.

$\therefore \quad BC = a, CA = b$ and $AB = c$

Again, AD, BE and CF are the medians of ΔABC

$$\therefore \quad AD = \frac{1}{2}\sqrt{2b^2 + 2c^2 - a^2}$$

$$BE = \frac{1}{2}\sqrt{2c^2 + 2a^2 - b^2}$$

and

$$CF = \frac{1}{2}\sqrt{2a^2 + 2b^2 - c^2}$$

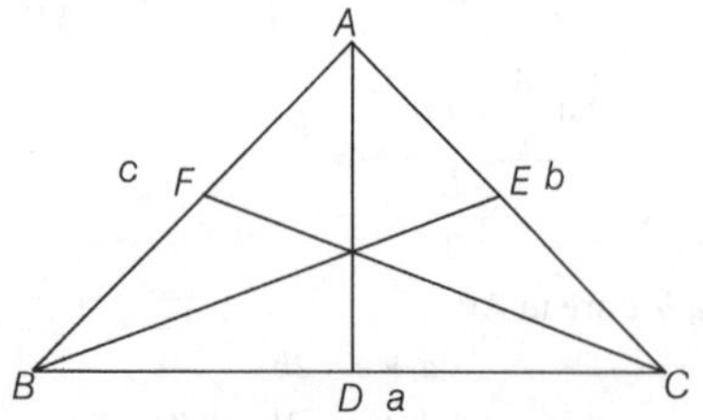

$$\therefore \quad AD^2 + BE^2 + CF^2 = \frac{1}{4}[2b^2 + 2c^2 - a^2 + 2c^2 + 2a^2 - b^2 + 2a^2 + 2b^2 - c^2]$$

$$= \frac{1}{4}[3a^2 + 3b^2 + 3c^2]$$

$$= \frac{3}{4}[a^2 + b^2 + c^2]$$

$$= \frac{3}{4}[BC^2 + CA^2 + AB^2]$$

$$\therefore \quad \frac{AD^2 + BE^2 + CF^2}{BC^2 + CA^2 + AB^2} = \frac{3}{4}$$

166. We know that, in ΔABC length of median from the $\angle A$ is $\frac{1}{2}\sqrt{2b^2 + 2c^2 - a^2}$

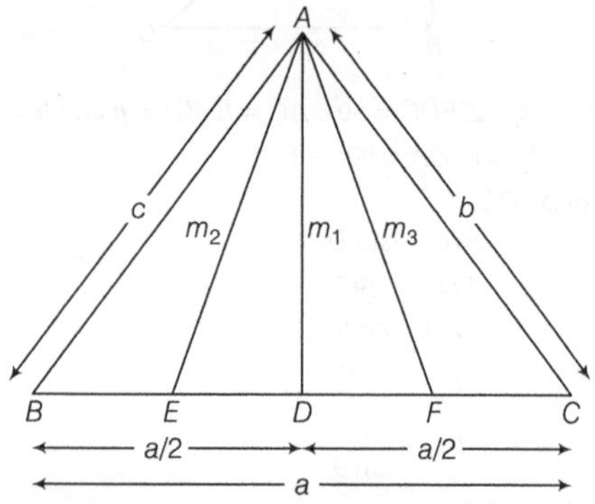

$$\therefore \quad AD^2 = m_1^2 = \frac{c^2 + b^2}{2} - \frac{a^2}{4} \quad \text{...(i)}$$

Similarly, in ΔABD,

$$AE^2 = m_2^2 = \frac{c^2 + m_1^2}{2} - \frac{\left(\frac{a}{2}\right)^2}{4} \quad \text{...(ii)}$$

and in ΔACD,

$$AF^2 = m_3^2 = \frac{b^2 + m_1^2}{2} - \frac{\left(\frac{a}{2}\right)^2}{4} \quad \text{...(iii)}$$

Now, $$m_2^2 + m_3^2 = m_1^2 + \left(\frac{b^2 + c^2}{2}\right) - 2\left(\frac{a^2}{16}\right)$$

$$= m_1^2 + m_1^2 + \frac{a^2}{4} - \frac{a^2}{8} \quad \text{[from Eq. (i)]}$$

$$= 2m_1^2 + \frac{a^2}{8}$$

$$\Rightarrow \quad \frac{a^2}{8} = m_2^2 + m_3^2 - 2m_1^2$$

167. We know from Napier's analogy that

$$\tan\left(\frac{B - C}{2}\right) = \frac{b - c}{b + c}\cot\frac{A}{2}$$

$$\tan\left(\frac{A - B}{2}\right) = \frac{a - b}{a + b}\cot\frac{C}{2}$$

$$\tan\left(\frac{C - A}{2}\right) = \frac{c - a}{c + a}\cot\frac{B}{2}$$

So, $$X = \tan\left(\frac{B-C}{2}\right)\cdot\tan\frac{A}{2}$$

$$X = \frac{b-c}{b+c}\cdot\cot\frac{A}{2}\cdot\tan\frac{A}{2} = \frac{b-c}{b+c}$$

Similarly, $$Y = \frac{a-b}{a+b},\ Z = \frac{c-a}{c+a}$$

Now, $$X + Y + Z + XYZ$$

$$= \frac{b-c}{b+c} + \frac{a-b}{a+b} + \frac{c-a}{c+a} + \left(\frac{b-c}{b+c}\right)\cdot\left(\frac{a-b}{a+b}\right)\cdot\left(\frac{c-a}{c+a}\right)$$

$$= \frac{(b-c)(a+b)(c+a) + (a-b)(b+c)(c+a) + (c-a)(b+c)(a+b)}{(b+c)(a+b)(c+a)} + \frac{(b-c)(a-b)(c-a)}{(b+c)(a+b)(c+a)}$$

After multiplication and solving, we get

$$X + Y + Z + XYZ = \frac{0}{(b+c)(b+a)(c+a)} = 0$$

168. Looking the figure, we see that $\angle T_1O_1R = 60°$ and it is the supplement of $\angle T_1AR = 120°$ {as an exterior angle for ΔABC}

Hence, $\angle AO_1R = 30°$

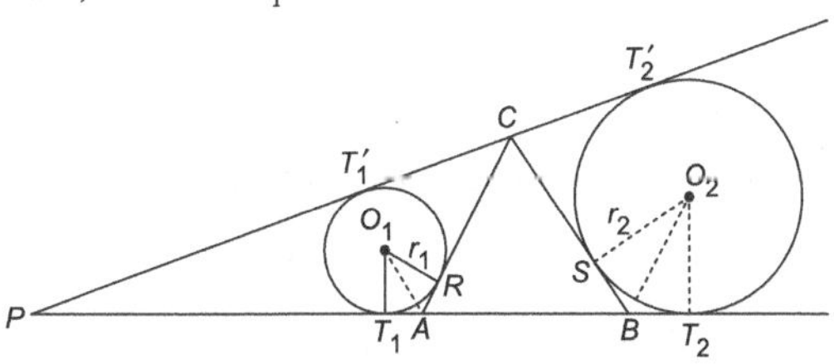

Similarly, we obtain $\angle BO_2S = 30°$

Since tangents drawn to a circle from external points are equal, we have

$$T_1T_2 = T_1A + AB + BT_2 = RA + SB + AB$$

$$= r_1\tan 30° + a + r_2\tan 30° = \frac{r_1 + r_2}{\sqrt{3}} + a.$$

and $$T_1'T_2' = T_2'C + CT_1'$$

$$= CR + CS = (a - RA) + (a - SB)$$

$$= 2a - \left(\frac{r_1 + r_2}{\sqrt{3}}\right)$$

Since, common external tangents to two circles are equal, $T_1T_2 = T_1'T_2'$

Hence, $$\frac{r_1 + r_3}{\sqrt{3}} + a = 2a - \frac{(r_1 + r_2)}{\sqrt{3}}$$

Hence, we find that $r_1 + r_2 = \frac{a\sqrt{3}}{2}$.

169. Equal chords subtend equal angles at the centre angles at the centre of circle; if each of sides of length i subtends an angles α_i $(i = 1, 2, 3)$ at the centre of the given circle, then $2\alpha_1 + 2\alpha_2 + 2\alpha_3 = 360°$

Hence $\frac{\alpha_1}{2} + \frac{\alpha_2}{2} = 90° - \frac{\alpha_3}{2}$,

and $\cos\left(\frac{\alpha_1}{2} + \frac{\alpha_2}{2}\right) = \cos\left(90° - \frac{\alpha_3}{2}\right) = \sin\frac{\alpha_3}{2}$

Next we apply the addition formula for the cosine;

$$\cos\frac{\alpha_1}{2}\cdot\cos\frac{\alpha_2}{2} - \sin\frac{\alpha_1}{2}\cdot\sin\frac{\alpha_2}{2} = \sin\frac{\alpha_3}{2}$$

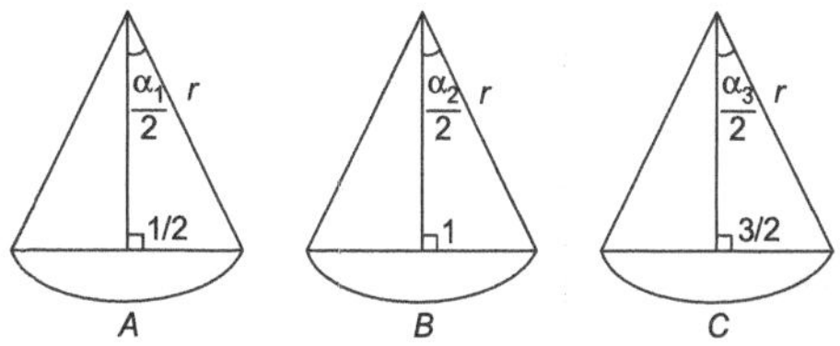

where, $\sin\frac{\alpha_1}{2} = \frac{1/2}{r}, \cos\frac{\alpha_1}{2} = \frac{\sqrt{4r^2 - 1}}{2r}$ [from (A)]

$\sin\frac{\alpha_2}{2} = \frac{1}{r},\ \cos\frac{\alpha_2}{2} = \frac{\sqrt{r^2 - 1}}{r}$ [from (B)]

$\sin\frac{\alpha_3}{2} = \frac{3/2}{r}$ [from (C)]

We substitute the expressions into Eq. (i) and obtain, after multiplying both sides by $2r^2$,

$$\sqrt{4r^2 - 1}.\sqrt{r^2 - 1} - 1 = 3r$$

Now, write it in the form;

$$\sqrt{(4r^2 - 1)(r^2 - 1)} = (3r + 1),$$

and square, obtaining.

$$(4r^2 - 1)(r^2 - 1) = 9r^2 + 6r + 1$$

which is equivalent to,

$$r(2r^3 - 7r - 3) = 0, \text{ since } r \neq 0 \text{ we have,}$$

$$2r^3 - 7r - 3 = 0.$$

170. Let, $BP = PQ = QC = x$

Also, let

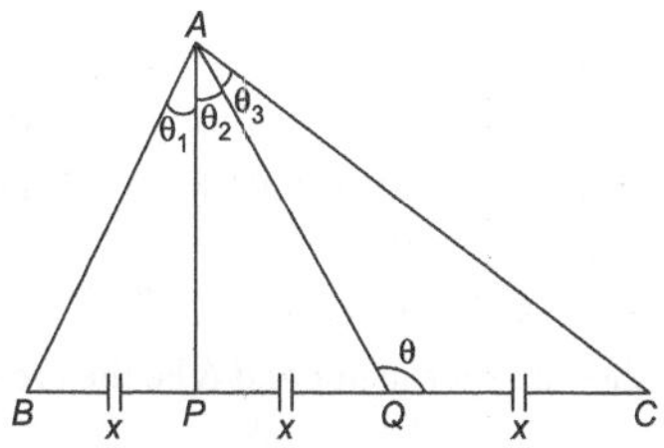

$$\angle BAP = \theta_1, \angle PAQ = \theta_2, \angle QAC = \theta_3$$

and let $\angle AQC = \theta$

Applying $m : n$ rule in ΔABC,

$$(2x + x)\cot\theta = 2x\cot(\theta_1 + \theta_2) - x\cot\theta_3$$

$$\Rightarrow \quad 3\cot\theta = 2\cot(\theta_1 + \theta_2) - \cot\theta_3 \quad \text{...(i)}$$

Applying $m : n$ rule in ΔAPC,

$$(x + x)\cot\theta = x\cot\theta_2 - x\cot\theta_3$$

$$2\cot\theta = \cot\theta_2 - \cot\theta_3 \quad \text{...(ii)}$$

On dividing Eq. (i) by Eq. (ii), and on solving we, get

$$\frac{3}{2} = \frac{2\cot(\theta_1 + \theta_2) - \cot\theta_3}{\cot\theta_2 - \cot\theta_3}$$

$$\Rightarrow \quad 4(1 + \cot^2\theta_2) = (\cot\theta_1 + \cot\theta_2)(\cot\theta_2 + \cot\theta_3)$$

$$\Rightarrow \quad 4\operatorname{cosec}^2\theta_2 = (\cot\theta_1 + \cot\theta_2)(\cot\theta_2 + \cot\theta_3)$$

171. Let the sides be $a, \frac{3a}{2}, \frac{9a}{4}$.

$$\text{Semi-perimeter} = \frac{a + \frac{3a}{2} + \frac{9a}{4}}{2} = \frac{19a}{8}$$

$$\text{Area of } \Delta = \sqrt{\frac{19a}{8}\cdot\frac{11a}{8}\cdot\frac{7a}{8}\cdot\frac{a}{8}} = \frac{a^2}{64}\sqrt{1463}.$$

$$\text{Now, circumradius} = \frac{\text{Product of sides}}{4\times \text{area of }\Delta}$$

$$\Rightarrow \quad \frac{54}{\sqrt{1463}} = \frac{a\times\frac{3a}{2}\times\frac{9a}{4}}{4\times\frac{a^2}{64}\times\sqrt{1463}}$$

$$\Rightarrow \quad a = 1.$$

Therefore, the sides of the triangles are $1, \frac{3}{2}$ & $\frac{9}{4}$.

172. Given $a + b + c = 2$

$$\Rightarrow \quad 1 - a + 1 - b + 1 - c = 1$$

or $\quad x + y + z = 1$

where $x = 1 - a, y = 1 - b, z = 1 - c$

Since, $\quad a + b > c$

$\Rightarrow \quad 0 < c < 1,$

similarly $\quad 0 < a, b < 1$

hence $\quad 0 < x, y, z < 1.$

Now, $a^2 + b^2 + c^2 + 2abc = (1 - x)^2 + (1 - y)^2 + (1 - z)^2 + 2(1 - x)(1 - y)(1 - z)$

$= 3 - 2(x + y + z) + (x^2 + y^2 + z^2) + 2[1 - (x + y + z) + (xy + yz + zx) - xyz]$

$= 1 + x^2 + y^2 + z^2 + 2(xy + yz + zx) - 2xyz$

$= 1 + (x + y + z)^2 - 2xyz$

$= 2 - 2xyz < 2$ as $0 < x, y, z < 1$

173. Let s be the semi-perimeter and Δ be the area of ΔABC. Then

$$r = \frac{\Delta}{s}, P = \frac{\Delta}{s - a}$$

$$\Rightarrow \quad \frac{r}{P} = \frac{s - a}{s}$$

Also, $\tan\frac{B}{2}\cdot\tan\frac{C}{2} = \frac{s - a}{s}$

$$\Rightarrow \quad \frac{r}{P} = \tan\frac{B}{2}\cdot\tan\frac{C}{2}$$

Let $\angle AP_1B = \alpha_1, \angle AP_2P_1 = \alpha_2, \angle AP_3P_2 = \alpha_3, \ldots,$

$\angle AP_{n-1}P_{n-2} = \alpha_{n-1}$

$$\Rightarrow \quad \frac{r_1}{P_1} = \tan\frac{B}{2}\cdot\tan\frac{\alpha_1}{2}$$

$$\frac{r_2}{P_2} = \tan\left(\frac{\pi - \alpha_1}{2}\right)\cdot\tan\frac{\alpha_2}{2}$$

$$\frac{r_3}{P_3} = \tan\left(\frac{\pi - \alpha_2}{2}\right)\cdot\tan\frac{\alpha_3}{2}$$

....

....

$$\frac{r_n}{P_n} = \tan\left(\frac{\pi - \alpha_{n-1}}{2}\right)\tan\frac{C}{2}.$$

$$\Rightarrow \quad \frac{r_1}{P_1}\cdot\frac{r_2}{P_2}\cdots\frac{r_n}{P_n} = \tan\frac{B}{2}\cdot\tan\frac{C}{2} = \frac{r}{P}$$

174. Let r be the radius of the circle.

The sides of the polygon, which subtends angle 2α at centre, has length $2r\sin\alpha$.

Hence, the area of this polygon,

$$= \frac{1}{2}(2r\sin\alpha).(r\cos\alpha) + \frac{1}{2}(2r\sin2\alpha)(r\cos2\alpha) + \ldots + \frac{1}{2}(2r\sin n\alpha)(r\cos n\alpha)$$

$$\Rightarrow \quad \frac{r^2}{2}[\sin2\alpha + \sin4\alpha + \ldots + \sin2n\alpha]$$

$$\Rightarrow \quad \frac{r^2}{4\sin\alpha}[(\cos\alpha - \cos3\alpha) + (\cos3\alpha - \cos5\alpha) + \ldots + (\cos(2n-1)\alpha - \cos(2n+1)\alpha)$$

$$\Rightarrow \quad A_1 = \frac{r^2}{4\sin\alpha}[\cos\alpha - \cos(2n + 1)\alpha]$$

$$= \frac{r^2}{4\sin(n)\alpha}[\sin\alpha.\sin(n + 1)\alpha]$$

Also, $\quad 2\alpha + 4\alpha + \ldots + 2n\alpha = 2\pi \Rightarrow \alpha(1 + 2 + 3\ldots + n) = \pi$

or $\quad \frac{\alpha n(n + 1)}{2} = \pi$ i.e. $(n + 1)\alpha = \frac{2\pi}{n}$

Also, $\quad A_2 =$ area of the regular polygon of n sides.

$$= \frac{1}{2}\left(2r\sin\frac{\pi}{4}\cdot r\cos\frac{\pi}{n}\right)n$$

$$= \frac{r^2 n}{2}\left(\sin\frac{2\pi}{n}\right) = \frac{r^2}{2}n\cdot\sin(n + 1)\alpha$$

Hence, $\quad \frac{A_1}{A_2} = \frac{\sin n\alpha}{n\sin\alpha}$

175. Let $M_1, M_2, \ldots, M_n$ are the foot of the altitudes drawn from the centre O to the sides of the polygon and $L_1, L_2, \ldots, L_n$ that of P.

$$PL_1 = OM_1 - OP\cos\theta$$

$$PL_1^2 = OM_1^2 + OP^2\cos^2\theta - 2OM_1OP\cos\theta$$

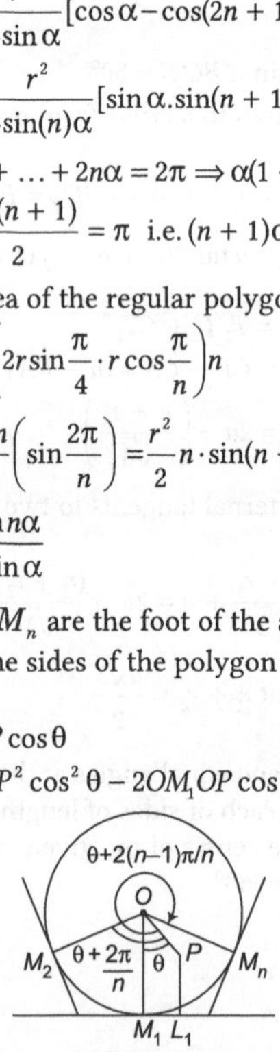

$$= OM_1^2 + \frac{1}{2}OP^2 + \frac{1}{2}OP^2\cos 2\theta - 2OM_1OP\cdot\cos\theta$$

$$\Sigma PL^2 = \Sigma OM_1^2 + \frac{1}{2}\Sigma OP^2 + \frac{1}{2}OP^2\Sigma\cos 2\theta - 2OM_1OP\,\Sigma\cos\theta$$

$$= na^2 + \frac{n}{2}c^2 + \frac{1}{2}c^2(0) - 2ac(0)$$

$$= n\left(a^2 + \frac{c^2}{2}\right)$$

Here, $\Sigma\cos\theta = \cos\theta + \cos\left(\theta + \frac{2\pi}{n}\right) + \cos\left(\theta + \frac{4\pi}{n}\right) + \ldots + \cos\left(\theta + \frac{(2n-1)\pi}{n}\right)$

$$= \sum_{r=0}^{n-1}\cos\left(\theta + \frac{2r\pi}{n}\right) = \text{real part of } \sum_{r=0}^{n-1} i^{(\theta + 2\pi/n)}$$

$$= \text{real part of } e^{i\theta}\sum_{r=0}^{n-1} e^{2\pi i/n} = 0$$

Similarly $\Sigma\cos 2\theta = 0$.

176. Let $c = a^3\cos 3B + 3a^2b\cos(2B - A) + 3ab^2\cos(B - 2A) + b^3\cos 3A$ and $s = a^3\sin 3B + 3a^2b\sin(2B - A) + 3ab^2\sin(B - 2A) + b^3\sin(-3A)$

Now, $c + is = \{a^3\cos 3B + 3a^2b\cos(2B - A) + 3ab^2\sin(B - 2A) + b^3\cos 3A\} + i\{a^3\sin 3B + 3a^2b\sin(2B - A) + 3ab^2\sin(B - 2A) + b^3\sin(-3A)\}$

$$= a^3e^{i3B} + 3a^2be^{i(2B-A)} + 3ab^2e^{i(B-2A)} + b^3e^{-i3A}\ (ae^{iB} + be^{-iA})^3$$

$$= (a\cos B + ai\sin B + b\cos A - bi\sin A)^3$$

$$= \{(a\cos B + b\cos A) + i(a\sin B - b\sin A)\}^3 = c^3$$

$\Rightarrow \{a^3\cos 3B + 3a^2b\cos(2B - A) + 3ab^2\cos(B - 2A) + b^3\cos 3A\} = c^3$ [by equating real parts]

177. $\Delta = \frac{1}{2}ah_1$

$\Rightarrow h_1 = \frac{2\Delta}{a}$, similarly, $h_2 = \frac{2\Delta}{b}, h_3 = \frac{2\Delta}{c}$

So, $\frac{h_1 + r}{h_1 - r} + \frac{h_2 + r}{h_2 - r} + \frac{h_3 + r}{h_3 - r}$

$$= \frac{\frac{2\Delta}{a} + \frac{\Delta}{s}}{\frac{2\Delta}{a} - \frac{\Delta}{s}} + \frac{\frac{2\Delta}{b} + \frac{\Delta}{s}}{\frac{2\Delta}{b} - \frac{\Delta}{s}} + \frac{\frac{2\Delta}{c} + \frac{\Delta}{s}}{\frac{2\Delta}{c} - \frac{\Delta}{s}}$$

$$= \frac{2s + a}{2s - a} + \frac{2s + b}{2s - b} + \frac{2s + c}{2s - c}$$

$$= \frac{4s}{2s - a} + \frac{4s}{2s - b} + \frac{4s}{2s - c} - 3$$

$$= 3\left[\frac{1}{3}\left\{\frac{4s}{2s - a} + \frac{4s}{2s - b} + \frac{4s}{2s - c}\right\}\right] - 3$$

$$\geq \left(\frac{3}{\frac{2s - a}{4s} + \frac{2s - b}{4s} + \frac{2s - c}{c}}\right) - 3 \text{ [since A.M.} \geq \text{H.M.]}$$

≥ 6

178. The given equation can be written as;

$$\frac{a\cos A + b\cos B + c\cos C}{a + b + c} = \frac{a\sin B + b\sin C + c\sin A}{9R}$$

$$\Rightarrow \frac{\sin A\cos A + \sin B\cos B + \sin C\cos C}{\sin A + \sin B + \sin C} = \frac{\frac{2\Delta}{c} + \frac{2\Delta}{a} + \frac{2\Delta}{b}}{9R}$$

$$\Rightarrow \frac{\sin 2A + \sin 2B + \sin 2C}{2(\sin A + \sin B + \sin C)} = \frac{2\Delta(ab + bc + ca)}{9abcR}$$

$$\Rightarrow \frac{2\sin C\{\cos(A - B) - \cos(A + B)\}}{4\cos\frac{C}{2}\left\{\cos\left(\frac{A + B}{2}\right) + \cos\left(\frac{A + B}{2}\right)\right\}} = \frac{2\Delta(ab + bc + ca)}{9abcR}$$

$$\Rightarrow \frac{4\sin A\sin B\sin C}{8\cos\frac{A}{2}\cos\frac{B}{2}\cos\frac{C}{2}} = 4\sin\frac{A}{2}\sin\frac{B}{2}\sin\frac{C}{2}$$

$$= \frac{2\Delta(ab + bc + ca)}{9abcR}$$

$$\Rightarrow \frac{9r}{2\Delta} = \frac{ab + bc + ca}{abc}$$

$$9abc = (a + b + c)(ab + bc + ca)$$

$$\Rightarrow a(b - c)^2 + b(c - a)^2 + c(a - b)^2 = 0 \Rightarrow a = b = c$$

Hence the result.

179. Area of ΔABC,

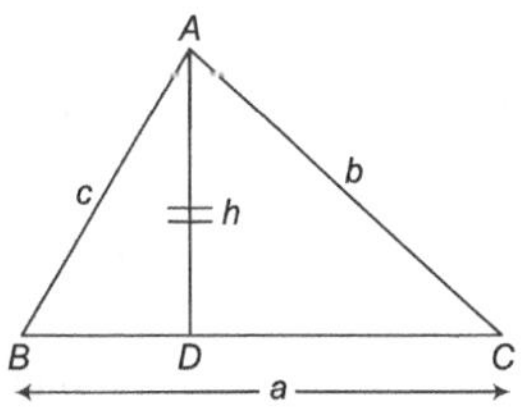

$$\Delta = \frac{1}{2}ah = \frac{1}{2}bc\sin A$$

$$\Rightarrow h = \frac{bc}{a}\sin A$$

$$\Rightarrow h^2 = \frac{b^2c^2}{a^2}(1 - \cos^2 A)$$

$$= \frac{b^2c^2}{a^2}(1 + \cos A)(1 - \cos A)$$

$$= \frac{b^2c^2}{a^2}\left(1 + \frac{b^2 + c^2 - a^2}{2bc}\right)\left(1 - \frac{b^2 + c^2 - a^2}{2bc}\right)$$

$$= \frac{1}{4a^2}((b + c)^2 + (a)^2)(a^2 - (b - c)^2)$$

$$\Rightarrow 4h^2 = \frac{1}{a^2}(a^2((b + c)^2 + (b - c)^2) - a^4 - (b^2 - c^2)^2)$$

$$= 2(b^2 + c^2) - a^2 - \frac{(b^2 - c^2)^2}{a^2}$$

$$\Rightarrow 4h^2 + a^2 = 2(b^2 + c^2) - \frac{(b^2 - c^2)^2}{a^2}$$

$$\Rightarrow 4h^2 + a^2 \leq 2(b^2 + c^2)$$

Clearly equality holds if $b = c$ (i.e., isosceles Δ)

180. Let, $\dfrac{B_1C_1}{BC}=\dfrac{A_1C_1}{AC}=\dfrac{A_1B_1}{AB}=\lambda$

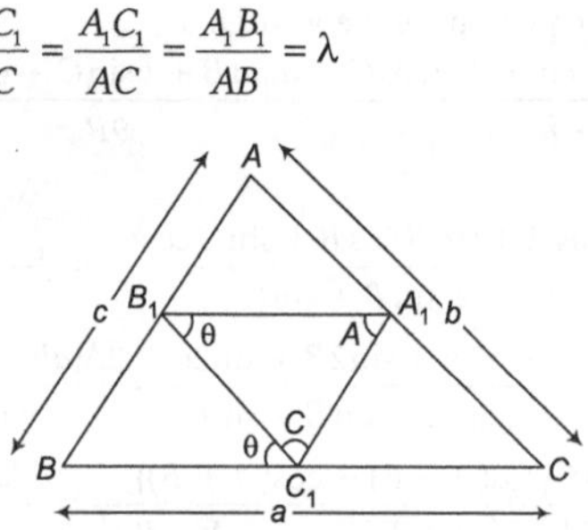

$\Rightarrow \quad B_1C_1=\lambda a,\ A_1C_1=\lambda b,\ A_1B_1=\lambda c$

$\Rightarrow \quad \dfrac{\text{Area of } \Delta A_1B_1C_1}{\text{Area of } DABC}=\lambda^2$

Let $\quad \angle B_1C_1B=\theta \Rightarrow \angle BB_1C_1=\pi-(\theta+B)$

$\angle A_1C_1C=\pi-(\theta+C)$

$\angle C_1A_1C=\pi-(C+\pi-\theta-C)=\theta$

using Sine rule in ΔBB_1C_1, we get :

$$\frac{B_1C_1}{\sin B}=\frac{BC_1}{\sin(\theta+B)}$$

$$\Rightarrow \frac{\lambda a.\sin(\theta+B)}{\sin B}=BC_1$$

Similarly using Sine rule in ΔC_1A_1C, we get

$$\frac{A_1C_1}{\sin C}=\frac{CC_1}{\sin\theta}\Rightarrow CC_1=\frac{\lambda b\cdot\sin\theta}{\sin C}$$

Now, $\quad a=BC_1+CC_1=\lambda\left(\dfrac{a\sin(\theta+B)}{\sin B}+\dfrac{b\sin\theta}{\sin C}\right)$

$$\Rightarrow \quad \lambda=\frac{\sin A\cdot\sin B\cdot\sin C}{\sin A\cdot\sin C\cdot\sin(\theta+B)+\sin\theta\cdot\sin^2 B}$$

$$=\frac{\sin A\sin B\sin C}{(\sin A\cdot\sin C\cdot\cos B+\sin^2 B)\sin\theta+\sin A\cdot\sin C\cdot\sin B\cdot\cos\theta}$$

$$\Rightarrow \lambda\ge\frac{\sin A\cdot\sin B\cdot\sin C}{\sqrt{(\sin A\cdot\sin C\cdot\cos B+\sin^2 B)^2+(\sin A\cdot\sin B\cdot\sin B)^2}}$$

$$\lambda^2\ge\frac{\sin^2 A\cdot\sin^2 B\cdot\sin^2 C}{\sin^2 A\cdot\sin^2 C+\sin^4 B+2\sin A\cdot\sin C\cdot\cos B\,.\,\sin^2 B}$$

$$\ge\frac{\sin^2 A\cdot\sin^2 B\cdot\sin^2 C}{\sin^2 A\cdot\sin^2 C+\sin^2 B\{\sin^2 B+\cos(A-C)\cdot\cos B-\cos(A+C)\cdot\cos B\}}$$

$$\ge\frac{\sin^2 A\cdot\sin^2 B\cdot\sin^2 C}{\sin^2 A\cdot\sin^2 C+\sin^2 B\{1-\cos^2 A+\sin^2 C\}}$$

$$\lambda^2_{\min}=\frac{\sin^2 A\cdot\sin^2 B\cdot\sin^2 C}{\sin^2 A\cdot\sin^2 C+\sin^2 B\cdot\sin^2 A+\sin^2 B\cdot\sin^2 C}$$

$$\therefore \quad \lambda^2_{\min}=\frac{1}{\operatorname{cosec}^2 A+\operatorname{cosec}^2 B+\operatorname{cosec}^2 C}$$

Thus, $\quad \dfrac{\text{ar}(\Delta A_1B_1C_1)}{\text{ar}(\Delta ABC)}\ge\dfrac{1}{\operatorname{cosec}^2 A+\operatorname{cosec}^2 B+\operatorname{cosec}^2 C}$

181. Let ABC be the triangle with D as the mid-point of AC.

$BD=l$ (given)

Area of $\Delta ABC=\dfrac{b^2}{c}\sin A$

Applying cosine rule to ΔABD, we get

$$\frac{b^2+\frac{b^2}{4}-l^2}{b^2}=\cos A\Rightarrow b^2=\frac{4l^2}{5-4\cos A}$$

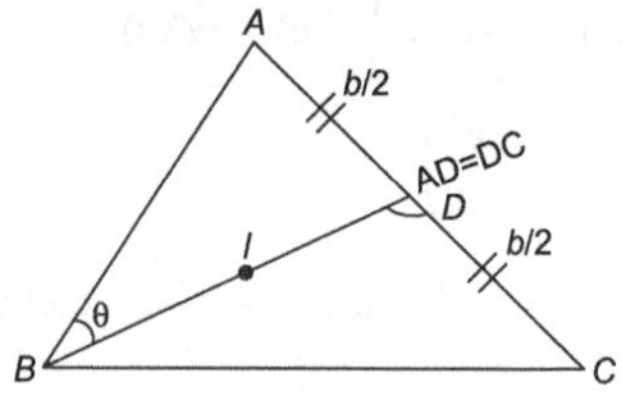

Hence, $\quad \Delta=2l^2\dfrac{\sin A}{5-4\cos A}$

$$\Rightarrow \quad \Delta=2l^2\cdot\frac{\frac{2t}{1-t^2}}{5-4\frac{(1-t^2)}{(1+t^2)}}$$

$$\Rightarrow \quad \Delta=\frac{4l^2t}{9t^2+1}=\frac{4l^2}{9t+\frac{1}{t}}$$

$\Rightarrow \quad \Delta\le\dfrac{2l^2}{3} \qquad \left[\because 9t+\dfrac{1}{t}\ge\left(\dfrac{1}{t}\cdot 9t\right)^{1/2}=6\right]$

Hence, maximum area $=\dfrac{2l^2}{3}$

Equality holds when, $9t=\dfrac{1}{t}\Rightarrow t=\dfrac{1}{3}$ [since $t>0$]

$\Rightarrow \quad \cos A=\dfrac{4}{5}$

182. Let $\angle BAA_1=\angle B_1AC=\theta$ and M and N be the points where incircles of Δ's BAA_1 and B_1AC touch the side BC.

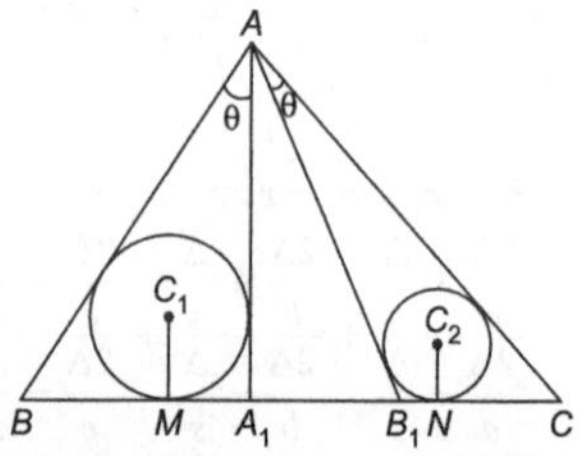

If R_1 and r_1 be the circum-radius and inradius of ΔBAA_1, then

$$r=4R_1\sin\frac{\theta}{2}\cdot\sin\frac{B}{2}\cdot\sin\left(\frac{\pi-(\theta+B)}{2}\right)$$

or $\quad r_1=4R_1\sin\dfrac{\theta}{2}\cdot\sin\dfrac{B}{2}\cdot\sin\dfrac{\alpha}{2},$

where $\alpha=\pi-(\theta+B)$

Now, $\quad BM=r_1\cos\dfrac{B}{2}=4R_1\sin\dfrac{\theta}{2}\cdot\sin\dfrac{\alpha}{2}\cdot\cos\dfrac{B}{2}$

and $\quad MA_1=r_1\cot\dfrac{\alpha}{2}=4R_1\sin\dfrac{\theta}{2}\cdot\sin\dfrac{B}{2}\cdot\cos\dfrac{\alpha}{2}$

Thus, $\dfrac{1}{BM}+\dfrac{1}{MA_1}=\dfrac{1}{4R_1\sin\theta/2}$

$$\left(\frac{1}{\sin\alpha/2\cdot\cos B/2}+\frac{1}{\sin B/2\cdot\cos\alpha/2}\right)$$

$$= \frac{1}{4R_1 \sin\theta/2} \cdot \frac{\sin\left\{\frac{\alpha + B}{2}\right\}}{\sin\alpha/2 \cdot \cos B/2 \cdot \sin B/2 \cdot \cos\alpha/2}$$

$$= \frac{\cos(\theta/2)}{R_1 \sin\theta/2 \sin\alpha \sin B} = \frac{2\cot\theta/2}{c \cdot \sin B}$$

Similarly, $\dfrac{1}{B_1N} + \dfrac{1}{NC} = \dfrac{2\cot\theta/2}{b\sin C}$

Hence, $\dfrac{1}{BM} + \dfrac{1}{MA_1} = \dfrac{1}{B_1N} + \dfrac{1}{NC}$ $\quad$ [as $b\sin C = c\sin B$]

183. Let the angle between PR and AC be θ

$\therefore$ In ΔAPC, from the sine rule

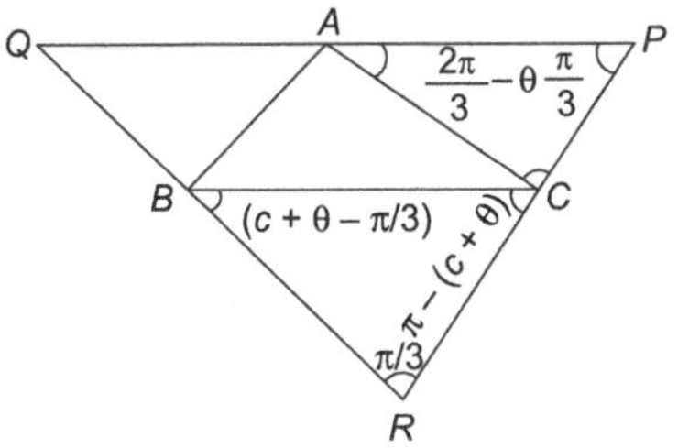

$$PC = \frac{b}{\sin\pi/3}\sin\left(\frac{2\pi}{3} - \theta\right)$$

In ΔBCR, from the sine rule.

$$CR = \frac{a\sin(c + \theta - \pi/3)}{\sin\pi/3}$$

$$\Rightarrow \quad PR = PC + CR = \frac{2}{\sqrt{3}}\left\{b\sin\left(\frac{\pi}{3} + \theta\right) + a\sin\left(c + \theta - \frac{\pi}{3}\right)\right\}$$

$$= \frac{2}{\sqrt{3}}$$

$$\left[\left(b\cos\frac{\pi}{3} + a\cos\left(c - \frac{\pi}{3}\right)\right)\sin\theta + \left(b\sin\frac{\pi}{3} + a\sin\left(c - \frac{\pi}{3}\right)\right)\cdot\cos\theta\right]$$

$\Rightarrow$ Max.

$$PR = \frac{2}{\sqrt{3}}\left[\sqrt{\left(b\cos\frac{\pi}{3} + a\cos\left(c - \frac{\pi}{3}\right)\right)^2 + \left(b\sin\frac{\pi}{3} + a\sin\left(c - \frac{\pi}{3}\right)\right)^2}\right]$$

$$= \frac{2}{\sqrt{3}}\sqrt{b^2 + a^2 + 2ab\cos\left(\frac{2\pi}{3} - c\right)}$$

Max. area of $\Delta PQR = \dfrac{\sqrt{3}}{4}[\text{max. } PR]^2$

$$= \frac{\sqrt{3}}{4}\cdot\frac{4}{3}[b^2 + a^2 - ab\cos C + \sqrt{3}ab\sin C]$$

$$= \frac{2b^2 + 2a^2 - 2ab\cos C}{2\sqrt{3}} + ab\sin C$$

$$= \frac{a^2 + b^2 + c^2}{2\sqrt{3}} + 2\Delta \qquad [\because 2ab\cos C = a^2 + b^2 - c^2]$$

184. Let the circle of radius r_A touch the sides AB and AC at D and E, where as the incircle touches their sides at D and E. Let O and O' be centres of the inscribed and that the circle with radius r_A. O and O' lie on the bisector of angle A.

Also, $AO' = OD'\operatorname{cosec}\dfrac{A}{2} = r_A\operatorname{cosec}\dfrac{A}{2}$

and $AO = OD\operatorname{cosec}\dfrac{A}{2} = r\operatorname{cosec}\dfrac{A}{2}$

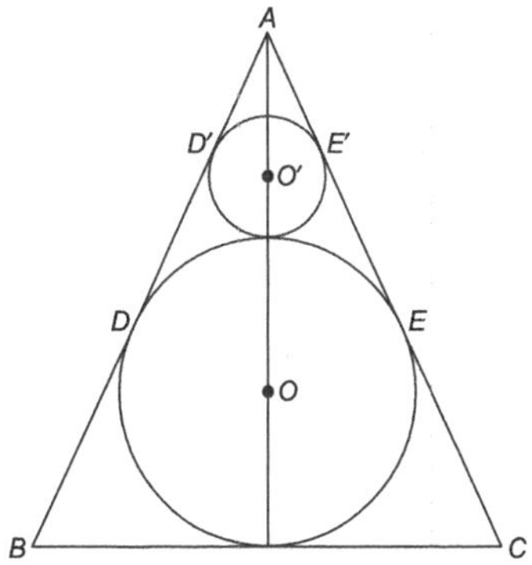

Hence, $OO' = r + r_A = AO - AO'$

$$= r\operatorname{cosec}\frac{A}{2} - r_A\operatorname{cosec}\frac{A}{2}$$

$$\Rightarrow \quad \frac{r_A}{r} = \frac{\operatorname{cosec}\frac{A}{2} - 1}{\operatorname{cosec}\frac{A}{2} + 1}$$

$$\Rightarrow \quad r_A = r\tan^2\left(\frac{\pi - A}{4}\right)$$

Similarly, $\quad r_B = r\tan^2\left(\dfrac{\pi - B}{4}\right), r_C = r\tan^2\left(\dfrac{\pi - C}{4}\right)$

Hence, $\sqrt{r_Ar_B} + \sqrt{r_Br_C} + \sqrt{r_Cr_A}$

$$= \left\{\tan\left(\frac{\pi - A}{4}\right) + \tan\left(\frac{\pi - B}{4}\right) + \tan\left(\frac{\pi - B}{4}\right)\cdot\tan\left(\frac{\pi - C}{4}\right) + \tan\left(\frac{\pi - C}{4}\right)\cdot\tan\left(\frac{\pi - A}{4}\right)\right\}$$

$$= \frac{r}{\cos\left(\frac{\pi - A}{4}\right)\cdot\cos\left(\frac{\pi - B}{4}\right)\cdot\cos\left(\frac{\pi - C}{4}\right)}$$

$$\left\{\sin\left(\frac{\pi - A}{4}\right)\sin\left(\frac{\pi - B}{4}\right)\sin\left(\frac{\pi - C}{4}\right) + \dots\right\}$$

$$= \frac{r}{\cos\left(\frac{\pi - A}{4}\right)\cdot\cos\left(\frac{\pi - B}{4}\right)\cdot\cos\left(\frac{\pi - C}{4}\right)}$$

$$\left\{\cos\left(\frac{\pi + A}{4}\right)\cdot\cos\left(\frac{\pi - A}{4}\right) + \sin\frac{\pi - B}{4}\cdot\sin\frac{\pi - C}{4}\cdot\cos\frac{\pi - A}{4}\right\}$$

$$= \frac{r\cos\left(\frac{\pi - A}{4}\right)}{\cos\left(\frac{\pi - A}{4}\right)\cdot\cos\left(\frac{\pi - B}{4}\right)\cdot\cos\left(\frac{\pi - C}{4}\right)}$$

$$= \frac{r\cos\left(\frac{\pi - A}{4}\right)\cdot\cos\left(\frac{\pi - B}{4}\right)\cdot\cos\left(\frac{\pi - C}{4}\right)}{\cos\left(\frac{\pi - A}{4}\right)\cdot\cos\left(\frac{\pi - B}{4}\right)\cdot\cos\left(\frac{\pi - C}{4}\right)}$$

$$\therefore \quad \sqrt{r_Ar_B} + \sqrt{r_Br_C} + \sqrt{r_Cr_A} = r$$

185. If a,b,c are the lengths of the sides of the Δ and x,y,z are the lengths of perpendicular from the points on the sides BC, CA, AB respectively, we have to minimise,

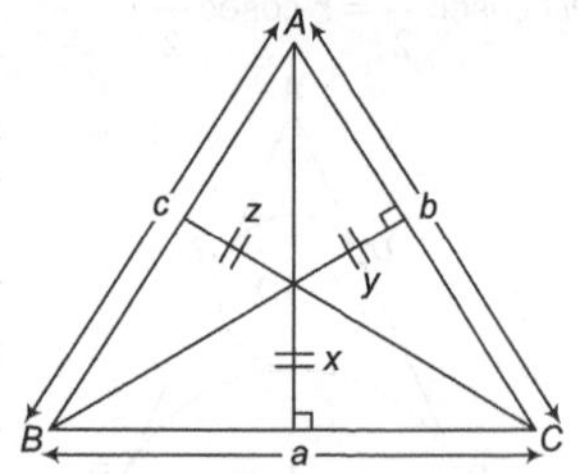

$$\Delta = x^2 + y^2 + z^2 \quad \text{...(i)}$$

We have,

$$\frac{1}{2}ax + \frac{1}{2}by + \frac{1}{2}cz = \Delta$$

$$\Rightarrow \quad ax + by + cz = 2\Delta \quad \text{...(ii)}$$

where, Δ is the area of the ΔABC

we have the identity;

$$\Rightarrow \quad (x^2 + y^2 + z^2)(a^2 + b^2 + c^2) - (ax + by + cz)^2$$
$$= (ax - by)^2 + (by - cz)^2 + (cz - ax)^2$$

$$\Rightarrow \quad (x^2 + y^2 + z^2)(a^2 + b^2 + c^2) \ge (ax + by + cz)^2$$

$$\Rightarrow \quad (x^2 + y^2 + z^2)(a^2 + b^2 + c^2) \ge 4\Delta^2 \quad \text{[using (ii)]}$$

$\Rightarrow x^2 + y^2 + z^2 \ge \dfrac{4\Delta^2}{a^2 + b^2 + c^2}$ and equality holds only when

$$\frac{x}{a} = \frac{y}{b} = \frac{z}{c} = \frac{ax + by + cz}{a^2 + b^2 + c^2} = \frac{2\Delta}{a^2 + b^2 + c^2}$$

$\therefore$ The minimum value of Δ is $\dfrac{4\Delta^2}{a^2 + b^2 + c^2}$.

$$\Delta_{\min} = \frac{4(s-a)(s-b)(s-c)s}{a^2 + b^2 + c^2}.$$

186. In the adjacent figure AA_1 and BB_1 are altitudes drawn from A and B on the sides BC and AC, respectively. Let P, O be the orthocentre and circum-centre of the Δ.

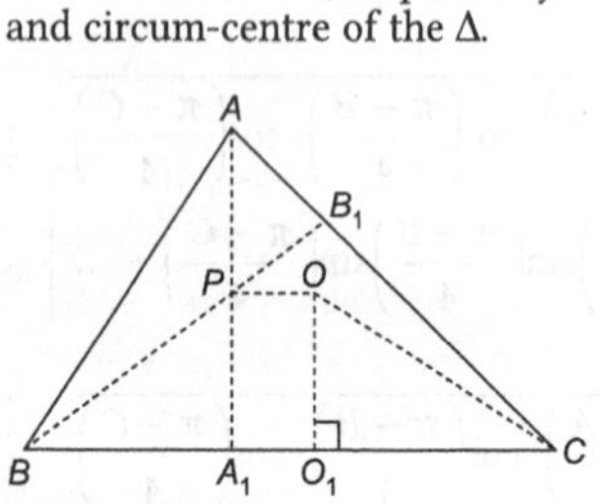

We have been given that $PA_1 = OO_1$.

In ΔOO_1C, $\angle O_1OC = A$

and $\quad OC = R \Rightarrow OO_1 = R\cos A$

Also, $\quad BA_1 = AB\cos B = c\cos B$

and $\quad PA_1 = BA_1 \cot C$

$$PA_1 = c\cos B \cdot \cot C = \frac{c\cos B\cos C}{\sin C}$$

$$= 2R\cos B\cos C$$

Thus, $\quad R\cos A = 2R\cos B\cos C$

$$\Rightarrow \quad 2\cos B\cos C + \cos(B + C) = 0$$

$$\Rightarrow \quad \sin B\sin C = 3\cos B\cos C$$

or $\quad \tan B\tan C = 3$

Now in any $\quad \Delta, \Sigma\tan A = 3\tan A$

$$\Rightarrow \quad \tan A + \tan B + \tan C = 3\tan A$$

$$\Rightarrow \quad \tan B + \tan C = 2\tan A$$

Now,

$$(\tan B - \tan C)^2 > 0 \text{ as } \angle B \ne \angle C$$

$$\Rightarrow \quad (\tan B + \tan C)^2 > 4\tan B\tan C$$

$$\Rightarrow \quad 4\tan^2 A > 12$$

$$\Rightarrow \quad \tan A > \sqrt{3} \quad \text{[as } \angle A \text{ is acute]}$$

$$\Rightarrow \quad A \in \left(\frac{\pi}{3}, \frac{\pi}{2}\right)$$

187. $BD = AB\cos B = c\cos B$

also $\quad \angle BHD = \dfrac{\pi}{2} - \angle EBC = \dfrac{\pi}{2} - \left(\dfrac{\pi}{2} - C\right) = C$

$$\Rightarrow \quad BH = \frac{BD}{\sin C} = \frac{c\cos B}{\sin C} = 2R\cos B$$

Now, points H, D, B and F are concyclic and BH is the diameter of the circle passing through these four points. In fact this circle is also the circum-circle of ΔBFD.

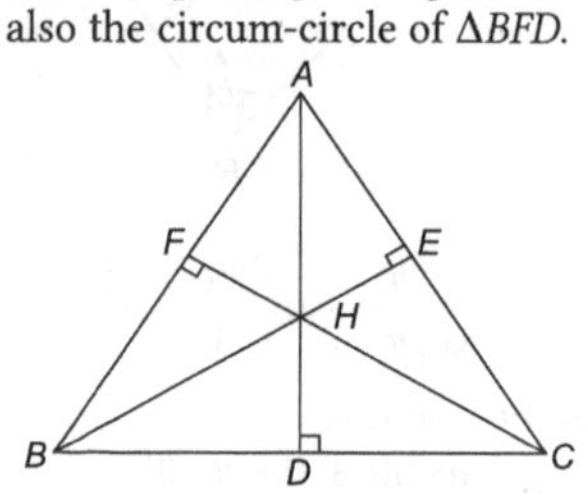

$$\Rightarrow \quad \frac{FD}{\sin B} = BH = 2R\cos B$$

$$\Rightarrow \quad FD = 2R\sin B\cos B = b\cos B$$

$$\Rightarrow \quad \frac{FD}{b} = \cos B$$

Similarly, $\quad \dfrac{EF}{a} = \cos A$ and $\dfrac{DE}{c} = \cos C$

Thus, $\quad \dfrac{EF}{a} + \dfrac{FD}{b} + \dfrac{DE}{c} = \cos A + \cos B + \cos C$

$$= 1 + 4\sin\frac{A}{2}\cdot\sin\frac{B}{2}\cdot\sin\frac{C}{2}$$

$$= 1 + \frac{r}{R} = \frac{R + r}{R}$$

$$\therefore \quad \frac{EF}{a} + \frac{FD}{b} + \frac{DE}{c} = \frac{R + r}{R}$$

188. Let the centres of circle be O_1 and O_2, and their radii be r_1 and r_2 respectively.

We have,

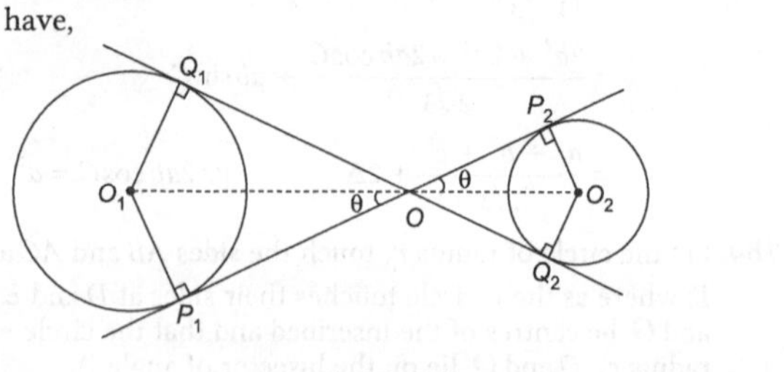

$$r_1 + r_2 = a \text{ and } O_1O_2 = 2a$$

Let, $\angle O_1OP_1 = \theta \Rightarrow \angle P_2OO_2 = \theta$

Now, $O_1O_2 = O_1O + OO_2 = r_1 \text{cosec}\,\theta + r_2 \text{cosec}\,\theta$

$$O_1O_2 = (r_1 + r_2)\,\text{cosec}\,\theta$$

$$\Rightarrow \quad \text{cosec}\,\theta = 2 \Rightarrow \theta = \frac{\pi}{6}$$

Now, $P_1O = r_1 \cot\theta$ and $OP_2 = r_2 \cot\theta$

$$\Rightarrow \quad P_1P_2 = (r_1 + r_2)\cot\theta = a\sqrt{3}$$

Central angle, $\angle Q_1O_1P_1 = 2\left(\frac{\pi}{2} - \theta\right) = \pi - 2\theta$

and $\angle P_2O_2Q_2 = \angle Q_1O_1P_1 = \pi - 2\theta$

$$\Rightarrow \quad \text{Total length of string} = 2P_1P_2 + (r_1 + r_2)(2\pi - (\pi - 2\theta))$$

$$= 2a\sqrt{3} + a\left(\pi + \frac{\pi}{3}\right)$$

$$= \left(\frac{4\pi}{3} + 2\sqrt{3}\right)a$$

189. Let A_1, B_1 and C_1 be the points of contact of the ex-circle opposite to vertex A, with the side AC, BC and AB, respectively.

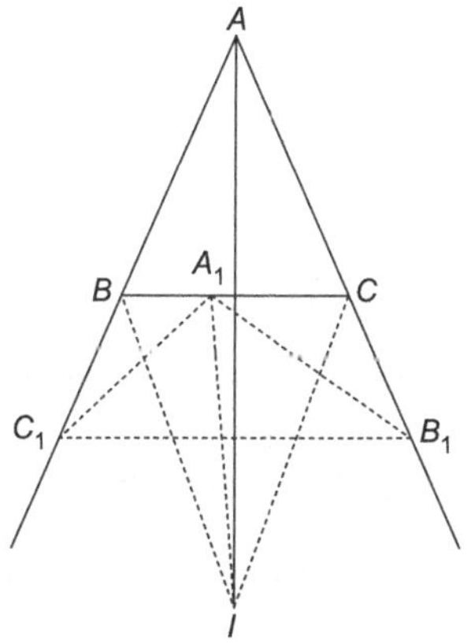

$$\Rightarrow \quad I_1A_1 = I_1B_1 = I_1C_1 = r_1$$

$$\angle B_1I_1A_1 = \pi - \angle B_1CA_1 = \pi - (\pi - C) = C$$

Similarly, $\angle A_1I_1C_1 = B$

$$\Rightarrow \quad \angle C_1I_1B_1 = (B + C)$$

Now, $\Delta_1 = \Delta A_1I_1B_1 + \Delta A_1I_1C_1 - \Delta B_1I_1C_1$

$$= \frac{1}{2}(r_1^2)(\sin C + \sin B - \sin A)$$

$$= \frac{1}{2}\frac{\Delta^2}{(s-a)^2}\left(\frac{c}{2R} + \frac{b}{2R} - \frac{a}{2R}\right)$$

$$= \frac{1}{2}\frac{\Delta^2}{(s-a)^2} \cdot \frac{(2s - 2a)}{2R} = \frac{\Delta^2}{2R(s-a)} = \frac{r_1\Delta}{2R}$$

$$\Rightarrow \quad \frac{\Delta_1}{\Delta} = \frac{r_1}{2R}$$

Similarly, $\frac{\Delta_2}{\Delta} = \frac{r_2}{2R}, \frac{\Delta_3}{\Delta} = \frac{r_3}{2R}$ and $\frac{\Delta_0}{\Delta} = \frac{r}{2R}$

Thus, $\frac{\Delta_1}{\Delta} + \frac{\Delta_2}{\Delta} + \frac{\Delta_3}{\Delta} - \frac{\Delta_0}{\Delta} = \frac{1}{2}R(r_1 + r_2 + r_3 - r)$

$$= \frac{1}{2R}(4R) = 2$$

190. Since, $\angle APB = \angle BPC = \angle CPA$

$\Rightarrow$ each of these angles is equal to $\frac{2\pi}{3}$

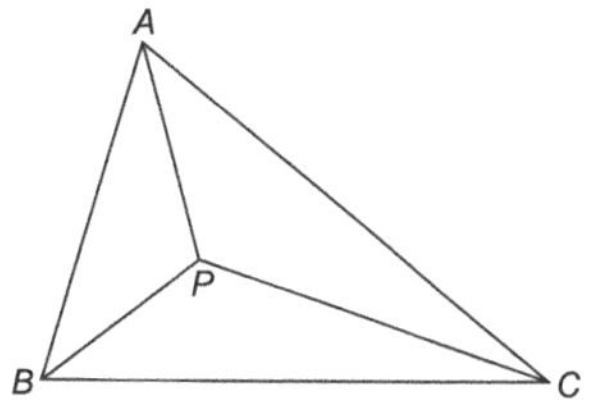

In ΔAPC we have,

$$PA^2 + PC^2 - 2PA.PC\cos\frac{2\pi}{3} = b^2$$

$$\Rightarrow \quad b^2 = PA^2 + PC^2 + PA \cdot PC$$

Similarly, in ΔBPC;

$$a^2 = PB^2 + PC^2 + PB \cdot PC$$

and in ΔAPB; $c^2 = AP^2 + PB^2 + PA \cdot PB$

Adding these results, we get

$$a^2 + b^2 + c^2 = 2(PA^2 + PB^2 + PC^2) + PA \cdot PB + PB \cdot PC + PC \cdot PA$$

$$\Rightarrow \quad a^2 + b^2 + c^2 = 2((PA + PB + PC)^2 - 2\Sigma PA \cdot PB) + \Sigma PA \cdot PB$$

Now, $\Delta = \Delta APC + \Delta BPC + \Delta APB$

$$= \frac{1}{2}\sin\frac{2\pi}{B}(AP \cdot PC + BP \cdot PC + PA \cdot PB) = \frac{\sqrt{3}}{4}(\Sigma PA \cdot PB)$$

Putting the value of $\Sigma PA \cdot PB$, we get

$$a^2 + b^2 + c^2 = 2(PA + PB + PC)^2 - 3 \cdot \frac{4\Delta}{\sqrt{3}}$$

$$\Rightarrow \quad (PA + PB + PC)^2 = \frac{a^2 + b^2 + c^2}{2} + 2\sqrt{3}\Delta$$

$$\Rightarrow \quad PA + PB + PC = \sqrt{\frac{a^2 + b^2 + c^2}{2} + 2\sqrt{3}\Delta}$$

191. In right angled Δ's $A'BD$ and $CA'D$

$$\angle A'BD = \angle CA'D$$

$$\angle BDA' = \angle CDA' = 90°$$

$\therefore$ The Δ's are equiangular and hence similar

So, $\frac{BD}{A'D} = \frac{A'D}{DC}$

So that $A'D^2 = BD.DC$

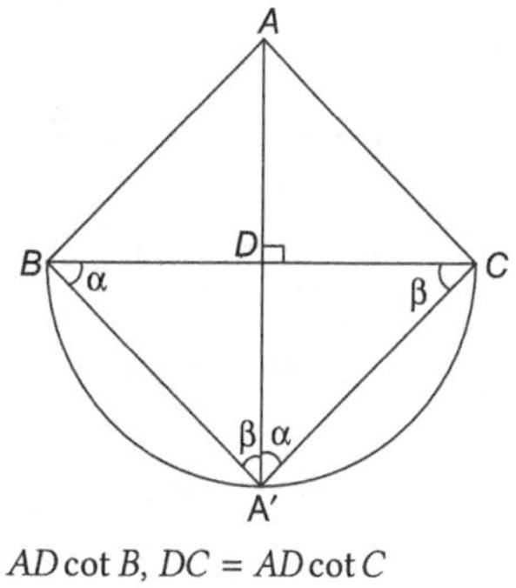

Since, $BD = AD\cot B$, $DC = AD\cot C$

$$\therefore \quad A'D^2 = AD^2 \cot B \cot C$$

or $\{\text{ar}\,(BCA')\}^2 = \left(\frac{1}{2}BC \cdot A'D\right)^2$

$$= \frac{1}{4} BC^2 \cdot AD^2 \cdot \cot B \cdot \cot C$$
$$= \{\text{ar}(ABC)^2\} \cdot \cot B \cdot \cot C \quad \text{...(i)}$$

Similarly,

$$\{\text{ar}(CAB')\}^2 = \{\text{ar}(ABC)\}^2 \cdot \cot C \cdot \cot A \quad \text{...(ii)}$$
$$\{\text{ar}(ABC')\}^2 = \{\text{ar}(ABC)\}^2 \cdot \cot A \cdot \cot B \quad \text{...(iii)}$$

On adding Eqs. (i), (ii) and (iii), we get the desired result because in any Δ;

$$\cot A \cot B + \cot B \cot C + \cot C \cdot \cot A = 1$$
$$\Rightarrow \{\text{ar}(BCA')\}^2 + \{\text{ar}(CAB')\}^2 + \{\text{ar}(ABC')\}^2 = \{\text{ar}\,(ABC)\}^2$$

192. Given a ΔXYZ, where $2s = x + y + z$

and $\frac{s-x}{4} = \frac{s-y}{3} = \frac{s-z}{2}$

$\therefore \quad \frac{s-x}{4} = \frac{s-y}{3} = \frac{s-z}{2} = \frac{3s-(x+y+z)}{4+3+2} = \frac{s}{9}$

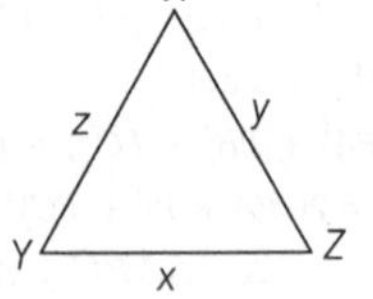

or $\quad \frac{s-x}{4} = \frac{s-y}{3} = \frac{s-z}{2} = \frac{s}{9} = \lambda$ (let)

$\Rightarrow \quad s = 9\lambda, s = 4\lambda + x, s = 3\lambda + y$ and $s = 2\lambda + z$

$\therefore \quad s = 9\lambda, x = 5\lambda, y = 6\lambda, z = 7\lambda$

Now, $\quad \Delta = \sqrt{s(s-x)(s-y)(s-z)}$ [heron's formula]

$$= \sqrt{9\lambda \cdot 4\lambda \cdot 3\lambda \cdot 2\lambda} = 6\sqrt{6}\lambda^2 \quad \text{...(i)}$$

Also, $\quad \pi r^2 = \frac{8\pi}{3} \Rightarrow r^2 = \frac{8}{3}$ ...(ii)

and $\quad R = \frac{xyz}{4\Delta} = \frac{(5\lambda)(6\lambda)(7\lambda)}{4 \cdot 6\sqrt{6}\lambda^2} = \frac{35\lambda}{4\sqrt{6}}$ ...(iii)

Now, $\quad r^2 = \frac{8}{3} = \frac{\Delta^2}{S^2} = \frac{216\lambda^4}{81\lambda^2} \Rightarrow \frac{8}{3} = \frac{8}{3}\lambda^2$ [from Eq. (ii)]

$\Rightarrow \quad \lambda = 1$

(a) $\Delta XYZ = 6\sqrt{6}\lambda^2 = 6\sqrt{6}$

$\therefore$ Option (a) is correct.

(b) Radius of circum-circle, $R = \frac{35}{4\sqrt{6}}\lambda = \frac{35}{4\sqrt{6}}$

$\therefore$ Option (b) is incorrect.

(c) Since, $\quad r = 4R \sin\frac{X}{2} \cdot \sin\frac{Y}{2} \cdot \sin\frac{Z}{2}$

$\Rightarrow \quad \frac{2\sqrt{2}}{\sqrt{3}} = 4 \cdot \frac{35}{4\sqrt{6}} \sin\frac{X}{2} \cdot \sin\frac{Y}{2} \cdot \sin\frac{Z}{2}$

$\Rightarrow \quad \frac{4}{35} = \sin\frac{X}{2} \cdot \sin\frac{Y}{2} \cdot \sin\frac{Z}{2}$

$\therefore$ Option (c) is correct.

(d) $\sin^2\left(\frac{X+Y}{2}\right) = \cos^2\left(\frac{Z}{2}\right)$, as $\frac{X+Y}{2} = 90^\circ - \frac{Z}{2}$

$$= \frac{s(s-z)}{xy} = \frac{9 \times 2}{5 \times 6} = \frac{3}{5}$$

$\therefore$ Option (d) is correct.

193. (i) $\cos C = \frac{a^2 + b^2 - c^2}{2ab}$

(ii) $R = \frac{abc}{4\Delta}, r = \frac{\Delta}{s}$

where, R, r, Δ denote the circum-radius, in-radius and area of triangle, respectively.

Let the sides of triangle be a, b and c.

Given, $\quad x = a + b$

$$y = ab$$
$$x^2 - c^2 = y$$

$\Rightarrow \quad a^2 + b^2 + 2ab - c^2 = ab$

$\Rightarrow \quad a^2 + b^2 - c^2 = -ab$

$\Rightarrow \quad \frac{a^2 + b^2 - c^2}{2ab} = -\frac{1}{2} = \cos 120^\circ$

$\Rightarrow \quad \angle C = \frac{2\pi}{3}$

$\because \quad R = \frac{abc}{4\Delta}, r = \frac{\Delta}{s}$

$\Rightarrow \quad \frac{r}{R} = \frac{4\Delta^2}{s(abc)} = \frac{4\left[\frac{1}{2}ab\sin\left(\frac{2\pi}{3}\right)\right]^2}{\frac{x+c}{2} \cdot y \cdot c}$

$\therefore \quad \frac{r}{R} = \frac{3y}{2c(x+c)}$

194. We know $\Delta = \frac{1}{2} ab \sin C \Rightarrow 15\sqrt{3} = \frac{1}{2} \times 6 \times 10 \times \sin C$

$\sin C = \frac{\sqrt{3}}{2}$ and C is given to be obtuse.

$\Rightarrow \quad C = \frac{2\pi}{3} c \Rightarrow = \sqrt{a^2 + b^2 - 2ab\cos C}$

$$C = \sqrt{6^2 + 10^2 - 2 \times 6 \times 10 \times \cos\frac{2\pi}{3}} = 14$$

$\therefore \quad r = \frac{\Delta}{s} \Rightarrow r^2 = \frac{225 \times 3}{\left(\frac{6+10+14}{2}\right)^2} = 3$

195. Whenever cosine of angle and sides are given or to find out, we should always use cosine law.

i.e. $\quad \cos A = \frac{b^2 + c^2 - a^2}{2bc}$,

$\cos B = \frac{a^2 + c^2 - b^2}{2ac}$ and $\cos C = \frac{a^2 + b^2 - c^2}{2ab}$

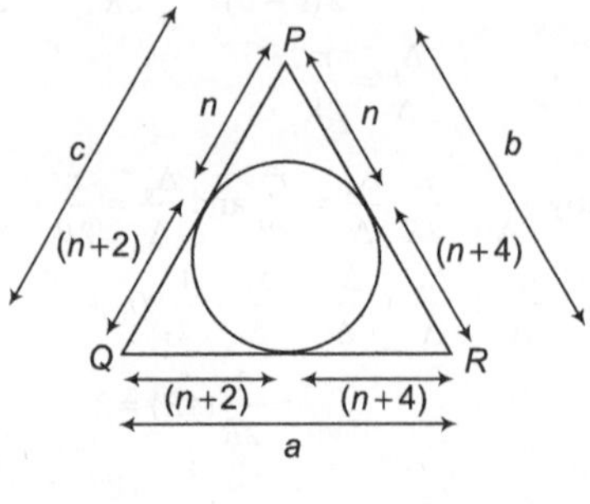

$\therefore \quad \cos P = \dfrac{b^2 + c^2 - a^2}{2bc}$

$\Rightarrow \quad \dfrac{1}{3} = \dfrac{(2n+4)^2 + (2n+2)^2 - (2n+6)^2}{2(2n+4)(2n+2)}$

$\left[\because \cos P = \dfrac{1}{3}, \text{ given}\right]$

$\Rightarrow \quad \dfrac{4n^2 - 16}{8(n+1)(n+2)} = \dfrac{1}{3} \Rightarrow \dfrac{n^2 - 4}{2(n+1)(n+2)} = \dfrac{1}{3}$

$\Rightarrow \quad \dfrac{(n-2)}{2(n+1)} = \dfrac{1}{3}$

$\Rightarrow \quad 3n - 6 = 2n + 2 \Rightarrow n = 8$

$\therefore$ Sides are $(2n+2), (2n+4), (2n+6)$, i.e. 18, 20, 22.

196. If ΔABC has sides a, b, c.

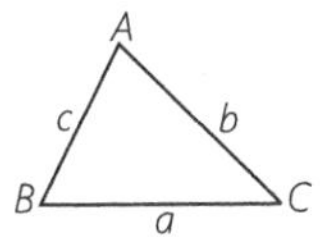

Then, $\tan(A/2) = \sqrt{\dfrac{(s-b)(s-a)}{s(s-a)}}$

where, $\quad s = \dfrac{a+b+c}{2} = \dfrac{2 + \frac{7}{2} + \frac{5}{2}}{2} = 4$

$\therefore \quad \dfrac{2\sin P - \sin 2P}{2\sin P + \sin 2P} = \dfrac{2\sin P(1 - \cos P)}{2\sin P(1 + \cos P)}$

$= \dfrac{2\sin^2(P/2)}{2\cos^2(P/2)} = \tan^2(P/2)$

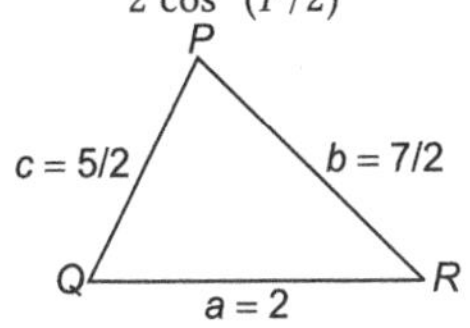

$\Rightarrow \quad \dfrac{(s-b)(s-c)}{s(s-a)} \times \dfrac{(s-b)(s-c)}{(s-b)(s-c)}$

$= \dfrac{[(s-b)^2(s-c)^2]}{\Delta^2} = \dfrac{\left(4 - \frac{7}{2}\right)^2\left(4 - \frac{5}{2}\right)^2}{\Delta^2}$

$= \left(\dfrac{3}{4\Delta}\right)^2$

197. Since, A, B, C are in AP.

$\Rightarrow \quad 2B = A + C$ i.e. $\angle B = 60°$

$\therefore \quad \dfrac{a}{c}(2\sin C\cos C) + \dfrac{c}{a}(2\sin A\cos A)$

$= 2k(a\cos C + c\cos A)$

$\left[\text{using, } \dfrac{a}{\sin A} = \dfrac{b}{\sin B} = \dfrac{c}{\sin C} = \dfrac{1}{k}\right]$

$= 2k(b) = 2\sin B \quad$ [using $b = a\cos C + c\cos A$]

$= \sqrt{3}$

198. Using, $\cos C = \dfrac{a^2 + b^2 - c^2}{2ab}$

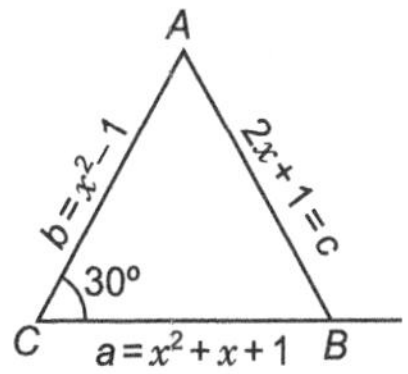

$\Rightarrow \quad \dfrac{\sqrt{3}}{2} = \dfrac{(x^2+x+1)^2 + (x^2-1)^2 - (2x+1)^2}{2(x^2+x+1)(x^2-1)}$

$\Rightarrow (x+2)(x+1)(x-1)x + (x^2-1)^2 = \sqrt{3}(x^2+x+1)(x^2-1)$

$\Rightarrow x^2 + 2x + (x^2 - 1) = \sqrt{3}(x^2 + x + 1)$

$\Rightarrow \quad (2-\sqrt{3})x^2 + (2-\sqrt{3})x - (\sqrt{3}+1) = 0$

$\Rightarrow \quad x = -(2+\sqrt{3})$ and $x = 1 + \sqrt{3}$

But $\quad x = -(2+\sqrt{3})$

$\Rightarrow$ c is negative.

$\therefore \quad x = 1 + \sqrt{3}$ is the only solution.

199. Given, $\cos B + \cos C = 4\sin^2\dfrac{A}{2}$

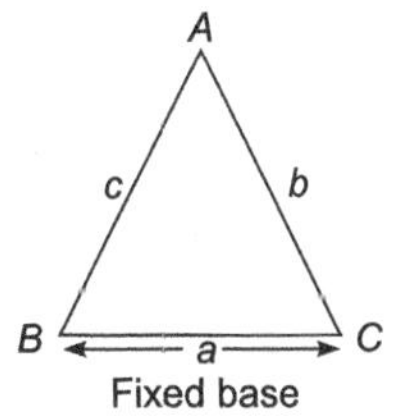

$\Rightarrow 2\cos\left(\dfrac{B+C}{2}\right)\cos\left(\dfrac{B-C}{2}\right) = 4\sin^2\dfrac{A}{2}$

$\Rightarrow 2\sin\dfrac{A}{2}\left[\cos\left(\dfrac{B-C}{2}\right) - 2\sin\dfrac{A}{2}\right] = 0$

$\Rightarrow \quad \cos\left(\dfrac{B-C}{2}\right) - 2\cos\left(\dfrac{B+C}{2}\right) = 0$

As $\quad \sin\dfrac{A}{2} \neq 0$

$\Rightarrow \quad -\cos\dfrac{B}{2}\cos\dfrac{C}{2} + 3\sin\dfrac{B}{2}\sin\dfrac{C}{2} = 0$

$\Rightarrow \quad \tan\dfrac{B}{2}\tan\dfrac{C}{2} = \dfrac{1}{3}$

$\Rightarrow \quad \sqrt{\dfrac{(s-a)(s-c)}{s(s-b)} \cdot \dfrac{(s-b)(s-a)}{s(s-c)}} = \dfrac{1}{3}$

$\Rightarrow \quad \dfrac{s-a}{s} = \dfrac{1}{3}$

$\Rightarrow \quad 2s = 3a \Rightarrow b + c = 2a$

$\therefore$ Locus of A is an ellipse.

200. In ΔABC, by sine rule, $\dfrac{a}{\sin A} = \dfrac{2\sqrt{2}}{\sin 30°} = \dfrac{4}{\sin C}$

$\Rightarrow \quad C = 45°, C' = 135°$

When, $\quad C = 45° \Rightarrow A = 180° - (45° + 30°) = 105°$

When, $\quad C' = 135° \Rightarrow A = 180° - (135° + 30°) = 15°$

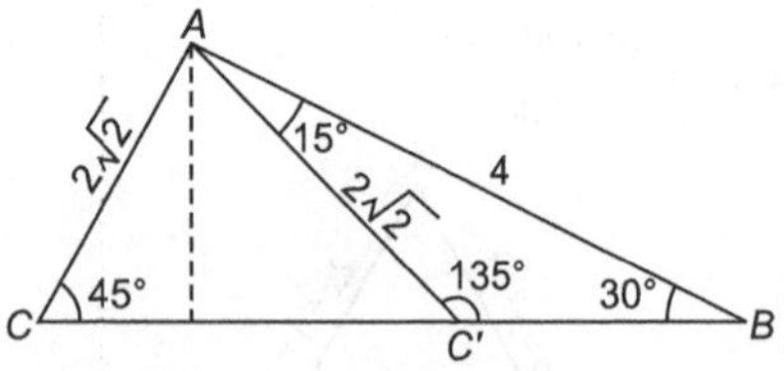

Area of $\Delta ABC = \frac{1}{2} AB \times AC \sin A = \frac{1}{2} \times 4 \times 2\sqrt{2} \sin (105°)$

$$= 4\sqrt{2} \times \frac{\sqrt{3}+1}{2\sqrt{2}} = 2(\sqrt{3}+1) \text{ sq units}$$

Area of $\Delta ABC' = \frac{1}{2} AB \times AC \sin A$

$$= \frac{1}{2} \times 4 \times 2\sqrt{2} \sin (15°) = 2(\sqrt{3}-1) \text{ sq units}$$

Difference of areas of triangle

$$= |2(\sqrt{3}+1) - 2(\sqrt{3}-1)| = 4 \text{ sq units}$$

Alternate Method

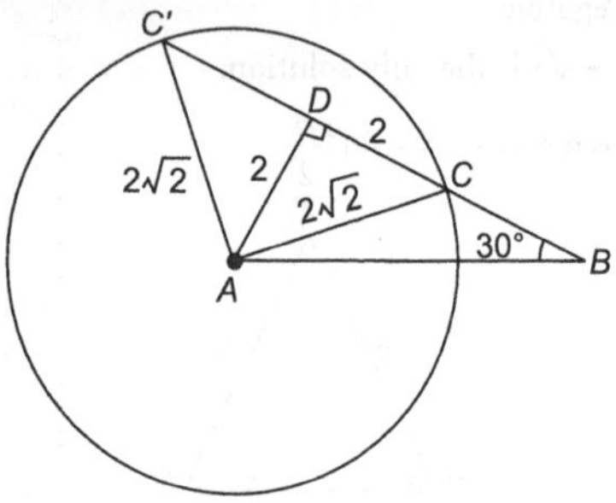

Here, $AD = 2$, $DC = 2$

Difference of areas of ΔABC and $\Delta ABC'$

$= $ Area of $\Delta ACC'$

$$= \frac{1}{2} AD \times CC' = \frac{1}{2} \times 2 \times 4 = 4 \text{ sq units}$$

201. Let a straight line through the vertex P of a given $\Delta\ PQR$ intersects the side QR at the point S and the circum-circle of $\Delta\ PQR$ at the point T.

Points P, Q, R, T are concyclic, then $PS \cdot ST = QS \cdot SR$

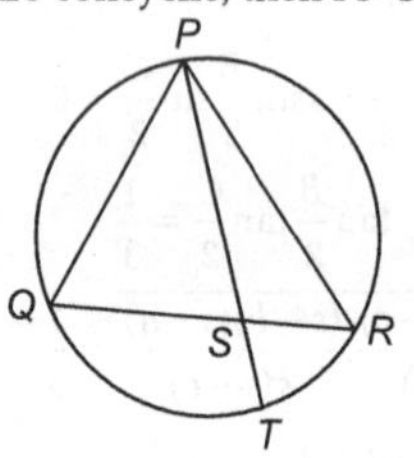

Now, $\frac{PS+ST}{2} > \sqrt{PS \cdot ST}$ [$\because$ AM > GM]

and $\frac{1}{PS} + \frac{1}{ST} > \frac{2}{\sqrt{PS \cdot ST}} = \frac{2}{\sqrt{QS \cdot SR}}$

Also, $\frac{SQ+QR}{2} > \sqrt{SQ \cdot SR}$

$\Rightarrow \frac{QR}{2} > \sqrt{SQ \cdot SR}$

$\Rightarrow \frac{1}{\sqrt{SQ \cdot SR}} > \frac{2}{QR} \Rightarrow \frac{2}{\sqrt{SQ \cdot SR}} > \frac{4}{QR}$

$\therefore \frac{1}{PS} + \frac{1}{ST} > \frac{2}{\sqrt{QS \cdot SR}} > \frac{4}{QR}$

202. Radius of in-circle is, $r = \frac{\Delta}{s}$

Since, $\Delta = 16\sqrt{2}$

Now, $s = \frac{6\sqrt{2}+6\sqrt{2}+4\sqrt{2}}{2} = 8\sqrt{2}$

$\therefore r = \frac{16\sqrt{2}}{8\sqrt{2}} = 2$

203. Equation of circum-circle of ΔPRS is

$$(x+1)(x-9) + y^2 + \lambda y = 0$$

It will pass through $(1, 2\sqrt{2})$, then $-16+8+\lambda \cdot 2\sqrt{2} = 0$

$\Rightarrow \lambda = \frac{8}{2\sqrt{2}} = 2\sqrt{2}$

$\therefore$ Equation of circum-circle is $x^2+y^2-8x+2\sqrt{2}y-9=0$

Hence, its radius is $3\sqrt{3}$.

Alternate Solution

Let $\angle PSR = \theta \Rightarrow \sin\theta = \frac{2\sqrt{2}}{2\sqrt{3}}$

$\therefore \sin\theta = \frac{PR}{2R}$

$\Rightarrow PR = 6\sqrt{2} = 2R \cdot \sin\theta$

$\Rightarrow R = 3\sqrt{3}$

204. Coordinates of P and Q are $(1, 2\sqrt{2})$ and $(1, -2\sqrt{2})$.

Now, $PQ = \sqrt{(4\sqrt{2})^2 + 0^2} = 4\sqrt{2}$

Area of $\Delta\, PQR = \frac{1}{2} \cdot 4\sqrt{2} \cdot 8 = 16\sqrt{2}$ sq units

Area of $\Delta\, PQS = \frac{1}{2} \cdot 4\sqrt{2} \cdot 2 = 4\sqrt{2}$ sq units

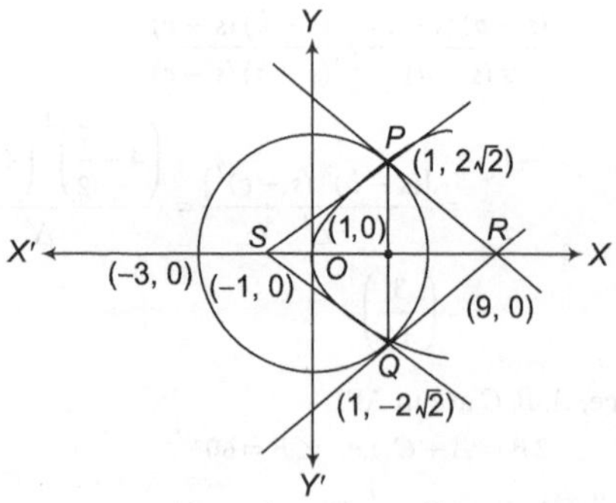

Ratio of areas of ΔPQS and ΔPQR is 1 : 4.

205. Since, $\Delta ABC = \Delta ABD + \Delta ACD$

$\Rightarrow \frac{1}{2} bc \sin A = \frac{1}{2} c\, AD \sin\frac{A}{2} + \frac{1}{2} b\, AD \sin\frac{A}{2}$

$\Rightarrow AD = \frac{2bc}{b+c} \cos\frac{A}{2}$

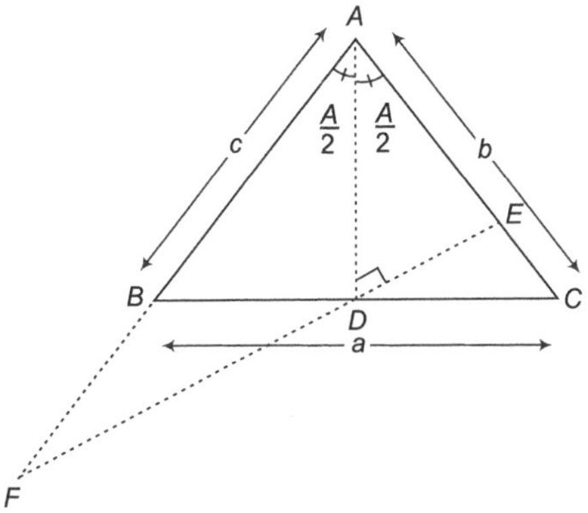

Again, $\quad AE = AD \sec \dfrac{A}{2} = \dfrac{2bc}{b+c}$

$\Rightarrow$ AE is HM of b and c.

$$EF = ED + DF = 2DE = 2AD \tan \frac{A}{2}$$

$$= 2\frac{2bc}{b+c}\cos \frac{A}{2} \tan \frac{A}{2} = \frac{4bc}{b+c} \sin \frac{A}{2}$$

Since, $AD \perp EF$ and $DE = DF$ and AD is bisector.

$\Rightarrow$ ΔAEF is isosceles.

Hence, (a), (b), (c), (d) are correct answers.

206. Let $AB = AC = a$ and $\angle A = 120°$.

$\therefore$ Area of triangle $= \dfrac{1}{2} a^2 \sin 120°$

where, $\quad a = AD + BD$

$$= \sqrt{3} \tan 30° + \sqrt{3} \cot 15°$$

$$= 1 + \frac{\sqrt{3}}{\tan(45° - 15°)}$$

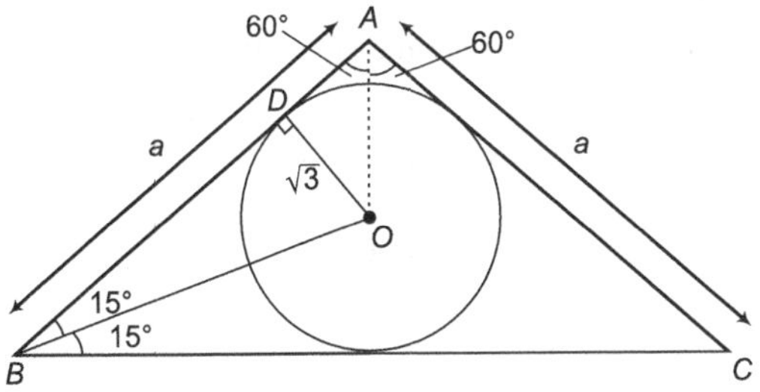

$$\Rightarrow \quad a = 1 + \sqrt{3}\left(\frac{1 + \tan 45° \tan 30°}{\tan 45° - \tan 30°}\right)$$

$$= 1 + \sqrt{3}\left(\frac{\sqrt{3}+1}{\sqrt{3}-1}\right)$$

$$= 4 + 2\sqrt{3}$$

$\therefore$ Area of a triangle $= \dfrac{1}{2}(4 + 2\sqrt{3})^2\left(\dfrac{\sqrt{3}}{2}\right)$

$$= (12 + 7\sqrt{3}) \text{ sq units}$$

207. Let a, b, c are the sides of ΔABC.

Now, $\quad \dfrac{b+c}{a} = \dfrac{k(\sin B + \sin C)}{k \sin A}$

$$= \frac{2\sin\left(\frac{B+C}{2}\right)\cos\left(\frac{B-C}{2}\right)}{2\sin\frac{A}{2}\cos\frac{A}{2}}$$

$$\Rightarrow \quad \frac{b+c}{a} = \frac{\cos\left(\frac{B-C}{2}\right)}{\sin\frac{A}{2}}$$

Also, $\quad \dfrac{b-c}{a} = \dfrac{\sin\left(\frac{B-C}{2}\right)}{\cos\frac{A}{2}}$

208. Let $AB = h$, then $AD = 2h$ and $AC = BC = \dfrac{h}{2}$

Again, let $\angle CPA = \alpha$

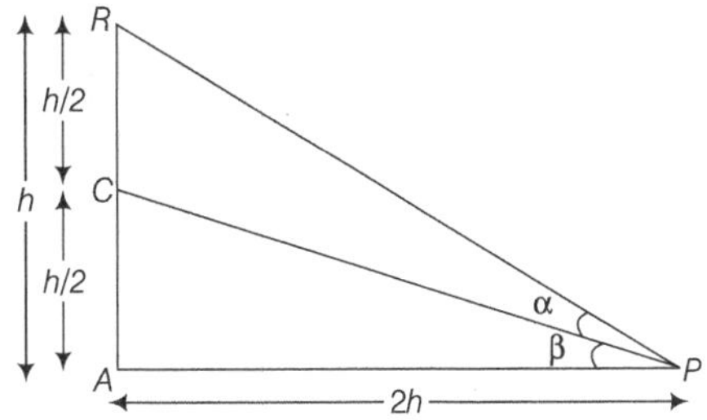

Now, in ΔABP, $\tan(\alpha + \beta) = \dfrac{AB}{AP} = \dfrac{h}{2h} = \dfrac{1}{2}$

Also, in ΔACP, $\tan\alpha = \dfrac{AC}{AP} = \dfrac{2}{2h} = \dfrac{1}{4}$

Now, $\tan\beta = \tan[(\alpha + \beta) - \alpha]$

$$= \frac{\tan(\alpha+\beta) - \tan\alpha}{1 + \tan(\alpha+\beta)\tan\alpha} = \frac{\frac{1}{2} - \frac{1}{4}}{1 + \frac{1}{2} \times \frac{1}{4}} = \frac{\frac{1}{4}}{\frac{9}{8}} = \frac{2}{9}$$

209. Let $AB = x$

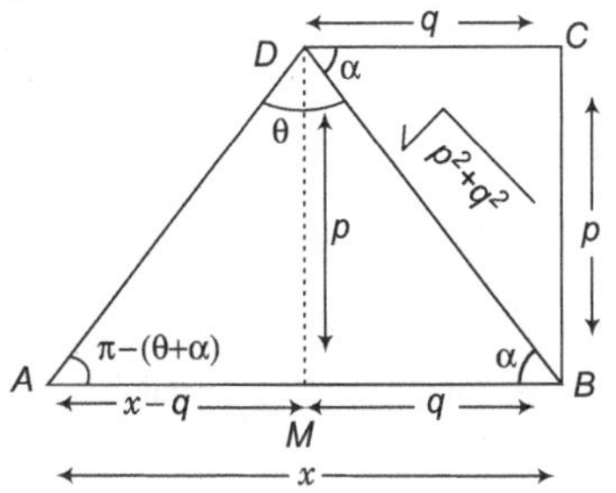

In ΔDAM, $\tan(\pi - \theta - \alpha) = \dfrac{p}{x-q}$

$$\Rightarrow \quad \tan(\theta + \alpha) = \frac{p}{q-x}$$

$$\Rightarrow \quad q - x = p\cot(\theta + \alpha)$$

$$\Rightarrow \quad x = q - p\cot(\theta + \alpha)$$

$$= q - p\left(\frac{\cot\theta\cot\alpha - 1}{\cot\alpha + \cot\theta}\right) \qquad \left[\because \cot\alpha = \frac{q}{p}\right]$$

$$= q - p\left(\frac{\frac{q}{p}\cot\theta - 1}{\frac{q}{p} + \cot\theta}\right) = q - p\left(\frac{q\cot\theta - p}{q + p\cot\theta}\right)$$

$$= q - p\left(\frac{q\cos\theta - p\sin\theta}{q\sin\theta + p\cos\theta}\right)$$

$$\Rightarrow \quad x = \frac{q^2\sin\theta + pq\cos\theta - pq\cos\theta + p^2\sin\theta}{p\cos\theta + q\sin\theta}$$

$$\Rightarrow \quad AB = \frac{(p^2 + q^2)\sin\theta}{p\cos\theta + q\sin\theta}$$

Alternate Solution

Applying sine rule in ΔABD,

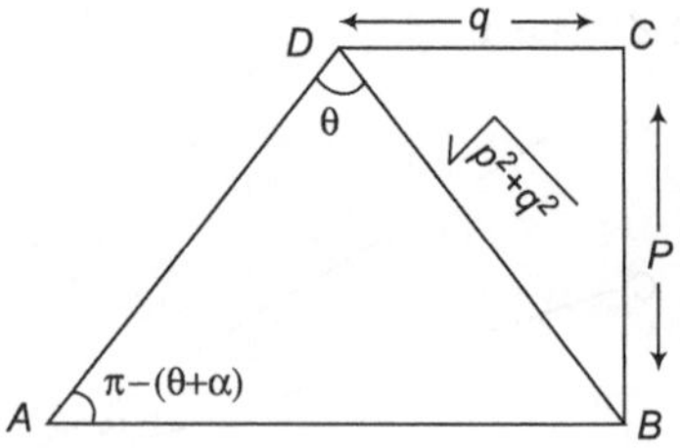

$$\frac{AB}{\sin\theta} = \frac{\sqrt{p^2 + q^2}}{\sin\{\pi - (\theta + \alpha)\}}$$

$$\Rightarrow \quad \frac{AB}{\sin\theta} = \frac{\sqrt{p^2 + q^2}}{\sin(\theta + \alpha)}$$

$$\Rightarrow \quad AB = \frac{\sqrt{p^2 + q^2}\sin\theta}{\sin\theta\cos\alpha + \cos\theta\sin\alpha} \quad \left[\because \cos\alpha = \frac{q}{\sqrt{p^2 + q^2}}\right.$$

$$= \frac{(p^2 + q^2)\sin\theta}{p\cos\theta + q\sin\theta} \quad \left.\text{and } \sin\alpha = \frac{p}{\sqrt{p^2 + q^2}}\right]$$

210.

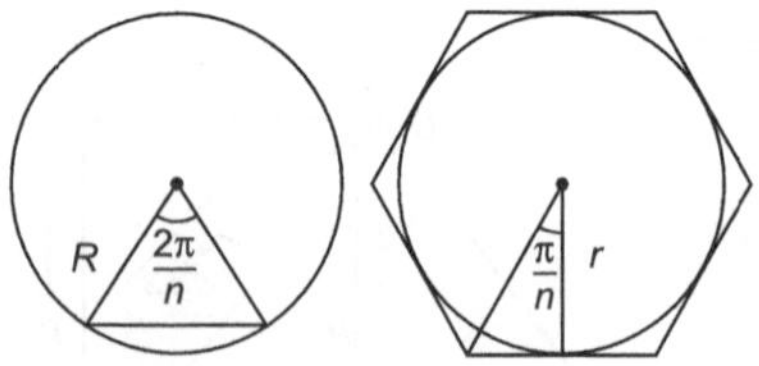

By formula of regular polygon,

$$\frac{a}{2R} = \sin\frac{\pi}{n} \text{ and } \frac{a}{2r} = \tan\frac{\pi}{n}$$

$$\therefore \quad \frac{r}{R} = \cos\frac{\pi}{n}$$

$n = 3$ gives $\frac{r}{R} = \frac{1}{2}$

$n = 4$ gives $\frac{r}{R} = \frac{1}{\sqrt{2}}$

$n = 6$ gives $\frac{r}{R} = \frac{\sqrt{3}}{2}$

211. We know that, $\frac{c}{\sin C} = 2R$

$$\Rightarrow \quad c = 2R \quad \text{...(i)} \quad [\because C = 90°]$$

and $\tan\frac{C}{2} = \frac{r}{s - c}$

$$\Rightarrow \tan\frac{\pi}{4} = \frac{r}{s - c}$$

$$\therefore \quad r = s - c$$

$$\Rightarrow \quad r = \frac{a + b + c}{2} - c$$

$$\Rightarrow a + b - c = 2r \quad \text{...(ii)}$$

On adding Eqs. (i) and (ii), we get

$$2(r + R) = a + b$$

212. In ΔBAD, $\cos(90° - B) = \frac{AD}{c}$

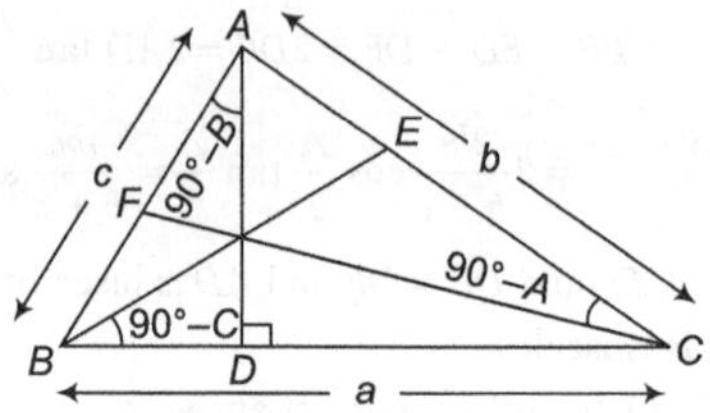

$$\Rightarrow \quad AD = c\sin B$$

Similarly, $BE = a\sin C$

and $CF = b\sin A$

Since, AD, BE and CF are in HP.

So, $c\sin B$, $a\sin C$ and $b\sin A$ are in HP.

$$\Rightarrow \quad \frac{1}{\sin C\sin B}, \frac{1}{\sin A\sin C} \text{ and } \frac{1}{\sin B\sin A}$$

are in AP.

Hence, $\sin A$, $\sin B$ and $\sin C$ are in AP.

Alternate Method

$$\text{ar}(\Delta ABC) = \frac{1}{2} \times BC \times AD$$

$$\Rightarrow \quad \Delta = \frac{1}{2} \times a \times AD$$

$$\Rightarrow \quad AD = \frac{2\Delta}{a}$$

Similarly, $BE = \frac{2\Delta}{b}$

and $CF = \frac{2\Delta}{c}$

Since, AD, BE and CF are in HP.

So, $\frac{1}{a}, \frac{1}{b}$ and $\frac{1}{c}$ are in HP.

Hence, a, b and c are in AP.

$\therefore$ $\sin A$, $\sin B$ and $\sin C$ are in AP.

CHAPTER

04

Inverse Trigonometric Functions

Learning Part

Practice Part

Session 1

Inverse Trigonometric Functions, Inverse Function, Domain and Range of Inverse Trigonometric Functions

Inverse Trigonometric Functions

Consider; $\sin\frac{\pi}{6}=\frac{1}{2}, \sin\frac{5\pi}{6}=\frac{1}{2}, \sin\frac{13\pi}{6}=\frac{1}{2},$

$\sin\frac{17\pi}{6}=\frac{1}{2}, \sin\left(-\frac{11\pi}{6}\right)=\frac{1}{2}, \sin\left(-\frac{7\pi}{6}\right)=\frac{1}{2},$ ... etc.

Now, if the question is "which is that angle or real number whose sine is $\frac{1}{2}$?"

Then there are infinite answers i.e. there is no unique answer.

Thus, the problem is, if the trigonometrical equation is $\sin y = x$, then for the given value of $x=\frac{1}{2}$, we have

$$y=\ldots,-\frac{7\pi}{6}, -\frac{11\pi}{6}, \frac{\pi}{6}, \frac{5\pi}{6}, \frac{13\pi}{6},\ldots$$

i.e.

for given value of x, we get infinite value of y.

$\therefore$ Correspondence from the set

$\{x \mid x \in R; -1 \le x \le 1\}$ to the set

$\{y \mid y \in R; \sin y = x\}$ is one to many correspondence

$\therefore$ It cannot be a function.

{$\because$ one-many and many-many are not function.}

However, if the question is "which is the numerically smallest angle or real number whose sine is $\frac{1}{2}$?"

Then the answer is $\frac{\pi}{6}$.

This is one and only one answer i.e. $\frac{\pi}{6}$ is the unique answer.

In this case, the relation $\sin\frac{\pi}{6}=\frac{1}{2}$ is also written as;

$$\frac{\pi}{6}=\sin^{-1}\left(\frac{1}{2}\right) \text{ or arc } \sin\left(\frac{1}{2}\right) \quad \{\text{read as sine inverse } \tfrac{1}{2}\}$$

It must therefore, be noted that $\sin^{-1} x$ is an angle and denotes the smallest numerical angle, whose sine is x. Similarly, $\cos^{-1} x$ and $\tan^{-1} x$, denotes an angle or real number; 'whose cosine is x' and 'whose tangent is x', respectively, provided that the answer given are numerically smallest available.

Note

(i) $\sin\frac{5\pi}{6}=\frac{1}{2}$ but $\frac{5\pi}{6} \ne \sin^{-1}(1/2)$

(ii) If there are two angles, one positive and the other negative having same numerical value. Then, we shall take the positive value.

e.g. $\cos\frac{\pi}{4}=\frac{1}{\sqrt{2}}$

and $\cos\left(-\frac{\pi}{4}\right)=\frac{1}{\sqrt{2}}$

But we write $\cos^{-1}\left(\frac{1}{\sqrt{2}}\right)=\frac{\pi}{4}$, not $\cos^{-1}\left(\frac{1}{\sqrt{2}}\right)=-\frac{\pi}{4}$

Inverse Function

Definition If a function, say f is one to one and onto from A to B, then function g which associates each element $y \in B$ to one and only one element $x \in A$, such that $y=f(x)$, then g is called the inverse function of f, denoted by $x=g(y)$.

Usually we denote $g=f^{-1}$ (Read as f inverse)

$$\therefore \quad x=f^{-1}(y)$$

Remark

If $y=f(x)$ and $x=g(y)$ are two functions such that $f\{g(y)\}=y$ and $g\{f(x)\}=x$ then f and g are said to be inverse functions of each other. Before we start with the definitions of $\sin^{-1} x, \cos^{-1} x$, $\tan^{-1} x$, etc., let us study the following discussion which will help us to define these inverse functions.

Domain and Range of Inverse Trigonometric Functions

1. If $\sin y = x$, then $y = \sin^{-1} x$, under certain conditions.

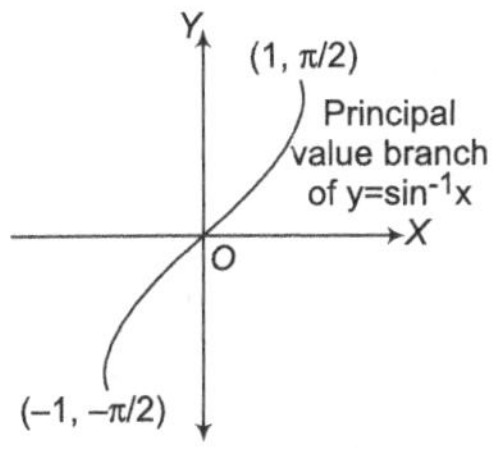

$\because \quad -1 \le \sin y \le 1$; but $\sin y = x$.

$\therefore \quad -1 \le x \le 1$

Also, $\quad \sin y = -1 \Rightarrow y = -\pi/2$

and $\quad \sin y = 1 \Rightarrow y = \pi/2$

Keeping in mind numerically smallest angles or real numbers, we have $-\pi/2 \le y \le \pi/2$

These restrictions on the values of x and y make the function $\sin y = x$, one-one and onto so that the inverse function exists, i.e. $y = \sin^{-1} x$ is meaningful.

Thus, $\quad$ Domain : $x \in [-1, 1]$

Range : $\quad y \in [-\pi/2, \pi/2]$

Note

(i) We can restrict the values of y in any of the intervals $[-3\pi/2, -\pi/2]$, $[-\pi/2, \pi/2]$, $[\pi/2, 3\pi/2]$ etc. Corresponding to each such interval, we get a branch of the function $y = \sin^{-1} x$. The branch with range $[-\pi/2, \pi/2]$ is called the **principal value branch**, whereas other intervals as range give different branches of $\sin^{-1}$.

(ii) The numerically least angle is called the **principal value** of the function.

2. Let $\cos y = x$ then $y = \cos^{-1} x$, under certain conditions

$\because \quad -1 \le \cos y \le 1.$

$\therefore \quad -1 \le x \le 1$

Also, $\quad \cos y = -1 \Rightarrow y = \pi$

and $\quad \cos y = 1 \Rightarrow y = 0$

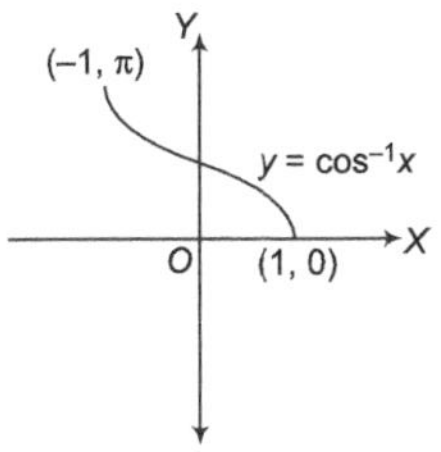

$\therefore \quad 0 \le y \le \pi$

These restrictions on the values of x and y make the function $\cos y = x$ one-one and onto so that the inverse function exists i.e. $y = \cos^{-1} x$ is meaningful.

Thus, $\quad$ Domain : $x \in [-1, 1]$

Range : $\quad y \in [0, \pi]$

3. If $\tan y = x$ then $y = \tan^{-1} x$, under certain conditions.

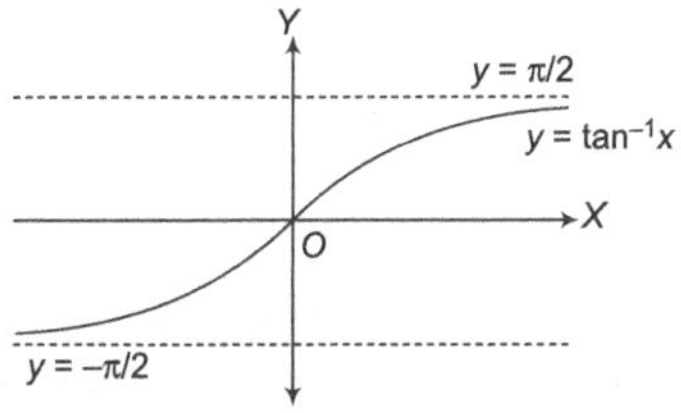

$\because \quad \tan y \in R$

$\therefore \quad x \in R$

Also, $\quad -\infty < \tan y < \infty \Rightarrow -\pi/2 < y < \pi/2$

These restrictions on x and y make the function, $\tan y = x$ one-one and onto so that the inverse function exists, i.e. $y = \tan^{-1} x$ is meaningful.

Thus, Domain : $x \subset R$, Range : $y \in (-\pi/2, \pi/2)$

4. If $\cot y = x$, then $y = \cot^{-1} x$, under certain conditions.

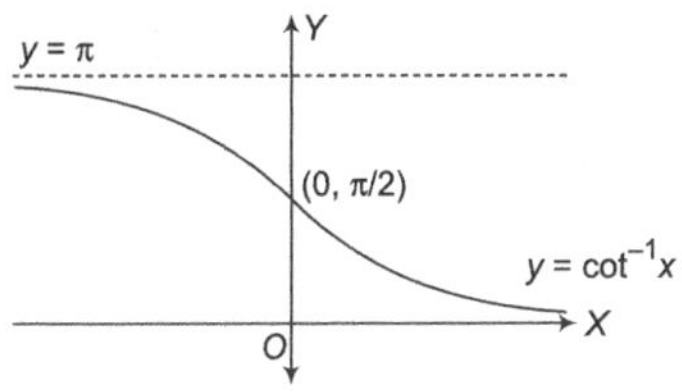

$\because \quad \cot y \in R \quad \therefore \quad x \in R$;

Also, $-\infty < \cot y < \infty \Rightarrow 0 < y < \pi$

These restrictions on x and y make the function, $\cot y = x$ one-one and onto so that the inverse function exists. i.e. $y = \cot^{-1} x$ is meaningful.

Thus, Domain : $x \in R$

Range : $y \in (0, \pi)$

5. If $\sec y = x$, then $y = \sec^{-1} x$, where $|x| \ge 1$ and $0 \le y \le \pi$, $y \ne \pi/2$

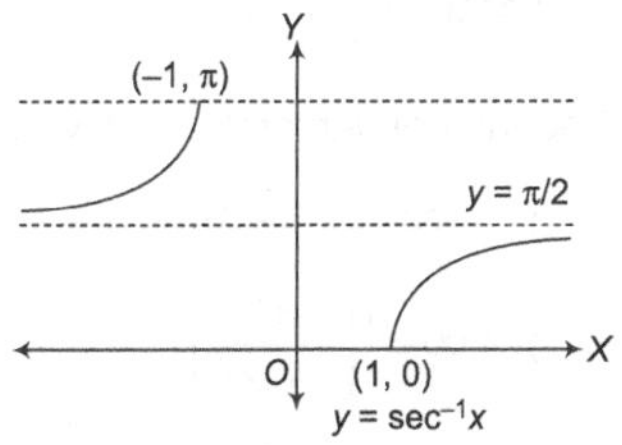

Here, Domain : $x \in R - (-1, 1)$

Range : $y \in [0, \pi] - \{\pi/2\}$

6. If $\text{cosec}\, y = x$ then $y = \text{cosec}^{-1} x$,
where $|x| \geq 1$ and $-\pi/2 \leq y \leq \pi/2,\ y \neq 0$

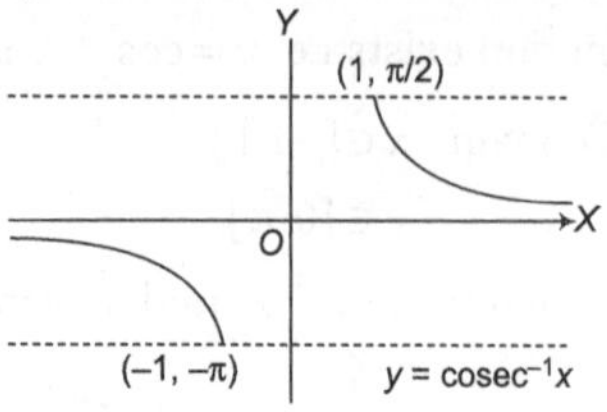

Here, Domain : $x \in R - (-1, 1)$

Range : $y \in [-\pi/2, \pi/2] - \{0\}$

Principal Values and Domains of Inverse Trigonometric/Circular Functions

	Function	Domain	Range (Principal Value Branch)
(i)	$y = \sin^{-1} x$	$-1 \leq x \leq 1$	$-\frac{\pi}{2} \leq y \leq \frac{\pi}{2}$
(ii)	$y = \cos^{-1} x$	$-1 \leq x \leq 1$	$0 \leq y \leq \pi$
(iii)	$y = \tan^{-1} x$	$x \in R$	$-\frac{\pi}{2} < y < \frac{\pi}{2}$
(iv)	$y = \text{cosec}^{-1} x$	$x \leq -1$ or $x \geq 1$	$-\frac{\pi}{2} \leq y \leq \frac{\pi}{2},\ y \neq 0$
(v)	$y = \sec^{-1} x$	$x \leq -1$ or $x \geq 1$	$0 \leq y \leq \pi;\ y \neq \frac{\pi}{2}$
(vi)	$y = \cot^{-1} x$	$x \in R$	$0 < y < \pi$

Note

(i) 1st quadrant is common to the range of all the inverse trigonometric functions.

(ii) 3rd quadrant is not used in inverse trigonometric functions.

(iii) 4th quadrant is used in the clockwise direction i.e. $-\frac{\pi}{2} \leq y \leq 0$.

(iv) No inverse function is periodic.

(v) If no branch of an inverse trigonometric function is mentioned, then it means the principal value branch of the function.

Example 1 Find the value of

$$\tan\left[\cos^{-1}\left(\frac{1}{2}\right) + \tan^{-1}\left(-\frac{1}{\sqrt{3}}\right)\right]$$

Sol. Let $y = \tan\left[\cos^{-1}\left(\frac{1}{2}\right) + \tan^{-1}\left(-\frac{1}{\sqrt{3}}\right)\right]$

$$\because \quad = \tan\left[\frac{\pi}{3} + \left(-\frac{\pi}{6}\right)\right] = \tan\left(\frac{\pi}{6}\right) = \frac{1}{\sqrt{3}}$$

Example 2 Find the value of $\cos\left[\cos^{-1}\left(\frac{-\sqrt{3}}{2}\right) + \frac{\pi}{6}\right]$

Sol. $\cos\left[\cos^{-1}\left(\frac{-\sqrt{3}}{2}\right) + \frac{\pi}{6}\right]$

$$= \cos\left\{\frac{5\pi}{6} + \frac{\pi}{6}\right\} = \cos(\pi) = -1$$

Example 3 Find domain of $\sin^{-1}(2x^2 - 1)$

Sol. Let $y = \sin^{-1}(2x^2 - 1)$

For y to be defined $-1 \leq (2x^2 - 1) \leq 1$

$$\Rightarrow 0 \leq 2x^2 \leq 2 \Rightarrow 0 \leq x^2 \leq 1$$

$$\Rightarrow x \in [-1, 1]$$

Exercise for Session 1

Find the value of the following

1. $\sin\left[\frac{\pi}{3} - \sin^{-1}\left(\frac{1}{2}\right)\right]$

2. $\text{cosec}[\sec^{-1}(-\sqrt{2}) + \cot^{-1}(-1)]$

Find the domain of the following

3. $y = \sec^{-1}(x^2 + 3x + 1)$

4. $y = \cos^{-1}\left(\frac{x^2}{1 + x^2}\right)$

5. $y = \tan^{-1}(\sqrt{x^2 - 1})$

Session 2

Property I of Inverse Trigonometric Functions

Property I

(i) $\sin^{-1}(\sin\theta)=\theta$; for all $\theta\in[-\pi/2,\pi/2]$

(ii) $\cos^{-1}(\cos\theta)=\theta$; for all $\theta\in[0,\pi]$

(iii) $\tan^{-1}(\tan\theta)=\theta$; for all $\theta\in(-\pi/2,\pi/2)$

(iv) $\text{cosec}^{-1}(\text{cosec}\,\theta)=\theta$; for all $\theta\in[-\pi/2,\pi/2], \theta\neq 0$

(v) $\sec^{-1}(\sec\theta)=\theta$; for all $\theta\in[0,\pi], \theta\neq\pi/2$

(vi) $\cot^{-1}(\cot\theta)=\theta$; for all $\theta\in(0,\pi)$

Proof We know that, if $f:A\to B$ is a bijection, then

$f^{-1}:B\to A$ exists such that

$f^{-1}of(x)=f^{-1}(f(x))=x$ for all $x\in A$

Clearly, all these results are direct consequences of this property.

Aliter For any $\theta\in[-\pi/2,\pi/2]$, let $\sin\theta=x$. Then, $\theta=\sin^{-1}x$

$$\therefore \quad \theta=\sin^{-1}(\sin\theta)$$

Hence, $\quad \sin^{-1}(\sin\theta)=\theta$ for all $\theta\in[-\pi/2,\pi/2]$

Similarly, we can prove other results.

Graphically they can be Shown As

(i) $y=\sin^{-1}(\sin x), x\in R, y\in\left[-\dfrac{\pi}{2},\dfrac{\pi}{2}\right]$, is periodic with period 2π.

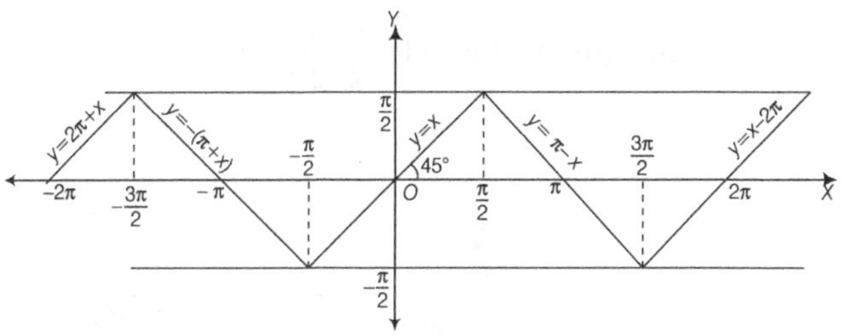

(ii) $y=\cos^{-1}(\cos x), x\in R, y\in[0,\pi]$, is periodic with period 2π.

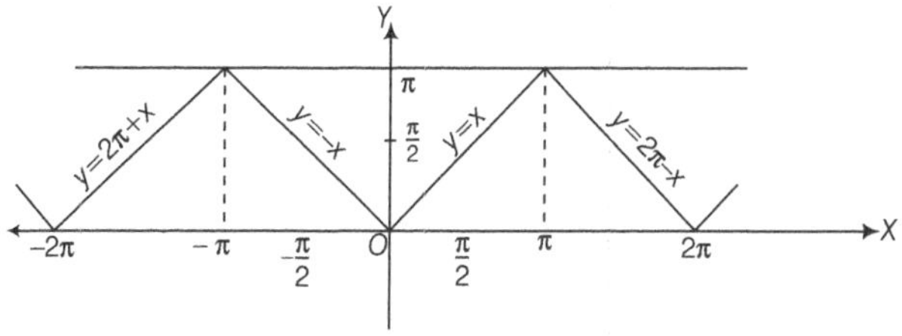

(iii) $y=\tan^{-1}(\tan x), x\in R-\left\{(2n-1)\dfrac{\pi}{2}, n\in I\right\}$,

$y\in\left(-\dfrac{\pi}{2},\dfrac{\pi}{2}\right)$ is periodic with period π.

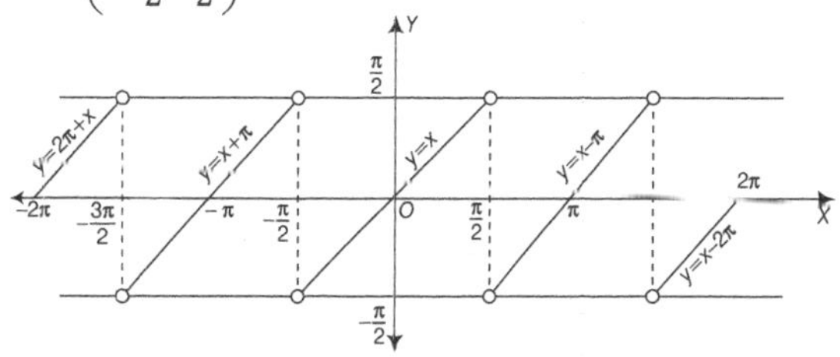

(iv) $y=\sec^{-1}(\sec x), x\in R-\left\{(2n-1)\dfrac{\pi}{2}, n\in I\right\}$,

$y\in\left[0,\dfrac{\pi}{2}\right)\cup\left(\dfrac{\pi}{2},\pi\right]$, is periodic with period 2π.

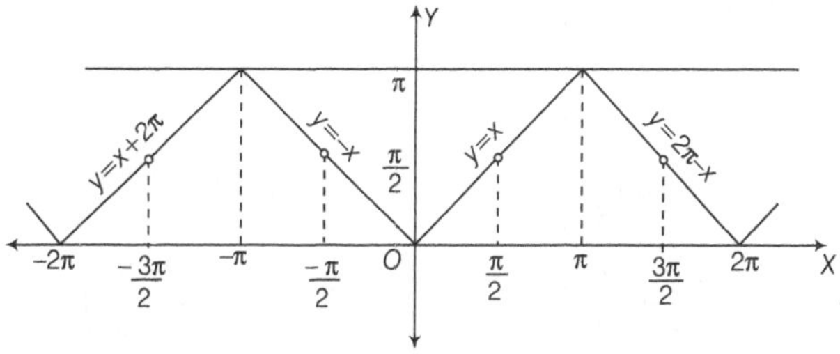

(v) $y=\cot^{-1}(\cot x), x\in R-\{n\pi, n\in I\}, y\in(0,\pi)$, is periodic with period π.

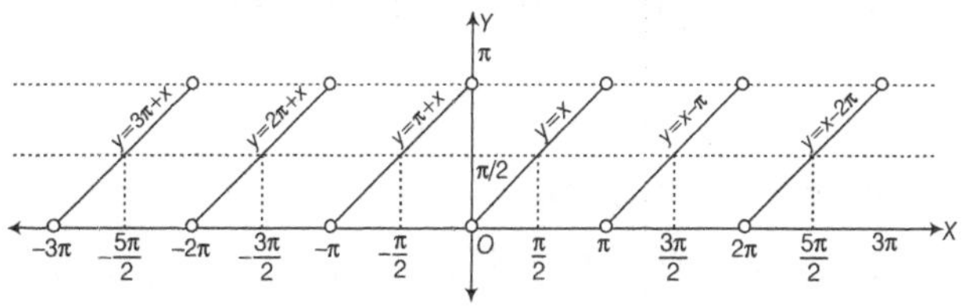

(vi) $y = \text{cosec}^{-1}(\text{cosec}\, x)$,

$x \in R - \{n\pi, n \in I\}, y \in \left[-\frac{\pi}{2}, \frac{\pi}{2}\right] - \{0\}$, is periodic with period 2π.

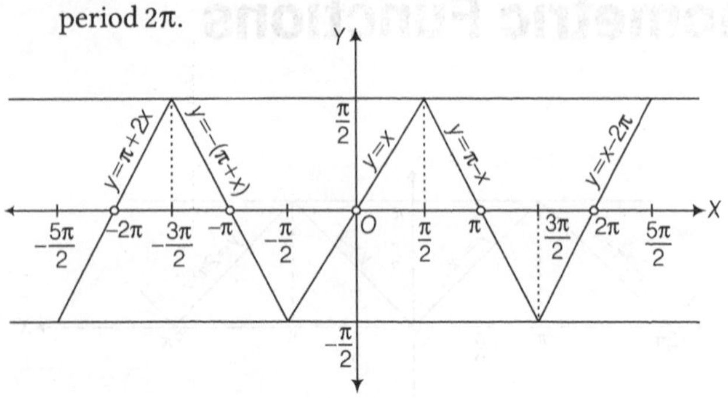

Remark

It should be noted that,

$\sin^{-1}(\sin\theta) \neq \theta$, if $\notin [-\pi/2, \pi/2]$. Infact, we have

$$\sin^{-1}(\sin\theta) = \begin{cases} -\pi-\theta, \text{ if } \theta \in [-3\pi/2, -\pi/2] \\ \theta, \text{ if } \theta \in [-\pi/2, \pi/2] \\ \pi-\theta, \text{ if } \theta \in [\pi/2, 3\pi/2] \\ -2\pi+\theta, \text{ if } \theta \in [3\pi/2, 5\pi/2] \text{ and so on} \end{cases}$$

$$\text{Similarly, } \cos^{-1}(\cos\theta) = \begin{cases} -\theta, \text{if } \theta \in [-\pi, 0] \\ \theta, \text{if } \theta \in [0, \pi] \\ 2\pi-\theta, \text{if } \theta \in [\pi, 2\pi] \\ -2\pi+\theta, \text{if } \theta \in [2\pi, 3\pi] \text{ and so on} \end{cases}$$

$$\tan^{-1}(\tan\theta) = \begin{cases} \theta+\pi, \text{if } \theta \in (-3\pi/2, -\pi/2) \\ \theta, \text{if } \theta \in (-\pi/2, \pi/2) \\ \theta-\pi, \text{if } \theta - \in (\pi/2, 3\pi/2) \\ \theta-2\pi, \text{if } \theta \in (3\pi/2, 5\pi/2) \text{ and so on} \end{cases}$$

Example 4. Evaluate the following

(i) $\sin^{-1}(\sin\pi/4)$ (ii) $\cos^{-1}\left(\cos\frac{2\pi}{3}\right)$

(iii) $\tan^{-1}\left(\tan\frac{\pi}{3}\right)$ (iv) $\sin^{-1}\left(\sin\frac{2\pi}{3}\right)$

(v) $\cos^{-1}\left(\cos\frac{7\pi}{6}\right)$ (vi) $\tan^{-1}\left(\tan\frac{2\pi}{3}\right)$

Sol. Recall that $\sin^{-1}(\sin\theta) = \theta$, if $-\pi/2 \le \theta \le \pi/2$

$\cos^{-1}(\cos\theta) = \theta$, if $0 \le \theta \le \pi$

and $\tan^{-1}(\tan\theta) = \theta$, if $-\pi/2 < \theta < \pi/2$

(i) $\sin^{-1}(\sin\pi/4) = \frac{\pi}{4}$ (ii) $\cos^{-1}\left(\cos\frac{2\pi}{3}\right) = \frac{2\pi}{3}$

(iii) $\tan^{-1}(\tan\pi/3) = \pi/3$

(iv) $\sin^{-1}\left(\sin\frac{2\pi}{3}\right) \neq \frac{2\pi}{3}$ as $\frac{2\pi}{3}$ does not lie between $-\pi/2$ and $\pi/2$.

Now, $\sin^{-1}\left(\sin\frac{2\pi}{3}\right) = \sin^{-1}(\sin(\pi - \pi/3))$ $\left\{\because \frac{2\pi}{3} = \pi - \frac{\pi}{3}\right\}$

$= \sin^{-1}(\sin\pi/3)$ $\{\because \sin(\pi-\theta) = \sin\theta\}$

$= \frac{\pi}{3}$.

(v) $\cos^{-1}\left(\cos\frac{7\pi}{6}\right) \neq \frac{7\pi}{6}$, because $\frac{7\pi}{6}$ does not lie between 0 and π.

Now, $\cos^{-1}\left(\cos\frac{7\pi}{6}\right) = \cos^{-1}\left(\cos\left(2\pi - \frac{5\pi}{6}\right)\right)$

$\left\{\because \frac{7\pi}{6} = 2\pi - \frac{5\pi}{6}\right\}$

$= \cos^{-1}\left(\cos\frac{5\pi}{6}\right) = \frac{5\pi}{6}$ $[\because \cos(2\pi-\theta) = \cos\theta]$

(vi) $\tan^{-1}\left(\tan\frac{2\pi}{3}\right) \neq \frac{2\pi}{3}$, because $\frac{2\pi}{3}$ does not lie between $-\frac{\pi}{2}$ and $\frac{\pi}{2}$.

Now, $\tan^{-1}\left(\tan\frac{2\pi}{3}\right) = \tan^{-1}(\tan(\pi - \pi/3))$

$\left\{\because \frac{2\pi}{3} = \pi - \frac{\pi}{3}\right\}$

$= \tan^{-1}\left(-\tan\frac{\pi}{3}\right)$ $\{\because \tan(\pi-\theta) = -\tan\theta\}$

$= \tan^{-1}\left(\tan\left(\frac{-\pi}{3}\right)\right)$ $[\because \tan(-\theta) = -\tan\theta]$

$= -\frac{\pi}{3}$.

Example 5. Evaluate the following

(i) $\sin^{-1}(\sin 7)$ (ii) $\sin^{-1}(\sin(-5))$

(iii) $\cos^{-1}(\cos 10)$ (iv) $\tan^{-1}(\tan(-6))$

Sol. (i) Let $y = \sin^{-1}(\sin 7)$

Note $\sin^{-1}(\sin 7) \neq 7$ as $7 \notin \left[-\frac{\pi}{2}, \frac{\pi}{2}\right]$

$\because$ $2\pi < 7 < \frac{5\pi}{2}$

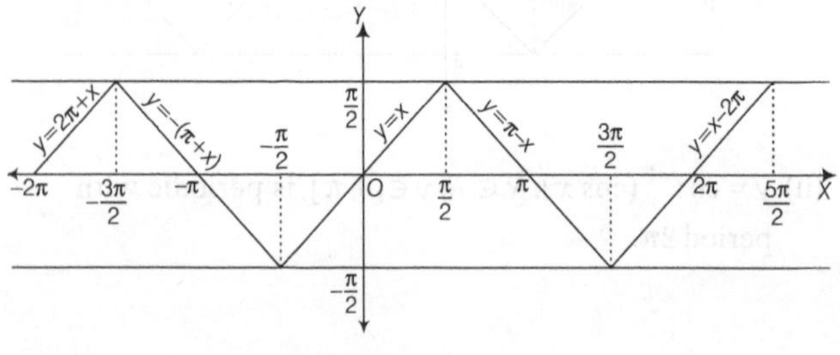

From the graph, we can see that if $2\pi \le x \le \frac{5\pi}{2}$, then

$y = \sin^{-1}(\sin x)$ can be written as

$$y = x - 2\pi$$

$\therefore \quad \sin^{-1}(\sin 7) = 7 - 2\pi$

Alternate Method

We know that, $\sin^{-1}(\sin\theta)=\theta$, if $-\pi/2 \le \theta \le \pi/2$

Here, $\theta = 7$ radians which does not lie between $-\pi/2$ and $\pi/2$.

But, $2\pi - 7$ and $7 - 2\pi$ both lie between $-\frac{\pi}{2}$ and $\frac{\pi}{2}$.

Also, $\sin(7 - 2\pi) = \sin(-(2\pi - 7)) = -\sin(2\pi - 7) = \sin 7$

$\therefore \quad \sin^{-1}(\sin 7) = \sin^{-1}(\sin(7 - 2\pi)) = 7 - 2\pi$

(ii) Here, $\theta = -5$ radians. Clearly, it does not lie between $-\frac{\pi}{2}$ and $\frac{\pi}{2}$. But $2\pi - 5$ lie between $-\frac{\pi}{2}$ and $\frac{\pi}{2}$.

Also, $\sin(2\pi - 5) = -\sin 5 = \sin(-5)$

$\therefore \quad \sin^{-1}(\sin(-5)) = \sin^{-1}(\sin(2\pi - 5)) = 2\pi - 5$

(iii) We know that, $\cos^{-1}(\cos\theta) = \theta$, if $0 \le \theta \le \pi$.

Here, $\theta = 10$ radians which does not lie between 0 and π. However, $(4\pi - 10)$ lies between 0 and π such that

$$\cos(4\pi - 10) = \cos 10$$

$\therefore \quad \cos^{-1}(\cos 10) = \cos^{-1}(\cos(4\pi - 10)) = 4\pi - 10$

(iv) We know that,

$\tan^{-1}(\tan\theta) = \theta$, if $-\pi/2 < \theta < \pi/2$.

Here, $\theta = -6$ radians which does not lie between $-\pi/2$ and $\pi/2$. But we find that $2\pi - 6$ lies between $-\pi/2$ and $\pi/2$ such that

$$\tan(2\pi - 6) = -\tan 6 = \tan(-6)$$

$\therefore \quad \tan^{-1}(\tan(-6)) = \tan^{-1}(\tan(2\pi - 6)) = 2\pi - 6$

Exercise for Session 2

1. Find the value of $\cos^{-1}\left(\cos\frac{2\pi}{3}\right) + \sin^{-1}\left(\sin\frac{2\pi}{3}\right)$.
2. Write the value of $\sin^{-1}\left(\sin\frac{3\pi}{5}\right)$.
3. Find the value of $\sin^{-1}\left(\cos\frac{33\pi}{5}\right)$.
4. Find the value of $\cos^{-1}(\cos 13)$.
5. Find $\sin^{-1}(\sin\theta)$, $\cos^{-1}(\cos\theta)$, $\tan^{-1}(\tan\theta)$, $\cot^{-1}(\cot\theta)$ for $\theta \in \left(\frac{5\pi}{2}, 3\pi\right)$.

Session 3

Property II of Inverse Trigonometric Functions

Property II

(i) $\sin(\sin^{-1} x) = x$,	for all $x \in [-1, 1]$
(ii) $\cos(\cos^{-1} x) = x$	for all $x \in [-1, 1]$
(iii) $\tan(\tan^{-1} x) = x$	for all $x \in R$
(iv) $\text{cosec}(\text{cosec}^{-1} x) = x$	for all $x \in (-\infty, -1] \cup [1, \infty)$
(v) $\sec(\sec^{-1} x) = x$,	for all $x \in (-\infty, -1] \cup [1, \infty)$
(vi) $\cot(\cot^{-1} x) = x$,	for all $x \in R$

Proof We know that, if $f: A \to B$ is a bijection, then $f^{-1}: B \to A$ exists such that $fof^{-1}(y) = f(f^{-1}(y)) = y$ for all $y \in B$.

Clearly, all these results are direct consequences of this property.

Aliter Let $\theta \in [-\pi/2, \pi/2]$ and $x \in [-1, 1]$ such that $\sin\theta = x$.

Then, $\theta = \sin^{-1} x \qquad \therefore\ x = \sin\theta = \sin(\sin^{-1} x)$

Hence, $\sin(\sin^{-1} x) = x$ for all $x \in [-1, 1]$

Similarly, we can prove other results.

Graphically they can be Shown As

(i) $y = \sin(\sin^{-1} x) = \cos(\cos^{-1} x) = x$,
$x \in [-1, 1], y \in [-1, 1]$; y is aperiodic

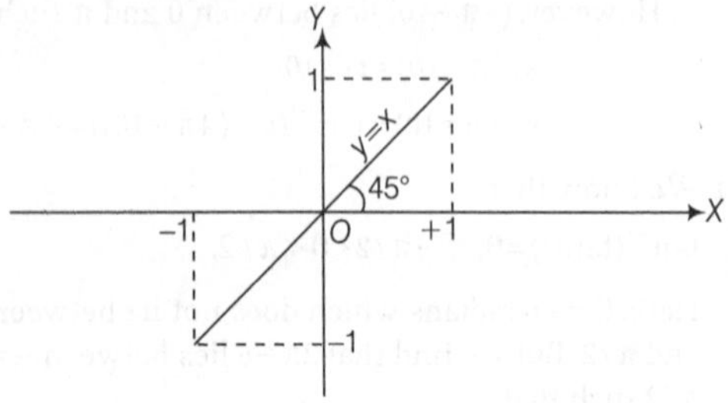

(ii) $y = \tan(\tan^{-1} x) = \cot(\cot^{-1} x) = x, x \in R, y \in R$; y is aperiodic

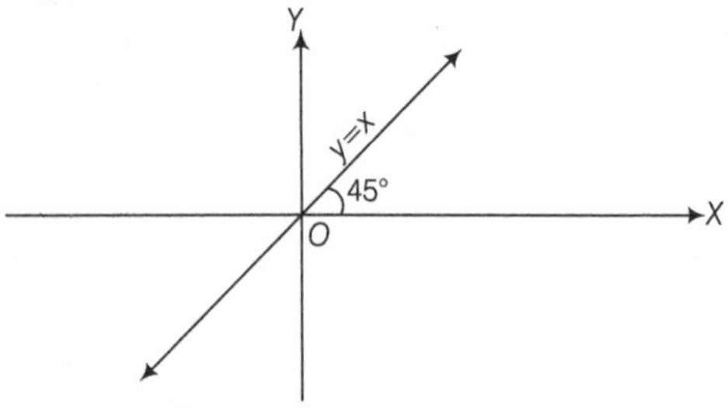

(iii) $y = \text{cosec}(\text{cosec}^{-1} x) = \sec(\sec^{-1} x) = x$,
$|x| \geq 1, |y| \geq 1$; y is aperiodic

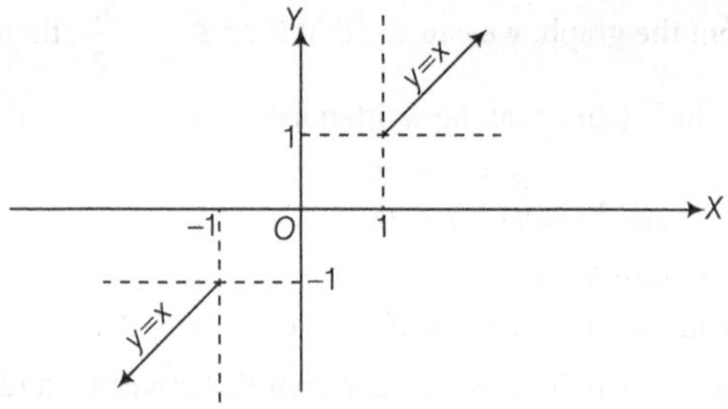

Example 6. Find the value of $\text{cosec}\left\{\cot\left(\cot^{-1}\frac{3\pi}{4}\right)\right\}$

Sol. Let $y = \text{cosec}\left\{\cot\left(\cot^{-1}\frac{3\pi}{4}\right)\right\}$ ...(i)

$\because \quad \cot(\cot^{-1} x) = x, \forall x \in R$

$\therefore \quad \cot\left(\cot^{-1}\frac{3\pi}{4}\right) = \frac{3\pi}{4}$

Now, from Eq. (i), we get $y = \text{cosec}\left(\frac{3\pi}{4}\right) = \sqrt{2}$

Example 7. Prove that
$$\sec^2(\tan^{-1} 2) + \text{cosec}^2(\cot^{-1} 3) = 15$$

Sol. $\sec^2(\tan^{-1}2) + \text{cosec}^2(\cot^{-1}3)$

$= \{1 + \tan^2(\tan^{-1}2)\} + \{1 + \cot^2(\cot^{-1}3)\}$

$= 1 + (\tan(\tan^{-1}2))^2 + 1 + (\cot(\cot^{-1}3))^2$

$= 2 + 2^2 + 3^2 = 15$

Exercise for Session 3

Evaluate the following :

1. $\cos\left\{\sin\left(\sin^{-1}\frac{\pi}{6}\right)\right\}$.
2. $\sin\left\{\cos\left(\cos^{-1}\frac{3\pi}{4}\right)\right\}$.
3. $\sin^2\left(\cos^{-1}\frac{1}{2}\right) + \cos^2\left(\sin^{-1}\frac{1}{3}\right)$.
4. $\tan^2(\sec^{-1}2) + \cot^2(\text{cosec}^{-1}3)$.
5. Find the solutions of the equation $\cos(\cos^{-1} x) = \text{cosec}(\text{cosec}^{-1} x)$.

Session 4

Property III, IV and V of Inverse Trigonometric Functions

Property III

(i) $\sin^{-1}(-x) = -\sin^{-1}(x)$, for all $x \in [-1, 1]$

(ii) $\cos^{-1}(-x) = \pi - \cos^{-1}(x)$, for all $x \in [-1, 1]$

(iii) $\tan^{-1}(-x) = -\tan^{-1} x$, for all $x \in R$

(iv) $\text{cosec}^{-1}(-x) = -\text{cosec}^{-1} x$, for all $x \in (-\infty, -1] \cup [1, \infty)$

(v) $\sec^{-1}(-x) = \pi - \sec^{-1} x$, for all $x \in (-\infty, -1] \cup [1, \infty)$

(vi) $\cot^{-1}(-x) = \pi - \cot^{-1} x$ for all $x \in R$

Proof (i) Clearly, $-x \in [-1, 1]$ for all $x \in [-1, 1]$

Let $\sin^{-1}(-x) = \theta$ ...(i)

Then, $-x = \sin\theta$

$\Rightarrow x = -\sin\theta$

$\Rightarrow x = \sin(-\theta)$

$\Rightarrow -\theta = \sin^{-1} x$

$\{\because x \in [-1, 1]$ and $-\theta \in [-\pi/2, \pi/2]$ for all $\theta \in [-\pi/2, \pi/2]$

$\Rightarrow \theta = -\sin^{-1} x$ ...(ii)

From Eqs. (i) and (ii), we get

$$\sin^{-1}(-x) = -\sin^{-1}(x)$$

(ii) Clearly, $-x \in [-1, 1]$ for all $x \in [-1, 1]$

Let $\cos^{-1}(-x) = \theta$ ...(i)

Then, $-x = \cos\theta$

$\Rightarrow x = -\cos\theta$

$\Rightarrow x = \cos(\pi - \theta)$

$\Rightarrow \cos^{-1} x = \pi - \theta$

$\{\because x \in [-1, 1]$ and $\pi - \theta \in [0, \pi]$ for all $\theta \in [0, \pi]\}$

$\Rightarrow \theta = \pi - \cos^{-1} x$ ...(ii)

From Eqs. (i) and (ii), we get

$$\cos^{-1}(-x) = \pi - \cos^{-1} x$$

Similarly, we can prove other results.

Example 8. Find the value of $\cos^{-1}\{\sin(-5)\}$

Sol. Let $y = \cos^{-1}\{\sin(-5)\}$

$= \cos^{-1}(-\sin 5)$

$\because \cos^{-1}(-x) = \pi - \cos^{-1} x, |x| \le 1$

$\therefore y = \pi - \cos^{-1}(\sin 5)$

$= \pi - \cos^{-1}\left\{\cos\left(\frac{\pi}{2} - 5\right)\right\}$ ... (i)

Here, $-2\pi < \left(\frac{\pi}{2} - 5\right) < -\pi$

and the graph of $\cos^{-1}(\cos x)$ is as

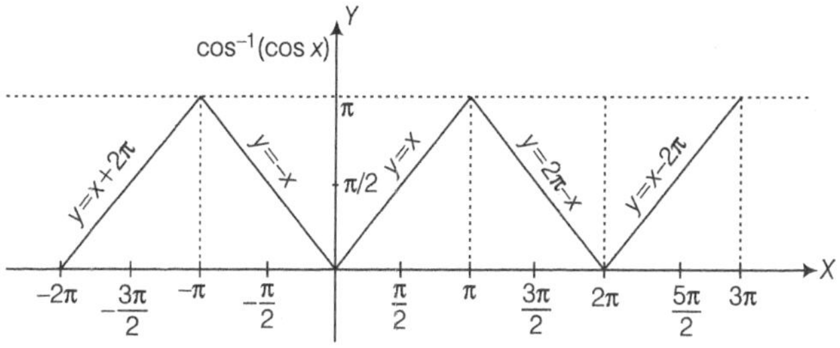

From the graph, we can see that if $-2\pi \le x \le -\pi$, then $y = \cos^{-1}(\cos x)$ can be written as $y = x + 2\pi$.

$\therefore$ From the graph

$$\cos^{-1}\left\{\cos\left(\frac{\pi}{2} - 5\right)\right\} = \left(\frac{\pi}{2} - 5\right) + 2\pi = \left(\frac{5\pi}{2} - 5\right)$$

Now, from Eq. (i), we get

$$y = \pi - \left(\frac{5\pi}{2} - 5\right) \Rightarrow y = 5 - \frac{3\pi}{2}$$

Example 9. Evaluate the following

(i) $\sin^{-1}\left(\sin\left(\frac{-3\pi}{4}\right)\right)$ (ii) $\cot^{-1}(\cot(-4))$

Sol. (i) $\sin^{-1}\left(\sin\left(\frac{-3\pi}{4}\right)\right) = \sin^{-1}\left(-\sin\left(\frac{3\pi}{4}\right)\right) = -\sin^{-1}\left(\sin\frac{3\pi}{4}\right)$

$= -\sin^{-1}\left(\sin\left(\pi - \frac{\pi}{4}\right)\right) = -\sin^{-1}\left(\sin\frac{\pi}{4}\right) = -\frac{\pi}{4}$

(ii) $\cot^{-1}(\cot(-4)) = \cot^{-1}(-\cot 4) = \pi - \cot^{-1}(\cot 4)$

$= \pi - \cot^{-1}(\cot(\pi + (4 - \pi)))$

$= \pi - \cot^{-1}(\cot(4 - \pi)) = 2\pi - 4$

Example 10. Evaluate the following

(i) $\sin\left(\frac{\pi}{2} - \sin^{-1}\left(\frac{-1}{2}\right)\right)$

(ii) $\sin\left(\frac{\pi}{2} - \sin^{-1}\left(-\frac{\sqrt{3}}{2}\right)\right)$

Sol. (i) $\sin\left(\frac{\pi}{2}-\sin^{-1}\left(-\frac{1}{2}\right)\right)=\sin\left(\frac{\pi}{2}+\sin^{-1}\left(\frac{1}{2}\right)\right)$

$$=\sin\left(\frac{\pi}{2}+\frac{\pi}{6}\right)=\sin\frac{2\pi}{3}=\frac{\sqrt{3}}{2}$$

(ii) $\sin\left(\frac{\pi}{2}-\sin^{-1}\left(-\frac{\sqrt{3}}{2}\right)\right)=\sin\left(\frac{\pi}{2}+\sin^{-1}\left(\frac{\sqrt{3}}{2}\right)\right)$

$$=\sin\left(\frac{\pi}{2}+\frac{\pi}{3}\right)=\sin\frac{5\pi}{6}=\frac{1}{2}$$

Property IV

(i) $\sin^{-1}\left(\frac{1}{x}\right)=\text{cosec}^{-1}x$, for all $x\in(-\infty,-1]\cup[1,\infty)$

(ii) $\cos^{-1}\left(\frac{1}{x}\right)=\sec^{-1}x$, for all $x\in(-\infty,-1]\cup[1,\infty)$

(iii) $\tan^{-1}\left(\frac{1}{x}\right)=\begin{cases}\cot^{-1}x, \text{for } x>0\\ -\pi+\cot^{-1}x, \text{for } x<0\end{cases}$

Proof

(i) Let $\text{cosec}^{-1}x=\theta$ ...(i)

Then, $x=\text{cosec}\,\theta$

$\Rightarrow \frac{1}{x}=\sin\theta$

$\because x\in(-\infty,-1]\cup[1,\infty) \Rightarrow \frac{1}{x}\in[-1,1]-\{0\}$

$\text{cosec}^{-1}x=\theta \Rightarrow \theta\in[-\pi/2,\pi/2]-\{0\}$

$\Rightarrow \theta=\sin^{-1}\left(\frac{1}{x}\right)$ ...(ii)

From Eqs. (i) and (ii), we get

$$\sin^{-1}\left(\frac{1}{x}\right)=\text{cosec}^{-1}x$$

(ii) Let $\sec^{-1}x=\theta$ ...(i)

Then, $x\in(-\infty,-1]\cup[1,\infty)$ and $\theta\in[0,\pi]-\{\pi/2\}$

Now, $\sec^{-1}x=\theta$

$\Rightarrow x=\sec\theta \Rightarrow \frac{1}{x}=\cos\theta$

$\Rightarrow \theta=\cos^{-1}\left(\frac{1}{x}\right)$ ...(ii)

$\left\{\begin{array}{l}\because x=(-\infty,-1]\cup[1,\infty)\\ \Rightarrow \frac{1}{x}\in[-1,1]-\{0\} \text{ and } \theta\in[0,\pi]-\left\{\frac{\pi}{2}\right\}\end{array}\right.$

From Eqs. (i) and (ii), we get

$$\cos^{-1}\left(\frac{1}{x}\right)=\sec^{-1}(x)$$

(iii) Let $\cot^{-1}x=\theta$. Clearly, $x\in R$, $x\neq 0$ and $\theta\in(0,\pi)$ ...(i)

Now, *two cases arises*

Case I When $x>0$

In this case, $\theta\in(0,\pi/2)$

$\therefore \cot^{-1}x=\theta$

$\Rightarrow x=\cot\theta$

$\Rightarrow \frac{1}{x}=\tan\theta$

$\theta=\tan^{-1}\left(\frac{1}{x}\right)$ ...(ii)

$\{\because \theta\in(0,\pi/2)\}$

From Eqs. (i) and (ii), we get

$$\tan^{-1}\left(\frac{1}{x}\right)=\cot^{-1}x, \text{ for all } x>0.$$

***Case* II** When $x<0$

In this case, $\theta\in(\pi/2,\pi)$ $\{\because x=\cot\theta<0\}$

Now, $\frac{\pi}{2}<\theta<\pi$

$\Rightarrow -\frac{\pi}{2}<\theta-\pi<0$

$\Rightarrow \theta-\pi\in(-\pi/2,0)$

$\therefore \cot^{-1}x=\theta$

$\Rightarrow x=\cot\theta$

$\Rightarrow \frac{1}{x}=\tan\theta$

$\Rightarrow \frac{1}{x}=-\tan(\pi-\theta)$ $\{\because \tan(\pi-\theta)=-\tan\theta\}$

$\Rightarrow \frac{1}{x}=\tan(\theta-\pi)$

$\Rightarrow \theta-\pi=\tan^{-1}\left(\frac{1}{x}\right)$ $\{\because \theta-\pi\in(-\pi/2,0)\}$

$\Rightarrow \tan^{-1}\left(\frac{1}{x}\right)=-\pi+\theta$ ...(iii)

From Eqs. (i) and (iii), we get

$$\tan^{-1}\left(\frac{1}{x}\right)=-\pi+\cot^{-1}x, \text{ if } x<0$$

Hence, $\tan^{-1}\left(\frac{1}{x}\right)=\begin{cases}\cot^{-1}x, \text{for } x>0\\ -\pi+\cot^{-1}x, \text{for } x<0\end{cases}$

Example 11. Evaluate the following

(i) $\sin\left(\text{cosec}^{-1}\frac{5}{3}\right)$ (ii) $\cot\left(\tan^{-1}\frac{3}{4}\right)$

Sol. (i) $\sin\left(\text{cosec}^{-1}\frac{5}{3}\right)=\sin\left(\sin^{-1}\frac{3}{5}\right)=\frac{3}{5}$

(ii) $\cot\left(\tan^{-1}\frac{3}{4}\right)=\cot\left(\cot^{-1}\frac{4}{3}\right)=\frac{4}{3}$

Example 12. Find the value of $\tan\left\{\cot^{-1}\left(\frac{-2}{3}\right)\right\}$

Sol. Let $y=\tan\left\{\cot^{-1}\left(\frac{-2}{3}\right)\right\}$ ...(i)

$\because \cot^{-1}(-x)=\pi-\cot^{-1}x,\ x\in R$

$\therefore$ Eq. (i) can be written as

$$y=\tan\left\{\pi-\cot^{-1}\left(\frac{2}{3}\right)\right\}$$

$$y=-\tan\left(\cot^{-1}\frac{2}{3}\right)$$

$\because \cot^{-1}x-\tan^{-1}\frac{1}{x}$ if $x>0$

$\therefore\quad y=-\tan\left(\tan^{-1}\frac{3}{2}\right)\Rightarrow y=-\frac{3}{2}$

Property V

(i) $\sin^{-1}x+\cos^{-1}x=\pi/2$, for all $x\in[-1,1]$

(ii) $\tan^{-1}x+\cot^{-1}x=\pi/2$, for all $x\in R$

(iii) $\sec^{-1}x+\text{cosec}^{-1}x=\frac{\pi}{2}$, for all $x\in(-\infty,-1]\cup[1,\infty)$

Proof

(i) Let, $\sin^{-1}x=\theta$ ...(i)

Then, $\theta\in[-\pi/2,\pi/2]$ $\{\because x\in[-1,1]\}$

$\Rightarrow -\pi/2\le\theta\le\pi/2$

$\Rightarrow -\pi/2\le-\theta\le\pi/2$

$\Rightarrow 0\le\frac{\pi}{2}-\theta\le\pi \Rightarrow \frac{\pi}{2}-\theta\in[0,\pi]$

Now, $\sin^{-1}x=\theta \Rightarrow x=\sin\theta$

$\Rightarrow x=\cos\left(\frac{\pi}{2}-\theta\right)$

$\Rightarrow \cos^{-1}x=\frac{\pi}{2}-\theta$

$\{\because x\in[-1,1]\text{ and }(\pi/2-\theta)\in[0,\pi]\}$

$\Rightarrow \theta+\cos^{-1}x=\pi/2$ ...(ii)

From Eqs. (i) and (ii), we get

$$\sin^{-1}x+\cos^{-1}x=\frac{\pi}{2}$$

(ii) Let, $\tan^{-1}x=\theta$ ...(i)

Then, $\theta\in(-\pi/2,\pi/2)$ $\{\because x\in R\}$

$\Rightarrow -\frac{\pi}{2}<\theta<\frac{\pi}{2}$

$\Rightarrow -\frac{\pi}{2}<-\theta<\frac{\pi}{2}$

$\Rightarrow 0<\frac{\pi}{2}-\theta<\pi$

$\Rightarrow \left(\frac{\pi}{2}-\theta\right)\in(0,\pi)$

Now, $\tan^{-1}x=\theta$

$\Rightarrow x=\tan\theta$

$\Rightarrow x=\cot(\pi/2-\theta)$

$\Rightarrow \cot^{-1}x=\frac{\pi}{2}-\theta$ $\{\because \pi/2-\theta\in(0,\pi)\}$

$\Rightarrow \theta+\cot^{-1}x=\frac{\pi}{2}$ ...(ii)

From Eqs. (i) and (ii), we get

$$\tan^{-1}x+\cot^{-1}x=\pi/2$$

(iii) Let, $\sec^{-1}x=\theta$ ...(i)

Then, $\theta\in[0,\pi]-\{\pi/2\}$ $\{\because x\in(-\infty,-1]\cup[1,\infty)\}$

$\Rightarrow 0\le\theta\le\pi,\ \theta\ne\pi/2$

$\Rightarrow -\pi\le-\theta\le0,\ \theta\ne\pi/2$

$\Rightarrow -\frac{\pi}{2}\le\frac{\pi}{2}-\theta\le\frac{\pi}{2},\ \frac{\pi}{2}-\theta\ne0$

$\Rightarrow \left(\frac{\pi}{2}-\theta\right)\in\left[-\frac{\pi}{2},\frac{\pi}{2}\right],\ \frac{\pi}{2}-\theta\ne0$

Now, $\sec^{-1}x=\theta$

$\Rightarrow x=\sec\theta$

$\Rightarrow x=\text{cosec}(\pi/2-\theta)$

$\Rightarrow \text{cosec}^{-1}x=\pi/2-\theta$

$\left\{\because\left(\frac{\pi}{2}-\theta\right)\in\left[-\frac{\pi}{2},\frac{\pi}{2}\right],\frac{\pi}{2}-\theta\ne0\right\}$

$\Rightarrow \theta+\text{cosec}^{-1}x=\pi/2$ ...(ii)

From Eqs. (i) and (ii), we get

$$\sec^{-1}x+\text{cosec}^{-1}x=\pi/2$$

Example 13. Find the value of of $\sin(2\cos^{-1} x + \sin^{-1} x)$ when $x = \frac{1}{5}$

Sol. Let $y = \sin[2\cos^{-1} x + \sin^{-1} x]$

$\because \quad \sin^{-1} x + \cos^{-1} x = \frac{\pi}{2}, |x| \le 1$

$\therefore \quad y = \sin\left[2\cos^{-1} x + \frac{\pi}{2} - \cos^{-1} x\right]$

$= \sin\left[\frac{\pi}{2} + \cos^{-1} x\right] = \cos(\cos^{-1} x)$

$\because \quad x = \frac{1}{5}$

$\therefore \quad y = \cos\left(\cos^{-1}\frac{1}{5}\right) \quad \text{... (i)}$

$\because \quad \cos(\cos^{-1} x) = x$ if $x \in [-1, 1]$

and $\quad \frac{1}{5} \in [-1,1]$

$\therefore \quad \cos\left(\cos^{-1}\frac{1}{5}\right) = \frac{1}{5}$

Now, from Eq. (i), we get $y = \frac{1}{5}$

Example 14. Solve $\sin^{-1} x - \cos^{-1} x = \cos^{-1}\frac{\sqrt{3}}{2}$.

Sol. We have, $\sin^{-1} x - \cos^{-1} x = \cos^{-1}\frac{\sqrt{3}}{2}$

$\Rightarrow \quad \sin^{-1} x - \cos^{-1} x = \frac{\pi}{6} \quad \text{...(i)}$

Also, $\sin^{-1} x + \cos^{-1} x = \frac{\pi}{2} \quad \text{...(ii)}$

On solving Eqs. (i) and (ii), we get $\sin^{-1} x = \frac{\pi}{3}$

and $\quad \cos^{-1} x = \frac{\pi}{6}$

$\Rightarrow \quad x = \frac{\sqrt{3}}{2}$ is the only solution.

Exercise for Session 4

Evaluate the following

1. $\tan^{-1}\left\{\tan\left(-\frac{7\pi}{8}\right)\right\}$.
2. $\tan^{-1}\left\{\cot\left(-\frac{1}{4}\right)\right\}$.
3. $\sec\left(\cos^{-1}\left(\frac{2}{3}\right)\right)$.
4. $\operatorname{cosec}\left(\sin^{-1}\left(-\frac{1}{\sqrt{3}}\right)\right)$.
5. $\cos\left[\cos^{-1}\left(\frac{-1}{3}\right) - \sin^{-1}\left(\frac{1}{3}\right)\right]$.
6. If $\sin^{-1} x = \frac{\pi}{5}$, for some $x \in (-1, 1)$, then find $\cos^{-1} x$.
7. If $\sec^{-1} x = \operatorname{cosec}^{-1} y$, then find the value of $\cos^{-1}\frac{1}{x} + \cos^{-1}\frac{1}{y}$.
8. Prove that $\tan^{-1} x + \tan^{-1}\frac{1}{x} = \begin{cases} \pi/2, \text{if } x > 0 \\ -\pi/2, \text{if } x < 0 \end{cases}$.

Solve the following

9. $5\tan^{-1} x + 3\cot^{-1} x = 2\pi$
10. $4\sin^{-1} x = \pi - \cos^{-1} x$

Session 5

Property VI, VII and VIII of Inverse Trigonometric Functions

Property VI

(i) $\tan^{-1} x + \tan^{-1} y$

$$= \begin{cases} \tan^{-1}\left(\dfrac{x+y}{1-xy}\right), \text{if } xy<1 \\ \pi + \tan^{-1}\left(\dfrac{x+y}{1-xy}\right), \text{if } x>0, y>0 \text{ and } xy>1 \\ -\pi + \tan^{-1}\left(\dfrac{x+y}{1-xy}\right), \text{if } x<0, y<0 \text{ and } xy>1 \end{cases}$$

(ii) $\tan^{-1} x - \tan^{-1} y$

$$= \begin{cases} \tan^{-1}\left(\dfrac{x-y}{1+xy}\right), \text{if } xy>-1 \\ \pi + \tan^{-1}\left(\dfrac{x-y}{1+xy}\right), \text{if } x>0, y<0 \text{ and } xy<-1 \\ -\pi + \tan^{-1}\left(\dfrac{x-y}{1+xy}\right), \text{if } x<0, y>0 \text{ and } xy<-1 \end{cases}$$

Proof

(i) Let $\tan^{-1} x = A$ and $\tan^{-1} y = B$. Then,

$x = \tan A$ and $y = \tan B$ and $A, B \in (-\pi/2, \pi/2)$

$\therefore \quad \tan(A+B) = \dfrac{\tan A + \tan B}{1 - \tan A \tan B} = \dfrac{x+y}{1-xy}$ ...(i)

Now, *the following cases arises.*

Case I When $x>0, y>0$ and $xy<1$

In this case, $x>0, y>0$ and $xy<1 \Rightarrow \dfrac{x+y}{1-xy}>0$

$\Rightarrow \quad \tan(A+B)>0$

$\Rightarrow A+B$ lies in I quadrant or in III quadrant.

$\Rightarrow \quad 0<A+B<\pi/2$

$$\left[\begin{matrix} \because x>0 \Rightarrow 0<A<\pi/2 \\ y>0 \Rightarrow 0<B<\pi/2 \end{matrix}\right\} \Rightarrow 0<A+B<\pi\bigg]$$

$\therefore \quad \tan(A+B) = \dfrac{x+y}{1-xy}$ {from Eq. (i)}

$\Rightarrow \quad A+B = \tan^{-1}\left(\dfrac{x+y}{1-xy}\right)$

$\Rightarrow \quad \tan^{-1} x + \tan^{-1} y = \tan^{-1}\left(\dfrac{x+y}{1-xy}\right)$

Case II When $x<0, y<0$ and $xy<1$.

In this case;

$x<0, y<0$ and $xy<1$

$\Rightarrow \quad \dfrac{x+y}{1-xy}<0$

$\Rightarrow \quad \tan(A+B)<0$ {from Eq. (i)}

$\Rightarrow A+B$ lies in II quadrant or in IV quadrant.

$\Rightarrow A+B$ lies in IV quadrant.

$$\left[\begin{matrix} \because x<0 \Rightarrow -\pi/2<A<0 \\ y<0 \Rightarrow -\pi/2<B<0 \end{matrix}\right\} \Rightarrow -\pi<A+B<0\bigg]$$

$\Rightarrow \quad \dfrac{-\pi}{2}<A+B<0$

$\therefore \quad \tan(A+B) = \dfrac{x+y}{1-xy}$ {from Eq. (i)}

$\Rightarrow \quad A+B = \tan^{-1}\left(\dfrac{x+y}{1-xy}\right)$

$\Rightarrow \quad \tan^{-1} x + \tan^{-1} y = \tan^{-1}\left(\dfrac{x+y}{1-xy}\right)$

Case III When $x>0$ and $y<0$ or $x<0$ and $y>0$

Consider,

$x>0$ and $y<0 \Rightarrow A \in (0, \pi/2)$

and $\quad B \in (-\pi/2, 0)$

$\Rightarrow \quad A+B \in (-\pi/2, \pi/2)$

$\therefore \quad \tan(A+B) = \dfrac{x+y}{1-xy}$ {from Eq. (i)}

$\Rightarrow \quad A+B = \tan^{-1}\left(\dfrac{x+y}{1-xy}\right)$

$\Rightarrow \quad \tan^{-1} x + \tan^{-1} y = \tan^{-1}\left(\dfrac{x+y}{1-xy}\right)$

Similarly, if $x<0$ and $y>0$, we have

$$\tan^{-1} x + \tan^{-1} y = \tan^{-1}\left(\dfrac{x+y}{1-xy}\right);$$

It follows from above three cases that

$$\tan^{-1} x+\tan^{-1} y=\tan^{-1}\left(\frac{x+y}{1-xy}\right) \text{ if } xy<1$$

Case IV $x>0, y>0$ and $xy>1$.

In this case, we have

$$x>0, y>0 \text{ and } xy>1$$

$$\Rightarrow \quad \frac{x+y}{1-xy}<0$$

$$\Rightarrow \quad \tan(A+B)<0$$

$$\left\{\text{from Eq. (i), } \tan(A+B)=\frac{x+y}{1-xy}\right\}$$

$\Rightarrow$ $A+B$ lies in II quadrant or in IV quadrant.

$\Rightarrow$ $A+B$ lies in II quadrant.

$$\{\because x>0, y>0 \Rightarrow A, B\in(0,\pi/2) \Rightarrow A+B\in(0,\pi)$$

$$\Rightarrow \quad \pi/2<A+B<\pi$$

$$\Rightarrow \quad \frac{\pi}{2}-\pi<(A+B)-\pi<0$$

$$\Rightarrow \quad -\pi/2<(A+B)-\pi<0$$

$$\therefore \tan(A+B)=\frac{x+y}{1-xy}$$

$$\Rightarrow \quad -\tan\{\pi-(A+B)\}=\frac{x+y}{1-xy}$$

$$\{\because \tan(\pi-(A+B))=-\tan(A+B)\}$$

$$\Rightarrow \quad \tan((A+B)-\pi)=\frac{x+y}{1-xy}$$

$$\Rightarrow \quad A+B-\pi=\tan^{-1}\left(\frac{x+y}{1-xy}\right)$$

$$\Rightarrow \quad A+B=\pi+\tan^{-1}\left(\frac{x+y}{1-xy}\right)$$

$$\Rightarrow \quad \tan^{-1} x+\tan^{-1} y=\pi+\tan^{-1}\left(\frac{x+y}{1-xy}\right)$$

Case V When $x<0, y<0$ and $xy>1$

In this case, we have

$$x<0, y<0 \text{ and } xy>1$$

$$\Rightarrow \quad \frac{x+y}{1-xy}>0$$

$$\Rightarrow \quad \tan(A+B)>0$$

$$\left\{\text{from Eq. (i), } \tan(A+B)=\frac{x+y}{1-xy}\right\}$$

$\Rightarrow \tan(A+B)$ lies either in I quadrant or III quadrant.

$\{\because x<0, y<0 \Rightarrow A, B\in(-\pi/2,0) \Rightarrow A+B\in(-\pi,0)\}$

$\Rightarrow A+B$ lies in III quadrant.

$$\Rightarrow \quad -\pi<A+B<-\pi/2$$

$$\Rightarrow \quad \pi-\pi<\pi+(A+B)<\pi-\pi/2$$

$$\Rightarrow \quad 0<\pi+(A+B)<\pi/2$$

Now, $$\tan(A+B)=\frac{x+y}{1-xy} \quad \{\text{from Eq. (i)}\}$$

$$\Rightarrow \quad \tan(\pi+(A+B))=\frac{x+y}{1-xy}$$

$$\Rightarrow \quad \pi+A+B=\tan^{-1}\left(\frac{x+y}{1-xy}\right)$$

$$\Rightarrow \quad A+B=-\pi+\tan^{-1}\left(\frac{x+y}{1-xy}\right)$$

$$\Rightarrow \quad \tan^{-1} x+\tan^{-1} y=-\pi+\tan^{-1}\left(\frac{x+y}{1-xy}\right)$$

(ii) This property can be proved by replacing y by $-y$ in above results.

Remark

If $x_1, x_2, x_3, \ldots, x_n \in R$, then

$$\tan^{-1} x_1+\tan^{-1} x_2+\ldots+\tan^{-1} x_n=\tan^{-1}\left(\frac{S_1-S_3+S_5\ldots}{1-S_2+S_4-S_6\ldots}\right)$$

where, S_k denotes the sum of the product of $x_1, x_2, \ldots, x_n$ taking k at a time.

Example 15. Prove that

(i) $\tan^{-1}\frac{1}{7}+\tan^{-1}\frac{1}{13}=\tan^{-1}\frac{2}{9}$

(ii) $\tan^{-1}2+\tan^{-1}3=\frac{3\pi}{4}$

(iii) $\tan^{-1}\frac{1}{2}+\tan^{-1}\frac{1}{5}+\tan^{-1}\frac{1}{8}=\frac{\pi}{4}$

(iv) $\tan^{-1}\frac{3}{4}+\tan^{-1}\frac{3}{5}-\tan^{-1}\frac{8}{19}=\frac{\pi}{4}$

(v) $\tan^{-1}\frac{1}{5}+\tan^{-1}\frac{1}{7}+\tan^{-1}\frac{1}{3}+\tan^{-1}\frac{1}{8}=\frac{\pi}{4}$

Sol. (i) LHS $= \tan^{-1}\frac{1}{7}+\tan^{-1}\frac{1}{13} = \tan^{-1}\left\{\frac{\frac{1}{7}+\frac{1}{13}}{1-\frac{1}{7}\times\frac{1}{13}}\right\}$

$$\left\{\because \tan^{-1} x+\tan^{-1} y=\tan^{-1}\left(\frac{x+y}{1-xy}\right); \text{if } xy<1\right\}$$

$$=\tan^{-1}\left(\frac{20}{90}\right)=\tan^{-1}\left(\frac{2}{9}\right)=\text{RHS}.$$

(ii) $\tan^{-1}2+\tan^{-1}3=\pi+\tan^{-1}\left(\frac{2+3}{1-2\times3}\right)$

$\left\{\text{using; } \tan^{-1}x+\tan^{-1}y=\pi+\tan^{-1}\left(\frac{x+y}{1-xy}\right), \text{if } xy>1\right\}$

$=\pi+\tan^{-1}(-1)$

$=\pi-\pi/4=3\pi/4$

(iii) $\tan^{-1}\frac{1}{2}+\tan^{-1}\frac{1}{5}+\tan^{-1}\frac{1}{8}$

$=\left(\tan^{-1}\frac{1}{2}+\tan^{-1}\frac{1}{5}\right)+\tan^{-1}\frac{1}{8}$

$=\tan^{-1}\left(\frac{\frac{1}{2}+\frac{1}{5}}{1-\frac{1}{2}\times\frac{1}{5}}\right)+\tan^{-1}\left(\frac{1}{8}\right)$

$\left\{\text{using; } \tan^{-1}x+\tan^{-1}y=\tan^{-1}\left(\frac{x+y}{1-xy}\right), \text{ if } xy<1\right\}$

$=\tan^{-1}\left(\frac{7}{9}\right)+\tan^{-1}\left(\frac{1}{8}\right)$

$=\tan^{-1}\left(\frac{\frac{7}{9}+\frac{1}{8}}{1-\frac{7}{9}\times\frac{1}{8}}\right)=\tan^{-1}\left(\frac{65}{65}\right)=\tan^{-1}(1)=\frac{\pi}{4}$

(iv) $\tan^{-1}\frac{3}{4}+\tan^{-1}\frac{3}{5}-\tan^{-1}\frac{8}{19}$

$=\left(\tan^{-1}\frac{3}{4}+\tan^{-1}\frac{3}{5}\right)-\tan^{-1}\frac{8}{19}$

$=\tan^{-1}\left(\frac{\frac{3}{4}+\frac{3}{5}}{1-\frac{3}{4}\times\frac{3}{5}}\right)-\tan^{-1}\left(\frac{8}{19}\right)$

$=\tan^{-1}\left(\frac{27}{11}\right)-\tan^{-1}\left(\frac{8}{19}\right)=\tan^{-1}\left(\frac{\frac{27}{11}-\frac{8}{19}}{1+\frac{27}{11}\times\frac{8}{19}}\right)$

$=\tan^{-1}\left(\frac{425}{425}\right)=\tan^{-1}1=\frac{\pi}{4}$

(v) $\left(\tan^{-1}\frac{1}{5}+\tan^{-1}\frac{1}{7}\right)+\left(\tan^{-1}\frac{1}{3}+\tan^{-1}\frac{1}{8}\right)$

$\Rightarrow \quad \tan^{-1}\left(\frac{\frac{1}{5}+\frac{1}{7}}{1-\frac{1}{5}\times\frac{1}{7}}\right)+\tan^{-1}\left(\frac{\frac{1}{3}+\frac{1}{8}}{1-\frac{1}{3}\times\frac{1}{8}}\right)$

$\Rightarrow \quad \tan^{-1}\left(\frac{6}{17}\right)+\tan^{-1}\left(\frac{11}{23}\right)$

$\Rightarrow \quad \tan^{-1}\left(\frac{\frac{6}{17}+\frac{11}{23}}{1-\frac{6}{17}\times\frac{11}{23}}\right)=\tan^{-1}\left(\frac{325}{325}\right)=\tan^{-1}(1)=\frac{\pi}{4}$

Example 16. Prove that $\tan^{-1}\frac{yz}{xr}+\tan^{-1}\frac{zx}{yr}+\tan^{-1}\frac{xy}{zr}=\frac{\pi}{2}$, where $x^2+y^2+z^2=r^2$

Sol. We know that,

$$\tan^{-1}x_1+\tan^{-1}x_2+\tan^{-1}x_3+\ldots+\tan^{-1}x_n$$
$$=\tan^{-1}\left(\frac{S_1-S_3+\ldots}{1-S_2+S_4\ldots}\right)$$

where, S_k denotes the sum of the product of $x_1, x_2, \ldots, x_n$ taking k at a time. Therefore,

$$\text{LHS}=\tan^{-1}\left(\frac{yz}{xr}\right)+\tan^{-1}\left(\frac{xz}{yr}\right)+\tan^{-1}\left(\frac{xy}{zr}\right)$$

$$=\tan^{-1}\left(\frac{\frac{yz}{xr}+\frac{xz}{yr}+\frac{xy}{zr}-\frac{xyz}{r^3}}{1-\left(\frac{x^2+y^2+z^2}{r^2}\right)}\right)$$

$$=\tan^{-1}\left(\frac{r^2\left(\frac{xy}{zr}+\frac{yz}{xr}+\frac{zx}{yr}-\frac{xyz}{r^3}\right)}{(x^2+y^2+z^2)-(x^2+y^2+z^2)}\right),$$

where $r^2=x^2+y^2+z^2$

$$=\tan^{-1}(\infty)=\frac{\pi}{2}=\text{LHS}$$

Example 17. Show that $(\tan^{-1}1+\tan^{-1}2+\tan^{-1}3)=\pi$

Sol. $\tan^{-1}2+\tan^{-1}3=\pi+\tan^{-1}\left(\frac{2+3}{1-2\times3}\right)$ {as (2) (3) >1}

$=\pi+\tan^{-1}(-1)=\pi-\tan^{-1}(1)$ ...(i)

$\therefore \quad \tan^{-1}(1)+\tan^{-1}(2)+\tan^{-1}(3)=\tan^{-1}(1)+\pi-\tan^{-1}(1)$ {using Eq. (i)}

$=\pi$

Example 18. Solve for x;

$\tan^{-1}(x+1)+\tan^{-1}x+\tan^{-1}(x-1)=\tan^{-1}3$

Sol. $\tan^{-1}(x+1)+\tan^{-1}(x-1)=\tan^{-1}3-\tan^{-1}x$

$\Rightarrow \tan^{-1}\left(\frac{x+1+x-1}{1-(x+1)(x-1)}\right)=\tan^{-1}\left(\frac{3-x}{1+3x}\right)$, when $x^2-1<1$ and $3x<1$

$\Rightarrow \quad \tan^{-1}\left(\frac{2x}{2-x^2}\right)=\tan^{-1}\left(\frac{3-x}{1+3x}\right)$, when $-\sqrt{2}<x<\sqrt{2}$

and $\quad x<\frac{1}{3}$

$\Rightarrow \quad \frac{2x}{2-x^2}=\frac{3-x}{1+3x}$, when $-\sqrt{2}<x<\frac{1}{3}$

$\Rightarrow \quad 2x(1+3x)=(2-x^2)(3-x)$, when $-\sqrt{2}<x<\frac{1}{3}$

$\Rightarrow \quad 2x+6x^2=6-2x-3x^2+x^3$

$\Rightarrow \quad x^3-9x^2-4x+6=0$

$\Rightarrow \quad (x+1)(x^2-10x+6)=0$, when $-\sqrt{2}<x<\frac{1}{3}$

$\Rightarrow \quad x=-1$ and neglecting $x^2-10x+6=0$ as its roots does not $\in\left(-\sqrt{2},\frac{1}{3}\right)$

$\therefore \quad x=-1$

Example 19. Find the sum of the series $\sum_{m=1}^{\infty}\tan^{-1}\left(\frac{2m}{m^4+m^2+2}\right)$

Sol. We have, $\sum_{m=1}^{\infty}\tan^{-1}\left(\frac{2m}{m^4+m^2+2}\right)$

$$=\sum_{m=1}^{\infty}\tan^{-1}\left(\frac{(m^2+m+1)-(m^2-m+1)}{1+\{(m^2+1)^2-m^2\}}\right)$$

$$=\sum_{m=1}^{\infty}\tan^{-1}\left(\frac{(m^2+m+1)-(m^2-m+1)}{1+(m^2+1+m)(m^2+1-m)}\right)$$

$$=\sum_{m=1}^{\infty}\tan^{-1}(m^2+m+1)-\tan^{-1}(m^2-m+1)$$

$$=\{\tan^{-1}(3)-\tan^{-1}(1)\}+\{\tan^{-1}(7)-\tan^{-1}(3)\}+\{\tan^{-1}(13)-\tan^{-1}(7)\}+\ldots+\infty$$

$$=\tan^{-1}\infty-\tan^{-1}1=\frac{\pi}{2}-\frac{\pi}{4}=\frac{\pi}{4}$$

Example 20. If $\tan^{-1}2$, $\tan^{-1}3$ are two angles of a triangle. Then, find the third angle.

Sol. Let the angle of Δ be,

$A=\tan^{-1}2$, $B=\tan^{-1}3$ and angle C

$\therefore \quad C=\pi-(\tan^{-1}2+\tan^{-1}3)$

$$=\pi-\left\{\pi+\tan^{-1}\left(\frac{2+3}{1-2\cdot 3}\right)\right\}$$

$$=\pi-\pi-\tan^{-1}(-1)=-\tan^{-1}(-1) \text{ or } C=\frac{\pi}{4}$$

$\therefore$ third angle is $\pi/4$

Example 21. Solve the following equations

(i) $\tan^{-1}\frac{x-1}{x-2}+\tan^{-1}\frac{x+1}{x+2}=\frac{\pi}{4}$

(ii) $\tan^{-1}2x+\tan^{-1}3x=\frac{\pi}{4}$

Sol. (i) $\tan^{-1}\left(\frac{x-1}{x-2}\right)+\tan^{-1}\left(\frac{x+1}{x+2}\right)=\frac{\pi}{4}$

$\Rightarrow \quad \tan^{-1}\left(\frac{x-1}{x-2}\right)=\tan^{-1}(1)-\tan^{-1}\left(\frac{x+1}{x+2}\right)$

{where, $\tan^{-1}(1)=\pi/4$}

$\Rightarrow \quad \tan^{-1}\left(\frac{x-1}{x-2}\right)=\tan^{-1}\left(\frac{1-\frac{x+1}{x+2}}{1+\frac{x+1}{x+2}}\right)$

$\Rightarrow \quad \tan^{-1}\left(\frac{x-1}{x-2}\right)=\tan^{-1}\left(\frac{x+2-x-1}{x+2+x+1}\right)$

$\Rightarrow \quad \tan^{-1}\left(\frac{x-1}{x-2}\right)=\tan^{-1}\left(\frac{1}{2x+3}\right)$

$\Rightarrow \quad \frac{x-1}{x-2}=\frac{1}{2x+3}$

$\Rightarrow \quad (2x+3)(x-1)=(x-2)$

$\Rightarrow \quad 2x^2-2x+3x-3=(x-2)$

$\Rightarrow \quad 2x^2+x-3=x-2 \Rightarrow 2x^2=1$

$\Rightarrow \quad x=\pm\frac{1}{\sqrt{2}}$

(ii) $\tan^{-1}2x+\tan^{-1}3x=\pi/4$

$\Rightarrow \quad \tan^{-1}\left\{\frac{2x+3x}{1-6x^2}\right\}=\frac{\pi}{4}$, if $6x^2<1$

$\Rightarrow \quad \frac{5x}{1-6x^2}=1$

$\Rightarrow \quad 6x^2+5x-1=0$, if $x^2<\frac{1}{6}$

$\Rightarrow \quad (6x-1)(x+1)=0$ and $-\frac{1}{\sqrt{6}}<x<\frac{1}{\sqrt{6}}$

$\Rightarrow \quad x=1/6, -1$ and $-\frac{1}{\sqrt{6}}<x<\frac{1}{\sqrt{6}}$

$\Rightarrow \quad x=1/6$, neglecting $x=-1$ as $x\in\left(-\frac{1}{\sqrt{6}},\frac{1}{\sqrt{6}}\right)$

Property VII

(1) $\sin^{-1}x+\sin^{-1}y$

$$=\begin{cases}\sin^{-1}\{x\sqrt{1-y^2}+y\sqrt{1-x^2}\}, & \text{if } -1\le x,y\le 1 \text{ and } x^2+y^2\le 1 \text{ or if } xy<0 \text{ and } x^2+y^2>1\\ \pi-\sin^{-1}\{x\sqrt{1-y^2}+y\sqrt{1-x^2}\}, & \text{if } 0<x,y\le 1 \text{ and } x^2+y^2>1\\ -\pi-\sin^{-1}\{x\sqrt{1-y^2}+y\sqrt{1-x^2}\}, & \text{if } -1\le x,y<0 \text{ and } x^2+y^2>1\end{cases}$$

(2) $\sin^{-1} x - \sin^{-1} y$

$$= \begin{cases} \sin^{-1}\{x\sqrt{1-y^2} - y\sqrt{1-x^2}\}, \text{if } -1 \le x, y \le 1 \\ \quad \text{and } x^2 + y^2 \le 1 \text{ or if } xy > 0 \text{ and } x^2 + y^2 > 1 \\ \pi - \sin^{-1}\{x\sqrt{1-y^2} - y\sqrt{1-x^2}\}, \text{if } 0 < x \le 1, \\ \quad -1 \le y < 0 \text{ and } x^2 + y^2 > 1 \\ -\pi - \sin^{-1}\{x\sqrt{1-y^2} - y\sqrt{1-x^2}\}, \text{if } -1 \le x < 0, \\ \quad 0 < y < 1 \text{ and } x^2 + y^2 > 1 \end{cases}$$

Proof Let, $\sin^{-1} x = A$ and $\sin^{-1} y = B$. Then,

$x = \sin A$, $y = \sin B$ and $A, B \in [-\pi/2, \pi/2]$

$\Rightarrow \quad \cos A = \sqrt{1-x^2}, \cos B = \sqrt{1-y^2}$

$\{\because \cos A, \cos B \ge 0 \text{ for } A, B \in [-\pi/2, \pi/2]\}$

$\therefore \quad \sin(A+B) = \sin A \cos B + \cos A \sin B$

$\Rightarrow \quad \sin(A+B) = x\sqrt{1-y^2} + y\sqrt{1-x^2} \quad \text{...(i)}$

$\sin(A-B) = x\sqrt{1-y^2} - y\sqrt{1-x^2} \quad \text{...(ii)}$

$\cos(A+B) = \sqrt{1-x^2}\sqrt{1-y^2} - xy \quad \text{...(iii)}$

$\cos(A-B) = \sqrt{1-x^2}\sqrt{1-y^2} + xy \quad \text{...(iv)}$

Case I When $-1 \le x, y \le 1$ and $x^2 + y^2 \le 1$

In this case, we have

$x^2 + y^2 \le 1 \Rightarrow 1 - x^2 \ge y^2$ and $1 - y^2 \ge x^2$

$\Rightarrow \quad (1-x^2)(1-y^2) \ge x^2y^2$

$\Rightarrow \quad \sqrt{1-x^2}\sqrt{1-y^2} - xy \ge 0$

$\Rightarrow \quad \cos(A+B) \ge 0 \quad \{\text{using Eq. (iii)}\}$

$\Rightarrow A + B$ lies either in I quadrant or in IV quadrant.

$\Rightarrow A + B \in \left[-\frac{\pi}{2}, \frac{\pi}{2}\right]$

$\left\{\because A, B \in \left[-\frac{\pi}{2}, \frac{\pi}{2}\right] \Rightarrow -\pi \le A + B \le \pi\right\}$

$\therefore \quad \sin(A+B) = x\sqrt{1-y^2} + y\sqrt{1-x^2}$

{from Eq. (i)}

$\Rightarrow \quad A + B = \sin^{-1}\{x\sqrt{1-y^2} + y\sqrt{1-x^2}\}$

$\{\because -\pi/2 \le A + B \le \pi/2\}$

$\Rightarrow \sin^{-1} x + \sin^{-1} y = \sin^{-1}\{x\sqrt{1-y^2} + y\sqrt{1-x^2}\}$

Case II When $xy < 0$ and $x^2 + y^2 > 1$.

In this case, we have

$xy < 0 \Rightarrow (x > 0 \text{ and } y < 0)$ or $(x < 0 \text{ and } y > 0)$

$\Rightarrow \left\{A \in \left(0, \frac{\pi}{2}\right] \text{ and } B \in \left[-\frac{\pi}{2}, 0\right)\right\}$ or $\left\{A \in \left[-\frac{\pi}{2}, 0\right)\right.$

and $\left.B \in \left(0, \frac{\pi}{2}\right]\right\} \Rightarrow -\frac{\pi}{2} \le A + B \le \frac{\pi}{2} \quad \text{...(v)}$

and $\quad x^2 + y^2 > 1$

$\Rightarrow 1 - x^2 < y^2$ and $1 - y^2 < x^2$

$\Rightarrow \quad (1-x^2)(1-y^2) < x^2y^2$

$\Rightarrow \quad (\sqrt{1-x^2}\sqrt{1-y^2})^2 < (|xy|)^2 \quad \{\because xy < 0\}$

$-|xy| < \sqrt{1-x^2}\sqrt{1-y^2} < |xy|$

$\Rightarrow \quad xy < \sqrt{1-x^2}\sqrt{1-y^2} < -xy \quad \{\because xy < 0\}$

$\Rightarrow \quad \sqrt{1-x^2}\sqrt{1-y^2} - xy > 0 \quad \{\therefore |xy| = -xy\}$

$\Rightarrow \quad \cos(A+B) > 0$

$\Rightarrow A + B$ lies either in I quadrant or in IV quadrant.

$\Rightarrow \quad A + B \in \left[-\frac{\pi}{2}, \frac{\pi}{2}\right] \quad \{\text{using Eq. (v)}\}$

$\therefore \quad \sin(A+B) = x\sqrt{1-y^2} + y\sqrt{1-x^2}$

$\Rightarrow \quad A + B = \sin^{-1}\{x\sqrt{1-y^2} + y\sqrt{1-x^2}\}$

$\left\{\because A + B \in \left[-\frac{\pi}{2}, \frac{\pi}{2}\right]\right\}$

$\Rightarrow \quad \sin^{-1} x + \sin^{-1} y = \sin^{-1}\{x\sqrt{1-y^2} + y\sqrt{1-x^2}\}$

Case III When $0 < x, y \le 1$ and $x^2 + y^2 > 1$.

In this case, we have

$0 < x, y \le 1 \Rightarrow A \in \left(0, \frac{\pi}{2}\right]$ and $B \in \left(0, \frac{\pi}{2}\right]$

$\Rightarrow \quad A + B \in (0, \pi] \quad \text{...(vi)}$

and, $\quad x^2 + y^2 > 1 \Rightarrow 1 - x^2 < y^2$ and $1 - y^2 < x^2$

$\Rightarrow \quad (1-x^2)(1-y^2) < x^2y^2$

$\Rightarrow \quad \sqrt{1-x^2}\sqrt{1-y^2} < xy \quad \{\because xy > 0\}$

$\Rightarrow \quad \sqrt{1-x^2}\sqrt{1-y^2} - xy < 0$

$\Rightarrow \quad \cos(A+B) < 0 \quad \{\text{using Eq. (iii)}\}$

$\Rightarrow (A + B)$ lies either in II quadrant or in III quadrant.

$\Rightarrow \quad \frac{\pi}{2} \le A + B \le \pi \quad \{\because A + B \in (0, \pi], \text{ from Eq. (vi)}\}$

$\Rightarrow \quad -\pi \le -(A+B) \le -\frac{\pi}{2} \Rightarrow 0 \le \pi - (A+B) \le \frac{\pi}{2}$

$\therefore \quad \sin(A+B) = x\sqrt{1-y^2} + y\sqrt{1-x^2} \quad \{\text{from Eq. (i)}\}$

$$\Rightarrow \quad \sin\{\pi-(A+B)\} = x\sqrt{1-y^2}+y\sqrt{1-x^2} \quad \{\because \sin(\pi-\theta)=\sin\theta\}$$

$$\Rightarrow \quad \pi-(A+B) = \sin^{-1}\{x\sqrt{1-y^2}+y\sqrt{1-x^2}\}$$

$$\Rightarrow \quad A+B = \pi-\sin^{-1}\{x\sqrt{1-y^2}+y\sqrt{1-x^2}\}$$

$$\Rightarrow \sin^{-1}x+\sin^{-1}y = \pi-\sin^{-1}\{x\sqrt{1-y^2}+y\sqrt{1-x^2}\}$$

Case IV When $-1 \le x, y < 0$ and $x^2+y^2 > 1$

In this case, we have

$$-1 \le x, y < 0 \Rightarrow A \in \left[-\frac{\pi}{2}, 0\right) \text{ and } B \in \left[-\frac{\pi}{2}, 0\right)$$

$$\Rightarrow \quad A+B \in [-\pi, 0) \quad \ldots\text{(vii)}$$

and $\quad x^2+y^2 > 1 \Rightarrow 1-x^2 < y^2$

and $\quad 1-y^2 < x^2 \Rightarrow (1-x^2)(1-y^2) < x^2y^2$

$$\Rightarrow \quad \sqrt{1-x^2}\sqrt{1-y^2} < xy \quad \{\because xy > 0\}$$

$$\Rightarrow \quad \sqrt{1-x^2}\sqrt{1-y^2} - xy < 0$$

$$\Rightarrow \quad \cos(A+B) < 0 \quad \{\text{using Eq. (iii)}\}$$

$\Rightarrow$ $A+B$ lies either in II quadrant or in III quadrant.

$$\Rightarrow \quad -\pi \le A+B \le -\frac{\pi}{2} \quad \{\text{using Eq. (vii)}\}$$

$$\Rightarrow \quad \frac{\pi}{2} \le -(A+B) \le \pi$$

$$\Rightarrow \quad -\frac{\pi}{2} \le -\pi-(A+B) \le 0$$

$$\therefore \quad \sin(A+B) = x\sqrt{1-y^2}+y\sqrt{1-x^2}$$

$$\Rightarrow -\sin\{\pi+(A+B)\} = x\sqrt{1-y^2}+y\sqrt{1-x^2}$$

$$\Rightarrow \sin\{-\pi-(A+B)\} = x\sqrt{1-y^2}+y\sqrt{1-x^2}$$

$$\Rightarrow \quad -\pi-(A+B) = \sin^{-1}\{x\sqrt{1-y^2}+y\sqrt{1-x^2}\}$$

$$\Rightarrow \quad A+B = -\pi-\sin^{-1}\{x\sqrt{1-y^2}+y\sqrt{1-x^2}\}$$

$$\Rightarrow \sin^{-1}x+\sin^{-1}y = -\pi-\sin^{-1}\{x\sqrt{1-y^2}+y\sqrt{1-x^2}\}$$

(2) Do yourself.

Property VIII

(1) $\cos^{-1}x+\cos^{-1}y$

$$= \begin{cases} \cos^{-1}\{xy-\sqrt{1-x^2}\sqrt{1-y^2}\}, \text{ if } -1 \le x, \\ \qquad y \le 1 \text{ and } x+y \ge 0 \\ 2\pi-\cos^{-1}\{xy-\sqrt{1-x^2}\sqrt{1-y^2}\}, \text{ if } -1 \le x, \\ \qquad y \le 1 \text{ and } x+y \le 0 \end{cases}$$

(2) $\cos^{-1}x-\cos^{-1}y$

$$= \begin{cases} \cos^{-1}\{xy+\sqrt{1-x^2}\sqrt{1-y^2}\}, \text{ if } -1 \le x, \\ \qquad y \le 1 \text{ and } x \le y \\ -\cos^{-1}\{xy+\sqrt{1-x^2}\sqrt{1-y^2}\}, \text{ if } -1 \le y \le 0, \\ \qquad 0 < x \le 1 \text{ and } x \ge y \end{cases}$$

Proof (1) Let $\cos^{-1}x = A$ and $\cos^{-1}y = B$. Then,

$$x = \cos A, \; y = \cos B \text{ and } A, B \in [0, \pi]$$

$$\Rightarrow \quad \sin A = \sqrt{1-x^2} \text{ and } \sin B = \sqrt{1-y^2}$$

$$\{\because \sin A, \sin B \ge 0 \text{ for } A, B \in [0, \pi]\}$$

$$\therefore \quad \cos(A+B) = xy-\sqrt{1-x^2}\sqrt{1-y^2} \quad \ldots\text{(i)}$$

$$\cos(A-B) = xy+\sqrt{1-x^2}\sqrt{1-y^2} \quad \ldots\text{(ii)}$$

Case I When $-1 \le x, y \le 1$ and $x+y \ge 0$

In this case,

$$-1 \le x, y \le 1 \Rightarrow A, B \in [0, \pi]$$

$$\Rightarrow \quad 0 \le A+B \le 2\pi \quad \ldots\text{(iii)}$$

and $\quad x+y \ge 0$

$$\Rightarrow \quad \cos A+\cos B \ge 0$$

$$\Rightarrow \quad \cos A \ge -\cos B$$

$$\Rightarrow \quad \cos A \ge \cos(\pi-B)$$

$$\Rightarrow \quad A \le \pi-B \quad \{\because \cos\theta \text{ is decreasing on } [0, \pi]\}$$

$$\Rightarrow \quad A+B \le \pi \quad \ldots\text{(iv)}$$

From Eqs. (iii) and (iv), we get

$$0 \le A+B \le \pi$$

$$\therefore \quad \cos(A+B) = xy-\sqrt{1-x^2}\sqrt{1-y^2}$$

$$\Rightarrow \quad A+B = \cos^{-1}\{xy-\sqrt{1-x^2}\sqrt{1-y^2}\}$$

$$\Rightarrow \quad \cos^{-1}x+\cos^{-1}y = \cos^{-1}\{xy-\sqrt{1-x^2}\sqrt{1-y^2}\}$$

Case II When $-1 \le x, y \le 1$ and $x+y \le 0$

In this case,

$$-1 \le x, y \le 1 \Rightarrow A, B \in [0, \pi]$$

$$\Rightarrow \quad 0 \le A+B \le 2\pi$$

and $\quad x+y \le 0 \Rightarrow \cos A+\cos B \le 0 \quad \ldots$(v)

$$\Rightarrow \quad \cos A \le -\cos B$$

$$\Rightarrow \quad \cos A \le \cos(\pi-B)$$

$$\Rightarrow \quad A \ge \pi-B$$

$$\{\because \cos\theta \text{ is decreasing on } [0, \pi]\}$$

$$\Rightarrow \quad A+B \ge \pi \quad \ldots\text{(vi)}$$

From Eqs. (v) and (vi), we get $\pi \le A+B \le 2\pi$

$$\Rightarrow \quad -\pi \ge -(A+B) \ge -2\pi$$

$$\Rightarrow \quad \pi \geq 2\pi - (A+B) \geq 0$$

$$\Rightarrow \quad 0 \leq 2\pi - (A+B) \leq \pi$$

$$\therefore \quad \cos(A+B) = xy - \sqrt{1-x^2}\sqrt{1-y^2}$$

$$\Rightarrow \quad \cos\{2\pi - (A+B)\} = xy - \sqrt{1-x^2}\sqrt{1-y^2}$$

$$\Rightarrow \quad 2\pi - (A+B) = \cos^{-1}\{xy - \sqrt{1-x^2}\sqrt{1-y^2}\}$$

$$\Rightarrow \quad A + B = 2\pi - \cos^{-1}\{xy - \sqrt{1-x^2}\sqrt{1-y^2}\}$$

$$\Rightarrow \quad \cos^{-1}x + \cos^{-1}y = 2\pi - \cos^{-1}\{xy - \sqrt{1-x^2}\sqrt{1-y^2}\}$$

(2) Do yourself.

Example 22. If $\cos^{-1}\frac{x}{a} + \cos^{-1}\frac{y}{b} = \alpha$, then show $\frac{x^2}{a^2} + \frac{y^2}{b^2} - \frac{2xy}{ab}\cos\alpha = \sin^2\alpha$

Sol. We have, $\cos^{-1}\left(\frac{x}{a}\right) + \cos^{-1}\left(\frac{y}{b}\right) = \alpha$

$$\Rightarrow \quad \cos^{-1}\left(\frac{x}{a}\cdot\frac{y}{b} - \sqrt{1-\frac{x^2}{a^2}}\sqrt{1-\frac{y^2}{b^2}}\right) = \alpha$$

$$\Rightarrow \quad \frac{xy}{ab} - \sqrt{\left(1-\frac{x^2}{a^2}\right)\left(1-\frac{y^2}{b^2}\right)} = \cos\alpha$$

$$\Rightarrow \quad \left(\frac{xy}{ab} - \cos\alpha\right)^2 = 1 - \frac{x^2}{a^2} - \frac{y^2}{b^2} + \frac{x^2y^2}{a^2b^2}$$

$$\Rightarrow \quad \cos^2\alpha + \frac{x^2y^2}{a^2b^2} - \frac{2xy\cos\alpha}{ab} = 1 - \frac{x^2}{a^2} - \frac{y^2}{b^2} + \frac{x^2y^2}{a^2b^2}$$

$$\Rightarrow \quad \frac{x^2}{a^2} + \frac{y^2}{b^2} - \frac{2xy}{ab}\cos\alpha = \sin^2\alpha$$

Example 23. If $\cos^{-1}\lambda + \cos^{-1}\mu + \cos^{-1}\gamma = 3\pi$, then find the value of $\lambda\mu + \mu\gamma + \gamma\lambda$.

Sol. We know, $0 \leq \cos^{-1}x \leq \pi$

Hence, from the question,

$$\cos^{-1}\lambda = \pi,\ \cos^{-1}\mu = \pi,\ \cos^{-1}\gamma = \pi$$

$$\therefore \quad \lambda = \mu = \gamma = -1$$

$\{\because \cos^{-1}\lambda + \cos^{-1}\mu + \cos^{-1}\gamma = 3\pi$ only when each value attains its maximum$\}$

$$\Rightarrow \quad \lambda\mu + \mu\gamma + \gamma\lambda = 3$$

Example 24. If $\sum_{i=1}^{2n} \cos^{-1} x_i = 0$, then find the value of $\sum_{i=1}^{2n} x_i$.

Sol. We know, $0 \leq \cos^{-1}x \leq \pi$.

$\therefore \cos^{-1}x_1, \cos^{-1}x_2, \cos^{-1}x_3, \ldots, \cos^{-1}x_{2n}$ are +ve and hence their sum is zero only when each value is zero.

i.e. $\quad \cos^{-1}x_i = 0$ for all i.

$$\Rightarrow \quad x_i = 1 \text{ for all } i \quad \ldots\text{(i)}$$

Now, $\sum_{i=1}^{2n} x_i = x_1 + x_2 + x_3 + \ldots + x_{2n}$

$$= \underbrace{\{1+1+1\ldots+1\}}_{2n \text{ times}} = 2n \quad \{\text{using Eq. (i)}\}$$

$$\Rightarrow \quad \sum_{i=1}^{2n} x_i = 2n$$

Example 25. If $\sum_{i=1}^{2n} \sin^{-1} x_i = n\pi$, then find the value of $\sum_{i=1}^{2n} x_i$.

Sol. We know, $-\frac{\pi}{2} \leq \sin^{-1}x \leq \frac{\pi}{2}$

Hence, from the question, each value $\sin^{-1}x_1, \sin^{-1}x_2, \sin^{-1}x_3, \ldots, \sin^{-1}x_{2n}$ is equal to $\frac{\pi}{2}$

i.e. $\quad \sin^{-1}x_i = \frac{\pi}{2}$ for all $i \Rightarrow x_i = 1$ for all i

Now, $\sum_{i=1}^{2n} x_i = x_1 + x_2 + \ldots + x_{2n} = \underbrace{\{1+1+\ldots+1\}}_{2n \text{ times}} = 2n$

$$\Rightarrow \quad \sum_{i=1}^{2n} x_i = 2n$$

Example 26. If $\sin^{-1}x + \sin^{-1}y + \sin^{-1}z = \frac{3\pi}{2}$, then find the value of $\Sigma\frac{(x^{101}+y^{101})(x^{202}+y^{202})}{(x^{303}+y^{303})(x^{404}+y^{404})}$

Sol. We have, $\sin^{-1}x + \sin^{-1}y + \sin^{-1}z = \frac{3\pi}{2}$

As we know that, $\sin^{-1}x \leq \frac{\pi}{2}$ for all $x \in [-1, 1]$

Hence, above result is possible only when

$$\sin^{-1}x = \frac{\pi}{2},\ \sin^{-1}y = \frac{\pi}{2},\ \sin^{-1}z = \frac{\pi}{2}$$

$$\Rightarrow \quad x = 1, y = 1, z = 1 \quad \ldots\text{(i)}$$

$$\therefore \quad \Sigma\frac{(x^{101}+y^{101})(x^{202}+y^{202})}{(x^{303}+y^{303})(x^{404}+y^{404})}$$

$$= \frac{(x^{101}+y^{101})(x^{202}+y^{202})}{(x^{303}+y^{303})(x^{404}+y^{404})} + \frac{(y^{101}+z^{101})(y^{202}+z^{202})}{(y^{303}+z^{303})(y^{404}+z^{404})} + \frac{(z^{101}+x^{101})(z^{202}+x^{202})}{(z^{303}+x^{303})(z^{404}+x^{404})} \quad \{\text{expanding summation}\}$$

$$= \frac{(1+1)(1+1)}{(1+1)(1+1)} + \frac{(1+1)(1+1)}{(1+1)(1+1)} + \frac{(1+1)(1+1)}{(1+1)(1+1)} \quad \{\text{using Eq. (i)}\}$$

$$= 1 + 1 + 1 = 3$$

Example 27. If $\cos^{-1}x+\cos^{-1}y+\cos^{-1}z=\pi$, prove that $x^2+y^2+z^2+2xyz=1$.

Sol. Here,

$$\cos^{-1}x+\cos^{-1}y+\cos^{-1}z=\pi$$

$$\Rightarrow \quad \cos^{-1}x+\cos^{-1}y=\pi-\cos^{-1}z$$

$$\Rightarrow \quad \cos^{-1}x+\cos^{-1}y=\cos^{-1}(-z)$$

$$\{\because \cos^{-1}(-x)=\pi-\cos^{-1}x\}$$

$$\Rightarrow \quad \cos^{-1}\{xy-\sqrt{1-x^2}\sqrt{1-y^2}\}=\cos^{-1}(-z)$$

$$\Rightarrow \quad xy-\sqrt{1-x^2}\sqrt{1-y^2}=-z$$

$$\Rightarrow \quad (xy+z)^2=(1-x^2)(1-y^2)$$

$$\Rightarrow \quad x^2y^2+z^2+2xyz=1-x^2-y^2+x^2y^2$$

$$\Rightarrow \quad x^2+y^2+z^2+2xyz=1.$$

Example 28. Find the sum of infinite series

$$s=\sin^{-1}\left(\frac{1}{\sqrt{2}}\right)+\sin^{-1}\left(\frac{\sqrt{2}-1}{\sqrt{6}}\right)+\sin^{-1}\left(\frac{\sqrt{3}-\sqrt{2}}{2\sqrt{3}}\right)+\ldots+\sin^{-1}\left(\frac{\sqrt{n}-\sqrt{n-1}}{\sqrt{n(n+1)}}\right)+\ldots\infty$$

Sol. Let, $T_n=\sin^{-1}\left(\frac{\sqrt{n}-\sqrt{n-1}}{\sqrt{n(n+1)}}\right)$

$$=\sin^{-1}\left(\frac{1}{\sqrt{n}}\sqrt{1-\frac{1}{n+1}}-\frac{1}{\sqrt{n+1}}\sqrt{1-\frac{1}{n}}\right)$$

$$\Rightarrow \quad T_n=\sin^{-1}\left(\frac{1}{\sqrt{n}}\right)-\sin^{-1}\left(\frac{1}{\sqrt{n+1}}\right)$$

$$\therefore \quad s=\sin^{-1}\left(\frac{1}{\sqrt{2}}\right)+T_2+T_3+T_4+\ldots\infty$$

$$=\sin^{-1}\left(\frac{1}{\sqrt{2}}\right)+\left\{\sin^{-1}\left(\frac{1}{\sqrt{2}}\right)-\sin^{-1}\left(\frac{1}{\sqrt{3}}\right)\right\}+\left\{\sin^{-1}\left(\frac{1}{\sqrt{3}}\right)-\sin^{-1}\left(\frac{1}{\sqrt{4}}\right)\right\}+\left\{\sin^{-1}\left(\frac{1}{\sqrt{4}}\right)-\sin^{-1}\left(\frac{1}{\sqrt{5}}\right)\right\}+\ldots\infty$$

$$=2\sin^{-1}\left(\frac{1}{\sqrt{2}}\right)$$

$$=2\cdot\frac{\pi}{4}=\frac{\pi}{2}.$$

Exercise for Session 5

1. Show that $\sin^{-1}\frac{3}{5}+\sin^{-1}\frac{15}{17}=\pi-\sin^{-1}\frac{84}{85}$
2. Evaluate $\sin^{-1}\frac{4}{5}+\sin^{-1}\frac{5}{13}+\sin^{-1}\frac{16}{65}$
3. If $\tan^{-1}4+\tan^{-1}5=\cot^{-1}\lambda$, then find '$\lambda$'.
4. Prove that $\cos^{-1}\left(\frac{7}{2}(1+\cos 2x)+\sqrt{(\sin^2 x-48\cos^2 x)}\sin x\right)=x-\cos^{-1}(7\cos x)$, where $x\in\left(0,\frac{\pi}{2}\right)$.

Solve the following

5. $\sin^{-1}\frac{1}{3}+\sin^{-1}\frac{2}{3}=\sin^{-1}x$
6. $\sin^{-1}x+\sin^{-1}2x=\frac{2\pi}{3}$

Session 6

Property IX of Inverse Trigonometric Functions

Property IX

(i) For $0 < x < 1$,

$$\sin^{-1} x = \cos^{-1} \sqrt{1-x^2} = \tan^{-1}\left(\frac{x}{\sqrt{1-x^2}}\right)$$

$$= \cot^{-1} \frac{\sqrt{1-x^2}}{x} = \sec^{-1}\left(\frac{1}{\sqrt{1-x^2}}\right)$$

$$= \operatorname{cosec}^{-1}\left(\frac{1}{x}\right)$$

(ii) For $0 < x < 1$,

$$\cos^{-1} x = \sin^{-1} \sqrt{1-x^2} = \tan^{-1}\left(\frac{\sqrt{1-x^2}}{x}\right)$$

$$= \cot^{-1}\left(\frac{x}{\sqrt{1-x^2}}\right) = \sec^{-1}\left(\frac{1}{x}\right)$$

$$= \operatorname{cosec}^{-1}\left(\frac{1}{\sqrt{1-x^2}}\right)$$

(iii) For $x > 0$, $\tan^{-1} x = \sin^{-1}\left(\frac{x}{\sqrt{1+x^2}}\right) = \cos^{-1}\left(\frac{1}{\sqrt{1+x^2}}\right)$

$$= \cot^{-1}\left(\frac{1}{x}\right) = \sec^{-1}(\sqrt{1+x^2})$$

$$= \operatorname{cosec}^{-1}\left(\frac{\sqrt{1+x^2}}{x}\right)$$

Proof

(i) Let, $\sin^{-1} x = \theta$. Then $x = \sin\theta$

Now, $\cos\theta = \sqrt{1-\sin^2\theta}$

$\Rightarrow \quad \cos\theta = \sqrt{1-x^2}$

$\Rightarrow \quad \theta = \cos^{-1}\sqrt{1-x^2}$

$\Rightarrow \quad \sin^{-1} x = \cos^{-1}\sqrt{1-x^2}$

$$= \sec^{-1}\left(\frac{1}{\sqrt{1-x^2}}\right) \qquad \{\because \theta = \sin^{-1} x\}$$

Again, $\tan\theta = \dfrac{\sin\theta}{\cos\theta} = \dfrac{\sin\theta}{\sqrt{1-\sin^2\theta}}$

$\Rightarrow \quad \tan\theta = \dfrac{x}{\sqrt{1-x^2}}$

$\Rightarrow \quad \theta = \tan^{-1}\left(\dfrac{x}{\sqrt{1-x^2}}\right)$

$\Rightarrow \quad \sin^{-1} x = \tan^{-1}\left(\dfrac{x}{\sqrt{1-x^2}}\right) \qquad \{\because \theta = \sin^{-1} x\}$

$\Rightarrow \quad \sin^{-1} x = \tan^{-1}\left(\dfrac{x}{\sqrt{1-x^2}}\right) = \cot^{-1}\left(\dfrac{\sqrt{1-x^2}}{x}\right)$

$$\left\{\because \tan^{-1} x = \cot^{-1}\frac{1}{x} \text{ for } x > 0\right\}$$

Hence,

$$\sin^{-1} x = \cos^{-1}\sqrt{1-x^2} = \sec^{-1}\left(\frac{1}{\sqrt{1-x^2}}\right)$$

$$= \tan^{-1}\left(\frac{x}{\sqrt{1-x^2}}\right)$$

$$= \cot^{-1}\left(\frac{\sqrt{1-x^2}}{x}\right)$$

$$= \operatorname{cosec}^{-1}\left(\frac{1}{x}\right)$$

Similarly, the other results can be proved.

Example 29. Evaluate the following

(i) $\sin\left(\tan^{-1}\frac{3}{4}\right)$ (ii) $\sin\left(\cos^{-1}\frac{3}{5}\right)$

(iii) $\cos\left(\tan^{-1}\frac{3}{4}\right)$ (iv) $\sin(\cot^{-1} x)$

Sol. (i) Let $y = \sin\left(\tan^{-1}\frac{3}{4}\right)$ ...(i)

Note

To find y we use $\sin(\sin^{-1} x) = x, -1, \le x \le 1$

For this we convert $\tan^{-1} x$ in $\sin^{-1} x$

Here, $\tan^{-1}\left(\frac{3}{4}\right) = \sin^{-1}\left(\frac{3/4}{\sqrt{1+(3/4)^2}}\right) = \sin^{-1}\left(\frac{3}{5}\right)$

Now, from Eq. (i), we get

or $\quad y = \sin\left(\sin^{-1}\frac{3}{5}\right)$

$\therefore \quad y = \frac{3}{5}$

(ii) $\cos^{-1}\frac{3}{5} = \sin^{-1}\sqrt{1-\left(\frac{3}{5}\right)^2} = \sin^{-1}\frac{4}{5}.$

$\therefore \quad \sin\left(\cos^{-1}\frac{3}{5}\right) = \sin\left(\sin^{-1}\frac{4}{5}\right) = \frac{4}{5}$

(iii) $\tan^{-1}\left(\frac{3}{4}\right) = \cos^{-1}\left(\frac{1}{\sqrt{1-\left(\frac{3}{4}\right)^2}}\right) = \cos^{-1}\frac{4}{5}$

$\therefore \quad \cos\left(\tan^{-1}\frac{3}{4}\right) = \cos\left(\cos^{-1}\frac{4}{5}\right) = \frac{4}{5}$

(iv) Let, $\cot^{-1} x = \tan^{-1}\left(\frac{1}{x}\right) = \sin^{-1}\left(\frac{1/x}{\sqrt{1+1/x^2}}\right)$

$= \sin^{-1}\left(\frac{1}{\sqrt{x^2+1}}\right)$

$\therefore \quad \sin(\cot^{-1} x) = \sin\left(\sin^{-1}\frac{1}{\sqrt{x^2+1}}\right) = \frac{1}{\sqrt{1+x^2}}$

| Example 30. Find the value of $\tan\left(\frac{1}{2}\cos^{-1}\frac{\sqrt{5}}{3}\right)$

Sol. Let $y = \tan\left(\frac{1}{2}\cos^{-1}\frac{\sqrt{5}}{3}\right)$ ...(i)

Let $\cos^{-1}\frac{\sqrt{5}}{3} = \theta \Rightarrow \theta \in \left(0, \frac{\pi}{2}\right)$ and $\cos\theta = \frac{\sqrt{5}}{3}$

$\therefore$ Eq. (i) becomes

$$y = \tan\left(\frac{\theta}{2}\right) \quad \text{... (ii)}$$

$\because \quad \tan^2\frac{\theta}{2} = \frac{1-\cos\theta}{1+\cos\theta} = \frac{1-\frac{\sqrt{5}}{3}}{1+\frac{\sqrt{5}}{3}}$

$= \frac{3-\sqrt{5}}{3-\sqrt{5}} = \frac{(3-\sqrt{5})^2}{4}$

$\tan\frac{\theta}{2} = \pm\left(\frac{3-\sqrt{5}}{2}\right)$ ... (iii)

$\because \quad \theta \in \left(0, \frac{\pi}{2}\right) \Rightarrow \frac{\theta}{2} \in \left(0, \frac{\pi}{4}\right)$

$\therefore \quad \tan\frac{\theta}{2} > 0$

So, from Eq. (iii), we get

$\tan\frac{\theta}{2} = \left(\frac{3-\sqrt{5}}{2}\right)$

Now, from Eq. (ii), we get

$y = \left(\frac{3-\sqrt{5}}{2}\right)$

| Example 31. Find the value of $\cos(2\cos^{-1} x + \sin^{-1} x)$ when $x = \frac{1}{5}$

Sol. Let $y = \cos[2\cos^{-1} x + \sin^{-1} x]$

$\because \quad \sin^{-1} x + \cos^1 x = \frac{\pi}{2}, |x| \le 1$

$\therefore \quad y = \cos\left[2\cos^{-1} x + \frac{\pi}{2} - \cos^{-1} x\right]$

$= \cos\left[\frac{\pi}{2} + \cos^{-1} x\right] = -\sin(\cos^{-1} x)$

$\because \quad x = \frac{1}{5}$

$\therefore \quad y = -\sin\left(\cos^{-1}\frac{1}{5}\right)$

$\Rightarrow \quad y = -\sin\left(\sin^{-1}\sqrt{1-\left(\frac{1}{5}\right)^2}\right)$

$= -\sin\left(\sin^{-1}\frac{\sqrt{24}}{5}\right) = -\frac{\sqrt{24}}{5}$

| Example 32. If $\sin^{-1} x + \sin^{-1} y + \sin^{-1} z = \pi$, prove that $x^4 + y^4 + z^4 + 4x^2y^2z^2 = 2(x^2y^2 + y^2z^2 + z^2x^2)$

Sol. We have

$\Rightarrow \quad \sin^{-1} x + \sin^{-1} y + \sin^{-1} z = \pi$

$\Rightarrow \quad \sin^{-1} x + \sin^{-1} y = \pi - \sin^{-1} z$

$\Rightarrow \quad \cos(\sin^{-1} x + \sin^{-1} y) = \cos(\pi - \sin^{-1} z)$

$\Rightarrow \quad \cos(\sin^{-1} x).\cos(\sin^{-1} y) - \sin(\sin^{-1} x).\sin(\sin^{-1} y)$

$= -\cos(\sin^{-1} z)$

$\Rightarrow \quad \sqrt{1-x^2}.\sqrt{1-y^2} - xy = -\sqrt{1-z^2}$

$\{\because \cos(\sin^{-1} x) = \cos\left(\cos^{-1}\sqrt{1-x^2}\right) = \sqrt{1-x^2}\}$

$\Rightarrow \quad \sqrt{1-x^2}.\sqrt{1-y^2} = xy - \sqrt{1-z^2}$

$\Rightarrow \quad (1-x^2)(1-y^2) = x^2y^2 + 1 - z^2 - 2xy\sqrt{1-z^2}$ {squaring both sides}

$\Rightarrow \quad 1 - x^2 - y^2 + x^2y^2 = x^2y^2 + 1 - z^2 - 2xy\sqrt{1-z^2}$

$\Rightarrow \quad x^2 + y^2 - z^2 = 2xy\sqrt{1-z^2}$

$\Rightarrow \quad (x^2+y^2-z^2)^2 = 4x^2y^2(1-z^2)$

$\Rightarrow \quad x^4 + y^4 + z^4 - 2x^2z^2 - 2y^2z^2 + 2x^2y^2 = 4x^2y^2 - 4x^2y^2z^2$

$\Rightarrow \quad x^4 + y^4 + z^4 + 4x^2y^2z^2 = 2(x^2y^2 + y^2z^2 + z^2x^2)$

Exercise for Session 6

Evaluate the following :

1. $\tan\left(\operatorname{cosec}^{-1}\frac{\sqrt{41}}{4}\right)$
2. $\sec\left(\cot^{-1}\frac{16}{63}\right)$
3. $\sin^2\left[\tan^{-1}\frac{3}{4}\right]$
4. $\sin\left\{\frac{1}{2}\cot^{-1}\left(\frac{-3}{4}\right)\right\}$
5. Show that $\cot\left[\sin^{-1}\sqrt{\frac{13}{17}}\right] = \sin\left[\tan^{-1}\frac{2}{3}\right]$

Session 7

Property X, XI, XII and XIII of Inverse Trigonometric Functions

Property X

(i) $2\sin^{-1}x = \begin{cases} \sin^{-1}(2x\sqrt{1-x^2}), & \text{if } -\frac{1}{\sqrt{2}} \le x \le \frac{1}{\sqrt{2}} \\ \pi - \sin^{-1}(2x\sqrt{1-x^2}), & \text{if } \frac{1}{\sqrt{2}} \le x \le 1 \\ -\pi - \sin^{-1}(2x\sqrt{1-x^2}), & \text{if } -1 \le x \le -\frac{1}{\sqrt{2}} \end{cases}$

(ii) $3\sin^{-1}x = \begin{cases} \sin^{-1}(3x-4x^3), & \text{if } -\frac{1}{2} \le x \le \frac{1}{2} \\ \pi - \sin^{-1}(3x-4x^3), & \text{if } \frac{1}{2} < x \le 1 \\ -\pi - \sin^{-1}(3x-4x^3), & \text{if } -1 \le x < -\frac{1}{2} \end{cases}$

Proof

(i) Let $\sin^{-1}x = \theta$. Then, $x = \sin\theta$

$\Rightarrow \quad \cos\theta = \sqrt{1-x^2}$ $\{\because \cos\theta > 0 \text{ for } \theta \in \left[-\frac{\pi}{2}, \frac{\pi}{2}\right]\}$

$\because \quad \sin 2\theta = 2\sin\theta\cos\theta$

$\therefore \quad \sin 2\theta = 2x\sqrt{1-x^2}$ ...(i)

Case I When $-\frac{1}{\sqrt{2}} \le x \le \frac{1}{\sqrt{2}}$

$-\frac{1}{\sqrt{2}} \le x \le \frac{1}{\sqrt{2}} \Rightarrow -\frac{1}{\sqrt{2}} \le \sin\theta \le \frac{1}{\sqrt{2}} \Rightarrow -\frac{\pi}{4} \le \theta \le \frac{\pi}{4}$

$\Rightarrow \quad -\frac{\pi}{2} \le 2\theta \le \frac{\pi}{2}$

Also, $-\frac{1}{\sqrt{2}} \le x \le \frac{1}{\sqrt{2}}$

$\Rightarrow \quad -1 \le 2x\sqrt{1-x^2} \le 1$

$\therefore \quad \sin 2\theta = 2x\sqrt{1-x^2}$ {from Eq. (i)}

$\Rightarrow \quad 2\theta = \sin^{-1}(2x\sqrt{1-x^2})$

$\Rightarrow \quad 2\sin^{-1} x = \sin^{-1}(2x\sqrt{1-x^2})$

Case II When $\frac{1}{\sqrt{2}} \le x \le 1$

$$\frac{1}{\sqrt{2}} \le x \le 1 \Rightarrow \frac{1}{\sqrt{2}} \le \sin\theta \le 1$$

$\Rightarrow \quad \frac{\pi}{4} \le \theta \le \frac{\pi}{2}$

$\Rightarrow \quad \frac{\pi}{2} \le 2\theta \le \pi$

$\Rightarrow \quad -\pi \le -2\theta \le -\frac{\pi}{2} \Rightarrow 0 \le \pi - 2\theta \le \frac{\pi}{2}$

Also, $\frac{1}{\sqrt{2}} \le x \le 1 \Rightarrow 0 \le 2x\sqrt{1-x^2} \le 1$

$\therefore \quad \sin 2\theta = 2x\sqrt{1-x^2}$ {from Eq. (i)}

$\Rightarrow \quad \sin(\pi - 2\theta) = 2x\sqrt{1-x^2}$

$\Rightarrow \quad \pi - 2\theta = \sin^{-1}(2x\sqrt{1-x^2})$

$\Rightarrow \quad \pi - 2\sin^{-1} x = \sin^{-1}(2x\sqrt{1-x^2})$

$\Rightarrow \quad 2\sin^{-1} x = \pi - \sin^{-1}(2x\sqrt{1-x^2})$

Case III When $-1 \le x \le -\frac{1}{\sqrt{2}}$

$$-1 \le x \le -\frac{1}{\sqrt{2}} \Rightarrow -1 \le \sin\theta \le -\frac{1}{\sqrt{2}}$$

$\Rightarrow \quad -\frac{\pi}{2} \le \theta \le -\frac{\pi}{4} \Rightarrow -\pi \le 2\theta \le -\frac{\pi}{2}$

$\Rightarrow \quad 0 \le \pi + 2\theta \le \frac{\pi}{2}$

Also, $\quad -1 \le x \le -\frac{1}{\sqrt{2}} \Rightarrow -1 \le 2x\sqrt{1-x^2} \le 0$

$\therefore \quad \sin 2\theta = 2x\sqrt{1-x^2}$ {from Eq. (i)}

$\Rightarrow \quad -\sin(\pi + 2\theta) = 2x\sqrt{1-x^2}$

$\Rightarrow \quad \sin(-\pi - 2\theta) = 2x\sqrt{1-x^2}$

$\Rightarrow \quad -\pi - 2\theta = \sin^{-1}(2x\sqrt{1-x^2})$

$\Rightarrow \quad -\pi - 2\sin^{-1} x = \sin^{-1}(2x\sqrt{1-x^2})$

$\Rightarrow \quad 2\sin^{-1} x = -\pi - \sin^{-1}(2x\sqrt{1-x^2})$

(ii) Let, $\sin^{-1} x = \theta$. Then, $x = \sin\theta$

$\therefore \quad \sin 3\theta = 3\sin\theta - 4\sin^3\theta$

$\Rightarrow \quad \sin 3\theta = 3x - 4x^3$

Case I When $-\frac{1}{2} \le x \le \frac{1}{2}$

$$-\frac{1}{2} \le x \le \frac{1}{2} \Rightarrow -\frac{1}{2} \le \sin\theta \le \frac{1}{2} \Rightarrow -\frac{\pi}{6} \le \theta \le \frac{\pi}{6}$$

$\Rightarrow \quad -\frac{\pi}{2} \le 3\theta \le \frac{\pi}{2}$

Also, $\quad -\frac{1}{2} \le x \le \frac{1}{2} \Rightarrow -1 \le 3x - 4x^3 \le 1$

$\therefore \quad \sin 3\theta = 3x - 4x^3$

$\Rightarrow \quad 3\theta = \sin^{-1}(3x - 4x^3)$

$\Rightarrow \quad 3\sin^{-1} x = \sin^{-1}(3x - 4x^3)$

Case II When $\frac{1}{2} < x \le 1$

$$\frac{1}{2} < x \le 1 \Rightarrow \frac{1}{2} < \sin\theta \le 1$$

$\Rightarrow \quad \frac{\pi}{6} < \theta \le \frac{\pi}{2}$

$\Rightarrow \quad \frac{\pi}{2} < 3\theta \le \frac{3\pi}{2}$

$\Rightarrow \quad -\frac{3\pi}{2} \le -3\theta < -\frac{\pi}{2}$

$\Rightarrow \quad -\frac{\pi}{2} \le \pi - 3\theta < \frac{\pi}{2}$

Also, $\frac{1}{2} < x \le 1 \Rightarrow -1 \le 3x - 4x^3 \le 1$

$\therefore \quad \sin 3\theta = (3x - 4x^3)$

$\Rightarrow \quad \sin(\pi - 3\theta) = (3x - 4x^3)$

$\Rightarrow \quad \pi - 3\theta = \sin^{-1}(3x - 4x^3)$

$\Rightarrow \quad \pi - 3\sin^{-1} x = \sin^{-1}(3x - 4x^3)$

$\Rightarrow \quad 3\sin^{-1} x = \pi - \sin^{-1}(3x - 4x^3)$

Case III When $-1 \le x < -\frac{1}{2}$

$$-1 \le x < -\frac{1}{2}$$

$\Rightarrow \quad -1 \le \sin\theta < -\frac{1}{2}$

$\Rightarrow \quad -\frac{\pi}{2} \le \theta < -\frac{\pi}{6}$

$\Rightarrow \quad -\frac{3\pi}{2} \le 3\theta < -\frac{\pi}{2}$

$\Rightarrow \quad -\frac{\pi}{2} \le \pi + 3\theta < \frac{\pi}{2}$

Also, $-1 \le x < -\frac{1}{2} \Rightarrow -1 \le 3x - 4x^3 \le 1$

$\therefore \quad \sin 3\theta = 3x - 4x^3$

$\Rightarrow \quad -\sin(\pi + 3\theta) = 3x - 4x^3$

$\{\sin(\pi + 3\theta) = -\sin 3\theta\}$

$\Rightarrow \quad \sin(-\pi - 3\theta) = 3x - 4x^3$

$\Rightarrow \quad -\pi - 3\theta = \sin^{-1}(3x - 4x^3)$

$\Rightarrow \quad -\pi - 3\sin^{-1} x = \sin^{-1}(3x - 4x^3)$

$\Rightarrow \quad 3\sin^{-1} x = -\pi - \sin^{-1}(3x - 4x^3)$

Property XI

(i) $2\cos^{-1} x = \begin{cases} \cos^{-1}(2x^2 - 1); & \text{if } 0 \le x \le 1 \\ 2\pi - \cos^{-1}(2x^2 - 1); & \text{if } -1 \le x \le 0 \end{cases}$

(ii) $3\cos^{-1} x = \begin{cases} \cos^{-1}(4x^3 - 3x), & \text{if } \frac{1}{2} \le x \le 1 \\ 2\pi - \cos^{-1}(4x^3 - 3x), & \text{if } -\frac{1}{2} \le x \le \frac{1}{2} \\ 2\pi + \cos^{-1}(4x^3 - 3x), & \text{if } -1 \le x \le -\frac{1}{2} \end{cases}$

Proof

(1) Let $\cos^{-1} x = \theta$. Then $x = \cos\theta$

$\therefore \quad \cos 2\theta = 2\cos^2\theta - 1 = 2x^2 - 1$

Case I When $0 \le x \le 1$

$\Rightarrow \quad 0 \le \cos\theta \le 1 \Rightarrow 0 \le \theta \le \frac{\pi}{2}$

$\Rightarrow \quad 0 \le 2\theta \le \pi$

Also, $0 \le x \le 1 \Rightarrow -1 \le 2x^2 - 1 \le 1$

$\therefore \quad \cos 2\theta = 2x^2 - 1$

$\Rightarrow \quad 2\theta = \cos^{-1}(2x^2 - 1)$

$\Rightarrow \quad 2\cos^{-1} x = \cos^{-1}(2x^2 - 1)$

Case II When $-1 \le x \le 0$

$-1 \le x \le 0 \Rightarrow -1 \le \cos\theta \le 0$

$\Rightarrow \quad \frac{\pi}{2} \le \theta \le \pi \Rightarrow \pi \le 2\theta \le 2\pi$

$\Rightarrow \quad -2\pi \le -2\theta \le -\pi$

$\Rightarrow \quad 0 \le 2\pi - 2\theta \le \pi$

Also, $-1 \le x \le 0 \Rightarrow -1 \le 2x^2 - 1 \le 1$

$\therefore \quad \cos 2\theta = (2x^2 - 1)$

$\Rightarrow \quad \cos(2\pi - 2\theta) = 2x^2 - 1$

$\Rightarrow \quad 2\pi - 2\theta = \cos^{-1}(2x^2 - 1)$

$\Rightarrow \quad 2\pi - 2\cos^{-1} x = \cos^{-1}(2x^2 - 1)$

$\Rightarrow \quad 2\cos^{-1} x = 2\pi - \cos^{-1}(2x^2 - 1)$

(ii) Let $\cos^{-1} x = \theta$. Then, $x = \cos\theta$

$\therefore \quad \cos 3\theta = 4\cos^3\theta - 3\cos\theta$

$\Rightarrow \quad \cos 3\theta = 4x^3 - 3x$

Case I When $\frac{1}{2} \le x \le 1$

$\frac{1}{2} \le x \le 1 \Rightarrow \frac{1}{2} \le \cos\theta \le 1$

$\Rightarrow \quad 0 \le \theta \le \frac{\pi}{3}$

$\Rightarrow \quad 0 \le 3\theta \le \pi$

Also, $\quad \frac{1}{2} \le x \le 1 \Rightarrow -1 \le 4x^3 - 3x \le 1$

$\therefore \quad \cos 3\theta = 4x^3 - 3x$

$\Rightarrow \quad 3\theta = \cos^{-1}(4x^3 - 3x)$

$\Rightarrow \quad 3\cos^{-1} x = \cos^{-1}(4x^3 - 3x)$

Case II When $-\frac{1}{2} \le x \le \frac{1}{2}$

$-\frac{1}{2} \le x \le \frac{1}{2}$

$\Rightarrow \quad -\frac{1}{2} \le \cos\theta \le \frac{1}{2}$

$\Rightarrow \quad \frac{\pi}{3} \le \theta \le \frac{2\pi}{3}$

$\Rightarrow \quad \pi \le 3\theta \le 2\pi$

$\Rightarrow \quad -2\pi \le -3\theta \le -\pi$

$\Rightarrow \quad 0 \le 2\pi - 3\theta \le \pi$

$\therefore \quad \cos 3\theta = 4x^3 - 3x$

$\Rightarrow \quad \cos(2\pi - 3\theta) = 4x^3 - 3x$

$\Rightarrow \quad 2\pi - 3\theta = \cos^{-1}(4x^3 - 3x)$

$\Rightarrow \quad 3\theta = 2\pi - \cos^{-1}(4x^3 - 3x)$

$\Rightarrow \quad 3\cos^{-1} x = 2\pi - \cos^{-1}(4x^3 - 3x)$

Case III When $-1 \le x \le -\frac{1}{2}$

$-1 \le x \le -\frac{1}{2} \Rightarrow -1 \le \cos\theta \le -\frac{1}{2}$

$\Rightarrow \quad \frac{2\pi}{3} \le \theta \le \pi$

$$\Rightarrow \quad 2\pi \le 3\theta \le 3\pi$$
$$\Rightarrow \quad -3\pi \le -3\theta \le -2\pi$$
$$\Rightarrow \quad -\pi \le 2\pi - 3\theta \le 0$$
$$\Rightarrow \quad 0 \le 3\theta - 2\pi \le \pi$$
$$\therefore \quad \cos 3\theta = 4x^3 - 3x$$
$$\Rightarrow \quad \cos(2\pi - 3\theta) = 4x^3 - 3x$$
$$\Rightarrow \quad \cos(3\theta - 2\pi) = 4x^3 - 3x$$
$$\Rightarrow \quad 3\theta - 2\pi = \cos^{-1}(4x^3 - 3x)$$
$$\Rightarrow \quad 3\theta = 2\pi + \cos^{-1}(4x^3 - 3x)$$
$$\Rightarrow \quad 3\cos^{-1} x = 2\pi + \cos^{-1}(4x^3 - 3x)$$

Property XII

$$\text{(i) } 2\tan^{-1} x = \begin{cases} \tan^{-1}\left(\dfrac{2x}{1-x^2}\right); & \text{if } -1 < x < 1 \\ \pi + \tan^{-1}\left(\dfrac{2x}{1-x^2}\right); & \text{if } x > 1 \\ -\pi + \tan^{-1}\left(\dfrac{2x}{1-x^2}\right); & \text{if } x < -1 \end{cases}$$

$$\text{(ii) } 3\tan^{-1} x = \begin{cases} \tan^{-1}\left(\dfrac{3x - x^3}{1-3x^2}\right); & \text{if } -\dfrac{1}{\sqrt{3}} < x < \dfrac{1}{\sqrt{3}} \\ \pi + \tan^{-1}\left(\dfrac{3x - x^3}{1-3x^2}\right); & \text{if } x > \dfrac{1}{\sqrt{3}} \\ -\pi + \tan^{-1}\left(\dfrac{3x - x^3}{1-3x^2}\right); & \text{if } x < -\dfrac{1}{\sqrt{3}} \end{cases}$$

Proof

(i) Let $\tan^{-1} x = \theta$. Then, $x = \tan\theta$

$$\therefore \quad \tan 2\theta = \frac{2\tan\theta}{1 - \tan^2\theta}$$

$$\text{or} \quad \tan 2\theta = \frac{2x}{1 - x^2}$$

Case I When $-1 < x < 1$

$$-1 < x < 1 \Rightarrow -1 < \tan\theta < 1$$
$$\Rightarrow \quad -\frac{\pi}{4} < \theta < \frac{\pi}{4} \Rightarrow -\frac{\pi}{2} < 2\theta < \frac{\pi}{2}$$
$$\therefore \quad \tan 2\theta = \frac{2x}{1-x^2}$$
$$\Rightarrow \quad 2\theta = \tan^{-1}\left(\frac{2x}{1-x^2}\right)$$
$$\Rightarrow \quad 2\tan^{-1} x = \tan^{-1}\left(\frac{2x}{1-x^2}\right)$$

Case II When $x > 1$

$$x > 1 \Rightarrow \tan\theta > 1$$
$$\Rightarrow \quad \frac{\pi}{2} > \theta > \frac{\pi}{4} \Rightarrow \pi > 2\theta > \frac{\pi}{2}$$
$$\Rightarrow \quad -\pi < -2\theta < -\frac{\pi}{2}$$
$$\Rightarrow \quad 0 < \pi - 2\theta < \frac{\pi}{2}$$
$$\Rightarrow \quad -\frac{\pi}{2} < -\pi + 2\theta < 0$$
$$\therefore \quad \tan 2\theta = \frac{2x}{1-x^2}$$
$$\Rightarrow \quad -\tan(\pi - 2\theta) = \frac{2x}{1-x^2}$$
$$\Rightarrow \quad \tan(-\pi + 2\theta) = \frac{2x}{1-x^2}$$
$$\Rightarrow \quad -\pi + 2\theta = \tan^{-1}\left(\frac{2x}{1-x^2}\right)$$
$$\Rightarrow \quad 2\theta = \pi + \tan^{-1}\left(\frac{2x}{1-x^2}\right)$$
$$\Rightarrow \quad 2\tan^{-1} x = \pi + \tan^{-1}\left(\frac{2x}{1-x^2}\right)$$

Case III When $x < -1$

$$x < -1 \Rightarrow \tan\theta < -1$$
$$\Rightarrow \quad -\frac{\pi}{2} < \theta < -\frac{\pi}{4}$$
$$\Rightarrow \quad -\pi < 2\theta < -\frac{\pi}{2}$$
$$\Rightarrow \quad 0 < \pi + 2\theta < \frac{\pi}{2}$$
$$\therefore \quad \tan 2\theta = \frac{2x}{1-x^2}$$
$$\Rightarrow \quad \tan(\pi + 2\theta) = \frac{2x}{1-x^2} \quad \{\because \tan(\pi + \alpha) = \tan\alpha\}$$
$$\Rightarrow \quad \pi + 2\theta = \tan^{-1}\left(\frac{2x}{1-x^2}\right)$$
$$\Rightarrow \quad 2\theta = -\pi + \tan^{-1}\left(\frac{2x}{1-x^2}\right)$$
$$\Rightarrow \quad 2\tan^{-1} x = -\pi + \tan^{-1}\left(\frac{2x}{1-x^2}\right)$$

(ii) Let $\tan^{-1} x = \theta$. Then $x = \tan\theta$

$$\therefore \quad \tan 3\theta = \frac{3\tan\theta - \tan^3\theta}{1 - 3\tan^2\theta}$$
$$\Rightarrow \quad \tan 3\theta = \frac{3x - x^3}{1 - 3x^2}$$

Case I When $-\frac{1}{\sqrt{3}} < x < \frac{1}{\sqrt{3}}$

$$-\frac{1}{\sqrt{3}} < x < \frac{1}{\sqrt{3}} \Rightarrow -\frac{1}{\sqrt{3}} < \tan\theta < \frac{1}{\sqrt{3}}$$

$$\Rightarrow \quad -\frac{\pi}{6} < \theta < \frac{\pi}{6} \Rightarrow -\frac{\pi}{2} < 3\theta < \frac{\pi}{2}$$

$$\therefore \quad \tan 3\theta = \frac{3x - x^3}{1 - 3x^2} \Rightarrow 3\theta = \tan^{-1}\left(\frac{3x - x^3}{1 - 3x^2}\right)$$

$$\Rightarrow \quad 3\tan^{-1} x = \tan^{-1}\left(\frac{3x - x^3}{1 - 3x^2}\right)$$

Case II When $x > \frac{1}{\sqrt{3}}$

$$x > \frac{1}{\sqrt{3}} \Rightarrow \tan\theta > \frac{1}{\sqrt{3}}$$

$$\Rightarrow \quad \frac{\pi}{2} > \theta > \frac{\pi}{6}$$

$$\Rightarrow \quad \frac{\pi}{2} < 3\theta < \frac{3\pi}{2}$$

$$\Rightarrow \quad -\frac{3\pi}{2} < -3\theta < -\frac{\pi}{2}$$

$$\Rightarrow \quad -\frac{\pi}{2} < \pi - 3\theta < \frac{\pi}{2}$$

$$\Rightarrow \quad -\frac{\pi}{2} < 3\theta - \pi < \frac{\pi}{2}$$

$$\therefore \quad \tan 3\theta = \frac{3x - x^3}{1 - 3x^2}$$

$$\Rightarrow \quad -\tan(\pi - 3\theta) = \frac{3x - x^3}{1 - 3x^2} \quad \{\because \tan(\pi - 3\theta) = -\tan 3\theta\}$$

$$\Rightarrow \quad 3\theta - \pi = \tan^{-1}\left(\frac{3x - x^3}{1 - 3x^2}\right)$$

$$\Rightarrow \quad 3\tan^{-1} x = \pi + \tan^{-1}\left(\frac{3x - x^3}{1 - 3x^2}\right)$$

Case III When $x < -\frac{1}{\sqrt{3}}$

$$x < -\frac{1}{\sqrt{3}} \Rightarrow \tan\theta < -\frac{1}{\sqrt{3}}$$

$$\Rightarrow \quad -\frac{\pi}{2} < \theta < -\frac{\pi}{6}$$

$$\Rightarrow \quad -\frac{3\pi}{2} < 3\theta < -\frac{\pi}{2}$$

$$\Rightarrow \quad -\frac{\pi}{2} < \pi + 3\theta < \frac{\pi}{2}$$

$$\therefore \quad \tan 3\theta = \frac{3x - x^3}{1 - 3x^2}$$

$$\Rightarrow \quad \tan(\pi + 3\theta) = \frac{3x - x^3}{1 - 3x^2} \qquad \{\because \tan(\pi + x) = \tan x\}$$

$$\Rightarrow \quad \pi + 3\theta = \tan^{-1}\left(\frac{3x - x^3}{1 - 3x^2}\right)$$

$$\Rightarrow \quad 3\tan^{-1} x = -\pi + \tan^{-1}\left(\frac{3x - x^3}{1 - 3x^2}\right)$$

Property XIII

(i) $$2\tan^{-1} x = \begin{cases} \sin^{-1}\left(\frac{2x}{1 + x^2}\right); & \text{if } -1 \le x \le 1 \\ \pi - \sin^{-1}\left(\frac{2x}{1 + x^2}\right); & \text{if } x > 1 \\ -\pi - \sin^{-1}\left(\frac{2x}{1 + x^2}\right); & \text{if } x < -1 \end{cases}$$

(ii) $$2\tan^{-1} x = \begin{cases} \cos^{-1}\left(\frac{1 - x^2}{1 + x^2}\right); & \text{if } 0 \le x < \infty \\ -\cos^{-1}\left(\frac{1 - x^2}{1 + x^2}\right); & \text{if } -\infty < x \le 0 \end{cases}$$

Proof

(i) Let $\tan^{-1} x = \theta$. Then, $x = \tan\theta$

$$\therefore \quad \sin 2\theta = \frac{2\tan\theta}{1 + \tan^2\theta} \Rightarrow \sin 2\theta = \frac{2x}{1 + x^2}$$

Case I When $-1 \le x \le 1$

$$-1 \le x \le 1 \Rightarrow -1 \le \tan\theta \le 1$$

$$\Rightarrow \quad -\frac{\pi}{4} \le \theta \le \frac{\pi}{4} \Rightarrow -\frac{\pi}{2} \le 2\theta \le \frac{\pi}{2}$$

$$\therefore \quad \sin 2\theta = \frac{2x}{1 + x^2} \Rightarrow 2\theta = \sin^{-1}\left(\frac{2x}{1 + x^2}\right)$$

$$\Rightarrow \quad 2\tan^{-1} x = \sin^{-1}\left(\frac{2x}{1 + x^2}\right)$$

Case II When $x > 1$

$$x > 1 \Rightarrow \tan\theta > 1$$

$$\Rightarrow \quad \frac{\pi}{4} < \theta < \frac{\pi}{2} \Rightarrow \frac{\pi}{2} < 2\theta < \pi$$

$$\Rightarrow \quad -\pi < -2\theta < -\frac{\pi}{2} \Rightarrow 0 < \pi - 2\theta < \frac{\pi}{2}$$

$$\therefore \quad \sin 2\theta = \frac{2x}{1 + x^2}$$

$$\Rightarrow \quad \sin(\pi - 2\theta) = \frac{2x}{1 + x^2}$$

$$\Rightarrow \quad \pi - 2\theta = \sin^{-1}\left(\frac{2x}{1 + x^2}\right)$$

$$\Rightarrow \quad \pi - 2\tan^{-1} x = \sin^{-1}\left(\frac{2x}{1+x^2}\right)$$

$$\Rightarrow \quad 2\tan^{-1} x = \pi - \sin^{-1}\left(\frac{2x}{1+x^2}\right)$$

Case III When $x < -1$

$$x < -1 \Rightarrow \tan\theta < -1$$

$$\Rightarrow \quad -\frac{\pi}{2} < \theta < -\frac{\pi}{4} \Rightarrow -\pi < 2\theta < -\frac{\pi}{2}$$

$$\Rightarrow \quad 0 < \pi + 2\theta < \frac{\pi}{2}$$

$$\Rightarrow \quad -\frac{\pi}{2} < -\pi - 2\theta < 0$$

$$\therefore \quad \sin 2\theta = \frac{2x}{1+x^2}$$

$$\Rightarrow \quad -\sin(\pi + 2\theta) = \frac{2x}{1+x^2}$$

$$\Rightarrow \quad \sin(-\pi - 2\theta) = \frac{2x}{1+x^2}$$

$$\Rightarrow \quad -\pi - 2\theta = \sin^{-1}\left(\frac{2x}{1+x^2}\right)$$

$$\Rightarrow \quad -\pi - 2\tan^{-1} x = \sin^{-1}\left(\frac{2x}{1+x^2}\right)$$

$$\Rightarrow \quad 2\tan^{-1} x = -\pi - \sin^{-1}\left(\frac{2x}{1+x^2}\right)$$

(ii) Let $\tan^{-1} x = \theta$

Then $x = \tan\theta$

$$\therefore \quad \cos 2\theta = \frac{1-\tan^2\theta}{1+\tan^2\theta} \Rightarrow \cos 2\theta = \frac{1-x^2}{1+x^2}$$

Case I When $0 \le x < \infty$

$$0 \le x < \infty$$

$$\Rightarrow \quad 0 \le \theta < \frac{\pi}{2}$$

$$\Rightarrow \quad 0 \le 2\theta < \pi$$

$$\therefore \quad \cos 2\theta = \frac{1-x^2}{1+x^2} \Rightarrow 2\theta = \cos^{-1}\left(\frac{1-x^2}{1+x^2}\right)$$

$$\Rightarrow \quad 2\tan^{-1} x = \cos^{-1}\left(\frac{1-x^2}{1+x^2}\right)$$

Case II When $-\infty < x \le 0$

$$-\infty < x \le 0$$

$$\Rightarrow \quad -\frac{\pi}{2} < \theta \le 0 \Rightarrow 0 \le -2\theta < \pi$$

$$\therefore \quad \cos 2\theta = \frac{1-x^2}{1+x^2}$$

$$\Rightarrow \quad \cos(-2\theta) = \frac{1-x^2}{1+x^2}$$

$$\Rightarrow \quad -2\theta = \cos^{-1}\left(\frac{1-x^2}{1+x^2}\right)$$

$$\Rightarrow \quad -2\tan^{-1} x = \cos^{-1}\left(\frac{1-x^2}{1+x^2}\right)$$

$$\Rightarrow \quad 2\tan^{-1} x = -\cos^{-1}\left(\frac{1-x^2}{1+x^2}\right)$$

Example 33. Define $y = \cos^{-1}(4x^3 - 3x)$ in terms of $\cos^{-1} x$ and also draw its graph.

Sol. We know, $3\cos^{-1} x = \begin{cases} \cos^{-1}(4x^3 - 3x), \text{ if } \frac{1}{2} \le x \le 1 \\ 2\pi - \cos^{-1}(4x^3 - 3x), \text{ if } -\frac{1}{2} \le x < \frac{1}{2} \\ 2\pi + \cos^{-1}(4x^3 - 3x), \text{ if } -1 \le x < -\frac{1}{2} \end{cases}$

$$\therefore \quad y = \cos^{-1}(4x^3 - 3x)$$

$$= \begin{cases} 3\cos^{-1} x & ; \quad \frac{1}{2} \le x \le 1 \\ 2\pi - 3\cos^{-1} x & ; \quad -\frac{1}{2} \le x < \frac{1}{2} \\ -2\pi + 3\cos^{-1} x & ; \quad -1 \le x < -\frac{1}{2} \end{cases}$$

Graph

For $y = \cos^{-1}(4x^3 - 3x)$

Domain : $[-1, 1]$

Range : $[0, \pi]$

(i) If $\frac{1}{2} \le x \le 1, y = 3\cos^{-1} x$

$$\Rightarrow \quad \frac{dy}{dx} = \frac{-3}{\sqrt{1-x^2}} = -3(1-x^2)^{-1/2} \quad \text{...(i)}$$

$$\Rightarrow \quad \frac{dy}{dx} < 0 \text{ if } x \in \left[\frac{1}{2}, 1\right)$$

$\Rightarrow$ Decreasing if $x \in \left[\frac{1}{2}, 1\right)$

Again if we differentiate Eq. (i) w.r.t 'x', we get

$$\frac{d^2y}{dx^2} = -\frac{3x}{(1-x^2)^{3/2}}$$

$$\Rightarrow \frac{d^2y}{dx^2} < 0 \text{ if } x \in \left[\frac{1}{2}, 1\right)$$

$$\Rightarrow \text{concavity downwards if } x \in \left[\frac{1}{2}, 1\right)$$

(ii) If $-\frac{1}{2} \le x < \frac{1}{2}$, $y = 2\pi - 3\cos^{-1} x$

$$\therefore \quad \frac{dy}{dx} = \frac{3}{\sqrt{1-x^2}} \Rightarrow \frac{dy}{dx} > 0 \text{ if } x \in \left[-\frac{1}{2}, \frac{1}{2}\right)$$

$$\Rightarrow \text{Increasing if } x \in \left[-\frac{1}{2}, \frac{1}{2}\right) \text{ and } \frac{d^2y}{dx^2} = \frac{3x}{(1-x^2)^{3/2}}$$

(a) If $x \in \left[-\frac{1}{2}, 0\right)$ then $\frac{d^2y}{dx^2} < 0$

$$\Rightarrow \text{Concavity downwards if } x \in \left[-\frac{1}{2}, 0\right)$$

(b) If $x \in \left[0, \frac{1}{2}\right)$ then $\frac{d^2y}{dx^2} > 0$

$$\Rightarrow \text{Concavity upwards if } x \in \left[0, \frac{1}{2}\right)$$

(iii) Similarly, if $-1 \le x < -\frac{1}{2}$ then $\frac{dy}{dx} < 0$ and $\frac{d^2y}{dx^2} > 0$

$\therefore$ the graph of $y = \cos^{-1}(4x^3 - 3x)$ is as

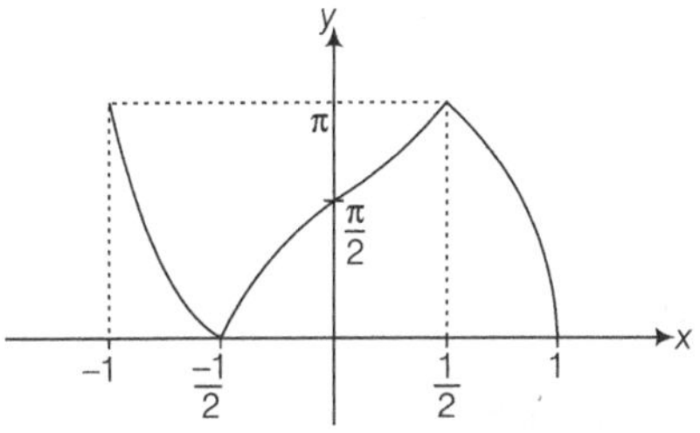

Example 34. Prove that

$$4\tan^{-1}\frac{1}{5} - \tan^{-1}\frac{1}{70} + \tan^{-1}\frac{1}{99} = \frac{\pi}{4}$$

Sol. $4\tan^{-1}\frac{1}{5} - \tan^{-1}\frac{1}{70} + \tan^{-1}\frac{1}{99}$

$$= 2\left(2\tan^{-1}\frac{1}{5}\right) - \tan^{-1}\frac{1}{70} + \tan^{-1}\frac{1}{99}$$

$$= 2\left\{\tan^{-1}\frac{2(1/5)}{1-(1/5)^2}\right\} - \tan^{-1}\frac{1}{70} + \tan^{-1}\frac{1}{99}$$

$$\left\{\because 2\tan^{-1}x = \tan^{-1}\left(\frac{2x}{1-x^2}\right), \text{ if } |x| < 1\right\}$$

$$= 2\tan^{-1}\frac{5}{12} - \left\{\tan^{-1}\frac{1}{70} - \tan^{-1}\frac{1}{99}\right\}$$

$$= \tan^{-1}\left(\frac{2(5/12)}{1-(5/12)^2}\right) - \tan^{-1}\left(\frac{\frac{1}{70} - \frac{1}{99}}{1 + \frac{1}{70} \times \frac{1}{99}}\right)$$

$$\left\{\because 2\tan^{-1}x = \tan^{-1}\left(\frac{2x}{1-x^2}\right) \text{ if } |x| < 1 \text{ and } \tan^{-1}x - \tan^{-1}y = \tan^{-1}\left(\frac{x-y}{1+xy}\right)\right\}$$

$$= \tan^{-1}\left(\frac{120}{119}\right) - \tan^{-1}\left(\frac{29}{6931}\right)$$

$$= \tan^{-1}\left(\frac{120}{119}\right) - \tan^{-1}\left(\frac{1}{239}\right)$$

$$= \tan^{-1}\left(\frac{\frac{120}{119} - \frac{1}{239}}{1 + \left(\frac{120}{119}\right)\left(\frac{1}{239}\right)}\right)$$

$$= \tan^{-1}(1) = \frac{\pi}{4}$$

Example 35. If $\sin^{-1}x + \sin^{-1}y + \sin^{-1}z = \pi$, prove that $x\sqrt{1-x^2} + y\sqrt{1-y^2} + z\sqrt{1-z^2} = 2xyz$.

Sol. Let $\sin^{-1}x = A$, $\sin^{-1}y = B$ and $\sin^{-1}z = C$.

Then, $x = \sin A$, $y = \sin B$, $z = \sin C$

We have,

$$\sin^{-1}x + \sin^{-1}y + \sin^{-1}z = \pi$$

$$\Rightarrow \quad A + B + C = \pi$$

$$\Rightarrow \quad \sin 2A + \sin 2B + \sin 2C = 4\sin A \sin B \sin C$$

$$\Rightarrow \quad 2\sin A \cdot \cos A + 2\sin B \cdot \cos B + 2\sin C \cdot \cos C = 4\sin A \sin B \sin C$$

$$\Rightarrow \quad x\sqrt{1-x^2} + y\sqrt{1-y^2} + z\sqrt{1-z^2} = 2xyz$$

Example 36. Solve $\sin[2\cos^{-1}\{\cot(2\tan^{-1}x)\}] = 0$

Sol. We have,

$$\sin[2\cos^{-1}\{\cot(2\tan^{-1}x)\}] = 0$$

$$\Rightarrow \quad \sin\left[2\cos^{-1}\left\{\cot\left(\tan^{-1}\left(\frac{2x}{1-x^2}\right)\right)\right\}\right] = 0$$

$$\left\{\because 2\tan^{-1}x = \tan^{-1}\left(\frac{2x}{1-x^2}\right)\right\}$$

$$\Rightarrow \quad \sin\left[2\cos^{-1}\left\{\cot\left(\cot^{-1}\left(\frac{1-x^2}{2x}\right)\right)\right\}\right] = 0$$

$$\left\{\because \cot^{-1}x = \tan^{-1}\frac{1}{x}\right]$$

$$\Rightarrow \quad \sin\left[2\cos^{-1}\left(\frac{1-x^2}{2x}\right)\right]=0$$

$$\Rightarrow \quad \sin\left[\sin^{-1}\left\{2\left(\frac{1-x^2}{2x}\right)\sqrt{1-\left(\frac{1-x^2}{2x}\right)^2}\right\}\right]=0$$

$$\{\because 2\cos^{-1}x=\sin^{-1}(2x\sqrt{1-x^2})\}$$

$$\Rightarrow \quad \left(\frac{1-x^2}{x}\right)\left[\sqrt{1-\left(\frac{1-x^2}{2x}\right)^2}\right]=0$$

$$\Rightarrow \quad \frac{1-x^2}{x}=0 \text{ or } 1-\left(\frac{1-x^2}{2x}\right)^2=0$$

$$\Rightarrow \quad 1-x^2=0 \text{ or } \left(\frac{1-x^2}{2x}\right)^2=1$$

$$\Rightarrow \quad x=\pm 1 \text{ or } (1-x^2)^2=4x^2$$

Now, $$(1-x^2)^2=4x^2$$

$$\Rightarrow \quad (1-x^2)^2-(2x)^2=0$$

$$\Rightarrow \quad (1-x^2-2x)(1-x^2+2x)=0$$

$$\Rightarrow \quad x^2+2x-1=0 \text{ or } x^2-2x-1=0$$

$$\Rightarrow \quad x=-1\pm\sqrt{2}$$

or $$x=1\pm\sqrt{2}$$

Hence, $x=\pm 1, -1\pm\sqrt{2}, 1\pm\sqrt{2}$ are the roots of the given equation.

Example 37. Find the value of

$$\tan\left\{\frac{1}{2}\sin^{-1}\left(\frac{2x}{1+x^2}\right)+\frac{1}{2}\cos^{-1}\left(\frac{1-y^2}{1+y^2}\right)\right\}$$

Sol. Let $x=\tan A$ and $y=\tan B$. Then,

$$\tan\left\{\frac{1}{2}\sin^{-1}\left(\frac{2x}{1+x^2}\right)+\frac{1}{2}\cos^{-1}\left(\frac{1-y^2}{1+y^2}\right)\right\}$$

$$=\tan\left\{\frac{1}{2}\sin^{-1}\left(\frac{2\tan A}{1+\tan^2 A}\right)+\frac{1}{2}\cos^{-1}\left(\frac{1-\tan^2 B}{1+\tan^2 B}\right)\right\}$$

$$=\tan\left\{\frac{1}{2}\sin^{-1}(\sin 2A)+\frac{1}{2}\cos^{-1}(\cos 2B)\right\}$$

$$=\tan\left(\frac{2A}{2}+\frac{2B}{2}\right)=\tan(A+B)$$

$$=\frac{\tan A+\tan B}{1-\tan A\tan B}=\frac{x+y}{1-xy}$$

Aliter

$$\tan\left[\frac{1}{2}\sin^{-1}\left(\frac{2x}{1+x^2}\right)+\frac{1}{2}\cos^{-1}\left(\frac{1-y^2}{1+y^2}\right)\right]$$

$$=\tan\left[\frac{1}{2}\cdot 2\tan^{-1}x+\frac{1}{2}\cdot 2\tan^{-1}y\right]$$

$$=\tan(\tan^{-1}x+\tan^{-1}y)$$

$$=\tan\left(\tan^{-1}\left(\frac{x+y}{1-xy}\right)\right)=\frac{x+y}{1-xy}$$

Exercise for Session 7

1. Define $y=\sin^{-1}(3x-4x^3)$ in terms of $\sin^{-1}x$ and also draw its graph.
2. Define $y=\tan^{-1}\left(\frac{3x-x^3}{1-3x^2}\right)$ in terms of $\tan^{-1}x$ and also draw its graph.

Draw the graph of the following

3. $y=\cos^{-1}(2x^2-1)$
4. $y=\sin^{-1}(2x\sqrt{1-x^2})$
5. $y=\tan^{-1}\frac{2x}{1-x^2}$
6. $y=\sin^{-1}\frac{2x}{1+x^2}$
7. $y=\cos^{-1}\frac{1-x^2}{1+x^2}$

JEE Type Solved Examples :
Single Option Correct Type Questions

Ex. 1. *Let* $f(x) = \sin x + \cos x + \tan x + \text{arc}\sin x + \text{arc}\cos x + \text{arc}\tan x$. *If M and m are maximum and minimum values of* $f(x)$, *then their arithmetic mean is equal to*

(a) $\frac{\pi}{2} + \cos 1$ (b) $\frac{\pi}{2} + \sin 1$

(c) $\frac{\pi}{4} + \tan 1 + \cos 1$ (d) $\frac{\pi}{4} + \tan 1 + \sin 1$

Sol. (*a*) Domain of f is $[-1,1]$; $f(x) = \sin x + \cos x + \tan x + \sin^{-1} x + \cos^{-1} x + \tan^{-1} x$

$$f'(x) = \cos x - \sin x + \underbrace{\sec^2 x}_{>1} + 0 + \underbrace{\frac{1}{1+x^2}}_{[1/2,1]}$$

Hence, $f'(x) > 0 \Rightarrow f$ is increasing

$\Rightarrow$ Range is $[f(-1), f(1)]$

$\therefore \quad f(x)|_{min} = f(-1) = -\sin 1 + \cos 1 - \tan 1$

$$-\frac{\pi}{2} + \pi - \frac{\pi}{4} = \frac{\pi}{4} + \cos 1 - \sin 1 - \tan 1$$

and $f(x)|_{max} = f(1) = \sin 1 + \cos 1 + \tan 1$

$$+\frac{\pi}{2} + \frac{\pi}{4} = \frac{3\pi}{4} + \cos 1 + \sin 1 + \tan 1$$

Now, $\quad \frac{M+m}{2} = \frac{\pi}{2} + \cos 1$

Ex. 2. *The value of* $5 \cdot \cot\left(\sum_{k=1}^{5} \cot^{-1}(k^2 + k + 1)\right)$ *is equal to*

(a) $\frac{5}{2}$ (b) 7

(c) −7 (d) $\frac{7}{2}$

Sol. (*b*) Consider

$$\sum_{k=1}^{5}\left(\tan^{-1}\left(\frac{(k+1)-k}{1+k(k+1)}\right)\right) = \sum_{k=1}^{5} \tan^{-1}(k+1) - \tan^{-1} k$$

Now, $\quad T_1 = \tan^{-1}(2) - \tan^{-1}(1)$;

$T_2 = \tan^{-1}(3) - \tan^{-1}(2)$ and so on

Hence, $\sum_{k=1}^{5} \cot^{-1}(k^2 + k + 1) = \tan^{-1}(6) - \tan^{-1}(1)$

$$= \tan^{-1}\left(\frac{5}{7}\right)$$

$$= \cot^{-1}\left(\frac{7}{5}\right)$$

$\therefore \quad 5\cot\left(\cot^{-1}\frac{7}{5}\right) = 7$

Ex. 3. *If the equation* $5 \text{ arc}\tan(x^2 + x + k) + 3 \text{ arc}\cot(x^2 + x + k) = 2\pi$, *has two distinct solutions, then the range of k, is*

(a) $\left(0, \frac{5}{4}\right]$ (b) $\left(-\infty, \frac{5}{4}\right)$ (c) $\left(\frac{5}{4}, \infty\right)$ (d) $\left(-\infty, \frac{5}{4}\right]$

Sol. (*b*) We have $2\pi = \frac{3\pi}{2} + 2\tan^{-1}(x^2 + x + k)$

$$\left(\text{As, } \tan^{-1}\alpha + \cot^{-1}\alpha = \frac{\pi}{2} \forall\, \alpha \in R\right)$$

$\Rightarrow \quad \tan^{-1}(x^2 + x + k) = \frac{\pi}{4} \Rightarrow x^2 + x + k = 1$

$\Rightarrow \quad x^2 + x + (k-1) = 0$

$\therefore$ For required condition, put $D > 0$

$\Rightarrow \quad 1 - 4(k-1) > 0 \Rightarrow 5 - 4k > 0 \Rightarrow k < \frac{5}{4}$

Ex. 4. *If* $f(x) = x^{11} + x^9 - x^7 + x^3 + 1$ *and* $f(\sin^{-1}(\sin 8)) = \alpha$, α *is constant, then* $f(\tan^{-1}(\tan 8))$ *is equal to*

(a) α (b) $\alpha - 2$

(c) $\alpha + 2$ (d) $2 - \alpha$

Sol. (*d*) $f(x) + f(-x) = 2$

Now, $\quad (\sin^{-1}(\sin 8)) = 3\pi - 8 = y$ (say)

and $\quad (\tan^{-1}(\tan 8)) = (8 - 3\pi) = -y$

Hence $\quad f(y) + f(-y) = 2$

Given $\quad f(y) = \alpha \Rightarrow f(-y) = 2 - \alpha$

Ex. 5. *If* $\sin^{-1}\left(x^2 - \frac{x^4}{3} + \frac{x^6}{9} \ldots\ldots\right) + \cos^{-1}\left(x^4 - \frac{x^8}{3} + \frac{x^{12}}{9} \ldots\ldots\right) = \frac{\pi}{2}$, *where* $0 \le |x| < \sqrt{3}$, *then number of values of 'x' is equal to*

(a) 1 (b) 2 (c) 3 (d) 4

Sol. (*c*) $\sin^{-1}\underbrace{\left(x^2 - \frac{x^4}{3} + \frac{x^6}{9} \ldots\ldots\right)}_{X} + \cos^{-1}\underbrace{\left(x^4 - \frac{x^8}{3} + \frac{x^{12}}{9} \ldots\ldots\right)}_{Y} = \frac{\pi}{2}$

$$\Rightarrow X = Y \Rightarrow \frac{x^2}{1 + \frac{x^2}{3}} = \frac{x^4}{1 + \frac{x^4}{3}} \Rightarrow \frac{3}{3 + x^2} = \frac{3x^2}{3 + x^4}$$

$\Rightarrow \quad 9 + 3x^4 = 9x^2 + 3x^4 \Rightarrow x^2 = 1$

Thus, $\quad x = 0, 1$ or -1

Hence, number of values is equal to 3.

Ex. 6. *Suppose* $3\sin^{-1}(\log_2 x)+\cos^{-1}(\log_2 y)=\pi/2$ *and* $\sin^{-1}(\log_2 x)+2\cos^{-1}(\log_2 y)=11\pi/6$, then the value of $x^{-2}+y^{-2}$ equals

(a) 6 (b) 7

(c) 5 (d) $\frac{7}{2}$

Sol. (a) Let $\sin^{-1}(\log_2 x)=a$ and $\cos^{-1}(\log_2 y)=b$

$$\therefore \quad 3a+b=\frac{\pi}{2} \text{ and } a+2b=\frac{11\pi}{6}$$

$$\Rightarrow \quad a=-\frac{\pi}{6} \text{ and } b=\pi$$

Hence, $x=\frac{1}{\sqrt{2}}$ and $y=\frac{1}{2} \Rightarrow \frac{1}{x^2}+\frac{1}{y^2}=6$

Ex. 7. *Range of* $f(x)=\sin^{-1}\log[x]+\log(\sin^{-1}[x])$, *where [ ] denotes GIF is*

(a) 1 (b) 2

(c) 0 (d) $\left\{\log\frac{\pi}{2}\right\}$

Sol. (d) Domain of $f(x)$ is [1, 2).

$\therefore$ Range is $\left\{\log\frac{\pi}{2}\right\}$.

Ex. 8. $\sum_{n=1}^{5}\sin^{-1}(\sin(2n-1))$ *is*

(a) 1 (b) 2

(c) 3 (d) 4

Sol. (a) $\sum_{n=1}^{5}\sin^{-1}(\sin(2n-1))$

$$=\sin^{-1}(\sin 1)+\sin^{-1}(\sin 3)+\sin^{-1}(\sin 5)+\sin^{-1}(\sin 7)+\sin^{-1}(\sin 9)$$

$$=1+\pi-3+5-2\pi+7-2\pi+3\pi-9=1$$

Ex. 9. *If* α *and* $\beta(\alpha>\beta)$ *are roots of the equation* $x^2-\sqrt{2}x+\sqrt{3-2\sqrt{2}}=0$, *then the value of* $(\cos^{-1}\alpha+\tan^{-1}\alpha+\tan^{-1}\beta)$ *is equal to*

(a) $\frac{3\pi}{8}$ (b) $\frac{5\pi}{8}$

(c) $\frac{7\pi}{8}$ (d) $\frac{\pi}{3}$

Sol. (a) $x^2-\sqrt{2}x+\sqrt{3-2\sqrt{2}}=0 \Rightarrow x^2-\sqrt{2}x+\sqrt{2}-1=0$

$$\Rightarrow x^2-1-\sqrt{2}(x-1)=0 \Rightarrow (x-1)(x+1-\sqrt{2})=0$$

$$x=1, \sqrt{2}-1$$

$$\therefore \quad \alpha=1 \text{ and } \beta=\sqrt{2}-1$$

Hence, $\cos^{-1}\alpha+\tan^{-1}\alpha+\tan^{-1}\beta=0+\frac{\pi}{4}+\frac{\pi}{8}=\frac{3\pi}{8}$

Ex. 10. *If the mapping* $f(x)=mx+c, m>0$ *maps* $[-1,1]$ *onto* $[0, 2]$, *then* $\tan\left(\tan^{-1}\frac{1}{7}+\cot^{-1}8+\cot^{-1}18\right)$ *is equal to*

(a) $f\left(\frac{2}{3}\right)$ (b) $f\left(\frac{1}{3}\right)$ (c) $f\left(\frac{-1}{3}\right)$ (d) $f\left(\frac{-2}{3}\right)$

Sol. (d) Clearly, $f(x)=x+1$ (As, $-1\le x\le 1$, $0\le x+1\le 2$) and $f(x)=mx+c$

Now, $\tan\left(\tan^{-1}\frac{1}{7}+\tan^{-1}\frac{1}{8}+\tan^{-1}\frac{1}{18}\right)$

$$=\tan\left(\tan^{-1}\left(\frac{\frac{1}{7}+\frac{1}{8}}{1-\frac{1}{7}\times\frac{1}{8}}\right)+\tan^{-1}\frac{1}{18}\right)$$

$$=\tan\left(\tan^{-1}\left(\frac{15}{55}\right)+\tan^{-1}\left(\frac{1}{18}\right)\right)$$

$$=\tan\left(\tan^{-1}\left(\frac{3}{11}\right)+\tan^{-1}\left(\frac{1}{18}\right)\right)$$

$$=\tan\left(\tan^{-1}\left(\frac{\frac{3}{11}+\frac{1}{18}}{1-\frac{3}{11}\times\frac{1}{18}}\right)\right)$$

$$=\tan\left(\tan^{-1}\frac{1}{3}\right)=\frac{1}{3}=f\left(\frac{-2}{3}\right)$$

Ex. 11. *If* $(\sin^{-1}a)^2+(\cos^{-1}b)^2+(\sec^{-1}c)^2+(\text{cosec}^{-1}d)^2=\frac{5\pi^2}{2}$, *then the value of* $(\sin^{-1}a)^2-(\cos^{-1}b)^2+(\sec^{-1}c)^2-(\text{cosec}^{-1}d)^2$

(a) $-\pi^2$ (b) $-\frac{\pi^2}{2}$

(c) 0 (d) $\frac{\pi^2}{2}$

Sol. (c) As $0\le(\sin^{-1}a)^2\le\frac{\pi^2}{4}$, $0\le(\cos^{-1}b)^2\le\pi^2$,

$0\le(\sec^{-1}c)^2\le\pi^2\left(\text{except }\frac{\pi^2}{4}\right)$ and $0<(\text{cosec}^{-1}d)^2\le\frac{\pi^2}{4}$

So, $0<(\sin^{-1}a)^2+(\cos^{-1}b)^2+(\sec^{-1}c)^2+(\text{cosec}^{-1}d)^2\le\frac{5\pi^2}{2}$

$\therefore (\sin^{-1}a)^2+(\cos^{-1}b)^2+(\sec^{-1}c)^2+(\text{cosec}^{-1}d)^2=\frac{5\pi^2}{2}$ (Given)

$$\Rightarrow \quad (\sin^{-1}a)^2=\frac{\pi^2}{4}$$

$$(\cos^{-1}b)^2=\pi^2$$

$$(\sec^{-1}c)^2=\pi^2$$

$$\text{and} \quad (\text{cosec}^{-1}d)^2=\frac{\pi^2}{4}$$

Hence, $(\sin^{-1}a)^2-(\cos^{-1}b)^2+(\sec^{-1}b)^2-(\text{cosec}^{-1}d)^2=0$

Ex. 12. *If* $f(x) = \sum_{r=1}^{n} \tan^{-1}\left(\frac{1}{x^2+(2r-1)x+(r^2-r+1)}\right)$, *then* $\left|\lim_{n\to\infty} f'(0)\right|$ *is*

(a) 1 (b) 2

(c) 3 (d) 4

Sol. (*a*) $f(x) = \sum_{r=1}^{n} (\tan^{-1}(x+r) - \tan^{-1}(x+r-1))$

$$\Rightarrow \quad f(x) = \tan^{-1}(x+n) - \tan^{-1}(x)$$

$$\Rightarrow \quad f'(x) = \frac{1}{1+(x+n)^2} - \frac{1}{1+x^2}$$

$$\Rightarrow \quad f'(0) = \frac{1}{1+n^2} - 1$$

Now, $\lim_{n\to\infty} f'(0) = -1 \Rightarrow \left|\lim_{n\to\infty} f'(0)\right| = 1$

Ex. 13. *The range of the function* $f(x) = \sec^{-1}(x) + \tan^{-1}(x)$, *is*

(a) $(0, \pi)$ (b) $\left(\frac{-\pi}{2}, \frac{3\pi}{2}\right)$

(c) $\left(0, \frac{3\pi}{4}\right]$ (d) None of these

Sol. (*a*) $D_f = (-\infty, -1] \cup [1, \infty)$

Also, f is an increasing function.

For, $\quad x \in (-\infty, -1], f(x) \in \left(0, \frac{3\pi}{4}\right]$ ... (i)

and for $\quad x \in [1, \infty), f(x) \in \left[\frac{\pi}{4}, \pi\right)$ ... (ii)

$\therefore$ For range of $f(x)$, (i) $\cup$ (ii) $\Rightarrow (0, \pi)$

Ex. 14. *The solution set of inequality* $(\cot^{-1}x)(\tan^{-1}x) + \left(2-\frac{\pi}{2}\right)\cot^{-1}x - 3\tan^{-1}x - 3\left(2-\frac{\pi}{2}\right) > 0$, *is*

(a) $x \in (\tan 2, \tan 3)$

(b) $x \in (\cot 3, \cot 2)$

(c) $x \in (-\infty, \tan 2) \cup (\tan 3, \infty)$

(d) $x \in (-\infty, \cot 3) \cup (\cot 2, \infty)$

Sol. (*b*) Given,

$$(\cot^{-1}x)(\tan^{-1}x) + \left(2-\frac{\pi}{2}\right)\cot^{-1}x - 3\tan^{-1}x - 3\left(2-\frac{\pi}{2}\right) > 0$$

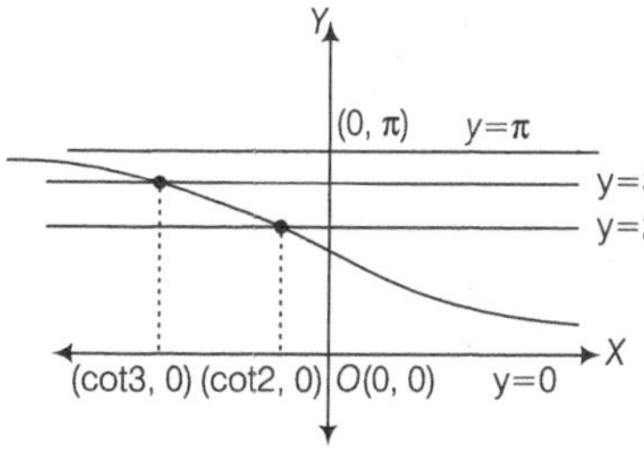

$$\Rightarrow \quad \cot^{-1}x\left(\tan^{-1}x + 2 - \frac{\pi}{2}\right) - 3\left(\tan^{-1}x + 2 - \frac{\pi}{2}\right) > 0$$

$$\Rightarrow \quad (\cot^{-1}x - 3)(2 - \cot^{-1}x) > 0$$

$$\left(\text{As } \tan^{-1}x - \frac{\pi}{2} = -\cot^{-1}x\right)$$

$$\Rightarrow \quad (\cot^{-1}x - 3)(\cot^{-1}x - 2) < 0$$

$$\Rightarrow \quad 2 < \cot^{-1}x < 3$$

$\Rightarrow \cot 3 < x < \cot 2$ (As $\cot^{-1}x$ is a decreasing function)

Hence, $x \in (\cot 3, \cot 2)$

Ex. 15. *Let* $f(x) = \sin(\sin^{-1}2x) + \text{cosec}(\text{cosec}^{-1}2x) + \tan(\tan^{-1}2x)$, *then which one of the following statement is/are incorrect?*

(a) $f(x)$ is odd function

(b) $f(x)$ is injective

(c) Range of $f(x)$ contains only two integers.

(d) The value of $f'\left(\frac{1}{2}\right)$ is equal to 6.

Sol. (*d*) Clearly $f(x) = 6x$ and domain $= \left\{\frac{-1}{2}, \frac{1}{2}\right\}$

JEE Type Solved Examples :
More than One Correct Option Type Questions

Ex. 16. *If* $f(x) = \cos^{-1}(\cos(x+1))$ *and* $g(x) = \sin^{-1}(\sin(x+2))$, *then*

(a) $f(1) + g(1) = (\pi - 1)$ (b) $f(1) > g(1)$

(c) $f(2) > g(2)$ (d) $f(2) < g(2)$

Sol. (a, b, c) $f(1) = \cos^{-1}(\cos 2) = 2$

$$g(1) = \sin^{-1}(\sin 3) = (\pi - 3)$$

$$f(1) + g(1) = (\pi - 1)$$

and $\quad f(1) > g(1)$

$$f(2) = \cos^{-1}(\cos 3) = 3$$

$$g(3) = \sin^{-1}(\sin 4) = (\pi - 4)$$

Ex. 17. $\sum_{r=1}^{\infty} \tan^{-1}\left(\dfrac{2}{(r+2)^2}\right)$ *is*

(a) $\pi - (\tan^{-1}2 + \tan^{-1}3)$ (b) $\dfrac{\pi}{4}$

(c) $\dfrac{7\pi}{4}$ (d) $\pi + (\tan^{-1}2 + \tan^{-1}3)$

Sol. *(a, b)* $T_r = \tan^{-1}\left(\dfrac{2}{(r^2+4r+3)+1}\right) = \tan^{-1}\left(\dfrac{(r+3)-(r+1)}{1+(r+3)(r+1)}\right)$

$$T_r = \tan^{-1}(r+3) - \tan^{-1}(r+1)$$

$$\text{Sum} = \pi - (\tan^{-1}(2) + an^{-1}(3))$$

$$= \pi - (\pi + \tan^{-1}(-1)) = \frac{\pi}{4}$$

Ex. 18. *If sides AB, BC and CA of a triangle ABC are represented by $x+2=0$, $3x+y=0$ and $x+3y+2=0$ respectively, then identify the correct statement.*

(a) $\Sigma \tan A = \dfrac{4}{3}$

(b) $\Pi \tan A = -\dfrac{4}{3}$

(c) $\Sigma \tan A \tan B = -\dfrac{41}{9}$

(d) $\sin^2(A+B) + \cos^2 C = \dfrac{5}{4}$

Sol. *(b, c)*

$$\angle A = \frac{\pi}{2} + \tan^{-1}\frac{1}{3}$$

$$\angle B = \frac{\pi}{2} - \tan^{-1}3$$

$$\angle C = \pi - (A+B)$$

$$= \tan^{-1}3 - \tan^{-1}\frac{1}{3}$$

$$= 2\tan^{-1}3 - \frac{\pi}{2}$$

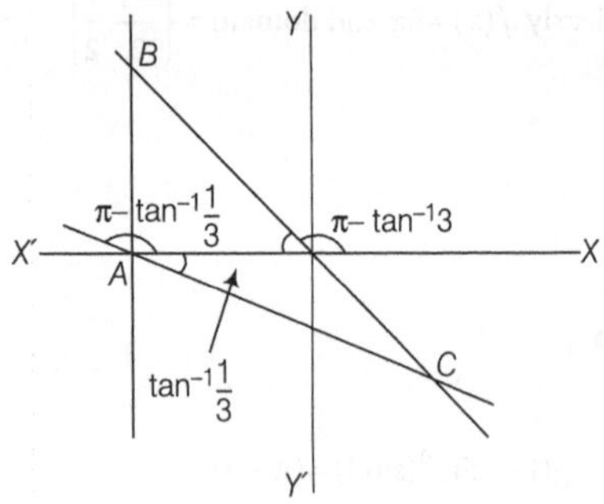

$$\tan A + \tan B + \tan C = \tan\left(\frac{\pi}{2} + \tan^{-1}\frac{1}{3}\right) + \tan\left(\frac{\pi}{2} - \tan^{-1}3\right) + \tan\left(\tan^{-1}3 - \tan^{-1}\frac{1}{3}\right)$$

$$= -\cot\left(\tan^{-1}\frac{1}{3}\right) + \cot(\tan^{-1}3) + \tan\left(\tan^{-1}\left(\frac{3-\frac{1}{3}}{1+3.\frac{1}{3}}\right)\right)$$

$$= -3 + \frac{1}{3} + \frac{4}{3} = -\frac{4}{3}$$

$$\Sigma \tan A = \Pi \tan A = -\frac{4}{3}$$

$$\Sigma \tan A \tan B = -3.\frac{1}{3} + \frac{1}{3}.\frac{4}{3} + \frac{4}{3}(-3) = -5 + \frac{4}{9} = -\frac{41}{9}$$

$$\sin^2(A+B) + \cos^2 C = \sin^2(\pi - C) + \cos^2 C = 1$$

Ex. 19. *Which of the following is/are correct?*

(a) $\cos(\cos(\cos^{-1}1)) < \sin(\sin^{-1}(\sin(\pi-1))) < \sin(\cos^{-1}(\cos(2\pi-2)))$

(b) $\cos(\cos(\cos^{-1}1)) < \sin(\cos^{-1}(\cos(2\pi-2))) < \sin(\sin^{-1}(\sin(\pi-1))) < \tan(\cot^{-1}(\cot 1))$

(c) $\sum_{t=1}^{5000} \cos^{-1}(\cos(2t\pi - 1)) = \sum_{t=1}^{2500} \cot^{-1}(\cot(t\pi+2))$, where $t \in I$

(d) $\cot^{-1}\cot\,\text{cosec}^{-1}\text{cosec}\,\sec^{-1}\sec\,\tan\,\tan^{-1}\cos\,\cos^{-1}\sin^{-1}\sin 4 = 4 - \pi$

Sol. *(a,b,c,d)* For (a) and (b)

$$\cos(\cos^{-1}1) = 1 \Rightarrow \cos(\cos(\cos^{-1}1)) = \cos 1$$

$$\sin^{-1}(\sin(\pi-1)) = \pi - (\pi - 1) = 1$$

$$\Rightarrow \quad \sin(\sin^{-1}(\sin(\pi-1))) = \sin 1$$

$$\cos^{-1}(\cos(2\pi-2)) = \cos^{-1}(\cos 2) = 2$$

$$\Rightarrow \quad \sin(\cos^{-1}(\cos(2\pi-2))) = \sin 2$$

$$\tan(\cot^{-1}(\cot 1)) = \tan 1$$

It is easy to compare $\cos 1, \sin 1, \sin 2, \tan 1$

$$\cos 1 < \sin 1 < \sin 2 < \tan 1 \Rightarrow \text{(a) is correct}$$

For (c)

$\because \cos^{-1}\cos x$ is periodic with period 2π

$$\therefore \quad \cos^{-1}\cos(2t\pi - 1) = \cos^{-1}(\cos 1) = 1 \;(t \in I)$$

$$\sum_{t=1}^{5000} \cos^{-1}\cos(2t\pi - 1) = 5000$$

Now, $\cot^{-1}\cot(t\pi + 2) = 2$ [$\cot^{-1}\cot x$ is periodic with period π]

$$\therefore \sum_{t=1}^{2500} \cot^{-1}\cot(t\pi+2) = 5000 \Rightarrow \text{(c) is correct}$$

(d) $\sin^{-1}\sin 4 = \pi - 4$

$$\cos\cos^{-1}(\pi - 4) = 4 - \pi$$

$$\tan\tan^{-1}(4-\pi) = \pi - 4$$

$$\sec^{-1}\sec(\pi - 4) = 4 - \pi$$

$$\text{cosec}^{-1}\text{cosec}(4-\pi) = 4 - \pi$$

$$\cot^{-1}\cot(4-\pi) = 4 - \pi$$

$\Rightarrow$ (d) is correct

Ex. 20. *Let x_1 and x_2 $(x_1 > x_2)$ be roots of the equation* $\sin^{-1}(\cos(\tan^{-1}(\text{cosec}(\cot^{-1} x)))) = \frac{\pi}{6}$, *then*

(a) $\sin^{-1}\frac{1}{x_1} + \cos^{-1}\frac{1}{x_2} = \pi$

(b) $\sin^{-1}\left(\frac{1}{x_1}\right) + \cos^{-1}\left(\frac{1}{x_2}\right) = 0$

(c) $\sin^{-1}\frac{1}{x_1} + \sin^{-1}\left(\frac{1}{x_2}\right) = 0$

(d) $\cos^{-1}\left(\frac{1}{x_1}\right) + \cos^{-1}\left(\frac{1}{x_2}\right) = \pi$

Sol. (a,c,d) Given, $\sin^{-1}(\cos(\tan^{-1}(\text{cosec}(\cot^{-1} x)))) = \frac{\pi}{6}$

$$\Rightarrow \quad \sin^{-1}(\cos(\tan^{-1}\sqrt{1+x^2})) = \frac{\pi}{6}$$

$$\Rightarrow \quad \sin^{-1}\left(\frac{1}{\sqrt{x^2+2}}\right) = \frac{\pi}{6}$$

$$\Rightarrow \quad \sqrt{x^2+2} = 2$$

$$\Rightarrow \quad x^2 = 2$$

$$\Rightarrow \quad x = \pm\sqrt{2}$$

So, $\quad x_1 = \sqrt{2}$ and $x_2 = -\sqrt{2}$

Now, verify alternatives.

JEE Type Solved Examples : Passage Based Questions

Passage I

(Ex. Nos. 21 to 22)

Suppose f, g and h be three real valued function defined on R.

Let $f(x) = 2x + |x|$, $g(x) = \frac{1}{3}(2x - |x|)$ *and* $h(x) = f(g(x))$

Ex. 21. *The range of the function* $k(x) = 1 + \frac{1}{\pi}(\cos^{-1}(h(x)) + \cot^{-1}(h(x)))$ *is equal to*

(a) $\left[\frac{1}{4}, \frac{7}{4}\right]$ (b) $\left[\frac{5}{4}, \frac{11}{4}\right]$

(c) $\left[\frac{1}{4}, \frac{5}{4}\right]$ (d) $\left[\frac{7}{4}, \frac{11}{4}\right]$

Ex. 22. *The domain of definition of the function* $l(x) = \sin^{-1}(f(x) - g(x))$ *is equal to*

(a) $\left(\frac{3}{8}, \infty\right]$ (b) $(-\infty, 1]$

(c) $[-1, 1]$ (d) $\left(-\infty, \frac{3}{8}\right]$

Sol. (Ex. Nos. 21 to 22)

We have $f(x) = \begin{cases} 3x, \geq 0 \\ x, x < 0 \end{cases}$ and $g(x) \begin{cases} \frac{x}{3}, x \geq 0 \\ x, x < 0 \end{cases}$

Clearly, f and g are inverse of each other.

Now, $h(x) = f(g(x)) = \begin{cases} 3\left(\frac{x}{3}\right) = x, x \geq 0 \\ x \quad , \quad x < 0 \end{cases}$

21. (b) As $h(x) = x, \forall\ x \in R$

$$\Rightarrow \quad k(x) = 1 + \frac{1}{\pi}(\cos^{-1} x + \cot^{-1} x)$$

Domain of $k(x) = [-1, 1]$ and $k(x)$ is decreasing function on $[-1, 1]$. As $k(x)$ is continuous function on $[-1, 1]$.

Now, $\quad k_{\text{min.}}(x = 1) = 1 + \frac{1}{\pi}(\cos^{-1} 1 + \cot^{-1} 1)$

$$= 1 + \frac{1}{\pi}\left(0 + \frac{\pi}{4}\right) = 1 + \frac{1}{4} = \frac{5}{4}$$

$$k_{\text{max.}}(x = -1) = 1 + \frac{1}{\pi}(\cos^{-1}(-1) + \cot^{-1}(-1))$$

$$= 1 + \frac{1}{\pi}\left[\pi + \frac{3\pi}{4}\right]$$

$$= 1 + \frac{7}{4} = \frac{11}{4}$$

$\Rightarrow$ Range of $k(x) = \left[\frac{5}{4}, \frac{11}{4}\right]$

22. (d) We have, $f(x) - g(x) = (2x + |x|) - \frac{1}{3}(2x - |x|)$

$$= \frac{4x}{3} + \frac{4}{3}|x|$$

$$= \begin{cases} \frac{8}{3}x; & x \geq 0 \\ 0\ ; & x < 0 \end{cases}$$

$\therefore$ For domain of function

$$0 \leq \frac{8x}{3} \leq 1 \Rightarrow 0 \leq x \leq \frac{3}{8}$$

$\Rightarrow \quad$ Domain of $l(x) = \left(-\infty, \frac{3}{8}\right]$

(**Note** Range of function $l(x) = \left[0, \frac{\pi}{2}\right]$)

Passage II

(Ex. Nos. 23 to 24)

In ΔABC, if $\angle B = \sec^{-1}\left(\frac{5}{4}\right) + \text{cosec}^{-1}\sqrt{5}$,

$$\angle C = \text{cosec}^{-1}\left(\frac{25}{7}\right) + \cot^{-1}\left(\frac{9}{13}\right) \text{ and } c = 3$$

(All symbols used have their usual meaning in a triangle.) On the basis of above information, answer the following questions.

Ex. 23. *$\tan A, \tan B, \tan C$ are in*

(a) AP (b) GP

(c) HP (d) neither AP, GP nor HP

Ex. 24. *The distance between orthocentre and centroid of triangle with sides $a^2, b^{\frac{4}{3}}$ and c is equal to*

(a) $\frac{5}{2}$ (b) $\frac{5}{3}$

(c) $\frac{10}{3}$ (d) $\frac{7}{2}$

Sol. (Ex. Nos. 23 to 24)

$$\angle B = \sec^{-1}\left(\frac{5}{4}\right) + \text{cosec}^{-1}\sqrt{5}$$

$$= \tan^{-1}\left(\frac{3}{4}\right) + \tan^{-1}\left(\frac{1}{2}\right)$$

$$= \tan^{-1}\frac{\frac{3}{4}+\frac{1}{2}}{1-\frac{3}{4}\cdot\frac{1}{2}} = \tan^{-1}2$$

$$\angle C = \text{cosec}^{-1}\left(\frac{25}{7}\right) + \cot^{-1}\left(\frac{9}{13}\right)$$

$$= \tan^{-1}\left(\frac{7}{24}\right) + \tan^{-1}\left(\frac{13}{9}\right)$$

$$= \tan^{-1}\left(\frac{\frac{7}{24}+\frac{13}{9}}{1-\frac{7}{24}\cdot\frac{13}{9}}\right) = \tan^{-1}3$$

$\tan B = 2$ and $\tan C = 3 \Rightarrow \tan A = 1$

$\because \quad \Sigma \tan A = \Pi \tan A$

$\therefore \quad \sin A = \frac{1}{\sqrt{2}}, \sin B = \frac{2}{\sqrt{5}}$ and $\sin C = \frac{3}{\sqrt{10}}$

$\therefore \quad \frac{a}{\sin A} = \frac{b}{\sin B} = \frac{c}{\sin C}$

$\Rightarrow \quad \sqrt{2}\,a = \frac{\sqrt{5}b}{2} = \frac{3\sqrt{10}}{3}$

Hence, $a = \sqrt{5}$ and $b = 2\sqrt{2}, c = 3$

23. (*a*) $\tan A = 1, \tan B = 2, \tan C = 3$ are in AP

24. (*b*) The triangle sides $a^2, b^{\frac{4}{3}}$ and c will have side-length 5, 4 and 3 respectively.

$\therefore$ Distance between orthocentre and centroid

$$= \frac{2}{3}\text{(circumradius)} = \frac{\text{hypotenuse}}{3} = \frac{5}{3}$$

JEE Type Solved Examples : Integer Answer Type Questions

Ex. 25. *Let $f(x) = x^2 - 2ax + a - 2$ and $g(x) = \left[2 + \sin^{-1}\frac{2x}{1+x^2}\right]$. If the set of real values of 'a' for which $f(g(x)) < 0, \forall x \in R$ is (k_1, k_2), then find the value of $(10k_1 - 3k_2)$.*

[**Note** : $[k]$ denotes greatest integer less than or equal to k.]

Sol. (8) We have $g(x) = \left[2 + \sin^{-1}\frac{2x}{1+x^2}\right]$

$$= 2 + \left[\sin^{-1}\frac{2x}{1+x^2}\right]$$

As, $\quad \sin^{1}\frac{2x}{1+x^2} \in \left[\frac{-\pi}{2}, \frac{\pi}{2}\right]$

$\therefore \quad \left[\sin^{-1}\frac{2x}{1+x^2}\right] = -2, -1, 0, 1$

Range of $g(x) = \{0, 1, 2, 3\}$ for $f(g(x)) < 0 \forall x \in R$

$\Rightarrow \quad f(0) < 0$ and $(3) < 0$

Now, $\quad f(0) < 0 \Rightarrow a - 2 < 0 \Rightarrow a < 2$

and $\quad f(3) < 0 \Rightarrow 9 - 6a + a - 2 < 0$

$a > \frac{7}{5}$

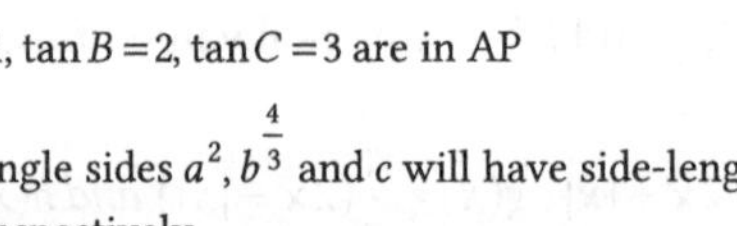

$\therefore \quad a \in \left(\frac{7}{5}, 2\right)$

Hence, $\quad k_1 = \frac{7}{5}, k_2 = 2$

$\therefore \quad (10k_1 - 3k_2) = 14 - 6 = 8$

Ex. 26. *Let x_1, x_2, x_3 be the solution of*

$$\tan^{-1}\left(\frac{2x+1}{x+1}\right)+\tan^{-1}\left(\frac{2x-1}{x-1}\right)=2\tan^{-1}(x+1) \text{ where}$$

$x_1 < x_2 < x_3$, then $2x_1 + x_2 + x_3^2$ is equal to

Sol. (1) Let $\alpha+\beta=2\gamma$ ...(i)

where $\tan\alpha=\frac{2x+1}{x+1}$, $\tan\beta=\frac{2x-1}{x-1}$ and

$\tan\gamma=x+1$

Taking tan in Eq. (i), we get

$$\frac{\tan\alpha+\tan\beta}{1-\tan\alpha\tan\beta}=\frac{2\tan\gamma}{1-\tan^2\gamma}$$

$$\Rightarrow \frac{\frac{2x+1}{x+1}+\frac{2x-1}{x-1}}{1-\frac{4x^2-1}{x^2-1}}=\frac{2(x+1)}{1-(x+1)^2}$$

$$\Rightarrow \frac{2x^2-x-1+2x^2+x-1}{x^2-1-4x^2+1}=\frac{2(x+1)}{-x^2-2x}$$

$$\Rightarrow \frac{2(2x^2-1)}{-3x^2}=\frac{2(x+1)}{-(x^2+2x)}$$

$$x=0$$

or $2x^3+4x^2-x-2=3x^2+3x$

$\Rightarrow 2x^3+x^2-4x-2=0$

$\Rightarrow x^2(2x+1)-2(2x+1)=0$

$\Rightarrow (x^2-2)(2x+1)=0$

$\Rightarrow x=\sqrt{2},-\sqrt{2}$ or $x=-\frac{1}{2}$

$x=-\sqrt{2}$ is rejected $\because$ it does not satisfy (i)

$\Rightarrow x_1=-\frac{1}{2}, x_2=0$ and $x_3=\sqrt{2}$

$\Rightarrow 2x_1+x_2+x_3^2=1$

Ex. 27. *If the range of function $f(x)=(\pi\sqrt{2}+\cos^{-1}\alpha)\,x^2+2(\cos^{-1}\beta)\,x+\pi\sqrt{2}-\cos^{-1}\alpha$ is $[0,\infty)$ then find the value of $|\alpha-\beta|+2\,\alpha\beta+1$*

Sol. (3) Given, $f(x)=(\pi\sqrt{2}+\cos^{-1}\alpha)\,x^2+2(\cos^{-1}\beta)x+\pi\sqrt{2}-\cos^{-1}\alpha$

Clearly, graph of $f(x)$ is parabola opening upward.

As, range of $f(x)$ is $[0,\infty)$, so discriminant $=0$

$\Rightarrow b^2-4ac=0$

$\Rightarrow 4(\cos^{-1}\beta)^2-4(\pi\sqrt{2}+\cos^{-1}\alpha)(\pi\sqrt{2}-\cos^{-1}\alpha)=0$

$\Rightarrow 4(\cos^{-1}\beta)^2-4(2\pi^2-(\cos^{-1}\alpha)^2)=0$

$\Rightarrow (\cos^{-1}\alpha)^2+(\cos^{-1}\beta)^2=2\pi^2$

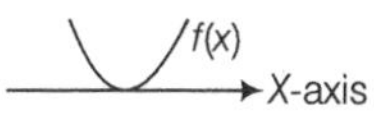

$\Rightarrow \cos^{-1}\alpha=\pi=\cos^{-1}\beta$

$\Rightarrow \alpha=\beta=-1$

Hence, $|\alpha-\beta|+2\alpha\beta+1=0+2+1=3$

Ex. 28. *Consider $f(x)=\sin^{-1}[2x]+\cos^{-1}([x]-1)$ (where [.] denotes greatest integer function.) If domain of $f(x)$ is $[a,b)$ and the range of $f(x)$ is $\{c,d\}$, then $a+b+\frac{2d}{c}$ is equal to (where $c<d$)*

Sol. (4) $f(x)=\sin^{-1}[2x]+\cos^{-1}([x]-1)$

$-1\le[2x]\le1$ and $-1\le[x]-1\le1$

$\Rightarrow -1\le 2x<2$ and $0\le[x]\le2$

$\Rightarrow -\frac{1}{2}<x<1$ and $0\le x<3$

$\Rightarrow 0\le x<1$ Domain

$\Rightarrow [x]=0$

$\Rightarrow 0\le 2x<2$

$\Rightarrow [2x]=0$ or 1

Now, $f(x)=\sin^{-1}[2x]+\cos^{-1}(-1)$

$$=\left(0 \text{ or } \frac{\pi}{2}\right)+\pi=\pi \text{ or } \frac{3\pi}{2}$$

$\Rightarrow u+b+\frac{2d}{c}-1+3=4$

Ex. 29. *Let $f(x)=\min(\tan^{-1}x,\cot^{-1}x)$ and $h(x)=f(x+2)-\pi/3$. Let x_1, x_2 (where $x_1<x_2$) be the integers in the range of $h(x)$, then the value of $(\cos^{-1}(\cos x_1)+\sin^{-1}(\sin x_2))$ is equal to*

Sol. (1) $f(x)=\begin{cases}\tan^{-1}x; & x\le1\\ \cot^{-1}x; & x>1\end{cases}$

$$\Rightarrow f(x+2)=\begin{cases}\tan^{-1}(x+2) & ; x\le-1\\ \cot^{-1}(x+2) & ; x>-1\end{cases}$$

$\therefore$ Range of $h(x)$ is $\left(-\frac{5\pi}{6},-\frac{\pi}{12}\right]$

$\therefore \cos^{-1}\cos(-2)+\sin^{-1}\sin(-1)=1$

Ex. 30. *If the area enclosed by the curves $f(x)=\cos^{-1}(\cos x)$ and $g(x)=\sin^{-1}(\cos x)$ in $x\in\left[\frac{9\pi}{4},\frac{15\pi}{4}\right]$ is $\frac{a\pi^2}{b}$ (where, a and b coprime), then find $(a-b)$.*

Sol. (1) We have $g(x)=\sin^{-1}(\cos x)=\frac{\pi}{2}-\cos^{-1}(\cos x)$

Both the curves bound the regions of same area in $\left[\frac{\pi}{4},\frac{7\pi}{4}\right],\left[\frac{9\pi}{4},\frac{15\pi}{4}\right]$ and so on

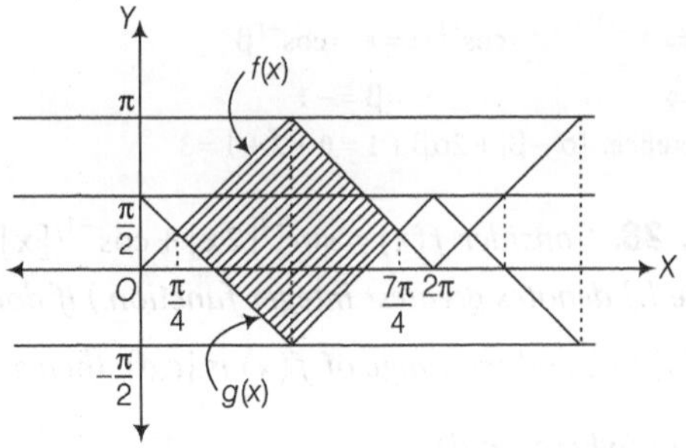

$\therefore$ Required area = area of shaded square

$$= \frac{9\pi^2}{8} = \frac{a\pi^2}{b}$$

$\therefore \quad a = 9$ and $b = 8$

Hence, $\quad a - b = 1$

Ex. 31. *Consider the curve $y = \tan^{-1} x$ and a point $A\left(1, \frac{\pi}{4}\right)$ on it. If the variable point $P_i(x_i, y_i)$ moves on the curve for $i = 1,2,3,...n(n \in N)$ such that $y_r = \sum_{m=1}^{r} \tan^{-1}\left(\frac{1}{2m^2}\right)$ and $B(x,y)$ be the limiting position of variable point P_n as $n \to \infty$, then the value of reciprocal of the slope of AB will be*

Sol. (2) $y = \lim_{n\to\infty} y_n = \lim_{n\to\infty} \sum_{m=1}^{n} \tan^{-1}\left(\frac{1}{2m^2}\right) =$

$$\lim_{n\to\infty} \sum_{m=1}^{n} \tan^{-1}\left\{\frac{2}{1 + (2m+1)(2m-1)}\right\}$$

$$= \lim_{n\to\infty} \sum_{m=1}^{n} \{\tan^{-1}(2m+1) - \tan^{-1}(2m-1)\}$$

$$= \lim_{n\to\infty} \{(\tan^{-1}3 - \tan^{-1}1) + (\tan^{-1}5 - \tan^{-1}3) + (\tan^{-1}7 - \tan^{-1}5).... + \tan^{-1}(2n+1) - \tan^{-1}(2n-1)\}$$

$$= \lim_{n\to\infty} \{\tan^{-1}(2n+1) - \tan^{-1}1\} \to \frac{\pi}{4}$$

$\because B \to \left(1, \frac{\pi}{4}\right)$ i.e. coordinates of B approach, towards those of 'A'.

$\therefore$ Chord AB approches to be the tangent to $y = f(x)$ at A

$\therefore$ (slope of $AB)^{-1} = \left[\frac{d}{dx}\tan^{-1} x\right]^{-1}_{\text{at } x=1}$

$$= (1 + x^2)_{x=1} = 2$$

Ex. 32. *If $\tan^{-1} x + \tan^{-1}\frac{\sqrt{1-y^2}}{y} = \frac{\pi}{3}$ and $\sin^{-1} y - \cos^{-1}\left(\frac{x}{\sqrt{1+x^2}}\right) = \frac{\pi}{6}$, then $\frac{5\sin^{-1} x}{\sin^{-1} y}$ is*

Sol. (6) $\tan^{-1} x + \cos^{-1} y = \frac{\pi}{3}$ and $\sin^{-1} y - \cot^{-1} x = \frac{\pi}{6}$

$\Rightarrow \quad \tan^{-1} x + \cos^{-1} y + \sin^{-1} y - \cot^{-1} x + \frac{\pi}{2}$

or $\quad \tan^{-1} x = \cot^{-1} x \Rightarrow x = 1$

Also, $\tan^{-1} x + \cos^{-1} y - \sin^{-1} y + \cot^{-1} x = \frac{\pi}{6}$

$\Rightarrow \pi - 2\sin^{-1} y = \frac{\pi}{6} \Rightarrow \sin^{-1} y = \frac{5\pi}{12}$

$\therefore \quad \frac{5\sin^{-1} x}{\sin^{-1} y} = 6$

Ex. 33. *If $A = \frac{1}{1}\cot^{-1}\left(\frac{1}{1}\right) + \frac{1}{2}\cot^{-1}\left(\frac{1}{2}\right) + \frac{1}{3}\cot^{-1}\left(\frac{1}{3}\right)$ and $B = 1\cot^{-1}(1) + 2\cot^{-1}(2) + 3\cot^{-1}(3)$ then $|B - A|$ is equal to $\frac{a\pi}{b} + \frac{c}{d}\cot^{-1}(3)$ where $a,b,c,d \in N$ are in their lowest form, find $(b - a - c - d)$*

Sol. (8) $B - A = (2\cot^{-1}(2) + 3\cot^{-1}(3)) - \left(\frac{1}{2}\cot^{-1}\left(\frac{1}{2}\right) + \frac{1}{3}\cot^{-1}\left(\frac{1}{3}\right)\right)$

$$= 2(\cot^{-1}2 + \cot^{-1}3) + \cot^{-1}3 - \left(\frac{1}{3}\left(\cot^{-1}\frac{1}{2} + \cot^{-1}\frac{1}{3}\right) + \frac{1}{6}\cot^{-1}\frac{1}{2}\right)$$

$$= \frac{\pi}{2} + \cot^{-1}3 - \left[\frac{\pi}{4} + \frac{1}{6}\tan^{-1}2\right]$$

$$= \frac{\pi}{4} + \cot^{-1}3 - \frac{1}{6}\left[\frac{3\pi}{4} - \tan^{-1}3\right]$$

$$= \frac{\pi}{8} + \cot^{-1}3 + \frac{1}{6}\tan^{-1}3$$

$$= \frac{\pi}{8} + \cot^{-1}3 - \frac{1}{6}\left(\frac{\pi}{2} - \cot^{-1}3\right)$$

$$= \frac{\pi}{8} + \frac{\pi}{12} + \cot^{-1}3 - \frac{1}{6}\cot^{-1}3$$

$$= \frac{5\pi}{24} + \frac{5}{6}\cot^{-1}3$$

Hence, $\quad a = 5; b = 24; c = 5; d = 6$

$b - a - c - d = 8$

JEE Type Solved Examples : Statement I and II Type Questions

- This section contains 2 questions. Each question contains **Statement I** (Assertion) and **Statement II** (Reason). Each question has 4 choices (a), (b), (c) and (d) out of which only one is correct. The choice are
  - (a) Both Statement I and Statement II are correct and Statement II is the correct explanation of Statement I
  - (b) Both Statement I and Statement II are correct but Statement II is not the correct explanation of Statement I.
  - (c) Statement I is correct but Statement II is incorrect.
  - (d) Statement II is correct but Statement I is incorrect.

Ex. 34. Statement I *If α, β are roots of $6x^2+11x+3=0$, then $\cos^{-1}\alpha$ exists but not $\cos^{-1}\beta\ (\alpha>\beta)$.*

Statement II *Domain of $\cos^{-1}x$ is $[-1,1]$.*

Sol. (*a*) Given, $6x^2+11x+3=0 \Rightarrow 6x^2+9x+2x+3=0$

$$\Rightarrow \quad 3x(2x+3)+1(2x+3)=0$$

$$\Rightarrow \quad (2x+3)(3x+1)-0 \Rightarrow x=\frac{-3}{2},\frac{-1}{3}$$

$$\therefore \quad \beta=\frac{-3}{2}, \alpha=\frac{-1}{3} \qquad \left(\because \frac{-1}{3}>\frac{-3}{2}\right)$$

$$\therefore \quad \cos^{-1}\left(\frac{-1}{3}\right) \text{ exists} \qquad \{\because \text{Domain of } \cos^{-1}x \text{ is } [-1,1]\}$$

Ex. 35. Statement I *If $\tan^{-1}x+\tan^{-1}y=\frac{\pi}{4}-\tan^{-1}z$ and $x+y+z=1$, then arithmetic mean of odd powers of x, y, z is equal to $1/3$.*

Statement II *For any x, y, z we have*

$$xyz-xy-yz-zx+x+y+z=1+(x-1)(y-1)(z-1)$$

Sol. (*b*) We have, $\tan^{-1}x+\tan^{-1}y+\tan^{-1}z=\frac{\pi}{4}$

Let $x=\tan A$, $y=\tan B$ and $z=\tan C$.

Then $\quad A+B+C=\frac{\pi}{4}$

Now, $\tan(A+B+C)=\dfrac{x+y+z-xyz}{1-(xy+yz+zx)}$

$$\Rightarrow \quad 1=\frac{x+y+z-xyz}{1-(xy+yz+zx)}$$

$$\Rightarrow \quad 1-(xy+yz+zx)=x+y+z-xyz$$

$$\Rightarrow \quad (x-1)(y-1)(z-1)=0$$

$\Rightarrow$ One of x,y,z is equal to 1

If $\quad z=1,\ x+y=0$

$\therefore \quad (x)^{odd}+(-x)^{odd}+1^{odd}=1$

Thus, AM of odd powers of x,y,z is equal to $\left(\frac{1}{3}\right)$.

JEE Type Solved Examples : Matching Type Questions

Ex. 36. *Match the principal values of $\cos^{-1}(8x^4-8x^2+1)$ given in column I with the corresponding intervals of x given in column II, for which it holds.*

	Column I		Column II
A	$4\cos^{-1}x$	p.	$0\le x\le\frac{1}{\sqrt{2}}$
B	$4\cos^{-1}x-2\pi$	q.	$\frac{1}{\sqrt{2}}\le x\le 1$
C	$2\pi-4\cos^{-1}x$	r.	$-1\le x\le-\frac{1}{\sqrt{2}}$
D	$4\pi-4\cos^{-1}x$	s.	$-\frac{1}{\sqrt{2}}\le x\le 0$

Sol. A→q, B→s, C→p, D→r

(A) $0\le\cos^{-1}(8x^4-8x^2+1)\le\pi$

$$\Rightarrow \quad 0\le 4\cos^{-1}x\le\pi$$

$$\Rightarrow \quad 0\le\cos^{-1}x\le\frac{\pi}{4} \Rightarrow \frac{1}{\sqrt{2}}\le x\le 1$$

(B) $0\le\cos^{-1}(8x^4-8x^2+1)\le\pi$

$$\Rightarrow \quad 0\le 4\cos^{-1}x-2\pi\le\pi$$

$$\Rightarrow \quad 2\pi\le 4\cos^{-1}x\le 3\pi$$

$$\Rightarrow \quad \frac{\pi}{2}\le\cos^{-1}x\le\frac{3\pi}{4} \Rightarrow -\frac{1}{\sqrt{2}}\le x\le 0$$

(C) $0\le 2\pi-4\cos^{-1}x\le\pi$

$$\Rightarrow \quad -2\pi\le-4\cos^{-1}x\le-\pi$$

$$\Rightarrow \quad 2\pi\ge 4\cos^{-1}x\ge\pi$$

$$\Rightarrow \quad \frac{\pi}{2}\ge\cos^{-1}x\ge\frac{\pi}{4} \Rightarrow 0\le x\le\frac{1}{\sqrt{2}}$$

(D) $0\le 4\pi-4\cos^{-1}x\le\pi$

$$\Rightarrow \quad -4\pi\le-4\cos^{-1}x\le-3\pi$$

$$\Rightarrow \quad 4\pi\ge 4\cos^{-1}x\ge 3\pi$$

$$\Rightarrow \quad \pi\ge\cos^{-1}x\ge\frac{3\pi}{4}$$

$$\Rightarrow \quad -1\le x\le-\frac{1}{\sqrt{2}}$$

Subjective Type Examples

Ex. 37. *If $A = 2\tan^{-1}(2\sqrt{2}-1)$ and $B = 3\sin^{-1}\left(\frac{1}{3}\right) + \sin^{-1}\left(\frac{3}{5}\right)$, then show $A > B$.*

Sol. We have,

$$A = 2\tan^{-1}(2\sqrt{2}-1) = 2\tan^{-1}(1.828)$$

$$\Rightarrow \quad A > 2\tan^{-1}(\sqrt{3})$$

$$\Rightarrow \quad A > \frac{2\pi}{3} \quad \text{...(i)}$$

Also we have,

$$\sin^{-1}\left(\frac{1}{3}\right) < \sin^{-1}\left(\frac{1}{2}\right) \Rightarrow \sin^{-1}\left(\frac{1}{3}\right) < \frac{\pi}{6}$$

$$\Rightarrow \quad 3\sin^{-1}\left(\frac{1}{3}\right) < \frac{\pi}{2}$$

Also,
$$3\sin^{-1}\left(\frac{1}{3}\right) = \sin^{-1}\left(3\cdot\frac{1}{3} - 4\left(\frac{1}{3}\right)^3\right)$$

$$= \sin^{-1}\left(\frac{23}{27}\right) = \sin^{-1}(0.852)$$

$$\Rightarrow \quad 3\sin^{-1}\left(\frac{1}{3}\right) < \sin^{-1}\left(\frac{\sqrt{3}}{2}\right)$$

$$\Rightarrow \quad 3\sin^{-1}\left(\frac{1}{3}\right) < \frac{\pi}{3}$$

Also,
$$\sin^{-1}\left(\frac{3}{5}\right) = \sin^{-1}(0.6) < \sin^{-1}\left(\frac{\sqrt{3}}{2}\right)$$

$$\Rightarrow \quad \sin^{-1}\left(\frac{3}{5}\right) < \frac{\pi}{3}$$

Hence,
$$B = 3\sin^{-1}\left(\frac{1}{3}\right) + \sin^{-1}\left(\frac{3}{5}\right) < \frac{2\pi}{3} \quad \text{...(ii)}$$

From (i) and (ii), we have

$$A > B.$$

Ex. 38. *Solve for x : $(\tan^{-1}x)^2 + (\cot^{-1}x)^2 = \frac{5\pi^2}{8}$.*

Sol. We have $(\tan^{-1}x)^2 + (\cot^{-1}x)^2 = \frac{5\pi^2}{8}$

$$\Rightarrow (\tan^{-1}x + \cot^{-1}x)^2 - 2\tan^{-1}x.\cot^{-1}x = \frac{5\pi^2}{8}$$

$$\Rightarrow \quad \left(\frac{\pi}{2}\right)^2 - 2\tan^{-1}x.\left(\frac{\pi}{2} - \tan^{-1}x\right) = \frac{5\pi^2}{8}$$

$$\left\{\because \tan^{-1}x + \cot^{-1}x = \frac{\pi}{2} \Rightarrow \cot^{-1}x = \frac{\pi}{2} - \tan^{-1}x\right\}$$

$$\Rightarrow \quad \frac{\pi^2}{4} - 2\cdot\frac{\pi}{2}\cdot\tan^{-1}x + 2(\tan^{-1}x)^2 = \frac{5\pi^2}{8}$$

$$\Rightarrow \quad 2(\tan^{-1}x)^2 - \pi\tan^{-1}x - \frac{3\pi^2}{8} = 0$$

$$\Rightarrow \quad \tan^{-1}x = -\frac{\pi}{4}, \frac{3\pi}{4}$$

$$\Rightarrow \quad \tan^{-1}x = -\frac{\pi}{4}$$

$$\Rightarrow \quad x = -1$$

$$\left\{\text{neglecting } \tan^{-1}x = \frac{3\pi}{4} \text{ as } \tan^{-1}x \in \left]-\frac{\pi}{2}, \frac{\pi}{2}\right[\right\}$$

Ex. 39. *Solve for x : If $[\sin^{-1}\cos^{-1}\sin^{-1}\tan^{-1}x] = 1$, where [.] denotes the greatest integer function.*

Sol. We have, $[\sin^{-1}\cos^{-1}\sin^{-1}\tan^{-1}x] = 1$

$$\Rightarrow \quad 1 \le \sin^{-1}.\cos^{-1}.\sin^{-1}.\tan^{-1}x \le \frac{\pi}{2}$$

$$\Rightarrow \quad \sin 1 \le \cos^{-1}.\sin^{-1}.\tan^{-1}x \le 1$$

$$\Rightarrow \quad \cos\sin 1 \ge \sin^{-1}.\tan^{-1}x \ge \cos 1$$

$$\Rightarrow \quad \sin\cos\sin 1 \ge \tan^{-1}x \ge \sin\cos 1$$

$$\Rightarrow \quad \tan\sin\cos\sin 1 \ge x \ge \tan\sin\cos 1$$

Hence, $x \in [\tan\sin\cos 1, \tan\sin\cos\sin 1]$

Ex. 40. *If $\tan^{-1}y = 4\tan^{-1}x \left(|x| < \tan\frac{\pi}{8}\right)$, find y as an algebraic function of x and hence prove that $\tan\frac{\pi}{8}$ is a root of the equation $x^4 - 6x^2 + 1 = 0$.*

Sol. We have,

$$\tan^{-1}y = 4\tan^{-1}x$$

$$\Rightarrow \tan^{-1}y = 2\tan^{-1}\frac{2x}{1-x^2} \quad (\text{as } |x| < 1)$$

$$= \tan^{-1}\frac{\frac{4x}{1-x^2}}{1 - \frac{4x^2}{(1-x^2)^2}}$$

$$= \tan^{-1}\frac{4x(1-x^2)}{x^4 - 6x^2 + 1} \quad \left(\text{as } \left|\frac{2x}{1-x^2}\right| < 1\right)$$

$$\Rightarrow \quad y = \frac{4x(1-x^2)}{x^4 - 6x^2 + 1}$$

If
$$x = \tan\frac{\pi}{8}$$

$$\Rightarrow \quad \tan^{-1}y = 4\tan^{-1}x = \frac{\pi}{2}$$

$$\Rightarrow \quad y = \infty \Rightarrow x^4 - 6x^2 + 1 = 0$$

Ex. 41. *If x_1, x_2, x_3, x_4 are the roots of the equation*

$$x^4 - x^3 \sin 2\beta + x^2 \cos 2\beta - x\cos\beta - \sin\beta = 0$$

Then show :

$$\tan^{-1} x_1 + \tan^{-1} x_2 + \tan^{-1} x_3 + \tan^{-1} x_4 = \frac{\pi}{2} - \beta$$

Sol. We have, $\Sigma x_1 = \sin 2\beta$,

$\Sigma x_1x_2 = \cos 2\beta$, $\Sigma x_1x_2x_3 = \cos\beta$

and $x_1x_2x_3x_4 = -\sin\beta$.

$$\left\{\begin{array}{l} \text{using; } ax^4 + bx^3 + cx^2 + dx + e = 0 \text{ has four root } x_1, x_2, x_3, x_4 \\ \Rightarrow \Sigma x_1 = x_1 + x_2 + x_3 + x_4 = -\frac{b}{a}, \\ x_1x_2 + x_2x_3 + x_1x_3 + x_1x_4 + x_2x_4 + x_1x_4 = \Sigma x_1x_2 = \frac{c}{a} \\ \Sigma x_1x_2x_3 = x_1x_2x_3 + x_1x_2x_4 + x_2x_3x_4 + x_1x_3x_4 = -\frac{d}{a} \\ \text{and } \quad x_1x_2x_3x_4 = \frac{e}{a} \end{array}\right\}$$

Let, $\alpha_1 = \tan^{-1} x_1$, $\alpha_2 = \tan^{-1} x_2$, $\alpha_3 = \tan^{-1} x_3$ and $\alpha_4 = \tan^{-1} x_4$

$\Rightarrow \quad \tan\alpha_1 = x_1$, $\tan\alpha_2 = x_2$, $\tan\alpha_3 = x_3$ and $\tan\alpha_4 = x_4$

$$\therefore \quad \tan(\alpha_1 + \alpha_2 + \alpha_3 + \alpha_4) = \frac{(s_1 - s_3)}{1 - s_2 + s_4} = \frac{\Sigma x_1 - \Sigma x_1x_2x_3}{1 - \Sigma x_1x_2 + x_1x_2x_3x_4}$$

$$= \frac{\sin 2\beta - \cos\beta}{1 - \cos 2\beta - \sin\beta} = \frac{\cos\beta(2\sin\beta - 1)}{\sin\beta(2\sin\beta - 1)}$$

$$= \cot\beta = \tan\left(\frac{\pi}{2} - \beta\right)$$

$$\Rightarrow \quad \alpha_1 + \alpha_2 + \alpha_3 + \alpha_4 = \frac{\pi}{2} - \beta$$

or $\tan^{-1} x_1 + \tan^{-1} x_2 + \tan^{-1} x_3 + \tan^{-1} x^4 = \frac{\pi}{2} - \beta$

Ex. 42. *Find the number of positive integral solutions of the equation :*

$$\tan^{-1} x + \cos^{-1} \frac{y}{\sqrt{1 - y^2}} = \sin^{-1} \frac{3}{\sqrt{10}}$$

Sol. Here,

$$\tan^{-1} x + \cos^{-1} \frac{y}{\sqrt{1 - y^2}} = \sin^{-1} \frac{3}{\sqrt{10}}$$

$$\Rightarrow \quad \tan^{-1} x + \tan^{-1}\left(\frac{1}{y}\right) = \tan^{-1}(3)$$

or $$\tan^{-1}\left(\frac{1}{y}\right) = \tan^{-1} 3 - \tan^{-1}(x)$$

or $$\tan^{-1}\left(\frac{1}{y}\right) = \tan^{-1}\left(\frac{3 - x}{1 + 3x}\right) \Rightarrow y = \frac{1 + 3x}{3 - x}$$

As x, y are positive integers, $x = 1, 2$ and correspondingly $y = 2, 7$.

$\therefore$ Solutions are $(x, y) = (1, 2), (2, 7)$ i.e. two solution.

Ex. 43. *If $\cot^{-1} \frac{n}{\pi} > \frac{\pi}{6}$, $n \in N$, then find the maximum value of n.*

Sol. Here, $\cot^{-1} \frac{n}{\pi} > \frac{\pi}{6}$

$$\Rightarrow \quad \cot\left(\cot^{-1} \frac{n}{\pi}\right) < \cot \frac{\pi}{6} \qquad \{\text{as } \cot x \text{ is decreasing in } [0, \pi/2]\}$$

$$\Rightarrow \quad \frac{n}{\pi} < \sqrt{3}$$

$$\Rightarrow \quad n < \sqrt{3}\pi = 5.5 \qquad \text{(approximately)}$$

So, the maximum value of n is 5, as $n \in N$.

Ex. 44. *If $\cot^{-1}\left(\frac{n^2 - 10n + 21 \cdot 6}{\pi}\right) > \frac{\pi}{6}$, $n \in N$, then find the minimum value of n.*

Sol. We have, $\cot^{-1}\left(\frac{n^2 - 10n + 21 \cdot 6}{\pi}\right) > \frac{\pi}{6}$

$$\Rightarrow \quad \frac{n^2 - 10n + 21 \cdot 6}{\pi} < \cot \frac{\pi}{6}$$

{as $\cot x$ is decreasing for $0 < x < \pi$}

$$\Rightarrow \quad n^2 - 10n + 21.6 < \pi\sqrt{3}$$

$$\Rightarrow \quad n^2 - 10n + 25 + 21 \cdot 6 - 25 < \pi\sqrt{3}$$

$$\Rightarrow \quad (n - 5)^2 < \pi\sqrt{3} + 3 \cdot 4$$

$$\Rightarrow \quad -\sqrt{\sqrt{3}\pi + 3 \cdot 4} < n - 5 < \sqrt{\sqrt{3}\pi + 3.4} \qquad \text{...(i)}$$

Since, $\sqrt{3}\pi = 5.5$ nearly

$$\therefore \quad \sqrt{\sqrt{3}\pi + 3 \cdot 4} \sim \sqrt{8 \cdot 9} \sim 2 \cdot 9$$

$$\Rightarrow \quad 2 \cdot 1 < n < 7 \cdot 9$$

$$\therefore \quad n = 3, 4, 5, 6, 7 \qquad \{\text{as } n \in N\}$$

or minimum value of $n = 3$.

Ex. 45. *Find the set of values of k for which $x^2 - kx + \sin^{-1}(\sin 4) > 0$ for all real x.*

Sol. We know,

$$\sin^{-1}(\sin 4) = \sin^{-1}(\sin(\pi - 4)) = \pi - 4$$

$$\left\{\because -\frac{\pi}{2} < \pi - 4 < \frac{\pi}{2}\right\}$$

$\therefore$ We have $x^2 - kx + \pi - 4 > 0$ for all $x \in R$

$$\therefore \quad D < 0, \text{ i.e. } k^2 - 4(\pi - 4) < 0$$

or $$k^2 + 4(4 - \pi) < 0$$

which is not true for any real k. {as $k^2 + 4(4 - \pi) > 0$}

Ex. 46. *Find the greatest and least value of;* $(\sin^{-1}x)^3+(\cos^{-1}x)^3$.

Sol. We have, $(\sin^{-1}x)^3+(\cos^{-1}x)^3=(\sin^{-1}x+\cos^{-1}x)$
$$[(\sin^{-1}x+\cos^{-1}x)^2-3\sin^{-1}x.\cos^{-1}x]$$
$$=\frac{\pi}{2}\left[\frac{\pi^2}{4}-3\sin^{-1}x\left(\frac{\pi}{2}-\sin^{-1}x\right)\right]$$
$$=\frac{3\pi}{2}\left[(\sin^{-1}x)^2-\frac{\pi}{2}\sin^{-1}x+\frac{\pi^2}{12}\right]$$
$$=\frac{3\pi}{2}\left[\left(\sin^{-1}x-\frac{\pi}{4}\right)^2+\frac{\pi^2}{12}-\frac{\pi^2}{16}\right]$$
$$=\frac{3\pi}{2}\left[\left(\sin^{-1}x-\frac{\pi}{4}\right)^2+\frac{\pi^2}{48}\right]$$

So, the least value is; $\frac{3\pi}{2}\cdot\frac{\pi^2}{48}=\frac{\pi^3}{32}\left\{\text{when}\left(\sin^{-1}x-\frac{\pi}{4}\right)=0\right\}$

Also, $\left(\sin^{-1}x-\frac{\pi}{4}\right)^2\le\left(\frac{3\pi}{4}\right)^2$

$\therefore$ The greatest value $=\frac{3\pi}{2}\left[\frac{9\pi^2}{16}+\frac{\pi^2}{48}\right]=\frac{7\pi^3}{8}$.

Ex. 47. *If x_r is given by,* $x_{r+1}=\sqrt{\frac{1}{2}(1+x_r)}$. *Then, show :* $\cos^{-1}x_0=\frac{\sqrt{1-x_0^2}}{x_1x_2x_3}$ *... up to infinity.*

Sol. Given : $x_{r+1}=\sqrt{\frac{1}{2}(1+x_r)}$

Let $x_0=\cos\theta$
$$x_1=\sqrt{\frac{1}{2}(1+x_0)}=\sqrt{\frac{1}{2}(1+\cos\theta)}=\cos\left(\frac{\theta}{2}\right)$$
$$x_2=\sqrt{\frac{1}{2}(1+x_1)}=\sqrt{\frac{1}{2}\left(1+\cos\frac{\theta}{2}\right)}=\cos\left(\frac{\theta}{2^2}\right)$$

Similarly, $x_3=\cos\left(\frac{\theta}{2^3}\right)$

....

....
$$x_n=\cos\left(\frac{\theta}{2^n}\right)$$
$$\Rightarrow\quad x_1x_2x_3\ldots x_n=\cos\frac{\theta}{2}.\cos\frac{\theta}{2^2}.\cos\frac{\theta}{2^3}\ldots\cos\frac{\theta}{2^n}$$
$$=\frac{\sin\theta}{2^n.\sin\frac{\pi}{2^n}}\quad\ldots\text{(i)}$$
$$\therefore\quad x_1x_2\ldots x_n\ldots\infty=\lim_{n\to\infty}(x_1x_2\ldots x_n)$$
$$=\lim_{n\to\infty}\left(\frac{\sin\theta}{2^n.\sin\frac{\theta}{2^n}}\right)\quad\{\text{using Eq. (i)}\}$$
$$=\lim_{n\to\infty}\frac{\sin\theta}{\theta\cdot\frac{\sin\frac{\theta}{2^n}}{\frac{\theta}{2^n}}}=\frac{\sin\theta}{\theta}\quad\ldots\text{(ii)}$$
$$\therefore\quad\frac{\sqrt{1-x_0^2}}{x_1x_2x_3\ldots x_\infty}=\frac{\sqrt{1-\cos^2\theta}}{\frac{\sin\theta}{\theta}}\quad\{\text{using Eq. (ii)}\}$$
$$=\theta=\cos^{-1}(x_0)\quad\{\because\cos\theta=x_0\}$$
$$\Rightarrow\quad\frac{\sqrt{1-x_0^2}}{x_1x_2x_3\ldots\infty}=\cos^{-1}x_0.$$

Ex. 48. *Express the equation;* $\cot^{-1}\frac{y}{\sqrt{1-x^2-y^2}}=2\tan^{-1}\sqrt{\frac{3-4x^2}{4x^2}}-\tan^{-1}\sqrt{\frac{3-4x^2}{x^2}}$ *as a rational integral equation in x and y.*

Sol. $\cot^{-1}\left(\frac{y}{\sqrt{1-x^2-y^2}}\right)=\tan^{-1}\left(\frac{\sqrt{1-x^2-y^2}}{y}\right)$

Also, $2\tan^{-1}\sqrt{\frac{3-4x^2}{4x^2}}=\tan^{-1}\frac{2\cdot\sqrt{\frac{3-4x^2}{4x^2}}}{1-\left(\frac{3-4x^2}{4x^2}\right)}$
$$=\tan^{-1}\left(\frac{4x\sqrt{3-4x^2}}{8x^2-3}\right)$$

Hence, the given equation is
$$\tan^{-1}\frac{\sqrt{1-x^2-y^2}}{y}=\tan^{-1}\left(\frac{4x\sqrt{3-4x^2}}{8x^2-3}\right)-\tan^{-1}\sqrt{\frac{3-4x^2}{x^2}}$$
$$=\tan^{-1}\left[\frac{\frac{4x\sqrt{3-4x^2}}{8x^2-3}-\sqrt{\frac{3-4x^2}{x^2}}}{1+\frac{4x\sqrt{3-4x^2}}{8x^2-3}\cdot\sqrt{\frac{3-4x^2}{x^2}}}\right]$$
$$=\tan^{-1}\left[\frac{(3-4x^2)^{3/2}}{9x-8x^3}\right]$$
$$\therefore\quad\frac{\sqrt{1-x^2-y^2}}{y}=\frac{(3-4x^2)^{3/2}}{9x-8x^3}$$

Squaring and simplifying
$$\frac{1-x^2}{y^2}=1+\frac{(3-4x^2)^3}{(9x-8x^3)^2}$$
$$=\frac{(9x-8x^3)^2+(3-4x^2)^3}{(9x-8x^3)^2}=\frac{27-27x^2}{(9x-8x^3)^2}$$
$$\therefore\quad y^2=\frac{x^2(9-8x^2)^2}{27}$$

Ex. 49. *Prove that*

$$\cos^{-1}\left(\frac{\cos x+\cos y}{1+\cos x\cdot\cos y}\right)=2\tan^{-1}\left(\tan\frac{x}{2}\tan\frac{y}{2}\right).$$

Sol. Let, $\tan\frac{x}{2}\cdot\tan\frac{y}{2}=\tan\theta$ ...(i)

Consider $\frac{\cos x+\cos y}{1+\cos x.\cos y}=\dfrac{\dfrac{1-\tan^2 x/2}{1+\tan^2 x/2}+\dfrac{1-\tan^2 y/2}{1+\tan^2 y/2}}{1+\dfrac{(1-\tan^2 x/2)}{(1+\tan^2 x/2)}\cdot\dfrac{(1-\tan^2 y/2)}{(1+\tan^2 y/2)}}$

$$=\frac{(1-\tan^2 x/2)(1+\tan^2 y/2)+(1+\tan^2 x/2)(1-\tan^2 y/2)}{(1+\tan^2 x/2)(1+\tan^2 y/2)+(1-\tan^2 x/2).(1-\tan^2 y/2)}$$

$$=\frac{1-\tan^2 x/2+\tan^2 y/2-\tan^2 x/2\tan^2 y/2+1-\tan^2 y/2+\tan^2 x/2-\tan^2 x/2\tan^2 y/2}{1+\tan^2 x/2+\tan^2 y/2+\tan^2 x/2\tan^2 y/2+1-\tan^2 x/2-\tan^2 y/2+\tan^2 x/2.\tan^2 y/2}$$

$$=\frac{2-2\tan^2 x/2\tan^2 y/2}{2+2\tan^2 x/2\tan^2 y/2}$$

$$=\frac{1-\tan^2\theta}{1+\tan^2\theta}$$ {from Eq. (i)}

$=\cos 2\theta$

$\therefore$ LHS $=\cos^{-1}(\cos 2\theta)=2\theta$

$$=2\left\{\tan^{-1}\left(\tan\frac{x}{2}\cdot\tan\frac{y}{2}\right)\right\}$$

Ex. 50. *If* $x=\text{cosec}(\tan^{-1}(\cos(\cot^{-1}(\sec(\sin^{-1}a)))))$ *and* $y=\sec(\cot^{-1}\sin(\tan^{-1}(\text{cosec}(\cos^{-1}a))))$, *where* $a\in[0,1]$. *Find the relationship between x and y in terms of 'a'.*

Sol. Here,

$$x=\text{cosec}(\tan^{-1}(\cos(\cot^{-1}(\sec(\sin^{-1}a)))))$$

$$\Rightarrow\quad x=\text{cosec}\left(\tan^{-1}\left(\cos\left(\cot^{-1}\left(\frac{1}{\sqrt{1-a^2}}\right)\right)\right)\right)$$

$$\Rightarrow\quad x=\text{cosec}\left(\tan^{-1}\left(\frac{1}{\sqrt{2-a^2}}\right)\right)$$

$$\Rightarrow\quad x=\sqrt{3-a^2}$$

and

$$y=\sec(\cot^{-1}\sin(\tan^{-1}(\text{cosec}(\cos^{-1}a))))$$

$$\Rightarrow\quad y=\sec\left(\cot^{-1}\left(\sin\left(\tan^{-1}\left(\frac{1}{\sqrt{1-a^2}}\right)\right)\right)\right)$$

$$\Rightarrow\quad y=\sec\left(\cot^{-1}\left(\frac{1}{\sqrt{2-a^2}}\right)\right)$$

$$\Rightarrow\quad y=\sqrt{3-a^2}$$ ...(ii)

From Eqs. (i) and (ii), $x=y=\sqrt{3-a^2}$.

Ex. 51. *Show that* $\tan^{-1}\left\{\tan\frac{\alpha}{2}\tan\left(\frac{\pi}{4}-\frac{\beta}{2}\right)\right\}=\tan^{-1}\left(\frac{\sin\alpha\cos\beta}{\cos\alpha+\sin\beta}\right)$ *where* $\alpha,\beta\in\left(0,\frac{\pi}{2}\right)$.

Sol. Since, $0<\alpha<\frac{\pi}{2}$

$\therefore\quad 0<\frac{\alpha}{2}<\frac{\pi}{4}\Rightarrow 0<\tan\frac{\alpha}{2}<1$ ...(i)

Similarly, $0<\beta<\frac{\pi}{2}$

$\Rightarrow\quad 0<\frac{\beta}{2}<\frac{\pi}{4}\Rightarrow -\frac{\pi}{4}<-\frac{\beta}{2}<0$

$\Rightarrow\quad 0<\frac{\pi}{4}-\frac{\beta}{2}<\frac{\pi}{4}\Rightarrow 0<\tan\left(\frac{\pi}{4}-\frac{\beta}{2}\right)<1$ ...(ii)

From Eqs. (i) and (ii), we get

$$0<\tan\frac{\alpha}{2}\tan\left(\frac{\pi}{4}-\frac{\beta}{2}\right)<1.$$

Since, $2\tan^{-1}x=\tan^{-1}\frac{2x}{1-x^2}$, when $x\in(0,1)$

$\therefore\quad 2\tan^{-1}\left\{\tan\frac{\alpha}{2}\tan\left(\frac{\pi}{4}-\frac{\beta}{2}\right)\right\}$

$$=\tan^{-1}\frac{2\tan\frac{\alpha}{2}\tan\left(\frac{\pi}{4}-\frac{\beta}{2}\right)}{1-\tan^2\frac{\alpha}{2}.\tan^2\left(\frac{\pi}{4}-\frac{\beta}{2}\right)}$$

$$=\tan^{-1}\frac{2\tan\frac{\alpha}{2}\cdot\dfrac{1-\tan\left(\frac{\beta}{2}\right)}{1+\tan\left(\frac{\beta}{2}\right)}}{1-\tan^2\frac{\alpha}{2}\left[\dfrac{1-\tan\frac{\beta}{2}}{1+\tan\frac{\beta}{2}}\right]^2}$$

$$=\tan^{-1}\frac{2\tan\frac{\alpha}{2}\left(1-\tan^2\frac{\beta}{2}\right)}{\left(1+\tan\frac{\beta}{2}\right)^2-\tan^2\frac{\alpha}{2}\left(1-\tan\frac{\beta}{2}\right)^2}$$

$$=\tan^{-1}\frac{2\tan\frac{\alpha}{2}\left(1-\tan^2\frac{\beta}{2}\right)}{\left(1+\tan^2\frac{\beta}{2}\right)\left(1-\tan^2\frac{\alpha}{2}\right)+2\tan\frac{\beta}{2}\left(1+\tan^2\frac{\alpha}{2}\right)}$$

$$=\tan^{-1}\frac{\dfrac{2\tan\frac{\alpha}{2}}{1+\tan^2\frac{\alpha}{2}}\cdot\dfrac{1-\tan^2\frac{\beta}{2}}{1+\tan^2\frac{\beta}{2}}}{\dfrac{1-\tan^2\frac{\alpha}{2}}{1+\tan^2\frac{\alpha}{2}}+\dfrac{2\tan\frac{\beta}{2}}{1+\tan^2\frac{\beta}{2}}}$$

$$=\tan^{-1}\left(\frac{\sin\alpha.\cos\beta}{\cos\alpha+\sin\beta}\right)$$

Ex. 52. *Solve :*

$$3\sin^{-1}\frac{2x}{1+x^2}-4\cos^{-1}\frac{1-x^2}{1+x^2}+2\tan^{-1}\frac{2x}{1-x^2}=\frac{\pi}{3}$$

Sol. Let $\tan^{-1}x=\theta$ for $x\ge 0$

Case I When $0\le x<1$, then $0\le\theta<\frac{\pi}{4}$ and so $0\le 2\theta<\frac{\pi}{2}$

$$\therefore\quad \sin^{-1}\frac{2x}{1+x^2}=\sin^{-1}(\sin 2\theta)=2\theta=2\tan^{-1}x$$

$$\cos^{-1}\frac{1-x^2}{1+x^2}=\cos^{-1}(\cos 2\theta)=2\theta=2\tan^{-1}x$$

$$\tan^{-1}\frac{2x}{1-x^2}=\tan^{-1}(\tan 2\theta)=2\theta=2\tan^{-1}x$$

Thus, $3\left(\sin^{-1}\frac{2x}{1+x^2}\right)-4\left(\cos^{-1}\frac{1-x^2}{1+x^2}\right)+2\left(\tan^{-1}\frac{2x}{1-x^2}\right)=\frac{\pi}{3}$

$$\Rightarrow\quad 3(2\tan^{-1}x)-4(2\tan^{-1}x)+2(2\tan^{-1}x)=\frac{\pi}{3}.$$

$$\Rightarrow\quad 2\tan^{-1}x=\frac{\pi}{3}$$

$$\Rightarrow\quad \tan^{-1}x=\frac{\pi}{6}\Rightarrow x=\frac{1}{\sqrt{3}}$$

as $\quad 0\le\frac{1}{\sqrt{3}}<1,\ x=\frac{1}{\sqrt{3}}$ is a solution

Case II When $x=1$, $\tan^{-1}\frac{2x}{1-x^2}$ is not defined.

$\therefore\quad x=1$, cannot be a solution.

Case III If $x>1$, then $\frac{\pi}{4}<\theta<\frac{\pi}{2}$ and so $\frac{\pi}{2}<2\theta<\pi$

$$\therefore\quad \sin^{-1}\left(\frac{2x}{1+x^2}\right)=\sin^{-1}(\sin 2\theta)=\pi-2\theta=\pi-2\tan^{-1}x.$$

$$\cos^{-1}\left(\frac{2x}{1+x^2}\right)=\cos^{-1}(\cos 2\theta)=2\theta=2\tan^{-1}x$$

and $\tan^{-1}\left(\frac{2x}{1-x^2}\right)=\tan^{-1}(\tan 2\theta)=2\theta-\pi=2\tan^{-1}x-\pi$

Thus, the given equation becomes

$$3(\pi-2\tan^{-1}x)-4(2\tan^{-1}x)+2(2\tan^{-1}x-\pi)=\frac{\pi}{3}$$

$$\Rightarrow\quad \pi-10\tan^{-1}x=\frac{\pi}{3}$$

or $\quad \tan^{-1}x=\frac{\pi}{15}$

i.e. $\quad x=\tan\frac{\pi}{15}<\tan\frac{\pi}{4}<1$

$\therefore\quad x=\tan\frac{\pi}{15}$ is not a solution.

Thus, $x=\frac{1}{\sqrt{3}}$ is the only solution for the given equation for $x\ge 0$.

Ex. 53. *Obtain the integral values of p for which the following system of equations possesses real solutions :*

$$\cos^{-1}x+(\sin^{-1}y)^2=\frac{p\pi^2}{4}\ \text{and}\ (\cos^{-1}x)(\sin^{-1}y)^2=\frac{\pi^4}{16}$$

Also, find these solution.

Sol. Let $\cos^{-1}x=a\Rightarrow a\in[0,\pi]$

and $\quad \sin^{-1}y=b\Rightarrow b\in\left[-\frac{\pi}{2},\frac{\pi}{2}\right]$

We have $\quad a+b^2=\frac{p\pi^2}{4}$ ...(i)

and $\quad ab^2=\frac{\pi^4}{16}$ ...(ii)

Since $\quad b^2\in\left[0,\frac{\pi^2}{4}\right]\Rightarrow a+b^2\in\left[0,\pi+\frac{\pi^2}{4}\right]$

So, Eq. (i) $\Rightarrow 0\le\frac{p\pi^2}{4}\le\pi+\frac{\pi^2}{4}$

i.e. $\quad 0\le p\le\frac{4}{\pi}+1$

Since $\quad p\in Z$, so $p=0,1$ or 2

But, if $p=0$, then $a=b=0$.

$\Rightarrow$ Equation (ii) will not be satisfied.

Now, substituting the value of b^2 from Eq. (i) in the Eq. (ii), we get

$$a\left(\frac{p\pi^2}{4}-a\right)=\frac{\pi^4}{16}\Rightarrow 16a^2-4p\pi^2a+\pi^4=0 \quad ...(iii)$$

Since, $\quad a\in R\Rightarrow D\ge 0$

i.e. $\quad 16p^2\pi^4-64\pi^4\ge 0\Rightarrow p^2\ge 4\Rightarrow p\ge 2$

Thus, we conclude that the only value of p that satisfies all conditions is $p=2$. Substituting $p=2$ in Eq. (iii), we get

$$16a^2-8\pi^2a+\pi^4=0$$

$$\Rightarrow\quad (4a-\pi^2)^2=0$$

$$\Rightarrow\quad a=\frac{\pi^2}{4}=\cos^{-1}x$$

$$\Rightarrow\quad x=\cos\frac{\pi^2}{4}$$

From Eq. (ii), we get $\frac{\pi^2}{4}\cdot b^2=\frac{\pi^4}{16}\Rightarrow b=\pm\frac{\pi}{2}=\sin^{-1}y$

$$\Rightarrow\quad y=\pm 1.$$

Ex. 54. *Solve the equation* $2(\sin^{-1}x)^2-(\sin^{-1}x)-6=0$

Sol. Let, $\sin^{-1}x=y$, we get

$$2y^2-y-6=0$$

$$2y^2-4y+3y-6=0$$

$$y=2 \text{ and } y=-1.5$$

$$\therefore\quad \sin^{-1}x=2 \text{ and } \sin^{-1}x=-1.5$$

Since $2>\frac{\pi}{2}$ and $|-1.5|<\frac{\pi}{2}$, the only solution is $x=\sin(-1.5)$.

Ex. 55. *Solve the equation* $\sin^{-1}6x+\sin^{-1}6\sqrt{3}x=-\frac{\pi}{2}$

Sol. Let us transfer $\sin^{-1}6\sqrt{3}x$ into the right hand side of the equation and calculate the sine of the both sides of the resulting equation

$$\sin(\sin^{-1}6x)=\sin\left(-\sin^{-1}6\sqrt{3}x-\frac{\pi}{2}\right)$$

$$\Rightarrow \quad 6x=-\sin(\sin^{-1}6\sqrt{3}x+\sin^{-1}1)$$

{using $\sin(-\theta)=-\sin(\theta)$}

$$\Rightarrow \quad 6x=-\sin(\sin^{-1}\sqrt{1-108x^2})$$

{using $\sin^{-1}x+\sin^{-1}y=\sin^{-1}\{x\sqrt{1-y^2}+y\sqrt{1-x^2}\}$}

$$\Rightarrow \quad 6x=-\sqrt{1-108x^2} \quad \text{...(i)}$$

Squaring both sides, we get

$$36x^2=1-108x^2 \Rightarrow 144x^2=1$$

whose roots are $x=\frac{1}{12}$ and $x=-\frac{1}{12}$.

Let us verify :

Substituting $x=\frac{-1}{12}$ in the given equation, we get

$$\sin^{-1}\left(-\frac{1}{2}\right)+\sin^{-1}\left(-\frac{\sqrt{3}}{2}\right)$$

$$=-\frac{\pi}{6}-\frac{\pi}{3}=-\frac{\pi}{2}$$

Thus, $x=\frac{-1}{12}$ is the root of given equation. But, when substituting $x=\frac{1}{12}$ in Eq. (i), we get

LHS $\quad 6x=\frac{1}{2}$

RHS $\quad -\sqrt{1-108x^2}=-1/2$

i.e. LHS $\neq$ RHS of Eq. (i).

Hence, $x=-\frac{1}{12}$ is a root of the given equation as it satisfy both given and Eq. (i).

Ex. 56. *Solve the equation :* $2\tan^{-1}(2x-1)=\cos^{-1}x$.

Sol. Here, $\quad 2\tan^{-1}(2x-1)=\cos^{-1}x$

or $\quad \cos(2\tan^{-1}(2x-1))=x \quad \left\{\text{We know } \cos 2\theta=\frac{1-\tan^2\theta}{1+\tan^2\theta}\right\}$

$$\Rightarrow \quad \frac{1-(2x-1)^2}{1+(2x-1)^2}=x$$

$$\Rightarrow \quad \frac{-2x^2+2x}{1-2x+2x^2}=x \Rightarrow 2x^3-x=0$$

$$\Rightarrow \quad x=0,\frac{\sqrt{2}}{2},-\frac{\sqrt{2}}{2}.$$

Now, to verify which of the following satisfy given equation,

Case I For $x=0$;

$$2\tan^{-1}(-1)=\cos^{-1}(0)$$

or $$-\frac{\pi}{2}=\frac{\pi}{2}$$

$\therefore x=0$ is not a solution of given equation.

Case II For $x=\frac{\sqrt{2}}{2}$

The right hand and the left hand side of the equation are equal to $\frac{\pi}{4}$ and $2\tan^{-1}(\sqrt{2}-1)$.

But $\quad \tan^{-1}(\sqrt{2}-1)=\frac{\pi}{8}$.

$\therefore \quad x=\frac{\sqrt{2}}{2}$ is a root of the given equation.

Case III For $x=-\frac{\sqrt{2}}{2}$

The left hand side of the equation is negative and the right hand side is positive.

Consequently, $x=-\frac{\sqrt{2}}{2}$ is not a root of the given equation.

Thus from above;

$x=\frac{\sqrt{2}}{2}$ is the only solution.

Inverse Trigonometric Functions Exercise 1 : Single Option Correct Type Questions

1. $\cot^{-1}\left(\sqrt{\dfrac{1-x^2}{1+x^2}}\right)$ is equal to

(a) $\cos^{-1}(x^2)$ (b) $\dfrac{\pi}{2}-\dfrac{1}{2}\cos^{-1}(x^2)$

(c) $\dfrac{\pi}{3}-\dfrac{1}{2}\cos^{-1}(x^2)$ (d) None of these

2. The value of $\cos\left(\dfrac{1}{2}\cos^{-1}\dfrac{1}{8}\right)$ is equal to

(a) 3/4 (b) $-3/4$

(c) 1/16 (d) 1/4

3. The inequality $\sin^{-1}(\sin 5) > x^2 - 4x$ holds if

(a) $x = 2-\sqrt{9-2\pi}$

(b) $x = 2+\sqrt{9-2\pi}$

(c) $x \in (2-\sqrt{9-2\pi}, 2+\sqrt{9-2\pi})$

(d) $x > 2+\sqrt{9-2\pi}$

4. The value of $\sin^{-1}\left\{\left(\sin\dfrac{\pi}{3}\right)\dfrac{x}{\sqrt{(x^2+k^2-kx)}}\right\}$

$-\cos^{-1}\left\{\left(\cos\dfrac{\pi}{6}\right)\dfrac{x}{\sqrt{(x^2+k^2-kx)}}\right\}$, where

$\left(\dfrac{k}{2}<x<2k, k>0\right)$ is

(a) $\tan^{-1}\left(\dfrac{2x^2+xk-k^2}{x^2-2xk+k^2}\right)$

(b) $\tan^{-1}\left(\dfrac{x^2+2xk-2k^2}{x^2-2xk+k^2}\right)$

(c) $\tan^{-1}\left(\dfrac{x^2+2xk-2k^2}{2x^2-2xk+2k^2}\right)$

(d) None of the above

5. If $a \le \tan^{-1}\left(\dfrac{1-x}{1+x}\right) \le b$ where $0 \le x \le 1$, then $(a, b) =$

(a) $\left(0, \dfrac{\pi}{4}\right)$ (b) $\left(0, \dfrac{\pi}{2}\right)$

(c) $\left(-\dfrac{\pi}{4}, \dfrac{\pi}{4}\right)$ (d) $\left(\dfrac{\pi}{4}, \dfrac{\pi}{2}\right)$

6. Sum of infinite terms of the series

$\cot^{-1}\left(1^2+\dfrac{3}{4}\right)+\cot^{-1}\left(2^2+\dfrac{3}{4}\right)+\cot^{-1}\left(3^2+\dfrac{3}{4}\right)+\dots$ is

(a) $\pi/4$ (b) $\tan^{-1}2$

(c) $\tan^{-1}3$ (d) $\tan^{-1}4$

7. Solution of equation $\cot^{-1}x+\sin^{-1}\dfrac{1}{\sqrt{5}}=\dfrac{\pi}{4}$ is

(a) $x=3$ (b) $x=1/\sqrt{5}$

(c) $x=0$ (d) None of these

8. Solution set of the inequality $(\cot^{-1}x)^2-(5\cot^{-1}x)+6>0$ is

(a) $(\cot 3, \cot 2)$

(b) $(-\infty, \cot 3) \cup (\cot 2, \infty)$

(c) $(\cot 2, \infty)$

(d) None of the above

9. Sum to infinite terms of the series $\tan^{-1}\left(\dfrac{1}{3}\right)+\tan^{-1}\left(\dfrac{2}{9}\right)$

$+\tan^{-1}\left(\dfrac{4}{33}\right)+\dots+\tan^{-1}\left(\dfrac{2^{n-1}}{1+2^{2n-1}}\right)+\dots$

(a) $\pi/4$ (b) $\pi/2$

(c) π (d) None of these

10. If $x+\dfrac{1}{x}=2$, the principal value of $\sin^{-1}x$ is

(a) $\dfrac{\pi}{4}$ (b) $\dfrac{\pi}{2}$

(c) π (d) $\dfrac{3\pi}{2}$

11. If $x \in \left(-\dfrac{\pi}{2}, \dfrac{\pi}{2}\right)$, then the value of

$\tan^{-1}\left(\dfrac{\tan x}{4}\right)+\tan^{-1}\left(\dfrac{3\sin 2x}{5+3\cos 2x}\right)$ is

(a) $x/2$ (b) $2x$

(c) $3x$ (d) x

12. If $\sin^{-1}x+\sin^{-1}y=\dfrac{2\pi}{3}$, then $\cos^{-1}x+\cos^{-1}y$

(a) $\dfrac{2\pi}{3}$ (b) $\dfrac{\pi}{3}$

(c) $\dfrac{\pi}{6}$ (d) π

13. $\cot\left(\dfrac{\pi}{4}-2\cot^{-1}3\right)$ is

(a) 1 (b) 7

(c) -1 (d) None of these

14. $\sin\left[\tan^{-1}\dfrac{1-x^2}{2x}+\cos^{-1}\dfrac{1-x^2}{1+x^2}\right]$ is

(a) 1 (b) 0

(c) -1 (d) None of these

15. If $\cos^{-1}\left(\frac{1-a^2}{1+a^2}\right)-\cos^{-1}\left(\frac{1-b^2}{1+b^2}\right)=2\tan^{-1}x$, then x is

(a) $\frac{a-b}{1+ab}$ (b) $\frac{b-a}{1+ab}$

(c) $\frac{a+b}{1-ab}$ (d) None of these

16. If $\left|\cos^{-1}\left(\frac{1-x^2}{1+x^2}\right)\right|<\frac{\pi}{3}$, then

(a) $x\in\left[-\frac{1}{\sqrt{3}},\frac{1}{\sqrt{3}}\right]$ (b) $x\in\left(-\frac{1}{\sqrt{3}},\frac{1}{\sqrt{3}}\right)$

(c) $x\in\left[0,\frac{1}{\sqrt{3}}\right]$ (d) None of these

17. The value of

$\cos^{-1}\left[\cot\left(\sin^{-1}\sqrt{\frac{2-\sqrt{3}}{4}}+\cos^{-1}\left(\frac{\sqrt{12}}{4}\right)+\sec^{-1}\sqrt{2}\right)\right]$ is

(a) 0 (b) $\frac{\pi}{4}$

(c) $\frac{\pi}{6}$ (d) $\frac{\pi}{2}$

18. If $\tan^{-1}\frac{x}{\pi}<\frac{\pi}{3}$, $x\in N$, then the maximum value of x is

(a) 2 (b) 5

(c) 7 (d) None of these

19. If $\tan^{-1}\frac{\sqrt{(1+x^2)}-\sqrt{(1-x^2)}}{\sqrt{(1+x^2)}+\sqrt{(1-x^2)}}=\alpha$, then x^2 is

(a) $\cos 2\alpha$ (b) $\sin 2\alpha$

(c) $\tan 2\alpha$ (d) $\cot 2\alpha$

20. The number of positive integral solutions of $\tan^{-1}x+\cot^{-1}y=\tan^{-1}3$ is

(a) one (b) two

(c) three (d) four

21. If $\operatorname{cosec}^{-1}(\operatorname{cosec} x)$ and $\operatorname{cosec}(\operatorname{cosec}^{-1}x)$ are equal functions, then the maximum range of value of x is

(a) $\left[-\frac{\pi}{2},-1\right]\cup\left[1,\frac{\pi}{2}\right]$

(b) $\left[-\frac{\pi}{2},0\right)\cup\left(0,\frac{\pi}{2}\right]$

(c) $(-\infty,-1]\cup[1,\infty)$

(d) $[-1,0)\cup(0,1]$

22. The value of $\lim_{|x|\to\infty}\cos(\tan^{-1}(\sin(\tan^{-1}x)))$ is equal to

(a) -1 (b) $\sqrt{2}$

(c) $-\frac{1}{\sqrt{2}}$ (d) $\frac{1}{\sqrt{2}}$

23. Complete solution set of $[\cot^{-1}x]+2[\tan^{-1}x]=0$, where [.] denotes the greatest integer function, is equal to

(a) $(0,\cot 1)$ (b) $(0,\tan 1)$

(c) $(\tan 1,\infty)$ (d) $(\cot 1,\tan 1)$

24. If $\sin^{-1}:[-1,1]\to\left[\frac{\pi}{2},\frac{3\pi}{2}\right]$ and $\cos^{-1}:[-1,1]\to[0,\pi]$ be two bijective function, respectively inverses of bijective functions $\sin:\left[\frac{\pi}{2},\frac{3\pi}{2}\right]\to[-1,1]$ and $\cos:[0,\pi]\to[-1,1]$, then $\sin^{-1}x+\cos^{-1}x$ is

(a) $\frac{\pi}{2}$ (b) π

(c) $\frac{3\pi}{2}$ (d) not a constant

25. If $a\sin^{-1}x-b\cos^{-1}x=c$, then $a\sin^{-1}x+b\cos^{-1}x$ is equal to

(a) 0 (b) $\frac{\pi ab+c(b-a)}{a+b}$

(c) $\frac{\pi}{2}$ (d) $\frac{\pi ab+c(a-b)}{a+b}$

26. The number of integer x satisfying $\sin^{-1}|x-2|+\cos^{-1}(1-|3-x|)=\frac{\pi}{2}$ is

(a) 1 (b) 2

(c) 3 (d) 4

27. The value of α such that $\sin^{-1}\frac{2}{\sqrt{5}}$, $\sin^{-1}\frac{3}{\sqrt{10}}$, $\sin^{-1}\alpha$ are the angles of a triangle is

(a) $\frac{-1}{\sqrt{2}}$ (b) $\frac{1}{2}$

(c) $\frac{1}{\sqrt{3}}$ (d) $\frac{1}{\sqrt{2}}$

28. Let $\begin{vmatrix}\tan^{-1}x & \tan^{-1}2x & \tan^{-1}3x\\ \tan^{-1}3x & \tan^{-1}x & \tan^{-1}2x\\ \tan^{-1}2x & \tan^{-1}3x & \tan^{-1}x\end{vmatrix}=0$, then the number of values of x satisfying the equation is

(a) 1 (b) 2

(c) 3 (d) 4

29. If the equation $x^3+bx^2+cx+1=0$, $(b<c)$, has only one real root α, then the value of $2\tan^{-1}(\operatorname{cosec}\alpha)+\tan^{-1}(2\sin\alpha\sec^2\alpha)$ is

(a) $-\pi$ (b) $-\frac{\pi}{2}$

(c) $\frac{\pi}{2}$ (d) π

30. Let $u = \cot^{-1}\sqrt{\cos 2\theta} - \tan^{-1}\sqrt{\cos 2\theta}$, then the value of $\sin u$ is

(a) $\cos 2\theta$ (b) $\sin 2\theta$
(c) $\tan^2\theta$ (d) $\cot^2\theta$

31. Let $f(x) = 2\tan^{-1} x = \begin{cases} \cos^{-1}\left(\dfrac{1-x^2}{1+x^2}\right), & x \ge 0 \\ -\cos^{-1}\left(\dfrac{1-x^2}{1+x^2}\right), & x < 0 \end{cases}$

The function $f(x)$ is continuous everywhere but not differentiable at x equals to

(a) 1 (b) -1
(c) 0 (d) $\dfrac{1}{\sqrt{2}}$

32. Let $f(x) = \sin^{-1}\left(\dfrac{2x}{1+x^2}\right) \forall x \in R$. The function $f(x)$ is continuous everywhere but not differentiable at x is/are

(a) 0, 1 (b) -1, 1
(c) -1, 0 (d) 0, 2

33. Let $f(x) = \tan^{-1}(x^2 - 18x + a) > 0 \; \forall x \in R$. Then the value of a lies in

(a) $(81, \infty)$ (b) $[81, \infty)$
(c) $(-\infty, 81)$ (d) $(-\infty, 81]$

34. Let $f(x) = \sin^{-1} 2x + \cos^{-1} 2x + \sec^{-1} 2x$. Then the sum of the maximum and minimum values of $f(x)$ is

(a) π (b) 2π
(c) 3π (d) $\dfrac{\pi}{2}$

35. If $\tan^{-1}\dfrac{b}{c+a} + \tan^{-1}\dfrac{c}{a+b} = \dfrac{\pi}{4}$, where a, b, c are the sides of ΔABC, then ΔABC is

(a) Acute-angled triangle
(b) Obtuse-angled triangle
(c) Right-angled triangle
(d) Equilateral triangle

36. Solutions of $\sin^{-1}(\sin x) = \sin x$ are, if $x \in (0, 2\pi)$

(a) 4 real roots
(b) 2 positive real roots
(c) 2 negative real roots
(d) 5 real roots

37. The equation $e^{\frac{2\sin^{-1} x}{\pi}} = \dfrac{y}{\log y}$, has

(a) Unique solution
(b) Infinite many solution
(c) $x = 1$
(d) $y = e$

38. Let $f(x) = 1 + 2\sin\left(\dfrac{e^x}{e^x + 1}\right), x \ge 0$, then $f^{-1}(x)$ is equal to (assuming f is bijective)

(a) $\log\left(\dfrac{\sin^{-1}\left(\dfrac{x-1}{2}\right)}{1 - \sin^{-1}\left(\dfrac{x-1}{2}\right)}\right)$ (b) $\log\left(\dfrac{\sin\left(\dfrac{x-1}{2}\right)}{1 - \sin\left(\dfrac{x-1}{2}\right)}\right)$

(c) $e^{\frac{\sin^{-1}\left(\frac{x-1}{2}\right)}{1-\sin^{-1}\left(\frac{x-1}{2}\right)}}$ (d) $e^{\frac{\sin\left(\frac{x-1}{2}\right)}{1-\sin\left(\frac{x-1}{2}\right)}}$

39. $\cos^{-1}(\cos(2\cot^{-1}(\sqrt{2} - 1)))$ is equal to

(a) $\sqrt{2} - 1$ (b) $\dfrac{\pi}{4}$
(c) $\dfrac{3\pi}{4}$ (d) None of these

40. The maximum value of $f(x) = \tan^{-1}\left(\dfrac{(\sqrt{12} - 2)x^2}{x^4 + 2x^2 + 3}\right)$ is

(a) 18° (b) 36°
(c) 22.5° (d) 15°

41. If $\tan^{-1}\dfrac{\sqrt{1+x^2} - 1}{x} = 4°$, then

(a) $x = \tan 2°$ (b) $x = \tan 4°$
(c) $x = \tan(1/4)°$ (d) $x = \tan 8°$

42. If $\tan^{-1}(\sin^2\theta - 2\sin\theta + 3) + \cot^{-1}(5^{\sec^2 y} + 1) = \dfrac{\pi}{2}$, then the value of $\cos^2\theta - \sin\theta$ is equal to

(a) 0
(b) -1
(c) 1
(d) None of the above

43. The number of solutions of the equation $|\tan^{-1}|x|| = \sqrt{(x^2+1)^2 - 4x^2}$ is

(a) 1 (b) 2
(c) 3 (d) 4

44. For any real number $x \ge 1$, the expression $\sec^2(\tan^{-1} x) - \tan^2(\sec^{-1} x)$ is equal to

(a) 1 (b) 2
(c) $2x^2$ (d) $2\sqrt{2}$

45. Let $f : R \to \left[0, \dfrac{\pi}{2}\right)$ be defined by $f(x) = \tan^{-1}(3x^2 + 6x + a)$. If $f(x)$ is an onto function, then the value of a is

(a) 1 (b) 2
(c) 3 (d) 4

46. The value of expression

$\tan^{-1}\left(\frac{\sqrt{2}}{2}\right)+\sin^{-1}\left(\frac{\sqrt{5}}{5}\right)-\cos^{-1}\left(\frac{\sqrt{10}}{10}\right)$ is

(a) $\cot^{-1}\left(\frac{1+\sqrt{2}}{1-\sqrt{2}}\right)$ (b) $\cot^{-1}\left(\frac{\sqrt{2}+1}{\sqrt{2}-1}\right)$

(c) $-\pi+\cot^{-1}\left(\frac{1+\sqrt{2}}{1-\sqrt{2}}\right)$ (d) $\pi-\cot^{-1}\left(\frac{1-\sqrt{2}}{1+\sqrt{2}}\right)$

47. The value of $\sec\left(2\cot^{-1}2+\cos^{-1}\frac{3}{5}\right)$ is equal to

(a) $\frac{25}{24}$ (b) $-\frac{24}{7}$

(c) $\frac{25}{7}$ (d) $-\frac{25}{7}$

48. Which one of the following statement is meaningless?

(a) $\cos^{-1}\left(\ln\left(\frac{2e+4}{3}\right)\right)$ (b) $\operatorname{cosec}^{-1}\left(\frac{\pi}{3}\right)$

(c) $\cot^{-1}\left(\frac{\pi}{2}\right)$ (d) $\sec^{-1}(\pi)$

49. The value of $\sec\left[\sin^{-1}\left(-\sin\frac{50\pi}{9}\right)+\cos^{-1}\cos\left(-\frac{31\pi}{9}\right)\right]$ is equal to

(a) $\sec\frac{10\pi}{9}$ (b) $\sec\frac{\pi}{9}$

(c) 1 (d) -1

50. The number k is such that $\tan\{\text{arc}\tan(2)+\text{arc}\tan(20k)\}=k$. The sum of all possible values of k is

(a) $-\frac{19}{40}$ (b) $-\frac{21}{40}$

(c) 0 (d) $\frac{1}{5}$

51. The value of $\sum_{r=2}^{\infty}\tan^{-1}\left(\frac{1}{r^2-5r+7}\right)$, is

(a) $\frac{\pi}{4}$ (b) $\frac{\pi}{2}$

(c) $\frac{3\pi}{4}$ (d) $\frac{5\pi}{4}$

52. If $x=\tan^{-1}1-\cos^{-1}\left(-\frac{1}{2}\right)+\sin^{-1}\frac{1}{2}$;

$y=\cos\left(\frac{1}{2}\cos^{-1}\left(\frac{1}{8}\right)\right)$, then

(a) $x=\pi y$

(b) $y=\pi x$

(c) $\tan x=-(4/3)y$

(d) $\tan x=(4/3)y$

53. The value of

$\tan^{-1}\left(\frac{1}{2}\tan 2A\right)+\tan^{-1}(\cot A)+\tan^{-1}(\cot^3 A)$ for $0<A<(\pi/4)$ is

(a) $4\tan^{-1}(1)$ (b) $2\tan^{-1}(2)$

(c) 0 (d) None of these

54. The sum $\sum_{n=1}^{\infty}\tan^{-1}\left(\frac{4n}{n^4-2n^2+2}\right)$ is equal to

(a) $\tan^{-1}\frac{1}{2}+\tan^{-1}\frac{2}{3}$ (b) $4\tan^{-1}1$

(c) $\frac{\pi}{2}$ (d) $\sec^{-1}(-\sqrt{2})$

55. Number of solution(s) of the equations

$\cos^{-1}(1-x)-2\cos^{-1}x=\frac{\pi}{2}$ is

(a) 3 (b) 2

(c) 1 (d) 0

56. There exists a positive real number x satisfying $\cos(\tan^{-1}x)=x$. The number value of $\cos^{-1}\left(\frac{x^2}{2}\right)$ is

(a) $\frac{\pi}{10}$ (b) $\frac{\pi}{5}$

(c) $\frac{2\pi}{5}$ (d) $\frac{4\pi}{5}$

57. The range of values of p for which the equation $\sin\cos^{-1}(\cos(\tan^{-1}x))=p$ has a solution is

(a) $\left(-\frac{1}{\sqrt{2}},\frac{1}{\sqrt{2}}\right]$ (b) $[0, 1)$

(c) $\left[\frac{1}{\sqrt{2}},1\right)$ (d) $(-1,1)$

58. Number of solutions of the equation

$\log_{10}(\sqrt{5\cos^{-1}x-1})+\frac{1}{2}\log_{10}(2\cos^{-1}x+3)$

$+\log_{10}\sqrt{5}=1$ is

(a) 0

(b) 1

(c) more than one but finite

(d) infinite

59. Which of the following is the solution set of the equations $\sin^{-1}x=\cos^{-1}x+\sin^{-1}(3x-2)$?

(a) $\left\{\frac{1}{2},1\right\}$ (b) $\left[\frac{1}{2},1\right]$

(c) $\left[\frac{1}{3},1\right]$ (d) $\left\{\frac{1}{3},1\right\}$

60. The set of values of x, satisfying the equation $\tan^2(\sin^{-1} x) > 1$ is

(a) $[-1,1]$ (b) $\left[-\frac{\sqrt{2}}{2}, \frac{\sqrt{2}}{2}\right]$

(c) $(-1,1) - \left[-\frac{\sqrt{2}}{2}, \frac{\sqrt{2}}{2}\right]$ (d) $[-1,1] - \left(-\frac{\sqrt{2}}{2}, \frac{\sqrt{2}}{2}\right)$

61. The solution set of the equation

$$\sin^{-1}\sqrt{1-x^2} + \cos^{-1} x = \cot^{-1}\left(\frac{\sqrt{1-x^2}}{x}\right) - \sin^{-1} x$$

(a) $[-1,1] - \{0\}$
(b) $(0,1] \cup \{-1\}$
(c) $[-1,0) \cup \{1\}$
(d) $[-1,1]$

62. The value of the angle $\tan^{-1}(\tan 65° - 2\tan 40°)$ in degrees is equal to

(a) $-20°$ (b) $20°$ (c) $25°$ (d) $40°$

63. If $\cos^{-1}\frac{x}{a} - \sin^{-1}\frac{y}{b} = \theta\,(a, b, \neq 0)$, then the maximum value of $b^2x^2 + a^2y^2 + 2ab\,xy\sin\theta$ equals

(a) ab (b) $(a+b)^2$
(c) $2(a+b)^2$ (d) a^2b^2

64. The value of $\sum_{r=1}^{\infty} \tan^{-1}\left(\frac{1}{r^2+5r+7}\right)$ is equal to

(a) $\tan^{-1}3$ (b) $\frac{\pi}{4}$

(c) $\sin^{-1}\frac{1}{\sqrt{10}}$ (d) $\cot^{-1}2$

65. The range of the function,

$f(x) = \tan^{-1}\left(\frac{1+x}{1-x}\right) - \tan^{-1} x$ is

(a) $\{\pi/4\}$ (b) $\{-(\pi/4), 3\pi/4\}$
(c) $\{\pi/4, -(3\pi/4)\}$ (d) $\{3\pi/4\}$

66. Let $g: R \to \left(0, \frac{\pi}{3}\right]$ is defined by $g(x) = \cos^{-1}\left(\frac{x^2-k}{1+x^2}\right)$

Then the possible values of 'k' for which g is surjective function, is

(a) $\left\{\frac{1}{2}\right\}$ (b) $\left(-1, -\frac{1}{2}\right]$

(c) $\left\{-\frac{1}{2}\right\}$ (d) $\left[-\frac{1}{2}, 1\right)$

67. Number of values of x satisfying simultaneously $\sin^{-1} x = 2\tan^{-1} x$ and

$\tan^{-1}\sqrt{x(x-1)} + \operatorname{cosec}^{-1}\sqrt{1+x-x^2} = \frac{\pi}{2}$, is

(a) 0 (b) 1
(c) 2 (d) 3

68. Number of values of x satisfying the equation $\cos(3\arccos(x-1)) = 0$ is equal to

(a) 0 (b) 1
(c) 2 (d) 3

69. Which one of the following function contains only one integer in its range?

[**Note** sgn (k) denotes the signum function of k.]

(a) $f(x) = \frac{1}{2}\cos^{-1}\left(\frac{1-x^2}{1+x^2}\right)$

(b) $g(x) = \text{sgn}\left(x + \frac{1}{x}\right)$

(c) $h(x) = \sin^2 x + 2\sin x + 2$

(d) $k(x) = \cos^{-1}(x^2 - 2x + 2)$

70. If range of the function $f(x) = \tan^{-1}(3x^2 + bx + 3)$, $x \in R$ is $\left[0, \frac{\pi}{2}\right)$, then square of sum of all possible values of b will be

(a) 0 (b) 18
(c) 72 (d) None of these

Inverse Trigonometric Functions Exercise 2 :
More than One Correct Type Questions

71. Let $\theta = \tan^{-1}\left(\tan\frac{5\pi}{4}\right)$ and $\phi = \tan^{-1}\left(-\tan\frac{2\pi}{3}\right)$ then

(a) $\theta > \phi$ (b) $4\theta - 3\phi = 0$

(c) $\theta + \phi = \frac{7\pi}{12}$ (d) None of these

72. Let $f(x) = e^{\cos^{-1}\sin(x+\pi/3)}$ then

(a) $f\left(\frac{8\pi}{9}\right) = e^{5\pi/18}$ (b) $f\left(\frac{8\pi}{9}\right) = e^{13\pi/18}$

(c) $f\left(-\frac{7\pi}{4}\right) = e^{\pi/12}$ (d) $f\left(-\frac{7\pi}{4}\right) = e^{11\pi/12}$

73. If the numerical value of $\tan\left\{\cos^{-1}\left(\frac{4}{5}\right)+\tan^{-1}\left(\frac{2}{3}\right)\right\}$ is $\frac{a}{b}$, HCF $(a, b) = 1$, then

(a) $a + b = 23$ (b) $a - b = 11$
(c) $3b = a + 1$ (d) $2a = 3b$

74. Let $f(x)=\sin^{-1} x+\cos^{-1} x$. Then $\frac{\pi}{2}$ is equal to

(a) $f\left(-\frac{1}{2}\right)$ (b) $f(k^2 - 2k + 3), k \in R$
(c) $f\left(\frac{1}{1+k^2}\right), k \in R$ (d) $f(-2)$

75. $\cos^{-1} x = \tan^{-1} x$, then

(a) $x^2 = \frac{\sqrt{5}-1}{2}$ (b) $x^2 = \frac{\sqrt{5}+1}{2}$
(c) $\sin(\cos^{-1} x) = \frac{\sqrt{5}-1}{2}$ (d) $\tan(\cos^{-1} x) = \frac{\sqrt{5}-1}{2}$

76. The value(s) of x satisfying the equation $\sin^{-1}|\sin x| = \sqrt{\sin^{-1}|\sin x|}$ is/are given by (n is any integer)

(a) $n\pi - 1$ (b) $n\pi$
(c) $n\pi + 1$ (d) $2n\pi + 1$

77. If $(\sin^{-1} x+\sin^{-1} w)(\sin^{-1} y+\sin^{-1} z)=\pi^2$, then

$$D=\begin{vmatrix} x^{N_1} & y^{N_2} \\ z^{N_3} & w^{N_4} \end{vmatrix} (N_1, N_2, N_3, N_4 \in N)$$

(a) has a maximum value of 2
(b) has a minimum value of 0
(c) 16 different D are possible
(d) has a minimum value of -2

78. Indicate the relation which can hold in their respective domain for infinite values of x.

(a) $\tan|\tan^{-1} x| = |x|$
(b) $\cot|\cot^{-1} x| = |x|$
(c) $\tan^{-1}|\tan x| = |x|$
(d) $\sin|\sin^{-1} x| = |x|$

79. To the equation $2^{2\pi/\cos^{-1} x} - \left(a+\frac{1}{2}\right) . 2^{\pi/\cos^{-1} x} - a^2 = 0$ has only one real root, then

(a) $1 \le a \le 3$ (b) $a \ge 1$
(c) $a \le -3$ (d) $a \ge 3$

80. $\sin^{-1}(\sin 3)+\sin^{-1}(\sin 4)+\sin^{-1}(\sin 5)$ when simplified reduces to

(a) an irrational number
(b) a rational number
(c) an even prime
(d) a negative integer

81. If $\cos^{-1} x + \cos^{-1} y + \cos^{-1} z = \pi$, then

(a) $x^2 + y^2 + z^2 + 2xyz = 1$
(b) $2(\sin^{-1} x + \sin^{-1} y + \sin^{-1} z) = \cos^{-1} x + \cos^{-1} y + \cos^{-1} z$
(c) $xy + yz + zx = x + y + z - 1$
(d) $\left(x+\frac{1}{x}\right)+\left(y+\frac{1}{y}\right)+\left(z+\frac{1}{z}\right) \ge 6$

82. $2\tan(\tan^{-1}(x)+\tan^{-1}(x^3))$ where $x \in R - \{-1, 1\}$ is equal to

(a) $\frac{2x}{1-x^2}$
(b) $\tan(2\tan^{-1} x)$
(c) $\tan(\cot^{-1}(-x) - \cot^{-1}(x))$
(d) $\tan(2\cot^{-1} x)$

83. Let $f(x)=\sin^{-1}|\sin x|+\cos^{-1}(\cos x)$. Which of the following statement(s) is/are TRUE?

(a) $f(f(3)) = \pi$
(b) $f(x)$ is periodic with fundamental period 2π
(c) $f(x)$ is neither even nor odd
(d) Range of $f(x)$ is $[0, 2\pi]$

84. If $f(x)=\sin^{-1} x . \cos^{-1} x . \tan^{-1} x . \cot^{-1} x . \sec^{-1} x \text{ cosec}^{-1} x$, then which of the following statement(s) hold(s) good?

(a) The graph of $y = f(x)$ does not lie above x axis
(b) The non-negative difference between maximum minimum value of the function $y = f(x)$ is $\frac{3\pi^6}{64}$
(c) The function $y = f(x)$ is not injective.
(d) Number of non-negative integers in the domain of $f(x)$ is two.

85. Let $\alpha = 3\cos^{-1}\left(\frac{5}{\sqrt{28}}\right)+3\tan^{-1}\left(\frac{\sqrt{3}}{2}\right)$ and $\beta = 4\sin^{-1}\left(\frac{7\sqrt{2}}{10}\right) - 4\tan^{-1}\left(\frac{3}{4}\right)$, then which of the following does not hold(s) good?

(a) $\alpha < \pi$ but $\beta > \pi$
(b) $\alpha > \pi$ but $\beta < \pi$
(c) Both α and β are equal
(d) $\cos(\alpha + \beta) = 0$

86. Let function $f(x)$ be defined as $f(x)=|\sin^{-1} x|+\cos^{-1}\left(\frac{1}{x}\right)$. Then which of the following is/are TRUE.

(a) $f(x)$ is injective in its domain.
(b) $f(x)$ is many-one in its domain.
(c) Range of f is singleton set
(d) $\text{sgn}(f(x)) = 1$, where sgn x denotes signum function of x.

87. Which of the following pairs(s) of function is(are) identical?

(a) $f(x) = \sin(\tan^{-1} x)$, $g(x) = \dfrac{x}{\sqrt{1 + x^2}}$

(b) $f(x) = \text{sgn}(\cot^{-1} x)$, $g(x) = \sec^2 x - \tan^2 x$, where sgn x denotes signum function of x.

(c) $f(x) = e^{\ln\left(\cos^{-1}\left(\frac{x^2-1}{x^2+1}\right)\right)}$, $g(x) = \cos^{-1}\left(\dfrac{x^2-1}{x^2+1}\right)$

(d) $f(x) = \sin^{-1}\left(\dfrac{2x}{1+x^2}\right)$, $g(x) = 2\tan^{-1} x$

88. The value of $\sum_{n=1}^{\infty} \cot^{-1}(n^2 + n + 1)$ is also equal to

(a) $\cot^{-1}(-1) + \sec^{-1}(1) - \text{cosec}^{-1}(1)$
(b) $\cot^{-1}(2) + \cot^{-1}(3)$
(c) minimum value of the function $f(x) = \tan^{-1}\left(\dfrac{1-x^2}{1+x^2}\right)$
(d) $\cos^{-1}\left(\cos 41\dfrac{\pi}{4}\right)$

89. Let $f : I - \{-1, 0, 1\} \to [-\pi, \pi]$ be defined as $f(x) = 2\tan^{-1} x - \tan^{-1}\left(\dfrac{2x}{1-x^2}\right)$, then which of the following statements(s) is (are) correct?

(a) $f(x)$ is bijective
(b) $f(x)$ is injective but not surjective
(c) $f(x)$ is neither injective nor surjective
(d) $f(x)$ is an odd function

[**Note** : I denotes the set of integers]

90. If $\log x = \dfrac{-1}{3}$, $\log y = \dfrac{2}{5}$ and $P = \log(\sin(\text{arc}\cos\sqrt{1-x^2}))$,

$Q = \log\left(\cos\left(\text{arc}\tan\dfrac{\sqrt{1-x^2y^2}}{xy}\right)\right)$, then

(a) $P = \dfrac{-1}{9}$ (b) $P + Q = \dfrac{-4}{15}$
(c) $P - Q = \dfrac{-2}{5}$ (d) $\dfrac{P}{Q} = -5$

Inverse Trigonometric Functions Exercise 3 : Passage Based Questions

Passage I

(Q. Nos. 91 to 93)

Let S denotes the set consisting of four functions and $S = \{[x], \sin^{-1} x, |x|, \{x\}\}$ where, $\{x\}$ denotes fractional part and $[x]$ denotes greatest integer function. Let A, B, C are subsets of S.

Suppose

A : consists of odd function(s)

B : consists of discontinuous function(s)

and C : consists of non-decreasing function(s) or increasing function(s).

If $f(x) \in A \cap C$; $g(x) \in B \cap C$; $h(x) \in B$ but not C and $l(x) \in$ neither A nor B nor C.

Then, answer the following.

91. The function $l(x)$ is

(a) periodic
(b) even
(c) odd
(d) neither odd nor even

92. The range of $g(f(x))$ is

(a) $\{-1, 0, 1\}$ (b) $\{-1, 0\}$
(c) $\{0, 1\}$ (d) $\{-2, -1, 0, 1\}$

93. The range of $f(h(x))$ is

(a) $\left(0, \dfrac{\pi}{2}\right)$ (b) $\left[0, \dfrac{\pi}{2}\right)$
(c) $\left(0, \dfrac{\pi}{2}\right]$ (d) $\left[0, \dfrac{\pi}{2}\right]$

Passage II

(Q. Nos. 94 to 96)

Let f be a real-valued function defined on R (the set of real numbers) such that $f(x) = \sin^{-1}(\sin x) + \cos^{-1}(\cos x)$

94. The value of $f(10)$ is equal to

(a) $6\pi - 20$ (b) $7\pi - 20$
(c) $20 - 7\pi$ (d) $20 - 6\pi$

95. The area bounded by curve $y = f(x)$ and x-axis from $\dfrac{\pi}{2} \le x \le \pi$ is equal to

(a) $\dfrac{\pi^2}{4}$ (b) $\dfrac{\pi^2}{2}$ (c) π^2 (d) $\dfrac{\pi^2}{8}$

96. Number of values of x in interval $(0, 3)$ so that $f(x)$ is an integer, is equal to

(a) 1 (b) 2
(c) 3 (d) 0

Passage III

(Q. Nos. 97 to 98)

Consider a real-valued function

$$f(x)=\sqrt{\sin^{-1}x+2}+\sqrt{1-\sin^{-1}x}$$

97. The domain of definition of $f(x)$ is

(a) $[-1,1]$ (b) $[\sin 1,1]$ (c) $[-1,\sin 1]$ (d) $[-1,0]$

98. The range of $f(x)$ is

(a) $[0,\sqrt{3}]$ (b) $[1,\sqrt{3}]$ (c) $[1,\sqrt{6}]$ (d) $[\sqrt{3},\sqrt{6}]$

Passage IV

(Q. Nos. 99 to 101)

Given that,

$$\tan^{-1}\left(\frac{2x}{1-x^2}\right)=\begin{cases}2\tan^{-1}x, |x|\le 1\\ -\pi+2\tan^{-1}x, x>1\\ \pi+2\tan^{-1}x, x<-1\end{cases}$$

$$\sin^{-1}\left(\frac{2x}{1+x^2}\right)=\begin{cases}2\tan^{-1}x, |x|\le 1\\ \pi-2\tan^{-1}x,\ x>1 \text{ and}\\ -(\pi+2\tan^{-1}x), x<-1\end{cases}$$

$\sin^{-1}x+\cos^{-1}x=\pi/2$ for $-1\le x\le 1$

99. $\sin^{-1}\left(\frac{4x}{x^2+4}\right)+2\tan^{-1}\left(-\frac{x}{2}\right)$ is independent of x, then

(a) $x\in[-3,4]$ (b) $x\in[-2,2]$
(c) $x\in[-1,1]$ (d) $x\in[1,\infty)$

100. If $\cos^{-1}\frac{6x}{1+9x^2}=-\frac{\pi}{2}+2\tan^{-1}3x$, then $x\in$

(a) $\left(\frac{1}{3},\infty\right)$ (b) $(-1,\infty)$
(c) $(-\infty, m-1)$ (d) None of these

101. If $(x-1)(x^2+1)>0$, then $\sin\left(\frac{1}{2}\tan^{-1}\frac{2x}{1-x^2}\tan^{-1}x\right)$ is equal to

(a) 1 (b) $\frac{1}{\sqrt{2}}$
(c) -1 (d) None of these

Passage V

(Q. Nos. 102 to 104)

For $x, y, z, t\in R$, $\sin^{-1}x+\cos^{-1}y+\sec^{-1}z\ge t^2-\sqrt{2\pi}t+3\pi$

102. The value of $x+y+z$ is equal to

(a) 1 (b) 0
(c) 2 (d) -1

103. The principal value of $\cos^{-1}(\cos 5t^2)$ is

(a) $\frac{3\pi}{2}$ (b) $\frac{\pi}{2}$
(c) $\frac{\pi}{3}$ (d) $\frac{2\pi}{3}$

104. The value of $\cos^{-1}(\min\{x, y, z\})$ is

(a) 0 (b) $\frac{\pi}{2}$ (c) π (d) $\frac{\pi}{3}$

Inverse Trigonometric Functions Exercise 4 :
Single Integer Answer Type Questions

105. Let $f(x)=\tan^{-1}\left(\frac{(x-2)}{x^2+2x+2}\right)$, then $26f'(1)$ is

106. Let $f(x)=(\text{arc}\tan x)^3+(\text{arc}\cot x)^3$. If the range of $f(x)$ is $[a,b)$, then find the value of $\frac{b}{7a}$.

107. If $\sum_{n=0}^{\infty} 2\,\text{arc}\cot\left(\frac{n^2+n+4}{2}\right)=k\pi$, then find the value of k.

108. Find the number of solutions of the equation $\tan\left(\sum_{r=1}^{5}\cot^{-1}(2r^2)\right)=\frac{5x+6}{6x+5}$.

109. $\lim_{z\to 0}\left[\{\max(\sin^{-1}x+\cos^{-1}x)^2, \min(x^2+4x+7)\}\cdot\frac{\sin^{-1}z}{z}\right]$ is equal to (where [.] denotes greatest integer function)

110. If $\sin(30°+\text{arc}\tan x)=\frac{13}{14}$ and $0<x<1$, the value of x is $\frac{a\sqrt{3}}{b}$, where a and b are positive integers with no common factors. Find the value of $\left(\frac{a+b}{2}\right)$.

111. Let $f:R\to\left(0,\frac{2\pi}{3}\right]$ defined as $f(x)=\cot^{-1}(x^2-4x+\alpha)$.

Find the smallest integral value of α such that $f(x)$ is into function.

112. Let L denotes the number of subjective functions $f: A \to B$, where set A contains 4 elements and set B contains 3 elements, M denotes number of elements in the range of the function $f(x) = \sec^{-1}(\text{sgn}\, x) + \text{cosec}^{-1}(\text{sgn}\, x)$, where sgn x denotes signum function of x and N denotes coefficient of t^5 in $(1+t^2)^5\,(1+t^3)^8$.

Find the value of $(N - LM)$

113. Number of solution(s) of the equations $\cos^{-1}(\cos x) = x^2$ is

114. If $\cos^{-1}(x) + \cos^{-1}(y) + \cos^{-1}(z) = \pi(\sec^2(u) + \sec^4(v) + \sec^6(w))$, where u, v, w are least non-negative angles such that $u < v < w$, then the value of $x^{2000} + y^{2002} + z^{2004} + \dfrac{36\pi}{u+v+w}$ is ....

115. Let $f(x) = \cos(\tan^{-1}(\sin(\cot^{-1} x)))$. The simplest form of $f(x)$ can be written as $\left(\dfrac{x^2 + A}{x^2 + B}\right)^{1/2}$. Then the value of $(A + B)$ is ....

116. Find the value of λ for which

$$\tan^{-1}\left\{\frac{\cos 2\alpha \sec 2\beta + \cos 2\beta \sec 2\alpha}{\lambda}\right\}$$
$$= \tan^{-1}\{\tan^2 \alpha + \beta) \tan^2(\alpha - \beta)\} + \tan^{-1} 1$$

117. The least value of n for which $(n-2)x^2 + 8x + n + 4 > \sin^{-1}(\sin 12) + \cos^{-1}(\cos 12)$, $\forall x \in R$, where $n \in N$, is ....

118. If $0 < \cos^{-1} x < 1$ and $1 + \sin(\cos^{-1} x) + \sin^2(\cos^{-1} x) + \sin^3(\cos^{-1} x) + \ldots \infty = 2$, then the value of $12x^2$ is ....

119. The number of real solutions of the equation $\sqrt{1 + \cos 2x} = \sqrt{2} \sin^{-1}(\sin x)$, $-\pi \le x \le \pi$, is

Inverse Trigonometric Functions Exercise 5 :
Statement I and II Type Questions

- This section contain 6 questions. Each question contains **Statement I** (Assertion) and **Statement II** (Reason). Each question has 4 choices (a), (b), (c) and (d) out of which only one is correct. The choices are
  (a) Statement I is True; Statement II is True; Statement II is a correct explanation for Statement I;
  (b) Statement I is True; Statement II is True; Statement II is NOT a correct explanation for Statement I;
  (c) Statement I is True; Statement II is False;
  (d) Statement I is False; Statement II is True.

120. Statement I $y = \tan^{-1}(\tan x)$ and $y = \cos^{-1}(\cos x)$ does not have any solution, if $x \in \left(\dfrac{\pi}{2}, \dfrac{3\pi}{2}\right)$

Statement II $y = \tan^{-1}(\tan x) = x - \pi$, $x \in \left(\dfrac{\pi}{2}, \dfrac{3\pi}{2}\right)$ and

$$y = \cos^{-1}(\cos x) = \begin{cases} 2\pi - x, & x \in \left[\pi, \dfrac{3\pi}{2}\right] \\ x, & x \in \left[\dfrac{\pi}{2}, \pi\right] \end{cases}$$

121. Statement I $\sin^{-1}\left(\dfrac{1}{\sqrt{e}}\right) > \tan^{-1}\left(\dfrac{1}{\sqrt{\pi}}\right)$

Statement II $\sin^{-1} x > \tan^{-1} y$ for $x > y$, $\forall\, x, y \in (0, 1)$

122. Statement I $\text{cosec}^{-1}\left(\dfrac{1}{2} + \dfrac{1}{\sqrt{2}}\right) > \sec^{-1}\left(\dfrac{1}{2} + \dfrac{1}{\sqrt{2}}\right)$

Statement II $\text{cosec}^{-1} x > \sec^{-1} x$, if $1 \le x < \sqrt{2}$

123. Let $f(x) = \sin^{-1}\left(\dfrac{2x}{1+x^2}\right)$

Statement I $f'(2) = -\dfrac{2}{5}$ and

Statement II $\sin^{-1}\left(\dfrac{2x}{1+x^2}\right) = \pi - 2\tan^{-1} x, \forall\, x > 1$

124. Statement I $\sin^{-1} 2x + \sin^{-1} 3x = \dfrac{\pi}{3}$

$\Rightarrow x = \sqrt{\dfrac{3}{76}}$ only.

and

Statement II Sum of two negative angles cannot be positive.

125. Statement I Number of roots of the equation $\cot^{-1} x \cos^{-1} 2x + \pi = 0$ is zero.

Statement II Range of $\cot^{-1} x$ and $\cos^{-1} x$ is $(0, \pi)$ and $[0, \pi]$, respectively.

Inverse Trigonometric Functions Exercise 6 : Matching Type Questions

126. Let $t_1 = (\sin^{-1} x)^{\sin^{-1} x}, t_2 = (\sin^{-1} x)^{\cos^{-1} x}$,

$t_3 = (\cos^{-1} x)^{\sin^{-1} x}, t_4 = (\cos^{-1} x)^{\cos^{-1} x}$,

Match the following items of **Column I** with **Column II**

	Column I		Column II
A.	$x \in (0, \cos 1)$	(p)	$t_1 > t_2 > t_4 > t_3$
B.	$x \in \left(\cos 1, \frac{1}{\sqrt{2}}\right)$	(q)	$t_4 > t_3 > t_1 > t_2$
C.	$x \in \left(\frac{1}{\sqrt{2}}, \sin 1\right)$	(r)	$t_2 > t_1 > t_4 > t_3$
D.	$x \in (\sin 1, 1)$	(s)	$t_3 > t_4 > t_1 > t_2$

127. Match the following items of **Column I** with **Column II**

	Column I		Column II
A.	$\sin^{-1} x + x > 0$, for	(p)	$x < 0$
B.	$\cos^{-1} x - x \geq 0$, for	(q)	$x \in (0, 1]$
C.	$\tan^{-1} x + x < 0$, for	(r)	$x \in [-1, 0)$
D.	$\cot^{-1} x + x > 0$, for	(s)	$x > 0$

Inverse Trigonometric Functions Exercise 7 : Subjective Type Questions

128. Solve $\sin^{-1}\left(\frac{5}{x}\right) + \sin^{-1}\left(\frac{12}{x}\right) = \frac{\pi}{2}$

129. Solve $\tan^{-1}\left(\frac{x+1}{x-1}\right) + \tan^{-1}\left(\frac{x-1}{x}\right) = \tan^{-1}(-7)$

130. If a, b, c are positive, show that

$$\tan^{-1}\sqrt{\frac{a(a+b+c)}{bc}} + \tan^{-1}\sqrt{\frac{b(a+b+c)}{ac}} + \tan^{-1}\sqrt{\frac{c(a+b+c)}{ab}} = \pi$$

131. Find the sum to the n term of the series

$$\operatorname{cosec}^{-1}\sqrt{10} + \operatorname{cosec}^{-1}\sqrt{50} + \operatorname{cosec}^{-1}\sqrt{170} + \ldots + \operatorname{cosec}^{-1}\sqrt{(n^2+1)(n^2+2n+2)}$$

132. If $x_i \in [0, 1] \, \forall \, i = 1, 2, 3, \ldots, 28$ then find the maximum value of

$$\sqrt{\sin^{-1} x_1}\sqrt{\cos^{-1} x_2} + \sqrt{\sin^{-1} x_2}\sqrt{\cos^{-1} x_3} + \sqrt{\sin^{-1} x_3}\sqrt{\cos^{-1} x_4} + \ldots + \sqrt{\sin^{-1} x_{28}}\sqrt{\cos^{-1} x_1}$$

133. Find the value of $\sum_{r=1}^{10}\sum_{s=1}^{10} \tan^{-1}\left(\frac{r}{s}\right)$.

134. Find the value

$$\lim_{n \to \infty} \sum_{k=2}^{n} \cos^{-1}\left(\frac{1+\sqrt{(k-1)k(k+1)(k+2)}}{k(k+1)}\right)$$

135. If $\frac{m\tan(\alpha-\theta)}{\cos^2\theta} = \frac{n\tan\theta}{\cos^2(\alpha-\theta)}$

Prove that : $\theta = \frac{1}{2}\left[\alpha - \tan^{-1}\left(\frac{n-m}{n+m}\tan\alpha\right)\right]$

136. Prove that :

$\tan^{-1}(e^{i\theta}) = \frac{n\pi}{2} + \frac{\pi}{4} - \frac{i}{2}\ln\tan\left(\frac{\pi}{4} - \frac{\theta}{2}\right)$, where '$n$' is an integer.

137. If the quadratic equation;

$4^{\sec^2\alpha}x^2 + 2x + \left(\beta^2 - \beta + \frac{1}{2}\right) = 0$ have meal roots, then find all the possible value of $\cos\alpha + \cos^{-1}\beta$.

Inverse Trigonometric Functions Exercise 8 : Questions Asked in Previous 10 Years Exam

(i) JEE Advanced & IIT-JEE

138. If $\alpha = 3\sin^{-1}\left(\frac{6}{11}\right)$ and $\beta = 3\cos^{-1}\left(\frac{4}{9}\right)$, where the inverse trigonometric functions take only the principal values, then the correct option(s) is/are

[More than one correct option, JEE 2015 Adv.]

(a) $\cos\beta > 0$ (b) $\sin\beta < 0$
(c) $\cos(\alpha+\beta) > 0$ (d) $\cos\alpha < 0$

139. If $0 < x < 1$, then $\sqrt{1+x^2}\,[\{x\cos(\cot^{-1}x) + \sin(\cot^{-1}x)\}^2 - 1]^{1/2}$ is equal to

[Single correct option, IIT-JEE 2008 3M]

(a) $\frac{x}{\sqrt{1+x^2}}$ (b) x
(c) $x\sqrt{1+x^2}$ (d) $\sqrt{1+x^2}$

(ii) JEE Main/AIEEE

140. If $\tan^{-1}y = \tan^{-1}x + \tan^{-1}\left(\frac{2x}{1-x^2}\right)$, where $|x| < \frac{1}{\sqrt{3}}$.

Then, the value of y is **[2015 JEE Main]**

(a) $\frac{3x - x^3}{1 - 3x^2}$ (b) $\frac{3x + x^3}{1 - 3x^2}$
(c) $\frac{3x - x^3}{1 + 3x^2}$ (d) $\frac{3x + x^3}{1 + 3x^2}$

141. The value of $\cot\left\{\sum_{n=1}^{23}\cot^{-1}\left(1 + \sum_{k=1}^{n} 2k\right)\right\}$ is **[2013 JEE Main]**

(a) $\frac{23}{25}$ (b) $\frac{25}{23}$ (c) $\frac{23}{24}$ (d) $\frac{24}{23}$

142. If x, y and z are in AP and $\tan^{-1}x$, $\tan^{-1}y$ and $\tan^{-1}z$ are also in AP, then **[2013 JEE Main]**

(a) $x = y = z$ (b) $2x = 3y = 6z$
(c) $6x = 3y = 2z$ (d) $6x = 4y = 3z$

143. The value of $\cot\left(\operatorname{cosec}^{-1}\frac{5}{3} + \tan^{-1}\frac{2}{3}\right)$ is **[AIEEE 2008]**

(a) $\frac{5}{17}$ (b) $\frac{6}{17}$
(c) $\frac{3}{17}$ (d) $\frac{4}{17}$

144. If $\sin^{-1}\left(\frac{x}{5}\right) + \operatorname{cosec}^{-1}\left(\frac{5}{4}\right) = \frac{\pi}{2}$, then the value of x is **[2007 AIEEE]**

(a) 1 (b) 3
(c) 4 (d) 5

145. If $\cos^{-1}x - \cos^{-1}\frac{y}{2} = \alpha$, then $4x^2 - 4xy\cos\alpha + y^2$ is equal to **[2005 AIEEE]**

(a) $-4\sin^2\alpha$ (b) $4\sin^2\alpha$
(c) 4 (d) $2\sin 2\alpha$

Answers

Exercise for Session 1

1. 1/2 **2.** –1 **3.** $(-\infty, -3] \cup [-2, -1] \cup [0, \infty)$
4. R **5.** $(-\infty, -1] \cup [1, \infty)$

Exercise for Session 2

1. π **2.** $\frac{2\pi}{5}$ **3.** $\frac{-\pi}{10}$
4. $13 - 4\pi$
5. $\sin^{-1}(\sin\theta) = 3\pi - \theta$; $\cos^{-1}(\cos\theta) = \theta - 2\pi$
$\tan^{-1}(\tan\theta) = \theta - 3\pi$; $\cot^{-1}(\cot\theta) = \theta - 2\pi$

Exercise for Session 3

1. $\frac{\sqrt{3}}{2}$ **2.** not defined **3.** $\frac{59}{36}$ **4.** 11 **5.** $x = \pm 1$

Exercise for Session 4

1. $\frac{\pi}{8}$ **2.** $\left(\frac{1}{4} - \frac{\pi}{2}\right)$ **3.** $\frac{3}{2}$ **4.** $-\sqrt{3}$
5. 0 **6.** $\frac{3\pi}{10}$ **7.** $\frac{\pi}{2}$ **9.** $x = 1$ **10.** $x = \frac{1}{2}$

Exercise for Session 5

2. $\frac{\pi}{2}$ **3.** $\lambda = \frac{-19}{9}$ **5.** $x = \frac{\sqrt{5} + 4\sqrt{2}}{9}$ **6.** $x = \frac{1}{2}$

Exercise for Session 6

1. $\frac{4}{5}$ **2.** $\frac{65}{16}$ **3.** $\frac{9}{25}$ **4.** $\frac{2\sqrt{5}}{5}$

Exercise for Session 7

1. $y = \sin^{-1}(3x - 4x^3) = \begin{cases} 3\sin^{-1}x & ; -\frac{1}{2} \le x \le \frac{1}{2} \\ \pi - 3\sin^{-1}x & ; \frac{1}{2} < x \le 1 \\ -\pi - 3\sin^{-1}x & ; -1 \le x < -\frac{1}{2} \end{cases}$

Graph of $y = \sin^{-1}(3x - 4x^3)$

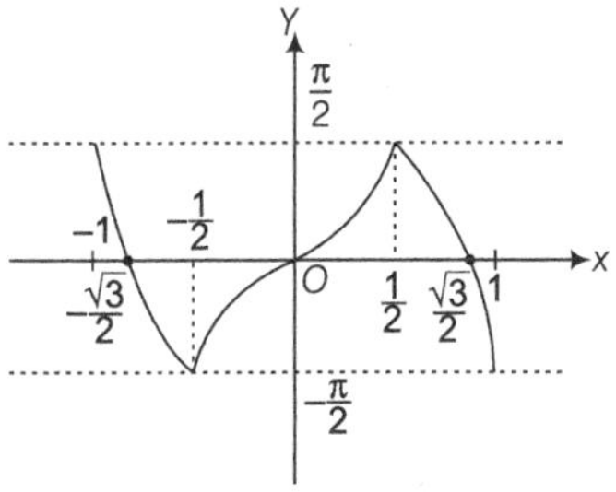

2. $y = \tan^{-1}\left(\dfrac{3x - x^3}{1 - 3x^2}\right) = \begin{cases} 3\tan^{-1}x & ; & -\dfrac{1}{\sqrt{3}} < x < \dfrac{1}{\sqrt{3}} \\ \pi + 3\tan^{-1}x & ; & -\infty < x < -\dfrac{1}{\sqrt{3}} \\ -\pi + 3\tan^{-1}x & ; & \dfrac{1}{\sqrt{3}} < x < \infty \end{cases}$

Graph of $y = \tan^{-1}\left(\dfrac{3x - x^3}{1 - 3x^2}\right)$

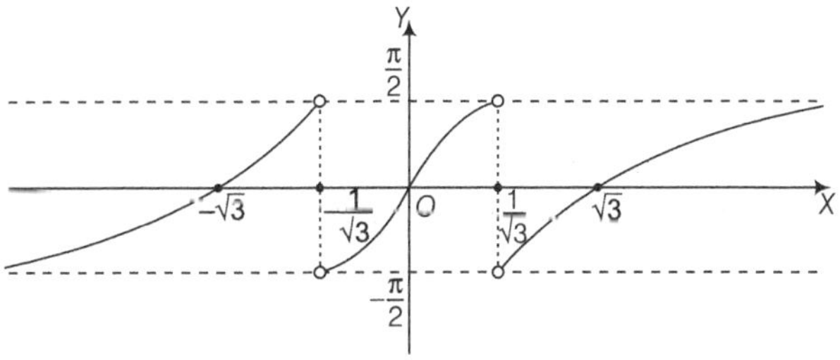

3.

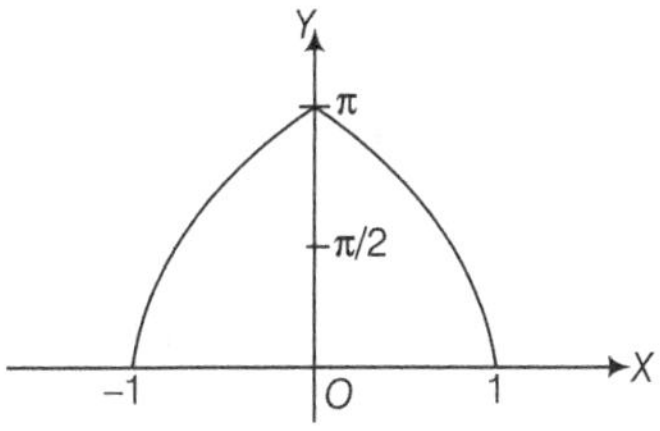

4.

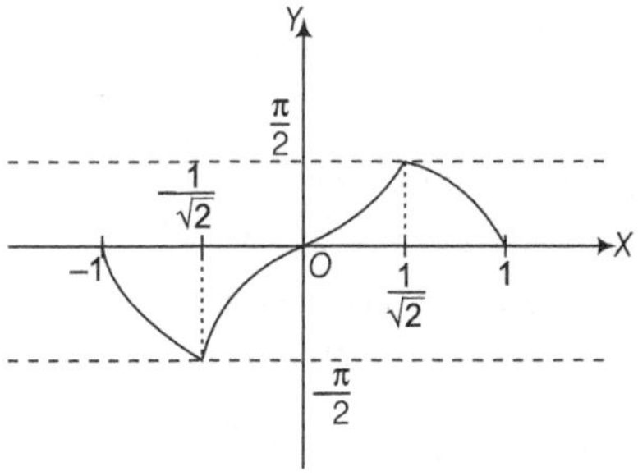

5.

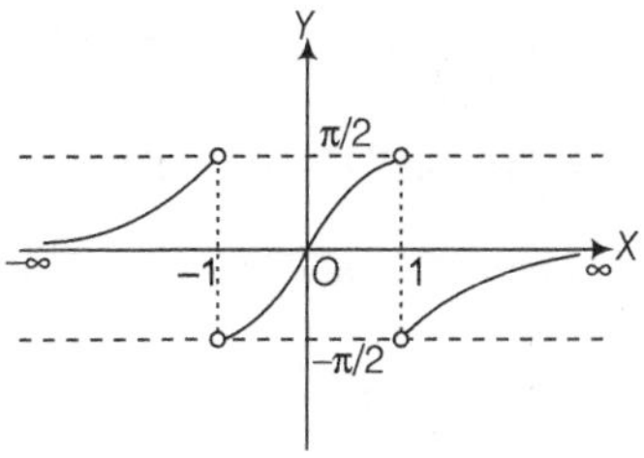

6.

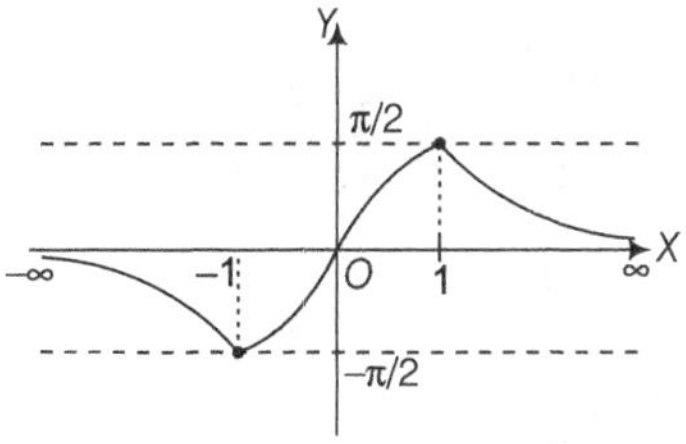

7.

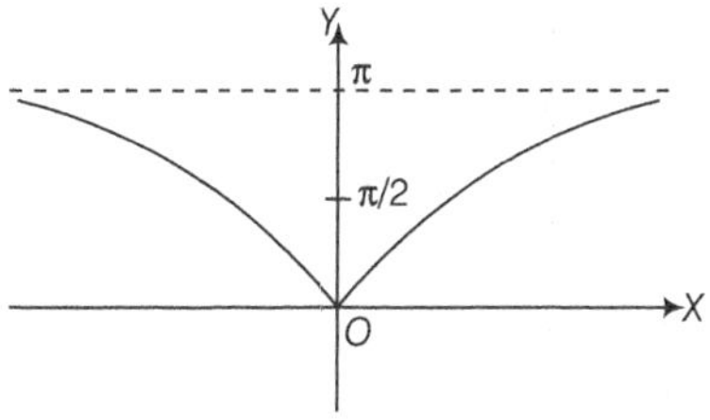

Chapter Exercises

1. (b)	2. (a)	3. (c)	4. (d)	5. (a)	6. (b)
7. (a)	8. (b)	9. (a)	10. (b)	11. (d)	12. (b)
13. (b)	14. (a)	15. (a)	16. (b)	17. (d)	18. (b)
19. (b)	20. (b)	21. (a)	22. (d)	23. (d)	24. (d)
25. (d)	26. (b)	27. (d)	28. (a)	29. (a)	30. (c)
31. (c)	32. (b)	33. (a)	34. (b)	35. (c)	36. (d)
37. (b)	38. (a)	39. (c)	40. (d)	41. (d)	42. (c)
43. (d)	44. (b)	45. (c)	46. (c)	47. (d)	48. (a)
49. (d)	50. (a)	51. (c)	52. (c)	53. (a)	54. (a)
55. (c)	56. (c)	57. (b)	58. (b)	59. (a)	60. (c)
61. (c)	62. (c)	63. (d)	64. (c)	65. (c)	66. (c)
67. (c)	68. (d)	69. (d)	70. (a)	71. (b,c)	72. (b,c)
73. (a,b,c)	74. (a,c)		75. (a,c)	76. (a,b,c)	77. (a,c,d)
78. (a,b,c,d)		79. (b,c)	80. (b,d)	81. (a,b)	
82. (a,b,c)		83. (a,b)	84. (a,b)		
85. (a,b,d)	86. (a,d)	87. (a,c)	88. (a,b,d)	89. (c,d)	90. (b,c,d)
91. (b)	92. (d)	93. (b)	94. (b)	95. (b)	96. (c)
97. (c)	98. (d)	99. (b)	100. (a)	101. (c)	102. (d)
103. (b)	104. (c)	105. (9)	106. (4)	107. (1)	108. (0)
109. (3)	110. (8)	111. (4)	112. (4)	113. (3)	114. (9)
115. (3)	116. (2)	117. (5)	118. (9)	119. (2)	120. (a)
121. (a)	122. (a)	123. (a)	124. (a)	125. (a)	

126. A → q; B → s; C → r; D → p;

127. A → q; B → r; C → p,r; D → q,r,s;

128. $x = 13$ 129. No Solution

131. $\tan^{-1}(n+1) - \dfrac{\pi}{4}$ 132. 7π 133. 25π 134. $\dfrac{\pi}{6}$

137. $\dfrac{\pi}{3} - 1$ and $\dfrac{\pi}{3} + 1$

138. (b,c,d)	139. (c)	140. (a)	141. (b)	142. (a)
143. (b)	144. (b)	145. (b)		

Solutions

1. Substituting $x^2 = \cos 2\theta$, we obtain

$$\cot^{-1}\left(\sqrt{\frac{1-x^2}{1+x^2}}\right) = \cot^{-1}\left(\sqrt{\frac{1-\cos 2\theta}{1+\cos 2\theta}}\right) = \cot^{-1}(\tan\theta)$$

$$= \cot^{-1}\left(\cot\left(\frac{\pi}{2}-\theta\right)\right)$$

$$= \frac{\pi}{2} - \theta$$

2. Let $\cos^{-1}\frac{1}{8} = \theta$, where $0 < \theta < \frac{\pi}{2}$. Then

$$\frac{1}{2}\cos^{-1}\frac{1}{8} = \frac{1}{2}\theta$$

$$\Rightarrow \quad \cos\left(\frac{1}{2}\cos^{-1}\frac{1}{8}\right) = \cos\frac{1}{2}\theta$$

Now $\quad \cos^{-1}\frac{1}{8} = \theta$

$$\Rightarrow \quad \cos\theta = \frac{1}{8} \Rightarrow 2\cos^2\frac{\theta}{2} - 1 = \frac{1}{8}$$

$$\Rightarrow \quad \cos^2\frac{\theta}{2} = \frac{9}{16}$$

$$\Rightarrow \quad \cos\frac{\theta}{2} = \frac{3}{4}$$

$$\left[\because 0 < \frac{\theta}{2} < \frac{\pi}{4}, \text{ so } \cos\frac{\theta}{2} \neq -\frac{3}{4}\right]$$

3. $\sin^{-1}(\sin 5) > x^2 - 4x$

$$\Rightarrow \quad \sin^{-1}(\sin(5-2\pi) > x^2 - 4x$$

$$\Rightarrow \quad 5 - 2\pi > x^2 - 4x$$

$$\Rightarrow \quad 9 - 2x > (x-2)^2$$

$$\Rightarrow \quad (x-2)^2 < 9 - 2\pi$$

$$\Rightarrow \quad -\sqrt{9-2\pi} < x - 2 < \sqrt{9-2\pi}$$

$$\Rightarrow \quad 2 - \sqrt{9-2\pi} < x < 2 + \sqrt{9-2\pi}$$

4. We have, $\sin^{-1}\left\{\left(\sin\frac{\pi}{3}\right)\frac{x}{\sqrt{(x^2+k^2-kx)}}\right\}$

$$- \cos^{-1}\left\{\left(\cos\frac{\pi}{6}\right)\frac{x}{\sqrt{(x^2+k^2-kx)}}\right\}$$

$$= \sin^{-1}\left\{\frac{\sqrt{3}x}{2\sqrt{x^2+k^2-kx}}\right\} - \cos^{-1}\left\{\frac{\sqrt{3}x}{2\sqrt{x^2+k^2-kx}}\right\}$$

$$= \frac{\pi}{2} - 2\cos^{-1}\left\{\frac{\sqrt{3}x}{\sqrt{4x^2-4kx+4k^2}}\right\}$$

$$= \frac{\pi}{2} - \cos^{-1}\left\{\frac{6x^2}{4x^2-4kx+4k^2} - 1\right\}$$

$$= \sin^{-1}\left\{\frac{2x^2+4xk-4k^2}{4x^2-4xk+4k^2}\right\}$$

$$= \sin^{-1}\left\{\frac{x^2+2xk-2k^2}{2x^2-2xk+2k^2}\right\}$$

5. $0 \le x \le 1 \Rightarrow 0 \le \tan^{-1}x \le \frac{\pi}{4}$

$$\Rightarrow \quad \frac{-\pi}{4} \le -\tan^{-1}x \le 0$$

$$\Rightarrow \quad 0 \le \frac{\pi}{4} - \tan^{-1}x \le \frac{\pi}{4}$$

6. $T_n = \cot^{-1}\left(n^2 + \frac{3}{4}\right) = \tan^{-1}\left(\frac{1}{n^2+\frac{3}{4}}\right)$

$$= \tan^{-1}\left(n+\frac{1}{2}\right) - \tan^{-1}\left(n-\frac{1}{2}\right)$$

$$S_n = \sum_{n=1}^{n} t_n = \tan^{-1}\left(n+\frac{1}{2}\right) - \tan^{-1}\frac{1}{2}$$

$$S_\infty = \tan^{-1}(\infty) - \tan^{-1}\frac{1}{2} = \frac{\pi}{2} - \tan^{-1}\frac{1}{2} = \cot^{-1}\frac{1}{2} = \tan^{-1}2$$

7. We have, $\cot^{-1}x + \sin^{-1}\frac{1}{\sqrt{5}} = \frac{\pi}{4}$

$$\Rightarrow \quad \tan^{-1}\frac{1}{x} + \tan^{-1}\frac{1/\sqrt{5}}{\sqrt{1-\frac{1}{5}}} = \frac{\pi}{4}$$

$$\Rightarrow \quad \tan^{-1}\frac{1}{x} + \tan^{-1}\frac{1}{2} = \tan^{-1}1$$

$$\Rightarrow \quad \tan^{-1}\frac{1}{x} = \tan^{-1}1 - \tan^{-1}\frac{1}{2}$$

$$\Rightarrow \quad \tan^{-1}\frac{1}{x} = \tan^{-1}\left(\frac{1-\frac{1}{2}}{1+1\cdot\frac{1}{2}}\right)$$

$$\Rightarrow \quad \tan^{-1}\frac{1}{x} = \tan^{-1}\left(\frac{1}{3}\right)$$

$$\Rightarrow \quad x = 3$$

8. $(\cot^{-1}x)^2 - (5\cot^{-1}x) + 6 > 0$

$$(\cot^{-1}x - 3)(\cot^{-1}x - 2) > 0$$

$$x > \cot 2 \text{ and } x < \cot 3$$

$$\therefore \quad x \in (-\infty, \cot 3) \cup (\cot 2, \infty)$$

9. $\lim_{n\to\infty}\sum_{r=1}^{n}\tan^{-1}\left(\frac{2^{r-1}}{1+2^{2r-1}}\right) = \lim_{n\to\infty}\sum_{r=1}^{n}\tan^{-1}\left(\frac{2^r - 2^{r-1}}{1+2^r\cdot 2^{r-1}}\right)$

$$= \lim_{n\to\infty}\sum_{r=1}^{n}\{\tan^{-1}(2^r) - \tan^{-1}(2^{r-1})\}$$

$$= \lim_{n\to\infty}(\tan^{-1}2^n - \tan^{-1}2^\circ)$$

$$= \tan^{-1}2^\infty - \tan^{-1}1$$

$$= \frac{\pi}{2} - \frac{\pi}{4} = \frac{\pi}{4}$$

10. $x + \frac{1}{x} = 2$

$$\left(\sqrt{x} - \frac{1}{\sqrt{x}}\right) = 0$$

$$\frac{x-1}{\sqrt{x}} = 0$$

$$x = 1$$

$$\therefore \quad \sin^{-1} x = \frac{\pi}{2}$$

11. $\tan^{-1}\left(\frac{\tan x}{4}\right) + \tan^{-1}\left(\frac{\frac{6\tan x}{1+\tan^2 x}}{5 + \frac{3(1-\tan^2 x)}{1+\tan^2 x}}\right)$

$$\Rightarrow \tan^{-1}\left(\frac{\tan x}{4}\right) + \tan^{-1}\left(\frac{3\tan x}{4+\tan^2 x}\right)$$

$$= \tan^{-1}\left(\frac{\frac{\tan x}{4} + \frac{3\tan x}{4+\tan^2 x}}{1 - \frac{3\tan^2 x}{4(4+\tan^2 x)}}\right)$$

$$= \tan^{-1}\left(\frac{16\tan x + \tan^3 x}{16 + \tan^2 x}\right)$$

$$= \tan^{-1}(\tan x) = x$$

12. $\left(\frac{\pi}{2} - \cos^{-1} x\right) + \left(\frac{\pi}{2} - \cos^{-1} y\right) = \frac{2\pi}{3}$

$$\Rightarrow \quad \cos^{-1} x + \cos^{-1} y = \frac{\pi}{3}$$

13. $\cot\left[\frac{\pi}{4} - \cot^{-1}\frac{3^2-1}{2\times 3}\right]$

$$= \cot\left[\frac{\pi}{4} - \cot^{-1}\frac{4}{3}\right]$$

$$= \frac{\cot\frac{\pi}{4}\cdot\frac{4}{3} + 1}{\frac{4}{3} - \cot\frac{\pi}{4}} = \frac{\frac{4}{3}+1}{\frac{4}{3}-1} = 7$$

14. $\sin\left[\tan^{-1}\frac{1-\tan^2\theta}{2\tan\theta} + \cos^{-1}\frac{1-\tan^2\theta}{1+\tan^2\theta}\right]$

[Put $x = \tan\theta$]

$$\Rightarrow \sin[\tan^{-1}(\cot 2\theta) + \cos^{-1}(\cos 2\theta)]$$

$$= \sin\left[\tan^{-1}\left(\tan\left(\frac{\pi}{2} - \theta\right)\right) + 2\theta\right]$$

$$= \sin\left(\frac{\pi}{2} - 2\theta + 2\theta\right) = 1$$

15. $2\tan^{-1} x = \cos^{-1}\frac{1-x^2}{1+x^2}$

$$\Rightarrow 2\tan^{-1} a - 2\tan^{-1} b = 2\tan^{-1} x$$

$$\Rightarrow \quad \frac{a-b}{1+ab} = x$$

16. $\frac{-\pi}{3} < \cos^{-1}\frac{1-x^2}{1+x^2} < \frac{\pi}{3}$

$$\Rightarrow \quad 0 \le \cos^{-1}\frac{1-x^2}{1+x^2} < \frac{\pi}{3}$$

$$\Rightarrow \quad \frac{1}{2} < \frac{1-x^2}{1+x^2} \le 1$$

$$\Rightarrow \quad 0 \le x^2 < \frac{1}{3}$$

$$\Rightarrow \quad x \in \left(\frac{-1}{\sqrt{3}} \cdot \frac{1}{\sqrt{3}}\right)$$

17. $\cos^{-1}\left[\cot\left(\sin^{-1}\sqrt{\frac{2-\sqrt{3}}{4}} + \cos^{-1}\frac{\sqrt{3}}{2} + \cos^{-1}\frac{1}{\sqrt{2}}\right)\right]$

$$\Rightarrow \cos^{-1}\left[\cot\left(\sin^{-1}\sqrt{\frac{2-\sqrt{3}}{4}} + \cos^{-1}\sqrt{\frac{2-\sqrt{3}}{4}}\right)\right]$$

$$= \cos^{-1}\left(\cot\frac{\pi}{2}\right) = \frac{\pi}{2}$$

18. $\tan\left(\tan^{-1}\frac{x}{\pi}\right) < \tan\frac{\pi}{3}$

$$\Rightarrow \quad \frac{x}{\pi} < \sqrt{3}$$

$\Rightarrow \quad x < \sqrt{3}\pi - 5.5$. Maximum value is 5.

19. $\frac{\sqrt{(1+x^2)} - \sqrt{(1-x^2)}}{\sqrt{(1+x^2)} + \sqrt{(1-x^2)}} = \frac{\tan\alpha}{1}$

Now, use 'C' and 'D' $\Rightarrow \frac{1+x^2}{1-x^2} = \frac{\tan^2\left(\frac{\pi}{4}+\alpha\right)}{1}$

Again, use 'C' and 'D' $= x^2 = -\cos\left(\frac{\pi}{2} + 2\alpha\right) = \sin 2\alpha$

20. $\tan^{-1} x + \cot^{-1} y = \tan^{-1} 3 \Rightarrow \cot^{-1} y = \tan^{-1} 3 - \tan^{-1} x$

$$\Rightarrow \quad \tan^{-1}\left(\frac{1}{y}\right) = \tan^{-1}\frac{3-x}{1+3x}$$

$$\Rightarrow \quad y = \frac{1+3x}{3-x}$$

Hence $x = 1, 2$, and $y = 2 \cdot 7$

21. $\text{cosec}^{-1}(\text{cosec } x) = x \;\forall\; x \in R = -(-1, 1)$

Also, range of $\text{cosec}^{-1}(\text{cosec } x) \in \left[-\frac{\pi}{2}, 0\right) \cup \left(0, \frac{\pi}{2}\right]$

So, combining these two, we get

$$x \in \left[-\frac{\pi}{2}, -1\right] \cup \left[1, \frac{\pi}{2}\right]$$

22. $\lim_{|x|\to\infty} \cos(\tan^{-1}(\sin(\tan^{-1} x)))$

$$= \cos(\tan^{-1}(\sin(\tan^{-1}\infty))) = \cos\left(\tan^{-1}\left(\sin\left(\frac{\pi}{2}\right)\right)\right)$$

$$= \cos(\tan^{-1}(1)) = \cos\left(\frac{\pi}{4}\right) = \frac{1}{\sqrt{2}}$$

23. $[\cot^{-1} x] + 2[\tan^{-1} x] = 0$

$\Rightarrow \quad [\cot^{-1} x] = 0, [\tan^{-1} x] = 0$

or $\quad [\cot^{-1} x] = 2, [\tan^{-1} x] = -1$

Now, $\quad [\cot^{-1} x] = 0 \Rightarrow x \in (\cot 1, \infty)$

$[\tan^{-1} x] = 0 \Rightarrow x \in (0, \tan 1)$

Therefore, $\quad [\cot^{-1} x] = [\tan^{-1} x] = 0, x \in (\cot 1, \tan 1)$

$[\cot^{-1} x] = 2 \Rightarrow x \in (\cot 3, \cot 2]$

$[\tan^{-1} x] = -1 \Rightarrow x \in [-\tan 1, 0)$

Hence, no such x exists.

Thus, the solution set is $(\cot 1, \tan 1)$

24. Let $\sin^{-1} x = \theta \Rightarrow x = \sin\theta, \frac{\pi}{2} \le \theta \le \frac{3\pi}{2}$

Now, $\cos^{-1} x = \cos^{-1}(\sin\theta) = \cos^{-1}\left(-\cos\left(\frac{3\pi}{2} - \theta\right)\right)$

$= \pi - \cos^{-1}\left(\cos\left(\frac{3\pi}{2} - \theta\right)\right)$

$= \pi - \left(\frac{3\pi}{2} - \theta\right)$, as $0 \le \frac{3\pi}{2} - \theta \le \pi$

$= \theta - \frac{\pi}{2} = \sin^{-1} x - \frac{\pi}{2}$

Hence, $\sin^{-1} x + \cos^{-1} x = 2\sin^{-1} x - \frac{\pi}{2}$

25. $a\sin^{-1} x - b\cos^{-1} x = c$

We have, $b\sin^{-1} x + b\cos^{-1} x = \frac{b\pi}{2}$

Adding $(a + b)\sin^{-1} x = \frac{b\pi}{2} + c$

or $\quad \sin^{-1} x = \dfrac{\left(\frac{b\pi}{2}\right) + c}{a + b} = \dfrac{b\pi + 2c}{2(a + b)}$

$\therefore \quad \cos^{-1} x = \frac{\pi}{2} - \frac{b\pi + 2c}{2(a + b)} = \frac{\pi a - 2c}{2(a + b)}$

$\Rightarrow a\sin^{-1} x + b\cos^{-1} x = \dfrac{\pi ab + c(a - b)}{a + b}$

26. $\sin^{-1}|x - 2| + \cos^{-1}(1 - |3 - x|) = \frac{\pi}{2}$

or $\quad |x - 2| = 1 - |3 - x|$

$|x - 2| + |3 - x| = 1$

or $\quad |x - 2| + |3 - x| = |(x - 2) + (3 - x)|$

or $\quad (x - 2)(3 - x) \ge 0$

or $\quad 2 \le x \le 3$

27. According to the question,

$\sin^{-1}\frac{2}{\sqrt{5}} + \sin^{-1}\frac{3}{\sqrt{10}} + \sin^{-1}\alpha = \pi$

$\Rightarrow \quad \tan^{-1}2 + \tan^{-1}3 + \tan^{-1}\left(\frac{\alpha}{\sqrt{1 - \alpha^2}}\right) = \pi$

$\Rightarrow \quad \pi + \tan^{-1}\frac{2 + 3}{1 - (2)(3)} + \tan^{-1}\left(\frac{\alpha}{\sqrt{1 - \alpha^2}}\right) = \pi$

$\Rightarrow \quad \tan^{-1}\left(\frac{\alpha}{\sqrt{1 - \alpha^2}}\right) = \frac{\pi}{4}$

$\Rightarrow \quad \frac{\alpha}{\sqrt{1 - \alpha^2}} = 1 \Rightarrow \alpha = \frac{1}{\sqrt{2}}$

28. Expanding, we have

$(\tan^{-1} x)^3 + (\tan^{-1} 2x)^3 + (\tan^{-1} 3x)^3$

$= 3\tan^{-1} x \tan^{-1} 2x \tan^{-1} 3x$

$\Rightarrow \quad x = 0$

29. Let $f(x) = x^3 + bx^2 + cx + 1$

$f(0) = 1 > 0, f(-1) = b - c < 0$

So, $\quad \alpha \in (-1, 0)$

So, $2\tan^{-1}(\operatorname{cosec}\alpha) + \tan^{-1}(2\sin\alpha\sec^2\alpha)$

$= 2\tan^{-1}\left(\frac{1}{\sin\alpha}\right) + \tan^{-1}\left(\frac{2\sin\alpha}{1 - \sin^2\alpha}\right)$

$= 2\left[\tan^{-1}\left(\frac{1}{\sin\alpha}\right) + \tan^{-1}(\sin\alpha)\right]$

$= 2\left(-\frac{\pi}{2}\right) = -\pi \qquad$ (as $\sin\alpha < 0$)

30. Given, $u = \cot^{-1}\sqrt{\cos 2\theta} - \tan^{-1}\sqrt{\cos 2\theta}$,

Put $\quad \cos 2\theta = \tan^2\theta$

$\Rightarrow \quad u = \cot^{-1}(\tan\theta) - \tan^{-1}(\tan\theta)$

$= \frac{\pi}{2} - \theta - \theta = \frac{\pi}{2} - 2\theta$

$\Rightarrow \quad \sin u = \sin\left(\frac{\pi}{2} - 2\theta\right) = \cos 2\theta = \tan^2\theta$

31.

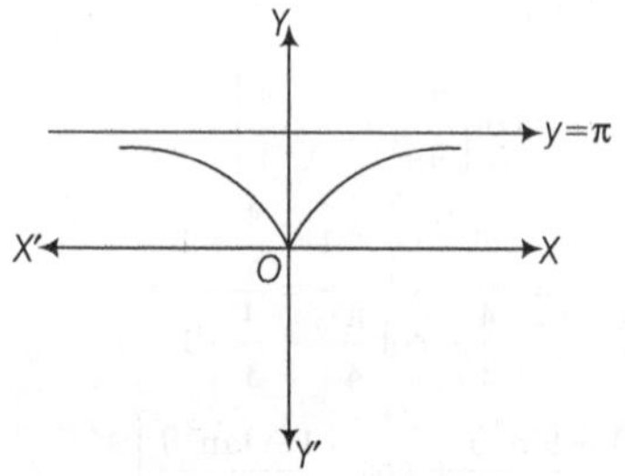

Thus, $f(x)$ is not differentiable at $x = 0$

32.

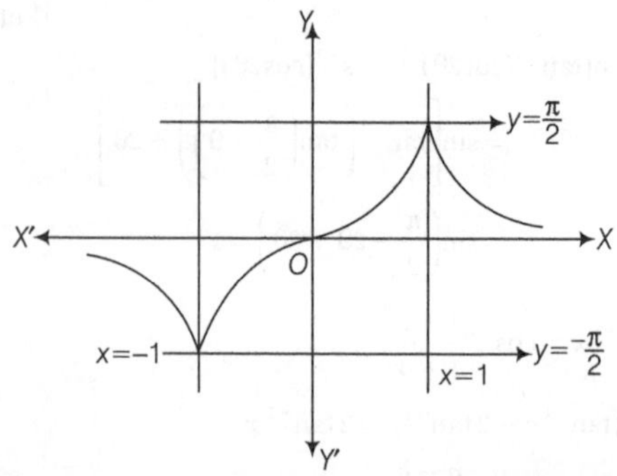

From the graph, it is clear that, $f(x)$ is not differentiable at $x = 1, -1$

33. Given, $f(x) = \tan^{-1}(x^2 - 18x + a) > 0$

$$\Rightarrow \quad \tan^{-1}(x^2 - 18x + a) > 0$$

$$\Rightarrow \quad x^2 - 18x + a > 0 \Rightarrow 18^2 - 4a < 0$$

$$\Rightarrow \quad a > \frac{18^2}{4} = \frac{18 \times 18}{4} = 81$$

$$\Rightarrow \quad a > 81 \Rightarrow a \in (81, \infty)$$

34. $f(x) = \sin^{-1}2x + \cos^{-1}2x + \sec^{-1}2x$

$$f(x) = \frac{\pi}{2} + \sec^{-1}2x$$

Graph of $\sec^{-1}2x$ is as following

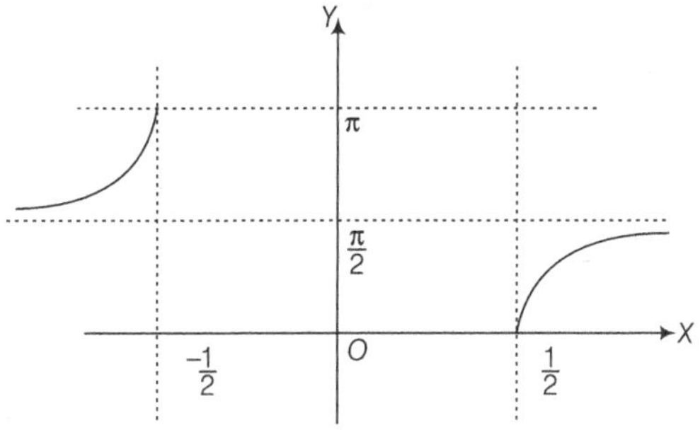

$$(f(x))_{\text{minimum}} = \frac{\pi}{2} + 0 = \frac{\pi}{2}$$

$$(f(x))_{\text{maximum}} = \frac{\pi}{2} + \pi = \frac{3\pi}{2}$$

$$\text{Sum} = \frac{\pi}{2} + \frac{3\pi}{2} = 2\pi$$

35. Given, $\tan^{-1}\dfrac{\left[\dfrac{b}{c+a} + \dfrac{c}{a+b}\right]}{1 - \dfrac{b}{c+a}\cdot\dfrac{c}{a+b}} = \dfrac{\pi}{4}$

$$\Rightarrow \quad \frac{ab + b^2 + c^2 + ac}{ac + bc + a^2 + ab - bc} = 1$$

$$\Rightarrow \quad ab + b^2 + c^2 + ac = ac + a^2 + ab$$

$$b^2 + c^2 = a^2$$

$\therefore \Delta ABC$ is right angled at A

36. By graph, clearly it has 5 real roots

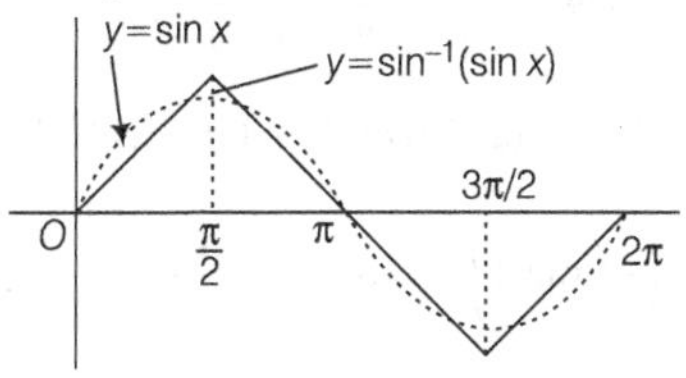

37. $-1 \le x \le 1, y > 0$

Both will be equated for infinite values of x and y. Therefore infinite many solutions.

38. $y = 1 + 2\sin\left(\dfrac{e^x}{e^x + 1}\right), x \ge 0$

$$\Rightarrow \quad y - 1 = 2\sin\left(\frac{e^x}{e^x+1}\right)$$

$$\Rightarrow \quad \frac{e^x}{e^x + 1} = \sin^{-1}\left(\frac{y-1}{2}\right)$$

$$\Rightarrow \quad e^x\left(\sin^{-1}\left(\frac{y-1}{2}\right) - 1\right) = -\sin^{-1}\left(\frac{y-1}{2}\right)$$

$$\Rightarrow \quad e^x = \frac{\sin^{-1}\left(\frac{y-1}{2}\right)}{1 - \sin^{-1}\left(\frac{y-1}{2}\right)}$$

$$x = \log\left(\frac{\sin^{-1}\left(\frac{y-1}{2}\right)}{1 - \sin^{-1}\left(\frac{y-1}{2}\right)}\right)$$

$$f^{-1}(x) = \log\left(\frac{\sin^{-1}\left(\frac{x-1}{2}\right)}{1 - \sin^{-1}\left(\frac{x-1}{2}\right)}\right)$$

39. $\cos^{-1}(\cos(2\cot^{-1}(\sqrt{2} - 1)))$

$$= \cos^{-1}(\cos(2(67\cdot 5°)))$$

$$= \cos^{-1}(\cos(135°))$$

$$= 135° = \frac{3\pi}{4}$$

40. $f(x) = \tan^{-1}\left(\dfrac{(\sqrt{12} - 2)x^2}{x^4 + 2x^2 + 3}\right) = \tan^{-1}\left(\dfrac{2(\sqrt{3} - 1)}{x^2 + \dfrac{3}{x^2} + 2}\right)$

As $x^2 + \dfrac{3}{x^2} \ge 2\sqrt{3}$ [using AM $\ge$ GM]

$$\Rightarrow \quad x^2 + \frac{3}{x^2} + 2 \ge 2 + 2\sqrt{3}$$

$$\therefore \quad (f(x))_{\max} = \tan^{-1}\left(\frac{2(\sqrt{3}-1)}{2(\sqrt{3}+1)}\right) = \frac{\pi}{12}$$

41. $\tan^{-1}\dfrac{\sqrt{1 + x^2} - 1}{x} = 4°$ $\qquad (x \ne 0)$

$$\Rightarrow \quad \frac{\sqrt{1+x^2} - 1}{x} = \tan 4°$$

$$\Rightarrow \quad \sqrt{1 + x^2} = 1 + x\tan 4°$$

$$\Rightarrow \quad 1 + x^2 = 2x\tan 4° + 1 + x^2\tan^2 4°$$

$$\Rightarrow \quad x = 0 \text{ or } x = \frac{2\tan 4°}{1 - \tan^2 4°} = \tan 8°$$

Since $x \ne 0$, we have $x = \tan 8°$

42. From the given equation $\sin^2\theta - 2\sin\theta + 3 = 5^{\sec^2 y} + 1$,

we get $\qquad (\sin\theta - 1)^2 + 2 = 5^{\sec^2 y} + 1$

LHS ≤ 6, RHS ≥ 6

Possible solution is $\sin\theta = -1$ when L.H.S. = R.H.S.

$$\Rightarrow \quad \cos^2\theta = 0$$

$$\Rightarrow \quad \cos^2\theta - \sin\theta = 1$$

43. $y = |x^2 - 1| = |\tan^{-1}|x||$

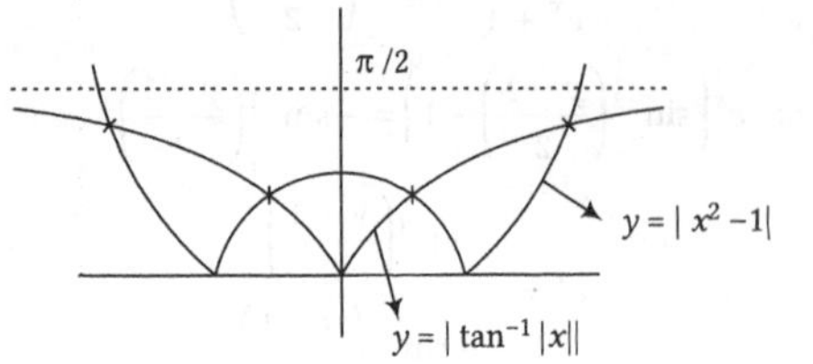

From graph, clearly it has 4 solutions.

44. $1 + \tan^2(\tan^{-1} x) - (\sec^2(\sec^{-1} x) - 1)$

$$= 1 + (\tan(\tan^{-1} x))^2 - (\sec(\sec^{-1} x))^2 + 1$$

$$= 1 + x^2 - x^2 + 1 = 2$$

45. The equation $3x^2 + 6x + a = 0$ must have equal roots

So, $D = 0$

$\Rightarrow \quad 36 - 12a = 0 \Rightarrow a = 3$

46. $\tan^{-1}\dfrac{1}{\sqrt{2}} + \sin^{-1}\dfrac{1}{\sqrt{5}} - \cos^{-1}\dfrac{1}{\sqrt{10}}$

$$= \tan^{-1}\frac{1}{\sqrt{2}} + \tan^{-1}\frac{1}{2} - \tan^{-1}3$$

$$= \tan^{-1}\frac{1}{\sqrt{2}} - \left[\tan^{-1}3 - \tan^{-1}\frac{1}{2}\right]$$

$$\Rightarrow \quad \tan^{-1}\frac{1}{\sqrt{2}} - \left[\tan^{-1}\frac{3 - \frac{1}{2}}{1 + \frac{3}{2}}\right] = \tan^{-1}\frac{1}{\sqrt{2}} - \tan^{-1}1$$

$$= -\left[\tan^{-1}1 - \tan^{-1}\frac{1}{\sqrt{2}}\right]$$

$$= -\left[\tan^{-1}\frac{1 - \frac{1}{\sqrt{2}}}{1 + \frac{1}{\sqrt{2}}}\right] = -\tan^{-1}\frac{\sqrt{2} - 1}{\sqrt{2} + 1}$$

$$= -\cot^{-1}\frac{\sqrt{2} + 1}{\sqrt{2} - 1} = -\left[\cot^{-1} - \left(\frac{1 + \sqrt{2}}{1 - \sqrt{2}}\right)\right]$$

$$= -\left[\pi - \cot^{-1}\frac{1 + \sqrt{2}}{1 - \sqrt{2}}\right] = -\pi + \cot^{-1}\frac{1 + \sqrt{2}}{1 - \sqrt{2}}$$

47. $\alpha = 2\tan^{-1}\dfrac{1}{2} + \tan^{-1}\dfrac{4}{3}$

$$= \tan^{-1}\frac{1}{1 - \frac{1}{4}} + \tan^{-1}\frac{4}{3} = 2\tan^{-1}\frac{4}{3}$$

$$\left(\text{using } 2\tan^{-1} x = \tan^{-1}\frac{2x}{1 - x^2}\right)$$

Let $\theta = \tan^{-1}\dfrac{4}{3} \Rightarrow \tan\theta = \dfrac{4}{3}$.

Now, $\quad \sec\alpha = \sec 2\theta = \dfrac{1 + \tan^2\theta}{1 - \tan^2\theta} = \dfrac{1 + (16/9)}{1 - (16/9)} = \dfrac{25}{-7}$

Hence, $\sec\theta = -\dfrac{25}{7}$

48. $\dfrac{2e + 4}{3} > e \Rightarrow \ln\left(\dfrac{2e + 4}{3}\right) > 1$

49. $\sin^{-1}\left(-\sin\dfrac{50\pi}{9}\right) = -\sin^{-1}\sin\left(\dfrac{50\pi}{9}\right) = -\sin^{-1}\sin\left(\dfrac{14\pi}{9}\right)$

$$= -\sin^{-1}\sin\left(2\pi - \frac{4\pi}{9}\right)$$

$$= -\sin^{-1}\left(-\sin\left(\frac{4\pi}{9}\right)\right) = \frac{4\pi}{9}$$

$$\cos^{-1}\cos\left(-\frac{31\pi}{9}\right) = \cos^{-1}\cos\left(\frac{31\pi}{9}\right)$$

$$= \cos^{-1}\cos\left(4\pi - \frac{5\pi}{9}\right) = \cos^{-1}\cos\frac{5\pi}{9} = \frac{5\pi}{9}$$

Hence, $\sec\left(\dfrac{4\pi}{9} + \dfrac{5\pi}{9}\right) = \sec\pi = -1$

50. $\tan\left\{\underbrace{\text{arc}\tan(2)}_{A} + \underbrace{\text{arc}\tan(20k)}_{B}\right\} = k;$

$\Rightarrow \quad \dfrac{\tan A + \tan B}{1 - \tan A \tan B} = k \Rightarrow \dfrac{2 + 20k}{1 - (2)(20k)} = k$

or $\quad 40k^2 + 19k + 2 = 0$

Now, sum of solutions, $k_1 + k_2 = -\dfrac{19}{40}$

51. $\displaystyle\sum_{r=2}^{\infty} \tan^{-1}\left(\frac{(r-2)-(r-3)}{1+(r-3)(r-2)}\right) = \sum_{r=2}^{\infty} (\tan^{-1}(r-2) - \tan^{-1}(r-3))$

Now, $\quad T_2 = \tan^{-1}0 - \tan^{-1}(-1)$

$T_3 = \tan^{1}1 - \tan^{-1}0$

$T_4 = \tan^{-1}2 - \tan^{-1}1$

$\vdots \qquad \vdots$

$T_n = \tan^{-1}(n-2) - \tan^{-1}(n-3)$

$\Rightarrow \quad S_n = \tan^{-1}(n-2) + \dfrac{\pi}{4}$

$\therefore \quad S_\infty = \dfrac{\pi}{2} + \dfrac{\pi}{4} = \dfrac{3\pi}{4}$

52. $x = -\pi/4;\ y = \cos\dfrac{\theta}{2};$ where $\cos\theta = \dfrac{1}{8}$

and $\quad \cos\dfrac{\theta}{2} = \sqrt{\dfrac{1 + \cos\theta}{2}} = \dfrac{3}{4}$

53. $\tan^{-1}\left(\dfrac{1}{2}\tan 2A\right) + \tan^{-1}(\cot A) + \tan^{-1}(\cot^3 A)$

$$= \tan^{-1}\left(\frac{1}{2}\tan 2A\right) + \tan^{-1}\left(\frac{\cot A + \cot^3 A}{1 - \cot^4 A}\right) + \pi$$

$$\left(0 < A < \frac{\pi}{4} \Rightarrow \cot A > 1\right)$$

$$= \tan^{-1}\left(\frac{\tan A}{1 - \tan^2 A}\right) + \pi + \tan^{-1}\frac{\cot A(1 + \cot^2 A)}{(1 - \cot^2 A)(1 + \cot^2 A)}$$

$$= \pi + \tan^{-1}\left(\frac{\tan A}{1 - \tan^2 A}\right) + \tan^{-1}\left(\frac{\cot A}{1 - \cot^2 A}\right)$$

$$= \pi = 4\tan^{-1}(1)$$

54. $T_n = \tan^{-1}\left(\frac{(n+1)^2-(n-1)^2}{1+(n^2-1)^2}\right) = \tan^{-1}(n+1)^2 - \tan^{-1}(n-1)^2$

$$S_n = \tan^{-1}(n+1)^2 + \tan^{-1}n^2 - \tan^{-1}1^2$$

$$\Rightarrow S_\infty = \frac{\pi}{2}+\frac{\pi}{2}-\frac{\pi}{4} = \pi - \frac{\pi}{4}$$

$$= \pi - \sec^{-1}(\sqrt{2})$$

$$= \sec^{-1}(-\sqrt{2})$$

55. $\frac{\pi}{2} - \cos^{-1}(1-x) + 2\cos^{-1}x = 0$

$$\sin^{-1}(1-x) + 2\cos^{-1}x = 0$$

Domain is [0, 1]

Now, in [0, 1], $\sin^{-1}(1-x) \in \left[0, \frac{\pi}{2}\right]$ and $2\cos^{-1}x \in [0, \pi]$

Hence, $\left.\begin{array}{l}\sin^{-1}(1-x) = 0\\ \text{and } \cos^{-1}x = 0\end{array}\right] \Rightarrow x = 1$

56. Let $\tan^{-1}(x) = \theta \Rightarrow x = \tan\theta$

$$\cos\theta = x \qquad \text{(given)}$$

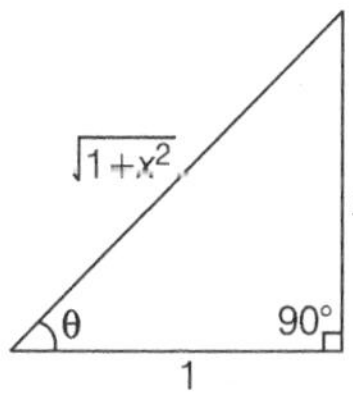

$$\frac{1}{\sqrt{1+x^2}} = x$$

$$x^2(1+x^2) = 1$$

$$\Rightarrow \quad x^2 = \frac{-1\pm\sqrt{5}}{2}$$

$$\Rightarrow \quad x^2 = \frac{\sqrt{5}-1}{2} \; (x^2 \text{ can not be} - \text{ve}) \Rightarrow \frac{x^2}{2} = \frac{\sqrt{5}-1}{4}$$

$$\cos^{-1}\left(\frac{\sqrt{5}-1}{4}\right) = \cos^{-1}\left(\sin\frac{\pi}{10}\right) = \cos^{-1}\left(\cos\frac{2\pi}{5}\right) = \frac{2\pi}{5}$$

57. $\sin\cos^{-1}(\cos(\tan^{-1}x)) = p$

For $x \in R \; \tan^{-1}x \in (-\pi/2, \pi/2)$

$\cos(\tan^{-1}x) \in (0, 1]$

$$\cos^{-1}\cos(\tan^{-1}x) \in [0, \pi/2)$$

$$\sin(\cos^{-1}(\cos(\tan^{-1}x))) \in [0, 1)$$

58. $\cos^{-1}x = t \Rightarrow x \in [-1, 1]$ and $t \in [0, \pi]$

Now, we have $\log_{10}\sqrt{5t-1} + \frac{1}{2}\log_{10}(2t+3) + \frac{1}{2}\log_{10}5 = 1;$

$$\left(t > \frac{1}{5} \text{ and } t > -\frac{3}{2}\right)$$

$$\Rightarrow \quad \log_{10}((5t-1)(2t+3).5) = 2$$

$$\Rightarrow \quad (5t-1)(2t+3).5 = 100$$

$$\Rightarrow \quad (5t-1)(2t+3) = 20$$

$$10t^2 + 13t - 3 = 20$$

$$\Rightarrow \quad 10t^2 + 23t - 10t - 23 = 0$$

$$t(10t+23) - (10t+23) = 0$$

$$(t-1)(10t+23) = 0 \Rightarrow t = 1 \text{ or } t = -\frac{23}{10} \text{ (rejected)}$$

$$\cos^{-1}x = 1 \Rightarrow x = \cos 1$$

59. $\sin^{-1}x + \cos^{-1}x = \frac{\pi}{2}$ ...(i)

and $\sin^{-1}x - \cos^{-1}x = \sin^{-1}(3x-2)$ ...(ii)

From Eqs. (i) and (ii), we get

$$2\cos^{-1}x = \cos^{-1}(3x-2)$$

Note that $(3x-2) \in [-1, 1]$ i.e. $-1 \le 3x-2 \le 1$

$$\Rightarrow \quad x \in \left[\frac{1}{3}, 1\right]$$

Now, $2\cos^{-1}x = \cos^{-1}(3x-2) \Rightarrow \cos^{-1}(2x^2-1) = \cos^{-1}(3x-2)$

$$\Rightarrow \quad 2x^2 - 3x + 1 = 0$$

$$\Rightarrow \quad x = 1 \text{ or } \frac{1}{2}$$

Hence, $x = 1$ or $\frac{1}{2}$

60. $\tan^2(\sin^{-1}x) > 1 \Rightarrow \tan(\sin^{-1}x) > 1$ or $\tan(\sin^{-1}x) < -1$

$$\sin^{-1}x > \frac{\pi}{4} \text{ or } \sin^{-1}x < \frac{\pi}{4}; x \in (-1, 1) - \left[-\frac{1}{\sqrt{2}}, \frac{1}{\sqrt{2}}\right]$$

Note that domain is $(-1, 1)$

61. $\sin^{-1}\sqrt{1-x^2} + \cos^{-1}x = \cot^{-1}\frac{\sqrt{1-x^2}}{x} - \sin^{-1}x$

or $\quad \frac{\pi}{2} + \sin^{-1}\sqrt{1-x^2} = \cot^{-1}\frac{\sqrt{1-x^2}}{x}$

$$\Rightarrow \tan^{-1}\frac{\sqrt{1-x^2}}{x} + \sin^{-1}\sqrt{1-x^2} = 0$$

$$\Rightarrow \quad -1 \le x < 0 \cup \{1\}$$

62. Consider $\tan 65° - 2\tan 40°$

$$= \tan(45° + 20°) - 2\tan 40°$$

$$= \frac{1+\tan 20°}{1-\tan 20°} - \frac{4\tan 20°}{1-\tan^2 20°}$$

$$= \frac{(1+\tan 20°)^2 - 4\tan 20°}{(1-\tan 20°)(1+\tan 20°)} = \frac{(1-\tan 20°)(1-\tan 20°)}{(1-\tan 20°)(1+\tan 20°)}$$

$$= \frac{(1-\tan 20°)}{(1+\tan 20°)} = \tan(45° - 20°) = \tan 25°$$

$\therefore \quad \tan^{-1}(\tan 25°) = 25°$

63. We have $\cos^{-1}\frac{x}{a} + \cos^{-1}\frac{y}{b} = \frac{\pi}{2} + \theta$

$$\Rightarrow \quad \frac{xy}{ab} - \sqrt{1-\frac{x^2}{a^2}}\sqrt{1-\frac{y^2}{b^2}} = -\sin\theta$$

$$\Rightarrow \quad \frac{xy}{ab} + \sin\theta = \sqrt{1-\frac{x^2}{a^2}}\sqrt{1-\frac{y^2}{b^2}}$$

On squaring both sides, we get

$$\frac{x^2y^2}{a^2b^2}+\sin^2\theta+\frac{2xy}{ab}\sin\theta=1-\frac{x^2}{a^2}-\frac{y^2}{b^2}+\frac{x^2y^2}{a^2b^2}$$

$$\Rightarrow \quad b^2x^2+a^2y^2+2ab\,xy\sin\theta=a^2b^2\cos^2\theta\le a^2b^2$$

64. $a_n=\tan^{-1}\left(\frac{1}{1+(n+3)(n+2)}\right)=\tan^{-1}\left(\frac{(n+3)-(n+2)}{1+(n+3)(n+2)}\right)$

$$=\tan^{-1}(n+3)-\tan^{-1}(n+2)$$

$$\therefore \quad S_n=f(n)-f(1)=\tan^{-1}(n+3)-\tan^{-1}3$$

$$\Rightarrow \quad S_\infty=\frac{\pi}{2}-\tan^{-1}3=\cot^{-1}3=\sin^{-1}\frac{1}{\sqrt{10}}$$

65. $\because \tan^{-1}\left(\frac{1+x}{1-x}\right)=\begin{cases}\tan^{-1}(1)+\tan^{-1}x\,,\text{ if } x<1\\ -\pi+\tan^{-1}(1)+\tan^{-1}x\,,\text{ if } x>1\end{cases}$

Now, $f(x)=\begin{cases}\tan^{-1}(1)\,,\text{ if } x<1\\ -\pi+\tan^{-1}(1)\,,\text{ if } x>1\end{cases}$

$\therefore$ Range $f=\left\{\frac{\pi}{4},\frac{-3\pi}{4}\right\}$

66. We have $\frac{1}{2}\le\frac{x^2-k}{1+x^2}<1\Rightarrow\frac{1}{2}\le1-\frac{(k+1)}{x^2+1}<1,\ \forall\ x\in R$

$$\Rightarrow \quad k+1>0 \text{ and } \frac{k+1}{x^2+1}\le\frac{1}{2}$$

So, $\quad k>-1$ and $x^2+1\ge 2k+2$

$\Rightarrow x^2-(2k+1)\ge0,\ \forall\ x\in R\Rightarrow 4(2k+1)\le0$

$$\therefore \quad k\le-\frac{1}{2}$$

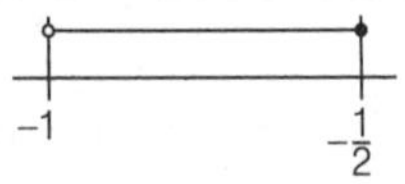

Hence, $k\in\left(-1,\frac{-1}{2}\right]$

67. Domain is $x\in[-1,1]$

Given, $\sin^{-1}x=2\tan^{-1}x$

$$\Rightarrow \quad \tan^{-1}\left(\frac{x}{\sqrt{1-x^2}}\right)=\tan^{-1}\left(\frac{2x}{1-x^2}\right)$$

$$\Rightarrow \quad \frac{x}{\sqrt{1-x^2}}=\frac{2x}{1-x^2}$$

$$\therefore \quad x=0 \text{ or } (1-x^2)^2=4(1-x^2)$$

$$\Rightarrow \quad (1-x^2)(3+x^2)=0$$

$$\Rightarrow \quad x=-1,0,1 \qquad \ldots \text{(i)}$$

$$\tan^{-1}\sqrt{x(x-1)}+\operatorname{cosec}^{-1}\sqrt{1+x-x^2}=\frac{\pi}{2}$$

$$x(x-1)\ge0\cap x-x^2\ge0\Rightarrow x(x-1)=0$$

$$\Rightarrow \quad x=0,1 \qquad \ldots \text{(ii)}$$

Now, Eqs. (i) $\cap$ (ii) gives $x=0,1$

Hence, number of common solution are 2.

68. Let $\theta=\text{arc}\cos(x-1)$

Now, $\quad \cos3\theta=4\cos^3\theta-3\cos\theta$

So, $\quad 4y^3-3y=0$, where $y=x-1$

$$\therefore \quad y=\pm\frac{\sqrt{3}}{2},0\Rightarrow x=1\pm\frac{\sqrt{3}}{2},1$$

Hence, three values of x.

Aliter :

$$\cos(3\cos^{-1}(x-1))=0$$

$$\Rightarrow 3\cos^{-1}(x-1)=(2n+1)\frac{\pi}{2},n\in Z$$

$$\Rightarrow \quad \cos^{-1}(x-1)=(2n+1)\frac{\pi}{6},n\in Z$$

$$\Rightarrow \quad \cos^{-1}(x-1)=\frac{\pi}{6},\frac{\pi}{2},\frac{5\pi}{6}$$

$$\Rightarrow \quad x-1=\frac{\sqrt{3}}{2},0,\frac{-\sqrt{3}}{2}$$

$$\therefore \quad x=1+\frac{\sqrt{3}}{2},1,1-\frac{\sqrt{3}}{2}$$

69. (a) $f(x)=\frac{1}{2}\cos^{-1}\left(\frac{1-x^2}{1+x^2}\right)$

$$D_f=R$$

As, $0\le\cos^{-1}\left(\frac{1-x^2}{1+x^2}\right)<\pi$

$$\therefore \quad R_f=\left[0,\frac{\pi}{2}\right)$$

(b) $g(x)=\text{sgn}\left(x+\frac{1}{x}\right)$

$$D_g=(-\infty,0)\cup(0,\infty)$$

$$R_g=\{-1,1\}$$

(c) $h(x)=\sin^2x+2\sin x+2$

$$D_h=R$$

Also, $h(x)=(\sin x+1)^2+1$

$$\therefore \quad R_h=[1,5]$$

(d) $k(x)=\cos^{-1}(x^2-2x+2)=\cos^{-1}((x-1)^2+1)$

$$\therefore \quad D_k=\{1\}$$

$$R_k=\{0\}$$

70. $\therefore 0\le\tan^{-1}(3x^2+bx+3)<\frac{\pi}{2}$

$$\Rightarrow \quad 0\le3x^2+bx+3<\infty$$

Thus, range of $3x^2+bx+3$ is $[0,\infty)$

Now, $D=b^2-4\cdot3\cdot3=0\Rightarrow b^2=36\Rightarrow b=\pm6$

Sum of values of $b=0$

$\therefore$Square of sum of values of $b=0$

71. $\theta=\tan^{-1}\left(\tan\frac{5\pi}{4}\right)$ and $\phi=\tan^{-1}\left(-\tan\frac{2\pi}{3}\right)$

$$=\tan^{-1}\tan\left(\pi+\frac{\pi}{4}\right) \text{ and } \phi=\tan^{-1}\left(-\tan\left(\pi-\frac{\pi}{3}\right)\right)$$

$= \tan^{-1}\tan\frac{\pi}{4}$ and $\phi = \tan^{-1}\tan\frac{\pi}{3}$

$= \frac{\pi}{4}$ and $\phi = \frac{\pi}{3}$

$\therefore \quad 4\theta - 3\phi = 0$ and $\theta + \phi = \frac{7\pi}{12}$

72. $f\left(\frac{8\pi}{9}\right) = e^{\cos^{-1}\sin\left(\frac{8\pi}{9}+\frac{\pi}{3}\right)} = e^{\cos^{-1}\sin\frac{11\pi}{9}}$

$= e^{\cos^{-1}\cos\frac{13\pi}{18}} = e^{13\pi/18}$

and $f\left(-\frac{7\pi}{4}\right) = e^{\cos^{-1}\sin\left(\frac{-7\pi}{4}+\frac{\pi}{3}\right)}$

$= e^{\cos^{-1}\sin\left(\frac{-17\pi}{12}\right)}$

$= e^{\cos^{-1}\cos\left(\frac{\pi}{12}\right)} = e^{\pi/12}$

73. Let $\cos^{-1}\left(\frac{4}{5}\right) = \alpha$, that is $\cos\alpha = \frac{4}{5}$, so that

$\tan\alpha = \sqrt{\left\{\left(\frac{5}{4}\right)^2 - 1\right\}} = \frac{3}{4}$

and $\tan\left(\cos^{-1}\frac{4}{5} + \tan^{-1}\frac{2}{3}\right) = \frac{\tan\alpha + \frac{2}{3}}{1 - \tan\alpha \cdot \frac{2}{3}}$

$= \frac{\frac{3}{4}+\frac{2}{3}}{1-\frac{2}{3}\cdot\frac{3}{4}} = \frac{17}{6} = \frac{a}{b}$ (given)

so, $\quad a = 17, b = 6, a + b = 23$

$a - b = 11$ and $3b = a + 1$

74. $\sin^{-1}x + \cos^{-1}x = \frac{\pi}{2}, \forall\ x \in [-1, 1]$

75. Given, $\cos^{-1}x = \tan^{-1}x$

$\Rightarrow \quad x = \cos\theta = \tan\theta$

$\Rightarrow \quad \cos^2\theta = \sin\theta$

$\Rightarrow \sin^2\theta + \sin\theta - 1 = 0$

$\Rightarrow \quad \sin\theta = \frac{-1 \pm \sqrt{1+4}}{2}$

$\Rightarrow \quad \sin\theta = \frac{\sqrt{5}-1}{2}$

$\Rightarrow \quad x^2 = \cos^2\theta = \frac{\sqrt{5}-1}{2}$

$\therefore$ Option (a) is correct.

and $\quad \sin(\cos^{-1}x) = \sin\theta = \frac{1}{2}(\sqrt{5}-1)$

Also, option (c) is correct

and $\quad \tan(\cos^{-1}x) = \tan\theta \neq \frac{1}{2}(\sqrt{5}-1)$

$\therefore$Option (d) is not correct.

76. The solution of $y = \sqrt{y}$ is $y = 0$ or $y = 1$

If $\sin^{-1}|\sin x| = 1 \Rightarrow x = 1$ or $x = \pi - 1$

But $y = \sin^{-1}|\sin x|$ is periodic with period π, so $x = n\pi + 1$ or $n\pi - 1$.

Again if $\sin^{-1}|\sin x| = 0$, $x = n\pi$

77. $(\sin^{-1}x + \sin^{-1}w)(\sin^{-1}y + \sin^{-1}z) = \pi^2$

$\therefore \quad \sin^{-1}x + \sin^{-1}w = \sin^{-1}y + \sin^{-1}z = \pi$

or $\quad \sin^{-1}x + \sin^{-1}w = \sin^{-1}y + \sin^{-1}z = -\pi$

$\therefore \quad x = y = z = w = 1$ or $x = y = z = w = -1$

Hence, the maximum value of $\begin{vmatrix} x^{N_1} & y^{N_2} \\ z^{N_3} & w^{N_4} \end{vmatrix} = \begin{vmatrix} 1 & -1 \\ 1 & 1 \end{vmatrix} = 2$ and

minimum value $\begin{vmatrix} -1 & 1 \\ 1 & 1 \end{vmatrix} = -2$ Also, there are 16 different determinants as each place value is either 1 or -1.

78. Since, $|\tan^{-1}x| = \begin{cases} \tan^{-1}x, & \text{if } x \geq 0 \\ -\tan^{-1}x, & \text{if } x < 0 \end{cases}$

$\Rightarrow \quad |\tan^{-1}x| = \tan^{-1}|x|\ \forall\ x \in R$

$\Rightarrow \quad \tan|\tan^{-1}x| = \tan\tan^{-1}|x| = |x|$

Also, $\quad |\cot^{-1}x| = \cot^{-1}x;\ \forall\ x \in R$

$\Rightarrow \quad \cot|\cot^{-1}x| = x,\ \forall\ x \in R$

$\tan^{-1}|\tan x| = \begin{cases} x, & \text{if } 0 < x < \frac{\pi}{2} \\ -x, & \text{if } -\frac{\pi}{2} < x < 0 \end{cases}$

$\sin|\sin^{-1}x| = \begin{cases} x, & x \in [0, 1] \\ -x, & x \in [-1, 0) \end{cases}$

79. $1 \leq \frac{\pi}{\cos^{-1}x} < \infty \Rightarrow 2 \leq \frac{\pi}{2^{\cos^{-1}x}} < \infty$

Hence, 2 should lie between or on the roots of

$t^2 - \left(a + \frac{1}{2}\right)t - a^2 = 0$, where $t = 2^{\pi/\cos^{-1}x}$

$\Rightarrow \quad f(2) \leq 0 \Rightarrow a^2 + 2a - 3 \geq 0$

$\Rightarrow a \in (-\infty, -3] \cup [1, \infty)$

80. $E = \pi - 3 + \pi - 4 + 5 - 2\pi = -2$

81. $\cos^{-1}x + \cos^{-1}y + \cos^{-1}z = \pi$

$\Rightarrow \quad \sin^{-1}x + \sin^{-1}y + \sin^{-1}z = \frac{\pi}{2}$

Also, $\quad \cos^{-1}x + \cos^{-1}y + \cos^{-1}z = \pi$

$\Rightarrow \quad \cos^{-1}x + \cos^{-1}y = \cos^{-1}(-z)$

$\Rightarrow \quad xy - \sqrt{1-x^2}\sqrt{1-y^2} = -z$

$\Rightarrow \quad x^2 + y^2 + z^2 + 2xyz = 1$

Option (d) can be true only if $x, y, z > 0$; for (c) put

$x = y = z = 1/2$

82. Let $\tan^{-1}x = \alpha$ and $\tan^{-1}x^3 = \beta$

$\tan\alpha = x$ and $\tan\beta = x^3$

$\therefore 2\tan(\alpha+\beta)=\dfrac{2(\tan\alpha+\tan\beta)}{1-\tan\alpha\tan\beta}=2\left[\dfrac{x+x^3}{1-x^4}\right]=\dfrac{2x}{1-x^2}$

Also, $\dfrac{2x}{1-x^2}=\dfrac{2\tan\alpha}{1-\tan^2\alpha}=\tan 2\alpha=\tan(2\tan^{-1}x)$

$=\tan\left(2\left(\dfrac{\pi}{2}-\cot^{-1}x\right)\right)=\tan(\pi-\cot^{-1}x-\cot^{-1}x)$

$=\tan(\cot^{-1}(-x)-\cot^{-1}(x))$

83. $f(x)=\begin{cases} 2x & ; \quad 0\le x\le \dfrac{\pi}{2} \\ \pi & ; \quad \dfrac{\pi}{2}<x\le\dfrac{3\pi}{2} \\ 4\pi-2x & ; \quad \dfrac{3\pi}{2}<x\le 2\pi \end{cases}$

Clearly, $f(x)$ is periodic function with period 2π. The graph of $f(x)$ is shown below.

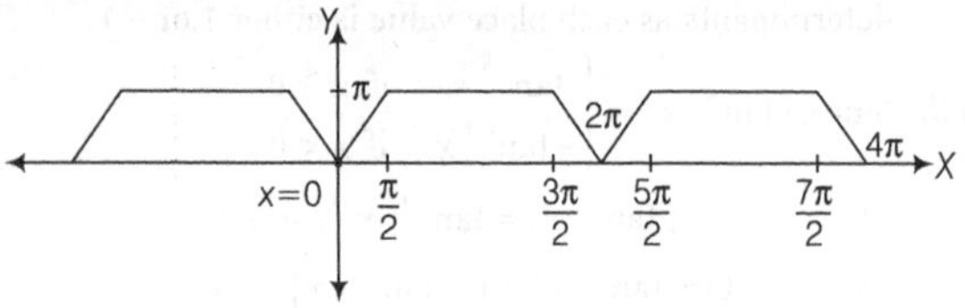

84. Domain of $\sin^{-1}x$ and $\cos^{-1}x$, each is $[-1,1]$ and that of $\sec^{-1}x$ and $\text{cosec}^{-1}x$, each is $(-\infty,-1]\cup[1,\infty)$

$\therefore$Domain of $f(x)$ must be $\{-1,1\}$

$\therefore$Range of $f(x)$ will be $\{f(-1),f(1)\}$

where, $f(-1)=\sin^{-1}(-1)\,.\cos^{-1}(-1)\,.\tan^{-1}(-1)\cot^{-1}(-1)$

$\cdot\sec^{-1}(-1)\cdot\text{cosec}^{-1}(-1)$

$=\left(\dfrac{-\pi}{2}\right).(\pi).\left(\dfrac{-\pi}{4}\right).\left(\dfrac{3\pi}{4}\right).(\pi).\left(\dfrac{-\pi}{2}\right)=\dfrac{-3\pi^6}{64}$

and $\quad f(1)=0$ {as $\cos^{-1}1=0$}

(i) Thus, the graph of $f(x)$ is a two point graph which doesn't lies above X-axis.

(ii) $f(x)_{\max}=0$ and $f(x)_{\min}=\dfrac{-3\pi^6}{64}$

Hence, $|f(x)_{\max}-f(x)_{\min}|=\dfrac{3\pi^6}{64}$

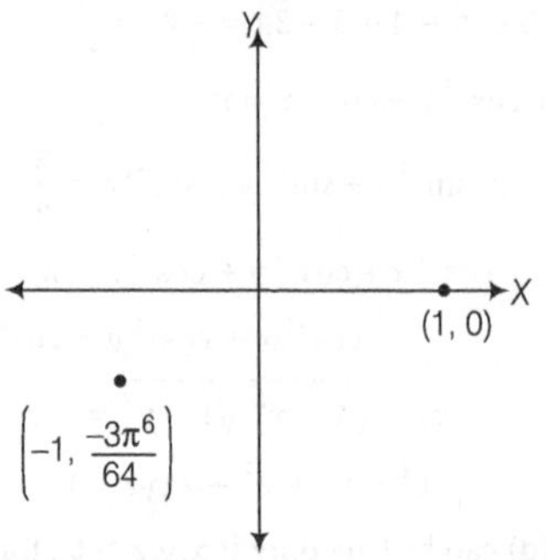

(iii) $f(x)$ is one-one. Hence, injective.

(iv) Domain is $\{-1,1\}$

$\therefore$Number of non-negative integers in the domain of $f(x)$ is one.

85. $\alpha=3\tan^{-1}\left(\dfrac{\sqrt3}{5}\right)+3\tan^{-1}\left(\dfrac{\sqrt3}{2}\right)=3$

$\left[\tan^{-1}\dfrac{\dfrac{\sqrt3}{5}+\dfrac{\sqrt3}{2}}{1-\dfrac{3}{10}}\right]=3\tan^{-1}\left(\dfrac{7\sqrt3}{7}\right)=\pi$

$\beta=4\left[\tan^{-1}7-\tan^{-1}\dfrac{3}{4}\right]=4\left[\tan^{-1}\dfrac{7-\dfrac{3}{4}}{1+\dfrac{21}{4}}\right]$

$=4\tan^{-1}\left(\dfrac{25}{25}\right)=\pi$

86. $f(x)=|\sin^{-1}x|+\cos^{-1}\left(\dfrac{1}{x}\right)$

Domain of $f(x)$ is $\{-1,1\}$

$f(1)=\dfrac{\pi}{2},\ f(-1)=\dfrac{3\pi}{2}$

So, function $f(x)$ is injective .

$\text{sgn}(f(x)=1\,(f(x)>0)$

Range of $\quad f(x)=\left\{\dfrac{\pi}{2},\dfrac{3\pi}{2}\right\}$

87. (a) $f(x)=\sin(\tan^{-1}x)$

Put $\tan^{-1}x=\theta\Rightarrow x=\tan\theta$

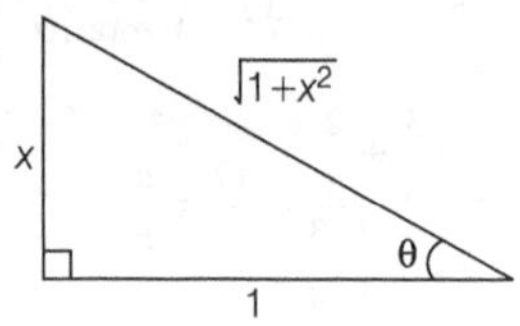

$\Rightarrow\ f(x)=\dfrac{x}{\sqrt{1+x^2}}=g(x)\Rightarrow$ identical functions

(b) $f(x)=\text{sgn}\,(\cot^{-1}x)=1,\forall\in R$

But $g(x)=\sec^2x-\tan^{-1}x\ne1,\forall\ x\in R$. (Think ? Domain of $g(x)$)

(c) As $\dfrac{x^2-1}{x^2+1}=\dfrac{x^2+1-2}{x^2+1}=1-\dfrac{2}{x^2+1}\in[-1,1)$

So, $\cos^{-1}\left(\dfrac{x^2-1}{x^2+1}\right)\in(0,\pi]$

So, $f(x)$ and $g(x)$ are identical functions.

(d) $f(x)=\sin^{-1}\left(\dfrac{2x}{1+x^2}\right)=\begin{cases}-(\pi+2\tan^{-1}x) & , \quad x\le-1 \\ 2\tan^{-1}x & , \quad -1<x<1 \\ \pi-2\tan^{-1}x & , \quad x\ge1\end{cases}$

So, $f(x)$ and $g(x)$ are not identical functions.

88. $T_n=\cot^{-1}(n^2+n+1)=\tan^{-1}\left(\dfrac{(n+1)-n}{1+n(n+1)}\right)$

$=(\tan^{-1}(n+1)-\tan^{-1}n)$

So, $\sum\limits_{n=1}^{n}T_n=(\tan^{-1}2-\tan^{-1}1)+$

$(\tan^{-1}3-\tan^{-1}2)+\ldots..+(\tan^{-1}(n+1)-\tan^{-1}n)$

$$= \tan^{-1}(n+1) - \tan^{-1}1$$

$$= \tan^{-1}\left[\frac{n+1-1}{1+(n+1)\cdot 1}\right] = \tan^{-1}\left(\frac{n}{n+2}\right)$$

$$\Rightarrow \quad \sum_{n=1}^{\infty} T_n = \tan^{-1}1 = \frac{\pi}{4}$$

(a) $\cot^{-1}(-1) + \sec^{-1}1 - \text{cosec}^{-1}1 = \frac{3\pi}{4} + 0 - \frac{\pi}{2}$

$$= \frac{3\pi - 2\pi}{4} = \frac{\pi}{4}$$

(b) $\cot^{-1}2 + \cot^{-1}3 = \tan^{-1}\left(\frac{1}{2}\right) + \tan^{1}\left(\frac{1}{3}\right) = \frac{\pi}{4}$

(c) $\because \quad -1 < \frac{1-x^2}{1+x^2} \le 1$

$$\therefore \quad \frac{-\pi}{4} < \tan^{-1}\left(\frac{1-x^2}{1+x^2}\right) \le \frac{\pi}{4}$$

Hence, minimum value of $f(x)$ does not exist and maximum value of $f(x)$ is $\frac{\pi}{4}$

(d) As $\frac{41\pi}{4} = \left(10\pi + \frac{\pi}{4}\right)$

So, $\quad \cos^{-1}\left(\cos\frac{41\pi}{4}\right) = \cos^{-1}\left(\cos\frac{\pi}{4}\right) = \frac{\pi}{4}$

89. We know that, $\tan^{-1}\left(\frac{2x}{1-x^2}\right)$

$$= \begin{cases} \pi + 2\tan^{-1}x &, \quad x < -1 \\ 2\tan^{-1}x &, \quad -1 < x \le 1 \\ -(\pi - 2\tan^{-1}x) &, \quad x > 1 \end{cases}$$

$$\therefore \quad f(x) = \begin{cases} -\pi, & x < -1 \\ \pi, & x > 1 \end{cases}$$

So, $\quad R_f = \{-\pi, \pi\}$

90. $P = \frac{-5}{15}$ and $Q = \frac{1}{15}$

Sol. (Q.Nos. 91-93)

$A = \{\sin^{-1}x\}$; $B = \{[x], \{x\}\}$; $C = \{[x], \sin^{-1}x\}$

$$f(x) = A \cap C = \sin^{-1}x$$

$$g(x) = B \cap C = [x]$$

$$h(x) = \{x\}$$

$|x|$ is a function which is neither odd, nor discontinuous, nor non-decreasing.

$\therefore \quad l(x)$ is $|x|$

91. $l(x) = |x|$ is an even function

92. $g(f(x)) = [\sin^{-1}x]$

$\because$ Range of $\sin^{-1}x$ is $\left[-\frac{\pi}{2}, \frac{\pi}{2}\right]$

$\therefore$ Range of $g(f(x)) = \{-2, -1, 0, 1\}$

93. $f(h(x)) = \sin^{-1}\{x\}$

Domain is R and range is $\left[0, \frac{\pi}{2}\right)$.

94. We have, $f(10) = \sin^{-1}(\sin 10) + \cos^{-1}(\cos 10)$

$$= (3\pi - 10) + (4\pi - 10) = (7\pi - 20)$$

95. Clearly, $f(x) = (\pi - x) + x = \pi, \forall \ x \in \left[\frac{\pi}{2}, \pi\right]$

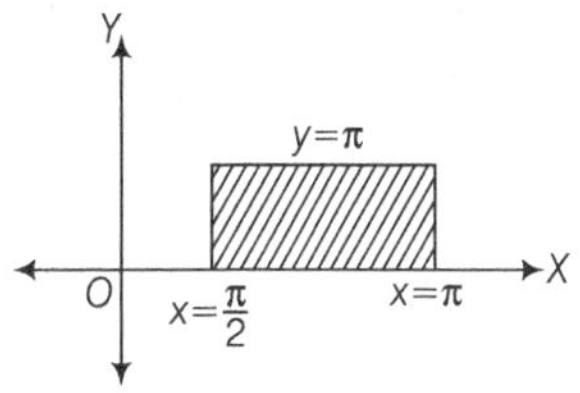

So, area $= \frac{\pi}{2} \times \pi = \frac{\pi^2}{2}$

96. As $\sin^{-1}(\sin x) = \begin{cases} x, & x \in \left[0, \frac{\pi}{2}\right) \\ \pi - x, & x \in \left[\frac{\pi}{2}, \pi\right) \end{cases}$

Also, $\quad \cos^{-1}(\cos x) = x, \ x \in [0, \pi)$

Now, $\quad f(x) = \begin{cases} 2x, & x \in \left[0, \frac{\pi}{2}\right) \\ \pi, & x \in \left[\frac{\pi}{2}, \pi\right) \end{cases}$

As, $\quad f(x) \in I$ in $x \in (0, 3) \Rightarrow x = \frac{1}{2}, 1, \frac{3}{2}$

So, number of values of x are three.

97. Given, $f(x) = \sqrt{\sin^{-1}x + 2} + \sqrt{1 - \sin^{-1}x}$

Clearly, for domain of $f(x)$, $1 - \sin^{-1}x \ge 0$

(As, $\sin^{-1}x + 2 > 0, \forall \ x \in [-1, 1]$)

$$\Rightarrow \quad \sin^{-1}x \le 1 \Rightarrow x \le \sin 1$$

$$\therefore \quad D_f = [-1, \sin 1]$$

98. Given, $f(x) = \sqrt{\sin^{-1}x + 2} + \sqrt{1 - \sin^{-1}x}$, where $x \in [-1, \sin 1]$

Let $y = \sqrt{\sin^{-1}x + 2} + \sqrt{1 - \sin^{-1}x}$

Then, $y > 0, \forall \ x \in [-1, \sin 1]$

Now, $\quad y^2 = (\sin^{-1}x + 2) + (1 - \sin^{-1}x) + 2\sqrt{(\sin^{-1}x + 2)(1 - \sin^{-1}x)}$

$$\Rightarrow \quad y^2 = 3 + 2\sqrt{\frac{9}{4} - \left(\sin^{-1}x + \frac{1}{2}\right)^2}$$

Clearly, $\quad y^2_{\max}\left(\sin^{-1}x = \frac{-1}{2}\right) = 3 + 2\sqrt{\frac{9}{4} - 0} = 3 + 3 = 6$

$$\Rightarrow \quad y_{\max}\left(x = -\sin\frac{1}{2}\right) = \sqrt{6}$$

Also, $\quad y^2_{\min}(\sin^{-1}x = 1) = 3 + 2\sqrt{\frac{9}{4} - \frac{9}{4}} = 3$

$$\Rightarrow \quad y_{\min}(x = \sin 1) = \sqrt{3}$$

Hence, range of $f = [\sqrt{3}, \sqrt{6}]$

99. $\sin^{-1}\left(\frac{4x}{x^2+4}\right)+2\tan^{-1}\left(-\frac{x}{2}\right)$

$$=\sin^{-1}\left(\frac{2\cdot\frac{x}{2}}{\left(\frac{x}{2}\right)^2+1}\right)-2\tan^{-1}\frac{x}{2}$$

$$=2\tan^{-1}\frac{x}{2}-2\tan^{-1}\frac{x}{2}=0$$

Here, $\left|\frac{x}{2}\right|\le 1$

$|x|\le 2 \Rightarrow -2\le x\le 2$

100. $\cos^{-1}\frac{6x}{1+9x^2}=-\frac{\pi}{2}+2\tan^{-1}3x$

$\Rightarrow \quad \frac{\pi}{2}-\sin^{-1}\frac{6x}{1-9x^2}=\frac{-\pi}{2}+2\tan^{-1}3x$

$\Rightarrow \quad \sin^{-1}\frac{6x}{1+9x^2}=\pi-2\tan^{-1}3x$

$\Rightarrow \quad \sin^{-1}\frac{2\cdot 3x}{1+(3x)^2}=\pi-2\tan^{-1}3x$

Above is true when $3x>1$

$\Rightarrow x>\frac{1}{3}$

$\Rightarrow x\in\left(\frac{1}{3},\infty\right)$

101. $(x-1)(x^2+1)>0 \Rightarrow x>1$

$\therefore \quad \sin\left[\frac{1}{2}\tan^{-1}\left(\frac{2x}{1-x^2}\right)-\tan^{-1}x\right]$

$$=\sin\left[\frac{1}{2}(-\pi+2\tan^{-1}x)-\tan^{-1}x\right]$$

$$=\sin\left(-\frac{\pi}{2}\right)=-1$$

Sol. (Q. Nos. 102 to 104)

$\sin^{-1}x\in\left[-\frac{\pi}{2},\frac{\pi}{2}\right]$

$\cos^{-1}y\in[0,\pi]$

$\sec^{-1}z\in\left[0,\frac{\pi}{2}\right)\cup\left(\frac{\pi}{2},\pi\right]$

$\Rightarrow \sin^{-1}x+\cos^{-1}y+\sec^{-1}z\le\frac{\pi}{2}+\pi+\pi=\frac{5\pi}{2}$

Also, $t^2-\sqrt{2\pi}t+3\pi=t^2-2\sqrt{\frac{\pi}{2}}t+\frac{\pi}{2}-\frac{\pi}{2}+3\pi$

$$=\left(t-\sqrt{\frac{\pi}{2}}\right)^2+\frac{5\pi}{2}\ge\frac{5\pi}{2}$$

The given inequation exists if equality holds, i.e.

LHS = RHS = $\frac{5\pi}{2}$

$\Rightarrow \quad x=1, y=-1, z=-1$ and $t=\sqrt{\frac{\pi}{2}}$

$\Rightarrow \quad \cos^{-1}(\cos 5t^2)=\cos^{-1}\left(\cos\left(\frac{5\pi}{2}\right)\right)=\frac{\pi}{2}$

$\cos^{-1}(\min\{x,y,z\})=\cos^{-1}(-1)=\pi$

105. $f(x)=\tan^{-1}\left(\frac{\left(\frac{x}{2}-1\right)}{1+x\left(\frac{x}{2}+1\right)}\right)$

$$=\tan^{-1}(x)-\tan^{-1}\left(\frac{x}{2}+1\right)$$

$$f'(x)=\frac{1}{(1+x^2)}-\frac{1}{\left(1+\left(\frac{x}{2}+1\right)^2\right)}\times\frac{1}{2}$$

$$f'(1)=\frac{1}{2}-\frac{1}{2\left(1+\frac{9}{7}\right)}=\frac{1}{2}-\frac{2}{13}=\frac{9}{26}$$

$\Rightarrow \quad 26f'(1)=9$

106. We have, $f(x)=(\tan^{-1}x)^3+(\cot^{-1}x)^3=(\tan^{-1}x+\cot^{-1}x)$ $((\tan^{-1}x)^2-(\tan^{-1}x)(\cot^{-1}x)+(\cot^{-1}x)^2)$

$$=\frac{\pi}{2}\left((\tan^{-1}x)^2-(\tan^{-1}x)\left(\frac{\pi}{2}-\tan^{-1}x\right)+\left(\frac{\pi}{2}-\tan^{-1}1x\right)^2\right)$$

$\left(\text{Using } \cot^{-1}x=\frac{\pi}{2}-\tan^{-1}x\right)$

$$=\frac{3\pi}{2}\left(\left(\tan^{-1}x-\frac{\pi}{4}\right)^2+\frac{\pi^2}{48}\right)$$

Clearly, $f(x)$ will be minimum when $\left(\tan^{-1}x-\frac{\pi}{4}\right)^2=0$

and $f(x)$ will be maximum when $\left(\tan^{-1}x-\frac{\pi}{4}\right)^2=\left(-\frac{\pi}{2}-\frac{\pi}{4}\right)^2$

$\therefore \quad a=f(x)_{\min}=\frac{3\pi}{2}\left(0+\frac{\pi^2}{48}\right)=\frac{\pi^3}{32}$

and $\quad b=f(x)_{\max}=\frac{3\pi}{2}\left(\left(\frac{-3\pi}{4}\right)^2+\frac{\pi^2}{48}\right)=\frac{7\pi^3}{8}$

Hence, $\quad \frac{b}{a}=\frac{\frac{7\pi^3}{8}}{\frac{7\pi^3}{32}}=4$

107. We have, $T_n=2\,\text{arc}\cot\left(\frac{n^2+n+4}{2}\right)=2\tan^{-1}\left(\frac{2}{n^2+n+4}\right)$

$$T_n=2\tan^{-1}\left(\frac{\frac{1}{2}}{1+\frac{n(n+1)}{2.2}}\right)=2\left(\tan^{-1}\left(\frac{n+1}{2}\right)-\tan^{-1}\left(\frac{n}{2}\right)\right)$$

Hence, $S_n=\sum_{n=0}^{n}T_n=2\sum_{n=0}^{n}\left(\tan^{-1}\left(\frac{n+1}{2}\right)-\tan^{-1}\left(\frac{n}{2}\right)\right)$

$$S_n = \tan^{-1}\left(\frac{1}{2}\right) - \tan^{-1}(0)$$

$$\tan^{-1}(1) - \tan^{-1}\left(\frac{1}{2}\right)$$

$$\tan^{-1}\left(\frac{n+1}{2}\right) - \tan^{-1}\left(\frac{n}{2}\right)$$

$$\Rightarrow \quad S_n = 2\tan^{-1}\left(\frac{n+1}{2}\right)$$

$$\therefore \quad \lim_{n\to\infty} S_n = 2.\left(\frac{\pi}{2}\right) = \pi = k\pi \text{ (given)} \Rightarrow k = 1$$

108. As, $\sum_{r=1}^{5} \cot^{-1}(2r^2) = \sum_{r=1}^{5} \tan^{-1}\left(\frac{2}{4r^2}\right)$

$$= \sum_{r=1}^{5} \tan^{-1}\left(\frac{(2r+1)(2r-1)}{1+(2r+1)(2r-1)}\right)$$

$$= \sum_{r=1}^{5} (\tan^{-1}(2r+1) - \tan^{-1}(2r-1))$$

$$= \tan^{-1}11 - \tan^{-1}1 = \tan^{-1}\left(\frac{11-1}{1+11\times 1}\right)$$

$$= \tan^{-1}\left(\frac{10}{12}\right) = \tan^{-1}\left(\frac{5}{6}\right)$$

$$\therefore \quad \tan\left(\sum_{r=1}^{5}\cot^{-1}(2r^2)\right) = \tan\left(\tan^{-1}\frac{5}{6}\right) = \frac{5}{6}$$

Now, $\tan\left(\sum_{r=1}^{5} \cot^{-1}(2r^2)\right) = \frac{5x+6}{6x+5} \Rightarrow \frac{5}{6} = \frac{5x+6}{6x+5}$

$\Rightarrow$ The given equations has no real solution.

109. $\min(x^2 + 4x + 7) = 3$

$$\max\left\{\frac{\pi^2}{4}, 3\right\} = 3$$

$$\lim_{z\to 0}\left[3.\frac{\sin^{-1} z}{z}\right] = 3$$

110. $\sin\left(\frac{\pi}{6} + \tan^{-1}x\right) = \frac{13}{14}$

$$\frac{\pi}{6} + \tan^{-1}x = \sin^{-1}\left(\frac{13}{14}\right) = \tan^{-1}\left(\frac{13}{3\sqrt{3}}\right)$$

$$\tan^{-1}x = \tan^{-1}\left(\frac{13}{3\sqrt{3}}\right) - \tan^{-1}\left(\frac{1}{\sqrt{3}}\right)$$

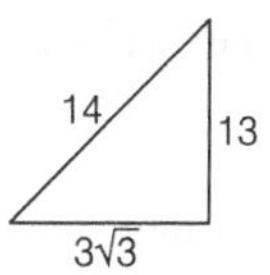

$$= \tan^{-1}\left[\frac{\frac{13}{3\sqrt{3}} - \frac{1}{\sqrt{3}}}{1+\frac{13}{9}}\right] = \tan^{-1}\left[\frac{10}{3\sqrt{3}}\left(\frac{9}{22}\right)\right]$$

$$\tan^{-1}x = \tan^{-1}\left[\frac{5\sqrt{3}}{11}\right]$$

$$\therefore \quad x = \frac{5\sqrt{3}}{11} \Rightarrow \frac{a\sqrt{3}}{b}$$

$$\Rightarrow \quad a + b = 16 \Rightarrow \left(\frac{a+b}{2}\right) = 8$$

Aliter : $\sin(30° + \tan^{-1}x) = \frac{13}{14}$

$$\Rightarrow \frac{1}{2}\cos(\tan^{-1}x) + \frac{\sqrt{3}}{2}\sin(\tan^{-1}x) = \frac{13}{14}$$

$$\Rightarrow \quad 7(1+\sqrt{3}x) = 13\sqrt{1+x^2}$$

$$\Rightarrow \quad 11x^2 - 49\sqrt{3}x + 60 = 0 \qquad \text{(On squaring)}$$

$$\therefore \quad x = \frac{49\sqrt{3} \pm 30\sqrt{3}}{22}$$

$$= \frac{5\sqrt{3}}{11}, 4\sqrt{3} \qquad \text{(Reject)}$$

$$\therefore \quad x = \frac{5\sqrt{3}}{11}$$

111. Clearly $x^2 - 4x + \alpha >$ or $\geq \frac{-1}{\sqrt{3}}, \forall\, x \in R$

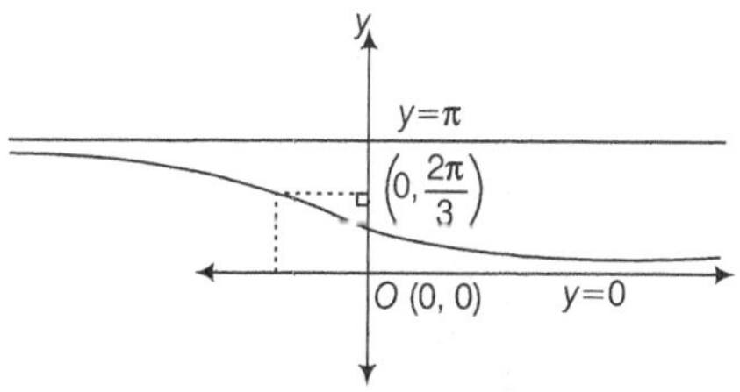

Graph of $y=\cot^{-1}x$

$$\Rightarrow \quad x^2 - 4x + \alpha + \frac{1}{\sqrt{3}} > 0, \forall\, x \in R$$

So, discriminant < 0 $\Rightarrow 16 - 4\left(\alpha + \frac{1}{\sqrt{3}}\right) < 0$

$$4 - \alpha - \frac{1}{\sqrt{3}} < 0 \Rightarrow \alpha > 4 - \frac{1}{\sqrt{3}}$$

$$\therefore \quad \alpha \in \left(4 - \frac{1}{\sqrt{3}}, \infty\right)$$

Hence, minimum integral $\alpha = 4$

112. **L** : $3^4 - [{}^3C_1(2^4 - 2) + {}^3C_2] = 36$

M : If $x > 0$, Sgn $(x) = 1$

$$f(x) = 0 + \frac{\pi}{2} = \frac{\pi}{2}$$

For $x = 0$ $f(x)$ is not defined

$\therefore$ for $x < 0$, $f(x) = \pi - \frac{\pi}{2} = \frac{\pi}{2}$

$$\therefore \quad M = 1$$

N: Coefficient of t^5 = coefficient of t^2 in $(1+t^2)^5 \times$ coefficient of t^3 in $(1+t^3)^8$

$$= 5 \times 8 = 40$$

Hence $\quad L = 36; M = 1$ and $N = 40$

$$\Rightarrow \quad N - LM = 40 - 36 = 4$$

113.

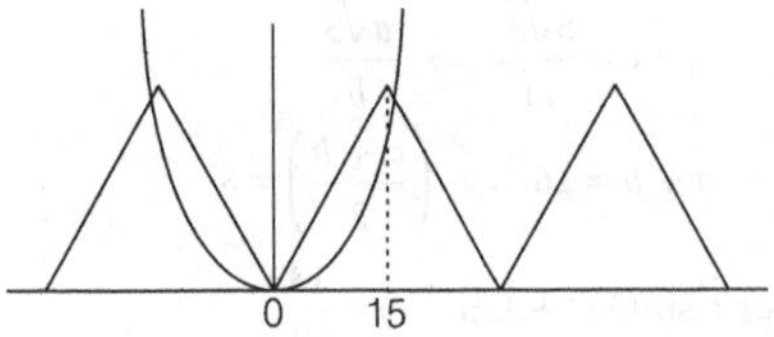

Number of solutions are three.

114. $\sec^2 u, \sec^4 v, \sec^6 w \in [1, \infty)$

$\therefore \quad \sec^2(u) + \sec^4(v) + \sec^6(w) \in [3, \infty)$

$\therefore \quad \pi(\sec^2 u + \sec^4 v + \sec^6 w) \in [3\pi, \infty)$

But $\quad \cos^{-1} x + \cos^{-1} y + \cos^{-1} z \in [0, 3\pi]$

So equation is possible of LHS = RHS = 3π

$\therefore \quad \cos^{-1} x = \cos^{-1} y = \cos^{-1} z = \pi$

$\therefore \quad x = y = z = -1$

and $\quad \sec^2 u = \sec^4 v = \sec^6 w = 1$

$\therefore \quad u = \pi, v = 2\pi, w = 3\pi$

$$\therefore x^{2000} + y^{2002} + z^{2004} + \frac{36\pi}{u + v + w} = 1 + 1 + 1 + \frac{36\pi}{6\pi} = 9$$

115. Given, $f(x) = \cos(\tan^{-1}(\sin(\cot^{-1} x)))$

$$= \cos\left(\tan^{-1}\left(\sin^{-1}\frac{1}{\sqrt{1 + x^2}}\right)\right)$$

$$= \cos\left(\tan^{-1}\left(\frac{1}{\sqrt{1 + x^2}}\right)\right)$$

$$= \cos\left(\cos^{-1}\left(\frac{\sqrt{1 + x^2}}{\sqrt{2 + x^2}}\right)\right)$$

$$= \frac{\sqrt{1 + x^2}}{\sqrt{2 + x^2}} = \left(\frac{x^2 + 1}{x^2 + 2}\right)^{1/2}$$

$\Rightarrow \quad A = 1, B = 2$

$\Rightarrow \quad A + B = 1 + 2 = 3$

116. We have,

RHS $= \tan^{-1}\{\tan^2(\alpha + \beta) \cdot \tan^2(\alpha - \beta)\} + \tan^{-1} 1$

$$= \tan^{-1}\left\{\frac{\tan^2(\alpha + \beta)\tan^2(\alpha - \beta) + 1}{1 - \tan^2(\alpha + \beta)\tan^2(\alpha - \beta)}\right\}$$

$$= \tan^{-1}\left\{\frac{\sin^2(\alpha + \beta)\sin^2(\alpha - \beta) + \cos^2(\alpha + \beta)\cos^2(\alpha - \beta)}{\cos^2(\alpha + \beta)\cos^2(\alpha - \beta) - \sin^2(\alpha + \beta)\sin^2(\alpha - \beta)}\right\}$$

$$= \tan^{-1}\left\{\frac{\{2\sin(\alpha + \beta)\sin(\alpha - \beta)\}^2 + \{2\cos(\alpha + \beta)\cos(\alpha - \beta)\}^2}{\{2\cos(\alpha + \beta)\cos(\alpha - \beta)\}^2 - \{2\sin(\alpha + \beta)\sin(\alpha - \beta)\}^2}\right\}$$

$$= \tan^{-1}\left\{\frac{(\cos 2\beta - \cos 2\alpha)^2 + (\cos 2\alpha + \cos 2\beta)^2}{(\cos 2\alpha + \cos 2\beta)^2 - (\cos 2\beta - \cos 2\alpha)^2}\right\}$$

$$= \tan^{-1}\left\{\frac{\cos 2\alpha + \cos^2 2\beta}{2\cos 2\alpha \cos 2\beta}\right\}$$

$$= \tan^{-1}\left\{\frac{\cos 2\alpha \sec 2\beta + \cos 2\beta \sec 2\alpha)}{2}\right\}$$

Hence, the value of λ is equal to 2.

117. We have, $\sin^{-1}(\sin 12) + \cos^{-1}(\cos 12)$

$$= \sin^{-1}(\sin(4\pi - (4\pi - 12))) + \cos^{-1}(\cos(4\pi - (4\pi - 12)))$$

$$= -(4\pi - 12) + (4\pi - 12) = 0$$

So that, $(n - 2)x^2 + 8x + n + 4 > 0, \forall x \in R$

$\Rightarrow \quad n - 2 > 0 \Rightarrow n \geq 3$

and $\quad 8^2 - 4(n - 2)(n + 4) < 0$

or, $\quad n^2 + 2n - 24 > 0 \Rightarrow n = 4$

$\Rightarrow \quad n \geq 5 \Rightarrow n = 5$

118. $1 + \sin(\cos^{-1} x) + \sin^2(\cos^{-1} x) + \ldots \infty = 2$

$$\Rightarrow \quad \frac{1}{1 - \sin(\cos^{-1} x)} = 2$$

or $\quad \frac{1}{2} = 1 - \sin(\cos^{-1} x)$

or $\quad \sin(\cos^{-1} x) = \frac{1}{2}$ or $\cos^{-1} x = \frac{\pi}{6}$

or $\quad x = \frac{\sqrt{3}}{2}$ or $12x^2 = 9$

119. Here $|\cos x| = \sin^{-1}(\sin x)$

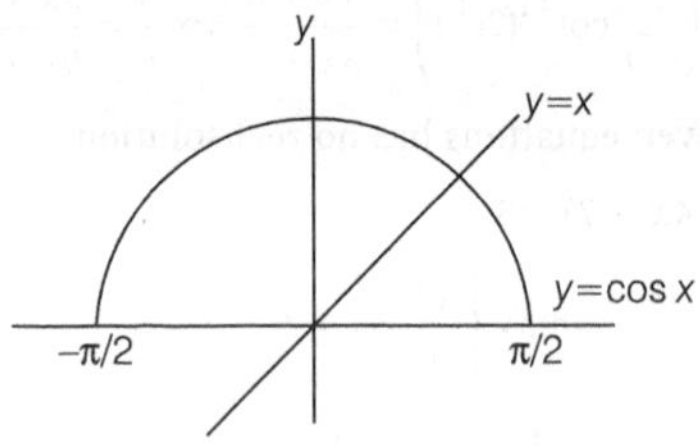

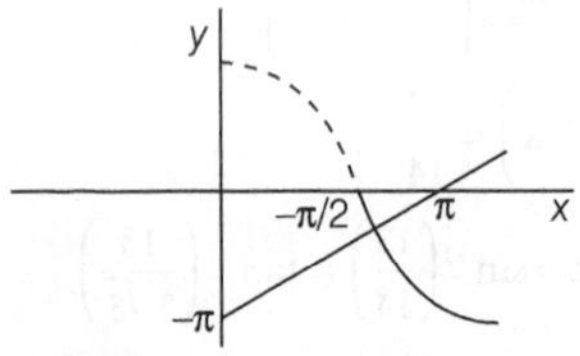

If $-\frac{\pi}{2} \leq x \leq \frac{\pi}{2}$ then $\cos x = x$

In the case there is one solution obtained graphically.

If $\frac{\pi}{2} < x \leq \pi$ then $-\cos x$

$$= \sin^{-1}\{\sin(\pi - x)\} = \pi - x$$

$\therefore \quad \cos x = x - \pi$

In this case there is one solution, obtained graphically.

If $-\pi \leq x < -\frac{\pi}{2}$ then

$$-\cos x = \sin^{-1}\{\sin(-\pi - x)\} = -x - \pi$$

i.e., $\quad \cos x = x + \pi.$

This gives no solution as can be seen from then graphs.

120. From graph

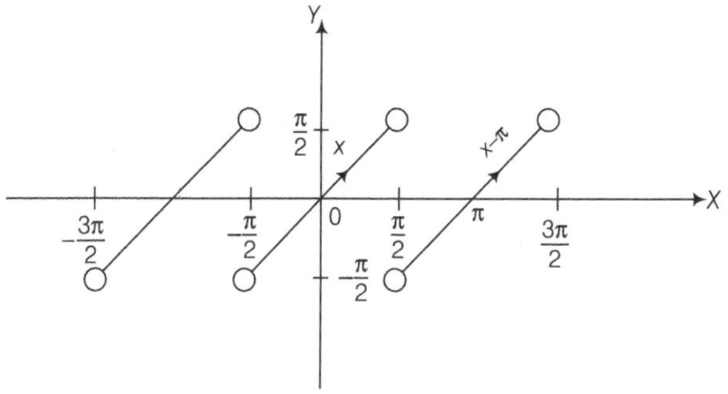

Graph for $\tan^{-1}(\tan x)$

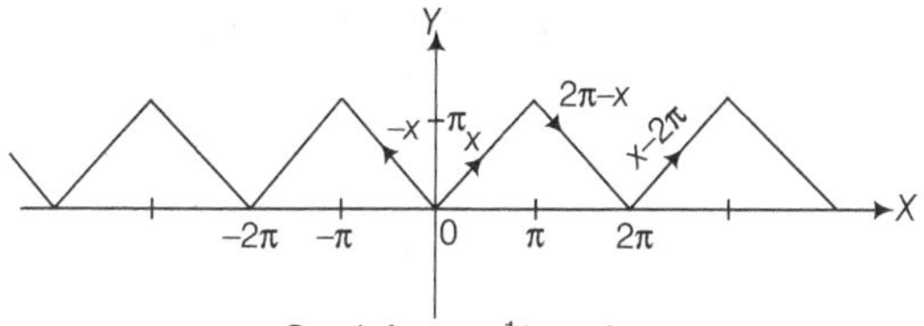

Graph for $\cos^{-1}(\cos x)$

Statement II is true and correct explanation for Statement 1.

121. $\sin^{-1} x = \tan^{-1} \dfrac{x}{\sqrt{1-x^2}} > \tan^{-1} x > \tan^{-1} y$

$$\left\{\because x > y, \frac{x}{\sqrt{1-x^2}} > x\right\}$$

$\therefore$ Statement II is true

$$e < \pi$$

$$\frac{1}{\sqrt{e}} > \frac{1}{\sqrt{\pi}}$$

by Statement II

$$\sin^{-1}\left(\frac{1}{\sqrt{e}}\right) > \tan^{-1}\left(\frac{1}{\sqrt{e}}\right) > \tan^{-1}\left(\frac{1}{\sqrt{\pi}}\right)$$

Statement I is true

122. $\text{cosec}^{-1} x > \sec^{-1} x$

$$\text{cosec}^{-1} x > \frac{\pi}{4} - \text{cosec}^{-1} x$$

$$\text{cosec}^{-1} x > \frac{\pi}{4}$$

$$1 \le x < \sqrt{2}$$

and $\left(\dfrac{1}{2} + \dfrac{1}{\sqrt{2}}\right) \in [1, \sqrt{2}]$

Statement II is true and explains Statement I.

123. $f(x) = \sin^{-1}\left(\dfrac{2x}{1+x^2}\right) = \pi - 2\tan^{-1} x, x \ge 1$

$$f'(x) = -\frac{2}{1+x^2}$$

$$\Rightarrow \quad f'(2) = -\frac{2}{5}$$

Statement I is true, Statement II is true, Statement II is a correct explanation for Statement I.

124. $\sin^{-1} 2x + \sin^{-1} 3x = \dfrac{\pi}{3}$

$$\therefore \quad \frac{\pi}{2} - \cos^{-1} 2x + \frac{\pi}{2} - \cos^{-1} 3x = \frac{\pi}{3}$$

$$\therefore \quad \cos^{-1} 2x + \cos^{-1} 3x = \frac{2\pi}{3}$$

$$6x^2 - \sqrt{1 - 13x^2 + 36x^4} = -\frac{1}{2}$$

$$\Rightarrow \quad \left\{6x^2 + \frac{1}{2}\right\}^2 = 1 - 13x^2 + 36x^4$$

$$\Rightarrow \quad 19x^2 = \frac{3}{4} \Rightarrow x = \pm\sqrt{\frac{3}{76}}$$

But sum of two negative angles cannot be $\dfrac{\pi}{3}$

$\therefore x = \sqrt{\dfrac{3}{76}}$ is only solution.

125. Statement 2 is correct, from which we can say $\cot^{-1} x + \cos^{-1} 2x = -\pi$ is not possible. Hence, both the statements are correct, and Statement 2 is the correct explanation of Statement 1.

126. (A) In $(0, \cos 1)$, we have $\cos^{-1} x > \sin^{-1} x$

Also, $\cos^{-1} x > 1$ and $\sin^{-1} x < 1$

The greatest is $(\cos^{-1} x)^{\cos^{-1} x} = t_4$ and the least is

$$(\sin^{-1} x)^{\cos^{-1} x} = t_2$$

and $\quad (\sin^{-1} x)^{\sin^{-1} x} < (\cos^{-1} x)^{\sin^{-1} x}$

$\Rightarrow \quad t_1 < t_3$

So, $\quad t_4 > t_3 > t_1 > t_2$

(B) Similarly, in $\cos 1 < x < \dfrac{1}{\sqrt{2}}$

$\cos^{-1} x > \sin^{-1} x$ and both are less than 1 So, greatest is t_3 and least is t_2 and $t_4 > t_1$.

Hence, $\quad t_3 > t_4 > t_1 > t_2$

(C) For $\dfrac{1}{\sqrt{2}} < x < \sin 1$

We have, $1 > \sin^{-1} x > \cos^{-1} x$

So, greatest is t_2 and least is t_3 also $t_1 > t_4$

Hence, $\quad t_2 > t_1 > t_4 > t_3$

(D) For $\sin 1 < x < 1$, we have

$$\sin^{-1} x > 1 > \cos^{-1} x$$

So, the greatest is t_1 and least is t_3 and $t_2 > t_4$

Hence, $\quad t_1 > t_2 > t_4 > t_3$

127. Follow the following graph for solution

(A)

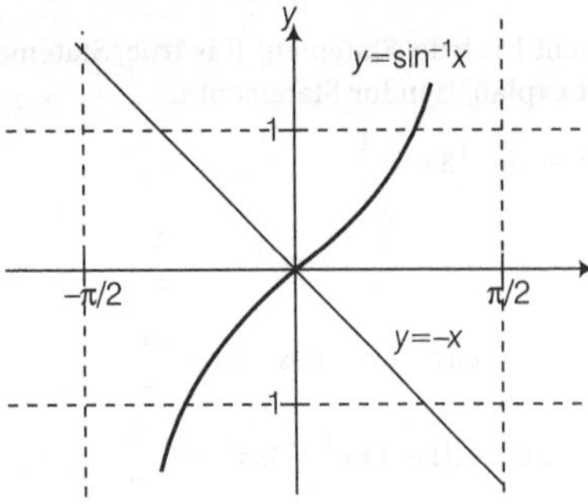

(B)

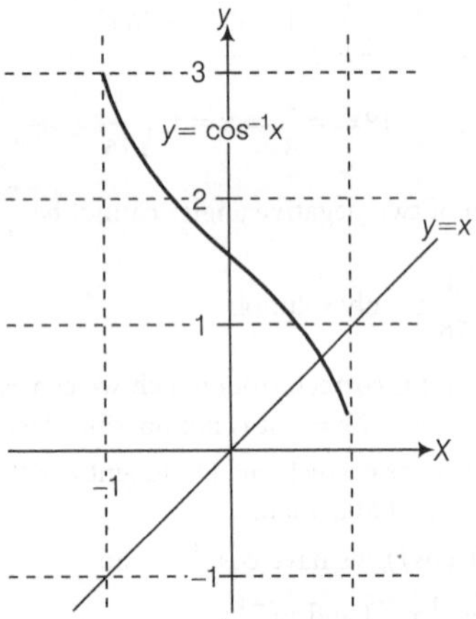

(A)

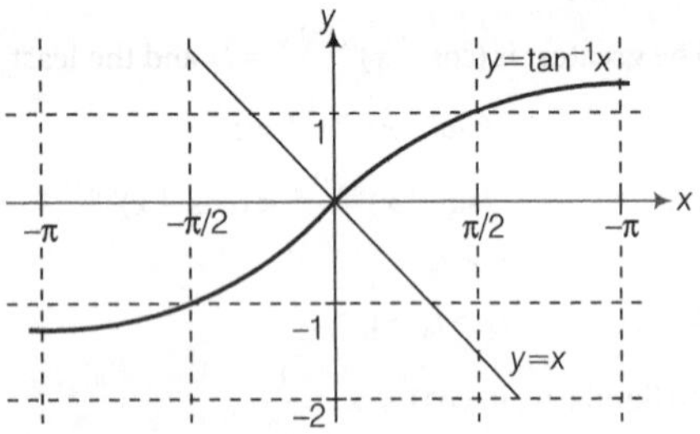

(C)

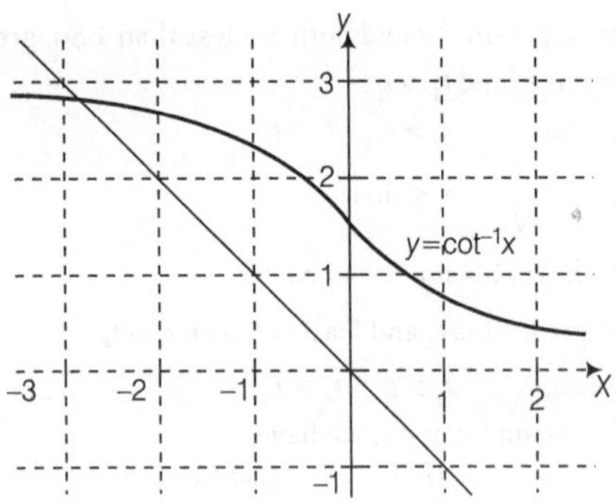

128. $\sin^{-1}\left(\frac{5}{x}\right) + \sin^{-1}\left(\frac{12}{x}\right) = \frac{\pi}{2}$

$$\Leftrightarrow \quad x > 12 \text{ and } \cos^{-1}\left(\frac{\sqrt{x^2 - 25}}{x}\right) + \sin^{-1}\left(\frac{12}{x}\right) = \frac{\pi}{2}$$

$$\Leftrightarrow \quad x > 12 \text{ and } \cos^{-1}\left(\frac{\sqrt{x^2 - 25}}{x}\right) = \cos^{-1}\left(\frac{12}{x}\right)$$

$$\Leftrightarrow \quad x > 12 \text{ and } \frac{\sqrt{x^2 - 25}}{x} = \frac{12}{x}$$

$$\Leftrightarrow \quad x > 12 \text{ and } x^2 - 25 = 12^2$$

$$\Leftrightarrow \quad x > 12 \text{ and } x^2 = 169$$

$$\Leftrightarrow \quad x > 12 \text{ and } x = \pm 13$$

$\therefore$ $x = 13$ is only solution.

129. Taking tan on both sides

$$\frac{\frac{x+1}{x-1} + \frac{x-1}{x}}{1 - \frac{x+1}{x-1} \cdot \frac{x-1}{x}} = -7 \text{ on simplification, we get } x = 2$$

Substituting, we get

$$\tan^{-1} 3 + \tan^{-1}\frac{1}{2} = \tan^{-1}(-7) \text{ but L.H.S is}$$

$\pi + \tan^{-1}(-7)$. Hence, no solution.

130. Let $X = \sqrt{\frac{a(a+b+c)}{bc}}$, $Y = \sqrt{\frac{b(a+b+c)}{ac}}$, $Z = \sqrt{\frac{c(a+b+c)}{ab}}$

$$XY = \sqrt{\frac{a(a+b+c)}{bc} \cdot \frac{b(a+b+c)}{ac}} = \frac{a+b+c}{c}$$

$$= \left(1 + \frac{a+b}{c}\right) > 1 \, (a, b, c > 0)$$

$$\therefore \quad \tan^{-1} X + \tan^{-1} Y = \pi + \tan^{-1}\left[\frac{X+Y}{1-XY}\right]$$

$$= \pi + \tan^{-1}\left(\frac{\sqrt{\frac{a(a+b+c)}{bc}} + \sqrt{\frac{b(a+b+c)}{ac}}}{1 - \frac{a+b+c}{c}}\right)$$

$$= \pi + \tan^{-1}\left(\frac{\sqrt{\frac{a+b+c}{c}}\left(\frac{\sqrt{a}}{\sqrt{b}} + \frac{\sqrt{b}}{\sqrt{a}}\right)}{-\frac{(a+b)}{c}}\right)$$

$$= \pi + \tan^{-1}\left(-\sqrt{\frac{a+b+c}{c}}\left(\frac{a+b}{\sqrt{ab}}\right)\left(\frac{c}{a+c}\right)\right)$$

$$= \pi - \tan^{-1}\frac{\sqrt{c(a+b+c)}}{ab} = \pi - \tan^{-1} Z$$

$$\therefore \quad \tan^{-1} X + \tan^{-1} Y + \tan^{-1} Z = \pi$$

131. Let $\theta = \operatorname{cosec}^{-1}\sqrt{(n^2+1)(n^2+2n+2)}$

$$\operatorname{cosec}^2\theta = (n^2+1)(n^2+2n+2)$$

$$= (n^2+1)^2 + 2n(n^2+1) + n^2 + 1 = (n^2+n+1)^2 + 1$$

$$\cot^2\theta = (n^2+n+1)^2$$

$$\tan\theta = \frac{1}{n^2+n+1} = \frac{(n+1)-n}{1+(n+1)n}$$

$$\theta = \tan^{-1}\left[\frac{(n+1)-n}{1+(n+1)n}\right]$$

$$= \tan^{-1}(n+1) - \tan^{-1} n$$

Thus, sum of n terms of the given series

$$= (\tan^{-1}2 - \tan^{-1}1) + (\tan^{-1}3 - \tan^{-1}2)$$
$$+ (\tan^{-1}4 - \tan^{-1}3) + \dots + (\tan^{-1}(n+1) - \tan^{-1}n)$$
$$= \tan^{-1}(n+1) - \pi/4$$

132. We have

$$E = \sqrt{\sin^{-1}x_1}\sqrt{\cos^{-1}x_2} + \sqrt{\sin^{-1}x_2}\sqrt{\cos^{-1}x_3}$$
$$+ \sqrt{\sin^{-1}x_3}\sqrt{\cos^{-1}x_4} + \dots + \sqrt{\sin^{-1}x_{28}}\sqrt{\cos^{-1}x_1}$$

$x_1 \in [0, 1]\ \forall i = 1, 2, 3, \dots, 28$

$\therefore \sin^{-1}x_1 > 0$

Now using A.M. $\geq$ G.M. we have $\frac{a^2+b^2}{2} \geq ab$, where $a, b > 0$

$$\therefore \quad \sqrt{\sin^{-1}x_1}\sqrt{\cos^{-1}x_2} \leq \left(\frac{\sin^{-1}x_1 + \cos^{-1}x_2}{2}\right)$$

$$\sqrt{\sin^{-1}x_2}\sqrt{\cos^{-1}x_3} \leq \left(\frac{\sin^{-1}x_2 + \cos^{-1}x_3}{2}\right)$$

$$\sqrt{\sin^{-1}x_{28}}\sqrt{\cos^{-1}x_1} \leq \left(\frac{\sin^{-1}x_{28} + \sin^{-1}x_1}{2}\right)$$

On adding all, we get

$$E \leq \sum_{i=1}^{28} \frac{\sin^{-1}x_i + \cos^{-1}x_i}{2}$$

$$\therefore \quad E \leq \frac{28\left(\frac{\pi}{2}\right)}{2}$$

$$\therefore \quad E_{\max} = 7\pi$$

133. $S = \sum_{r=1}^{10}\sum_{s=1}^{10} \tan^{-1}\left(\frac{r}{s}\right)$

$$\therefore \quad S = \sum_{r=1}^{10}\sum_{s=1}^{10} \tan^{-1}\left(\frac{s}{r}\right) \text{ (As } r \text{ and } s \text{ are independent)}$$

On adding, we get

$$2S = \sum_{r=1}^{10}\sum_{s=1}^{10}\left(\tan^{-1}\left(\frac{r}{s}\right) + \tan^{-1}\left(\frac{s}{r}\right)\right)$$

$$\Rightarrow \quad 2S = \sum_{r=1}^{10}\sum_{s=1}^{10}\frac{\pi}{2} = \frac{\pi}{2}\sum_{r=1}^{10}10 = \frac{100\pi}{2}$$

$$S = 25\pi$$

134. $S_{n-1} = \sum_{k=2}^{n} \cos^{-1}\left(\frac{1+\sqrt{(k-1)k(k+1)(k+2)}}{k(k+1)}\right)$

$$= \cos^{-1}\left(\frac{1}{k}\cdot\frac{1}{(k+1)} + \frac{1+\sqrt{(k-1)k(k+1)(k+2)}}{k(k+1)}\right)$$

$$= \cos^{-1}\left(\frac{1}{k}\cdot\frac{1}{k+1} + \sqrt{1-\frac{1}{k^2}}\sqrt{1-\frac{1}{(k+1)^2}}\right)$$

$$= \cos^{-1}\frac{1}{k+1} - \cos^{-1}\frac{1}{k}$$

Substituting $k = 2, 3, 4 \dots$, we get

$$t_2 = \cos^{-1}\left(\frac{1}{3}\right) - \cos^{-1}\left(\frac{1}{2}\right)$$

$$t_3 = \cos^{-1}\left(\frac{1}{4}\right) - \cos^{-1}\left(\frac{1}{3}\right)$$

$$t_n = \cos^{-1}\left(\frac{1}{n+1}\right) - \cos^{-1}\left(\frac{1}{n}\right)$$

$$S_{n-1} = \cos^{-1}\left(\frac{1}{n+1}\right) - \cos^{-1}\left(\frac{1}{2}\right)$$

$$\therefore \quad S = \lim_{n\to\infty}\left[\cos^{-1}\left(\frac{1}{n+1}\right) - \cos^{-1}\left(\frac{1}{2}\right)\right]$$

$$= \cos^{-1}(0) - \cos^{-1}\left(\frac{1}{2}\right)$$

$$= \frac{\pi}{2} - \frac{\pi}{3} = \frac{\pi}{6}$$

135. Given, $\frac{m\tan(\alpha-\theta)}{\cos^2\theta} = \frac{n\tan\theta}{\cos^2(\alpha-\theta)}$

$$\Rightarrow \quad \frac{n}{m} = \frac{\tan(\alpha-\theta).\cos^2(\alpha-\theta)}{\cos^2\theta.\tan\theta}$$

$$= \frac{2\sin(\alpha-\theta).\cos(\alpha-\theta)}{2\sin\theta.\cos\theta}$$

$$\Rightarrow \quad \frac{n}{m} = \frac{\sin(2\alpha-2\theta)}{\sin(2\theta)}$$

using componendo and dividendo; we get.

$$\Rightarrow \quad \frac{n-m}{n+m} = \frac{\sin(2\alpha-2\theta)-\sin2\theta}{\sin(2\alpha-2\theta)+\sin2\theta} = \frac{2\sin(\alpha-2\theta).\cos\alpha}{2\sin\alpha\cos(\alpha-2\theta)}$$

$$\Rightarrow \quad \left(\frac{n-m}{n+m}\right)\tan\alpha = \tan(\alpha-2\theta)$$

$$\Rightarrow \quad \alpha - 2\theta = \tan^{-1}\left\{\left(\frac{n-m}{n+m}\right)\tan\alpha\right\}$$

$$\Rightarrow \quad 2\theta = \alpha - \tan^{-1}\left\{\left(\frac{n-m}{n+m}\right)\tan\alpha\right\}$$

$$\Rightarrow \quad \theta = \frac{1}{2}\left[\alpha - \tan^{-1}\left\{\left(\frac{n-m}{n+m}\right)\tan\alpha\right\}\right]$$

136. Let $\tan^{-1}(e^{i\theta}) = A + iB$ or $e^{i\theta} = \tan(A+iB)$

$$\Rightarrow \quad \tan^{-1}(e^{-i\theta}) = A - iB. \text{ or } e^{-i\theta} = \tan(A-iB)$$

Now, $\cos 2A = \cos(A+iB+A-iB)$

$$\therefore \ \cos2A = \cos(A+iB).\cos(A-iB) - \sin(A+iB)\sin(A-iB)$$
$$= \cos(A+iB).\cos(A-iB)\{1-\tan(A+iB).\tan(A-iB)\}$$
$$= \cos(A+iB).\cos(A-iB)\{1-e^{i\theta}.e^{-i\theta}\}$$
$$= \cos(A+iB).\cos(A-iB)\ \{1-1\} = 0$$

$$\Rightarrow \quad 2A = (2n+1)\frac{\pi}{2} \text{ or } A = \frac{n\pi}{2} + \frac{\pi}{4}$$

Also $\tan 2iB = \tan(A+iB-A+iB)$

$$= \frac{\tan(A+iB)-\tan(A-iB)}{1+\tan(A+iB).\tan(A-iB)}$$

$$= \frac{e^{i\theta}-e^{-i\theta}}{1+e^{i\theta}.e^{-i\theta}} = i\sin\theta$$

$$\Rightarrow \quad \frac{e^{-2B} - e^{2B}}{i(e^{-2B} + e^{2B})} = i\sin\theta$$

or $$\frac{e^{2B} - e^{-2B}}{e^{2B} + e^{-2B}} = \sin\theta$$

or $$\frac{e^{2B}}{e^{-2B}} = \frac{1 + \sin\theta}{1 - \sin\theta} = \frac{1 + \cos\left(\frac{\pi}{2} - \theta\right)}{1 - \cos\left(\frac{\pi}{2} - \theta\right)}$$

$$e^{4B} = \frac{\cos^2\left(\frac{\pi}{4} - \frac{\theta}{2}\right)}{\sin^2\left(\frac{\pi}{4} - \frac{\theta}{2}\right)}$$

$$\Rightarrow \quad e^{2B} = \cot\left(\frac{\pi}{4} - \frac{\theta}{2}\right)$$

$$\Rightarrow \quad B = -\frac{1}{2}\ln\tan\left(\frac{\pi}{4} - \frac{\theta}{2}\right)$$

137. The quadratic equation $4^{\sec^2\alpha}.x^2 + 2x + \left(\beta^2 - \beta + \frac{1}{2}\right) = 0$ have real roots.

$$\Rightarrow \text{discriminant} = 4 - 4.4^{\sec^2\alpha}\left(\beta^2 - \beta + \frac{1}{2}\right) \ge 0$$

$$\Rightarrow \quad 4^{\sec^2\alpha}\left(\beta^2 - \beta + \frac{1}{2}\right) \le 1 \quad \text{...(i)}$$

But $4^{\sec^2\alpha} \ge 4$, $\beta^2 - \beta + \frac{1}{2} = \left(\beta - \frac{1}{2}\right)^2 + \frac{1}{4} \ge \frac{1}{4}$

i.e. equation will be satisfied only when

$$4^{\sec^2\alpha} = 4 \quad \text{and } \beta^2 - \beta + \frac{1}{2} = \frac{1}{4}$$

$$\Rightarrow \quad \sec^2\alpha = 1 \text{ and } \left(\beta - \frac{1}{2}\right)^2 = 0$$

$$\Rightarrow \quad \cos^2\alpha = 1 \text{ and } \beta = \frac{1}{2}$$

$$\Rightarrow \quad \alpha = n\pi \text{ and } \beta = \frac{1}{2}$$

$$\therefore \quad \cos\alpha + \cos^{-1}\beta = \cos n\pi + \cos^{-1}\frac{1}{2}$$

$$= \begin{cases} 1 + \frac{\pi}{3}, & \text{when } n \text{ is even integer} \\ -1 + \frac{\pi}{3}, & \text{when } n \text{ is odd integer} \end{cases}$$

i.e. the value of;

$$\cos\alpha + \cos^{-1}\beta \text{ is } \frac{\pi}{3} - 1 \text{ and } \frac{\pi}{3} + 1.$$

138. Here, $\alpha = 3\sin^{-1}\left(\frac{6}{11}\right)$ and $\beta = 3\cos^{-1}\left(\frac{4}{9}\right)$ as $\frac{6}{11} > \frac{1}{2}$

$$\Rightarrow \quad \sin^{-1}\left(\frac{6}{11}\right) > \sin^{-1}\left(\frac{1}{2}\right) = \frac{\pi}{6}$$

$$\therefore \quad \alpha = 3\sin^{-1}\left(\frac{6}{11}\right) > \frac{\pi}{2} \Rightarrow \cos\alpha < 0$$

Now, $\beta = 3\cos^{-1}\left(\frac{4}{9}\right)$

As $\frac{4}{9} < \frac{1}{2} \Rightarrow \cos^{-1}\left(\frac{4}{9}\right) > \cos^{-1}\left(\frac{1}{2}\right) = \frac{\pi}{3}$

$$\therefore \quad \beta = 3\cos^{-1}\left(\frac{4}{9}\right) > \pi$$

$$\therefore \quad \cos\beta < 0 \text{ and } \sin\beta < 0$$

Now, $\alpha + \beta$ is slightly greater than $\frac{3\pi}{2}$.

$$\therefore \quad \cos(\alpha + \beta) > 0$$

139. We have, $0 < x < 1$

Let $\cot^{-1} x = \theta \Rightarrow \cot\theta = x$

$$\Rightarrow \quad \sin\theta = \frac{1}{\sqrt{1 + x^2}} = \sin(\cot^{-1} x)$$

and $$\cos\theta = \frac{x}{\sqrt{1 + x^2}} = \cos(\cot^{-1} x)$$

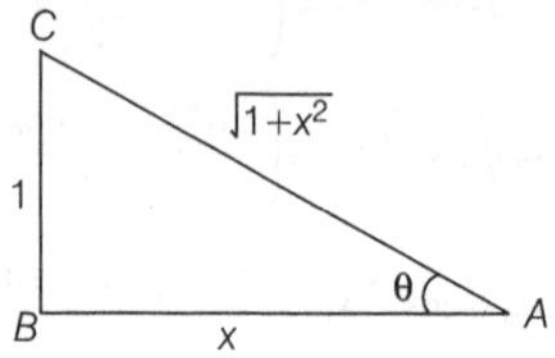

Now, $\sqrt{1 + x^2}\,[\{x\cos(\cot^{-1} x) + \sin(\cot^{-1} x)\}^2 - 1]^{1/2}$

$$= \sqrt{1 + x^2}\left[\left(x\frac{x}{\sqrt{1 + x^2}} + \frac{1}{\sqrt{1 + x^2}}\right)^2 - 1\right]^{1/2}$$

$$= \sqrt{1 + x^2}\left[\left(\frac{1 + x^2}{\sqrt{1 + x^2}}\right)^2 - 1\right]^{1/2}$$

$$= \sqrt{1 + x^2}[1 + x^2 - 1]^{1/2} = x\sqrt{1 + x^2}$$

140. Given, $\tan^{-1} y = \tan^{-1} x + \tan^{-1}\left(\frac{2x}{1 - x^2}\right)$, where $|x| < \frac{1}{\sqrt{3}}$

$$\Rightarrow \quad \tan^{-1} y = \tan^{-1}\left\{\frac{x + \frac{2x}{1 - x^2}}{1 - x\left(\frac{2x}{1 - x^2}\right)}\right\}$$

$$\left[\because \tan^{-1} x + \tan^{-1} y = \tan^{-1}\left(\frac{x + y}{1 - xy}\right),\right.$$

$$\left.\text{where } x > 0, y > 0 \text{ and } xy < 1\right]$$

$$= \tan^{-1}\left(\frac{x - x^3 + 2x}{1 - x^2 - 2x^2}\right)$$

$$\tan^{-1} y = \tan^{-1}\left(\frac{3x - x^3}{1 - 3x^2}\right)$$

$$\Rightarrow \quad y = \frac{3x - x^3}{1 - 3x^2}$$

Aliter

$$|x| < \frac{1}{\sqrt{3}} \Rightarrow -\frac{1}{\sqrt{3}} < x < \frac{1}{\sqrt{3}}$$

Let $x = \tan\theta$

$$\Rightarrow \quad -\frac{\pi}{6} < \theta < \frac{\pi}{6}$$

$$\therefore \quad \tan^{-1} y = \theta + \tan^{-1}(\tan 2\theta)$$

$$= \theta + 2\theta = 3\theta$$

$$\Rightarrow \quad y = \tan 3\theta$$

$$\Rightarrow \quad y = \frac{3\tan\theta - \tan^3\theta}{1 - 3\tan^2\theta}$$

$$\Rightarrow \quad y = \frac{3x - x^3}{1 - 3x^2}$$

141. We have, $\cot\left[\sum_{n=1}^{23} \cot^{-1}\left(1 + \sum_{k=1}^{n} 2k\right)\right]$

$$\Rightarrow \cot\left[\sum_{n=1}^{23} \cot^{-1}(1 + 2 + 4 + 6 + 8 + \ldots + 2n)\right]$$

$$\Rightarrow \cot\left[\sum_{n=1}^{23} \cot^{-1}\{1 + n(n+1)\}\right]$$

$$\Rightarrow \quad \cot\left[\sum_{n=1}^{23} \tan^{-1}\frac{1}{1 + n(n+1)}\right]$$

$$\Rightarrow \quad \cot\left[\sum_{n=1}^{23} \tan^{-1}\left\{\frac{(n+1) - n}{1 + n(n+1)}\right\}\right]$$

$$\Rightarrow \quad \cot\left[\sum_{n=1}^{23} (\tan^{-1}(n+1) - \tan^{-1} n)\right]$$

$$\Rightarrow \cot[(\tan^{-1}2 - \tan^{-1}1) + (\tan^{-1}3 - \tan^{-1}2) + (\tan^{-1}4 - \tan^{-1}3)] + \ldots + (\tan^{-1}24 - \tan^{-1}23)]$$

$$\Rightarrow \cot(\tan^{-1}24 - \tan^{-1}1)$$

$$\Rightarrow \cot\left(\tan^{-1}\frac{24 - 1}{1 + 24\cdot(1)}\right) = \cot\left(\tan^{-1}\frac{23}{25}\right)$$

$$= \cot\left(\cot^{-1}\frac{25}{23}\right) = \frac{25}{23}$$

142. Since, x, y and z are in an AP.

$$\therefore \quad 2y = x + z$$

Also, $\tan^{-1} x$, $\tan^{-1} y$ and $\tan^{-1} z$ are in an AP.

$$\therefore \quad 2\tan^{-1} y = \tan^{-1} x + \tan^{-1}(z)$$

$$\Rightarrow \quad \tan^{-1}\left(\frac{2y}{1 - y^2}\right) = \tan^{-1}\left(\frac{x + z}{1 - xz}\right)$$

$$\Rightarrow \quad \frac{x + z}{1 - y^2} = \frac{x + z}{1 - xz}$$

$$\Rightarrow \quad y^2 = xz$$

Since, x, y and z are in an AP as well as in a GP.

$$\therefore \quad x = y = z$$

143. Since, $\operatorname{cosec}^{-1}\left(\frac{5}{3}\right) = \tan^{-1}\left(\frac{3}{4}\right)$

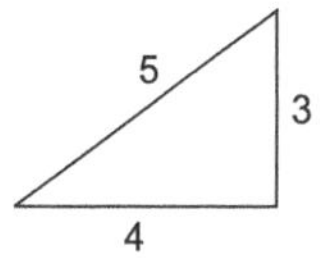

$$\therefore \quad \cot\left(\tan^{-1}\frac{3}{4} + \tan^{-1}\frac{2}{3}\right)$$

$$= \cot\tan^{-1}\left[\frac{\frac{3}{4} + \frac{2}{3}}{1 - \frac{1}{2}}\right] = \cot\tan^{-1}\left[\frac{\left(\frac{17}{12}\right)}{\left(\frac{1}{2}\right)}\right]$$

$$= \cot\left[\tan^{-1}\left(\frac{17}{6}\right)\right] = \frac{6}{17}$$

144. We have, $\sin^{-1}\left(\frac{x}{5}\right) + \operatorname{cosec}^{-1}\left(\frac{5}{4}\right) = \frac{\pi}{2}$

$$\Rightarrow \quad \sin^{-1}\left(\frac{x}{5}\right) + \sin^{-1}\left(\frac{4}{5}\right) = \frac{\pi}{2}$$

$$\Rightarrow \quad \sin^{-1}\left(\frac{x}{5}\right) = \frac{\pi}{2} - \sin^{-1}\left(\frac{4}{5}\right)$$

$$\Rightarrow \quad \sin^{-1}\left(\frac{x}{5}\right) = \cos^{-1}\left(\frac{4}{5}\right)$$

$$\Rightarrow \quad \sin^{-1}\left(\frac{x}{5}\right) = \sin^{-1}\left(\frac{3}{5}\right)$$

$$\therefore \quad x = 3$$

145. Given that, $\cos^{-1} x - \cos^{-1}\frac{y}{2} = \alpha$

$$\Rightarrow \quad \cos^{-1}\left(\frac{xy}{2} + \sqrt{1 - x^2}\sqrt{1 - \frac{y^2}{4}}\right) = \alpha$$

$$\Rightarrow \quad \frac{xy}{2} + \sqrt{1 - x^2}\sqrt{1 - \frac{y^2}{4}} = \cos\alpha$$

$$\Rightarrow \quad 2\sqrt{1 - x^2}\sqrt{1 - \frac{y^2}{4}} = 2\cos\alpha - xy$$

On squaring both sides, we get

$$\frac{4(1 - x^2)(4 - y^2)}{4} = 4\cos^2\alpha + x^2y^2 - 4xy\cos\alpha$$

$$\Rightarrow \quad 4 - 4x^2 - y^2 + x^2y^2 = 4\cos^2\alpha + x^2y^2 - 4xy\cos\alpha$$

$$\therefore \quad 4x^2 - 4xy\cos\alpha + y^2 = 4\sin^2\alpha$$

Previous Years' Questions

JEE Main and Advanced (2021-18)

Trigonometric Functions and Identities

1. cosec 18° is a root of the equation **[JEE Main 2021]**

(a) $x^2+2x-4=0$ (b) $4x^2+2x-1=0$

(c) $x^2-2x+4=0$ (d) $x^2-2x-4=0$

Ans. *(d)*

$$\text{cosec}18° = \frac{1}{\sin 18°} = \frac{4}{\sqrt{5}-1} = \sqrt{5}+1$$

If $x=\sqrt{5}+1$, then

$$(x-1)^2 = 5 \Rightarrow x^2-2x-4=0$$

2. The value of

$2\sin\left(\frac{\pi}{8}\right)\sin\left(\frac{2\pi}{8}\right)\sin\left(\frac{3\pi}{8}\right)\sin\left(\frac{5\pi}{8}\right)\sin\left(\frac{6\pi}{8}\right)\sin\left(\frac{7\pi}{8}\right)$ is **[JEE Main 2021]**

(a) $\frac{1}{4\sqrt{2}}$ (b) $\frac{1}{4}$

(c) $\frac{1}{8}$ (d) $\frac{1}{8\sqrt{2}}$

Ans. *(c)*

$$2\sin\left(\frac{\pi}{8}\right)\sin\left(\frac{2\pi}{8}\right)\sin\left(\frac{3\pi}{8}\right)\sin\left(\frac{5\pi}{8}\right)\sin\left(\frac{6\pi}{8}\right)\sin\left(\frac{7\pi}{8}\right)$$

$$=2\sin\frac{\pi}{8}\sin\frac{2\pi}{8}\sin\frac{3\pi}{8}\sin\left(\pi-\frac{3\pi}{8}\right)\sin\left(\pi-\frac{2\pi}{8}\right)\sin\left(\pi-\frac{\pi}{8}\right)$$

$$=2\sin\frac{\pi}{8}\sin\frac{2\pi}{8}\sin\frac{3\pi}{8}\sin\frac{3\pi}{8}\sin\frac{2\pi}{8}\sin\frac{\pi}{8}$$

$$=2\sin^2\left(\frac{\pi}{8}\right)\sin^2\left(\frac{2\pi}{8}\right)\sin^2\left(\frac{3\pi}{8}\right)$$

$$2\sin^2\frac{\pi}{8}\left(\frac{1}{\sqrt{2}}\right)^2\sin^2\left(\frac{\pi}{2}-\frac{\pi}{8}\right)$$

$$=2\sin^2\frac{\pi}{8}\times\frac{1}{2}\times\cos^2\frac{\pi}{8}$$

$$=\sin^2\left(\frac{\pi}{8}\right)\cos^2\left(\frac{\pi}{8}\right)$$

$$=\frac{1}{4}\left(2\sin\frac{\pi}{8}\cos\frac{\pi}{8}\right)^2\frac{1}{4}\sin^2\left(\frac{\pi}{4}\right)$$

$$=\frac{1}{4}\left(\frac{1}{\sqrt{2}}\right)^2=\frac{1}{8}$$

3. Let $f: R \to R$ be defined as

$f(x+y)+f(x-y)=2f(x)f(y), f\left(\frac{1}{2}\right)=-1.$

Then, the value of

$\sum_{k=1}^{20}\frac{1}{\sin(k)\sin[k+f(k)]}$ is equal to **[JEE Main 2021]**

(a) $\text{cosec}^2(21)\cos(20)\cos(2)$

(b) $\sec^2(1)\sec(21)\cos(20)$

(c) $\text{cosec}^2(1)\,\text{cosec}(21)\sin(20)$

(d) $\sec^2(21)\sin(20)\sin(2)$

Ans. *(c)*

$f: R \to R$ is defined as

$$f(x+y)+f(x-y)=2f(x)f(y)$$

Let $f(x)=\cos\lambda x$

$\because$ $f\left(\frac{1}{2}\right)=-1 \Rightarrow \cos\frac{\lambda}{2}=-1$

$\Rightarrow$ $\lambda=2\pi$

Thus, $f(x)=\cos 2\pi x$

Now, k is a natural number.

Thus, $f(k)=1$ $(\because \cos 2n\pi=1, n\in Z)$

Now, $\sum_{k=1}^{20}\frac{1}{\sin k\cdot\sin[k+f(k)]}$

$$=\sum_{k=1}^{20}\frac{1}{\sin k\cdot\sin(k+1)}$$

Multiply and divide by $\sin 1$,

$$=\frac{1}{\sin 1}\sum_{k=1}^{20}\frac{\sin 1}{\sin k\cdot\sin(k+1)}$$

$$=\frac{1}{\sin 1}\sum_{k=1}^{20}\frac{\sin[(k+1)-k]}{\sin k\cdot\sin(k+1)}$$

$$=\frac{1}{\sin 1}\sum_{k=1}^{20}[\cot k-\cot(k+1)]$$

$$=\frac{1}{\sin 1}(\cot 1-\cot 21)$$

$$=\frac{1}{\sin 1}\left(\frac{\cos 1}{\sin 1}-\frac{\cos 21}{\sin 21}\right)$$

$$=\frac{1}{\sin 1}\left(\frac{\cos 1\sin 21-\cos 21\sin 1}{\sin 1\cdot\sin 21}\right)$$

$$=\frac{1}{\sin 1}\left(\frac{\sin 20}{\sin 1\cdot\sin 21}\right)$$

$$=\text{cosec}^2 1\cdot\text{cosec}\,21\cdot\sin 20$$

4. The value of $\cot\frac{\pi}{24}$ is **[JEE Main 2021]**

(a) $\sqrt{2}+\sqrt{3}+2-\sqrt{6}$

(b) $\sqrt{2}+\sqrt{3}+2+\sqrt{6}$

(c) $\sqrt{2}-\sqrt{3}-2+\sqrt{6}$

(d) $3\sqrt{2}-\sqrt{3}-\sqrt{6}$

Ans. *(b)*

We know that,

$$\cot\theta=\frac{1+\cos 2\theta}{\sin 2\theta}$$

When $\theta=\frac{\pi}{24}$,

$$\cot\left(\frac{\pi}{24}\right)=\frac{1+\cos\left(\frac{\pi}{12}\right)}{\sin\left(\frac{\pi}{12}\right)}$$

$$=\frac{1+\left(\frac{\sqrt{3}+1}{2\sqrt{2}}\right)}{\frac{\sqrt{3}-1}{2\sqrt{2}}}$$

$$=\frac{2\sqrt{2}+\sqrt{3}+1}{\sqrt{3}-1}\times\frac{\sqrt{3}+1}{\sqrt{3}+1}$$

$$=\frac{2\sqrt{6}+2\sqrt{2}+3+\sqrt{3}+\sqrt{3}+1}{2}$$

$$=\frac{2(\sqrt{6}+\sqrt{2}+\sqrt{3}+2)}{2}$$

$$=\sqrt{6}+\sqrt{2}+\sqrt{3}+2$$

5. If $15\sin^4\alpha+10\cos^4\alpha=6$, for some $\alpha\in R$, then the value of $27\sec^6\alpha+8\text{cosec}^6\alpha$ is equal to **[JEE Main 2021]**

(a) 350 (b) 500 (c) 400 (d) 250

Ans. *(d)*

Given, $15\sin^4\alpha+10\cos^4\alpha=6$

$\Rightarrow$

$15\sin^4\alpha+10\cos^4\alpha=6(\sin^2\alpha+\cos^2\alpha)^2$

$\Rightarrow$ $15\sin^4\alpha+10\cos^4\alpha=6$

$(\sin^4\alpha+\cos^4\alpha+2\sin^2\alpha\cos^2\alpha)$

$\Rightarrow 9\sin^4\alpha+4\cos^4\alpha-12\sin^2\alpha\cos^2\alpha=0$

$\Rightarrow$ $(3\sin^2\alpha-2\cos^2\alpha)^2=0$

$\Rightarrow$ $3\sin^2\alpha-2\cos^2\alpha=0$

$\Rightarrow$ $3\sin^2\alpha=2\cos^2\alpha$

$\Rightarrow \quad \tan^2 \alpha = \frac{2}{3}$

$\therefore \quad \cot^2 \alpha = \frac{3}{2}$

Now, $27\sec^6 \alpha + 8\text{cosec}^6\alpha$

$= 27(\sec^2 \alpha)^3 + 8(\text{cosec}^2\alpha)^3$

$= 27(1+\tan^2 \alpha)^3 + 8(+\cot^2 \alpha)^3$

$= 27\left(1+\frac{2}{3}\right)^3 + 8\left(1+\frac{3}{2}\right)^3 = 250$

6. If for $x \in \left(0, \frac{\pi}{3}\right)$, $\log_{10} \sin x + \log_{10} \cos x = -1$ and $\log_{10}(\sin x + \cos x) = \frac{1}{2}(\log_{10} n - 1), n > 0$, then the value of n is equal to **[JEE Main 2021]**

(a) 20 (b) 12

(c) 9 (d) 16

Ans. *(b)*

$\log_{10}\sin x + \log_{10}\cos x = -1, x \in (0, \pi/2)$

$\log_{10}(\sin x \cos x) = -1$

$\Rightarrow \quad \sin x \cos x = 10^{-1} = 1/10$

$\log_{10}(\sin x + \cos x) = 1/2(\log_{10} n - 1), n > 0$

$2\log_{10}(\sin x + \cos x)$

$= (\log_{10} n - \log_{10} 10)$

$\Rightarrow \quad \log_{10}(\sin x + \cos x)^2 = \log_{10}(n/10)$

$\Rightarrow \quad (\sin x + \cos x)^2 = n/10$

$\Rightarrow \sin^2 x + \cos^2 x + 2\sin x \cos x = \frac{n}{10}$

$\Rightarrow 1 + 2(1/10) = n/10 \Rightarrow 12/10 = n/10$

$\therefore \quad n = 12$

7. If $0 < a, b < 1$ and $\tan^{-1} a + \tan^{-1} b = \frac{\pi}{4}$, then the value of $(a+b) - \left(\frac{a^2+b^2}{2}\right) + \left(\frac{a^3+b^3}{3}\right) - \left(\frac{a^4+b^4}{4}\right) + \ldots$ is **[JEE Main 2021]**

(a) $\log_e 2$ (b) $e^2 - 1$

(c) e (d) $\log_e\left(\frac{e}{2}\right)$

Ans. *(a)*

Given, $\tan^{-1} a + \tan^{-1} b = \frac{\pi}{4}$

$\Rightarrow \quad \tan^{-1}\left[\frac{a+b}{1-ab}\right] = \frac{\pi}{4}$

$\Rightarrow \quad \frac{a+b}{1-ab} = \tan\frac{\pi}{4} = 1$

$\Rightarrow \quad a + b = 1 - ab$

$\Rightarrow \quad (1+a)(1+b) = 2 \quad \ldots(i)$

Now,

$(a+b) - \left(\frac{a^2+b^2}{2}\right) + \left(\frac{a^3+b^3}{3}\right) - \left(\frac{a^4+b^4}{4}\right) + \ldots$

$= \left[a - \frac{a^2}{2} + \frac{a^3}{3} - \frac{a^4}{4} + \ldots\right] + \left[b - \frac{b^2}{2} + \frac{b^3}{3} - \frac{b^4}{4} + \ldots\right]$

$= \log_e(1+a) + \log_e(1+b)$ {Using expansion of $\log_e(1+x)$}

$= \log_e(1+a)(1+b)$ $\{\because \log a + \log b = \log ab\}$

$= \log_e 2$ [use Eq. (i)]

8. If $0 < x, y < \pi$ and $\cos x + \cos y - \cos(x+y) = \frac{3}{2}$, then $\sin x + \cos y$ is equal to **[JEE Main 2021]**

(a) $\frac{1}{2}$ (b) $\frac{\sqrt{3}}{2}$

(c) $\frac{1-\sqrt{3}}{2}$ (d) $\frac{1+\sqrt{3}}{2}$

Ans. *(d)*

Given, $\cos x + \cos y - \cos(x+y) = \frac{3}{2}$

$\Rightarrow 2\cos\left(\frac{x+y}{2}\right)\cos\left(\frac{x-y}{2}\right) - \left[2\cos^2\left(\frac{x+y}{2}\right) - 1\right] = \frac{3}{2}$

[Use formula,

$\cos a + \cos b = 2\cos\left(\frac{a+b}{2}\right)\cos\left(\frac{a-b}{2}\right)$,

$\cos 2x = 2\cos^2 x - 1$]

$\Rightarrow$

$2\cos\left(\frac{x+y}{2}\right)\cos\left(\frac{x-y}{2}\right) - 2\cos^2\left(\frac{x+y}{2}\right)$

$= \frac{3}{2} - 1 = \frac{1}{2}$

$\Rightarrow$

$4\cos\left(\frac{x+y}{2}\right)\cos\left(\frac{x-y}{2}\right) - 4\cos^2\left(\frac{x+y}{2}\right)$

$= \frac{1}{2} \times 2 = 1$

$= \cos^2\frac{(x-y)}{2} + \sin^2\frac{(x-y)}{2}$

$\Rightarrow \left[\cos\left(\frac{x-y}{2}\right) - 2\cos\left(\frac{x+y}{2}\right)\right]^2 + \sin^2\left(\frac{x-y}{2}\right) = 0$

$\Rightarrow \sin\left(\frac{x-y}{2}\right) = 0$

and $\cos\left(\frac{x-y}{2}\right) - 2\cos\left(\frac{x+y}{2}\right) = 0$

$\Rightarrow x = y$ and $\cos 0 - 2\cos x = 0$ (use $x = y$)

Given, $\cos x = \frac{1}{2} = \cos y$

$\therefore \quad \sin x = \sqrt{1-\cos^2 x}$

$= \sqrt{1-\frac{1}{4}} = \frac{\sqrt{3}}{2}$

$\therefore \sin x + \cos y = \frac{\sqrt{3}}{2} + \frac{1}{2}$

$= \frac{1+\sqrt{3}}{2}$

9. Let $\alpha = \max_{x \in R}\{8^{2\sin 3x} \cdot 4^{4\cos 3x}\}$ and $\beta = \max_{x \in R}\{8^{2\sin 3x} \cdot 4^{4\cos 3x}\}$. If $8x^2 + bx + c = 0$ is a quadratic equation whose roots are $\alpha^{1/5}$ and $\beta^{1/5}$, then the value of $c - b$ is equal to **[JEE Main 2021]**

(a) 42 (b) 47

(c) 43 (d) 50

Ans. *(a)*

Let $\alpha = \max(8^{2\sin 3x} \cdot 4^{4\cos 3x})$

$= \max(2^{6\sin 3x} \cdot 2^{8\cos 3x})$

$= \max(2^{6\sin 3x + 8\cos 3x})$

and $\beta = \max(8^{2\sin 3x} \cdot 4^{4\cos 3x})$

$= \max(2^{6\sin 3x} \cdot 2^{8\cos 3x})$

$= \max(2^{6\sin 3x + 8\cos 3x})$

Now, determine the range of $6\sin 3x + 8\cos 3x$

($\because$ Range of $a\sin x + b\cos x$

$= [-\sqrt{a^2+b^2}, \sqrt{a^2+b^2}]$

$= [-\sqrt{6^2+8^2}, \sqrt{6^2+8^2}]$

$= [-10, 10]$

So, $\quad \alpha = 2^{10}$ and $\beta = 2^{-10}$

Now, $\quad \alpha^{1/5} = (2^{10})^{1/5} = 4$

$\beta^{1/5} = (2^{-10})^{1/5} = \frac{1}{4}$

Quadratic equation with roots 4 and $\frac{1}{4}$ is

$x^2 - \left(4 + \frac{1}{4}\right)x + 4 \times \frac{1}{4} = 0$

$\Rightarrow \quad x^2 - \frac{17}{4}x + 1 = 0$

Multiplying both sides by 8,

$8x^2 - 34x + 8 = 0$

On comparing, $8x^2 + bx + c$, we get

$b = -34$ and $c = 8$

So, $\quad c - b = 8 - (-34) = 42$

10. If $L=\sin^2\left(\frac{\pi}{16}\right)-\sin^2\left(\frac{\pi}{8}\right)$ and $M=\cos^2\left(\frac{\pi}{16}\right)-\sin^2\left(\frac{\pi}{8}\right)$, then

[JEE Main 2020]

(a) $L=-\frac{1}{2\sqrt{2}}+\frac{1}{2}\cos\frac{\pi}{8}$

(b) $L=\frac{1}{4\sqrt{2}}-\frac{1}{4}\cos\frac{\pi}{8}$

(c) $M=\frac{1}{4\sqrt{2}}+\frac{1}{4}\cos\frac{\pi}{8}$

(d) $M=\frac{1}{2\sqrt{2}}+\frac{1}{2}\cos\frac{\pi}{8}$

Ans. *(d)*

Given, $L=\sin^2\frac{\pi}{16}-\sin^2\frac{\pi}{8}$

$=\sin\left(\frac{\pi}{16}+\frac{\pi}{8}\right)\sin\left(\frac{\pi}{16}-\frac{\pi}{8}\right)$

$=\sin\frac{3\pi}{16}\sin\left(-\frac{\pi}{16}\right)=\frac{1}{2}\left[2\sin\frac{3\pi}{16}\sin\left(-\frac{\pi}{16}\right)\right]$

$=\frac{1}{2}\left[\cos\frac{\pi}{4}-\cos\frac{2\pi}{16}\right]=\frac{1}{2\sqrt{2}}-\frac{1}{2}\cos\frac{\pi}{8}.$

and, $M=\cos^2\frac{\pi}{16}-\sin^2\frac{\pi}{8}$

$=\cos\left(\frac{\pi}{16}+\frac{\pi}{8}\right)\cos\left(\frac{\pi}{16}-\frac{\pi}{8}\right)$

$=\cos\frac{3\pi}{16}\cos\left(-\frac{\pi}{16}\right)$

$=\frac{1}{2}\left[2\cos\frac{3\pi}{16}\cos\frac{\pi}{16}\right]$

$=\frac{1}{2}\left[\cos\frac{\pi}{4}+\cos\frac{2\pi}{16}\right]$

$=\frac{1}{2\sqrt{2}}+\frac{1}{2}\cos\frac{\pi}{8}.$

11. If $\frac{\sqrt{2}\sin\alpha}{\sqrt{1+\cos 2\alpha}}=\frac{1}{7}$ and $\sqrt{\frac{1-\cos 2\beta}{2}}=\frac{1}{\sqrt{10}}$, $\alpha,\beta\in\left(0,\frac{\pi}{2}\right)$, then $\tan(\alpha+2\beta)$ is equal to ....

[JEE Main 2020]

Ans. *(1)*

It is given that, for $\alpha,\beta\in\left(0,\frac{\pi}{2}\right)$

$$\frac{\sqrt{2}\sin\alpha}{\sqrt{1+\cos 2\alpha}}=\frac{1}{7}\Rightarrow\frac{\sqrt{2}\sin\alpha}{\sqrt{2}|\cos\alpha|}=\frac{1}{7}$$

$\Rightarrow\quad \tan\alpha=\frac{1}{7}\ \because\ \alpha\in\left(0,\frac{\pi}{2}\right)$ ...(i)

and $\sqrt{\frac{1-\cos 2\beta}{2}}=\frac{1}{\sqrt{10}}$

$\Rightarrow\quad |\sin\beta|=\frac{1}{\sqrt{10}}$

$\Rightarrow\quad \sin\beta=\frac{1}{\sqrt{10}}\ \because\ \beta\in\left(0,\frac{\pi}{2}\right)$

$\therefore\quad \tan\beta=\frac{1}{3}$ ...(ii)

$\therefore\quad \tan 2\beta=\frac{2\tan\beta}{1-\tan^2\beta}=\frac{\frac{2}{3}}{1-\frac{1}{9}}=\frac{6}{8}=\frac{3}{4}$

Now, $\tan(\alpha+2\beta)=\frac{\tan\alpha+\tan 2\beta}{1-\tan\alpha\tan 2\beta}$

$=\frac{\frac{1}{7}+\frac{3}{4}}{1-\left(\frac{1}{7}\times\frac{3}{4}\right)}=\frac{4+21}{28-3}=\frac{25}{25}=1$

Hence, answer is 1.

12. The value of $\cos^3\left(\frac{\pi}{8}\right)\cdot\cos\left(\frac{3\pi}{8}\right)+\sin^3\left(\frac{\pi}{8}\right)\cdot\sin\left(\frac{3\pi}{8}\right)$ is

[JEE Main 2020]

(a) $\frac{1}{4}$ (b) $\frac{1}{2\sqrt{2}}$ (c) $\frac{1}{2}$ (d) $\frac{1}{\sqrt{2}}$

Ans. *(b)*

Given trigonometric expression

$$\cos^3\left(\frac{\pi}{8}\right)\cos\left(\frac{3\pi}{8}\right)+\sin^3\left(\frac{\pi}{8}\right)\sin\frac{3\pi}{8}$$

$$\cos^3\left(\frac{\pi}{8}\right)\cos\left(\frac{\pi}{2}-\frac{\pi}{8}\right)+\sin^3\left(\frac{\pi}{8}\right)\sin\left(\frac{\pi}{2}-\frac{\pi}{8}\right)$$

$\left[\because\frac{\pi}{2}-\frac{\pi}{8}=\frac{3\pi}{8}\right]$

$=\cos^3\left(\frac{\pi}{8}\right)\sin\left(\frac{\pi}{8}\right)+\sin^3\left(\frac{\pi}{8}\right)\cos\left(\frac{\pi}{8}\right)$

$=\sin\left(\frac{\pi}{8}\right)\cos\left(\frac{\pi}{8}\right)\left[\cos^2\left(\frac{\pi}{8}\right)+\sin^2\left(\frac{\pi}{8}\right)\right]$

$=\frac{1}{2}\times 2\sin\left(\frac{\pi}{8}\right)\cos\left(\frac{\pi}{8}\right)=\frac{1}{2}\sin\frac{\pi}{4}=\frac{1}{2\sqrt{2}}$

Hence, option (b) is correct.

13 If $x=\sum_{n=0}^{\infty}(-1)^n\tan^{2n}\theta$ and $y=\sum_{n=0}^{\infty}\cos^{2n}\theta$, for $0<\theta<\frac{\pi}{4}$, then

[JEE Main 2020]

(a) $y(1+x)=1$ (b) $y(1-x)=1$

(c) $x(1+y)=1$ (d) $x(1-y)=1$

Ans. *(b)*

It is given that $x=\sum_{n=0}^{\infty}(-1)^n\tan^{2x}\theta$

$=1-\tan^2\theta+\tan^4\theta-\tan^6\theta+....+\text{upto}\ \infty$

$=\frac{1}{1+\tan^2\theta}\qquad\{\because\ \theta\in\left(0,\frac{\pi}{4}\right)$

$\Rightarrow\quad \tan^2\theta\in(0,1)\}$

$x=\cos^2\theta$...(i)

and $\quad y=\sum_{n=0}^{\infty}\cos^{2n}\theta$

$=1+\cos^2\theta+\cos^4\theta+\cos^6\theta+\ldots+\text{upto}\ \infty$

$=\frac{1}{1-\cos^2\theta}\ \{\because\ \theta\in\left(0,\frac{\pi}{4}\right)$

$\Rightarrow\cos^2\theta\in\left(0,\frac{1}{2}\right)\}$

$=\frac{1}{\sin^2\theta}\Rightarrow\ \sin^2\theta=\frac{1}{y}$ ...(ii)

On adding Eqs. (i) and (ii), we get

$$1=x+\frac{1}{y}$$

$\Rightarrow\quad y(1-x)=1$

Hence, option (b) is correct.

14. Let $f:[0,2]\to\mathbf{R}$ be the function defined by

$$f(x)=(3-\sin(2\pi x))\sin\left(\pi x-\frac{\pi}{4}\right)-\sin\left(3\pi x+\frac{\pi}{4}\right).$$

If $\alpha,\beta\in[0,2]$ are such that $\{x\in[0,2]:f(x)\geq 0\}=[\alpha,\beta]$, then the value of $\beta-\alpha$ is ......

[JEE Advanced 2020]

Ans. (1.00)

The given function $f:[0,2]\to R$ defined by

$f(x)=(3-\sin(2\pi x))\sin\left(\pi x-\frac{\pi}{4}\right)-\sin\left(3\pi x+\frac{\pi}{4}\right)$

$=(3-\sin(2\pi x))\left[\frac{\sin\pi x}{\sqrt{2}}-\frac{\cos\pi x}{\sqrt{2}}\right]-\left\{\frac{\sin 3\pi x}{\sqrt{2}}+\frac{\cos(3\pi x)}{\sqrt{2}}\right\}$

$=(3-\sin(2\pi x))\frac{[\sin(\pi x)-\cos(\pi x)]}{\sqrt{2}}-\frac{1}{\sqrt{2}}[3\sin(\pi x)-4\sin^3(\pi x)+4\cos^3(\pi x)-3\cos(\pi x)]$

$=\frac{\sin(\pi x)-\cos(\pi x)}{\sqrt{2}}[3-\sin(2\pi x)-3+4\{\sin^2(\pi x)+\cos^2(\pi x)+\sin(\pi x)\cos(\pi x)\}]$

$=\frac{\sin(\pi x)-\cos(\pi x)}{\sqrt{2}}[4+\sin(2\pi x)]$

As, $f(x)\geq 0\ \forall\in[\alpha,\beta]$, where $\alpha,\beta\in[0,2]$, so $\sin(\pi x)-\cos(\pi x)\geq 0$ as $4+\sin(2\pi x)>0$ $\forall x\in R$.

$\Rightarrow\quad \pi x\in\left[\frac{\pi}{4},\frac{5\pi}{4}\right]\Rightarrow x\in\left[\frac{1}{4},\frac{5}{4}\right]$

$\therefore\quad \alpha=\frac{1}{4}$ and $\beta=\frac{5}{4}$

Therefore the value of $(\beta-\alpha)=1$

15. If $\cos(\alpha+\beta)=\frac{3}{5}$, $\sin(\alpha-\beta)=\frac{5}{13}$ and $0<\alpha,\beta<\frac{\pi}{4}$, then $\tan(2\alpha)$ is equal to

[JEE Main 2019]

(a) $\frac{63}{52}$ (b) $\frac{63}{16}$
(c) $\frac{21}{16}$ (d) $\frac{33}{52}$

Ans. *(b)*

Given, $\sin(\alpha-\beta)=\frac{5}{13}$ and $\cos(\alpha+\beta)=\frac{3}{5}$,

where $\alpha,\beta\in\left(0,\frac{\pi}{4}\right)$

Since, $0<\alpha<\frac{\pi}{4}$ and $0<\beta<\frac{\pi}{4}$

$\therefore\ 0<\alpha+\beta<\frac{\pi}{4}+\frac{\pi}{4}=\frac{\pi}{2}$

$\Rightarrow\quad 0<\alpha+\beta<\frac{\pi}{2}$

Also, $-\frac{\pi}{4}<-\beta<0$

$\therefore\quad 0-\frac{\pi}{4}<\alpha-\beta<\frac{\pi}{4}+0\Rightarrow$

$-\frac{\pi}{4}<\alpha-\beta<\frac{\pi}{4}$

$\therefore\ \alpha+\beta\in\left(0,\frac{\pi}{2}\right)$ and $\alpha-\beta\in\left(-\frac{\pi}{4},\frac{\pi}{4}\right)$

But $\sin(\alpha-\beta)>0$, therefore $\alpha-\beta\in\left(0,\frac{\pi}{4}\right)$.

Now, $\sin(\alpha-\beta)=\frac{5}{13}$

$\Rightarrow\quad \tan(\alpha-\beta)=\frac{5}{12}$ ...(i)

and $\cos(\alpha+\beta)=\frac{3}{5}$

$\Rightarrow\quad \tan(\alpha+\beta)=\frac{4}{3}$ ...(ii)

Now, $\tan(2\alpha)=\tan[(\alpha+\beta)+(\alpha-\beta)]$

$$=\frac{\tan(\alpha+\beta)+\tan(\alpha-\beta)}{1-\tan(\alpha+\beta)\tan(\alpha-\beta)}$$

$$=\frac{\frac{4}{3}+\frac{5}{12}}{1-\frac{4}{3}\times\frac{5}{12}}$$

[from Eqs. (i) and (ii)]

$$=\frac{48+15}{36-20}=\frac{63}{16}$$

16. The value of $\sin 10^\circ\sin 30^\circ\sin 50^\circ\sin 70^\circ$ is

[JEE Main 2019]

(a) $\frac{1}{36}$ (b) $\frac{1}{32}$
(c) $\frac{1}{16}$ (d) $\frac{1}{18}$

Ans. *(c)*

We have, $\sin 10^\circ\sin 30^\circ\sin 50^\circ\sin 70^\circ$

$=\sin(30^\circ)[\sin(10^\circ)\sin(50^\circ)\sin(70^\circ)]$

$=\frac{1}{2}[\sin(10^\circ)\sin(60^\circ-10^\circ)\sin(60^\circ+10^\circ)]$

$=\frac{1}{2}\left[\frac{1}{4}\sin(3(10^\circ))\right]$

$\left[\because \sin\theta\sin(60^\circ-\theta)\sin(60^\circ+\theta)=\frac{1}{4}\sin 3\theta\right]$

$=\frac{1}{8}\sin 30^\circ=\frac{1}{8}\times\frac{1}{2}=\frac{1}{16}$

17 The equation $y=\sin x\sin(x+2)-\sin^2(x+1)$ represents a straight line lying in

[JEE Main 2019]

(a) second and third quadrants only
(b) first, second and fourth quadrants
(c) first, third and fourth quadrants
(d) third and fourth quadrants only

Ans. *(d)*

Key Idea Use formulae :
$2\sin A\sin B=\cos(A-B)-\cos(A+B)$
and $\cos 2\theta=1-2\sin^2\theta$

Given equation is
$y=\sin x\sin(x+2)-\sin^2(x+1)$
$=\frac{1}{2}[\cos 2-\cos(2x+2)]-\frac{1}{2}$
$[1-\cos(2x+2)]$

$[\because 2\sin A\sin B=\cos(A-B)-\cos(A+B)$ and $\cos 2\theta=1-2\sin^2\theta$
$\Rightarrow 2\sin^2\theta=1-\cos 2\theta]$

$=\frac{1}{2}\cos 2-\frac{1}{2}\cos(2x+2)-\frac{1}{2}+\frac{1}{2}\cos(2x+2)$

$=\frac{1}{2}(\cos(2)-1)=-\frac{1}{2}(2\sin^2(1))$

$=-\sin^2(1)<0\Rightarrow y<0$

and as we know that $y<0$, is in third and fourth quadrants only.

18. For any $\theta\in\left(\frac{\pi}{4},\frac{\pi}{2}\right)$, the expression $3(\sin\theta-\cos\theta)^4+6(\sin\theta+\cos\theta)^2+4\sin^6\theta$ equals **[JEE Main 2019]**

(a) $13-4\cos^4\theta+2\sin^2\theta\cos^2\theta$
(b) $13-4\cos^2\theta+6\cos^4\theta$
(c) $13-4\cos^2\theta+6\sin^2\theta\cos^2\theta$
(d) $13-4\cos^6\theta$

Ans. *(d)*

Given expression

$=3(\sin\theta-\cos\theta)^4+6(\sin\theta+\cos\theta)^2+4\sin^6\theta$

$=3((\sin\theta-\cos\theta)^2)^2+6(\sin\theta+\cos\theta)^2+4(\sin^2\theta)^3$

$=3(1-\sin 2\theta)^2+6(1+\sin 2\theta)+4(1-\cos^2\theta)^3$

$[\because 1+\sin 2\theta=(\cos\theta+\sin\theta)^2$
and $1-\sin 2\theta=(\cos\theta-\sin\theta)^2]$

$=3(1^2+\sin^2 2\theta-2\sin 2\theta)+6(1+\sin 2\theta)+4(1-\cos^6\theta-3\cos^2\theta+3\cos^4\theta)$

$[\because (a-b)^2=a^2+b^2-2ab$
and $(a-b)^3=a^3-b^3-3a^2b+3ab^2]$

$=3+3\sin^2 2\theta-6\sin 2\theta+6+6\sin 2\theta+4-4\cos^6\theta-12\cos^2\theta+12\cos^4\theta$

$=13+3\sin^2 2\theta-4\cos^6\theta-12\cos^2\theta+12\cos^4\theta$

$=13+3(2\sin\theta\cos\theta)^2-4\cos^6\theta-12\cos^2\theta(1-\cos^2\theta)$

$=13+12\sin^2\theta\cos^2\theta-4\cos^6\theta-12\cos^2\theta\sin^2\theta$

$=13-4\cos^6\theta$

19. The value of $\cos\frac{\pi}{2^2}\cdot\cos\frac{\pi}{2^3}\,.......\cos\frac{\pi}{2^{10}}\cdot\sin\frac{\pi}{2^{10}}$ is **[JEE Main 2019]**

(a) $\frac{1}{1024}$ (b) $\frac{1}{2}$ (c) $\frac{1}{512}$ (d) $\frac{1}{256}$

Ans. *(c)*

We know that,

$\cos\alpha\cdot\cos(2\alpha)\cos(2^2\alpha)...\cos(2^{n-1}\alpha)$

$=\frac{\sin(2^n\alpha)}{2^n\sin\alpha}$

$\therefore \cos\frac{\pi}{2^2}\cdot\cos\frac{\pi}{2^3}...\cos\frac{\pi}{2^{10}}\cdot\sin\frac{\pi}{2^{10}}$

$$=\left\{\frac{\sin\left(\frac{\pi}{2^{10}}2^9\right)}{2^9\sin\left(\frac{\pi}{2^{10}}\right)}\right\}\sin\frac{\pi}{2^{10}}$$

$\left[\because \text{here, } \alpha=\frac{\pi}{2^{10}} \text{ and } n=9\right]$

$$=\frac{1}{2^9}\sin\left(\frac{\pi}{2}\right)=\frac{1}{2^9}=\frac{1}{512}$$

20. The maximum value of $3\cos\theta+5\sin\left(\theta-\frac{\pi}{6}\right)$ for any real value of θ is **[JEE Main 2019]**

(a) $\frac{\sqrt{79}}{2}$ (b) $\sqrt{34}$
(c) $\sqrt{31}$ (d) $\sqrt{19}$

Ans. *(d)*

Given expression $3\cos\theta+5\sin\left(\theta-\frac{\pi}{6}\right)$

$=3\cos\theta+5\left(\sin\theta\cos\frac{\pi}{6}-\sin\frac{\pi}{6}\cos\theta\right)$

$=3\cos\theta+5\left(\frac{\sqrt{3}}{2}\sin\theta-\frac{1}{2}\cos\theta\right)$

$=3\cos\theta-\frac{5}{2}\cos\theta+\frac{5\sqrt{3}}{2}\sin\theta$

$=\frac{1}{2}\cos\theta+\frac{5\sqrt{3}}{2}\sin\theta$

$\because$ The maximum value of $a\cos\theta + b\sin\theta$ is $\sqrt{a^2+b^2}$

So, maximum value of $\frac{1}{2}\cos\theta + \frac{5\sqrt{3}}{2}\sin\theta$ is

$$=\sqrt{\left(\frac{1}{2}\right)^2+\left(\frac{5\sqrt{3}}{2}\right)^2}=\sqrt{\frac{1}{4}+\frac{75}{4}}$$

$$=\sqrt{\frac{76}{4}}=\sqrt{19}.$$

21. Let a, b, c be three non-zero real numbers such that the equation

$\sqrt{3}\,a\cos x + 2b\sin x = c, x \in \left[-\frac{\pi}{2}, \frac{\pi}{2}\right]$, has two distinct real roots α and β with $\alpha+\beta=\frac{\pi}{3}$. Then, the value of $\frac{b}{a}$ is .... **[JEE Advanced 2018]**

Ans. *(0.5)*

We have, α, β are the roots of

$$\sqrt{3}\,a\cos x + 2b\sin x = c$$

$\therefore$ $\sqrt{3}\,a\cos\alpha + 2b\sin\alpha = c$ ...(i)

and $\sqrt{3}\,a\cos\beta + 2b\sin\beta = c$ ...(ii)

On subtracting Eq. (ii) from Eq. (i), we get

$$\sqrt{3}a(\cos\alpha - \cos\beta) + 2b(\sin\alpha - \sin\beta) = 0$$

$$\Rightarrow \sqrt{3}a\left(-2\sin\left(\frac{\alpha+\beta}{2}\right)\right)\sin\left(\frac{\alpha-\beta}{2}\right) + 2b\left(2\cos\left(\frac{\alpha+\beta}{2}\right)\right)\sin\left(\frac{\alpha-\beta}{2}\right)=0$$

$$\Rightarrow \sqrt{3}\,a\sin\left(\frac{\alpha+\beta}{2}\right) = 2b\cos\left(\frac{\alpha+\beta}{2}\right)$$

$$\Rightarrow \tan\left(\frac{\alpha+\beta}{2}\right) = \frac{2b}{\sqrt{3}a}$$

$$\Rightarrow \tan\left(\frac{\pi}{6}\right) = \frac{2b}{\sqrt{3}a} \quad \left[\because \alpha+\beta=\frac{\pi}{3}, \text{given}\right]$$

$$\Rightarrow \frac{1}{\sqrt{3}} = \frac{2b}{\sqrt{3}a} \Rightarrow \frac{b}{a}=\frac{1}{2} \Rightarrow \frac{b}{a}=0.5$$

Trigonometric Equations and in Equations

1. If n is the number of solutions of the equation $2\cos x\left(4\sin\left(\frac{\pi}{4}+x\right)\sin\left(\frac{\pi}{4}-x\right)-1\right)=1$, $x\in[0,\pi]$ and S is the sum of all these solutions, then the ordered pair (n, S) is **[JEE Main 2021]**

(a) $(3, 13\pi/9)$ (b) $(2, 2\pi/3)$
(c) $(2, 8\pi/9)$ (d) $(3, 5\pi/3)$

Ans. *(a)*

$$2\cos x\left(4\sin\left(\frac{\pi}{4}+x\right)\sin\left(\frac{\pi}{4}-x\right)-1\right)=1$$

$$\Rightarrow 2\cos x\left(2\cos(2x)-2\cos\left(\frac{\pi}{2}\right)-1\right)=1$$

$$\Rightarrow 2\cos x(4\cos^2 x-3)=1$$

$$\Rightarrow \cos 3x = \frac{1}{2}$$

$$\Rightarrow 3x=\frac{\pi}{3},\frac{5\pi}{3},\frac{7\pi}{3} \Rightarrow x=\frac{\pi}{9},\frac{5\pi}{9},\frac{7\pi}{9}$$

Number of solutions $= n = 3$

Sum of solutions $= S = \frac{13\pi}{9}$

2. The number of solutions of the equation $32^{\tan^2 x}+32^{\sec^2 x}=81$, $0\le x\le\frac{\pi}{4}$ is **[JEE Main 2021]**

(a) 3 (b) 1 (c) 0 (d) 2

Ans. *(b)*

$$32^{\tan^2 x}+32^{\sec^2 x}=81$$

$$\Rightarrow 32^{\tan^2 x}+32^{1+\tan^2 x}=81$$

$$\Rightarrow 33\times 32^{\tan^2 x}=81$$

$$\Rightarrow 32^{\tan^2 x}=\frac{27}{11}$$

$$\Rightarrow \tan^2 x = \ln_{32}\left(\frac{27}{11}\right)$$

$$\Rightarrow \tan x = \sqrt{\ln_{32}\left(\frac{27}{11}\right)} \in (0,1)$$

$\Rightarrow$ One solution in $\left[0,\frac{\pi}{4}\right]$.

3. Let S be the sum of all solutions (in radians) of the equation $\sin^4\theta+\cos^4\theta-\sin\theta\cos\theta=0$ in $[0, 4\pi]$. Then, $\frac{8S}{\pi}$ is equal to **[JEE Main 2021]**

Ans. *(56)*

Given, $\sin^4\theta+\cos^4\theta-\sin\theta\cos\theta=0$

$\theta\in[0,4\pi]$

$$\Rightarrow 1-2\sin^2\theta\cos^2\theta-\sin\theta\cos\theta=0$$

$$\Rightarrow 2-\sin^2 2\theta-\sin 2\theta=0$$

$$\Rightarrow \sin^2 2\theta+\sin 2\theta-2=0$$

$$\Rightarrow (\sin 2\theta+2)(\sin 2\theta-1)=0$$

$$\Rightarrow \sin 2\theta=1, 2\theta\in[0,8\pi]$$

$$\theta=\frac{\pi}{4},\frac{5\pi}{4},\frac{9\pi}{4},\frac{13\pi}{4}$$

Sum of solutions $S=\frac{28\pi}{4}$

Then, $\frac{8S}{\pi}=\frac{8}{\pi}\times\frac{28\pi}{4}=56$

4. The number of solutions of $\sin^7 x+\cos^7 x=1$, $x\in[0,4\pi]$ is equal to **[JEE Main 2021]**

(a) 11 (b) 7
(c) 5 (d) 9

Ans. *(c)*

$\sin^7 x+\cos^7 x=1$ ...(i)

As,

$\sin^2 x+\cos^2 x=1$ ...(ii)

Now, $\sin^7 x\le\sin^2 x$ and $\cos^7 x\le\cos^2 x$.

But according to question, Eqs. (i) and (ii), it is only possible, when

$\sin^7 x=\sin^2 x$ and $\cos^7 x=\cos^2 x$

So, $\sin^2 x+\cos^2 x=1$

When $\sin x=0$ and $\cos x=1$

or $\sin x=1$ and $\cos x=0$

$$\Rightarrow x=0, 2\pi, 4\pi, \frac{\pi}{2},\frac{5\pi}{2}$$

$\therefore$ 5 solutions.

5. The number of solutions of the equation $|\cot x|=\cot x+\frac{1}{\sin x}$ in the interval $[0, 2\pi]$ is .... **[JEE Main 2021]**

Ans. *(1)*

Given,

$|\cot x|=\cot x+\frac{1}{\sin x}$ ...(i)

If $\cot x>0$, then $|\cot x|=\cot x$

From Eq. (i), $\cot x=\cot x+\frac{1}{\sin x}$

$\Rightarrow \frac{1}{\sin x}=0$ (not possible)

If $\cot x<0$, then $|\cot x|=-\cot x$

From Eq. (ii), $-\cot x=\cot x+\frac{1}{\sin x}$

$$\Rightarrow 2\cot x+\frac{1}{\sin x}=0\Rightarrow 2\cos x=-1$$

$$\Rightarrow x=\frac{2\pi}{3} \text{ or } \frac{4\pi}{3}$$

Here, $x=\frac{4\pi}{3}$ rejected because $\frac{4\pi}{3}\in$ third quadrant and in third quadrant $\cot x$ is positive. Since, we considered $\cot x<0$.

$\therefore x=2\pi/3$ is the only one solution.

6. The number of roots of the equation, $(81)^{\sin^2 x}+(81)^{\cos^2 x}=30$ in the interval $[0, \pi]$ is equal to **[JEE Main 2021]**

(a) 3 (b) 4 (c) 8 (d) 2

Ans. *(b)*

Given, $81^{\sin^2 x}+81^{\cos^2 x}=30$

$\Rightarrow 81^{\sin^2 x}+81^{(1-\sin^2 x)}=30$

$\Rightarrow 81^{\sin^2 x}+\dfrac{81}{81^{\sin^2 x}}=30$

Let $81^{\sin^2 x}=y$.

$\therefore \quad y+\dfrac{81}{y}=30 \Rightarrow y^2-30y+81=0$

$\Rightarrow \quad (y-27)(y-3)=0$

$\Rightarrow \quad y=3$ or $y=27$

$81^{\sin^2 x}=3$ or $81^{\sin^2 x}=27$

$3^{4\sin^2 x}=3$ or $3^{4\sin^2 x}=3^3$

$\Rightarrow \quad 4\sin^2 x=1$ or $4\sin^2 x=3$

$\Rightarrow \quad \sin^2 x=1/4$ or $\sin^2 x=3/4$

$\Rightarrow \quad \sin^2 x=\sin^2(\pi/6)$

or $\quad \sin^2 x=\sin^2(\pi/3)$

$\Rightarrow \quad x=n\pi\pm\pi/6$ or $x=n\pi\pm\pi/3$

From $[0, \pi]$,

$x=\pi/6, 5\pi/6$ or $x=\pi/3, 2\pi/3$

Hence, the total number of solutions $=4$

7. If $\sqrt{3}(\cos^2 x)=(\sqrt{3}-1)\cos x+1$, the number of solutions of the given equation when $x\in\left[0, \dfrac{\pi}{2}\right]$ is .... **[JEE Main 2021]**

Ans. *(1)*

Given, $\sqrt{3}\cos^2 x=(\sqrt{3}-1)\cos x+1$, $x\in[0, \pi/2]$

Let $\cos x=t$, then

$\sqrt{3}t^2=(\sqrt{3}-1)t+1$

$\Rightarrow \quad \sqrt{3}t^2-\sqrt{3}t+t-1=0$

$\Rightarrow \quad (\sqrt{3}t^2-\sqrt{3}t)+(t-1)=0$

$\Rightarrow \quad \sqrt{3}t(t-1)+1(t-1)=0$

$\Rightarrow \quad (t-1)(\sqrt{3}t+1)=0$

This gives $t=1$ and $t=\dfrac{-1}{\sqrt{3}}$

Put, $t=\cos x$, then

$\cos x=1$ and $\cos x=\dfrac{-1}{\sqrt{3}}$

$\cos x=-1/\sqrt{3}$ is rejected as $x\in[0, \pi/2]$

$\therefore \cos x=1$

Since, $x\in\left[0, \dfrac{\pi}{2}\right]$, then $\cos x=\cos 0$

This gives $x=0$ is only solution.

Therefore, number of solution when $x\in[0, \pi/2]$.

8. The number of integral values of k for which the equation $3\sin x+4\cos x=k+1$ has a solution, $k\in R$ is .... **[JEE Main 2021]**

Ans. *(11)*

Given, $3\sin x+4\cos x=k+1$ ... (i)

Multiply and divide LHS of Eq. (i) by $\sqrt{3^2+4^2}=5$

i.e. $5\left(\dfrac{3}{5}\sin x+\dfrac{4}{5}\cos x\right)=k+1$

$\Rightarrow 5(\cos\alpha\sin x+\sin\alpha\cos x)=k+1$

[Let $\cos\alpha=3/5$ then $\sin\alpha=\sqrt{1-(3/5)^2}=\dfrac{4}{5}$]

$\Rightarrow 5\sin(x+\alpha)=k+1$

[Use $\sin(a+b)=\sin a\cos b+\cos a\sin b$]

$\Rightarrow \sin(x+\alpha)=\dfrac{k+1}{5}$

Let, $x+\alpha=\theta$

Then, $\sin\theta=\dfrac{k+1}{5}$

$\because \quad -1\le\sin\theta\le 1 \Rightarrow -1\le\dfrac{k+1}{5}\le 1$

$\Rightarrow \quad -5\le k+1\le 5 \Rightarrow -6\le k\le 4$

$\therefore$ Possible integral values of k are $-6, -5, -4, -3, -2, -1, 0, 1, 2, 3$ and 4.

i.e. Total 11 integral values of k are possible for which Eq. (i) has solution.

9. If the equation $\cos^4\theta+\sin^4\theta+\lambda=0$ has real solutions for θ, then λ lies in the interval **[JEE Main 2020]**

(a) $\left(-\dfrac{5}{4}, -1\right)$ (b) $\left[-1, -\dfrac{1}{2}\right]$

(c) $\left(-\dfrac{1}{2}, -\dfrac{1}{4}\right]$ (d) $\left[-\dfrac{3}{2}, -\dfrac{5}{4}\right]$

Ans. *(b)*

The expression, $\cos^4\theta+\sin^4\theta$

$=(\cos^2\theta+\sin^2\theta)^2-2\sin^2\theta\cos^2\theta$

$=1-\dfrac{1}{2}\sin^2(2\theta)$

$\because \quad \sin^2(2\theta)\in[0, 1]$

$\Rightarrow \quad -\dfrac{1}{2}\sin^2(2\theta)\in\left[-\dfrac{1}{2}, 0\right]$

$\Rightarrow \quad 1-\dfrac{1}{2}\sin^2(2\theta)\in\left[\dfrac{1}{2}, 1\right]$

Now, as $\cos^4\theta+\sin^4\theta+\lambda=0$

$\Rightarrow \lambda=-(\cos^4\theta+\sin^4\theta)$

$\Rightarrow \lambda\in[-1, -1/2]$ for real solution of the given equation $\cos^4\theta+\sin^4\theta+\lambda=0$ for θ.

Hence, option (b) is correct.

10. The number of distinct solutions of the equation, $\log_{1/2}|\sin x|=2-\log_{1/2}|\cos x|$ in the interval $[0, 2\pi]$, is **[JEE Main 2020]**

Ans. *(8)*

Given equation,

$\log_{1/2}|\sin x|=2-\log_{1/2}|\cos x|$

$\Rightarrow \quad -\log_2|\sin x|=2+\log_2|\cos x|$

$\Rightarrow \quad \log_2|\sin x|+\log_2|\cos x|+\log_2 4=0$

$\Rightarrow \quad \log_2(4|\sin x||\cos x|)=0$

$\Rightarrow \quad 4|\sin x||\cos x|=1$

$\Rightarrow \quad \sin 2x=\pm\dfrac{1}{2}$

$\because \quad x\in[0, 2\pi]\Rightarrow 2x\in[0, 4\pi]$

$\therefore$ For $2x\in[0, 2\pi]$, $\sin 2x=\pm\dfrac{1}{2}$ has four solutions and $2x\in[2\pi, 4\pi]$, $\sin 2x=\pm\dfrac{1}{2}$ has four more solutions.

$\therefore$ Total number of solutions are 8.

Hence, answer is 8.

11. All the pairs (x, y) that satisfy the inequality $2^{\sqrt{\sin^2 x-2\sin x+5}}\cdot\dfrac{1}{4^{\sin^2 y}}\le 1$ also satisfy the equation **[JEE Main 2019]**

(a) $2|\sin x|=3\sin y$ (b) $\sin x=|\sin y|$

(c) $\sin x=2\sin y$ (d) $2\sin x=\sin y$

Ans. *(b)*

Given, inequality is

$2^{\sqrt{\sin^2 x-2\sin x+5}}\cdot\dfrac{1}{4^{\sin^2 y}}\le 1$

$\Rightarrow \quad 2^{\sqrt{(\sin x-1)^2+4}}\cdot 2^{-2\sin^2 y}\le 1$

$\Rightarrow \quad 2^{\sqrt{(\sin x-1)^2+4}}\le 2^{2\sin^2 y}$

$\Rightarrow \quad \sqrt{(\sin x-1)^2+4}\le 2\sin^2 y$

[if $a>1$ and $a^m\le a^n\Rightarrow m\le n$]

$\because$ Range of $\sqrt{(\sin x-1)^2+4}$ is $[2, 2\sqrt{2}]$ and range of $2\sin^2 y$ is $[0, 2]$.

$\therefore$ The above inequality holds, if

$\sqrt{(\sin x-1)^2+4}=2=2\sin^2 y$

$\Rightarrow \quad \sin x=1$ and $\sin^2 y=1$

$\Rightarrow \quad \sin x=|\sin y|$ [from the options]

12. Let $S=\{\theta\in[-2\pi, 2\pi]: 2\cos^2\theta+3\sin\theta=0\}$, then the sum of the elements of S is **[JEE Main 2019]**

(a) 2π (b) π

(c) $\dfrac{5\pi}{3}$ (d) $\dfrac{13\pi}{6}$

Ans. *(a)*

We have, $\theta\in[-2\pi, 2\pi]$

and $\quad 2\cos^2\theta+3\sin\theta=0$

$\Rightarrow \quad 2(1-\sin^2\theta)+3\sin\theta=0$

$\Rightarrow \quad 2-2\sin^2\theta+3\sin\theta=0$

$\Rightarrow \quad 2\sin^2\theta - 3\sin\theta - 2 = 0$

$\Rightarrow \quad 2\sin^2\theta - 4\sin\theta + \sin\theta - 2 = 0$

$\Rightarrow 2\sin\theta(\sin\theta - 2) + 1(\sin\theta - 2) = 0$

$\Rightarrow \quad (\sin\theta - 2)(2\sin\theta + 1) = 0$

$\Rightarrow \quad \sin\theta = \frac{-1}{2} \quad [\because (\sin\theta - 2) \neq 0]$

$\Rightarrow \quad \theta = 2\pi - \frac{\pi}{6}, -\pi + \frac{\pi}{6}, -\frac{\pi}{6}, \pi + \frac{\pi}{6}$

$[\because \theta \in [-2\pi, 2\pi]]$

Now, sum of all solutions

$= 2\pi - \frac{\pi}{6} - \pi + \frac{\pi}{6} - \frac{\pi}{6} + \pi + \frac{\pi}{6} = 2\pi$

13. The number of solutions of the equation $1 + \sin^4 x = \cos^2 3x, x \in \left[-\frac{5\pi}{2}, \frac{5\pi}{2}\right]$ is **[JEE Main 2019]**

(a) 3 (b) 5 (c) 7 (d) 4

Ans. *(b)*

Given equation is $1 + \sin^4 x = \cos^2(3x)$

Since, range of $(1 + \sin^4 x) = [1, 2]$

and range of $\cos^2(3x) = [0, 1]$

So, the given equation holds if

$1 + \sin^4 x = 1 = \cos^2(3x)$

$\Rightarrow \quad \sin^4 x = 0$ and $\cos^2 3x = 1$

Since, $\quad x \in \left[-\frac{5\pi}{2}, \frac{5\pi}{2}\right]$

$\therefore \quad x = -2\pi, -\pi, 0, \pi, 2\pi.$

Thus, there are five different values of x is possible.

14. Let S be the set of all $\alpha \in R$ such that the equation, $\cos 2x + \alpha \sin x = 2\alpha - 7$ has a solution. Then, S is equal to **[JEE Main 2019]**

(a) R (b) $[1, 4]$ (c) $[3, 7]$ (d) $[2, 6]$

Ans. *(d)*

The given trigonometric equation is

$\cos 2x + \alpha \sin x = 2\alpha - 7$

$\Rightarrow \quad 1 - 2\sin^2 x + \alpha \sin x = 2\alpha - 7$

$[\because \cos 2x = 1 - 2\sin^2 x]$

$\Rightarrow \quad 2\sin^2 x - \alpha \sin x + 2\alpha - 8 = 0$

$\Rightarrow \quad 2(\sin^2 x - 4) - \alpha(\sin x - 2) = 0$

$\Rightarrow 2(\sin^2 x - 2)(\sin x + 2) - \alpha(\sin x - 2) = 0$

$\Rightarrow \quad (\sin x - 2)(2\sin x + 4 - \alpha) = 0$

$\Rightarrow \quad 2\sin x + 4 - \alpha = 0 \quad [\because \sin x + 2 \neq 0]$

$\Rightarrow \quad \sin x = \frac{\alpha - 4}{2} \quad \text{...(i)}$

Now, as we know $-1 \leq \sin x \leq 1$

$\therefore \quad -1 \leq \frac{\alpha - 4}{2} \leq 1$ [from Eq. (i)]

$\Rightarrow \quad -2 \leq \alpha - 4 \leq 2$

$\Rightarrow \quad 2 \leq \alpha \leq 6 \Rightarrow \alpha \in [2, 6]$

15. If $0 \leq x < \frac{\pi}{2}$, then the number of values of x for which $\sin x - \sin 2x + \sin 3x = 0$, is **[JEE Main 2019]**

(a) 2 (b) 3 (c) 1 (d) 4

Ans. *(a)*

We have, $\sin x - \sin 2x + \sin 3x = 0$

$\Rightarrow \quad (\sin x + \sin 3x) - \sin 2x = 0$

$\Rightarrow 2\sin\left(\frac{x + 3x}{2}\right)\cos\left(\frac{x - 3x}{2}\right) - \sin 2x = 0$

$\left[\because \sin C + \sin D = 2\sin\left(\frac{C+D}{2}\right)\cos\left(\frac{C-D}{2}\right)\right]$

$\Rightarrow 2\sin 2x \cos x - \sin 2x = 0$

$[\because \cos(-\theta) = \cos\theta]$

$\Rightarrow \quad \sin 2x(2\cos x - 1) = 0$

$\Rightarrow \quad \sin 2x = 0$ or $2\cos x - 1 = 0$

$\Rightarrow \quad 2x = 0, \pi, \ldots$ or $\cos x = \frac{1}{2}$

$\Rightarrow \quad x = 0, \frac{\pi}{2} \ldots$ or $x = \frac{\pi}{3}$

In the interval $\left[0, \frac{\pi}{2}\right)$ only two values satisfy, namely $x = 0$ and $x = \frac{\pi}{3}$.

16. The sum of all values of $\theta \in \left(0, \frac{\pi}{2}\right)$ satisfying $\sin^2 2\theta + \cos^4 2\theta = \frac{3}{4}$ is **[JEE Main 2019]**

(a) $\frac{3\pi}{8}$ (b) $\frac{5\pi}{4}$ (c) $\frac{\pi}{2}$ (d) π

Ans. *(c)*

Given, $\sin^2 2\theta + \cos^4 2\theta = \frac{3}{4}$

$\Rightarrow \quad (1 - \cos^2 2\theta) + \cos^4 2\theta = \frac{3}{4}$

$(\because \sin^2 x = 1 - \cos^2 x)$

$\Rightarrow 4\cos^4 2\theta - 4\cos^2 2\theta + 1 = 0$

$\Rightarrow \quad (2\cos^2 2\theta - 1)^2 = 0$

$\Rightarrow \quad 2\cos^2 2\theta - 1 = 0$

$\Rightarrow \quad \cos^2 2\theta = \frac{1}{2} \Rightarrow \cos 2\theta = \pm\frac{1}{\sqrt{2}}$

If $\theta \in \left(0, \frac{\pi}{2}\right)$, then $2\theta \in (0, \pi)$

$\therefore \quad \cos 2\theta = \pm\frac{1}{\sqrt{2}}$

$\Rightarrow \quad 2\theta = \frac{\pi}{4}, \frac{3\pi}{4},$

$\left[\because \cos\left(\frac{3\pi}{4}\right) = \cos\left(\pi - \frac{\pi}{4}\right) = -\cos\frac{\pi}{4} = -\frac{1}{\sqrt{2}}\right]$

$\Rightarrow \quad \theta = \frac{\pi}{8}, \frac{3\pi}{8}$

Sum of values of $\theta = \frac{\pi}{8} + \frac{3\pi}{8} = \frac{\pi}{2}$

Properties and Solutions of Triangles

1. A vertical pole fixed to the horizontal ground is divided in the ratio 3 : 7 by a mark on it with lower part shorter than the upper part. If the two parts subtend equal angles at a point on the ground 18 m away from the base of the pole, then the height of the pole (in m) is **[JEE Main 2021]**

(a) $12\sqrt{15}$ (b) $12\sqrt{10}$ (c) $8\sqrt{10}$ (d) $6\sqrt{10}$

Ans. *(b)*

Let height of pole $= 10h$

$\tan\alpha = \frac{3H}{18}$

$\tan 2\alpha = \frac{10H}{18} = \frac{2\tan\alpha}{1 - \tan^2\alpha}$

$\frac{5H}{9} = \frac{2\cdot\left(\frac{H}{6}\right)}{1 - \frac{H^2}{36}}$

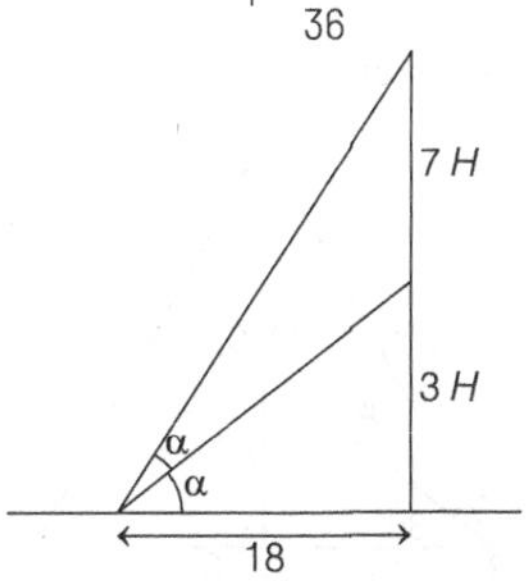

$1 - \frac{H^2}{36} = \frac{3}{5} \Rightarrow \frac{H^2}{36} = \frac{2}{5}$

$H^2 = \frac{72}{5} \Rightarrow H = \sqrt{\frac{72}{5}} \Rightarrow 10H = 12\sqrt{10}$

2. Let $\frac{\sin A}{\sin B} = \frac{\sin(A - C)}{\sin(C - B)}$, where A, B and C are angles of a ΔABC. If the lengths of the sides opposite these angles are a, b and c respectively, then **[JEE Main 2021]**

(a) $b^2 - a^2 = a^2 + c^2$

(b) b^2, c^2 and a^2 are in AP

(c) c^2, a^2 and b^2 are in AP

(d) a^2, b^2 and c^2 are in AP

Ans. *(b)*

Given, $\frac{\sin A}{\sin B} = \frac{\sin(A-C)}{\sin(C-B)}$

$\Rightarrow \frac{\sin(B+C)}{\sin(A+C)} = \frac{\sin(A-C)}{\sin(C-B)}$

$[\because A+B+C=\pi]$

$\Rightarrow \sin(C+B)\cdot\sin(C-B)$
$=\sin(A+C)\cdot\sin(A-C)$

$\Rightarrow \sin^2 C - \sin^2 B = \sin^2 A - \sin^2 C$

$\Rightarrow 2\sin^2 C = \sin^2 A + \sin^2 B$

$\Rightarrow 2(2R\sin C)^2 = (2R\sin A)^2 + (2R\sin B)^2$

$\Rightarrow 2c^2 = a^2 + b^2$

$\left[\because \frac{a}{\sin A} = \frac{b}{\sin B} = \frac{c}{\sin C} = 2R\right]$

$\Rightarrow a^2, c^2, b^2$ are in AP.

$\Rightarrow b^2, c^2, a^2$ are in AP

3. A spherical gas balloon of radius 16 m subtends an angle 60° at the eye of the observer *A* while the angle of elevation of its center from the eye of *A* is 75°. Then the height (in metre) of the top most point of the balloon from the level of the observer's eye is **[JEE Main 2021]**

(a) $8(2+2\sqrt{3}+\sqrt{2})$ (b) $8(\sqrt{6}+\sqrt{2}+2)$
(c) $8(\sqrt{2}+2+\sqrt{3})$ (d) $8(\sqrt{6}-\sqrt{2}+2)$

Ans. *(b)*

According to the question,

$XY = OX + OY$

$\Rightarrow XY = 16 + OA\sin 75°$

$\Rightarrow XY = 16 + OQ\ \text{cosec}\,30°\sin 75°$

$\Rightarrow XY = 16 + 16\frac{\sin 75°}{\sin 30°}$

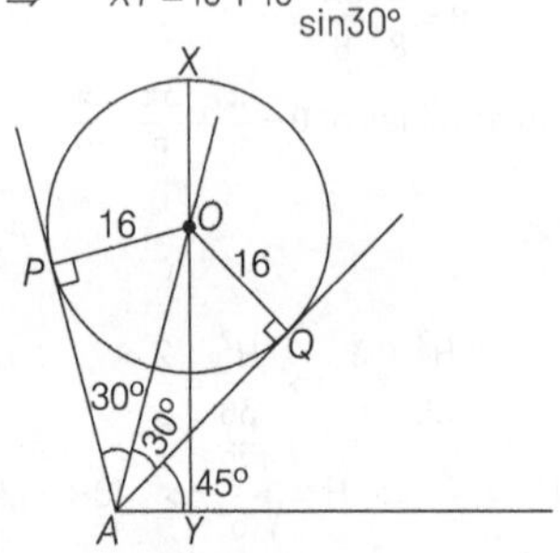

$\Rightarrow XY = 16(1 + 2\sin 75°)$

$\Rightarrow XY = 16\left[1 + 2\left(\frac{\sqrt{3}+1}{2\sqrt{2}}\right)\right]$

$\Rightarrow XY = 16\left(\frac{\sqrt{2}+\sqrt{3}+1}{\sqrt{2}}\right)$

$XY = 8(2+\sqrt{6}+\sqrt{2})$

4. If in a triangle *ABC*, *AB* = 5 units, $\angle B = \cos^{-1}\left(\frac{3}{5}\right)$ and radius of circumcircle of ΔABC is 5 units, then the area (in square units) of ΔABC is **[JEE Main 2021]**

(a) $10+6\sqrt{2}$ (b) $8+2\sqrt{2}$
(c) $6+8\sqrt{3}$ (d) $4+2\sqrt{3}$

Ans. *(c)*

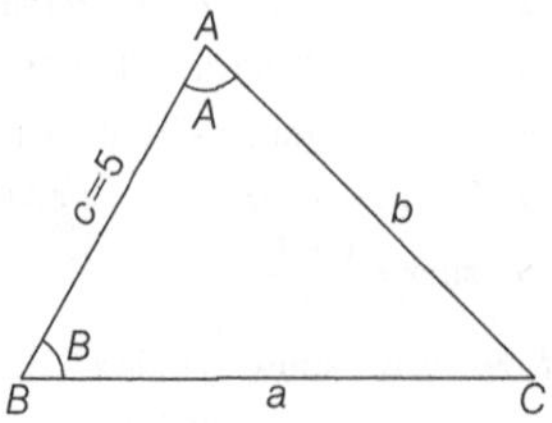

$\cos B = \frac{3}{5}$

$\Rightarrow \sin B = \frac{4}{5} \Rightarrow R = 5$

$\frac{a}{\sin A} = \frac{b}{\sin B} = \frac{c}{\sin C} = 2R$

$\Rightarrow \frac{b}{4/5} = 10 \Rightarrow b = 8$

$\cos B = \frac{a^2+c^2-b^2}{2ac}$

$\Rightarrow \frac{3}{5} = \frac{a^2+25-64}{2\cdot a\cdot 5}$

$\Rightarrow 6a = a^2 - 39$

$\Rightarrow a^2 - 6a - 39 = 0 \Rightarrow a = \frac{6 \pm 8\sqrt{3}}{2}$

$\therefore a = 3 + 4\sqrt{3}$

$\Rightarrow \text{Area} = \frac{abc}{4R} = \frac{(3+4\sqrt{3})8\cdot 5}{4\cdot 5}$

$\therefore \text{Area} = 6 + 8\sqrt{3}$

5. Let the centroid of an equilateral triangle ΔABC be at the origin. Let one of the sides of the equilateral triangle be along the straight line $x+y=3$. If *R* and *r* be the radius of circumcircle and incircle, respectively of ΔABC, then $(R+r)$ is equal to **[JEE Main 2021]**

(a) $\frac{9}{\sqrt{2}}$ (b) $7\sqrt{2}$
(c) $2\sqrt{2}$ (d) $3\sqrt{2}$

Ans. *(a)*

Let equation of *BC* is $x+y=3$

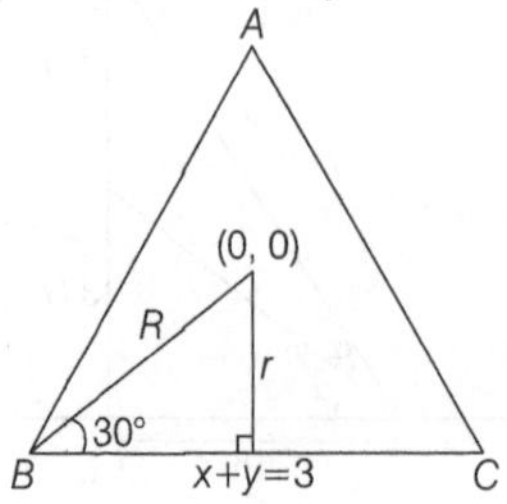

$\therefore r = \frac{|0+0-3|}{\sqrt{2}} = \frac{3}{\sqrt{2}}$

In equilateral triangle,

$\sin 30° = \frac{r}{R} \Rightarrow \frac{1}{2} = \frac{r}{R} \Rightarrow R = 2r$

$\therefore r + R = r + 2r = 3r = 3.\frac{3}{\sqrt{2}} = \frac{9}{\sqrt{2}}$

6. A pole stands vertically inside a triangular park *AB(c)* Let the angle of elevation of the top of the pole from each corner of the park be $\frac{\pi}{3}$. If the radius of the circumcircle of ΔABC is 2, then the height of the pole is equal to **[JEE Main 2021]**

(a) $\frac{2\sqrt{3}}{3}$ (b) $2\sqrt{3}$
(c) $\sqrt{3}$ (d) $\frac{1}{\sqrt{3}}$

Ans. *(b)*

Let $PO = h$, $CO = 2 =$ radius and $\angle PCO = \frac{\pi}{3}$

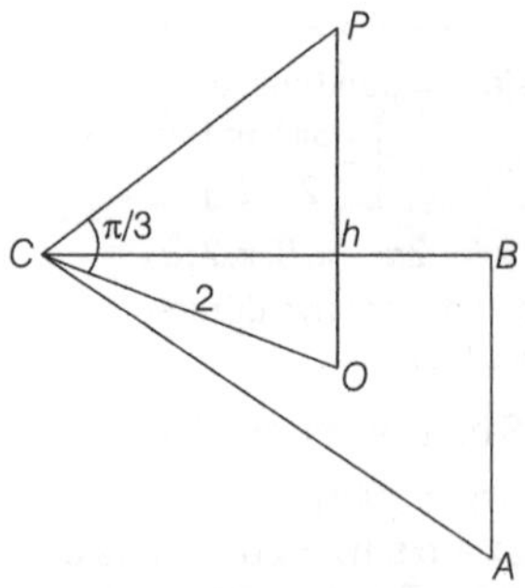

$\tan\frac{\pi}{3} = \frac{PO}{OC}$

$\Rightarrow \sqrt{3} = \frac{h}{2} \Rightarrow h = 2\sqrt{3}$

7. In ΔABC, the lengths of sides *AC* and *AB* are 12 cm and 5 cm, respectively. If the area of ΔABC is 30 cm^2 and *R* and *r* are respectively the radii of circumcircle and incircle of ΔABC, then the value of $2R+r$ (in cm) is equal to .... **[JEE Main 2021]**

Ans. *(15)*

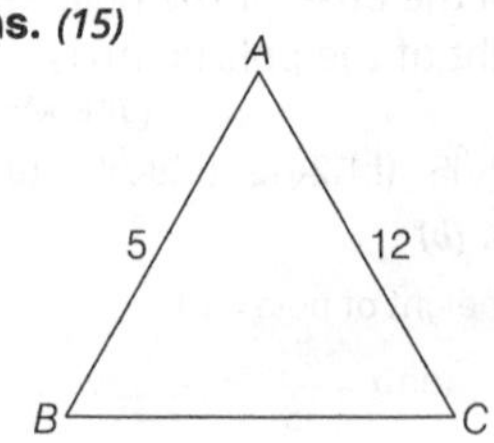

Area of $\Delta ABC = 30\text{cm}^2$

$\frac{1}{2}\cdot 5\cdot 12\sin A = 30\text{cm}^2$

$\Rightarrow \sin A = 1 \Rightarrow A = 90°$

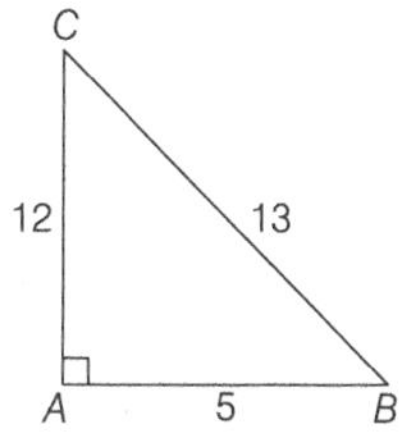

Here, $AB^2 + AC^2 = BC^2$

$\Rightarrow \quad 12^2 + 5^2 = BC^2$

$\Rightarrow \quad BC = 13$

$\therefore \quad R = \frac{a}{2\sin A} = \frac{13}{2\cdot\sin 90} = \frac{13}{2}$

$r = \frac{\Delta}{s} = \frac{30}{\left(\frac{12+5+13}{2}\right)} = \frac{30}{15} = 2$

$\therefore \quad 2R + r = 2\cdot\frac{13}{2} + 2 = 15$

8. In a ΔABC, let $AB = \sqrt{23}$, $BC = 3$ and $CA = 4$. Then, the value of $\frac{\cot A + \cot C}{\cot B}$ is **[JEE Advanced 2021]**

Ans. *(2)*

Given, $AB = \sqrt{23} = c$

$BC = 3 = a$

$CA = 4 = b$

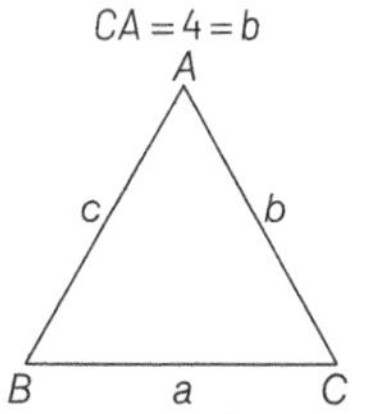

Now, $\frac{\cot A + \cot C}{\cot B} = \frac{\frac{\cos A}{\sin A} + \frac{\cos C}{\sin C}}{\frac{\cos B}{\sin B}}$

$= \frac{\frac{b^2+c^2-a^2}{2bc(\sin A)} + \frac{a^2+b^2-c^2}{2ab(\sin C)}}{\frac{c^2+a^2-b^2}{2ac(\sin B)}}$

$= \frac{\frac{b^2+c^2-a^2}{4\Delta} + \frac{a^2+b^2-c^2}{4\Delta}}{\frac{c^2+a^2-b^2}{4\Delta}}$

$= \frac{b^2+c^2-a^2+a^2+b^2-c^2}{c^2+a^2-b^2}$

$= \frac{2b^2}{a^2+c^2-b^2} = \frac{2\times 16}{9+23-16} = \frac{32}{16} = 2$

9. Two vertical poles $AB = 15$ m and $CD = 10$ m are standing apart on a horizontal ground with points A and C on the groun(d) If P is the point of intersection of BC and AD, then the height of P (in meters) above the line AC is **[JEE Main 2020]**

(a) 20/3 (b) 6 (c) 10/3 (d) 5

Ans. *(b)*

The diagram of given situation is

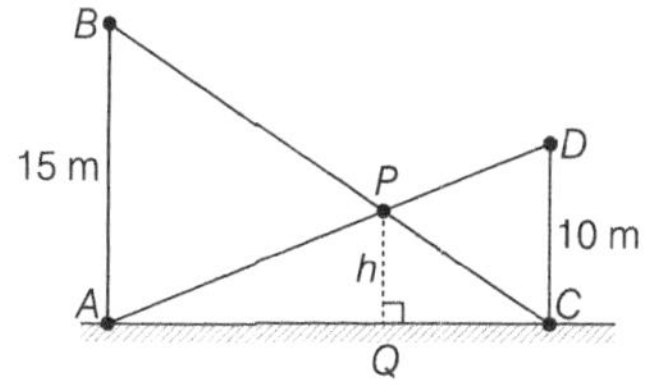

Let height of P above the line AC is h m and $AQ = x$ m and $CQ = y$ m, so

$\frac{10}{x+y} = \frac{h}{x}$ ...(i)

and $\frac{15}{x+y} = \frac{h}{y}$ ...(ii)

From Eqs. (i) and (ii), we get

$\frac{10}{15} = \frac{y}{x} \Rightarrow 2x = 3y$ ...(iii)

Now, from Eqs. (i) and (iii), we get

$\frac{10}{x+\frac{2x}{3}} = \frac{h}{x} \Rightarrow \frac{10}{\frac{5}{3}} = h \Rightarrow h = 6\text{ m}$

Hence, option (b) is correct.

10. The angle of elevation of a cloud C from a point P, 200 m above a still lake is 30°. If the angle of depression of the image of C in the lake from the point P is 60°, then PC (in m) is equal to : **[JEE Main 2020]**

(a) 100 (b) 400
(c) $200\sqrt{3}$ (d) $400\sqrt{3}$

Ans. *(b)*

According to given information in question, from the diagram, where $PQ = MN = 200$m and let $CM = l$ m. So, $NC' = (l + 200)$m

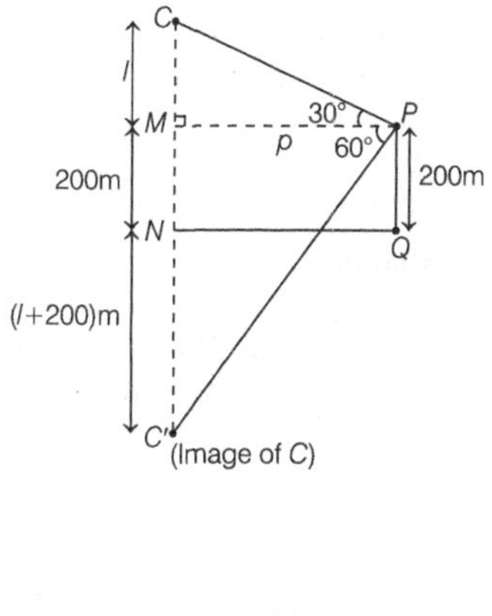

So, in right angled triangle CMP, we have

$\tan 30° = \frac{l}{p} \Rightarrow p = \sqrt{3}l$ ...(i)

and in right angle triangled PMC', we have

$\tan 60° = \frac{200 + l + 200}{p}$

$\Rightarrow \quad \sqrt{3}p = 400 + l$ ...(ii)

From Eqs. (i) and (ii), we get

$2l = 400 \Rightarrow l = 200\text{m}$ and $p = 200\sqrt{3}\text{m}$

$\therefore \quad PC = \sqrt{l^2 + p^2} = 400\text{m}.$

11. The angle of elevation of the top of a hill from a point on the horizontal plane passing through the foot of the hill is found to be 45°. After walking a distance of 80 m towards the top, up a slope inclined at an angle of 30° to the horizontal plane, the angle of elevation of the top of the hill becomes 75°. Then, the height of the hill (in meters) is .... **[JEE Main 2020]**

Ans. *(80.00)*

According to the given situation if we draw the diagram.

$\because \quad \angle ACB = 45°$

$\therefore \quad AB = BC = h$ [let]

Now, from the diagram

$\sin 30° = \frac{x}{80}$ (in ΔCDE) $\Rightarrow x = 40\text{m}$

and $\cos 30° = \frac{y}{80} \Rightarrow y = 40\sqrt{3}$ m

Now, in ΔAEF,

$\tan 75° = \frac{h-x}{h-y} \Rightarrow \frac{\sqrt{3}+1}{\sqrt{3}-1} = \frac{h-40}{h-40\sqrt{3}}$

$\Rightarrow \sqrt{3}h + h - 120 - 40\sqrt{3}$

$= \sqrt{3}h - h - 40\sqrt{3} + 40$

$\Rightarrow \quad 2h = 160 \Rightarrow h = 80\text{m}$

12. Let AD and BC be two vertical poles at A and B respectively on a horizontal groun(d) If $AD = 8$ m, $BC = 11$ m and $AB = 10$ m; then the distance (in meters) of a point M on AB from the point A such that $MD^2 + MC^2$ is minimum is .... **[JEE Main 2020]**

Ans. *(5.00)*

On drawing the diagram of given situation

$MD^2 + MC^2 = 64 + x^2 + 121 + (10 - x)^2$

$= f(x)$ [say]

$\Rightarrow \quad f(x) = 2x^2 - 20x + 285$

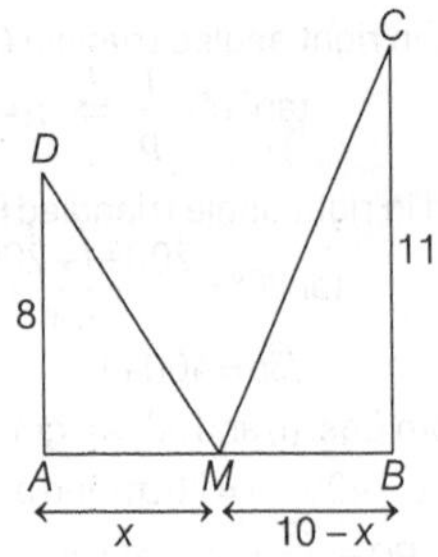

For minima, $f'(x)=0 \Rightarrow 4x-20=0$

$\Rightarrow x=5$

As $f''(x)=4>0$

$\therefore AM=5\,\text{m}$

13. The angle of elevation of the summit of a mountain from a point on the ground is 45°. After climbing up one km towards the summit at an inclination of 30° from the ground, the angle of elevation of the summit is found to be 60°. Then, the height (in km) of the summit from the ground is **[JEE Main 2020]**

(a) $\dfrac{\sqrt{3}-1}{\sqrt{3}+1}$ (b) $\dfrac{\sqrt{3}+1}{\sqrt{3}-1}$

(c) $\dfrac{1}{\sqrt{3}-1}$ (d) $\dfrac{1}{\sqrt{3}+1}$

Ans. *(c)*

From the diagram as $\angle PRQ=45°$

$\therefore PQ=RQ$

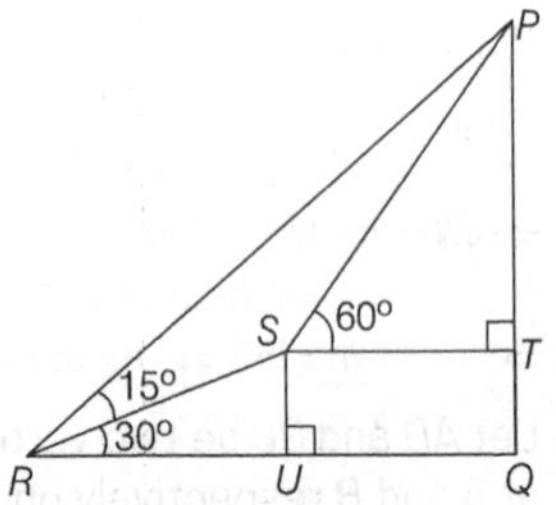

and $\angle SPT=30°$ and $\angle RPQ=45°$

$\therefore \angle PRS=\angle SPR \Rightarrow SR=SP=1\,\text{km}$

In ΔRSU,

$\sin 30°=\dfrac{SU}{RS} \Rightarrow SU=\dfrac{1}{2}\,\text{km}$

$\therefore$ Height of mountain $=PT+TQ=PT+SU$

$=\dfrac{\sqrt{3}}{2}+\dfrac{1}{2}=\dfrac{\sqrt{3}+1}{2}\,\text{km}$ or $\dfrac{1}{\sqrt{3}-1}\,\text{km}$

14. Let x, y and z be positive real numbers. Suppose x, y and z are the lengths of the sides of a triangle opposite to its angles X, Y and Z, respectively. If

$$\tan\frac{X}{2}+\tan\frac{Z}{2}=\frac{2y}{x+y+z},$$

then which of the following statements is/are TRUE? **[JEE Advanced 2020]**

(a) $2Y=X+Z$

(b) $Y=X+Z$

(c) $\tan\dfrac{X}{2}=\dfrac{x}{y+z}$

(d) $x^2+z^2-y^2=xz$

Ans. *(b, c)*

For a ΔXYZ, it is given that

$$\tan\frac{X}{2}+\tan\frac{Z}{2}=\frac{2y}{x+y+z}$$

$$\Rightarrow \frac{\Delta}{s(s-x)}+\frac{\Delta}{s(s-z)}=\frac{y}{s}$$

$$\Rightarrow \Delta\frac{(s-z+s-x)}{(s-x)(s-z)}=y$$

$$\Rightarrow \Delta=(s-x)(s-z)$$

$$\Rightarrow s(s-x)(s-y)(s-z)=(s-x)^2(s-z)^2$$

$$\Rightarrow s^2-sy=s^2-(x+z)s+xz$$

$$\Rightarrow s(x+z-y)=xz$$

$$\Rightarrow (x+z)^2-y^2=2xz$$

$$\Rightarrow x^2+z^2=y^2$$

$$\therefore y=\frac{\pi}{2}$$

$$\because X+Y+Z=\pi$$

$$\Rightarrow X+Z=\frac{\pi}{2}=Y$$

$$\Rightarrow X+Z=Y$$

$$\because \tan\frac{X}{2}=\sqrt{\frac{1-\cos X}{1+\cos X}}=\sqrt{\frac{1-\frac{z}{y}}{1+\frac{z}{y}}}$$

$$=\sqrt{\frac{y-z}{y+z}}=\frac{\sqrt{y^2-z^2}}{y+z}$$

$$\tan\frac{X}{2}=\frac{x}{y+z} \quad \{\because x^2+z^2=y^2\}$$

15. Two vertical poles of heights, 20 m and 80 m stand apart on a horizontal plane. The height (in m) of the point of intersection of the lines joining the top of each pole to the foot of the other, from this horizontal plane is **[JEE Main 2019]**

(a) 15 (b) 16

(c) 12 (d) 18

Ans. *(b)*

Let a first pole AB having height 20 m and second pole PQ having height 80 m and $\angle PBQ=\alpha$, $\angle AQB=\beta$

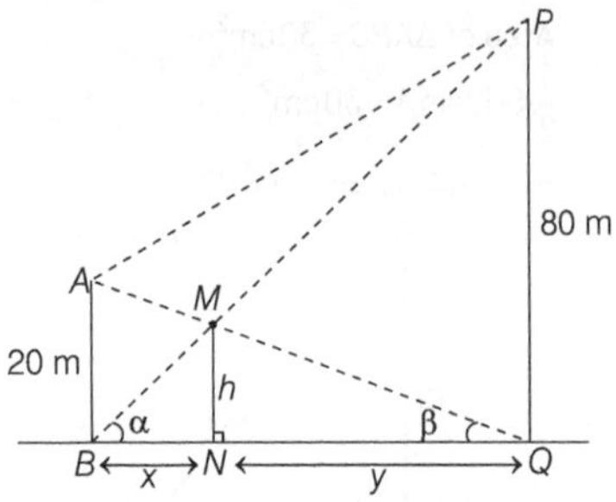

and $MN=h\,\text{m}$ is the height of intersection point from the horizontal plane

$\because \tan\alpha=\dfrac{h}{x}=\dfrac{80}{x+y}$ [in ΔMNB and ΔPQB] ...(i)

and $\tan\beta=\dfrac{h}{y}=\dfrac{20}{x+y}$ [in ΔMNQ and ΔABQ] ...(ii)

From Eqs. (i) and (ii), we get

$\dfrac{y}{x}=4 \Rightarrow y=4x$ ...(iii)

From Eqs. (i) and (iii), we get

$$\frac{h}{x}=\frac{80}{x+4x}$$

$$\Rightarrow h=\frac{80}{5}=16\,\text{m}$$

16. Two poles standing on a horizontal ground are of heights 5 m and 10 m, respectively. The line joining their tops makes an angle of 15° with ground. Then, the distance (in m) between the poles, is **[JEE Main 2019]**

(a) $5(\sqrt{3}+1)$ (b) $\dfrac{5}{2}(2+\sqrt{3})$

(c) $10(\sqrt{3}-1)$ (d) $5(2+\sqrt{3})$

Ans. *(d)*

Given heights of two poles are 5 m and 10 m.

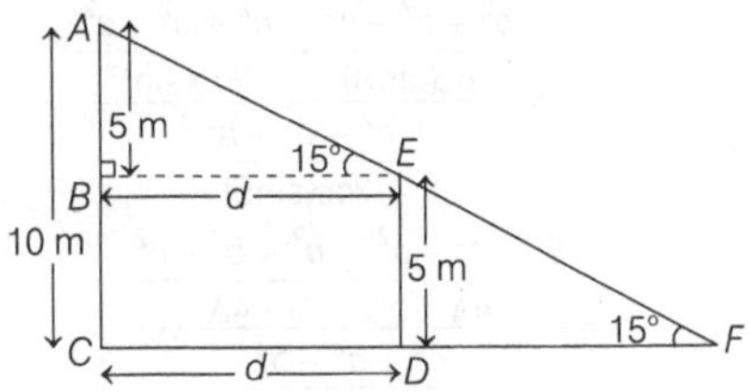

i.e. from figure $AC=10\,\text{m}$, $DE=5\,\text{m}$

$\therefore AB=AC-DE=10-5=5\,\text{m}$

Let d be the distance between two poles.

Clearly, $\Delta ABE \sim \Delta ACF$ [by AA-similarity criterion]

$\therefore \angle AEB=15°$

In ΔABE, we have

$$\tan 15° = \frac{AB}{BE} \Rightarrow \frac{\sqrt{3}-1}{\sqrt{3}+1} = \frac{5}{d}$$

$$\left[\because \tan 15° = \frac{\sqrt{3}-1}{\sqrt{3}+1}\right]$$

$$\Rightarrow \quad d = \frac{5(\sqrt{3}+1)}{(\sqrt{3}-1)}$$

$$\Rightarrow \quad d = 5\frac{\sqrt{3}+1}{\sqrt{3}-1} \times \frac{\sqrt{3}+1}{\sqrt{3}+1}$$

$$= \frac{5(3+2\sqrt{3}+1}{3-1} = \frac{5(2\sqrt{3}+4)}{2}$$

$$= \frac{2\times 5(\sqrt{3}+2)}{2} = 5(2+\sqrt{3})\text{ m}$$

17. The angle of elevation of the top of a vertical tower standing on a horizontal plane is observed to be 45° from a point A on the plane. Let B be the point 30 m vertically above the point A. If the angle of elevation of the top of the tower from B be 30°, then the distance (in m) of the foot of the tower from the point A is **[JEE Main 2019]**

(a) $15(3+\sqrt{3})$ (b) $15(5-\sqrt{3})$

(c) $15(3-\sqrt{3})$ (d) $15(1+\sqrt{3})$

Ans. *(a)*

According to the question, we have the following figure.

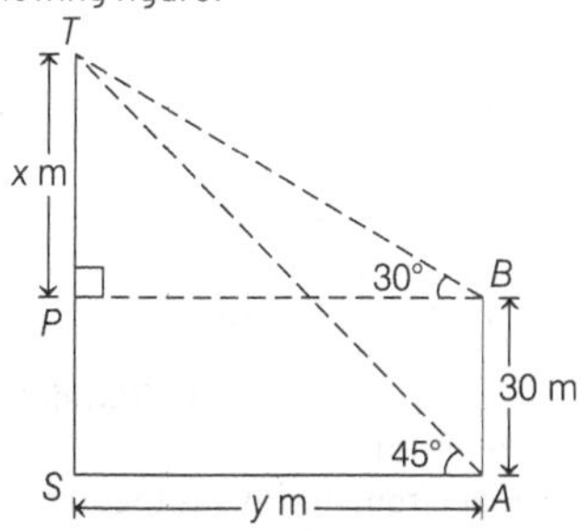

Now, let distance of foot of the tower from the point A is y m.

Draw $BP \perp ST$ such that $PT = x$ m.

Then, in ΔTPB, we have

$$\tan 30° = \frac{x}{y}$$

$$\Rightarrow \quad x = \frac{1}{\sqrt{3}}y \quad \text{...(i)}$$

and in ΔTSA, we have

$$\tan 45° = \frac{x+30}{y}$$

$$\Rightarrow \quad y = x + 30 \quad \text{...(ii)}$$

On the elimination of quantity x from Eqs. (i) and (ii), we get

$$y = \frac{1}{\sqrt{3}}y + 30$$

$$\Rightarrow \quad y\left(1 - \frac{1}{\sqrt{3}}\right) = 30$$

$$\Rightarrow \quad y = \frac{30\sqrt{3}}{\sqrt{3}-1} = \frac{30\sqrt{3}(\sqrt{3}+1)}{3-1}$$

$$= \frac{30}{2}\sqrt{3}(\sqrt{3}+1) = 15(3+\sqrt{3})$$

18. Consider a triangular plot ABC with sides $AB = 7$ m, $BC = 5$ m and $CA = 6$ m. A vertical lamp-post at the mid-point D of AC subtends an angle 30° at B. The height (in m) of the lamp-post is **[JEE Main 2019]**

(a) $\frac{2}{3}\sqrt{21}$ (b) $2\sqrt{21}$

(c) $7\sqrt{3}$ (d) $\frac{3}{2}\sqrt{21}$

Ans. *(a)*

According to given information, we have the following figure.

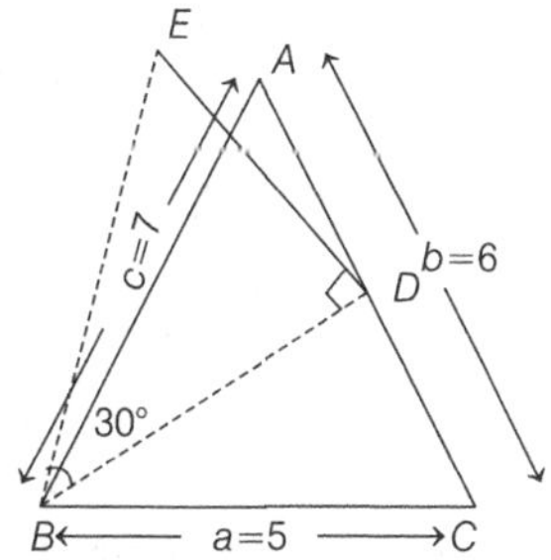

Clearly, length of $BD = \frac{1}{2}\sqrt{2a^2 + 2c^2 - b^2}$,

(using Appollonius theorem)

where, $c = AB = 7, a = BC = 5$

and $b = CA = 6$

$$\therefore \quad BD = \frac{1}{2}\sqrt{2\times 25 + 2\times 49 - 36}$$

$$= \frac{1}{2}\sqrt{112} = \frac{1}{2}4\sqrt{7} = 2\sqrt{7}$$

Now, let $ED = h$ be the height of the lamp post.

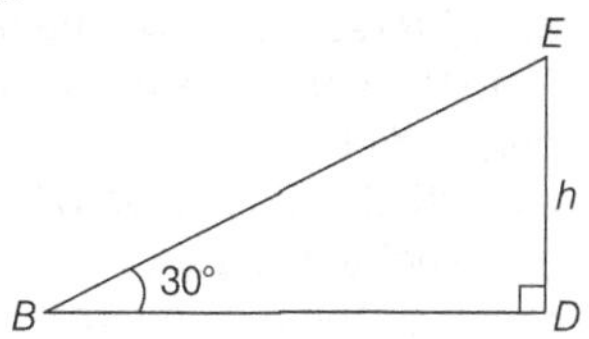

Then, in ΔBDE, $\tan 30° = \frac{h}{BD}$

$$\Rightarrow \quad \frac{1}{\sqrt{3}} = \frac{h}{2\sqrt{7}} \Rightarrow h = \frac{2\sqrt{7}}{\sqrt{3}} = \frac{2}{3}\sqrt{21}$$

19. With the usual notation, in ΔABC, if $\angle A + \angle B = 120°$, $a = \sqrt{3}+1$ and $b = \sqrt{3}-1$, then the ratio $\angle A : \angle B$, is **[JEE Main 2019]**

(a) 7 : 1 (b) 3 : 1

(c) 9 : 7 (d) 5 : 3

Ans. *(a)*

For a ΔABC, it is given that $a = \sqrt{3}+1$, $b = \sqrt{3}-1$ and $\angle A + \angle B = 120^\circ$

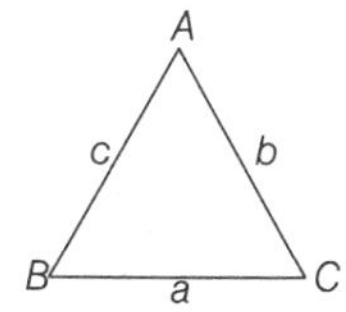

Clearly, $\angle C = 60^\circ$ [$\because \angle A + \angle B + \angle C = 180^\circ$]

Now, by tangent law, we have

$$\tan\frac{A-B}{2} = \frac{a-b}{a+b}\cot\frac{C}{2}$$

$$= \frac{(\sqrt{3}+1)-(\sqrt{3}-1)}{(\sqrt{3}+1)+(\sqrt{3}-1)}\cot\left(\frac{60^\circ}{2}\right)$$

$$= \frac{2}{2\sqrt{3}}\cot(30^\circ) = \frac{1}{\sqrt{3}} \times \sqrt{3} = 1$$

$$\Rightarrow \tan\left(\frac{A-B}{2}\right) = 1 = \tan 45^\circ$$

$$\Rightarrow \quad \frac{A-B}{2} = 45^\circ \Rightarrow \angle A - \angle B = 90^\circ$$

On solving $\angle A - \angle B = 90^\circ$ and $\angle A + \angle B = 120^\circ$, we get

$$\angle A = 105^\circ \text{ and } \angle B = 15^\circ$$

So, $\angle A : \angle B = 7 : 1$

20. If the area of an equilateral triangle inscribed in the circle,

$$x^2 + y^2 + 10x + 12y + c = 0$$

is $27\sqrt{3}$ sq units, then c is equal to **[JEE Main 2019]**

(a) 20 (b) −25 (c) 13 (d) 25

Ans. *(d)*

Clearly, centre of the circumscribed circle is the centroid (G) of the equilateral ΔABC. [$\because$ in an equilateral triangle circumcentre and centroid coincide]

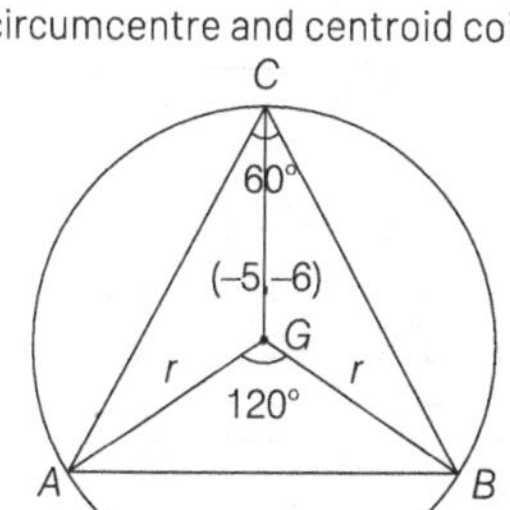

Also, we know that

$$\Delta AGB \cong \Delta BGC \cong \Delta CGA$$

[by SAS congruence rule]

$\therefore \; ar(\Delta ABC) = 3\, ar(\Delta AGB)$

$= 3\left(\frac{1}{2} r^2 \sin 120^\circ\right)$

[$\because$ area of triangle $\Delta = \frac{1}{2} ab \sin(\angle C)$]

$\because \quad ar(\Delta ABC) = 27\sqrt{3}$ [given]

$\therefore \quad \frac{3}{2} r^2 \frac{\sqrt{3}}{2} = 27\sqrt{3}$

[$\sin 120^\circ = \sin(180^\circ - 60^\circ) = \sin 60^\circ = \frac{\sqrt{3}}{2}$]

$\Rightarrow \quad r^2 = 4 \times 9 \Rightarrow r = 6$

Now, radius of circle,

$r = \sqrt{g^2 + f^2 - c}$

$\Rightarrow \quad 6 = \sqrt{25 + 36 - c}$

[$\because$ in the given equation of circle $2g = 10$ and $2f = 12 \Rightarrow g = 5$ and $f = 6$]

$\Rightarrow \quad 36 = 25 + 36 - c \Rightarrow c = 25$

21. In a triangle, the sum of lengths of two sides is x and the product of the lengths of the same two sides is y. If $x^2 - c^2 = y$, where c is the length of the third side of the triangle, then the circumradius of the triangle is **[JEE Main 2019]**

(a) $\frac{c}{3}$ (b) $\frac{c}{\sqrt{3}}$ (c) $\frac{3}{2}y$ (d) $\frac{y}{\sqrt{3}}$

Ans. *(b)*

We know that,

$\frac{a}{\sin A} = \frac{b}{\sin B} = \frac{c}{\sin C} = 2R$

and given that, $a + b = x$, $ab = y$ and $x^2 - c^2 = y$

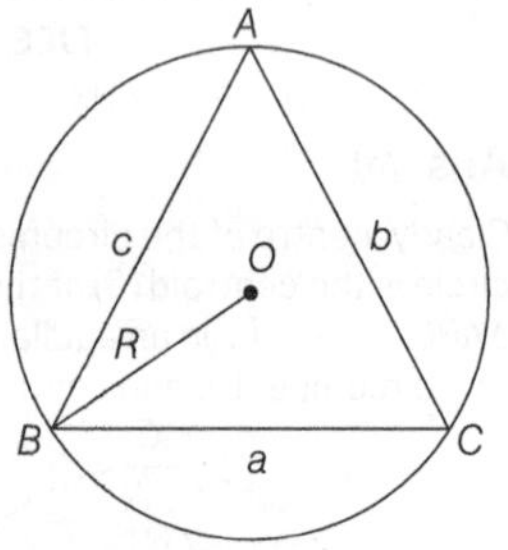

$\therefore \quad (a+b)^2 - c^2 = ab$

$\Rightarrow \quad a^2 + b^2 - c^2 = -2ab + ab$

$\Rightarrow \quad a^2 + b^2 - c^2 = -ab$

$\Rightarrow \quad \frac{a^2 + b^2 - c^2}{2ab} = \frac{-ab}{2ab} = -\frac{1}{2}$

$\therefore \quad \cos C = -\frac{1}{2} \Rightarrow C = 120^\circ$

[using cosine rule, $\cos C = \frac{a^2 + b^2 - c^2}{2ab}$]

Now, $\frac{c}{\sin C} = 2R$

$\Rightarrow \quad R = \frac{1}{2} \frac{c}{\sin(120^\circ)} = \frac{c}{2} \frac{2}{\sqrt{3}}$

$\therefore \quad R = \frac{c}{\sqrt{3}}$

22. If the angle of elevation of a cloud from a point P which is 25 m above a lake be 30º and the angle of depression of reflection of the cloud in the lake from P be 60º, then the height of the cloud (in meters) from the surface of the lake is **[JEE Main 2019]**

(a) 50 (b) 60 (c) 45 (d) 42

Ans. *(a)*

According to given information, we have the following figure,

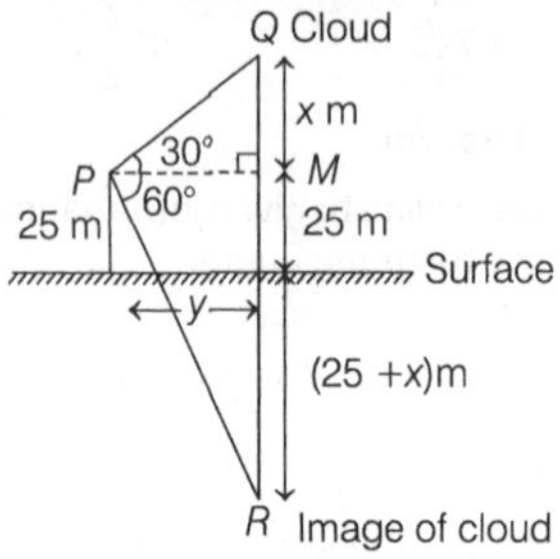

In ΔPQM, $\quad \tan 30^\circ = \frac{x}{y}$ ...(i)

In ΔPRM, $\quad \tan 60^\circ = \frac{25 + (25 + x)}{y}$ ...(ii)

On eliminating 'y' from Eqs. (i) and (ii), we get

$\sqrt{3} = \frac{25 + (25 + x)}{\sqrt{3}x}$

$\Rightarrow \quad 3x = 50 + x$

$\Rightarrow \quad 2x = 50 \Rightarrow x = 25\,m.$

$\therefore$ Height of cloud from surface

$= x + 25 = 50\,m$

23. In a non-right-angled triangle ΔPQR, let p, q, r denote the lengths of the sides opposite to the angles at P, Q, R respectively. The median from R meets the side PQ at S, the perpendicular from P meets the side QR at E, and RS and PE intersect at O. If $p = \sqrt{3}$, $q = 1$, and the radius of the circumcircle of the ΔPQR equals 1, then which of the following options is/are correct? **[JEE Advanced 2019]**

(a) Length of $OE = \frac{1}{6}$

(b) Length of $RS = \frac{\sqrt{7}}{2}$

(c) Area of $\Delta SOE = \frac{\sqrt{3}}{12}$

(d) Radius of incircle of $\Delta PQR = \frac{\sqrt{3}}{2}(2 - \sqrt{3})$

Ans. *(a,b,d)*

Let a non-right angled ΔPQR.

Now, by sine rule

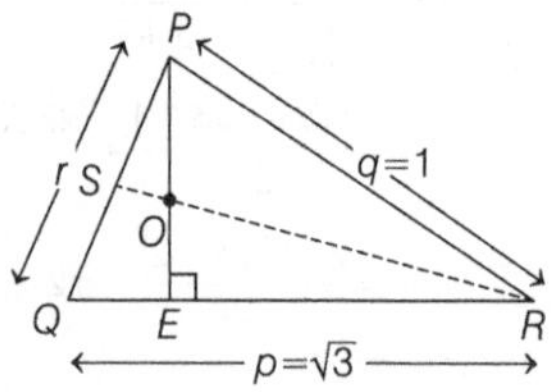

$\frac{p}{\sin P} = \frac{q}{\sin Q} = \frac{r}{\sin R} = 2 \times \text{circumradius}$

$\Rightarrow \frac{\sqrt{3}}{\sin P} = \frac{1}{\sin Q} = \frac{r}{\sin R} = 2 \times 1$

[circumradius = 1 unit]

$\Rightarrow \sin P = \frac{\sqrt{3}}{2}$ and $\sin Q = \frac{1}{2}$

$\Rightarrow P = 120º$ and $Q = 30º$

($\because \Delta PQR$ is non-right angled triangle)

So, $R = 30º$

$\Rightarrow r = 1$, so ΔPQR is an isosceles triangle. And, since RS and PE are the median of ΔPQR, so 'O' is centroid of the ΔPQR.

Now,

Option (a),

From Apollonius theorem,

$2(PE^2 + QE^2) = PQ^2 + PR^2$

$\Rightarrow 2\left(PE^2 + \frac{3}{4}\right) = 1 + 1$

$\Rightarrow PE^2 = 1 - \frac{3}{4} \Rightarrow PE^2 = \frac{1}{4} \Rightarrow PE = \frac{1}{2}$ units

and $OE = \frac{1}{3} PE = \frac{1}{6}$ units

[$\because O$ divides PE is 2 : 1]

Option (b),

Again from Apollonius theorem,

$2(PS^2 + RS^2) = PR^2 + QR^2$

$\Rightarrow \quad 2\left(\frac{1}{4} + RS^2\right) = 1 + 3$

$\Rightarrow RS^2 = 2 - \frac{1}{4} \Rightarrow RS^2 = \frac{7}{4}$

$\Rightarrow \quad RS = \frac{\sqrt{7}}{2}$ units

Option (c),

Area of $\Delta SOE = \frac{1}{2}(OE)(ST)$

$= \frac{1}{2} \times \frac{1}{6}[(PS)\sin 60º]$

$= \frac{1}{12} \times \frac{1}{2} \times \frac{\sqrt{3}}{2}$

$= \frac{\sqrt{3}}{48}$ square units

Option (d),

$\because$ Inradius of

$$\Delta PQR = \frac{\Delta}{s} = \frac{\frac{1}{2}pq\sin R}{\frac{1}{2}(p+q+r)} = \frac{\frac{1}{2}(\sqrt{3})(1)\frac{1}{2}}{\frac{1}{2}(\sqrt{3}+1+1)}$$

$$= \frac{\sqrt{3}}{2}(2-\sqrt{3}) \text{ units}$$

Hence, options (a), (b) and (d) are correct.

24. PQR is a triangular park with $PQ = PR = 200$ m. A TV tower stands at the mid-point of QR. If the angles of elevation of the top of the tower at P, Q and R are respectively $45°$, $30°$ and $30°$, then the height of the tower (in m) is **[JEE Main 2018]**

(a) 100 (b) 50

(c) $100\sqrt{3}$ (d) $50\sqrt{2}$

Ans. *(a)*

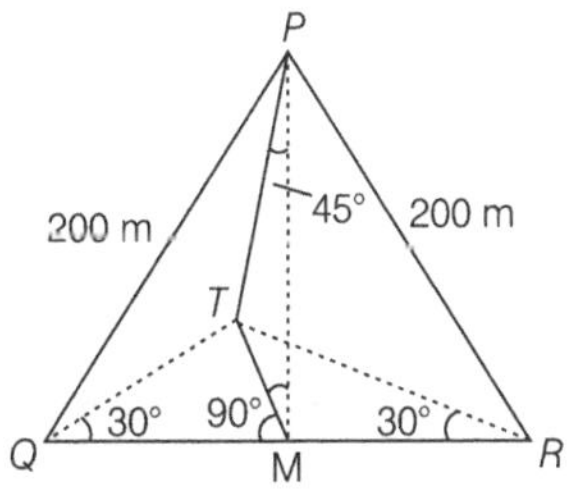

Let height of tower TM be h.

In ΔPMT, $\tan 45° = \frac{TM}{PM} \Rightarrow 1 = \frac{h}{PM}$

$\Rightarrow \quad PM = h$

In ΔTQM, $\tan 30° = \frac{h}{QM}$;

$$QM = \sqrt{3}h$$

In ΔPMQ, $PM^2 + QM^2 = PQ^2$

$$h^2 + (\sqrt{3}h)^2 = (200)^2$$

$\Rightarrow \quad 4h^2 = (200)^2 \Rightarrow h = 100$ m

25. In a ΔPQR, let $\angle PQR = 30°$ and the sides PQ and QR have lengths $10\sqrt{3}$ and 10, respectively. Then, which of the following statement(s) is (are) TRUE? **[JEE Advanced 2018]**

(a) $\angle QPR = 45°$

(b) The area of the ΔPQR is $25\sqrt{3}$ and $\angle QRP = 120°$

(c) The radius of the incircle of the ΔPQR is $10\sqrt{3} - 15$

(d) The area of the circumcircle of the ΔPQR is 100π

Ans. *(b, c, d)*

We have,

In ΔPQR

$$\angle PQR = 30°$$
$$PQ = 10\sqrt{3}$$
$$QR = 10$$

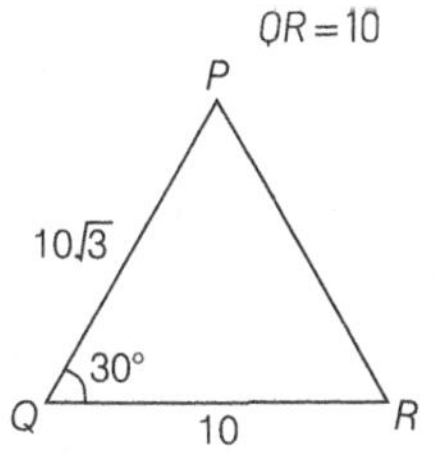

By cosine rule

$$\cos 30° = \frac{PQ^2 + QR^2 - PR^2}{2PQ \cdot QR}$$

$$\Rightarrow \quad \frac{\sqrt{3}}{2} = \frac{300 + 100 - PR^2}{200\sqrt{3}}$$

$$\Rightarrow \quad 300 = 300 + 100 - PR^2$$

$$\Rightarrow \quad PR = 10$$

Since, $PR = QR = 10$

$\therefore \quad \angle QPR = 30°$ and $\angle QRP = 120°$

Area of $\Delta PQR = \frac{1}{2} PQ \cdot QR \cdot \sin 30°$

$$= \frac{1}{2} \times 10\sqrt{3} \times 10 \times \frac{1}{2} = 25\sqrt{3}$$

Radius of incircle of

$$\Delta PQR = \frac{\text{Area of } \Delta PQR}{\text{Semi-perimetre of } \Delta PQR}$$

i.e. $$r = \frac{\Delta}{s} = \frac{25\sqrt{3}}{\frac{10\sqrt{3} + 10 + 10}{2}} = \frac{25\sqrt{3}}{5(\sqrt{3} + 2)}$$

$$\Rightarrow \quad r = 5\sqrt{3}(2 - \sqrt{3}) = 10\sqrt{3} - 15$$

and radius of circumcircle $(R) = \frac{abc}{4\Delta}$

$$= \frac{10\sqrt{3} \times 10 \times 10}{4 \times 25\sqrt{3}} = 10$$

$\therefore$ Area of circumcircle of

$$\Delta PQR = \pi R^2 = 100\pi$$

Hence, option (b), (c) and (d) are correct answer.

Inverse Trigonometric Functions

1. $\cos^{-1}(\cos(-5)) + \sin^{-1}(\sin(6)) - \tan^{-1}(\tan(12))$ is equal to (The inverse trigonometric functions take the principal values) **[JEE Main 2021]**

(a) $3\pi - 11$ (b) $4\pi - 9$

(c) $4\pi - 11$ (d) $3\pi + 1$

Ans. *(c)*

$\cos^{-1}(\cos(-5)) + \sin^{-1}(\sin(6)) - \tan^{-1}(\tan(12))$

$$= 2\pi - 5 + (-2\pi + 6) - (12 - 4\pi) = 4\pi - 11$$

2. If $(\sin^{-1} x)^2 - (\cos^{-1} x)^2 = a$, $0 < x < 1$, $a \neq 0$, then the value of $2x^2 - 1$ is **[JEE Main 2021]**

(a) $\cos\left(\frac{4a}{\pi}\right)$ (b) $\sin\left(\frac{2a}{\pi}\right)$

(c) $\cos\left(\frac{2a}{\pi}\right)$ (d) $\sin\left(\frac{4a}{\pi}\right)$

Ans. *(b)*

Given, $(\sin^{-1} x)^2 - (\cos^{-1} x)^2 = a$

$\Rightarrow$

$(\sin^{-1} x + \cos^{-1} x)(\sin^{-1} x - \cos^{-1} x) = a$

$$\Rightarrow \quad \frac{\pi}{2}(\sin^{-1} x - \cos^{-1} x) = a$$

$$\Rightarrow \quad \frac{\pi}{2} - 2\cos^{-1} x = \frac{2a}{\pi}$$

$$\Rightarrow \quad 2\cos^{-1} x = \frac{\pi}{2} - \frac{2a}{\pi}$$

$$\Rightarrow \quad \cos^{-1}(2x^2 - 1) = \frac{\pi}{2} - \frac{2a}{\pi}$$

$$\Rightarrow \quad 2x^2 - 1 = \cos\left(\frac{\pi}{2} - \frac{2a}{\pi}\right)$$

$$\Rightarrow \quad 2x^2 - 1 = \sin\left(\frac{2a}{\pi}\right)$$

3. Let M and m respectively be the maximum and minimum values of the function $f(x) = \tan^{-1}(\sin x + \cos x)$ in $\left[0, \frac{\pi}{2}\right]$, then the value of $\tan(M - m)$ is equal to **[JEE Main 2021]**

(a) $2 + \sqrt{3}$ (b) $2 - \sqrt{3}$

(c) $3 + 2\sqrt{2}$ (d) $3 - 2\sqrt{2}$

Ans. *(d)*

We have, $f(x) = \tan^{-1}(\sin x + \cos x)$

$$\because \quad x \in \left[0, \frac{\pi}{2}\right]$$

$\Rightarrow 1 \le \sin x + \cos x \le \sqrt{2}$

$[\because -\sqrt{A^2 + B^2} \le A\sin x + B\cos x \le \sqrt{A^2 + B^2}]$

$\Rightarrow \tan^{-1}(1) \le \tan^{-1}(\sin x + \cos x) \le \tan^{-1}(\sqrt{2})$

$\therefore \quad m = \tan^{-1}(1)$ and $M = \tan^{-1}(\sqrt{2})$

$\therefore \quad M - m = \tan^{-1}\sqrt{2} - \tan^{-1}(\sqrt{1})$

$$= \tan^{-1}\left(\frac{\sqrt{2} - 1}{1 + \sqrt{2}}\right) = \tan^{-1}(3 - 2\sqrt{2})$$

$\Rightarrow \tan(M - m) = 3 - 2\sqrt{2}$

4. If $y(x)=\cot^{-1}\left(\dfrac{\sqrt{1+\sin x}+\sqrt{1-\sin x}}{\sqrt{1+\sin x}-\sqrt{1-\sin x}}\right)$, $x\in\left(\dfrac{\pi}{2},\pi\right)$, then $\dfrac{dy}{dx}$ at $x=\dfrac{5\pi}{6}$ is

[JEE Main 2021]

(a) $-\dfrac{1}{2}$ (b) -1 (c) $\dfrac{1}{2}$ (d) 0

Ans. *(a)*

$$y(x)=\cot^{-1}\left(\frac{\sqrt{1+\sin x}+\sqrt{1-\sin x}}{\sqrt{1+\sin x}-\sqrt{1-\sin x}}\right),$$

$$x\in\left(\frac{\pi}{2},\pi\right)$$

$$=\cot^{-1}\left(\frac{\left|\cos\frac{x}{2}+\sin\frac{x}{2}\right|+\left|\cos\frac{x}{2}-\sin\frac{x}{2}\right|}{\left|\cos\frac{x}{2}+\sin\frac{x}{2}\right|-\left|\cos\frac{x}{2}-\sin\frac{x}{2}\right|}\right)$$

$$=\cot^{-1}\left(\frac{\cos\frac{x}{2}+\sin\frac{x}{2}+\sin\frac{x}{2}-\cos\frac{x}{2}}{\cos\frac{x}{2}+\sin\frac{x}{2}-\sin\frac{x}{2}+\cos\frac{x}{2}}\right)$$

$$\left[\because \frac{x}{2}\in\left(\frac{\pi}{4},\frac{\pi}{2}\right)\right]$$

$$=\cot^{-1}\left(\frac{\sin\frac{x}{2}}{\cos\frac{x}{2}}\right)=\cot^{-1}\left(\tan\frac{x}{2}\right)$$

$$=\cot^{-1}\left(\cot\left(\frac{\pi}{2}-\frac{x}{2}\right)\right)$$

$$\Rightarrow\quad y(x)=\frac{\pi}{2}-\frac{x}{2}$$

$$\therefore\quad y'(x)=\frac{-1}{2}$$

5. A 10 inches long pencil AB with mid-point C and a small eraser P are placed on the horizontal top of a table such that $PC=\sqrt{5}$ inches and $\angle PCB=\tan^{-1}(2)$. The acute angle through which the pencil must be rotated about C so that the perpendicular distance between eraser and pencil becomes exactly 1 inch is

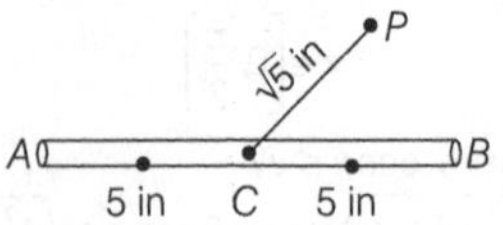

[JEE Main 2021]

(a) $\tan^{-1}\left(\dfrac{3}{4}\right)$ (b) $\tan^{-1}(1)$

(c) $\tan^{-1}\left(\dfrac{4}{3}\right)$ (d) $\tan^{-1}\left(\dfrac{1}{2}\right)$

Ans. *(a)*

Let the angle before rotation be θ, then

$$\theta=\angle PCB=\tan^{-1}(2)$$

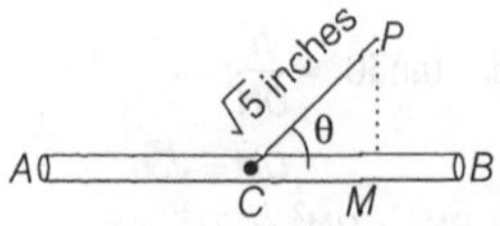

Now, after rotation let angle become α.

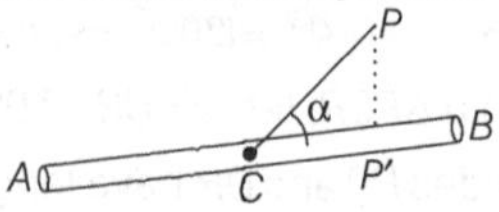

InΔPCM

$$PC=\sqrt{5}$$

$$PM=\sqrt{5}\sin\theta=\sqrt{5}\left(\frac{2}{\sqrt{5}}\right)=2$$

After rotation perpendicular distance becomes $PP'=1$

$$PC\sin\alpha=1$$

$$\Rightarrow\quad \sqrt{5}\sin\alpha=1$$

$$\Rightarrow\quad \sin\alpha=\frac{1}{\sqrt{5}}$$

$$\Rightarrow\quad \tan\alpha=\frac{\sin\alpha}{\sqrt{1-\sin^2\alpha}}=\frac{1}{2}$$

$\therefore$Rotated angle

$$=\theta-\alpha=\tan^{-1}(2)-\tan^{-1}\left(\frac{1}{2}\right)$$

$$=\tan^{-1}\left(\frac{2-1/2}{1+2\times 1/2}\right)=\tan^{-1}\left(\frac{3}{4}\right)$$

6. The number of real roots of the equation

$$\tan^{-1}\sqrt{x(x+1)}+\sin^{-1}\sqrt{x^2+x+1}=\frac{\pi}{4}$$

is **[JEE Main 2021]**

(a) 1 (b) 4 (c) 3 (d) 0

Ans. *(d)*

$$\tan^{-1}\sqrt{x(x+1)}+\sin^{-1}\sqrt{x^2+x+1}=\frac{\pi}{4}$$

Domain, $x(x+1)\geq 0$

$$0\leq x^2+x+1\leq 1$$

So, only when $x^2+x=0$, equation will be define(d)

$x=0,-1$

At $x=0$, $\tan^{-1}0+\sin^{-1}1=\dfrac{\pi}{2}$

$x=-1$, $\tan^{-1}0+\sin^{-1}1=\dfrac{\pi}{2}$

$\therefore$ No solution.

7. The value of

$\tan\left(2\tan^{-1}\left(\dfrac{3}{5}\right)+\sin^{-1}\left(\dfrac{5}{13}\right)\right)$ is equal to **[JEE Main 2021]**

(a) $\dfrac{-181}{69}$ (b) $\dfrac{220}{21}$

(c) $\dfrac{-291}{76}$ (d) $\dfrac{151}{63}$

Ans. *(b)*

Let

$$A=\tan\left(2\tan^{-1}\left(\frac{3}{5}\right)+\sin^{-1}\left(\frac{5}{13}\right)\right)\quad\text{...(i)}$$

Now, using $2\tan^{-1}x=\tan^{-1}\dfrac{2x}{1-x^2}$

$$2\tan^{-1}\left(\frac{3}{5}\right)=\tan^{-1}\left(\frac{2\left(\frac{3}{5}\right)}{1-\left(\frac{3}{5}\right)^2}\right)$$

$$=\tan^{-1}\left(\frac{\frac{6}{5}}{1-\frac{9}{25}}\right)$$

$$=\tan^{-1}\left(\frac{30}{16}\right)=\tan^{-1}\left(\frac{15}{8}\right)$$

Let $\sin^{-1}\left(\dfrac{5}{13}\right)=\theta$, then $\sin\theta=\dfrac{5}{13}$

$$\therefore\quad \tan\theta=\frac{\sin\theta}{\sqrt{1-\sin^2\theta}}=\frac{\frac{5}{13}}{\sqrt{1-\left(\frac{5}{3}\right)^2}}=\frac{\frac{5}{13}}{\frac{12}{13}}$$

$$\Rightarrow\quad \theta=\tan^{-1}\left(\frac{5}{12}\right)=\sin^{-1}\left(\frac{5}{13}\right)$$

From Eq. (i),

$$A=\tan\left(\tan^{-1}\left(\frac{15}{8}\right)+\tan^{-1}\left(\frac{5}{12}\right)\right)$$

$$=\tan\left[\tan^{-1}\left(\frac{\frac{15}{8}+\frac{5}{12}}{1-\left(\frac{15}{8}\right)\left(\frac{5}{12}\right)}\right)\right]$$

$$=\tan\left[\tan^{-1}\left(\frac{\frac{55}{24}}{\frac{21}{96}}\right)\right]=\tan\left[\tan^{-1}\frac{55\times 4}{21}\right]$$

$$=\tan\left(\tan^{-1}\frac{220}{21}\right)=\frac{220}{21}$$

8. If

$\cot^{-1}(\alpha)=\cot^{-1}2+\cot^{-1}8+\cot^{-1}18$ $+\cot^{-1}32+\ldots$.. upto 100 terms, then α is **[JEE Main 2021]**

(a) 1.01 (b) 1.00

(c) 1.02 (d) 1.03

Ans. *(a)*

$\cot^{-1}\alpha=\cot^{-1}2+\cot^{-1}8+\cot^{-1}18$ $+\cot^{-1}32\ldots$ upto 100 terms

Let T_n be the nth term of $\cot^{-1}\alpha$.

$$T_n=\cot^{-1}(2n^2)=\tan^{-1}\left(\frac{1}{2n^2}\right)$$

$$=\tan^{-1}\left[\frac{(2n+1)-(2n-1)}{1+(2n+1)(2n-1)}\right]$$

$$\because\quad \tan^{-1}x-\tan^{-1}y=\tan^{-1}\left(\frac{x-y}{1+xy}\right)$$

$\therefore \quad \tan^{-1}\left\{\frac{(2n+1)-(2n-1)}{1+(2n+1)(2n-1)}\right\}$

$= \tan^{-1}(2n+1) - \tan^{-1}(2n-1)$

$T_1 = \tan^{-1}3 - \tan^{-1}1$

$T_2 = \tan^{-1}5 - \tan^{-1}3$

$T_3 = \tan^{-1}7 - \tan^{-1}5$

$\vdots$

$T_{99} = \tan^{-1}199 - \tan^{-1}197$

$T_{100} = \tan^{-1}201 - \tan^{-1}199$

$\sum T_r = \tan^{-1}201 - \tan^{-1}1$

$= \tan^{-1}\left(\frac{201-1}{1+201\cdot 1}\right)$

$= \tan^{-1}\left(\frac{200}{202}\right)$

$\Rightarrow \cot^{-1}(\alpha) = \tan^{-1}\left(\frac{200}{202}\right)$

$\Rightarrow \cot^{-1}(\alpha) = \cot^{-1}\left(\frac{202}{200}\right)$

$\therefore \alpha = \frac{202}{200} = \frac{101}{100} = 1.01$

9. The sum of possible values of x for $\tan^{-1}(x+1) + \cot^{-1}\left(\frac{1}{x-1}\right) = \tan^{-1}\left(\frac{8}{31}\right)$ is **[JEE Main 2021]**

(a) $\frac{-32}{4}$ (b) $-\frac{31}{4}$

(c) $-\frac{30}{4}$ (d) $-\frac{33}{4}$

Ans. ***(a)***

$\tan^{-1}(x+1) + \cot^{-1}\left(\frac{1}{x-1}\right) = \tan^{-1}\left(\frac{8}{31}\right)$

$\Rightarrow \tan^{-1}(x+1) + \tan^{-1}(x-1) = \tan^{-1}(8/31)$

$\Rightarrow \quad \tan^{-1}\left[\frac{(x+1)+(x-1)}{1-(x+1)(x-1)}\right] = \tan^{-1}\left(\frac{8}{31}\right)$

$\Rightarrow \quad \tan^{-1}\left[\frac{2x}{1-(x^2-1)}\right] = \tan^{-1}\left(\frac{8}{31}\right)$

$\Rightarrow \quad \frac{2x}{2-x^2} = \frac{8}{31}$

$\Rightarrow \quad 62x = 16 - 8x^2$

$\Rightarrow \quad 8x^2 + 62x - 16 = 0$

$\Rightarrow \quad 2(x+8)(4x-1) = 0$

$\Rightarrow \quad x = -8, 1/4$

But at $x = 1/4$

$\text{LHS} \Rightarrow \tan^{-1}\left(1+\frac{1}{4}\right) + \cot^{-1}\left(\frac{1}{\frac{1}{4}-1}\right)$

$= \tan^{-1}\frac{5}{4} + \cot^{-1}\left(\frac{-4}{3}\right)$

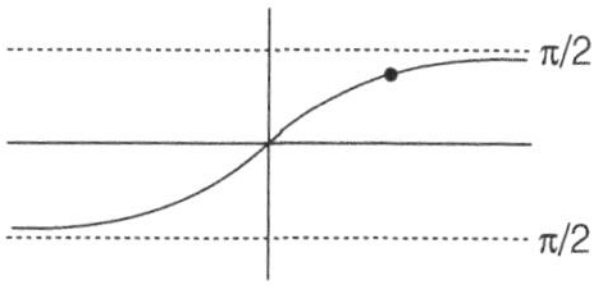

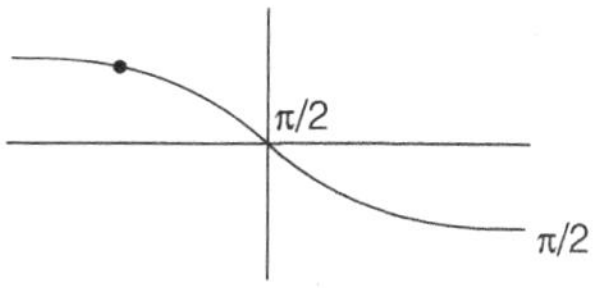

$= \underbrace{\underbrace{\tan^{-1}\frac{5}{4}}_{<\pi/2} + \underbrace{\cot^{-1}\left(\frac{-4}{3}\right)}_{>\pi/2}}_{>\pi/2}$

$\therefore \text{LHS} > \pi/2$

$\text{RHS} = \underbrace{\tan^{-1}\left(\frac{8}{31}\right)}_{<\pi/2}$

As, LHS $> \pi/2$ and RHS $< \pi/2$.

So, $x = -8$ is the only solution.

10. The number of solutions of the equation $\sin^{-1}\left[x^2+\frac{1}{3}\right] + \cos^{-1}\left[x^2-\frac{2}{3}\right] = x^2$, for $x \in [-1, 1]$, and $[x]$ denotes the greatest integer less than or equal to x, is **[JEE Main 2021]**

(a) 2 (b) 0

(c) 4 (d) infinite

Ans. ***(b)***

Given, $\sin^{-1}\left[x^2+\frac{1}{3}\right] + \cos^{-1}\left[x^2-\frac{2}{3}\right] = x^2$

$\Rightarrow$

$\sin^{-1}\left[\left(x^2-\frac{2}{3}+1\right)\right] + \cos^{-1}\left[x^2-\frac{2}{3}\right] = x^2$

$\Rightarrow \sin^{-1}\left[\left(x^2-\frac{2}{3}\right)+1\right] + \cos^{-1}\left[x^2-\frac{2}{3}\right]$

$= x^2 \quad \text{...(i)}$

$(\because [x+n] = [x]+n, n \in I)$

$\because \left[x^2-\frac{2}{3}\right]$ gives always integral values.

$\therefore \left[x^2-\frac{2}{3}\right] = 0, -1$ are possible values for $x \in [-1, 1]$ $\quad [\because -1 \le x \le 1$

$\therefore \quad 0 \le x^2 \le 1$

$\Rightarrow \quad -\frac{2}{3} \le x^2 - \frac{2}{3} \le 1 - \frac{2}{3}$

$\Rightarrow \quad -\frac{2}{3} \le x^2 - \frac{2}{3} \le \frac{1}{3}$

$\Rightarrow \quad -0.66 \le x^2 - \frac{2}{3} \le 0.33$

$\therefore \quad \left[x^2-\frac{2}{3}\right] = -1, 0$ are possibilities.]

Case I If $\left[x^2-\frac{2}{3}\right] = 0$

Then, Eq. (i) becomes,

$\sin^{-1}(1) + \cos^{-1}(0) = x^2$

$\Rightarrow \quad x^2 = \pi$

$\Rightarrow \quad x = \pm\sqrt{\pi}$

But at this value of

$x^2, \left[x^2-\frac{2}{3}\right] \ne 0$

$\therefore \quad x^2 = \pi$ (Rejected)

Case II If $\left[x^2-\frac{2}{3}\right] = -1$

Then, Eq. (i) becomes,

$\sin^{-1}(0) + \cos^{-1}(-1) = x^2$

$\Rightarrow \quad x^2 = \pi$

$\Rightarrow \quad x = \pm\sqrt{\pi}$

But at this value of x^2, $\left[x^2-\frac{2}{3}\right] \ne -1$

$\therefore \quad x^2 = \pi$

(Rejected also)

Hence, there is no solution for Eq. (i).

$\therefore$ Total number of solution(s) = 0

11. Let $S_k = \sum_{r=1}^{k} \tan^{-1}\left(\frac{6^r}{2^{2r+1}+3^{2r+1}}\right)$. Then, $\lim_{k\to\infty} S_k$ is equal to **[JEE Main 2021]**

(a) $\tan^{-1}\left(\frac{3}{2}\right)$ (b) $\frac{\pi}{2}$

(c) $\cot^{-1}\left(\frac{3}{2}\right)$ (d) $\tan^{-1}(3)$

Ans. ***(c)***

$S_K = \sum_{r=1}^{k} \tan^{-1}\left(\frac{6^r}{2^{2r+1}+3^{2r+1}}\right)$

$= \sum \tan^{-1} \frac{2^r 3^{r+1} - 3^r 2^{r+1}}{2^{2r+1}\left[1+\left(\frac{3}{2}\right)^{2r+1}\right]}$

$= \sum \tan^{-1} \frac{2^{2r+1}\left[\left(\frac{3}{2}\right)^{r+1} - \left(\frac{3}{2}\right)^r\right]}{2^{2r+1}\left[1+\left(\frac{3}{2}\right)^{2r+1}\right]}$

$= \sum_{r=1}^{k} \tan^{-1} \frac{\left(\frac{3}{2}\right)^{r+1} - \left(\frac{3}{2}\right)^r}{1+\left(\frac{3}{2}\right)^{2r+1}}$

$= \sum_{r=1}^{k} \tan^{-1}\left(\frac{3}{2}\right)^{r+1} - \tan^{-1}\left(\frac{3}{2}\right)^r$

$= \tan^{-1}\left(\frac{3}{2}\right)^2 - \tan^{-1}\left(\frac{3}{2}\right)$

$$\tan^{-1}\left(\frac{3}{2}\right)^3 - \tan^{-1}\left(\frac{3}{2}\right)^2$$

$$\vdots$$

$$\tan^{-1}\left(\frac{3}{2}\right)^{k+1} - \tan^{-1}\left(\frac{3}{2}\right)^k$$

$$\Rightarrow \quad S_k = \tan^{-1}\left(\frac{3}{2}\right)^{k+1} - \tan^{-1}\left(\frac{3}{2}\right)$$

When $k \to \infty$, $\tan^{-1}\left(\frac{3}{2}\right)^{k+1} \to \pi/2$

$$\lim_{k\to\infty} S_k = \frac{\pi}{2} - \tan^{-1}\left(\frac{3}{2}\right) = \cot^{-1}\left(\frac{3}{2}\right)$$

12. Given that the inverse trigonometric functions take principal values only. Then, the number of real values of x which satisfy $\sin^{-1}\left(\frac{3x}{2}\right) + \sin^{-1}\left(\frac{4x}{5}\right) = \sin^{-1} x$ is equal to **[JEE Main 2021]**

(a) 2
(b) 1
(c) 3
(d) 0

Ans. *(c)*

Given, $\sin^{-1}\left(\frac{3x}{5}\right) + \sin^{-1}\left(\frac{4x}{5}\right) = \sin^{-1} x$

$$\Rightarrow \sin^{-1}\left(\frac{3x}{5}\sqrt{1-\frac{16x^2}{25}} + \frac{4x}{5}\sqrt{1-\frac{9x^2}{25}}\right) = \sin^{-1} x$$

$$\Rightarrow 3x\sqrt{25-16x^2} + 4x\sqrt{25-9x^2} = 25x$$

$$\Rightarrow \sqrt{225x^2 - 114x^4} + \sqrt{400x^2 - 144x^4} = 25x \quad \text{...(i)}$$

$$(400x^2 - 144x^4) - (225x^2 - 144x^4) = 175x^2 \quad \text{...(ii)}$$

On dividing Eq. (ii) by Eq. (i),

$$\sqrt{400x^2 - 144x^4} - \sqrt{225x^2 - 144x^2} = 7x \quad \text{...(iii)}$$

Now, adding Eqs. (i) and (iii),

$$2\sqrt{400x^2 - 144x^4} = 32x$$

$$\Rightarrow \quad 400x^2 - 144x^4 = 256x^2$$

$$\Rightarrow \quad 144x^2 - 144x^4 = 0$$

$$\Rightarrow \quad 144x^2(1 - x^2) = 0$$

$$x = 0, -1, 1$$

Hence, 3 real values for x satisfies the equation.

13. If $\frac{\sin^{-1}(x)}{a} = \frac{\cos^{-1} x}{b} = \frac{\tan^{-1} y}{c}$, $0 < x < 1$, then the value of $\cos\left(\frac{\pi c}{a+b}\right)$ is **[JEE Main 2021]**

(a) $\frac{1-y^2}{y\sqrt{y}}$
(b) $1 - y^2$
(c) $\frac{1-y^2}{1+y^2}$
(d) $\frac{1-y^2}{2y}$

Ans. *(c)*

$$\frac{\sin^{-1} x}{a} = \frac{\cos^{-1} x}{b} = \frac{\tan^{-1} y}{c} \quad \text{...(i)}$$

Take first two terms of Eq. (i)

$$\frac{\sin^{-1} x}{a} = \frac{\cos^{-1} x}{b}$$

$$\Rightarrow \quad \frac{\sin^{-1} x}{a} = \frac{\cos^{-1} x}{b} = \frac{\sin^{-1} x + \cos^{-1} x}{a+b}$$

$$\Rightarrow \quad \frac{\sin^{-1} x}{a} = \frac{\cos^{-1} x}{b} = \frac{\pi/2}{a+b}$$

$$\left[\because \sin^{-1} x + \cos^{-1} x = \frac{\pi}{2}\right]$$

$$\Rightarrow \quad \frac{\sin^{-} x}{a} = \frac{\cos^{-1} x}{b} = \frac{\pi/2}{a+b} = \frac{\tan^{-1} y}{c}$$

Using last two terms,

$$\frac{\tan^{-1} y}{c} = \frac{\pi/2}{a+b}$$

$$\Rightarrow \quad \tan^{-1} y = \frac{\pi c}{2(a+b)}$$

$$\Rightarrow \quad 2\tan^{-1} y = \frac{\pi c}{(a+b)}$$

$$\Rightarrow \quad \cos^{-1}\left(\frac{1-y^2}{1+y^2}\right) = \frac{\pi c}{a+b}$$

$$\left[\because 2\tan^{-1} y = \cos^{-1}\left(\frac{1-y^2}{1+y^2}\right)\right]$$

$$\Rightarrow \quad \frac{1-y^2}{1+y^2} = \cos\left(\frac{\pi c}{a+b}\right)$$

$$\therefore \quad \cos\left(\frac{\pi c}{a+b}\right) = \frac{1-y^2}{1+y^2}$$

14. $\operatorname{cosec}\left[2\cot^{-1}(5) + \cos^{-1}\left(\frac{4}{5}\right)\right]$ is equal to **[JEE Main 2021]**

(a) $\frac{56}{33}$ (b) $\frac{65}{33}$ (c) $\frac{65}{56}$ (d) $\frac{75}{56}$

Ans. *(c)*

$\operatorname{cosec}[2\cot^{-1}(5) + \cos^{-1}(4/5)]$

$$= \operatorname{cosec}\left[2\tan^{-1}\left(\frac{1}{5}\right) + \cos^{-1}\left(\frac{4}{5}\right)\right]$$

$$\left[\because \tan^{-1} x = \cot^{-1}\left(\frac{1}{x}\right)\right]$$

$$= \operatorname{cosec}\left[\tan^{-1}\left(\frac{2\left(\frac{1}{5}\right)}{1-\left(\frac{1}{5}\right)^2}\right) + \cos^{-1}\left(\frac{4}{5}\right)\right]$$

$$\left[\therefore 2\tan^{-1}\theta = \tan^{-1}\left(\frac{2\theta}{1-\theta^2}\right)\right]$$

$$= \operatorname{cosec}\left(\tan^{-1}\frac{5}{12} + \cos^{-1}\frac{4}{5}\right)$$

Let $\tan^{-1}\left(\frac{5}{12}\right) = x$, then $\tan x = \frac{5}{12}$ gives

$\sin x = \frac{5}{13}$, $\cos x = \frac{12}{13}$

Let $\cos^{-1}\left(\frac{4}{5}\right) = y$, then $\cos y = \frac{4}{5}$ gives,

$\sin y = \frac{3}{5}$

$$\text{Now, } \operatorname{cosec}(x+y) = \frac{1}{\sin(x+y)}$$

$$= \frac{1}{\sin x \cos y + \cos x \sin y}$$

$$= \frac{1}{\left(\frac{5}{13}\right)\left(\frac{4}{5}\right) + \left(\frac{12}{13}\right)\left(\frac{3}{5}\right)} = \frac{65}{56}$$

15. A possible value of $\tan\left(\frac{1}{4}\sin^{-1}\frac{\sqrt{63}}{8}\right)$ is **[JEE Main 2021]**

(a) $\frac{1}{\sqrt{7}}$
(b) $2\sqrt{2} - 1$
(c) $\sqrt{7} - 1$
(d) $\frac{1}{2\sqrt{2}}$

Ans. *(a)*

Given, $\tan\left(\frac{1}{4}\sin^{-1}\frac{\sqrt{63}}{8}\right)$

Let $\sin^{-1}\frac{\sqrt{63}}{8} = \theta$

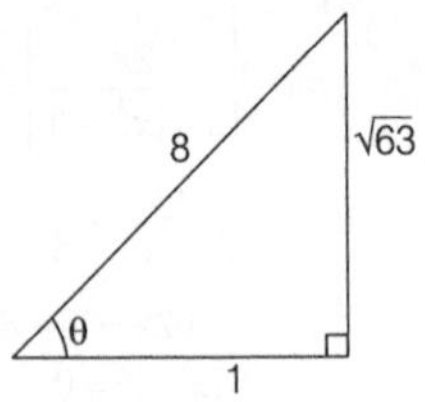

$$\Rightarrow \quad \sin\theta = \frac{\sqrt{63}}{8}$$

$$\Rightarrow \quad \cos\theta = \frac{1}{8}$$

$$\text{Also, } \cos\frac{\theta}{2} = \sqrt{\frac{1+\cos\theta}{2}}$$

$$= \sqrt{\frac{1+\frac{1}{8}}{2}} = \sqrt{\frac{9}{16}} = \frac{3}{4}$$

$\therefore \tan\left(\frac{1}{4}\sin^{-1}\frac{\sqrt{63}}{8}\right) = \tan\left(\frac{\theta}{4}\right)$

$$= \sqrt{\frac{1-\cos\frac{\theta}{2}}{1+\cos\frac{\theta}{2}}}$$

$$= \sqrt{\frac{1-\frac{3}{4}}{1+\frac{3}{4}}} = \frac{1}{\sqrt{7}}$$

16. For any positive integer n, let $S_n : (0, \infty) \to R$ be defined by $S_n(x) = \sum_{k=1}^{n} \cot^{-1}\left(\frac{1+k(k+1)x^2}{x}\right)$, where for any $x \in R$, $\cot^{-1}(x) \in (0, \pi)$ and $\tan^{-1}(x) \in \left(-\frac{\pi}{2}, \frac{\pi}{2}\right)$. Then, which of the following statements is (are) TRUE? **[JEE Advanced 2021]**

(a) $S_{10}(x) = \frac{\pi}{2} - \tan^{-1}\left(\frac{1+11x^2}{10x}\right)$, for all $x > 0$

(b) $\lim_{n\to\infty} \cot(S_n(x)) = x$, for all $x > 0$

(c) The equation $S_3(x) = \frac{\pi}{4}$ has a root in $(0, \infty)$

(d) $\tan(S_n(x)) \leq \frac{1}{2}$, for all $n \geq 1$ and $x > 0$

Ans. *(a and b)*

For option (a)

$S_n(x) = \sum_{k=1}^{n} \cot^{-1}\left[\frac{1+k(k+1)x^2}{x}\right]$ can be written as

$$S_n(x) = \sum_{k=1}^{n} \tan^{-1}\left[\frac{(k+1)x - kx}{1 + kx\cdot(k+1)x}\right]$$

$$= \sum_{k=1}^{n}[\tan^{-1}(k+1)x - \tan^{-1}(kx)]$$

$$= \tan^{-1}(n+1)x - \tan^{-1}x$$

$$= \tan^{-1}\left(\frac{nx}{1+(n+1)x^2}\right)$$

Now, $S_{10}(x) = \tan^{-1}\left(\frac{10x}{1+11x^2}\right)$

$$= \frac{\pi}{2} - \cot^{-1}\left(\frac{10x}{1+11x^2}\right)$$

$$= \frac{\pi}{2} - \tan^{-1}\left(\frac{1+11x^2}{10x}\right)$$

Option (a) is correct.

For option (b)

$$\lim_{n\to\infty} \cot(S_n(x)) = \cot\left(\tan^{-1}\left(\frac{x}{x^2}\right)\right)$$

$$= \cot\left(\tan^{-1}\left(\frac{1}{x}\right)\right) = \cot(\cot^{-1}x) = x, x > 0$$

Option (b) is correct.

For option (c)

$S_3(x) = \frac{\pi}{4} \Rightarrow \frac{3x}{1+4x^2} = 1$

$\Rightarrow 4x^2 - 3x + 1 = 0$ has no real root.

Option (c) is incorrect.

For option (d)

For $x = 1$, $\tan(S_n(x)) = \frac{n}{n+2}$ which is greater than $\frac{1}{2}$ for $n \geq 3$.

Option (d) is incorrect.

17. The domain of the function $f(x) = \sin^{-1}\left(\frac{|x|+5}{x^2+1}\right)$ is $(-\infty, -a] \cup [a, \infty)$. Then a is equal to **[JEE Main 2020]**

(a) $\frac{\sqrt{17}}{2}+1$ (b) $\frac{\sqrt{17}-1}{2}$

(c) $\frac{1+\sqrt{17}}{2}$ (d) $\frac{\sqrt{17}}{2}$

Ans. *(c)*

Given function, $f(x) = \sin^{-1}\left(\frac{|x|+5}{x^2+1}\right)$

For, domain of 'f'

$\frac{|x|+5}{x^2+1} \leq 1$, as $\frac{|x|+5}{x^2+1} > 0 \; \forall x \in R$

$\Rightarrow |x| + 5 \leq x^2 + 1$ as $x^2 + 1$ is positive $\forall x \in R$

$\Rightarrow x^2 - |x| - 4 \geq 0$

$\Rightarrow |x|^2 - |x| - 4 \geq 0$

$[\because \forall x \in R, x^2 = |x|^2]$ ... (i)

$\because$ For $|x|^2 - |x| - 4 = 0 \Rightarrow |x| = \frac{1 \pm \sqrt{1+16}}{2}$

$\Rightarrow |x| = \frac{1 \pm \sqrt{17}}{2} \Rightarrow |x| = \frac{1+\sqrt{17}}{2}$ as $|x| \geq 0$

$\therefore |x| \geq \frac{\sqrt{17}+1}{2}$

$\Rightarrow x \in \left(-\infty, -\frac{\sqrt{17}+1}{2}\right] \cup \left[\frac{\sqrt{17}+1}{2}, \infty\right)$

$\therefore a = \frac{\sqrt{17}+1}{2}$

18. $2\pi - \left(\sin^{-1}\frac{4}{5} + \sin^{-1}\frac{5}{13} + \sin^{-1}\frac{16}{65}\right)$ is equal to **[JEE Main 2020]**

(a) $\frac{\pi}{2}$ (b) $\frac{5\pi}{4}$ (c) $\frac{3\pi}{2}$ (d) $\frac{7\pi}{4}$

Ans. *(c)*

We have,

$$2\pi - \left(\sin^{-1}\frac{4}{5} + \sin^{-1}\frac{5}{13} + \sin^{-1}\frac{16}{65}\right)$$

$$= 2\pi - \left(\tan^{-1}\frac{4}{3} + \tan^{-1}\frac{5}{12} + \tan^{-1}\frac{16}{63}\right)$$

$$= 2\pi - \left(\tan^{-1}\frac{\frac{4}{3}+\frac{5}{12}}{1-\frac{5}{9}} + \tan^{-1}\frac{16}{63}\right)$$

$$= 2\pi - \left(\tan^{-1}\left(\frac{21\times3}{4\times4}\right) + \tan^{-1}\frac{16}{63}\right)$$

$$= 2\pi - \left(\tan^{-1}\frac{63}{16} + \tan^{-1}\frac{16}{63}\right)$$

$$= 2\pi - \left(\cot^{-1}\frac{16}{63} + \tan^{-1}\frac{16}{63}\right)$$

$$= 2\pi - \frac{\pi}{2} = \frac{3\pi}{2}$$

Hence, option (c) is correct.

19. If S is the sum of the first 10 terms of the series $\tan^{-1}\left(\frac{1}{3}\right) + \tan^{-1}\left(\frac{1}{7}\right) + \tan^{-1}\left(\frac{1}{13}\right) + \tan^{-1}\left(\frac{1}{21}\right) + \ldots$, then $\tan(S)$ is equal to **[JEE Main 2020]**

(a) $\frac{6}{5}$ (b) $\frac{5}{6}$

(c) $\frac{10}{11}$ (d) $\frac{5}{11}$

Ans. *(b)*

Given, series is

$$S = \tan^{-1}\left(\frac{1}{3}\right) + \tan^{-1}\left(\frac{1}{7}\right) + \tan^{-1}\frac{1}{13} + \tan^{-1}\left(\frac{1}{21}\right) + \ldots + \text{ up to 10 terms}$$

$$= \tan^{-1}\frac{1}{1+2} + \tan^{-1}\frac{1}{1+6} + \tan^{-1}\frac{1}{1+12} + \tan^{-1}\frac{1}{1+20} + \ldots + \text{ up to 10 terms}$$

$$= \tan^{-1}\frac{2-1}{1+(1\cdot2)} + \tan^{-1}\frac{3-2}{1+(2\cdot3)} + \tan^{-1}\frac{4-3}{1+(3\cdot4)} + \tan^{-1}\frac{5-4}{1+(4\cdot5)} + \ldots + \text{ up to 10 terms}$$

$$= [\tan^{-1}2 - \tan^{-1}1] + [\tan^{-1}3 - \tan^{-1}2] + [\tan^{-1}4 - \tan^{-1}3] + [\tan^{-1}5 - \tan^{-1}4] + \ldots\ldots + \text{ upto 10 terms}$$

$$= \tan^{-1}11 - \tan^{-1}1$$

$$= \tan^{-1}\left(\frac{11-1}{1+(11\times1)}\right)$$

$$\Rightarrow S = \tan^{-1}\left(\frac{10}{12}\right)$$

$$\Rightarrow \tan(S) = \frac{5}{6}$$

20. If $f'(x)=\tan^{-1}(\sec x+\tan x)$, $\frac{\pi}{2}<x<\frac{\pi}{2}$, and $f(0)=0$, then $f(1)$ is equal to

(a) $\frac{\pi+1}{4}$ (b) $\frac{\pi+2}{4}$ (c) $\frac{1}{4}$ (d) $\frac{\pi-1}{4}$

[JEE Main 2020]

Ans. *(a)*

Given $f'(x)=\tan^{-1}(\sec x+\tan x)$, $-\frac{\pi}{2}<x<\frac{\pi}{2}$

$$\Rightarrow f'(x)=\tan^{-1}\left(\frac{1+\sin x}{\cos x}\right)$$

$$\Rightarrow f'(x)=\tan^{-1}\left(\frac{1-\cos\left(\frac{\pi}{2}+x\right)}{\sin\left(\frac{\pi}{2}+x\right)}\right)$$

$$=\tan^{-1}\left(\frac{2\sin^2\left(\frac{\pi}{4}+\frac{x}{2}\right)}{2\sin\left(\frac{\pi}{4}+\frac{x}{2}\right)\cos\left(\frac{\pi}{4}+\frac{x}{2}\right)}\right)$$

$$=\tan^{-1}\tan\left(\frac{\pi}{4}+\frac{x}{2}\right)$$

$$f'(x)=\frac{\pi}{4}+\frac{x}{2}$$

$$\left[\because x\in\left(-\frac{\pi}{2},\frac{\pi}{2}\right),\ \text{so } \frac{x}{2}\in\left(-\frac{\pi}{4},\frac{\pi}{4}\right)\right]$$

Now, $f(x)=\frac{\pi x}{4}+\frac{x^2}{4}+C$ [on integration]

$\because\quad f(0)=0\Rightarrow C=0$

$\therefore\quad f(x)=\frac{\pi x}{4}+\frac{x^2}{4}$

So, $\quad f(1)=\frac{\pi+1}{4}$

Hence, option (a) is correct.

21. If $\alpha=\cos^{-1}\left(\frac{3}{5}\right)$, $\beta=\tan^{-1}\left(\frac{1}{3}\right)$, where $0<\alpha,\beta<\frac{\pi}{2}$, then $\alpha-\beta$ is equal to

[JEE Main 2019]

(a) $\tan^{-1}\left(\frac{9}{5\sqrt{10}}\right)$ (b) $\cos^{-1}\left(\frac{9}{5\sqrt{10}}\right)$

(c) $\tan^{-1}\left(\frac{9}{14}\right)$ (d) $\sin^{-1}\left(\frac{9}{5\sqrt{10}}\right)$

Ans. *(d)*

Given, $\alpha=\cos^{-1}\left(\frac{3}{5}\right)$ and $\beta=\tan^{-1}\left(\frac{1}{3}\right)$

where $0<\alpha,\beta<\frac{\pi}{2}$

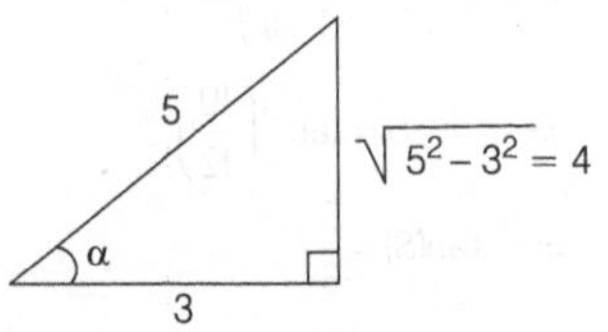

Clearly, $\alpha=\tan^{-1}\frac{4}{3}$

So, $\alpha-\beta=\tan^{-1}\frac{4}{3}-\tan^{-1}\frac{1}{3}$

$$=\tan^{-1}\left(\frac{\frac{4}{3}-\frac{1}{3}}{1+\left(\frac{4}{3}\times\frac{1}{3}\right)}\right)$$

$$\left[\because \tan^{-1}x-\tan^{-1}y=\tan^{-1}\frac{x-y}{1+xy},\ \text{if } xy>-1\right]$$

$$=\tan^{-1}\frac{1}{1+\frac{4}{9}}=\tan^{-1}\frac{9}{13}$$

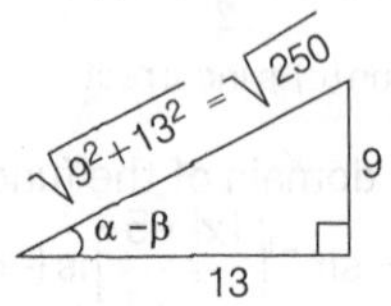

$$=\sin^{-1}\frac{9}{\sqrt{9^2+13^2}}$$

$$=\sin^{-1}\frac{9}{\sqrt{250}}=\sin^{-1}\left(\frac{9}{5\sqrt{10}}\right)$$

22. If $\cos^{-1}x-\cos^{-1}\frac{y}{2}=\alpha$, where $-1\le x\le 1$, $-2\le y\le 2$, $x\le\frac{y}{2}$, then for all x,y, $4x^2-4xy\cos\alpha+y^2$ is equal to

[JEE Main 2019]

(a) $2\sin^2\alpha$ (b) $4\cos^2\alpha+2x^2y^2$

(c) $4\sin^2\alpha$ (d) $4\sin^2\alpha-2x^2y^2$

Ans. *(c)*

Given equation is

$\cos^{-1}x-\cos^{-1}\frac{y}{2}=\alpha$, where $-1\le x\le 1$,

$-2\le y\le 2$ and $x\le\frac{y}{2}$

$$\therefore \cos^{-1}\left(x\frac{y}{2}+\sqrt{1-x^2}\ \sqrt{1-(y/2)^2}\right)=\alpha$$

$$\left[\because \cos^{-1}x-\cos^{-1}y=\cos^{-1}(xy+\sqrt{1-x^2}\sqrt{1-y^2})\ \ |x|,|y|\le 1 \text{ and } x+y\ge 0\right]$$

$$\Rightarrow \frac{xy}{2}+\sqrt{1-x^2}\ \sqrt{1-(y/2)^2}=\cos\alpha$$

$$\Rightarrow \sqrt{1-x^2}\ \sqrt{1-(y/2)^2}=\cos\alpha-\frac{xy}{2}$$

On squaring both sides, we get

$$(1-x^2)\left(1-\frac{y^2}{4}\right)=\cos^2\alpha+\frac{x^2y^2}{4}-2\frac{xy}{2}\cos\alpha$$

$$\Rightarrow 1-x^2-\frac{y^2}{4}+\frac{x^2y^2}{4}$$

$$=\cos^2\alpha+\frac{x^2y^2}{4}-xy\cos\alpha$$

$$\Rightarrow x^2+\frac{y^2}{4}-xy\cos\alpha=1-\cos^2\alpha$$

$$\Rightarrow 4x^2-4xy\cos\alpha+y^2=4\sin^2\alpha$$

23. The value of $\sin^{-1}\left(\frac{12}{13}\right)-\sin^{-1}\left(\frac{3}{5}\right)$ is equal to

[JEE Main 2019]

(a) $\pi-\sin^{-1}\left(\frac{63}{65}\right)$ (b) $\frac{\pi}{2}-\sin^{-1}\left(\frac{56}{65}\right)$

(c) $\frac{\pi}{2}-\cos^{-1}\left(\frac{9}{65}\right)$ (d) $\pi-\cos^{-1}\left(\frac{33}{65}\right)$

Ans. *(b)*

Key Idea Use formulae

(i) $\sin^{-1}x-\sin^{-1}y$

$=\sin^{-1}(x\sqrt{1-y^2}-y\sqrt{1-x^2})$

if $x^2+y^2\le 1$

or if $xy>0$ and $x^2+y^2>1$

$\forall x,y\in[-1,1]$

(ii) $\sin^{-1}x=\cos^{-1}\sqrt{1-x^2}$ and

(iii) $\sin^{-1}\theta+\cos^{-1}\theta=\frac{\pi}{2}$

We have,

$$\sin^{-1}\left(\frac{12}{13}\right)-\sin^{-1}\left(\frac{3}{5}\right)$$

$$=\sin^{-1}\left(\frac{12}{13}\sqrt{1-\left(\frac{3}{5}\right)^2}-\frac{3}{5}\sqrt{1-\left(\frac{12}{13}\right)^2}\right)$$

$$\left[\because \sin^{-1}x-\sin^{-1}y=\sin^{-1}(x\sqrt{1-y^2}-y\sqrt{1-x^2}),\ \text{if } x^2+y^2\le 1 \text{ or if } xy>0 \text{ and } x^2+y^2>1\ \forall x,y\in[-1,1]\right]$$

$$=\sin^{-1}\left(\frac{12}{13}\times\frac{4}{5}-\frac{3}{5}\times\frac{5}{13}\right)$$

$$=\sin^{-1}\left(\frac{48-15}{65}\right)$$

$$=\sin^{-1}\left(\frac{33}{65}\right)=\cos^{-1}\sqrt{1-\left(\frac{33}{65}\right)^2}$$

$$=\cos^{-1}\sqrt{\frac{3136}{4225}}$$

$$[\because \sin^{-1}x=\cos^{-1}\sqrt{1-x^2}]$$

$$=\cos^{-1}\left(\frac{56}{65}\right)=\frac{\pi}{2}-\sin^{-1}\left(\frac{56}{65}\right)$$

$$\left[\because \sin^{-1}\theta+\cot^{-1}\theta=\frac{\pi}{2}\right]$$

24. If $\cos^{-1}\left(\frac{2}{3x}\right)+\cos^{-1}\left(\frac{3}{4x}\right)=\frac{\pi}{2}\left(x>\frac{3}{4}\right)$, then x is equal to

[JEE Main 2019]

(a) $\frac{\sqrt{145}}{10}$ (b) $\frac{\sqrt{146}}{12}$

(c) $\frac{\sqrt{145}}{12}$ (d) $\frac{\sqrt{145}}{11}$

Ans. *(c)*

Key Idea Use the formula,

$\cos^{-1}x+\cos^{-1}y$

$=\cos^{-1}(xy-\sqrt{1-x^2}\sqrt{1-y^2})$

We have, $\cos^{-1}\left(\frac{2}{3x}\right)+\cos^{-1}\left(\frac{3}{4x}\right)=\frac{\pi}{2}$

$\Rightarrow \cos^{-1}\left(\frac{2}{3x}\cdot\frac{3}{4x}-\sqrt{1-\frac{4}{9x^2}}\sqrt{1-\frac{9}{16x^2}}\right)=\frac{\pi}{2}$

$[\because \cos^{-1}x+\cos^{-1}y$
$=\cos^{-1}(xy-\sqrt{1-x^2}\sqrt{1-y^2})]$

$\Rightarrow \cos^{-1}\left(\frac{1}{2x^2}-\frac{\sqrt{9x^2-4}\sqrt{16x^2-9}}{12x^2}\right)=\frac{\pi}{2}$

$\Rightarrow \frac{6-\sqrt{9x^2-4}\sqrt{16x^2-9}}{12x^2}=\cos\frac{\pi}{2}=0$

$\Rightarrow \sqrt{9x^2-4}\sqrt{16x^2-9}=6$

On squaring both sides, we get

$(9x^2-4)(16x^2-9)=36$

$\Rightarrow 144x^4-81x^2-64x^2+36=36$

$\Rightarrow 144x^4-145x^2=0$

$\Rightarrow x^2(144x^2-145)=0$

$\Rightarrow x=0 \text{ or } x=\pm\sqrt{\frac{145}{144}}=\pm\frac{\sqrt{145}}{12}$

But $x>\frac{3}{4}$ $\therefore$ $x=\frac{\sqrt{145}}{12}$

25. If $x=\sin^{-1}(\sin 10)$ and $y=\cos^{-1}(\cos 10)$, then $y-x$ is equal to **[JEE Main 2019]**

(a) 0 (b) 10 (c) 7π (d) π

Ans. *(d)*

The graph of $y=\sin^{-1}(\sin x)$ is

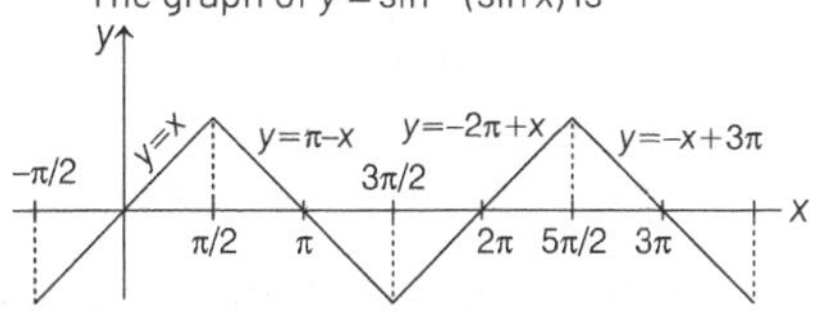

$\therefore$ $x=\sin^{-1}(\sin 10)=-10+3\pi$ ...(i)

and the graph of $y=\cos^{-1}(\cos x)$ is

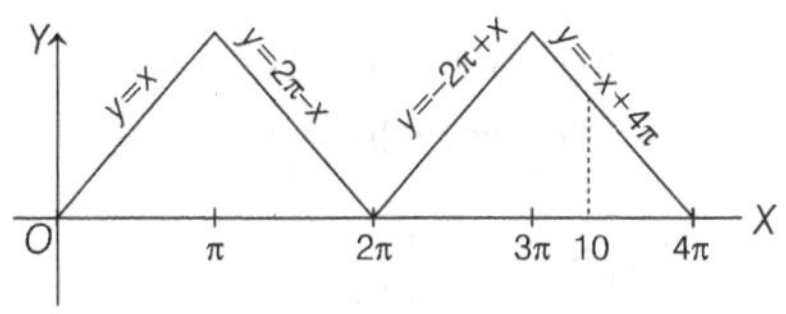

$\therefore$ $y=\cos^{-1}(\cos 10)=-10+4\pi$ ...(ii)

Now, from Eqs. (i) and (ii),

$y-x=(-10+4\pi)-(-10+3\pi)=\pi$

26. The value of $\cot\left(\sum_{n=1}^{19}\cot^{-1}\left(1+\sum_{p=1}^{n}2p\right)\right)$ is **[JEE Main 2019]**

(a) $\frac{23}{22}$ (b) $\frac{21}{19}$ (c) $\frac{19}{21}$ (d) $\frac{22}{23}$

Ans. *(b)*

Consider, $\cot\left(\sum_{n=1}^{19}\cot^{-1}\left(1+\sum_{p=1}^{n}2p\right)\right)$

$=\cot\left(\sum_{n=1}^{19}\cot^{-1}(1+n(n+1))\right)$

$\left[\because \sum_{p=1}^{n}p=\frac{n(n+1)}{2}\right]$

$=\cot\left(\sum_{n=1}^{19}\cot^{-1}(1+n+n^2)\right)$

$=\cot\left(\sum_{n=1}^{19}\tan^{-1}\frac{1}{1+n(n+1)}\right)$

$[\because \cot^{-1}x=\tan^{-1}\frac{1}{x}, \text{if } x>0]$

$=\cot\left(\sum_{n=1}^{19}\tan^{-1}\left(\frac{(n+1)-n}{1+n(n+1)}\right)\right)$

$[\text{put } 1=(n+1)-n]$

$=\cot\sum_{n=1}^{19}(\tan^{-1}(n+1)-\tan^{-1}n)$

$\left[\because \tan^{-1}\frac{x-y}{1+xy}=\tan^{-1}x-\tan^{-1}y\right]$

$=\cot[(\tan^{-1}2-\tan^{-1}1)+(\tan^{-1}3-\tan^{-1}2)+$
$......+(\tan^{-1}20-\tan^{-1}19)]$

$=\cot(\tan^{-1}20-\tan^{-1}1)$

$=\cot\left[\left(\frac{\pi}{2}-\cot^{-1}20\right)-\left(\frac{\pi}{2}-\cot^{-1}1\right)\right]$

$[\because \tan^{-1}x+\cot^{-1}x=\pi/2]$

$=\cot(\cot^{-1}1-\cot^{-1}20)$

$=\frac{\cot(\cot^{-1}1)\cot(\cot^{-1}20)+1}{\cot(\cot^{-1}20)-\cot(\cot^{-1}1)}$

$\left[\because \cot(A-B)]=\frac{\cot A\cot B+1}{\cot B-\cot A}\right]$

$=\frac{(1\times 20)+1}{20-1}$ $[\because \cot(\cot^{-1}x)=x]$

$=\frac{21}{19}$

27. All x satisfying the inequality $(\cot^{-1}x)^2-7(\cot^{-1}x)+10>0$, lie in the interval **[JEE Main 2019]**

(a) $(-\infty, \cot 5)\cup(\cot 2, \infty)$

(b) $(\cot 5, \cot 4)$

(c) $(\cot 2, \infty)$

(d) $(-\infty, \cot 5)\cup(\cot 4, \cot 2)$

Ans. *(c)*

Given, $(\cot^{-1}x)^2-7(\cot^{-1}x)+10>0$

$\Rightarrow (\cot^{-1}x-2)(\cot^{-1}x-5)>0$

(by factorisation)

$\Rightarrow \cot^{-1}x<2 \text{ or } \cot^{-1}x>5$

By wavy curve method,

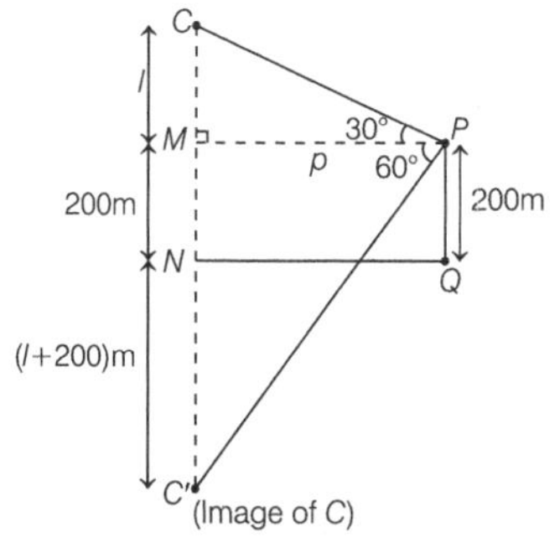

$\cot^{-1}x\in(-\infty,2)\cup(5,\infty)$

$\cot^{-1}x\in(0,2)$

$[\because \text{range of } \cot^{-1}x \text{ is } (0,\pi)]$

$\therefore$ $x\in(\cot 2,\infty)$

28. Considering only the principal values of inverse functions, the set $A=\left\{x\geq 0: \tan^{-1}(2x)+\tan^{-1}(3x)=\frac{\pi}{4}\right\}$ **[JEE Main 2019]**

(a) is an empty set

(b) is a singleton

(c) contains more than two elements

(d) contains two elements

Ans. *(b)*

Given equation is

$\tan^{-1}(2x)+\tan^{-1}(3x)=\frac{\pi}{4}, x\geq 0$

$\Rightarrow \tan^{-1}\frac{5x}{1-6x^2}=\frac{\pi}{4}, 6x^2<1$

$\left[\because \tan^{-1}x+\tan^{-1}y=\tan^{-1}\left(\frac{x+y}{1-xy}\right), xy<1\right]$

$\Rightarrow \frac{5x}{1-6x^2}=1, x^2<\frac{1}{6}$

$\Rightarrow 6x^2+5x-1=0, 0\leq x<\frac{1}{\sqrt{6}}$

$[\because x\geq 0]$

$\Rightarrow 6x^2+6x-x-1=0,$ $0\leq x<\frac{1}{\sqrt{6}}$

$\Rightarrow 6x(x+1)-1(x+1)=0,$ $0\leq x<\frac{1}{\sqrt{6}}$

$\Rightarrow (6x-1)(x+1)=0,$ $0\leq x<\frac{1}{\sqrt{6}}$

$\Rightarrow x=\frac{1}{6},-1,$ $0\leq x<\frac{1}{\sqrt{6}}$

$\Rightarrow x=\frac{1}{6},$ $[\because 0\leq x<\frac{1}{\sqrt{6}}]$

So 'A' is a singleton set.

29. The value of $\sec^{-1}\left(\frac{1}{4}\sum_{k=0}^{10}\sec\left(\frac{7\pi}{12}+\frac{k\pi}{2}\right)\sec\left(\frac{7\pi}{12}+\frac{(k+1)\pi}{2}\right)\right)$ in the interval $\left[-\frac{\pi}{4},\frac{3\pi}{4}\right]$ equals...... **[JEE Advanced 2019]**

Ans. *(0)*

$$\because \sum_{k=0}^{10} \sec\left(\frac{7\pi}{12}+\frac{k\pi}{2}\right)\sec\left(\frac{7\pi}{12}+\frac{(k+1)\pi}{2}\right)$$

$$=\sum_{k=0}^{10}\frac{1}{\cos\left(\frac{7\pi}{12}+\frac{k\pi}{2}\right)\cos\left(\frac{7\pi}{12}+\frac{(k+1)\pi}{2}\right)}$$

$$=\sum_{k=0}^{10}\frac{\sin\left[\left(\frac{7\pi}{12}+\frac{(k+1)\pi}{2}\right)-\left(\frac{7\pi}{12}+\frac{k\pi}{2}\right)\right]}{\cos\left(\frac{7\pi}{12}+\frac{k\pi}{2}\right)\cos\left(\frac{7\pi}{12}+\frac{(k+1)\pi}{2}\right)}$$

$$\left[\because \frac{7\pi}{12}+\frac{(k+1)\pi}{2}-\left(\frac{7\pi}{12}+\frac{k\pi}{2}\right)=\frac{\pi}{2}\text{ and }\sin\frac{\pi}{2}=1\right]$$

$$=\sum_{k=0}^{10}\frac{\sin\left(\frac{7\pi}{12}+\frac{(k+1)\pi}{2}\right)\cos\left(\frac{7\pi}{12}+\frac{k\pi}{2}\right)-\sin\left(\frac{7\pi}{12}+\frac{k\pi}{2}\right)\cos\left(\frac{7\pi}{12}+\frac{(k+1)\pi}{2}\right)}{\cos\left(\frac{7\pi}{12}+\frac{k\pi}{2}\right)\cos\left(\frac{7\pi}{12}+\frac{(k+1)\pi}{2}\right)}$$

$$=\sum_{k=0}^{10}\left[\tan\left(\frac{7\pi}{12}+\frac{(k+1)\pi}{2}\right)-\tan\left(\frac{7\pi}{12}+\frac{k\pi}{2}\right)\right]$$

$$=\tan\left(\frac{7\pi}{12}+\frac{\pi}{2}\right)-\tan\left(\frac{7\pi}{12}\right)$$
$$+\tan\left(\frac{7\pi}{12}+\frac{2\pi}{2}\right)-\tan\left(\frac{7\pi}{12}+\frac{\pi}{2}\right)$$
$$\vdots$$
$$+\tan\left(\frac{7\pi}{12}+\frac{11\pi}{2}\right)-\tan\left(\frac{7\pi}{12}+\frac{10\pi}{2}\right)$$

$$=\tan\left(\frac{7\pi}{12}+\frac{11\pi}{2}\right)-\tan\frac{7\pi}{12}$$

$$=\tan\frac{\pi}{12}+\cot\frac{\pi}{12}$$

$$=\frac{1}{\sin\frac{\pi}{12}\cos\frac{\pi}{12}}=\frac{2}{\sin\frac{\pi}{6}}=4$$

So,

$$\sec^{-1}\left(\frac{1}{4}\sum_{k=0}^{10}\sec\left(\frac{7\pi}{12}+\frac{k\pi}{2}\right)\sec\left(\frac{7\pi}{12}+\frac{(k+1)\pi}{2}\right)\right)$$
$$=\sec^{-1}(1)=0$$

30. The number of real solutions of the equation $\sin^{-1}\left(\sum_{i=1}^{\infty}x^{i+1}-x\sum_{i=1}^{\infty}\left(\frac{x}{2}\right)^i\right)$ $=\frac{\pi}{2}-\cos^{-1}\left(\sum_{i=1}^{\infty}\left(-\frac{x}{2}\right)^i-\sum_{i=1}^{\infty}(-x)^i\right)$ lying in the interval $\left(-\frac{1}{2},\frac{1}{2}\right)$ is ....

(Here, the inverse trigonometric functions $\sin^{-1}x$ and $\cos^{-1}x$ assume values in $\left[-\frac{\pi}{2},\frac{\pi}{2}\right]$ and $[0,\pi]$, respectively.)

[JEE Advanced 2018]

Ans. *(2)*

We have, $\sin^{-1}\left(\sum_{i=1}^{\infty}x^{i+1}-x\sum_{i=1}^{\infty}\left(\frac{x}{2}\right)^i\right)$

$$=\frac{\pi}{2}-\cos^{-1}\left(\sum_{i=1}^{\infty}\left(\frac{-x}{2}\right)^i-\sum_{i=1}^{\infty}(-x)^i\right)$$

$$\Rightarrow \sin^{-1}\left[\frac{x^2}{1-x}-\frac{x\cdot\frac{x}{2}}{1-\frac{x}{2}}\right]=\frac{\pi}{2}-\cos^{-1}\left[\frac{\frac{-x}{2}}{1+\frac{x}{2}}-\frac{(-x)}{1+x}\right]$$

$$\left[\because \sum_{i=1}^{\infty}x^{i+1}=x^2+x^3+x^4+\ldots=\frac{x^2}{1-x}\right.$$
using sum of infinite terms of GP $\Big]$

$$\Rightarrow \sin^{-1}\left[\frac{x^2}{1-x}-\frac{x^2}{2-x}\right]=\frac{\pi}{2}-\cos^{-1}\left[\frac{x}{1+x}-\frac{x}{2+x}\right]$$

$$\Rightarrow \sin^{-1}\left[\frac{x^2}{1-x}-\frac{x^2}{2-x}\right]=\sin^{-1}\left(\frac{x}{1+x}-\frac{x}{2+x}\right)$$

$$\left[\because \sin^{-1}x=\frac{\pi}{2}-\cos^{-1}x\right]$$

$$\Rightarrow \frac{x^2}{1-x}-\frac{x^2}{2-x}=\frac{x}{1+x}-\frac{x}{2+x}$$

$$\Rightarrow x^2\left(\frac{2-x-1+x}{(1-x)(2-x)}\right)=x\frac{(2+x-1-x)}{(1+x)(2+x)}$$

$$\Rightarrow \frac{x}{2-3x+x^2}=\frac{1}{2+3x+x^2}\text{ or }x=0$$

$$\Rightarrow x^3+3x^2+2x=x^2-3x+2$$

$$\Rightarrow x^3+2x^2+5x-2=0\text{ or }x=0$$

Let $f(x)=x^3+2x^2+5x-2$

$f'(x)=3x^2+4x+5$

$f'(x)>0, \forall x\in R$

$\therefore$ x^3+2x^2+5x-2 has only one real roots

Therefore, total number of real solution is 2.

31. For any positive integer n, define $f_n:(0,\infty)\to R$ as

$$f_n(x)=\sum_{j=1}^{n}\tan^{-1}\left(\frac{1}{1+(x+j)(x+j-1)}\right)$$

for all $x\in(0,\infty)$. (Here, the inverse trigonometric function $\tan^{-1}x$ assumes values in $\left(-\frac{\pi}{2},\frac{\pi}{2}\right)$. Then, which of the following statement(s) is (are) TRUE? **[JEE Advanced 2018]**

(a) $\sum_{j=1}^{5}\tan^2(f_j(0))=55$

(b) $\sum_{j=1}^{10}(1+f'_j(0))\sec^2(f_j(0))=10$

(c) For any fixed positive integer n, $\lim_{x\to\infty}\tan(f_n(x))=\frac{1}{n}$

(d) For any fixed positive integer n, $\lim_{x\to\infty}\sec^2(f_n(x))=1$

Ans. *(d)*

We have,

$$f_n(x)=\sum_{j=1}^{n}\tan^{-1}\left(\frac{1}{1+(x+j)(x+j-1)}\right)\text{ for all }x\in(0,\infty)$$

$$\Rightarrow f_n(x)=\sum_{j=1}^{n}\tan^{-1}\left(\frac{(x+j)-(x+j-1)}{1+(x+j)(x+j-1)}\right)$$

$$\Rightarrow f_n(x)=\sum_{j=1}^{n}[\tan^{-1}(x+j)-\tan^{-1}(x+j-1)]$$

$$\Rightarrow f_n(x)=(\tan^{-1}(x+1)-\tan^{-1}x)+(\tan^{-1}(x+2)-\tan^{-1}(x+1))+(\tan^{-1}(x+3)-\tan^{-1}(x+2))+\ldots+(\tan^{-1}(x+n)-\tan^{-1}(x+n-1))$$

$$\Rightarrow f_n(x)=\tan^{-1}(x+n)-\tan^{-1}x$$

This statement is false as $x\neq 0$, i.e., $x\in(0,\infty)$

(b) This statement is also false as $0\notin(0,\infty)$

(c) $f_n(x)=\tan^{-1}(x+n)-\tan^{-1}x$

$$\lim_{x\to\infty}\tan(f_n(x))=\lim_{x\to\infty}\tan(\tan^{-1}(x+n)-\tan^{-1}x)$$

$$\Rightarrow \lim_{x\to\infty}\tan(f_n(x))=\lim_{x\to\infty}\tan\left(\tan^{-1}\frac{n}{1+nx+x^2}\right)$$

$$=\lim_{x\to\infty}\frac{n}{1+nx+x^2}=0$$

$\therefore$ (c) statement is false.

(d) $\lim_{x\to\infty}\sec^2(f_n(x))=\lim_{x\to\infty}(1+\tan^2 f_n(x))$

$$=1+\lim_{x\to\infty}\tan^2(f_n(x))$$
$$=1+0=1$$

$\therefore$ statement (d) is true.